Textbook of MICROBIOLOGY

Textbook of MICROBIOLOGY

Surinder Kumar MD DNB MNAMS
Director Professor
Department of Microbiology
Maulana Azad Medical College
New Delhi, India

Foreword
Vishwa Mohan Katoch

JAYPEE BROTHERS MEDICAL PUBLISHERS
The Health Sciences Publisher
New Delhi | London

Jaypee Brothers Medical Publishers (P) Ltd

Headquarters
EMCA House
23/23-B, Ansari Road, Daryaganj
New Delhi 110 002, India
Landline: +91-11-23272143, +91-11-23272703
+91-11-23282021, +91-11-23245672
E-mail: jaypee@jaypeebrothers.com

Corporate Office
4838/24, Ansari Road, Daryaganj
New Delhi 110 002, India
Phone: +91-11-43574357
Fax: +91-11-43574314
E-mail: jaypee@jaypeebrothers.com

Overseas Offices
J.P. Medical Ltd.
83 Victoria Street, London
SW1H 0HW (UK)
Phone: +44-2031708910
E-mail: info@jpmedpub.com

EU GPSR Authorised Representative
Logos Europe, 9 rue Nicolas Poussin
17000, La Rochelle, France
Phone: +33 (0) 6 67 93 73 78
E-mail: contact@logoseurope.eu

Website: www.jaypeebrothers.com
Website: www.jaypeedigital.com

Inquiries for bulk sales may be solicited at: jaypee@jaypeebrothers.com

Textbook of Microbiology

First Edition: 2012
Reprint: **2025**

ISBN : 978-93-5025-510-0

Printed in India

Dedicated to

My father
Late Shri Lachhman Das

My mother
Smt Bal Kaur

My wife
Dr (Prof) Savita Kumari

and

My sons
Sourabh Kumar and Sanchit Kumar
whose love and energy make everything I do possible.

Dr. Vishwa Mohan Katoch
MD, FNASc, FAMS, FASc, FNA
Secretary to the Government of India
(Department of Health Research)
Ministry of Health & Family Welfare &
Director-General, ICMR

Indian Council of Medical Research
(Department of Health Research)
Ministry of Health & Family Welfare
V Ramalingaswami Bhawan, Ansari Nagar
New Delhi - 110 029 (INDIA)

Foreword

It is my privilege to write this foreword for *Textbook of Microbiology* by Professor Surinder Kumar. All of us know that Microbiology is an extremely diverse discipline and is undergoing continuous evolution as technology changes and new microbes are identified by use of modern molecular techniques. In medicine, microbiology's impact stays, whether it be emerging diseases, the development of new vaccines, drugs, and bioengineered organisms, the roles of viruses in cancer or the use of microbes to clean-up toxic wastes. An observation made by the renowned microbiologist Louis Pasteur, about 120 years ago, "Life would not long remain possible in the absence of microbes" seems to be even truer than ever. The threat of many infectious diseases is still a fact of life in spite of developments in economy, better drugs to treat the infections caused by them and vaccines for prevention. New pathogens are constantly being discovered and incurable infectious diseases continue to haunt us. New and incurable infectious diseases remain a worldwide problem.

Microbiology is an inherently valuable and useful discipline that offers an intimate view of an invisible world. The amount of information on microbiology is so vast that microbiology books have generally followed one of the two main tracks. The traditional books are usually very exhaustive encyclopedias of microbiologic facts which may serve as excellent reference works but are too long and detailed to be read by the typical medical student, who is trying to keep up with several classes simultaneously of other subjects within a limited period. Short concise microbiology books may not serve the purpose of a student because the amount of material covered in them may be even less than that provided in the class lectures of a typical medical microbiology course.

An overview of the book, written by Professor Surinder Kumar shows that, it has potential of emerging as a comprehensive review for graduate students, residents, and health professionals interested in infectious diseases. Coverage of fundamental aspects such as General Bacteriology, Immunology, Systemic Bacteriology, Virology, Medical Mycology, Miscellaneous and Diagnostic Medical Microbiology, appears to be well balanced and aptly illustrated with tables, illustrated figures, and photographs have been added throughout the text to help readers understand and retain information. I hope that the book will be received with the enthusiasm and will fulfill the needs of medical students, teachers, microbiologists and health professionals.

Vishwa Mohan Katoch
MD FNASc FAMS FASc FNA
Secretary to the Government of India
(Department of Health Research)
Ministry of Health and Family Welfare and
Director-General, ICMR

Preface

Microbiology is an extremely diverse discipline and can be a bewildering field to the novice. The traditional books seem to be exhaustive of microbiologic facts, are too long and detailed to be read by the typical medical student who is trying to keep up with several classes simultaneously of other subjects. On the contrary, some of the books are in brevity, too cursory and with insufficient information on microbiology not useful to a typical medical student. The microbiology text presented here, was written after 25 years of teaching medical students and searching for a book that was both readable and complete enough to meet their needs. It contains all of the information that is pertinent to medical students who are studying microbiology keeping in mind their examination. It also provides a solid background of microbiology while describing the organisms in a manner that is clinically relevant.

Although, the text was designed to teach undergraduate and postgraduate medical students, it should also serve as a review tool for individuals who are taking medical examinations and persons working in health-related professions, physicians and infectious disease scientists. The mass of material may appear overwhelming for undergraduate students but that is real life.

Microbiology has expanded beyond recognition with various medical specialty and it is not possible for any textbook to cover all aspects of medical microbiology in depth. The textbook is divided into seven sections, based on the major disciplines included within microbiology: General Bacteriology, Immunology, Systemic Bacteriology, Virology, Medical Mycology, Miscellaneous and Diagnostic Medical Microbiology. The chapters themselves are comprehensive yet free of unnecessary detail and provide the reader with a framework for understanding. Mycology and parasitology have continued to flourish and have blossomed into fields of study of their own rights. Therefore, parasitology has not been included in the book which has a sturdy independence. I shall be thankful for any comment or suggestions from students, teachers and all the readers of the book for further improvements.

Surinder Kumar

Acknowledgments

This book took years for writing but is a lifetime preparation. I would like to thank those who set example in teaching and prodding which helped me to develop a thirst for knowledge as well as methods for quenching that thirst. I am greatly indebted to a variety of mentors, friends and colleagues who encouraged me and gave valuable suggestions for improving the text. I am always indebted to my late brother Sant Swaran Dev whose advice, guidance and true life philosophy gave me strength and courage to continue my work. I would like to thank my family for patiently enduring the writing of this book, which seemed at times to be an endless process. I am especially grateful to my wife Dr Savita Kumari, Professor, Department of Internal Medicine, Postgraduate Institute of Medical Education and Research (PGIMER), Chandigarh, India, for her support and encouragement and to two dear children, Sourabh Kumar and Sanchit Kumar (both are now medical students), who gave me their childhood moments without any complaint for completing the book so I could retain my sanity.

I am particularly indebted to Dr Vishwa Mohan Katoch, Secretary to the Government of India (Department of Health Research), Ministry of Health and Family Welfare and Director-General, Indian Council of Medical Research (ICMR), New Delhi, India, who generously agreed to write a foreword for this book. My special thanks goes to Dr Sanjeev R Saigal, my PhD student and now my Research Associate ICMR, who stood by me every time. My sincere appreciation also goes to Mr Tarun Duneja (Director-Publishing), Mr KK Raman (Production Manager), Mr Sunil Kumar Dogra (Production Executive), Mr Neelambar Pant (Production Coordinator), Mr Ravinder Kumar, Mr Sanjeev Kumar, Mr Akhilesh Kumar Dubey, Mr Gyanendra Kumar and the entire team of M/s Jaypee Brothers Medical Publishers (P) Ltd, New Delhi, India, for their support in this project.

Contents

Section One: General Bacteriology

Section Two: Immunology

Section Three: Systemic Bacteriology

Section Four: Virology

Section Five: Medical Mycology

Section Six: Miscellaneous

Section Seven: Diagnostic Medical Microbiology

SECTION ONE

GENERAL BACTERIOLOGY

CHAPTER 1 Historical Development of Microbiology

History is bunk—(Henry Ford 1863-1947)
In the field of observation, chance favors only prepared minds –Louis Pasteur

LEARNING OBJECTIVES

After reading and studying this chapter, you should be able to:

- Discuss contributions of Antony van Leeuwenhoek.
- List contributions of Louis Pasteur.
- Describe contributions of Robert Koch.
- Discuss Koch's postulates.
- Describe contributions of Paul Ehrlich.

INTRODUCTION

Microbiology is the study of living organisms of microscopic size. **Medical microbiology** is the subdivision concerned with the causative agents of infectious disease of man, the response of the host to infection and various methods of diagnosis, treatment and prevention. The term microbe was first used by Sedillot in 1878, but now is commonly replaced by microorganisms.

INFECTION AND CONTAGION

Ancient Belief

Among ancient peoples, epidemic and even endemic diseases were believed to be supernatural in origin, sent by the gods as punishment for the sins of human kind. Sacrifices and lustrations to appease the anger of the gods were sought for the treatment and, more important, the prevention of disease. There was never any difficulty in finding particular sets of sins to justify a specific epidemic since humans are wilful and wantom by nature.

Concept of Contagion

Long before microbes had been seen, observations on communicable diseases had given rise to the concept of contagion: The spread of disease by contact, direct or indirect. This idea was implicit in the laws enacted in early biblical times to prevent the spread of leprosy.

Invisible Living Creatures Produced Disease

Varro in the second century BC later recorded the principle of contagion by invisible creatures. Roger Bacon, in the thirteenth century more than a millennium later, postulated that invisible living creatures produced disease. Fracastorius (1546), a physician of Verona concluded that communicable diseases were caused by living agents (germs) 'seminaria' or 'seeds'. Kircher (1659) reported finding minute worms in the blood of plague victims, but with the equipment available to him, it is more likely that what he observed were only blood cells. von Plenciz (1762) suggested that each disease was caused by a separate agents.

DISCOVERY OF MICROORGANISMS

Even before microorganisms were seen, some investigators suspected their existence and responsibility for disease. Among others, the Roman philospher **Lucretius (about 98-55 BC**) and the physician **Girolamo Fracastoro** (1478-1553) suggested that disease was caused by invisible living creatures.

First Observation of Microorganisms

As microbes are invisible to the unaided eye, direct observation of microorganisms had to await the development of the microscope.

Antony van Leeuwenhoek (1632-1723)

The credit for having first observed and reported bacteria belongs to Antony van Leeuwenhoek. Antoni van Leeuwenhoek, the Dutchman, was a draper and haberdasher in Delft, Holland. He had little education, but great patience and curiosity. His hobby was grinding lenses and observing diverse materials through them. He was the amateur microscopist and was the first person to observe microorganisms (1673) using a simple microscope. In 1683 he made accurate descriptions of various types of bacteria and communicated them to the

Royal Society of London. Their importance in medicine and in other areas of biology came to be recognized two centuries later.

Contributions of Antony von Leeuwenhoek

1. **He constructed the first microscope:** Consisted of a single biconvex lens that magnified about x200.
2. **The first person to observe microorganisms:** Microorganisms were first seen by Antony van Leeuwenhoek (1673) and he found many microorganisms in materials such as water, mud, saliva and the intestinal contents of healthy subjects, and he recognized them as living creatures (animalcules) and to Leeuwenhoek the world of "little animalcules" represented only a curiosity of nature.
3. **Accurate description of bacteria:** He first accurately described the different shapes of bacteria as cocci (spheres), bacilli (rods) and spirochetes (spiral filaments) and communicated them to Royal Society of London in 1683.

CONFLICT OVER SPONTANEOUS GENERATION

Spontaneous Generation (Abiogenesis)

From earliest times, people had believed in **spontaneous generation (abiogenesis)** that living organisms could develop from nonliving matter. Even great **Aristotle (384-322 BC)** thought animal could originate from the soil. This view was finally challenged by the Italian physician **Francesco Redi** (1626-1697) and proved that gauze placed over jar containing meat prevented maggots forming the meat. Similar experiments by others helped discredit the theory for larger organisms.

Evidence Pro

Some proposed that microorganisms arose by spontaneous generation though larger organisms did not. **John Needham (1713-1781** the English priest) **in 1745,** published experiments purporting the spontaneous generation (abiogenesis) of microorganisms in putrescible fluids. **Felix Pouchet (1859),** the French naturalist, claimed to have carried out experiments conclusively proving that microbial growth could occur without air contamination. This claim provoked **Louis Pasteur** (1822-1895) to settle the matter once and for all.

Evidence Con

Spontaneous Generation Experiment: Lazzaro Spallanzani (1729-1799), an Italian priest and naturalist opposed this view who boiled beef broth for an hour, sealed the flasks, and observed no formation of microbes. **Franz Schulze (1815-1873), Theodore Schwann (1810-1882), Georg Friedrich Schroder** and **Theodor von Dusch** attempted to counter such arguments.

Louis Pasteur (1822-1895) settled the matter once and for all. In a series of classic experiments, Pasteur proved conclusively that all forms of life, even microbes, arose only from their like and not *de novo*. Pasteur was able to filter microorganisms from the air and concluded that this was the source of contamination and finally, in **1859,** in public controversy with **Pouchet,** prepared boiled broth in flasks with long narrow gooseneck tubes that were open to the air. Air could pass but microorganisms settled in the gooseneck, and no growth developed in any of the flasks. If the necks were broken, growth commenced immediately. Pasteur had not only resolved the controversy by 1861 but also had shown how to keep solution sterile.

Tyndallization—John Tyndall (1820-1893): John Tyndall (1820-1893), the English physicist finally, dealt a final blow to spontaneous generation **in 1877.** He completed the story by proving that dust did not indeed carry germs and that if dust was absent, broth remained sterile if directly exposed to air. He was able to explain satisfactorily the need for prolonged heating to eliminate microbial life from infusions.

Heat stable form and **a heat-sensitive form**—He exposed infusions to heat for varying time and concluded that bacteria existed in two forms: **a heat stable form** and **a heat-sensitive form**. Heat-stable forms were destroyed either by prolonged or intermittent heating. Intermittent heating, now called **tyndallization,** killed both forms since the heat-stable forms changed to heat-sensitive forms between periods of heat treatment. This method of 'tyndallization' served to eliminate many of anomalies reported by the advocates of heterogenesis.

Heat-Resistant Forms as Spores

Ferdinand Cohn (1828-1898), the German botanist, discovered the evidence of heat-resistant forms as spores. Spores as well as vegetative forms were responsible for the appearance of microbial life in inadequately heated infusions.

ROLE OF MICROORGANISMS IN DISEASE

A firm basis for the casual nature of infectious disease was established only in the latter half of the nineteenth century. Fungi, being larger than bacteria, were the first agents to be recognized. **Agostino Bassi (1773-1856)** demonstrated in 1835 that a silkworm disease called muscardine was due to a fungal infection. **MJ Berkeley (1845)** proved that the great potato blight of Ireland was caused by a fungus. Following his success with the study of fermentation, Pasteur was asked by French government to investigate the **pebrine disease of silkworm** that was disrupting the silk industry. He showed that the disease was due to a protozoan parasite after several years of work.

Empirical Observations

The etiologic role of bacteria was first established with anthrax. **Pollender** (1849) and **Davaine** (1850) observed anthrax bacilli in the blood of animals dying of the disease.

Indirect transmission was recognized in the 1840s, when American poet-physician **Oliver Wendell Holmes** (1843) in Boston, USA and **Ignaz Semmelweis** in **Vienna (1846)** had independently concluded that puerperal sepsis was contagious. They blamed obstetricians moving with unwashed hands from one patient to the next for the prevalence of puerperal sepsis in hospitals. **Semmelweis** also identified its mode of transmission by doctors and medical students attending on women in labor in the hospital and had prevented it by the simple measure of washing hands in an antiseptic solution. But those pioneers encountered enormous resistance from the insulted physicians. Semmelweis was persecuted by medical orthodoxy and driven insane for the service to medicine and humanity.

Relationship of a Spirillum to relapsing fever—Obermeier (1872) discovered the relationship of a *Spirillum* to relapsing fever and demonstrated for the first time the presence of a pathogenic microorganism in the blood of a human being.

SCIENTIFIC DEVELOPMENT OF MICROBIOLOGY

The development of microbiology as a scientific discipline dates from Louis Pasteur, perfection on microbiological studies by Robert Koch, the introduction of antiseptic surgery by Lord Lister and contributions of Paul Ehrlich in chemotherapy.

Louis Pasteur (1822-95)

Louis Pasteur (1822-95) was born in the village of Dole, France on December 27, 1822 the son of humble parents. His father was a tanner. He was originally trained as a chemist, but his studies on fermentation led him to take interest in microorganisms. His discoveries revolutionized medical practice, although he never studied medicine.

Father of microbiology—Louis Pasteur (Fig. 1.1) is known as "**Father of microbiology**" because his contribution led to the development of microbiology as a separate scientific discipline.

Fig. 1.1: Louis Pasteur

Contributions of Louis Pasteur in Microbiology (Box 1.1)

1. Coined the term microbiology.
2. Proposed germ theory of disease.
3. Disapproved theory of spontaneous generation.
4. Developed sterilization techniques.
5. Developed methods and techniques for cultivation of microorganisms.
6. Studies on pebrine (silk worm disease), anthrax, chicken cholera and hydrophobia.
7. Pasteurization.
8. Coined the term vaccine.
9. Discovery of attenuation and chicken cholera vaccine.
10. Developed live attenuated anthrax vaccine.
11. Developed rabies vaccine.
12. Noticed Pneumococci.

Joseph Lister (1827-1912)

Joseph Lister was a professor of Surgery in Glasgo Royal Infirmary. He was impressed with Pasteur's study on the involvement of microorganisms in fermentation and putrefaction.

- **Developed a system of antiseptic surgery**—He developed a system of antiseptic surgery designed to prevent microorganisms from entering wounds. The approach was remarkably successful and transformed surgery after Lister published his findings in 1867. It also provided strong evidence for the role of microorganism in disease because phenol, which killed bacteria, also prevented wound infections.
- **Father of modern surgery**—He established the guiding principle of antisepsis for good surgical practice and was milestone in the evolution of surgical practice from the era of 'laudable pus' to modern aseptic techniques. For this work he is called the "Father of modern surgery"

Robert Koch (1843-1910)

Robert Koch (Fig. 1.2) was the German physician. The first direct demonstration of the role of bacteria in carrying disease came by the study of anthrax by Koch. Winner of the Nobel Prize in 1905, Robert Koch is known as "**Father of bacteriology**".

Contributions of Robert Koch

1. **Staining techniqes:** He described **methods for the easy microscopic examination of bacteria** in dried, fixed films stained with aniline dyes (1877).
2. **Hanging drop method:** He was the first to use **hanging drop method** by studying bacterial motility.

Box 1.1: Contributions of Louis Pasteur in microbiology

1. **Coined the term Microbiology**—Pasteur coined the term microbiology for the study of living organisms of microscopic size.
2. **Proposed germ theory of disease**—He established that putrefaction and fermentation was the result of microbial activity and that different types of fermentations were associated with different types of microoganisms (1857).
3. **Disapproved theory of spontaneous generation**—He disapproved the theory of spontaneous generation in 1860-61. In a series of classic experiments, Pasteur proved conclusively that all forms of life, even microbes, arose only from their like and not *de novo*.
4. **Developed sterilization techniques**—He introduced sterilization techniques and developed the steam sterilizer, hot-air oven and autoclave in the course of these studies.
5. **Developed methods and techniques for cultivation of microorganisms**. He showed that for successful cultivation it was necessary to discover a suitable growth medium and to establish optimal conditions of temperature, acidity or alkalinity, and oxygen tension.
6. **Studies on pebrine (silkworm disease), anthrax, chicken cholera and hydrophobia**.
7. **Pasteurization**—He devised the process of destroying bacteria, known as pasteurization (1863-65). This process (pasteurization) is employed to preserve milk and certain other perishable foods throughout the civilized world today.
8. **Coined the term vaccine**—It was Pasteur who coined the term vaccine for such prophylactic preparations to commemorate the first of such preparations namely cowpox, employed by Jenner for protection against smallpox.
9. **Discovery of the process of attenuation and chicken cholera vaccine**—An accidental observation that chicken cholera bacillus cultures left on the bench for several weeks lost their pathogenic, property but retained their ability to protect the birds against subsequent infection by them, led to the discovery of the process of attenuation and the development of live vaccines.
10. **Developed live attenuated anthrax vaccine**—He attenuated cultures of the anthrax bacillus by incubation at high temperature (42-43°C) and proved that inoculation of such cultures in animals induced specific protection against anthrax. The success of such immunization was dramatically demonstrated by a public experiment on a farm at Pouilly-le-Fort (1881) during which vaccinated sheep, goats and cows were challenged with a virulent anthrax bacillus culture. All the vaccinated animals survived the challenge, while an equal number of unvaccinated control animals succumbed to it.
11. **Developed rabies vaccine**—The crowning achievement of Pasteur was the successful application of the principle of vaccination to the prevention of rabies, or hydrophobia, in human beings and developed Pasteur rabies vaccine in 1885. He did not know that rabies was caused by a virus, but he managed to develop a live attenuated vaccine for the disease.
12. **Noticed Pneumococci**—Pneumococci were first noticed by Pasteur and Sternberg independently in 1881.

3. **Methods for isolating pure cultures of bacteria:** He devised a **simple method for isolating pure cultures of bacteria** by plating out mixed material on a solid culture medium and to isolate pure cultures of pathogens.
4. Discoveries of the causal agents of anthrax (1876), tuberculosis (1882), and cholera (1883).
5. **Koch's postulates:** It was necessary to introduce criteria for proving the claims that a microorganism isolated from a disease was indeed causally related to it. Robert Koch proved that microorganisms cause disease. Koch used the criteria proposed by his former teacher, Jacob Henle (1809-1885), to establish the relationship between *Bacillus anthracis* and anthrax and published his findings in 1876. (Fig. 1.3). His criteria for proving the causal relationship between a microorganism and a specific disease are known as Koch's postulates (1876), which are used today to prove that a particular microorganism causes a particular disease (Box 1.2).
6. **Koch's phenomenon:** Koch (1890) observed that a guinea pig already infected with the bacillus responded with an exaggerated response when injected with the tubercle bacillus or its protein. This hypersensitivity reaction is known as Koch's phenomenon.

Fig. 1.2: Robert Koch

Important Discoveries by other Scientists

Koch began to gather round him the group of followers who were destined to introduce his methods into many laboratories throughout the world. **Hansen** (1874) described the leprosy bacillus; **Neisser (1879)** discovered the gonococcus in the pus discharge from urethra; **Eberth (1880)** observed the typhoid bacillus; **Alexander Ogston (1881)** described the staphylococci in abscess and suppurative lesions; **Loeffler (1884)** observed and described the diphtheria bacillus; **Nicolaier (1884)** observed the tetanus bacillus in soil; **Rosenbach (in 1886)** demonstrated the tetanus bacillus with round terminal spore; **Fraenkel (1886)** described the pneumococcus; in **1887 Weichselbaum** described and isolated the meningococcus from the spinal fluid of a patient; in 1887 **Bruce** identified the causative agent of malta fever; in 1905 **Schaudin and Hoffman** discovered the syphilis.

Box 1.2: Koch's postulates

Koch's postulates are a series of guidelines for the experimental study of infectious disease. According to these, a microorganism can be accepted as the causative agent of an infectious disease only (Fig. 1.3) if the following conditions are satisfied:

Postulate 1: The organism should be regularly found in the lesions of the disease.

Postulate 2: It should be possible to isolate the organism in pure culture from the lesions.

Postulate 3-Inoculation of the pure culture into suitable laboratory animals should reproduce the lesion of the disease.

Postulate 4: It should be possible to reisolate the organism in pure culture from the lesions produced in the experimental animals.

Subsequently an additional fifth criterion introduced states that specific antibodies to the organism should be demonstrable in the serum of patients suffering from the disease.

Limitations of Koch's postulates: Even today Koch's postulates are considered whenever a new infectious disease arises. These criteria have proved invaluable in identifying pathogens, but they cannot always be met, for example, some organisms (including all viruses) cannot be grown on artificial media, and some are pathogenic only for man. *Mycobacterium leprae*, a causative agent of leprosy, has not been cultured on artificial medium so far and not fulfilling Koch's postulates.

GOLDEN ERA OF MEDICAL BACTERIOLOGY

Koch's postulates permitted Koch and his students to identify many of the causes of the most infectious diseases of humans and animals. Koch had now assembled the techniques needed to investigate the bacterial causes of many communicable diseases. The powerful methodology developed by Koch introduced the **"Golden era of medical bacteriology"**. By 1882 Koch had used these techniques to isolate bacillus of tuberculosis. There followed a golden era of 30 to 40 years in which most of the major bacterial pathogens were isolated.

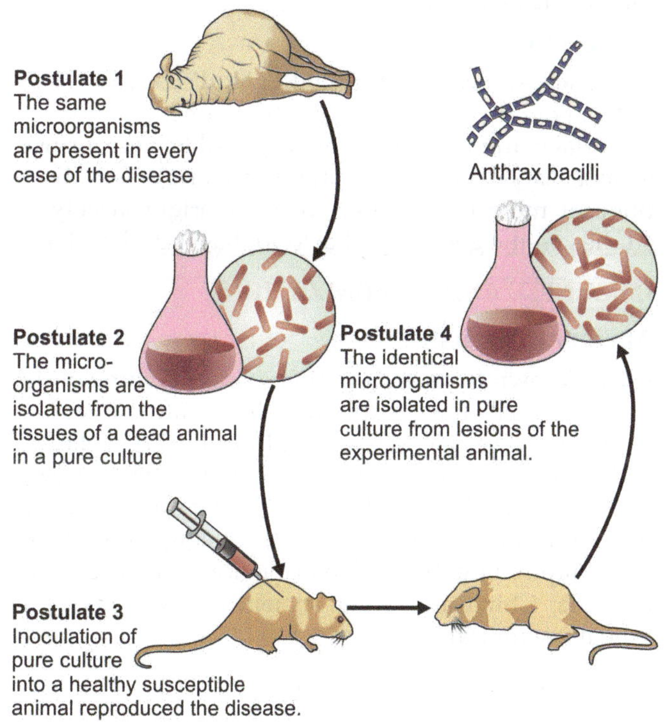

Fig. 1.3: Demonstration of Koch's postulates

PAUL EHRLICH (1854-1915)

Paul Ehrlich, an outstanding German Scientist and genius of extraordinary activity also known as **"Father of chemotherapy"**.

Contributions of Paul Ehrlich

1. **Stains to cells and tissues:** He applied stains to cells and tissues for the purpose of revealing their function.
2. **Acid-fastness of tubercle bacillus:** He reported the acid-fastness of tubercle bacillus.
3. **Methods of standardizing toxin and antitoxin:** He introduced methods of standardizing toxin and antitoxin and coined the term minimum lethal dose.
4. **Side chain theory of antibody production:** He proposed side chain theory of antibody production.
5. **Salvarsan introduction:** He introduced *salvarsan*, an arsenical compound, sometimes called the **'magic bullet'**. It was capable of destroying the spirochete of syphilis with only moderate toxic effects. He continued his experimentation until 1912 when he announced the discovery of **neosalvarsan.** Thus

he created a new branch of medicine known as **chemotherapy.**

GOLDEN AGE OF MICROBIOLOGY (1854-1914)

For about 60 years, beginning with the work of Pasteur, there was an explosion of discoveries in microbiology. The period from **1854 to 1914** has been appropriately named the **Golden Age of Microbiology**. During this period, rapid advances, spearheaded mainly by Pasteur and Robert Koch, led to the establishment of microbiology as a science.

DISCOVERY OF VIRUSES

As a science, virology evolved later than bacteriology. Although the physical nature of viruses was not fully revealed until the invention of the electron microscope, the infections they cause have been known and feared since the dawn of history.

Infectious Agents Smaller than Bacteria

The existence of viruses became evident during the closing years of the nineteenth century, the infectious agents of numerous diseases were being isolated and many infectious diseases had been proved to be caused by bacteria. But there remained a large number of diseases for which no bacterial cause could be established until it was realized that the responsible agents were smaller than bacteria.

Various Infections

Rabies in Dogs

Pasteur had suspected that **rabies in dogs** could be caused by a microbe too small to be seen under the microscope.

Tobacco Mosaic Disease

Iwanowski (1892), Russian scientist and **Martinus Beijrinck** (1898) in Holland, attributed the cause of tobacco- mosaic disease to the infectious agents in bacteria-free filtrates to be living, but fluid—contagium vivum fluidum and introduced the term virus (Latin for '*poison*') for such filterable infectious agents.

Foot and Mouth Disease of Cattle

Friedrich Loeffler and **Paul Frosch** at the same time in 1898 in Germany found that foot and mouth disease of cattle was also caused by a similar filter-passing virus.

Yellow Fever

The discovery of first human disease proved to have a viral etiology was **yellow fever**. The US Army Yellow Fever Commission under **Walter Reed** in Cuba (1902) showed that this human disease (yellow fever) was not only a filterable virus, but also transmitted through the bite of infected mosquitos.

Electron Microscope

Viruses could not be visualized under the light microscope or grown in culture media so investigation of viruses and the disease caused by them was rendered difficult. Larger viruses could be seen under light microscope after appropriate staining, but their detailed morphology could only be studied by electron microscope by **Ruska (1934)**.

Cultivation of Viruses

The technique of growing them on chick embryos was developed by **Goodpasture in 1930s**. The use of living human and animal tissue cells for the *in vitro* culture of viruses was developed by **John Enders (1949) and others.**

Virus infection and malignancy

Leukemia

Vilhelm Ellerman and **Oluf Bang (1908)** in Copenhagen put forth the possibility that virus infection could lead to malignancy by reporting that leukemia could be transmitted between chickens by cell-free filtrates.

Sarcoma in Fowls

Peyton Rous (1911) three years later isolated a virus causing sarcoma in fowls. Several viruses have been blamed to cause natural and experimental tumors in birds and animals. Viruses also cause malignant transformation of infected cells in tissue culture.

Viral Oncogenesis

The discovery of viral and cellular oncogenes have put forth the possible mechanisms of viral oncogenesis. Positive proof a virus causing of human malignancy was established when the virus of human T-cell leukemia was isolated in 1980.

Bacteriophages

Frederick W Twort (1915) and **Felix d' Herelle (1917)** independently discovered a lytic phenomenon in bacterial cultures. The agents responsible were termed *bacteriophages* (virus that attack bacteria). The discipline of molecular biology owes it origin largely to studies on the genetics of bacteriophages and bacteria.

IMMUNITY AND IMMUNIZATION

Ancient Knowledge

It was known from ancient times since the time of the ancient Greeks that people who have suffered from a distinctive disease, such as smallpox, measles, plague, yellow fever and various other infectious diseases, resisted it on subsequent exposures and rarely contract it second time. The practice of producing a mild form of smallpox intentionally (variolation) was prevalent in India, China and other ancient civilizations from time immemorial.

Edward Jenner (1749-1823)

The first scientific attempts at artificial immunizations in the late eighteenth century by **Edward Jenner (1749-1823)** from England. He observed the immunity to smallpox in milkmaids who were exposed to occupational cowpox infection, introduced the technique of vaccination using cowpox material (1796). It was on May 14, 1796, that Jenner extracted the contents of a pustule from the arm of a cowpox-infected milkmaid, **Sarah Nelmes**, and injected it into the arm of eight- year-old James Phipps. Jenner's vaccination paved the way for the ultimate eradication of smallpox. Edward Jenner is known as the "**Father of immunology**".

Live Vaccines

Further work on immunization was carried out by Louis Pasteur and derived attenuated (reduced in virulence) live *vaccines* for **fowl cholera, anthrax, swine erysipelas and rabies.** He called them **vaccine** in honor of Jenner's work with cowpox or vaccinia (Latin *vacca* for cow). In 1881 he made a convincing controlled trial of his anthrax vaccine. Today, attenuated live vaccines are used with outstanding success against such diseases as **tuberculosis, poliomyelitis, measles and yellow fever.**

Vaccine for Hydrophobia

Pasteur's development of a vaccine for hydrophobia made the greatest impact in medicine. This was acclaimed throughout the world. The Pasteur Institute, Paris was built by public contributions and similar institutions were established soon in many other countries for the preparation of vaccines and for the investigation of infectious diseases.

Antibodies and Complement

Nuttal (1888) observed that defibrinated blood had a bactericidal effect and **Buchner (1889)** noticed that this effect was abolished by heating the sera for one hour at 55°C. This heat-labile bactericidal factor was termed '*alexine*'. The first step in elucidating the mechanisms of acquired immunity was the discovery of antibodies by **Emil von Behring and Shibasaburo Kitasato (1890)** in the sera of animals which had received sublethal dose of diphtheria or tetanus toxoid. **Pfeiffer (1893)** demonstrated bactericidal effect *in vivo* by injecting live cholera vibrios intraperitoneally in guinea pigs previously injected with killed vibrios. **Bordet (1895)** proved the humoral nature of such activity. He defined two components in this reaction, the first being heat stable substance *'antibody'* found in the immune sera and the second being heat labile identical with **Buchner's** alexine, subsequently named *'complement'*.

Cellular Concept of Immunity

Elie Metchnikoff (1883) discovered the phenomenon of phagocytosis and developed the cellular concept of immunity. **Paul Ehrlich** hypothesized that immunity could be explained by presence of noncellular components of blood. **Wright (1903)** discovered opsonization, in which antibodies and phagocytic cells act in conjunction. Both Metchnikoff and Ehrlich shared a Nobel Prize in 1908 for their contributions to the emerging science of immunology. The pioneering work of **Landsteiner** laid the foundation of immunochemistry.

Allergy

Koch's phenomenon: Koch (1890) had noticed that when the tubercle bacillus or its protein was injected into a guinea pig already infected with the bacillus, an exaggerated response took place—a hypersensitivity reaction known as **Koch's phenomenon.**

Anaphylaxis: Portier and Richet (1902), studying the effect of the toxic extracts of sea anemones in dogs made the paradoxical observation that dogs which had prior contact with the toxin were, abnormally sensitive to even minute quantities of it subsequently. This phenomenon was termed **anaphylaxis** and led to the development of the discipline of allergy.

Selection Theory of Antibody

In 1955, Jerne proposed the **'natural selection theory'** of antibody synthesis. **Burnet (1957)** modified this into *clonal selection theory*.

Immunological Surveillance

Burnet (1967) developed the concept of *immunological surveillance* based on the original suggestion of **Thomas (1959)**, according to which the primary function of the immune system is to preserve the integrity of the body, seeking and destroying all 'foreign antigens'. Malignancy was thought to be a failure of this function.

Transplantation

Another aspect of this role of immunity is in the rejection of homografts. Understanding of the immunological basis of transplantation, largely due to the work of **Medawar and Burnet**, made successful transplants possible by elective immunosuppression and proper selection of donors based on histocompatibility.

SEROTHERAPY AND CHEMOTHERAPY

Antisera

The work of **Behring and Kitasato** led to the successful use of antisera raised in animals for the treatment of patients with diphtheria, tetanus, pneumonia and other diseases. Antisera were the only specific therapeutic agents available for the management of infectious diseases till **Domagk (1935)** initiated scientific chemotherapy with the discovery of prontosil.

Magic Bullet

Ehrlich (1909) discovered **salvarsan (arsphenamine),** sometimes called the **'magic bullet'** was capable of destroying the spirochete of syphilis with only moderate toxic effects. In 1912 he announced the discovery of **neosalvarsan.** This gave him the title, "**Father of chemotherapy**".

Antibiotics—A Fortunate Accident

The modern era of antibiotics developed only after **Gerhard Domagk (1895-1964)** found that prontosil (the forerunner of sulfonamides) had a dramatic effect on streptococcal infection in 1935. **Sir Alexander Fleming (1881-1955)** made accidental discovery that the fungus *Penicillium notatum* produces a substance which destroys staphylococci. In the 1940s, **Florey and Chain** and their associates demonsrates its clinical value. This was the beginning of the antibiotics era. **Selman Waksman** exploited the potential for antibiotic production among soil microorganisms in the 1940s. Within 25 years of these discoveries, most of the major groups of antimicrobial agents had been recognized and more recent developments have chiefly involved chemical alteration of existing molecules.

Microbes Control is Far more Difficult

The global eradication of smallpox inspired visions of similar campaigns against other major pestilences. But it was realized that controlling microbes was a far more difficult than was imagined when new infectious diseases began to appear. Unceasing vigilance appears essential to protect man from microbes because of problems of emergence of drug resistance and appearance of new agents of infectious disease. The most notorious of these is undoubtedly the human immunodeficincy virus (HIV), the causative agent of acquired immune deficiency sndrome (AIDS). The rise and fall of AIDS produces a sobering reminder of the potential impact of microbial disease.

DEVELOPMENT OF MOLECULAR BIOLOGY AND MOLECULAR GENETICS

Molecular Biology

Oswald Avery with Colin Macleod and Maclyn McCarty in 1944 showed that deoxyribonucleic acid (DNA) transformed nonvirulent pneumococci to virulent organisms. This discovery of chemical nature of hereditary material heralded the beginning of the merger of microbiology and molecular biology

Recombinant DNA Technology

In the 1970s new discoveries in microbiology led to the development of recombinant DNA technology and genetic engineering from work in microbial genetics and molecular biology.

NOBEL PRIZES AWARDED FOR RESEARCH IN MICROBIOLOGY

The number of Nobel laureates in Medicine and Physiology for their contribution in microbiology is evidence of the positive contribution made to human health by the science of microbiology. About one-third of these have been awarded to scientists working on microbiological problems (Table 1.1).

KNOW MORE

- Microbes, also called microorganisms, are minute living things that individually are usually too small to be seen with the unaided eye. The group includes bacteria, fungi (yeasts and molds), protozoa, and microscopic algae). It also includes viruses, those noncellular entities sometimes regarded as being at the border between life and nonlife.

Antony von Leeuwenhoek (1632-1723): His observational report were enthusiastic and accurate and created some interest at the time, but unfortunately Leeuwenhoek treated these investigations as a hobby and did not really found as a science because he kept his methods secret and left no students to continue his work.

- Although **Fracastoro** and others had suggested that invisible organisms produced disease, most believed that disease was due to causes such as supernatural forces, poisonous vapors called miasmas, and imbalances between the four humors thought to be present in the body.

☞ KEY POINTS

- Microbiology is the study of living organisms of microscopic size.
- Antony van Leeuwenhoek was the first person to describe microorganisms.
- The spontaneous generation of microorganisms was disproved by Spallanzani, Pasteur, Tyndall, and others.
- The work of Bassi, Pasteur, Koch, and others supported the germ theory of disease. Lister provided indirect evidence with his development of antiseptic surgery.
- **Louis Pasteur:** Pasteur showed that fermentations were caused by microorganisms and that some microorganisms could live in the absence of oxygen. He is known as **"Father of Microbiology". Contributions of Louis Pasteur in Microbiology** are very important.
- **Joseph Lister** developed a system of antiseptic surgery. For this work he is called the **"Father of modern surgery"**
- **Robert Koch**
  a. Koch developed the techniques required to grow bacteria on solid media and to isolate pure cultures of pathogens.
  b. Koch's postulates are used to prove a direct relationship between a suspected pathogen and a disease.
- **Paul Ehrlich** is known as **"Father of chemotherapy"**.
- The existence of viruses became evident during the closing years of the nineteenth century.

Table 1.1: Nobel laureates for research in microbiology		
Year	*Nobel laureates*	*Contribution*
1901	Emil A von Behring	Developed a diphtheria antitoxin.
1902	Ronald Ross	Discovered how malaria is transmitted.
1905	Robert Koch	Tuberculosis—discovery of causative agent.
1907	CLA Laveron	Discovery of malaria parasite in an unstained preparation of fresh blood.
1908	Paul Ehrlich and Elie Metchnikoff	Developed theories on immunity. Described phagocytosis, the intake of solid materials by cells.
1913	Charles Richet	Anaphylaxis.
1919	Jules Bordet	Discovered roles of complement and antibody in cytolysis, developed complement fixation test.
1928	Charles Nicolle	Typhus exanthematicus
1930	Karl Landsteiner	Described ABO blood groups; solidified chemical basis for antigen-antibody reactions.
1939	Gerhardt Domagk	Antibacterial effect of prontosil.
1945	Alexander Fleming, Ernst Chain, and Howard Florey	Discovered penicillin.
1951	Max Theiler	Yellow fever vaccine
1952	Selman A Waksman	Development of streptomycin. He coined the term 'antibiotic'.
1954	John F Enders, Thomas H Weller, and Frederick C Robbins	Cultured poliovirus in cell cultures.
1960	Sir Macfarlane Burnet and Sir Peter Brian Medawar	Immunological tolerance, clonal selection theory
1962	James D Watson, Frances HC Crick, And Maurice AF Wilkins	Double helix structure of deoxyribonucleic acid (DNA).
1966	Francois Jacob, Andre Lwoff and Jacques Monod	Regulatory mechanisms in microbial genes (concept of 'lac operon').
1966	Peyton Ross	Viral oncogenes (avian sarcoma)
1968	Robert Holley, Har Gobind Khorana, and Marshall W Nirenberg	Genetic code
1969	Max Delbruck, AD Hershey and Salvador Luria	Mechanism of virus infection in living cells
1972	Gerald M Edelman and Rodney R Porter	Described the nature and structure of antibodies.
1975	David Baltimore, Renato Dulbecco and Howard M Temin	Interactions between tumor viruses and genetic material of the cells.
1977	Rosalyn Yalow	Developed inmmunoassay
1980	Baruj Benacerraf, Jean Dausset and George Snell	HLA antigens
1984	Cesar Milstein, Georges Kohler Neils Jerne	Developed hybridoma technology for production of monoclonal antibodies.
1987	S Tonegawa	Described the genetics of antibody production.
1989	J Michael Bishop and Harold E Varmus	Discovered cancer-causing genes called oncogenes.
1990	Joseph E Murray and E Donnall Thomas	Performed the first successful organ transplants by using immunosuppressive agents.
1993	Kary B Mullis	Discovered the polymerase chain reaction (PCR) to amplify DNA.
1996	Peter C Doherty and Rolf M Zinkernagel	Cell mediated immune defences
1997	Stanley B Prusiner	Prion discovery
2001	Leland H Hartwell, Paul M Nurse, and R Timothy Hunt	Discovered genes that encode proteins regulating cell division
2005	Barry J Marshall and J Robin Warren	*Helicobacter pylori* and its role in gastritis and peptic ulcer disease

Contd....

Contd....

2007	Mario R Capecchi, Oliver Smithies and Sir Martin J Evans	Creation of knockout mice for stem cell research
2008	Luc Montagnier and Francoise Barre-Sinoussi	Discovery of human immunodeficiency virus
	Herald zur Hausen	Human papillomaviruses causing cervical cancer

- Vaccines against anthrax and rabies were made by Pasteur; von Behring and Kitasato prepared antitoxins for diphtheria and tetanus.
- **Edward Jenner** is known as the **"Father of immunology"**.
- Ehrlich is given the title, **"Father of chemotherapy"**.
- In the twentieth century microbiology contributed greatly to the fields of biochemistry and genetics. It also helped stimulate the rise of molecular biology.
- The positive contribution has been made to human health by the science of microbiology.

IMPORTANT QUESTIONS

I. Write short notes on:
   (a) Contributions of Antony van Leeuwenhoek
   (b) Contributions of Louis Pasteur
   (c) Contributions of Robert Koch
   (d) Koch's postulates
   (e) Contributions of Edward Jenner
   (f) Contributions of Paul Ehrlich
   (g) Name four Nobel laureates in Microbiology

FURTHER READING

Benacerraf B, et al. A history of bacteriology and immunology. London: William Heinemann, 1980.

Brock TD. Milestones in microbiology. American Society for Microbiology, Washington, DC, 1999.

Bulloch W. The history of bacteriology, London Oxford University Press, 1938.

Collard P. The development of microbiology. Cambridge University Press, 1976

deKruif P. Microbe Hunters. London: Hutchison, 1958.

Foster WD. A history of medical bacteriology and immunology. London: Cox and Wyman, 1970.

Mann J. The elusive magic bullet: the search for the perfect drug. Oxford, Oxford University Press, Oxford 1999.

Parish HJ. Victory with vaccines—The story of immunisation. London: Livingstone, 1968.

Waterson AP, Wilkinson L. An introduction to the history of virology. London: Cambridge University Press, 1978.

Williams G. Virus Hunters. London: Hutchison, 1960.

CHAPTER 2

Microscopy

Learning Objectives

After reading and studying this chapter, you should be able to:
- Discuss microscopic methods.
- Explain the principle and describe uses of the following: darkfield microscopy; phase-contrast microscopy; fluorescent microscopy and electron microscopy.

INTRODUCTION

Microscope is an optical instrument used to magnify (enlarge) minute objects or microorganisms which cannot be seen by naked eye. The study of the morphology of bacteria requires the use of microscopes. In general, microscopy is used in microbiology for two basic purposes: the **initial detection of microbes** and **the preliminary or definitive identification of microbes**.

MICROSCOPY: INSTRUMENTS

Microscopic Methods

A. Light microscopy
1. Bright-field (light) microscope
2. Dark-field microscopy
3. Phase-contrast microscopy
4. Fluorescent microscopy
5. Confocal

B. Electron microscopy
1. Transmission
2. Scanning

C. Scanning probe microscopy

A. Light Microscopy

Light microscopy refers to the use of any kind of microscope that uses visible light to observe specimens. Here we examine several types of light microscopy. A modern **compound light microscope (LM)** has a series of lenses and uses visible light as its source of illumination (Fig. 2.1).

Principles of Light Microscopy

In light microscopy, light typically passes through a specimen and then through a series of magnifying lenses.

1. The Bright-Field Microscope

The most common type of light microscope, and the easiest to use, is the **bright-field microscope,** which evenly illuminates the field of view.

Compound Light Microscopy

A series of finely ground lenses (Fig. 2.1) forms a clearly focused image that is many times larger than the specimen itself. This magnification is achieved when light rays from an **illuminator,** the **light source** used to illuminate the specimen positioned on a stage, pass through a **condenser,** which has lenses that direct the light rays through the specimen. From here, light rays pass into the **objective lenses,** the lenses closest to the specimen. The image of the specimen is magnified again by the **ocular lens, or *eyepiece*.** In bright-field microscopy, the

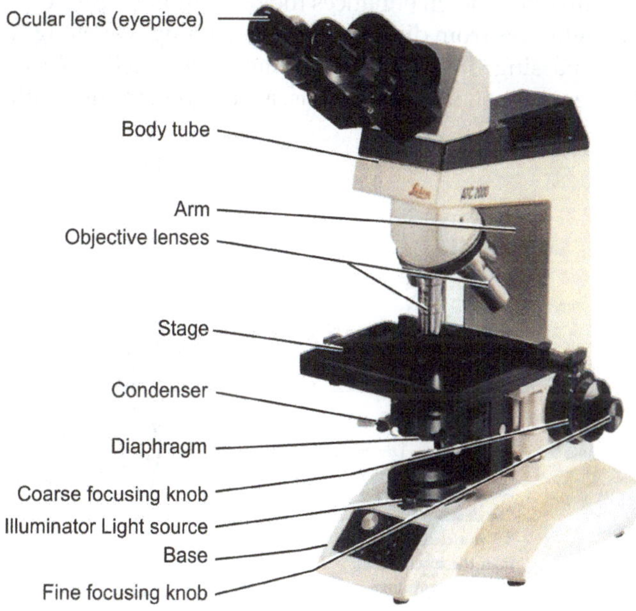

Fig. 2.1: Principal parts of compound light microscope

specimen is visualized by transillumination, with light passing up through the condenser to the specimen. The image is then magnified first by the objective lens and then by the ocular lens.

Magnification

We can calculate the **total magnification** of a specimen by multiplying the objective lens magnification (power) by the ocular lens magnification (power). Three different objective lenses are commonly used: **low power × 10); high dry × 40); and oil immersion × 100).** Most oculars magnify specimens by a factor of 10. Multiplying the magnification of a specific objective lens with that of the ocular, we see that the total magnifications would be 100 × for low power, 400 × for high power, and 1000 × for oil immersion.

Resolution

The limitation of bright-field microscopy is the resolution (also called **resolving power)** of the image (i.e. the ability to distinguish that two objects are separate and not one). The resolving power of a microscope is determined by the wavelength of light used to illuminate the subject and the angle of light entering the objective lens (referred to as the numerical aperture). A general principle of microscopy is that the shorter the wavelength of light used in the instrument, the greater the resolution.

Immersion Oil

The white light used in a compound light microscope has a relatively long wavelength and cannot resolve structures smaller than about 0.2 µm. To achieve high magnification (100 ×) with good resolution, the objective lens must be small. Immersion oil is placed between the glass slide and the oil immersion objective lens. The immersion oil has the same refractive index as glass, so the oil becomes part of the optics of the glass of the microscope. The oil enhances the resolution by preventing light rays from dispersing and changing wavelength after passing through the specimens. A specific objective lens, the oil immersion lens, is designed for use with oil; this lens provides 100 × magnification on most light microscopes.

Uses

Bacteria may be examined under light microscope, either in the **living state or after fixation and staining.** The arrangement, motility and approximate size of the organisms can be observed by the examination of wet films or 'hanging drops'.

2. Dark-Ground (Dark-Field) Microscope

Dark-field microscopy is frequently performed on the same microscope on which bright-field microscopy is performed. Instead of the normal condenser, a dark-field microscope uses a dark-field condenser that contains an opaque disk. The disk blocks light that would enter the objective lens directly. Only light that is reflected off (turned away from) the specimen enters the objective lens (Fig. 2.2B). Because there is no direct background light, the specimen appears light against a dark background. This creates a "dark-field" that contrasts against the highlighted edge of the specimens and results when the oblique rays are reflected from the edge of the specimen upward into the objective of the microscope.

Use

This technique is particularly valuable for observing organisms such as *Treponema pallidum,* a spirochete which is less than 0.2 µm in diameter and therefore cannot be observed with direct light.

Disadvantage

Internal structure of organisms cannot be studied because light passes around rather than through organisms.

3. Phase-Contrast Microscopy (Fig. 2.2C)

Principle

The principle of phase-contrast microscopy is based on the wave nature of light rays, and the fact that light rays can be *in phase* (their peaks and valleys match) or out of phase. In a phase-contrast microscope, one set of light

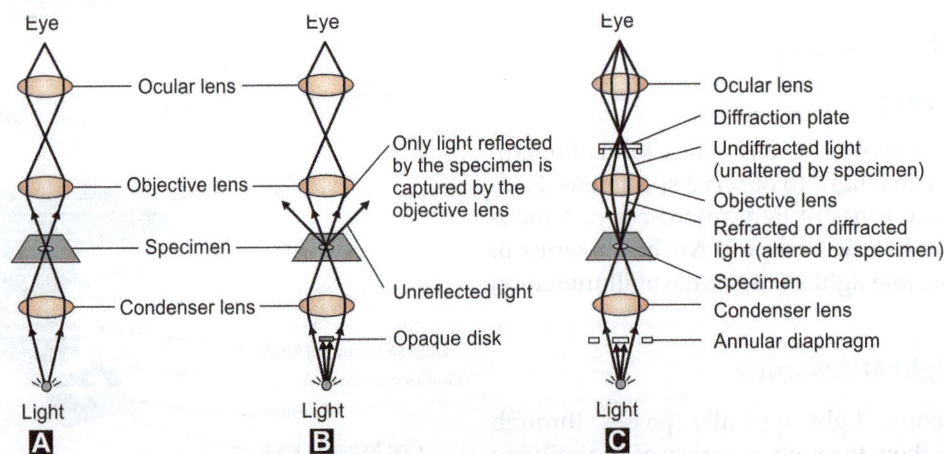

Fig. 2.2: (A) Bright-field, (B) dark-field, and (C) phase-contrast microscopy. The illustrations show the contrasting light pathways of each of these types of microscopy

rays comes directly from the light source. The other set comes from light that is reflected or diffracted from a particular structure in the specimen. (*Diffraction* is the scattering of light rays as they "touch" a specimen's edge. The diffracted rays are bent away from the parallel light rays that pass farther from the specimen). When the two sets of light rays-direct rays and reflected or diffracted rays are brought together, they form an image of the specimen on the ocular lens, containing areas that are relatively light (in-phase), through shades of gray, to black (out of phase). Through the use of annular rings in the condenser and the objective lens, the differences in phase are amplified so that in-phase light appears brighter than out-of-phase light. The special phase condenser consists of annular diaphragms on a rotating disk fitted to the bottom of the condenser.

Advantage

Phase-contrast microscopy improves the contrast, the internal structures of a cell become more sharply defined and makes evident the structures within cells that differ in thickness or refractive index.

Uses

1. To study unstained living cells
2. Detailed examination of internal structures in living microorganisms.
3. To study flagellar movements and motility of bacteria and protozoans
4. To study intestinal and other live protozoa such as amoebae and Trichomonas
5. To examine fungi grown in culture.

Differential Interference Contrast (DIC) Microscopy

Differential interference contrast (DIC) microscopy is similar to phase-contrast microscopy in that it uses differences in refractive indexes. However, in a DIC microscope two beams of light are used instead of one. In addition, prisms split each light beam, adding contrasting colors to the specimen. Therefore, the resolution of a DIC microscope is higher than that of a standard phase-contrast microscope. Also, the image is brightly colored and appears nearly three-dimensional.

4. Fluorescent Microscopy

Fluorescence microscopy takes advantage of fluorescence, the ability of substances to absorb short wavelengths of light (ultraviolet) and give off light at a longer wavelength (visible). If tissues, cells or bacteria are stained with a fluorescent dye and are examined under the microscope with ultraviolet light instead of ordinary visible light, they become luminous and are seen as bright objects against a dark background. The color that the cells will appear depends on the type of dye and the light filters. The fluorescence microscope is used to observe cells or other material that are either naturally fluorescent or have been stained or tagged with fluorescent dyes.

Principal Use

The principal use of fluorescence microscopy is a diagnostic technique called the **fluorescent antibody (FA) technique**, or **immunofluorescence** employed for detection of antigen **(direct fluorescent antibody technique)** and antibodies **(indirect fluorescent antibody methods).**

Epifluorescence Microscope

A common variation of the standard fluorescence microscope is the epifluorescence microscope, which projects the ultraviolet light through the objective lens and onto the specimen. Because the light is not transmitted through the specimen, cells can be observed attached to soil particles or other opaque materials.

5. Confocal Microscopy

In confocal microscopy, lenses focus a laser beam to illuminate a given point on one vertical plane of a specimen. Like fluorescent microscopy, specimens are stained with fluorochromes so they will emit, or return light. Mirrors then scan the laser beam across the specimen, illuminating successive regions and planes until the entire specimen has been scanned. Each plane corresponds to an image of one fine slice of the specimen. A computer then assembles the data and constructs a three-dimensional image, which is displayed on a screen. In effect, this microscope is a miniature CAT scan for cells.

Frequently, the specimens are first stained or tagged with a fluorescent dye. By using certain fluorescent tags that bind specifically to a given protein or other compound, the precise cellular location of that compound can be determined. In some cases, multiple different tags that bind to specific molecules are used, each having a distinct color.

Uses

i. To obtain three-dimensional images of entire cells and cellular components
ii. **To evaluate cellular physiology—**by monitoring the distributions and concentrations of substances such as ATP and calcium ions.

B. Electron Microscopy

- Electron microscopy is in some ways comparable to light microscopy. Rather than using glass lenses, visible light, and the eye to observe the specimen, the electron microscope uses electromagnetic lenses, electrons, and a fluorescent screen to produce the magnified image (figure). That image can be captured on photographic film to create an electron photomicrograph.
- The superior resolution of the electron microscope is due to the fact that electrons have a much shorter wavelength than the photons of white light.

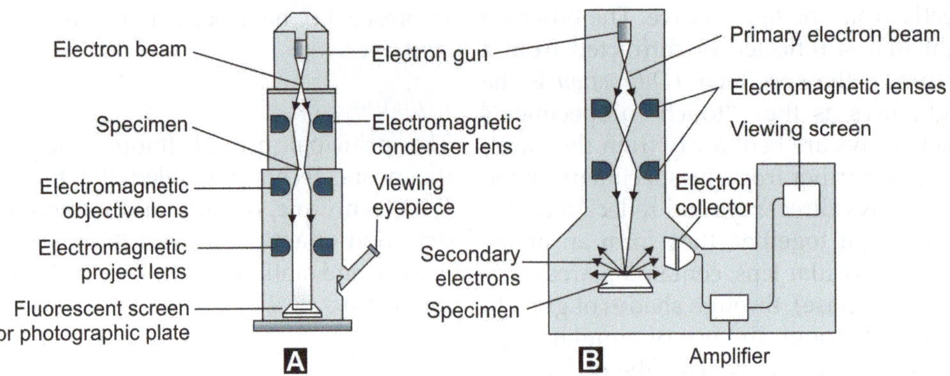

Fig. 2.3: (A) Transmission and (B) scanning electron microscopy. The illustrations show the pathways of electron beams used to create images of the specimens

- The electron beam is focused by circular electromagnets, which are analogous to the lenses in the light microscope. The object which is held in the path of the beam scatters the electrons and produces an image which is focused on a fluorescent viewing screen.
- The wavelength of electrons used in an EM is 0.005 nm, as compared to 500 nm with visible light, i.e. about 100,000 times shorter than that of ordinary light. Theoretically, if conditions were identical in the optical and electron microscopes, the resolving power of the EM should be 100,000 times (resolution down to 0.0025 nm). However, the numerical aperture of an EM lens is very small (the diameter of the aperture is only a few micrometers) and does not approach the width of that of an optical microscope objective. In practice, the best resolution that can be obtained is 0.3 to 0.5 nm, a hundred times better than that of the light microscope.

Types of electron microscopes (Figs 2.3A and B)

The illustrations show the pathways of electron beams used to create images of the specimens. (A) Transmission: In a transmission electron microscope, electrons pass through the specimen and are scattered. Magnetic lenses focus the image onto a fluorescent screen or photographic plate; (B) Scanning: In a scanning electron microscope, primary electrons sweep across the specimen and knock electrons from its surface. These secondary electrons are picked up by a collector, amplified, and transmitted onto a viewing screen or photographic plate

There are two types of electron microscopes in general use:

i. Transmission Electron Microscope (TEM)

In transmission electron microscope (TEM) electrons like light pass directly through the specimen that has been prepared by thin sectioning, freeze fracturing, or freeze etching. It is used to observe fine details of cell structure.

ii. Scanning Electron Microscope (SEM)

Scanning electron microscopes scan a beam of electrons back and forth over the surface of a specimen producing three-dimensional views of the surfaces of whole microorganisms.

C. Scanning-Probe Microscopy

Scanning probe microscopes map the bumps and valleys of a surface on an atomic scale. Their resolving power is much greater than the electron microscope, and the samples do not need special preparation as they do for electron microscopy. Among the new scanned-probe microscopes are:

1. **Scanning tunneling microscopy (STM):** They are used to provide incredibly detailed views of molecules such as DNA.
2. **Atomic force microscopy (AFM):** They produce three-dimensional images of the surface of a molecule.

☞ KEY POINTS

Microscopy: A simple microscope consists of one lens; a compound microscope has multiple lenses.

Compound Light Microscopy: The most common microscope used in microbiology is the compound microscope/bright-field microscope/light microscope; Immersion oil is used with the oil immersion lens to reduce light loss between the slide and the lens.

- **Dark-field Microscopy:** The dark-field microscope directs light toward a specimen at an angle. It is most useful for detecting the presence of extremely small organisms.
- **Phase-Contrast Microscopy:** A phase-contrast microscope brings direct and reflected or diffracted light rays together (in phase) and amplifies differences in refraction to form an image of the specimen on the ocular lens. It allows the detailed observation of living organisms.

- **Differential Interference Contrast (DIC) Microscopy:** It allows detailed observations of living cells.
- **Confocal Microscopy:** In confocal microscopy, a specimen is stained with a fluorescent dye and illuminated one plane at a time.
- **Fluorescence Microscopy:** In fluorescence microscopy, specimens are first stained with fluorochromes and then viewed through a compound microscope by using an ultraviolet light source. The microorganisms appear as bright objects against a dark background.
- **Electron Microscopy:** Electron microscopes use electromagnetic lenses, instead of glass lenses, electrons, and fluorescent screens to produce a magnified image. Two types of EM: (i) Transmission electron microscopes (TEMs). (ii) Scanning electron microscopes

Scanned-Probe Microscopy

Scanning probe microscopes map the bumps and valleys of a surface on an atomic scale. Their resolving power is much greater than the electron microscope, and the samples do not need special preparation as they do for electron microscopy.

KNOW MORE

The most common type of microscope is the bright-field microscope.

- Brightfield illumination is used for stained smears.
- Unstained cells are more productively observed using dark-field, phase-contrast, or DIC microscopy.

Scanning-Probe Microscopes

Since the early 1980s, several new types of microscopes, called **scanned-probe microscopes**, have been developed Scanning probe microscopes make it possible to view images at an atomic scale. Among the new scanned-probe microscopes are the scanning tunneling microscope and the atomic force microscope, discussed next.

IMPORTANT QUESTIONS

1. Name various microscopic instruments used in microbiology and describe the working principle of compound microscope.
2. Describe the principles of bright-field, dark-field, phase-contrast, fluorescent, and electron microscopy. Give one example in which each method would be used.
3. Describe the differences between the principles of a light microscope and an electron microscope.
4. Describe the principles involved in the light microscope, phase-contrast microscope and electron microscope.
5. Write short notes on:
   Compound microscope/bright-field microscope.
   Resolution.
   Dark-field microscope
   - Fluorescent microscope
   - Phase-contrast microscope

   Confocal Microscopy
   Electron microscope.
   Transmission electron microscope (TEM).
   Scanning electron microscope (SEM)
   Scanned-probe microscopy

FURTHER READING

Tortora GJ, FunkeBR, Case CL. Microbiology an introduction observing microorganisms through a microscope: 2004. 8(E): Perason Education.

CHAPTER 3

Morphology of Bacteria

Learning Objectives

After reading and studying this chapter, you should be able to:

- Differentiate between prokaryotes and eukaryotes.
- Describe anatomy of bacterial cell.
- Describe cell envelope.
- Describe bacterial cell wall.
- Discuss capsule or bacterial capsule.
- Describe bacterial flagellae.
- Describe fimbriae or pili.
- Discuss bacterial spores or endospores.
- Explain L-forms of bacteria.

INTRODUCTION

Microorganisms are generally regarded as living forms that are microscopic in size and relatively simple, usually unicellular, in structure. The bacteria are single-celled organisms that reproduce by simple division, i.e. binary fission. Most are free living and contain the genetic information, energy-producing and biosynthetic systems necessary for growth and reproduction.

Microorganisms are a heterogeneous group of several distinct classes of living beings. Whittaker's system recognizes five-kingdoms of living things—**Monera (bacteria), Protista, Fungi, Plantae, and Animalia**. Five kingdoms have been modified further by the development of **three domains, or Superkingdoms** system—the **Bacteria,** the **Archaea** (meaning ancient), and the **Eucarya.**

COMPARISON OF PROKARYOTIC CELLS-EUKARYOTIC CELLS

All living organisms on earth are composed of one or the other of two types of cells: ***prokaryotic*** cells and **eukaryotic** cells based on differences in cellular organization and biochemistry.

- Prokaryotes: Bacteria and Archaea.
- Eukaryotes: Eucarya.

1. *Prokaryotes:* Prokaryotic cells (*pro* or primitive nucleus) do not have a membrane-bound nucleus. All bacteria and blue-green algae are prokaryotes.

   Bacteria are prokaryotic microorganisms and don't contain chlorophyll but in contrast, blue-green algae possess chlorophyll. They are unicellular and do not show true branching, except in the so called 'higher bacteria' (Actinomycetales).
2. *Eukaryotes*: Eukaryotic cells (*eu* or true nucleus), have a membrane-bound nucleus. Other algae (excluding blue-green algae), fungi, slime moulds, protozoa, higher plants, and animals are eukaryotic.

   Three fundamental characteristics are often considered to distinguish prokaryotes from eukaryotes: small size, absence of a complex, organelle-containing cytoplasm, and the absence of a nuclear membrane (Table 3.1).

Size of Bacteria

Bacteria are very small in size. The unit of measurement in bacteriology is the micron or micrometer (μm).

1 micron (μ) or micrometer (μm) = a millionth part of a meter or a thousandth of a millimeter.

1 millimicron (mμ) or nanometer (nm) = one thousandth of a micron or one millionth of a millimeter.

1 Angstrom unit (Å) − one tenth of a nanometer.

The diameter of the smallest body that can be resolved and seen clearly with naked eye is 200 μm. Bacteria of medical importance generally measure 0.2-1.5 μm in diameter and about 3-5 μm in length. To see bacteria, a light microscope must be used. The best light microscope, using the most advanced optics, are capable of magnifications of 1000 to 2000 times. The electron microscope provides superb resolving power. Many types of microscopes are used for examination of bacteria.

STUDY OF BACTERIA

Stained Preparations

Because most microorganisms appear almost colorless when viewed through a standard light microscope, we often must prepare them for observation. Live bacteria

Table 3.1: Principle differences between prokaryotic and eukaryotic cells		
Characteristics	*Prokaryotic*	*Eukaryotic*
Size (approximate)	0.5 to3 μm	>5 μm
Nucleus		
Nuclear membrane	Absent	Present
Nucleolus	Absent	Present
Chromosome	One (circular)	More than one (linear)
Deoxyribonucleoprotein	Absent	Present
Division	By binary fission	By mitosis
Cytoplasm		
Cytoplasmic streaming	Absent	Present
Mitochondria	Absent	Present
Golgi apparatus	Absent	Present
Lysosomes	Absent	Present
Pinocytosis	Absent	Present
Endoplasmic reticulum	Absent	Present
Chemical composition		
Sterol	Absent	Present
Muramic acid	Present	Absent
Examples	Eubacteria, Archaea All bacteria and blue-green algae	Fungi, slime moulds, protozoa, higher plants, and animals including humans

do not show much structural detail under the light microscope due to lack of contrast. Hence it is customary to use staining techniques to produce color contrast. Staining simply means coloring the microorganisms with a dye that emphasizes certain structures.

Bacteria may be stained in the living state, but this type of staining is employed only for special purposes. Routine methods for staining of bacteria involve drying and fixing smears, procedures that kill them. Fixing simultaneously kills the microorganisms and attaches them to the slide. It also preserves various parts of microbes in their natural state with only minimal distortion. Various staining techniques are commonly used in bacteriology.

Shape of Bacteria

Depending on their shape, bacteria are classified into several varieties (Fig. 3.1):

1. **Cocci:** Cocci (from kokkos meaning berry) are spherical, or nearly spherical.
2. **Bacilli:** Bacilli (from baculus meaning rod) are relatively straight, rod shaped (cylindrical) cells. In some of the bacilli, the length of the cells may be equal to width. Such bacillary forms are known as coccobacilli and have to be carefully differentiated from cocci.
3. **Vibrios:** Vibrios are curved or comma-shaped rods and derive the name from their characteristic vibratory motility.
4. **Spirilla:** Spirilla are rigid spiral or helical forms.
5. **Spirochetes:** Spirochetes (from speira meaning coil and chaite meaning hair) are flexuous spiral forms.

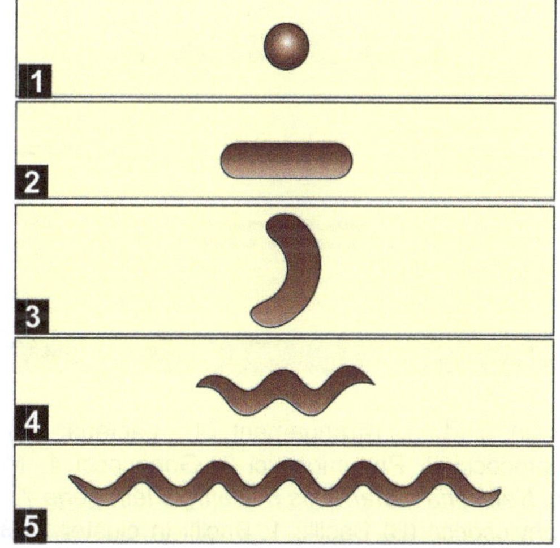

Fig. 3.1: Shape of bacteria: 1. Coccus; 2. Bacillus; 3. Vibrio; 4. Spirillum; 5. Spirochete

6. **Mycoplasma:** Mycoplasma are cell wall deficient bacteria and hence do not possess a stable morphology. They occur as round or oval bodies and interlacing filaments.

ARRANGEMENT OF BACTERIAL CELLS

Pathogenic bacterial species appear as sphere (cocci), rods (bacilli), and spirals. Bacteria sometimes show characteristic cellular arrangement or grouping (Figs 3.2A and B). The type of cellular arrangement is determined by the plane through which binary fission takes place and by the tendency of the daughter cells to remain attached even after division.

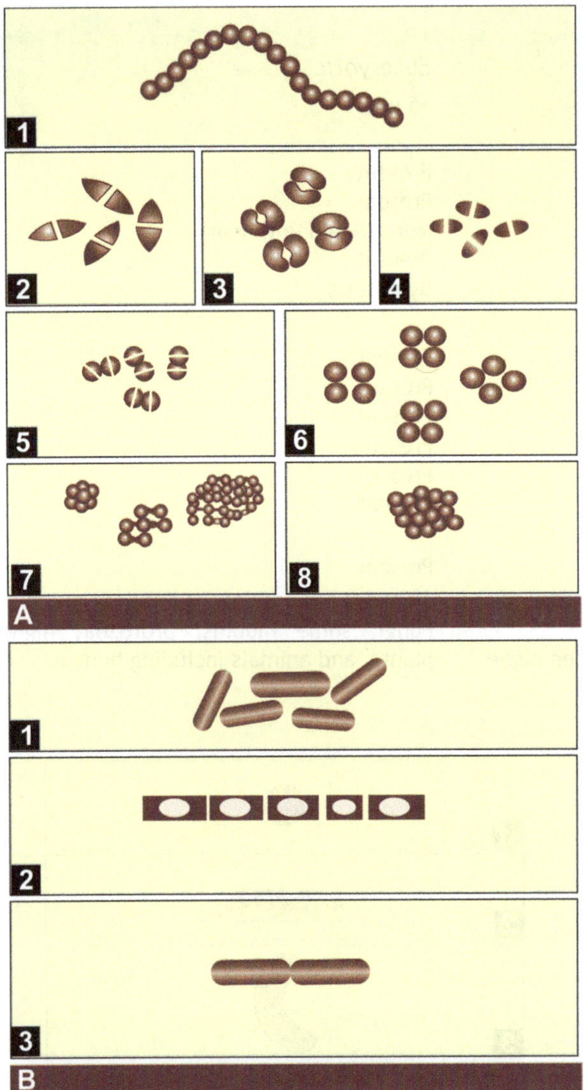

Figs 3.2A and B: Arrangement of bacteria: (A) Cocci: 1. Streptococci 2. Pneumococci 3. Gonococci 4. Meningococci 5. *Neisseria catarrhalis* 6. *Gaffkya tetragena* 7. *Sarcina* 8. *Staphylococci;* (B) Bacilli: 1. Bacilli in cluster 2. Bacilli in chains (*B. anthrax*) 3. Diplobacilli (*K. pneumoniae*)

Cocci Arrangement

i. **Diplococci:** Cocci may be arranged in pairs (diplococci) when cocci divide and remain together.
ii. **Long chains:** Long chains (Streptococcus, Enterococcus, and Lactococcus) when cells adhere after repeated divisions in one plane.
iii. **Grape like clusters:** Grape like clusters (staphylococci) when cocci divide in random planes.
iv. **Tetrads:** Square groups of four cells (tetrads) when cocci divide in two planes as in members of the genus *Micrococcus*.
v. **Cubical packets:** Cubical packets of eight of cells (genus *Sarcina*) when cocci divide in three planes.

Bacilli Arrangement

Bacilli split only across their short axes, therefore, the patterns formed by them are limited. The shape of the rod's end often varies between species and may be flat, rounded, cigar-shaped, or bifurcated. Some bacilli too may be arranged in **chains** (streptobacilli). Others are arranged at various angles to each other, resembling the letter V presenting a **cuneiform** or **Chinese letter arrangement** and is characteristic of *Corynebacterium diphtheriae*.

ANATOMY OF THE BACTERIAL CELL

The principal structures of the bacterial cell are shown in (Fig. 3.3). Bacterial Cell Components can be divided into:

a. **The outer layer or cell envelope** consists of two components:
   1. Cell wall.
   2. Cytoplasmic or plasma membrane—beneath cell wall.
b. **Cellular appendages—**Besides these essential components, some bacteria may possess additional structures such as **capsule, fimbriae, and flagella.**

Capsule: Some bacteria produce a protective gelatinous covering layer called a capsule outside the cell wall. If the capsule is too thin to be seen with light microsope (<0.2 µm) it is called **microcapsule**.

Loose slime: Soluble, large-molecular, amorphous, viscid colloidal material may be dispersed by the bacterium into the environment as loose slime.

Flagella: Some bacteria carry external filamentous appendages protruding from the cell wall; **flagella,** which are organs of locomotion; **fimbriae**, which appear to be organs of adhesion; and **pili**, which are involved in the transfer of genetic material.

B. Cell Interior

Those structures and substances that are bounded by the cytoplasmic membrane, compose the cell interior and include cytoplasm, cytoplasmic inclusions (mesosomes, ribosomes, inclusion granules, vacuoles) and a single circular chromosome of deoxyribonucleic acid (DNA).

A. Cell envelope and its Appendages

a. The Outer Layer or Cell Envelope

1. Cell Wall

The cell wall is the layer that lies just outside the plasma membrane. It is 10-25 nm thick, strong and relatively rigid, though with some elasticity, and openly porous, being freely permeable to solute molecules smaller than 10 kDa in mass and 1 nm in diameter.

Functions of the cell wall:

1. To impart **shape and rigidity** to the cell.
2. It **supports the weak cytoplasmic membrane**

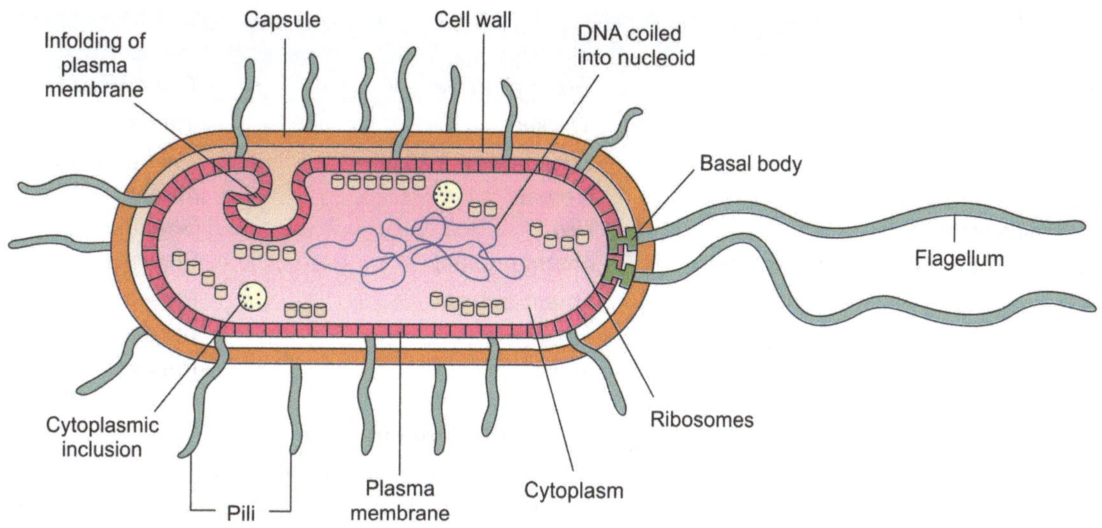

Fig. 3.3: Anatomy of a bacterial cell

against the high internal osmotic pressure of the protoplasm (ranges from 5 and 25 atm).

3. Maintains the characteristic shape of the bacterium.
4. It takes part in **cell division.**
5. Also functions in **interactions (e.g. adhesion) with other bacteria** and with mammalian cells.
6. Provide **specific protein and carbohydrate receptors** for the attachment of some bacterial viruses.

Chemical Structure of Bacterial Cell Wall

Chemically the cell wall is composed of mucopeptide (peptidoglycan or murein) scaffolding formed by N-acetyl glucosamine and N-acetyl muramic acid molecules alternating in chains, which are crosslinked by peptide bonds (Fig. 3.4).

Peptidoglycan consists of three parts (Fig. 3.4):

1. A backbone—composed of alternating N-acetylglucosamine and N-acetylmuramic acid.
2. A set of identical tetrapeptide side chains attached to N-acetylmuramic acid.
3. A set of identical pentapeptide cross-bridges.

In all bacterial species, the backbone is the same, however, tetrapeptide side chains and pentapeptide cross-bridges vary from species to species (Fig. 3.4). Several antibiotics interfere with construction of the cell wall peptidoglycan.

In gram-positive bacteria, the cell wall consists mainly of peptidoglycan and teichoic acids, whereas in gram-negative bacteria, the cell wall is more complex in both the anatomical and chemical sense and includes the thinner peptidoglycan and outer membrane.

Difference Between Cell Wall of Gram-positive and Gram-negative Bacteria

The difference between gram-positive and gram-negative bacteria has been shown to reside in the cell wall. In general, the walls of the gram-positive bacteria have simpler chemical nature than those of gram-negative bacteria (Table 3.2). The integrity of the cell wall is essential to the viability of the bacterium. The protoplasm may swell from osmotic inflow of water and burst the weak cytoplasmic membrane if the wall is weakened or ruptured. This process of lethal disintegration and dissolution is termed **lysis.**

Gram-positive Bacterial Cell Wall

The gram-positive bacterial cell wall (Fig. 3.5) is about 80 nm thick and is composed mostly of several layers of peptidoglycan.

Peptidoglycan: It constitutes 50-90 percent of the dry weight of the wall and are thicker and stronger (more extensively cross-linked) than those of gram-negative bacteria (peptidoglycan comprising 5-10 percent of the

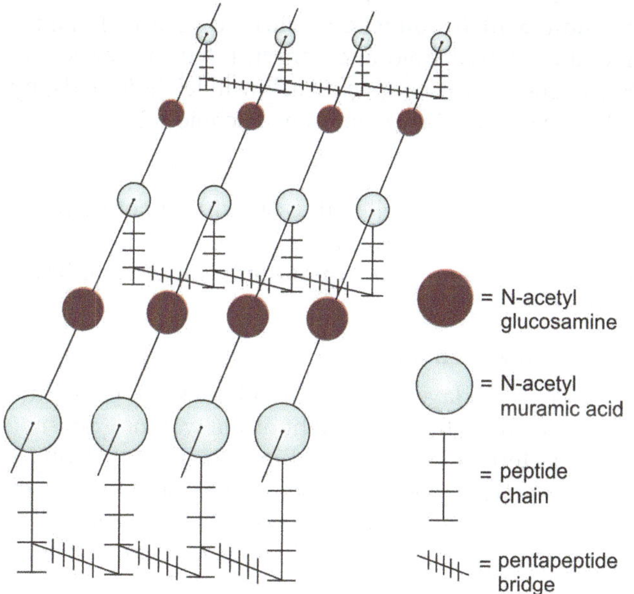

Fig. 3.4: Chemical structure of bacterial cell wall

Table 3.2: Comparison of cell walls of gram-positive and gram-negative bacteria

Characteristic	*Gram-positive*	*Gram-negative*
1. Thickness	Thicker	Thinner
2. Peptidoglycan	Thick layer (16-80 nm)	2 nm (thin layer)
3. Teichoic acid	Present	Absent
4. Variety of amino acids	Few	Several
5. Aromatic and sulfur containing amino acids	Absent or scant	Present
6. Lipids	Absent or scant	Present
7. Porin proteins	Absent	Present
8. Periplasmic region	Absent	Present

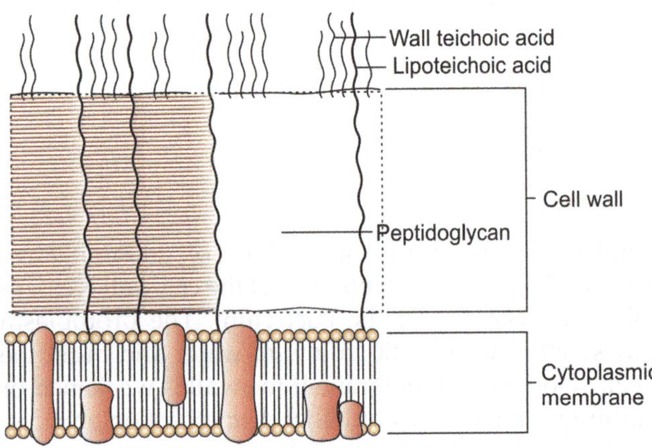

Fig. 3.5: Gram-positive cell wall

wall material). By contrast, gram-negative cell walls contain only a thin layer of peptidoglycan.

Teichoic acid: In addition, the cell walls of gram-positive bacteria contain **teichoic acids**, which consist primarily of an alcohol (such as glycerol or ribitol) and phosphate. There are two types of teichoic acids: **wall teichoic acid**, covalently linked to peptidoglycan, and **membrane teichoic acid (lipoteichoic acid)**, covalently linked to membrane glycolipid and concentrated in mesosomes. Some gram-positive species lack wall teichoic acids, but all appear to contain membrane teichoic acids.

Gram-negative Cell Wall

The gram-negative cell wall is structurally quite different from that of gram-positive cells (Fig. 3.6). It consists of **peptidoglycan and, lipoprotein, outer membrane, and lipopolysaccharide**.

i. Peptidoglycan layer

Peptidoglycan layer is a single-unit thick and constitutes 5-10 percent of the dry weight of the wall of gram-negative bacteria. It is bonded to lipoproteins covalently in the outer membrane and plasma membrane and is in the periplasmic, a gel like fluid between the outer membrane and plasma membrane. The periplasmic contains a high concentration of degradive enzymes and transport proteins. The periplasmic space is approximately 20 to 40 percent of the cell volume, which is far from insignificant.

ii. Lipoprotein

Lipoprotein, or murein lipoproteins seemingly attach (both covalently and noncovalently) to the peptidoglycan by their protein portion, and to the outer membrane by their lipid component.

Function: To stabilize the outer membrane and anchor it to the peptidoglycan layer.

iii. Outer membrane

External to the peptidoglycan, and attached to it by lipoproteins is the outer membrane. It is a bilayered structure. Its inner leaflet is composed of phospholipid while phospholipids of the outer leaflet are replaced by lipopolysaccharide (LPS) molecules.

Functions:

a. **A protective barrier:** A most important function is to serve as a protective barrier. It prevents or slows the entry of the salts, antibiotics and other toxic substances that might kill or injure the bacterium.
b. **Porins or transmembrane proteins:** In addition to LPS, the outer membrane also contains several important proteins that function in the selective transport of the nutrients into the cell. **Porins** or **transmembrane proteins**, traverse the outer membrane and form trimeric channels that permit the passage of molecules such as nucleotides, disaccharides, peptides, amino acids, vitamin B12, and iron.

iv. Lipopolysaccharide (LPS)

A structural component that is unique to the gram negative outer membrane is lipopolysaccharide (LPS). It is a large complex molecule that contains lipids and carbohydrates and consists of three components:

a. **Lipid A** is the lipid portion of LPS and is embedded in the top layer of the outer membrane. When gram-negative bacteria die, they release Lipid A, which functions as an **endotoxin**. All the toxicity of the endotoxin is due to lipid A which is responsible for the endotoxic activities, that is, pyrogenicity, lethal effect, tissue necrosis, anticomplementary activity, B-cell mitogenicity, immunoadjuvant property and anti-tumor activity.
b. **The core polysaccharide** is attached to lipid A and a terminal series of repeat unit contains unusual sugars. Its role is to provide stability.

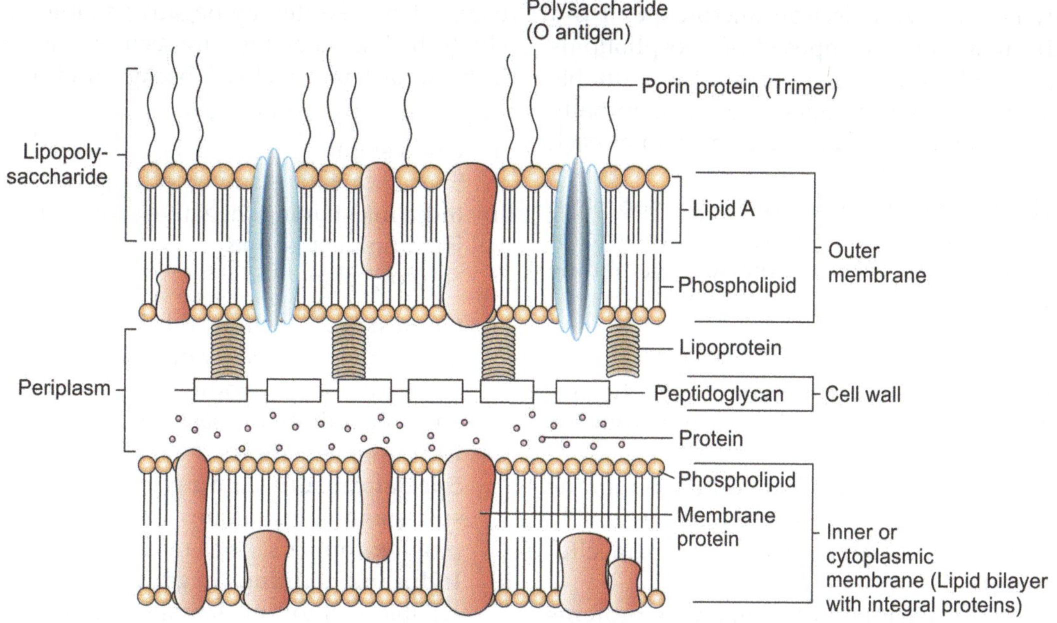

Fig. 3.6: Gram-negative cell wall

c. **O polysaccharide:** Extends outward from the core polysaccharide and is composed of sugar molecules. O polysaccharides function as antigens and are useful for distinguishing species of gram-negative bacteria. The role is comparable to that of teichoic acids in gram-positive cells. Polysaccharide represents a major surface antigen of the bacterial cell. It is known as **O antigen.** Bacteria carrying LPS containing O antigen form smooth colonies in bacteriological media in contrast to those lacking the O antigen, which form rough colonies.

Demonstration of Cell Wall

The cell wall cannot be seen by light microscopy and does not stain with simple dyes. It can be demonstrated by:

i. **Plasmolysis:** When placed in a hypertonic solution, the cytoplasm loses water by osmosis and shrinks, while the cell wall retains its original shape and size (bacterial ghost).
ii. Microdissection.
iii. Exposure to specific antibody.
iv. Mechanical rupture of the cell.
v. Differential staining procedures.
vi. Electron microscopy.

Enzymes that Attack Cell Walls

Various enzymes attack cell walls:

i. **Lysozyme:** The enzyme **lysozyme**, which is found in animal secretions (tears, saliva, nasal secretions) as well as in egg white, is a natural body defence substance which lyses bacteria of many species. It acts by hydrolyzing linkage of the peptidoglycan backbone. Gram-positive bacteria treated with lysozyme in low osmotic strength media lye. The outer membrane of the gram-negative cell wall prevents excess of lysozyme unless disrupted by an agent such as ethylenediaminetetraacetic acid (EDTA), a chelating agent.
ii. **Autolysins:** Bacteria themselves possess enzymes, called **autolysins,** able to hydrolyze their own cell wall substances. These enzymes presumably play essential functions in cell growth and division, but their activity is most apparent during the dissolution of dead cells **(autolysis).**

Protoplasts and Spheroplasts

Protoplasts

The gram-positive cell wall is almost completely destroyed by lysozyme. The cellular contents that remain surrounded by the plasma membrane may remain intact if lysis does not occur. This wall-less cell is termed a protoplast. They contain cytoplasmic membrane and cell wall is totally lacking. Typically, a protoplast is spherical and is still capable of carrying on metabolism.

Spheroplasts

When lysozyme is applied to gram-negative cells, usually the wall is not destroyed to the same extent as in gram-positive cells. Some of the outer membrane also remains. In this case, the cellular contents, plasma membrane, and remaining outer wall layer are called a **spheroplast.** It is also a spherical structure. They are capable of reverting to parent bacterial form when cell wall inhibitor is removed from the culture medium.

2. Cytoplasmic (Plasma) Membrane

Structure: The cytoplasmic (plasma) membrane limits the bacterial protoplast. It is thin (5-10 nm thick), elastic and can only be seen with electron microscope. It is a typical "**unit membrane**", composed of phospholipids and proteins. Lipid molecules are arrayed in a double layer with their hydrophilic polar regions externally aligned and in contact with a layer of protein at each surface.

Chemically, the membrane consists of lipoprotein with small amounts of carbohydrate. With the exception of *Mycoplasma,* bacterial cytoplasmic membrane lacks sterols.

Functions of Cytoplasmic Membrane

i. **Semipermeable membrane**—controlling the inflow and outflow of metabolites to and from the protoplasm.
ii. **Housing enzymes**—involved in outer membrane synthesis, cell wall synthesis, and in the assembly and secretion of extracytoplasmic and extracellular substances.
iii. **Housing many sensory and chemotaxis proteins** that monitor chemical and physical changes in the environment.
iv. **Generation of chemical energy (i.e, ATP).**
v. **Cell motility.**
vi. **Mediation of chromosomal segregation during replication.**

b. Cellular Appendages

i. Capsule or Slime Layer

Structure: Many bacteria synthesize large amount of extracellular polymer in their natural environments. When the polymer forms a condensed, well defined layer closely surrounding the cell, it is called the **capsule** as in the pneumococcus. If the polymer is easily washed off and does not appear to be associated with the cell in any definite fashion, it is referred as a **slime layer** as in *Leuconostoc*. A **glycocalyx** is a network of polysaccharide extending from the surface of bacteria and other cells. Capsules too thin to be seen under the light microscope are called **microcapsules**.

Composition of capsules and slime layers: Cap sules and slime layers usually are composed of polysaccharide (for example pneumococcus) or of polypeptide in some bacteria (for example, *Bacillus anthracis* and *Yersinia pestis*). Some bacteria may have both a capsule and a slime layer (for example, *Streptococcus salivarius).* Bacteria secreting large amounts of slime produce mucoid growth on agar, which is of a stringy consistency when touched with the loop.

Capsulated bacteria: *Streptococcus pneumoniae,* several groups of streptococci, *Neisseria meningitidis, Klebsiella, Haemophilus influenzae, Yersinia* and *Bacillus.*

Staining techniques: Slime has little affinity for basic dyes and is not visible in Gram stained smears. Special capsule staining techniques are available, usually employing copper salts as mordants. Capsules may be readily demonstrated by negative staining in wet films with India ink, when they are seen as clear halos around the bacteria, against a black background (Fig. 3.7).

Demonstration of Capsule

i. Gram stain.
ii. Special capsule staining techniques.
iii. India ink staining (negative staining).
iv. Electron microscope.
v. Serological methods.

i. **Gram stain:** Capsules and slime are usually not visible in films stained by ordinary methods except as clear haloes surrounding the stained smear. Slime has little affinity for basic dyes and is not visible in Gram stained smears.
ii. **Special capsule staining techniques:** Special capsule staining techniques are available, usually employing copper salts as mordants.
iii. **India ink staining (negative staining):** The most reliable method of demonstration is by 'negative' staining in wet films with India ink and the capsule appears as a clear halo around the bacterium, against a dark background in the film.
iv. **Electron microscope:** They can also be studied with electron microscope.
v. **Serological methods:** Capsular material is antigenic and may be demonstrated by serological methods. Capsules may also be visualized by reaction with capsule-specific antibodies which causes a characteristic swelling of the capsule. When a suspension of a capsulated bacterium is mixed with its specific anticapsular serum and examined under the microscope, the capsule becomes very prominent and appears 'swollen' due to an increase in its refractivity. It is known as the **capsule-swelling reaction** or **Quellung reaction** (Quellung—(*Ger*) *swelling*), described by Neufeld (1902) was widely

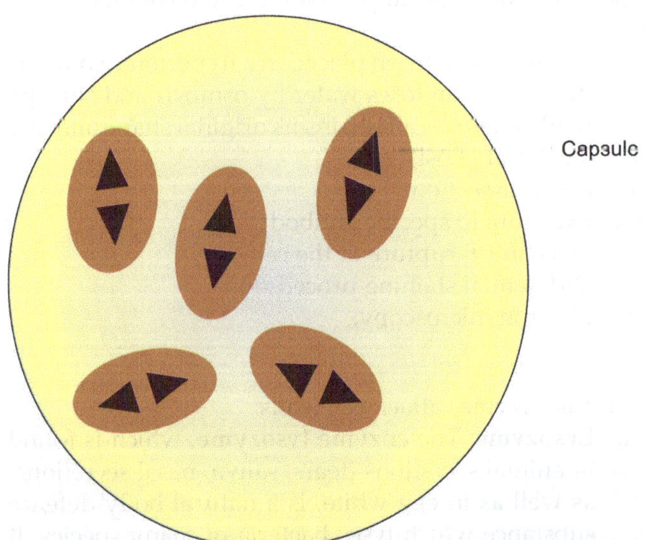

Fig. 3.7: Pneumococci negatively stained with India ink to show capsule

employed for the typing of pneumococci in the pre-sulphonamide days when lobar pneumonia used to be treated with specific anticapsular sera.

Use of capsule-swelling reaction: This phenomenon is seen in and allows rapid identification of capsular serotypes of *Streptococcus pneumoniae, Neisseria meningitidis*, several groups of streptococci, *Klebsiella, Haemophilus influenzae, Yersinia* and *Bacillus.*

Functions of Capsule

i. **Virulence factor:** Capsules often act as a virulence factor by protecting the bacterium from ingestion by phagocytosis, and noncapsulate mutant of these bacteria are nonvirulent. Repeated subcultures *in vitro* lead to the loss of capsule and also of virulence.
ii. **Protection of the cell wall:** In protecting the cell wall attack by various kinds of antibacterial agents, e.g. bacteriophages, colicines, complement, lysozyme and other lytic enzymes.
iii. **Identification and typing of bacteria:** Capsular antigen is specific for bacteria and can be used for identification and typing of bacteria.

ii. Flagella

Motile bacteria, except spirochetes, possess one or more unbranched, long, sinuous filaments called **flagella**, which are the organs of locomotion.

Structure: They are long, hollow, helical filaments, usually several times the length of the cell. They are 3-20 µm long and are of uniform diameter (0.01-0.013 µm) and terminate in a square tip. It originates in the bacterial protoplasm and is extruded through the cell wall. Flagella consists of largely or entirely of a protein, **flagellin**, belonging to the same chemical group as myosin, the contractile protein of muscle.

Though flagella of different genera of bacteria have the same chemical composition, they are antigenically different. Flagella are highly antigenic and flagellar antigens induce specific antibodies in high titers. Several of the immune responses mounted by host systems are directed against these flagellar antigen. Flagellar antibodies are not protective but are useful in serodiagnosis.

Flagella can be found on both gram-positive and gram-negative bacilli. Most cocci are immotile, whereas about half of the bacilli and almost all of the spirilla possess flagella.

Parts and Composition

Each flagellum consists of three parts (Fig. 3.8).

i. Filament
ii. Hook
iii. Basal body.

   i. **Filament:** The filament is the longest and most obvious portion which extends from the cell surface to the tip.
   ii. **Hook:** The hook is a short, curved segment which links the filament to its basal body and functions as universal joint between the basal body and the filament.
   iii. **Basal body:** The basal body is embedded in the cell (cytoplasmic membrane). In the gram-negative bacteria, the basal body has four rings connected to a central rod (L, P, S and M). The outer L and P rings associated with the lipopolysaccharide and peptidoglycan layers respectively. S ring is located just above the cytoplasmic membrane and the inner M ring contacts the cytoplasmic membrane.

   Gram-positive bacteria have only two basal body rings, an inner ring connected to the cytoplasmic membrane and an outer one probably attached to peptidoglycan.

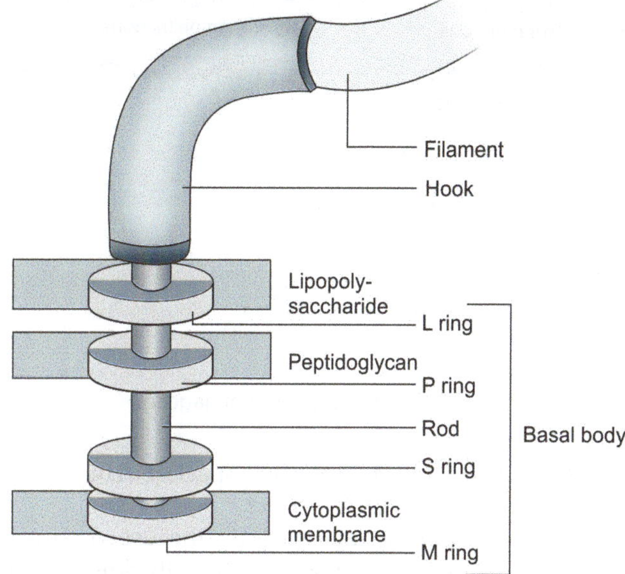

Fig. 3.8: The structure of bacterial flagellum

Arrangement/Types (Fig. 3.9)

The number and location of flagella are distinctive for each genus. There are four types of flagella arrangement:

- **Monotrichous**—Single polar flagellum (e.g. Cholera vibrio).
- **Amphitrichous**—Single flagellum at both ends (e.g. *Alcaligenes faecalis).*
- **Lophotrichous**—Tuft of flagella at one or both ends (e.g. spirilla).
- **Peritrichous**—Flagella surrounding the cell (e.g. Typhoid bacilli).

Demonstration of Flagella

Flagella are about 0.02 µm in thickness and hence beyond the resolution limit of the light microscope. The following methods are used for its demonstration:

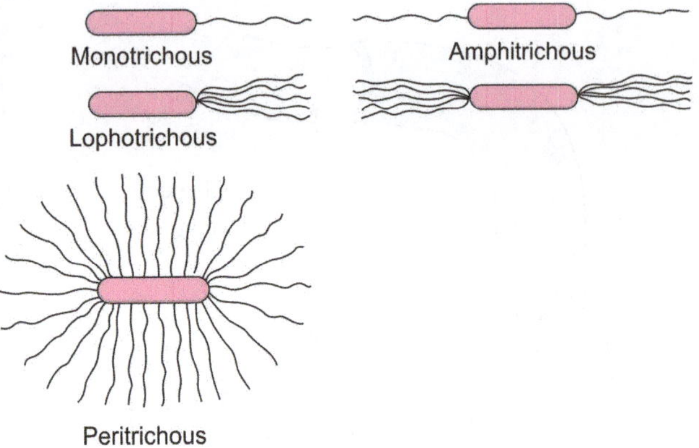

Fig. 3.9: Arrangement of flagella

i. **Dark ground illumination**: In special circumstances by dark ground illumination.
ii. **Special staining methods:** By the use of special staining methods in which their thickness is increased by mordanting.
iii. **Electron microscopy.**
iv. **Indirect methods:** Indirect methods by which motility of bacteria can be seen or demonstrated due to the difficulty of demonstrating flagella directly.
   a. **Microscopically in fluid suspensions (in a hanging drop or under a coverslip):** On microscopic examination of wet films, motile bacteria are seen swimming in different directions across the field, with a darting, wriggling or tumbling movement.
   b. **By spread of bacterial growth as a film over agar**, e.g. swarming growth of *Proteus sp*.
   c. **Turbidity spreading through semisolid agar,** e.g. Craigie's tube method.

iii. Fimbria or Pili

Structure and synthesis: Many gram-negative bacteria have short, fine, hair like surface appendages called fimbriae or pili depending on their function. They are shorter and thinner than flagella (0.1 to 1.5 μm in length and uniform width between 4 and 8 nm) and emerge from the cell wall. Single cells have been seen to be covered with as few as 10 fimbriae to as many as 1000. They originate in the **cytoplasmic membrane** and are composed of structural protein subunits termed **pilins** like flagella.

They occur in non-motile, as well as in motile strains. Fimbriae are antigenic. It is necessary to ensure that the bacterial antigens employed for serological tests and preparation of antisera are devoid of fimbriae as members of different genera may possess the same fimbrial antigen.

Demonstration of Fimbriae

1. **Electron microscopy:** They cannot be seen with the light microscope but are only visible in an electron microscope due to their smaller size.
2. **Hemagglutination:** Most fimbriate bacteria bear fimbriae of a type that enables them to adhere to, among other kinds of tissue cells, the red blood cells of many animal tissues. They adhere to guinea pig, fowl, horse and pig red cells very strongly, to human cells moderately strongly, to sheep cells weakly and to ox cells scarcely at all. The adherence functions can be detected by the ability of bacteria to adhere to epithelial cells or to cause hemagglutination. Therefore, a simple hemagglutination test can be used to determine whether a culture contains fimbriate bacilli or not. The adherence of bacteria to red cells, cultured cells, or tissue surfaces can be competitively blocked by the fimbriae.

Types of Fimbriae

Different types of fimbriae have different adhesive properties and can be divided into four types: Type 1 fimbriae (mannose-sensitive), Type 2 fimbriae, Type 3 fimbriae (mannose-resistant) and Type 4 fimbriae.

Functions of Pili

Two classes can be distinguished on the basis of their function: **ordinary (common) pili** and **sex pili.**

A. **Ordinary (common) pili:** Fimbriae probably function as **organs of adhesions** that allow attachment of a bacterial cell to other cells or surfaces. The adhesive property may be of value to the bacteria in holding them in nutritionally favorable-microenvironments.
B. **Sex pili:** Sex pili are similar to fimbriae, about 1-10 per cell, but are functionally different. These are longer and fewer in number than other fimbriae. They are genetically determined by **sex factors or conjugative plasmids** and appear to be involved in the transfer of DNA during conjugation. They are found on **'male' bacteria** and help in the attachment of those cells to 'female' bacteria, forming hollow conjugation tubes through which, it is assumed, genetic material is transferred from the donor to the recipient cell. Some bacterial viruses attach specifically to receptors on sex pili at the start of their reproductive cycle. Pili are classified into different types (for example, F, I) based on their susceptibility to specific bacteriophage.

B. Cell interior

1. Cytoplasm

The cytoplasm of the bacterial cell is a viscous watery solution or soft gel, containing a variety of organic or inorganic solutes, and numerous ribosomes and polysomes. The cytoplasm of bacteria differs from that of the higher eukaryotic organisms in not containing an endoplasmic reticulum or membrane-bearing microsomes, mitochondria, lysomes and in showing signs of internal mobility, e.g. cytoplasmic streaming, the formation, migration and disappearance of vacuoles, and amoeboid movement.

The cytoplasm may contain granules or inclusions such as starch, glycogen, poly-β-hydroxy/alkanoates, sulphur globules, magnetosomes, parasporal bodies, gas vesicles and endospores. The cytoplasm stains uniformly with basic dyes in young cultures.

2. Ribosomes

The cytoplasmic region of a well studied bacterium consists predominantly of small, electron-dense particles called **ribosomes**. The ribosomes are the location for all bacterial protein synthesis. Bacterial ribosomes are slightly smaller (10-20 nm) than eukaryotic ribosomes and they have a sedimentation rate of 70S (S or Svedberg unit), being composed of a 30S and 50S subunit. The prokaryotic ribosomes differ from their eukaryotic counterparts in that the latter are heavier and categorized as 80S (composed of 40S and 60S subunits).

Bacterial and host cell ribosomes have some considerable differences which allows us to use antibacterial agents such as streptomycin which interferes with bacterial metabolism at the ribosomal level without unduly upsetting human ribosomal function.

Types of RNA: Ribosomes are composed of different proteins associated with three types of ribonucleic acid (RNA)—**messenger RNA (mRNA), ribosomal RNA (rRNA),** and **transfer RNA (tRNA).**

i. **Messenger (m) RNA:** Most genes encode proteins and are transcribed into **messenger RNA (mRNA).** These molecules are translated during protein synthesis.
ii. **Ribosomal RNA (rRNA):** Ribosomes, composed of rRNA and proteins, also are central to translation and provide the site at which translation occurs.
iii. **Transfer RNA (tRNA):** Transfer RNA (tRNA) specifically transfers the genetic information carried in the mRNA into functional proteins.

3. Mesosomes (Chondroids)

Strucure: These are convoluted or multilaminated membranous bodies formed as invaginations of the plasma membrane into the cytoplasm. Mesosomes develop sometimes in relation to the nuclear body and often from the sites of cross-wall formation in gram-positive bacteria but not so common in gram-negative bacteria.

Types of Mesosomes

These are of two types:

i. **Septal mesosomes:** They function in the formation of cross-walls during cell division.
ii. **Lateral mesosomes:** They are present in a more random fashion.

Functions of Mesosomes

i. **Compartmenting of DNA:** These are thought to be involved in mechanisms responsible for the compartmenting of DNA at cell division and sporulation.
ii. **Sites of the respiratory enzymes:** They provide increased membrane surface and are the principle sites of the respiratory enzymes in bacteria and are analogous to mitochondria of eukaryotes.

4. Intracytoplasmic Inclusions

The synthesis and accumulation of intracellular inclusions in different prokaryotes is usually dependent upon both nutritional availability and environmental growth conditions. These are not permanent or essential structures, and may be absent under certain conditions of growth. These bodies are usually for storage, and reduce osmotic pressure by tying up molecules in particulate form. They consist of **volutin (polyphosphate), lipid, glycogen, starch or sulfur.**

Metachromatic or volutin granules: Volutin granules (*metachromatic or Babes-Ernst granules*) function as storage reservoirs for phosphate, are highly refractive, strongly basophilic bodies consisting of polymetaphosphate. They are characteristically found in diphtheria bacillus and also occur in the plague bacillus (*Yersinia pestis*), mycobacteria (e.g. *Mycobacterium tuberculosis*) and others.

Staining

i. **Methylene blue or toluidine blue staining:** With the blue dyes—methylene blue or toluidine blue, they stain a red violet color, contrasting with the blue staining of the bacterial protoplasm and thus show the metachromatic effect (reflecting the fact they take on a red appearance when stained with blue dye).
ii. **Neisser and Albert staining:** The granules can be demonstrated with even greater color contrast by special methods such as those of Neisser and Albert.
iii. **Acid-fast staining:** Volutin granules are slightly acid-fast resisting decoloration by 1 percent sulfuric acid.
iv. **Wet films:** They are more refractile than the protoplasm and are sometimes distinguishable in unstained wet films.
v. **Electron microscopy:** They appear as very opaque, clear demarcated bodies by electron microscopy.

5. Bacterial Nucleus

Structure: The genetic material of a bacterial cell is contained in a single, long molecule of double-stranded deoxyribonucleic acid (DNA) which can be extracted in the form of a closed circular thread about 1 mm (1000 µm) long, about 1000 times the length of the cell. Therefore, it occurs tightly coiled like a skein of woollen-thread. The nuclear deoxyribonucleic acid (DNA) is not associated with basic protein. The bacterial chromosome is haploid and replicates by simple fission instead of by mitosis as in higher cells.

Bacterial nucleus does not possess nuclear membrane (separating them from the cytoplasm), nucleolus, and deoxyribonucleoprotein. The differences between the nuclei of bacteria and higher organisms form the main basis for classifying them as prokaryotes and eukaryotes.

Demonstration

i. **Acid or ribonuclease hydrolysis:** By acid or ribonuclease hydrolysis and subsequent staining for nuclear material.
ii. **By electron microscopy.**
iii. **Light microscope:** The nucleoid also is visible in the light microscope after staining with the Feulgan stain, which specifically reacts with DNA. They appear as oval or elongated bodies, generally one per cell.

Plasmids

Bacteria may possess extranuclear genetic elements in the cytoplasm consisting of DNA, termed *plasmids* or *episomes* (see Chapter 10). These can exist and replicate independently of the chromosome or may be integrated with it. In either case they normally are inherited or passed on the progeny either through conjugation or the agency of bateriophages.

Function of plasmids: Plasmids are not essential for host growth and reproduction they inhabit, but may **confer on it certain properties** such as drug resistance, bacteriocin production, resistance to toxic metal ions, enterotoxin production, enhanced pathogenicity, reduced sensitivity to mutagens, or to the ability to degrade complex organic molecules which may constitute a survival advantage.

6. Bacterial Spore

A number of gram-positive bacteria, such as those of the genera *Clostridium* and *Bacillus* can form a special resistant dormant structure called an **endospore** or, simply, **spores.** Endospore develop when essential nutrients are depleted. In sporulation, each vegetative cell forms only one spore, and in subsequent germination, each spore gives rise to a single vegetative cell. Sporulation in bacteria, therefore, is not a method of reproduction but of preservation.

Sporulation: Spore formation, sporogenesis or sporulation normally commences when growth ceases due to lack of nutrients, depletion of the nitrogen or carbon source (or both) being the most significant factor. New antigens appear on sporulation that are not found in the vegetative cell.

Stages: It is a complex process and may be divided into several stages (Fig. 3.10).

- **Spore septum:** In the first observable stage of sporulation, a newly replicated bacterial chromosome and a small portion of cytoplasm are isolated by an ingrowth of the plasma membrane called a **spore septum**.

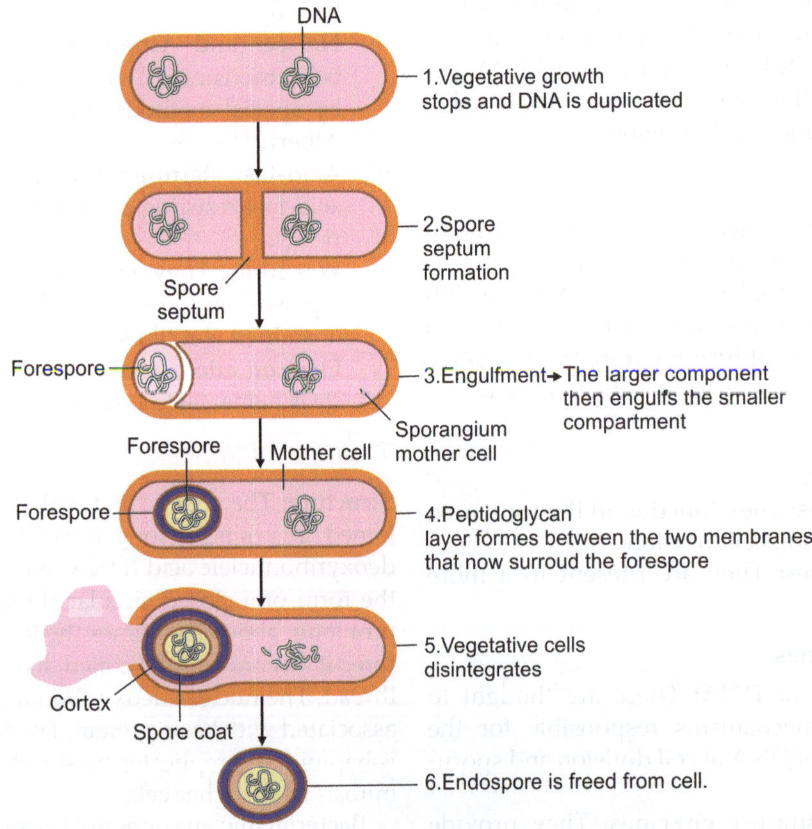

Fig. 3.10: The stages of endospore formation

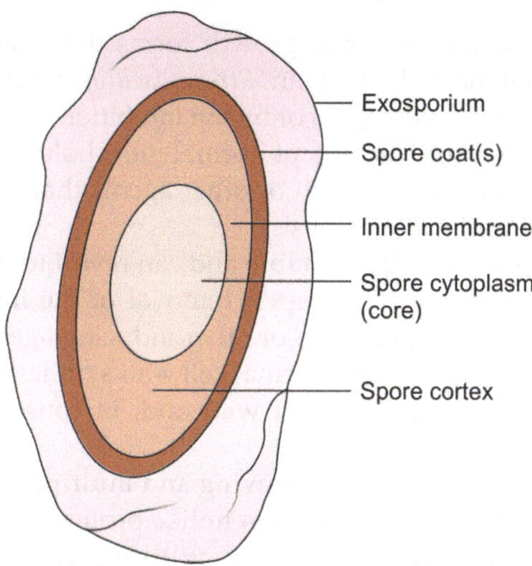

Fig. 3.11: Bacterial spore (cross-section)

- **Forespore:** The spore septum becomes a **double-layered membrane** that surrounds the chromosome and cytoplasm. Structure, entirely enclosed within the original cell, is called a **forespore.**
- **Spore coat:** The forespore is subsequently completely encircled by dividing septum as a double layered membrane. The two spore membranes now engage in active synthesis of various layers of the spore. The inner layer becomes the **inner membrane**. Between the two layers is laid **spore cortex** and outer layer is transformed into **spore coat** which consists of several layers. In some species from outer layer also develops **exosporium** which bears ridges and folds (Fig. 3.10).
- **Free endospore:** Finally exosporium disintegrates and the endospore is freed.

Properties of Endospores (Fig. 3.11)

1. **Core:** The fully developed spore has the **core** which is the spore protoplast containing the normal cell structures but is metabolically inactive.
2. **Spore wall:** The innermost layer surrounding the inner spore membrane is called the spore wall. It contains normal peptidoglycan and becomes the cell wall of the germinating vegetative cell.
3. **Cortex:** The cortex is the thickest layer of the spore envelope. Cortex peptidoglycan is extremely sensitive to lysozyme, and its autolysins plays a role in spore germination.
4. **Spore coat: Cortex**, in turn, is enclosed by fairly thick spore coat.
5. **Exosporium:** Spores of some species have an additional, apparently rather loose covering known as the exosporium, which may have distinctive ridges and grooves.

Germination

Germination is the process of conversion of a spore into vegetative cells under suitable conditions. It occurs in three stages: activation, initiation and outgrowth. Once activated, a spore will initiate germination if the environmental conditions are favorable. The spore loses its refractility and swells. The spore wall is shed and the germ cell appears by rupturing the spore coat. The germ cell elongates to form the vegetative bacterium.

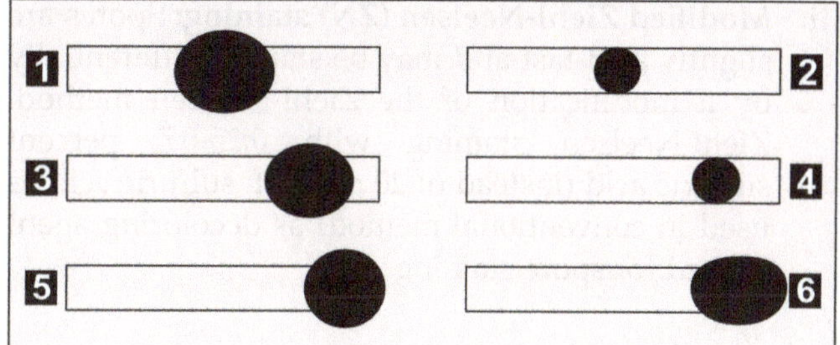

Fig. 3.12: Types of spores: 1. Central, bulging, 2. Central, not bulging, 3. Subterminal, bulging, 4. Subterminal, not bulging, 5. Terminal, spherical, 6. Terminal, oval

Shape and Position of Spores

The shape and position of the spore and its size relative to the parent cell are species characteristics. Spores may be central (equatorial), subterminal (close to one end), or terminal (Fig. 3.12). The appearance may be spherical, ovoid or elongated, and being narrower than the cell, or broader and bulging it. The diameter of spore may be same or less than the width of bacteria *(Bacillus)*, or may be wider than the bacillary body producing a distension or bulge in the cell *(Clostridium)*.

Resistance

Bacterial spores constitute some of the most resistant forms of life. They may remain viable for centuries. Though some spores may resist boiling for prolonged periods, spores of all medically important species are destroyed by autoclaving at 120°C for 15 minutes. Methods of sterilization and disinfection should ensure that spores also are destroyed. Sporulation helps bacteria survive for long periods under unfavorable environments.

Endospore heat resistance probably is due to several factors: calcium-dipicolinate and acid-soluble protein stabilization of DNA, protoplast dehydration, the spore coat, DNA repair, the greater stability of the cell proteins in bacteria adapted to growth at high temperatures and others.

Demonstration

Endospores can be examined with both light and electron microscope. In unstained preparations, the spore is recognized within the parent cell by its greater refractibility.

i. **Gram staining:** Spores appear as an unstained refractile body within the cell. When mature, the spore resists coloration by simple stains, appearing as a clear space within the stained cell protoplasm.

ii. **Modified Ziehl-Neelsen (ZN) staining:** Spores are slightly acid-fast and may be stained differentially by a modification of the Ziehl-Neelsen method. Ziehl-Neelsen staining with 0.25-0.5 percent sulfuric acid (instead of 20 percent sulfuric acid as used in conventional method) as decoloring agent is used for spore staining.

Uses of Spores

1. **Importance in food, industrial, and medical microbiology:** Endospores are of great practical importance in food, industrial, and medical microbiology because of their resistance and the fact that several species of endospore forming bacteria are dangerous pathogens. Endospore often survive boiling for an hour or more. Therefore, autoclaves must be used to sterilize many materials.
2. **Sterilization control:** For proper sterilization, spores of certain species of bacteria are employed as indicator, e.g. *Bacillus stearothermophilus* which is destroyed at a temperature of 121°C for 10 to 20 minutes (same temperature and time as used in autoclaving). Prior to its use, these spores may be kept in autoclave. Proper sterilization is indicated by the absence of the spores after autoclaving.
3. **Research:** Spore formation is well suited for research on the construction of complex biological structures.

PLEOMORPHISM AND INVOLUTION FORM

During growth, bacteria of a single strain may show considerable variation in size and shape, or form a proportion of cells that are swollen, spherical, elongated or pear shaped. This is known as **pleomorphism** and occurs most readily in certain species (e.g. *Streptobacillus moniliformis* and *Yersinia pestis)* in aging cultures on artificial medium and especially in the presence of antagonistic substances such as penicillin, glycine, lithium chloride, sodium chloride in high concentration, and organic acids at low pH. The abnormal cells are generally regarded as **degenerate** or **involution forms**. Many of cells may be non viable, whilst others may grow and revert to normal form when transferred to a suitable environment.

Pleomorphism and involution forms are often due to defective cell wall synthesis. Involution forms may also develop due to the activity of autolytic enzymes.

L-FORMS OF BACTERIA (CELL-WALL-DEFECTIVE ORGANISMS)

- The first isolation of a naturally occurring L-form took place when Kleineberger-Nobel, studying *Streptobacillus moniliformis* in the Lister Institute, London, observed abnormal forms of the bacteria and named them **L-forms** after the Lister Institute, London (hence, the "L").
- These are abnormal growth forms that may arise spontaneously (e.g. in *Streptobacillus moniliformis* and *Bacteriodes* spp.) or by the inhibition of cell wall synthesis in bacteria of normal morphology (in the presence of penicillin or other agents that interfere with cell wall synthesis).
- L-forms may be **unstable** and can revert to the normal bacillary form upon removal of the inducing stimulus as penicillin or other inducing agents and are able to resume normal cell wall synthesis.
- They **lack a rigid cell wall** and, in consequence, regular shape and size.
- They are capable of **growing and multiplying** on a suitable nutrient medium unlike protoplasts.
- L-forms are **difficult to cultivate** and usually require a medium that is solidified with agar as well as having the right osmotic strength.
- Colonies of L-phase organisms on agar media are small and have a characteristic **'fried egg'** appearance, rather like *Mycoplasma,* which usually lack peptidoglycan.
- **L-forms resemble *Mycoplasma* in several ways,** including morphology, type of growth on agar and filterability. It is possible that mycoplasmas represent stable L-forms of as yet unidentified parent bacteria. L-forms should probably be regarded as laboratory artifacts that do not occur or survive to any important extent in natural habitat.
- L-forms are **non-pathogenic to laboratory animals**.
- It has been proposed that the formation of L-forms during antibiotic therapy could explain **relapses after treatment** because, theoretically, such L-forms could revert to their original bacterial forms after cessation of therapy, resulting in a relapse of disease. L-forms in the host may produce **chronic infections** and are relatively resistant to antibiotic treatment, they present special problems in chemotherapy. L-forms are more stable than protoplasts and spheroplasts.

KNOW MORE

Protoplasts

Protoplasts metabolize and grow in size but they do not multiply if maintained on osmotically protective medium. Protoplasts cannot revert to normal bacterial morphology. These are unstable. Hypertonic condition is necessary for their maintenance. These are derived from gram-positive bacteria.

Spheroplasts

Spheroplasts retain outer membrane and entrapped peptidoglycan from gram-negative cells. It differs from the protoplast in that some cell wall material is retained.

They are produced in presence of penicillin. Because spheroplasts retain a residual cell wall, therefore, they are osmotically less sensitive than protoplasts and are often capable of growing on an ordinary agar medium. They are capable of reverting to parent bacterial form when cell wall inhibitor is removed from the culture medium. Spheroplasts may be called unstable L-forms because of their resemblance with L-forms of bacteria. If such cells are able to grow and divide, they are called **L-forms**.

Antibiotics such as penicillin interfere with cell wall synthesis.

☞ KEY POINTS

- Bacteria are unicellular, and most of them multiply by binary fission. They are proparyotes
- **Eukaryotes:** have a membrane-bound nucleus.
- **Shapes:** Most common prokaryotes are either cocci or bacilli, virios, spirilla, spirochetes, mycoplasma.
- **The structure of the prokaryotic cell:** Bacterial cell consists of outer layer (cell wall and plasma membrane) and cellular appendages (capsule, fimbriae and flagella). The cell wall surrounds the plasma membrane and protects the cell from changes in water pressure.
- **Gram-positive cell wall:** The gram-positive cell wall contains a relatively thick layer of peptidoglycan. Teichoic acids stick out of the peptidoglycan layer.
- **Gram-negative cell wall**
    1. Gram-negative bacteria have a lipopolysaccharide lipoprotein-phospholipid outer membrane surrounding a thin peptidoglycan layer.
    2. Periplasm contains a variety of proteins, including those involved in nutrient degradation and transport.
    3. Porins form small channels that permit small molecules to pass through the outer membrane.
- L-forms are gram-positive or gram-negative bacteria that do not make a cell wall.
- The plasma (cytoplasmic) membrane: The plasma membrane encloses the cytoplasm and is selectively permeable. **Mesosomes**, irregular infoldings of the plasma membrane, are artifacts, not true cell structures.
- **Glycocalyx:** The glycocalyx (capsule, slime layer, or extracellular polysaccharide) is a gelatinous polysaccharide and/or polypeptide covering.
- **Flagella:** Flagella are relatively long filamentous appendages consisting of a filament, hook, and basal body that is responsible for most types of bacterial motility.
- **Pili** are short hair-like appendages found on some gram-negative bacteria. Pili are structures used for attachment and facilitate attachment to a surface.
- **Cytoplasm-**The cytoplasm is mostly water, with inorganic and organic molecules, DNA, ribosomes, and inclusions.
- **Ribosomes** carry out protein synthesis; it can be inhibited by certain antibiotics.
- **Inclusions:** Inclusions are reserve deposits found in prokaryotic and eukaryotic cells.
- **Nucleus-**The nuclear area contains the DNA of the bacterial chromosome.
- **Endospores** are a dormant stage produced by members of *Bacillus* and *Clostridium* for survival during adverse environmental conditions.

IMPORTANT QUESTIONS

1. Describe briefly the anatomy of bacterial cell.
2. Draw a labeled diagram of bacterial cell. Write briefly on cell wall of bacteria.
3. Write short notes on:
    - Bacterial cell wall.
    - Cytoplasmic membrane.
    - Plasmids or episomes.
    - Capsule or bacterial capsule.
    - Bacterial flagellae.
    - Fimbriae or pili.
    - Bacterial spores or endospores.
    - L-forms of bacteria.

FURTHER READING

Balows A, Truper HG, Dworkin M, Harder W, Schleifer KH. 1992. The prokaryotes, 2nd edn. New York: Springer-Verlag.

Driks A. Bacillus subtilis spore coat. Microbiol. Mol Biol Rev 1999;63(1):1-20.

Hoppert M, Matyer F. Prokaryotes. American Scientist 1999; 87:518-25.

Moat AG, Foster JW. Microbiology physiology, 3rd edn. Wiley-Liss 1995.

Trun NJ, Marko JF. Architecture of a bacterial chromosome. ASM News 1998;64(5):276-83.

CHAPTER 4

Physiology of Bacteria

Learning Objectives

After reading and studying this chapter, you should be able to:

- Explain generation time of bacteria.
- Describe and draw bacterial growth curve.
- Define the atmospheric requirement of microaerophilic bacteria and capnophilic bacteria.
- Explain redox potential.

PRINCIPLES OF BACTERIAL GROWTH

Growth may be defined as the orderly increase of all of the chemical constituents of the cell. Bacterial growth involves both an increase in the size of individuals and increase in the number of individuals. Whatever the balance between these two processes, the net effect is an increase in the total mass **(biomass).**

A. Bacterial Division

Bacteria divide by **binary fission** where individual cells enlarge and divide to yield two progeny of approximately equal size (Fig. 4.1). Nuclear division precedes cell division and, therefore, in a growing population, many cells carrying two nuclear bodies can be seen. The cell division occurs by a constrictive or pinching process, or by the ingrowth of a transverse septum across the cell. The daughter cells may remain partially attached after division in some species.

Generation Time or Doubling Time

The size of a population of bacteria in a favorable growth medium increases exponentially. The interval of time between two cell divisions, or the time required for a bacterium to give rise to two daughter cells under optimum conditions, is known as the **generation time or doubling time.**

Examples: In coliform bacilli and many other medically important bacteria, it about 20 minutes, in tubercle bacilli it is about 20 hours and in lepra bacilli it is about 20 days. It is often difficult to grasp fully the scale of exponential microbial growth.

As bacteria reproduce so rapidly and by geometric progression, a single bacterial cell can theoretically give rise to 10^{21} progeny in 24 hours, with a mass of approximately 4,000 tonnes. In actual practice, exponential growth cannot be sustained indefinitely in a closed system (batch culture) with limited available nutrients.

Colonies: Bacteria growing on solid media form **colonies.** Each colony represents a clone of cells derived from a single parent cell. In contrast to growth in broth, far less is known about the state of the bacteria in a mature macroscopic colony on an agar plate. It is likely that all phases of growth are represented in colonies,

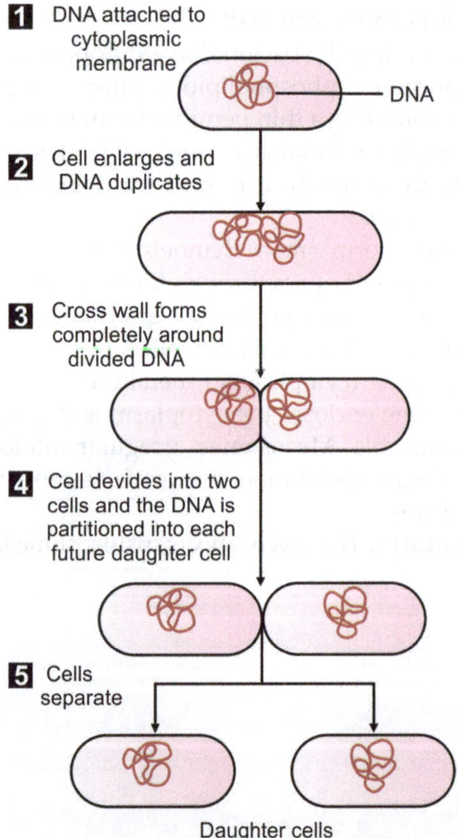

Fig. 4.1: Binary fission in bacteria

depending on the location of a particular cell and age of the culture. In liquid media, growth is diffuse.

Bacterial count: Growth of bacteria is diffuse in liquid media and they form colonies on solid media. Each colony consists of a clone of cells derived from a single parent cell. Bacteria in a culture medium or clinical specimen can be counted by two methods:

1. Total count

This is total number of bacteria present in a specimen irrespective of whether they are living or dead. Total count is done by counting the bacteria under microscope using counting chamber and by comparing the growth with standard opacity tubes.

2. Viable count

This measures only viable (living) cells which are capable of growing and producing a colony on a suitable medium. The *viable count* measures the number of living cells, that is, cells capable of multiplication.

Determination of Viable Counts—Dilution or Plating Methods

1. Dilution method

In **dilution method**, the suspension is diluted to a point beyond which unit quantities do not yield growth when inoculated into suitable liquid media. With varying dilutions several tubes are inoculated and the viable counts calculated statistically from the number of tubes showing growth. The method does not give accurate values, but is used in the 'presumptive coliform count' in drinking water.

2. Plating method

In the plating method, appropriate dilutions are inoculated on solid media by:

i. **Pour-plate method**
ii. **Spread plate method**—in which serial dilutions are dropped on the surface of dried plates and colony counts obtained.

Detection and Measurement of Bacterial Growth

1. Direct cell counts
2. Viable cell counts
3. Measuring biomass
4. Measuring cell products

B. Bacterial Growth Curve

If a suitable liquid medium is inoculated with bacterium and incubated, its growth follows a definitive course. Small samples are taken at regular intervals after inoculation and plotted in relation to time. A plotting of the data will yield a characteristic growth curve (Fig. 4.2). The changes of slope on such a graph indicate the transition from one phase of development to another.

Phases of Bacterial Growth Curve

The bacterial growth curve can be divided into four major phases: (i) lag phase (ii) exponential or log (logarithmic) phase (iii) stationary phase, and (iv) decline phase. These phases reflect the physiologic state of the organisms in the culture at that particular time.

1. Lag phase

When microorganisms are introduced into fresh culture medium, usually no immediate increase in cell number occurs, and therefore this period is called the **lag phase.** After inoculation, there is an increase in cell size at a time when little or no cell division is occurring. During this time, however, the cells are not dormant. This initial period is the time required for adaptation to the new environment, during which the necessary enzymes and metabolic intermediates are built up in adequate quantities for multiplication to proceed. (The situation is analogous to a factory being equipped to produce automobiles; there is considerable tooling-up activity but no immediate increase in the automobile population).

The lag phase varies considerably in length with the species, nature of the medium, size of inoculum and environmental factors such as temperature and nutrients present in the new medium.

2. Log (logarithmic) or exponential phase

Following the lag phase, the cells start dividing and their numbers increase exponentially or by geometric progression with time. If the logarithm of the viable count is plotted against time, a straight line will be obtained. The population is most uniform in terms of chemical and physiological properties during this phase. Therefore, exponential phase cultures are usually used in biochemical and physiological studies.

Exponential phase is of limited duration because of (i) exhaustion of nutrients; (ii) accumulation of toxic metabolic end products; (iii) rise in cell density, (iv) change in pH; and (v) decrease in oxygen tension (in case of aerobic organisms).

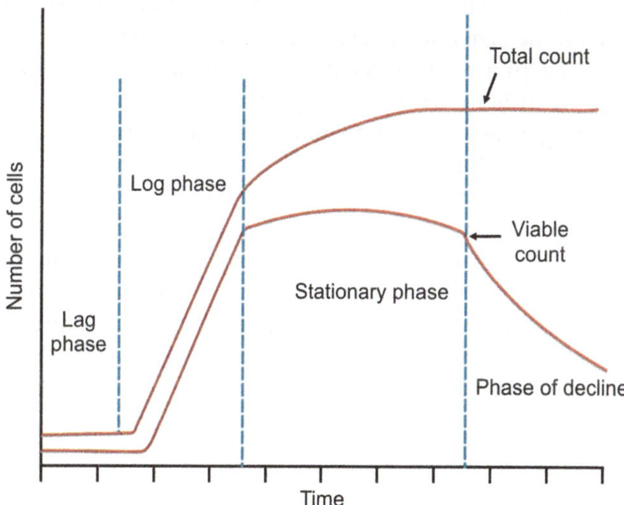

Fig. 4.2: Bacterial growth curve. The viable count shows lag, log, stationary and decline phases. In the total count, the phase of decline is not evident

The log phase is the time when cells are most active metabolically and is preferred for industrial purposes. However, during their log phase of growth, microorganisms are particularly sensitive to adverse conditions. Radiation and many antimicrobial drugs, e.g. the antibiotic penicillin—exert their effect by interfering with some important step in the growth process and are, therefore, most harmful to cells during this phase.

3. Stationary phase

After a varying period of exponential growth, cell division stops due to depletion of nutrients and accumulation of toxic products. Eventually growth slows down, and the total bacterial cell number reaches a maximum and stabilizes. The number of progeny cells formed is just enough to replace the number of cells that die. The growth curve becomes horizontal. The viable count remains stationary as an equilibrium exists between the dying cells and the newly formed cells.

4. Decline or death phase

The death phase is the period when the population decreases due to cell death. Eventually the rate of death exceeds the rate of reproduction, and the number of viable cells declines. Like bacterial growth, death is exponential cell death may also be caused by autolysis besides nutrient deprivation and buildup of toxic wastes. Finally, after a variable period, all the cells die and culture becomes sterile.

When the total count is plotted, it parallels the viable count up to the stationary phase, but it continues steadily without any phase of decline. Even the total count shows a phase of decline with autolytic enzymes.

Association of Growth Curve and Cell Changes

The various stages of the growth curve are associated with morphological and physiological alterations of the cells. It has been possible to define the effect of growth rate on the size and specialized growth techniques.

1. **Lag phase:** Bacteria have the maximum cell size towards the end of the lag phase.
2. **Log phase:** Cells are smaller and stain uniformly in the log phase.
3. **Stationary phase:** In the stationary phase, cells frequently are Gram variable and show irregular staining due to the presence of intracellular storage granules. Sporulation occurs at this stage and also many bacteria produce secondary metabolic products such as exotoxins and antibiotics.
4. **Decline phase:** In the phase of decline, involution forms are common.

Batch Culture or Closed System

In the laboratory, bacteria are typically grown in broth contained in a tube or flask, or on an agar plate. These are considered **batch or closed systems.**

Continuous Culture

To maintain cells in a state of continuous growth, nutrients must be continuously added and waste products removed. This is called **continuous culture** or **open system.**

Continuous culture is using a **chemostat** in which cells of a growing culture are continuously harvested and nutrients continuously replenished. The second type of continuous culture system, the **turbidostat,** has a photocell that measures the absorbance or turbidity of the culture in the growth vessel. The flow rate of media through the vessel is automatically regulated to maintain a predetermined turbidity or cell density.

BACTERIAL NUTRITION

The minimum nutritional requirements for growth and multiplication of bacteria are water, a source of carbon, a source of nitrogen and some inorganic salts. The water content of bacterial cells can vary from 75 to 90 percent of the total weight and is the vehicle for the entry of all cells and for the elimination of all waste products. It participates in the metabolic reactions and also forms an integral part of the protoplasm.

Categories of Requirements for Microbial Growth

The requirements for microbial growth can be divided into two main categories: (i) chemical and (ii) physical.

1. Chemical Requirements

Chemical requirements include sources of carbon, nitrogen, sulfur, phosphorus, trace elements, oxygen, and organic growth factors.

A. Major elements (Macroelements or macronutrients)

Elements that make up cell constituents are called **major elements (macroelements or macronutrients) and o**ver 95 percent of cell dry weight is made up of a few major elements. These include carbon, oxygen, hydrogen, nitrogen, sulfur, phosphorus, potassium, magnesium, calcium, and iron.

B. Trace elements

Some elements, termed **trace elements or micronutrients** are required in very minute amounts by all cells. They include **cobalt, zinc, copper, molybdenum,** and **manganese.** These elements form parts of enzymes or may be required for enzyme function. They aid in the catalysis of reactions and maintenance of protein structure. Very small amounts of these trace elements are found in most natural environments, including water.

2. Growth Factors

Some bacteria require certain organic compounds in minute quantities known as growth factors or bacterial vitamins. A **growth factor** is an organic compound which a cell must contain in order to grow, but which it is unable to synthesize. These low molecular weight compounds that must be provided to a particular bacterium are called **growth factors or bacterial vitamins**. Growth factors are called **'essential'** when growth does not occur in their absence, or **'accessory'** when they enhance growth without being absolutely necessary for it. In many cases, bacterial vitamins are identical with the vitamins necessary for mammalian nutrition, particularly those belonging to the B group, thiamine, riboflavin, nicotinic acid, pyridoxine, folic acid and vitamin B_{12}.

Microbiological Assays

Ideally the amount of growth resulting is directly proportional to the quantity of growth factor present. This is the principle of **microbiological assays** which are specific, sensitive, and simple. They still are used in the assay of substances like **vitamin B_{12} and biotin.**

3. Energy Sources

Organisms derive energy either from sunlight or metabolizing chemical compounds.

Phototrophs: Organisms that gain energy from light are called **phototrophs** (*photo* means "light").

Chemotrophs: Organisms that obtain energy by metabolizing chemical compounds are called **chemotrophs** (*chemo* means "chemical").

2. Physical Factors Influencing Microbial Growth

1. **Temperature**
2. **Oxygen**
3. **Carbon dioxide**
4. **Moisture and drying**
5. **pH**
6. **Light**
7. **Osmotic effect**
8. **Mechanical and sonic stresses**

1. Temperature

Optimum Temperature

Each bacterial species has an optimal temperature for growth and a temperature range above and below which growth is blocked. The temperature at which growth occur best is known as the **'optimum temperature'**. Thus, bacteria pathogenic for humans usually grow at 37°C (our body temperature). Bacteria are divided into three groups on the basis of temperature ranges through which they grow:

i. **Mesophilic:** Bacteria which grow between **10°C and 45°C**, with optimal growth between 20-40°C.
   Examples: All parasites of warm blooded animals are mesophilic and it is the largest group of bacteria including all of the pathogenic forms.
ii. **Psychrophilic:** Psychrophilic bacteria (cold-loving) are organisms that grow between –5 to 30°C, optimum at **10 to 20°C.**
   Examples: They are soil and water saprophytes and though not of direct medical importance, may cause spoilage of refrigerated food. These organisms may be capable of growth in food and pharmaceuticals stored at normal refrigeration temperatures (0-8°C).
iii. **Thermophilic:** Thermophiles (heat-loving) have growth range 25-80°C, optimum at 50-60°C. They may cause spoilage of underprocessed canned food and can be a source of proteins with remarkable thermotolerant properties such as taq polymerase, the key enzyme used in the polymerase chain reaction.
   Examples: Some thermophiles (like *Bacillus stearothermophilus*) form spores that are exceptionally thermotolerant.

2. Oxygen

Based on their O_2 requirements, prokaryotes can be separated into aerobes and anaerobes.

A. Aerobic bacteria

Require oxygen for growth and may be:

i. **Obligate aerobes:** They have an absolute or obligate requirement for oxygen (O_2), like the cholera vibrio.
ii. **Facultative anaerobes:** They are ordinarily aerobic but can also grow in the absence of oxygen, though less abundantly, e.g. *Staphylococcus* spp.; *Escherichia coli*, etc. Most bacteria of medical importance are facultative anaerobes.
iii. **Microaerophilic organisms:** They grow best at low oxygen tension (~5%) e.g. *Campylobacter spp., Helicobacter spp.*

B. Anaerobic bacteria

Grow in absence of oxygen.

Obligate anaerobes: They may even die on exposure to oxygen, e.g. *Clostridium tetani,* Bacteroides fragilis.

3. Carbon Dioxide

All bacteria require small amount of carbon dioxide for growth. Thus, this requirement is usually met by the carbon dioxide present in the atmosphere, or produced endogenously by cellular metabolism. Some organisms such as *Brucella abortus*, require much higher levels of carbon dioxide (5-10%) for growth, especially on fresh isolation **(capnophilic)**. Pneumococci and gonococci are other capnophilic which grow better in air supplemented with 5 to 10 percent CO_2.

4. Moisture and Drying

Moisture is very essential for the growth of the bacteria because water is essential ingredient of bacterial protoplasm and hence drying is lethal to cells. However, the effect of drying varies in different species

Examples:

i. *Treponema pallidum* are highly sensitive, while others like **staphylococci** withstand resistant to desiccation for months.
ii. **Bacterial spores: Bacterial spores** are particularly resistant to desiccation and survive in the dry state for several decades.
iii. **Freeze drying or lyophilization:** Drying in vacuum in the cold **(freeze drying or lyophilization)** is a method of preservation of bacteria, viruses and many labile biological materials. On a larger scale, it is used for preserving therapeutic antisera, human plasma, antibiotics and vaccines.

5. pH

Most bacteria can live and multiply within the range of pH 5 (acidic) to pH 8 (basic) and have a pH optimum near neutral (pH 7).

Most pathogenic bacteria grow best at a neutral or slightly alkaline pH (7.2 to 7.6). Some **acidophilic** bacteria such as **lactobacilli** grow under acidic conditions while *cholerae vibrio,* grow at high degrees of alkalinity (well above pH 8), whereas most other bacteria grow in a range of 6 to 7.5. Numerous fungi grow well at pH 4 or 5.

6. Light

Darkness provides a favorable condition for growth and viability of bacteria. Bacteria are sensitive to ultraviolet light and other radiations as ultraviolet rays from direct sunlight or a mercury lamp are bactericidal. Bacteria are also killed by ionizing radiations.

Exposure to light may influence pigment production. Photochromogenic mycobacteria form pigment only on exposure to light.

7. Osmotic Effect

Tolerance to osmotic variation: Bacteria are more tolerant to osmotic variation because of the mechanical strength of the cell wall. Except for the mycoplasma and other cell wall defective organisms, the majority of the bacteria are osmotically tolerant.

Plasmolysis: When microorganisms with rigid cell walls are placed in a hypertonic environment, water leaves and the plasma membrane shrinks away from the wall, a process known as **plasmolysis.** This occurs more readily in gram-negative bacteria than in gram-positive bacteria.

Plasmoptysis: Sudden transfer of bacteria from concentrated solution to distilled water may also cause **plasmoptysis** due to excessive osmotic imbibition of water leading to swelling and rupture of the cell.

8. Mechanical and Sonic Stresses

In spite of tough walls of bacteria, they may be ruptured by mechanical stress such as grinding or vigorous shaking with glass beads. Exposure to ultrasonic vibration may also disintegrate bacteria.

BACTERIAL METABOLISM

Metabolism of the substance is defined as the series of changes of substance (carbohydrate, protein or fat) within the bacterial cell from absorption to elimination. Metabolism may be **aerobic** or **anaerobic** and can be divided into two classes of chemical reactions: those that release energy and those that require energy.

Energy Production

There are three critical processes of bacterial energy production; **aerobic respiration, anaerobic respiration and fermentation (Fig. 4.3).** There are two general aspects of energy production: the concept of **oxidation-reduction** and the **mechanisms of ATP generation**

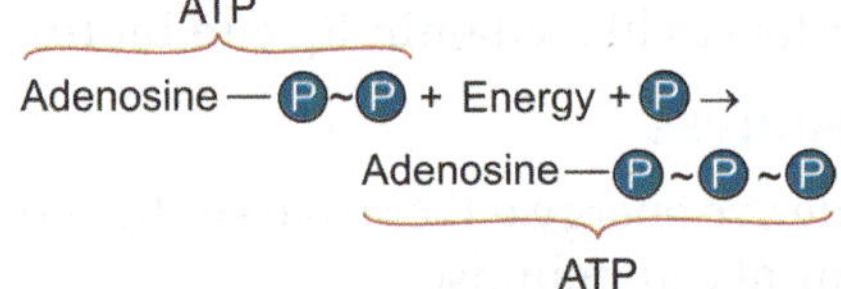

Oxidation-reduction (Redox) Reactions

Oxidation is the removal of electrons, and reduction is the addition of electrons. In other words, each time one substance is oxidized, another is simultaneously reduced. The pairing of these reactions is called **oxidation-reduction or a redox reaction** (Fig. 4.4).

Generation of ATP

Much of the energy released during oxidation-reduction reactions is trapped within the cell by the formation of ATP. Specifically, a phosphate group, P, is added to ADP with the input of energy to form ATP:

The symbol ~ designates a "high-energy" bond—that is, one that can readily be broken to release usable energy. The high-energy bond that attaches the third

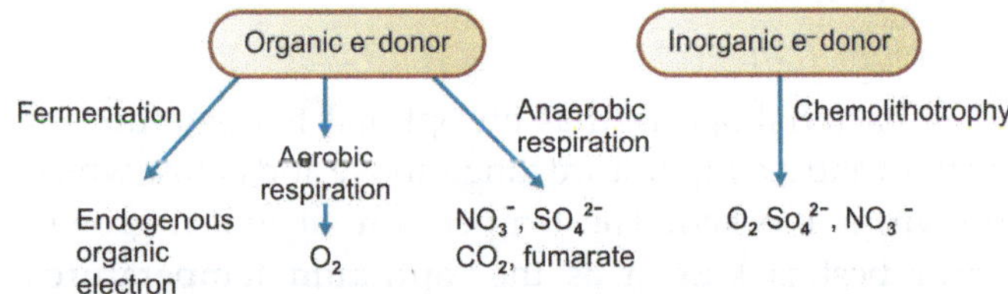

Fig. 4.3: Patterns of energy release

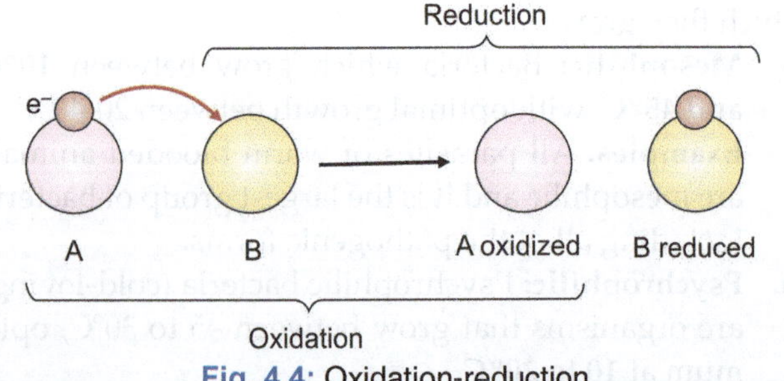

Fig. 4.4: Oxidation-reduction

P (phosphate group) in a sense contains the energy stored in this reaction. When this P (phosphate group) is removed, usable energy is released. **The addition of P** (phosphate group) **to a chemical compound is called phosphorylation.**

Mechanisms of phosphorylation: There are two general mechanisms for ATP production in bacterial cells.

i. Substrate-level phosphorylation

In substrate level phosphorylation, high energy phosphate bonds produced by the central pathways are donated to adenosine diphosphate (ADP) to form ATP. Additionally, pyruvate, a primary intermediate in the central pathways, serves as the initial substrate for several other pathways that also can generate ATP by substrate level phosphorylation. These other pathways constitute **fermentative metabolism,** which does not require oxygen and produces various end products, including alcohols, acids, carbon dioxide, and hydrogen. The specific fermentative pathways used, and hence end products produced, vary with different bacterial species. Detecting these products serves as an important basis for laboratory identification of bacteria. Fermentation is carried out by both obligate and facultative anaerobes.

ii. Oxidative phosphorylation

In oxidative phosphorylation, an electron transport system is involved that conducts a series of electron transfers from reduced carrier molecules such as NADH2 and NADPH2, produced in the central pathways, to a terminal electron acceptor. The energy produced by the series of oxidation-reduction reactions is used to generate ATP from ADP. The process is known as **aerobic respiration** when oxidative phosphorylation uses oxygen as the terminal electron acceptor. **Anaerobic respiration** refers to processes that use final electron acceptors other than oxygen.

OXIDATION-REDUCTION (O-R) POTENTIAL (REDOX POTENTIAL)

The ability of a substance to take up or part with electrons is known as **oxidation-reduction or redox or Eh potential.** When an unattackable electrode is immersed in a solution, electrical potential is set up between the electrode and the solution. This depends upon the state of oxidation or reduction of the solution. This electrode potential (Eh) can be measured in millivolts. The more oxidized the system, the higher the potential and in reduced system potential is lower. Obligate anaerobes are unable to grow unless the Eh is at least as low as 0.2 volt.

Toxic Derivatives of Oxygen

i. Obligate aerobes and facultative anaerobes

Some oxidation-reduction (redox) reactions occurring in the presence of oxygen commonly result in the formation of the reactive **superoxide (O_2^-)** and **hydroxyl ($OH^\cdot$)** radicals as well as **hydrogen peroxide (H_2O_2).** These products of oxygen reduction are extremely toxic because they are powerful oxidizing agents and rapidly destroy cellular constituents. To cope with this, **obligate aerobes** and facultative anaerobes usually contain the enzymes **superoxide dismutase (SOD)** and **catalase,** which catalyzes the destruction of superoxide radical and hydrogen peroxide respectively. Peroxidase also can be used to destroy hydrogen peroxide.

$$2O_2^- + 2H^+ \xrightarrow{\text{superoxide dismutase}} O_2 + H_2O_2$$

$$2H_2O_2 \xrightarrow{\text{catalase}} 2H_2O + O_2$$

ii. Obligate anaerobes

All strict anaerobes lack both enzymes or have them in very low concentrations and therefore cannot tolerate O_2. Another reason might be that anaerobes possess essential enzymes that are active only in the reduced state.

KNOW MORE

Detection and Measurement of Bacterial Growth

A. Direct cell counts

1. **Direct microscopic count:** In this method known as the direct microscopic count, a measured volume of a bacterial suspension is placed within a defined area on a microscope slide. *Breed count method* is used to count the number of bacteria in milk. A specially designed slide called a *Petroff-Hausser cell counter* is also used in direct microscopic counts.
2. **Cell-counting instruments:** Coulter counters and flow cytometers count total cells in dilute solutions.

B. Viable cell counts

The *viable count* measures the number of living cells, that is, cells capable of multiplication.

C. Measuring biomass

1. Turbidity; 2. Dry weight; 3. Chemical Constitents.

D. Measuring cell products

1. Acid; 2. Gases; 3. ATP (Firefly luciferase catalyzes light-emitting reaction when ATP is present).

Groups of Proparyotes

1. Autotrophs

Organisms that can use **inorganic carbon** in the form of carbon dioxide as their carbon source are called **autotrophs (auto means self).** They are capable of independent existence in water and soil and are of no medical importance, though they are of vital concern in agriculture and the maintenance of soil fertility.

Role of autotrophs

Carbon fixation: They play a critical role in the cycling of carbon in the environment because they can convert inorganic carbon (CO_2) to an organic form, the process of **carbon fixation.** The earth would quickly run out of organic carbon without carbon fixation, which is essential to humans and other animals.

2. Heterotrophs

Organisms that use **organic carbon** are called **heterotrophs (hetero—different; troph—nourishment).** They are unable to utilize carbon dioxide as the sole source of carbon and use reduced, preformed organic molecules as carbon sources.

KEY POINTS

- The growth of a population is an increase in the number of cells. Most bacteria multiply by binary fission.
- The time required for a population to double in number is the **generation time**.
  1. **Total count** is total number of bacteria present in a specimen irrespective of whether they are living or dead.
  2. **Viable count:** This measures only viable (living) cells which are capable of growing and producing a colony on a suitable medium.
- **Bacterial growth curve:** The bacterial growth curve can be divided into four major phases: lag phase, exponential or log (logarithmic) phase, stationary phase, and decline phase.

The requirements for microbial growth—chemical and physical:

Chemical Requirements

All organisms require a carbon source, nitrogen, and other chemicals required for microbial growth include sulfur, phosphorus, trace elements, and for some microorganisms, organic growth factors.

Physical Factors Influencing Microbial Growth

- **Temperature:** Organisms can be grouped as psychrophiles, psychrotrophs, mesophiles, thermophiles, or hyperthermophiles based on their optimum growth temperatures.
- **Oxygen (O_2) Requirements:** Organisms can be grouped as obligate aerobes, obligate anaerobes, facultative anaerobes, microaerophiles based on their oxygen (O_2) requirements.
- **CO_2:** Microbes that grow better at high CO_2 concentrations are called ***capnophiles***.
- **pH:**
- **Light:**
- **Osmotic effect:**
- **Mechanical and sonic stresses:**
  Bacterial metabolism: Metabolism may be **aerobic** or **anaerobic** and can be divided into two classes of chemical reactions: those that release energy and those that require energy.
- Oxidation is the removal of electrons, and reduction is the addition of electrons.The pairing of these reactions is called **oxidation-reduction or a redox reaction.**

IMPORTANT QUESTIONS

1. Draw a typical bacterial growth curve and describe it.
2. What are the heterotrophic bacteria? Discuss the nutritional and physical requirements for the growth of the bacteria.
3. Write short notes on:
   Bacterial growth curve/growth phases of bacteria.
   Redox potential

FURTHER READING

Dawes IW, Sutherland W. Microbial Physiology, 2nd edn., Oxford: Blackwell 1991.

Ingraham JL, Maaloe O, Neidhardt FE. Growth of Bacterial Cell. Sinauer 1983.

Moat AO, Foster JW. Microbial Physiology, 3rd edn; New York: Wiley-Liss, 1995.

Stanier RY, et al. The Microbial World, 5th edn. New Jersey: Prentice-Hail, 1986.

CHAPTER 5

Sterilization and Disinfection

Learning Objectives

After reading and studying this chapter, you should be able to:

- Define the following terms: sterilization, disinfection and antisepsis.
- Describe various agents used in sterilization.
- Describe sterilization by moist heat.
- Describe the different heat methods and their respective applications.
- Describe the following: Pasteurization, tyndallization or intermittent sterilization or fractional sterilization, inspissation or serum inspissator, hot air oven.
- Explain principle and functioning of autoclave.
- Describe filtration—uses and types.
- Discuss types of radiation and their uses.
- Give the mechanism of action for each type of chemical agent commonly used in antiseptics and disinfectants.
- Describe the following: Aldehydes as disinfectants, uses of formaldehyde and glutaraldehyde as disinfectants.
- Explain vapor-phase disinfectants or gaseous sterilization and discuss the role of ethylene oxide in sterilization of disposable items.
- Describe various tests used for testing of disinfectants.

INTRODUCTION

From the beginning of recorded history, people have practiced disinfection and sterilization, even though the existence of microorganisms was long unsuspected. The Egyptians used fire to sterilize infectious material and disinfectants to embalm bodies, and the Greeks burned sulfur to fumigate buildings. Mosaic law commanded the Hebrews to burn any clothing suspected of being contaminated with the leprosy bacterium.

Applications of Sterilization and Disinfection

Important applications in practical microbiology and in the practice of medicine and surgery.

1. **Aseptic techniques:** Used in microbiological research, the preservation of food and the prevention of the disease.
2. **Sterile apparatus and culture media: Laboratory work with pure cultures** requires the use of sterile apparatus and culture media.
3. **The need to avoid infecting the patient:** Requires the use of equipment, instruments, dressings and parentral drugs that are free from living microorganisms, or at least from those which may give rise to infection.

DEFINITIONS OF FREQUENTLY USED TERMS

Sterilization

Sterilization [Latin *sterilis*, unable to produce offspring or barren] is defined as the process by which an article, surface, or medium is freed of all living microorganisms either in the vegetative or spore state. When sterilization is achieved by a chemical agent, the chemical is called **sterilant**.

Disinfection

Disinfection is the killing, inhibition, or removal of microorganisms that may cause disease.

Antiseptics

Antiseptics are chemical agents applied to tissue to prevent infection by killing or inhibiting pathogen growth; they also reduce the total microbial population. Antiseptics are generally not as toxic as disinfectant because they must not destroy too much host tissue.

METHODS OF STERILIZATION AND DISINFECTION (TABLE 5.1)

A. Physical agents
B. Chemical agents.

Table 5.1: Methods of sterilization and disinfection

A. Physical agents
- 1. Sunlight
- 2. Drying
- 3. Heat
  - a. Dry heat
    - I. Incineration
    - II. Red heat
    - III. Flaming
    - IV. Hot air sterilizer
    - V. Microwave ovens.
  - b. Moist heat
    - I. Pasteurization
    - II. Boiling
    - III. Steam under normal pressure
    - IV. Steam under pressure.
- 4. Filtration
- 5. Radiation
- 6. Ultrasonic and sonic vibrations.

B. Chemical agents
- 1. Agents that damage the cell membranes
  - a. Surface-active disinfectants
  - b. Phenolic compounds
  - c. Alcohols.
- 2. Agents that damage proteins
  - a. Acids and alkalies
  - b. Alcohols.
- 3. Agents that modify functional groups of proteins and nucleic acids
  - a. Heavy metals
  - b. Oxidizing agents
  - c. Dyes
  - d. Alkylating agents.

a. Physical Agents

1. Sunlight

Sunlight has appreciable bactericidal activity and plays an important role in the spontaneous sterilization that occurs under natural conditions. Its disinfectant action is due primarily to its content of ultraviolet rays. It is a natural method of sterilization in cases of water in tanks, rivers and lakes. Direct sunlight, as in tropical countryside where it is not filtered off by impurities in the atmosphere, has an active germicidal effect due to the combined effect of ultraviolet and heat rays.

2. Drying

Water constitutes four-fifths of the weight of the bacterial cell and is essential for the growth of bacteria. Therefore, drying in air has a deleterious effect on many bacteria. However, this method is unreliable and is only of theoretical interest. Spores are unaffected by drying.

3. Heat

Heat is the most reliable and universally applicable method of sterilization and, wherever possible, should be methods of choice. Either **dry** or **moist heat** may be applied. Materials that may be damaged by heat can be sterilized at lower temperature, for longer periods or by repeated cycles.

Factors influencing sterilization by heat:

1. **Nature of heat**- Dry heat or moist heat
2. **Temperature and time:** The time required for sterilization is inversely related to the temperature of exposure. This relationship may be expressed by the term **thermal death point (TDT)**, which refers to the minimum time required to kill a suspension of organisms at a predetermined temperature in a specified environment.
3. **The number of microbes:** The sterilization time is related to the number of organisms in the suspension. The more microbes there are to begin with, the longer it takes to eliminate the entire population.
4. **Characteristics of the organisms:** The sterilization time is also related to the presence or absence of **spores,** the **strain** and **characteristics of organism**. In general, vegetative bacteria and viruses are more susceptible and bacterial spores the more resistant, to sterilizing and disinfecting agents.
5. **The nature of contaminated material:** The nature of the material in which the organisms are heated affects the rate of killing. The presence of organic substances, proteins, nucleic acids, starch, gelatin, sugar, fats and oils, increase the thermal death time. The presence of disinfectants and high acid or alkaline pH hasten bacterial killing.

Mechanism of Action

Dry heat: The lethal effect of dry heat, or dessication in general, is usually due to **protein denaturation, oxidative damage,** and **toxic effects of elevated levels of electrolytes**.

Moist heat: Kills microorganisms by **coagulation and denaturation of their enzymes** and **structural proteins.**

a. Dry Heat Sterilization

- i. Red heat
- ii. Flaming
- iii. Incineration
- iv. Hot air sterilizer
- v. Microwave ovens

i. Red Heat

Inoculating wires loops and points of forceps are sterilized by holding them almost vertically in a Bunsen flame until red hot along their whole length, almost up to the tip of their metal holder. The points of forceps and the surface of searing spatulae may also be heated until red.

ii. Flaming

Scalpel blades, glass slides, mouth of culture tubes and bottles are exposed to a flame for a few seconds without heating them to become red hot.

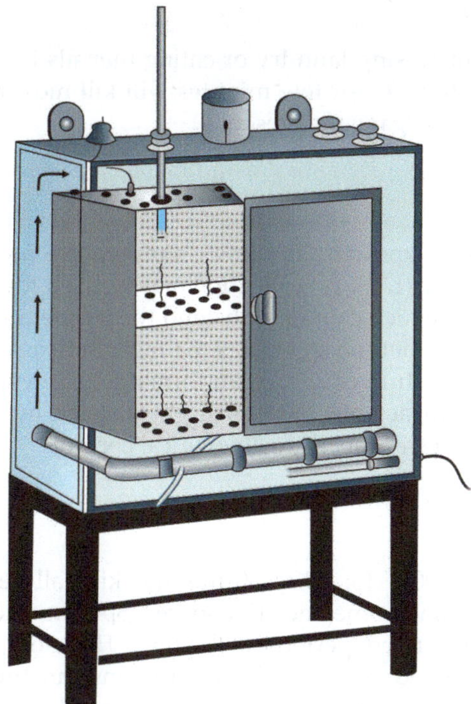

Fig. 5.1: Hot air oven

iii. Incineration

This is an efficient method for the sterilization and disposal of contaminated materials at a high temperature. By this method, infective material is reduced to ashes, such as pathological waste materials, surgical dressings, contaminated material, animal carcasses and other clinical wastes are safely destroyed by incineration.

iv. Hot Air Oven

Hot air sterilizer is the most widely used method of sterilization by dry heat. It is used to process materials which can withstand high temperatures for length of time needed for sterilization by dry heat, but which are likely to be affected by contact with steam. Hot air oven is electrically heated, with heating (Fig. 5.1). It should be fitted with a fan to provide forced air circulation throughout the oven chamber, a temperature indicator, a control thermostat and timer, open mesh shelving and adequate wall insulation.

Preparation of Load

1. **No overloading:** It must **not be overloaded** and the individual articles or packs of the load are positioned to allow free circulation of hot air between and around the item.
2. **Articles:** Should be thoroughly **clean and dry.**
3. **Glassware:** Should be perfectly **dry before being placed in the oven.**
4. **Test tubes and flasks:** Should be **wrapped in paper.**
5. **Rubber materials**, except silicon rubber, will not withstand the sterilizing temperature.
6. **Cotton plugs:** May get charred at 180°C.
7. **Heat-sensitive materials:** Dry heat sterilization is slow and not suitable for heat-sensitive materials like many plastic and rubber items.

Sterilizing Cycle

i. The sterilization hold time

It is set to **160°C for 2 hours or 170°C for 1 hour, or 180°C for 30 minutes (Table 5.2).**

ii. Cutting instruments

Such as those used in ophthalmic surgery, should ideally be sterilized at **150°C for two hours**.

iii. Oils, glycerol and dusting powder

The British Pharmacopoeia recommends a holding time of **one hour at 150°C** for oils, glycerol and dusting powder.

Cooling

Cooling may take up to several hours and, therefore do not attempt to open the chamber door until the chamber and load have **cooled below 80°C**. Glassware is liable to crack if cold air is admitted suddenly while it is still very hot.

Uses of Hot Air Oven

It is a method of choice for sterilization of:

1. **Glassware:** Such as tubes, flasks, measuring cylinders, all-glass syringes, glass petri dishes and glass pipettes.
2. **Metal instruments:** Such as forceps, scissors and scalpels.
3. **Nonaqueous materials and powders, oils and greases** in sealed containers and swab sticks packed in test tubes.

Sterilization Controls

Two types of controls are available:

A. Biological control

An envelope containing a filter paper strip impregnated with 10^6 spores of *Bacillus subtilis* subsp *niger* is inserted into suitable packs. After sterilization is over, the strips are removed and inoculated into tryptone soy broth and incubated anerobically at 37ºC for five days. No growth of *Bacillus subtilis* subp *niger* indicates proper sterilization.

Table 5.2: Minimum recommended times for heat sterilization

Process	*Temperature (C)*	*Hold time (min)*
Dry heat	160	120
	170	60
	180	30
Moist heat	121	15
	126	10
	134	3

B. Chemical indicator

A chemical indicator such as **Browne's tubes No. 3** containing red solution is inserted in each load and a color change from red to green is observed which indicates proper sterilization.

C. Thermocouples

Thermocouples may also be used periodically.

v. Microwave Ovens

In microwave ovens, the heating effect is not uniform and no reliable sterilization process using microwaves is presently available.

b. Moist heat

Moist heat is divided into three forms:

A. **At temperature below 100°C**
B. **At a temperature of 100°C**
C. **At temperature above 100°C.**

A. At a Temperature Below 100°C

Heat labile fluids may be disinfected, though not sterilized, by heating at 56°C for 1 hour. Such treatment is sufficient to kill vegetative bacteria but not spores. It includes:

i. Pasteurization of milk

Disinfection by moist heat at temperature below 100°C is termed **pasteurization.** Milk can be pasteurized in two ways. The temperature is employed either **63°C for 30 minutes** (holder **method**) or **72°C for 15-20 seconds** (the **flash method**) followed by **rapid cooling to 13°C or lower.** Large quantities of milk are now usually subjected **to flash pasteurization** or **high-temperature short-term (HTST) pasteurization.**

The dairy industry also sometimes uses **ultrahigh-temperature (UHT) sterilization.** Milk and milk products are heated at 140 to 150°C for 1 to 3 seconds.

All nonsporing pathogens such as mycobacteria, brucellae and salmonellae are destroyed by these processes. *Coxiella burnetii* is relatively heat resistant and may survive the holder method.

ii. Vaccine preparation

Vaccines prepared from nonsporing bacteria may be inactivated in a water bath at 60°C for one hour as most vegetative bacteria are killed at this tempratuture and time. Serum or body fluids containing coagulable proteins can be sterilized by heating for one hour at 56°C in a water bath on several successive days.

iii. Inspissation

Media such as Lowenstein-Jensen and Loeffler's serum are rendered sterile by heating at 80 to 85°C for half an hour on three successive days **(fractional sterilization).** This process is called **inspissation** and instrument used is called *inspissator.*

iv. Water bath

Washing or rinsing laundry or eating utensils in water bath at 70 to 80°C for few minutes will kill most nonsporing microorganisms present.

Low Temperature Steam Formaldehyde (LTSF) Sterilization

A method known as low temperature steam-formaldehyde (LTSF) sterilization may be used for sterilizing items which cannot withstand the temperature of 100°C. In this method steam at subatmospheric pressure at the temperature of 75°C with formaldehyde vapor is used. The efficacy of LTSF sterilizers is tested by using *Bacillus stearothermophilus* as biological control.

B. Temperature at 100°C

a. Boiling

Boiling at 100°C for 10 to 30 minutes kills all vegetative spores and some bacterial spores. Sporing bacteria require prolonged periods of boiling. Therefore, it is not recommended for sterilization of instruments for surgical procedures and should be regarded only as a means of disinfection. Hard water should not be used. Addition of 2 percent sodium bicarbonate may promote sterilization. The procedure is not to be used if better methods are available.

Uses

i. For the **disinfection of surgical instruments prior to processing.**
ii. **For** the disinfection of medical and surgical equipment—when sterility is not essential in emergency or under field conditions.

b. Steam at atmospheric pressure at 100°C for 90 minutes

Pure steam in equilibrium with boiling water at normal atmospheric pressure, i.e. 'free steam', has a temperature of 100°C and is used to disinfect selective heat-labile culture media in the laboratory which may decompose if subjected to higher temperatures. This can be provided by the traditional **Koch and Arnold steamer** (or by the multipurpose autoclave).

Koch and Arnold Steamer

A Koch and Arnold steamer consists of upright metal tank with a removable lid incorporating a chimney, which allows the free escape of steam. Water is added on the bottom and there is a perforated shelf above water level. Articles to be sterilized are placed on this perforated shelf just above the level of water. Water in the bottom of the tank is heated by gas or electricity (Fig. 5.2). They are exposed to steam at atmospheric pressure for 90 minutes. One single exposure to steam for 90 minutes ensures complete sterilization. Most of the vegetative forms are killed by this method except thermophiles.

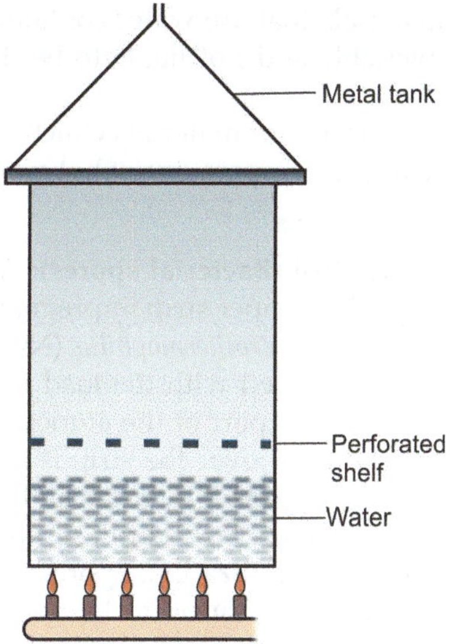

Fig. 5.2: Steamer

c. Tyndallization

An exposure of steam at 100°C for **20** minutes on three successive days is called **tyndallization or intermittent sterilization.** This is a fractional method of sterilization. The instrument commonly used is Koch and Arnold steamer.

Principle

Vegetative cells and some spores are killed during the first heating and that the more resistant spores subsequently germinate and are killed during either the second or the third heating. Though generally adequate, this method may fail with spores of certain anaerobes and thermophiles.

Uses

This method is useful in sterilizing heat-sensitive culture media containing such materials as carbohydrates, egg or serum. which are damaged by higher temperature of autoclave.

C. Moist Heat at Temperatures Above 100°C

Steam Under Pressure

Steam above 100°C or saturated steam is more efficient sterilizing agent than hot air because:

1. It provides greater lethal action of moist heat.
2. It is quicker in heating up articles to be sterilized.
3. It can penetrate easily porous material such as cotton wool stoppers, paper and cloth wrappers, bundles of surgical linen and hollow apparatus.

Autoclave

Autoclaving is the process of sterilization by saturated steam under high pressure above 100°C. Steam sterilization is carried out in a pressure chamber called an **autoclave** (a device somewhat like a fancy pressure cooker).

Various Components of Autoclave

In its simplest form, the laboratory autoclave consists of a **vertical or horizontal cylinder of gunmetal or stainless steel**, in a supporting sheet iron case. **The lid** or **door** is fastened by screw clamps and made airtight by a suitable washer. The autoclave has on its lid or upper side a discharge tap for air and steam, a pressure gauge and a safety valve that can be set to blow off at any desired pressure. Heating is done by gas or electricity (Fig. 5.3). The domestic pressure cooker serves as a miniature autoclave and may be used for sterilizing small articles in clinics and similar establishments.

Principle of Autoclave

The principle of the autoclave or steam sterilizer is that water boils when its vapor pressure equals that of the surrounding atmosphere. When pressure inside a closed vessel increases, the temperature at which water boils also increases. Saturated steam has penetrative power and is a better sterilizing agent than dry heat.

Steam condenses to water and gives up its latent heat to that surface when it comes into contact with a cooler surface. The energy available from this latent heat is considerable. For example, 1600 ml steam at 100°C and at atmospheric pressure condenses into one ml of water at 100°C and releases 518 calories of heat). The large reduction in volume, sucks in more steam to the area and the process continues till the temperature of that surface is raised to that of the steam. The water of condensation ensures moist conditions for killing of the exposed microorganisms.

Bacteria are intrinsically more susceptible to moist heat as bacterial protein coagulates rapidly and condensed water ensures moist conditions for killing the microbes present.

Types of Steam Sterilizers

Several types of steam sterilizers are available (Table 5.3). Even the domestic pressure cooker can be used as a sterilizer.

Procedure

1. **Water:** Sufficient water is put in the cylinder. Above this is a perforated shelf on which articles to be sterilized are placed, and the autoclave is heated.
2. **Lid:** The lid is screwed tight with the discharge tap open and the safety valve is adjusted to the required pressure.

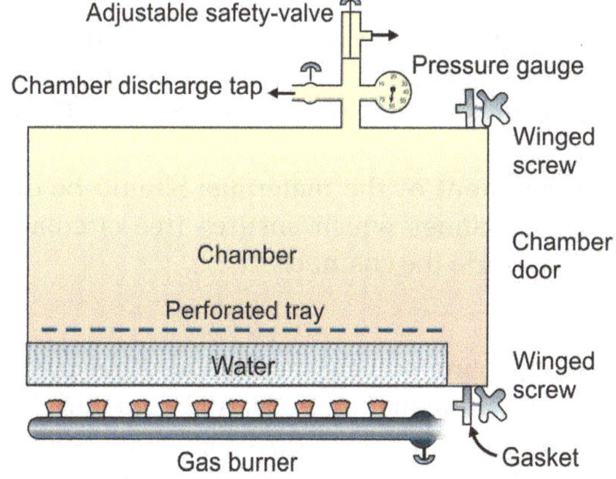

Fig. 5.3: A simple autoclave

Table 5.3: Types of steam sterilizer

1. Simple laboratory autoclave.
2. Transportable bench top autoclaves.
3. Large simple autoclaves.
4. Downward displacement laboratory autoclaves.
5. Media preparators.
6. Multipurpose laboratory autoclaves.
7. Other autoclaves.
   - High security autoclave
   - Porous load sterilizer
   - Low temperature steam.

3. **Air removal:** The steam-air mixture is allowed to escape freely till all the air has been displaced. To know when all the air inside the autoclave has escaped, the discharge tap is connected with one end of a rubber tube and the other end of it is placed in water. When the air bubbles stop coming, it indicates that all the air from inside the autoclave has been removed.
4. **The discharge tap is now closed.**
5. **Holding period:** The steam pressure rises inside and when it reaches the desired set level (15 psi), the safety valve opens and the excess steam escapes. From this point, the holding period (15 minutes) is calculated.
6. **Autoclave cooling:** When the holding period is over, the heater is turned off and the autoclave allowed to cool till the pressure gauge indicates that the pressure inside is equal to the atmospheric pressure.
7. **Air entry in the autoclave:** The discharge tap is opened slowly and air is allowed to enter the autoclave.
8. **Removal of articles:** The lid is now opened and the sterilized articles removed.

Note: If the tap is opened when the pressure inside is high, liquid media will tend to boil violently and spill from the container and sometimes an explosion may occur. If opened after the pressure inside has fallen below atmospheric pressure, an excessive amount of water would have evaporated and lost from the media.

Precautions

i. **Air escape from the chamber:** Since temperature of air-steam mixture is lower than that of pure steam, **the air must be allowed to escape from the chamber.**
ii. **Arrangement of the materials:** Should be done in such a manner which ensures free circulation of steam inside the chamber.

Uses

i. For sterilizing culture media and other laboratory supplies, aqueous solutions, rubber material, dressing materials, gowns, dressing, linen, gloves, instruments and pharmaceutical products.
ii. For all materials that are water containing, permeable or wettable and not liable to be damaged by the process.
iii. Particularly useful for materials which cannot withstand the higher temperature of hot air oven.

Sterilization Controls

A. **Biological control (Bacterial spores):** An envelope containing a filter paper strip impregnated with 10^6 spores of *Bacillus stearothermophilus* (NCTC 10003 or ATCC 7953) is placed with the load in the coolest and least accessible part of the autoclave chamber. After sterilization is over the strip is removed and inoculated into tryptone soy broth and incubated at 56°C for 5 days. No growth of *B. stearothermophilus* indicates proper sterilization. Spores of this organism withstand 121°C for up to 12 minutes and this has made the organism ideal for testing autoclaves.
B. **Chemical control:** A Browne's tube containing red solution changes to green when exposed to temperature of 121°C for 15 minutes in autoclave. It indicates proper sterilization.
C. **Autoclave tapes.**
D. **Thermocouples:** May also be used which records the temperature by a potentiometer.

4. Filtration

Filtration is the principal method used in the laboratory for the sterilization of heat labile materials, e.g. sera, solutions of sugars or antibiotics used for preparation of culture media.

Uses of Filtration

1. **Heat sensitive solutions:** For sterilization of pharmaceuticals, ophthalmic solutions, culture media, oils, antibiotics, and other heat sensitive solutions.
2. **For separation of bacteriophages and bacterial toxins from bacteria.**
3. **Isolation of organisms which are scanty in fluids.**
4. **Concentration of bacteria from liquids:** The filter disks concentrate bacteria from liquids, e.g. in testing water samples for cholera vibrios or typhoid bacilli.
5. **For virus isolation:** By obtaining bacteria-free filtrates of clinical samples as viruses pass through ordinary filters. Most viruses and certain bacteria such as Mycoplasmas can pass through filters with a pore size as low as 0.22 µm and cannot be kept back by the bacterial filters, therefore, serum sterilized by filtration cannot be employed for clinical use.

Types of Filters

i. Earthware filters
ii. Asbestos filters
iii. Sintered glass filters
iv. Membrane filters
v. Syringe filters

vi. Vacuum and 'in-line' filters
vii. Pressure filtration
viii. Air filters.

i. **Earthware filters:** These are manufactured in several different grades of porosity and have been used widely for purification of water for industrial and drinking purposes. The fluid to be sterilized is forced by suction or pressure from inside to outside or vice versa. After use they can be sterilized by scrubbing with stiff brush followed by boiling and autoclaving. They are of two types:
   a. Unglazed ceramic filters, e.g. Chamberland and Doulton filters.
   b. Compressed diatomaceous earth filters, e.g. the Berkefeld and Mandler filters.
ii. **Asbestos filters (Seitz filter):** They are made up of a disk of asbestos (magnesium trisilicate). It is supported on a perforated metal disk within a metal funnel. The funnel is loosely assembled with asbestos in position and sterilized in the autoclave. It is then fitted on to a sterile flask through a silicone rubber bung. The fluid to be sterilized is put into the funnel and flask connected to the exhaust pump through its side tap. Sterilized fluid is collected from the flask and filter disk is discarded after use. These disks are available with different grades of porosity.
   Examples: Seitz filter, Carlson and Sterimat filters.
iii. **Sintered glass filters:** They are prepared by size grading powdered glass followed by heating. The pore size can be controlled by the general particle size of the glass powder. The filters are easily cleaned, have low absorption properties and do not shed particles, but they are fragile and relatively expensive.
iv. **Membrane filters:** Membrane filters consist of a variety of polymeric materials such as cellulose nitrate, cellulose diacetate, polycarbonate and polyester. They are manufactured as disks from 13 to 293 mm diameter and with porosities from 0.015 to 12 μm. They come in a wide range of average pore diameters (APD), the 0.22 μm size being most widely used for sterilization because the pore size is smaller than that of bacteria.

Uses

1. They are used routinely in **water analysis and purification.**
2. Sterilization and sterility testing.
3. For preparing sterile solutions for parenteral use.
4. **Bacterial counts of water:** They can also be used for bacterial counts of water. A known amount of water is filtered through the membrane filter disk. The upper side of the disk is then placed on an appropriate moist culture medium and incubated. The colonies that develop can be counted and viable count calculated.

v. **Syringe filters:** Membrane of 13 and 25 mm diameter are commonly fitted in syringe-like holders of stainless steel or polycarbonate. For sterilization, the fluid is forced through the disk (membrane) by pressing the piston of the syringe.
vi. **Vacuum and 'in-line' filters:** Membranes of 25 and 45 mm diameter are used, either with in-line filter holders of Teflon or stainless steel and aluminium, or with vacuum holders of borosilicate glass, polycarbonate or stainless steel. They are suitable for the sterilization or disinfection of large volumes of liquid or air.
vii. **Pressure filtration:** Large membranes, 100 to 293 mm in diameter, or filter cartridges housed in pressure filter holders, may be used for the production of very pure water for laboratory use. They may be fitted with a Teflon filter so that the assembly is autoclavable with the filter *in situ*.
viii. **Air filters:** Air can also be sterilized by filtration. Large volumes of air may be rapidly freed from infection by passage through **high efficiency particulate air (HEPA)** filters which remove particles of 0.3 μm or larger (and probably particles <0.1 μm) are one of the most important air filtration systems. They are widely used in air filtration, especially that type incorporating **laminar air flow (LAF).** Laminar air flow (LAF) units are of two types, horizontal and vertical, depending upon the direction of the air flow.

5. Radiation

Two types of radiations are used:
I. Nonionizing
II. Ionizing

I. Nonionizing

Infrared and ultraviolet rays are of non-ionizing type. The effectiveness of UV light as a lethal and mutagenic agent is closely correlated with its wavelength. The most effective bactericidal wavelength is in the 240 to 280 nm range, with the optimum of about 260 nm, the wavelength most effectively absorbed by DNA and this infers with DNA replication.

Ultraviolet radiation can be produced artificially by mercury vapor lamps. Unlike ionizing radiation, the energy of UV radiation is lower and its power of penetration is poor. UV radiation around 260 nm is quite lethal but does not penetrate glass, dirt films, water and other substances very effectively. Because of this disadvantage, UV radiation is used as a sterilizing agent only in a few specific situations.

Microbial Sensitivity to UV Radiation

i. Bacterial spores are generally more resistant to UV light than are vegetative cells.
ii. Viruses are also inactivated and tend to be more sensitive than bacterial spores.

iii. Human immunodeficiency virus (HIV) is not inactivated by UV radiation.

Practical Applications of UV Radiation

1. **To disinfect drinking water.**
2. **Disinfection of enclosed areas**—such as entryways, hospital wards, operating theaters, laboratories and in ventilated safety cabinets in which dangerous microorganisms are being handled.

Precautions

Because UV radiation burns skin and damages eyes, people working in such areas must be certain that the UV lamps are off when the areas are in use.

II. Ionizing Radiation

These include X-rays, γ(gamma) rays and cosmic rays. These have very high penetrative power and are highly lethal to all cells including bacteria. Ionizing radiations damage the DNA by various mechanisms and include structural defects in microbial DNA synthesis, leading to cell death. Bacterial spores are generally more resistant than vegetative cells, and spores are among the most radiation resistant microorganisms known.

Applications

i. **For sterilization in pharmacy and medicine.**
ii. **Sterilization of packaged disposable articles:** Such as plastic syringes, intravenous lines, catheters and gloves that are unable to withstand heat.

Cold Sterilization

Since there is no appreciable increase in temperature in this method it is known as **cold sterilization.** Large commercial plants use gamma radiation emitted from a radioactive element, usually cobalt 60 for this type of sterilization.

iii. **Use for antibiotics, hormones, sutures, and vaccines** and to prevent food spoilage.

B. Chemical Agents

Germicidal chemicals can be used to disinfect and, in some cases, sterilize.

Characteristics of a Disinfectant

An ideal antiseptic or disinfectant should:

- Have a wide spectrum of activity and must be effective against a wide variety of infectious agents (gram-positive and gram-negative bacteria, acid-fast bacteria, bacterial endospores, fungi, and viruses)
- Be active at high dilutions and in the presence of organic matter
- Be effective in acid as well as alkaline media
- Have speedy action
- Have high penetrating power
- Be stable
- Be compatible with other antiseptics and disinfectants
- Not corrode metals
- Not cause local irritation or sensitization
- Not interfere with healing
- Not be toxic if absorbed into circulation
- Be cheap and easily available
- Be safe and easy to use

Such an ideal chemical is yet to be found.

Factors that Determine the Potency of Disinfectants

1. The concentration and stability of the agent.
2. Nature of the organism.
3. Time of action.
4. pH.
5. Temperature.
6. The presence of organic (especially protein) or other interfering substances.
7. Nature of the item to be disinfected.

Category of Disinfectant

Disinfection processes have been categorized as high level, intermediate level, and low level.

1. High-level Disinfection

High-level disinfection can generally approach sterilization in effectiveness, whereas spore forms can survive intermediate-level disinfection, and many microbes can remain viable when exposed to low-level disinfection.

High-level disinfectants are used for items involved with invasive procedures that cannot withstand sterilization procedures (e.g. certain types of endoscopes, surgical instruments with plastic or other components that cannot be autoclaved).

Examples

i. Treatment with moist heat
ii. Use of liquids such as glutaraldehyde, hydrogen peroxide, peracetic acid, chlorine dioxide, and other chlorine compounds.

2. Intermediate-level Disinfectants

Intermediate-level disinfectants are used to clean surfaces or instruments in which contamination with bacterial spores and other highly resilient organisms is unlikely. These include flexible fiberoptic endoscopes, laryngoscopes, vaginal specula, anesthesia breathing circuits, and other items. These have been referred to as semicritical instruments and devices.

Examples: Alcohols, iodophor compounds, phenolic compounds

3. Low-level Disinfectants

Low-level disinfectants are used to treat noncritical instruments and devices such as blood pressure cuffs, electrocardiogram electrodes, and stethoscopes. They do not penetrate through mucosal surfaces or into sterile tissues although these items come into contact with patients.

Examples: Quaternary ammonium compounds.

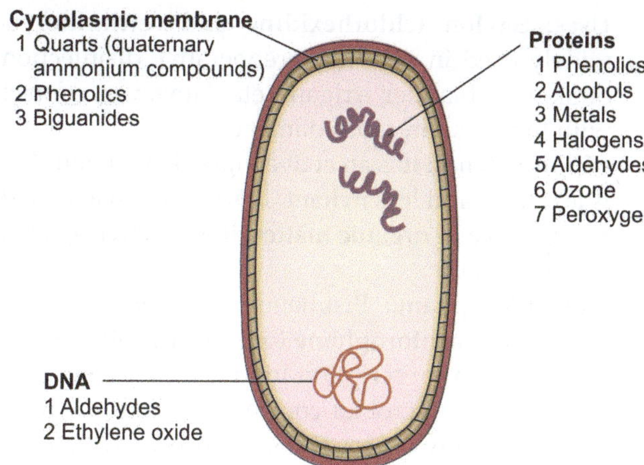

Fig. 5.4: Sites of action of chemical agents

Mechanisms of Antimicrobial Action (Fig. 5.4)

Most chemical germicides react irreversibly with vital enzymes and other proteins, the cytoplasmic membrane, or viral envelopes, although their precise mechanisms of action are often not completely understood (Fig. 5.4). The main modes of action are:

A. Agents that Damage the Cell Membrane

1. Surface active disinfectants
2. Phenolic compounds
3. Alcohols.

B. Agents that Denature Proteins

1. Acids and alkalies
2. Alcohols

C. Agents that Modify Functional Groups of Proteins and Nucleic Acids

1. Heavy metals and their compounds
2. Oxidizing agents: Halogens
   - Hydrogen peroxide.
3. Dyes: Aniline dyes
   - Acridine dyes
4. Alkylating agents: Aldehydes—Formaldehyde, Glutaraldehyde
   - Ethylene oxide.

A. Agents that damage the cell membrane

1. Surface-active Agents

Substances that alter the energy relationships at interfaces, producing a reduction of surface or interfacial tension, are referred to as **surface-active agents or surfactants**. They possess both hydrophobic (water-repelling) and hydrophilic (water-attracting) groups.

Classification

These surfactants are classified into anionic, cationic, nonionic and ampholytic (amphoteric). Of these, the cationic and anionic compounds have been the most useful antibacterial agents.

a. Cationic agents

These act on phosphate groups of cell membrane phospholipids and also enter the cell. This leads to loss of membrane semipermeability and Ieakage from the cell of nitrogen and phosphorus containing compounds. The agent which enters the cell, denatures its proteins. **Quaternary ammonium compounds (quats)** are the most important cationic compounds.

Examples of quaternary ammonium compounds include cetrimide (cetavalon), benzalkonium chloride (Zephiran, a brand name) and Cetylpyrimidium chloride (Cepacol, a brand name). Antimicrobial activity is affected greatly by organic matter and by pH, most active at alkaline pH and acid inactivates them. They are inactivated by hard water and soap. Disinfection may be enhanced by the appropriate combination of a surface-active agent with disinfectant to improve contact spread and cleansing properties.

Uses

i. They are primarily active against gram-positive, nonsporing bacteria; lethal to gram-negative organisms at high concentrations.
ii. They are also fungistatic and active against viruses with lipid envelopes (e.g. herpes and influenza) and much less against nonenveloped viruses (e.g. enteroviruses).

b. Anionic agents

These include soap and fatty acids. Anionic surfactants such as common soaps usually have strong detergent but weak antimicrobial properties. These agents are most active at acid pH and are active against gram-positive organisms but are relatively ineffective against gram-negative species.

c. Ampholytic (amphoteric) compounds

Known as **'Tego'** compounds, possess detergent properties of anionic and antimicrobial activity of cationic compounds. They are active over a wide range of pH but organic matter markedly reduces their activity.

Uses

They are effective against a wide range of gram-positive and gram-negative organisms and some viruses at a concentration of 1 percent in water.

2. Phenols and Phenolics

Phenol

In 1867, Joseph Lister, the father of antiseptic surgery, first introduced **phenol** (carbolic acid) to control surgical infections in the operating room. Phenols are obtained by distillation of coal tar between temperatures of 170°C and 270°C. It is now rarely used as an antiseptic or disinfectant because it irritates the skin and has a disagreeable odor. Phenol is bactericidal at a concentration of 1 percent.

Phenolics

Derivatives of phenol are called **phenolics.** Phenolics exert antimicrobial activity by injuring lipid-containing plasma membranes, which results in leakage of cellular contents.

Uses

i. **Use in hospitals:** Phenolic disinfectants derived from coal tar are widely used as disinfectants for **various purposes in hospitals.**
ii. They are active against **gram-positive and gram-negative bacteria,** moderately active against **Mycobacteria** and have little activity against **spores and viruses.**
iii. **Disinfection of pus, saliva, and feces:** A useful property of phenolics as disinfectants is that they remain active in the presence of organic compounds and are resistant to inactivation by organic matter, are stable, and persist for long periods after application. For these reasons, phenolics are suitable agents for disinfecting pus, saliva, and feces.
iv. They are used mainly for discarded cultures, contaminated pipettes and other infected material.
v. It is used for preservation of sera and vaccines at a concentration of 0.5 percent.

Disadvantages of Phenolics

i. Disagreeable odor.
ii. Skin irritation.
iii. It is readily absorbed by skin and causes **toxicity.**

Phenol Derivatives

Certain phenol derivatives like **cresol, chlorhexidine, chloroxylenol** and **hexachlorophane** are commonly used as antiseptics.

i. **Cresols:** Cresols, obtained industrially by the distillation of coal tar, are emulsified with green soap and sold under the trade names of *Lysol* and *Creolin*. '**White fluids'** such as Lysol are effective but are irritant to the skin. They are active against a wide range of organisms. They are most commonly used for sterilization of infected glasswares, cleaning floors, disinfection of excreta. They are not readily inactivated by the presence of organic matter.
ii. **Chlorhexidine:** *Chlorhexidine* is a member of the biguanide group with a broad spectrum of activity. It is frequently used for microbial control on skin and mucous membranes. They are bactericidal at a high dilution. Its killing effect is related to the injury it causes to the plasma membrane.

    They are more active against gram-positive than gram-negative bacteria. They are biocidal against most vegetative bacteria and fungi *Mycobacteria* are relatively resistant, endospores and protozoan cysts are not affected. The only viruses affected are certain enveloped (lipophilic) types.

    Uses: Savlon (chlorhexidine and cetrimide) is widely used in wounds, preoperative disinfection of skin, as bladder irrigant, etc. However, contact with the eyes can cause damage.
iii. **Chloroxylenol:** It is an active ingredient of dettol. It is less toxic and less irritant. It is readily inactivated by presence of organic matter. It is inactive against *Pseudomonas*.
iv. **Hexachlorophane:** Bisphenols are derivatives of phenol. **Hexachlorophene (**one bisphenol) is an ingredient of a prescription lotion, used for surgical and hospital microbial control procedures. Gram-positive staphylococci and streptococci, which can cause skin infections in newborns, are particularly susceptible to hexachlorophane so it is used notably for prophylaxis against staphylococcal infection in nurseries. However, it can cause **neurotoxicity (brain damage),** especially in infants, and its use is now severely restricted.

B. Agents that Denature Proteins

Among the chemical agents that denature cellular proteins are the acids, alkalies, alcohols, acetone and other organic solvents.

Acids and Alkalies

Acids and alkalies exert their antibacterial activity through their free H^+ and OH^- ions, through the undissociated molecules, or by altering the pH of the organism's environment. Many aliphatic and aromatic acids are employed as preservatives, especially in the food industry, and to some extent in pharmaceutical and cosmetic products. They are not sporicidal and the activity of the acids, but not the esters, is very pH dependent. The antimicrobial action of alkalies as sodium hydroxide, calcium hydroxide, sodium carbonate is related to hydroxyl ion concentration.

Alcohols

Ethyl alcohol (ethanol) and **isopropyl alcohol** are the most frequently used. They rapidly kill bacteria including tubercle bacilli but they have no action on spores and viruses. However, human immunodeficiency virus (HIV) is susceptible to ethyl alcohol (70%) and isopropyl alcohol (35%) in the absence of organic matter. They must be used at a concentration of 60 to 70% in water to be effective. They are most frequently used as skin disinfectants and act by **denaturing bacterial proteins.**

Isopropyl alcohol is preferred as it is better fat solvent, more bactericidal and less volatile. It has been recommended as a replacement for ethanol for the sterilization of thermometers.

Methyl alcohol is effective against fungal spores and is used for treating cabinets and incubators affected by them. Methyl alcohol vapor is toxic and inflammable.

C. Agents that Modify Functional Groups of Proteins and Nucleic Acids

1. Heavy Metals

For many years the ions of heavy metals such as **mercury, silver, arsenic, zinc and copper** were used as germicides. Heavy metals combine with proteins, often with their sulphahydryl groups, and inactivate them. They may also precipitate cell proteins.

i. **Mercuric chloride:** It is very toxic and at present has limited use. Mercury compounds in the form of organic derivatives, e.g. phenylmercuric nitrate, and acetate, thiomersal and mercurochrome are less toxic and are used as mild antiseptic. They are active against both gram-positive and gram-negative bacteria but are sporistatic at ambient temperatures and have limited fungicidal action.

ii. **Silver nitrate:** The most commonly employed of the silver salts is **silver nitrate.**

Use

1. Highly **bactericidal for the gonococcus.**
2. Routinely used for the **prophylaxis of the ophthalmic neonatorum in newborn infants** in a 1 percent solution.
3. **To prevent infection of burns:** Topical application of silver nitrate or silver sulphadiazine in cream has significantly reduced the mortality in these patients

iii. **Copper derivatives:** Are used as algicides, fungicides, wood, paint, cellulose and fabric preservation.

2. Oxidizing Agents

The most useful antimicrobial agents in this group are the **halogens** and **hydrogen peroxide**. They inactivate enzymes by converting functional-SH groups to the oxidized S-S form.

i. Halogens

Chlorine and **iodine** are among our most useful disinfectants. They are bactericidal and sporicidal. They are active in very high dilutions and their action is very rapid.

a. Iodine

Iodine compounds are the most effective halogens available for disinfection. It is actively bactericidal, with moderate action against spores. It is active against the tubercle bacteria and viruses.

Uses

i. **Skin disinfectant:** Iodine in aqueous and alcoholic solution has been used widely as a skin disinfectant. Iodine often has been applied as **tincture of iodine**, 2 percent or more iodine in a water-ethanol solution of potassium iodide. Although it is an effective antiseptic, the skin may be damaged, a stain is left, and iodine allergies can result.

ii. **Iodophors:** Mixtures of iodine with various surface-active agents that act as carriers for iodine are known as *iodophors* (*iodo*, "iodine"; *phor* "carrier"). They are used in hospitals for preoperative skin degerming and in hospitals and laboratories for disinfecting. **Povidine-iodine (Betadine)** for wounds and *Wescodyne* for skin and laboratory disinfection are some popular brands.

b. Chlorine

In addition to **chlorine** itself, there are three types of chlorine compounds—**hypochlorites** and **inorganic and organic chloramines**. The disinfectant action of all chlorine compounds is due to the liberation of free chlorine. When elemental chlorine or hypochlorites are added to water, the chlorine reacts with water to form hypochlorous acid (HOCL), which in neutral or acidic solution is a strong oxidizing agent and an effective disinfectant.

The activity of chlorine is markedly influenced by the presence of organic matter. Chlorine has a special place in the treatment of water supply and combinations of hypochlorite. Organic compounds and alkaline detergents can reduce the effectiveness of chlorine compounds.

Uses

The usual disinfectant for water supplies, swimming pools, dairy and food industries.

Hypochlorites

The hypochlorites have a bactericidal, fungicidal, virucidal and rapidly sporicidal action. It should not be used in the presence of formaldehyde as one of the reaction products is found to be carcinogenic.

Uses

i. Widely used in the **food and dairy industries** for sanitizing dairy and food processing equipments.
ii. As sanitizers in most households, hospitals, and public buildings.
iii. Widespread application as laboratory disinfectants on bench surfaces and in discard pots.
iv. Bleaching powder or hypochlorite solution is the most widely used for human immunodeficiency virus (HIV) infected material. Hypochlorite solution decays rapidly and should be prepared daily.

Chloramines are used as antiseptics for dressing wounds.

ii. Hydrogen Peroxide

Hydrogen peroxide (H_2O_2) effectively kills most bacteria at a concentration of 3 to 6 percent and kills all organisms, including spores, at higher concentrations (10-25%). The active oxidant form is not hydrogen peroxide but rather the free hydroxyl radical formed by the decomposition of hydrogen peroxide.

Uses

It is used to disinfect plastic implants, contact lenses, and surgical prostheses.

3. Dyes

Aniline dyes and **acridine dyes** are two groups of dyes which are used extensively as skin and wound antiseptic. Both are **bacteriostatic in high dilution** but are of low bactericidal activity.

i. **Aniline dyes:** Of the aniline dyes, derivatives of triphenylmethane, especially **brilliant green, malachite green and crystal violet** have many uses.

   They are highly selective for gram-positive than against gram-negative organisms and have been used in the laboratory in the formulation of selective culture media. They have no activity against tubercle bacilli, and hence the use of malachite green in the Lowenstein-Jenson medium.

   They are considerably inhibited by organic material such as pus though they are nonirritant to the tissues and nontoxic. These dyes are more active at alkaline pH. Their lethal effects on bacteria are believed to due to their reaction with acid groups in the cell.

ii. **Acridine dyes:** The acridine dyes, often referred to as "**flavines**" due to their yellow color, exert a bactericidal and bacteriostatic effect on a number of organisms. They are more active against gram-positive organisms than against gram-negative but are not as selective as the aniline dyes. Unlike the aniline dyes, antimicrobial activity is retained in the presence of serum or pus. The more important dyes are **proflavine, acriflavine, euflavine and aninacrine**. They show no significant differences in potency.
   - **Proflavine and acriflavine:** Proflavine and acriflavine are among the compounds of clinical use and have been employed in **wound antiseptics**. If impregnated in gauze, they are slowly released in a moist environment, and hence their advantage and use in clinical medicine. They interfere with the synthesis of nucleic acids and proteins in both bacterial and mammalian cells.

4. Alkylating Agents

The lethal effects of aldehydes (formaldehyde and glutaraldehyde) and ethylene dioxide result from their alkylating action on proteins.

i. Formaldehyde

Formaldehyde is active against the amino group in the protein molecules. It is lethal to bacteria and their spores (but less than glutaraldehyde), viruses and fungi. It is employed in the liquid and vapor states. It combines readily with proteins and is less effective than in the presence of organic matter. Formaldehyde is commercially available in aqueous solutions containing 37 percent formaldehyde (formalin) or as paraformaldehyde, a solid polymer that contains 91 to 99 percent formaldehyde. Formalin destroys all organisms, including spores when used as a sufficiently high concentration. Exposure of skin or mucous membranes to formaldehyde can be toxic and the gas is irritant and toxic when inhaled.

Uses

a. **Formalin:**
   i. Used for preserving fresh tissues and is the major component of embalming fluids.
   ii. Formalin (generally from 0.2-04%) has been used extensively to inactivate viruses in the **preparation of vaccines**. 10 percent formalin containing 0.5 percent sodium tetraborate is used to sterilize clean metal instruments.

b. **Formaldehyde use:**
   i. It is used to preserve anatomical specimens.
   ii. For destroying anthrax spores in hair.
   iii. As an antiseptic mouthwash.
   iv. For the disinfection of membranes in dialysis equipment.
   v. A preservative in hair shampoos.

c. **Formaldehyde gas**
   i. Used for sterilizing instruments, heat sensitive catheters and for fumigating wards, sick rooms and laboratories.
   ii. Clothing, bedding, furniture and books can be satisfactorily disinfected under properly controlled conditions.

ii. Glutaraldehyde

This has an action similar to formaldehyde. It has a broad-spectrum action against vegetative bacteria including mycobacteria, fungi and viruses, but acts more slowly against spores. It is more active and less toxic than formaldehyde. It is used as 2 percent buffered solution. It can be used for delicate instruments having lenses. A 2 percent buffered solution of glutaraldehyde is an effective disinfectant. It is available commercially as 'cidex'. It usually disinfects objects within about 10 minutes but may require as 12 hours to destroy all spores.

Disadvantages of Glutaraldehyde

i. It is irritant to the eyes, skin and respiratory mucosa.
ii. It is more active at alkaline pH levels ("activated" by sodium hydroxide) but is less stable.
iii. It is also inactivated by organic material, so items to be treated must be cleaned.

Uses

i. **Cold sterilant:** It has been used increasingly as a **cold sterilant** for **surgical instruments and endoscopes.** It has no deleterious effect on the cement or lenses of instruments such as **cystoscopes and endoscopes**
ii. Used safely to sterilize corrugated rubber anesthetic tubes and face masks, plastic endotracheal tubes, metal instruments and polythene tubing.

Vapor-Phase Disinfectants

1. Ethylene oxide

This is a colorless liquid with a boiling point of 10.7°C and is a highly penetrating gas with a sweet ethereal smell. It is highly inflammable and in concentrations in air greater than 3 percent, highly explosive. By mixing it with inert gases such as carbon dioxide or nitrogen, to a concentration of 10 percent, its explosive tendency is eliminated. It diffuses through many types of porous materials and readily penetrates some plastics. It is unsuitable for fumigating rooms because of its explosive property.

It is highly lethal to all kinds of microbes including spores and tubercle bacilli. Its action is due to its alkylating the amino, carboxyl, hydroxyl and sulfhydryl groups in protein molecules. In addition, it reacts with DNA and RNA. It is effective against all types of microorganisms including viruses and spores.

Uses

a. **Sterilization of articles liable to be damaged by heat:** It is specially used for sterilizing heart-lung machines, respirators, sutures, dental equipment, books and clothing.
b. **Sterilization of a wide range of materials-** It has been successfully used to sterilize a wide range of materials such as glass, metal and paper surfaces, clothing, plastics, soil, some foods and tobacco.

Disadvantages of Ethylene Oxide

i. It is an irritant, and personnel working with it have to take strict precautions.
ii. Its use as a disinfectant presents a potential toxicity to human beings, including mutagenicity and carcinogenicity. *Bacillus globigi,* a red-pigmented variant of *B. subtilis,* has been used to test ethylene oxide sterilizers.

2. Formaldehyde Gas

It is used for fumigation of complex heat-sensitive equipment, including anesthetic machine and baby incubators and for periodic decontamination of laboratory safety cabinets.

- **Fumigation of operation theaters and other rooms:** This is also widely employed for fumigation of operation theaters and other rooms (such as isolation rooms). After sealing the windows and other outlets, formaldehyde gas is generated by adding 150 g of $KMnO_4$ to 280 ml formalin for every 1000 cu. ft (28.3 cu. m) of room volume. The reaction produces considerable heat, and so heat resistant vessels should be used. After starting generation of formaldehyde vapor, the doors should be sealed and left unopened for 48 hours.

  Formaldehyde has an extremely unpleasant odor and is irritant to mucous membranes. The effect of irritant vapors should be nullified by exposure to ammonia vapor after completion of disinfection.

3. Betapropiolactone (BPL)

This is a condensation product of ketane and formaldehyde with a boiling point of 163°C. It also destroys microorganisms more readily than ethylene oxide but does not penetrate materials well and may be carcinogenic. For sterilization of biological products, 0.2 percent BPL is used. It is capable of killing all microorganisms and is very active against viruses.

Use

In the liquid form it has been used to sterilize vaccines and sera.

> **Note:** Recently vapor-phase hydrogen peroxide has been used to decontaminate biological safety cabinets. Peracetic acid is other vapor-phase sterilant.

RECOMMENDED CONCENTRATIONS OF VARIOUS DISINFECTANTS

The recommended concentrations of various disinfectants commonly used in the hospitals are given in **Table 5.4. Table 5.5** shows various methods of sterilization/disinfection of some important materials.

TESTING OF DISINFECTANTS

There is no single reliable test available to determine the efficiency of disinfectant due to a number of parameters, which influence disinfectant activity. The following tests are used for testing disinfectants:

1. Phenol coefficient test
   - Rideal-Walker test
   - Chick-Martin test
2. Minimum inhibitory concentration (MIC)
3. Kelsey-Sykes capacity test
4. In-use test

1. **Phenol coefficient test:** These tests are irrevalent. The test organism is inappropriate and the test irreproducible.

   Rideal Walker test: The best-known disinfectant screening test is the phenol coefficient test in which potency of a disinfectant is compared with that of phenol.

   Procedure: A series of dilutions of phenol and the experimental disinfectant are inoculated with the test bacteria such as typhoid bacilli. Suspensions containing equal numbers of typhoid bacilli are submitted to the action of varying concentrations of phenol and of the disinfectant to be tested. The dilution of the test disinfectant which sterilizes the suspension in a given time, divided by the corresponding dilution of phenol is stated as the **phenol coefficient** (Phenol = 1) of the disinfectant.

   Interpretation: A phenol coefficient of 1.0 means

Table 5.4: List of the recommended concentrations of disinfectants commonly used in the hospitals	
Disinfectant	*Concentration*
Betadine (Iodophore)	2%
Bleching powder (calcium hypochlorite)	14 gm in one liter of water
Dettol (chloroxylenol)	4%
Ethyl alcohol	70%
Glutaraldehyde	2%
Lysol	2.5%
Savlon (chlorhexidine and cetrimide)	2%, 5%
Sodium hypochlorite	1%, 0.1%

Table 5.5: Various procedures of sterilization/disinfection of some important materials.	
Materials	*Methods*
1. Metallic inoculating wires	Red heat
2. Infective materials like soiled dressings, bed	Burning (incineration)
3. Glasswares: Syringes, petridishes, test tubes, flasks, universal containers, oily fluids (paraffin)	Hot air oven
4. Metal instruments	Autoclaving, hot air oven, infrared radiation
5. Serum, body fluids, bacterial vaccines	Waterbath, at 56°C × 1 hour, vaccine bath at × 1 hour
6. Gloves, aprons, dressings, catheters, surgical instruments except sharp instruments.	Autoclaving
Sharp instruments	5% cresol
7. Suture materials except catgut	Autoclaving
Catgut	Ionizing radiation
8. Milk	Pasteurization
9. Most of the culture media	Autoclaving
10. Culture media containing egg, serum or sugar	Tyndallization
11. Toxin, ascitic fluid, serum, sugar and antibiotic solutions	Filtration
12. Rubber, plastic and polythene tubes including disposable syringes	Gamma radiation, ethylene oxide gas
13. Feces and urine, vomitus, sputum	Bleaching powder, cresols, formalin, burning, autoclaving
14. Disposable syringes, rubber or plastic disposable goods, bone and tissue grafts, adhesive dressings	Ionizing radiation
15. Antitoxic sera, serum, urine	Merthiolate (1:10,000)
16. Operation theater, wards and laboratory or floor space	Formaldehyde gas and cresols (Lysol) Sodium hypochlorite (1%)
17. Polythene tubing, fabrics, machine	Ethylene oxide
18. Water	Chlorine as hypochlorites 0.2%
19. Skin	Tincture iodine, spirit (70% ethanol), savlon (phenol derivative).
20. Woollen blankets, wool and hides	Formaldehyde gas
21. Sterilization of operation theater	Formaldehyde gas (50 cc formalin and 25 gm $KMnO_4$ per 100 cu. ft. space)

that the disinfectant in question has the same effectiveness as phenol and a coefficient of less than 1.0 means it is less effective and more than 1.0 means it is more effective. Higher the phenol coefficient, more effective is the disinfectant. This test does not reflect natural conditions as the bacteria and the disinfectant react directly without any organic matter being present. Modifications have therefore been suggested.

Chick-Martin test: Chick-Martin test is modification of Rideal-Walker test. In this test, the disinfectant acts in the presence of organic matter (dried yeast or feces) to simulate natural situations. Even this modification falls short of simulating natural conditions. Various other modifications have been introduced, but no test is entirely satisfactory.

2. **Minimum inhibitory concentration (MIC):** This test measures the lowest concentration of the disinfectant that inhibits the growth of *S. typhi* in a nutrient medium. Tests for the MIC of the disinfectant are also irrelevant as the number of organisms exposed to the disinfectant is too low and the time of exposure too long.
3. **Kelsey-Sykes test (Capacity test):** The main feature of the test is that instead of one addition of a large inoculum of the test organisms, the additions are made in increments with or without organic matter and this gives a measure of the capacity of the disinfectant to cope with successive bacterial invasions. Capacity test is designed to simulate the natural conditions under which the disinfectants are used in the hospitals.

   Procedure: Test organisms (*Staph. aureus, E. coli, Ps. aeruginosa* and *Proteus vulgaris*) are added to the disinfectant in three successive lots at 0, 10 and 20 minutes in both clean and dirty conditions. Each addition is in contact with the disinfectant for 8 minutes, therefore, samples are transferred at 8, 18 and 28 minutes respectively to a recovery medium. The disinfectant is judged by its ability to kill bacteria as judged by recovery on subculture (growth or no growth in recovery media).
4. **In-use test:** The in-use test should only be performed to confirm that the chosen disinfectant has been effective under the conditions and period of use. For example, this test should be performed on the diluted disinfectant in discard jars after they have been used and left overnight. The liquid phase of disinfectant solutions is examined quantitatively for viable organisms in actual use hospital practice. The ability of a disinfectant to inactivate a known number of a standard strain of a pathogenic staphylococcus on a given surface within a certain time determines its efficiency. Generally, the results of such tests are more useful than those of phenol coefficient test and its modifications.

STERILIZATION OF PRIONS

Prions are infectious proteins without any detectable nucleic acid. They have properties distinct from other infectious agents and are highly resistant to physical and chemical agents, particularly in their resistance to conventional inactivation methods. They are nonconventional transmissible agents that cause transmissible degenerative encephalopathies (TDE).

1. **Dry heat:** Prions are extremely resistant to dry heat. A temperature of 360°C for one hour has been reported not to be completely effective.
2. **Wet heat:** They are more resistant to steam sterilization than conventional transmissible agents (bacteria and their spores, fungi and viruses). Steam is found to be effective if a temperature of 134-138°C is maintained for 18 minutes.
3. **Chemicals:** These agents are inactivated by sodium hypochlorite (25% available chlorine) treatment at room temperature for one hour. They are also sensitive to phenol (90%), household bleach, ether, acetone, urea (6 mol/L), sodium dodecyl sulphate (10%) and iodine disinfection. Other chemicals like aldehydes, potassium permanganate, hydrogen peroxide, ethylene oxide, β-propiolactone, ethanol, proteases and ionizing radiations have been found to be ineffective.

KNOW MORE

Oxidizing agents: For certain purposes—iodine as a skin disinfectant and chlorine as a water disinfectant—are unequated. They are unique among disinfectants in that their activity is almost exclusively bactericidal and in that they are effective against sporulating organisms.

Disinfection of Skin

Washing with soap and water removes most of the transient surface contaminants and some of the resident bacteria. Chlorhexidine or iodine detergent scrub should be used if the hands are likely to have become contaminated with pathogens. As an alternative, the hands may be rubbed with a solution of 0.5 percent chlorhexidine and 1 percent glycerol in 70 percent isopropylalcohol.

The wound should be cleaned and irrigated with a mild disinfectant such as chlorhexidine with cetrimide, e.g. savlon if as a result of a laboratory accident the skin is broken. If **the unbroken skin** is grossly contaminated with bacterial or viral pathogens, it should at once be rinsed with phenolic or hypochlorite disinfectant available for normal use on the bench, e.g. 2 percent hycolin and then immediately be washed with water.

☞ KEY POINTS

- **Sterilization** is the process by which an article, surface, or medium is freed of all living microorganisms either in the vegetative or spore state.
- **Disinfection** is the killing, inhibition, or removal of microorganisms that may cause disease.
- **Antisepsis** is the prevention of sepsis or putrefaction either by killing microorganisms or by preventing their growth.

METHODS OF STERILIZATION AND DISINFECTION

A. Physical Agents

Dry Heat: (a) Flaming; (b) Incineration; (c) Hot-air sterilizatio

Moist Heat: (a) Pasteurization—Heat treatment for milk (72°C for about 15 sec) that kills all pathogens and most nonpathogens; **(b) Boiling; (c) Steam under normal pressure**

Tyndallization: An exposure of steam at 100°C for **20** minutes on three successive days is called **tyndallization or intermittent sterilization.**

Steam Under Pressure

Autoclaving: Very effective method of sterilization; at about 15 psi of pressure (121°C).

Filtration

Various types of filters are now available: (i) Earthware filters; (ii) Asbestos filters; (iii) Sintered glass filters; (iv) Membrane filters; (v) Syringe filters; (vi) Vacuum and 'in-line' filters; (vii) Pressure filtration; (viii) Air filters.

Radiation

1. Ionizing; (2) Nonionizing.

Chemical Agents: Although generally less reliable than heat, these chemicals are suitable for treating large surfaces and many heat-sensitive items.

1. Surface-active Agents: (a) Cationic agents; (b) Anionic agents; (c) Ampholytic (amphoteric) agents:

2. Phenol derivatives—certain phenol derivatives like **cresol, chlorhexidine, chloroxylenol** and **hexachlorophane** are commonly used as antiseptics.

3. Alcohols: Ethyl alcohol (ethanol) and isopropyl alcohol are the most frequently used.

4. Heavy metals: Mercuric chloride, silver nitrate

5. Oxidizing agents

Halogens

- **a. Iodine:** Mixtures of iodine with various surface-active agents that act as carriers for iodine are known as *iodophors*
- **b. Chlorine (i) Hypochlorites (ii) Hydrogen peroxide**

3. Dyes: Aniline dyes and acridine dyes are two groups of dyes which are used extensively as skin and wound antiseptic
4. Alkylating agents: The lethal effects of aldehydes (formaldehyde and glutaraldehyde) and ethylene dioxide result from their alkylating action on proteins.

Vapor-phase Disinfectants: (i) Ethylene oxide; (ii) Formaldehyde gas; (iii) Betapropiolactone (BPL).

Testing of Disinfectants

1. **Phenol coefficient test** (Rideal-Walker method, Chick-Martin method)
2. **Minimum inhibitory concentration (MIC)**
3. **Kelsey-Sykes capacity test**
4. **In-use test**

IMPORTANT QUESTIONS

1. Define sterilization and disinfection. Classify the various agents used in sterilization. Add a note on the principle and functioning of autoclave.
2. Define the terms sterilization, disinfection and antisepsis. Name various agents used for sterilization and discuss the role of hot air oven in sterilization.
3. Write short notes on:
   - Hot air oven.
   - Inspissation or serum inspissator
   - Autoclave.
   - Sterilization by radiation and its practical applications.
4. Write briefly about:
   - Sterilization by moist heat
   - Pasteurization.
   - Tyndallization or intermittent sterilization or fractional sterilization.
   - Filtration.
5. Name various types of disinfectants and discuss the role of halogens in chemical disinfection.
6. Write short notes on:
   a. Vapor-phase disinfectants or gaseous sterilization.
   b. Surface active disinfectants
   c. Quaternary ammonium compounds
   d. Oxidizing agents.
   e. Testing of disinfectants.
   f. Sterilization of prions.

FURTHER READING

Alder VG, Simpson RA. Heat sterilization. A. Sterilization and disinfection by heat methods. Principles and Practice of Disinfection and Sterilization, 2nd edn, Russell AD, Hugo WB, Ayliffe GAJ, (Eds) Oxford: Blackwell Science, Oxford 1992: 483-98.

Ayliffe GAJ, Coats D, Hoffman PN. Chemical Disinfection in Hospitals, 2nd edn, Public Health Laboratory Service, London 1993.

Denyer SP. Filtration sterilization. Principles and Practice of Disinfection, Preservation and Sterilization, 2nd edn, eds Russell AD, Hugo WB, Ayliffe GAJ (Eds), Blackwell Science Oxford 1992: 573-604.

Gardner JF, Peel MM. Introduction to Sterilization and Disinfection, Edinburg: Churchill Livingstone, Edinburg 1991.

Russell AD, et al. Principles and Practice of Disinfection, Sterilisation and Preservation. 2nd edn. Oxford: Blackwell Scientific 1992.

6 Culture Media

Learning Objectives

After reading and studying this chapter, you should be able to:

- Describe classification of media.
- Differentiate between the following: Enriched media and enrichment media; Indicator media and differential media; selective media and differential media with suitable examples.
- Discuss liquid media and composition and uses.

INTRODUCTION

Culture medium: A nutrient material prepared for the growth of microorganisms in a laboratory is called a **culture medium**.

Inoculum: When microbes are introduced into a culture medium to initiate growth, they are called an **inoculum**.

Culture: The microbes that grow and multiply in or on a culture medium are referred to as a **culture.**

Cultivation is the process of growing microorganisms in culture by taking bacteria from the infection site (i.e. *in vivo* environment) by some means of specimen collection and growing them in the artificial environment of the laboratory (i.e. the *in vitro* environment). By appropriate procedures they have to be grown separately *(isolated)* on *culture media* and obtained as *pure cultures* for study. Once grown in culture, most bacterial populations are easily observed without microscopy and are present in sufficient quantities to allow laboratory testing to be performed.

MAIN PURPOSES OF BACTERIAL CULTIVATION

Bacterial cultivation has three main purposes:

i. **To grow and isolate** all bacteria present in an infection.
ii. **Infection and contaminants or colonizers:** To determine which of the bacteria that grow are most likely causing infection and which are likely contaminants or colonizers.
iii. **Identification and characterization:** To obtain sufficient growth of clinically relevant bacteria to allow identification and characterization.

COMMON INGREDIENTS OF CULTURE MEDIA

Some of the components of culture media are as follows:

1. **Water:** Tap water is often suitable for culture media, particularly if it has a low mineral content, but if the local supply is found unsuitable, glass-distilled or demineralized water must be used instead.
2. **Agar:** Agar (or agar-agar) is prepared from a variety of seaweeds and is now universally used for preparing solid media. The chief component of agar is a long-chain polysaccharide. It also contains a variety of impurities including inorganic salts, a small amount of protein like material and sometimes traces of long-chain fatty acids which are inhibitory to growth. The minerals present are mainly magnesium and calcium.

   Agar does not add to the nutritive properties of a medium and is not affected by the growth of bacteria. The melting and solidifying points of agar solutions are not the same. At the concentrations normally used, most bacteriological agars melt **at about 95°C** and solidify only when cooled to **about 42°C**. There are considerable differences in the properties of the agars manufactured in different places, and even between different batches from the same source. A concentration of 1-2 percent usually yields a suitable gel, but the manufacturer's instructions should be followed. The jellifying property varies in different brands of agar; for example, New Zealand agar has more jellifying capacity than Japanese agar. Agar is manufactured either in long shreds or as powder. There may be variations between different batches of the same brand, apart from the differences due to a different source of agar.

3. **Peptone:** Another almost universal ingredient of common media is peptone. It is a complex mixture of partially digested proteins. The important constituents are peptones, proteoses, amino acids, a variety of inorganic salts including phosphates, potassium and magnesium, and certain accessory growth factors such as nicotinic acid and riboflavin. The brands of peptone supplied by different manufacturers show appreciable differences in composition and growth-promoting properties. Moreover, variations may occur between different batches of one brand.

   Apart from the standard grades of bacteriological peptone, some manufacturers supply special grades of peptone recommended for particular purposes, e.g. Neopeptone, proteose peptone, mycological peptone, etc. Commercially available peptones or digest broth can be used. Meat extract is also available commercially and is known as Lab-Lemco.
4. **Yeast extract:** It contains a wide range of amino acids, growth factors and inorganic salts. Yeast extract is used mainly as a comprehensive source of growth factors and may be substituted for meat extract in culture media.
5. **Malt extract:** It consists mainly of maltose (about 50%), starch, dextrins and glucose, and contains about 5 percent of proteins and protein breakdown products, and a wide range of mineral salts and growth factors.
6. **Blood and serum:** Thsese are used for enriching culture media. Either human or animal blood can be used. Usually 5-10 percent blood is used and the most usual concentration is 10 percent. Serum is used in certain media.

CLASSIFICATION OF MEDIA

Media have been classified in many ways (Table 6.1).

Phases of Growth Media

Growth media are used in either of two phases: liquid (broth) or solid (agar). In some instances (e.g. certain blood culture methods), a biphasic medium that contains both a liquid and a solid phase is used.

A. Liquid (broth) media: In broth media nutrients are dissolved in water, and bacterial growth is indicated by a change in broth's appearance from clear to turbid, (i.e. cloudy).

Numerous culture media have been devised. The earliest culture media were liquid, which made the isolation of bacteria to prepare pure cultures extremely difficult. The original media used by Louis Pasteur were liquids such as urine or meat broth.

Table 6.1: Classification of media

A. Based on phases of growth media	**C. Special media**
1. Liquid (broth) media	(i) Enriched media
2. Solid (agar) media	(ii) Enrichment media
3. Semisolid media	(iii) Selective media
	(iv) Indicator or differential media
	(v) Transport media
	(vi) Sugar media
B. Based on nutritional factors	**D. Reducing media**
1. Simple media (Basal media)	1. Liquid (broth) media
2. Complex media	2. Solid (agar) media
3. Synthetic or defined media	3. Semisolid media

Disadvantages

i. Growths usually do not exhibit specially characteristic appearances in them and they are of only limited use in identifying species.
ii. Also, organisms cannot be separated with certainty from mixtures by growth in liquid media.

   If solid media are used, these disadvantages are overcome.

Uses

i. For obtaining bacterial growth from blood or water when large volumes have to be tested,
ii. For preparing bulk cultures of antigens or vaccines.

B. Solid (agar) media: Solid media are made by adding a solidifying agent to the nutrients and water. **Agarose** is the most common solidifing agent. The petri dish containing the agar is referred to as **agar**.

Colonies: While bacteria grow diffusely in liquids, they produce discrete visible growth on solid media. If inoculated in suitable dilutions, bacteria form **colonies**, which are clones of cells originating from a single bacterial cell. Bacterial cultures derived from a single colony or clone are considered "**pure**". On solid media, bacteria have distinct colony morphology and exhibit many other characteristic features such as pigmentation or hemolysis, making identification easy.

History of solid medium: The earliest solid medium was cooked cut potato used by Robert Koch. In 1881 Koch published an article describing the use of boiled potatoes. Later he introduced gelatin to solidify liquid media but it was not satisfactory as gelatin is liquefied at 24°C and also by many proteolytic bacteria. About a year later, in 1882, agar was first used as a solidifying agent. Agar had been used by the East Indies Dutch to make jellies and jams. Fannie Eilshemius Hesse, the New Jersey-born wife of Walther Hesse, one of Koch's assistants, had learned of agar from a Dutch acquaintance and suggested its use when she heard of the difficulties with gelatin. Agar solidified medium was an instant success and continues to be essential in all areas of microbiology.

Table 6.2: Characteristics of culture media	
Type	*Characteristics*
1. Complex media	Composed of ingredients such as peptones and extracts, which may vary in their chemical composition.
2. Chemically defined media	Composed of precise mixtures of pure chemicals such as ammonium sulfate
3. Special media	
i. Enriched media	They are used to grow bacteria which are more exacting in their nutritional needs. Substances such as blood, serum, or egg are added to a basal medium
ii. Enrichment media	Similar to selective media but designed to increase numbers of desired microbes to detectable levels.
iii. Selective media	Suppression of unwanted microbes and encouraging desired microbes
iv. Indicator media	Contain an indicator which changes color when a bacterium grows in them
v. Differential media	Medium that contains an ingredient that can be changed by certain bacteria in a recognizable way. Differentiation of colonies of desired microbes from others
vi. Sugar media	Sugar fermentation reactions are carried out for the identification of most of the organisms
vii. Transport media	Transport medium is a holding medium designed to preserve the viability of microorganisms in the specimen but not allow multiplication
4. Reducing media	Growth of obligate anaerobes

C. Semisolid media: For special purposes where agar is added to media in concentrations that are too low to solidify them. At 0.2 to 0.5 percent it yields a semisolid medium through which motile, but not nonmotile, bacteria may spread.

Based on Nutritional Factors (Table 6.2)

1. Simple media
2. Complex media
3. Synthetic or chemically defined media.

1. Simple media (Basal media)

Simple media are those which contain only basic nutrients required for the growth of ordinary organisms, and used as a general purpose media, e.g. peptone water, nutrient broth and nutrient agar (Table 6.3 and 6.4). These simple media are generally used as the basis to prepare enriched media; hence known as basal media.

2. Complex media

Media that contain some ingredients of unknown chemical composition are called **complex media** (Table 6.2). A complex medium contains a variety of ingredients such as meat juices and digested proteins. The exact chemical composition of these ingredients can be highly variable although a specific amount of each ingredient is in the medium. One common ingredient is **peptone.** This is protein taken from any of a variety of sources that has been hydrolyzed to amino acids and short peptides by treatment with enzymes, acids, or alkali. **Extracts,** which are the water-soluble components of a substance, are also used.

Nutrient broth: A commonly used complex medium, **nutrient broth (in liquid form)**. It is a simple basal liquid medium, supports growth of many organisms (Table 6.3).

Types of Nutrient Broth

There are **three types** of nutrient broth

1. **Meat infusion broth:** It is an aqueous extract of lean meat to which peptone is added.
2. **Meat extract broth:** It is prepared as a mixture of commercial peptone and meat extract.
3. **Digest broth:** It consists of a watery extract of lean meat that has been digested with a proteolytic enzyme so that additional peptone need not be added. Digest broths are economical and are good for obtaining luxuriant growths of exacting organisms.

Uses of Nutrient Broth

i. It is used **as a base** to prepare many other culture media.
ii. It is also used **to study growth curve.**

Nutrient Agar (Table 6.4, Fig. 6.1)

Nutrient agar is prepared by adding agar at a concentration of 2 percent to the nutrient broth. It is the simplest and commonest medium used routinely in microbiology laboratories, to grow nonfastidious bacteria. Nutrient agar is commonly referred to as "agar medium". Nutrient agar is used in the form of slopes, butts in test tubes or plates in petridishes.

Uses of Nutrient Agar

1. For routine laboratory work.
2. To study colony characteristics.
3. To detect pigment formation.
4. For antibiotic sensitivity testing.

Table 6.3: Representative types of liquid media

Medium	Composition	Characteristics
1. **Peptone water**	Peptone—10 gm Sodium chloride (NaCl)—5 gm Water—1 Liter (pH 7.4-7.5)	i. Basis for carbohydrate fermentation media ii. For testing the formation of indole
2. **Nutrient broth**	Peptone water Meat extract	i. For routine culture ii. As a base to prepare many other culture media iii. To study growth curve
3. **Glucose broth**	Nutrient broth + Glucose (1% commonest)	Blood culture Promotes luxuriant growth of many organisms Glucose acts as reducing agent
4. **Enrichment media**		
i. Tertrathinate broth	Nutrient broth Sodium thiosulfate Calcium carbonate Iodine solution Phenol red	Enriches salmonellae and sometimes shigellae
ii. Senite F broth	Sodium selenite Peptone Lactose	It inhibits coliform bacilli while permitting salmonellae and many shigellae to grow
iii. Alkaline peptone water	Peptone—10 gm Sodium chloride(NaCl)—10gm Distilled water—1 Liter	Excellent medium for enriching the number of *V. cholerae* and other vibrio species in a fecal specimen
5. **Anaerobic media**		
i. Thioglycollate broth	Yeast extract, Casein hydrolysate, pancreatic digest Glucose, L-cysteine, Agar, Sodium chloride, Sodium thioglycollate, Resazurin sodium solution, Water	Supports growth of anerobes, aerobes, microaerophilic, and fastidious microorganisms
ii. Roberson's cooked meat broth (RCM)	Nutrient broth, Predigested cooked meat of ox heart	Culture of anaerobic bacteria Preservaton of stock cuture of aerobic bacteria

5. To maintain stock cultures.
6. For slide agglutination and other tests. Pure growth of the bacterium on nutrient agar medium is used.

Semisolid agar: If the concentration of agar is reduced to 0.2 to 0.5 percent, **semisolid or sloppy agar** is obtained which enables motile organisms to spread but not non-motile bacteria. When agar concentration is further decreased to 0.05 to 0.1 percent, agar prevents convection currents and prevents the diffusion of air in thioglycollate media, used for the growth of anaerobic and microaerophilic organisms.

Firm agar: If the concentration of agar is increased to 6 **percent** it is called **firm agar**. It prevents spreading or swarming by organisms such as *Proteus* species and *Clostridium tetani* and enable them to form discrete colonies.

3. Synthetic or chemically defined media

They are prepared exclusively from pure chemical substances and their exact composition is known. Chemically defined media are used for various experimental purposes. These are used for various special studies such as metabolic requirements of the experimental microorganisms. Simple peptone water medium, 1 percent peptone with 0.5 percent NaCl in water, may be considered a semidefined medium since its composition is approximately known.

C. Special Media (Table 6.2)

i. Enriched media (Table 6.2, 6.4)

These are prepared to meet the nutritional requirements of more exacting bacteria by the addition of substances such as blood, serum or egg to a basal medium.

Table 6.4: Representative types of agar media		
Medium	*Composition*	*Characteristics*
A. Simple medium		
Nutrient agar	Nutrient broth, agar (2-3%)	Complex medium used for **routine laboratory work**. Supports the growth of a variety of **nonfastidious bacteria**
B. Enriched media		
i. Blood agar	Nutrient agar. Sheep blood (5-10%)	In addition to being **enriched medium**, it is an **indicator medium** showing the hemolytic properties of bacteria such as *Streptococcus pyogenes*
ii. Chocolate agar	Heated blood agar (55°C × 2 hr)	This medium is used to culture **fastidious bacteria**, such as *Haemophilus influenzae*, the neisseriae and pnemocococcus
iii. Loeffler's serum slope (LSS)	Nutrient broth serum (Ox, sheep or horse) glucose	Culture of *Corynebactreim diphtheriae*
C. Selective media		
Deoxycholate citrate agar (DCA)	Nutrient agar, sodium deoxycholate, sodium citrate, lactose, neutral red	Suitable for the isolation of **Salmonella and Shigella.**
Thayer Martin	Blood agar base Hemoglobin and supplement B Colistin, nystatin, vancomycin, trimethoprim	Selective for *Neisseria* species, which are fastidious. Selective-contains antibiotics that inhibit most organisms except *Neisseria* species
D. Indicator media		
MacConkey agar	Peptone, Sodium taurocholate, agar, neutral red lactose neutral red	Isolation and differentiation of **lactose fermenting (LF)** and **nonlactose fermenting (NLF)** enteric bacilli. **Selective** because bile salts inhibits nonintestinal bacteria and lactose with and the neutral red to distinguish lactose fermenting (LF) coliforms from the non-lactose fermenting (NLF) salmonella and shigella groups. **Differential** because the pH indicator turns red when the sugar in the medium, lactose, is fermented.

Fig. 6.1: Nutrient agar

Examples of Enriched Media

- **Blood agar (Fig. 6.2):** Many medically important bacteria are fastidious, requiring a medium that is even richer than nutrient agar commonly used in clinical laboratories is **blood agar.** Blood agar is used for isolation of **Streptococci, pneumococci, Hemophilus.**
- **Chocolate agar (Fig. 6.3):** A medium used to culture even more fastidious bacteria is **chocolate agar,** named for its appearance rather than the ingredients. It is used for isolation of *Neisseria* (meningococci and gonococci) and Haemophilus.
- **Bordet-Gengou agar:** For isolation of *Bordetella.*
- **Loeffler's serum slope (Fig. 6.4):** It is used for the isolation of *Corynebacterium diphtheriae.*
- **Dorset's egg medium:** It is used for the cultivation of *mycobacterium tuberculosis* and *corynebacterium diphtheriae.*

ii. Enrichment media (Table 6.2, 6.3)

When a substance is added to a liquid medium which inhibits the growth of unwanted bacteria and favors the growth of wanted bacteria is known as **enrichment medium**. This medium for an enrichment culture is **usually liquid** and provides nutrients and environmental conditions that favor the growth of a particular microbe but not others. The result is an absolute increase in the numbers of the wanted bacterium relative to the other bacteria.

This is often the case for soil or fecal samples. In mixed cultures or in materials containing more than

Fig. 6.2: Blood agar

Fig. 6.3: Chocolate agar

one bacterium, the bacterium to be isolated is often overgrown by the unwanted bacteria. Usually the nonpathogenic or commensal bacteria tend to overgrow the pathogenic ones, for example, *S.Typhi*, being overgrown by *E. coli* in cultures from feces. In such situations, substances which have a stimulating effect on the bacteria to be grown or an inhibitory effect on those to be suppressed are incorporated in the medium.

Examples

a. **Tetrathionate broth:** Tetrathionate inhibits coliforms while allowing typhoid-paratyphoid bacilli to grow freely in fecal sample.
b. **Selenite F (F for Feces) broth:** It is used for dysentery bacilli.
c. **Alkaline peptone water:** It is used for *V. cholerae* from feces.

iii. Selective media (Table 6.2, 6.4)

When a substance is added to a solid medium which inhibits the growth of unwanted bacteria but favors the growth of wanted bacteria it is known as **selective media.** Inhibitory agents used for this purpose include dyes, bile salts, alcohols, acids, and antibiotics. These

Fig. 6.4: Loeffler's serum slope (LSS)

media are used to isolate particular bacteria from specimens where mixed bacterial flora is expected.

Examples

a. **Deoxycholate citrate agar (DCA):** Addition of deoxycholate acts as a selective agent for dysentery bacilli (isolation of shigellae).
b. **Wilson Blair's brilliant green bismuth sulfite agar medium.** This medium is used for the isolation of typhoid and paratyphoid bacilli.
c. **Lowenstein-Jensen medium (Fig. 6.5):** This medium is used for *Mycobacterium tuberculosis*.
d. **Thayer-Martin medium:** The isolation of gonococci and meningococci.
e. **Potassium tellurite medium:** The isolation of diphtheria bacilli.
f. **TCBS (Thiosulphate, citrate, bilesalts, sucrose) agar** for the isolation of *Vibrio cholerae.*

iv. Indicator media

These media contain an indicator which changes color.

Examples

a. **Wilson-Blair medium:** There is incorporation of sulfite in Wilson-Blair medium. *S. Typhi* reduces sulfite to sulfide in the presence of glucose and the colonies of *S. Typhi* have a black metallic sheen.
b. **Potassium tellurite:** Potassium tellurite in McLeod's medium is reduced to metallic tellurium by the diphtheria bacillus to produce black colonies.
c. **MacConkey agar:** MacConkey agar indicates lactose fermenting property. LF produce pink colonies and Nonlactose fermenters (NLF) produce colorless colonies due to neutral red indicator (Fig. 6.6).

v. Differential media (Table 6.2)

A medium which has substances incorporated in it, enabling it to bring out differing characteristics of bacteria and thus helping to distinguish between them, is called a **differential medium.**

Fig. 6.5: Lowenstein-Jensen medium

Fig. 6.6: MacConkey agar

Examples

a. **Blood agar**: Blood agar is both a differential medium and an enriched one. It distinguishes between hemolytic and nonhemolytic bacteria. Hemolytic bacteria (e.g. many streptococci and staphylococci isolated from throats) produce clear zones around their colonies because of red blood cell destruction (Fig. 6.2).
b. **Nagler's medium:** There are many special media for demonstrating particular characteristics, like Nagler's medium which enables us to view lecithinase activity.
c. **MacConkey agar (Fig. 6.6):** MacConkey agar is both differential and selective. It contains peptone, meat extracat, NaCI, bile salt, lactose and neutral red indicator. Bile salt inhibits the growth of gram-positive bacteria. Some bacteria that ferment the sugar produce acid, which decreases medium pH and causes the neutral red indicator to give a pink or red color to colonies which are known as lactose fermenters (LF). Nonlactose fermenters (NLF) form colorless, pale colonies. This may also be termed indicator medium. Some organisms ferment lactose late, and called as Late-Lactose fermenters (LLF).
LF: For example, *Esch. coli, Klebsiellae* sp.
NLF: For example, *Salmonella, Shigella, proteus*
LLF: For example, *Shigella sonnei.*

vi. Sugar media (Table 6.2)

For the identification of most of the organisms, sugar fermentation reactions are carried out. Carbohydrate fermentation is used 'for characterization and identification of bacteria, particularly important in the study of gram negative bacilli. Sugar media are used to test fermentation.

Sugar used for sugar media: The term 'sugar' in microbiology denotes any fermentable substance. They may be:

1. **Monosaccharides:**
   a. Pentoses, e.g. arabinose, xylose, rhamnose.
   b. Hexoses, e.g. glucose, fructose, mannose, sorbose, galactose.
2. **Disaccharides**, e.g. Sucrose, maltose, lactose, trehalose, cellobiose.
3. **Trisaccharides**, e.g. raffinose.
4. **Polysaccharides**, e.g. starch, inulin, dextrin, glycogen.
5. **Polyhydric a1cohols**, e.g. glycerol, erythritol, adonitol, mannitol, dulcitol, sorbitol, inositol.
6. **Glycosides**, e.g. salicin, aesculin.
7. **Organic acids**, e.g. tartarate, citrate, mucate, gluconate, malonate.

Usual sugar media: The usual sugar media consist of 1 percent of the sugar in peptone water along with an appropriate indicator (Anrade's indicator–0.005 percent acid fuchsin in NaOH). A small tube (Durham's tube) is kept inverted in the sugar tube to detect gas production. The color of the medium is light yellow. The test bacterium is inoculated and incubated. Acid production is indicated by the development of pink color. Gas accumulates in the inner Durham's tube (Fig. 6.7).

Hiss serum sugars: Hiss serum sugars are used for organisms which are exacting in their growth requirements (fastidious organisms) like Streptococci, pneumococci, gonococci, meningococci and diphtheria bacilli. Hiss' serum water consists of 25 percent serum in distilled water.

vii. Transport media (Table 6.2)

A transport medium is a holding medium designed to preserve the viability of microorganisms in the specimen but not allow multiplication.

Delicate organisms (like gonococci) which may not survive the time taken for transporting the specimen to the laboratory or the normal flora may overgrow pathogenic flora *(Salmonella, Shigella* and *V. cholerae),* such special media are devised to maintain the viability of the pathogen termed as "transport media".

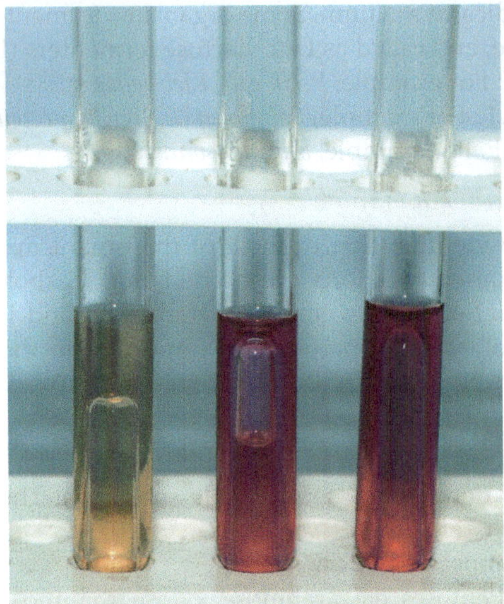

Fig. 6.7: Usual sugar media

Examples

i. Stuart's transport medium and Amies transport medium for gonococci.
ii. Buffered glycerol saline for enteric bacilli.
iii. Pike's medium for *Streptococcus pyogenes*, pneumococci and *Haemophilus influenzae* in nose and throat swabs.
iv. Cary-Blair medium for *Vibrio cholerae* and *Campylobacter*.
v. Venkatraman-Ramakrishnan (V-R) fluid (pH 9.2) for *Vibrio cholerae*.
vi. Alkaline peptone water (pH 8.6) for *Vibrio cholerae*.

5. Anaerobic Media (Table 6.2, 6.3)

These media are used to grow anaerobic organisms, and contain reducing substances. These include:

i. Thioglycollate broth.
ii. Cooked meat broth.

i. Thioglycollate broth

Thioglycollate broth contains reducing agents such as sodium thioglycollate, glucose, vitamin C (ascorbic acid), cysteine and agar (concentration of 0.05%), with methylene blue. Thioglycollate acts as a reducing agent and creates an anaerobic environment deeper in the tube, allows anaerobic bacteria to grow. Glucose and thioglycollate maintain the anaerobic condition. Agar (0.05%) prevents convection currents of air. Methylene blue or resazurin act as an oxidation-reduction potential indicator, which should show that the medium is anaerobic except in the surface layer in addition to a reducing agent and semisolid agar.

ii. Cooked meat broth or RCM broth (Robertson's Cooked meat broth)

The medium contains meat infusion or nutrient broth, minced, cooked ox heart tissue. Liquid paraffin is layered over broth to prevent the entry of air. Unsaturated fatty

Fig. 6.8: Cooked meat broth

acids in chopped meat act as reducing agent and absorb oxygen. In addition to its reducing effect, the meat provides a variety of nutritional substances for bacterial growth. Hematin catalyzes this reducing action and sulfhydryl groups create a low redox potential. All these factors favors the growth of anaerobic bacteria. The specimen is inoculated deep in the medium in contact with the meat (Fig. 6.8).

RCM broth detects proteolytic and saccharolytic activities of anaerobic bacteria. Proteolytic bacteria blacken the meat with the formation of foul smelling sulfur compounds. Saccharolytic bactreia turn meat red slightly with sour smell.

Preserving Bacterial Cultures

1. Refrigeration
2. Deep freezing
3. Lyophilization (freeze drying)
4. Cold storage
5. Drying methods.

Media to Test Special Properties

Various media to test special properties like urease production, and composite media for simultaneous demonstration of different features have been devised. They are dealt with in the appropriate chapters.

Color Code

A color code is usually adopted for identifying prepared media, This depends on the laboratory or group of laboratories. One color or a mixture of colors is used on the cotton stopper, or color paints are used on the caps.

Process of Media Making

It is essential to monitor the quality of culture media at all stages in their preparation and use. Culture media used to be prepared in laboratories themselves, starting with basic ingredients. Not only was this laborious but it also led to considerable batch variation in the quality of media. The process of media making has become simpler and its quality more uniform with the ready availability of commercial dehydrated culture media.

KNOW MORE

Preserving Bacterial Cultures

1. **Refrigeration:** Refrigeration can be used for the short-term storage of bacterial cultures.
2. **Deep freezing: Deep freezing** is a process in which a pure culture of microbes is placed in a suspending liquid and quick-frozen at temperatures ranging from - 50° to -95°C. The culture can usually be thawed and cultured even several years later.
3. **Lyophilization (freeze drying):** During **lyophilization (freeze drying),** a suspension of microbes is quickly frozen at temperatures ranging from -54° to -95°C, and the water is removed by a high vacuum (sublimation). The organisms can be revived at any time by hydration with a suitable liquid nutrient medium.
4. **Cold storage:** Cold storage is a common method of preserving strains of bacteria.
5. **Drying methods:** A number of drying methods for drying suspensions of bacteria for preservation purposes have been developed.

KEY POINTS

- A culture medium is any material prepared for the growth of bacteria in a laboratory.
- Microbes that grow and multiply in or on a culture medium are known as a **culture.**
- Agar is a common solidifying agent for a culture medium, a complex polysaccharide from red algae.
- **Simple media** include the nutrient broth and peptone water, which form the basis of other media.
- **Complex Media:** A complex medium is one in which the exact chemical composition varies slightly from batch to batch.
- **A chemically defined medium** is one in which the exact chemical composition is known.
- **Enriched media** are solid media supplemented with blood, serum, etc. (e.g. blood agar, chocolate agar, Loffler's serum slope, Lowenstein-Jensen medium, etc.).
- **Enrichment media:** An enrichment culture is used to encourage the growth of a particular microorganism in a mixed culture and are the liquid media (e.g. selenite F broth or tetrathionate broth).
- **Selective Media:** By inhibiting unwanted organisms with salts, dyes, or other chemicals, selective media allow growth of only the desired microbes (e.g. TCBS, DCA, LJ, MSA, XLD, etc).
- **Differential media or indicator media** distinguish one microorganism from one another growing on the same media (e.g. eosin methylene blue medium, MacConkey medium, mannitol salt agar, etc).
- **Sugar media** usually consist of 1 percent sugar in peptone water along with appropriate indicator.
- **Transport media** are used to maintain the viability of certain delicate organisms during their transport to the laboratory (e.g. Stuart's transport medium, alkaline peptone water, etc).
- **Anaerobic media:** Reducing media chemically remove molecular oxygen that might interfere with the growth of anaerobes.
- **Preserving bacterial cultures:** Microbes can be preserved for long periods of time by deep freezing or lyophilization (freeze drying).

IMPORTANT QUESTIONS

1. What are culture media? Classify and discuss them briefly.
2. Distinguish between a selective medium and a differential medium.
3. Write short notes on:
   - Liquid media
   - Solid media
   - Simple media/Basal media.
   - Defined media/synthetic media.
   - Agar/Agar-agar.
   - Complex media
   - Nutrient broth.
   - Nutrient agar.
   - Enriched media
   - Loeffler's serum slope.
   - Blood agar
   - Enrichment media
   - Selenite F broth.
   - Tetrathionate broth
   - Selective media.
   - Lowenstein-Jensen (LJ) medium.
   - MacConkey agar.
   - Indicator media.
   - Differential media
   - Transport media.
   - Sugar media.
   - Anaerobic media.
   - Thioglycollate medium/thioglycollate broth.
   - Cooked meat broth (CMB) or RCM broth.

FURTHER READING

Collee JC, et al. Mackie and McCartney Practical Medical Microbilology, 14th edn. London: Churchill Livingstone 1996:95-111.

CHAPTER 7

Culture Methods

Learning Objectives

After reading and studying this chapter, you should be able to:

- Discuss anaerobic culture methods.
- Explain the principle and describe uses of the following: McIntosh and Filde's anaerobic jar; cooked meat broth (CMB).

INTRODUCTION

Culture methods employed depend on the purpose for which they are intended.

Indications for Culture

In the clinical laboratory, the indications for culture are mainly to:

1. Isolate bacteria in pure culture
2. Demonstrate their properties
3. Obtain sufficient growth for preparation of antigens and for other tests
4. Type isolates by methods such as bacteriophage and bacteriocin susceptibility
5. Determine sensitivity to antibiotics
6. Estimate viable counts
7. Maintain stock cultures.

METHODS OF BACTERIAL CULTURE

The methods of bacterial culture used in the clinical laboratory include **streak culture, lawn culture, stroke culture, stab culture, pour-plate culture, shake culture and liquid culture**. Special methods are employed for culturing **anaerobic bacteria**. The **sweep plate method** is used for estimating bacteria in the dust on clothing.

1. Streak Culture (Surface Plating)

This method is routinely employed for the isolation of bacteria in pure culture from clinical specimens. A platinum loop No. 23 SWG, 6.5 cm long, is charged with the specimen to be cultured. Owing to the high cost of platinum, loops for routine work are made of nichrome resistance wire, No. 24 SWG. The loop is flat, circular and completely closed with 2-4 mm internal diameter mounted on a handle.

One loopful of the specimen is smeared thoroughly over area A (Fig. 7.1), on the surface of a well dried plate, to give a **well-inoculum or 'well'**. The loop is re-sterilized and drawn from the well in two or three parallel lines on to the fresh surface of the medium (B). This process is repeated as shown (C, D, E), care being taken to sterilize the loop, and cool it on unseeded medium, between each sequence. At each step the inoculum is derived from the most distal part of the immediately preceding strokes.

Plates are incubated in the inverted position with the lid underneath. On incubation, growth may be confluent at the site of original inoculation (well), but becomes progressively thinner, and well separated colonies are obtained over the final series of streaks.

2. Lawn Culture or Carpet Culture

Lawn cultures are prepared by flooding the surface of the plate with a liquid culture or suspension of the bacterium, pipetting off the excess inoculum and incubating the plate. Alternatively, the surface of the plate may

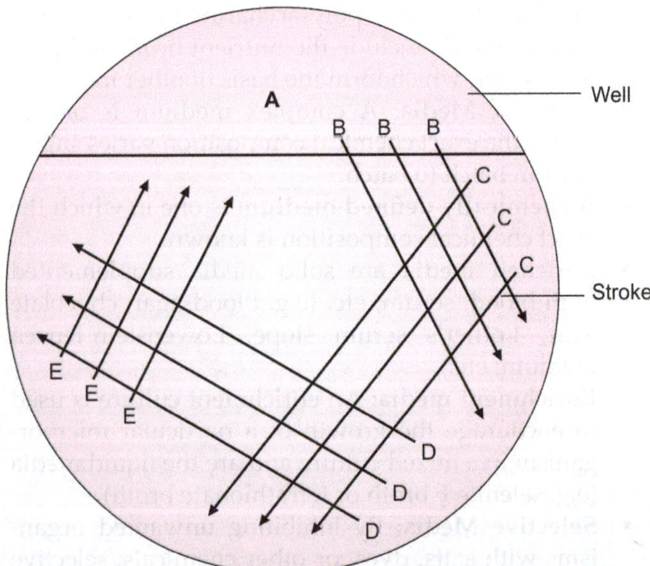

Fig. 7.1: Streak culture (streak plating) on solid media

be inoculated by applying a swab soaked in the bacterial culture or suspension. After incubation, lawn culture provides a uniform growth of the bacterium.

Uses

i. **Antibiotic susceptibility testing:** It is useful for antibiotic susceptibility testing by disk diffusion method
ii. **Bacteriophage typing.**
iii. **For preparation of bacterial antigens and vaccines:** It may also be employed when a large amount of growth is required on solid media as, for instance, in the preparation of bacterial antigens and vaccines.

3. Stroke Culture

Stroke culture is made in tubes containing agar slope or slant. Slopes are seeded by lightly smearing the surface of agar with loop in a zig-zag pattern taking care not to cut the agar.

It is employed for providing pure growth of the bacterium for **slide agglutination** and **other diagnostic tests.**

4. Stab Culture

The preparation of the stab cultures, a suitable medium such as nutrient gelatin or glucose agar is punctured with a long, straight, charged wire into the center of the medium and withdrawing it in the same line to avoid splitting the medium. The medium is allowed to set, with the tube in the upright position, providing a flat surface at the top of the medium.

Uses

i. Mainly for demonstration of gelatin liquefaction.
ii. Demonstration of oxygen requirement of the bacterium under study.
iii. For the maintenance of stock cultures.
iv. To study motility of bacteria in semisolid agar.

Pour-Plate Culture

This method is used for counting the number of living bacteria or groups of bacteria in a liquid culture or suspension. A measured amount of the suspension is mixed with molten agar medium in a Petri dish. Either 1.0 ml or 0.1 ml of dilutions of the bacterial suspension is introduced into a Petri dish. The nutrient medium, in which the agar is kept liquid by holding it in a water bath at 45-50°C, is poured over the sample, which is then mixed into the medium by gentle agitation of the plate. When the agar solidifies, the plate is incubated inverted at 37°C for 48 hours, or most suitable for the species examined. After incubation, colonies will grow within the nutrient agar (from cells suspended in the nutrient medium as the agar solidifies) as well as on the surface of the agar plate and can be enumerated using colony counters.

Uses

i. Gives an estimate of the **viable bacterial count** in a suspension.
ii. Recommended method for **quantitative urine cultures.**

5. Shake Culture

It is made by melting nutrient agar in a test tube, cooling it to 45°C and inoculating it while molten from a liquid medium with a drop from a capillary pipette or a wetted straight wire, depending on the desired size of the inoculum. Withdraw the pipette or wire and flame the mouth of the tube if it has a cotton-wool plug. Replace the cap or plug and discard the pipette into disinfectant or flame the wire. Mix the contents of the tube by rotation between the palms of the hands before the agar solidifies. Incubate it at 37°C for 24 hours and look for the growth of the organisms.

6. Liquid Cultures

Liquid cultures in tubes, bottles or flasks may be inoculated by touching with a charged loop or by adding the inoculum with pipettes or syringes.

Uses

i. **Blood culture and for sterility:** Large inocula can be employed and hence this method is adopted for blood culture and for sterility tests, where the concentration of bacteria in the inocula are expected to be small.
ii. **Dilution in the medium** For inocula containing antibiotics and other inhibitory substances liquid cultures are preferable, as these are rendered ineffective by dilution in the medium.
iii. **Large yields:** Liquid cultures are also preferred when large yields are desired, the yield being enhanced by agitation, aeration, addition of nutrients and removal of toxic metabolites (continuous culture methods).

Disadvantages

i. It does not provide a pure culture from mixed inocula—the major disadvantage.
ii. Identification of bacteria is not possible.

AEROBIC CULTURE

For cultivation of aerobes the incubation is done in an incubator under normal atmospheric condition. Incubation of cultures at 37°C is standard practice in the culture of bacteria pathogenic to man.

Culture in an Atmosphere with Added Carbon Dioxide

Some organisms, such as *Brucella abortus* and capnophilic streptococci, require extra CO_2 in the air in which they are grown and others, such as the pneumococcus and gonococcus, grow better in air supplemented with 5 to 10 percent CO_2.

ANAEROBIC CULTURE METHODS

Anaerobic bacteria require incubation without oxygen and differ in their requirement and sensitivity to oxygen. Obligate anaerobes will not grow from small inocula unless oxygen is absent and the Eh of the medium is low.

METHODS OF ANAEROBIOSIS

Anaerobiosis can be achieved by a number of methods such as:

A. Production of a vacuum.
B. Displacement of oxygen by other gases.
C. By absorption of oxygen by chemical or biological methods.
D. By displacement and combustion of oxygen.
E. By reducing agents.
F. Other anaerobic culture systems.

A. By Production of a Vacuum

This was attempted by incubating cultures in a vacuum desiccator, but proved to be unsatisfactory because some oxygen is always left behind. In the vacuum produced, fluid cultures may boil over and the media may get detached from the plates. This method is not in use now. Displacement of oxygen with gases such as hydrogen, nitrogen, helium or carbon dioxide is sometimes employed.

Drawbacks

i. Repeated evacuation and refilling.
ii. Moreover, oxygen can never be removed completely and this method rarely produces complete anaerobiosis.

B. Displacement of Oxygen by other Gases

Candle Jar

Candle jar is a popular, but ineffective method. Here inoculated plates are placed inside a large airtight container and a lighted candle kept in it before the lid is sealed. Although it is expected that the burning candle will use all the available oxygen inside before it gets extinguished, but in practice some amount of oxygen is always left behind. The candle jar provides a concentration of carbon dioxide which stimulates the growth of most bacteria.

C. By Absorption of Oxygen by Chemical or Biological Methods

i. Chemical methods
   a. Pyrogallic acid
   b. Mixture of powdered chromium and sulfuric acid
ii. Biological methods.

i. Chemical Methods

a. Pyrogallic Acid

Buchner (1888) first introduced alkaline pyrogallol for anaerobiosis which absorbs oxygen. This method has subsequently been used with different modifications. In a large tube (Buchner's tube) containing solution of sodium hydroxide; pyrogallic acid is added. The Buchner's tube is then placed inside an air-tight jar and provides anaerobiosis but a small amount of carbon monoxide, which is formed during the reaction, may be inhibitory to some bacteria.

The method has been applied to single tube and plate cultures.

The Spray anaerobic dish is a glass dish with its bottom partitioned into two halves, the top accommodating half of a Petri dish carrying the medium. Pyrogallic acid and sodium hydroxide are placed in the separate halves at the bottom of the dish. The inoculated culture plate is inverted on the top of the dish and is sealed completely. The dish is then rocked to mix the reagents, producing anaerobiosis.

The anaerobic dish is not in use now.

b. Mixture of powdered chromium and sulfuric Acid

A mixture of chromium and sulfuric acid can also be used for producing anaerobiosis within jars (Rosenthal method) or with yellow phosphorous. The two chemicals react in presence of available oxygen and produce chromous sulfate.

Gaspak

The Gaspak is now the method of choice for preparing anaerobic jars. The Gaspak is commercially available in the form of a disposable packet of aluminum foil containing pellets of sodium borohydride and cobalt chloride and of citric acid and sodium bicarbonate.

After the inoculated plates are kept in the jar, water is added to the disposable aluminum foil packet and the packet is immediately put in the jar and its lid is screwed tight. Reactions then take place to supply hydrogen and carbon dioxide. Hydrogen combines with oxygen in the presence of a catalyst (e.g. alumina pellets coated with palladium) present in the under surface of the lid of the jar to produce an anaerobic environment.

The Gaspak is simple and effective, eliminating the need for drawing a vacuum and adding hydrogen. As the standard gaspak jar is not evacuated before use a relatively large volume of water is formed during catalysis.

The Gaskit (Unipath) is a similar development and is said to yield a more assured volume of carbon dioxide.

ii. Biological Methods

Absorption of oxygen from small closed systems has been attempted by incubation along with aerobic bacteria, germinating seeds or chopped vegetables. With aerobic bacteria, this has been attempted by incubating aerobic organisms along with anaerobic bacteria. Two blood agar plates are taken one is inoculated with aerobic bacteria (Pseudomonas aeruginosa) and the other with specimen of anaerobic bacteria. Then these two plates are placed one over the other and sealed along the rims and are incubated. Anaerobiosis produced by such biological methods is slow and ineffective.

D. By Displacement and Combustion of Oxygen

Anaerobic Jars

Anaerobic jars provide the method of choice when an oxygen-free or anaerobic atmosphere is required for obtaining surface growths of anaerobes.

McIntosh and Filde's Anaerobic Jar

Anaerobiosis obtained by **McIntosh and Filde's anaerobic jar** (Fig. 7.2) is the most dependable and widely used method.

The jar (c. 20 × 12.5 cm) should be made of metal or robust plastic with a lid that can be clamped' down on a gasket to make it air tight. Glass jars have been used in the past, but, as explosions occasionally occur, their use is not justified. The lid is furnished with two tubes with valves, one acting as gas inlet and the other as the outlet. The lid also has two terminals which can be connected to an electrical supply. On its undersurface it carries a gauze sachet carrying alumina pellets coated with palladium (palladinized alumina). It acts as a room temperature catalyst for the conversion of hydrogen and oxygen into water.

Procedure

Inoculated culture plates are placed inside the jar with the medium uppermost and lid downwards and the lid clamped tight. The outlet tube is connected to a vacuum pump and the air inside is evacuated. Approximately 6/7 of the air is evacuated (pressure reduced to 100 mm Hg, i.e. 660 mm below atmospheric) and this is monitored on a vacuum gauge. The outlet tap is then closed and the inlet tube connected to a hydrogen supply. Hydrogen is drawn in rapidly. As soon as this inrush of gas has ceased the inlet tap is also closed and the jar is held on the bench for 10 minutes. If the catalyst is normally active, a decrease in pressure of at least 20 mm Hg occurs within the 10 minutes period after the admission of hydrogen. This can be detected on the gauge when the valve is opened.

Catalyst

After the jar is filled with hydrogen, the electrical terminals are connected to a current supply to heat the catalyst and if room temperature catalyst is used, heating is not required. Catalysis will continue until all the oxygen in the jar has been used up. The catalyst will help to combine hydrogen and residual oxygen to form water. The jar is then incubated at 37°C.

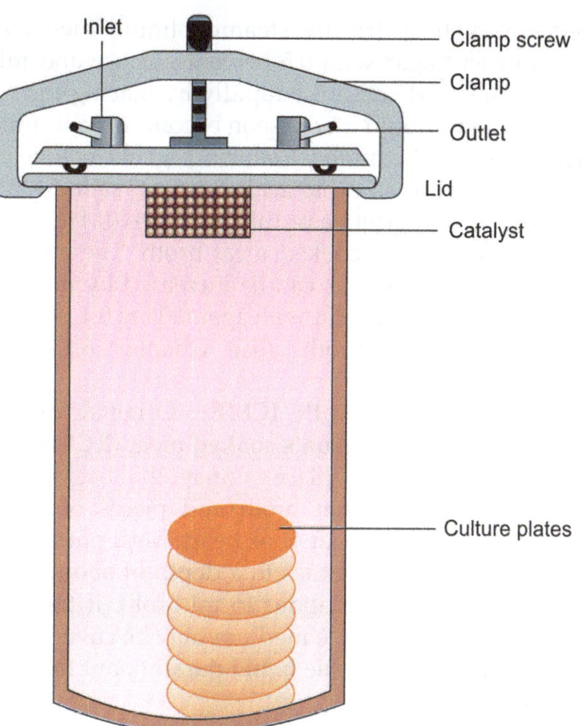

Fig. 7.2: McIntosh and Filde's anaerobic jar

Indicator

An indicator should be employed for verifying the anaerobic condition in the jars. **Reduced methylene blue** is generally used as indicator (mixture of NaOH, methylene blue and glucose). It becomes colorless anaerobically but regains blue color on exposure to oxygen.

In addition to, or instead of, using a chemical indicator, some workers include in the jar a plate inoculated with a known strict anaerobe such as *Clostridium tetani* or *Bacteroides fragilis*, and of a strict aerobe, such as *Pseudomonas aeruginosa*. This method is quite reliable if the indicator anaerobe grows and the aerobe does not.

Disadvantage

Any anaerobic jar system has the major disadvantage that the plates have to be removed from the jar to be examined and this, of course, exposes the colonies to oxygen, which is especially hazardous to the anaerobes during their first 48 hours of growth. For this reason, a suitable holding system always should be used in conjunction with anaerobic jars, placed in an oxygen-free holding system, removed one by one for rapid microscopic examination of colonies, and then quickly returned to the holding system. Plates never should remain in room air on the open bench.

E. By Reducing Agents

In the clinical laboratory a variety of methods are available for the culture of anaerobic organisms. The simplest method is exclusion of oxygen from the medium, and is effected by growing the organisms within the culture medium such as freshly steamed liquid media and deep nutrient agar with 0.5 percent glucose and minimal shaking and solidified rapidly by placing the tube in cold water. Liquid media soon become aerobic unless a reducing agent is added. Reducing agents include glucose (0.5 to 1%), ascorbic acid (0.1%), cysteine (0.1%), sodium mercaptoacetate or thioglycollate (0.1%), or the particles of meat in cooked meat broth. The effectiveness of reducing agents can be increased by making a liquid medium semisolid with agar 0.05 to 0.1 percent.

i. Thioglycollate broth: (See Chapter 6, Culture Media).
ii. **Cooked meat broth (CMB):** Original medium known as **'Robertson's cooked meat (RCM) medium**) has a special place in anaerobic bacteriology. It contains nutrient broth and pieces of fat-free minced cooked meat of ox heart. Meat particles are placed in 30 ml bottles to a depth of about 2.5 cm and covered with about 15 ml broth. If test tubes are used the surface medium may be covered with a 1 cm layer of sterile liquid paraffin, but this is not essential.

Principle

i. Unsaturated fatty acids present in meat utilize oxygen for auto-oxidation, the reaction is being catalyzed by hematin in the meat.
ii. Certain reducing substances such as glutathione and cysteine present in meat also utilize oxygen.
iii. Sulphydryl compounds (present in cysteine) also contributes for a reduced oxidation-reduction (OR) potential.

Method

Liquid media should be prereduced by holding in a boiling water bath for 10 minutes to drive off dissolved oxygen, then quickly cooled to 37°C just before use. The surface of the CMB medium may be covered with a layer of sterile liquid paraffin for strict anaerobiosis.

Interpretation

It permits the growth of even strict anaerobes and indicates their saccharolytic or proteolytic activities. With growth of saccharolytic anaerobes (*C. welchii*), color of meat pieces turns red while it becomes black in case of proteolytic an aerobes (*C. tetani*).

Uses of CMB

i. CMB is suitable for growing anaerobes in air.
ii. Also for the preservation of stock cultures of aerobic organisms.

The inoculum is introduced deep in the medium in contact with the meat.

F. Other Anaerobic Culture Systems

a. **Hungate procedure:** More meticulous anaerobic methods have been developed to prevent access of air to media during preparation, inoculation and incubation. One of these methods was pioneered in 1950 by Hungate, the manipulation of all these procedures under oxygen-free gas. Surface colonies are grown in roll tubes in which a thin layer of agar coats the inside of the tube. The medium must be transparent for surface colonies to be visible, and this precludes the use of blood agar.
b. **Anaerobic cabinets:** Anaerobic cabinets are commercially produced for the processing of specimens and the subsequent incubation of cultures and subcultures in an oxygen free atmosphere enriched with 5 to 10 percent CO_2. The advantage of anaerobic cabinets is that all of the processing, including periodic examination of plates and preparation of subcultures, can be done without exposure to oxygen.
c. **Anaerobic bags or pouches:** Anaerobic bags are available commercially. One or two inoculated plates are placed into a bag and an oxygen removal system is activated and the bag is sealed and incubated. Plates can be examined for growth without removing the plates from bag, thus without exposing the colonies to oxygen.

METHODS OF ISOLATING PURE CULTURES

Most studies and tests of the physiological, serological and other characters of bacteria are valid only when made on a pure culture. The following methods may be employed for isolating pure cultures of bacteria from mixtures:

1. **Surface plating:** Surface plating is the method routinely employed in clinical bacteriology and enables the isolation of distinct colonies which may be picked out, if necessary for further purification and study.
2. **Use of selective, enrichment or indicator media:** Enrichment, selective and indicator media are widely used for the isolation of pathogens from specimens such as feces, with varied flora.
   i. **Selective media:** Selective media such as tellurite media for the diphtheria bacillus, have been devised so that, the majority of organisms likely to be associated with those for which the media are used will not grow, and the isolation of pure cultures is thus facilitated.
   ii. **Enrichment media:** Enrichment media such as selenite broth for *Salmonella* sp, favor the multiplication of particular species as a step towards their isolation in pure culture.
   iii. **Indicator media:** Indicator media, such as Willis and Hobbs medium for *Clostridium* sp, contain ingredients that change in appearance with particular organisms and so assist their isolation.

3. **Selective treatment of the specimen before culture:**
   i. **Heating at 65°C for 30 minutes or at higher temperatures for shorter periods:** can be used to separate spores from vegetative bacilli but there is no guarantee that spores will germinate under subsequent cultural conditions. For example, by heating a mixture containing vegetative and spore forming bacteria at 80°C the former can be eliminated. This method is useful for the isolation of tetanus bacilli from dust and similar sources.
   ii. **Pretreatment of specimens with appropriate bactericidal substances:** Pure cultures may be obtained by pretreatment of specimens with appropriate bactericidal substances which destroy the unwanted bacteria. This method is the standard practice for the isolation of tubercle bacilli from sputum and other clinical specimens, by treatment with alkali, acid or other substances to which most commensals are susceptible but tubercle bacilli are resistant.
4. Use of selective growth conditions:
   i. **Separation of bacteria with different temperature optima:** The temperature and atmosphere chosen for a culture automatically preclude the growth of many bacteria. Incubation at 37°C, used for most medically important bacteria, is too warm for some air contaminants, which subsequently appear as colonies when plates are kept at room temperature. Some pathogens are selectively favoured by growth at temperatures above 37ºC. Only thermophilic bacteria grow at 60ºC. A mixture containing *Neisseria meningitidis* and *Neisseria catarrhalis* can be purified by incubation at 22ºC when only the latter grows.
   ii. **Cultivation under aerobic or anaerobic conditions:** Obligate aerobes and anaerobes may be separated by cultivation under aerobic or anaerobic conditions. Strict anaerobes will not grow in air and most facultative anaerobes grow less vigorously under anaerobic than under aerobic conditions. Shake cultures in Veillon tubes were in use formerly but are now obsolete.
5. **Separation of motile from nonmotile bacteria can be effected using Cragie's tube:** This consists of a tube of semisolid agar, with a narrow tube open at both ends placed in the center of the medium in such a way that it projects above the level of the agar. Inoculation of the mixture is made into the central tube. On incubation, subculture is taken from the surface of the medium in the outer tube because the motile bacteria alone traverse the agar and appear at the top of the medium outside the central tube.

   A **U-tube** also serves the same purpose, inoculation being performed in one limb and the subculture taken from the other. This method can also be used to obtain phase variants in Salmonella species.
6. **Animal inoculation:** Pathogenic bacteria may be isolated from mixtures by inoculation into appropriate animals due to the fact that laboratory animals are highly susceptible to certain organisms for example, the mouse to the pneumococcus. Advances in the development of culture media have now restricted the requirement for animal work to specialist centers.

Examples

i. **Pneumococcus:** If a mixture of organisms containing the pneumococcus, e.g. in sputum, is inoculated subcutaneously into a mouse, the animal dies of pneumococcal septicemia in 12 to 48 hours and the organism can be obtained in pure culture from the heart blood.
ii. **Anthrax bacilli: Anthrax bacilli** can be distinguished from other aerobic sporulating bacilli by inoculation into mice or guinea pigs. Anthrax bacilli produce a fatal septicemia and may be cultured pure from the heart blood.
iii. **Tubercle bacillus:** Similarly, the tubercle bacillus can be isolated from contaminating organisms by inoculation of an infected specimen into a guinea-pig. The tubercle bacillus is found in a pure state in the resulting lesions.

7. **Filters:** Bacteria of differing sizes may be separated by the use of selective filters. Filters are widely used for separating viruses from bacteria.

KNOW MORE

Culture in an Atmosphere with Added Carbon Dioxide

Some organisms, such as *Brucella abortus* and capnophilic streptococci, require extra CO_2 in the air in which they are grown. Others, such as the pneumococcus and gonococcus, grow better in air supplemented with 5 to 10 percent CO_2. For this dedicated CO_2 jars should be used. For small numbers of cultures an alternative approach is to use a CO_2 generating kit.

KEY POINTS

- The methods of bacterial culture used in the clinical laboratory include **streak culture, lawn culture, stroke culture, stab culture, pour-plate culture, shake culture and liquid culture.** Special methods are employed for culturing **anaerobic bacteria.**

- **Streak culture (surface plating):** This method is routinely employed for the isolation of bacteria in pure culture from clinical specimens.

  Anaerobic culture methods: Anaerobic bacteria require incubation without oxygen and differ in their requirement and sensitivity to oxygen.
- **Anaerobic jars:** Anaerobiosis obtained by **Mclntosh and Filde's anaerobic jar** is the most dependable and widely used method.
- Gaspak is now the method of choice for preparing anaerobic jars.

  Cooked meat broth (CMB): Original medium known as **'Robertson's cooked meat (RCM) medium'**) has a special place in anaerobic bacteriology.
- **Methods of isolating pure cultures:** Various methods may be employed for isolating pure cultures of bacteria from mixtures

IMPORTANT QUESTIONS

1. Discuss in detail anaerobic culture methods.
2. Enumerate various methods of bacterial culture. Describe in detail the anaerobic methods of cultivation.
3. Write short notes on:
   Streak culture
   Lawn culture
   Stroke culture.
   Stab culture.
   Pour-plate culture
   Anaerobic culture methods/culture of anaerobic bacteria.
   McIntosh and Fildes anaerobic jar.
   Gas pak.
   Cooked meat broth (CMB)/RCM broth.
   Methods of isolating pure cultures.

FURTHER READING

Collee JC, et al. Mackie and McCartney Practical Medical Microbilology. London: Churchill Livingstone 1996;113-129.

CHAPTER 8

Identification of Bacteria

Learning Objectives

After reading and studying this chapter, you should be able to:

- **Explain principle and discuss interpretation of the following biochemical reactions:** 1. Sugar fermentation; 2. Indole production; 3. Methyl red (MR) test; 4. Voges-Proskauer (VP) test for acetoin production; 5. Citrate utilization; 6. Nitrate reduction test; 7. Urease test; 8. Catalase production; 9. Oxidase test; 10. Phenylalanine deaminase test; 11. Hydrogen sulfide production; 12. Triple-sugar iron (TSI).
- List various examples of bacteria **giving positive biochemical reactions as mentioned above**.

METHODS USED TO IDENTIFY BACTERIA

Once a bacterium has been obtained in pure culture, it has to be identified. Identification schemes are classified into one of the two categories (Table 8.1):

1. Phenotypic characteristics
2. Genotypic characteristics.

1. Phenotypic Characteristics

Until recently, identification of bacteria of only pure culture (single strain) was made by phenotypic methods.

A. Microscopic Morphology

The morphology of the bacterium depends on a number of factors such as the strain studied, nature of the culture medium, temperature and time of incubation, age of the culture and the number of subcultures it has undergone. The characteristics noted are shown in Table 8.2.

B. Staining Reactions

A number of staining techniques for the identification of bacteria, are available. Of these, *Gram stain* and *Ziehl-Neelsen* stain are most important.

The Gram stain divides bacteria into **gram-positive** and **gram-negative.** With Ziehl-Neelsen staining bacteria divides into **acid fast** and **non-acid fast.** Numerous other stains are used for special purposes, such as demonstration of flagella, capsule, spores, and metachromatic granules. The fluorescent antibody technique enables one to identify them according to their surface antigens.

The age of the culture is important. Staining characteristics either vary or are not brought out well in older cultures. Simple stains bring out the morphology best.The study of morphology and staining characteristics helps in preliminary'identification of the isolate.

Table 8.1: Methods used to identify bacteria

Phenotypic characteristics	*Genotypic characteristics*
A. Microscopic morphology	Nucleic acid hybridization
B. Staining reactions	PCR-Amplifying specific DNA sequences
C. Metabolism differences	Sequencing rRNA genes
i. Macroscopic morphology or cultural characteristics	
ii. Fermentation and other biochemical reactions.	
D. Serology	
E. Antibiotic tolerance (resistance) tests, dye tolerance, and other inhibition tests	
F. Bacteriophage and bacteriocin typing	
G. Pathogenicity.	

Table 8.2: Morphological characteristics of bacteria

1. **Shape:** Cocci, spherical, oval or lanceolate, short rods, long rods, filaments, commas or spirals.
2. **Axis:** Straight or curved.
3. **Size:** Length and breadth. Considerable variations in shape and size leading to club, navicular and swollen or shadow or giant forms may be seen.
4. **Sides:** Parallel, bulging, concave or irregular.
5. **Ends:** Rounded, truncate, concave or pointed.
6. **Arrangement:** They may be arranged singly, in pairs, in tetrads or in packets of eight, or in chains, short or long, in the case of cocci; bacilli may be arranged at random, in short or long chains, in Chinese letter patterns, as palisades or in bundles; vibrios may be single or in S or spiral forms.
7. **Irregular forms:** Variations in shape and size; club, filamentous, branched, navicular, citron, fusiform, giant swollen.
8. **Motility:** Nonmotile, sluggishly motile, actively motile or may exhibit darting motility.
9. **Flagella:** Without flagella, that is atrichate, or monotrichate, lophotrichate, amphitrichate or peritrichate.
10. **Fimbriae:** In electron micrographs. approximate number and size, polar or peritrichate.
11. **Spores:** The spores, when present may be oval or spherical or ellipsoidal and may be of the same width or wider than that of the bacillary body. The spores may be equatorial, subterminal or terminal.
12. **Capsules:** Present or absent, indefinite mucoid sheath or envelope.
13. **Staining:** Even, irregular, unipolar, bipolar, beaded, barred; and variations in depth between different organisms; presence of metachromatic granules; reaction to Gram and to Ziehl-Neelsen stains.

NOTE: Hanging drop preparations, dark ground illumination, phase contrast or electron microscopy, all help in these studies.

Table 8.3: Description of appearance of growth in solid or in liquid media

1. **Shape:** Circular, irregular, or rhizoid.
2. **Size:** In millimeters.
3. **Elevation:** Effuse, elevated, convex, concave, umbonate or umbilicate (Fig. 8.1).
4. **Margins:** Bevelled or otherwise.
5. **Surface:** Smooth, wavy, rough, granular, papillate or glistening.
6. **Edges:** Entire, undulate, crenated, fimbriate or curled (Fig. 8.2).
7. **Color:** By reflected and transmitted light; fluorescent, iridescent, opalescent, self-luminous.
8. **Structure:** Opaque, translucent or transparent.
9. **Consistency:** Membranous, friable, butyrous or viscid.
10. **Emulsifiability:** Easy or difficult.
11. **Differentiation:** Differentiated into a central and a peripheral portion.

In a stroke culture, note:

1. The degree of growth—scanty, moderate, or profuse.
2. Their nature-discrete or confluent, filiform, spreading or rhizoid.
3. Their elevation, surface, edges, color, structure, odor, emulsifiability, consistency and changes in the medium.

Growth in liquid medium:

Degree: None, scanty, moderate, abundant or profuse.

Turbidity: Present or absent; if present, slight, moderate or dense; uniform, granular or flocculent.

Deposit: Present or absent: if present, slight, moderate or abundant; powdery, granular, flocculent, membranous or viscid; disintegrating completely or incompletely on shaking.

Surface growth: Present or absent; if present, ring growth around wall of tube; or surface pellicle, which is thin or thick; with a smooth, granular or rough surface and which disintegrates completely or incompletely on shaking. All aerobes have tendency to grow on surface of media due to more content of oxygen present on the surface, e.g. *Pseudomonas* sp.

C. Metabolic Differences

The requirements of oxygen, the need for carbon dioxide, the capacity to form pigments, and the production of hemolysis help in classification.

i. Cultural characteristics or macroscopic morphology

These provide additional information for the identification of the bacterium. The characteristics revealed in different types of media are noted. While studying colonies on solid media, the various features are noted (Table 8.3).

ii. Biochemical reactions (Table 8.4)

A large number of biochemical tests can be employed for the identification of different bacteria. These include:

1. Sugar Fermentation

This is tested in sugar media.

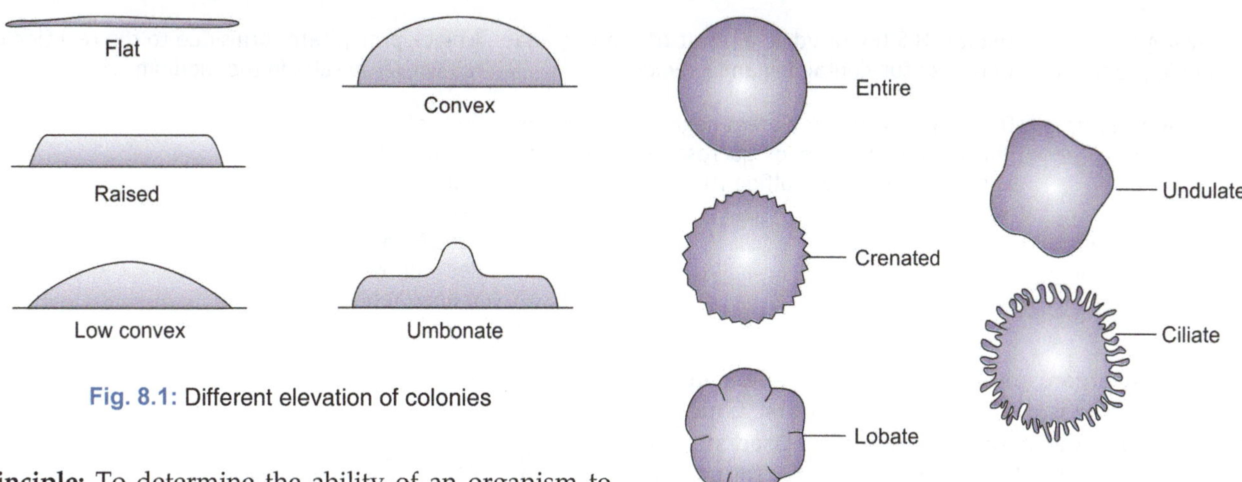

Fig. 8.1: Different elevation of colonies

Fig. 8.2: Description of edges of colonies

Principle: To determine the ability of an organism to ferment a specific carbohydrate (sugar) incorporated in a medium producing acid or acid with gas.

Table 8.4: Some important biochemical tests

Biochemical test	*Principle of the test*	*Positive reaction*
1. Sugar fermentation	Detects the acidity resulting from fermentation of the sugar incorporated into the medium producing acid or acid with gas.	Acid production is shown by change in the color of the medium to pink or red. An inverted tube traps any gas that is made.
2. Indole	Detects the ability of certain bacteria to decompose the amino acid tryptophan to indole by an enzyme tryptophanase indicates a positive reaction.	The product, indole, reacts with a chemical reagent that is added, turning the reagent a deep red color.
3. Methyl red (MR) test	Detects mixed acids, the characteristic end products of a particular fermentation pathway. 8 mixed adds, p. 153	The medium becomes acidic (pH < 4.5); a red color develops upon the addition of a pH indicator.
4. Voges-Proskauer (VP) test for acetoin production	Detects acetoin, an intermediate of the fermentation pathway that leads to the production of a 2, 3-butandiol.	A red color develops upon addition of chemicals that detect acetoin.
5. Citrate utilization	Determines whether or not citrate can be used as a sole carbon source.	Growth, which is usually accompanied by the color change of a pH indicator.
6. Nitrate reduction test	This is a test for the presence of the enzyme nitrate reductase which causes the reduction of nitrate to nitrite which can be tested for by an appropriate colorimetric reagent.	Red color develops within few minutes, indicates the presence of nitrite and hence the ability of the organism to reduce nitrate. Almost all Enterobacteriaceae reduce nitrate.
7. Urease test	Detects the ability of an organism to produce an enzyme urease which splits urea to carbon dioxide and ammonia.	Ammonia makes the medium alkaline and thus phenol red indicator changes to pink/red in color.
8. Catalase test	Detects the activity of the enzyme catalase, which causes the breakdown of hydrogen peroxide to produce O_2 and water.	Bubbles
9. Oxidase test	Detects the activity of cytochrome c oxidase, a component of the electron transport chain of specific organisms	A dark color develops upon the addition of a specific reagent
10. Phenylalanine deaminase test	Detects the enzymatic removal of the amino group from phenylalanine.	The product of the reaction, phenylpyruvic acid, reacts with ferric chloride to give the medium a green color.

11.	Hydrogen sulphide production	Detects H_2S liberated as a result of the degradation of sulfur-containing amino acids.	A black precipitate forms due to the reaction of H_2S with iron salts in the medium.
12.	Triple-sugar iron (TSI) agar	Determines whether a gram-negative rod utilizes glucose and lactose or sucrose fermentatively and forms hydrogen sulfide (H_2S)	Red color—Alkaline Yellow color—Acidic Slant/Butt Red/Red Red/Yellow Red/Yellow H_2S Yellow/Yellow

Method: Test organism is inoculated in a sugar medium and incubated at 37°C for 18-24 hours. **Glucose, lactose, sucrose and mannitol** are widely used sugars. Sugar media contains **1 percent sugar.** Indicator used is **Andrade's indicator** (a solution of acid fuchsin to which is added sodium hydroxide).

Interpretation (Figs 8.3 and 8.4)

Positive—Pinkish-red (acidic)—acid production is shown by change in the color of the medium to pink or red.

Negative—Yellow to colorless (alkaline).

Gas production can be seen as bubbles in Durham's tube.

Examples of Fermentative Bacteria

Glucose fermenters: All members of the family Enterobacteriaceae

Glucose and lactose fermenters: *Escherichia coli, Klebsiella* sp.

Glucose and mannitol fermenters: *Salmonella* sp.

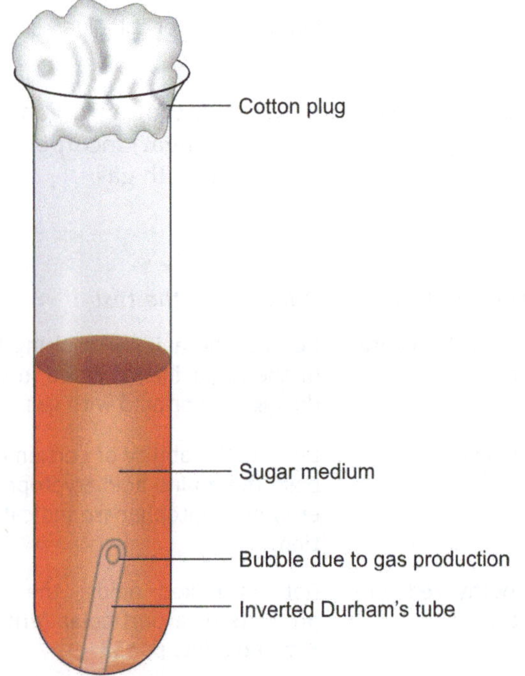

Fig. 8.3: Inverted Durham's tube showing gas production

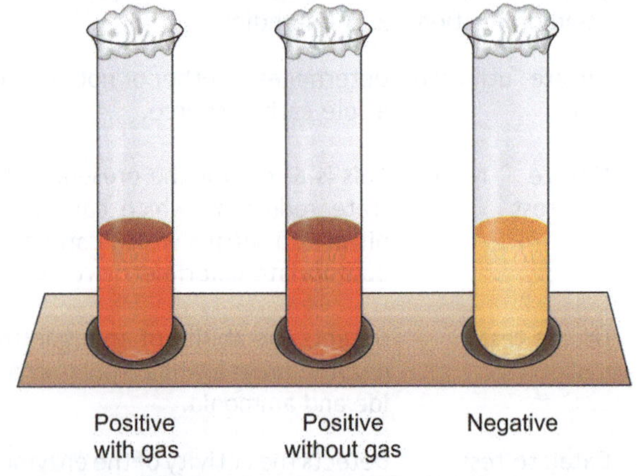

Fig. 8.4: Sugar fermentation tests

2. Indole Production

Principle: This test demonstrates the ability of certain bacteria to decompose the amino acid tryptophan to indole, which accumulates in the medium. Tryptophan is decomposed by an enzyme tryptophanase produced by certain bacteria.

$$\textbf{Tryptophan} \xrightarrow{\text{Tryptohanase}} \textbf{Indole}$$

Method: Indole production is detected by inoculating the test bacterium into peptone water (tryptophan rich) and incubating it at 37°C for 48-96 hours. Add 0.5 ml Kovac's reagent and shake gently.

Kovac's reagent consists of:

Paradimethylaminobenzaldehyde	10 g
Amyl or isoamyl alcohol	150 ml
Conc. hydrochloric acid	50 ml

Interpretation (Fig. 8.5)

Indole positive—A **red color** in the alcohol layer indicates a positive reaction.

Indole negative—Yellow colored ring (color of Kovac's reagent) near the surface of the medium.

Positive and Negative Bacteria

Indole positive: *Esch. coli Shigella, Edwardsiella, Proteus* sp. other than *P. mirabilis.*

Indole negative: *Klebsiella* sp. *Enterobacter* sp. *Serratia., Hafnia* spp., *P. mirabilis*

3. Methyl Red (MR) Test

Principle: The methyl red test is employed to detect the production of sufficient acid during the fermentation of glucose so that pH of the medium falls and it is maintained below 4.5 (Fig. 8.6).

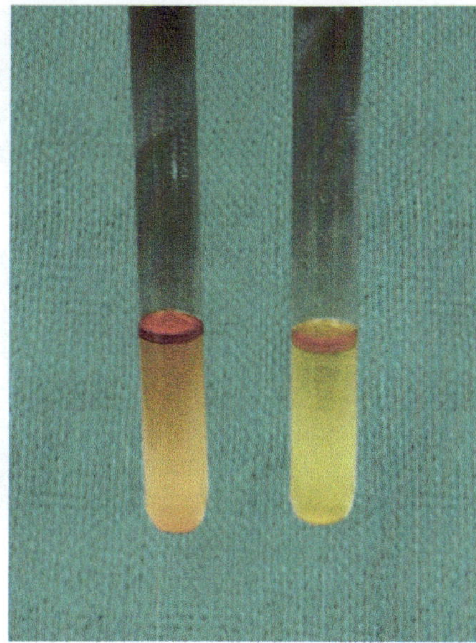

Fig. 8.5: Indole test

Medium

Method: Inoculate the test organism in liquid medium (glucose phosphate broth) and incubate at 37°C for 2 to 5 days. Then add five drops of 0.04 percent solution of methyl red. Mix well and read the result immediately.

Interpretation (Fig. 8.7)

Positive: Bright red color.
Negative: Yellow color.
Note: If the results after 48 hours are equivocal, the test should be repeated with cultures that have been incubated for 5 days.

MR-positive and -negative Bacteria

MR-positive: *Esch. coli*, *Shigella* sp., *Edwardsiella* sp, *Yersinia* sp., *Listeria monocytogenes.*

MR-negative: *Klebsiella*, *Enterobacter* sp. *Serratia* sp, *Hafnia* sp.

4. Voges-Proskauer (VP) Test for Acetoin Production

Principle: Many bacteria ferment carbohydrates with the production **of acetyl methyl carbinol (acetoin)** or its reduction product **2, 3 butylene glycol.** The substances can be tested for by a colorimetric reaction between **diacetyl** formed during the test by oxidation of acetyl methyl carbinol or 2, 3 butylene glycol and a guanidino group under alkaline conditions. In the presence of **potassium hydroxide** and **atmospheric oxygen,** acetoin is converted to diacetyl, and α-naphthol serves as a catalyst to form a red complex (Fig. 8.6).

Medium: Glucose phosphate peptone water, as for the methyl red test.

Method: Inoculate test organism in glucose phosphate broth and incubate at 37°C or 30°C for 48 hours only.

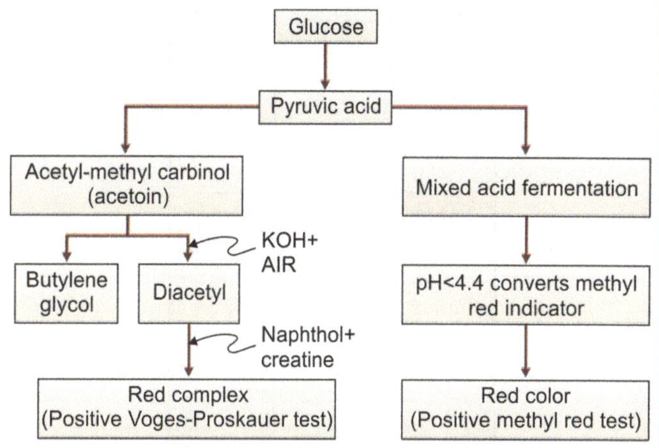

Fig. 8.6: Mixed-acid and butylene glycol pathway of dextrose fermentation

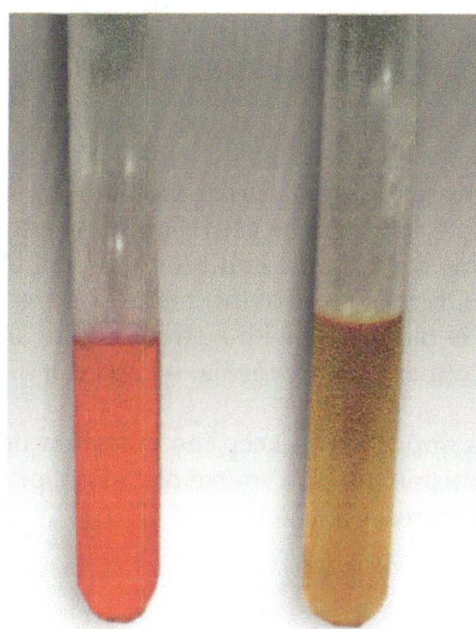

Fig. 8.7: Methyl red (MR) test

Add 1 ml of 40 percent potassium hydroxide and 3 ml of a 5 percent solution of α-naphthol in absolute ethanol.

Interpretation of VP Test (Fig. 8.8)

Positive reaction—Development of a pink color in 2-5 minutes, becoming crimson in 30 minutes.
Negative reaction—Colorless for 30 minutes.
Note: The tube can be shaken at intervals to ensure maximum aeration. Naphthol is a carcinogen.

VP-positive and -negative Bacteria

VP-positive: *Klebsiella* sp., *Enterobacter* sp. *Serratia* sp. *Eltor vibrios*, *Staphylococcus.*
VP-negative: *Esch. coli*, *Shigella* sp, *Edwardsiella*, *Micrococcus.*

5. Citrate Utilization

Principle: This is a test for the ability of an organism to utilize citrate as the sole carbon and energy source for growth and an ammonium salt as the sole source of nitrogen with resulting alkalinity.

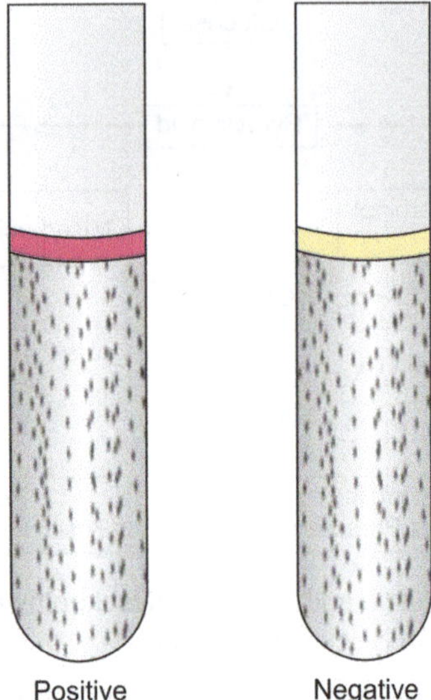

Fig. 8.8: Voges-Proskauer (VP) test

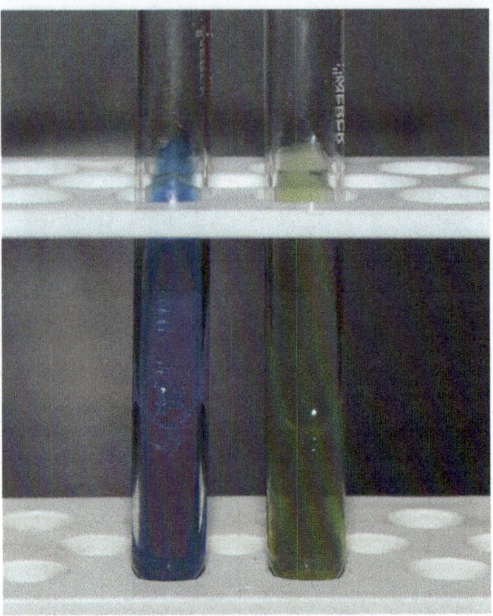
Fig. 8.9: Citrate utilization test

Method: Koser's liquid citrate medium or solid Simmons' citrate agar may be used. Simmon's citrate medium contains agar, citrate and bromothymol blue as an indicator. Original color of the medium is green. A part of colony is picked up with a straight wire and inoculated into either of these media. Incubate at 37°C for 96 hours.

Note: It is important to keep the inoculum light, since dead organisms can be a source of carbon, producing a false positive reaction.

Interpretation (Fig. 8.9)

1. Simmons' citrate medium

Positive: Blue color and streak of growth. Blue color is due to the alkaline pH which results from utilization of citrate. Bromothymol blue (indicator) is blue in alkaline conditions.

Negative—Original green color and no growth.

2. Koser's citrate medium

Positive—Turbidity, i.e. growth.

Negative—No turbidity.

Citrate Positive and Negative Bacteria

Citrate-positive: *Klebsiella* sp., *Enterobacter* sp. *Serratia* sp., *Hafnia* spp, *Salmonella* sp. *except S. typhi*, *Citrobacter* sp.,

Citrate negative: *Esch. coli*, *Edwardsiella*, *Shigella* sp., *Salmonella typhi.*

IMViC Tests for the Family Enterobacteriaceae (Table 8.5)

Indole, MR, VP and citrate tests are very useful in the identification and classification of enteric gram-negative bacteria. These tests are commonly referred to by the sigla 'IMViC' tests. 'IMViC' tests for the family Enterobacteriacae are shown in Table 8.5. in general (with exceptions).

6. Nitrate Reduction Test

Principle: This is a test for the presence of the enzyme **nitrate reductase** which causes the reduction of nitrate to nitrite which can be tested for by an appropriate colorimetric reagent. Almost all Enterobacteriaceae reduce nitrate.

$$\text{Nitrate} \xrightarrow{\text{Nitrate reductase}} \text{Nitrite}$$

Method: Inoculate test organism in 5 ml medium containing potassium nitrate, peptone and distilled water. Incubate it at 37°C for 96 hours. Add 0.1 ml of, the test reagent to the test culturewhich consists of equal volumes of 0.8 percent sulphanilic acid and 0.5 percent α-naphthylamine in 5 N acetic acid mixed just before use.

Interpretation (Fig. 8.10) of Nitrate Reduction Test

Positive—Red color develops within few minutes (indicates the presence of nitrite and hence the ability of the organism to reduce nitrate).

Negative—No color development.

Nitrate reduction positive and negative bacteria

Nitrate reduction positive—All members of Enterobacteriaceae, *Branhamella catarrhalis.*

Nitrate reduction negative—*Haemophilus ducreyi.*

7. Urease Test

Principle

To determine the ability of an organism to produce an enzyme urease which splits urea to ammonia. Ammonia makes the medium alkaline and thus phenol red indicator changes to pink/red in color.

$$\text{Urea} \xrightarrow{\text{Urease}} CO_2 + H_2O + 2NH_3 \text{--} (NH_4)_2CO_3$$

Procedure

The test organism is inoculated on the entire slope of **Christensen's medium** which contains urea and phenol red indicator in addition to other constituents including agar. It is incubated at 37°C and examined after 4 hours and after overnight incubation. The test should not be considered negative till after four days of incubation.

Christensen's urease medium contains:

- Peptone water
- Urea (20%)
- Agar
- Phenol red.

Interpretation (Fig. 8.11) of Urease Test

Urease-positive—Purple-pink color
Urease-negative—Pale yellow color (Fig. 8.11).

Note: Development of purple-pink color indicates production of urease. The latter in the presence of water converts urea into ammonia and carbon dioxide. Ammonia makes the medium alkaline and phenol red indicator changes to purple-pink in color.

Urease-positive and -negative Bacteria

Urease-positive: *Klebsiella* sp., *Proteus* sp., *Yersinia enterocolitica, Helicobacter pylori*

Urease-negative: *Esch. coli, Providencia* sp., *Yersinia pestis.*

8. Catalase Production

Principle

This demonstrates the presence of catalase, an enzyme that catalyzes the release of oxygen from hydrogen peroxide.

$$H_2O_2 \xrightarrow{\text{Catalase}} H_2O + O_{\text{(Air bubbles)}}$$

Method

1. One ml of hydrogen peroxide solution, H_2O_2 (10 vol), is poured over a 24 hours nutrient agar slope culture of the test organism and the tube is held in a slanting position.
2. Alternatively, a small amount of the culture to be tested is picked from a nutrient agar slope with a clean sterile platinum loop or a clean, thin glass rod and dip it in a drop of 10 percent hydrogen peroxide on a clean glass slide.

Interpretation

Positive test—Immediate bubbling, easily observed (O_2 formed).

Negative test—No bubbling (no O_2 formed).

Note: A false positive reaction may be obtained if the culture medium contains catalase (e.g. blood agar), or if an iron wire loop is used.

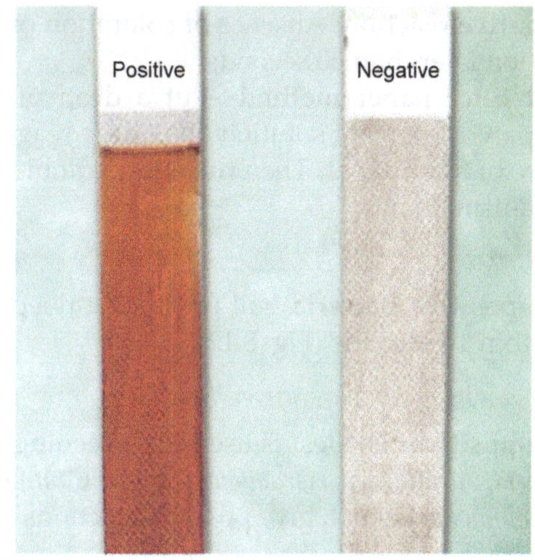

Fig. 8.10: Nitrate reduction test

Catalase-positive and -negative Bacteria

Catalase-positive: All members of Enterobacteriaceae except *Shigella dysentriae* type 1; *Staphylococcus, Micrococcus, Bacillus.*

Catalase-negative: *Shigella dysenteriae* type 1, *Streptococcus, Clostridium.*

9. Oxidase Test

Principle

To determine the presence of bacterial **cytochrome oxidase** using the oxidation of the substrate, a redox dye, **tetramethyl-p-phenylene-diamine dihydrochloride** (oxidase reagent). The dye is reduced to a deep purple color. A positive test (presence of oxidase) is indicated by the development of a dark purple color. No color development indicates a negative test and the absence of the enzyme.

Method

1. **Plate method:** Cultures are made on a suitable solid growth medium. A freshly prepared 1 percent solution of **tetramethyl-p-phenylene-diamine dihydrochloride** is poured on to the plate so as to cover the surface and is then decanted. The colonies of oxidase-positive organisms rapidly develop a **purple color.**
2. **Dry filter paper method:** A strip of filter paper soaked in the oxidase reagent is removed, laid in a Petri dish and moistened with distilled water. The colony to be tested is picked up with a platinum loop and smeared over the moist area.

   Interpretation: Positive reaction is indicated by an intense deep-purple hue, appearing within 5-10 seconds.

Negative reaction: Absence of coloration or by coloration later than 60 seconds.

3. **Wet filter paper method**—Put a drop of freshly prepared 1 percent solution of oxidase reagent on a piece of filter paper. Then rub a few colonies of test organism on it.

Interpretation

Oxidase-positive bacteria will produce a deep purple color within 10 seconds (Fig. 8.12).

Oxidase-positive and -negative Bacteria

Oxidase-positive: The test is used for screening species of *Neisseria, Alcaligenes, Aeromonas, Vibrio, Campylobacter* and *Pseudomonas,* which give positive reactions.

Oxidase negative: Family **Enterobacteriaceae**—All the members of the family Enterobacteriaceae are oxidase-negative.

10. Phenylalanine Deaminase Test

Phenylalanine deaminase determines whether the organism possesses the enzyme phenylalanine deaminase that deaminates phenylalanine to phenylpyruvic acid, which reacts with ferric salts to give a green color. This test is also commonly called as **PPA test**. This test is useful in initial differentiation of *Proteus, Morganella,* and *Providencia* from the rest of the Enterobacteriaceae.

Phenylalanine—Phenylalanine deaminase—Phenylpyruvic acid—$FeCl_3$—Green color.

Procedure

Agar slants of the medium containing phenylalanine is inoculated with a fairly heavy inoculum and incubated at 37°C for overnight. A few drops of 10 percent ferric chloride solution is added directly to the surface of the agar. If the test is positive, a green color will develop in the fluid and in the slope.

Interpretation

Positive—Green color (Fig. 8.13).
Negative—No color change.

Positive and Negative Bacteria

PPA-positive: *Proteus* sp., *Morganella* sp., *Providencia* sp.
PPA-negative: All members of the rest of Enterobacteriaceae.

11. Hydrogen Sulfide Production

Principle

Some organisms decompose sulphur-containing amino acids producing H_2S among the products. It detects H_2S liberated as a result of the degradation of sulphur containing amino acids.

Procedure

The organisms can be grown in culture tubes. Between the cotton plug and the tube, insert a filter paper strip soaked in **10 percent lead acetate solution** and dried.

Fig. 8.11: Urease test

Browning of the paper indicates H_2S production. It has variable sensitivity. When cultured in media containing lead acetate or ferric ammonium citrate or ferrous acetate they turn them black or brown. This method is more sensitive than lead acetate strip method.

Interpretation

Positive—Black color.
Negative—No change in color.

Positive and Negative Bacteria

H_2S-positive: *Proteus mirabilis, Proteus vulgaris, Salmonella* sp. with some exceptions.
H_2S-negative: *Morganella* sp., *Salmonella paratyphi A, S. cholerae-suis.*

12. Potassium Cyanide Test

Principle

This tests the ability of an organism to grow in the presence of cyanide. In view of the toxicity of potassium cyanide, the use of this test is limited and it must be strictly controlled.

Method

Inoculate buffered peptone water medium, containing 1 in 13,000 concentration of potassium cyanide, with test organism. Incubate at 37°C for 24 to 48 hours.

Interpretation

Positive—Turbidity due to growth.
Negative—Clear (no growth).

Positive and Negative Bacteria

Positive KCN test: *Klebsiella* sp., *Citrobacter freundii, Pseudomonas aeruginosa*
Negative KCN test: *Salmonella* sp., *Esch. coli, Alcaligenes faecalis*

13. Triple-Sugar Iron (TSI) Agar

Principle

Triple sugar iron (TSI) agar is used to determine whether a gram-negative rod utilizes glucose and lactose or sucrose fermentatively and forms hydrogen sulfide (H_2S). TSI medium facilitates preliminary identification of gram-negative bacilli.

Media

TSI is a composite medium and contains 10 parts lactose: 10 parts sucrose: 1 part glucose and peptone. Phenol red and ferrous sulfate serve as indicators of acidification and H_2S formation respectively. The medium is distributed in tubes, with a **butt** and **slant**.

Procedure

Medium is inoculated with bacterial culture by a straight wire pierced deep in the **butt** (stab culture). Incubate the tube at 35°C in ambient air for 18 to 24 hours.

Glucose Fermenting Organism

If the tube is inoculated with a **glucose-fermenting organism** that cannot utilize lactose, only a relatively small quantity of acid can be obtained from the 0.1 percent concentration of glucose in the medium. Initially the entire medium becomes **acidic (yellow)** in **8 to 12 hours.** However, within the next few hours, the glucose supply is completely exhausted and the bacteria begin oxidative degradation of the amino acids within the slant portion of the tube where oxygen is present. This results in the release of **amines** that soon counteract the small quantities of acid present in the slant. The entire slant reverts to an **alkaline pH** (color returns to **red**) by 18 to 24 hours. The **deep (anaerobic portion)** of the tube remains **acidic (yellow)** because amino acid degradation is insufficient to counteract the acid formed.

Lactose-fermenting organism: If the tube is inoculated with a lactose-fermenting organism, then fermentation continues as the organism is able to use lactose (present in 10 times the concentration of glucose), even though the glucose is completely used up after the first 8 to 12 hours. Therefore, when, in addition to glucose, lactose and/or sucrose are fermented, the large amount of fermentation products formed on the slant will more than neutralize the alkaline amines and render the slant **acidic (yellow),** provided the reaction is read in 18 to 24 hours. Consequently, when the tube is examined at the end of 18 to 24 hours, acid production from fermentation of lactose is still occurring and both the **slant and the deep appear yellow,** resulting in an **acid-slant-acid deep reaction.** Reactions in TSI should not be read beyond 24 hours of incubation, because aerobic oxidation of the fermentation products from lactose and/or sucrose does proceed and the slant will eventually revert to the alkaline state.

H_2S prodcing bacteria: Certain bacteria produce H_2S which is a colorless gas. H_2S combines with ferric ions (from ferric salts) to form ferous sulphide as black precipitate and is manifested by blackening of the butt of the medium in the tube

Interpretation (Table 8.5, Fig. 8.12)

Red color (alkaline)—No fermentation

Yellow color (acidic)—Fermentation of carbohydrate

Bubbles in the butt—Gas is also produced during fermentation of carbohydrate.

Blackening of the medium—H_2S production

To determine the ability of an organism to attack specific carbohydrates incorporated in a growth medium, with or without the production of gas, along with the determination of possible hydrogen sulphide (H_2S) production.

Table 8.5: Expected results of TSI agar test

Slant/butt	*Color*	*Utilization and Interpretation*
1. Alkaline slant / alkaline butt (K/K)	Red/Red	No fermentation of glucose, lactose or sucrose. This is characteristic of nonfermentive bacteria, such as *Pseudomonas aeruginosa*
2. Alkaline slant / acid butt (K/A)	Red/Yellow	Glucose fermented; lactose (or sucrose for TSI medium) not fermented. This is characteristic of non-lactose-fermenting bacteria, such as *Shigella* species
3. Alkaline slant acid (Black) deep (K / A / H_2S)	Red/Yellow H_2S	Glucose fermented; lactose not fermented, hydrogen sulfide produced. This is characteristic of non-lactose-fermenting, hydrogen sulfide-producing bacteria, such as *Salmonella* species, *Citrobacter* species, and *Proteus* species
4. Acid slant/Acid deep (A/A)	Yellow/Yellow	Glucose and lactose (or sucrose with TSI) fermented. This is characteristic of lactose-fermenting coliforms, such as *Escherichia coli* and the *Klebsiella-Enterobacter* species

K—Alkaline; A—Acidic; H_2S—Hydrogen sulfide

D. Serology

By using specific sera we can identify organisms by agglutination or other suitable serological reactions. Unknown bacterial antigen is identified by known specific antisera by slide agglutination test. Immunofluorescence test is useful in some cases.

E. Antibiotic Tolerance Tests, Dye Tolerance, and Other Inhibition Tests

These range from disc tests of resistance to antibiotics such as penicillin, bacitracin, gentamicin, novobiocin and metronidazole, to special tests that demonstrate tolerance of dyes and other chemicals such as gentian violet, sodium taurocholate and optochin. The concentrations of antibiotics used for characterization tests often exceed those that would be used in conventional sensitivity test discs. In this context, *resistance* or *tolerance tests* should not be confused with *sensitivity tests*.

F. Bacteriophage and Bacteriocin Typing

These enable intraspecies typing of some bacteria.

G. Pathogenicity

Pathogenicity tests by inoculation of the test organism into laboratory animals like the guinea pig, rabbit, rat and mouse were common procedures for identification of isolates in the past. The animals may be inoculated by intradermal, subcutaneous, intramuscular, intraperitoneal, intracerebral or intravenous, or by oral or nasal spray. They are rarely used now because simpler in vitro tests are available.

2. Genomic Characterization

An alternative approach to identification involves the application of modern, rapid molecular methods and identification of a bacterial species from a complex population is now possible.Molecular methods such as polymerase' chain reaction and other amplification procedures coupled with nucleic acid probes carrying specific DNA or RNA base sequences are now widely used for identifying microbes.

Characterization of Strain Differences

1. Biochemical typing.
2. Serological typing.
3. Antibiogram.
4. Phage typing.
5. Genomic typing: (i) Pulsed-field gel electrophoresis (ii) Ribotyping.

KNOW MORE

Automated Methods

While classical phenotypic characterization of isolates takes days, automated methods are now available which only take hours. Identification is simplified by the detection of specific enzymes, toxins, antigens or metabolic end products of the isolates. For example, many obligate anaerobes can be identified rapidly by gas liquid chromatography of the short chain fatty acids produced by them during glucose fermentation.

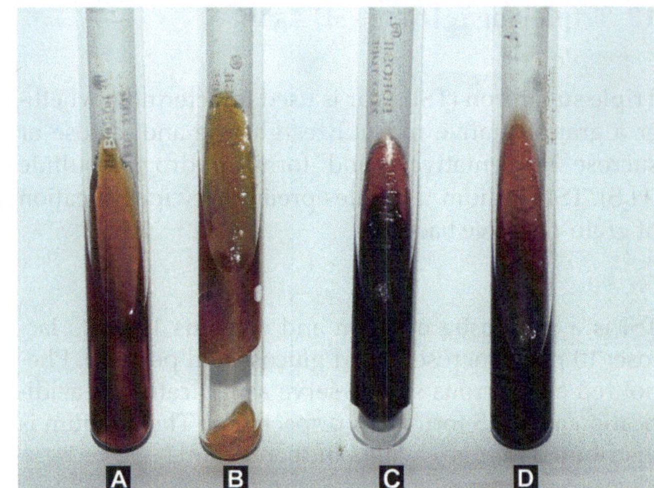

Fig. 8.12: Triple sugar iron agar. (A) Uninoculated; (B) Acid slant/acid butt with gas, no H_2S (A/A); (C) Alkaline slant/acid butt, no gas, H_2S-positive (K/A H_2S^+); (D) Alkaline slant/no change butt, no gas, no H_2S (K/NC)

Characterization of Strain Differences

1. **Biochemical typing:** Biochemical tests are most commonly used to identify various species of bacteria, but in some cases they can be used to distinguish different strains. A strain that has a characteristic biochemical pattern is called a biovar or a biotype.
2. **Serological typing:** Proteins and carbohydrates that vary among strains can be used to differentiate strains. A strain that has a characteristic serological type is called a **serovar** or a **serotype**.
3. **Antibiogram:** Antibiotic susceptibility patterns, or antibiograms, can also be used to distinguish among different strains. Again, this method has now largely been replaced by molecular techniques.
4. **Phage typing:** Strains of a given species sometimes differ in their susceptibility to various types of bacteriophage.
5. **Genomic typing:** Molecular methods can be used to detect restriction fragment length polymorphisms (RFLPs).
   i. **Pulsed-field gel electrophoresis:** The genomic DNA is digested with an enzyme that cuts the chromosome into 10 to 20 large fragments. These are then separated using a special modification of gel electrophoresis. The pattern of sizes can

be determined directly by looking at the stained gel.

ii. **Ribotyping:** The genomic DNA is digested into many small fragments. These are then separated by gel electrophoresis and transferred to a membrane, which is then probed with labeled rDNA, resulting in a distinct pattern of bands.

☞ KEY POINTS

Methods used to identify bacteria

1. **Phenotypic characteristics:**
   A. Microscopic morphology.
   B. Staining reactions.
   C. Metabolism differences
      i. Cultural characteristics.
      ii. Fermentation and other biochemical reactions.
   D. Serology—By using specific sera we can identify organisms by agglutination or other suitable serological reactions.
   E. Antibiotic tolerance (resistance) tests, dye tolerance, and other inhibition tests.
   F. Bacteriophage and bacteriocin typing.
   G. Pathogenicity.

IMPORTANT QUESTIONS

Write short notes on:

a. Indole production
b. Methyl red test
c. Voges-Proskauer (VP) test
d. Citrate utilization test
e. Nitrate reduction test
f. Urease test
g. Catalase test
h. Oxidase test
i. Phenylalanine deaminase test.

FURTHER READING

Collee JC, et al. Mackie & McCartney Practical Medical Microbilology. London: Churchill Livingstone 1996;131-149.

Duerden BI, Towner KJ, Magee JT. Isolation, description and identification of bacteria. Topley and Wilson's Microbiology and Microbial Infections, Balows A, Sussaman M (Eds). vol 2, 9 edn, Arnold, London, Sydney, Auckland 1998;65-84.

CHAPTER 9

Bacterial Taxonomy

Learning Objectives

After reading and studying this chapter, you should be able to:

- Discuss bacterial classifications.

TAXONOMY

Taxonomy is the science that studies organisms in order to arrange them into groups; those organisms with similar properties are grouped together and separated from those that are different. Taxonomy can be viewed as three separate but interrelated areas:

1. **Classification:** The process of arranging organisms into similar or related groups, primarily to provide easy identification and study.
2. **Identification:** The process of characterizing organisms.
3. **Nomenclature:** The system of assigning of names to organisms.

BACTERIAL CLASSIFICATION

Three-domain System

The classification scheme currently favored by most microbiologists is the three-domain system. This designates all organisms as belonging to one of the three *domains—**Bacteria, Archaea,*** **and** ***Eucarya.*** The system is based on the work of Carl Woese and colleagues who compared the sequences of nucleotide bases in ribosomal RNA from a wide variety of organisms.

Five-kingdom System

The most widely accepted system was the five-kingdom system, proposed by RH Whittaker in 1969 before the three-domain classification system was introduced. The five kingdoms in this system are the **Plantae, Animalia, Fungi, Protista** (mostly single-celled eukaryotes) and **Prokaryotae.** While the five-kingdom system recognizes the obvious morphological differences between plants and animals, it does not reflect the recent genetic insights of the ribosomal RNA data, which indicates that plants and animals are more closely related to each other than *Archaea* are to *Bacteria.*

Taxonomic Hierarchies

Taxonomic classification categories are arranged in a hierarchical order, with the **species** being the basic unit. The species designation gives a formal taxonomic status to a group of **related isolates or strains**, which, in turn, permits their identification.

Kingdoms are divided successively into **Division, class, order, family, tribe, genus** and **species.**

Example

The full taxonomical position of the **typhoid bacillus** is as follows:

Division *Protophyta*
Class *Schizomycetes*
Order *Eubacteriales*
Family *Enterobacteriaceae*
Tribe *Salmonellae*
Genus *Salmonella*
Species *Salmonella Typhi*

Species Concept in Bacteria

Species is the standard taxonomical unit in biology. In higher organisms such as plants and animals, the basic taxonomic unit, a species is generally considered to be a group of morphologically similar organisms that are capable of interbreeding to produce fertile offspring. But in bacteria, the species concept is vague and ill defined.

In spite of these difficulties, the concept of species provides a convenient unit in bacterial taxonomy. Besides morphological features, criteria useful for the definition of bacterial species are physiological, biochemical, antigenic and pathogenic properties. As 'species' is a genetic concept, definitive information can be obtained by comparison of the nucleotide base ratios, which are constant for any one species but may be different in different species. Historically, taxonomists have relied heavily on phenotypic attributes to classify prokaryotes; however, the development and application of molecular techniques such as nucleotide sequencing is finally making it possible to determine the genetic relatedness of microorganisms.

CLASSIFICATION SYSTEMS

There are two general ways in which classification systems can be constructed.

Phenetic System

Organisms can be grouped together based on overall similarity to form a **phenetic system.** Computers may be used to analyze data for the production of phenetic classifications. The process is called **numerical taxonomy.**

Phylogenetic System

They can be grouped based on probable evolutionary relationships to produce a **phylogenetic system.**

Adansonian or Numerical Classification

The Adansonian or numerical classification, so called after Michael Adanson who introduced it in the eighteenth century, avoids the use of weighted characteristics. It merely takes into account all the characteristics expressed at the time of study and makes no phylogenetic assumption. Hence it is called a *phenetic system.* It gives equal weight to all measurable features, and groups organisms on the basis of similarities of several characteristics.

The development of computers has extended the scope of phenetic classification by permitting comparisons of very large numbers of properties of several organisms at the same time. This is known as ***numerical taxonomy.*** Information about the properties of organisms is converted into a form suitable for numerical analysis and then compared by means of a computer. The resulting classification is based on general similarity as judged by comparison of many characteristics, each given equal weight. This approach was not feasible before the advent of computers because of the large number of calculations involved.

Phylogenetic Classification

There are two approaches to bacterial classification. The hierarchical classification represents a branching tree like arrangement, one characteristic being employed for division at each branch or level. These are systems based on evolutionary relationships rather than general resemblance. Here some characteristics are arbitrarily given special weightage. Depending on the characteristic so chosen, the classification would give different patterns. For example, the intestinal gram-negative bacilli have been traditionally classified depending on whether they ferment lactose or not. While this provides a useful distinction between the pathogenic and nonpathogenic groups of these bacilli, a different but useful classification could be obtained using fermentation of sucrose as the criterion. While classification based on weighted characteristic is a convenient method, it has the serious drawback that the characters used may not be valid. This has proven difficult for procaryotes and other microorganisms, primarily because of the lack of a good fossil record. While there is no "official" classification of prokaryotes, microbiologists generally rely on the reference text **Bergey's Manual of Systematic Bacteriology** as a guide.

Molecular or Genetic Classification

This is based on the degree of genetic relatedness of different organisms. This classification is said to be the most natural or fundamental method since all properties are ultimately based on the genes present. DNA relatedness can be tested by studying the nucleotide sequences of DNA and by DNA hybridization or recombination methods. The nucleotide base composition and base ratio (Adenine- Thymine: Guanine-Cytosine ratio) varies widely among different groups of microorganisms, though it is constant for members of the same species. Molecular classification has been employed more with viruses than with bacteria.

At present no standard classification of bacteria is universally accepted and applied, although *Bergey's Manual of Determinative Bacteriology* is widely used as an authoritative source.

Intraspecies Classification

It is often necessary to subclassify bacterial species for diagnostic or epidemiological purposes on the basis of biochemical properties (biotypes), antigenic features (serotypes), bacteriophage susceptibility (phage types) or production of bacteriocins (colicin types). A species may be divided first into **groups** and then into **types**.

The application of newer techniques from immunology, biochemistry and genetics has led to much greater discrimination in intraspecies typing. The methods used are of two types: phenotypic and genotypic. Phenotypic methods include electrophoretic typing of bacterial proteins and immunoblotting. Genotypic methods include plasmid profile analysis, restriction endonuclease analysis of chromosomal DNA with Southern blotting, PCR and nucleotide sequence analysis. Some of these techniques are considered in the chapter 10 (Bacterial Genetics).

Nomenclature

Nomenclature, the naming of microorganisms according to established rules and guidelines, provides the accepted labels by which organisms are universally recognized. Bacteria are given names according to an official set of internationally recognized rules, the *International Code for the Nomenclature of Bacteria.* Bacterial nomenclature is governed by an international code prepared by the *International Committee on Systematic Bacteriology* and published as *Approved Lists of Bacterial Names* in the *International Journal of Systematic and Evolutionary Microbiology;* most new species are also first described in this journal, and a species is only considered to be validly published if it appears on a validation list in this journal.

Casual or Common Name

Two kinds of names are given to bacteria. The first is the ***casual* or common name** which varies from country to country and is in the local language. Names such as 'typhoid bacillus' and 'gonococcus' are casual names. Such names are useful for communication at the local level.

Scientific or International Name

The second is the scientific or international name which is the same throughout the world. The scientific name consists usually of two words, the first being the name of the **genus** and the second the specific epithet. In this **binomial (two-name) system** of nomenclature, every organism is assigned **a genus** and **species** name. The generic name is usually a Latin noun. The specific epithet is an adjective or noun and indicates some property of the species (for example *albus,* meaning white), the animal in which it is found (for example *suis,* means pig), the disease it causes *(tetani,* of tetanus), the person who discovered it *(welchii,* after Welch) or the place of its isolation *(london).* The genus designation, which is always **capitalized,** and the species designation, which is **never capitalized,** even if it refers to a person or place (for example *Salmonella london).*

Type Cultures

As a point of reference, type cultures of bacteria are maintained in international reference laboratories. The type cultures contain representatives of all established species. The original cultures of any new species described are deposited in type collections. They are made available by the reference laboratories to other workers for study and comparison.

☞ KEY POINTS

- Bacterial taxonomy—It comprises three components; (1) Identification of an unknown isolate with a defined and named unit; (2) Classification of organism; (3) Nomenclature or naming of the microbial isolates.
- Bacterial classifications include Adansonian, phylogenetic, and genetic classifications.
- The phylogenetic classification is a type of hierarchical classification that represents a branching tree-like arrangement, one characteristic being employed for divisions at each branch or level.
- The Adansonian classification gives equal weight to all measurable features and groups of bacteria on the basis of similarities of several characteristics.
  - The genetic or molecular classification is based on homology of the DNA base sequences of the microorganisms.
- Nomenclature refers to the naming of microorganisms.

IMPORTANT QUESTION

1. Write briefly about bacterial taxonomy.

FURTHER READING

Cowan ST. Dictionary of Medical Taxonomy. London: Oxford University Press, 1978.

Cowan ST, Steel KJ. Manual for the Identification of Medical Bacteria. London: Cambridge University Press.

Finegold $M, Baron EJ. Bailey and Scott's Diagnostic Bacteriology. 7th edn. St. Louis: CV. Mosby 1986.

Goodfellow M, Board RG, (Ed). Microbiological Classification and Identification. London: Academic Press.

Murray Pet AI, (Eds) 1999. Manual of Clinical Microbiology. 7th edn, Washington: American Society for Microbiology, 1965.

CHAPTER 10

Bacterial Genetics

Learning Objectives

After reading and studying this chapter, you should be able to:

- Explain Lac operon.
- Discuss regulations or control of gene expression.
- Discuss structures and functions of the plasmids.
- Describe mutations.
- Discuss various methods of gene transfer.
- Differentiate among the mechanisms of transformation, transduction, lysogenic conversion and conjugation in the transfer of genetic material from one bacterium to another.
- Discuss resistance transfer factor (RTF).
- Differentiate between mutational and plasmid mediated drug resistance.
- Discuss transposons.
- Describe principle and clinical applications of the following; nucleic and probes; genetic engineering; polymerase chain reaction.
- Describe gene therapy.

INTRODUCTION

Bacteria also breed true and maintain their characteristics from generation to generation like other organisms. However, a small proportion of their progeny exhibits variations in particular properties. It was only since 1940s that principles of genetics were applied to bacteria and their viruses. This has led not merely to a better understanding of the genetic processes but also to fundamental advances in biology and biochemistry and to the birth of a new branch of science, molecular biology.

Genetics: Genetics is the study of genes, their structure and function, heredity and variation.

Genomics: The study and analysis of the nucleotide sequence of DNA is called genomics.

Genome: The complete set of genetic information for a cell is referred to as its genome.

STRUCTURE AND FUNCTIONS OF THE GENETIC MATERIAL

Nucleic Acid Structure

A substance called deoxyribonucleic acid (DNA) is the substance of which genes are made. DNA and another substance called ribonucleic acid (RNA), are together referred to as nucleic acids because they were first discovered in the nuclei of cells.

Nucleotides: Nucleotides are the structural units of nucleic acids. Nucleotides are named according to their nitrogenous base.

Parts of Nucleotide

Each nucleotide has three parts:

i. **A nitrogen-containing base**

Purines and pyrimidines: The nitrogen-containing bases are cyclic compounds made up of carbon, hydrogen, oxygen, and nitrogen atoms. The bases are named adenine (A), thymine (T), cytosine (C), guanine (G), and uracil (U). A and G are double-ring structures called purines, whereas T, C, and U are single ring structures referred to as pyrimidines.

ii. **A pentose (five-carbon) sugar called deoxyribose or ribose**

iii. **A phosphate group (phosphoric acid).**

Nucleoside: The term nucleoside refers to the combination of a purine or pyrimidine plus a pentose sugar; it does not contain a phosphate group.

A. Deoxyribonucleic Acid (DNA) Structure

Double helix: According to the model proposed by Watson and Crick, a DNA molecule consists of two long strands wrapped around each other to form a **double helix** (Fig. 10.1). The double helix looks like a twisted ladder, and each strand is composed of **many nucleotides.** The two strands are held together by **weak hydrogen bonds** between the nitrogenous bases of the opposing strands.

Sugar-phosphate backbone: Every strand of DNA composing the double helix has a "backbone" consisting of

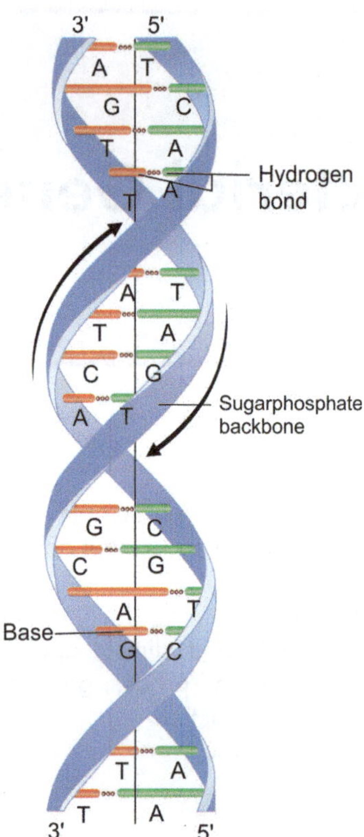

Fig. 10.1: A schematic drawing of the Watson-Crick structure of DNA, showing helical sugar phosphate backbone of the two strands held together by hydrogen bonding between the bones

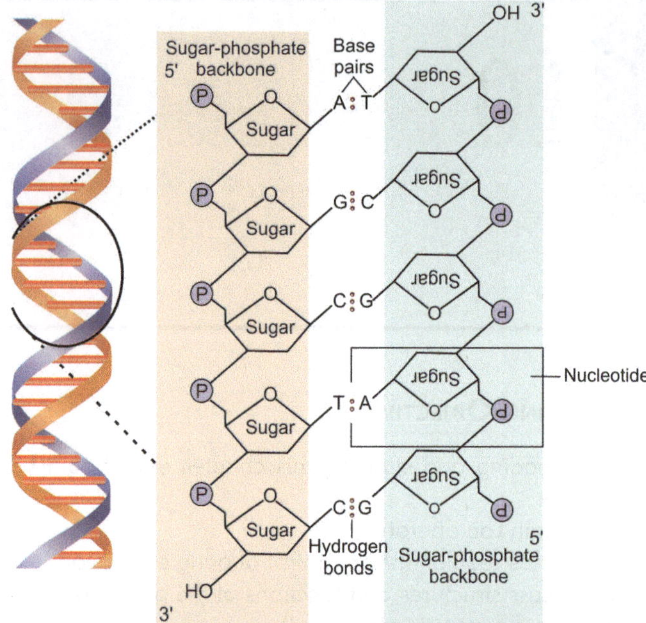

Fig. 10.2: Chemical structure of segment of double stranded DNA. The two stands of DNA are antiparallel; one is oriented in the 5' or 3' direction, and its complement is oriented in the 3' or 5' direction. Hydrogen bonding occurs between the complementary base pairs, three bonds form between a G-C base pair, and two bonds form between an A-T base pair

alternating deoxyribose **sugar** and **phosphate groups**. The deoxyribose of one nucleotide is joined to the phosphate group of the next. The nitrogen containing bases make up the rungs of the ladder. **Purine A** is always paired with the **pyrimidine T** and that the purine G is always paired with the **pyrimidine C**. The bases are held together by hydrogen bonds; A and T are held by two hydrogen bonds, and G and C are held by three (Fig. 10.2).

Base pairing: The characteristic bonding of A to T and G to C is called **base pairing** and is fundamental to the remarkable functionality of DNA. Because of the rules of base pairing, one strand can always be used as a **template** for the synthesis of the complementary opposing strand.

Complementary: Because the sequence of bases of one strand is determined by the sequence of bases of the other, the bases are said to be **complementary.** While the two strands of DNA in the double helix are complementary, they are also **antiparallel**. That is, they are oriented in opposite directions. One strand is oriented in the 5′ to 3′ direction and its complement is oriented in the 3′ to 5′ direction. This also has important implications in the function and synthesis of nucleic acids.

Code and Codon

Genetic information is stored in DNA as a code. The unit of code is known as codon. It consists of a sequence of three bases. Therefore, code is triplet. Each codon specifies or codes for a single amino acid, but more than one codon may exist for a single amino acid. Therefore, code is degenerate. Thus, the triplet AGA codes for arginine but the triplets AGG, CGU, CGC, CGA and CGG also code for the same amino acid. Three codons UAA, UAG and UGA do not code for any amino acid and are known as nonsense codons. They act as punctuation marks, terminating the message for synthesis of a polypeptide.

Cistron or Gene

A segment of DNA carrying a number of codons specifying for a particular polypeptide is known as cistron or gene. A large number of genes constitute a locus and a large number of loci constitute cell genome. DNA can be compared with a book of information. Letters represent nucleotides, words represent codons, sentences represent genes, paragraphs represent loci and entire book as DNA molecule or cell genome.

A DNA molecule consists of a large number of genes, each of which contains hundreds of thousands of nucleotides. The bacterial chromosome consists of a double-stranded molecule of DNA arranged in a circular form. When straightened, it is about 1,000 µm in length. The length of DNA is usually expressed as kilobases (1 kb = 1,000 base pairs). Bacterial DNA is about 4,000 kb and the human genome about 3 million kb long.

Introns and exons: In higher forms of life, several stretches of DNA that do not appear to function as codons occur between the coding sequences of genes.

These apparently useless noncoding intrusions are called introns, while the stretches of coded genes are called exons. During transcription, the genome is copied in its entirety, both introns and exons. The introns are then excised from the RNA copy before being translated by the ribosomes into proteins.

B. Ribonucleic Acid (RNA) Structure

Differences between DNA and RNA: RNA, the second principal kind of nucleic acid, differs from DNA in several respects.

Types of RNA: Three major kinds of RNA have been identified in cells. These are referred to as messenger RNA (mRNA), ribosomal RNA (rRNA), and transfer RNA (tRNA). Each type of RNA has a specific role in protein synthesis.

i. **Messenger RNA (mRNA):** Most genes encode proteins and are transcribed into messenger RNA (mRNA). These molecules are translated during protein synthesis. Alternatively, the gene product can be a ribosomal RNA (rRNA) or transfer RNA (tRNA), each of which plays a different but critical role in protein synthesis.
ii. **Ribosomal RNA (rRNA):** Ribosomes, composed of rRNA and proteins, also are central to translation and provide the site at which translation occurs.

Gene Expression

Gene expression involves two separate but inter-related processes, transcription and translation.

A. **Transcription:** Transcription is the process of synthesizing RNA from a DNA template. The DNA acts as a template for the transcription of RNA by RNA polymerase for subsequent protein production within the cell. RNA polymerase attaches itself to the beginning of a gene on DNA and synthesises mRNA, using one of the strands in DNA as a template. This process is known as **transcription**. The bases in mRNA will be complementary to one strand of DNA since DNA acts as a template for synthesis of mRNA.
B. **Translation:** Translation is the process of decoding the information carried on the mRNA to synthesize the specified protein.

Process of translation: The process of translation requires three major components **mRNA, ribosomes,** and **tRNAs** in addition to various accessory proteins.

Messenger RNA (mRNA): The **mRNA** is a temporary copy of genetic information. It carries the coded information for making specific proteins from DNA to ribosomes, where proteins are synthesized.

Ribosomes: Serve as the sites of translation, and their structure facilitates the joining of one amino acid to another.

Transfer RNA (tRNA): The **tRNA** molecule contains a **triplet** at one end and amino acid at the other end. The ribosome moves along the mRNA until the entire mRNA molecule has been translated into corresponding sequences of amino acids. Finally, the sequence of amino acids in the resulting polypeptide chain determines the configuration into which the polypeptide chain folds itself, which in many cases determines the enzymic properties of the completed protein.

Central dogma of molecular biology: The flow of information from DNA to RNA to protein is often referred to as the **central dogma of molecular biology** (DNA→RNA→ polypeptide) was once believed that information flow proceeded only in this direction. These processes are illustrated schematically in (Fig. 10.3).

EXTRACHROMOSOMAL GENETIC ELEMENTS

Plasmids

Most bacteria possess extrachromosomal genetic in addition to chromosomal DNA elements known as **plasmid.** It consists of a circular piece of double-stranded DNA, can replicate autonomously **(independent replicons)** and can maintain in the cytoplasm of a bacterium for many generations. They are found mainly in bacteria but also in some eukaryotic microorganisms. Plasmids DNA sometimes may be integrated with chromosomal DNA. The name **episome** was employed for such integrated forms, though this distinction is not usually made now. These are not essential for the normal life and functioning of the host bacterium. They may confer on it properties leading to survival advantage under appropriate conditions such as resistance to antibiotics, bacteriocin production.

Conjugative or nonconjugative plasmids: Such plasmid that contains the information for self-transfer to another cell by conjugation is known as **conjugative or self-transmissible plasmid.** Those plasmids which do not possess information for self transfer to another cell are known as **nonconjugative or nonself-transmissible plasmids.** They can, however, be transferred with the help of transfer factor such as colicin plasmid (Col ~) and through the agency of bacteriophages (transduction).

Plasmid Classification

Two methods: Physical and genetic methods.

1. **Physical methods: Restriction endonuclease fingerprints** are generating from purified plasmid DNA. Closely related plasmids will produce the same, or very similar, pattern (fingerprints), while unrelated plasmids will produce different fingerprints.
2. **Genetic methods:** Incompatibility typing—Plasmids may be classified by **incompatibility testing.** This method relies on the fact that closely related plasmids are unable to coexist stably in the same bacterial cell, while unrelated plasmids can. Plasmids which are sufficiently closely related to interfere with each other's replication in this manner are said to be incompatible and to belong to the same

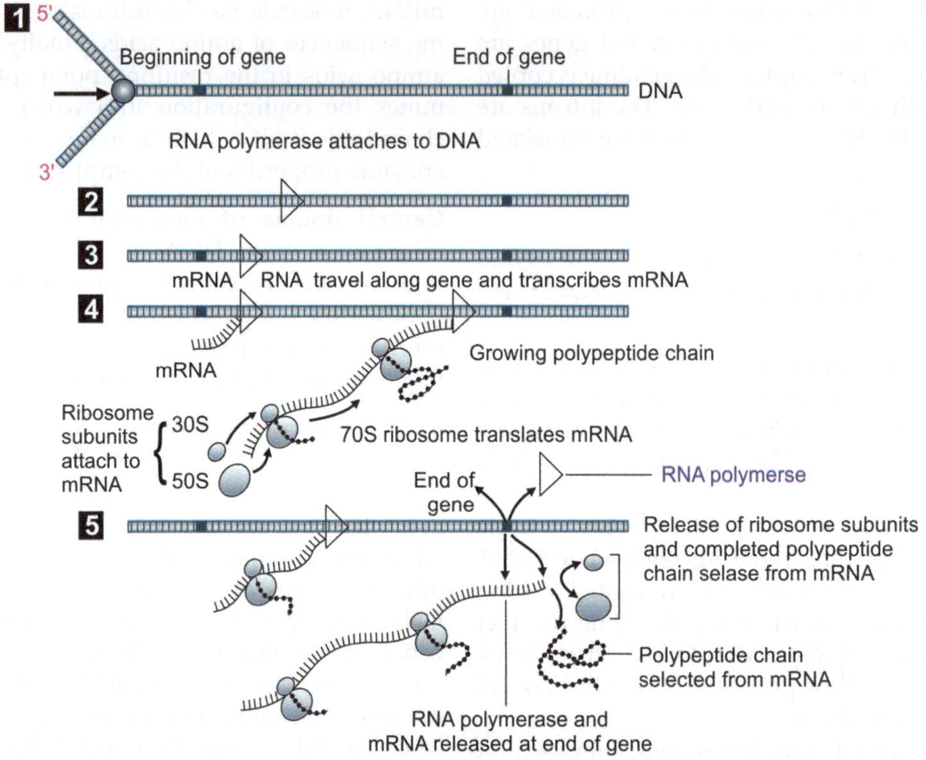

Fig. 10.3: Synthesis of polypeptide

incompatibility group. In contrast, unrelated plasmids can coexist stably and are therefore said to belong to different incompatibility groups. They have been classified based on the types of conjugation tube induced, which determine the susceptibility of the host bacterium to lysis by some virulent bacteriophages. All members of the same plasmid incompatibility group produce the same type of pilus for conjugation.

Uses of Plamids

1. **Confer on properties leading to survival advantage to bsctreia:** Such as antibiotic resistance, bacteriocin production, resistance to toxic metal ions, enterotoxin production, enhanced pathogenicity, reduced sensitivity to mutagens, or the ability to degrade complex organic molecules.
2. **Vectors in genetic engineering:** Plasmids have become important vectors in genetic engineering by their ability to transfer genes from one cell to another.
3. **Clean up of environmental wastes:** Because of their ability to degrade and detoxify a variety of unusual compounds, many of them are being investigated for possible use in the cleanup of environmental wastes.

GENOTYPIC AND PHENOTYPIC VARIATIONS

There are two types of variation
A. Phenotypic variation
B. Genotypic variation

A. Phenotypic Variation

The phenotype ('Pheno'; display) is the physical expression of various characters by bacterial cells in a given environment. These properties are determined not only by its genome (genotype), but also by its environment. Phenotypic variations are reversible.

Examples of environmental influence on bacteria:

1. **Synthesis of flagella:** The typhoid bacillus is normally flagellated. But the flagella are not synthesized when grown in phenol agar. This is only a phenotypic variation determined by the environment and is reversed when subcultured from phenol agar into broth.
2. **Synthesis of enzyme:** Another example of environmental influence is the synthesis by *Escherchia coli* of the enzyme beta-galactosidase, necessary for lactose fermentation. *E. coli* possesses the genetic information for the synthesis of the enzyme but the actual synthesis takes place only when it is grown in a medium containing lactose. When grown in a medium containing glucose only, the enzyme is not synthesized.

Such enzymes which are synthesized only when induced by the substrate are called ***induced enzymes***. The enzymes which are synthesized irrespective of the presence or absence of the substrate are called ***constitutive enzymes***. Regulation of enzyme induction illustrates the economy of nature, the enzyme being produced only when appropriate substrates are present.

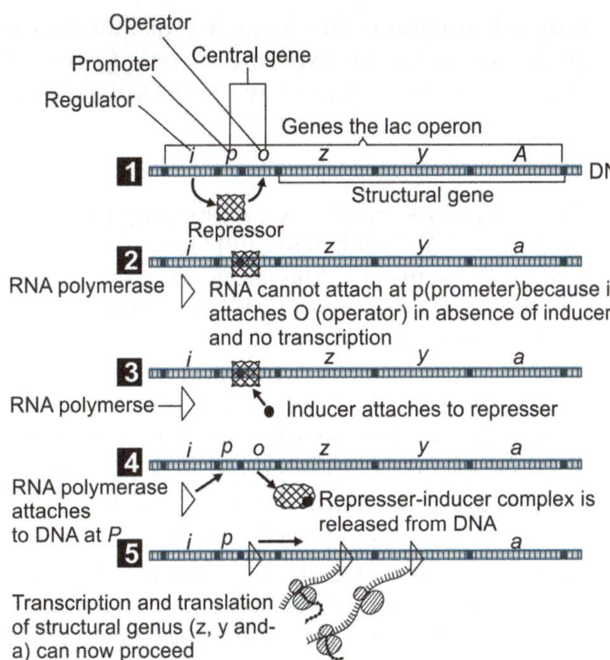

Fig. 10.4: Lac operon of *Escherichia coli*

Regulation of Gene Expression

In prokaryotic bacteria the process of gene expression is regulated mainly at the transcriptional level, thereby conserving the energy supply and the transcription-translation apparatus.

Lac Operon

Originally It was proposed in the early 1960s by Jacob and Monod. They suggested that segments of bacterial DNA are organized into functional units called **operons** of which the most well known example is the lactose operon of *E. coli* (Fig. 10.4).

Lactose fermentation requires three enzymes: beta-galactosidase, galactoside permease and transacetylase coded by **structural genes** *Lac* Z, *Lac* Y and *Lac A* of *Lac* operon respectively (Fig. 10.4). Adjacent to the structural gene is the operator, which is a sequence of bases that controls the expression (transcription) of the structural genes and a **promoter** next to the operator where the RNA polymerase binds that will transcribe the structural genes. Also important, but not part of the operon is a distant **Regulatory gene** (in this case is Lac I) which codes for a ***repressor protein***. It is a protein molecule which can combine with either operator region on the chromosome or with the inducer (lactose).

Effect of Lactose on the Control of the Lactose Operon

For transcription to occur as the first stage in protein synthesis, **RNA polymerase** has to attach to DNA at a specific **promoter** region and transcribe the DNA in a fixed direction. In resting stage when lactose (inducer) is not present in the medium, **repressor molecule** is bound to the operator, preventing the passage of RNA polymerase from **promoter** to the structural genes. The repressor molecule has an affinity for lactose, in the presence of which it leaves the operator region free enabling the transcription to take place. When lactose present is completely metabolized, the repressor again attaches to the operator, switching off transcription. Lactose acts both as inducer and the substrate for the enzyme.

Table 10.1: Distinction between genotypic and phenotypic variation

Phenotypic variation	*Genotypic variation*
1. Reversible	1. Not reversible
2. Temporary	2. Stable
3. Not heritable	3. Heritable
4. Dependent on environmental conditions and altering when these change	4. Not influenced by the environment

B. Genotypic Variation

The sum total of the genes that make up the genetic apparatus of a cell (genome) establishes its genotype. It is the hereditary constitution of the cell that is transmitted to its progeny. The genotype includes the complete genetic potential of the cell, all of which may or may not be expressed in a given environmental situation.

Genotypic variations are stable, heritable and not influenced by the environment. They may occur by mutation or by one of the mechanisms of genetic transfer or exchange, such as transformation, transduction, lysogenic conversion and conjugation. The distinction is between phenotypic and genotypic variation is important (Table 10.1).

Mutation

It is a random, undirected, heritable variation caused by an alteration in the nucleotide sequence at some point of the DNA of the cell. It may be due to addition, deletion or substitution of one or more bases (Fig. 10.5). Since mutation may occur in any of the several thousand genes of the cell, and different mutations in the same gene may produce different effects in the cell, the number of possible mutations is very large.

Molecular mechanism of mutation: The molecular mechanism of mutation is that during DNA replication, some 'error' creeps in while the progeny strands are copied. For instance, instead of thymine bonding with adenine, it may, due to tautomerism, sometimes bond with guanine.

Frequency of Mutation

Mutation is a natural event, taking place all the time at its particular frequency in all the dividing forms of life. Particular mutations occur at fairly constant rates, normally between once per 10^4 and once per 10^{10} cell divisions. Some of these mutations will be viable and could be selected by particular environmental conditions during subculture.

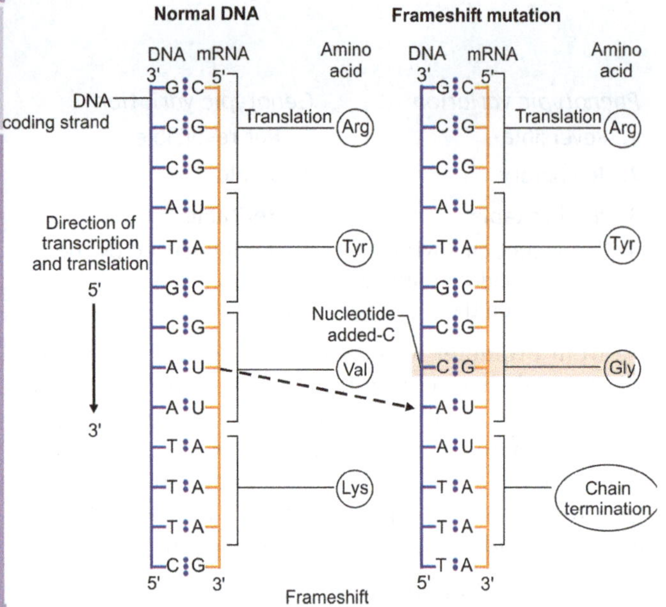

Fig. 10.5: Frameshift mutation

Though mutations are taking place all the time, most mutants go unrecognized as the mutation may involve minor function or it may be lethal. Mutation is best appreciated when it involves a function which can be readily observed. Mutants can be detected by selecting or testing for an altered phenotype. For example, an *E. coli* mutant that loses its ability to ferment lactose can be readily detected on MacConkey agar but is unrecognisable on nutrient agar.

Lethal mutation: Some mutations involve vital functions, and such mutants are **nonviable *(lethal mutation)*.** An important type of lethal mutation is **conditional mutation.**

Conditional mutation: A conditional lethal mutant may be able to live under certain permissive conditions but not under other or nonpermissive conditions. The commonest type of conditional mutant is the **temperature sensitive(ts) mutant**, which is able to live at the permissive temperature (say, 35°C), but not at the restrictive temperature (say, 39°C).

Recognition of mutation: Mutation can be best recognized when it involves a function which can be readily observed by experimental methods like alteration in colonial morphology, pigmentation, alteration in cell surface antigens, sensitivity to bacteriophages or bacteriocins, loss of ability to produce capsule or flagella, loss of virulence and change in biochemical characters.

Types of Mutation

Mutations can be divided conveniently into:

A. **Spontaneous mutation:** Many mutations occur spontaneously in nature and these spontaneous mutations apparently occur in the absence of any mutation-causing agents.

B. **Induced mutation:** The frequency of mutation is greatly enhanced by exposure of cells to several agents (mutagens) which may be physical or chemical.

Mutagens

1. **Physical agents:** (i) UV rays; (ii) Ionising radiation, e.g. X-rays; (iii) Visible light; (iv) Heat.
2. **Chemical agents:** (i) Alkylating agents; (ii) Acridine dyes; (iii) 5-Bromouracil; (iv) 2-aminopurine; (v) Nitrous acid.

C. **Point mutations:** As the name suggests, point mutations affect just one point (base pair) in a gene. Such mutations may be a change to or substitution of a different base pair. Alternatively, a point mutation can result in the deletion or addition of a base pair. It is in general, reversible and is of two classes.

1. **Base pair substitution:** This comprises those mutants in which a single base pair (nucleotide) has been substituted for another pair, and can be subdivided into **transition** (one purine is replaced by other purine or a pyrimidine is replaced by other pyrimidine) and **tranversion** (substitution .of a purine for a pyramidine and vice versa in base pairing).

Example

Normal sequence: *THE FAT CAT ATE THE RAT*
Substitution: *THE FAT CAN ATE THE RAT*

Depending on the placement of the substituted base, when the mRNA is translated this may cause no change **(silent mutation)**, lead to the insertion of the wrong amino acid **(missense mutation)** or generate a stop codon **(nonsense mutation)**, prematurely terminating the polypeptide.

Missense mutation: If the triplet code is altered so as to specify an amino acid different from that normally located at a particular position in the protein. This is called missense mutation.

Nonsense mutation: Deletion of a nucleotide within a gene may cause premature polypeptide chain termination by generating a nonsense codon *(UAG, UAA or UGA)*. This is known as nonsense mutation.

2. Base pair deletion or insetion

Frameshift mutations: If the number of bases inserted or deleted is not a multiple of three, there will be shift in the reading frame, i.e. frameshift mutations. This shifts the normal **'reading frame'** of the coded message forming newest of triplet codon. The coded message is read correctly up to the point of addition or deletion, but the subsequent codons will specify the incorrect amino acids (Fig. 10.6).

For the same reason, in an infected patient, a variety of mutants will appear spontaneously in the population that grows from the few bacteria originally enter-

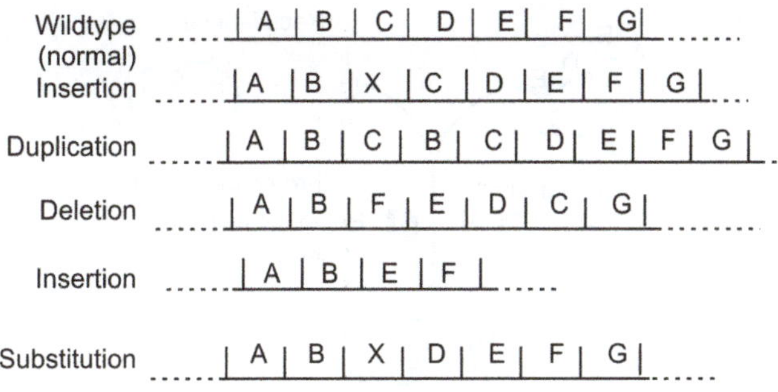

Fig. 10.6: Examples of types of mutation. A portion of the wild type chromosome is shown on the top sequence from which different mutational rearrangements are derived

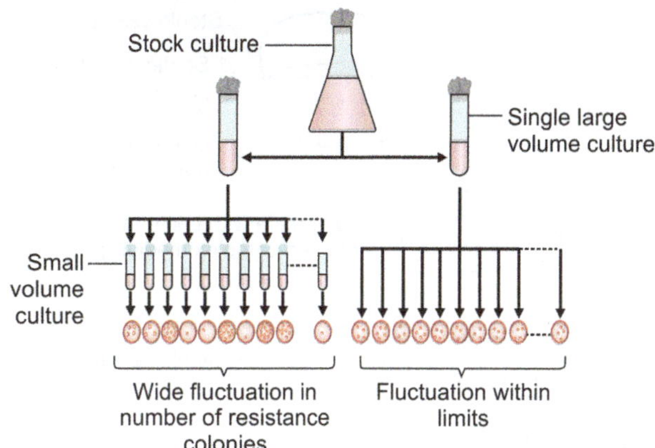

Fig. 10.7: Fluctuation test

ing the body. Such mutations may enhance the ability of an organism to grow in the body, e.g. by conferring antibiotic resistance, enhanced virulence, or altered surface antigens. In such a situation, cells with the mutation will rapidly outgrow cells without the mutation, so that selection of the mutant cells occurs and they soon become the predominant type.

Survival advantage of mutation: When mutation confers a survival advantage it is of vital importance. For example, if a streptomycin resistant mutant of the tubercle bacillus develops in a patient under treatment with the drug, it multiplies selectively and ultimately replaces the original drug sensitive population of bacteria. But the mutation confers no survival advantage in a patient who is not on treatment, so preferential multipication of the mutant does not occur.

Importance of Bacterial Mutation

Drug resistance and development of live vaccines: The practical importance of bacterial mutation is mainly in the field of drug resistance and development of live vaccines. Some organisms have been subcultured in the laboratory for many generations until they lost their virulence for man (e.g. BCG vaccine). This is known as live attenuated vaccine.

Mutant Selection in Laboratory

1. **Fluctuation test—Luria and Delbruck (1943)** provided the proof that bacteria undergo **spontaneous mutation** independent of the environment by the **'fluctuation test'**. They found that very wide fluctuations occurred in the numbers of bacteriophage resistant *E. coli* colonies when samples were plated from several separate small volume cultures, as compared to samples tested from a single large volume culture (Fig. 10.7). Statistically, this indicated that mutations occurred randomly in the separate small volume cultures, some early and some late, resulting in the wide fluctuation. In the large volume cultures, fluctuations were within limits of sampling error. However, the logic of this experiment was not widely appreciated by microbiologists, probably due to the complicated statistical interpretation.

2. Other Tests

A. Direct (positive) slection
B. Indirect (negative) selection
C. Conditional lethal mutants
D. The Ames test.

A. **Direct selection:** Direct selection involves inoculating cells onto a medium on which the mutant, but not the parent, can grow. For example, mutants resistant to the antibiotic streptomycin can be easily selected directly by inoculating cells onto a medium containing streptomycin. Only the rare resistant cells in the population will form a colony. Antimicrobial-resistant mutants are usually very easy to isolate by direct selection.

B. **Indirect selection:** Indirect selection is required to isolate an auxotrophic mutant, one that requires a growth factor which the parent strain does not. There is no medium on which the mutant will grow and the parent will not.

1. **Replica Plating:** An ingenious technique for indirect selection of auxotrophic mutants, replica plating, was devized by the husband and wife team of Joshua and Esther Lederberg in the early 1950s (Fig. 10.8). In this technique, a master plate containing isolated colonies of all cells growing on an enriched medium is pressed onto sterile velvet, a fabric with tiny threads that stand on end like tiny bristles (velvet template). This operation transfers some cells of every bacterial colony onto the velvet. Next, two sterile plates, one containing a glucose-salts (minimal) medium and the second an enriched, complex medium, are pressed in succession onto the same velvet. This procedure transfers cells imprinted on the velvet from the master plate to both the glucose-salts medium and the enriched medium.

All cells that do not have a nutritional requirement will form colonies on both the enriched and the glucose-salts medium, but auxotrophs will only

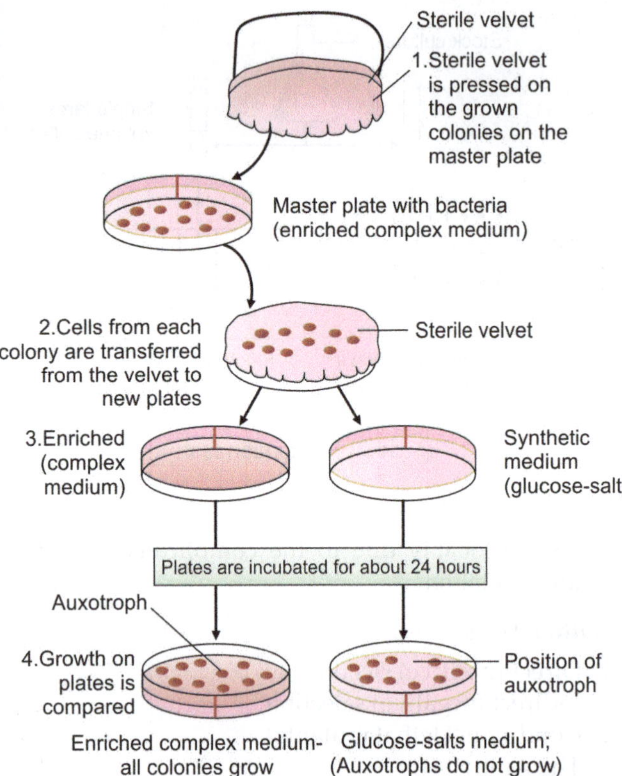

Fig. 10.8: Replica plating method

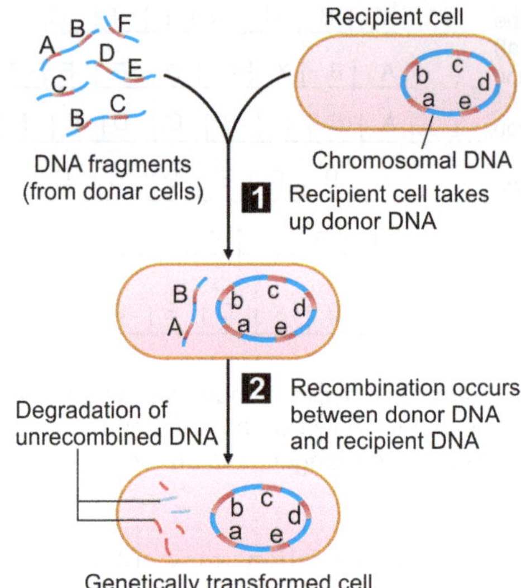

Fig. 10.9: Mechanism of genetic transformation in bacteria

form colonies on the enriched medium glucose-salts medium.

2. **Penicillin Enrichment:** In cases where the parent cell is sensitive to penicillin, the proportion of auxotrophic mutant in the population can be increased by a technique called **penicillin enrichment**.

C. **Conditional lethal mutants:** The name conditional lethal mutants are given to mutants in essential genes. They can be isolated since mutant proteins often function at a low temperature such as 25°C but not at a higher temperature such as 37°C. One class of conditional lethal mutants are called **temperature-sensitive mutant** that can grow only at their lower range of temperature as a result of defective proteins.

D. **Ames Test:** The test is based on the ability of a potential mutagen to revert an auxotrophic mutant to its prototrophic form. Specifically, the Ames test measures the effect of a test chemical on the rate of reversion of a specific Salmonella strain, a histidine requiring auxotroph, to one that no longer requires histidine. If the chemical is mutagenic, it will increase the reversion rate of the strain relative to that observed when no chemical is added (the control).

TRANSMISSION OF GENETIC MATERIAL (GENE TRANSFER)

A change in the genome of a bacterial cell may be caused either by a mutation in the DNA of the cell or result from the acquisition of additional DNA from an external source. DNA may be transferred between bacteria by the following mechanisms:

A. Transformation
B. Transduction
C. Lysogenic conversion
D. Conjugation.

A. Transformation

Transformation is the transfer of genetic information through the agency of free (naked) DNA. (Fig. 10.9). It was the first example of genetic exchange in bacteria to have been discovered. The initial experiment on transformation was performed by Frederick Griffith in England in 1928.

Griffith's Experiment Demonstrating Genetic Transformation

Streptococcus pneumoniae in capsulated form is an extremely virulent organism for mice, whereas non capsulated variants are avirulent.

i. Griffith (1928) found that injections of **living encapsulated bacteria** killed the mouse (Fig. 10.10).
ii. Injections of **live nonencapsulated bacteria or dead encapsulated bacteria** did not kill the mouse.
iii. When the **dead encapsulated bacteria** were mixed with live **nonencapsulated bacteria** and injected into the mice, many of the mice died. In the blood of the dead mice, Griffith found living, encapsulated bacteria. Hereditary material (genes) from the dead bacteria had entered the live cells and changed them genetically so that, their progeny were encapsulated and therefore virulent.

Such transformation was subsequently demonstrated *in vitro* also. The nature of the transforming principle was identified as DNA by Oswald T Avery and his associates Colin M MacLeod and Maclyn McCarty in the United States in 1944.

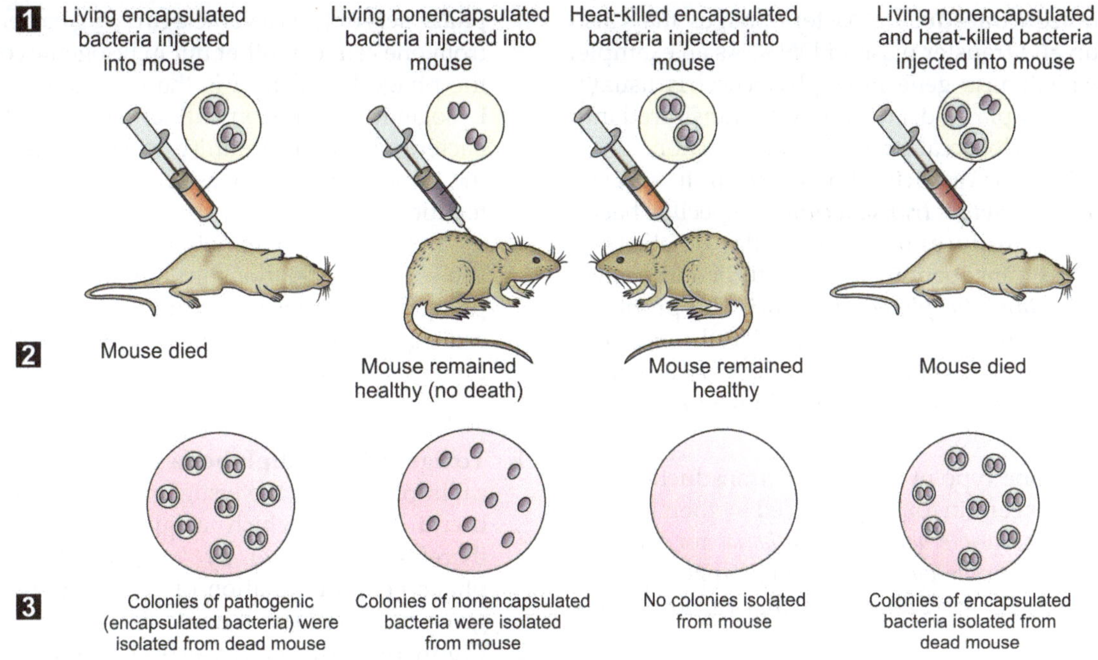

Fig. 10.10: Transformation experiment of Griffith

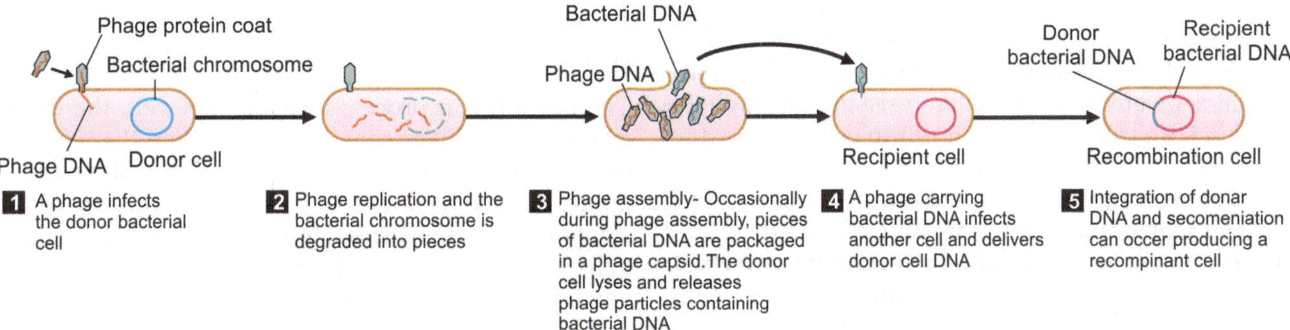

Fig. 10.11: Transduction by a bacteriophage (showing generalized transduction)

Transformation and bacteria: Transformation occurs naturally among very few genera of bacteria, including *Bacillus, Haemophilus, Neisseria, Acinetobacter,* and certain strains of the genera *Streptococcus* and *Staphylococcus.*

B. Transduction

The transfer of a portion of the DNA from one bacterium to another by a bacteriophage is known as transduction. Bacteriophages are viruses that parasitise bacteria and consist of a nucleic acid core and a protein coat. Most bacteriophages carry their genetic information (the phage genome) as a length of double-stranded DNA coiled up inside a protein coat. When bacteriophages multiply inside an infected bacterial cell, each phage head is normally filled with a copy of the replicated phage genome. During the assembly of bacteriophage progeny inside infected bacteria, **'packaging errors'** may occur occasionally. A phage particle may have at its core a segment of the host DNA besides its own nucleic acid. When this particle infects another bacterium, DNA transfer is effected and the recipient cell acquires new characteristic coded by the donor DNA. Bacterial genes have been transduced by the phage into the second cell (Fig. 10.11).

Types of Transduction

Two major types of transduction are known to occur in bacteria: generalized transduction and specialized transduction.

1. **Generalized transduction:** Since phages of this type pick up any portion of the bacterial chromosome at random are termed **generalized transducing phages**. Transduction of cellular DNA by a virus can lead to recombination between the DNA of the donor host cell and the DNA of the recipient host cell. The process of generalized transduction is typical of bacteriophages. Genes can be transduced only between fairly closely related strains as particular phages usually attack only a limited range of bacteria. As well as chromosomal genes,

generalized transducing bacteriophages may also pick up and transfer plasmid DNA. As an example, the penicillinase gene in staphylococci is usually located on a plasmid, and it may be transferred into other staphylococcal strains by transduction.

2. **Specialized or restricted transduction:** In ***specialized or restricted transduction***, a specific bacteriophage transduces only a particular genetic trait. Restricted transduction has been studied intensively in the ***'lambda' phage of E. coli***. The prophage lambda is inserted in the bacterial chromosome only between the genes determining galactose utilisation (gal) and biotin synthesis (bio) and therefore it transduces only either of these.

Exampels: In one type of specialized transduction, the phage codes for certain toxins produced by their bacterial hosts, such as *Corynebacterium diphtheriae* for diphtheria toxin, *Streptococcus pyogenes* for erythrogenic toxin, and *E. coli* 0157:H7 for Shiga toxin that causes bloody diarrhea.

Role of Transduction

1. **In episomes and plasmids:** Transduction is not only confined to transfer of chromosomal DNA but episomes and plasmids may also be transduced.
2. **Penicillin resistance in staphylococci:** The plasmids determining penicillin resistance in staphylococci are transferred from cell to cell by transduction.
3. **Genetic mapping of bacteria:** It provides an excellent tool for the genetic mapping of bacteria.
4. **Treatment of some inborn metabolic defects:** It has also been proposed as a method of genetic engineering in the treatment of some inborn metabolic defects.

C. Lysogenic Conversion

Bacteriophages exhibit two types of life cycle.
(1) Virulent or lytic cycle; (2) Temperate or nonlytic cycle:

i. **Virulent or lytic cycle:** In the virulent or lytic cycle, large numbers of progeny phages are built up inside the host bacterium, which ruptures to release them.
ii. **Temperate or nonlytic cycle:** In the temperate or nonlytic cycle, the host bacterium is unharmed. The phage DNA becomes integrated with the bacterial chromosome as the prophage, and is replicated stably as part of the host cell chromosome and is transferred to the daughter cells. This process is called lysogeny and bacteria harbouring prophages are called **lysogenic bacteria**. In lysogenic bacteria, the prophage behaves as an additional segment of the bacterial chromosome, coding for new characteristic. This process by which the prophage DNA confers genetic information to a bacterium is called **lysogenic or phage conversion**. In transduction, the phage acts only as a vehicle carrying bacterial genes from one cell to another but in lysogenic conversion the phage DNA itself is the new genetic element. Lysogenic conversion influences susceptibility to bacteriophages (immunity to superinfection with the same or related phages) and antigenic characteristics.

Lysogeny is extremely frequent in nature. This is a symbiotic relationship in which the integrated phage DNA imparts immunity to the lysogenized cell against superinfection by genetically related phages, as well as certain unrelated phages.

Medical Importance

1. **Toxigenicity in diphtheria bacilli:** Of great medical importance is the lysogenic conversion in **diphtheria bacilli**, which acquire toxigenicity (and therefore virulence) by lysogenization with the phage beta. Elimination of the phage from a toxigenic strain renders it nontoxigenic.
2. **Production of staphylococci, streptococci and clostridia toxins:** It is probable that the production of many toxins by **staphylococci, streptococci** and **clostridia** is also dependent upon lysogenic conversion by specific bacteriophages.

D. Conjugation

Conjugation is a process in which one cell, **the donor or male cell**, makes contact with another, the recipient or female cell, and DNA is transferred directly from the donor into the recipient (Fig. 10.12). This has been considered to be the bacterial equivalent of sexual mating in higher organisms but the analogy is irrelevant as, following conjugation, the female bacterium is in turn converted into a male cell. Conjugation, in contrast to transformation and transduction, requires contact between donor and recipient cells.

Lederberg and Tatum (1946) first described bacterial conjugation in a strain of *E. coli* called K12 and has been most extensively studied in this strain. Certain types of plasmids, known as **transfer factors or sex factors**, carry the genetic information necessary for conjugation to occur. Only cells that contain such a plasmid can act as donors; those lacking a sex factor act as recipients.

Donor cells must carry the plasmid, and recipient cells usually do not. Conjugation takes place between a male cell and a female cell. Transfer of DNA between cells by conjugation requires direct contact between the donor and recipient cells. The maleness or donor status of a cell is determined by the presence in it of a plasmid, which codes for specialized fimbria (sex pilus) which projects from the surface of the cell. The tip of the pilus attaches to the surface of a recipient cell and holds the two cells together so that DNA can then pass into the recipient cell. It is probable, but not absolutely certain, that transfer actually occurs through the pilus. Alternatively, the pilus could act simply as a mechanism by which the donor and female cells are drawn together.

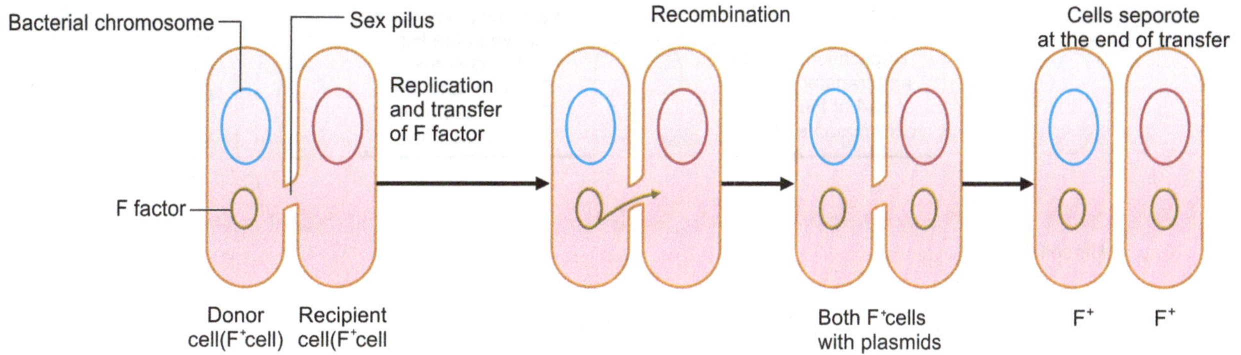

Fig. 10.12: Process of conjugation

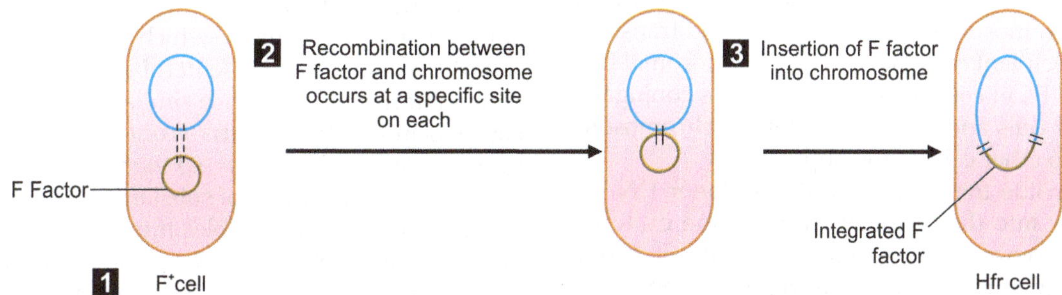

Fig. 10.13: High-frequency recombination (HfR)

Different types of pilus are specified by different types of plasmid and can therefore be used as an aid to plasmid classification.

Types of Conjugation

Three types of conjugation:

1. Plasmid transfer
2. Chromosomal gene transfer
3. Plasmid and chromosomal gene transfer.

1. *Plasmid transfer*

Populations of *E. coli* can be divided into two types of cells. One, the donor cell, contains an **F or fertility plasmid** and is designated **F$^+$**. The other, the recipient cell, does not contain this plasmid and is called **F$^-$**. DNA is transferred only in one direction, from F$^+$ to F$^-$ that is, in a polar fashion. Consequently, F$^+$ cells are often referred to as **males** and the F$^-$cells as **females**. The F plasmid codes for the synthesis of a structure, the **sex or F pilus *(Conjugation tube)***, the protein appendage that attaches the donor to the recipient cell (Fig. 10.12). The tip of the pilus attaches to the surface of a recipient cell and holds the two cells together so that DNA can then pass into the recipient cell. Donor cells can transfer their F plasmid but not their chromosome into recipient cells. In this way a sex factor may spread rapidly through a whole population of recipient cells. This process is sometimes described as infectious spread of a plasmid. The maleness in bacteria is thus a **transmissible or 'infectious'** characteristic. The plasmid also carries information required for its own transfer.

F Factor (Fertility Factor)

The F factor is a transfer factor that contains the genetic information necessary for the synthesis of the sex pilus and for self-transfer but is devoid of other identifiable genetic markers such as drug resistance. Cells that contain the F plasmid free in the cytoplasm (F$^+$ cells) have no unusual characteristics apart from the ability to produce F pili and to transfer the F plasmid to F$^-$ cells by conjugation. It was in *E. coli* K12 that the role of plasmids in conjugation was first recognized. When other similar plasmids were also discovered, the name '**transfer factor**' came to be used for all such plasmids which conferred on their host cells the ability to act as donors in conjugation.

2. *Chromosomal transfer*

High-frequency recombination (Hfr) (Fig. 10.13): The bacterial chromosome is sometimes transferred to F– cells by a few cells in the F population. The F factor is, actually an episome and has the ability to exist in some cells in the 'integrated state' or inserted into the host chromosome in a very small proportion of F+ cells. Once inserted, the entire chromosome behaves like an enormous F plasmid, and hence, chromosomal genes can be transferred in the normal sex factor manner to a recipient cell at a relatively high frequency. Cultures of cells in which the F plasmid has inserted into the chromosome are consequently termed high-frequency recombination (Hfr) strains because such cells are able to transfer chromosomal genes to recipient cells with high frequency.

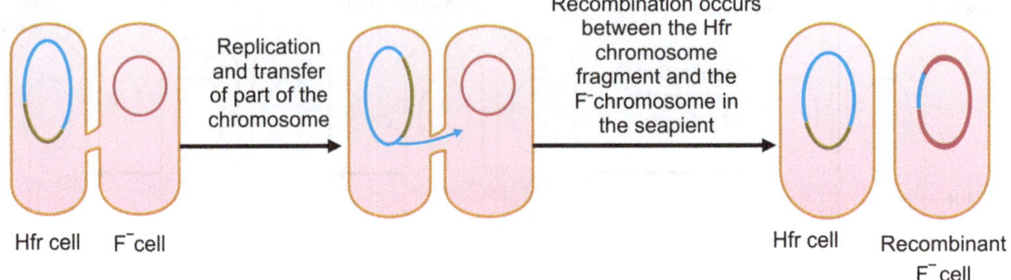

Fig. 10.14: Sexduction

Differences Between Chromosome and Plasmid Transfer

There are, however, two major differences in the transfer of the chromosome compared to plasmid transfer.

i. Most recipient cells that receive chromosomal DNA remain F^-, in contrast to recipient cells conjugating with F^+ cells and cannot transfer the chromosomal DNA to other cells in the population.
ii. Once inside the recipient cell, the donor DNA integrates into the recipient chromosome by replacing homologous genes in the recipient cell through recombination and it cannot replicate and will be destroyed.

3. *Plasmid and Chromosomal Transfer*

There is an additional mechanism by which chromosomal genes may be mobilized by conjugation to a recipient cell. This conversion of an F^+ cell into the Hfr state is reversible. When the F factor reverts from the integrated state to the free state, the F plasmid is not always excised accurately, and occasionally an F plasmid is excised together with some of the neighboring chromosomal genes. An F plasmid that has picked up a small portion of the chromosome in this way is known as an ***F-prime* (F′)**. When an F′ cell mates with a recipient, it transfers along with the F factor, the host genes incorporated with it. This process of transfer of host genes through the F′ factor resembles transduction and has therefore been called **sexduction** (Fig. 10.14).

It is important to emphasize that the F plasmid system is confined to *Esch. coli* and other closely related enteric bacteria. Many other plasmids are capable of mediating conjugation, and sometimes chromosome mobilization, not only in *Esch. coli* but also in other bacteria, e.g. plasmid RP4 and its relatives have been used to mediate conjugation in a wide range of gram-negative bacteria. In gram-positive bacteria, such as *Enterococcus faecalis*, and several *Streptomyces species*, there have been reports of conjugation systems. Conjugation is used to map the location of genes on a bacterial chromosome.

Medically Important Factors Transferred by Conjugation

Colicogenic (Col) factor and resistance transfer factor (RTF) are two medically important factors which can be transferred by conjugation.

A. Colicinogenic (Col) Factor

Several strains of coliform bacteria produce colicins—antibiotic-like substances which are specifically and selectively lethal to other enterobacteria. Bacteria other than coliforms also produce similar kind of substances, e.g. pyocin by *Pseudomonas pyocyanea*, diphthericin by *Corynebacterium diphtheriae*, so the name bacteriocin has been given to this group of substances. The specificity of action of bacteriocins enables intraspecies classification of certain bacteria, e.g. *Shigella sonnei*, *Ps. aeruginosa*.

Colicin production is determined by a plasmid called the Col factor, which resembles the F factor in promoting conjugation, leading to self-transfer and, at times, transfer of chromosomal segments.

B. Resistance Factors or R Plasmids

Resistance factors (R factors) are plasmids that have significant medical importance as it leads to the spread of multiple drug resistance among bacteria. They were first discovered in Japan in 1959 after several dysentery epidemics by the Shigella strains, resistant simultaneously to usual four drugs. In addition, they observed that patients excreting such Shigella strains also shed in their feces other normal bacteria from the patients such as *E. coli* strains resistant to the same drugs. Transfer of multiple drug resistance was demonstrated between *E. coli* and Shigella strains both in vitro and in vivo. The resistance is plasmid mediated and is transferred by conjugation. This mechanism of drug resistance is known as **transferable, episomal or infectious drug resistance.**

Characteristics of R factor: This R plasmid consists of two components: **RTF+r determinants.** The whole plasmid **(RTF+r determinants)** is known as the **R factor**. An R factor can have several r determinants, and resistance to as many as eight or more drugs can be transferred simultaneously (Fig. 10.15).

Resistance transfer factor (RTF): The transfer factor called the **resistance transfer factor (RTF)** is responsible for conjugal transfer.

Resistance determinant (r): A **resistance determinant (r)** code for resistance against various drugs.

Nonconjugative and conjugative plasmids: Sometimes the RTF may dissociate from the r determinants, the

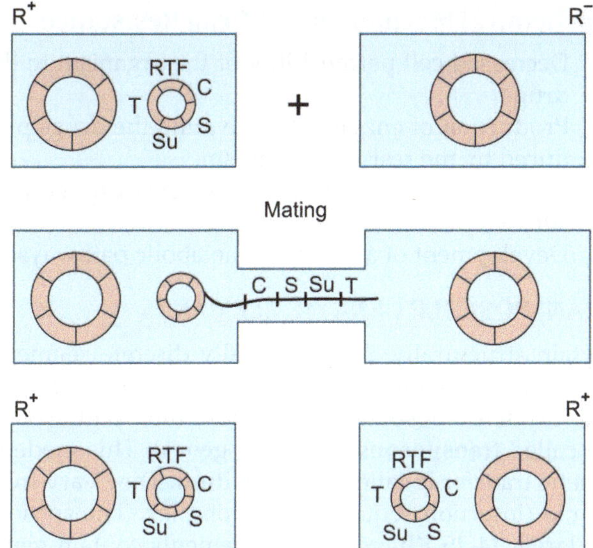

Fig. 10.15: Transferable drug resistance. Two regions of a R plasmid are RTF+r. The RTF contains genes needed for plasmid replication and transfer of the plasmid by conjugation to other bacteria; r determination carries genes for reisistance to different antibiotics

two components existing as separate plasmids. In such cases, though the host cell remains drug resistant, the resistance is not transferable. Those plasmids which lack RTF, i.e. possess only r determinants are known as **nonconjugative or nonself-transmissible plasmids** but they still code for drug resistance. Those plasmids which possess both RTF and r determinants are known as **conjugative or self-transmissible plasmids**. The RTF can have attached to it determinants other than those for drug resistance. In some enteropathogenic *E. coli* enterotoxin and hemolysin production are transmitted by this transfer factor.

Factors influencing R factor transfer: The transfer can be effected readily *in vitro*. It also occurs *in vivo* but in the normal gut, it is inhibited by several factors such as anaerobic conditions, bile salts, alkaline pH and the abundance of anaerobic gram-positive bacteria minimizing the chances of contact between donor cells and suitable recipient cells. But in the intestines of persons on oral antibiotic therapy, transfer occurs readily due to the destruction of the sensitive normal flora and the selection pressure produced by the drug.

Features of R Factor

1. **Transfer to antimicrobial and heavy metal-sensitive bacteria:** They can be transferred to antimicrobial and heavy metal-sensitive bacteria, thereby conferring simultaneous resistance to several antimicrobials and heavy metals encoded by the R genes.
2. **Wide host ranges:** Many R plasmids have wide host ranges and can multiply in a wide variety of different gram-negative genera such as *Shigella, Salmonella, Escherichia, Yersinia, Klebsiella, Vibrio,* and *Pseudomonas.*
3. **Serious problems for the treatment of infectious diseases:** R factors present very serious problems for the treatment of infectious diseases with antibiotics.
4. **Transmission from animals to man:** Bacteria carrying R factors can be transmitted from animals to man. Hence, indiscriminate use of antibiotics in veterinary practice or in animal feeds can also lead to an increase of multiple drug resistance in the community. The addition of antibiotics in animal feeds has for this reason been prohibited by legislation in some countries. Widespread resistance has considerably diminished the clinical efficacy of most antibiotics.

GENETIC MECHANISMS OF DRUG RESISTANCE IN BACTERIA

All of the properties of a microorganism are determined ultimately by genes located either on the chromosome or on plasmids or on lysogenic bacteriophages. It is important to distinguish between intrinsic and acquired resistance with regard to antibiotic resistance. Intrinsic resistance is dependent upon the natural insusceptibility of an organism. In contrast, acquired resistance involves changes in the DNA content of a cell, such that the cell acquires a phenotype (i.e. antibiotic resistance) which is not inherent in that particular species.

Mechanisms of Drug Resistance in Bacteria

A. By mutation (chromosomal)
B. By genetic exchange.

A. By mutation (chromosomal)

Mutational resistance is mainly of two types:

1. **Stepwise mutation:** The stepwise mutation, as seen with penicillin, where high levels of resistance are achieved only by a series of small-step mutations. The target is altered so that, it can no longer bind a drug as efficiently, although it still has some residual affinity. This is also called the multistep pattern of resistance.
2. **Single large-step mutation or 'one-step' mutation:** As seen with streptomycin, the drug target is altered by mutation so that, it is totally unable to bind a drug, where the mutants differ widely in the degree of resistance, some exhibiting low resistance, while others may be highly resistant, and some even streptomycin dependent. This type of mutation occurs with streptomycin, but is otherwise not very common clinically.

Clinical importance in tuberculosis: Mutational resistance is of clinical importance in tuberculosis. If a patient is treated with streptomycin alone initially the bacilli die in large numbers but soon resistant mutants appear and multiply unchecked. If only one drug is given to the patient, the few resistant mutant bacteria will multiply and eventually cause a relapse of the disease. If two or more antituberculous, drugs are used for combined treatment, repopulation by resistant mutants does not occur, as a mutant resistant to one drug will be destroyed by the other drug. The frequency with which double or triple mutations occur spontaneously in the same cell is so low as to be clinically insignificant.

The possibility of a mutant exhibiting resistance to multiple drugs simultaneously is so remote as to be virtually nonexistent. However, inspire of this knowledge, inadequate or inappropriate treatment over the years has caused extensive resistance in tubercle bacilli, leading to a pandemic of multidrug resistant tuberculosis (MDR TB) across the world.

B. By Genetic exchange

Transferable antibiotic resistance:

1. **Transformation:** Resistance transfer by transformation can be demonstrated experimentally but its significance in nature is not known.
2. **Transduction:** Acquisition of resistance by transduction is common in staphylococci. The penicillinase plasmids, which are transmitted by transduction, may also carry determinants for resistance to mercuric chloride and erythromycin.
3. **Conjugation:** Of the three modes of gene transfer in bacteria, it is plasmid-mediated conjugation that is of greatest significance in terms of drug resistance. Plasmids conferring resistance to one or more unrelated groups of antibiotics (R plasmids) can be transferred rapidly by conjugation throughout the population.

   Transferable drug resistance mediated by the R factor is the most important method of drug resistance. Acquisition of an R factor simultaneously confers resistance to several drugs and therefore treatment with a combination of drugs is not useful. Multidrug resistant strains of plague bacilli have been isolated from patients in Madagascar in 1995. Table 10.2 shows the comparison of mutational and transferable drug resistance.

Biochemical Mechanisms of Drug Resistance

1. Decreased cell permeability of the organism to the drug.
2. Production of enzymes inactivating the drugs produced by the resistant organism.
3. Modification of the properties of the drug receptor site.
4. Development of alternative metabolic pathways.

TRANSPOSABLE GENETIC ELEMENTS

Certain structurally and genetically discrete segments of DNA which move around between chromosomal and extrachromosomal DNA molecules within cells are called **transposons (jumping genes)**. This mode of genetic transfer is called **transposition**. They vary from simple (insertion sequences) to complex. Transposons are larger (4-25 Kb) composite elements contain genes for movement as well as genes that encode for various functions such as drug resistance and toxin production.

Unlike other processes that reorganize DNA, transposition does not require extensive areas of homology between the transposon and its destination site. Transposons provide a natural mechanism for the movement of genes from one chromosome to another. In the 1950s, American geneticist Barbara McClintock discovered transposons in corn during her studies on maize genetics, for which she was awarded the Nobel prize for Medicine in 1983.

Structure of Transposons

The simplest transposable elements are **insertion sequences (IS)**, or **IS elements.** It is a short sequence of DNA (around 750 to 1,600 base pairs in length contain contain a gene that codes for enzyme (tranposase which catalyzes transposition). The tranposase gene is bounded at each end by **inverted repeats (IR)** sequences that function as recognition sites for tranposon.

The more complex is called a **composite trasposon.** All transposons contain the information for their own transposition. A transposon is a segment of DNA with one or more genes in the center, and the two ends carrying **'inverted repeat'** sequences of nucleotides-nucleotide sequences complementary to each other but in the

Table 10.2: Comparison of mutational and transferable drug resistance

Mutational drug resistance	*Transferable drug resistance*
1. Resistance to one drug at a time	1. Multiple drug resistance at one time
2. Low degree resistance	2. High degree resistance
3. Can overcome by high drug dose	3. High dose ineffective
4. Development of drug resistance can be prevented by treatment with combination of drug	4. Development of drug resistance cannot be prevented treatment with combination of drugs
5. Resistance is not transferable	5. Resistance is transferable to other organisms to other organisms
6. Mutants may be metabolically defective	6. Not defective
7. Virulence may be low	7. Virulence not decreased

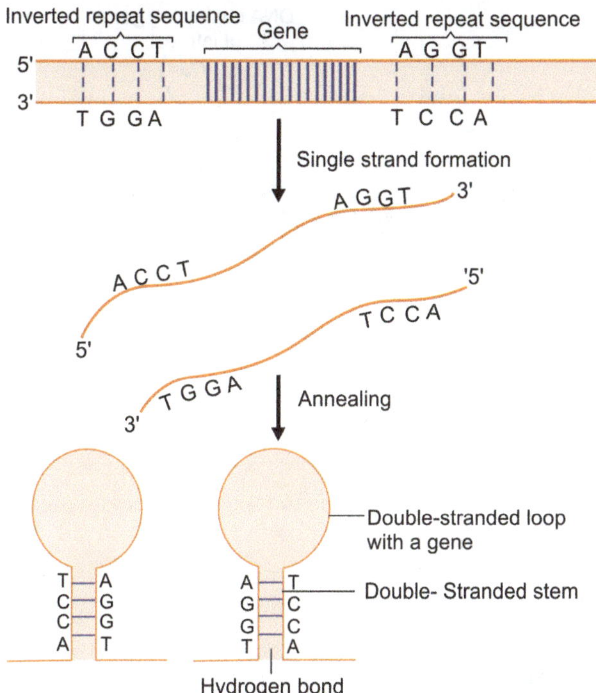

Fig. 10.16: Structure of transposon

reverse order. Because of this feature, each strand of the transposon can form a single-stranded loop carrying the gene, and a double-stranded stem formed by hydrogen bonding between the terminal inverted repeat sequences (Fig. 10.16).

Significance

1. **Mechanism for amplifying genetic transfers:** Transposition is a mechanism for **amplifying genetic transfers** in nature and has been identified in microorganisms, plants and animals.
2. **Spread from one organism or species to another:** Furthermore, because they may be carried between cells on plasmids or viruses, they can also spread from one organism-or even species-to another.
3. **Antibiotic resistance:** Many transoposons contain genes for antibiotic resistance.
4. **Gene manipulations:** Transposons appear to accomplish in nature, gene manipulations similar to the laboratory manipulations that have been called 'genetic engineering'. Transposons are thus, a potentially powerful mediator of evolution in organisms.

MOLECULAR GENETICS

Discoveries in microbial genetics have provided the basis for the discipline of molecular genetics, which is concerned with the analysis and manipulation of DNA using biochemical and microbiological techniques. It has been stated that these techniques have revolutionized the study of biology and medicine, probably more than any technique since the development of the light microscopes. Some techniques and applications of molecular genetics are discussed below.

Genetic Engineering

Genetic engineering is a method of inserting foreign genes into a bacterium and obtaining chemically useful products. Genetic engineering, also known as **recombinant DNA (rDNA) technology**, uses the techniques and tools developed by the bacterial geneticists to purify, amplify, modify, and express specific gene sequences. The use of genetic engineering and "cloning" has revolutionized biology and medicine.

Genetic Engineering Procedure (Fig. 10.17)

This consists of isolation of the genes coding for any desired protein from microorganisms or from cells of higher forms of life including human beings, and their **introduction** into suitable microorganisms, in which the genes would be functional, directing the production of the **specific protein.** Such cloning of genes in microorganisms enables the preparation of the desired protein in pure form, in large quantities and at a reasonable cost.

Basic Tools of Genetic Engineering

1. Cloning vectors
2. Restriction enzymes
3. DNA ligase.

1. **Cloning vectors:** Cloning vectors can be used to deliver the DNA sequences into receptive bacteria and amplify the desired sequence. Many types of vectors are currently used such as plasmid vectors, bacteriophages and other viruses, cosmid vectors, and artificial chromosomes.
2. **Restriction endonucleases (restriction enzymes):** These are used to cleave DNA reproducibly at defined sequences. **Restriction enzymes** are microbial enzymes which cleave double-stranded DNA at specific oligonucleotide sequences. Restriction enzyme recognizes and cuts, or digests, only one particular sequence of nucleotide bases in DNA, and it cuts this sequence in the same way each time. These enzymes are present in many prokaryotes organisms. e.g. restriction endonuclease Eco RI, Hind Ill, Bgl H, Pst 1 and Sma 1 are obtained from *E. coli, H. influenzae, B. globigii, Provindentia stuarti* and *Serratia marcescens* respectively.
3. **DNA ligase:** The enzyme that links the fragment to the cloning vector.

Application of Genetic Engineering

Genetic engineering has been used to isolate and express the genes for useful proteins in bacteria, yeast, or even insect cells and has become an established branch of biotechnology with great scope for commercial exploitation.

1. **Production of vaccines:** Foot and mouth disease, hepatitis B and rabies viruses.
   DNA vaccine: In the future, it may be sufficient to inject plasmid DNA capable of expressing the

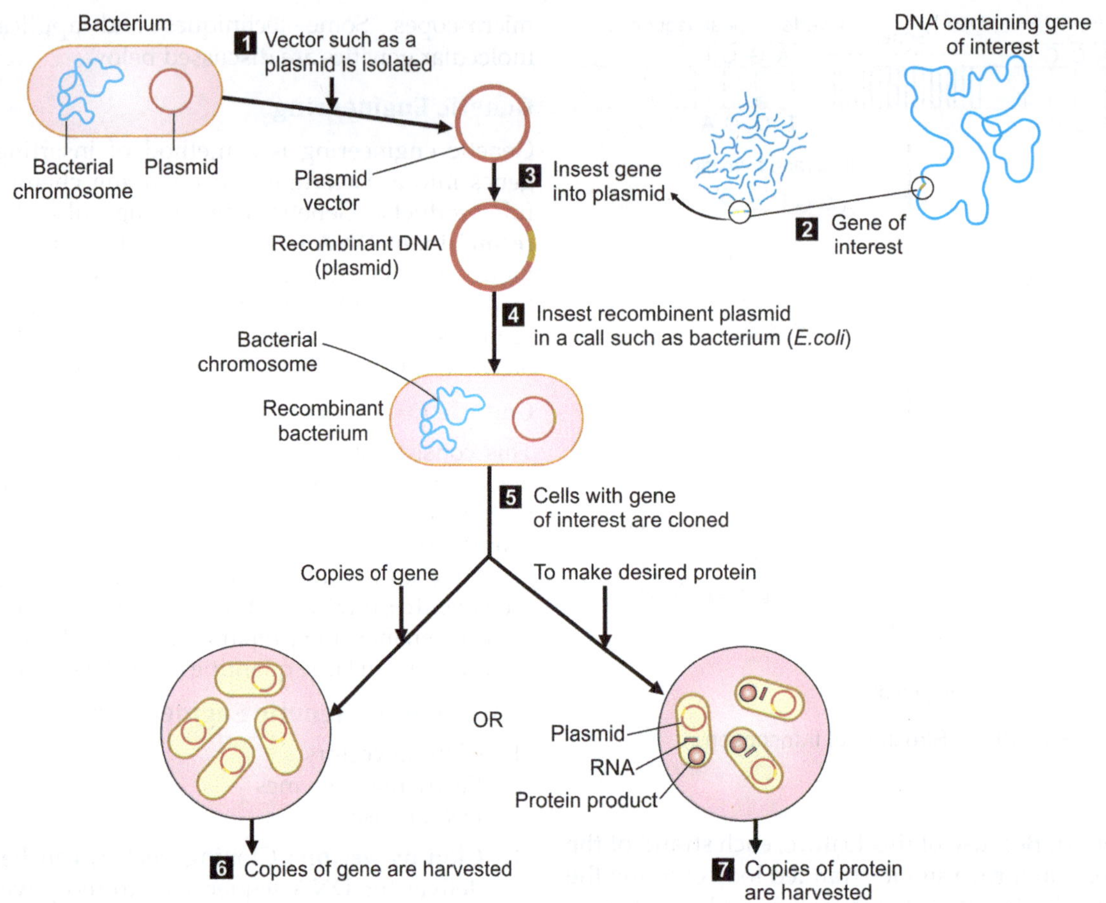

Fig. 10.17: Genetic engineering procedure

desired immunogen (DNA vaccine) into an individual to let the host cells express the immunogen and generate the immune response.

2. **Production of proteins of therapeutic interest:** Human growth hormones, human insulin, erythropoietin, blood clotting factor VIII, tissue plasminogen activator, interferons, tumor necrosis factor, interleukin-1, 2 and 3, granulocyte colony stimulating factor, epidermal growth factor, fibroblast growth factor, somatostatin, growth hormones are proteins of therapeutic interest.

   Cloned human insulin, interferons, somatostatin, growth hormones and many other biologicals have already been marketed.
3. **Gene therapy:** Genetic diseases can be cured by introducing normal genes into the patient.
4. **Others:** It has also become essential to laboratory diagnosis, forensic science, agriculture, and many other disciplines.

GENETIC PROBES

DNA Probes

DNA probes are pieces of radiolabelled or chromogenically labeled pieces of single-stranded DNA that will bind to DNA that is complementary to the probe using hybridizing technique. The specificity of the interaction in base pairing during DNA or RNA synthesis enables the production of specific DNA probes.

DNA probes can be used like antibodies as sensitive and specific tools to detect, locate and quantitate specific nucleic acid sequences in clinical specimens. Individual species or strains of an infectious agent can be detected even if they are not growing or replicating because of the specificity and sensitivity of DNA probe techniques.

Development of nucleic acid probe: All microorganisms, simple or complex, contain some unique sequences of DNA or RNA within their genome that 'distinguish them from all other organisms. DNA probes are chemically synthesized or obtained by cloning specific genomic fragments or an entire viral genome into bacterial vectors **(plasmids, cosmids).**

Hybridization: Hybridization is the process whereby two single-strands of nucleic acid come together to form a stable double-stranded molecule.

Detection of hybridization: Hybridization assays require that one nucleic acid strand (probe) originates from an organism of known identity and the other strand **(the target)** originates from unknown organisms

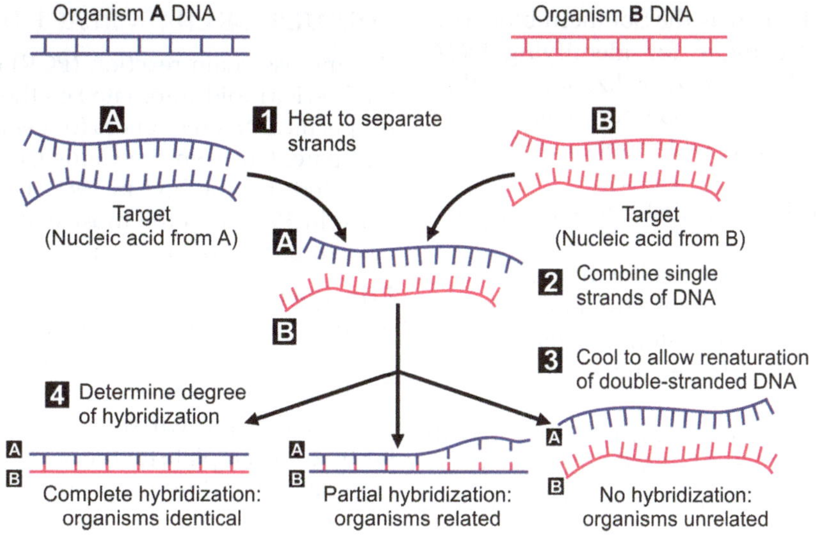

Fig. 10.18: Principle of nucleic acid hybridization

to be detected or identified. The sample serves as a source of nucleic acid to be analyzed and can consist of a suspension of an unknown organism (for culture confirmation) or a clinical specimen such as sputum or stool

Identification of unknown organism is established by positive hybridization (i.e. duplex formation) between a probe nucleic acid strand (from unknown organisms) and a target nucleic acid strand from the organism to be identified. Failure to hybridize indicates lack of homology between probe and target nucleic acid. Positive hybridization identifies the unknown organism as being the same as the probe-source organisms. With a negative hybridization test, the organism remains undetected or unidentified (Fig. 10.18).

Applications of Nucleic Acid Probes

DNA probes have already been used successfully to identify a wide variety of pathogens, from simple viruses to pathogenic bacteria and parasites.

1. **Antibiotic resistance:** Probes have also been developed which can recognize specific antibiotic resistance genes, so that, antimicrobial susceptibility of an infecting organism can be determined directly without primary isolation and growth.
2. **Culture confirmation:** DNA probes are being used for culture confirmation as an alternative to conventional, time-consuming or labor-intensive methods in the diagnostic laboratory. For example, DNA hybridization makes it possible to rapidly identify *Mycobacterium tuberculosis, M kansasii, M avium complex,* and *M gordonae* isolated in culture, significantly reducing the time for reporting of the species of the isolate Table 10.3.
3. **Direct detection in clinical specimen:** For detection of fastidious organisms directly in clinical specimens probe technology may also be used. Examples are *Neisseria gonorrhoeae* and *Chlamydia trachomatis*.

List of microorganims where nucleic acid probes have been applied to diagnostic microbiology is given in Table 10.3. Commercial kits incorporating DNA probes are now available to detect a range of bacteria and viruses.

Table 10.3: Microorganisms where nucleic acid probes have been applied to diagnostic microbiology

A. Culture confirmation and identification	*B. Direct detection In clinical specimen*
1. Bacteria	1. Bacteria
• *Mycobacterium tuberculosis*	• *Mycobacterium tuberculosis*
• *M. avium*	• *Neisseria gonorrhoeae*
• *M. avium complex*	• *Chlamydia trachomatis*
• *Neisseria gonorrhoeae*	• *Legionella pneumophila*
• *Chlamydia trachomatis*	• *Mycoplasma pneumoniae*
• *Haemophilus influenzas*	• *Bordetella pertussis*
• *Listeria monocytogenes*	• *Gardnerella vaginalis*
• *Campylobacter spp.*	• *Streptococcus pyogenes*
• *Enterococcus spp.*	2. Viruses
• *Streptococcus atalactic*	• Enteroviruses
2. Viruses	• Human papillomavirus
• Human papillomavirus	3. Fungi
3. Fungi	• *Candida* spp.
• *Blastomyces dermatitidis*	4. Protozoa
• *Coccidioides immitis*	• *Trichomonas vaginalis*
Histoplasma capsulatum	
• *Cryptococcus neoformans*	

BLOTTING TECHNIQES

1. **Southern blotting:** Very specific DNA sequences can be detected using hybridization techniques

with a technique known as the Southern blot. This highly sensitive technique for identifying DNA fragments by DNA: DNA hybridization is called Southern blotting, after EM Southern who devised it. This technique has very wide application in DNA analysis:

i. DNA to be hybridized is first cleaved by restriction endonucleases.
ii. Then the pieces are separated on the basis of size and charge by agarose gel electrophoresis.
iii. These fragments are transferred (adsorbed) to a nitrocellulose or nylon membrane that is laid over the gel.
iv. When this membrane is allowed to react with labeled probe (radioactive single-stranded DNA probes), only the fragment containing the specific sequence of DNA that hybridizes the probe is detected.
v. These will hybridize with homologous DNA to form radioactive double-stranded segments, which can be detected on X-ray film.

2. **Northern blotting:** An analogous procedure for the analysis of RNA has been called northern blotting (as opposed to southern blotting). Here the RNA mixture is separated by gel electrophoresis, blotted and identified using labeled DNA or RNA probes.
3. **Western blotting (Western Blot Assay)**: A similar technique for the ldentificaion of proteins (antigens) is called *immunoblotting* (or, in conformity with other blotting technique, *western blotting*). The Western blot detects antibody instead of DNA.
   i. The DNA (or RNA) of a particular etiologtc agent is treated with endonucleases, or the protein components of an agent are treated with proteinases to create fragments of different size.
   ii. The nucleic acid or protein fragments are separated by agarose gel electrophoresis, i.e. PAGE (sodium dodecyl sulfate-polyacrylamide gel electrophoresis).
   iii. The patterns are then blotted onto nitrocellulose as in the Southern hybridization assay.
   iv. The filter paper containing specific nucleic acids or proteins is then allowed to react with antiserum from a patient or animal suspected of containing antibodies against the agent. If present, antibodies will bind to the protein or nucleic acid against which they were created, and they can then be detected by visual methods for the detection of antibodies (e.g. enzyme-labeled probes, fluorescent markers).

Use: Diagnosis of HIV antibody—The western blot test has received wide publicity as the confirmatory test for the diagnosis of HIV antibody in sera. The specificity of the test depends on its ability to separately identify antibodies directed against different antigens of the pathogen (for example, against the surface, core and reverse transcriptase antigens of HIV).

POLYMERASE CHAIN REACTION (PCR FIG. 10.19)

Polymerase chain reaction (PCR) or gene amplification method is a rapid automated method for the amplification of specific DNA sequences (or genes). PCR was invented and patented from the Cetus Corporation. Kary Mulis invented this method in 1989. He was awarded Nobel Prize in 1993. It is the most widely used target nucleic acid amplification method.

Principle: PCR is based on repeated cycles of high temperature template denaturation, oligonucleotide primer annealing, and polymerase mediated extension Table 10.4. PCR consists of several cycles of sequential DNA replication where the products of the first cycle become the template for the next cycle. It makes available abundant quantities of specific DNA sequences starting from sources containing minimal quantities of the same.

Procedure: The reaction consists of three essential steps:

1. **Denaturation:** Heat at 94°C is applied to the target DNA, breaking the bonds that hold the strands together. This is known as denaturation.
2. **Primer annealing:** The temperature is reduced to 50-60°C, then oligonucleotide primers attach to target DNA. This temperature is called annealing temperature and the process is known as annealing of primers.
3. **Extension of primer target duplex:** Then polymerase enzyme triggers the formation of new DNA strand from the nucleotides. Extension of the primers is done by a thermostable *Taq polymerase* (purified from *Thermus aquaticus*, a thermophilic bacterium that lives in hot springs at temperatures of 70-75°C). This is known as **primer extension**. When, the temperature is again raised the new strands separate and the process begins again.

   All of the necessary reagents are added to a tube, which is placed in a **thermocycler**. These programmable thermal cyders are used to maintain continuous reaction cycles. As shown in Fig. 10.18, for each **target sequence** originally present in the PCR mixture, two double-stranded fragments containing the target sequence are produced after one cycle. Therefore, with completion of each cycle there is a doubling of target nucleic acid and after 30 to 40 cycles 10^7 to 10^8 target copies will be present in the reaction mixture, a sharp contrast to the days required by conventional amplification method (culture).

Detection of PCR Products

The specific PCR amplification product containing the target nucleic acid of interest is referred to as the amplicon. Amplified sequences of target DNA can be detected by a variety of methods.

1. Gel electrophoresis and ethidium bromide staining.
2. Southern blot and dot-blot analysis with either radioactive or nonradioactive probes.

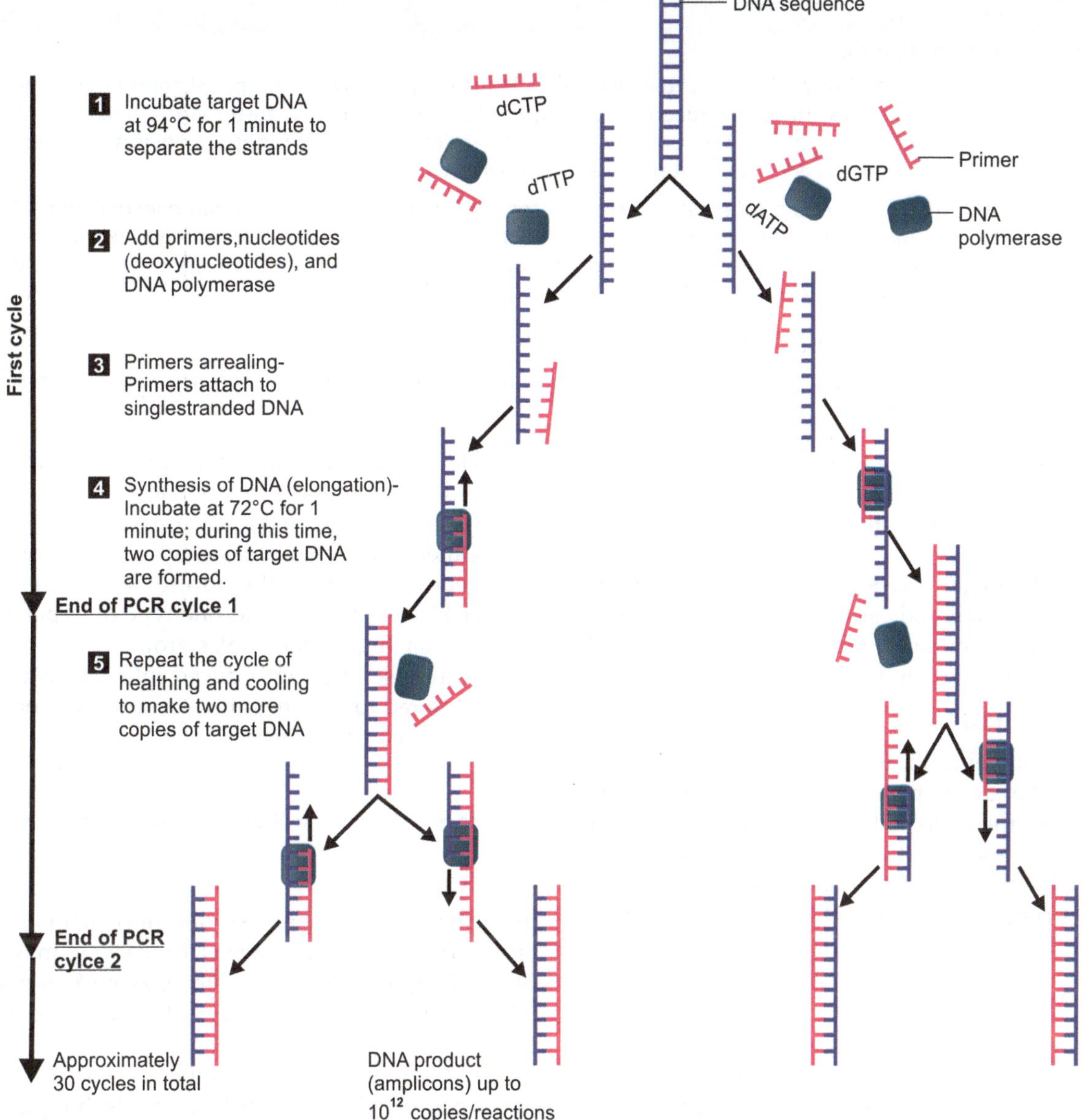

Fig. 10.19: Polymerase chain reaction

3. Oligomer restriction.
4. Oligomer hybridization.
5. Reverse dot-blot.

Table 10.4: Programming of cycles in polymerase chain reaction

Step	Temperature	Time	Cycle
1. Initial denaturation	94°C	5 minutes	First cycle
Denaturation	94°C	30 seconds	
2. Annealing of primers	55°C	1 minute	30 cycles
3. Primer extension	72°C	1 minute	
Final primer extension	72°C	7 minutes	Last cycle

Applications of PCR

The development of PCR or gene amplification method is a major methodological breakthrough in molecular biology. Within a short span, this method has found its way into nearly every type of laboratory from forensic to ecology and from diagnosis to pure research.

1. **In clinical laboratory:** Applications of PCR in clinical laboratory aregiven in Table 10.5.
2. **In diagnosis of inherited disorders:** The PCR technology is being widely used to amplify gene segments that contain known mutations for diagnosis of inherited diseases such as sickle cell anemia, β-thalassemia, cystic fibrosis, etc.

   PCR is especially useful for prenatal diagnosis of inherited diseases, where cells obtained from the fetus by amniocentesis are very few.

Table.10.5: Applications of PCR in clinical laboratory	
Diagnosis of infections due to:	
A. Viruses	**HIV-1, HIV-2, HTLV-1**, cytomegalovirus, human papillomavirus, herpes simplex viruses, hepatitis B virus, HCV, HDV, HEV, rubella virus, Epstein-Barr virus, varicella-zoster virus, human herpes virus-6 and 7, parvovirus B19, enteroviruses, coxsackieviruses/echoviruses, rhinoviruses, measles virus, rotavirus, l adenovirus, respiratory syncytial virus.
B. Bacteria	*Mycobacterium tuberculosis, Mycobacterium avium complex, Legionella pneumophila, Chlamydia trachomatis, Mycoplasma pneumoniae, Helicobacter pylori, Burkholderia pseudomallei, Campylobacter spp., Corynebacterium diphtheriae, Leptospira interrogans, Streptococcus pyogenes, Streptococcus pneumoniae, Yersinia enterocolitica.*
C. Fungi	*Candida* spp., *Cryptococcus neoformans, Aspergillus* spp. *Pneumocystis carinii.*
D. Protozoa	*Toxoplasma gondii, Trypanosoma cruzi, Enterocytozoon bieneusi, Encephalitozoon hellem, Plasmodiumspp.*

3. **In cancer detection:** Identification of mutations in oncosuppressor genes such as retinoblastoma gene can help to identify individuals at high-risk of cancer.
4. **In medicolegal cases:** PCR allows DNA in a single cell, hair follicle or sperm to be amplified enormously and analyzed. The pattern obtained is then compared with that of various suspects.

Derivations of the PCR Method

The powerful amplification capacity of PCR has prompted the development of several modifications that enhance the utility of this methodology, particularly in the diagnostic setting. Specific examples include multiplex PCR, nested PCR, quantitative PCR, RT-PCR, arbitrary primed PCR, and PCR for nucleotide sequencing.

GENE THERAPY

Gene therapy is inserting a missing gene or replacing a defective one in human cells. This technique uses a harmless virus to carry the missing or new gene into certain host cells, where the gene is picked up and inserted into the appropriate chromosome. It is possible to imagine removing some cells from a person and transforming them with a normal gene to replace a defective or mutated gene. When these cells are returned to the person, they should function normally. The benefit of this therapy is to permanently cure the physiological dysfunction by repairing the genetic defect.

Adenoviruses and retroviruses are used most often to deliver genes; however, some researchers are working with plasmid vectors.

Applications

1. Since 1990, gene therapy has been used to treat patients with aenosine deaminase (ADA) deficiency, a cause of severe combined immunodeficiency disease (SCID); **Duchenne's muscular dystrophy**, a muscle-destroying disease; cystic fibrosis; and **LDL-receptor deficiency**. Results are still being evaluated.
2. Hemophilia: The first gene therapy to treat hemophilia in humans was done in 1999. An attenuated retrovirus was used as the vector.
3. Other genetic diseases may also be treatable by gene therapy in the future, including diabetes, sickle cell disease, and one type of hypercholesterolemia, high blood cholesterol.
4. Antisense DNA introduced into cells is also being explored to treat hepatitis, cancers, and one type of coronary artery disease.

KNOW MORE

Reverse Polymerase Chain Reactions

Instead of Taq polymerase described above, *Tth polymerase* obtained from *Thermus thermophilus* may be used. At high temperature, the Tth enzyme possesses both DNA polymerase and reverse transcriptase activities. This allows both cDNA synthesis from mRNA followed by PCR amplification. A number of biotech companies have commercialized amplification techniques for use in the clinical microbiology laboratory.

KEY POINTS

- Genetics is the study of genes, their structure and function, heredity and variation.
- **Characteristics of DNA:** A single strand of DNA has a 5′ end and a 3′ end; the two strands of DNA in the double helix are antiparallel; they are oriented in opposite directions. A gene is a segment of DNA, a sequence of nucleotides, that codes for a functional product, usually a protein.
- **Characteristics of RNA:** A single-stranded RNA fragment is transcribed from one of the two strands of DNA. There are three different functional groups of RNA molecules: messenger RNA (mRNA), ribosomal RNA (rRNA), and transfer RNA (tRNA).
- **Protein synthesis:** The Central Dogma Identifies the flow of genetic Information. The anabolism of proteins takes place by a complex mechanism in which

the genetic information in DNA is first transcribed to a genetic message in RNA and then translated to a sequence of amino acids in the protein.

- **Regulating the expression of genes:** Protein synthesis is generally controlled by regulating the synthesis of mRNA molecules. The best understood is the operon model where binding of a repressor protein to the operon represses transcription. Other factors, such as inducers or corepressors, influence the ability of the repressor protein to bind to the operon.
- **Plasmids:** Plasmids carry nonessential information. Bacteria often have one or more small closed loops of DNA called plasmids. These structures carry information that can confer selective advantages (e.g. antibiotic resistance, protein toxins) to the bacteria that contain them.
- **Mutation:** Mutation is a permanent change in the cellular DNA. This can occur spontaneously in nature resulting from a replication error or the effects of natural radiation.
- **Gene mutation:**
  - Spontaneous mutations occur without the presence of any mutagen and are the result of natural processes.
  - Mutagens are agents in the environment that cause permanent changes in DNA.
  - Induced mutations are deliberate mutations resulting from a mutagen.
  - Point mutations affect one base pair in a gene. Base pairs in the DNA can change in one of two ways.
    i. **Base-pair substitution:** The result can be a silent, missense, or nonsense mutation (which create stop codons).

       A point mutation also can occur from the loss or gain of a base pair. Such mutations change the reading frame and often lead to loss of function by the protein being synthesized.

    ii. **Removal or addition of nucleotides:** Frame shift mutations involve the addition or deletion of nucleotides, rendering all genes downstream of a stop codon and in the same operon non-functional.
- **Mutant selection:**
  - Indirect selection is required when the desired mutant does not grow on a medium on which the parent grows.
  - Replica plating involves the simultaneous transfer of all the colonies on one plate to another and the comparison of the growth of individual colonies on both plates.
  - Penicillin enrichment increases the proportion of mutants in a population by killing growing bacteria in a medium on which only nonmutants will grow.
- **Mechanisms of gene transfer**
  1. **Transformation:** DNA-mediated transformation involves the transfer of "naked" DNA.
  2. **Transduction:** Transduction involves the transfer of bacterial DNA by a bacteriophage.
  3. Lysogenic conversion has two types of life cycles
  4. Conjugation
- This process requires contact between living cells.
- One type of genetic donor cell is an F^+; recipient cells are F^-. F cells contain plasmids called F factors; these are transferred to the F^- cells during conjugation.
- When the plasmid becomes incorporated into the chromosome, the cell is called an Hfr (high-frequency of recombination) cell.
- During conjugation, an Hfr cell can transfer chromosomal DNA to an F^- cell. Usually, the Hfr chromosome breaks before it is fully transferred. R plasmids
- **Mechanisms of Drug Resistance in Bacteria:** (A) By mutation (chromosomal); (B) By genetic exchange.
- **Transposons** are small segments of DNA that can move from one region to another region of the same chromosome, or to a different chromosome or a plasmid. They vary from simple (insertion sequences) to complex.
- **Genetic engineering techniques** can be used to isolate virtually any gene with a biochemically recognizable property. Human genes can be cloned in bacteria. Bacteria are used as the biochemical factories for the synthesis of such proteins as insulin, interferon, and human growth hormone.
- **Nucleic acid probes** are segments of DNA and RNA labeled with radioisotopes or enzymes that can hybridize to complementary nucleic acids with high degree of specificity.
- **Blotting techniqes:** 1. Southern blotting is used for identifying DNA fragments by DNA/DNA hybridization; 2. Northern blotting is used for the analysis of RNA; 3. Western blotting (Western Blot Assay) is used for the identificaion of proteins antigens) is called immunoblotting
  - **Polymerase chain reaction (PCR):** PCRs are amplified methods. PCR is used to make multiple copies of a desired piece of DNA enzymatically. It can be used to increase the amounts of DNA in samples to detectable levels.

IMPORTANT QUESTIONS

1. Describe the structure and functions of the plasmids.
2. Define mutation. Discuss various types of mutations.
3. Name the various methods of gene transfer. Discuss anyone of these in detail.
4. Write short notes on

- Extrachromosomal genetic elements
- Lac operon
- Transformation
- Transduction
- Lysogenic conversion
- Conjugation
- Fertility factor (F-factor)
- High-frequency recombination (Hfr)
- R plasmids
- Resistance transfer factor (RTF).

5. Differentiate between mutational and plasmid mediated drug resistance in a tabulated form.
6. Write short notes on:
   - Transposable genetic elements.
   - Genetic engineering.
   - Restriction endonucleases.
7. Discuss polymerase chain reaction. What are the applications of this reaction in clinical practice?
8. Write short notes on:
   - DNA probes and their applications
   - Blotting techniques.

FURTHER READING

Bennett PM. The spread of drug resistance. Poplation Genetics of Bacteria. In: Baumberg S, Young JPW, Wellington EMH, Saunders JR (Eds), Cambridge. Cambridge University Press 1995;317-44.

Berg CM, Berg DE, Groisman EA. Transposable elements and the genetic engineering of bacteria, Mobile DNA. In: Berg DE, Howe MM (Eds)., ASM Press Washington. DC 1989;879-925.

Chapman-Hall. Watson JD, et al. Recombinant DNA. New York: Scientific American Books 1983.

Erlich HA. PCR Technology - Principles and Applications of DNA Amplification. New York: Stockton Press, New York 1989.

Innis MA, Gelfand DH, Sninsky JJ, White T. PCR Protocols - A Guide to Methods and Applications. San Academic Press, Diego 1990.

Mazel D, Davies J. Antibiotic resistance in microbes. Cellular and Molecular Life Sciences 1999;56:742-754.

Mazodier P, Davies J. Gene transfer between distantly related bacteria. Annu Rev Genet 1991;25:147-71.

CHAPTER 11

Infection

LEARNING OBJECTIVES

After reading and studying this chapter, you should be able to:

- Define the terms saprophytes, parasite, commensal, pathogen.
- Describe classification of infections.
- Define and differentiate primary, secondary, opportunistic and reinfections.
- Name and define various types of carriers.
- Discuss modes of spread of infection giving suitable examples.
- List the differences between exotoxins and endotoxins.

Infection and immunity involve interaction between the animal body (host) and the infecting microorganisms.

MICROORGANISMS AND HOST

Based on their relationship to their host they can be divided into saprophytes and parasites.

A. Saprophytes
B. Parasites

A. Saprophytes

Saprophytes (from Greek sapros decayed; and phyton plant) are free-living microbes that live on dead or decaying organic matter. They are found in soil and water and play an important role in the degradation of organic materials in nature. They are of little relevance in infectious disease because they are generally incapable of multiplying on living tissues. However, saprophytes like *Bacillus subtilis* may cause infection sometimes when host resistance is lowered.

B. Parasites

Parasites are microbes that can establish themselves and multiply in the hosts. There are many parasitic agents or organisms among the viruses, bacteria, fungi, plants, and animals. By convention, when the word parasite is used without quantification, it refers specifically to a protozoan or helminthic (nematode, trematode, cestode) organisms. Parasite mirobes may be either pathogens or commensals:

Pathogens

Pathogens (from Greek *pathos*, disease, and *gen*, to produce) are the microorganisms or agents, which are capable of producing disease in the host. Its ability to cause disease is called *pathogenicity*.

Types of Pathogens

They are two types: **Primary** and **opportunist pathogens**.

a. Primary (frank) pathogens

Primary (frank) pathogens are the organisms, which are capable of producing disease in previously healthy individuals with intact immunological defences. However, these bacteria may more readily cause disease in individuals with impaired defences.

b. Opportunist pathogens

Opportunist pathogens rarely cause disease in individuals with intact immunological and anatomical defences. These bacteria are able to cause disease only when such defences are impaired or compromised, as a result of congenital or acquired disease or by the use of immunosuppressive therapy or surgical techniques. Many opportunistic pathogens are part of the normal human flora, e.g. coagulase-negative staphylococci and *Escherichia coli.*

Commensals

Commensals (organisms of normal flora) are the micro organisms that live in complete harmony with the host without causing any damage to it. Skin and mucous membranes are sterile at birth. The normal bacterial flora of the body consist largely of commensals. Many commensals behave as facultative pathogens in that they can produce disease when the host resistance is lowered.

INFECTION AND INFECTIOUS DISEASE

It is necessary to distinguish between the term 'infection' and 'infectious disease'.

Infection

The lodgement and multiplication of a parasite in or on the tissues of a host constitute infection. It does not invariably result in disease. In fact, disease is but a rare consequence of infection, which is a common natural event.

Infectious disease

An **infectious disease** is any change from a state of health in a part or all of the host body is not capable of carrying on its normal functions due to the presence of an organism or its products.

CLASSIFICATION OF INFECTIONS

Infections may be classified in various ways.

1. **Primary infection:** Initial infection with a parasite in a host is termed *primary infection.*
2. **Reinfections:** Subsequent infections by the same parasite in the host are termed *reinfections.*
3. **Secondary infection:** When a new parasite sets up an infection in a host whose resistance is lowered by a preexisting infectious disease, this is termed *secondary infection.*
4. **Local infection:** The term *Local infection* (more appropriately *local sepsis)* indicates a condition where, due to infection or sepsis at localized sites such as appendix or tonsils, generalized effects are produced.
5. **Cross infection:** When in a patient already suffering from a disease a new infection is set up from another host or another external source, it is termed *cross infection.*
6. **Nosocomial infections:** Cross infections occurring in hospitals are called *nosocomial infections* (from Greek *nosocomion* hospital).
7. **Iatrogenic infection:** The term *iatrogenic infection* refers to physician induced infections resulting from investigative, therapeutic or other procedures. Depending on whether the source of infection is from the host's own body or from external sources, infections are classified as *endogenous* or *exogenous,* respectively. Based on the clinical effects of infections, they may be classified into different varieties.
8. **Inapparent infection:** Inapparent infection is one where clinical effects are not apparent.
9. **Subclinical infection:** The term *subclinical infection* is often used as a synonym to inapparent infection.
10. **Atypical infection:** Atypical infection is one in which the typical or characteristic clinical manifestations of the particular infectious disease are not present.
11. **Latent infection:** Some parasites, following infection, may remain in the tissues in a latent or hidden form proliferating and producing clinical disease when the host resistance is lowered. This is termed *latent infection.*

SOURCES OF INFECTION

A. Human beings
B. Animals
C. Insects
D. Soil and water
E. Food

A. Human Beings

The commonest source of infection for human beings is human beings themselves. The parasite may originate from a patient or carrier. Humans play a substantial role as microbial reservoirs.

Humans serving as the microbial reservoir

i. Indeed, the passage of a neonate from the sterile environment of the mother's womb through the birth canal, which is heavily colonized with various microbial agents, is a primary example of one human directly acquiring microorganisms from another human serving as the reservoir.
ii. Acquisition of "strep" throat through touching.
iii. Hepatitis by blood transfusions.
iv. Gonorrhea, syphilis, and (AIDS) acquired immune deficiency syndrome by sexual contact.
v. Tuberculosis by coughing; and the common cold through sneezing.

Carrier

A **carrier** is person who harbors the microorganisms without suffering from any ill effect, because of it. There are several types of carriers.

1. **Convalescent carrier:** A **convalescent carrier** is an individual who has recovered from the infectious disease but continues to harbor large numbers of the pathogen.
2. **Healthy carrier:** A **healthy carrier** is an individual who harbors the pathogen but is not ill.
3. **Incubatory carrier:** An **incubatory carrier** is an individual who is incubating the pathogen in large numbers but is not yet ill.
4. **Temporary carriers:** Convalescent, healthy, and incubatory carriers may harbor the pathogen for only a brief period (hours, days, or weeks) and lasts less than six months and then called **casual, acute, transient or temporary carriers.**

   Chronic carriers: They harbor the pathogen for long periods (months, years, or life).
6. **Contact carriers:** The term **contact carrier** is applied to a person who acquires the pathogen from a patient.
7. **Paradoxical carrier: Paradoxical carrier** refers to a carrier who acquires the pathogens from another carrier.

Carriers may be classified according to **portal of exit** of the infectious agent such as **urinary carriers, intestinal carriers, respiratory carriers, nasal carriers,** etc.

B. Animals

Reservoir Hosts

Many pathogens are capable of causing infections in both human beings and animals. Therefore, animals may act as a source of infection of such organisms. These, animals serve to maintain the parasite in nature and act as **reservoir** and they are, therefore, called **reservoir hosts**.

Zoonosis

The diseases and infections, which are transmissible to man from animals are called **zoonosis**. Humans contact the pathogens by several mechanisms.

Examples of zoonotic diseases:

Bacterial

Anthrax, brucellosis, Q fever, leptospirosis, bovine tuber-culosis, bubonic plague, Salmonella food poisoning.

Viral

Rabies, yellow fever, cowpox, monkeypox

Protozoal

Leishmaniasis, toxoplasmosis, trypanosomiasis, babesiosis.

Helminthic

Echinococcosis, teniasis, trichinellosis

Fungal

Microsporum canis, Trichophyton verrucosum.

C. Insects

Arthropodborne Diseases

Blood-sucking insects such as mosquitoes, ticks, mites, flies, and lice may transmit pathogens to human beings and diseases so caused are called *arthropodborne diseases.*

Vectors

Insects that transmit infections are called *vectors*. Vector-borne transmission can be of two types either mechanical (external) or biological (internal).

i. **Mechanical vector:** The disease agent is transmitted mechanically by the arthropod. Carriage is passive, with no growth of the pathogen during transmission.

   Examples: Transmission of diarrhea, dysentery, typhoid, food poisoning and trachoma by the housefly.

ii. **Biological vectors:** Biological vectors are those in whom the pathogens multiply or undergo developmental changes with or without multiplication. Biological vectors transmit infection only after the pathogen has multiplied in them sufficiently or has undergone a developmental cycle. The interval between the time of entry of the pathogen into the vector and the vector becoming infective is called the *extrinsic incubation period.*

   Examples: *Aedes aegypti* mosquito in yellow fever, Anopheles mosquito in malaria.

Reservoir Hosts

Besides acting as vectors, some insects may also act as reservoir hosts (for example, ticks in relapsing fever and spotted fever). Infection is maintained in such insects by transovarial or transstadial passage.

D. Soil and Water

i. Soil

Some pathogens can survive in the soil for long periods.

Examples

a. **Spores of tetanus and gas gangrene:** Spores of tetanus and gas gangrene remain viable in the soil for several decades and serve as source of infection. The human and animal intestine is the normal habitat of these organisms and they enter the soil through their feces.
b. **Fungi and parasites:** Fungi (causing mycetoma, sporotrichosis, histoplasmosis) and parasites such as roundworms and hookworms also survive in the soil and cause human infection.

ii. Water

Water may act as the source of infection either due to contamination with pathogenic microorganisms (*Shigella, Salmonella, Vibrio cholerae,* poliomyelitis virus, hepatitis virus) or due the presence of aquatic vector (cyclops containing larvae of guinea worm infection).

E. Food

Contaminated food may act as source of infection of organisms causing food poisoning, gastroenteritis, diarrhea and dysentery. There are two primary types of food-related diseases: foodborne infections and food intoxicants.

MODES OF TRANSMISSION OF INFECTION

The human host may acquire microbial agents by various means referred to as the *modes of transmission*. Pathogenic organisms can spread from one host to another by a variety of mechanisms. These include:

1. Contact
2. Inhalation
3. Ingestion
4. Inoculation
5. Insects
6. Congenital
7. Iatrogenic and laboratory infections

1. Contact

Infection may be acquired by contact, which may be direct or indirect.

a. Direct contact

Direct contact implies an actual physical interaction with the infectious source. Diseases transmitted by direct contact include STD (sexually transmitted diseases such as syphilis, gonorrea, lymphogranuloma venereum, lymphogranuloma inguinale, trichomoniasis, herpes simplex type 2 hepatitis B and acquired immuno deficiency syndrome (AIDS), leprosy, leptospirosis, skin and eye infections.

The term *contagious disease* had been used for diseases transmitted by direct contact and *infectious disease* signifying all other modes of transmission.

b. Indirect contact—fomites

Indirect contact may be through the agency of fomites, which are inanimate objects such as clothing, pencils or toys which may be contaminated by a pathogen from one person and act as a vehicle for its transmission to another. Pencils shared by school children may act as fomites in the transmission of diphtheria, and face towels in trachoma.

Common examples of intermediary inanimate objects include thermometers, eating utensils, drinking cups, and bedding. This embraces a variety of mechanisms including the traditional 5 F's - "flies, fingers, fomites, food and fluid".

2. Inhalation

Droplet nuclei

Respiratory infections such as common cold, influenza, measles, mumps, tuberculosis and whooping cough are acquired by inhalation. Such microbes are shed by the patients into the environment, in secretions from the nose or throat during sneezing, speaking , coughing and other forceful expiratory activities. Large droplets more than 0.1 mm in diameter fly forwards and downwards from the mouth to the distance of a few feet and they reach the floor within a few seconds or they may fall on the eyes, face, mouth and clothes of the person standing in front of the producer of the spray. Small droplets, under 0.1 mm in diameter, evaporate immediately to become minute particles or *droplet nuclei* (usually 1-10 μm in diameter) which remain suspended in the air for long periods, acting as sources of infection. Particles of 10 μm or greater in diameter are filtered off by nose. Particles in the 1-5 μm range are liable to be easily drawn into the alveoli of the lungs and may be retained there.

Diseases spread by droplet nuclei: These include tuberculosis, influenza, chickenpox, measles, Q fever and many respiratory infections.

Dust

Some of the larger droplets which are expelled during talking, coughing or sneezing, settle down by their sheer weight on the floor, carpets, furniture, clothes, bedding, linen and other objects in the immediate environment and become part of the dust.

The diseases carried by infected dust: Include streptococcal and staphylococcal infection, pneumonia, tuberculosis, Q fever and psittacosis. Airborne dust is primarily inhaled, but may settle on uncovered food and milk. This type of transmission is most common in hospital-acquired (nosocomial) infection.

3. Ingestion

Intestinal infections are generally acquired by the ingestion of food or drink contaminated by pathogens. Infection transmitted by ingestion may be waterborne (cholera), foodborne (food poisoning) or handborne (dysentery). Diseases transmitted by water and food include chiefly infections of the alimentary tract, e.g. acute diarrheas, typhoid fever, cholera, polio, hepatitis A, food poisoning and intestinal parasites.

4. Inoculation

The disease agent may be inoculated directly into the skin or mucosa, e.g. rabies virus deposited subcutaneously by dog bite, tetanus spores implanted in deep wounds, and arboviruses injected by insect vectors.

Infection by inoculation may be iatrogenic when unsterile syringes and surgical equipment are employed. Hepatitis B and the human immunodeficiency virus (HIV) may be transmitted through transfusion of infected blood, or the use of contaminated syringes and needles, particularly among addicts of injectable drugs.

5. Insects

Vectorborne

In infectious disease epidemiology, *vector* is defined as an arthropod or any living carrier (e.g. snail) that transports an infectious agent to a susceptible individual. In some diseases, blood-sucking insects play an important role in the spread of infection from one individual to another. Table 11.1 shows common arthropods and diseases transmitted by them. Transmission by a vector may be mechanical or biological.

6. Congenital

Vertical Transmission

Some pathogens are able to cross the placental barrier and reach the fetus in utero. This is known as **vertical transmission**. This is another form of direct transmission. Vertical transmission may result in abortion, miscarriage or stillbirth. Live infants may be born with manifestations of a disease, as in congenital syphilis. Intrauterine infection with the rubella virus, especially in the first trimester of pregnancy, may interfere with oncogenesis and lead to congenital malformation. Such infections are known as teratogenic infections.

Examples: So-called TORCH agents (*Toxoplasma gondii*, rubella virus, cytomegalovirus and herpes virus), varicella virus, syphilis, hepatitis B, coxsackie B and AIDS.

7. Iatrogenic and Laboratory Infections

If meticulous care in asepsis is not taken, infections like AIDS and hepatitis B may sometimes be transmitted during administration of injections, lumber puncture and catheterization. Modern methods of treatment

Table 11.1: Arthropodborne diseases

Arthropod	*Diseases Transmitted*
1. Mosquito	Malaria, filaria, viral encephalitis (e.g. Japanese encephalitis), viral fevers (e.g. dengue, West Nile, viral hemorrhagic fevers (e.g. yellow fever, dengue hemorrhagic fever)
2. Housefly	Typhoid and paratyphoid fever, diarrhea, dysentery, cholera, gastroenteritis, amebiasis, helminthic infestations, poliomyelitis, conjunctivitis, trachoma, anthrax, yaws, etc.
3. Sand fly	Kala-azar, oriental sore, sand fly fever, oraya fever
4. Tsetse fly	Sleeping sickness
5. Louse	Epidemic typhus, relapsing fever, trench fever, pediculosis
6. Rat flea	Bubonic plague, endemic typhus, chiggerosis, Hymenolepis diminuta
7. Blackfly	Onchocerciasis
8. Reduviid bug	Chagas disease
9. Hard tick	Tick typhus, viral encephalitis, viral fevers, viral hemorrhagic fever, (e.g. Kyasanur forest disease), tularemia, tick paralysis, human babesiosis
10. Soft tick	Q fever, relapsing fever
11. Trombiculid mite	Scrub typhus, Rickettsialpox
12. Itch-mite	Scabies
13. Cyclops	Guinea worm disease, fish tapeworm (*D. latum*)
14. Cockroaches	Enteric pathogens

such as exchange transfusion, dialysis, and heart and transplant surgery have increased the possibilities for iatrogenic infections. These are known as iatrogenic or physician-induced infections. Laboratory personnel handling infectious material and doing mouth-pipetting are particularly at risk.

FACTORS PREDISPOSING TO MICROBIAL PATHOGENICITY

Pathogenicity and Virulence

Pathogenicity

Denotes the ability of a microbial species to cause disease. The term virulence (Latin *virulentia,* from *virus,* poison) denotes the ability of a strain of a species to produce disease.

Virulence

Virulence provides a quantitative measure of pathogenicity, or the likelihood of causing disease. For example, encapsulated pneumococci are more virulent than nonencapsulated pneumococci, and *Escherichia coli* that express Shiga-like toxins are more virulent than those that do not express these toxins. The virulence of a strain is not constant and may undergo spontaneous or induced variation.

Exaltation

Enhancement of virulence is known as *exaltation*. This can be induced by serial passage of a strain in an experimental animal.

Attenuation

Reduction of virulence is known as *attenuation* and can be induced by passage through unfavorable hosts, repeated culture in artificial media, growth under high temperature or in the presence of weak antiseptics, desiccation, or prolonged storage in culture.

Virulence Factors

Virulence factors refer to the properties (i.e. gene products) that enable a microorganism to establish itself on or within a host of a particular species and enhance its potential to cause disease. Virulence is determined by three characteristics of the pathogens: invasiveness, infectivity, and pathogenic potential. A major aspect of pathogenic potential is toxigenicity.

Determinants of Virulence

1. Transmissibility
2. Adhesion
3. Invasiveness
4. Toxigenicity
5. Avoidance of host defence mechanisms
6. Enzymes
7. Plasmids
8. Bacteriophages
9. Communicability
10. Infecting dose
11. Route of infection

1. Transmissibility

The first step of the infectious process is the entry of the microorganism into the host by one of several ports: the respiratory tract, gastrointestinal tract, urogenital tract, or through skin that has been cut, punctured, or burned. Once entry is achieved, the pathogen must overcome a diversity of host defenses before it can establish itself. These include phagocytosis, the acidic environments of the stomach and urogenital tract, and various hydrolytic and proteolytic enzymes found in saliva, in stomach, and in the small intestine.

2. Adhesion

Adhesins

The initial event in the pathogenesis is the attachment of the bacteria to body surfaces. This attachment is not a chance event but a specific reaction between surface *receptors* on host cells and adhesive structures *(ligands)* on the surface of bacteria. These adhesive structures are called **adhesins.**

Adhesions may occur as organized structures, such as **fimbriae or fibrillae and pili, or as colonization factors.** Nonspecific surface properties of the bacterium, including **surface charge** and **hydrophobicity,** also contribute to the initial stages of the adhesion process.

Some bacteria (for example, *Escherichia coli*) use their pili to adhere to the surface of host cells. Group A streptococci have similar structures (fimbriae). A striking example of the importance of adhesion is that of *Neisseria gonorrhoeae* in which strains that lack pili are not pathogenic. If bacterium is invasive in nature, adherence helps in penetrating host cells.

Adhesins as Virulence Factors

Adhesins serve as virulence factors, and loss of adhesins often renders the strain avirulent. Adhesins are usually made of protein and are antigenic in nature. Specific immunization with adhesins has been attempted as a method of prophylaxis in some infections, as for instance against *E. coli* diarrhea in calves and piglets, and gonorrhea in human beings.

3. Invasiveness

Invasiveness signifies the ability of a pathogen to spread in the host tissues after establishing infection. For many disease-producing bacteria, invasion of the host's epithelium is central to the infectious process. Highly invasive pathogens characteristically produce spreading or generalized lesions (e.g. streptococcal septicemia following wound infection), while less invasive pathogens cause more localized lesions (e.g. staphylococcal abscess). Some pathogens though capable of causing serious or even fatal diseases lack invasiveness (e.g. the tetanus bacillus which remains confined to the site of entry and produces the disease by elaborating a potent toxin).

4. Toxigenicity

Some bacteria cause disease by producing toxins, of which there are two general types: the exotoxins and the endotoxins. Both gram-positive and gram-negative bacteria secrete the exotoxins, which are proteins. In contrast, the endotoxins, which are lipopolysaccharides, are not secreted, but instead are integral components of the cell walls of gram-negative bacteria (Table 11.2).

Exotoxins

Exotoxins are soluble, heat-labile proteins inactivated at 60 to 80°C which are secreted by certain species of bacteria and diffuse readily into the surrounding medium.

These are highly potent in minute amounts and include some of the most poisonous substances known. It is estimated that as little as one microgram of tetanus exotoxin can kill an adult human, One mg of tetanus or botulinum toxin is sufficient to kill more than one million guinea pigs and 3 kg of botulinum toxin can kill all the inhabitants of the world.

Treatment with dilute formaldehyde destroys the toxic activity of most exotoxins, but does not affect their antigenicity. Formaldehyde—inactivated toxins, called toxoids, are thus useful in preparing vaccines.

Exotoxin proteins are in many cases encoded by genes carried on plasmids or temperate bacteriophage.

They exhibit specific tissue affinity and pharmacological activities, each toxin producing a typical effect which can be made out by characteristic clinical manifestations or autopsy appearances.

They are associated with specific diseases and have specific mechanisms of action.

They are easily inactivated by formaldehyde, iodine, and other chemicals to form immunogenic **toxoids.**

They are unable to produce a fever in the host directly and often given the name of the disease they produce (e.g. the diphtheria toxin).

Exotoxins are generally formed by gram-positive bacteria but may also be produced by some gram-negative organisms such as Shiga's dysentery bacillus, cholera vibrio and enterotoxigenic *E. coli.*

Endotoxins

These are heat-stable, lipopolysaccharide (LPS) components of the outer membranes of gram-negative bacteria but not gram-positive bacteria. Their toxicity depends upon the the component (lipid A).

They are released into the host's circulation following bacterial cell lysis.

They are toxic only at high doses (milligram per kilogram amounts).

They cannot be toxoided.

They are poor antigens and weakly immunogenic and their toxicity is not completely neutralized by the homologous antibodies.

They do not exhibit specific pharmacological activities. They are generally similar, despite source.

All endotoxins, produce similar effects whether isolated from pathogenic or nonpathogenic bacteria.

Administration of small quantities of endotoxin in susceptible animals causes an elevation of body temperature manifested within 15 minutes and lasting for several hours. The pyrogenic effect of fluids used for intravenous administration is usually due to the presence of endotoxins from contaminant bacteria. They are usually capable of producing general systematic effects. Intravenous injections of large doses of endotoxin and massive gram-negative septicemias cause endotoxic shock marked by fever, leukopenia, thrombocytopenia, significant fall in blood pressure, circulatory collapse and bloody diarrhea leading to death (Table 11.2).

Table 11.2: Differences between exotoxins and endotoxins

Exotoxins	*Endotoxins*
1. Proteins	1. Lipopolysaccharide on outer membrane. Lipid A portion is toxic
2. Heat-labile (inactivated at 60°-80°C)	2. Heat-stable
3. Actively secreted by the cells; diffuse into the surrounding medium	3. Form integral part of the cell wall; do not diffuse into surrounding medium
4. Readily separable from cultures by physical means such as filtration	4. Obtained only by cell lysis
5. Action often enzymic	5. No enzymic action
6. Specific pharmacological effect for each exotoxin	6. Nonspecific action of all endotoxins
7. Specific tissue affinities	7. No Specific tissue affinities
8. Highly toxic and fatal in microgram quantities	8. Moderate toxicity. Active only in very large doses
9. Highly antigenic	9. Weakly antigenic
10. Action specifically neutralized by antibody	10. Neutralization by antibody ineffective
11. Usually do not produce fever	11. Usually produce fever by release of interlukin-1
12. Produced by both gram-positive bacteria and gram-negative bacteria	12. Produced by gram-negative bacteria only
13. Frequently controlled by extrachromosomal genes (e.g. plasmids)	13. Synthesized directly by chromosomal genes
14. Disease, e.g. Botulism, diphtheria, tetanus	14. Gram-negative infections, meningococcemia

5. Avoidance of Host Defence Mechanisms

Bacteria also have evolved many mechanisms to evade host defenses. Several of these evasive mechanisms are now discussed.

a. Capsules

Some bacteria such as *Streptococcus pneumoniae, Neisseria meningitidis,* and *Haemophilus influenzae* can produce a slippery mucoid capsule that prevents the phagocyte from effectively contacting the bacterium, and noncapsulate variants usually exhibit much reduced virulence.

b. Streptococcal M protein

Other bacteria evade phagocytosis by producing specialized surface proteins such as the M protein on *Streptococcus pyogenes.*

c. Resistance to killing by phagocytic cells

Some bacteria have evolved the ability to survive inside neutrophils, monocytes, and macrophages. These pathogens not only survive within macrophages and other phagocytes, but may actually multiply intracellulary.

Different organisms use different strategies for survival:

i. To escape from the phagosome before it merges with the lysosome, as seen with *Listeria monocytogenes, Shigella,* and *Rickettsia.*
ii. To prevent phagolysosome fusion, e.g. in *Mycobacterium tuberculosis,* probably at least partly because of its waxy external layer.
iii. Production of catalase by *Staph. aureus* and *N. gonorrhoeae* is thought to protect these organisms from such toxic products.

d. Antigenic variation

Variation in surface antigen composition during the course of infection provides a mechanism of avoidance of specific immune responses directed at those antigens.

Examples

i. Pathogenic *Neisseria.*
ii. The borrelia generate antigenic variation.
iii. Other bacteria show strain-specific antigenic variability such as group A streptococci 75 antigenically distinct serotypes of M proteins.

e. Serum resistance

To survive in the blood, bacteria must be able to resist lysis as a result of deposition of complement on the bacterial surface. Some gram-negative bacteria can lengthen the 0 chains in their lipopolysaccharide to prevent complement activation.

f. Siderophore and iron acquisition

Many bacteria produce these low molecular weight compounds called siderophores that can acquire iron from the host's iron binding proteins. This property enhances the virulence.

6. Enzymes

Many species of bacteria produce tissue-degrading enzymes that play important roles in the infection process.

i. **Coagulase:** Coagulase is produced by *S. aureus.* This thrombin-like enzyme prevents phagocytosis by forming a fibrin barrier around the bacteria and walling off the lesion.
ii. **Lecithinase-C and collagenase:** *C. perfringens* produces lecithinase-C and collagenase. Lecithinase-

C damages cell membranes by splitting lecithin to phosphorylcholine and diglyceride while collagenase degrades collagen, the major protein of fibrous connective tissue thus promoting spread of infection in tissue.

iii. **Hyaluronidases:** Hyaluronidases split hyaluronic acid which is a component of intercellular connective tissue and thus facilitate the spread of infection along tissue spaces.

iv. **Streptokinase (fibrinolysin):** Many hemolytic streptococci produce streptokinase (fibrinolysin). Fibrinolysins promote the spread of infections by breaking down the fibrin barrier in tissues.

v. **Cytolysins:** These include, hemolysins capable of destroying erythrocytes and leukocidins damage polymorphonuclear leukocytes.

   a. **Streptolysin O and streptolysin S** are produced by group A streptococci.
   b. Most strains of ***Staphylococcus aureus*** **produce hemolysins.**
   c. ***Escherichia coli*** strains that cause urinary tract infections produce hemolysins whereas those strains that are part of normal gastrointestinal flora may or may not produce hemolysins.

vi. **IgA 1 proteases:** These enzymes specifically cleave immunoglobulin IgA which protects at mucosal surfaces. It is an important virulence factor of *N. gonorrhoeae, N. meningitidis, H. influenzae, S. pneumoniae,* some strains of *Prevotella melaninogenica* and some streptococci associated with dental caries.

7. Plasmids

Plasmids are extrachromosomal DNA segments that carry genes for antibiotic resistance known as R-factors. Multiple drug resistance (R) plasmids increase the severity of clinical disease by their resistance to antibiotic therapy.

Genes coding for some virulence characteristics may be plasmidborne. Surface antigens responsible for the colonization of intestinal mucosa by *E. coli* and enterotoxin production by *E. coli* and *Staph. aureus* are examples of plasmidborne virulence factors.

8. Bacteriophages

The classical example of phage directed virulence is seen in diphtheria. All the strains of *C. diphtheriae* produce exotoxin only when they are lysogenized with a bacteriophage called betaphage. In diphtheria bacilli, the gene for toxin production is present in beta or other *tox+* corynephages. The elimination of this phage abolishes the toxigenicity of the bacillus.

9. Communicability

The ability of a microbe to spread from one host to another is known as communicability. This property determines the survival and distribution of a parasite in a community but does not influence the production of disease in an individual host. A correlation need not exist between virulence and communicability. In fact, a high degree of communicability may not be exhibited by a highly virulent parasite due to its rapidly lethal effect on the host.

In general, infections in which the pathogen is shed in secretions, as in respiratory or intestinal-diseases, are highly communicable. In some instances, infection represents a dead end, as in hydrophobia, there being an interruption in the spread of the pathogen to other hosts.

Occurrence of epidemic and pandemic diseases requires that the pathogen should possess high degree of virulence and communicability.

10. Infecting Dose

Adequate number of bacteria is required for successful infections. The dosage may be estimated as the minimum infecting dose (MID) or minimum lethal dose (MLD).

Minimum Infecting Dose (MID)

Minimum number of bacteria required to produce clinical evidence of infection in a susceptible animal under standard conditions is called minimum infective dose (MID).

Minimum Lethal Dose (MLD)

MLD is a minimum number of bacteria that produce death in the animal under standard conditions.

As animals exhibit considerable individual variation in susceptibility, these doses are more correctly estimated as statistical expressions, ID50 and LD50 as the dose required to infect or kill 50 percent of the animals tested under standard conditions.

11. Route of Infection

Some bacteria can initiate infection whatever be the mode of entry such as streptococci. Certain bacteria are infective when introduced through optimal route, for example, cholera vibrios can produce lesion only when administered by oral route, but unable to cause infection when introduced subcutaneously. However, *Staphylococcus aureus* can cause lesion whatever may be the portal of entry. Probably this difference is related to modes by which different bacteria are able to initiate tissue damage and establish themselves.

Bacteria after introduction into tissues also differ in their sites of election in the host body. They also differ in the ability to produce damage of different organs in different species of animals. Lesions are caused mainly in the kidney and infrequently in the liver and spleen when tubercle bacilli injected into rabbits but in guinea pigs the lesions are mainly in the liver and spleen, the kidneys being spared. The reasons are largely obscure for such selective multiplication in tissues, though they may be related to the presence in tissues of substances that may selectively hinder or favor their multiplication.

TYPES OF INFECTIOUS DISEASES

Infectious diseases may be localized or generalized.

A. Localized

Localized infections may be superficial or deep-seated.

B. Generalized

Generalized infection involves the spread of the infecting agent from the site of entry by contiguity, through tissue spaces or channels, along the lymphatics or through the bloodstream (bacteremia) which leads to dissemination of organisms.

1. Bacteremia

Circulation of bacteria in the blood is known as *bacteremia.* Transient bacteremia is a frequent event even in healthy individuals and may occur during chewing, brushing of teeth or straining at stools. The bacteria are immediately mopped up by phagocytic cells and are unable to initiate infection. Bacteremia of greater severity and longer duration is seen during generalized infections as in typhoid fever.

2. Septicemia

It is the condition where bacteria circulate and multiply in the blood, form toxic products and cause high, swinging type of fever.

3. Pyemia

It is condition where pyogenic bacteria produce septicemia with multiple abscesses in internal organs such as the spleen, liver and kidney.

EPIDEMIOLOGICAL TERMINOLOGY

Depending on the spread of infectious diseases in community, they may be classified as endemic, epidemic, and pandemic.

1. Endemic

The disease which is constantly present in a particular area, e.g. typhoid fever is endemic in most parts of India.

2. Epidemic

The disease that spreads rapidly, involving many persons in a particular area at the same time, is called epidemic disease, e.g. meningococcal meningitis. In the cold countries influenza causes annual winter epidemics.

3. Pandemic

It is an epidemic that spreads through many areas of the world involving very large number of persons within a short period, e.g. cholera, influenza and enteroviral conjunctivitis.

Epidemics vary in the rapidity of spread. Waterborne diseases such as cholera and hepatitis may cause explosive outbreaks, while diseases which spread by person-to-person contact evolve more slowly. Such creeping or smouldering epidemics, as that of cerebrospinal fever, are termed *prosodemic* diseases.

KNOW MORE

Infection

Humans and microorganisms inhabit the same planet, and their paths cross in many and varied ways so that interactions are inevitable.

Serum Resistance

In the enterobacteriaceae, resistance is primarily due to the composition of the lipopolysaccharide (LPS) present in the bacterial outer membrane. Others such as *Neisseria gonorrhoeae* generate serum resistance.

KEY POINTS

- Infection and immunity involve interaction between the animal body (host) and the infecting microorganisms.
- Parasites are microbes that can establish themselves and multiply in the hosts. Parasite mirobes may be either pathogens or commensals:

Infection

Infections may be classified in various ways:

Sources of Infection: 1. Human beings—from a patient or carrier; 2. Animals; 3. Insects; 4. Soil and water; 5. Water. 6. Food

Modes of Transmission of Infection

1. Contact; 2. Inhalation; 3. Ingestion; 4. Inoculation; 5. Insects; 6. Congenital; 7. Iatrogenic and laboratory infections.

Determinants of Virulence

1. Transmissibility; 2. Adhesion; 3. Invasiveness; 4. Toxigenicity; 5. Enzymes; 6. Plasmids; 7. Bacteriophages; 8. Communicability; 9. Infecting dose; 10. Route of infection.

Types of Infectious Diseases

A. Localized infections may be superficial or deep-seated.

B. Generalized infection

1. **Bacteremia;** 2. **Septicemia; 3. Pyemia**

IMPORTANT QUESTIONS

1. Describe in detail the sources of infections to humans beings.
2. What are the various modes of spread of infection? Describe each in brief giving suitable examples.
3. Describe the factors determining microbial pathogenicity and virulence.
4. Distinguish between exotoxins and endotoxins in a tabulated form.

FURTHER READING

Mims CA. The Pathogenesis of Infectious Disease, 3rd edn. London: Academic Press 1987.

Poxton IR, Arbuthnot JP. Determinants of bacterial virulence, Chap. 13 In: Topley and Wilson's Principles of Bacteriology Virology and Immunity, 8th edn. Vol. 1. London: Edward Arnold 1990.

Relman DA, Falkow S: A molecular perspective of microbial pathogenicity. In: Mandell, Donglas and practice of intertious diseases, 5th ed. Mandell GL, Bennett JE, Dolan R (Editos) Churchill Livingstone 2000.

Salyers AA, Whitt DD. Bacterial Pathogenesis: A Molecular Approach. American Society for Microbiology Press, 1994.

SECTION TWO

IMMUNOLOGY

CHAPTER

12 Immunity

LEARNING OBJECTIVES

After reading and studying this chapter, you should be able to:

- Describe innate immunity, artificial active immunity, natural passive immunity, herd immunity.
- Differentiate between active and passive immunity.

DEFINITION

Immunity Latin *immunis,* free of burden] refers to the resistance exhibited by the host towards injury caused by microorganisms and their products.

The complex reaction a host animal undergoes after contact with microorganisms can be grouped under the broadly defined heading of **resistance.** Protection against infectious diseases is only one of the consequences of the immune system, which in its entirety is concerned with the reaction of the body against any foreign antigen.

CLASSIFICATION

Immunity against infectious diseases is of different types. The discrimination between self and nonself, and the subsequent destruction and removal of foreign material, is accomplished by two arms of immune system, the **innate** (or **"natural") immune system,** and the **adaptive** (or **"acquired"), specific immune system**.

Immunity

i. Innate (or natural) immunity		
	a. Nonspecific	Species
		Racial
		Individual
	b. Specific	Species
		Racial
		Individual
ii. Acquired (or adaptive) immunity		
	a. Active	Natural
		Artificial
	b. Passive	Natural
		Artificial

1. Innate or natural immunity

The body's first line of defense against invasion by microorganisms is the innate immunity or "natural" immune system, which is essential for the health of an organism. It is the resistance to infections which an individual possesses by virtue of his genetic or constitutional make up. Repeated exposure to a pathogen does not enhance the innate immune system.

a. Nonspecific and Specific Immunity

It may be **nonspecific**, when it indicates a degree of resistance to infections in general, or **specific** where resistance to a particular pathogen is concerned. Innate immunity may be considered at the level of **species, race or individual.**

i. Species Immunity

Resistance or susceptibility (lack of resistance) to infections can vary from one species of animal to other. It refers to the total or relative refractoriness to a pathogen, shown by all members of a species.

Examples

i. **Mice** are extremely susceptible to infection by *Streptococcus pneumoniae*. **Humans**, on the other hand, are relatively resistant to *Streptococcus pneumoniae* infection.
ii. **The rat** is strikingly resistant to diphtheria whilst the guinea pig and humans are highly susceptible.
iii. **All human beings** are totally unsusceptible to plant pathogens and to many animal pathogens such as rinderplast or distemper.

Mechanisms of Species Immunity

Physiological and biochemical differences: The mechanisms of species immunity are not clearly understood but may be due to **physiological** and **biochemical differences** between the tissues of the different host species, which determine whether or not a pathogen can multiply in them.

2. Racial Immunity

Within a species, different races may show differences in susceptibility to infections. This is known as **racial immunity.** Such racial differences are known to be **genetic in origin**, and by selection and inbreeding.

Examples

i. **High resistance of Algerian sheep to anthrax:** It is the classic example.
ii. **Susceptibility to tuberculosis:** The people of Negroid origin in the USA are more susceptible than the Caucasians to tuberculosis. But such comparisons are vitiated by external influences such as differences in socioeconomic levels.
iii. **Genetic resistance to *Plasmodium falciparum* malaria:** It is seen in some parts of Africa and the Mediterranean coast and is attributed to the hereditary abnormality of the red blood cells (sickling) prevalent in the area. These red blood cells cannot be parasitized by malarial parasite. It confers immunity to infection by the malarial parasite and may have evolved from the survival advantage conferred by it in a malarial environment.

3. Individual Immunity

The differences in innate immunity exhibited by different individuals in a race is known as **individual immunity**. The role of heredity in determining resistance to infection is well illustrated by studies on tuberculosis in twins. If one homozygous twin develops tuberculosis, the other twin has a 3 to 1 chance of developing the disease compared with a 1 in 3 chance if twins are heterozygous.

Factors Influencing the Level of Immunity

1. Age

i. Fetus in Utero

The two extreme of life carry higher susceptibility to infectious diseases as compared with adults. **The fetus in utero** is protected from maternal infection by the placental barrier. But some pathogens cross this barrier causing overwhelming infection leading to **fetal death**, while others such as *Toxoplasma gondii,* rubella, herpes, cytomegaloviruses lead to **congenital malformations.** The higher susceptibility of the young appears to associate with immaturity of immune system.

ii. Newborn Animals

Newborn animals are more susceptible to experimental infection than adult animals, e.g. coxsackievirus causes fatal infection in suckling mice but not in adult mice. Some infections like measles, mumps, poliomyelitis and chickenpox tend to be more severe in adults than in young children. This may be due to more active immune response producing greater tissue damage.

iii. In the Elderly

It besides a general waning of the activities of the immune system, physical abnormalities (e.g. prostatic enlargement leading to stasis of urine) or long-term exposure to environmental factors (e.g. smoking) are common causes of increased susceptibility to infection.

2. Hormonal Influences and Sex

i. Endocrine Disorders

There is an increased susceptibility to infection in endocrine disorders such as diabetes mellitus, hypothyroidism and adrenal dysfunction (increased corticoids secretion). The reason for this disease have not yet been clarified but may be related to enzyme or hormone activities.

Glucocorticoids are anti-inflammatory agents, decreasing the ability of phagocytes to ingest material. They also have beneficial effect by interfering in some way with toxic effects of bacterial products such as endotoxin. In diabetics, staphylococcal, streptococcal and certain fungal infections such as candidiasis, aspergillosis and mucormycosis occur more frequently. Pregnant women are more susceptible to microbial infection due to increased steroid levels during pregnancy.

ii. Sex

There is no marked difference in susceptibility to infections between the sexes. In general, incidence and death rate from infectious diseases are greater in males than in females. However, infectious hepatitis and whooping cough have a higher morbidity and mortality in females.

3. Nutrition

In general, both humoral and cell mediated immune processes are reduced in malnutrition although the adverse effects of poor nutrition on susceptibility to certain infectious agents are not now seriously questioned. **Protein calorie malnutrition** lowers C3 and factor B of the complement system, decreases the interferon response, and inhibits neutrophil activity.

Experimental evidence in animals has shown that inadequate diet may be correlated with increased susceptibility of a variety of bacterial diseases, associated with decreased phagocytic activity and leucopenia. Viruses are intracellular parasites and malnutrition might have an effect on virus production, but the usual outcome is enhanced disease due to impaired immune responses, especially the cytotoxic response.

4. Stress

A growing body of evidence has demonstrated an inverse relation between stress and immune function. The end result is an increased susceptibility to infection.

Mechanisms of innate immunity

1. Mechanical Barriers and Surface Secretions

The first defenses are the external and internal body surfaces that are in relatively direct contact with the external environment and as are the body areas with which microorganisms will initially associate. These surfaces include:

A. Skin (including conjuctival epithelium covering the eye)
B. Mucous membranes that line the mouth or oral cavity, the respiratory tract, and the genitourinary tract.

A. Skin

The intact skin and the mucous membranes provide mechanical barriers that prevent the entrance of most microbial species. In conditions where the skin is damaged, such as in burns patients and after traumatic injury or surgery, infections can be a serious problem.

Even though the structure of the skin itself undoubtedly gives a great deal of protection, considerably more important are the fatty acids secreted by the sebaceous glands and the propionic acid by the normal flora of the skin. Secretions from the sebaceous glands contain both saturated and unsaturated fatty acids that kill many bacteria and fungi.

A striking example of this type of infection is seen in the case of the fungi causing ringworm of the scalp (species of *Microsporum* and *Trichophyton*). This infection is difficult to cure in children, but after puberty it disappears without treatment, presumably as a result of a change in the amount and kinds of fatty acids secreted by the sebaceous glands.

The bactericidal activity of skin secretions is illustrated by the frequent mycotic and pyogenic infections seen in persons who immerse their hands in soapy water for long periods occupationally.

B. Mucous Membrane

General protective mechanisms: A major protective component of mucous membranes is the **mucus** itself. This substance serves to trap bacteria before they can reach the outer surface of the cells, lubricates the cells to prevent damage that may promote microbial invasion, and contain numerous specific (i.e. antibodies) and nonspecific antibacterial substances. In addition to mucous activity and flow mediated by cilia action, rapid cellular shedding and tight intercellular connections provide effective barriers. As is the case with skin, specific cell clusters, known as **mucosa-associated lymphoid tissue,** exist below the outer cell layer and mediate specific protective mechanisms against microbial invasion.

Specific protective characteristics: Besides the general protective properties of mucosal cells, the lining of the different body tracts has other characteristics specific to each anatomic site.

a. Mouth or Oral Cavity

The mouth or oral cavity is protected by the flow of saliva that physically carries microorganisms away from the cell surfaces and also contains the lysozyme, which destroys bacterial cell walls, and antibodies. Harmful agents also heavily colonize the mouth with microorganisms that contribute to protection by producing substances that hinder successful invasion. Particles deposited in the mouth are swallowed and subjected to the action of the digestive juices.

b. Gastrointestinal Tract

i. **Stomach:** In the gastrointestinal tract, several systems function to inactivate bacteria. The low pH and proteolytic enzymes of the stomach help keep the numbers of microorganisms low. The high acidity of the stomach destroys most microorganisms. The pH becomes progressively alkaline from the duodenum to the ileum.
ii. **Small intestine:** In the **small intestine**, protection is provided by the presence of bile salts that disrupt bacterial membranes and the fast flow of the intestinal contents that hinders microbial attachment to mucosal cells.
iii. **Ileum: The ileum** contains a rich and varied flora and in the large intestine, the bulk of the contents is composed of bacteria. Abundant resident microflora in the large bowel also contributes significantly to protection.

c. Upper Respiratory Tract

i. **Architecture of the nose:** In the upper respiratory tract, nasal hairs keep out large airborne particles that may contain microorganisms. The very architecture of the nose prevents entry of microorganisms to a large extent, the inhaled particles being arrested at or near the nasal orifices. Those that pass beyond are held by the mucus lining the epithelium, and are swept back to the pharynx where they tend to be swallowed or coughed out.
ii. **Sticky mucus:** The sticky mucus covering the respiratory tract acts as a trapping mechanism for inhaled particles.
iii. **Ciliary motion:** Ciliary motion transports the trapped organisms back up the respiratory tract to the external openings.
iv. **Cough reflex:** Cough reflex is an important defence mechanism of the respiratory tract and propels the organisms away from the lungs.
v. **Mucopolysaccharide:** Nasal and respiratory secretions contain **mucopolysaccharide** capable of combining with influenza and certain other viruses. When organisms enter the body via mucus membrane, they tend to be taken up by phagocytes and are transported into regional lymphatic channels that carry them to the lymph nodes. Particles that manage to reach the pulmonary alveoli are ingested by the phagocytic cells present there.

d. Genitourinary Tract

i. **Normal flow of urine: The normal flow of urine** flushes the urinary system, carrying microorganisms away from the body.

ii. **Spermine and zinc:** Spermine and zinc present in the semen carry out antibacterial activity. Because of short urethra, bladder infection is more common in females.

iii. **Acidity of the adult vagina:** The low pH (acidity) of the adult vagina, due to fermentation of glycogen in the epithelial cells by the resident aciduric bacilli, provides an inhospitable environment for colonization by pathogens. A thick mucus plug in the cervical opening is a substantial barrier.

e. Conjunctiva

i. ***Lachrymal* fluid:** Conjunctiva is continually being assaulted by microbe-laden dust and is kept moist by the continuous flushing action of tears (lachrymal fluid). The eyes become susceptible to infection when lachrymal secretions are absent. Tears contain large amounts of lysozyme, lactoferrin, and sIgA and thus provide mechanical as well as physical protection.

ii. **Lysozyme:** Tears contain the antibacterial substance lysozyme, first described by Fleming (1922). It is a basic protein of low molecular weight which acts as a muraminidase. Lysozyme is present in tissue fluids and in nearly all secretions except cerebrospinal fluid, sweat and urine. It acts by splitting certain polysaccharide components of the cell walls of susceptible bacteria. In the concentrations seen in tears and other secretions, lysozyme is active only against some nonpathogenic gram-positive bacteria. However, it occurs in phagocytic cells in concentrations high enough to be lethal to many pathogens.

2. Antibacterial Substances in Blood and Tissues

Many microbial substances are present in the tissue and body fluids. These are nonspecific. These molecules all show the characteristics of innate immunity—there is no specific recognition of the microorganisms and the response is not enhanced on re-exposure to the same antigen.

a. Complement System

The complement system possesses bactericidal activity and plays an important role in the destruction of pathogenic bacteria that invade the blood and tissues.

b. Other Substances

Several substances possessing antibacterial properties have been described in blood and tissues. These substances possess antibacterial properties demonstrable experimentally but their relevance in the natural context is not clearly understood.

These include:

1. **Beta lysine:** A relatively thermostable substance active against anthrax and related bacilli.
2. **Basic polypeptides** such as leukins extracted from leukocytes and plakins from platelets.
3. **Acidic substances**, such as lactic acid found in muscle tissue and in the inflammatory zones; and
4. **lactoperoxidase in milk.**

C. Interferon

The production of interferon is a method of defence against viral infections. These are a family of antiviral agents produced by live or killed viruses and certain other inducers. Interferon has been shown to be more important than specific antibodies in protection against and recovery from certain acute viral infections. Tissues and body secretions contain other antiviral substances.

3. Microbial Antagonisms

The skin and mucous surfaces have resident bacterial flora which prevent colonization by pathogens. Invasion by extraneous microbes may be due to alteration of normal resident flora, causing serious diseases such as staphylococcal or clostridial enterocolitis or candidiasis following oral antibiotics. The extreme susceptibility of germ free animals of all types of infections is an example of the importance of normal bacterial flora in native immunity.

4. Cellular Factors in Innate Immunity

Natural defense against the invasion of blood and tissues by microorganisms and other foreign particles is mediated to a large extent by phagocytic cells which ingest and destroy them. **Phagocytic cells,** originally discovered by Metchnikoff (1883), were classified by him into **microphages (polymorphonuclear leukocytes) and macrophages.**

Macrophages

Macrophages consist of histiocytes which are the wandering ameboid cells seen in tissues, fixed reticuloendothelial cells and monocytes of blood. Monocytes enter the blood, and differentiate after they reach the capillaries of a particular tissue. In connective tissue they are known as **histiocytes,** in kidneys as **mesangial cells,** in bones as **osteoclast,** in brain as **microglial cells,** in lungs as **alveolar macrophages,** in liver as **Kupffer cells,** and in spleen, lymph nodes and thymus as **sinus lining macrophages.**

Phagocytic cells reach the sites of inflammation in large numbers, attracted by chemotactic substances, and ingest particulate materials. Capsulated bacteria, such as pneumococci, are not readily phagocytosed except in the presence of opsonins. They are more readily phagocytosed when trapped against a firm surface such as the alveolar wall than when they are free in tissue fluids.

Phagocytosis

Phagocytosis is part of the innate immune response, during which microorganisms, foreign particles, and cellular debris are engulfed by phagocytic cells such as neutrophils and monocytes in the circulation, and macrophages and neutrophils in interstitial spaces.

Stages of Phagocytosis

Phagocytosis consists of three stages:

1. Attachment of the microorganisms or foreign material to the phagocytic cell.
2. Ingestion by the cell and drawn into the cell by **endocytosis**. Once internalized, the bacteria are trapped within phagocytic vacuoles (**phagosomes**) in the cytoplasm.
3. Degradation of the foreign material or microorganism within the phagocytic cells.

The phagosome containing the material to be destroyed fuses with a granule-containing lysosome, generating a **phagolysosome.** The bacteria are subjected to the action of the lytic enzymes in the phagolysosome and are destroyed.

Some bacteria, such as mycobacteria and brucellae, resist intracellular digestion and may actively multiply inside the phagocytic cells. Phagocytosis in such instances may actually help to disseminate infection to different parts of the body. The importance of phagocytosis in protection against infection is evidenced by the enhanced susceptibility to infection seen either when the phagocytic cells are depleted, as in agranulocytosis, or when they are functionally deficient, as in chronic granulomatous disease.

In nonspecific defence against viral infections and tumors a class of lymphocytes called natural killer (NK) cells are important. They selectively kill virus infected cells and tumor cells. NK cells are activated by interferons.

5. Inflammation

If the surface chemical and physiologic defences of the body are breached by a pathogen, inflammation can result, which is an important, nonspecific defence mechanism. Sequences of events in acute inflammation in response to an injury will be:

1. Vasodilation.
2. Increased vascular permeability.
3. Emigration of leukocytes.
4. Chemotaxis.
5. Phagocytosis.

Vasodilation

The inflammatory response causes the normally tight junctions between endothelial cells lining the capillaries, and between epithelial cells of the mucosal surface, to reversibly separate. Increased blood flow to injured area provides increased delivery of plasma proteins, neutrophils, and phagocytes (**vasodilation**).

Increased Vascular Permeability

Protein-rich exudates containing immunoglobulins and complement moves into injured area (**increased permeability)**. Neutrophils and macrophages adhere to endothelial cells of capillaries.

Emigration of Leukocytes

Leukocytes squeeze through gaps created by contraction of endothelial cells (**emigration of leukocytes**).

Chemotaxis

Neutrophils and macrophages move to site of injury in response to gradient of chemotactic mediators released by injured tissue (**chemotaxis)**.

Phagocytosis

Phagocyte attaches to the microorganisms and engulfs it by endocytosis and microorganisms are degraded by oxygen radicals and digestive enzymes (**phagocytosis**), whereas others (natural killer cells) limit the infection by releasing the compounds toxic to microorganisms. Inflammation is also accompanied by an increased concentration of serum proteins called acutephase proteins.

6. Fever

Following infection a rise of temperature is a natural defense mechanism. It not merely helps to accelerate the physiological processes but may, in some cases, actually destroy the infecting pathogens. For example, antibody production and T cell proliferation are more efficient at higher body temperature than at normal levels. Before penicillin era, therapeutic induction of fever was employed for the destruction of *Treponema pallidum* in patients suffering from syphilis. Fever aids recovery from viral infections by stimulating the production of interferon.

7. Acute Phase Proteins

A sudden increase in the plasma concentration of certain proteins, collectively termed '**acute phase proteins**' occurs as a result of infection or tissue injury. These include **C reactive protein (CRP), mannose binding protein, alpha-1-acid glycoprotein, serum amyloid P component and many others**. The alternative pathway of complement is activate by CRP and some other acute phase proteins. They are believed to enhance host resistance, prevent tissue injury and promote repair of inflammatory lesions.

ii. Acquired Immunity

Acquired immunity refers to the resistance that an individual acquires during his lifetime. Acquired immunity can be obtained by natural or artificial means and actively or passively. Acquired immunity is of two types: active immunity and passive immunity. (Fig. 12.1, Table 12.1).

a. Active Immunity

Active immunity is induced after contact with foreign antigens. It is also known as ***adaptive immunity*** as it represents an adaptive response of the host to a specific pathogen or other antigen. This involves the active functioning of the host's immune apparatus leading to the synthesis of antibodies and/or the production of immunologically active cells.

Immune Response

a. Primary Response

Active immunity sets in only after a **latent period** which is required for the immunological machinery to be set in motion. There is often a **negative phase** during the development of active immunity during which the level of measurable immunity may actually be lower than it was before the antigenic stimulus. This is because the antigen combines with any pre-existing antibody and lowers its level in circulation. Once developed, the active immunity is **long lasting**.

b. Secondary Response

If an individual who has been actively immunized against an antigen, experiences the same antigen subsequently, the immune response occurs more quickly and abundantly than during the first encounter. This is known as **secondary response**. This implies that the immune system is able to retain for long periods the memory of a prior antigenic exposure and to produce a secondary type of response when it encounters the same antigen again. This is known as **immunological memory.** Active immunization is more effective and confers better protection than passive immunization.

Types of Active Immunity

1. Natural active immunity
2. Artificial active immunity

1. Natural Active Immunity

Natural active immunity results from either a clinical or an inapparent infection by a microbe. The immune system responds by producing antibodies and activated lymphocytes that inactivate or destroy the antigen. Such immunity is usually **long lasting** but the duration varies with the type of pathogen. The immunity is **life-long** following many viral diseases such as chickenpox or measles. A large majority of adults in the developing countries possess natural active immunity to poliomyelitis due to repeated inapparent infections with the polioviruses during childhood.

The immunity appears to be **short-lived** in some viral diseases, such as **influenza or common cold**. In case of **influenza**, short-lived immunity is due to the ability of the virus to undergo antigenic variation, so that immunity following first infection is not effective against second infection due to an antigenically different virus. In **common cold,** the apparent lack of immunity is because the same clinical picture can be caused by infection with a large number of different viruses. In general, the immunity following bacterial infection is generally less permanent than that following viral infections. Some, such as typhoid fever, induce durable protection.

In some infections like **syphilis and malaria**, the immunity to reinfection lasts only as long as the original infection remains active. Once the disease is cured, the patient becomes susceptible to the infection again. This special type of immunity known as **'premunition'** or **infection-immunity**. In **chancroid,** another venereal disease, caused by *Haemophilus ducreyi,* there does not appear to be any effective immunity as the patient may develop lesions following reinfection even while the original infection is active.

A. Active immunity

1. Natural—Clinical
   - Subclinical infection
2. Artificial
   Vaccination—Live
   - Killed

B. Passive immunity

1. Natural—Through placenta
   - Through breast milk
2. Artificial—Immune serum
   - Immune cells

Fig. 12.1: Types of immunity

2. Artificial Active Immunity

Artificial active immunity is the resistance induced by vaccines. **Vaccines** are preparations of live or killed microorganisms or their products used for immunization. Vaccines are made with either: 1. Live, attenuated microorganisms, 2. Killed microorganisms, 3. Microbial extract, 4. Vaccine conjugates, 5. Inactivated toxoids.

Table 12.1: Comparison of active and passive immunity

Active immunity	*Passive immunity*
1. Produced actively by host's immune system	1. Received passively. No active host participation
2. Induced by infection or by immunogens	2. Readymade antibody transferred
3. Durable effective protection	3. Transient, less effective
4. Immunity effective only after lag period, i.e. time required for generation of antibodies and immunocompetent cells.	4. Immediate immunity
5. Immunological memory present	5. No memory
6. Booster effect on subsequent dose	6. Subsequent dose less effective
7. 'Negative phase' may occur	7. No negative phase
8. Not applicable in the immunodeficient	8. Applicable in immunodeficient

Both bacterial and viral pathogens are targeted by these diverse means.

Examples of Vaccines

Bacterial vaccines

a. Live (BCG vaccine for tuberculosis)
b. Killed (Cholera vaccine)
c. Subunit (Typhoid Vi-antigen)
d. Bacterial products (Tetanus toxoid)

Viral Vaccines

a. Live
  - Oral polio vaccine—Sabin
  - 17D vaccine for yellow fever
  - MMR vaccine for measles, mumps, rubella.

b. Killed
  - Injectable polio vaccine—Salk
  - Neural and non-neural vaccines for rabies
  - Hepatitis B vaccine

c. **Subunit—**Hepatitis B vaccine

Live Vaccines

Live vaccines initiate an infection without causing any injury or disease. The immunity following live vaccine administration, therefore, parallels that following natural infection. However, it may be of a **lower order** than induced by infection. In, general, live vaccines are **more potent** immunizing agents than killed vaccines The **immunity lasts for several years** but booster doses may be necessary. Live vaccines may be administered **orally** (as with the Sabin vaccine for poliomyelitis) **or parenterally** (as with the measles vaccine).

Killed Vaccines

Killed vaccines are usually **safe** and generally **less immunogenic** than live vaccines, and protection lasts only for a **short period**. They have, therefore, to be administered repeatedly, generally at least two doses being required for the production of immunity. The first is known as the **primary dose** and the subsequent doses as **booster doses**. Killed vaccines are usually administered by **subcutaneous or intramuscular route.** Parenteral administration provides humoral antibody response. Antibody response to killed vaccines is improved by the addition of '**adjuvants**', for example, aluminum phosphate adjuvant vaccine for cholera.

b. Passive Immunity

The immunity that is transferred to a recipient in a 'readymade' form is known as **passive immunity**. Here the recipient's immune system plays no active role. There is no lag or latent period in passive immunity, protection being effective immediately after passive immunization. There is no negative phase. The immunity is transient, usually lasting for days or weeks, only till the passively transmitted antibodies are metabolized and eliminated. There is no secondary type response in passive immunity. Rather, subsequent administration of antibodies is less effective due to immune elimination. When a foreign antibody is administered a second time, it is eliminated more rapidly than initially. Following the first injection of an antibody (such as horse serum), its elimination is only by metabolic breakdown but during subsequent injections its elimination is much quicker because it combines with antibodies to horse serum that would have been produced following its initial injection. The usefulness of repeated passive immunization is limited by this factor of immune elimination. This happens when foreign (horse) serum is used and when human serum is used immune elimination is not a problem.

Main Advantage of Passive Immunity

i. **The prompt availability** of large amount of antibody.
ii. It is employed where **instant immunity** is required as in case of **diphtheria, tetanus, botulism, rabies, hepatitis A and B** following exposure because of its immediate action

1. Natural Passive Immunity

This is the resistance passively transferred from mother to baby through the placenta. After birth, immunoglobulins are passed to the newborn through the **breast milk.** The **human colostrum**, is rich in IgA antibodies which are resistant to intestinal digestion, gives protection to the neonate up to three months of age.

The human fetus acquires some ability to synthesize antibodies (lgM) from about the twentieth week of life but its immunological capacity is still inadequate at birth. It is only by about the age of three months that the infant acquires a satisfactory level of immunological independence. Until then, maternal antibodies give passive protection against infectious diseases to the infant.

Transport of antibodies across the placenta is an active process and, therefore, the concentration of antibody in the fetal blood may sometimes be higher than that seen in the mother. Protection so afforded will ordinarily be adequate against all the common infectious diseases in the locality. Therefore, most pediatric infections are more common after the age of three months when maternal immunoglobulins disappear than in younger infants.

By active immunization of mothers during pregnancy, it is possible to improve the quality of passive immunity in the infants because pregnant woman's antibodies pass across the placenta to her fetus. Immunization of pregnant women with tetanus toxoid is recommended for this purpose in countries where neonatal tetanus is common.

2. Artificial Passive Immunity

Artificial passive immunity is the resistance passively transferred to a recipient by the administration of antibodies. Although this type of immunity is immediate, it is short lived and lasts only a few weeks to a few months. The agents used for this purpose are pooled human gamma globulin, hyperimmune sera of animal or human origin and convalescent sera. These are used for prophylaxis and therapy.

Types of immunoglobulin preparations: Two types of immunoglobulin preparations are available for passive immunization.

A. Human Immunoglobulins

a. Human normal immunoglobulin
b. Human specific immunoglobulin

a. Human Normal Immunoglobulin

Human normal immunoglobulin (HNIG) is used to provide temporary protection against **hepatitis A** infection for travelers to endemic areas and to control institutional and hosehold outbreaks of hepatits A and to prevent **measles** in highly susceptible individuals. HNIG also protects those with **agammaglobulinemia.**

b. Specific (Hyperimmune) Human Immunoglobulin

These preparations are made from the plasma of patients who have recovered recently from an infection or are obtained from individuals who have been immunized against a specific infection. Preparations of specific immunoglobulins are available for passive immunization against tetanus (human tetanus immunoglobulin; HTIG), hepatitis B (HBIG), human rabies immunoglobulin (HRIG), varicella-zoster immunoglobulin (ZIG) and Antivaccinia immunoglobulin (AVIG).

Human immune serum does not lead to any hypersensitivity reaction, therefore, there is no immune elimination and its half-life is more than that of animal sera. It has to be ensured that all preparations from human sera are free from the risk of human immunodeficiency virus (HIV), hepatitis B, hepatitis C and other viruses.

B. Nonhuman (Antisera)

The term **aniserum** is applied to materials prepared in animals. **Equine hyperimmune sera** such as antitetanus serum (ATS) prepared from hyperimmunized horses used to be extensively employed. They gave temporary protection but disadvantage is that may give rise to hypersensitivity and immune elimination. Since human immunoglobulin prepararations exist only for a small number of diseases, antitoxins prepared from nonhuman sources (against tetanus, diphtheria, botulism, gas gangrene and snake bite) are still the main stay of passive immunization.

Indications of Passive Immunization

1. **To provide immediate protection** to a nonimmune host exposed to an infection and lack active immunity to that pathogen and when there is insufficient time for active immunization to take effect, e.g. exposure to a toxin or poison.
2. **Treatment of some infections.**
3. **For the suppression of active immunity** when it may be injurious, e.g. administration of anti-Rh (D) IgG to Rh-negative mother, bearing Rh-positive baby at the time of delivery to prevent isoimmunization.
4. **Immunocompromised or immunodeficient individuals,** e.g. children with hypogammaglobulinemia, individuals with AIDS, patients receiving chemotherapy, organ transplant recipients receiving immunosuppressive therapy.

Combined Immunization

Combined immunization is a combination of active and passive methods of immunization which is sometimes employed. For example, it is often undertaken in some diseases such as tetanus, diphtheria, rabies. Ideally, whenever passive immunization is employed for immediate protection, combined immunization is to be preferred, as in the protection of a nonimmune individual with a tetanus prone wound. The person exposed to tetanus may be injected ATS on one arm and tetanus toxoid on the other arm with separate syringe followed by full course of tetanus toxoid. Similarly, AIDS and diphtheria toxoid can also be practiced.

Adoptive Immunity

Injection of immunologically competent lymphocytes is known as *adoptive immunity* and does not have general application. Instead of whole lymphocytes, an extract of immunologically competent lymphocytes, known as the **'transfer factor'**, can be used. This has attempted in the treatment of certain types of diseases for example, lepromatous leprosy.

MEASUREMENT OF IMMUNITY

It is not possible to measure accurately the level of immunity in an individual. Estimates of immunity are generally made by statistical methods using large numbers of individuals.

Demonstration of the Specific Antibody

Immunity can be tested by a simple method by relating its level to some convenient indicator, such as demonstration of the specific antibody which is not always reliable because the immune response to a pathogen consists of the formation of antibodies to several antigens present in it, as also to the production of cellular immunity.

A variety of techniques can be used to demonstrate antibodies such as agglutination, precipitation, complement fixation, hemagglutination inhibition, neutralization, ELISA and others. In the absence of exact information as to which antigen of the pathogen constitutes the *'protective antigen''* serological attempts to measure immunity are at best only approximations.

In Vitro or *In Vivo* Methods

In some instances, as in diphtheria where pathogenesis is due to a well defined antigen (the toxin), the level of immunity can be assayed by *in vitro* or *in vivo* (Schick test) methods. Skin tests for delayed hypersensitivity and *in vitro* tests for CMI give an indication of immunity where protection is associated with cell mediated immunity.

LOCAL IMMUNITY

Besredka (1919-24), proposed the concept of local immunity and it has gained importance in the treatment of infections which are localized or where it is operative in combating infection at the site of primary entry of the pathogen. Local immunity is conferred by secretory immunoglobulin A (*secretory IgA*) produced locally by plasma cells present on mucosal surfaces or in secretory glands. There appears to be a selective transport of such antibodies between the various mucosal surfaces and secretory glands.

Examples

1. Poliomyelitis immunization

In poliomyelitis, active immunization provides systemic immunity with the killed vaccine. The antibodies neutralize the virus when it enters the bloodstream. But it does not prevent multiplication of the virus at the site of entry, the gut mucosa, and its fecal shedding. Natural infection or immunization with the live oral vaccine provide local intestinal immunity.

2. Influenza immunization

Similarly, in influenza, immunization with the killed vaccine evokes humoral antibody response and the antibody titer in respiratory secretions is often not high enough to prevent infection. Natural infection or the live influenza vacicine administered intranasally provides local immunity.

HERD IMMUNITY

It is the level of resistance of a community or a group of people to a particular disease and is relevant in the control of epidemic diseases. When a large number of individuals in a community *(herd)* are immune to a pathogen the herd immunity is said to be satisfactory. Herd immunity provides a human barrier to the spread of the disease in the human herd. It can be affected by several factors such as the environment and the strength of an individual's immune system.

Low Herd Immunity

Epidemics are likely to occur on the introduction of a suitable pathogen when herd immunity is low which is due to the presence of large numbers of susceptible individuals in the community.

High Level of Herd Immunity

Eradication of communicable diseases depends on the development of a high level of herd immunity rather than on the development of a high level of immunity in individuals. For herd immunity to operate well in a community or a country, vaccine uptake rates must exceed 90 percent.

KNOW MORE

Lysozyme

A number of enzymes cleave peptidoglycan and cause bacterial lysis. These enzymes fall into three general categories:

1. Endo-N-acetyl hexosamines, known as lysozymes.
2. Endopeptidases.
3. Amidases.

Acute Phase Proteins

Certain proteins in the plasma, collectively termed 'acute phase proteins', increase in concentration in response to early 'alarm' mediators such as interleukin-1(IL-1), IL-6 and tumor necrosis factor (TNF), released as a result of infection or tissue injury.

KEY POINTS

- Immunity refers to the resistance exhibited by the host towards injury caused by microorganisms and their products.
- Innate or natural immunity is the resistance to infections which an individual possesses by virtue of his genetic or constitutional make up.
- Factors influencing the level of immunity are age, hormonal influences and sex, nutrition and stress
- Mechanisms of innate immunity
  - Mechanical barriers and surface secretions.
  - Antibacterial substances in blood and tissues.
  - Microbial antagonisms.
  - Cellular factors in innate immunity.
  - Inflammation.
  - Fever.
  - Acute phase proteins.
- Acquired immunity is of two types: (i) active immunity and (ii) passive immunity
- Natural active immunity results from either a clinical or an inapparent infection by a microbe.
- Artificial active immunity is the resistance induced by vaccines.
- The immunity that is transferred to a recipient in a 'readymade' form is known as passive immunity.
- Herd immunity is the level of resistance of a community or a group of people to a particular disease.

IMPORTANT QUESTIONS

1. Define and classify immunity. Discuss mechanisms of innate immunity.
2. Tabulate the differences between active and passive immunity.
3. Write short notes on:
   a. Innate immunity
   b. Active immunity
   c. Passive immunity
   d. Herd immunity.

FURTHER READING

Gabay C, Kushner I. Acute phase proteins. N Engl J Med 1999; 340:448.

Kwaitkowsky D. Susceptibility to infection. Br Med J 2000; 321:1061.

Roitt IM. Essential Immunology. London 8th (Edn): Blackwell Scientific 1994.

Weir DM, Stewart J. Immunology, 8th (Edn). Edinburgh: Churchill Livingstone, 1997.

CHAPTER 13

Antigens

Learning Objectives

After reading and studying this chapter, you should be able to:

- Describe haptens, heterophile antigens and superantigens.

INTRODUCTION

Antigens *(antibody generator)* are the substances that can stimulate an immune response and, given the opportunity, react specifically by binding with the effector molecules (antibodies) and effector cells (lymphocytes). The ability of a molecule to function as an antigen depends on its size and structural complexity.

Immunogen: A protein or carbohydrate that challenges the immune system and that can initiate an immune response is called an **immunogen**.

TYPES OF ANTIGEN

Two attributes of antigenicity are (1) induction of an immune response (immunogenicity), and (2) specific reaction with antibodies or sensitized cells (immunological reactivity). Based on the ability of antigens to carry out these two functions, they may be classified into different types.

A. Complete Antigen

Complete antigen is able to induce antibody formation and produce a specific and observable reaction with antibody so produced.

B. Haptens (Incomplete Immunogen)

Haptens (Latin *haptein,* to grasp) are low-molecular-weight molecules which cannot induce an immune response when injected by themselves (i.e. they are not *immunogenic)* but can do so when covalently coupled to a large protein molecule, called the *carrier* molecule. Conjugation of a hapten to a carrier protein makes the hapten immunogenic because the carrier protein can be processed and presented to specific T cells, as a result, both hapten-specific and carrier-specific antibodies can be made.

Example

Penicillin

One example of a hapten is penicillin. By itself penicillin is not antigenic. However, when it combines with certain serum proteins of sensitive individuals, the resulting molecule does initiate a severe and sometimes fatal allergic immune reaction. In these instances the hapten is acting as an antigenic determinant on the carrier molecule.

Types of Haptens

Haptens may be simple or complex.

1. **Simple haptens:** *Simple haptens* are nonprecipitating. They can inhibit precipitation of specific antibodies by the corresponding antigen or complex hapten. Simple haptens are univalent, since it is assumed that precipitation requires the antigen to have or more antibody combining sites.
2. **Complex haptens:** *Complex haptens* can precipitate with specific antibodies, complex haptens are polyvalent.

ANTIGENIC DETERMINANT OR EPITOME

The smallest unit of antigenicity is known as the *antigenic determinant* or *epitope* (Fig. 13.1). Each antigen can have several **antigenic determinant sites** or **epitopes**. Chemically, determinants include sugars, organic acids and bases, amino acid side chains, hydrocarbons and aromatic groups. Within a protein, an epitope may be formed by a specific sequence (**sequential or linear epitope**) and may be present as a single linear segment of the primary sequence or a three dimentional structure (**configurational epitope**) formed by bringing together on the surface residues from different sites of the peptide chain during its folding into the tertiary structure.

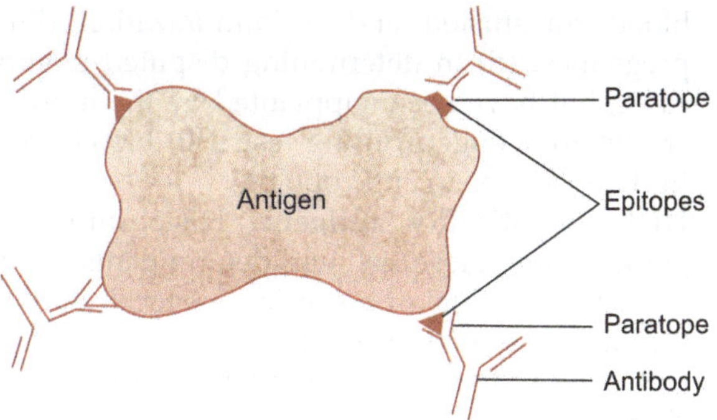

Fig. 13.1: Epitopes of antigen and paratopes of antibody

T cells recognize sequential epitopes, while B cells identify the tertiary configuration of the conformational epitopes. The combining area on the antibody molecule, corresponding to the epitope, is called the *paratope* (Fig. 13.1).

Valence

The number of antigenic determinant sites on the surface of an antigen is its **valence.** The valence determines the number of antibody molecules that can combine with the antigen at one time. The antigen is **monovalent** or **multivalent**.

a. Monovalent

If one determinant site is present, the antigen is **monovalent**.

b. Multivalent

Most antigens, however, have more than one determinant site or more than one copy of the same epitope and are termed **multivalent**. Multivalent antigens generally elicit a, stronger immune response than do monovalent antigens.

DETERMINANTS OF ANTIGENICITY

A number of properties have been identified which make a substance antigenic but the exact basis of antigenicity is still not clear.

1. Size
2. Chemical nature
3. Foreignness
4. Susceptibility to tissue enzymes
5. Antigenic specificity
6. Species specificities
7. Isospecificities
8. Autospecifity
9. Organ specificity
10. Heterogenetic (heterophile) specificity

1. Size

Molecular Weight

Antigenicity depends upon the molecular weight. Very large molecules such as the crustacean respiratory pigment hemocyanin are very powerful antigens and are widely used in experimental immunology. Particles with low molecular weight (less than 5000) are nonantigenic or feebly so. Low molecular weight substances may be rendered antigenic by adsorbing them on large inert particles such as **bentonite or kaolin.**

Some low molecular weight chemical substances such as picryl chloride, formaldehyde and drugs such as aspirin, penicillin and sulphonamides may be antigenic and appear to contradict the requirement that an antigen be large. These substances are highly antigenic, particularly, if applied to the skin. The reason for this appears to be that the complex of such a substance, acting as hapten, with a tissue protein acting as a carrier, forms a complete antigen. This phenomenon has important implication in the development of certain types of hypersensitivity.

2. Chemical Nature

Not all molecules are immunogens. In general, proteins are the best immunogens and carbohydrates are weaker immunogens. Lipids and nucleic acids are poor immunogenic. Their antigenicity is enhanced by combination with proteins. A certain amount of chemical complexity is required, e.g. amino acid homopolymers are less immunogenic than heteropolymers containing two or three different amino acids. That probably explains why proteins which are composed of about 20 different amino acids are better antigens than polysaccharides which have only four or five monosaccharide units. However, all proteins are not antigenic. A well known exception is gelatin, which is nonimmunogenic due to its structural unstability.

3. Foreignness

Some antigens are more effective in inducing an immune response than others. Only antigens which are '**foreign**' to the individual **(nonself)** induce an immune response because host distinguishes self from nonself and normally does not respond to self. In general, the greater the difference between the antigen (Ag) and similar molecules in the host's body, the greater the immune response that is generated.

4. Susceptibility to Tissue Enzymes

Only those substances which can be metabolized and susceptibility to the tissue enzymes behave as antigens. Antigens introduced into the body are degraded by the host into the fragments of appropriate size containing antigenic determinants. Phagocytes and intracellular enzymes appear to play an essential role in breaking down antigens into immunologic fragments.

Substances unsusceptible to the tissue enzymes such as polystyrene latex are not antigenic. Substances that are insoluble in body fluid and not metabolized are not antigenic. Substances very rapidly broken down by tissue enzymes are also not antigenic. Polypeptides consisting of L-amino acids are antigenic while synthetic polypeptides composed of D-amino acids which are not metabolized in the body are not antigenic.

5. Antigenic Specificity

Chemical Groupings

Foreignness of a substance to an animal can depend on the presence of chemical groupings that are not normally found in the animal's body. It was first demonstrated by Obermayer and Pick and confirmed by Karl Landsteiner that the basis of antigenic specificity is stereochemical. The classic contributions of Landsteiner and his cotemporaries demonstrated the specificity of antibodies produced to hapten-protein conjugates. Antigenic specificity varies with the position of antigenic determinant, i.e. whether it is in *ortho, meta* or *para* positions. The influence of spatial configuration of the determining group was shown by differences in antigenic specificity of the *dextro, levo* and *meso* isomers of substances such as tartaric acid. Thus, both the chemical nature and the three-dimensional structure of the hapten play major roles in determining the specificity of antigen-antibody interaction.

Antigenic specificity is not absolute. In many cases an antibody specific for one antigen may display significant cross-reactivity for an apparently unrelated antigen.

6. Species Specificities

Tissues of all individuals in a species possess species specific antigens. Thus, human blood proteins can be differentiated from animal proteins by specific antigen-antibody reaction. However, some degree of cross-reactivity is seen between antigens from related species. This immunological relationship parallels phylogenetic relationships and it has been (i) used in tracing relationships between species. (ii) Species- specific antigens also possess forensic applications in the identification of species of blood and of seminal stains (iii) Phylogenetic relationships are reflected in the extent of cross-reaction between antigens from different species that cause hypersensitivity. An individual sensitized to horse serum will react with serum from other equines but may not do so with bovine serum.

7. Iso-specificities

Isoantigens or alloantigens are antigens found in some but not all members of a species. These are able to produce alloantibodies or isoantibodies in individuals who are free from the antigens. On the basis of isoantigens a species may be divided into different groups.

Examples of Isoantigens

i. **Human erythrocytes antigens:** The best example of isoantigens is human erythrocytes antigens on the basis of which all humans can be divided into different blood groups; A, B, AB and O. Each of these groups may be further divided into Rh-positive or Rh-negative. This carries clinical importance (i) in blood transfusion and isoimmunization during pregnancy, (ii) in determining disputed paternity cases, but have been supplanted by the more discriminatory fingerprinting tests (iii) blood groups find application in anthropology.
ii. **Histocompatibility antigens:** Histocompatibility antigens are those cellular determinants specific to each individual of a species. Histocompatibility typing is essential in organ/tissue transplantation from one individual to another within a species. These antigens are associated with plasma membrane of tissue cells and are responsible for evoking immunological response against graft unless it is antigenically identical to that of the recipient.These antigens are encoded by genes known as histocompatibility genes which collectively constitute major histocompatibility complex (MHC). MHC products present on the surface of leukocytes are known as human leucokyte-associated (HLA) antigens. These have been studied extensively in organ transplantation. Major histocompatibility antigens in man and mouse are known as HLA and H2 respectively.

8. Autospecificity

Sequestrated antigens: Autologous or self-antigens are ordinarily nonantigenic but there are exceptions. Certain self-antigens are present in closed system and are not accessible to the immune apparatus and these are known as **sequestrated antigens.**

Example of Sequestrated Antigens

1. **Lens protein:** Sequestrated antigens that are not normally found free in circulation or tissue fluids (such as **lens protein** normally confined within its capsule) are not recognized as self-antigens. When the antigen leaks out, following penetrating injury, it may induce an immune response causing damage to the lens of the other eye.
2. **Sperm:** Similarly, antigens that are absent during embryonic life and develop later (such as **sperm**) are also not recognized as self-antigens. When the sperm antigen enters the circulation, it is immunogenic and is believed to be the pathogenesis of orchitis following mumps. This is one of the mechanisms of pathogenesis of autoimmune diseases.

9. Organ Specificity

Some organs such as brain, kidney and lens protein of different species, share the same antigens. These antigens are known as organ-specific antigens, characteristic of an organ or tissue and found in different species. Injection of heterologous organ-specific antigens may induce an immune response damaging the particular organs or tissue in the host.

Example

Neuroparalytic complications: The neuroparalytic complications following antirabic vaccination using sheep brain vaccines are a consequence of brain-specific antigens shared by sheep and man. The sheep brain antigens induce an immunological response in the vaccines, damaging their nervous tissue due to the cross-reaction between human and sheep brain antigens.

10. Heterogenetic (Heterophile) Specificity

Same or closely related antigens occurring in different biological species, classes and kingdoms are known as heterogenetic or heterophile antigens. Examples of such heterophile antigens.

Examples of Heterophile Antigens

i. **Forssman antigen:** Forssman antigen which is a lipid carbohydrate complex widely distributed in man, animals, birds, plants and bacteria. It is absent in rabbits, so anti-Forssman antibody can be prepared in these animals.
ii. **Weil-Felix reaction:** It is an agglutination test in which patient sera are tested for agglutinins to the O antigens of certain nonmotile Proteus strains OX19, OX2 and OXK. Cross-reaction between O antigen of these strains of *Proteus* and certain rickettsial antigens is the basis of this test.
iii. **Paul-Bunnell test:** During infectious-mononucleosis heterophile antibodies appear in the serum of the patient. These antibodies agglutinate sheep erythrocytes. This test is known as Paul-Bunnel test.
iv. **Cold agglutinin test:** Agglutination of human O group erythrocytes at 4°C by the sera of patients suffering from primary atypical pneumonia.
v. **Agglutination of Streptococcus MG:** Agglutination of *Streptococcus* MG by the sera of the patients of primary atypical pneumonia.

TOLEROGENS

Antigens do not always exhibit immunogenicity or evoke antibody formation. An antigen presented at one concentration might induce specific immunological unresponsiveness or tolerance in some instances, while at another concentration it might promote immunity. An antigen that induces tolerance is referred to as **tolerogen**.

Types of Tolerance

Two forms of tolerance can be defined-

A. Natural tolerance
B. Acquired tolerance

A. Natural Tolerance

The nonresponse to self-molecules is due to natural tolerance and it appears during fetal development when the immune system is being formed. Therefore, an individual's immune system does not normally, react against self-antigen. If this tolerance breaks down and the body responds to self-molecules then an autoimmune disease will develop.

B. Acquired Tolerance

Acquired tolerance arises when a potential immunogen induces a state of unresponsiveness to itself.

BIOLOGICAL CLASSES OF ANTIGENS

Depending on their ability to induce antibody formation, antigens are classified as **T cell dependent (TD)** and **T cell independent (TI) antigens**.

Antibodies are produced by B cells and their derivatives, plasma cells. For the full expression of this function, however, the cooperation of T lymphocytes is necessary.

1. T Cell Independent (TI) Antigens

A number of antigens will stimulate specific immunoglobulin production directly, without the apparent participation of T cells. Such antigens are called **T cell independent (TI) antigens**. TI antigens are present on the surface of infectious organisms. Mitogens and T independent antigens have an inherent ability to drive B cells into division and differentiation. An example is the lipopolysaccharide of gram-negative organisms or the pneumococcal polysaccharide of the various species of *S. pneumoniae*. The immune response generated to these antigens tends to be similar on each exposure, i.e. IgM is the antibody and the response shows little memory.

2. T Cell Dependent (TD) Antigens

Many antigens do not stimulate antibody production without the help of T lymphocytes are called **T cell dependent (TD) antigens** TD antigens are structurally more complex, such as erythrocytes, serum proteins and a variety of protein-hapten complex. These antigens first bind to the B cell, which must then be exposed to T cell derived lymphokines, i.e. helper factors, before antibody can be produced. They are immunogenic over a wide range and do not cause tolerance readily.

T dependent responses rely on T cells and their products to control the antibody class, affinity and memory. They induce the full gamut of immunoglobulin isotypes—IgM, IgG, IgA and IgE. They produce immunological memory, require preliminary processing and are rapidly metabolized in the body.

SUPERANTIGENS

Superantigens are bacterial proteins which can interact with antigen-presenting cells (APCs) and T cells in nonspecific manner. This activity does not involve the endocytic processing required for typical antigen presentation but instead occurs by concurrent association with MHC class II molecules of the APCs and the Vβ domain of the T cell receptor. This interaction activates a large number of T cells (10%) than conventional antigens (about 1%), explaining the massive cytokine expression and immunomodulation. The antigens which provoke such a drastic immune response are termed **superantigen.**

Examples

1. **Staphylococcal enterotoxins:** Toxic shock syndrome toxin, exfoliative toxin and some viral proteins.
2. **Association with diseases:** Superantigens should be considered possible chronic associates in such diseases as rheumatic fever, arthritis, Kawasaki syndrome, atopic dermatitis, and one type of psoriasis.

KNOW MORE

Antigens

Determinants of Antigenicity

Foreignness: In general, the antigenicity of a substance is related to the degree of its foreignness. Antigens from other individuals of the same species are less antigenic than those from other species. Antigens from related species are less antigenic than those from distant species.

This was first recognized by Ehrlich that an individual normally does not mount an immune response against his own normal constituent antigens, who postulated the concept of 'horror autotoxicus' (which means fear of self-poisoning). The theory of clonal selection, first proposed by Sir Frank Macfarlane Burnet, postulated that during the prenatal period, the developing lymphoid system learns to distinguish "self" from "nonself". Several mechanisms may be responsible for this self-nonself discrimination. As a result, individuals normally, develop a long-term tolerance to molecules that are components of their own cells. Breakdown of this homeostatic mechanism results in autoimmunization and autoimmune disease.

☞ KEY POINTS

- Antigens are the substances that can stimulate an immune response and given the opportunity, react specifically by binding with the effector molecules (antibodies) and effector cells (lymphocytes).
- Types of antigen are: (A) Complete antigen; (B) Haptens (incomplete immunogen).
- Determinants of antigenicity are (1) Size; (2) Chemical nature; (3) Foreignness; (4) Susceptibility to tissue enzymes; (5) Antigenic specificity; (6) Species; pecificities; (7) Isospecifities; (8) Autospecificity; (9) Organ specificity; 10. Heterogenetic (heterophile) specificity.
- Tolerogens: An antigen that induces tolerance is referred to as tolerogen.
- Biological classes of antigens are T cell dependent (TD) and T cell independent (TI) antigens.
- Superantigens are bacterial proteins which can interact with antigen-presenting cells (APCs) and T cells in nonspecific manner. The antigens which provoke such a drastic immune response are termed superantiges .

IMPORTANT QUESTIONS

1. What is an antigen? Discuss briefly various determinants of antigenicity.
2. Write short notes on:
   i. Haptens
   ii. Heterophile antigens

FURTHER READING

Herman A, Kappler JW, et al. Superantigens: mechanisms of T cell stimulation and role in immune responses. Annu Rev Immunol 1991;9:745-72.

Roitt I, et al. Immunology, 6th edn Mosby, 2001.

CHAPTER 14

Antibodies—Immunoglobulins

LEARNING OBJECTIVES

After reading and studying this chapter, you should be able to:

- Define antibody and draw labeled diagram of immunoglobulin.
- Describe structure and functions of IgG, IgA and IgM.
- Discuss properties of IgM, IgG, IgA, IgD, and IgE.
- Draw labeled diagram of IgG, IgM and IgA.

INTRODUCTION

Towards the end of the 19th century, von Behring and Kitasato in Berlin found that the serum of an approximately immunized animal contained specific neutralizing substances or antitoxins. This was the first demonstration of the activity of what are now known as antibodies or immunoglobulins.

Antibody or immunoglobulin (Ig): An **antibody or immunoglobulin (Ig)** is a glycoprotein that is made in response to an antigen, and can recognize and bind to the antigen that caused its production.

Immunoglobulins provide a structural and chemical concept, while the term **'antibody'** is a biological and functional concept. **All antibodies are immunoglobulins, but all immunoglobulins may not be antibodies.**

Physiochemical Properties of Antibodies

1. **Electrophoretic mobility:** Serum proteins can be separated into **soluble albumins** and insoluble globulins by fractionation of immune sera by half saturation with ammonium sulfate. Globulins could be separated into water soluble **pseudoglobulins** and **insoluble euglobulins**. Most antibodies were found to be euglobulins. Serum glycoproteins can be separated according to their charge by movement through a gel in an electric field and classified as **albumin** and **globulin** (alpha-1, alpha-2, beta, and gamma globulin) (Fig. 14.1). Tiselius and Kabat (1938) showed that antibody activity was associated with the gammaglobulin fraction. The term **gammaglobulin** thereafter became synonymous with **'antibody'.**
2. **Sedimentation and molecular weight:** Most antibodies are sedimented at 7S (MW 150,000-180,000) based on sedimentation studies. Some heavier antibodies-19S globulins (MW about 900,000) were designated as M or macroglobulins.
3. **Physicochemical and antigenic structure:** On the basis of physicochemical and antigenic structure Igs can be divided into five distinct classes or isotypes namely IgG, IgA, IgM, IgD and IgE. Within certain classes there are subclasses that show slight differences in structure and function from other.

ANTIBODY STRUCTURE

Papain Digestion

Antibody structure monomers have a Y shape with antigen-binding site at the end of each arm of the Y. is FC region.

Porter, Edelman, Nisonoff and their colleagues poineered studies involving the cleavage of the immunoglobulin molecule, which have led to a detailed picture of its structure. Rabbit IgG antibody to egg albumin was digested by **papain** in the presence of cysteine, each molecule of immunoglobulin is split by papain into three parts (Fig. 14.2). Two of the fragments are identical and are referred to as **Fab (Fragment antigen-binding)** because they retain the immunoglobulin's ability to bind specifically to an antigen. The third fragment **Fc (Fragment crystallizable)** which can be crystallized, does not bind antigen, but contributes to the biologic activity.

Pepsin Digestion

When treated with **pepsin,** a 5S fragment is obtained, which is composed essentially of **two Fab fragments**

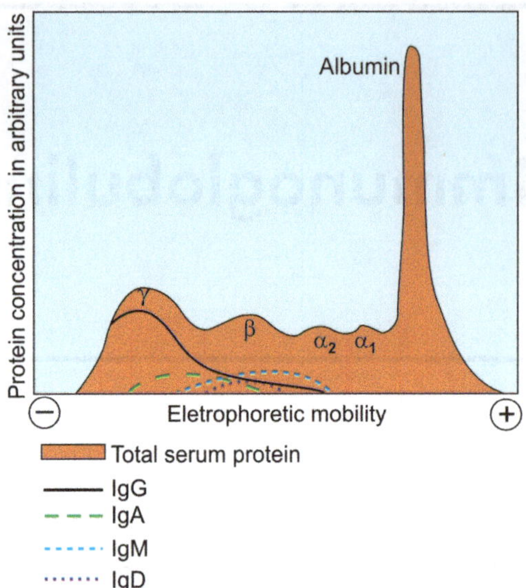

Fig. 14.1: Electrophoretic pattern of human serum showing the main components

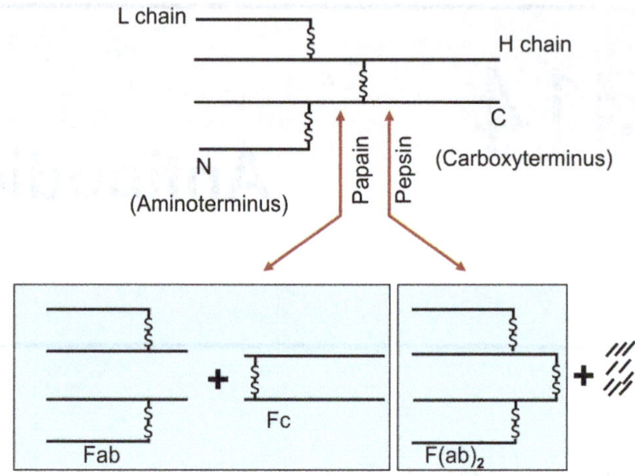

Fig. 14.2: Basic structure of an immunoglobulin molecule and the fragments obtained by the cleavage by papain and pepsin

held together in position. It is bivalent and precipitates with the antigen. This fragment is called **F(ab)$_2$**. The Fc portion is digested by pepsin into smaller fragments (Fig. 14.2).

Light and Heavy Chains

All immunoglobulins are composed of the same basic units consisting of four chains: **two identical 'light' (L)** chains and **two identical heavy chains**. The L chain is attached to the H chain by a disulfide bond. The two H chains are joined together by 1-5 S-S bonds, depending on the class of immunoglobulins (Figs 14.3A and B). The smaller chains are called **'light' (L)** chains and the larger ones **'heavy' (H)** chains. The variable region contains the **antigen-binding site;** the **constant region** encompasses the entire Fc region as well as part of the Fab regions.

1. Classes of L Chains

The L chains are similar in all classes of immunoglobulins. **The L chain** has a molecular weight of approximately 25,000 and the H chain of 50,000. There are two classes of L chains, designated kappa (κ) and lambda (λ). Both are normally expressed in every individual. However, each B cell expresses (and each antibody contains) only one type of L chain—either κ or λ but not both. Kappa (κ) and lambda (λ) are named after Korngold and Lapari who originally described them. Kappa (κ) and lambda (λ) chains occur in a ratio of about 2:1 in human sera.

2. Classes of H Chains

The H chains are structurally and antigenically distinct for each class and are designated by the Greek letter corresponding to the immunoglobulin class as shown below:

Immunoglobulin class	*H chain*
IgG	gamma (γ)
IgA	alpha (α)
IgM	mu (μ)
IgD	delta (δ)
IgE	epsilon (ε)

In humans there are five classes of heavy chains designated by lower case Greek letters: gamma (γ), alpha (α), mu (μ), delta (δ), and epsilon (ε). The properties of these heavy chains determine respectively the five immunoglobulin classes—IgG, IgA, IgM, IgD. Each immunoglobulin class differs in its general properties, half-life, distribution in the body, and interaction with other components of the host's defensive systems.

3. Constant and Variable Regions

The antigen-combining site of the molecule is at its amino terminus. All immunoglobulin chains possess a **constant ("C") region** that determines the biologic action of an antibody, and a **variable ("V") region** that binds a unique epitope. Both light and heavy chains contain two different regions. Of the 220 amino acids, those constitute carboxy-terminal half of L chain occurs in a constant sequence. This part of the chain is called **Constant regions (C$_L$)**. Only two sequence patterns are seen in the constant region—those determining *kappa (κ) and lambda (λ)* specificities. On the other hand, the amino acid sequence in the amino terminal half of the chain is highly variable, the variability determining the immunological specificity of the antibody molecules. It is, therefore, called **variable region (V$_L$)**. Within the light chain variable domain are **hypervariable regions** or **complementarity-determining regions (CDRs)** that differ in amino acid sequence more frequently than the rest of the variable domain.

The H chain also has **Constant regions (C_H)** and **variable regions (V_H)**. While in the L chain the two regions are of equal length, in the H chains the variable region constitutes approximately only a fifth of the chain and is located at its amino terminus. The **variable regions (V_L and V_H)** from different antibodies do have different sequences. It is the variable regions (V_L, V_H) that when folded together form the **antigen binding sites**. The other domains of the heavy chains are termed constant domains and are numbered CH_1, CH_2, CH_3, and sometimes CH_4, starting with the domain next to the variable domain. It is the portion of H chains present in Fab fragment. H chains carry a carbohydrate moiety, which is distinct for each class of immunoglobulins.

4. Fc Fragment

The Fc fragment is composed of the carboxy terminal portion of the H chains. It can be crystallized, and is therefore called **Fc (fragment-crystallizable)**. This third fragment does not bind antigen, but contributes to the antibody's biologic activity, such as binding complement, transplacental passage, skin fixation and catabolic rate.

Functions of Fc

Binds complement leading to complement fixation.

- Binds to cell receptors (FcRs).
- Determines passage of IgG across the placental barrier.
- Determines skin fixation and catabolic rate.

Antigenic determinants that distinguish one class of antibody from another are also located on Fc fragment.

5. Immunoglobulin Domain

Each immunoglobulin peptide chain has internal disulfide links in addition to interchain disulfide bonds which bridge the H and L chains. These interchain disulfide bonds form loops in the peptide chain, and each of the loops is compactly folded to form a globular domain, each domain having a separate function. The variable region domains, V_L and V_H, are responsible for the formation of a specific antigen binding site. The CH_2 region in IgG binds $C1_q$ in the classical complement sequence, and the CH_3 domain mediates adherence to the monocyte surface. The areas of the H chain in the C region between the first and second C region domains (CH_1 and CH_2) is the **hinge region**. It is more flexible and is more exposed to enzymes and chemicals. Papain acts here to produce one Fc and two Fab fragments (Figs 14.3A and B).

IMMUNOGLOBULIN CLASSES

Human serum contain five classes of immunoglobulins—IgG, IgA, IgM, IgD and IgE in the descending order of the concentration. Table 14.1 shows their diferent features.

1. Immunoglobulin G (IgG)

1. This is the **major immunoglobulin** in human serum, accounting for about 80 percent of the total immunoglobulin pool.
2. It has a sedimentation coefficint of **7S** and a molecular weight of **150,000**.
3. It contains less carbohydrate than other immunoglobulins.

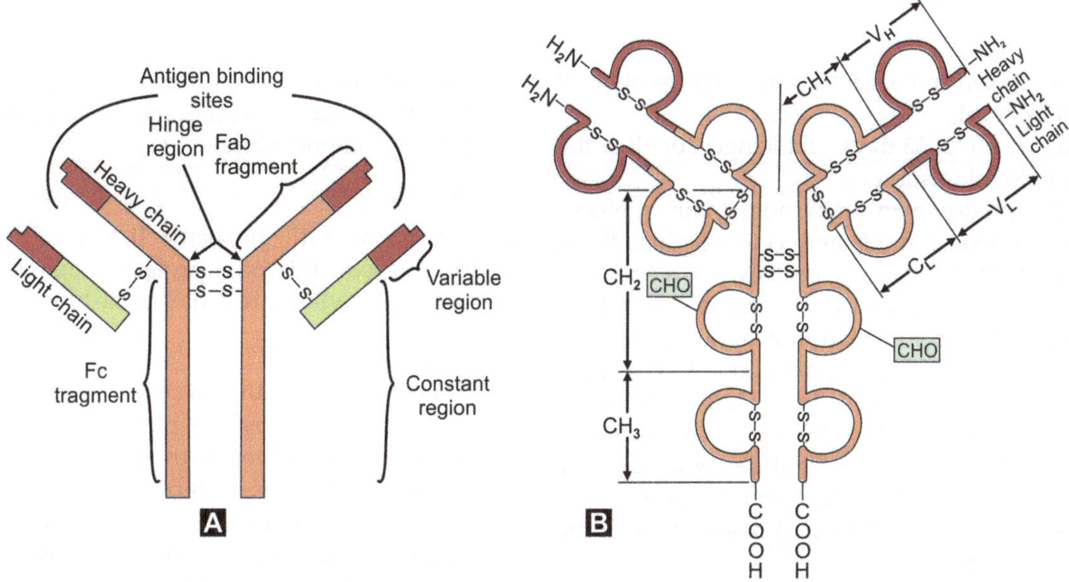

Figs 14.3A and B: (A) General structure of the IgG class antibody. The four-peptide chain structure of the IgG molecule composed of two identical heavy (H) and two identical light (L) chains linked by interchains disulfide bonds. (B) Loops formed by intrachain disulfide bonds are domains. Each chain has one domain in the variable region (V_H and V_L). Each light chain has one domain in the constant region (C_L) while each heavy chain has three domains in the constant region (CH_1, CH_2 and CH_3). Between CH_1 and CH_2 is the Hinge region

Property	IgG	IgA*	IgM	IgD	IgE
A. Physical properties					
1. Sedimentation coefficient(S)	7	7	19	7	8
2. Molecular weight in kilodaltons	150,000	160,000	900,000	180,000	190,000
3. Carbohydrate (%)	3	8	12	13	12
4. Number of four-chain units per molecule	1	1-3	5-6	1	1
B. Physiologic properties					
1. Normal adult serum concentration (mg/ml)	12	2	1.2	0.03	0.00004
2. Half-life (in days)	23	6	5	2-8	1-5
3. Daily production (mg/kg)	34	24	3.3	0.4	0.0023
4. Intravascular distribution (%)	45	42	80	75	50
C. Biologic properties					
1. Complement-fixation					
Classical	++	-	+++	-	-
Alternative	-	+	-	-	-
2. Placental transport to fetus	+	-	-	-	-
3. Present in milk	+	+	-	-	-
4. Selective selection by submucus glands	-	+	-	-	-
5. Anaphylactic hypersensitivity	-	-	-	-	++++
6. Heat stabilty	+	+	+	+	-
D. Major characterstics	Most abundant Ig; Longest half life: Crosses placenta; opsonizes antigen.	Protects mucosal surfaces	Very efficienct against bacteremia	Mainly lymphocyte receptor; major surface components of B cells	Initiates inflammation; raised in helminthic infections; causes allergy symptoms

Table14.1: Physical, physiologic and biologic properties of human serum immunoglobulins

* IgA may occur in 7S, 9S and 11S forms.

4. The normal serum concentration of IgG is about 8 **to 16 mg per ml.**
5. It has a half-life of **23 days**—the longest of all of the immunoglobulin isotypes.
6. IgG is the predominant immunoglobulin in blood, lymph, peritoneal fluid, and cerebrospinal fluid, and it is distributed nearly equally between extra- and intravascular spaces. Therefore, IgG is particularly suitable for passive immunization done by the transfer of serum containing antibodies (antiserum).
7. Four subclasses of IgG (Ig1, Ig2, Ig3, Ig4) have been recognized. Each subclass possesses a distinct type of γ chain which can be identified with specific antiserum. They constitute about 65 percent, 23 percent, 8 percent and 4 percent respectively of the total human IgG.
8. **Catabolism** of IgG is unique in that it varies with its serum concentration. When its level is raised, as in chronic malaria, kala-azar or myeloma, the IgG synthesized against a particular antigen will be catabolized rapidly and may result in the particular antibody deficiency. Conversely, in hypogammaglobulilnemia, the IgG given for treatment will be catabolized only slowly.

Functions of IgG

IgG is a very versatile molecule. It may be considered a general purpose antibody, protective against those infectious agents which are active in the blood and tissues.

Examples of its Functions

i. **Transfer from mother to fetus:** IgG is the only class of Igs that can cross the placenta and is responsible for the protection of the infant during first few months of life, thus conferring the mother's humoral immunity to infection to the fetus and neonate. However, 'subclass IgG2 does not cross the placenta. It is not synthesized by the fetus in any significant amount.
ii. **Opsonization:** IgG binds to microorganisms and enhances their phagocytosis. Extracellular killing of target cells coated with IgG antibody is mediated

through recognition of the surface Fc fragment by K cells bearing the appropriate receptors. IgG can aid natural killer (NK) cells in finding their targets. Interaction of IgG complexes with platelet Fc receptors probably leads to aggregation and vasoactive amine release.

iii. **Fixing to guinea pig skin:** IgG is the only Ig which has the property of fixing to guinea pig skin, but the significance of this property is not known.

iv. **Immunological reactions:** IgG participates in most immunological reactions such as complement fixation, precipitation and neutralization of toxins and viruses.

v. **Immobilize bacteria:** IgG can also immobilize bacteria by binding to their cilia or flagella.

vi. **Suppresses the homologous antibody synthesis:** Passively administered IgG suppresses the homologous antibody synthesis by a feedback process. This property is utilized for prevention of isoimmunization of Rh-negative mother bearing Rh-positive baby by administration of anti-Rh (D) IgG at the time of delivery.

2. Immunoglobulin A (IgA)

1. It is the second most abundant class, constituting about 10 to 13 percent of serum immunoglobulins.
2. The normal serum level is 0.6 to 4.2 mg per ml.
3. It has a half-life of 6-8 days.
4. IgA is the primary immunoglobulin found in external secretions, such as mucus, tears, saliva, gastric fluid, colostrum and sweat. It exists in different forms in these various solutions.
5. IgA occurs in two forms. **Serum IgA and secretory IgA (SIgA).**

Serum IgA

Serum IgA is monomeric (one four-chain unit) 7S molecule (MW about 160,000).

Secretory IgA (SIgA)

In contrast, IgA found on mucosal surfaces and in secretions is a dimer formed by two monomer units joined together at their carboxy terminals by a glycopeptide termed the **J chain (J for joining)**. This dimeric form is more important form, known as **secretory IgA (SIgA)**. Dimeric SIgA is synthesized by plasma cells situated near the mucosal or glandular epithelium. J chains are also found in other polymeric immunoglobulins such as IgM.

Secretory Component

Secretory IgA (SIgA) contains another glycine rich polypeptide called the **secretory component** or **secretory piece** (Fig. 14.4). This is not produced by lymphoid cells but by mucosal or glandular epithelial cells. The dimeric IgA molecules are released by plasma cells (mature B cells), and then bind to a receptor on the basal membranes of adjacent epithelial cells. This receptor is called the **poly-Ig receptor**. This receptor binds tightly to the IgA dimers, and transports them through the epithelial cells to extracellular fluids such as the mucus of the respiratory and digestive tracts. When the poly-Ig receptor-IgA dimmer complex arrives on the exterior surface of the mucosal cell, the poly-Ig receptor is cleaved. The portion of the receptor that stays attached to the IgA dimer is called the **secretory piece or secretory (S) component**.

The secretory component is resistant to degradation by digestive enzymes. The secretory piece is believed to **protect IgA from denaturation by bacterial proteases** in sites such as the intestinal mucosa which have a rich and varied bacterial flora. SIgA is a much larger molecule than serum IgA (11S; MW about 400.000).

Subclasses

There are two subclasses of IgA in humans: IgA_1 and IgA_2.

IgA_1: In serum, IgA_1 constitutes 80 percent to 90 percent of IgA while SIgA consists of about equal amounts of the two subclasses. Certain streptococci and pathogenic Neisseria produce proteases that specifically cleave the heavy chain of IgA_1.

IgA_2: IgA_2 lacks interchain disulfide bonds between the heavy and light chains. Though IgA_2 is a minor component of serum IgA, it is the dominant form in the secretions. It is resistant to such cleavage because it has a shorter hinge region and lacks the proline-rich site cleaved by the proteases.

Functions of IgA

i. **Local immunity:** Secretory IgA (SIgA) is selectively concentrated in secretions and on mucus surfaces forming an **'antibody paste'** and is believed to play an important role in local immunity against respiratory and intestinal pathogens.

ii. **Prevention of organisms entry into body tissues:** IgA antibodies may function by inhibiting the adherence of microorganisms to the surface of mucosal cells by covering the organisms and thereby preventing their entry into body tissues.

iii. **Newborn protection:** IgA present in breast milk provides the newborn with protection against infection.

iv. **Agglutination:** It can cause agglutination, and can also prevent viruses from entering cells.

v. **Alternative pathways activation:** IgA does not fix complement but it can activate the alternative pathways.

vi. **Phagocytosis and intracellular killing:** It promotes phagocytosis and intracellular killing of microorganisms.

3. Immunoglobulin M (IgM)

1. About 10 percent of normal serum Igs consists of this class.

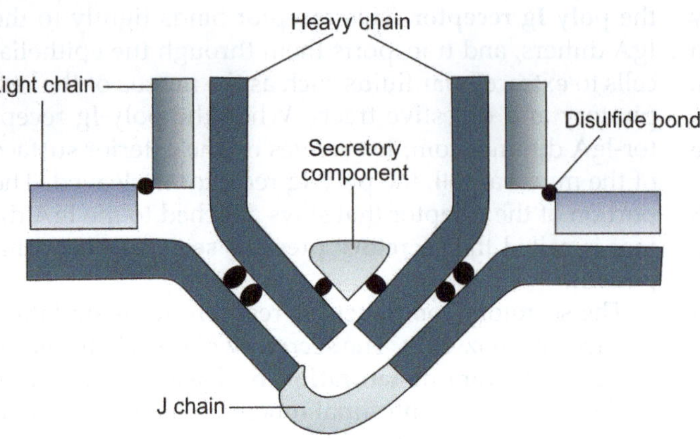

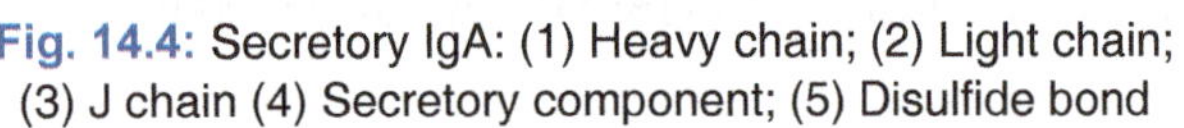
Fig. 14.4: Secretory IgA: (1) Heavy chain; (2) Light chain; (3) J chain (4) Secretory component; (5) Disulfide bond

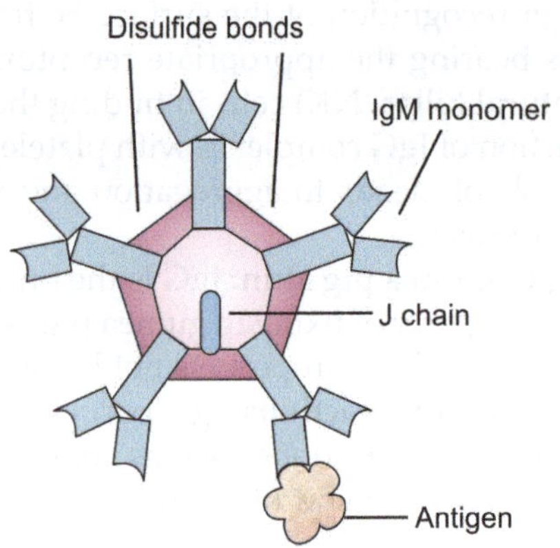

Fig. 14.5: IgM molecule, pentameric molecule, composed of five identical monomers

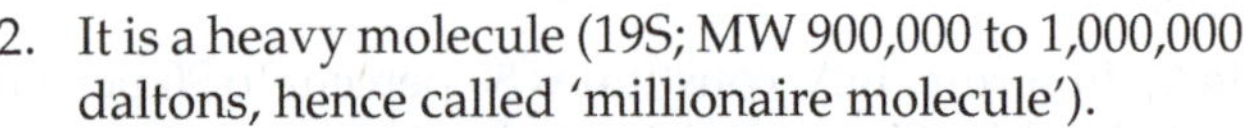

2. It is a heavy molecule (19S; MW 900,000 to 1,000,000 daltons, hence called 'millionaire molecule').
3. The normal serum level of IgM is 1.2 mg/ml.
4. It has a half-life of about 5 days.
5. IgM is the first immunoglobulin to appear after exposure to an antigen.
6. In the circulation, IgM exists as a pentamer of five four-chain units. The five identical IgM monomers are connected to each other by a polypeptide joining J chain. Polymerization of the subunits depends upon the presence of the J chain as with IgA. Monomeric IgM and IgD are present on the surface of mature, naive B cells.
7. IgM contains 10 Fab fragments, and thus 10 antigen-binding sites. Though the theoretical valency is ten, this is observed only with small haptens. The effective valency falls to five with larger antigens, probably due to steric hindrance.
   Most of IgM (80 percent) is intravascular in distribution.
8. Phylogenetically IgM is the oldest Ig class. IgM is the first class of antibody produced during the primary immune response. It is also the earliest to be synthesized by fetus beginning by about 20 weeks of age. As it cannot cross the placental barrier, the presence of IgM in the fetus or newborn indicates intrauterine infection. Its detection is, therefore, useful for the diagnosis of congenital infections such as syphilis, rubella, human immunodeficiency virus (HIV) infection and toxoplasmosis.
9. They are relatively short-lived hence their demonstration in the serum indicates recent infection.
10. Treatment of serum with 0.12 M 2-mercaptoethanol selectively destroys IgM without affecting IgG antibodies.This provides a simple method for differential estimation of IgG and IgM antibodies.
11. Isohemagglutinins (anti-A and anti-B) and antibodies to *S. Typhi* O antigen and Wassermann reaction antibodies in syphilis are usually IgM.
12. IgM agglutinates bacteria, activates complement by the classical pathway, and enhances the ingestion of pathogens by phagocytic cells. IgM is normally restricted to the intravascular space because of its high molecular weight.

4. Immunoglobulin D (IgD)

1. IgD has a monomer structure similar to IgG.
2. Its molecular weight is 180,000 daltons.
3. IgD is an immunoglobulin found in trace amounts in the blood serum (0.03 mg/ml).
4. Half-life is about 3 days.
5. IgD antibodies are abundant in combination with IgM on the surface of B cells and bind antigens, thus signaling the B cell to start antibody production.
6. Two subclasses of IgD (IgD1 and IgD2) are known.

5. Immunoglobulin E (IgE)

1. It resembles IgG structurally and also known as reagin antibody.
2. IgE is an 8S molecules (MW 19,000) and half-life of two days.
3. It is present in extremely low amounts in serum.
4. It exhibits unique properties such as **heat lability** (inactivated at 56°C in one hour).
5. It is susceptible to mercaptoethanol.
6. It does not pass the placental barrier.
7. IgE does not activate complement nor agglutinate antigens.
8. **Allergic reactions:** IgE molecules bind tightly by their Fc (stem) regions to receptors on mast cells and basophils, specialized cells that participate in allergic reactions. IgE may be elevated in allergic (atopic) individuals, and is responsible for many of the symptoms of allergies, bronchial asthma and even systemic anaphylaxis. Allergy mediated by IgE is termed a type I hypersensitivity response.
9. **Immunity against helminthic parasites:** This may have evolved to provide immunity against

helminthic parasites. Children living in insanitary conditions, with a high load of intestinal parasites, have high serum levels of IgE.

10. **Extravascular:** It is mostly found **extravascularly** in the lining of the respiratory and intestinal tracts.
11. **Protection against pathogens:** The physiological role of IgE appears to be protection against pathogens by mast cell degranulation and release of inflammatory mediators.

Role of Different Immunoglobulin Classes

IgG: Protects the body fluids
IgA: Protects the body surfaces
IgM: Protects the blood stream
IgE: Mediates type I hypersensitivity
IgD: Role not known.

ANTIGENIC DETERMINANTS ON IMMUNOGLOBULINS

Since Igs are high molecular weight glycoproteins, they can serve as very potent antigens. Differences in amino acid sequences between different Ig classes, subclasses and types determine the antigenic specificity of Igs. Igs carry three major categories of antigenic determinants.

1. Isotypes

These determinants are shared by all members of the same species ***(Iso is the same for each person)*** referring to the variations in the **heavy chain constant regions (CH)** associated with the different classes. Thus, different classes of immunoglobulins are differentiated on the basis of isotypic markers on H chains. Various H chain markers are gamma, mu, alpha, delta and epsilon in immunoglobulins IgG, IgM, IgA, IgD and IgE respectively. and subclasses (IgGI, IgG2, IgG3, IgG4, IgAI and IgA2) Similarly, light chains are also distinguished by isotypic markers into κ and λ.

2. Allotypes

Allotypes are individual specific determinants within a species. Some but not all individuals of a particular species possesses them. **(Every one [allo] of them cannot have the same IgG)**. These are distinct amino acid residues located primarily in γ and α H chains and κ L chains. **Allotypes are** the genetically controlled allelic forms of immunoglobulin molecules. Antibodies can be formed against an allotypic determinant by injecting Igs into another member of the species that does not possess the antigen. In humans a number of allotypic markers have been discovered on Igs.

Allotypic systems in humans are the **Gm system** (for gamma marker), Am on alpha heavy chains, and Km (originally designated as **In V system) on kappa lights chains**. Allotypic markers are useful in testing paternity and population genetics.

Gm system: The Gm is associated with the Fc portion of the IgG heavy chain. More than 25 Gm types have been identified so far.

Am: Genetic markers associated with IgA are called **Am.**

InV system: The InV system is associated with the **kappa light chain** and so has been renamed Km. Three **Km** allotypes have been identified.

To date, in the human system no allotypic markers have been found for lambda light chains or μ, δ or ε, heavy chains.

3. Idiotypes

Idiotypes refer to individual specific immunoglobulin molecules that differ in the hypervariable region of the Fab portion. ***(There are many different idiots)***. These are located on the V regions of L and H chains at or near the antigen-combining sites. These antigenic determinants are also called idiotopes analogus to epitope of classical antigens.

Abnormal Immunoglobulins

Other structurally similar proteins are seen in serum in many pathological processes, and sometimes even in healthy persons apart from antibodies in following pathological conditions.

A. Multiple myeloma
B. Heavy chain disease
C. Cryoglobulinemia

A. Multiple Myeloma

The abnormal plasma cells are **myeloma cells** which also collect in the solid part of the bone. The disease is called **"multiple myeloma".** Myeloma is a plasma cell dyscrasia in which there is unchecked proliferation of one clone of plasma cells, resulting in the excessive production of the particular immunoglobulin synthesised by the clone. Such immunoglobulins are, therefore, called **monoclonal.**

Waldenstrom's macroglobulinemia: Multiple myeloma may affect plasma cells synthesizing IgG, IgA, IgD or IgE. Myeloma involving IgM producing cells (lymphoplasmacytoid cells) is known as **Waldenstrom's macroglobulinemia**. In this condition, there occurs excessive production of the respective myeloma proteins (M proteins) and of their light chains (Bence-Jones proteins).

Bence-Jones proteins: In most patients, the myeloma cells also secrete excessive amounts of light chains. These excess light chains were first discovered in the urine of myeloma patients and were named **Bence-Jones proteins**, for their discoverer Bence-Jones (1847). Bence-Jones proteins can be identified in urine by its characteristic property of coagulation when heated to 50°C but redissolving at 70°C. These proteins are the light chains of immunoglobulins and so may occur as the kappa or lambda forms. But in anyone patient, the chain is either kappa or lambda only, and never both, being uniform in all other respects also.

B. Heavy Chain Disease

It is a lymphoid neoplasia characterized by the overproduction of the Fc parts of the immunoglobulin heavy chains.

C. Cryoglobulinemia

It is the presence in blood of cryoglobulin, which is precipitated in the microvasculature on exposure to cold. It is a condition in which there is the formation of a gel or a precipitate on cooling the serum, which redissolves on warming. It may not always be associated with disease but is often found in **myelomas, macroglobulinemias and autoimmune conditions such as systemic lupus erythematosus.** Most cryoglobulins consist of either IgG, IgM or mixture of the two.

Because of the monoclonal nature of Bence-Jones and other M proteins, they have been valuable models for the understanding of immunoglobulin structure and function.

KNOW MORE

Antibodies—Immunoglobulins

WHO (1964) endorsed the generic term **'immunoglobulin'** and was internationally accepted for **'proteins of animal origin endowed with known antibody activity and for certain other proteins related to them by chemical structure'.** The definition includes, besides antibody globulins, the abnormal proteins found in myeloma, macroglobulinemia, cryoglobulinemia and the naturally occurring subunits of immunoglobulins. Immunoglobulins constitute 20 to 25 percent of the total serum.

☞ KEY POINTS

- An antibody or immunoglobulin (Ig) is a glycoprotein that is made in response to an antigen, and can recognize and bind to the antigen that caused its production.
- An antibody molecule consists of two identical light chains and two identical heavy chains, which are linked by disulfide bonds. Each heavy chain has an amino-terminal variable region followed by a constant region.
- The heavy chain isotype determines the class of an antibody.
- Each of the domains in the immunoglobulin molecule has a characteristic tertiary structure called the **immunoglobulin fold.**
- Within the amino-terminal variable domain of each heavy and light chain are three complementarity-determining regions (CDRs). These polypeptide regions contribute the antigen-binding site of an antibody, determining its specificity.
- Antigenic determinants on immunoglobulins are: (1) Isotypes; (2) Allotypes; (3) Idiotypes
- Abnormal immunoglobulins are multiple myeloma, heavy chain disease C and cryoglobulinemia.

IMPORTANT QUESTIONS

1. Name various classes of immunoglobulins and describe structure and functions of IgG, IgA and IgM.
2. Write shot notes on:
   a. Immunoglobulin G (IgG)
   b. Immunoglobulin M (IgM)
   c. Immunoglobulin A (IgA)

FURTHER READING

Burton DR. Antibody: the flexible adapter molecule. Trends Biochem Sci 1990;15:64-9.

Goodman JW. Immunoglobulin Structure and Function. 7(Eds) In Basic and Clinical Immunology 1991;.

Roitt IM, Delves O. Encyclopedia of Immunology. London: Academic Press 1992.

CHAPTER 15

The Complement System

Learning Objectives

After reading and studying this chapter, you should be able to:

- Describe sequence of events when the classical pathway and the alternative pathway of the complement system is activated.
- Discuss biological effects of complement.
- Describe complement deficiencies and associated diseases.

Complement: The term **'complement' (C)** refers to a system of factors which occur in normal serum and are activated characteristically by antigen-antibody interaction and subsequently mediate a number of biologically significant consequences.

COMPLEMENT SYSTEM

The complement system is an alarm and a weapon against infection, especially bacterial infection. The complement system includes serum and membrane-bound proteins that function in both acquired and constitutive (natural) host defence system. These proteins are highly regulated and interact via a series of proteolytic cascades.

The complement system belongs to the group of biological effector mechanisms (called ***triggered enzyme cascades***) which also includes coagulation, fibrinolytic and kinin systems. Such biological cascades have distinct advantages. For example, each enzyme in the cascade is able to activate many molecules of the succeeding component providing for amplification of the response at each step. Every step has its own control mechanisms so that, the cascade can be regulated with precision.

Pfeiffer phenomenon: It was discovered by Pfeiffer (1894) that cholera vibrios were lysed when injected intraperitoneally into specifically immunized guinea pigs (bacteriolysis *in vivo* or Pfeiffer phenomenon). Bacteria were similarly lysed *in vitro* when added to the cell-free serum of immunized animals.

General Properties of Complement

1. **Present in the sera of all mammals and of most lower animals:** Complement is present in the sera of all mammals and also in that of most lower animals, including birds, amphibians and fishes. Hepatocytes, blood monocytes, epithelial cells of the gastrointestinal tract, and tissue macrophages synthesize complement proteins.
2. **Nonspecific serological reagent:** It is a nonspecific serological reagent in that complement from one species can react with antibodies from other species, though the efficiency of reaction is influenced by the taxonomic distance between the species.
3. **Serum molecules:** The complement system consists of approximately 30 serum molecules constituting nearly 10 percent of the total serum proteins and forming one of the major defence systems of the body. A series of circulating and self-cell surface regulatory proteins keep the complement system in check.
4. **Heat labile:** Complement as a whole is heat labile, its cytolytic activity undergoing spontaneous denaturation slowly at room temperature and being destroyed in 30 minutes at 56°C though some of its components are heat stable. A serum deprived of its complement activity by heating at 56°C for 30 minutes, is then said to be **'inactivated'**.
5. **Complement fixation, binding or consumption:** Complement (C) ordinarily does not bind to free antigen or antibody but only to antibody which has combined with its antigen. Various terms such as fixation, binding or consumption have been used to refer to the combination of C with bound immunoglobulin, leading to the activation of the classical C pathway.

   All classes of Ig do not fix complement. Only IgM, IgG3, 1 and 2 (in that order) fix complement, but not IgG4, IgA, IgD or IgE.
6. **Site of complement binding:** The site of complement binding is located on the Fc piece of the Ig

molecule (CH_2 — domain on IgG, CH_4 on IgM), and is expressed only when Ig is combined with its antigen. The fixation of complement is not influenced by the nature of antigens, but only by the class of immunoglobulins.

Components of Complement

The complement system comprises a group of serum proteins, many of which exist in inactive forms. The complement system consists of at least twenty chemically and immunologically distinct serum proteins comprising the complement components, the properdin system and the control proteins.

There are nine components of complement called C1 to C9. The fraction C1 occurs in serum as a calcium ion dependent complex, which on chelation with EDT A yields three protein subunits called C1q, r, and s. Thus, C is made up of a total of 11 different proteins. C fractions are named C1 to C9 in the sequence of the cascading reaction, except that C4 comes after C1, before C2. In normal serum C3 is present in the highest concentration (1.2 mg/ml) whereas C2 in the lowest concentration (0.015 mg/ml). Activation of the complement system can be initiated either by antigen-antibody complexes or by a variety of nonimmunologic molecules.

Complement components react in a specific sequence as a cascade either through the classical or alternative pathway. Both have same result, i.e. lysis or damage of target cell. Classical pathway is triggered by specific antigen-antibody complex; the alternative pathway can be initiated by endotoxin, lipopolysaccharides or zymosan (yeast cell wall).

Proteins unique to the alternative pathway are identified by capital letters other than "C" (for example, B, D, and P).

Complement is normally present in circulation in inactive form, but when its activity is induced by antigen-antibody reaction or other stimuli, complement components react in a specific sequence as a cascade either through the classical or alternative pathway.

Complement (C) is normally present in the body in an inactive form. Activation of the complement system can be initiated either by antigen-antibody complexes or by a variety of nonimmunologic molecules. Basically, the complement cascade is a series of reactions in which the preceding components act as enzymes on the succeeding components, cleaving them into dissimilar fragments—**larger** and **smaller. The larger fragments** usually join the cascade. **The smaller fragments** which are released often possess biological effects which contribute to defence mechanisms by amplifying the inflammatory process, increasing vascular permeability, inducing smooth muscle contraction, causing chemotaxis of leukocytes, promoting virus neutralization, detoxifying endotoxins and effecting the release of histamine from mast cells.

PRINCIPLE PATHWAYS OF COMPLEMENT ACTIVATION

Three principle pathways are involved in complement activation (classical pathway, alternate or properdin pathway, and lectin pathway) all of which converge on the activation of the third component C3. Sequential activation of complement components occurs via two main pathways, i.e. classical pathway, alternate or properdin pathway). The final steps that lead to a membrane attack are the same in all pathways.

A. Classical Complement Pathway

The classical pathway is so called because it was the first one identified. The chain of events in which C components react in a specific sequence following activation of C1 and typically culminate in immune cytolysis is known as the classical pathway (Fig. 15.1). It consists of the following steps:

1. **Antigen-antibody binding:** The first step is the binding of C1 to the antigen-antibody complex. The recognition unit of C1 is C1q, which reacts with the Fc piece of bound IgM or IgG. Following binding of antigen to antibody, the C1 complement component, which is composed of three proteins (q, r and s), attaches to the Fc portion of the antibody molecule through its C1q subcomponent. C1q has six combining sites. Effective activation occurs only when C1q is attached to immunoglobulins by at least two of its binding sites. One molecule of IgM or two molecules of IgG can therefore initiate the process. C1q binding in the presence of calcium ions leads to sequential activation of C1r and s. In the presence of calcium ions, a trimolecular complex (C1 qrs–Ag–Ab) that has esterase activity is rapidly formed.
2. **Production of C3 convertase:** Activated Cls is an esterase (Cls esterase), one molecule of which can cleave several molecules of C4–an instance of amplification. Activated C_1 cleaves C4 into two pieces C4a and C4b (C4 $\rightarrow$ C4a + C4b). C4a is an anaphylatoxin and C4b which binds to cell membrane along with C1.

   C14b in the presence of magnesium ions cleaves C2 into two pieces (C2 $\rightarrow$ C2a + C2b). C2a remains linked to cell bound C4b, and C2b which is released into fluid phase. The pieces recombine, forming C4b2a has enzymatic activity and is referred to as the classical pathway C3 convertase.
3. **Production of C5 convertase:** C3 convertase cleaves C3 into two fragments (C3 $\rightarrow$ C3a + C3b). C3a is soluble, and is an anaphylatoxin, and C3b which remains cell-bound along with C4b2a to form a trimolecular complex C4b2a3b which has enzymatic activity and is called C5 convertase of the classic pathway.
4. **Formation of the membrane attack complex (MAC):** The terminal stage of the classic pathway

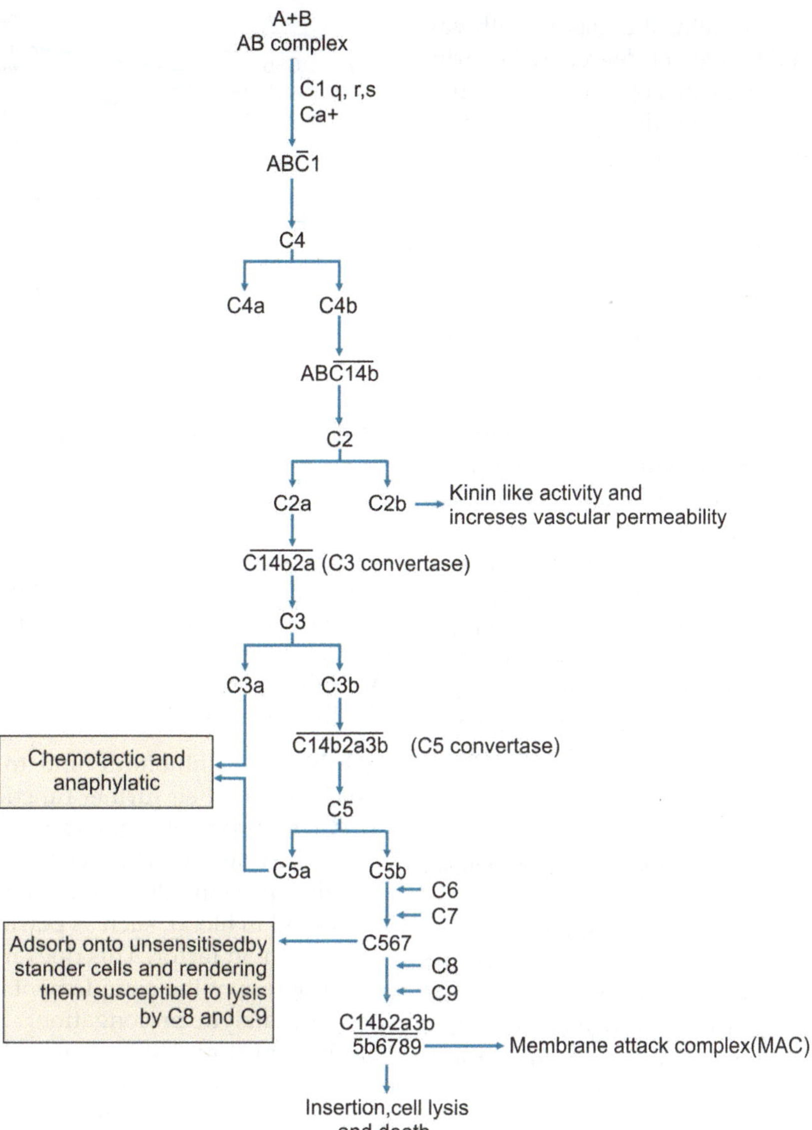

Fig. 15.1: Complement cascade—the classical pathway

involves creation of membrane attack complex (MAC), which is also called the lytic unit.

Initiation of membrane attack complex (MAC) assembly begins with cleavage of C5 by C5 convertase into C5a and C5b fragments (C5 → C5a + C5b). The C5a the most potent anaphylatoxin in the body and C5b, which continues with the cascade. C6 and C7 then join together. A heat stable trimolecular complex C567 is formed part of which binds to the cell membrane and prepares it for lysis by C8 and C9 which join the reaction subsequently. This complex (C5b67) inserts itself into the plasma membrane of the target cell. Most of C567 escape and serve to amplify the reaction by adsorbing onto unsensitized 'bystander cells' and rendering them susceptible to lysis by C8 and C9.

The unbound C567 has chemotactic activity, though the effect is transient due to its rapid inactivation. C8 and C9 then bind, forming the membrane attack complex (C5b6789) that creates a pore in the plasma membrane of the target cell. The mechanism of complement mediated cytolysis is the production of 'holes', approximately 100 A in diameter on the cell membrane. This disrupts the osmotic integrity of the membrane, leading to the release of the cell contents.

Lysozyme from the blood enters through the pore and digests the peptidoglycan cell wall causing the bacterium to lyse osmotically if the cell is a gram-negative bacterium. In contrast, gram-positive bacteria are resistant to the cytolytic action of the membrane attack complex because they lack an exposed outer membrane and the thick peptidoglycan prevents an attack on the plasma membrane.

Although the classical pathway is generally activated by the antigen-antibody complex or aggregated immunoglobulin, the classic pathway can also be activated to a lesser degree by heparin, DNA, certain retroviruses, mycoplasma, C-reactive protein (CRP), mannose binding protein (MBP), and certain "trypsin-like" proteases.

B. Alternative Complement Pathway

In the complement cascade the central process is the activation of C3, which is the major component of C. In the classical pathway, activation of C3 is achieved by C42 (classical C3 convertase). The activation of C3 without prior participation of C142 is known as the 'alternative pathway' (Fig. 15.2).

The alternate pathway of complement activation (the properdin pathway) does not require the formation of antigen-antibody complexes for activation. The first example of the alternative pathway was the demonstration by Pillemer (1954) of the 'properdin system' as a group of serum proteins contributing to antimicrobial defence without requiring specific antibodies. The activator in this system was zymosan, a polysaccharide from the yeast cell wall, but many other substances can also activate the pathway. These activators include bacterial endotoxins, IgA and D, the cobra venom factor and the nephritic factor (a protein present in the serum of glomerulonephritis patients).

1. **Production of alternative pathway C3 convertase. C5 convertase and MAC** (Fig. 15.2).

This pathway bypasses both the recognition unit and the assembly of the activation unit as described for the classic pathway. Instead, there are at least three normal serum proteins that, when activated together with C3, form a functional **C3 convertase** and a **C5 convertase**. These are factor B, factor D, and properdin (P).

The binding of C3b to an activator is the first step in the alternative pathway. Although C3b is present in the circulation but in the free state it is rapidly inactivated by the serum protein factors H and I. However, bound C3b is protected from such inactivation. The bound C3b, in the presence of Mg^{++}, interacts with plasma protein factor B forming C3bB which is also known as 'C3 proactivator convertase' to form a magnesium dependent complex 'C3b, B'. Factor B in the complex is cleaved by serum factor D (also called 'C3 proactivator convertase') into two fragments; Ba and Bb. Fragment Ba is released into the medium. Fragment Bb remains bound to C3b producing C3bBb. C3bBb acts as the alternate pathway C3 convertase, capable of producing more C3b. This enzyme C3bBb is extremely labile. The function of properdin (also called Factor P) a serum protein is to stabilize the C3 convertase, which hydrolyzes C3, leading to further steps in the cascade, as in the classical pathway (Fig. 15.1).

Regulation of the Complement System

Unchecked complement activity can cause not only exhaustion of the complement system but also serious damage to tissues. A series of circulating and self-cell surface regulatory proteins keep the complement system in check. Several in built control mechanisms regulate the complement cascade at different steps. These are mainly of two kinds:

A. Inhibitors
B. Inactivators

A. **Inhibitors:** Inhibitors bind to complement components and halt their further function:

1. **Cl esterase:** Normal serum contains an inhibitor of Cl esterase. This heat labile alpha neuraminoglycoprotein also inhibits many other esterases found in blood, such as plasmin, kininogen and the Hageman factor. This does not prevent the normal progress of the complement cascade but checks its autocatalytic prolongation.
2. **Vitronectin:** Vitronectin, also known as the S protein. The S protein present in normal serum binds to C567 and modulates the cytolytic action of the membrane attack complex.

B. **Inactivators:** Inactivators are enzymes that destroy complement proteins:

1. **Factor 1:** Control of C3b is necessary for the regulation of both the classic and the alternate pathways. Normal serum contains an endopeptidase, called Factor 1 which cleaves C3b and possibly C4b.
2. **Factor H:** Another beta globulin factor H acts in concert with factor I modulating C3 activation. It has a strong affinity for C3b, and, after binding C3b, it exerts its control.
3. **Anaphylatoxin inactivator:** Anaphylatoxin inactivator is an alpha globulin that enzymatically degrades C3a, C4a and C5a which are anaphylatoxins released during the C cascade.
4. **C4 binding protein:** It is a normal human serum protein that binds tightly to activated C4 and enhances C4b degradation.

Many other regulators of C have been reported such as the cell surface proteins, decay-accelerating factor (DAF), homologous restriction factor, complement receptor-1 (CR-1).

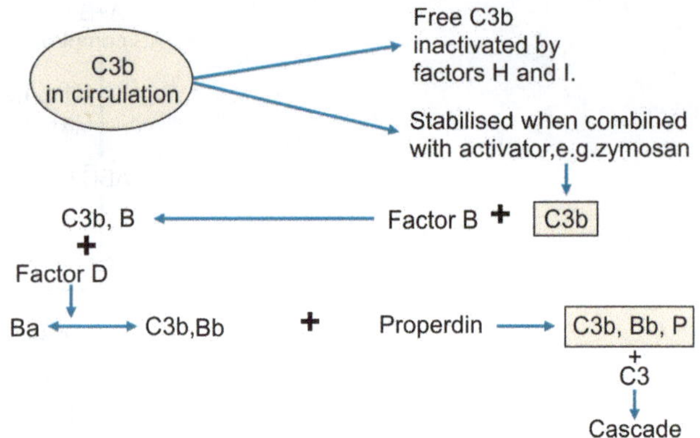

Fig. 15.2: Complement cascade—the alternative pathway

Biological Effects of Complement (C)

1. **Bacteriolysis and cytolysis:** Complement mediates immunological membrane damage. This results in bacteriolysis and cytolysis. Cells vary in their susceptibility to complement mediated lysis.
2. **Virus neutralization:** Neutralization of certain viruses requires the participation of C, e.g. neutralization of herpes virus by IgM antibody requires the participation of C1, C4 and possibly C3.
3. **Anaphylatoxins:** C fragments released during the cascade reaction help in amplifying the inflammatory response. The cleavage products of both pathways of complement activation, C3a and C5a are anaphylatoxic (histamine releasing) and chemotactic. C4a also has anaphylatoxin activity but is less potent even than C3a. C567 is chemotactic and also brings about reactive lysis.
4. **Immune adherence and opsonization:** An important function of C is to facilitate the uptake and destruction of pathogens by phagocytic cells. This opsonic effect is based on the presence on the surface of phagocytic cells, (macrophages, monocytes, neutrophils and others) of complement receptors or CRs. If immune complexes have activated the complement system, the C3b bound to them stimulate phagocytosis and removal of immune complexes. This facilitated phagocytosis is referred to as opsonization. C3b can act as a bridge to bring antibody-coated material into intimate contact with phagocytic cells, inducing their phagocytosis and destruction.
5. **Chemotaxis:** Any substance that attracts leukocytes to an area of inflammation is a chemotactic agent. Factors Ba (the split product from the alternate pathway) and C5a are both chemotactic for PMNs and macrophages, thus contributing to local inflammation. C5b67, the partially formed attack complex, also has been implicated as a chemotactic agent.
6. **Hypersensitivity reaction:** Complement participates in
   i. Type II hypersensitivity (cytotoxic) reactions.
   ii. Type III (immune complex) hypersensitivity reactions.
7. **Autoimmune diseases:** Serum C (complement) components are decreased in many autoimmune diseases such as systemic lupus erythromatosus and rheumatoid arthritis. They may, therefore, be involved in the pathogenesis of autoimmune diseases.

   C plays a major role in the pathogenesis of autoimmune hemolytic anemia, paroxysmal nocturnal hemoglobinuria and hereditary angioneurotic edema.
8. **Endotoxic shock:** Endotoxins can efficiently activate the alternative pathway of the complement cascade. There is massive C3 fixation and platelets adherence in endotoxic shock. Large scale platelet lysis and release of large amounts of platelet factor lead to disseminated intravascular coagulation (DIC) and thrombocytopenia. In endotoxic shock with gram-negative septicemia or dengue hemorrhagic fever may have a similar pathogenesis. Depletion of C protects against the Schwartzman reaction. Schwartzman reaction produced in rabbits by intravenous injection of endotoxin is a good model of excessive C3 activation.

QUANTITATION OF COMPLEMENT (C) AND ITS COMPONENTS

The lysis of RBCs provides the basis of measurement of 'C' activity. Measurement of the complement levels in the serum can be accomplished by estimating the highest dilution of the serum lysing sheep erythrocytes sensitized with antierythrocytic antibody. The hemolytic unit of C (CH50) may be defined as that amount of complement that lyses 50 percent of sensitized erythrocytes under defined conditions.

BIOSYNTHESIS OF COMPLEMENT

Various complement components are synthesized in different parts of the body, e.g. intestinal epithelium (C1), macrophages (C2, C4), spleen (C5, C8) and liver (C3, C6, C9). The site of synthesis of C7 is not known. The control mechanism that controls the synthesis of the complement component is not known.

COMPLEMENT DEFICIENCIES

Some animal and humans have been found to possess genetic defects that either result in a deficiency in a complement component or in a deficient regulatory system for the control of the activated components of complement. These generally lead to enhanced susceptibility to infectious diseases. Indeed, complement deficiencies have been associated with recurrent bacterial and fungal infections as well as with collagen-vascular inflammatory diseases.

A deficiency or dysfunction of C1 esterase inhibitor results in hereditary angiedema, an autosomal dominant heritable disease. Deficiency of C3b inactivator (factor 1), factor D and properdin predispose to recurrent infections.

Individuals with defects in the complement components C1q, C1r, C1s are predisposed to develop SLE and lupus nephritis. Deficiency in C3 leads to an increased susceptibility to bacterial infections and a predisposition to immune complex disease as deficiency in C2 and C4, both of which are located within the MHC region. The development of SLE-like symptoms in C1q knockout mice parallels the human situation.

Complement deficiency	Association with disease
C1 inhibitor	Hereditary angioneurotic edema
C1r	Systemic lupus erythematosus-like disease, frequently ftal from overewhelming infection
C2	Increased susceptibility to infections
C3	Recurrent bacterial infections
C4	Systemic lupus erythematosus-like disease
C5	Recurrent infections—lupus-like disease
C6,C7,C8	Recurrent infections: Disseminated gonococcal infections
C9	Not more susceptible to disease than other individuals in the general population`
Factor 1	Low C3 levels with recurrent bacterial infections

Table 15.1: Human genetic deficiencies of complement components and associated diseases

KNOW MORE

Complement Nomenclature

Several complement components are proenymes, which must be cleaved to form active enzymes. The complement system utilizes a unique nomenclature. Most complement plasma proteins are named with a capital "C," followed by a number (for example, C3). C is an abbreviation for the complement system. The components of the classical pathway are numbered from C1 to C9.

KEY POINTS

- The complement system comprises a group of serum proteins, many of which exist in inactive forms. It is present in all normal individuals in their blood.
- Complement activation occurs by the classical, alternative, or lectin pathways, each of which is initiated differently.
- The classical pathway is activated with the formation of soluble antigen-antibody complexes (immune complexes) or the binding of antibody to antigen on a suitable target, such as a bacterial cell.
- Activation of the alternative and lectin pathways is antibody-independent. These pathways are initiated by reaction of complement proteins with surface molecules of microorganisms. The lectin pathway is activated by lectins.
- Clinical consequences of inherited complement deficiencies range from increases in susceptibility to infection to tissue damage caused by immune complexes.

IMPORTANT QUESTIONS

1. Define complement. What is the sequence of events when the classical pathway of the complement system is activated?
2. Write short notes on:
   i. Alternative pathway of complement.
   ii. Biological effects of complement.

FURTHER READING

Clas F, Loos M. Complement in Health and Disease, In: Whaley K Lancaster: MTP Press 1987;201-25.

Colten HR, Rosen FS. Complement deficiencies. Annu Rev Immunol 1992;10:809-34.

Janeway CA, Travers P. Immunobiology, 2nd Edn. London: Current Biology 1996.

Liszewski MK, Ferris TC, et al. Control of the complement system. Adv Immuno1 1996;61:201-83.

Morgan BP, Walport MJ. Complement deficiency and disease. Immunol Today 1991;I2:301-6.

Tomlinson S. Complement defense mechanisms. Cure Opin Immunol 1993;5:83.

CHAPTER

16 Antigen-Antibody Reactions

Learning Objectives

After reading and studying this chapter, you should be able to:

- Differentiate between precipitation and agglutination
- Describe prozone phenomenon.
- Discuss mechanism and applications of precipitation reactions giving suitable examples.
- Describe types of precipitation reactions.
- Describe principle and applications of immunoelectrophoresis, radial immunodiffusion, counterimmunoelectrophoresis (CIE), rocket electrophoresis.
- Describe applications of agglutination reactions and their uses.
- Discuss principle and applications of agglutination reactions.
- Describe principle of complement fixation test.
- Discuss principle and clinical applications of immunofluorescence technique.
- Discuss principle, various types and clinical applications of ELISA technique.

INTRODUCTION

The antigen-antibody interaction is a bimolecular association that exhibits exquisite specificity. It is similar to an enzyme-substrate interaction, with an important distinction: it does not lead to an irreversible chemical alteration in either the antibody or the antigen and is therefore is reversible.

Uses

1. ***In vivo* or in the body**
   i. **Protection:** The *in vivo* interaction that occur in vertebrate animals are antibody absolutely essential in protecting the animal against the continuous onslaught of viruses, microorganisms and their products, certain macromolecules, and cancer cells.
   ii. **Basis of antibody-mediated immunity:** In the body, they form the basis of antibody-mediated immunity in infectious diseases, or of tissue injury in some types of hypersensitivity and autoimmune diseases.
2. ***In vitro* or in the laboratory:** Many of these same reactions can take place outside the animal body (*in vitro*) under controlled laboratory conditions and extensively used in diagnostic testing.
   i. These assays can be used **to detect the presence of either antibody or antigen**
   ii. **Vital roles in diagnosing diseases.**
   iii. **Monitoring the level of the humoral immune response.**
   iv. **Identifying molecules of biological or medical interest.**

ANTIGEN-ANTIBODY INTERACTIONS

The reactions between antigen and antibody occurs in three stantigenes:

1. **Primary stage:** In primary or initial interaction there is no visible effect and the reaction is rapid, occurs even at low temperatures and obeys the laws of physical chemistry and thermodynamics. The reaction is reversible, the combination between antigen and antibody being effected by the weaker intermolecular forces such as van der Waals' forces, ionic bonds and hydrogen bonding, rater than by the firmer covalent bonding. Free and bound antigen or antibody can be estimated separately in the reaction mixture by using a number of physical and chemical methods.
2. **Secondary stage:** The primary interaction in most instances, but not all, is followed by the secondary stage, leading to demonstrable events such as precipitation, agglutination, lysis of cells, killing of live antigens, neutralization of toxins and other biologically active antigens, fixation of complement, immobilization of motile organisms and enhancement of phagocytosis. When such reactions were discovered, it was believed that a different type of antibody was responsible for each type of reaction and the antibodies were designated by the reactions they were thought to produce.

The antibody causing agglutination was called *agglutinin,* that causing precipitation *precipitin* and the corresponding antigen, *agglutinogen, precipitinogen,* and so on. Zinsser's unitarian hypothesis (by the 1920s) replaced this view which held that an antigen gave rise to only one antibody, which was capable of producing all the different reactions depending on the nature of the antigen and the conditions of the reaction. Both these views are extreme and fallacious. The truth is that a single antibody can cause precipitation, agglutination and most of the other serological reactions but it is also true that an antigen can stimulate the production of different classes of immunoglobulins which are different in their reaction capacities as well as in other properties (Table 16.1).

3. **Tertiary reactions:** Some antigen-antibody reactions occurring *in vivo* initiate chain reactions that lead to neutralization or destruction of injurious antigens, or to tissue damage. These are tertiary reactions and include humoral immunity against infectious diseaseas well as clinical allergy and other immunological diseases.

GENERAL CHARACISTICS OF ANTIGEN-ANTIBODY REACTIONS

1. **Highly specific:** Antigen-antibody reaction is highly specific. In some cases antibody elicited by one antigen can cross-react with an unrelated antigen due to antigenic similarity or relatedness.
2. **Lock and key arrangement:** Both antigens and antibodies participate in the formation of agglutinates or precipitates. The molecules are held together in lock and key arrangement.
3. **No denaturation:** There is no denaturation of antigen or antibody during the reaction.
4. **Surface antigens:** During combination only surface antigens participate. Therefore, surface antigens are more relevant.
5. **Entire molecules react:** Entire molecules react and not fragments.
6. **Combination is firm and reversible:** The combination is firm but reversible. The firmness of the union is influenced by the **affinity** and **avidity** of the reaction.

   Affinity denotes the intensity of attraction between the antigen and antibody molecules. **Avidity** is the strength of the bond after the formation of the antigen-antibody complexes.

Table 16.1: Efficacy of the different immunoglobulin classes in various serological reactions

Serological reaction	*IgM*	*IgG*	*IgA*
1. Precipitation	Weak	*Strong*	Variable
2. Agglutination	*Strong*	Weak	Moderate
3. Complement fixation	Weak	*Strong*	Negative

7. **Combination in varying proportions:** Unlike chemicals with fixed valencies, antigens and antibodies can combine in varying proportions. Both antigens and antibodies are multivalent. Antibodies are generally bivalent, though IgM molecules may have five or ten combining sites. Antigens may have valencies up to hundreds.

ANTIGEN AND ANTIBODY MEASUREMENT

Numerous methods can be used to determine the presence or the amount of antigens and antibodies. Measurement may be in terms of **mass** (e.g. nitrogen) or more commonly as units or titer.

Titer

Antibody titer of a serum is the highest dilution of the serum which shows an observable reaction with the antigen in the particular test. It is usually expressed as the reciprocals of the dilution of the serum.

PARAMETERS OF SEROLOGICAL TESTS

Two important parameters of serological tests are sensitivity and specificity:

Sensitivity

Sensitivity is defined as the ability of a test to identify correctly all those who have the disease, i.e. "**true positives**". A 90 percent sensitivity means that 90 percent of the diseased people screened by the test will give a "true positive" result and the remaining 10 percent a "false negative" result.

Specificity

Specificity has been defined as the ability of a test to identify correctly those who do not have the disease i.e. "**true negatives**". A 90 percent specificity means that 90 percent of nondiseased persons will give "true negative" and remaining 10 percent false positive result, 10 percent of nondiseased persons screened by the test will be wrongly classified as "diseased" when they are not.

SEROLOGICAL REACTIONS

The study of antigen-antibody reactions *in vitro* is called ***serology*** **[serum and–ology].** Antigen-antibody reactions *in vitro* are known as **serological reactions**.

Types of Antigen and Antibody Reactions

A. Precipitation reactions
B. Agglutination reactions
C. Complement fixation test (CFT)
D. Neutralization tests
E. Opsonization
F. Immunofluorescence
G. Radioimmunoassay (RIA)
H. Enzyme-linked immunosorbent assay (ELISA)
I. Immunoelectroblot techniques

J. Immunochromatographic tests
K. Immunoelectronmicroscopic tests

A. Precipitation Reactions

Precipitation

When a soluble antigen combines with its antibody in the presence of electrolytes (NaCl) at a suitable temperature and pH, the antigen-antibody complex forms an insoluble **precipitate** and is called **precipitation.** Antibodies that thus aggregate soluble antigens are called **precipitins**. Precipitation is relatively less sensitive for the detection of antibodies. Precipitation can take place in liquid media or in gels such as agar, agarose or polyacrylamide.

Flocculation

When instead of sedimenting, the precipitate remains suspended as floccules, the reaction is known as **flocculation.**

Zone Phenomenon

A quantitative precipitation reaction can be performed by placing a constant amount of antibody in a series of tubes and adding increasing amounts of antigen to the tubes. This plot of the amount of antibody precipitated versus increasing antigen concentration (at constant total antibody) reveals three zones (Fig. 16.1): This is called **zone phenomenon**. Zoning occurs in agglutination and some other serological reactions.

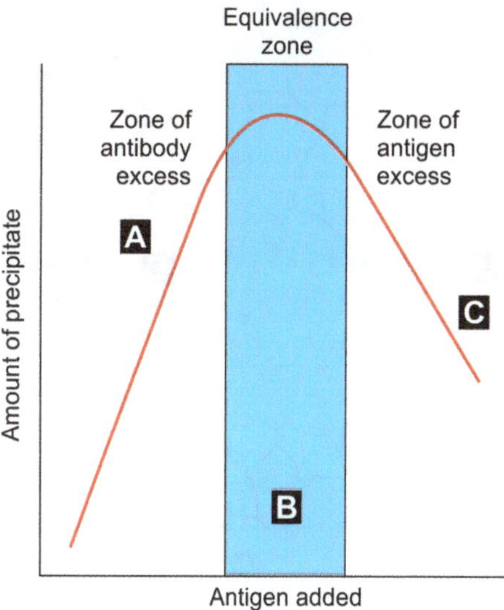

Fig. 16.1: A quantitative precipitation test showing: A. prozone (zone of antibody excess), B. zone of equivalence, C. zone of antigen excess

Three zones

1. **Zone of antibody excess or prozone (ascending part):** Only a small amount of precipitate is formed in the first few tubes, which contain relatively little antigen, and hence, an excess of antibody.
   Importance of prozone: The prozone is of importance in clinical serology, as sera rich in antibody or may sometimes give a false negative precipitation or agglutination result, unless several dilutions are tested.
2. **Equivalence zone (peak):** The so-called **equivalence zone** occurs when the ratio of antibody to antigen is optimal.
3. **Zone of antigen excess or postzone (descending part):** In the tubes containing more antigen, the amount of precipitate increases up to a point, after which it decreases as a result of smaller complexes being formed in the zone of antigen excess.

Mechanism of Precipitation

The **lattice hypothesis** was proposed by Marrack (1934) to explain the mechanism of precipitation. According to this concept, multivalent antigens combine with bivalent antibodies in varying proportions, depending on the antigen-antibody ratio in the reacting mixture, and is now widely accepted. Precipitation results when a large lattice is formed consisting of alternating antigen and antibody molecules. This is possible only in the **zone of equivalence.** The antigen must be either bivalent or polyvalent. In the zones of antigen or antibody excess, the lattice does not enlarge, as the valencies of the antibody and the antigen, respectively are fully satisfied. In either case, extensive lattices cannot be formed and precipitation is inhibited (Figs 16.2A to C). The lattice hypothesis holds good for agglutination also.

Applications of Precipitation Reaction

The precipitation test may be carried out as either a qualitative or quantitative test. As little as 1 µg of protein can be detected by precipitation tests so it is very sensitive in the detection of antigens. It has the following applicationas:

1. Forensic application in the identification of blood and seminal stains.
2. In testing for food adulterants.
3. To standardize toxins and antitoxins.

Types of Precipitation and Flocculation Tests

The following types of precipitation and flocculation tests are in common use:

Ring Test

This consists of layering the antigen solution over a column of antiserum in a narrow tube and is the simplest type of precipitation test. A precipitate forms at the junction of the two liquids. Ring tests have only a few clinical applications now.

Examples

i. Ascoli's thermoprecipitin test
ii. The grouping of streptococci by the Lancefield technique.

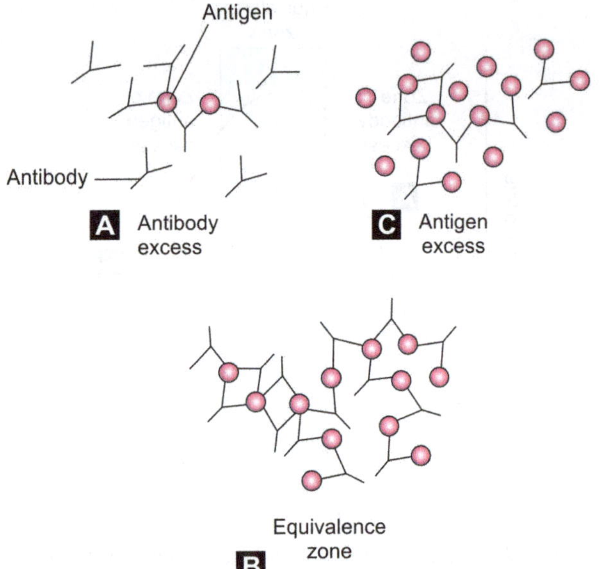

Figs 16.2A to C: Mechanism of precipitation by lattice formation. In A (antibody excess) and C (antigen excess), lattice formation does not occur. In B (zone of equivalence), lattice formation and precipitation occur optimally

Slide Test (Slide Flocculation Test)

When a drop each of the antigen and antiserum are placed on a slide and mixed by shaking, floccules appear.

Example

The VDRL test for syphilis is an example of slide flocculation.

Tube Test (Tube Flocculation Test)

i. **The Kahn test for syphilis** is an example of a tube flocculation test.
ii. **Standardization of toxins and toxoids:** A quantitative tube flocculation test is employed for the standardization of toxins and toxoids.

Immunodiffusion (Precipitation Reactions in Gels)

Several modifications of the classic precipitin reaction have been developed. Such changes are designed either to increase the sensitivity of the reaction or to identify specific antigen-antibody reactions occurring in a system containing multiple antigens and antibodies.

Immunodiffusion refers to a precipitation reaction that occurs between an antibody and antigen in an agar gel medium. Immunodiffusion is usually performed in a soft (1%) agar gel.

Advantages

There are several advantages in allowing precipitation to occur in a gel rather than in a liquid medium.

1. The reaction is visible as a distinct band of precipitation, which is stable and can be stained for preservation, if necessary.
2. These human chorionic gonadotropin (hCG) can be used to determine relative concentrations of antibodies or antigens, to compare antigens, or to determine the relative purity of an antigen preparation.
3. It also indicates identity, cross reaction and nonidentity between different antigens.

Types of Immunodiffusion Tests

1. **Single diffusion in one dimension (Oudin procedure)**
   Antibody is incorporated in agar gel in a test tube. Antigen solution is then layered over it. The antigen diffuses downward through the agar gel and wherever it reaches in optimum concentration with antibody, a line of precipitation is formed (Fig. 16.3). The line of precipitation moves downwards as more antigen diffuses. The number of bands indicates the number of different antigens present.
2. **Double diffusion in one dimension (Oakley-Fulthorpe procedure)**
   Here, the antibody is incorporated in gel in a test tube, above which is placed a column of plain agar and antigen is layered on surface of this. Antigen and antibody diffuse (double diffusion) towards each other (in one dimension) through the intervening column of plain agar and forms a band of precipitate where they meet at optimum proportion (Fig. 16.3).
3. **Single diffusion in two dimensions (Radial immunodiffusion—Mancini method)**
   The assay is carried out by incorporating monospecific antiserum (antiserum containing only antibody to the antigen being assayed for) into melted agar and allowing the agar to solidify on a glass plate in a thin layer. Wells then are punched into the agar, and different dilutions of the antigen are placed into the various wells. As the antigen diffuses into the agar, the region of equivalence is established and a ring of precipitation, a precipitin ring, forms around the well (Fig. 16.4). The area of the precipitin ring is proportional to the concentration of antigen.

Uses

i. To quantitate serum immunoglobulins, complement proteins, other substances—this method is commonly used.

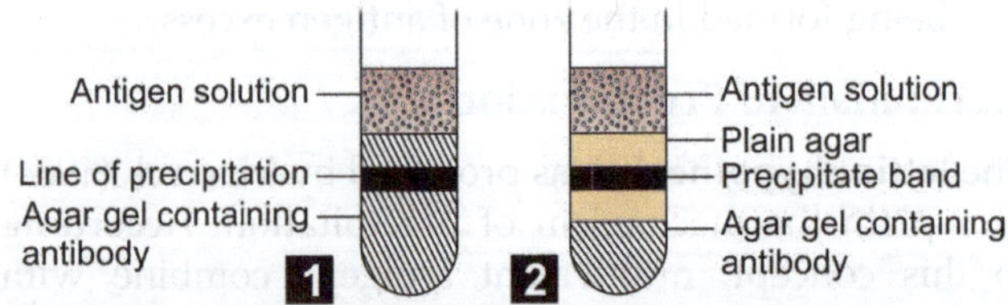

Fig. 16.3: 1. Single diffusion in one dimension (Oudin procedure) 2. Double diffusion in one dimension (Oakley-Fulthorpe procedure)

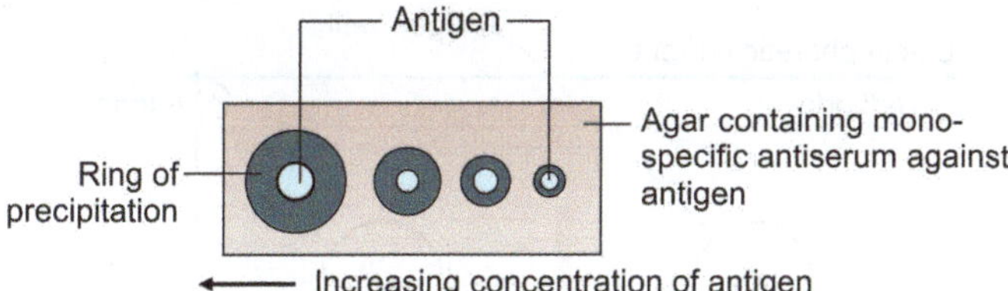

Fig. 16.4: Single diffusion in two dimensions (Radial immunodiffusion—Mancini method)

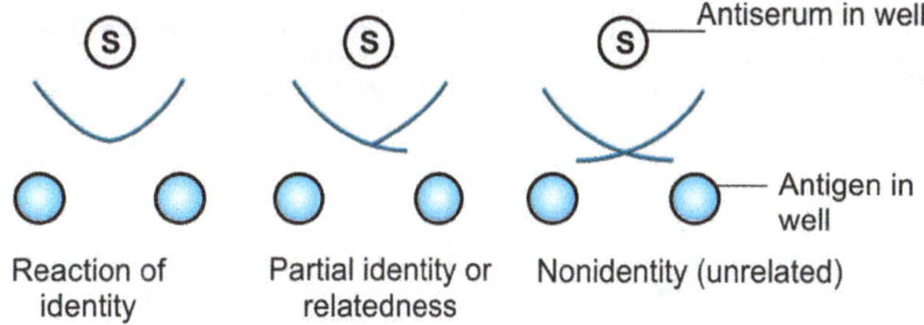

Fig. 16.5: Double diffusion in two dimensions (Ouchterlony technique)

ii. For screening sera for antibodies to influenza viruses, among others.
iii. For the laboratory diagnosis of multiple myeloma or agammaglobulinemia—it is used routinely.

Double Diffusion in Two Dimensions (Ouchterlony Technique)

In the Ouchterlony method, both antigen and antibody diffuse (hence, double diffusion) radially from wells toward each other, thereby establishing a concentration gradient. When soluble antigen and soluble antibody are placed in separate small wells punched into agar that has solidified on a slide or glass plate, the antigen and the antibody will diffuse through the agar. The holes are located only a few millimeters apart, and antibody and antigen will interact to form a line of precipitate in the area in which they are in optimal proportions.

In specimens containing several soluble antigens, multiple precipitin lines are observed, each occurring between the wells at a position that depends on the concentration of that particular antigen and its antibody (Fig. 16.5). This simple technique is an effective **qualitative** tool for determining the relationship between antigens and the number of different antigen-antibody systems present. The visible line of precipitation permits a comparison of antigens for identity, partial identity or nonidentity against a given selected antibody.

Example

Elek test for toxigenicity in diphtheria bacilli: A special variety of double diffusion in two dimensions is the **Elek test for toxigenicity** in diphtheria bacilli. When diphtheria bacilli are streaked at right angles to a filter paper strip carrying the antitoxin implanted on a plate of suitable medium, arrowhead shaped lines of precipitation appear on incubation, if the bacillus is toxigenic

Immunoelectrophoresis

Immunoelectrophoresis is a procedure that combines electrophoresis and double diffusion for the separation of antigen-antibody reactions in gels.

In this procedure, antigens are separated by electrophoresis in an agar gel. A small drop of solution containing the antigens (usually proteins) is placed into a small well, punched out of solidified agar on a small glass plate. The plate then is placed in an electric field to allow for the electrophoretic migration of the antigens. Positively charged proteins move to the negative electrode, and negatively charged proteins move to the positive electrode Different antigens migrate at different rates or even in different directions, depending on their size and charge, and on the conditions of electrophoresis. A trough is then cut next to the wells and filled with antibody and diffusion allowed to proceed for 18-24 hours. The antibody and the antigens diffuse toward each other, resulting in the formation of precipitin bands or arcs wherever they are in optimal proportions. The resulting precipitin lines can be photographed and the slides dried, stained and preserved for record.

Uses

i. **To separate many antigens:** This procedure can be used to separate many antigens, as well as to indicate the potential purity of an antigen. Over 30 different antigens can be identified in human serum by this method.
ii. **To separate the major blood proteins in serum for certain diagnostic tests:** this assay is used.
iii. **For testing for normal and abnormal proteins in serum and urine (Fig. 16.6).**

Electroimmunodiffusion

Immunodiffusion is a slow process. The development of precipitin lines can be speeded up by electrically driving the antigen and antibody in a gel, rather than simply allowing them to come in contact by diffusion. Various methods have been described combining electrophoresis with diffusion. Of these, frequently used methods in the clinical laboratory are:

1. One-dimensional double electoimmunodiffusion (counterimmunoelectrophoresis)
2. One-dimensional single electroimmunodiffusion (rocket electrophoresis)

1. Counterimmunoelectrophoresis (CIE)) or Countercurrent electrophoresis (CIEP)

Counterimmunoelectrophoresis can be used only for antigens and antibody that migrate in opposite directions in the electric field. The test is set up similarly to that for double diffusion in agar. Two wells are punched about 1 cm apart in an agar slab on a glass plate. The antigen and antibody solutions are placed in these wells in such a way that when the electric field

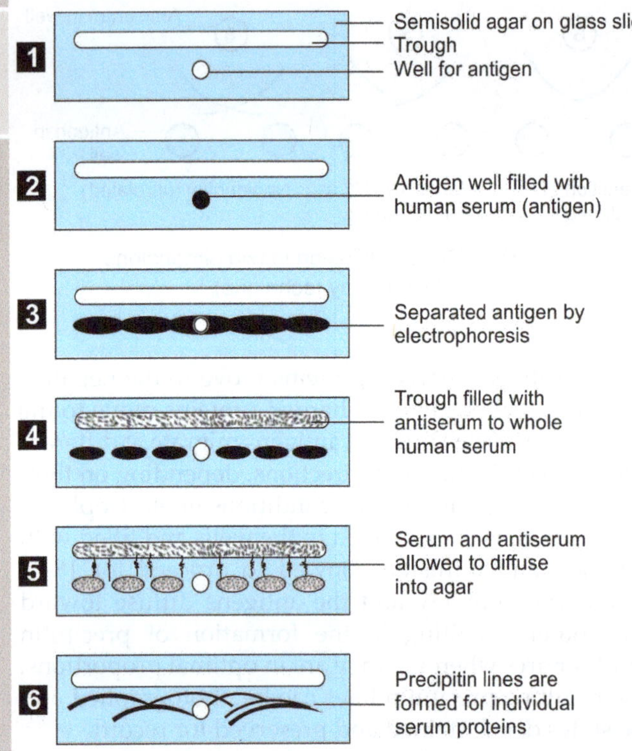

Fig. 16.6: Immunoelectrophoresis

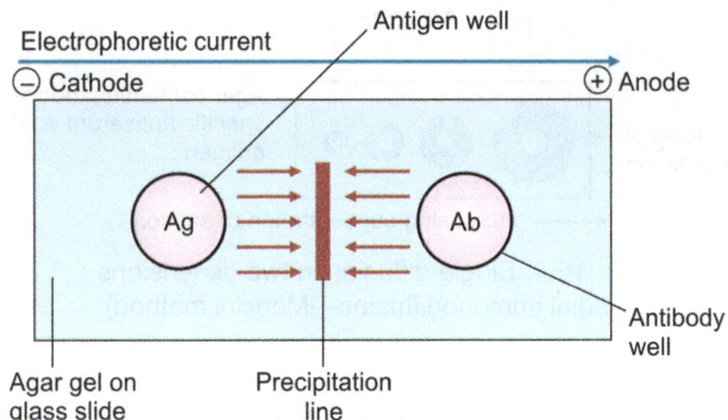

Fig. 16.7: Counterimmunoelectrophoresis (CIE)

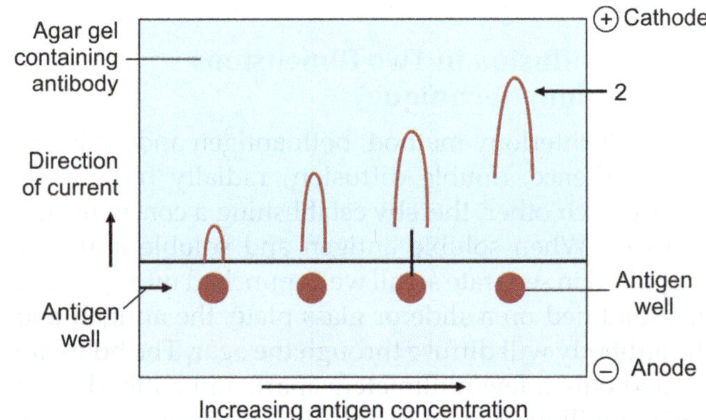

Fig 16.8: Rocket electrophoresis

is applied across the plate, the antigen will migrate toward the antibody, and the antibody will migrate toward the antigen. A precipitin line will form when the antigen and antibody meet in optimal proportions resulting in precipitation at a point between them (Fig. 16.7).

Advantages of Counterimmunoelectrophoresis (CIE)

1. **More sensitive:** It is 10 times more sensitive than simple diffusion in agar.
2. **More rapid assay:** It is a much more rapid assay (as little as 30 minutes for some antigens) than is double diffusion in gel.

Clinical Applications

i. **For detection of various antigens:** Such as hepatitis B surface antigen (HBs antigen) and alpha-fetoprotein in serum and meningococcal and cryptococcal antigens in CSF.
ii. **To detect the presence of anti-DNA antibody in the serum of patients with several autoimmune disorders.**

2. One-dimensional Single Electroimmunodiffusion (Rocket Electrophoresis)

As in radial immunodiffusion, small wells are cut in an agarose gel slab on a glass plate. The agar contains antibody to one or more antigens of interest. The antigen, in increasing concentration, is placed in wells, and an electric potential is applied across the plate. A negatively charged antigen is electrophoresed in a gel containing antibody. The antigens migrate into the gel and precipitate with the antibody in the gel at the appropriate position. The precipitate formed between antigen and antibody has the shape of a rocket. (Fig. 16.8). The height (distance from the antigen well to the top of the precipitin band) of which is proportional to the concentration of antigen in the well. It is so named because the precipitin bands resemble rockets.Thus, this method is primarily a quantitative technique.

Main Application

Quantitative estimation of antigens.

Laurell's Two dimensional Electrophoresis

A variant of rocket electrophoresis is Laurell's two dimensional electrophoresis. In this technique, the antigen mixture is first electrophoretically separated in a direction perpendicular to that of the final rocket stage. By this method one can quantitate each of several antigens in a mixture (Figs 16.9A and B).

B. Agglutination Reactions

When a particulate antigen is mixed with its antibody in the presence of electrolytes at a suitable temperature and pH, the particles are **clumped or agglutinated.** When antigen is present on the surface of a cell or particle, the addition of antibody causes a **clumping or agglutination** of the cells. Antibodies that produce such reactions are called **agglutinin.** This reaction is analogous to the

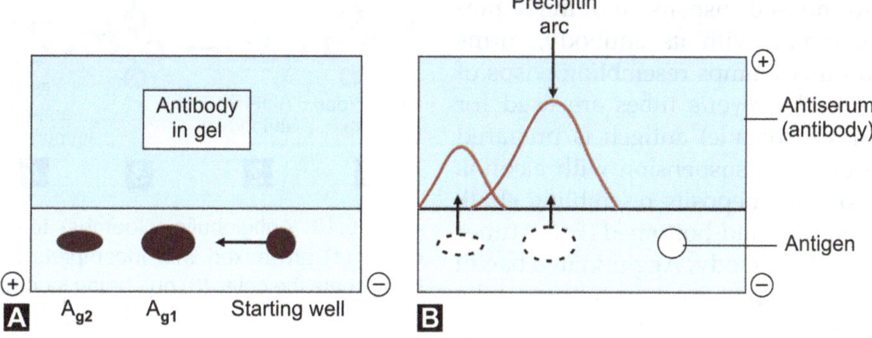

Figs 16.9A and B: Laurell's variant of rocket electrophoresis (two-dimensional electrophoresis): (A) First run-antigens are separated by electrophoretic mobility; (B) The second run is done at right angles to the first which drives the antigens into the antiserum containing gel to form precipitation peaks. The area of the peak is proportional to the concentration of the antigen

precipitin reaction, in that antibody acts as a bridge to form a lattice network of antibody and cells. Agglutination occurs optimally when antigen and antibodies react in equivalent proportion.

Agglutinations reactions are more sensitive than precipitin reactions because of the direct nature of the antigen/carrier/antibody interaction.

Prozone Phenomenon

The zone phenomenon may be seen when either an antibody or an antigen is in excess. Just an excess of antibody inhibits precipitation reactions, such excess can also inhibit reactions. This inhibition is called the **prozone phenomenon.** Several mechanisms can cause the prozone effect.

Blocking Antibodies

Occasionally antibodies are formed that will react with the antigenic determinants on a cell but will not cause agglutination (e.g. anti-Rh and anti-Brucella). These antibodies are called **blocking antibodies** because they inhibit agglutination by the complete antibody added subsequently. At one time blocking antibodies were thought to be monovalent (possessing only one antigen-binding site) and thus unable to form a lattice. However, this is now known not to be true.

Applications of Agglutination Reaction

1. **Slide agglutination**
2. **Tube agglutination**
3. **Antiglobulin (Coombs') test**
4. **Passive (indirect) agglutination test**

1. Slide Agglutination

A drop of sterile saline is placed on one of the divisions of the slide or plate and is emulsified in it, with an inoculating loop, just sufficient bacterial culture to make it visibly milky. The drop is examined with a hand lens or low power microscope to ascertain that the bacteria are well separated and not in visible clumps. With a small loop of thin platinum wire, a bulging loopful of the diagnostic serum is taken and is placed on the side just beside the bacterial suspension and is then mixed into the latter. The mixing is completed by tilting the slide to and fro for 30-60 seconds while viewing under a good light against a dark backround with the naked eye. Distinct clumping within 60 seconds is a positive result.

On another area of the slide or plate, in parallel with the test, a control test is performed adding only saline, instead of serum, to the bacterial suspension. If clumping takes place, the bacterial suspension is autoagglutinable and the reaction with the serum must be disregarded.

Slide agglutination is rapid and convenient, but in order to obtain rapid agglutination, the antiserum is used undiluted or in low dilution.

Uses

i. **For the identification of unknown bacterial cultures:** This method is widely used for the identification of unknown bacterial cultures isolated from clinical specimens such as *Salmonella* and *Shigella*.
ii. **Very rapid:** It is very rapid and requires only small quantities of culture and serum.
iii. **Also as the method for blood grouping and cross matching.**

2. Tube Agglutination

The usual agglutination test actually is only a semiquantitative measure of antibody. Serum from a patient thought to be infected with a given bacterium is serially diluted in a series of tubes to which the bacteria is added. The last tube showing visible agglutination will reflect the serum antibody **titer** of the patient. The reciprocal of the greatest serum dilution that elicits a positive agglutination is known as the agglutinin titer.

Uses of Tube Agglutination

Tube agglutination is routinely employed for the serological diagnosis of typhoid, brucellosis and typhus fever.

i. **Widal test:** In the Widal test used for the diagnosis of enteric fever, two types of antigens are used: the flagellar antigens (H) and somatic (O) antigen.

H antigen is a formolised suspension of the organisms which combining with its antibody, forms large, loose and fluffy clumps resembling wisps of cotton-wool. Conical Dreyer's tubes are used for H agglutination. O (somatic) antigen is prepared by treating the bacterial suspension with alcohol. It forms tight, compact deposits resembling chalk powder at the base of round-bottomed (Felix) tubes on combination with antibody. Agglutinated bacilli spread out in a disk like pattern at the bottom of the tube, whereas, negative reaction shows a compact button like deposit.

ii. **Tube agglutination test for brucellosis:** The tube agglutination test for brucellosis may be complicated by **prozone phenomenon** and the presence of **blocking antibodies**. To prevent false negative results due to prozone, several dilution of the serum should be tested. Incomplete or blocking antibodies may be detected by performing the test in hypertonic (5%) saline or albumin saline or more reliably by the antiglobulin (Coombs) test.

iii. **Weil-Felix reaction:** Weil-Felix reaction for serodiagnosis of **typhus fevers** is heterophil agglutination test. This is based on heterophil antibody that reacts with microorganisms or cells of unrelated species due to common antigen sharing. Weil-Felix reaction for serodiagnosis of typhus fevers is based on the sharing of common antigen between typhus rickettsiae and some strains of Proteus bacilli (OX 19, OX2 and OXK).

iv. **Paul-Bunnell test:** Red blood cells are used as antigens in Paul-Bunnell test and cold agglutination test. Paul-Bunnell test is based on the presence of sheep cell agglutinins in the sera of infectious mononucleosis patients, which are adsorbed by ox red cells but not by guinea pig kidney extract. IgM antibodies capable of agglutinating human red cells at 0-4°C (cold agglutinins) are sometimes found in certain human diseases including mycoplasmal (primary atypical pneumonia) pneumonia, malaria, trypanosomiasis and aquired hemolytic anemia.

3. Antiglobulin (Coombs') Test

Anti-Rh antibodies are of the IgG type, but they normally will not agglutinate Rh-positive RBCs. The antiglobulin test for Rh *antibodies* that he originally devised is the British immunologist, RRA. Coombs, still used today.

Principle of the Antiglobulin Test

When sera containing incomplete anti-Rh antibodies are mixed with Rh positive red cells, the antibody coats the surface of the erythrocytes but they are not agglutinated when such antibody-coated erythrocytes are washed to free all unattached protein and are treated with a *rabbit* antiserum against human gammaglobulin (antiglobulin or Coombs serum), the cells are agglutinated. This is the principle of the antiglobulin test (Fig. 16.10).

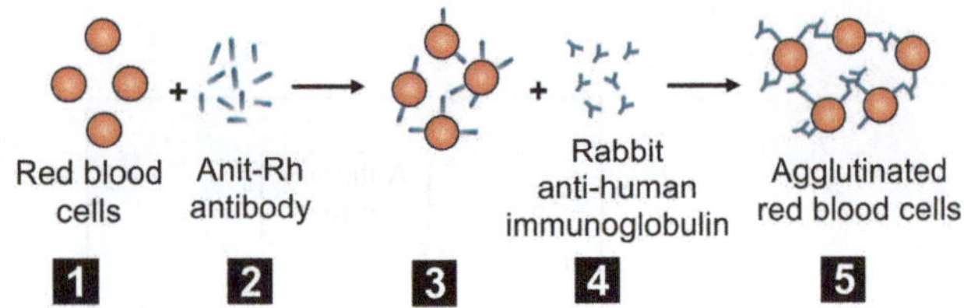

Fig. 16.10: Antiglobulin (Coombs) test. Rh positive erythrocytes; (1) are mixed with incomplete antibody; (2) The antibody coats the cells; (3) but, being incomplete, cannot produce agglutination. On addition of antiglobulin serum; (4) which is complete antibody to immunoglobulin, agglutination takes place (5)

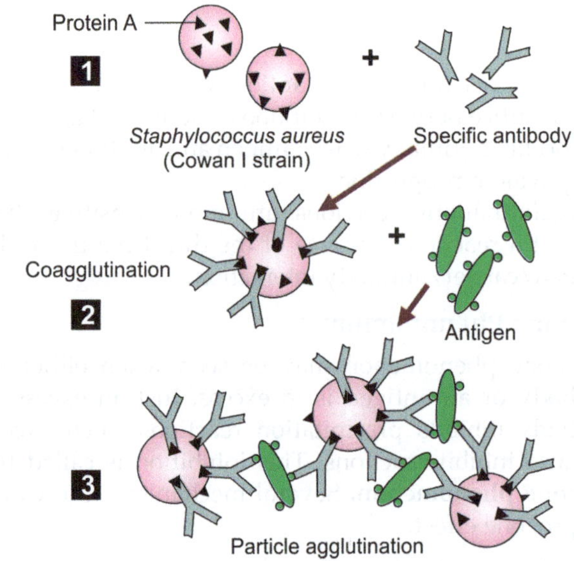

Fig. 16.11: Coagglutination

Types of Coombs Test

Coombs test is of two types: direct and indirect.

Direct Coombs Test

In the most direct, coombs test, the sensitization of the erythrocytes with incomplete antibodies takes place *in vivo* as in case of hemolytic disease of the newborn due to Rh incompatibility. Agglutination results when the red cells of erythroblastic infants are washed free of unattached protein and then a drop of combs serum is mixed.

Indirect Coombs Test

In the indirect Coombs test, sensitization of RBCs with incomplete antibodies is performed *in vitro*. It is accomplished by first reacting standard Rh-positive RBCs with the serum to be tested or reacting RBCs of an unknown Rh type with a standard anti-Rh antibody, and then, after a short incubation, adding antihuman Ig (immunoglobulin). If the serum contains Rh antibodies or the RBCs are Rh-positive, agglutination will occur, as shown schematically in Figure 16.11.

Uses

For demonstrating any type of incomplete or nonagglutinating antibody, e.g. in brucellosis.

4. Passive (Indirect) Agglutination Test

The difference between a precipitin reaction and agglutination is based on whether the antigen is soluble or particulate. A precipitation reaction can be converted into agglutination reaction by coating soluble antigen on to the surface of **carrier particles.** Such test is more convenient and more sensitive for detection of antibodies. Such tests are known as passive agglutination tests.

The commonly used carrier particles are red cells, latex particles or bentonite. Human or sheep erythrocytes adsorb a variety of antigens. Polysaccharide antigens may be adsorbed by simple mixing with the cells. For adsorption of protein antigens, tanned red cells are used.

Reversed Passive Agglutination

When instead of antigen, the antibody is adsorbed to carrier particles in tests for estimation of antigens, the technique is known as **reversed passive agglutination.**

Examples of Passive Agglutination

Hemagglutination Test

Rose-Waaler Test

A special type of passive agglutination test is the Rose-Waaler test. In rheumatoid arthriris, RA factor (an anti-gammaglobulin autoantibody) appears in the serum. It acts as antibody to human IgG. The RA factor is able to agglutinate red cells coated with globulins. The antigen used for the test is a suspension of sheep erythrocytes sensitized with a subagglutinating dose of rabbit antisheep erythrocyte antibody (amboceptor).

Treponema Pallidum Hemagglutination (TPHA)

One of the most widely used passive agglutination tests employing erythrocytes is *Treponema pallidum* hemagglutination (TPHA) for serological diagnosis of treponemal infection.

Latex Agglutination Test

Polystyrene latex, which can be manufactured as uniform spherical particles, 0.8 µm in diameter, can adsorb several types of antigens.

Latex agglutination tests (latex fixation tests) are widely employed in clinical laboratory for the detection of antistreptolysin-O (ASO), C reactive protein (CRP), RA factor, human chorionic gonadotropin (hCG) and many other antigens.

Coagglutination

Principle

Similar to latex agglutination, coagglutination uses antibody bound to a particle to enhance visibility of the agglutination reaction between antigen and antibody. Certain strains of (the Cowan strain, ATCC 12498) have a high content of surface protein A. Protein A on the *Staph. aureus* cell wall binds the Fc portion of the immunoglobulin molecule, leaving the Fab portion free to bind antigen. Visible agglutination of the staphylococcal cells serves as a positive test to indicate antigen-antibody binding.

Uses

Coagglutination can be used for detecting the presence of antigens in serum, urine and CSF of various streptococcal groups, including Lancefield groups A, B, C, D, F, G, and N; *Streptococcus pneumoniae; Neisseria meningitidis; N. gonorrhoeae;* and *Haemophilus influenzae* types A to F grown in culture.

Agglutination Inhibition

A modification of the agglutination reaction, called **agglutination inhibition,** provides a highly sensitive assay for small quantities of an antigen.

C. Complement Fixation Test (CFT)

When complement binds to an antigen-antibody complex it becomes "fixed" and "used up.' Complement fixation tests are very sensitive and can be used to detect extremely small amounts of an antibody for a suspect microorgamism in an individual's serum. This can detect as little as 0.04 µg of antibody nitrogen and 0.01 µg of antigen.

CFT is a complex procedure consisting of **two steps** and **five reagents**—antigen, antibody, complement, sheep erythrocytes and amboceptor (rabbit antibody to sheep red cells). Each of these reagents has to be separately standardized.

This test consists **of two separate systems** and these two systems are tested in sequence (Figs 16.12A and B). Appropriate controls should be used, including the following: antigen and serum controls to ensure that they are not anticomplementary; complement control to ensure that the desired amount of complement is added, and cell control to see that sensitized erythrocytes do not undergo lysis in the absence of complement.

Procedure

Test system consists of

i. **Antigen**—suspected of causing the patient's disease;
ii. **Patient's serum (Antibody);**
iii. **Complement.**

A known **antigen** is mixed with **test serum** lacking complement. When immune complexes have had time to form, **complement** is added to the mixture. If the patient's serum contains antibody to the antigen, the resulting antigen-antibody complexes will bind all of the complement added. In most of the cases, fixation of complement with antigen-antibody complex causes in itself no visible effect. (Antigen+Antibody+ Complement).

Indicator system consists of **sheep red cells (antigen) coated with anti-sheep-red cell antibody and complement**—an exogenous source of **complement** (usually guinea pig serum). Afterwards, sensitized indicator cells, usually sheep red blood cells previously coated with complement-fixing antibodies are added to the mixture. Complement lyses antibody coated red cells.

Interpretation

Positive CF Test—Absence of Lysis

The specific antibodies are present in the test serum and complement is consumed by the immune complexes insufficient amount of complement will be available to lyse the indicator cells.

Negative CF Test—*Lysis of the Indicator Cells*

Lysis of the indicator cells indicates lack of antibody and a negative CF test. Lysis of the indicator cells (Figs 16.12A and B) results if immune complexes do not form in part of the test because the antibodies are not present in the test serum, complement remains and lyses the indicator cells.

Uses

1. **Diagnosis of syphilis:** Complement fixation was once used in the diagnosis of syphilis (the Wassermann test).
2. In the diagnosis of certain viral, fungal, rickettsial, chlamydial, and protozoan diseases.
3. For diagnosing infection caused by fungi, respiratory viruses, and arboviruses, as well as to diagnose Q fever. This test is still probably the most common method.

Anticomplementary Effect

Nonspecific adsorption of complement may give false positive results. Some sera may develop anticomplementary properties on aging, and after bacterial contamination. Hemolyzed blood serum also has anticomplementary effect.

Other Complement Dependent Serological Tests

Complement takes part in many immunological reactions and is absorbed during the combination of antigens with their antibodies. In the presence of the appropriate antibodies, complement lyses erythrocytes, kills and, in some cases, lyses bacteria, immobilizes motile organisms, promotes phagocytosis and immune adherence and contributes to tissue damage in certain types of hypersensitivity.

i. **Immune adherence:** Some bacteria, like *Vibrio cholerae* and *Treponema pallidum,* react with specific antibody in the presence of complement and particulate material such as erythrocytes or platelets. The

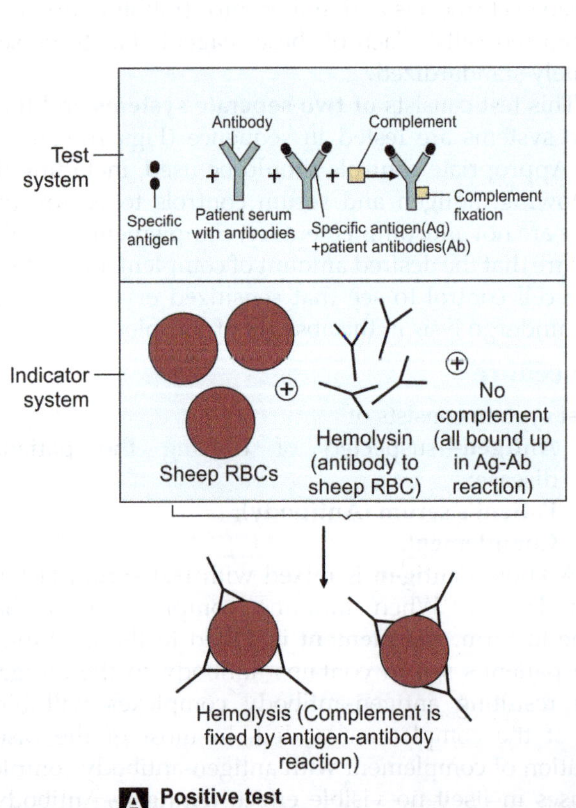

A Positive test

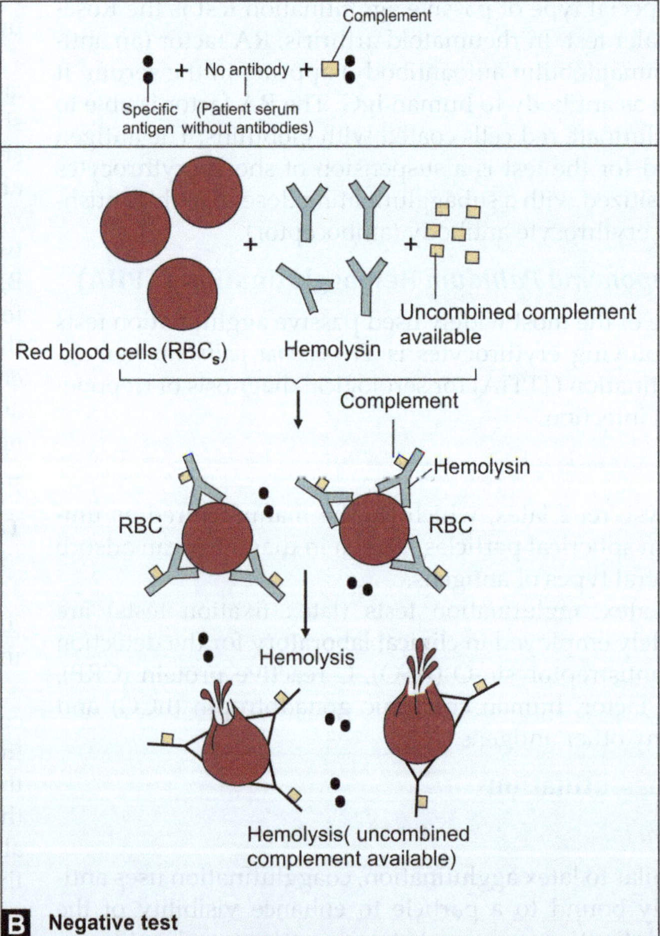

B Negative test

Figs 16.12A and B: Complement fixation test

bacteria are aggregated and adhere to the cells. This is known as immune adherence. Adherence occurs through the activated C3b component of complement.

ii. **Immobilization test:** In the *Treponema pallidum* immobilization test, a highly specific test, formerly considered the 'gold standard' for the serological diagnosis of syphilis, the test serum is incubated anaerobically with a suspension of live treponemes and complement. If antibodies are present, the treponemes will be found to be immobilized.

iii. **Cytolytic or cytocidal tests:** These are also complement dependent. When *V. cholerae* is mixed with its antibody in the presence of complement, the bacterium is killed and lysed. This forms the basis of the vibriocidal antibody test for the measurement of anticholera antibodies.

Indirect Complement Fixation Test

Certain avian (for example, duck, turkey, parrot) and mammalian (for example, horse, cat) sera do not fix guinea pig complement. The indirect complement fixation test may be employed when such sera are to be tested. Here the test is set up in duplicate and after the first step, the standard antiserum known to fix complement is added to one set. If the test serum contained antibody, the antigen would have been used up in the first step and therefore the standard antiserum added subsequently would not be able to fix complement. Therefore, hemolysis indicates a positive result in the indirect test.

Conglutinating Complement Absorption Test

An alternative method is the conglutinating complement absorption test for systems which do not fix guinea pig complement. This uses horse complement which is nonhemolytic. The indicator system is sensitized sheep erythrocytes mixed with bovine serum. Bovine serum contains a beta globulin component called conglutinin, which acts as antibody to complement. Therefore, conglutinin causes agglutination of sensitized sheep erythrocytes (conglutination) if they have combined with complement. If the horse complement had been used up by the antigen-antibody interaction in the first step, agglutination of sensitized cells will not occur.

D. Neutralization Tests

These are of two types: viral neutralization tests and toxin neutralization tests.

1. Viral Neutralization

Neutralization of viruses by their antibodies can be demonstrated in various systems. IgG, IgM, and IgA antibodies can bind to some viruses during their extracellular phase and inactivate them. This antibody mediated viral inactivation is called **viral neutralization**. Fixation of the classical pathway complement component C3b to the virus aids the neutralization process. Viral neutralization prevents a viral infection due to the inability of the virus to bind to its target cell.

Indicator Systems

Laboratory animals or tissue culture cells are used as "indicator systems" in these tests. The test serum is diluted serially, incubated with a known amount of virus and the mixture is then added to indicator cultures: animals, embryonated hen's egg and tissue culture. The highest dilution of serum ablating infectivity in 50 percent of virus-serum mixtures tested is taken as the titer. When bacteriophages are seeded in appropriate dilution on lawn cultures, plaques of lysis are produced. Specific antiphage serum inhibits plaque formation.

2. Toxin Neutralization

Bacterial exotoxins are good antigens and their activity may be completely neutralized by appropriate concentrations of specific antibody. Antibody to bacterial exotoxin is usually referred to as antitoxin which are important clinically, in protection against and recovery from diseases such as diphtheria and tetanus.

Bacterial endotoxins are poorly antigenic and their toxicity is not neutralized by antisera.

Toxin neutralization can be tested *in vivo* or *in vitro*.

Toxin Neutralization In Vivo

i. **Toxigenicity test:** The neutralizing capacity of an antitoxin can be assayed by neutralization test in which mixture of toxin and antitoxin is injected into a susceptible animal and the least amount of antitoxin that prevents death or disease in the animal is estimated.

ii. **Schick test:** With the diphtheria toxin, which in small doses causes cutaneous reaction, neutralization test can be carried out on the human skin. The Schick test is based on the ability of circulating antitoxin to neutralize the diphtheria toxin given intradermally. Neutralization (no reaction) indicates immunity and redness and erythema indicates susceptibility to diphtheria.

Toxin Neutralization In Vitro

If a toxin has a demonstrable *in vitro* effect, this effect can be neutralized by specific antitoxin.

Examples

i. **Antistreptolysin O (ASO) test:** Antistreptolysin O (ASO)—antitoxin, present in the serum of the patient suffering from *Strep. pyogenes* infection, neutralizes the hemolytic activity of the streptococcal O hemolysin (toxin).

ii. **Nagler's reaction:** Another example of *in vitro* toxin—antitoxin neutralization is Nagler's reaction. *Cl. perfringens* produces α-toxin which is a phospholipase (lecithinase-C). This produces opalescence in serum or egg yolk media. This reaction is specifically neutralized by the antitoxin.

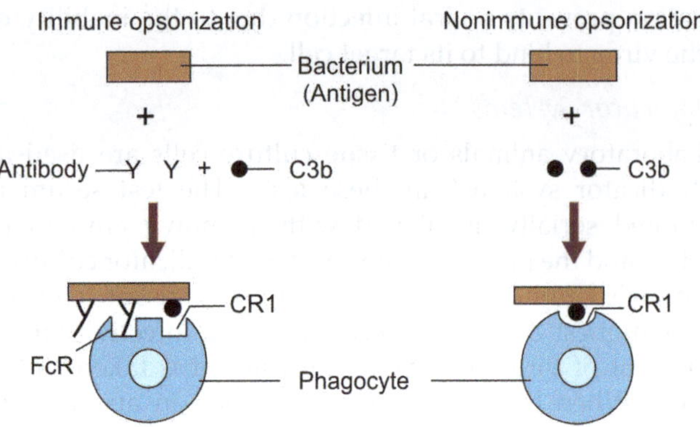

Fig. 16.13: Opsonization

iii. **Agar gel precipitation test:** Agar gel precipitation test is employed to detect production of toxin by *Corynebacterium diphtheriae.*

E. Opsonization

Opsonization is the process in which microorganisms or other particles are coated by antibody and/or complement, and thus prepared for "recognition" and ingestion by phagocytic cells. Opsonizing antibodies, especially IgG1 and IgG3, bind to Fc receptors on the surface of macrophages and neutrophils. This binding forms a bridge between the phagocyte and the antigen. This phagocytic process can be greatly enhanced by opsonization.

Immune Opsonization

Phagocytes such as macrophages, monocytes and neutrophils possess surface receptors (CR1) for C3b and Fc receptors for antibody. If immune complexes have activated the complement system, then Fc and CR1 receptors, present on the phagocyte, bind Fc region of antibody and C3b bound on immune complexes respectively thus facilitating their phagocytosis. This facilitated phagocytosis by antibody and complement is known as **immune opsonization.**

Nonimmune Opsonization

On the contrary, nonimmune opsonization requires only C3b (opsonin) for opsonization. The alternative pathway can be activated by bacteria in the blood stream and generate C3b, which coats the bacteria. C3b binds to CR1 receptors present on the phagocytes thus facilitating their phagocytosis. Soluble immune complexes, viruses and tumor cells are also opsonized and removed by the same mechanism (Fig. 16.13).

F. Immunofluorescence

Immunofluorescence is a process in which dyes called fluorochromes are exposed to UV, violet, or blue light to make them fluoresce or emit visible light. Fluorescent molecules absorb light of one wavelength (excitation) and emit light of another wavelength (emission). Albert Coons and his colleagues (1942) showed that fluorescent dyes can be conjugated to antibodies and that such 'labelled' antibodies can be used and identify antigens in tissues. Antibody molecules bound to antigens in cells or tissue sections can similarly be visualized. The emitted light can be viewed with a fluorescence microscope, which is equipped with a UV light source.

Fluorescent Dyes

Rhodamine B or **fluorescein isothiocyanate (FITC)** are the most commonly used fluorescent dyes. **Fluorescein** emits an intense yellow-green fluorescence**. Rhodamine** emits a deep red fluorescence. **Phycoerythrin and phycobiliproteins** are other highly fluorescent substances and have also come into use.

The most commonly used fluorescent dyes such as rhodamine B or fluorescein isothiocyanate (FITC) can be coupled to antibody molecules without changing the antibody's capacity to bind to a specific antigen. The dyes can be conjugated to the Fc region of an antibody molecule without affecting the specificity of the antibody. For detection of antibodies by immunofluorescence, the *sandwich* technique can be employed. The antibody is first allowed to react with unlabelled antigen, which is then treated with fluorescent labelled antibody. A sandwich is thus formed—the antigen being in the middle and labelled and unlabelled antibodies on either side.

Fluorescent dyes may also be conjugated with complement. Labeled complement is a versatile tool and can be employed for the detection of antigen or antibody.

Types of Immunofluorescence

There are two main kinds of fluorescent antibody assays: direct and indirect.

1. Direct Immunofluorescence (Fig. 16.14A)

In **direct staining,** the specific antibody (the primary antibody) is directly conjugated with fluorescein. It involves fixing the specimen (cell or microorganism) containing the antigen of interest onto a slide. Fluorescein-labeled antibodies are then added to the slide and incubated. The slide is washed to remove any unbound antibody and examined with the fluorescence microscope for a yellow-green fluorescence. The pattern of fluorescence reveals the antigen's location.

Uses

i. It is used to identify antigens such as those found on the surface of group A streptococci.
ii. To diagnose enteropathogenic *Escherichia coli, Neisseria meningitidis, Salmonella typhi, Shigella sonnei, Listeria monocytogenes, Haemophilus influenzae* type b.
iii. To diagnose rabies virus.

2. Indirect Immunofluorescence (Fig. 16.14B)

In **indirect staining,** the primary antibody is unlabeled and is detected with an additional fluorochrome-labeled

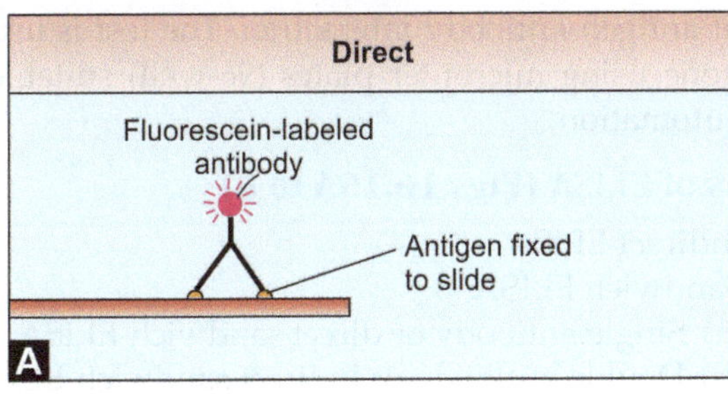

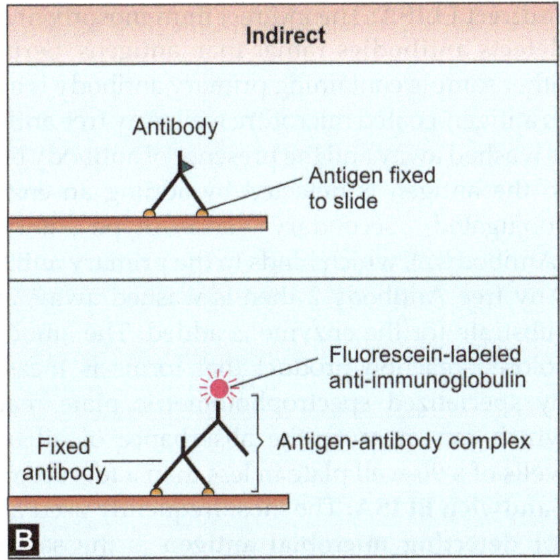

Figs 16.14A and B: Direct and indirect immunofluorescence

reagent. Indirect immunofluorescence is used to detect the presence of antibodies in serum following an individual's exposure to microorganisms.

In this technique, a known antigen is fixed onto a slide. The test antiserum is then added, and if the specific antibody is present, it reacts with antigen to form a complex. When fluorescein-labeled anti-immunoglobulin is added, it reacts with the fixed antibody. After incubation and washing, the slide is examined with the fluorescence microscope. The occurrence of fluorescence shows that antibody specific to the test antigen is present in the serum.

Uses

Diagnosis of Syphilis

It is used to identify the presence of *Treponema pallidum* antibodies in the diagnosis of syphilis (Fluorescent treponemal antibody absorption (FTA-ABS) test as well as antibodies produced in response to other microorganisms.

Advantages

Indirect immunofluorescence staining has two advantages over direct staining.

i. The primary antibody does not need to be conjugated with a fluorochrome.
ii. Increase the sensitivity of staining.

G. Radioimmunoassay

One of the most sensitive techniques for detecting antigen or antibody is radioimmunoassay (RIA). The technique was first developed by two endocrinologists, SA Berson and Rosalyn Yalow, in 1960 to determine levels of insulin—anti-insulin complexes in diabetics. Although their original attempts to publish a report of this research met with some resistance from immunologists, the technique soon proved its own value for measuring hormones, serum proteins, drugs, and vitamins at concentrations of 0.001 micrograms per milliliter or less. The significance of the technique was acknowledged by the award of a Nobel Prize to Yalow in 1977, some years after Berson's death.

Several different versions of the radioimmunoassay (RIA) are available. Although RIA has sensitivity and specificity, the problem associated with radioactive waste disposal and the instability of certain radionucleotide has limited the expansion of RIA techniques.

Principle of Radioimmunoassay (RIA)

The principle of RIA is based on competitive binding of radiolabeled antigen (e.g. ^{125}I and unlabeled antigen to a high-affinity antibody. The labeled and unlabelled (test) antigens compete for the limited binding sites on the antibody. This competition is determined by the level of the unlabelled (test) antigen present in reacting system.

Procedure

1. Measured quantities of **labeled antigen** (radiolabeled) antigen (of the same kind being tested) and antibody (specific to the antigen being tested) are mixed and incubated, one mixture with and one without added test sample.
   The **labeled antigen** is mixed with **antibody** at a concentration that saturates the antigen-binding sites of the antibody molecule.
2. Then increasing amounts of the **test sample (unlabeled antigen)** of unknown concentration are added. The antibody does not distinguish labeled from unlabeled antigen, and so the two kinds of antigen compete for available binding sites on the antibody.
3. With increasing concentrations of test antigen (unlabeled antigen), more labeled antigen will be displaced from the binding sites.
4. The antigen-antibody complex is washed to remove unbound radiolabeled antigen from the mixture. The antigen is separated into '**free**' and **'bound' fractions** after the reaction and their radioactive counts measured.
5. The radioactivity associated with the antibody is then detected by means of radioisotope analyzers and autoradiography (photographic emulsions that show areas of radioactivity). A little amount of bound radioactivity indicates that there is large amount of antigen; and a large amount of bound

radioactivity indicates that there is little antigen in the sample.

6. The **standard dose response or reference curve** has to be prepared first for any reacting system. This is plotted on graph by taking the ratio of bound radiolabeled antigen to free radiolabeled antigen against varying known amounts of unlabelled antigen and the antigen concentration of the test sample can be read from this curve. This curve will allow the measurement of antigen present by merely counting the radioactivity present in the test antigen-antibody precipitated complexes.

Uses

i. **To measure the concentration of certain hormones** such as insulin, testosterone, growth hormone, and glucagon because of its extreme sensitivity (measuring picograms of antigen per milliliter).
ii. **To determine the presence of the hepatitis B antigen** which can be present in the serum of asymptomatic blood donors.

h. Enzyme-linked Immunosorbent Assay (ELISA)

Enzyme-linked immunosorbent assay, commonly known as ELISA or EIA), is similar in principle to RIA but the radioactive tag used in RIA techniques can be replaced with an enzyme. When this enzyme is linked to an antibody and used to detect and measure other antibodies or antigens, the assay is called the enzyme-linked immunosorbent assay (ELISA). An enzyme conjugated with antibody reacts with a colorless substrate to generate a colored reaction product. Such a substrate is called a chromogenic substrate.

Enzyme-linked immunosorbent assay is highly sensitive, highly specific and less expensive technique used in serology to detect antigens or antibodies.

Principle

ELISA is based on two principles:

1. Solid phase immunoassay are more widely used. It refers to the binding of either antigen or antibody to a variety of solid materials, such as polyvinyl or polycarbonate wells or membranes (disks) of polyacrylamide, paper or plastic or metal beads or some other solid matrix.
2. Antigens and antibodies can be covalently attached to an active enzyme (such as alkaline phosphatase or horseradish peroxidase and β-galactosidase), with the resulting complexes still fully functional, both immunologically and enzymatically. Enzyme activity is used to measure the quantity of antigen or antibody present in the test sample. After all unreacted material is washed away, the substrate for the enzyme is added (usually one that will yield a colored product, such as p-nitrophenol phosphate for alkaline phosphatase), and the conversion of the substrate from colorless to color is a measure of antigen-antibody interaction. The test is usually done using microtiter plates (96-well) suitable for automation.

Types of ELISA (Figs 16.15A to C)

1. Indirect ELISA
2. Sandwich ELISA
   (a) Single antibody or direct sandwich ELISA
   (b) Double-antibody or indirect sandwich ELISA
3. Competitive ELISA

1. **Indirect ELISA:** The indirect immunosorbent assay **detects antibodies** rather than antigens. Serum or other sample containing primary antibody is added to antigen-coated microtitre well. Any free antibody is washed away and the presence of antibody bound to the antigen is detected by adding an enzyme-conjugated secondary anti-isotype antibody (Antibody 2), which binds to the primary antibody. Any free Antibody 2 then is washed away, and a substrate for the enzyme is added. The amount of colored reaction product that forms is measured by specialized spectrophotometric plate readers, which can measure the absorbance of all of the wells of a 96-well plate in less than a few seconds.
2. **Sandwich ELISA:** The most frequently used ELISA for **detecting microbial antigen** is the sandwich solid-phase ELISA. It is of two types:
   a. **Single antibody or direct sandwich ELISA:** In this technique, the antibody (rather than the antigen) is immobilized on a microtiter well. The test sample is then exposed to the solid-phase antibody, to which the antigen, if present, will bind. After the well is washed, a second enzyme-linked antibody specific for test antigen is added and allowed to react with the bound antigen. The conjugated antibody will react with the antigen held to the solid-phase by the first antibody, forming an **antibody-antigen-antibody sandwich** on the solid phase. After any free second antibody is removed by washing, substrate is added, and the colored reaction product is measured.
   b. **Double-antibody or indirect sandwich ELISA (Fig. 16.15)**
   It is used for the **detection of antigens**. In this assay, specific antibody is placed in wells of a microtiter plate (or it may be attached to a membrane). The antibody is absorbed onto the walls, coating and sensitizing the plate. A test antigen is then added to each well. If the antigen reacts with the antibody, the antigen is retained when the well is washed to remove unbound antigen. An antibody enzyme conjugate specific for the antigen is then added to each well. The final complex is formed of an outer antibody-enzyme, middle antigen, and inner antibody—that is, it

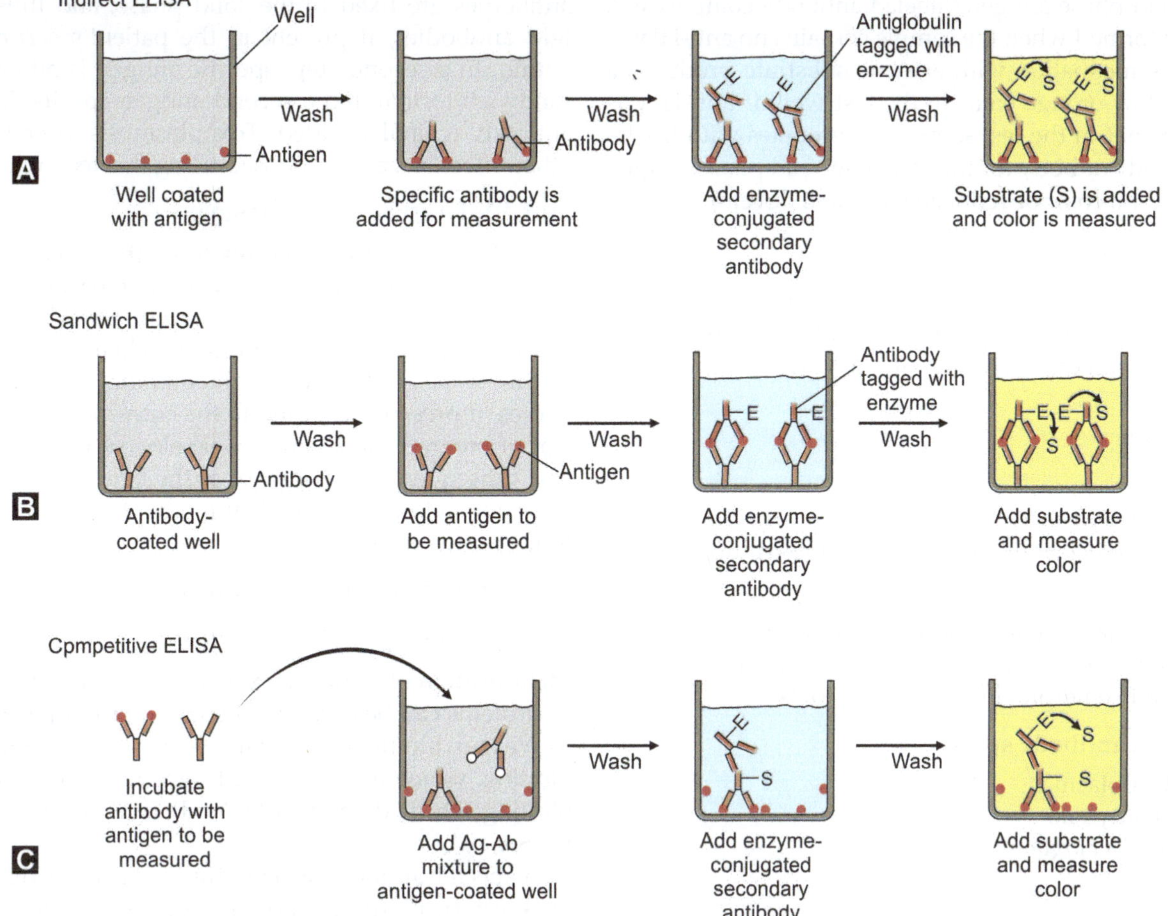

Figs 16.15A to C: Enzyme-linked immunosorbent assay (ELISA)

is a layered (Antibody-Antigen-Antibody) sandwich.

In the double antibody ELISA, the second antibody of the sandwich must be from a different species than the solid-phase antibody, otherwise, the anti-immunoglobulin conjugate reacts with the solid phase antibody, producing high background activity. A substrate that the enzyme will convert to a colored product is then added, and any resulting product is quantitatively measured by optical density scanning of the plate.

If the antigen has reacted with the absorbed antibodies in the first step, the ELISA test is positive. If the antigen is not recognized by the absorbed antibody, the ELISA test is negative because the unattached antigen has been washed away, and no antibody-enzyme is bound.

3. **Competitive ELISA:** The test can be made more specific by making serum antibody and enzyme labeled antibody compete for the binding sites on the antigen. In this technique, antibody is first incubated in solution with a sample containing antigen. The antigen-antibody mixture is then added to an antigen coated microtiter well. The more antigen present in the sample, the less free antibody will be available to bind to the antigen-coated well. Addition of an enzyme-conjugated, secondary antibody specific for the isotype of the primary antibody can be used to determine the amount of primary antibody bound to the well as in an indirect ELISA. In the competitive assay, however, the higher the concentration of antigen in the original sample, the lower the absorbance.

Illustration

The principle of this test too can be illustrated by outlining its application for detection of anti-HIV antibodies in the patient serum.

i. The wells of the polystyrene microtiter plate are coated with HIV antigens which constitutes the solid-phase antigen.
ii. The test sample and human anti-HIV labelled with the enzyme horse-radish peroxidase are incubated in such a well.

iii. Solid phase antigen/labeled antibody complex will be formed when the sample contains no anti-HIV.
iv. The incubation with enzyme substrate produces a yellow-orange color in the test well. If anti-HIV is present in the test sample, it competes with the labeled antibody for the available solid phase antigen and no color or reduced color will develop.

USES OF ELISA

ELISA is a simple and versatile technique. It needs only microliter quantities of reactants. ELISA has been used to detect antigens and antibodies of various microorganisms.

Examples

Parasites

- *Entamoeba histolytica* antigens in feces
- *Toxoplasma* antigens in the patient serum.

Bacteria

- *Haemophilus influenzae* antigens in spinal fluid
- β-haemolytic streptococcal antigen in spinal fluid
- Labile enterotoxin of *E. coli* in stools.

To detect antibody specific for:

- Mycoplasmas
- Chlamydiae
- *Borrelia burgdorferi.*

Viruses

To detect antibody specific for:

- Hepatitis virus antigens
- Herpes simplex viruses 1 and 2
- Respiratory syncytial virus (RSV)
- Cytomegalovirus
- Human immunodeficiency virus (HIV)
- Rubella virus (both IgG and IgM)
- Adenovirus antigens—in nasopharyngeal specimens.

Cassette-Based Membrane-Bound ELISA Assays

The introduction of membrane-bound ELISA components has improved sensitivity and ease of use dramatically. **Cassette-based membrane-bound ELISA assays**, designed for testing a single serum, can be performed rapidly (often within 10 minutes) as compared with the 2-4 hours taken for microplate ELISA. There is no need for microplate washers or readers. The result is read visually. Inbuilt positive and negative controls are usually provided for validation of the test procedure.

Examples of Cassette ELISA

i. Used for the detection of HIV type 1 and 2 antibodies.
ii. To detect antibodies to *Helicobacter pylori, T. gondii,* and some other infectious agents are available.

Antibody Capture ELISAs

Antibody capture ELISAs are particularly valuable for detecting IgM in the presence of IgG. Anti-IgM antibodies are fixed to the solid phase, and thus only IgM antibodies, if present in the patient's serum, are bound. In a second step, specific antigen is added in a sandwich format and a second antigen-specific labeled antibody is finally added. Toxoplasmosis, rubella, and other infections are diagnosed using this technology.

Slot-Blot and Dot-Blot Assays

Slot-blot and **dot-blot assays** force the target antigen through a membrane filter, causing it to become affixed in the shape of the hole (a dot or a slot). Several antigens can be placed on one membrane. When test (patient) serum is layered onto the membrane, specific antibodies, if present, will bind to the corresponding dot or slot of antigen. Addition of a labeled second antibody and subsequent development of the label allows visual detection of the presence of antibodies based on the pattern of antigen sites.

I. Immunoelectroblot Techniqus

Western Blotting (Immunoblotting)

Identification of a specific protein in a complex mixture of proteins can be accomplished by a technique known as Western blotting, named for its similarity to Southern blotting, which detects DNA fragments, and Northern blotting which detects mRNAs. It is a variation of an ELISA.

A specific antibody in a mixture can also be identified by Western blotting. Known antigens of well-defined molecular weight are separated by **SDS-polyacrylamide gel (SDS-PAGE)**, and blotted onto nitrocellulose in this case. The separated bands of known antigens are then probed with the sample (test sample) suspected of containing antibody specific for one or more of these antigens. Reaction of an antibody with a band is detected by using either radiolabeled or enzyme-linked secondary antibody that is specific for the species of the antibodies in the test sample. The enzyme substrate is subsequently added which indicates positive test. The substrate changes color in the presence of enzyme and permanently stains the nitrocellulose paper. The position of the band on the paper indicates the antigen with which the antibody has reacted.

Application

i. **Confirmatory testing for human immunodeficiency virus (HIV).**
ii. **Antibodies against microbes with numerous cross-reacting antibodies,** such as *T. pallidum, B. burgdorferi,* Herpes simplex virus types 1 and 2, and Human immunodeficiency virus (HIV) are identified more specifically using this technology than a method that tests for only one general antibody type.

J. Immunochromatographic Tests

A one-step qualitative immunochromatographic technique has found wide application in serodiagnosis due to its simplicity, economy and reliability. A description of its use for HBs antigen detection illustrates the method. The test system is a small cassette containing a membrane impregnated with anti-HBs antigen antibody colloidal gold dye conjugate. The membrane is exposed at three windows on the cassette. The test serum is dropped into the first window. As the serum travels upstream by capillary action, a colored band appears at the second window (test site) if the serum contains HBs antigen, due the formation of a HBs antigen-antibody conjugate complex. This is the positive reaction. Absence of a colored band at the test site indicates a negative reaction. Simultaneously, a colored band should appear in every case at the third window, which forms an inbuilt control, in the absence of which the test is invalid. The test is claimed to be nearly as sensitive and specific as EIA tests.

K. Immunoelectronmicroscopic Tests

1. **Immunoelectronmicroscopy:** When viral particles mixed with specific antisera are observed under the electron microscope, they are seen to be clumped. This is known as immune electron microscopy. This finds application in the study of some viruses such as the hepatitis A virus and the viruses causing diarrhea.
2. **Immunoferritin test:** Ferritin (an electron-dense substance from horse spleen) can be conjugated with antibody, and such labelled antibody reacting with an antigen can be visualised under the electron microscope.
3. **Immunoenzyme test:** Some stable enzymes, such as peroxidase, can be conjugated with antibodies. Tissue sections carrying the corresponding antigens are treated with peroxidase labeled antisera. If the tissue section possesses specific antigen then peroxidase bound to the antigen can be visualized under the electron microscope, by microhistochemical methods. In immunoenzyme tests, some other enzymes, such as glucose oxidase, phosphatases and tyrosinase, may also be included.

KNOW MORE

The source of complement for laboratory use is guinea pig serum. Complement is unstable and deteriorates on keeping at ordinary temperatures. As complement activity is heat labile, the serum should be freshly drawn, or preserved by applying two methods;

1. Rapid drying of the serum from the frozen state in vacuo ('freeze drying')
2. With special preservatives as in Richardson's method (addition to the liquid serum of hypertonic sodium chloride or other salts).

Standardization of Complement and Amboceptor

Guinea pig serum is first titrated for complement activity. One unit or minimum hemolytic dose (MHD) of complement is the highest dilution of guinea pig serum that lyses one unit volume of washed sheep RBCs in the presence of excess of hemolysin (amboceptor) within a fixed time (usually 30 to 60 minutes) at a fixed temperature (37°C). Similarly, amboceptor should also be titrated. One MHD of hemolysin is defined as highest dilution of the serum that lyses one unit volume of washed sheep RBCs in the presence of excess complement within a fixed time (usually 30 to 60 minutes at 37°C).

☞ KEY POINTS

- The antigen-antibody reaction depend on four types of noncovalent interactions: hydrogen bonding, iononic bonds, van der Waals bonds, and hydrophobic interactions
- Various types of antigen-antibody reactions are:
  - **Precipitation test** is a type of antigen-antibody reaction, in which the antigen occurs in a soluble form.. Immunodiffusion procedures are precipitation reactions carried out in an agar gel medium. Electrophoresis can be combined with precipitation in gels in a technique called **immunoelectrophoresis.** Immunoelectrophoresis combines electrophoresis with immunodiffusion for the analysis of serum proteins.
  - **Agglutination reactions:** The interaction of particulate antigens with antibodies leads to agglutination reactions. Direct agglutination reactions can be: Slide agglutination test, Tube agglutination test, Heterophile agglutination tests and Antiglobulin (Coombs) test. Passive agglutination reaction depending on the carrier particles used.
  - Complement-fixation reactions are serological tests based on the depletion of a fixed amount of complement in the presence of an antigen-antibodyreactionandare:(i)Complementfixationtest; (ii) Immune adherence test; (iii) Immobilization test; (iv) Cytolytic or cytocidal reactions.
  - **Neutralization reactions:** In neutralization reactions, the harmful effects of a bacterial exotoxin or virus are eliminated by a specific antibody.
  - **Opsonization** is immune opsonization or nonimmune opsonization.
  - **Immunofloroscence:** Direct immunofluorescence test is used to detect unknown antigen in a cell or tissue. Indirect immunofluorescence test is used for detection of specific antibodies in the serum and other body fluids.
  - **Radioimmunoassay (RIA)** is a highly sensitive and quantitative procedure that utilizes radioactively labeled antigen or antibody.

- Various types of **enzyme-linked immunosorbent assay (ELISA)** depends on an enzyme-substrate reaction and are direct ELISA, indirect ELISA, sandwich ELISA and competitive ELISA.
- **Western blotting** is done for the detection of proteins.
- **Immunoelectronmicroscopic tests** include immunoelectronmicroscopy, immunoenzyme test, and immunoferritin test.

IMPORTANT QUESTIONS

1. Name various antigen-antibody reactions. Describe the principle and applications of precipitation reactions giving suitable examples.
2. Define agglutination reaction. Discuss the principle and applications of agglutination reactions giving suitable examples.
3. Write short notes on:
   - Zone phenomenon (or) prozone
   - Immunodiffusion (or) gel diffusion
   - Immunoelectrophoresis
   - Counterimmunoelectrophoresis (CIE) or countercurrent eletrophoresis (CIEP)
   - Rocket electrophoresis
   - Coagglutination
   - Neutralization tests
   - Opsonization
   - Complement fixation test (CFT).
   - Immunofluorescence tests

   Radioimmunoaasay (RIA)

   ELISA—its principle and application.

FURTHER READING

Balows A, et a1 1991. Manual of Clinical Microbiology.Washington DC: American Society of Microbiology.

Hudson L, Hay FC. Practical Immunology, 3rd edn. Oxford: Blackwell 1989.

Stites DP, Terr AI. Basic and Clinical Immunology, 7th edn. Norwalk: Appleton-Lange 1991.

CHAPTER 17

Structures and Functions of the Immune System

Learning Objectives

After reading and studying this chapter, you should be able to:
- Differentiate between T and B cells in a tabulated form.
- Describe cluster of differentiation (CD).
- Discuss the following: Natural killer cells or NK cells; killer cells or K cells or ADCC cells; human leukocyte antigen (HLA).

INTRODUCTION

The lymphoreticular system is a complex organization of cells of diverse morphology distributed widely in different organs and tissues of the body responsible for immunity. **Lymphoreticular cells** consist of lymphoid and reticuloendothelial components. The lymphoid cells (lymphocytes and plasma cells) are primarily concerned with the specific immune response. The phagocytic cells (polymorphonuclear leucocytes and macrophages), forming part of the reticuloendothelial system, are primarily concerned with the 'scavenger' functions of eliminating effete cells and foreign particles, thus contributing to nonspecific immunity by removing microorganisms from blood and tissues. They also play a role in specific immunity, both in the afferent and efferent limbs of the immune response.

TYPES OF IMMUNE RESPONSE

The functional anatomy of the lymphoid system can be appreciated only against the background of the 'two component concept' of immunity. The immune response to an antigen, whatever its nature, can be of two broad types:

1. Humoral or Antibody Mediated Immunity (HMI or AMI)

Humoral immunity is mediated by antibodies produced by plasma cells and present in blood and other body fluids (hence the name 'humoral' from 'humor', the old term for body fluids).

2. Cellular or Cell Mediated Immunity (CMI)

Cellular immunity is mediated directly by sensitized lymphocytes. Cells for each of these components develop through separate channels and remain independent, though they may also interact in some instances (Fig. 17.1).

ORGANS AND TISSUES OF THE IMMUNE SYSTEM

A number of morphologically and functionally diverse organs and tissues have various functions in the development of immune responses. Based on function, the organs and tissues of the immune system can be divided into primary or secondary lymphoid organs or tissues.

A. Primary (Central) Organs or Tissues

The primary organs or tissues are where immature lymphocytes mature and differentiate into antigen-sensitive mature B and T cells. These are:

i. Thymus: Primary lymphoid organ.
ii. Bone marrow: Primary lymphoid tissue.

B. Secondary (Peripheral) Organs and Tissues

The secondary organs and tissues serve as areas where lymphocytes may encounter and bind antigen, whereupon they proliferate and differentiate into fully mature, antigen-specific effector cells.

Examples: Lymph nodes, spleen, various mucosal associated lymphoid tissues (MALT)—such as gut-associated lymphoid tissue (GALT), SALT (skin associated lymphoid tissues).

A. Central (Primary) Lymphoid Organs

The thymus and bone marrow are the primary (or central) lymphoid organs, where maturation of lymphocytes takes place. The thymus is the primary lymphoid organ and bone marrow is the primary lymphoid tissue.

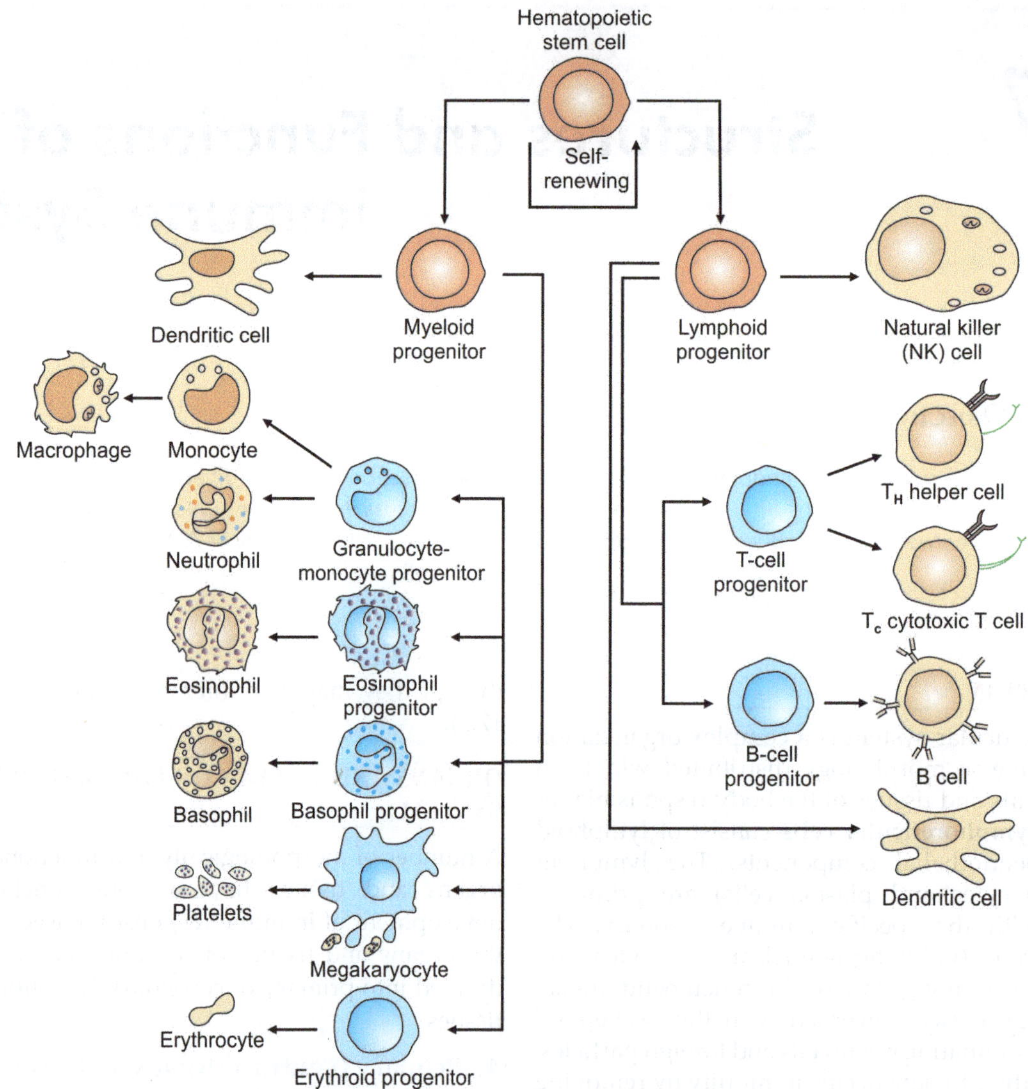

Fig. 17.1: Development of T and B cell systems

1. Thymus: A T Cell Factory

Thymus is a flat, bilobed organ situated above the heart and is the site of T cell development and maturation.

Function of the Thymus—Immunological Competence On The Lymphocytes

The primary function of the thymus is the production of thymic lymphocytes. Precursor cells from the bone marrow migrate into the thymus to the outer cortex where they proliferate. As they mature and acquire T cell surface markers, they move to the inner cortex where approximately 90 percent die, possibly as part of the acquisition of immune tolerance. The other 10 percent move into the medulla, become mature T cells, and enter the bloodstream. In the thymus, the lymphocytes acquire new surface antigens ('Thy' antigens).

Thymus (T) dependent lymphocytes: Lymphocytes produced in the thymus are called 'thymus (T) dependent lymphocytes' or 'T cells'. Lymphocyte proliferation in the thymus is not dependent on antigenic stimulation, unlike in the peripheral organs. Antigen introduced directly into the thymus may lead to a local immunity. Differentiation and maturation are under the influence of hormones such as **thymopoietin** and **thymosin**, produced by the epithelial elements in the thymus.

Thymus dependent antigens: T lymphocytes are selectively seeded into certain sites in the peripheral lymphatic tissues, being found in the white pulp of the spleen, around the central arterioles, and in the paracortical areas of lymph nodes. These regions have been termed **'thymus dependent'** as they are found grossly depleted after neonatal thymectomy.

Thymus and immune function: The importance of thymus in lymphocyte proliferation and development of CMI is evident from the lymphopenia, deficient graft rejection and the so called **'runt disease'** seen in neonatal thymectomized mice. These thymectomized mice show a dramatic decrease in circulating lymphocytes of the T cell lineage and an absence of cell-mediated immunity.

A **congenital birth defect in humans (DiGeorge's syndrome)** and in certain mice **(nude mice)** in which the thymus fails to develop are other evidence of the importance of the thymus. In both cases, there is an absence of circulating T cells and of cell-mediated immunity and an increase in infectious disease.

2. Bursa of Fabricius

In birds, undifferentiated lymphocytes move from the bone marrow to the **Bursa of Fabricius** where B cells mature. Stem cells from the yolk sac, fetal liver and bone marrow enter the bursa, proliferate and develop into immunocompetent **"bursal dependent"** or **B cells** (to designate their origin in the **bursa**). These migrate and seed selective areas in the peripheral lymphoid organs—the mantle, germinal follicles and perifollicular regions of the spleen, and the far cortical areas and medullary cords of lymphocytes. These are called **'bursa dependent'** or **'thymus independent areas'**. B lymphocytes transform into plasma cells and secretes antibodies following antigenic stimulation.

3. Bone Marrow

In humans and other mammals, the bone marrow acts as the bursa equivalent and is the site of B cell origin and development. Arising from lymphoid progenitors, immature B cells proliferate and differentiate within the bone marrow, and stromal cells within the bone marrow interact directly with the B cells and secrete various cytokines that are required for development. A selection process within the bone marrow eliminates B cells with self-reactive antibody receptors like thymic selection during T cell maturation. Bone marrow is not the site of B cell development in all species.

B. Peripheral (Secondary) Lymphoid Organs

1. Lymph Nodes

They are encapsulated bean-shaped structures containing a reticular network packed with lymphocytes, macrophages, and dendritic cells. Clustered at junctions of the lymphatic vessels, lymph nodes are the first organized lymphoid structure to encounter antigens that enter the tissue spaces.

Structure of lymph node: Lymph nodes have an indentation called the **hilus** where blood vessels enter and leave the node. A typical lymph node is surrounded by a fibrous **capsule** from which **trabeculae** penetrate into the nodes.

Morphology: Morphologically, a lymph node can be divided into three roughly concentric regions, each of which supports a distinct *microenvironment* (Fig. 17.2).

i. *Cortex:* The outermost layer, the cortex (B cell area), contains lymphocytes (mostly B cells), macrophages,

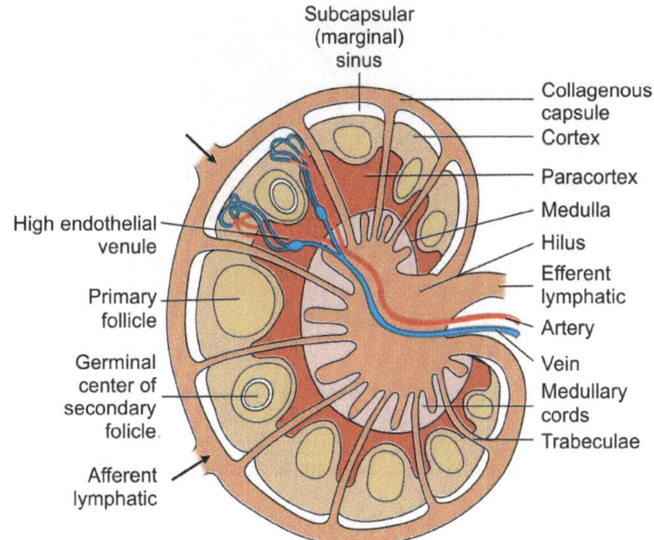

Fig. 17.2: Diagrammatic section of lymph node

and follicular dendritic cells arranged in **primary follicles**. After antigenic challenge, the primary follicles enlarge into **secondary follicles**, each containing a germinal center. The cortex lacks primary follicles and germinal centers in children with B cell deficiencies.

ii. *Paracortex (T cell area):* Beneath the cortex is the **paracortex**, which is populated largely by T lymphocytes and also contains interdigitating dendritic cells and thought to have migrated from tissues to the node. The paracortex is sometimes referred to as a **thymus-dependent area** in contrast to the cortex, which is a **thymus-independent area.**

iii. *Medulla:* The innermost layer of a lymph node is the **medulla** (consisting of cellular cords containing T cells, B cells, abundant plasma cells and macrophages). In the medulla, the lymphocytes, plasma cells and macrophages are arranged as elongated branching bands **(medullary cords).** The cortical follicles and medullary cords contain B lymphocytes and constitute the **bursa dependent areas.** Paracortical area contains T lymphocytes and constitutes the thymus dependent area.

Functions of Lymph Node

i. Lymph nodes act as a filter for lymph, each group of nodes draining a specific part of the body.
ii. They phagocytose foreign materials including microorganisms.
iii. They help in the proliferation and circulation of T and B cells.
iv. They enlarge following local antigenic stimulation.

2. Spleen

The spleen is the large secondary lymphoid organ located in the abdominal cavity.

Structure of spleen: It has a **capsule** from which extends a number of projections **(trabeculae)** into the interior

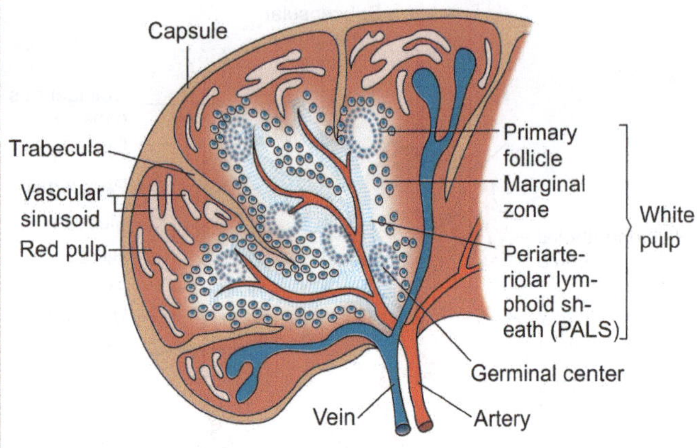

Fig. 17.3: Schematic diagram of splenic architecture

to form a compartmentalized structure. The compartments are of two types, **white pulp** and the **red pulp (Fig. 17.3)**.

White pulp: The branches of the splenic artery travel along the trabeculae, and leaving them branch again to form the **central arterioles**, which are surrounded by a sheath of lymphoid tissues. This part is known as the **white pulp**. The central arterioles proceed onto the red pulp, so called because of the abundance of red blood cells in it. The splenic **white pulp** surrounds the branches of the splenic artery, forming a **periarteriolar lymphoid sheath (PALS)** populated mainly by T lymphocytes. Primary lymphoid follicles are attached to the PALS. These follicles are rich in B cells and some of them contain **germinal centers** which develop following antigenic stimulation. Upon antigenic challenge, these **primary follicles** develop into characteristic **secondary follicles** containing germinal centers, where rapidly dividing B cells (centroblasts) and plasma cells are surrounding by dense clusters of concentrically arranged lymphocytes. Outside the germinal center is a **'mantle layer'** of lymphocytes. The **marginal zone**, is located peripheral to the PALS, is populated by lymphocytes and macrophages. The lymphatic sheath surrounding the central arterioles is the thymus dependent area of the spleen. The perifollicular region, germinal center and mantle layer form the bursa dependent (thymus independent) areas (Fig. 17.4).

Red pulp: The splenic **red pulp** consists of a network of sinusoids populated by macrophages and numerous red blood cells (erythrocytes) and few lymphocytes.

Functions of Spleen

i. Functions as the graveyard for blood cells.
ii. Mounting immune responses to antigens in the blood stream is a major role. The spleen specializes in filtering the blood and trapping blood-borne microorganisms and antigens.

3. Mucosa-Associated Lymphoid Tissue (MALT)

The mucosa lining the alimentary, respiratory, genitourinary and other lumina and surfaces are constantly exposed to numerous antigens. These vulnerable membrane surfaces are defended by a group of organized lymphoid tissues known collectively as **mucosal-associated lymphoid tissue (MALT)**. Like the skin, mucous membranes also have this specialized immune barrier.

Structure of MALT: Structurally, these tissues range from loose, barely organized clusters of lymphoid cells in the lamina propria of intestinal villi to well-organized structures such as the familiar tonsils and appendix, as well as Peyer's patches, which are found within the submucosal layer of the intestinal lining.

Gut-associated Lymphoid tissue (GALT): The system most studied is the **gut-associated lymphoid tissue (GALT)**. GALT includes the tonsils, adenoids, and Peyer's patches in the intestine.

Bronchial-associated lymphoid tissue (BALT): Less well-organized mucosal-associated lymphoid tissue also occurs in the respiratory system and is called **bronchial-associated lymphoid tissue** (BALT). The diffuse MALT in the urogenital system does not have a specific name.

Mucosal or secretory immune system: Mucous membranes are an effective barrier to the entrance of most pathogens, which contributes to nonspecific immunity. This indicates the existence of a common **mucosal** or **secretory immune system** and explains the superiority of oral or nasal immunization over the parenteral route for many enteric and respiratory infections

CELLS OF THE LYMPHORETICULAR SYSTEM

Lymphoid Cells: Cells of the Immune System

The cells responsible for both nonspecific and specific immunity are the white blood cells called **leukocytes**. All of the leukocytes originate from pluripotent stem cells in the fetal liver and in the bone marrow of the animal host from which they migrate to other body sites, undergo further development, and perform their various functions.

Lymphocytes: Lymphocytes constitute 20-40 percent of the body's white blood cells and 99 percent of the cells in the lymph. These lymphocytes continually circulate in the blood and lymph and are capable of migrating into the tissue spaces and lymphoid organs, thereby integrating the immune system to a high degree. Many mature lymphoid cells are long-lived, and persist as memory cells for many years.

Lymphocytes are now recognized as the major cellular elements responsible for immunological responses. The lymphocytes can be broadly subdivided into three populations—B cells, T cells, and natural killer cells. According to their size, lymphocytes can be classified into small (5-8 µm), medium (8-12 µm) and large (12-15 µm) lymphocytes.

Short-lived and long-lived lymphocytes: Depending on their lifespan, they can be classified as short-lived and long-lived lymphocytes. In human beings, the short-lived lymphocytes have a lifespan of about two weeks, while the long lived cells may last for three years or more, or even for life. Short-lived lymphocytes are the effector cells in immune response, while the long-lived cells act as the storehouse for immunological memory. Long lived cells are mainly thymus derived.

Lymphatic recirculation: Lymphopoiesis takes place mainly in the central lymphoid organs where they differentiate and mature before entering the circulation and then the peripheral lymphoid organs and tissues. These populations of lymphocytes do not remain distinct but mix together in a process known as **'lymphocyte recirculation'**. There is a constant traffic of lymphocytes through the blood, lymph, lymphatic organs and tissues. This recirculation ensures that following introduction of antigen into any part of the body, lymphocytes of appropriate specificity would reach the site during their ceaseless wandering and mount an immune response. A lymphocyte completes one cycle of recirculation in about one or two days. Recirculating lymphocytes can be recruited by the lymphoid tissues whenever necessary. Recirculating lymphocytes are mainly T cells. B cells tend to be more sessile. Chronic thoracic duct drainage will therefore result in selective T cell depletion.

Differences Between T and B Cells

Many tests help in differentiation of T and B cells (Table 17.1). These include:

1. **Thymus-specific antigens:** T cells have thymus-specific antigens, which are absent on B cells.
2. **T cell receptor (TCR):** All T cells express an antigen-binding **T cell receptor (TCR),** which resembles but differs from antibody, and **CD2-** and **CD3- associated** proteins on their surface.
3. **Surface immunoglobulins:** B cells have immunoglobulin on their surface.
4. **SRBC rosette:** T cells bind to sheep erythrocytes forming rosettes (SRBC or E rosette through CD2 molecule. B cells do not.
5. **EAC rosettes:** B cells bind to sheep erythrocytes coated with antibody and complement, forming EAC rosettes, due to the presence of a C3 receptor (CR2) on the B cell surface. This receptor (CR2) also acts as a receptor for the Epstein-Barr virus. T cells do not possess this.
6. **Microvilli:** Viewed under the scanning microscope, T cells are generally free of cytoplasmic surface projections, while B cells have an extensively filamentous surface, with numerous microvilli.
7. **Blast transformation:** T cells undergo blast transformation, evidenced by enhanced DNA synthesis, on treatment with mitogens such as phytohemagglutinin (PHA) or Concanavalin A (Con A), while B cells undergo similar transformation with bacterial endotoxins, *Staphylococcus aureus* (Cowan 1 strain) or EB virus.

Table 17.1: Comparison of T cells and B cells

Property	*T cell*	*B cell*
A. Origin	Bone marrow	Bone marrow
B. Maturation	Thymus	Bursal equivalent: Bone marrow, Payer's patches
C. Location		
1. Peripheral blood	65-85%	15-25%
2. Lymph node	60-75%	30-35%
3. Spleen	25-45%	55-60%
4. Thoracic duct	80-90%	10-20%
5. Thymus	96%	Negligible
D. Thymus specific antigens	+	-
E. CD3 receptor	+	-
F. Surface immunoglobulins	-	+
G. Receptor for Fc piece of IgG	-	+
H. SRBC rosette (CD2; measles receptor)	+	-
I. EAC rosette (C3 receptor; CR2; EBV receptor)	-	+
J. Numerous microvilli, on surface	-	+
K. Blast transformation with:		
1. anti-CD3	+	-
2. anti-Ig	-	+
3. PHA	+	-
4. Concanavalin A	+	-
5. Endotoxins	-	+

T Lymphocytes

T Cell Maturation (Fig. 17.4)

T lymphocytes derive their name from their site of maturation in the **thymus**. T cell precursors from the yolk sac, fetal liver and bone marrow migrate to the thymus during the embryonic and postnatal stages. Once in the thymus, the stem cells begin to differentiate into thymic lymphocytes (called **thymocytes**). T cells change their phenotype during maturation. The cortex contains, mostly, rapidly dividing immature thymocytes. The earliest identifiable cells of T lineage are the $CD7^+$ pro-T cells, which acquire CD2 on entering the thymus. They synthesize CD3 in the cytoplasm and become pre-T cells. T cell receptor (TCR) synthesis also takes place. These earliest cells lack the two major T cell markers (CD4 and CD8) and are termed **double negative cells.** They express cytoplasmic but not surface TCR-associated CD3 and now committed to become T cells. They continue to express CD7 together with CD2 and CD5. T cell receptor (TCR) synthesis also takes place.

Immature T cells in the thymus exhibit CD7, 2, 3, I, 4 and 8, besides TCR. On functional maturity, they lose CDI and differentiate into the two major subsets $CD8^-4^+$ or $CD4^-8^+$. The CD4 and CD8 proteins are coreceptors for the TCR, because they facilitate the interaction of the TCR with the antigen-presenting MHC molecule and can enhance the activation response.

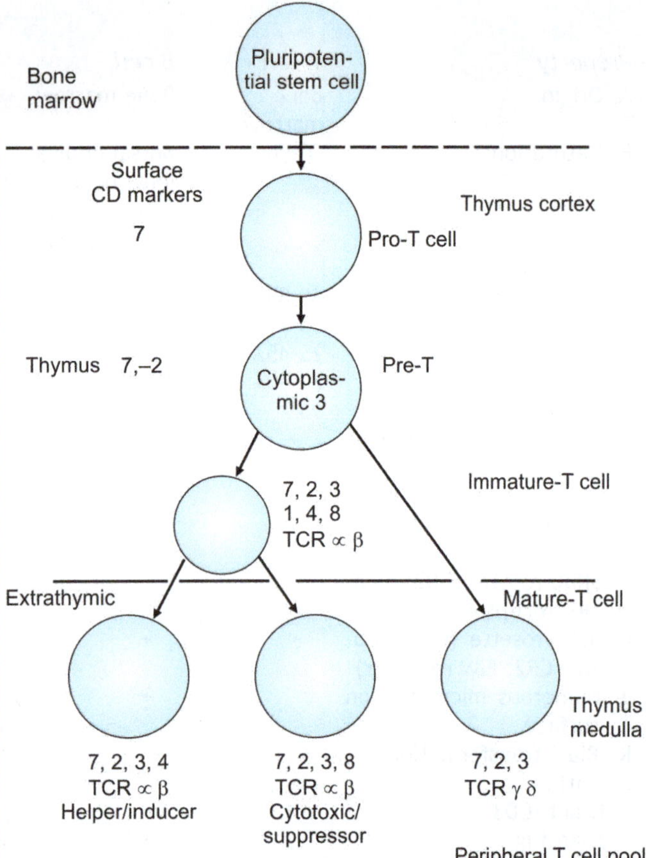

Fig. 17.4: T cell maturation pathway

Mature $CD8^-4^+$ TCR αβ cells are helper/inducer cells, inducing B cell differentiation, stimulating proliferation of $CD8^+$ cytotoxic cells, producing lymphokines and regulating certain stages of erythropoiesis. $CD4^-8^+$ TCR αβ cells are suppressor cytotoxic cells, inhibiting B cell antibody synthesis and acting as cytotoxic effector cells. Minor subsets of $CD4^+$ cells and $CD8^+$ cells also exist. Small numbers of $CD4^+8^+$ and $CD4^-8^-$ cells are also present in circulation.

Application: Sequential antigenic changes characterizing T cell maturation enable their easy identification. This has application in defining T cell malignancies. Acute T cell malignancies such as lymphoblastic leukemia and lymphomas involve early T cells, Pro-T cells and other immature forms. Chronic T cell malignancies like mycosis fungoides, peripheral T cell lymphomas and HTLV-1 associated adult I cell leukemias involve mature T cells, mainly CD^{4+} cells.

MHC restriction: T cells also develop MHC restriction. Therefore, $CD8^+$ cells respond only to foreign antigens presented along with HLA Class I, and $CD4^+$ cells to those presented with HLA Class II molecules. Approximately two-thirds of the peripheral blood T cells are CD4 T helper cells, and approximately one-third are CD8 cytotoxic cells. This is the basis for the CD4/CD8 ratio used to assess HIV infection.

T cell receptor (TCR): TCR is a heterodimer of glycoprotein chains expressed on the T cell surface, which in association with CD3 acts as the antigen recognition unit, analogous to the Ig on the surface of B cells. These transmembrane receptors, found only on the surface of T cells, consist of two different polypeptide chains—either a plus b, or g plus d, that are attached to each other by disulfide bonds. Large majority of T cells carry ab TCR.

The structure of these heterodimers is similar to those of the immunoglobulins (antibodies), and the TCRs are therefore referred to as members of the **immunoglobulin gene superfamily**.

The function of TCR γδ cells is not well understood, but they are believed to be immune surveillance cells on epithelial surfaces and a form of defense against intracellular bacteria.

Types of T Cells

Based on their surface markers, MHC restriction, target cells and function, the following T cells subtypes are recognized:

a. On the Basis of Surface Markers

Classification of lymphocytes on the basis of surface markers makes use of two important characteristics:

1. Cluster of differentiation or cluster determinant (CD).
2. Antigen recognition receptors.

1. Cluster of Differentiation or Cluster Determinant (CD)

The term **cluster of differentiation (CD)** refers families of surface glycoprotein antigens that can be recognized by specific antibodies produced against them. These markers reflect the stage of differentiation and functional properties of the cells. Thus a cell displaying CDI is identified by the binding of antibodies against CDI. Each class of leucocyte displays a diagnostic pattern of CDs. Over 200 CD markers have been identified so far.

Examples

- CD3 is expressed only by T cells.
- CD19 is expressed only by B cells.
- CD64 is expressed only by monocytes.
- CD66 is expressed only by granulocytes.
- CD68 is expressed only by macrophages.
- On the other hand, CD18 and CD45 are expressed by a variety of leucocyte types.

2. Antigen Recognition Receptors

These include membrane-bound (surface) immunoglobulins (mIgs or sIgs) in B cells and T cell receptors (TCRs) in T cells. In contrast to CDs, which can serve as diagnostic feature for all leucocytes, antigen recognition receptors are limited to B and T lymphocytes only. Both mIgs and TCRs serve as specific surface receptors, recognizing and interacting with only single antigenic determinant on the antigen. Reaction of antigens with mIgs and TCRs activates B cells and T cells respectively, leading to proliferation and differentiation.

The antigen specificity of the mIgs in B cells and TCRs in T cells is predetermined, and the sole effect of antigen is to select out a cell with appropriate surface receptor and induce it to clonally expand and differentiate into a cell that will produce the antibody it has been predetermined to make or produce specific clones of effector T cells respectively.

b. On the Basis of Functions

A. Regulatory T Cells

1. **Helper/inducer cell (Th)**—Helper/inducer cell (Th), with CD4 surface marker, MHC class II restriction is generally stimulating and promoting the growth of T cells and macrophages. They help B cells make antibody in response to antigenic challenge; stimulate cell mediated immunity. Based on the different profiles of cytokines produced, two subsets are identified Th1 and Th2.
   Th1 cells: Th1 cells produce mainly the cytokines interferon gamma (IFN-γ) and interleukin-2 (IL-2) which activate macrophages and T cells promoting CMI, destruction of target cells and killing of intracellular microbes, such as tubercle and lepra bacilli.
   Th2 cells: Th2 cells produce mainly the cytokines IL4, 5 and 6 which stimulate B cells to form antibodies.
2. **Suppressor T cells (Ts cells):** These have CD8 surface marker and MHC class I restriction. They can suppress B cell and T cell response.

B. Effector Cells

1. **Delayed type—Hypersensitivity T cells (Td cells)**—They are involved in delayed hypersensitivity and cell-mediated immune response.
2. **Cytotoxic T cells (Tc cells):** They are also called $CD8^+$ cells with CD8 surface marker and MHC class I restriction. They can kill and lyse target cells carrying new or foreign antigens, including tumor, allograft and virus infected cells.
   Memory cells (Tm)—both CD4 and CD8 cells provide memory and an amnestic immune response.

Regulation of Immune Response

Immune response is regulated by mutually opposing influence of Th cells and Tc cells. Their balanced activity results in optimal response. Overactivity of helper cells and decreased suppressor cell activity may lead to conditions like autoimmunity. Diminished helper cell function or increased suppressor activity causes immunodeficiency. Helper cells constitute about 65 percent and suppressor cells about 35 percent of circulating T lymphocytes.

B cells and Plasma Cells

B cell Maturation

The B lymphocyte derived its letter designation from its site of maturation, in the bursa of fabricius in birds; and the bone marrow of mammals, including humans and mice. B cell differentiation also takes place in the fetal liver and fetal spleen. About 5-15 percent of the circulating lymphoid pool are B cells defined by the presence of surface immunoglobulin.

During ontogeny, the first recognizable B cell possesses only surface IgM. As embryonic development proceeds, these surface immunoglobulin (sIg)-bearing B cells acquire Fc receptors, surface IgD, class 11 molecules (Ia antigens), and complement receptors, in essentially in that order. Thus, most B cells in mature individuals have the following cell-surface phenotype: IgM IgD FcR Ia CR. Plasma cells, which are descendants of B cells, lack all these markers.

Mature B cell undergoes **clonal proliferation** on contact with its appropriate antigen. Interaction between antigen and the membrane-bound antibody on a mature naive B cell, as well as interactions with T cells and macrophages, selectively induces the activation and differentiation of B cell clones of corresponding specificity. In this process, the B cell divides repeatedly and differentiates over a 4- to 5-day period, generating a population of **plasma cells** and **memory cells.** Plasma cells, which have lower levels of membrane-bound antibody than B cells, synthesize and secrete antibody. **Memory cells** circulate until activated by specific antigen. All clonal progeny from a given B cell secrete antibody molecules with the same antigen-binding specificity.

Plasma Cell

Plasma cells are fully differentiated antibody-secreting effector cells of the B cell lineage. Antigenically stimulated B cells undergo **blast transformation**, becoming successively **plasmoblasts**, intermediate transitional cells and **plasma cells**. Plasma cells are factories of antibody production. They are terminally differentiated cells, and many die in 1 or 2 weeks. A plasma cell makes an antibody of a single specificity, of a single immunoglobulin class and allotype and a single light chain type only. An exception is found in primary antibody response, when a plasma cell producing IgM initially and later it may switch to IgG production. Lymphocytes, lymphoblasts and transitional cells may also synthesize Ig to some extent, while plasma cell is the best antibody producing cell.

Null Cells

A small proportion (5-10%) of lymphocytes that lack distinguishing phenotypic markers characteristic of T or B lymphocytes are called null cells. Because of their morphology, they are also known as large granular lymphocytes (LGL). LGL are a heterogeneous group of cells with differences in their functional and surface marker features. The member of this group is the:

a. **Natural killer (NK) cells:** The most important member. The term NK cell is sometimes used as a common name for all null cells.

b. **Antibody-dependent cellular cytotoxic (ADCC) cells.**
c. **Lymphokine activated killer (LAK) cells.**

a. Natural Killer (NK) Cells

Natural killer (NK) cells are derived from large granular lymphocytes (LGL). They possess spontaneous cytotoxicity towards various target cells, mainly malignant and virus infected cells. They differ from killer cells in being independent of antibody. NK activity can be enhanced by lymphokines, specifically interferon-γ. NK cells differ from Tc cells in other properties as well-NK activity is 'natural' or 'nonimmune' as it does not require sensitization by prior antigenic contact. NK cells therefore form part of the innate immune setup.

Function: They are considered to be important in:

i. Immune surveillance.
ii. Natural defence against virus infected and maligant mutant cells.
iii. NK cells are capable of nonspecific killing of virus-transformed target cells and are involved in allograft and tumor rejection.

b. Antibody-dependent Cell Mediated Cytotoxicity (ADCC)

A subpopulation of LGLs possesses surface receptors for the *Fc* part of Ig. They are capable of lysing or killing target cells sensitized with IgG antibodies. This is an example of a process known as antibody-dependent cell mediated cytotoxicity (ADCC). This antibody dependent cellular cytotoxicity is distinct from the action of cytotoxic T cells, which is independent of antibody. ADCC cells were formerly called killer (K) cells but are now classified with NK cells.

c. Lymphokine Activated Killer (LAK) Cells

Lymphokine activated killer (LAK) cells are NK lymphocytes treated with interleukin-2 (lL-2), which are cytotoxic to a wide range of tumor cells without affecting normal cells. LAK cells have shown promise in the treatment of some tumors such as renal cell carcinoma. IL-2 also acts as a growth factor for NK cells.

Phagocytic Cells

Phagocytic cells are the **mononuclear macrophages** (of blood and tissues) and the **polymorphonuclear microphages**.

i. **Mononuclear cells:** The mononuclear phagocytic system consists of monocytes circulating in the blood and macrophages in the tissues. Both types are highly phagocytic and make up the monocyte-macrophage system. The blood macrophages (monocytes) are the largest of the lymphoid cells found in peripheral blood (12-15 µm). The tissue macrophages (histiocytes) are larger (15-20 µm). Mononuclear macrophage cells originate in the bone marrow from precursor cells and become monocytes in about six days. Monocytes in circulation have an approximate half-life of three days.

Monocytes: Monocytes (Greek *monos*, single, and *cyte*, cell) are mononuclear phagocytic leukocytes with an ovoid or kidney-shaped nucleus and granules in the cytoplasm that stain gray-blue. They are produced in the bone marrow and enter the blood, circulate for about eight hours, enlarge, migrate to the tissues, and mature into macrophages.

Macrophages: Macrophages [Greek *macros*, large, and phage, to eat] are derived from monocytes and are also classified as mononuclear phagocytic leukocytes.

Monocytes leave the circulation and reach various tissues to become transformed into macrophages, with morphological and functional features characteristic of the tissues. Macrophages spread throughout the animal body and take up residence in specific tissues where they are given special names, e.g. alveolar macrophages in the lung, histiocytes in connective tissues, Kupffer cells in the liver, Mesangial cells in the kidney, microglial cells in the brain and osteoclasts in the bone.

Functions of Macrophages

1. ***Phagocytosis:*** The primary function of macrophages is phagocytosis. Since macrophages are highly phagocytic, their function in nonspecific resistance will be discussed in more detail, in the context of phagocytosis.
2. Antigen presentation to T cells to initiate immune responses.
3. Secretion of cytokines to activate and promote innate immune response.

Although normally in a resting state, macrophages are activated by a variety of stimuli in the course of an immune response. Phagocytosis of particulate antigens serves as an initial activating stimulus. However, macrophage activity can be further enhanced by cytokines secreted by activated Th cells, by mediators of the inflammatory response, and by components of bacterial cell walls. Macrophages secrete interleukin-1, interleukin-6, tumor necrosis factor, and interleukin-12, in response to bacterial interaction, which stimulate immune and inflammatory responses, including fever. One of the most potent activators of macrophages is interferon gamma (IFN-γ) secreted by activated Th cells.

Polymorphonuclear microphages: Microphages are the polymorphonuclear leucocytes of the blood. Because of the irregular-shaped nuclei, granulocytes are also called **polymorphonuclear leukocytes or PMNs**. Three types of granulocytes exist: basophils, eosinophils, and neutrophils.

a. **Neutrophils:** Neutrophils are actively phagocytic and form the predominant cell type in acute inflammation. The phagocytic property of neutrophils is nonspecific, except for its augmentation by opsonins.

b. **Eosinophils:** They primarily inhabit tissues rather than the bloodstream. Their distinctive feature is the presence of two types of granules—the small, and the large ones. The granules contain a variety of hydrolytic enzymes which bring about extracellular killing of large parasites. Eosinophils possess phagocytic activity but only to a limited degree. They are found in large numbers in allergic inflammation, parasitic infections and around antigen-antibody complexes.
c. **Basophils:** Basophils [Greek *basis*, base, and *philein*, to love] have an irregular-shaped nucleus with two lobes, and the granules stain bluish-black with basic dyes. Basophil leukocytes are found in the blood and tissues (mast cells). Their cytoplasm has large numbers of prominent basophilic granules containing heparin, histamine, serotonin and other hydrolytic enzymes. Degranulation of mast cells, with release of pharmacologically active agents, constitutes the effector mechanism in anaphylactic and atopic allergy.

Dendritic cells: While macrophages are the major antigen presenting cells, another type of cell known as the dendritic cell also performs this function. Dendritic cells are bone marrow derived cells of a lineage different from the macrophages and T or B lymphocytes. Four types of dendritic cells are known: Langerhans cells, interstitial dendritic cells, myeloid cells, and lymphoid dendritic cells. They possess MHC class II antigens but not Fc or sheep RBC receptors or surface immunoglobulins.

They are more potent antigen-presenting cells than macrophages and B cells, both of which need to be activated before they can function as antigen-presenting cells (APCs). Dendritic cells are specially involved in the presentation of antigens to T cells during the primary immune response. The B cell is another antigen presenting cell, particularly during the secondary immune response.

MAJOR HISTOCOMPATIBILITY COMPLEX

The major histocompatibility complex (MHC) is a remarkable cluster of genes that control T cell recognition of self and nonself. The complex was named *histocompatibility* because it was first detected through the rejection of tissue grafts exchanged between different strains of mice. It was a *major* system because rejection was rapid and difficult to control, and it was *complex* because many genes of differing function were clustered together. No less than eight Nobel prizes have been awarded to pioneers in transplantation and immunologic tolerance.

MHC proteins play a pivotal role in "presenting" antigens to T cells. In humans, the MHC is called the human leukocyte antigen (HLA) complex. Unlike most proteins that have a fixed, defined structure, key amino acid sequences in the MHC proteins vary widely from person to person, and thus they act as "identity markers" on the surface of cells. Interestingly, these proteins were first detected by their effect on transplant rejection (that is, tissue incompatibility).

Classes of MHC Molecules

The major histocompatibility complex is a collection of genes arrayed within a long continuous stretch of DNA on chromosome 6 in humans and chromosome 17 in mice. Early studies on MHC were carried out in mice. However, all species of animals (including human beings) examined subsequently were found to possess a similar complex of genes on a segment of one chromosome pair, coding for three different classes of proteins. These proteins are grouped into three classes.

They are the principal antigens involved in graft rejection and cell mediated cytolysis. Class I molecules may function as components of hormone receptors.

HLA Complex

The name 'histocompatibility complex' arose because its discovery was based on transplantation experiments, and only later were the other two components of the complex identified. The major antigens determining histocompatibility in human beings are alloantigens, characteristically found on the surface of leucocytes. Human MHC antigens are therefore synonymous with **human leucocyte antigens (HLA)**, and the MHC complex of genes with the HLA complex.

The HLA complex of genes is located on the short arm of chromosome 6 (Fig. 17.5). It consists of three separate clusters of genes:

1. **HLA class I comprising A, B and C loci:** HLA class I comprising A, B and C loci. Class I proteins are encoded by the HLA-A, HLA-B, and HLA-C genes in humans. HLA class I antigens (A, B and C) are found on the surface of virtually all nucleated cells and, in some species (i.e. mice, but not humans), on red blood cells as well.

   They are the principal antigens involved in graft rejection and cell mediated cytolysis. Class I molecules may function as components of hormone receptors.
2. **Class II or the D region consisting of DR, DQ and DP loci:** Class II proteins are encoded by the HLA-D region. There are three main sets: the DP, DQ and DR encoded molecules. HLA class II antigens are more restricted in distribution, being found only on cells of the immune system—macrophages,

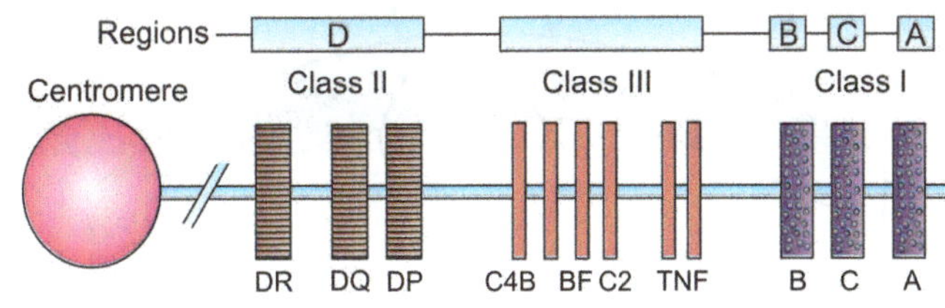

Fig. 17.5: HLA complex loci on chromosome

dendritic cells, activated T cells, and particularly on B cells.

Class II genes are also involved in graft rejection and in the regulation of immune responses to exogenously produced but processed soluble antigen.

3. **Class III or the complement region:** Class III or the complement region containing genes for complement components C2 and C4 of the classical pathway, as well as properdin factor B of the alternative pathway, heat shock proteins and tumor necrosis factors—alpha and beta.

HLA loci are multiallelic, that is, the gene occupying the locus can be any of several alternative forms (alleles). As each allele determines a distinct product (antigen), the HLA system is very pleomorphic. For example, at least 24 distinct alleles have been identified at HLA locus A and 50 at B.

HLA Molecules

HLA antigens are two-chain glycoprotein molecules anchored on the surface membrane of cells. Class I and class II MHC are membrane-bound glycoproteins that are closely related in both structure and function.

Class I MHC molecules: Class I MHC molecules consist of a heavy peptide chain (alpha chain) noncovalently linked to a much smaller peptide called beta 2-microglobulin (beta chain) (Fig. 17.6). The alpha chain is a transmembrane glycoprotein encoded by polymorphic genes within the A, B and C regions of the human HLA complex. The beta chain has a constant aminoacid sequence and is coded for by a gene on chromosome 15. Association of the alpha chain with β_2-microglobulin is required for expression of class I molecules on cell membranes. The α chain is anchored in the plasma membrane by its hydrophobic transmembrane segment and hydrophilic cytoplasmic tail.

The alpha chain of class I MHC molecules consists of three globoid domains (alpha 1, alpha 2, alpha 3) which protrude from the cell membrane and a small length of transmembrane C terminus reaching into the cytoplasm. The distal domains (alpha 1 and alpha 2) have highly variable aminoacid sequences and are folded to form a cavity or groove between them. Class I MHCs bind small peptides produced in the cytoplasm by degradation of self-proteins, nonself proteins synthesized in bacteria- or virus-infected cells, or tumor cells. Protein antigens processed by macrophages or dendritic cells to form small peptides are bound to this groove for presentation to CD8 T cells. The T cell will recognize the antigen only when presented as a complex with the MHC class I molecule and not otherwise (**MHC restriction**). When so presented, the CD8 cytotoxic killer cell destroys the target cell (for example, a virus infected cell).

Class II MHC molecules: Class II MHC molecules are heterodimers, consisting of an alpha and a beta chain (Fig. 17.7) which associate by noncovalent interactions. Like class I alpha chains, class II MHC molecules are membrane-bound glycoproteins that contain external domains, a transmembrane segment, and a cytoplasmic anchor segment. Each chain has two domains, the proximal domain being the constant region and the distal, the variable. Each chain in a class II molecule contains two external domains: $\alpha 1$ and $\alpha 2$ domains in one chain and $\beta 1$ and $\beta 2$ domains in the other. The two distal domains (alpha 1, beta 1) constitute the antigen-binding site for recognition by CD4 lymphocytes, in a fashion similar to the recognition of the Class I antigen peptide complex

Fig. 17.6: Structure of HLA class I molecule

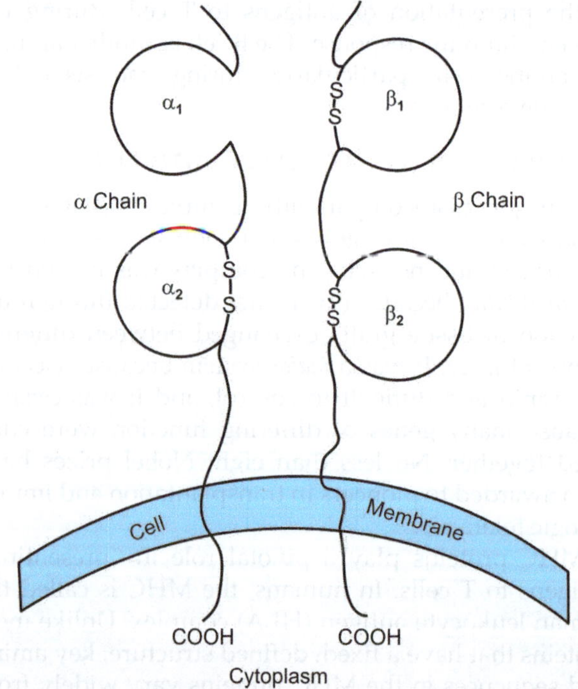

Fig. 17.7: Structure of HLA class II molecule

by CD8 T cells. Class II MHCs bind peptides that are derived from antigens brought into antigen-presenting cells (APCs) by endocytosis or phagocytosis, and then degraded in vesicles by lysosomal acid hydrolases. As with the class I MHCs, the class II MHC-peptide fragment complex is presented on the surface of the APC (for example, macrophages, dendritic cells, and B cells), and this complex is recognized by the appropriate T cell receptor of CD4 T cells. Antigen-driven clonal expansion of the specific T cells occurs, resulting in cytokine-mediated helper T cells functions.

HLA class II molecules are primarily responsible for the graft-versus-host response and the mixed leukocyte reaction (MLR).

Both class I and II molecules are members of the **immunoglobulin supergene family**. These two types of molecule are folded into domains of a similar overall structure to immunoglobulin and, along with other molecules of the immune system involved in recognition process, are thought to have evolved from a common ancestral molecule. The immune response (Ir) genes which control immunological responses to specific antigens are believed to be situated in the HLA class I region, probably associated with the DR locus. Ir genes have been studied extensively in mice and located in the I region of mouse MHC. They code for Ia (I region associated) antigens consisting of IA and IE proteins. However, the relevance of Ir genes in humans is not clear.

Role of MHC Diversity

1. **Transplantation:** The MHC system was originally identified in the context of transplantation, which is an artificial event.
2. **For protecting the species from the broadest possible number of pathogens:** The reason for the enormous diversity found in the MHCs is presumably a mechanism that is essential for protecting the species from the broadest possible number of pathogens. The primary aim of the MHC may be defense against microbes and not against the graft.
3. **Nonimmnological phenomena:** MHC has been implicated in a number of nonimmnological phenomena such as individual odor, body weight in mice and egg laying in chickens.

HLA Typing

Antisera for HLA typing were obtained principally from multiparous women. These multiparous women tend to have antibodies to the HLA antigens of their husbands, due to sensitization during pregnancy. However, monoclonal antibodies to HLA antigens have been developed.

Typing is done serologically by **microcytotoxicity**. It tests for complement mediated lysis of peripheral blood lymphocytes with a standard set of typing sera. However, serological typing is not possible for HLA-DR antigens, which are detected by the **mixed leucocyte reaction (MLR) and primed lymphocyte typing (PLT)**, respectively. **Genetic methods** are being used increasingly for HLA-typing in advanced centers which employ restriction fragment length polymorphism (RFLP) and gene sequence specific oligonucleotide probe typing.

Uses of HLA Typing

1. **Tissue transplantation:** HLA typing is used primarily for testing compatibility between recipients and potential donors before tissue transplantation.
2. **Disputed paternity:** It also has applications in disputed paternity.
3. **Anthropological studies:** As the prevalence of HLA types varies widely between different human races and ethnic groups, HLA typing is used in anthropological studies.
4. **An association between HLA types and diseases:** An association has been observed between HLA types and certain diseases. For example, strong association has been found between ankylosing spondylitis and HLA-B27, rheumatoid arthritis and HLA-DR4, and many autoimmune conditions and HLA-DR3.

MHC Restriction

Both $CD4^+$ and $CD8^+$ T cells can recognize antigen only when it is presented by a self-MHC molecule, an attribute called *self-MHC restriction*. Both class I and class II antigens operate in this phenomenon. Cytotoxic T lymphocytes from immunized mice are able to kill and lyse virus infected target cells only when the T cells and target cells are of the same MHC type, so that the T cells can recognize class I MHC antigens on the target cells. Helper T cells can accept antigen presented by macrophages only when the macrophages bear the same class II MHC molecules on the surface. For T cells participating in delayed type hypersensitivity, the antigen has to be presented along with class II MHC.

In 1996, Peter Doherty and Rolf Zinkernagel were awarded the Nobel Prize for Medicine for their seminal contributions in this area in view of the great importance of MHC restriction in immunological control.

KNOW MORE

- **Role of MHC diversity:** The reason for the enormous diversity found in the MHCs is presumably a mechanism that is essential for protecting the species from the broadest possible number of pathogens. For example, if there were only one type of MHC on all human cells, the probability that all invading organisms have epitopes that could be recognized by the single MHC peptide binding cleft would presumably be small. Therefore, in the absence of a MHC-peptide complex, no T cell response could be instituted, and a large segment of the human population could be killed by the resulting massive infection by increasing the specificity of self antigens, the MHC prevents microbes with related antigenic make up sneaking past host immune defences by molecular mimicry. The primary aim of the MHC may be defense against microbes and not against the graft.

☞ KEY POINTS

- The immune system is organized into several special tissues which are collectively termed as lymphoid or immune tissues.
- The lymphoid organs, based on their function, are classified into primary (central) and secondary (peripheral) lymphoid organs.
- The primary lymphoid organs provide sites where lymphocytes mature and become antigenically committed. T lymphocytes mature within the thymus, and B lymphocytes arise and mature within the bone marrow of humans, mice, and several other animals, but not all vertebrates.
- Secondary lymphoid organs capture antigens and provide sites where lymphocytes become activated by interaction with antigens.
- There are three types of lymphocytes: B cells, T cells, and natural killer cells (NK cells). The three types of lymphoid cells are best distinguished on the basis of function and the presence of various membrane molecules.
- Lymphocytes can be classified as thymus-derived cells or bone marrow-derived cells.
- T cells perform two important functions: cytotoxicity and delayed hypersensitivity.
- T cells play key role in regulating antibody production and CMI, and in suppression of certain immune responses.
- Bone marrow-derived lymphocytes are known as B lymphocytes or B cells and perform two important functions.
- NK cells have the ability to kill certain virally infected cells and tumor cells without prior sensitization.
- Antigen-presenting cells (APCs) include macrophages and dendritic cells.
- The MHC in humans is known as HLA complex. In humans, the HLA complex of genes is located on short arm of chromosome 6 containing several genes that are critical to immune function.
- The genes encode MHC proteins that are classified into three groups or classes known as the class I, class II, and class III molecules.
- HLA typing or tissue typing are usually performed for 1. Tissue transplantation; 2. Disputed paternity; 3. Anthropological studies; 4. An association between HLA types and diseases.

IMPORTANT QUESTIONS

1. Differentiate between T and B cells in a tabulated form.
2. Give an account of lymphocytes.
3. Write short notes on:
   - Mucosa-associated lymphoid tissue
   - Subsets of T lymphocytes
   - B lymphocytes
   - Null cells (or) large granular lymphocytes (LGL)

   Write briefly on:
   - Major histocompatibilty complex (MHC)
   - Human leukocyte antigen (HLA)
   - HLA typing
   - MHC restriction.

FURTHER READING

Abbas AK, Lichman AH, Prober JS. Cellular and Molecular Immunology, 3rd edn, Saunders 1997.

Chapel H, Haeney M. Essentials of Clinical Immunology. 5th edn. London: Blackwell Science.

Peakman M, Vergani D. Basic and Clinical Immunology. Edinburgh Churchill-Livingstone 1997.

Klein JK, Sato A. The HLA system. New Eng J Med 2000; 343:702.

CHAPTER 18

Immune Response

Learning Objectives

After reading and studying this chapter, you should be able to:

- Differentiate between primary and secondary humoral immune responses.
- Discuss monoclonal antibodies—principle, technique and applications.
- Describe the following: Cytokines; immunological tolerance.

DEFINITION

The immune response is the specific reactivity induced in a host by an antigenic stimulus. For the generation of immune response, antigen must interact with and activate a number of different cells. In addition, these cells must interact with each other. It is generally equated with protection against invading microorganisms in infectious disease. But the immune response has a much wider scope and includes reactions against any antigen, living or nonliving. It may lead to consequences that are beneficial, indifferent or injurious to the host. The state of specific nonreactivity (tolerance) induced by certain types of antigenic stimuli is also included in it.

TYPE OF IMMUNE RESPONSE

The immune response can be divided into two types—the humoral (antibody mediated) and the cellular (cell mediated) types. The generation and control of immune responses are a consequence of a complex series of antigen-mediated interactions between various cell types. The two are usually developed together, though at times one or the other may be predominant or exclusive. They usually act in conjunction but may sometimes act in opposition.

a. Antibody Mediated Immunity (AMI)

1. Provides primary defense against most extracellular bacterial pathogens
2. Helps in defense against viruses that infect through the respiratory or intestinal tracts
3. Prevents recurrence of virus infections
4. It also participates in the pathogenesis of immediate (types 1, 2 and 3) hypersensitivity and certain autoimmune diseases.

b. Cell-Mediated Immunity (CMI)

i. Protects against fungi, viruses and facultative intracellular bacterial pathogens like *Mycobacterium tuberculosis, Mycobacterium leprae, Brucella* and *Salmonella,* and parasites like *Leishmania* and trypanosomes.

ii. It also participates in the rejection of homografts and graft-versus-host reaction.

iii. It mediates the pathogenesis of delayed (type 4) hypersensitivity and certain autoimmune diseases.

iv. It provides immunological surveillance and immunity against cancer.

HUMORAL IMMUNITY

Synthesis of Antibody

On exposure to antigen, antibody production follows a characteristic pattern (Fig. 18.1). The production of antibodies consists of three steps:

1. Lag Phase

A lag phase, the immediate stage following antigenic stimulus during which no antibody is detec in circulation. This is the time taken for the interactions described above to take place and antibody to reach a level that can be measured.

2. Log Phase

A log phase in which there is steady rise in the titer of antibodies. There is an exponential rise in the antibody level or titer.

3. A Plateau or Steady State

This log phase is followed by a plateau with a constant level of antibody when there is equilibrium between antibody synthesis and catabolism.

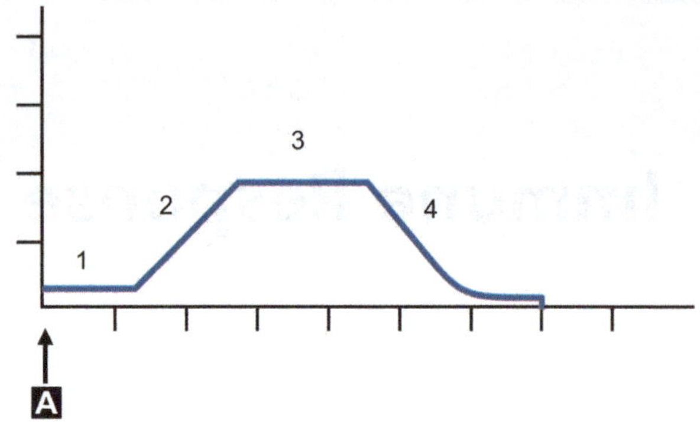

Fig. 18.1: Primary immune response. An antigenic stimulus; 1. Latent period; 2. Log phase (rise in titer of serum antibody); 3. Steady state of antibody titer; 4. Decline of antibody titer

4. The Phase of Decline

The amount of antibody then declines due to the clearing of antigen-antibody complexes and the natural catabolism of the immunoglobulin, i.e. the catabolism exceeds the production and the titer falls (Fig 18.1).

Primary and Secondary Responses

The kinetics and other characteristics of the humoral response differ considerably depending on whether the humoral response results from activation of naive lymphocytes (primary response) or memory lymphocytes (secondary response). In both cases, activation leads to production of secreted antibodies of various isotypes, which differ in their ability to mediate specific effector functions.

a. Primary Humoral Response

The first contact of an exogenous antigen with an individual generates a primary humoral response, characterized by the production of antibody-secreting plasma cells and memory B cells. The kinetics of the primary response, as measured by serum antibody level, depend on the nature of the antigen, the route of antigen administration, the presence or absence of adjuvants, and the species or strain being immunized.

When first introduced, the antigen selects the cells that can react with it. In all cases, however, a primary response to antigen is characterized by a **lag phase**, during which naive B cells undergo clonal selection, subsequent clonal expansion, and differentiation into memory cells or plasma cells. The duration of the lag phase varies with the nature of the antigen.

During a primary humoral response, IgM is secreted initially, often followed by a switch to an increasing proportion of IgG. Depending on the persistence of the antigen, a primary response can last for various periods, from only a few days to several weeks.

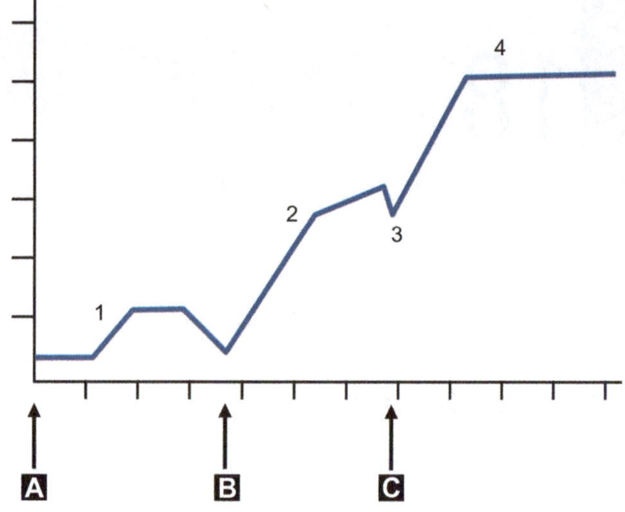

Fig. 18.2: Effect of repeated antigenic stimulus. A, B, C antigenic stimuli; 1. Primary immune response; 2. Secondary immune response; 3. Negative phase; 4. High level of antibody following booster injection

b. Secondary Humoral Response (Fig. 18.2)

On subsequent exposure, the responding cells, i.e. memory cells, are at a different level of activation and are present at an increased frequency. Therefore, there is a shorter lag before antibody can be detected: the main isotype is IgG.

Activation of memory cells by antigen results in a secondary antibody response that can be distinguished from the primary response in several ways. The secondary response has a shorter lag period, reaches a greater magnitude, lasts longer and is also characterized by secretion of antibody with a higher affinity for the antigen, and isotypes other than IgM predominate. The level of antibody produced is 10 or more times greater than during the primary response.

Priming dose and booster doses: A single injection of an antigen helps more in sensitizing or priming the immunocompetent cells producing the particular antibody than in the actual elaboration of high levels of antibody. Only by subsequent injections of the antigen, effective levels of antibody are usually induced. The first injection is known as the **'priming' dose** and subsequent injections as **'booster' doses.** With live vaccines, a single dose is sufficient as multiplication of the organism in the body provides a continuing antigenic stimulus that acts as both the priming and booster dose.

Negative phase: If the same animal is subsequently exposed to the same antigen already carrying the specific antibody in circulation, a temporary fall in the level of circulating antibody occurs due to the combination of the antigen with the pre-existing antibody. This is known as the **'negative phase'**. It is followed by an increase in the titer of the antibody exceeding the initial level (Fig. 18.2).

FATE OF ANTIGEN IN TISSUES

The manner in which an antigen is dealt within the body depends on factors such as the physical and chemical nature of the antigen, its dose and route of entry, and whether the antigenic stimulus is primary or secondary.

Physical and chemical nature of the antigen: Particulate antigens are removed from circulation in two phases. The first is the **nonimmune phase** during which the antigen is engulfed by the phagocytic cells, broken down and eliminated. The phase of immune elimination begins with the appearance of the specific antibody, during which antigen-antibody complexes are formed and are rapidly phagocytosed, resulting in an accelerated disappearance of the antigen from circulation.

Soluble antigens: Three phases can be recognized with **soluble antigens:** Equilibration, metabolism and immune elimination. The phase of equilibration consists of diffusion of the antigen to the extravascular spaces. During the metabolic phase, the level of the antigen falls due to catabolic decay. During the phase of immune elimination, there is rapid elimination of the antigen with the formation of antigen-antibody complexes. Such complexes can cause tissue damage and may be responsible for 'immune complex diseases' such as serum sickness.

Route of entry: Antigens introduced intravenously are rapidly localized in the spleen, liver, bone marrow, kidneys and lungs. They are broken down by the reticuloendothelial cells and excreted in the urine, about 70-80 percent being thus eliminated within one or two days. In contrast, antigens introduced subcutaneously are mainly localized in the draining lymph nodes, only small amounts being found in the spleen.

Speed of elimination: The speed of elimination of an antigen is related to the speed at which it is metabolized. Protein antigens are generally eliminated within days or weeks, whereas polysaccharides which are metabolized slowly, persist for months or years. Pneumococcal polysaccharide, for instance, may persist up to 20 years in human beings, following a single injection.

PRODUCTION OF ANTIBODIES

The majority of antigens will stimulate B cells only if they have the assistance of T lymphocyte helper (T_H) cells.Antigens can be divided into two categories based on their apparent need for for T_H cells for the induction of antibody synthesis (1) those that require Th cells, referred To as T-dependent antigens (TD-antigens) such as proteins and erythrocytes; and (2) those that do not require T_H cells, called **T-independent antigens (TI-antigens)** such as polysaccharides and other structurally simple molecules with repeating epitopes. Immune response to an antigen is brought about by three types of cells - antigen processing cells (APC- principally macrophages and dendritic cells), T cells and B cells. The sequence of events is as follows:

1. Antigen Processing and Presentation

For successful development of antibody response to a T-dependent antigen, the antigen must be associated with MHC class II molecules on the surface of an antigen-presenting cell (APC). APC can ingest antigen, degrade it and present it to T cells. T cell is able to recognise only when the processed antigen is presented on the surface of APC, in association with MHC molecules to the T cell carrying the receptor (TCR) for the epitope . The antigen has to be presented complexed with MHC Class II in the case of CD4 (Helper T/T_H) cells, and for CD8 (cytotoxic T/Tc) cells with MHC Class I molecules. B cells, which possess surface Ig and MHC Class II molecules, can also present antigens to T cells, particularly during the secondary response.

2. T cell and B cell Activation

The activation of resting CD4 (Helper T/T_H) cells require two signals. The first signal is a combination of the T cell receptor (TCR) with the MHC class II complexed antigen. The second is the co-stimultory signal i.e. **interleukin-1 (IL-1)** which is produed by the APC. The activated T_H cell forms interleukin-2 and other cytokines such as interleukin-4, IL-5 and **IL-6** required for B cell activation. These **interleukin-4** (formerly known as B cell stimulatory factor I), **IL-5 (B-cell growth factor, BCGF)** and **IL-6** (formerly called **B-cell stimulatory factor, BCSF-2)** activate B cells which have combined with their respective antigens to clonally proliferate and differentiate into antibody-secreting **plasma cells.**

B cells carry surface receptors which consist of IgM or other immunoglobulin classes. A plasma cell secretes an antibody of a single specificity of a single antibody class (IgM, IgG or any other single class) depending upon these receptors,. However, primary antibody response is characterized by the initial production of IgM, and and later switching over to form IgG. Class switching is influenced by different combinations of lyniphokines produced by helper T (T_H) cells. Under the direction of cytokines produced by effector T-helper (T_H) cells, some B cells become programmed to produce antibodies other than IgM. Following antigenic stimulus, not all B lymphocytes are converted into plasma cells. A small proportion of activated B cells become long lived memory cells which produce a secondary type of response to subsequent contact with the antigen. During secondary antigenic stimulus,the increased antibody response is due to the memory cells induced by the primary contact with the antigen.

Cytotoxic T (CD8/Tc) cells are activated when they contact antigens presented along with MHC Class I molecules. Again, these cells require two signals to be activated (1) antigen fragment in association with MHC class1is the first signal; (2)s co-stimulatory signal IL2, which is secreted by activated TH cells is the second signal. On contact with a target cell carrying the antigen on its surface, the activated Tc cells release cytotoxins that destroy the target, which may be virus infected or tumour cells. Some Tc cells also become memory cells.

Monoclonal Antibodies

Monoclonal Antibodies

When a clone of lymphocytes or plasma cells undergoes selective proliferation, as in multiple myeloma, antibodies with a single antigenic specificity accumulate. Such antibodies produced by a single clone and directed against a single antigenic determinant are called **monoclonal antibodies,** e.g. plasma cell tumor (myeloma). In myeloma, antibodies are produced by a single clone of plasma cells directed against a single antigenic determinant and hence the antibodies are homogeneous and monoclonal.

Polyclonal Antibodies

Antibodies that are produced ordinarily by infection or immunization in an animal in response to a single antigen are heterogeneous as they are synthesised by several different clones of B cells against different epitopes of the same antigen, i.e. they are **polyclonal**. Such antisera contain immunoglobulins of different classes with specificities against different epitopes of the antigen.

Hybridoma Technology

By laboratory manipulation, Kohler and Milstein (1975) prepared a hybrid cell line (hybridoma) by fusion of a mouse myeloma cell with an antibody producing lymphocyte from spleen or lymph node of the same inbred strain of mouse. In recognition of the great importance of this hybridoma technology, the Nobel Prize for Medicine was awarded to them in 1984. Such hybrid cells can produce virtually unlimited quantities of monoclonal antibody of any required specificity indefinitely in cell culture conditions.

Procedure (Fig. 18.3)

The production of monoclonal antibodies involves the following steps listed below:

1. Selection of Antigen

Monoclonal antibodies can be produced against any substance recognized as an antigen by the immune system of the animal being injected. Using a pure antigen is ideal.

2. Animal Immunization

Animal (usually mouse) is immunized with a pure selective antigen and when good immune response occurs, it is killed and B lymphocytes are harvested from the spleen or lymph node. A suspension of spleen cells (B cells) is prepared.

3. Fusion of Splenic Lymphocytes and Myeloma Cells

A suspension of splenic cells is then fused with a myeloma cell-line by incubating in presence of polyethylene glycol.

Because cells cannot remain viable in cell culture for very long, they must be fused together with cells that are able to survive and multiply in tissue culture, that is, the continuously propagating, or immortal cells, of multiple myeloma (a malignant tumor of antibody producing plasma cells).

4. Selection of Hybrid Lymphocyte-Myeloma Cells

Lymphocytes from the spleen of mice immunized with the desired antigen are fused with mouse myeloma cells grown in culture which do not form immunoglobulins and are deficient in the enzyme hypoxanthine phosphoribosyl transferase (HPRT). Antibody-producing spleen cells, however, possess the enzyme. The fused cells are placed in basal culture medium (HAT medium containing hypoxanthine, aminopterin and thymidine) which does not permit the growth of the enzyme deficient myeloma cells. Thus, fused hybridoma cells survive in the selective medium and can be recognized by their ability to grow indefinitely in the medium. Unfused antibody-producing lymphoid cells die after several multiplications *in vitro* because they are not immortal, and unfused myeloma cells die in the presence of the toxic enzyme substrates. The only surviving cells will be true hybrids. These hybrid cells are called **hybridomas** (they are hybrids of the two cells).

5. Cloning the Hybridoma Cells

The single hybrid cells producing the desired antibody must be isolated and grown as a clone.

6. Screening for Desired Antibodies

The growth medium supernatant from the microdilution tray wells in which the hybridoma cells are growing is then tested for the presence of the desired antibody.

7. Mass Production of Monoclonal Antibodies

Clones producing antibodies against the desired antigen are selected for continuous cultivation. Such hybridomas can be maintained indefinitely in culture and will continue to form monoclonal antibodies. They may be also injected intraperitoneally in mice. Within days, a tumor known as a hybridoma develops and monoclonal antibodies may be obtained by harvesting the ascitic fluid produced. Ascitic fluid can be removed from mice many times over the animals' lifetimes.

Uses of Monoclonal Antibody

1. They are routinely used **in the typing of tissue**.
2. Use in the **identification and epidemiological study of infectious microorganisms.**

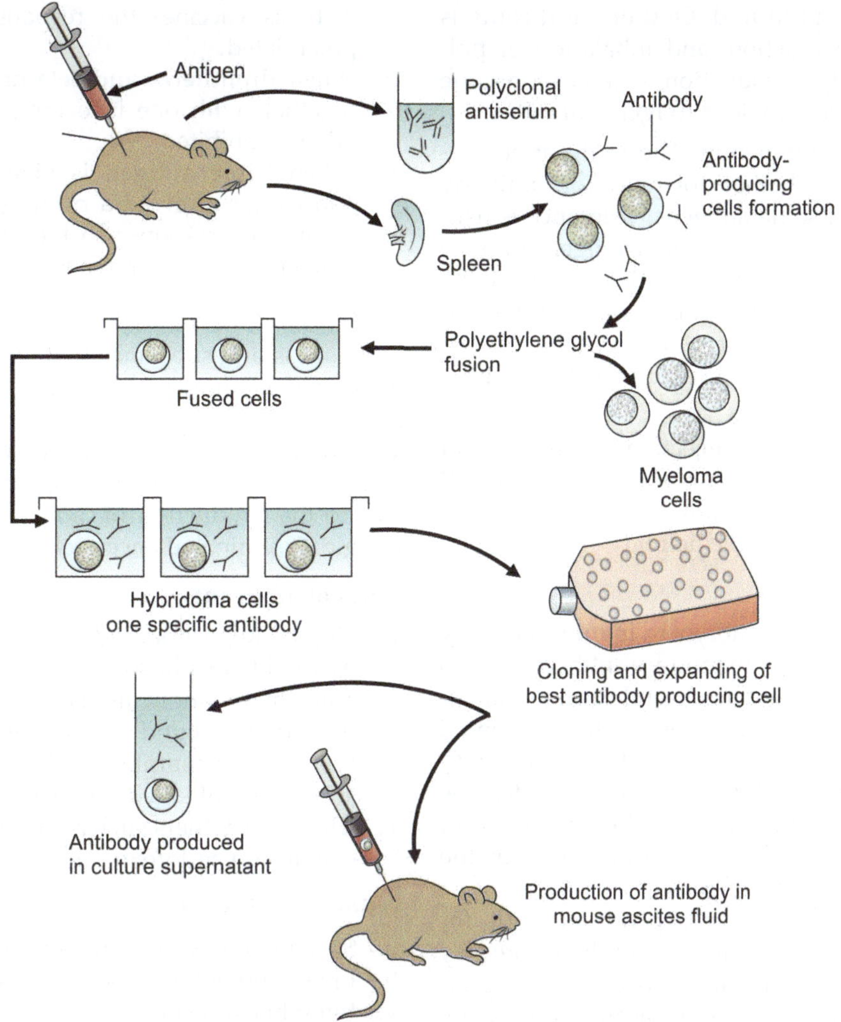

Fig. 18.3: Production of a monoclonal antibody

3. Use in the **identification of tumor and other surface antigens.**
4. Use in the **classification of leukemias.**
5. Use in the **identification of functional populations of different types of T cells.**
6. **Anticipated future uses:**
   i. Passive immunizations against infectious agents and toxic drugs.
   ii. Tissue and organ graft protection.
   iii. Stimulation of tumor rejection and elimination.
   iv. Manipulation of the immune response.
   v. Preparation of more specific and sensitive diagnostic procedures.
   vi. Delivery of antitumor agents (immunotoxins) to tumor cells.

Factors Influencing Antibody Production

1. Genetic Factors

The immune response is under genetic control. Level of immune response to a partcular antigen is controlled by the MHC class II molecules. Response in different individuals to same antigen varies due to genetic factor. Persons capable of responding to a particular antigen are called *responder* and those who cannot respond are termed *nonresponder*. The Ir (immune response) genes control this property.

2. Nutritional Status

Malnutrition affects host susceptibility to certain micro-organisms, especially bacteria.

Examples:

i. **Protein calorie malnutrition** suppresses both humoral and cellular immune responses, the latter more severely.
ii. **Deficiences of amino acids and vitamins:** Deficiences of amino acids have been shown to cause a decrease in antibody synthesis.

3. Route of Administration

There is better humoral immune response following parenteral administration of antigen than oral or nasal routes. Large particulate antigens, such as bacteria or erythrocytes, are more effective when injected into tissues. The route of administration may also influence the type of antibody produced. Oral or nasal route is most sui for IgA production, and inhalation of pollen antigens induce IgE production, whereas the same antigens given parenterally lead to IgG antibodies.

Sulzberger-Chase phenomenon: The route of administration determines whether tolerance or antibody response results with some antigens. Tolerance is usually induced by the injection of protein antigens into the mesenteric vein or intrathymically. Sulzberger (1929) and Chase (1959) showed that guinea pigs can be rendered specifically tolerant if certain antigens are fed before a parenteral challenge **(Sulzberger-Chase phenomenon)**.

Site of injection: The site of injection seems relevant with some antigens. It has been reported that hepatitis B vaccine is less immunogenic following gluteal injection than following injection into the deltoid.

4. Age

i. **Embryonic life:** The embryo is immunologically immature. The capacity to produce antibodies starts only with the development and differentiation of lymphoid organs. When the potential immunocompetent cell comes into contact with its specific antigen during embryonic life, the response is elimination of the cell or induction of tolerance. This is believed to be the basis for the nonantigenicity of self antigens.
ii. **Infant:** The infant has to depend on itself for antibody production from 3-6 months of age, by which time the maternal antibodies disappear. However, full competence is acquired only by about 5-7 years for IgG and 10-15 years for IgA.
iii. **Antigens concerned:** The ontogeny of antibody response also depends on the antigens concerned. B cell responses to most proteins and other T cell dependent antigens develop early, while responses to polysaccharides and other T cell independent antigens develop only later, usually by two years.

5. Multiple Vaccines

Antigenic competition: The effects may vary when two or more antigens are administered simultaneously, Antibodies may be produced against the different antigens, or antibody response to one or the other of the antigens may be enhanced, or the response to one or more of them may be diminished **(antigenic competition).** Such antigenic competition is important, from a practical point of view, in immunization with polyvalent antigens. Examples:

i. When **two bacterial vaccines** (for example, typhoid and cholera vaccines) are given in a mixed form, the antibody response to each is not influenced by the other.
ii. When **toxoids are given along with bacterial vaccines** (for example, triple vaccine containing diphtheria and tetanus toxoids along with *Bordetella pertussis* vaccine) the response to the toxoid is potentiated.
iii. When **diphtheria and tetanus toxoids** are given together, with one in excess, the response to the other is inhibited.
iv. When **triple antigen** is given to a person who had earlier received a primary dose of diphtheria toxoid, the response to the tetanus and pertussis antigens will be diminished.

6. Adjuvant

Adjuvant is any substance that enhances the immunogenicity of an antigen. Adjuvants may confer immunogenicity on nonantigenic substances, increase the concentration and persistence of the circulating antibody, induce or enhance the degree of cellular immunity and lead to the production of 'adjuvant diseases' such as allergic disseminated encephalomyelitis.

Types of adjuvants

a. **Depot:** Repository adjuvants such as aluminum hydroxide or phosphate, and Freund's incomplete adjuvant (water in archis oil).
b. **Bacterial:** Freund's complete adjuvant is the Freund's incomplete adjuvant along with a suspension of killed tubercle bacilli.
c. **Chemical:** Others such as silica particles, beryllium sulfate and endotoxins activate macrophages.

Action of Adjuvants

a. Sustained release of antigen from depot
b. Liberation of lymphocytes activating factor
c. Lymphocytes stimulation: B cell, T cell or both.
d. Stimulate CMI

Freund's Complete Adjuvant

The most potent adjuvant is Freund's complete adjuvant, which is the incomplete adjuvant along with a suspension of killed tubercle bacilli. Besides increasing the humoral immune response, it induces a high degree of cellular immunity (delayed hypersensitivity) as well. It is unsui for human use as it produces a local granuloma.

7. Immunosuppressive Agents

These inhibit the immune response. They are useful in certain situations like transplantation, when it becomes necessary to prevent graft rejection.

Examples: X-irradiation, radiomimetic drugs, corticosteroids, antimetabolites and other cytotoxic chemicals, and antilymphocytic serum.

i. **X-irradiation:** Antibody response is suppressed by sublethal whole body irradiation. Antibody production does not occur when antigenic stimulus follows 24 hours after irradiation. The primary response is more radiosensitive than the secondary response.
ii. **Radiomimetic drugs:** Radiomimetic drugs are agents with an action resembling that of X-rays.

They belong in general to the class of alkylating agents (for example, cyclophosphamide, nitrogen mustard). In human beings, cyclophosphamide is given for three days after the antigen, completely suppresses the antibody response. It is much less effective when given before the antigen.

iii. **Corticosteroids:** Corticosteroids cause depletion of lymphocytes from the blood and lymphoid organs. They also stabilize the membranes of cells and lysosomes, inhibiting histamine release and the inflammatory response. Therapeutic doses have little effect on the antibody formation in human beings.

iv. **Antimetabolites:** They include folic acid antagonists (methotrexate), alkylating agents (cyclophosphamide) and analogs of purine (6-mercaptopurine, azathioprine), cytosine (cytosine arabinoside) and uracil (5-fluorouracil).

   Antimetabolites are substances that interfere with the synthesis of DNA, RNA or both and thus inhibit cell division and differentiation necessary for humoral and cellular immune responses. Many antimetabolites find clinical application in the prevention of graft rejection.

v. **Cyclosporine:** The drug most widely used now for immunosuppression is cyclosporine. It is not cytotoxic for lymphocytes and has no antimitotic activity. It selectively inhibits helper T cell activity. A related drug is rapamycin.

vi. **Antilymphocyte serum (ALS):** Antilymphocyte serum (ALS) is a heterologous antiserum raised against lymphocytes. Antibody prepared against thymus cells is called antithymocyte serum (ATS). The corresponding globulin preparations are called ALG and ATG. They were used to prevent graft rejection. ALS is devoid of any action other than that on lymphocytes while all other immunosuppressive agents have undesirable side effects.

ALS acts primarily against T lymphocytes and therefore specifically on cell mediated immunity. Humoral antibody response to thymus independent antigens is unaffected and may even be enhanced. ALS acts only against lymphocytes in circulation and not cells in lymphoid organs. As ALS is a foreign protein, its effect is decreased on repeated administration, which may also lead to serum sickness and other hypersensitivity reactions. Monoclonal antibodies against specific lymphocyte membrane antigens have been prepared.

8. Effect of Antibody

Passive administration of the homologous antibody will suppress specifically the humoral immune response to an antigen. The action appears to be by a feedback mechanism. The primary response is more susceptible to inhibition than the secondary response.

The inhibitory effect of a passively administered antibody on the humoral immune response has been applied in the prevention of Rh sensitization in Rh negative women carrying Rh positive fetuses. This is achieved by the administration of anti-Rh globulin immediately following delivery (within 72 hours).

This effect is also relevant in the practice of combined immunization as in diphtheria and tetanus. Adsorbed toxoid should be used as the inhibitory effect is much less than with fluid toxoid.

Intravenous administration of immune globulin has been shown to have immunomodulatory effects. It has been used in the treatment of many diseases of presumed immunopathologic etiology, such as thrombocytopenias and autoimmune hemolytic anemias.

9. Superantigens

Superantigens are certain protein molecules that activate very large numbers of T cells irrespective of their antigenic specificities such as staphylococcal enterotoxins. Super antigens bypass antigen processing and presentation. And so named because they bind directly. Such superantigens have beeen found in staphylococci, streptococci, mycoplasma and other species. This exaggerated T cell activation leads to massive outpouring of T cell cytokines, causing multisystem dysfunctions, such as seen in staphylococcal toxic shock syndrome.

10. Mitogens

Mitogens are certain substances that induce division of lymphocytes and other cells. Some of these, like the lectin glycoproteins bind to sugars on the surface of responsive cells and activate them, causing a polyclonal reaction. At low concentrations, they stimulate B cells without polyclonal activation. Lipopolysaccharide is such a B cell mitogen.

CELL-MEDIATED IMMUNE RESPONSES

The term 'cell-mediated immunity' (CMI) refers to the specific immune responses which involve T lymphocyte-mediated functions that do not involve antibodies. This form of immunity can be transferred from donor to recipient with intact lymphocytes, but not with antisera, hence it is called **cell-mediated immune reaction.**

Scope of CMI

CMI participates in the following immunological functions:

1. Delayed hypersensitivity (type IV hypersensitivity).
2. Immunity in infectious diseases caused by obligate and facultative intracellular parasites. These include infections with bacteria (for example, tuberculosis, leprosy, listeriosis, brucellosis), fungi (for example, histoplasmosis, coccidioidomycosis, blastomycosis), protozoa (for example, leishmaniasis, trypanosomiasis) and viruses (for example, measles, mumps).
3. Transplantation immunity and graft-versus-host reaction.
4. Immunological surveillance and immunity against cancer.

5. Pathogenesis of certain autoimmune diseases (for example, thyroiditis, encephalomyelitis).

Induction of CMI

Antigen-specific cell-mediated immunity is mediated by T lymphocytes. A second, smaller population of cells (which have the morphologic features of lymphocytes but lack T or B cell markers, mediate cellular cytotoxicity, and are not antigen specific) includes natural killer (NK) and killer (K) cells. As in case of antibody-mediated immune response, cell-mediated immune response can also be divided into primary and secondary cell-mediated immune responses.

Primary Cell-mediated Immune Response

This is produced by initial contact with a foreign antigen. Foreign antigen is presented by antigen presenting cells (APCs) to T cells leading to their activation. Each T cell bears on its surface a specific receptor (TCR) for one epitope and combines only with antigens carrying that epitope. On contact with the appropriate antigen, T cells undergo blast transformation, clonal proliferation and differentiation into memory cells and effector cells providing CMI. T cells recognize antigens only when presented with MHC molecules.

The T cell group is composed of:

1. **Helper T (Th) cells** which react with antigens presented on the surface of macrophages or other cells, complexed with MHC class II molecules. They then release biological mediators (lymphokines) which activate macrophages, enabling them to kill intracellular parasites.
2. **T cytotoxic cells (Tc cells)**, which recognize antigen on the surface of cells (such as virus infected, tumor or allograft cells), in association with MHC class I molecules, secrete lymphokines and destroy the target cells.

The NK group kills target cells by mechanisms similar to those that the Tc cell uses. The NK cells kill some tumor cells but lack antigen-specific receptors. K cells recognize antibody-coated target cells by means of a cell-surface Fc receptor on the K cell.

Secondary Cell-mediated Immune Response

If the same host is subsequently exposed to the same antigen, then the secondary cell-mediated immune response is usually more pronounced and occurs more rapidly. Because of the availability of specific memory cells, an increased number of effector cells are produced.

CYTOKINES

Cytokine (Greek *cyto*, cell, and *kinesis*, movement) are biologically active substances produced by cells that influence other cells. Biologically active substances released by activated T lymphocytes were called **lymphokines.** When released from mononuclear phagocytes, these proteins are called **monokines.**

The term **interleukin** was introduced for those products of leucocytes which exert a regulatory influence on other cells; and if their effect is to stimulate the growth and differentiation of immature leukocytes in the bone marrow, they are called **colony-stimulating factors (CSFs).**

Interferons, growth factors and others were found to have similar effects. Therefore, all of them have been grouped under the term **cytokines.**

Recently cytokines have been grouped into the following categories or families: chemokines, hematopoietins, interleukins, and members of the tumor necrosis factor (TNF) family. Some examples of these cytokine families are provided in Table 18.1.

Characteristics of Cytokines

i. Cytokines are peptide mediators or intercellular messengers which regulate immunological, inflammatory and reparative host responses.
ii. They are highly potent hormone-like substances, active even at femtomolar (l0 -15 M) concentrations.
iii. They differ from endocrine hormones in being produced not by specialized glands but by widely distributed cells (such as lymphocytes, macrophages, platelets and fibroblasts).
iv. Cytokines can affect the same cell responsible for their production (an autocrine function), nearby cells (a paracrine function), or can be distributed by the circulatory system to their target cells (an endocrine function).
v. Their production is induced by nonspecific stimuli such as a viral, bacterial, or parasitic infection; cancer; inflammation; or the interaction between a T cell and antigen. Some cytokines also can induce the production.
vi. Cytokines have been named based on the biological effects they produce.

Features of Important Cytokines

Various cytokines are given in Table 18.1.

A. Interleukins

Interleukin-1

Originally described as the leucocyte activating factor (LAF) in 1972 and as the B cell activating factor (BAF) in 1974, this cytokine was renamed interleukin-1 (IL-1) in 1979. It is a stable polypeptide retaining its activity up to 56°C and between pH 3-11. IL-1 occurs in two molecular forms, IL-1 alpha and beta. IL-1 is principally secreted by macrophages and monocytes but can be produced by most other nucleated cells also. Its production is stimulated by antigens, toxins, injury and inflammatory processes and inhibited by cyclosporin A, corticosteroids and prostaglandins.

Immunological Effects of IL-1

The immunological effects of IL-l include:

i. Stimulation of T cells for the production of IL-2 and other lymphokines.
ii. B cell proliferation and antibody synthesis.
iii. Neutrophil chemotaxis and phagocytosis.
iv. It mediates a wide range of metabolic, physiological, inflammatory and hematological effects by acting on bone marrow, epithelial and synovial cells, fibroblasts, osteoclasts, hepatocytes, vascular endothelium and other targets.

IL-1 is an important endogenous pyrogen. Together with the tumor necrosis factor (TNF), it is responsible for many of the hematological changes in septic shock and also enhances the initial meningeal inflammation in bacterial meningitis. Cytokine inhibitors such as dexamethasone have been found to protect against the sequelae of such excessive meningeal inflammation. On the other hand, IL-1 has a beneficial effect in severe infections in immunocompromised hosts.

Interleukin-2

A T cell growth factor (TCGF) produced by activated T cells was discovered in 1976, which induced T cell proliferation and enabled their maintenance in continuous culture, contributed greatly to the understanding of T cell functions. This cytokine, renamed IL-2, is a powerful modulator of the immune response. It is the major activator of T and B cells and stimulates cytotoxic T cells and NK cells. It converts some null cells (LGL) into lymphokine activated killer (LAK) cells which can destroy NK resistant tumor cells. This property has been used in the treatment of certain types of cancer.

Interleukin-3

IL-3 is a growth factor for bone marrow stem cells. It stimulates multilineage hematopoiesis and is therefore known also as the multicolony stimulating factor (multi-CSF).

Interleukin-4

IL-4 activates resting B cells and acts as a B cell differentiating factor. It also acts as a growth factor for T cells and mast cells. It enhances the action of cytotoxic T cells. Formerly it was known as the B cell growth factor-1 (BCGF-1). It may have a role in atopic hypersensitivity as it augments IgE synthesis.

Interleukin-5

IL-5 causes proliferation of activated B cells. It also induces maturation of eosinophils. Formerly, it was known as the B cell growth factor-II.

Interleukin-6

IL-6 is produced by stimulated T and B cells, macrophages and fibroblasts. It induces immunoglobulin synthesis by activated B cells and formation of IL2 receptors on T cells. It has a stimulatory effect on hepatocytes, nerve cells and hematopoietic cells. It acts as an inflammatory response mediator in host defense against infections.

B. Colony Stimulating Factors (CSF)

These cytokines stimulate the growth and differentiation of pluripotent stem cells in the bone marrow. They have been named after the types of cell colonies they induce in soft agar culture—for example, granulocyte (G), or mononuclear (M) CSF (Table 18.1). IL-3 which induces growth of all types of hematopoietic cells is known as multi-CSF. In the body they cause other effects also, presumably by inducing cascades of other cytokines. They are responsible for adjusting the rate of production of blood cells according to requirements, for example, the massive granulocyte response seen in pyogenic infections.

Colony stimulating factors have clinical applications for treating hematopoietic dysfunctions in infections and malignancies.

C. Tumor Necrosis Factors (TNF)

The tumor necrosis factor occurs as two types, alpha and beta.

TNF-α: A serum factor found to induce hemorrhagic necrosis in certain tumors was named the tumor necrosis factor. The same substance was independently described as cachectin, a serum factor causing the wasting syndrome (cachexia) during chronic infections. This has been renamed TNFα. It is formed principally by activated macrophages and monocytes. It resembles IL-l in possessing a very wide spectrum of biological activities such as participation in the manifestations of endotoxic other cytokines.

TNF-β: TNF-β is produced principally by T helper cells and formerly known as lymphotoxin. Its effects are similar to those of TNF-α.

D. Interferons (IFNs)

Interferons (IFNs) are a group of related low molecular weight, regulatory cytokines produced by certain eukaryotic cells in response to a viral infection. Besides defending against viruses, they also help regulate the immune response. Originally identified as antiviral agents, interferons are now classified as cytokines. Interferons usually are species specific but virus nonspecific.

There are three classes of IFNs, alpha produced by leukocytes, beta produced by fibroblasts and gamma by T cells activated by antigens, mitogens or exposure to IL-2. IFN-γ causes many immunological effects, such as macrophage activation, augmentation of neutrophil and monocyte functions, and antitumor activity.

E. Other Cytokines

Transforming growth factor beta (TGF β): The transforming growth factor beta (TGF β) was so named because of its ability to transform fibroblasts. Besides acting as a growth factor for fibroblasts and promoting

Table 18.1: Cytokines: Hormones of the immune system

Cytokine	*Main sources*	*Major functions*
A. Interleukins		
IL-1 (α and β)	Macrophages and other cell type	Proliferation and differentiation of T, B and other cells; pyrogenic; induce acute phase proteins; bone marrow cell proliferation
IL-2	T cells	Promote growth and differentiation of T and B cells, cytotoxicity of T and NK cells, secretion of other lymphokines
IL-3	T cells	Multi-CSF
IL-4	Th cells	Proliferation of B and cytotoxic T cells; increase IgG1 and IgE production; enhance MHC class II and IgE receptors
IL-5	Th cells	Proliferation of eosinophils, stimulate IgA and IgM production
IL-6	Th cells	Promote B cell differentiation; IgG production, acute phase proteins
IL-7	Bone marrow, spleen, stromal cells	B and T cell growth factor
IL-8	Macrophages, others	Neutrophil chemotactic factor
IL-9	T cell	T cell growth and proliferation
IL-I0	T, B cells, macrophages	Inhibit IFN production and mononuclear cell functions
IL-11	Bone marrow stromal cells	Induce acute phase proteins
IL-12	T cells	Activate NK cells
IL-13	T cells	Inhibit mononuclear cell functions
B. Colony-stimulating factors		
GM-CSF	T cells, macrophages, fibroblasts	T cell and macrophage growth stimulation
G-CSF	Fibroblasts, endothelium	Granulocyte growth stimulation
M-CSF	Fibroblasts, endothelium	Macrophages growth stimulation
C. Tumor necrosis factors		
TNF-α	Macrophages, monocytes	Tumor cytotoxicity, lipolysis, wasting, acute phase proteins, phagocytic cell activation, antiviral and antiparasitic effects, endotoxic shock
TNF-β	T cells	Induce other cytokines
D. Interferons		
IFN-α	Leukocytes	
IFN-β	Fibroblasts	Antiviral activity
IFN-γ	T cells	Antiviral, macrophage activation; MHC class I and II expression on cells
E. Others		
TGF-β	T and B cells	Inhibit T and B cell proliferation and hematopoiesis; promote wound healing
LIF	T cells	Proliferation of stem cells; eosinophil chemotaxis

wound healing, it also acts as a down regulator of some immunological and hematological processes.

Leukemia inhibitory factor (LIF): The leukemia inhibitory factor (LIF), produced by T cells, helps stem cell proliferation and eosinophil chemotaxis.

Cytokine production is regulated by exogenous stimuli such as antigens and mitogens, as well as by endogenous factors such as neuroendocrine hormonal peptides (corticosteroids, endorphins) and products of lipoxygenase and cyclo-oxygenase pathways. They also regulate each other by positive and negative feedbacks.

A number of cytokines (for example, IL-1, 2, 3, colony stimulating factors, interferons) have already found therapeutic application. With better understanding of their properties, it is possible that many cytokines, their agonists and antagonists could eventually be used in the management of inflammatory, infectious, autoimmune and neoplastic conditions.

Detection of CMI

Development of CMI can be detected by following methods:

1. **Skin test for DH:** The original method for detecting CMI was the skin test for delayed hypersensitivity (for example, the tuberculin test).
2. **Lymphocytes transformation test:** Transformation of cultured sensitized T lymphocytes on contact with the antigen
3. **Target cell destruction:** Killing of cultured cells by T lymphocytes sensitized against them.
4. **Migration inhibiting factor (MIF) test:** It is commonly employed. As originally described, this consisted of incubating in a culture chamber, packed peritoneal macrophages in a capillary tube. The macrophages migrate to form a lacy, fan-like pattern. If the macrophages are from a guinea pig sensitized to tuberculoprotein, addition of tuberculin to the culture chamber will inhibit the migration.

This has been adapted for clinical use by incubating human peripheral leukocytes in capillary tubes in culture chambers. The leukocytes are prevented from migrating when an antigen to which the individual has CMI is introduced into the culture medium. It is possible to make a semiquantitative assessment of the migration inhibition by comparison with the control.

Transfer Factor

Lawrence (1954) reported transfer of CMI in man by injection of extract from the leukocytes from immunized individual. This extract is known as the 'transfer factor' (TF). The transferred immunity is specific in that CMI can be transferred only to those antigens to which the donor is sensitive.

Properties of Transfer Factor

TF is a dialysable, low molecular weight substance (MW 2000 to 4000), resistant to trypsin, DNAase, RNAase and freeze thawing. It is stable for several years at –20°C and in the lyophilized form at 4°C. It is inactivated at 56°C in 30 minutes. It is not antigenic. Chemically, it appears to be a polypeptide-polynucleotide.

TF is highly potent, an extract from 0.1 ml of packed leucocytes being sufficient for transfer. The transferred CMI is systemic and not local at the injected site alone. Delayed hypersensitivity and various *in vitro* correlates of CMI can be demonstrated in the recipient following TF injection. Humoral immunity is not transmitted by TF; TF transfers CMI to all the antigens to which the donor is sensitive, *en bloc*. It is possible to transfer CMI from the recipient to another in a serial fashion.

Mode of Action of Transfer Factor

The mode of action of TF is not known. It appears to stimulate the release of lymphokines from sensitized T lymphocytes. It does not promote antibody synthesis. TF could be an informational molecule or a specific gene derepressor capable of inducing antigenically uncommitted lymphocytes to produce antigen specific receptors. TF activity was till recently demonstrable only in human beings but it has now been reported in monkeys, guinea pigs and mice.

Applications of Transfer Factor

1. It has been used to **restore immune capacity** in patients with T cell deficiency (Wiskott-Aldrich syndrome).
2. It has also been used in **the treatment of disseminated infections associated with deficient CMI** (lepromatous leprosy, tuberculosis, mucocutaneous candidiasis).
3. It has been employed in the treatment of malignant melanoma and may be beneficial in other types of cancer as well.
4. Its use has been suggested in some **autoimmune diseases** (systemic lupus erythematosus, rheumatoid arthritis) and diseases of unknown etiology (sarcoidosis, multiple sclerosis).

IMMUNOLOGICAL TOLERANCE

Immunologic tolerance is defined as the absence of a specific immune response resulting from a previous exposure to the inducing antigen. This nonreactivity is specific to the particular antigen, immune reactivity to other antigens being unaffected. The most no example is immunologic tolerance to self. Any antigen that comes into contact with the immunological system during embryonic life would be recognized as a self antigen and would not induce any immune response.

Burnet and Fenner (1949) suggested that the unresponsiveness of individuals to self antigens was due to the contact of the immature immunological system with self antigens during embryonic life. Any antigen that comes into contact with the immunological system during embryonic life would be recognized as a self antigen and would not induce any immune response. They postulated that tolerance could be induced against foreign antigens if they were administered during embryonic life. Medawar and his colleagues (1953) proved this experimentally using two strains of syngeneic mice. When a skin graft from one inbred strain of mice (CBA) is applied on a mouse of another strain (A), it is rejected. If CBA cells are injected into fetal or newborn strain A mice, however, the latter when they grow up will freely accept skin grafts from CBA mice. The content of the self-antigen appears to have been enlarged by contact with a foreign antigen during embryonic life. This phenomenon is called 'specific immunological tolerance'.

Types of Immunological Tolerance

Two forms of tolerance are known:

1. Natural
2. Acquired

1. Natural Tolerance

The body develops immune tolerance towards self- molecules due to natural tolerance during growth of fetus. Autoimmune disease will develop in the event of breakdown of tolerance to self-tissues. Foreign antigens that may be introduced during growth of foetus, the foetus will develop immune tolerance.

Naturally occurring tolerance is found in certain viral infections such as congenital rubella and cytomegalovirus infections.

2. Acquired Tolerance

Tolerance in later life may occur under certain special circumstances, e.g. ingestion of bovine myelin in high concentrations can make an individual tolerant to myelin. This concept is now being considered in the tolerant treatment for the autoimmunopathogenesis that causes multiple sclerosis.

Parameters that Affect the Induction of Tolerance

These include age, dose of antigen, route of tolerogen administration, physical nature of the antigen, and various treatments that reduce activation of positive regulatory cells of the immune response.

1. Age

The younger the recipient, the easier it is to induce tolerance, and, of course, it is easiest *in utero*. Development of tolerance is not confined to the embryo or newborn but can occur in adults also. This undoubtedly reflects both the number and immunologic maturity of the lymphocytes that constitute the immune system.

2. Dose of Antigen

The induction of tolerance is dose dependent. There is a threshold dose, below which tolerance is not induced. Further increase in dose increases the duration of tolerance. With certain antigens, tolerance can be induced by two types of doses, one high and the other low, with intermediate doses producing immunity instead of tolerance. These are are known as 'high zone' and 'low zone' tolerance respectively. A special type of high zone tolerance is **Felton's immunological paralysis.**

3. Route of Tolerogen Administration

The route of tolerogen contact is also important, that is, intravenously administered antigens make faster contact with more cells at higher concentrations, compared with antigens administered subcutaneously or intraperitoneally. Certain haptens that are immunogenic in guinea pigs by the intradermal route are tolerogenic orally or intravenously.

4. Nature of Tolerogen

Soluble antigens and haptens are more tolerogenic than particulate antigens.

5. Various Treatments

When human gammaglobulin is heat aggregated, it is highly immunogenic in mice but is tolerogenic when deaggregated.

6. Genetic Background

Rabbits and mice can be rendered tolerant more rapidly than guinea pigs and chickens.

Mechanism of Tolerance

Tolerance can arise through three possible mechanisms:

1. Clonal deletion
2. Clonal anergy
3. Suppression.

1. Clonal Deletion

In embryonic life, clones of B and T cells possessing receptors that recognize self-antigens are selectively deleted or eliminated and, therefore, no longer available to respond upon subsequent exposure to that antigen. This is known as clonal deletion.

2. Clonal Anergy

Clones of B and T cells expressing receptors that recognize self-antigen might remain but they cannot be activated. This is known as clonal anergy.

3. Suppression

Clones of B and T cells expressing receptors that recognize self-antigens are preserved. Antigen recognition might be capable of causing activation, however, expression of immune response might be inhibited or blocked through active suppression.

THEORIES OF IMMUNE RESPONSE

Theories of immunity fall into two categories: instructive and selective.

A. Instructive theories
   1. Direct template theories
   2. Indirect template theory.
B. Selective theories
   1. Side chain theory.
   2. Natural selection theory
   3. Clonal selection theory.

A. Instructive Theories

The instructive theories postulate that an immunocompetent cell is capable of synthesizing antibodies of any specificity. The antigen encounters an immunocompetent cell and instructs it to produce the complementary antibody.

1. **Direct template theories:** These theories were proposed by Brein I and Haurowitz (1930), Alexander (1931), and Mudd (1932). According to these, the antigen or the antigenic determinant enters the antibody-forming cell and serves as a template against which antibody molecules are synthesized so that they have combining sites complementary

to the antigenic determinants. These are therefore known as 'direct template' theories. A more detailed model was presented by Pauling (1940).

2. **Indirect template theory:** This theory was proposed by Burnet and Fenner (1949). According to this theory, the entry of the antigenic determinant into the antibody producing cell induced in it a heri change. A 'genocopy' of the antigenic determinant was thus incorporated in its genome and transmitted to the progeny cells (indirect template). This theory explained specificity and the secondary response but became untenable with advances in the molecular biology of protein synthesis.

Burnet and Fenner were the first to explain the non-antigenicity of self antigens by postulating the embryonic recognition of 'self-markers'.

B. Selective Theories

1. **Side chain theory:** This theory was proposed by Ehrlich (1898). According to side chain theory, immunocompetent cells (ICCs) have surface receptors capable of reacting with antigens which have complementary side chains. When foreign antigens are introduced into the body, they combine with those cell receptors which have a complementary fit. This inactivates the receptors. There is an over production of the same type of receptors which circulate as antibodies as a compensatory mechanism.
2. **Natural selection theory:** This theory was proposed by Jerne (1955) which postulates that about a million globulin (antibody) molecules were formed in embryonic life, which covered the full range of antigenic specificities. These globulins were the 'natural antibodies'. When an antigen was introduced, it combined selectively with the globulin that had the nearest complementary 'fit'. The globulin, with the combined antigen, homed in on the antibody forming cells and stimulated them to synthesize the same kind of antibody. Here, selection was postulated at the level of the antibody molecule. It did not explain the fact that immunological memory resides in the cells, and not in serum.
3. **Clonal selection theory:** This theory was proposed by Burnet (1957) which shifted immunological specificity to the cellular level. This theory states that during immunological development, cells capable of reacting with different antigens were formed by a process of somatic mutation. Clones of cells that had immunological reactivity with self antigens were eliminated during embryonic life. Such clones are called **forbidden clones**. Their persistence or development in later life by somatic mutation could lead to autoimmune processes. Each immunocompetent cell was capable of reacting with one antigen (or a small number of antigens) which could recognize and combine with antigens introduced into the body. The result of the contact with the specific antigen was cellular proliferation to form clones synthesizing the antibody. This theory is more widely accepted than other theories.

Jerne has postulated the *network hypothesis* as an explanation for the mechanism of regulation of antibody response. Niels K Jerne was awarded the Nobel Prize for Medicine in 1984 for his theoretical contribution to antibody formation and regulation of the immune system. Recently the genetic basis of antibody diversity has been clarified. The discovery of *split genes* for immunoglobulins demolished the long standing dogma of *'one gene-one protein'* and has important implications in biology, beyond immunology. Susumu Tonegawa was awarded the Nobel Prize for Medicine in 1987 for this discovery.

KNOW MORE

Immunological Tolerance

The idea of tolerance stems from the observation of Owen (1945) that **nonidentical** (dizygotic) twin cattle share a common placental circulation *in utero* and the blood of these genetically dissimilar twins contained a mixture of erythrocytes. In these **red blood chimeras,** a s mixture of red blood cells is maintained throughout life.

- Strain differences in tolerance induction are seen within species. The higher the degree of immunocompetence of the host, the more difficult it is to induce tolerance.

KEY POINTS

- The immune response is the specific reactivity induced in a host by an antigenic stimulus and can be divided into two types: the humoral (antibody mediated) and the cellular (cell mediated) types.
- The humoral response results from activation of naive lymphocytes (primary response) or memory lymphocytes (secondary response).
- **Monoclonal antibodies:** Such antibodies produced by a single clone and directed against a single antigenic determinant are called monoclonal antibodies, e.g. plasma cell tumor (myeloma).
- **'Cell-mediated immunity' (CMI):** The term 'cell mediated immunity' (CMI) refers to the specific immune responses which involve T-lymphocyte-mediated functions that do not involve antibodies.
- **Primary cell-mediated immune response** is produced by initial contact with a foreign antigen.
- **Secondary cell-mediated immune response:** If the same host is subsequently exposed to the same antigen, then the secondary cell-mediated immune response is usually more pronounced and occurs more rapidly.

- **Cytokines:** Cytokines are biologically active substances produced by cells that influence other cells. Interferons, growth factors and others were found to have similar effects. Therefore all of them have been grouped under the term cytokines.
- **Transfer factor:** Transfer of CMI in man by injection of extract from the leukocytes from immunized individual.
- **Immunologic tolerance:** is defined as the absence of a specific immune response resulting from a previous exposure to the inducing antigen. Two forms of tolerance are known: 1. Natural; 2. Acquired.
- Theories of immunity fall into two categories: instructive and selective.

IMPORTANT QUESTIONS

I. Discuss primary and secondary humoral immune responses.
2. Discuss briefly about:
   Monoclonal antibodies—production and applications.
   – Adjuvants.
   – Cytokines.
   – Theories of antibody production.
3. Write short notes on:
   – Transfer factor.
   – Burnet's clonal selection theory.
   – Immunological tolerance.

FURTHER READING

Dekker, Tonegawa S. Somatic generation of antibody diversity, Nature 1983;302:575.

Dinarello CA, Mier JW. The lymhokines. N Engl J Med 1987; 317:940-5.

Gershon RK, Hondo K. Infectious immunological tolerance. Immunology 1971;21:903-14.

Jerne NK. The generative grammar of the immune system. Science 1985;29,1057.

Llewellyn MB, et al. Monoclonal antibodies, BMJ 1992;305: 1269,1348,1424.

Remick DG, Friedland S. Cytokines in health and disease. 2nd edn. Basel: Marcel 1997.

Roitt I. Essential Immunology, 9th edn. Blackwell Scientific Publications 1996.

Weir DM, Stewart. Immunology. 8th edn. Edinburgh: Churchill-Livingstone 1997.

Schwartz RS. Diversity of the immune response. New Eng J Med 2003;348:1017.

CHAPTER 19

Immunodeficiency Diseases

Learning Objectives

After reading and studying this chapter, you should be able to:

- Classify and enumerate of immunodeficiency diseases.
- List primary and secondary immunodeficiency syndromes.

INTRODUCTION

Immunodeficiency diseases are conditions where the defence mechanisms of the body are impaired, leading to repeated microbial infections of varying severity and sometimes enhanced susceptibility to malignancies.

Immunodeficiency disease results from the absence, or failure of normal function, of one or more elements of the immune system. Specific immunodeficiency diseases involve abnormalities of T or B cells of the adaptive immune system. Nonspecific immunodeficiency diseases involve abnormalities of elements such as complement or phagocytes, which act nonspecifically in immunity.

CLASSIFICATION OF IMMUNODEFICIENCY DISEASES

Immunodeficiencies may be classified as **primary** or **secondary**.

Primary: A condition resulting from a genetic or developmental defect in the immune system is called a primary immunodeficiency.

Secondary: Secondary immunodeficiency, or acquired immunodeficiency is the loss of immune function and results from exposure to various agents.

PRIMARY IMMUNODEFICIENCIES

The established types of primary immunodeficiency syndromes are listed in Table 19.1. The lymphoid cells disorders may affect T cells, B cells, or both B and T cells. Deficiencies involving components of nonspecific mediators of innate immunity, such as phagocytes or complement, are impaired. The myeloid cell disorders affect **phagocytic function**.

A. DISORDERS OF SPECIFIC IMMUNITY

1. Humoral Immunodeficiencies (B Cell Defects)

a. X-linked Agammaglobulinemia (XLA)

The model B cell deficiency is X-linked agammaglobulinemia or Bruton's hypogammaglobulinemia. This rare disorder, which appears to selectively involve very early B lineage cells, was the first recognized of all of the primary immunodeficiencies, and was discovered in 1952 by Colonel Ogden Bruton. It is seen only in male infants. The incidence of this condition has been reported as one in a hundred thousand population in the United Kingdom. Such children frequently have a family history of brothers or maternal uncles with recurring infections.

Manifestations

Manifestations are not apparent till about six months of age due to the passive protection afforded by maternal antibodies. The disease presents as recurrent serious infections with pyogenic bacteria, particularly with pneumococci, streptococci, meningococci, *Pseudomonas* and *H. influenzae.* Patients respond normally to viral infections such as measles and chickenpox.

All classes of immunoglobulins are grossly depleted in the serum, the IgG level being less than a tenth, and IgA and IgM less than a hundredth of the normal level. Tonsils and adenoids are atrophic. Lymph node biopsy reveals a depletion of cells of the bursa-dependent areas. Plasma cells and germinal centers are absent even after antigenic stimulation. There is a marked decrease in the proportion of B cells in circulation. Antibody formation does not occur even after injections of antigens.

Table 19.1: Classification of primary immunodeficiency syndromes

A. Disorders of specific immunity

I. Humoral immunodeficiencies (B cell defects)
- a. X-linked agammaglobulinemia
- b. Transient hypogammaglobulinemia of infancy
- c. Common variable immunodeficiency (late onset hypogammaglobulinemia)
- d. Selective immunoglobulin deficiencies (IgA, IgM or IgG subclasses)
- e. Immunodeficiencies with hyper-IgM
- f. Transcobalamin II deficiency

II. Cellular immunodeficiencies (T cell defects)
- a. Thymic hypoplasia (Digeorge's syndrome)
- b. Chronic mucocutaneous candidiasis
- c. Purine nucleoside phosphorylase (PNP) deficiency

III. Combined immunodeficiencies (B and T cell defects)
- a. Cellular immunodeficiency with abnormal immunoglobulin synthesis (Nezelof syndrome)
- b. Ataxia telangiectasia
- c. Wiskott-Aldrich syndrome
- d Immunodeficiency with thymoma
- e. Immunodeficiency with short-limbed dwarfism
- f Episodic lymphopenia with lymphocytotoxin
- g. Severe combined immunodeficiencies
  1. 'Swiss type' agammaglobulinemia
  2. Reticular dysgenesis of de Vaal
  3. Adenosine deaminase (ADA) deficiency

B. Disorders of complement
- a. Complement component deficiencies
- b. Complement inhibitor deficiencies

C. Disorders of phagocytosis
- a. Chronic granulomatous disease
- b. Myeloperoxidase deficiency
- c. Chediak-Higashi syndrome
- d. Leukocyte G6PD deficiency
- e. Job's syndrome
- f. Tuftsin deficiency
- g. Lazy leukocyte syndrome
- h. Hyper-IgE syndrome
- i. Actin-binding protein deficiency
- j. Shwachman's disease

Cell-mediated Immunity

Cell-mediated immunity is not affected. Delayed hypersensitivity of tuberculin and contact dermatitis types can be demonstrated. Allograft rejection is normal. Arthritis, hemolytic anemia and atopic manifestations are frequently observed. However, the wheal-and-flare response of atopic hypersensitivity cannot be demonstrated.

Management

Its management consists of the maintenance of an adequate level of immunoglobulins. To provide these, whole plasma infusions have been employed, the donors being tested for hepatitis and other transmissible infections.

b. Transient Hypogammaglobulinemia of Infancy

This is due to an abnormal delay in the initiation of IgG synthesis in some patients. By 3 months of age, normal infants begin to synthesize their own IgG. When there is a delay, immunodeficiency occurs. Recurrent otitis media and respiratory infections are the common diseases found in these conditions. Spontaneous recovery occurs between 18 and 30 months of age. It may be found in infants of both sexes.

Treatment

Treatment with gammaglobulin may be required in some cases but it is not recommended prophylactically, as it may contribute to prolongation of immunodeficiency by a negative feedback inhibition of IgG synthesis.

c. Common Variable Immunodeficiency (CVID)

CVID is characterized by a profound decrease in numbers of antibody-producing plasma cells, low levels of most immunoglobulin isotypes (hypogammaglobulinemia), and recurrent infections. The condition is usually manifested later in life (in the second or third decade of life) than other deficiencies and is sometimes called *late onset hypogammaglobulinemia* or incorrectly, *acquired hypogammaglobulinemia*. Patients with CVID, are very susceptible to pyogenic organisms and to the intestinal protozoan, *Giardia lamblia,* which cause severe diarrhea. The B cells are not defective; instead, they fail to receive proper signals from the T cells. B cells fail to mature into plasma cells and secrete immunoglobulins.

Treatment

Patients with CVID should be treated with intravenous gamma globulin as it provides protection against recurrent pyogenic infections. Many patients develop autoimmune diseases, most prominently pernicious anemia, and the reason for this is not known.

d. Selective Immunoglobulin Deficiencies

In these conditions, there is selective deficiency of one or more immunoglobulin classes, while the others remain normal or elevated. These '**dysgammaglobulinemias**' are common and have been reported in about one percent of all patients with recurrent infections.

i. Selective IgA Deficiency

IgA deficiency is by far the most common, with a reported incidence of about 0.2 percent in normal populations. Recurrent respiratory and genitourinary tract infections resulting from lack of secreted IgA on mucosal surfaces are common. In addition, problems such as intestinal malabsorption, allergic disease, and autoimmune disorders may also be associated with low IgA levels.

ii. Selective IgM Deficiency

Selective IgM deficiency is a rare condition and has been found to be associated with septicemia, may be accompanied by various malignancies or by autoimmune disease. IgG deficiencies are also rare.

Treatment

These are often not noticed until adulthood and have been observed in relation with chronic progressive bronchiectasis. These can be effectively treated by administration of immunoglobulin.

e. X-linked Hyper-IgM Syndrome (XHM)

X-linked hyper-IgM syndrome (XHM) is characterized by a deficiency of IgG, IgA, and IgE, and elevated levels of IgM (hyper M). XHM syndrome is generally inherited as an X-linked recessive disorder. Childern with XHM suffer recurrent infections, especially respiratory infections. Patients show enhanced susceptibility to infections and autoimmune processes such as thrombocytopenia, neutropenia, hemolytic anemia and renal lesions. Elevated IgM level with immunodeficiency is sometimes seen in congenital rubella.

f. Transcobalamin II Deficiency

In this disorder, patients show metabolic effects of vitamin B_{12} deficiency including megaloblastic anemia and intestinal villus atrophy and is inherited as autosomal recessive. The associated immunological defects are depleted plasma cells, diminished immunoglobulin levels and impaired phagocytosis.

Treatment with vitamin B_{12} has been reported to restore hematopoietic, gastrointestinal and B cells functions but not phagocytic activity.

2. Cellular Immunodeficiencies (T Cell Defects)

a. Thymic Hypoplasia (DiGeorge's Syndrome)

Thymic hypoplasia results from dysmorphogenesis of the third and fourth pharyngeal pouches during early embryogenesis, leading to hypoplasia or aplasia of the thymus and parathyroid glands. It is probably due to some intrauterine infection or other complication.

The stage at which the causative developmental defect occurs has been determined, and the syndrome is sometimes called the *third and fourth pharyngeal pouch syndrome* to reflect its precise embryonic origin.

The immunodeficiency primarily involves cell-mediated immunity. The thymus dependent areas of lymph nodes and spleen are depleted of lymphocytes. Delayed hypersensitivity and graft rejection are depressed. Although B cells are present in normal numbers, affected individuals do not produce antibody in response to immunization with specific antigens. Thymic transplantation is of some value for correcting the T cell defects, but many DiGeorge patients have such severe heart disease that their chances of survival are poor, even if the immune defects are corrected.

b. Purine Nucleoside Phosphorylase (PNP) Deficiency

The enzyme purine nucleoside phosphorylase (PNP) is involved in the sequential degradation of purines to hypoxanthine and finally to uric acid. The lack of enzyme PNP because of a gene defect in chromosome 14 results in impaired metabolism of cytosine and inosine to purine. These patients show decreased T cell proliferation leading to decreased T cell-mediated immunity and recurrent or chronic infections. They usually present with hypoplastic anemia and recurrent pneumonia, diarrhea and candidiasis. A low serum uric acid may point to the diagnosis.

3. Combined Immunodeficiencies (B and T Cell Deficiency)

a. Cellular Immunodeficiency with Abnormal Immunoglobulin Synthesis (Nezelof Syndrome)

The term Nezelof syndrome has been rather loosely applied to a group of disorders, probably of varied etiology, where depressed cell-mediated immunity is associated with selectively elevated, decreased or normal levels of immunoglobulin. The consistent features are a marked deficiency of T cell immunity and varying degrees of deficiency of B cell immunity. Affected individuals suffer from chronic diarrhea, viral and fungal infections, and a general failure to thrive. Autoimmune processes such as hemolytic anemia are common. antigenic stimuli do not induce antibody formation in spite of normal levels of immunoglobulins.

Treatment

Histocompatible bone marrow transplantation, transfer factor and thymus transplantation have been used for treatment, with success in some cases. Adequate antimicrobial therapy is essential for the treatment of microbial infection.

b. Ataxia Telangiectasia

Ataxia telangiectasia is a disease syndrome that includes deficiency of IgA and sometimes of IgE. It is inherited as an autosomal recessive trait. The most prominent clinical features are progressive cerebellar ataxia, oculocutaneous telangiectasias, chronic sinopulmonary disease, a high incidence of malignancy, and a variable humoral and cellular immunodeficiency. Affected infants develop a wobbly gait (ataxia) at about 18 months. Telangiectasia (dilated capillaries) involving the conjunctiva, face and other parts of the body usually appears at five or six years of age. Death occurs due to sinopulmonary infection early in life, or malignancy in the second or third decade.

The defective cell-mediated immunity results in an impairment of delayed hypersensitivity and graft rejection. The disease is progressive, with both neurological defects and immunodeficiency becoming more severe with time.

Transfer factor therapy and fetal thymus transplants have been tried with some benefit.

c. Wiskott-Aldrich Syndrome (WAS)

This is an X-linked recessive syndrome that is characterized clinically by the triad of **eczema, thrombocytopenic purpura, and undue susceptibility to infection**. Affected boys rarely survive the first decade of life, death being due to infection, hemorrhage or lymphoreticular malignancy.

Cell-mediated immunity undergoes progressive deterioration associated with cellular depletion of the thymus and the paracortical areas of lymph nodes. Serum IgM level is low but IgG and IgA levels are normal or elevated. Isohemagglutinins are absent in the serum. The humoral defect appears to be a specific inability to respond to polysaccharide antigens.

Bone marrow transplantation and transfer factor therapy have been found beneficial.

d. Immunodeficiency with Thymoma

This syndrome, occurring usually in adults, consists of thymoma, impaired cell-mediated immunity and agammaglobulinemia. This association of hypogammaglobulinemia with spindle cell thymoma usually occurs relatively late in adult life. It is frequently accompanied by aplastic anemia.

e. Severe Combined Immunodeficiency (SCID)

The family of disorders termed SCID stems from the defects in lymphoid development that affect either T cells or both T and B cells. The SCID syndrome is characterized by gross functional impairment of both humoral and cell-mediated immunity and by susceptibility to devastating fungal, bacterial and viral infections. It is usually congenital, may be inherited either as an X-linked or autosomal recessive defect, or may occur sporadically. Affected infants rarely survive beyond 1 year without treatment.

i. Swiss Type Agammaglobulinemia

Agammaglobulinemia with lymphocytopenia and severe defect in cell-mediated immunity was reported by Swiss workers in 1958. This has been referred to as *Swiss type agammaglobulinemia.* The basic defect is presumed to be at the level of the lymphoid stem cell.

ii. Adenosine Deaminase (ADA) Deficiency

Adenosine deaminase (ADA) *deficiency* is the first immunodeficiency disease associated with an enzyme deficiency. ADA catalyzes the conversion of adenosine to inosine, an important step in the purine metabolic pathway. Its deficiency results in accumulation of adenosine, which interferes with purine metabolism and DNA synthesis. The range of immunodeficiency varies from complete absence to mild abnormalities of B and T cell functions. The condition is associated with chondrocyte abnormalities which can be discerned radiologically.

iii. Reticular Dysgenesis

This is the most serious form of SCID. Here the defect is in the development of the multipotent bone marrow stem cell, as a result of which there is a total failure of myelopoiesis leading to lymphopenia, neutropenia, thrombocytopenia, anemia and bone marrow aplasia. A baby with this disorder usually dies within the first year of life from recurrent, intractable infections.

iv. Recombinase Activating Gene (RAG 1/2) Deficiency

It is an autosomal recessive form of SCID results from a mutation in either of the genes encoding RAG-1 and RAG-2.

v. Interleukin Receptor γ Chain (γ_c) Deficiency

The defective gene encodes a common γ chain of the receptors for IL-2,-4,-7,-9, and 15, thus disrupting the action of this important set of lymphokines.

vi. Janus-associated Kinase 3 (JAK3) Deficiency

The same $T^- NK^- B^+$ SCID phenotype seen in X-linked SCID can be inherited as an autosomal recessive disease due to mutations in the gene for *JAK3 protein kinase deficiency*. This enzyme associates with the common γ chain of the receptors for IL-2,-4,-7,-9 and 15 to serve as a key element in their signal transduction pathways.

B. DISORDERS OF COMPLEMENT

a. Complement Component Deficiencies

Genetic deficiencies have been detected for almost all the complement components in human beings. The defects are transmitted as autosomal recessive traits. Hemolytic and other functional activities are completely restored by supplying the deficient factor. Deficiency of Clr and C4 is associated with systemic lupus erythematosus. Recurrent pyogenic infections were found associated with C3 deficiency and neisserial infections with deficiency of C6, C7 and C8. So far, there is no known disease with deficiency of C9.

b. Complement Inhibitor Deficiencies

i. C1 Inhibitor

Clinically, the most important deficiency of the complement system is that of the **C1 inhibitor**. **Hereditary angioneurotic edema (HAE)** is due to a genetic deficiency of C1 inhibitor and is transmitted as an autosomal dominant condition. It manifests clinically as localized edema of the tissue, often following trauma, but sometimes with no known cause.

Management

Androgens, aminocaproic acid and its analogue tranexamic acid have been found useful in the management of this condition.

ii. Deficiency of C3b Inactivator

The rare deficiency of C3b inactivator has been associated with chronic recurrent pyogenic lesions.

C. DISORDERS OF PHAGOCYTSIS

Phagocytosis may be impaired by either intrinsic or extrinsic defects. Intrinsic disorders may be due to defects within the phagocytic cell, such as enzyme deficiencies. Extrinsic disorders may be due to a deficiency of opsonic antibody, complement or other factors promoting phagocytosis, or to the effects of drugs or antineutrophil autoantibodies. Phagocytic dysfunction leads to increased susceptibility to infection, ranging from mild recurrent skin infections to overwhelming systemic infection.

a. Chronic Granulomatous Disease (CGD)

Chronic granulomatous disease (CGD) is a group of disorders, most of which are X-linked recessive, with some that are autosomal recessive. Individuals with CGD possess polymorphonuclear leukocytes that phagocytize invading bacteria normally but are unable to kill many of the ingested microorganisms.

Bacterial Infection

The bacteria involved in the recurrent infections are catalase positive pyogenic pathogens such as staphylococci and coliforms. Catalase negative pathogens are handled normally. Leukocytes from the patients are unable to kill catalase positive bacteria following phagocytosis. The bacteria multiply in the cells and, being protected from antibodies and antibiotics by their intracellular position, set up chronic suppurative infection.

Bactericidal Defect

The diminished bactericidal capacity of the phagocytic cells is associated with a decrease of some metabolic processes like oxygen consumption, hexose monophosphate pathway activity and production of hydrogen peroxide. The diminished H_2O_2 production appears to be the major reason for the bactericidal defect.

Manifestation

This familial disease manifests itself as recurrent infection with low grade pathogens, starting early in life. The progress is chronic and the outcome fatal. Symptoms occur by two years of age, with patients suffering from recurrent opportunistic infections, primarily involving bacteria and fungi.

Leukocytes from the patients fail to reduce nitroblue tetrazolium (NBT) during phagocytosis. This property has been used as a screening method (NBT test) for the diagnosis of chronic granulomatous disease.

Treatment

Treatment consists of broad-spectrum antibiotics and antifungal agents. Interferon-γ injections to activate phagocytes improve the clinical course of CGD.

b. Myeloperoxidase (MPO) Deficiency

In this rare disease, leukocytes are deficient in myeloperoxidase (MPO). They generally remain well and free from recurrent infections. However, they are liable to develop recurrent *Candida albicans* infection.

c. Chediak-Higashi Syndrome

This autosomal recessive disease is characterized by decreased pigmentation of the skin, eyes and hair, photophobia, nystagmus and giant peroxidase positive inclusions in the cytoplasm of leukocytes. The inclusions may be the result of autophagocytic activity. The leukocytes possess diminished phagocytic activity. The individuals with this syndrome suffer from recurrent infections similar to those seen in persons with CGD.

d. Leukocyte G-6-PD Deficiency

In this rare disease leukocytes are deficient in glucose-6-phosphate dehydrogenase. These patients show diminished bactericidal activity after phagocytosis leading to repeated bacterial infections. The condition resembles chronic granulomatous disease in reduced myeloperoxidase activity and susceptibility to microbial agents, but the NBT test may be normal.

e. Job's Syndrome

This is characterized by multiple large 'cold' staphylococcal abscesses containing large quantities of pus, occurring inflammatory response. Atopic eczema, chronic nasal discharge and otitis media are common features. The serum immunoglobulins are normal, except for elevated IgE. The pathogenesis of the syndrome is not clear but it is probably a primary defect in phagocytic function.

f. Tuftsin Deficiency

A leukokinin capable of stimulating phagocytosis, discovered at Tufts University, Boston, has been designated 'tuftsin'. Patients with tuftsin deficiency have been reported to be prone to local and systemic bacterial infections.

g. Lazy Leukocyte Syndrome

The basic defect here is in chemotaxis and neutrophil mobility. The bone marrow has a normal number of neutrophils but there is a peripheral neutropenia, with poor leukocyte response to chemical and inflammatory stimulation. Patients show an increased susceptibility to bacterial infection, with recurrent stomatitis, gingivitis and otitis.

h. Hyper-IgE Syndrome

These patients, of both sexes, have an early onset of eczema and recurrent bacterial infections such as abscesses, pneumonia and secondary infection of eczema. The organisms responsible include *Staphylococcus aureus* and *Streptococcus pyogenes.* Cellular and humoral immune mechanisms are normal but serum IgE levels are usually more than ten times the normal level.

i. Actin-binding Protein Deficiency

Frequent infection and slow mobility of leukocytes result from the defective actin-binding protein in these patients.

j. Shwachman's Disease

In this condition frequent infections are found together with decreased neutrophil mobility, pancreatic malfunction and bone abnormalities.

k. Leukocyte Adhesion Deficiency (LAD)

LAD is due to integrin gene defects. This receptor, an integrin called complement receptor 3 (CR3), is deficient in patients with LAD and consequently they develop severe bacterial infections, particularly of the mouth and gastrointestinal tract.

SECONDARY IMMUNODEFICIENCIES

Secondary or acquired deficiencies of immunological mechanisms can occur secondarily to a number of disease states such as metabolic disorders, malnutrition, malignancy and infections or after exposure to drugs and chemicals. AIDS is a secondary immunodeficiency is far more common than primary immunodeficiency.

Deficiencies of Humoral and Cellular Immune Response

Deficiencies of humoral and cellular immune response may occur secondarily during the course of many disease processes.

A. Humoral Immunity Depression

Humoral deficiency results when B cells are depleted as in lymphoid malignancy, particularly in chronic lymphatic leukemia; when immunoglobulin catabolism is increased as in the nephrotic syndrome; when excessive loss of serum protein occurs as in exfoliative skin disease and in protein-losing enteropathies; and when excessive production of abnormal immunoglobulins occurs as in multiple myeloma.

B. Cell-mediated Immunity Depression

Cell-mediated immunity is depressed in lymphoreticular malignancies like Hodgkin's lymphoma, obstruction to lymph circulation and infiltration of the thymus-dependent area of lymph nodes with nonlymphoid cells as in lepromatous leprosy; and transiently, following certain viral infections such as measles.

C. Humoral and Cell-mediated Immunity Depression

Both types of immune responses are adversely affected in nutritional deprivation. Aging also causes waning in the efficiency of acquired immunity. Immunodeficiency follows the intentional or unintentional administration of immunosuppressive agents.

KNOW MORE

C1 Inhibitor

Normally, Cl inhibitor is involved in inactivation of complement system and other pathways, like clotting, fibrinolytic and kinin. These systems may be activated in Cl inhibitor deficiency. Formation of bradykinin and C2 kinin is due to activation of factor XII (Hageman factor). These kinins act on the postcapillary venules and cause contraction of endothelial cells and formation of gaps which allow the plasma leakage and production of edema.

☞ KEY POINTS

- Immunodeficiency results from the failure of one or more components of the immune system. Immunodeficiency disorders can occur in T cells, B cells, complement and phagocytes.
- These can be classified as primary or secondary immunodeficiencies.
- **A primary immunodeficiency** may affect either adaptive or innate immune functions.
- **Examples of complement deficiencies**—patients with deficiencies in C2 and C4 components, paroxysmal nocturnal hemoglobinuria, and hereditary angioedema
- **Disorders of phagocytosis**—include chronic granulomatous disease (CGO), myeloperoxidase deficiency Chediak-Higashi syndrome, leukocyte adhesion deficiency, Job's syndrome.
- **Secondary immunodeficiencies** occur secondary to numerous diseases or conditions and can result from malnutrition, immunosuppressive agents, infections (such as AIDS), and malignancies.

IMPORTANT QUESTIONS

1. What are immunodeficiency diseases? Classify and enumerate immunodeficiency diseases.
2. Write short notes on:
   a. DiGeorge's syndrome
   b. B cell defect
   c. T cell defect
   d. Disorders of complement

FURTHER READING

Fischer A, Amaiz-Villena A. Immunodeficiencies of genetic origin. Immunol Today 1995;16:510.

Roitt I, Brostoff J, Male D (Eds). Primary immunodeficiency. In: Immunology, 5 (Edn) Harcourt Brace & Company Asia PTE Ltd 1998;285-92.

Rosen FS, et al. The primary immunodeficiencies. New Engl J Med 1984;11:235,300.

Rosen FS, Seligman M (Eds). Immunodeficiencies, Switzerland: Harwood Academic Publishers GmbH 1993.

CHAPTER 20

Hypersensitivity Reactions

Learning Objectives

After reading and studying this chapter, you should be able to:

- Compare major types of hypersensitivity reactions.
- Differentiate between immediate and delayed hypersensitivity.
- Discuss type I, type II, type-III type IV hypersensitivity reactions—Mechanism and examples.

Hypersensitivity: Hypersensitivity is an exaggerated immune response that results in tissue damage and is manifested in the individual on second or subsequent contact with an antigen.

Immune responses to foreign antigens are, for the most part, beneficial to the responding individual. Nevertheless, at times the response to a seemingly innocuous antigen can result in tissue damage and even death. This inappropriate immune response is termed hypersensitivity or allergy.

Although, the word hypersensitivity implies an increased response, the response is not always heightened but may, instead, be an inappropriate immune response to an antigen. Hypersensitive reactions may develop in the course of either humoral or cell-mediated responses.

Allergy: The term allergy (from the Greek word meaning altered reaction) was coined by von Pirquet merely to indicate an altered reactivity on second contact with an antigen. In time, however, the term allergy has become synonymous with hypersensitivity. Increased resistance, called **immunity**, and increased susceptibility, called **hypersensitivity**, were regarded as opposite forms of allergy.

CLASSIFICATION OF HYPERSENSITIVITY REACTIONS

Hypersensitivity reactions have been classified traditionally into 'immediate' and 'delayed' types, based on the time required for a sensitized host to develop clinical reactions on re-exposure to the antigen.

I. **Immediate hypersensitivity:** In **immediate hypersensitivity** the symptoms are manifest within minutes or hours after a sensitized recipient encounters antigen.

II. **Delayed hypersensitivity: Delayed type hypersensitivity (DTH)** is so named in recognition of the delay of symptoms until days after exposure.

The major differences between the immediate and delayed types of hypersensitivity reactions are shown in (Table 20.1).

Clinical Types

The immediate and delayed reactions are subdivided into several distinct clinical types:

I. Immediate hypersensitivity (B cell or antibody mediated)
   - Anaphylaxis
   - Atopy
   - Antibody mediated cell damage
   - Arthus phenomenon
   - Serum sickness.

II. Delayed hypersensitivity (T cell mediated)
   - Infection (tuberculin) type
   - Contact dermatitis type

Gell and Coombs Classification

Peter Gell and Robert Coombs developed a classification system for reactions responsible for hypersensitivities in 1963. Their system correlates cliniclal symptoms with information about immunologic events that occur during hypersensitive reactions. The Gell-Coombs classification system divides hypersensitivity into four types: I, II, III, and IV.

Type I (Anaphylactic, IgE or Reagin Dependent)

Type I or immediate hypersensitivity is characterized by the production of antibodies ('cytotropic' IgE antibodies) which bind specifically to a high-affinity receptor on mast cells and basophils in sensitized individuals. Subsequent exposure to the same antigen will combine

Table 20.1: Distinguishing features of immediate and delayed types of hypersensitivity		
Characteristic	*Immediate hypersensitivity*	*Delayed hypersensitivity*
1. Time of reaction after challenge with antigen	1. Reaction appears and recedes rapidly.	1. Appears slowly, lasts longer.
2. Induction	2. Induced by antigens or haptens.	2. Antigen or hapten intradermally or with by any route.
3. Immune response	3. Circulating antibodies present and responsible for reaction; 'antibody mediated' reaction.	3. Circulating antibodies may be absent and are not responsible for reaction; 'cell-mediated' reaction.
4. Transfer of hypersensitivity	4. Passive transfer possible with serum.	4. Cannot be transferred with serum; but possible with T cells or transfer factor.
5. Desensitization	5. Desensitization easy, but short-lived.	5. Difficult, but long-lasting.

Table 20.2: Comparison of major types of hypersensitivity reactions			
Type of reaction	*Time required for manifestation*	*Mediators*	*Clinical syndrome*
Type I: IgE type	Minutes	IgE: Histamine and other pharmacological agents	1. Anaphylaxis 2. Atopy
Type II: Cytolytic and cytotoxic	Variable: Hours to days	IgG: IgM, complement	1. Transfusion reactions 2. Rh incompatibility
Type III: Immune complex	Variable: Hours to days	IgG: IgM, C, leukocytes	1. Arthus reaction 2. Serum sickness
Type IV: Delayed hypersensitivity	Hours to days	T cells; Lymphokines; macrophages	1. Tuberculin test 2. Contact dermatitis 3. Graft rejection 4. Tumor immunity

with the cell fixed antibody, leading to release of pharmacologically active substances (vasoactive amines) which produce the clinical reaction.

Type II (Cytotoxic or Cell Stimulating)

Type II reaction is initiated by IgG (or rarely IgM) antibodies that react either with cell surface or tissue antigens. Cell or tissue damage occurs in the presence of complement or mononuclear cells. Type II reactions are intermediate between hypersensitivity and autoimmunity. Combination with antibody may, in some instances, cause stimulation instead of damage. An example is the 'long acting thyroid stimulator' (LATS), an antibody against some determinant on thyroid cells, which stimulates excessive secretion of thyroid hormone. (Such antibody mediated cell stimulation has also been called type V hypersensitivity).

Type III (Immune Complex or Toxic Complex Disease)

Type III or immune complex disease occurs when excess complexes are formed in the circulation. These may precipitate in and around small blood vessels, causing damage to cells secondarily, or on membranes, interfering with their function.

Type IV (Delayed or Cell-Mediated Hypersensitivity)

Type IV or cell mediated reactions are those in which specific T cells are the primary effector cells. The antigen activates specifically sensitized CD4 and CD8 T cells, leading to the secretion of lymphokines, with fluid and phagocyte accumulation. The classification and some of the features of hypersensitivity reactions are shown in (Table 20.2).

TYPE I HYPERSENSITIVITY (IGE DEPENDENT)

A type I hypersensitive reaction is induced by certain types of antigens referred to as ***allergens***, and has all the hallmarks of a normal humoral response. Allergic reactions occur when an individual who has produced IgE antibody in response to an innocuous antigen (allergen) subsequently encounters the same allergen. Type I, or anaphylactic, reactions often occur within 2 to 30 minutes after a person sensitized to an antigen is re-exposed to that antigen.

Anaphylaxis

Anaphylaxis means "the opposite of protected", from the prefix ana-, against, and the Greek *phylaxis*, protection. Anaphylaxis is an inclusive term for the reactions caused when certain antigens combine with IgE antibodies. Anaphylactic responses can be:

A. **Systemic reactions:** Systemic reactions produce shock and breathing difficulties and are sometimes fatal.

B. **Localized reactions:** Localized reactions is chronic or recurrent, nonfatal, typically localized form called atopy.

A. Systemic Anaphylaxis (or Anaphylactic Shock)

Systemic anaphylaxis is a generalized response that occurs when an individual sensitized to an allergen receives a subsequent exposure to it. Antigen enters the bloodstream and becomes widespread. Instead of being localized in area of the skin or respiratory tract, the reaction affects almost the entire body.

Systemic anaphylaxis is a shock-like and often fatal state whose onset occurs within minutes of a type I hypersensitive reaction. This was the response observed by Portier and Richet in dogs after antigenic challenge. Theobald Smith (1902) had noticed a similar phenomenon in guinea pigs, following widely spaced injections of toxin-antitoxin mixtures. Ehrlich named this the **'Theobald Smith phenomenon'** and showed that it was independent of the toxin and antitoxin used, since the phenomenon could be induced with normal serum also.

Sensitizing Dose and the Shocking Dose

Sensitization is most effective when the antigen is introduced parenterally but may occur by any route, including ingestion or inhalation. In susceptible species, very minute doses can sensitize the host. Antigens as well as haptens can induce anaphylaxis. There should be an interval of at least 2 to 3 weeks between the **sensitizing dose** and the **shocking dose**. Once sensitized, the individual remains so for long periods. The shocking dose is most effective when injected intravenously, less effective intraperitoneally or subcutaneously and least effective intradermally. The shocking antigen must be identical or immunologically closely related to the sensitizing antigen. The clinical features of anaphylaxis are the same with any antigen but vary between species.

Target Tissues or 'Shock Organs'

The clinical effects are due to smooth muscle contraction and increased vascular permeability. The organs affected vary with the species. Tissues or organs predominantly involved in the anaphylactic reaction are known as **'target tissues'** or **'shock organs'**. Other changes seen in anaphylaxis are edema, decreased coagulability of blood, fall in blood pressure and temperature, leukopenia and thrombocytopenia.

Experimental animals: Systemic anaphylaxis can be induced in a variety of experimental animals and is seen occasionally in humans. There is considerable species variation in susceptibility to anaphylaxis. Guinea pigs are highly susceptible and rats are very resistant. Rabbits, dogs and human beings are of intermediate susceptibility.

i. **Guinea pigs:** Active sensitization in **guinea pigs** is induced by a single injection of a foreign protein such as egg albumin. The animal is usually challenged with an intravenous injection of the same protein after an incubation period of about 2 weeks. **Sequence of events:** The guinea pig will exhibit a dramatic sequence of events. Within 1 minute, the animal becomes restless, its respiration becomes labored, and its blood pressure drops. As the smooth muscles of the gastrointestinal tract and bladder contract, the guinea pig defecates and urinates. Finally bronchiole constriction results in death by asphyxiation within 2 to 4 minutes of the injection. The heart continues to beat for sometime after the respiration has stopped. These events all stem from the systemic vasodilation and smooth-muscle contraction brought on by mediators released in the course of the reaction. Postmortem examination reveals that massive edema, shock, and bronchiole constriction are the major causes of death.

ii. **Rabbits:** In rabbits, death in anaphylactic shock is due to constriction of the pulmonary artery and its branches, leading to extreme dilatation of the right side of the heart. Respiratory movements continue after the cessation of the heartbeat.

iii. **Dogs:** In dogs, the reaction is slower and takes 1 to 2 hours. There is constriction of the hepatic venous system with gross engorgement of the liver and profound fall of blood pressure.

Systemic Anaphylaxis in Humans

Systemic anaphylaxis in humans is characterized by a similar sequence of events that occurs when an individual sensitized to an allergen receives a subsequent exposure to it. In human beings, fatal anaphylaxis is fortunately rare. The reaction is immediate due to the large amount of mast cell mediators released over a short period.

Symptoms and signs: Symptoms and signs of anaphylactic shock begin with itching of the scalp and tongue, flushing of the skin over the whole body and difficulty in breathing due to bronchial spasm. There may be nausea, vomiting, abdominal pain and diarrhea, sometimes with blood in the stool. Acute hypotension, loss of consciousness and death follow. The arterioles dilate, which greatly reduces arterial blood pressure and increases capillary permeability with rapid loss of fluid into the tissue spaces. Because of these reactions the individual can die within a few minutes from reduced venous return, asphyxiation, reduced blood pressure, and circulatory shock.

Antigens: A wide range of antigens have been shown to trigger this reaction in susceptible humans, including the venom from bee, wasp, hornet, and ant stings; drugs, such as penicillin, insulin, and antitoxins; and seafood and nuts. If not treated quickly, these reactions can be fatal.

Treatment

Epinephrine (adrenaline) is the drug of choice for systemic anaphylactic reactions, a drug that constricts blood vessels and raises the blood pressure. Prompt treatment with adrenaline can be life saving. Adrenaline is to be administered, 0.5 ml of a 1 in 1000 solution, subcutaneously or intramuscularly, the dose being repeated up to a total of 2 ml over 15 minutes, if necessary.

Cutaneous Anaphylaxis

If a person is suspected to be allergic to a particular substance, a skin test can be done. Small amounts of potential allergens are introduced at specific skin sites either by intradermal injection or by superficial scratching. If a person is allergic to the allergen, there is a wheal and flare (local anaphylaxis) within 30 min. Wheal is a pale central area of puffiness due to edema caused by increased capillary permeability which is surrounded by flare (caused by hyperemia and subsequent erythema). **Use:** Cutaneous anaphylaxis (skin test for type I hypersensitivity) is useful in testing for hypersensitivity and in identifying the allergen responsible in atopic diseases. A number of tests can be applied to sites on the forearm or back of an individual at one time.

Precaution: In highly sensitized individuals, even the skin test may lead to serious and even fatal reactions. Hence, a syringe loaded with adrenaline should always be kept ready whenever, a skin test is performed to detect anaphylactic hypersensitivity.

Passive Cutaneous Anaphylaxis (PCA)

This test is an extremely sensitive *in vivo* method for detection of antibodies developed by Zoltan Ovary (1952). PCA is fundamentally the same as the human PK reaction. In this procedure, a small volume of the antibody is injected intradermally into a normal animal. The corresponding antigen is injected intravenously 4 to 24 hours afterwards along with a dye such as Evans blue that is strongly bound to serum albumin. There will be an immediate blueing at the site of intradermal injection due to vasodilatation and increased capillary permeability (wheal-and-flare reaction).

Use of PCA: PCA can be used to detect human IgG antibody which is heterocytotropic (capable of fixing to cells of other species) but not IgE which is homocytotropic (capable of fixing to cells of homologous species only).

Anaphylaxis *In Vitro* (Schultz-Dale Phenomenon)

Schultz and Dale (1910) demonstrated that isolated tissues, such as intestinal or uterine muscle strips from sensitized guinea pigs, held in a bath of Ringer's solution will contract vigorously on addition of the specific antigen to the bath. This is known as the Schultz-Dale phenomenon. The reaction is specific and will be elicited only by the antigen to which the animal is sensitive. By treatment with serum from sensitised animals, tissues from normal animals can be passively sensitized. The actual test is done by adding the antigen and observing subsequent contractions.

Mechanism of Anaphylaxis

The immunologic basis for hypersensitivity is cytotropic IgE antibody. Free IgE antibody in circulation is not relevant in anaphylaxis. Thus, an animal with a high titer of circulating antibody may be refractory to shock, while anaphylaxis may be caused by cell fixed antibody, even in the absence of detectable circulating antibody. While in human beings, IgE is the cytophilic antibody, in the guinea pig and mouse the analogous cytophilic antibody is IgG 1.

After an initial contact with an antigen, the individual produces IgE antibodies. IgE molecules are bound to surface receptors on mast cells and basophils. These cells carry large numbers of such receptors called FcER receptors, analogous to TCR receptors on T cell surface. IgE molecules attach to these receptors by their Fc end, leaving two antigen-binding sites free. Mast cells and basophils coated by IgE are said to be sensitized, making the individual allergic to the allergen.

Following exposure to the shocking dose, the antigen molecules combine with the cell bound IgE, bridging the gap between adjacent antibody molecules. This cross-linking increases the permeability of the cells to calcium ions and leads to degranulation, with release of biologically active substances contained in the granules. The pharmacologically active mediators released from the granules act on the surrounding tissues. The manifestations of anaphylaxis are due to pharmacological mediators, which can be classified as two types either primary or secondary. *The primary mediators* are the preformed contents of mast cells and basophile granules produced before degranulation and are stored in the granules (histamine, proteases, eosinophil chemotactic factor, neutrophil chemotactic factor, and heparin). *Secondary mediators* are newly formed upon stimulation by mast cells, basophils and other leukocytes (slow reacting substance of anaphylaxis, prostaglandins and platelet activating factor, and cytokines such as IL-3, IL-4, IL-5, IL-6; GM-CSF). These mediators trigger smooth muscle contractions, vasodilation, increased vascular permeability, and mucus secretion (Fig. 20.1).

A. Primary Mediators of Anaphylaxis

The primary mediators are produced before degranulation and are stored in the granules. The most significant primary mediators are histamine, proteases, eosinophil chemotactic factor, neutrophil chemotactic factor and heparin.

1. **Histamine:** Histamine, which is formed by decarboxylation of the amino acid histidine and is localized in the granules of mast cells and basophils and in the platelets of some species. Its biological effects are observed within minutes of mast cell activation because it is stored preformed in the granules. Released into the skin, histamine stimulates sensory nerves, producing burning and itching sensations. It causes vasodilatation and hyperemia by an

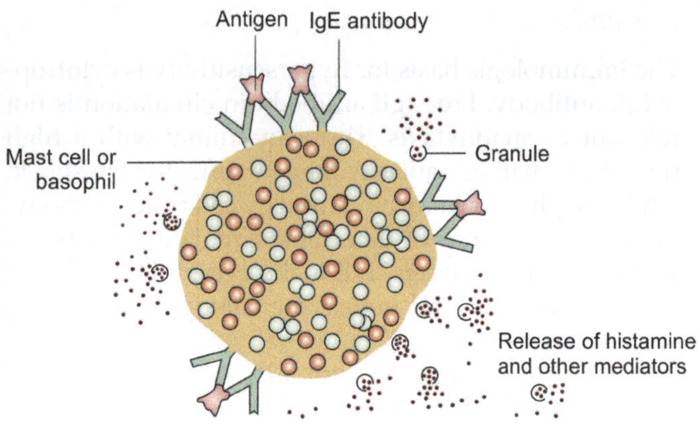

Fig. 20.1: Antigen-induced mediator release from mast cell

axon reflex (flare effect) and edema by increasing capillary permeability (wheal effect). Histamine induces smooth muscle contraction in diverse tissues and organs, including vasculature, intestines, uterus and especially the bronchioles. It also stimulates secretions (secretagogue effect).

2. **Heparin:** Heparin is an acidic mucopolysaccharide. It contributes to anaphylaxis in dogs, but apparently not in human beings.
3. **Serotinin (5-hydroxytryptamine):** This is a base derived from decarbxylation of tryptophan. It is found in the intestinal mucosa, brain tissue and platelets.It is present in the platelets of most species. It induces contraction of smooth muscle, increased vascular permeability, and capillary dilatation. Its is important in anaphylaxis in rats and mice but its role in human beings is uncertain.
4. **Chemotactic factors**
   i. *Eosinophil chemotactic factor (ECF-A):* Eosinophil chemotactic factors of anaphylaxis (ECF-A) are acidic tetrapeptides released from mast cell granules which are strongly chemotactic for eosinophils. These probably contribute to the eosinophilia accompanying many hypersensitivity states.
   ii. *Neutrophil chemotactic factors (NCF):* A high molecular weight chemotactic factor has been identified, which attracts neutrophils (NCF).
5. **Proteases:** Enzymatic mediators such as proteases and hydrolases are also released from mast cell granules. These enzymes most probably are involved in the digestion of blood vessel basement membranes, resulting in increased permeability to a variety of cell types.

B. Secondary Mediators of Anaphylaxis

The secondary mediators either are synthesized after target-cell activation or are released by the breakdown of membrane phospholipids during the degranulation process.

1. **Platelet activating factor (PAF):** Platelet activating factor (PAF) is a low molecular weight lipid released from basophils which causes aggregation of platelets and release of their vasoactive amines. It is a low-molecular-weight lipid, which induces a rapid wheal and flare reaction when injected into human skin.
2. **Leukotrienes and prostaglandin:** They are derived by two different pathways from arachidonic acid, which is formed from disrupted cell membranes of mast cells and other leukocytes. The lipoxygenase pathway leads to the formation of leukotrienes, while the cyclo-oxygenase pathway leads to formation of prostaglandins and thromboxane. As secondary mediators, the leukotrienes and prostaglandins are not formed until the mast cell undergoes degranulation and the enzymatic breakdown of phospholipids in the plasma membrane. Their effects are more pronounced and longer lasting, however, than those of histamine.
   A substance originally demonstrated in lungs, producing slow, sustained contraction of smooth muscles, and therefore termed slow reacting substance of anaphylaxis (SRS-A) has since been identified a family of leukotrienes (LTB4, C4, D4, lM). The leukotrienes mediate bronchoconstriction, increased vascular permeability, and mucus production. In humans, the leukotrienes are thought to contribute to the prolonged bronchospasm and buildup of mucus seen in asthmatics.
   Prostaglandin F2-α and thromboxane A2 are powerful, but transient, bronchoconstrictors. Prostaglandins also affect secretion by mucus glands, platelet adhesion, permeability and dilatation of capillaries and the pain threshold.
3. **Cytokines:** Human mast cells secrete IL-4, IL-5, IL-6, and TNF-α. These cytokines alter the local microenvironment, eventually leading to the recruitment of inflammatory cells such as neutrophils and eosinophils. IL-4 increases 1gE production by B cells. IL-5 is especially important in the recruitment and activation of eosinophils. The high concentrations of TNF-α secreted by mast cells may contribute to shock in systemic anaphylaxis.

Other Mediators of Anaphylaxis

Besides the products of mast cells and other leukocytes, several other biologically substances are known to induce mast cell degranulation and have been implicated in anaphylaxis. All do so by bypassing the IgE receptor and interacting with other receptors on the mast cell membrane. These include split products from complement activation, C3a and C5a and bradykinin and other kinins formed from plasma kininogens.

Anaphylactoid Reaction

Intravenous injection of peptone, trypsin and certain other substances provokes a clinical reaction resembling anaphylactic shock. This is termed 'anaphylactoid reaction'. The clinical resemblance is due to the same chemical mediators participating in both reactions.

Anaphylactoid shock has no immunological basis and, It is a nonspecific mechanism involving the activation of complement and the release of anaphylatoxins which is the only difference.

B. Localized Anaphylaxis (Atopy)

The term 'atopy' (literally meaning out of place or strangeness) refers to naturally occurring familial hypersensitivities of human beings. It was introduced by Coca (1923) and typified by hay fever and asthma. The antigens commonly involved in atopy are **inhalants** (for example, plant pollen, fungal spores, animal dander, and house dust mites or other types of fine particles suspended in air) or **ingestants** (for example, milk, milk products, eggs, meat, fish or cereal). Some of them are **contact allergens**, to which the skin and conjunctiva may be exposed. The symptoms depend primarily on the route by which the antigen enters the body. These atopens are generally not good antigens when injected parenterally but induce IgE antibodies, formerly termed as 'reagin' antibodies. Atopic sensitization is developed spontaneously following natural contact with atopens. It is difficult to induce atopy artificially.

Predisposition to atopy is genetically determined, probably linked to MHC genotypes. Atopy therefore runs in families. What is inherited is not sensitivity to a particular antigen, or a particular atopic syndrome but the tendency to produce IgE antibodies in unusually large quantities. All individuals are capable of forming IgE antibodies in small amounts but in atopics IgE response is preponderant. About 10 percent of persons have this tendency to overproduce IgE. It has been reported that bottle fed infants tend to develop atopy in later life more often than breastfed babies.

Mechanism of Atopy

The mechanism of development of atopy is essentially the same as that of systemic anaphylaxis. The symptoms of atopy are caused by the release of pharmacologically active substances following the combination of the antigen and the cell fixed IgE. Atopy is likely to develop, when allergen is localized or absorbed slowly, on the other hand systemic anaphylaxis is likely to develop if large quantities of allergen are quickly distributed throughout the body.

Atopic sensitivity is due to an overproduction of IgE antibodies. This is often associated with a deficiency of IgA. This association has led to the suggestion that IgA deficiency may predispose to atopy. The distribution of lymphocytes capable of synthesizing IgA and IgE is closely parallel, especially in the submucosa. In normal individuals, the inhalant and ingestant antigens are dealt with by IgA lining the respiratory and intestinal mucosa and therefore they do not come into contact with the potential IgE producing cells. When IgA is deficient, the antigens cause massive stimulation of IgE forming cells, leading to overproduction of IgE.

Clinical Expression of Atopic Reactions

The clinical expression of atopic reactions is usually determined by the portal of entry of the antigen-conjunctivitis, rhinitis, gastrointestinal symptoms and dermatitis following exposure through the eyes, respiratory tract, intestine or skin, respectively. Sometimes the effects may be at sites remote from the portal of entry, for example, urticaria following ingestion of the allergen. Specific desensitization (hyposensitization) is often practised in the treatment of atopy. Atopic allergies, which afflict at least 20 percent of the population in developed countries, include a wide range of IgE-mediated disorders, including **allergic rhinitis (hay fever), asthma, food allergies and atopic dermatitis (eczema).**

Methods to Detect Type I Hypersensitivity Reactions

Formerly, measurement of reaginic antibody could be done only by an *in vivo* assay or by skin testing.

1. **Prausnitz-Kustner (PK) reaction:** Prausnitz and Kustner in 1921, demonstrated transmission of IgE-mediated type I hypersensitivity by injecting serum containing IgE antibodies from allergic person into the skin of a normal or nonallergic person. Serum from Kustner, who was hypersensitive to certain species of cooked fish, was injected intracutaneously in Prausnitz (normal) followed 24 hours later by an intracutaneous injection of cooked fish, to which Kustner was sensitive, into the same site in Prausnitz. This led to wheal and flare reaction within 20 minutes. As IgE antibody is homocytotropic, the test has to be carried out on human skin, therefore, there is risk of transmission of hepatitis B virus and human immunodeficiency virus.
2. **Skin testing:** It is done by injecting small amounts of allergen into the skin of an allergic individual and looking for a wheal and flare reaction.
3. **Radioimmunosorbent test (RIST):** Another method of assessing type I hypersensitivity is to determine the serum level of total IgE antibody by the radioimmunosorbent test (RIST). This highly sensitive technique, based on the radioimmunoassay, can detect nanomolar levels of total IgE.
4. **Radioallergosorbent test (RAST):** The similar radioallergosorbent test (RAST) detects the serum level of IgE specific for a given allergen.

TYPE II HYPERSENSITIVITY: CYTOLYTIC AND CYTOTOXIC

These reactions involve a combination of **IgG (or IgM) antibodies** with an antigenic determinants on the surface of cells. Antibody can activate the **complement** system, creating pores in the membrane of a foreign cell, or it can mediate cell destruction by **antibody dependent cell-mediated cytotoxicity (ADCC)**. Type II hypersensitivity is generally, called **cytolytic or cytotoxic reactions** because it results in the destruction of host cells, either

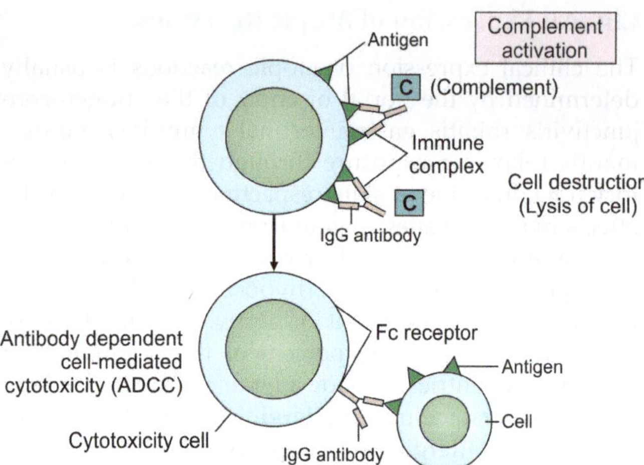

Fig. 20.2: Type II hypersensitivity

by lysis or toxic mediators. Type II hypersensitive reactions involve antibody-mediated destruction of cells (Fig. 20.2).

Examples

1. **Transfusion reactions:** A **transfusion reaction** can occur if a patient receives erythrocytes differing antigenically from his or her own during blood transfusion. In transfusion reactions red blood cells are destroyed as a result of reacting with circulating antibodies. These involve blood group systems that include the ABO and Rh antigens.
2. **Hemolytic disease of the newborn:** If the child is Rh+, Rh– mother can become sensitized to this antigen during birth causing the mother's body to produce anti-Rh antibodies of the IgG type. If the fetus in a subsequent pregnancy is Rh+, her anti-Rh antibodies will cross the placenta and destroy fetal RBCs. The fetal body responds to this immune attack by producing large numbers of immature RBCs call erythroblasts. Thus, the term erythroblastosis fetalis was used to describe what is now called hemolytic disease of the newborn (HDNB). The clinical features of HDN may vary from a mere accentuation of the physiological jaundice in the newborn to erythroblastosis fetalis or intrauterine death due to hydrops fetalis.
3. **Drug-induced cytotoxic reactions:** Some persons develop antibodies against their blood elements, resulting in autoimmune hemolytic anemia, agranulocytosis or thrombocytopenia. Blood platelets (thrombocytes) that are destroyed by drug-induced cytotoxic reactions in the disease called thrombocytopenic purpura (quinine is a familiar example). Drugs may bind similarly to white or red blood cells (RBCs), causing local hemorrhaging and yielding symptoms described as "blueberry muffin" skin mottling. Immune-caused destruction of granulocytic white cells is called agranulocytosis, and it affects the body's phagocytic defenses. When RBCs are destroyed in the same manner, the condition is termed hemolytic anemia.
4. **Anemia due to infectious diseases:** A variety of infectious diseases due to salmonella organisms and mycobacteria are associated with hemolytic anemia. due to an immune reaction against a lipopolysaccharide bacterial endotoxin that becomes coated on to the erythrocytes of the patient.

TYPE III HYPERSENSITIVITY—IMMUNE COMPLEX-MEDIATED

Type III reactions involve antibodies against soluble antigens circulating in the serum. The antigen-antibody complexes are deposited in organs and cause inflammatory damage. The tissue damage that results from the deposition of immune complexes is caused by the activation of complement, platelets and phagocytes; in essence, an acute inflammatory response (Fig. 20.3).

Models of immune complex-mediated disease: Two basic models of immune complex-mediated disease have been well-characterized: the *Arthus reaction* and *serum sickness*. These differ somewhat in both their mode of development and their clinical manifestations.

A. Arthus Reaction (Local Immune Complex Disease)

A localized form of experimental immune complex-mediated vasculitis is called **Arthus reaction**. Arthus (1903), observed that when rabbits were repeatedly injected subcutaneously with normal horse serum, the initial injections were without any local effect, but with later injections, there occurred intense local reaction consisting of edema, induration and hemorrhagic necrosis. This is known as **Arthus reaction** and is a local manifestation of generalized hypersensitivity. The tissue damage is due to formation of local precipitating immune complexes which are deposited on the endothelial lining of the blood vessels. Antigen-antibody complexes can then trigger and activate complement leading to release of inflammatory molecules. This leads to increased vascular permeability and infiltration of the site with neutrophils. Leukocyte-platelet thrombi are formed that reduce the blood supply and lead to tissue necrosis. The Arthus reaction can be passively transferred with sera containing precipitating antibodies (IgG, IgM) in high titers.

Clinical syndrome: Arthus reaction forms a pathogenic component of many clinical syndromes. Examples:

i. **Farmer's lung:** Intrapulmonary Arthus-like reaction to inhaled antigens, such as thermophilic actinomycetes from moldy hay or grain causes Farmer's lung.
ii. **"Pigeon fancier's disease"** results from inhalation of a serum protein in dust derived from dried pigeon feces and other types of hypersensitivity pneumonitis.

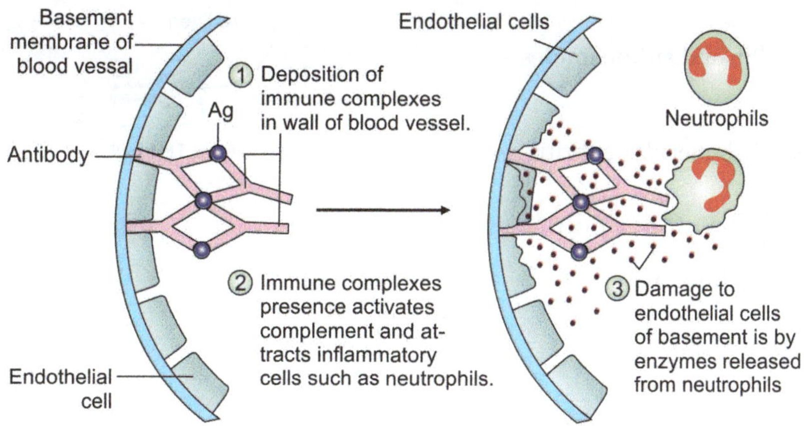

Fig. 20.3: Immune complex-mediated hypersensitivity. (1) Immune complexes on the basement membrane of the wall of a blood vessel, where they; (2) activate complement and attract inflammatory cells such as neutrophils to the site. (3) The neutrophilis discharge enzymes as they react with the immune complexes, resulting in damage to tissue cells

Passive Cutaneous Form of The Arthus Reaction

In passive cutaneous form of the Arthus reaction an antiserum is first injected intravenously into a nonsensitive recipient and the corresponding antigen is then injected into the skin.

Reversed Passive Arthus Reaction

In this the antiserum is injected into the recipient's skin and the antigen is then injected into the same dermal site or intravenously. Both injections are made at about the same time, so that sufficient amount of antibody will remain near the injection site to precipitate with the antigen and cause local inflammation and necrosis.

B. Serum Sickness (Systemic Immune Complex Disease)

This is a systemic form of type III hypersensitivity. When large amounts of antigen enter the bloodstream and bind to antibody, circulating immune complexes can form. If antigen is in excess, small complexes form because these are not easily cleared by the phagocytic cells, they can cause tissue-damaging type III reactions at various sites.

Historically, generalized type III reactions were often observed after the administration of antitoxins containing foreign serum, such as horse antitetanus or antidiphtheria serum. The occurrence of disease caused by immune complexes was suspected by von Pirquet (1911). This appeared 7 to 12 days following a single injection of a high concentration of foreign serum such as the diphtheria antitoxin. The clinical syndrome consists of fever, weakness, lymphadenopathy, splenomegaly, arthritis, glomerulonephritis, endocarditis, vasculitis, urticarial rashes, abdominal pain, nausea and vomiting.

Pathogenesis: The pathogenesis is the formation of immune complexes (consisting of the foreign serum and antibody to it that reaches high enough titers by 7 to 12 days) and the circulating immune complexes deposit in the blood vessel walls and tissues, leading to increased vascular permeability and thus, to inflammatory diseases such as glomerulonephritis and arthritis.

Antigen-antibody aggregates can fix **complement** leading to **inflammation** and **tissue damage**. The plasma concentration of complement falls due to massive complement activation and fixation by antigen-antibody complexes. The disease is self-limited. Initially, the circulating immune complexes are in antigen excess and produce inflammatory lesions, but as antibody production rises, the immune complexes increase in size as zone of equivalence is reached. These larger immune complexes are more susceptible to phagocytosis and immune elimination and cleared by the cells of reticulo endothelial system of liver and spleen. When all foreign antigen is thus eliminated and free antibody appears, the symptoms clear without any sequlae.

The latent period of 7 to 12 days is required only for serum sickness following a single injection. With subsequent injections, the disease manifests earlier. Serum sickness differs from other types of hypersensitivity reaction in that a single injection can serve as both as sensitizing dose and and the shocking dose. A heterologous serum injections are not used often now and serum sickness is now commonly studied in rabbits by giving them an intravenous injection of a soluble protein such as bovine serum albumin (BSA). The syndrome is currently more commonly seen following injections of penicillin or other antibiotics.

Diseases associated with immune complexes: Complexes of antibody with various bacterial, viral, and parasitic antigens have been shown to induce a variety of type III hypersensitive reactions. Formation of immune complexes contributes to the pathogenesis of a number of conditions other than serum sickness. Raised levels of immune complexes have been demonstrated in many infective and other conditions. These include the following:

1. Autoimmune diseases
   - Systemic lupus erythematosus
   - Rheumatoid arthritis
   - Goodpasture's syndrome.

2. Drug reactions
   – Allergies to penicillin and sulfonamides
3. Infectious diseases
   – Bacterial infections
     - Poststreptococcal glomerulonephritis
     - Endocarditis
     - Lepromatous leprosy
     - Secondary syphilis
     - Infected shunts in children
   – Viral infections
     - Dengue hemorrhagic fever
     - Hepatitis B
     - Cytomegalovirus
     - Infectious mononucleosis
     - Subacute sclerosing panencephalitis
   – Parasitic infections
     - Malaria
     - Toxoplasmosis
     - Trypanosomiasis
     - Schistosomiasis
     - Filariasis

TYPE IV HYPERSENSITIVITY—DELAYED HYPERSENSITIVITY

Type IV hypersensitivity reactions (delayed hypersensitivity) constitute one aspect of cell-mediated immune response and are caused mainly by T cells. These are typically provoked by intracellular microbial infections or haptens like simple chemicals applied on the skin, evolve slowly and consist of a mixed cellular reaction involving lymphocytes and macrophages in particular. It is named delayed hypersensitivity because it appears in 24 to 48 hours after the presensitized host encounters the antigen, while immediate hypersensitivity reactions develop in 1/2 to 12 hours. A major factor in the delay is the time required for the participating T cells and macrophages to migrate to and accumulate near the foreign antigens. The T cells involved in delayed type hypersensitivity reactions are primarily TD cells. In some types of hypersensitivities resulting in tissue damage, Tc cells may also participate.

Causes of Type IV Reactions

Type IV reactions occur when antigens, especially those binding to tissue cells, are phagocytosed by macrophages and then presented to receptors on the Th1 cell surface in the context of class I MHC. Contact between the antigen and Th1 cell causes the cell to proliferate and release cytokines. Cytokines attract lymphocytes, macrophages, and basophils to the affected tissue. Extensive tissue damage may result (Fig. 20.4).

Delayed hypersensitivity cannot be passively transferred by serum but can be transferred by lymphocytes or the transfer factor.

Types of delayed hypersensitivity: Two types of delayed hypersensitivity are recognized the tuberculin (infection) type and the contact dermatitis type.

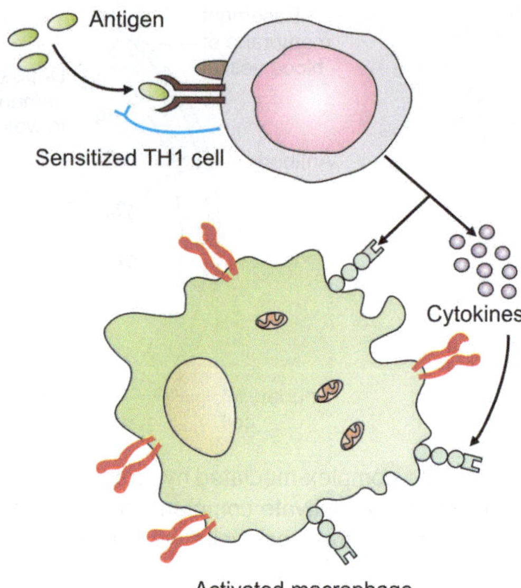

Fig. 20.4: Type IV (delayed or cell-mediated) hypersensitivity

1. Tuberculin (Infection) Type

This form of hypersensitivity was originally described by Koch. In tuberculin hypersensitivity tuberculin or purified protein derivative (PPD) is injected into the skin of the forearm. intradermally in an individual sensitized to tuberculoprotein by prior infection or immunization. An indurated (firm and hard) inflammatory reaction, 10 mm or more in diameter, develops at the site of injection within 48 to 72 hours. It is characterized by erythema due to increased blood flow to the damaged area and the infiltration with a large number of mononuclear cells, mainly T lymphocytes and about 10 to 20 percent macrophages into the injection site are responsible for the induration. The size of the induration is directly related to the amount of antigen that was introduced and to the degree of hypersensitivity of the tested individual. In unsensitized individuals, the tuberculin injection provokes no response.

The tuberculin test therefore provides useful indication of the state of delayed hypersensitivity (cell-mediated immunity) to the bacilli. The tuberculin test differs from the skin test for type I hypersensitivity not only in the longer interval for appearance but also in its morphology and histology.

Tuberculin type hypersensitivity develops in many infections with bacteria, fungi, viruses and parasites, especially when the infection is subacute or chronic and the pathogen intracellular. A similar hypersensitivity is developed in allograft reaction and in many autoimmune diseases.

2. Contact Dermatitis Type

Allergic contact dermatitis is caused by haptens that combine with proteins in the skin to form the allergen that elicits the immune response. The haptens are the antigenic determinants, and the skin proteins are the carrier molecules for the haptens. The substances involved are in themselves not antigenic but may acquire antigenicity on combination with skin proteins. Subsequent contact with allergen in a sensitized individual leads to contact dermatitis. As most of the substances involved are fat soluble, passage along sebaceous glands may be the method of entry of the allergens.

Examples: Examples of these haptens include cosmetics, plant materials (catechol molecules from poison ivy and poison oak), topical chemotherapeutic agents, metals (nickel and chromium), and chemicals like dyes, picryl chloride and dinitrochlorobenzene, drugs such as penicillin and toiletries and jewelry (especially jewelry containing nickel).

Mechanism of action: Most of these substances are small molecules that can complex with skin proteins. This complex is internalized by antigen—presenting cells in the skin (e.g. Langerhans' cells), then processed and presented together with class II MHC molecules, causing activation of sensitized Th1 cells. The sensitized T cells travel to the skin site, where on contacting the antigen they release various lymphokines. Approximately 48 to 72 hours after the second exposure, the secreted cytokines cause macrophages to accumulate at the site. Th1 cells secrete IFNγ and IL-2 which activate macrophages and other lymphocytes. Th-2 cells release IL-4, IL-5, GM-CSF and other factors that lead to an influx of eosinophils and tissue damage. Activated Tc cells mediate killing of target cells. Tissue damage results from lytic enzymes released from activated macrophages. Contact with the allergen in a sensitized individual leads to 'contact dermatitis'. The lesions varying from macules and papules to vesicles that break down, leaving behind raw weeping areas typical of acute eczematous dermatitis.

Detection by 'patch test': Hypersensitivity is detected by the **'patch test'**. The allergen is applied to the skin under an adherent dressing. Sensitivity is indicated by itching appearing in 4 and 5 hours, and local reaction which may vary from erythema to vesicle or blister formation, after 24 to 28 hours.

3. Granulomatous Hypersensitivity

Granulomatous hypersensitivity reactions develop over a period of 21 to 28 days; the granulomas are formed by the aggregation and proliferation of macrophages, and may persist for weeks. In terms of its clinical consequences, this is by far the most serious type of type IV hypersensitivity response.

Several important chronic diseases involve cell and tissue destruction by type IV hypersensitivity reactions. These diseases are caused by viruses, mycobacteria, protozoa, and fungi that produce chronic infections in which the macrophages and T cells are continually stimulated. Examples are leprosy, tuberculosis, leishmaniasis, candidiasis, and herpes simplex lesions.

Type V: Hypersensitivity (Stimulatory Type) Jones-Mote Reaction (or) Cutaneous Basophil Hypersensitivity

This is an antibody-mediated hypersensitivity and is a modification of type II hypersensitivity reaction. Antibodies interact with antigens on cell surface which leads to cell proliferation and differentiation instead of inhibition or killing. Antigen-antibody reaction enhances the activity of affected cell.

Example of Grave's disease: Thyroid hormones are produced in excess quantity in grave's disease. **Long acting thyroid stimulating (LATS)** antibody is an autoantibody to thyroid membrane antigen. It is presumed that LATS combines with a TSH receptor on thyroid cell surface and brings about the the same effect as TSH resulting in excessive secretion of thyroid hormone.

SHWARTZMAN REACTION

This is not an immune reaction but rather a perturbation in factors affecting intravascular coagulation. It is, however, traditionally described along with hypersensitivity reactions because of superficial resemblance.

Shwartzman (1928) observed that when a culture filtrate of *S. typhi* (endotoxin) is injected intradermally in a rabbit, followed by the same filtrate (endotoxin) intravenously 24 hours later, a hemorrhagic necrotic lesion develops at the site of the intradermal injection. The intradermal and intravenous injections need not be of the same or even related endotoxins. Culture suspensions or filtrates of a variety of bacteria will sensitize the skin to intravenous injection by an equally wide variety of cultures or filtrates. This absence of specificity and the short interval between the two doses preclude any immunological basis for the reaction.

Mechanism

The first dose is called **preparatory dose.** The second dose is called **provocative dose.** This intravenous (provocative) injection can be a variety of substances—bacterial endotoxins, antigen-antibody complexes, starch, kaolin and others. The preparatory injection causes accumulation of leukocytes which condition the site by release of lysosomal enzymes. These enzymes damage capillary walls. Following provocative dose, there occurs intravascular clotting, the thrombi leading to necrosis of vessel walls and hemorrhage. This is called **local form of Shwartzman reaction.**

When both injections are given intravenously the animal dies 12 to 24 hours after the second dose. The animal develops disseminated intravascular coagulation (DIC). Autopsy reveals bilateral cortical necrosis of kidneys and patchy necrosis of the liver, spleen and other organs. Sanarelli (1924) described an essentially similar phenomenon in experimental cholera. The reaction is therefore called the **Sanarelli-Shwartzman reaction or the generalized Shwartzman reaction.**

Clinical Conditions

1. **Waterhouse Friderichsen syndrome:** Purpuric rashes of meningococcal septicemia and the acute hemorrhagic adrenal necrosis found in overwhelming infections (Waterhouse-Friderichsen syndrome) mechanisms similar to the Shwartzman reaction may operate.
2. **Septic shock syndrome:** Many infections, particularly Gram-negative septicemias can lead to a septic shock syndrome with profound hypotension, hypoxia and oliguria. gram-negative bacteria are more likely to cause fatal septicemias than other infectious agents. This may sometimes be accompanied by the **adult respiratory distress syndrome (ARDS)** with overwhelming neutrophil invasion of lungs. Although multiple organs are affected by endotoxemia, the lung is particularly vulnerable to serious, irreversible damage because of its high concentrations of tissue macrophages.

KNOW MORE

The immune response can be likened to fire. Fire is essential for warmth and cooking and in many ways it is beneficial, but exactly the same fire is destructive if it starts in the wrong place or becomes uncontrolled. Similarly, the immune response is essential for protection, but it can be destructive if it out of control.

Desensitization or Hyposensitization

One way of preventing immediate hypersensitivity reactions is to inject the person with extremely dilute solutions of the antigen, paradoxical in that even tiny amounts of antigen can severe generalized anaphylaxis. This form of immunotherapy called desensitization or hyposensitization. The concentration of antigen in the injected solution is very gradually increased over a period of months in a series of injections. The individual gradually becomes less and less sensitive to the antigen and may even lose the hypersensitivity entirely.

☞ KEY POINTS

- **Hypersensitivity** is an exaggerated immune response that results in tissue damage and is manifested in the individual on second or subsequent contact with an antigen. Hypersensitivity reactions are of 5 types: Types I, II, III, IV and V.
- **Type I hypersensitivity reaction** is mediated by IgE antibodies.
- Clinical manifestations of type I reactions include generalized or systemic anaphylaxis or localized anaphylactic (type I) reactions.
- **Type II hypersensitivity reactions, or cytotoxic reactions** are caused by antibodies that can destroy normal cells by complement lysis or by antibody-dependent cellular cytotoxicity (ADCC). Transfusion reactions and hemolytic disease of the newborn are type II reactions.
- **Type III hypersensitivity reactions** are mediated by small antigen-antibody complexes that activate complement and other inflammatory systems, attract neutrophils, and contribute to inflammation. Deposition of immune complexes near the site of antigen entry can induce an **Arthus reaction**, (localized reaction) and serum sickness.
- The small immune complexes are often deposited in small blood vessels in organs, where they cause inflammatory disease—for example, glomerulonephritis in the kidney or arthritis in the joints.
- **Type IV hypersensitivity reaction** involves the cell-mediated branch of the immune system. Antigen activation of sensitized Th1 cells induces release of various cytokines that cause macrophages to accumulate and become activated. The net effect of the activation of macrophages is to release lytic enzymes that cause localized tissue damage. Two types of delayed hypersensitivity are: tuberculin type and contact dermatitis.

IMPORTANT QUESTIONS

1. What is hypersensitivity? How do you classify various types of hypersensitivity reactions? Describe type I hypersensitivity reactions.
2. Write short notes on:
   - Anaphylaxis
   - Atopy
   - Type III hypersensitivity (or) immune complex diseases
   - Arthus reaction
   - Serum sickness
   - Type IV hypersensitivity (or) delayed hypersensitivity (DTH).

FURTHER READING

Blackwell. Evan PW. Anaphylaxis, BMJ 1998;316:1442.

Davies KA, Hird V, et al. A study of in vivo immune complex formation and clearing in man. J Immunol 1990; 44:4613-20.

Friedman PS. The immunology of allergic contact dermatitis: the DNCB story. Adv Dermatol 1991;5:175-96.

Holgate ST, Church MK. Allergy. London: Gower 1993.

Howarth PH. Allergy: Pathogenic mechanism. New Engj Med 1998;343:712.

Roitt IM. Essential Immunology. 8th Edn. Oxford: Blackwell; 1994.

Williams RC. Immune Complexes in Clinical and Experimental Medicine, Massachusetts: Harvard University Press; 1980.

CHAPTER 21

Autoimmunity

Learning Objectives

After reading and studying this chapter, you should be able to:

- Describe the mechanisms of autoimmunity.
- Classify autoimmune diseases.
- List autoimmine diseases.

INTRODUCTION

One of the classically accepted features of the immune system is the capacity to distinguish self from nonself. Normally, a person is tolerized to self-antigens during the development of the immune system as a fetus and later in life by other mechanisms (e.g. oral tolerization). Early in the last century, Paul Ehrlich realized that the immune system could go away and, instead of reacting against foreign antigens, could focus its attack on self-antigens. He termed this condition "horror autotoxicus". But he did not regard autoimmunization as an impossibility and even envisaged its pathogenic possibility.

Definition

Autoimmunity is a condition in which structural or functional damage is produced by the action of immunologically competent cells or antibodies against the normal components of the body. Autoimmunity is due to copious production of autoantibodies and autoreactive T cells. Autoimmunity literally means 'protection against self' but it actually implies 'injury to self' and therefore it has been criticized as a contradiction in terms. 'Autoallergy' has been suggested as an acceptable alternative but the term autoimmunity has the sanction of wide use.

We now understand that, while mechanisms of self-tolerance normally protect an individual from potentially self-reactive lymphocytes, there are failures. They result in an inappropriate response of the immune system against self-components termed autoimmunity. Autoimmunity represents the end result of the breakdown of one or more of the basic mechanisms regulating immune tolerance.

FEATURES OF DISEASES OF AUTOIMMUNE ORIGIN

Diseases of autoimmune origin usually exhibit the following features:

1. An elevated level of immunoglobulins.
2. Demonstrable autoantibodies.
3. Deposition of immunoglobulins or their derivatives at sites of election, such as renal glomeruli.
4. Accumulation of lymphocytes and plasma cells at the sites of lesion.
5. Benefit from corticosteroids or other immunosuppressive therapy.
6. The occurrence of more than one type of autoimmune lesion in an individual.
7. A genetic predisposition towards autoimmunity.
8. Incidence higher among females.
9. Chronicity: Usually nonreversible.

MECHANISMS OF AUTOIMMUNITY

A variety of mechanisms have been proposed for induction of autoimmunity:

1. Forbidden clones
2. Neoantigens or altered antigens
3. Molecular mimicry
4. Polyclonal B cell activation
5. Activity of helper and suppressor T cells
6. Sequestered antigens
7. Defects in the idiotype-anti-idiotype network
8. Genetic factor

1. Forbidden Clones

Antibody-forming lymphocytes capable of reacting with different antigens are formed according to clonal selection theory. During embryonic life, clones of cells that have immunological reactivity with self-antigens are eliminated. Such clones are called forbidden clones. Their persistence or development in later life by somatic mutation can lead to autoimmunity

2. Neoantigens or Altered Antigens

Cells or tissues may undergo antigenic alteration as a result of physical, chemical or biological influences. Such altered or 'neoantigens' may elicit an immune response. Cells or tissues may undergo antigenic alteration by physical agents such as irradiation. Photosensitivity and cold allergy may represent sensitization to self-antigens, altered by light and cold, respectively.

Several chemical agents including drugs can combine with cells and tissues and alter their antigenic structure. Skin contact with a variety of chemicals may lead to contact dermatitis. Drug-induced anemia, leukopenia and thrombocytopenia often have autoimmune basis. Viruses and other intracellular pathogens may induce alterations of cell antigens leading to autoimmunity. Alteration of cell antigens is also induced by bacterial enzymes. Viral infections, such as infectious mononucleosis, are known to often precede autoimmune diseases. Neuraminidases formed by myxoviruses and many bacteria act on erythrocytes releasing the T antigen. The almost universal occurrence of T agglutinins in human sera is believed to represent a harmless autoimmune response following infections. Neoantigens may also arise by mutation. Such mutant cells may be immunogenic.

3. Molecular Mimicry

A number of viruses and bacteria have been shown to possess antigenic determinants that are identical or similar to normal host cell components. The fortuitous similarity between some foreign and self-antigens is the basis of the 'cross reacting antigens' theory of autoimmunity. Immunological damage may result from immune responses induced by cross reacting antigens. Molecular mimicry by cross-reactive microbial antigens can stimulate autoreactive B and T cells.

Organ specific antigens are present in several species. Injections of heterologous organ specific antigens may induce an immune response damaging the particular organ or tissue in the host. One of the best examples of this type of autoimmune reaction is **postrabies encephalitis**. This is the neurological injury that used to be the complication of antirabic immunization in human beings with the neural vaccine of infected sheep brain tissue partially denatured by treatment with phenol. In a vaccinated person, these sheep brain cell antigens could induce formation of antibodies and activated T cells, which cross-react with the recipient's own brain cells, leading to encephalitis.

Infection

Immunological injury due to cross reacting antigens can also follow infection. Streptococcal M proteins and the heart muscle share antigenic characteristics. The immune response induced by repeated streptococcal infection can therefore damage the heart. Nephritogenic strains of streptococci possess antigens found in the renal glomeruli. Infection with such strains may lead to the glomerulonephritis due to antigenic sharing.

There may also be cross-reactivity between HLA-B27 and certain strains of klebsiella in connection with ankylosing spondilytis, and cross-rectivity between bacterial heat-shock proteins and DR4 in relationship to rheumatoid arthritis. There is evidence for mimicry between myelin basic protein (MBP) and viral peptides.

4. Polyclonal B cell Activation

While an antigen generally, activates only its corresponding B cell, certain stimuli nonspecically turn on multiple B clones. Such stimuli include chemicals (for example, 2-mercaptoethanol), bacterial products (PPD, lipopolysaccharides), enzymes (trypsin), antibiotics (nystatin) and infection with some bacteria (mycoplasma), viruses (EB virus, CM virus) and parasites (malaria). Multiple nonspecific antibodies form during some infectious diseases, such as autoantibodies reactive to T and B cells, rheumatoid factor, and antinuclear antibodies in infectious mononucleosis, antihuman erythrocyte cold antibodies in mycoplasma pneumonia. Many AIDS patients also show high levels of nonspecific antibody and autoantibodies to RBCs and platelets. These patients are often coinfected with other viruses such as EBV and cytomegalovirus, which may induce the polyclonal B cell activation that results in autoantibody production.

5. Activity of Helper and Suppressor T Cells

Enhanced helper T cell and decreased suppressor T cells functions have been suggested as causes of autoimmunity. Defects in the thymus, in stem cell development and macrophages function have also been postulated as causes.

6. Sequestered Antigens

Certain self-antigens are present in closed systems and are not accessible to the immune apparatus. These are known as sequestered antigens.

Examples

i. Lens Antigen of the Eye

The lens protein is enclosed in its capsule and does not circulate in the blood. Hence, immunological tolerance against this antigen is not established during fetal life. The release of lens protein after eye damage has been shown to lead on occasion to the formation of autoantibodies. When the antigen leaks out, following penetrating injury, it may induce an immune response causing damage to the lens of the other eye.

ii. Sperm Antigens

An example of 'sequestration in time' is seen with sperm antigens. Sperm arise late in development and sequestered from the circulation. As spermatozoa develop only with puberty, the antigen can not induce tolerance during fetal life. However, after a vasectomy, some sperm antigens are released into the circulation and can induce autoantibody formation in some men. This is also believed to be the pathogenesis of orchitis following mumps. The virus damages the basement membrane of seminiferous tubules leading to the leakage of sperms and initiation of an immune response resulting in orchitis.

iii. Heart Muscle Antigens

Similarly, release of heart muscle antigens after myocardial infarction has been shown to lead to the formation of autoantibodies.

7. Defects in the Idiotype-Anti-idiotype Network

It is possible that abnormalities in the generation of appropriate anti-idiotype antibodies either at the B or T cell level, are responsible for the development of autoimmunity in certain circumstances.

Additional factors that appear to be important determinants in the induction of autoimmunity include age, sex, genetic background, exposure to infectious agents, and environmental contacts.

8. Genetic Factors

Genetic factors such as defective *Ir* or immunoglobulin genes have also been postulated. In human autoimmune diseases and in animal models, genetic factors appear to influence the development and fate of autoimmune state. In spite of so many different possible mechanisms proposed, their actual role in autoimmunity, if any, has not been established.

CLASSIFICATION OF AUTOIMMUNE DISEASES

Based on the site of involvement and nature of lesions, autoimmune diseases may be classified as:

A. Localized (or organ-specific)
B. Systemic (or nonorgan-specific)

A. Localized (Organ-Specific) Autoimmune Diseases (Table 21.1)

The immune response is directed to a target antigen unique to a single organ or gland in an organ-specific

Table 21.1: Some autoimmune diseases in humans

	Organ-specific autoimmune diseases	
Disease	**Self-antigen**	**Immune response**
Hashimoto's thyroiditis	Thyroid proteins and cell	Autoantibodies
Graves' disease	Thyroid-stimulating hormone receptor	Autoantibodies
Goodpasture's syndrome	Renal and lung basement membranes	Autoantibodies
Autoimmune hemolytic anemia	RBC membrane proteins	Autoantibodies
Addison's disease	Adrenal cells	Autoantibodies
Idiopathic thrombocytopenia purpura	Platelet membrane proteins	Autoantibodies
Insulin-dependent diabetes mellitus Myasthenia gravis	Pancreatic beta cells	T_{DTH} autoantibodies
Myocardial infarction	Heart	Autoantibodies
Pernicious anemia	Gastric parietal cells; intrinsic factor	Autoantibodies
Poststreptococcal glomerulonephritis	Kidney	Antigen-antibody complexes
Myasthenia gravis	Acetylcholine receptors	Autoantibodies (blocking)
Spontaneous infertility	Sperm	Autoantibodies
	Systemic autoimmune diseases	
Disease	**Self-antigen**	**Immune response**
Systemic lupus erythematosus (SLE)	DNA, nuclear protein, RBC and platelet membranes	Autoantibodies, immune complexes
Multiple sclerosis	Brain or white matter	Th1 cells and Tc cells, autoantibodies
Rheumatoid arthritis	Connective tissue,IgG	Autoantibodies, immune complexes
Scleroderma	Nuclei, heart, lungs, gastrointestinal tract, kidney	Autoantibodies
Sjögren's syndrome	Salivary gland, liver, kidney, thyroid	Autoantibodies
Ankylosing spondylitis	Vertebrae	Immune complexes

autoimmune disease, so that the manifestations are largely limited to that organ.

Diseases

1. Autoimmune Diseases of the Thyroid Gland

i. Hashimoto's Thyroiditis (Lymphadenoid Goiter)

This is the most typical and best studied of organ-specific autoimmune diseases. In 1956, Roitt and Doniach in England, demonstrated antithyroglobulin antibodies in the sera of patients by precipitation in gel, and Witebsky and Rose in the USA by the more sensitive passive hemagglutination test. The latter workers also reproduced the disease in rabbits by immunization with autologous thyroid tissue obtained by hemithyroidectomy.

In Hashimoto's thyroiditis, which is most frequently seen in middle-aged women and is associated with an enlargement of the thyroid gland and symptoms of hypothyroidism or frank myxedema. Histologically, the glandular structure is replaced by lymphoid tissue consisting of lymphocytes, histiocytes and plasma cells. Antibodies are formed to a number of thyroid proteins, including thyroglobulin and thyroid peroxidase, both of which are involved in the uptake of iodine. Binding of the autoantibodies to these proteins interferes with iodine uptake and leads to decreased production of thyroid hormones (hypothyroidism).

ii. Thyrotoxicosis (Grave's Disease)

Thyroid changes in Graves' disease are autoimmune in origin and initiated by IgG antibodies against specific domains of the thyroid stimulating hormone (TSH). Unlike TSH, however, the autoantibodies are not regulated, and consequently they overstimulate the thyroid. These autoantibodies are called long-acting thyroid-stimulating (LATS) antibodies for this reason. The result is that the thyroid gland is stimulated to produce increased amounts of thyroid hormones and becomes greatly enlarged.

2. Addison's Disease

Two important causes of Addison's disease are autoimmune adrenalitis and tuberculous adrenalitis. The immunological basis of Addison's disease is suggested by lymphocytic infiltration of the adrenal glands and the presence of circulating antibodies directed against the cells of the zona glomerulosa. Similar lesions can be produced in experimental animals by immunization with adrenal tissue in Freund's adjuvant.

3. Autoimmune Orchitis

In guinea pigs, experimental allergic orchitis with progressive damage to germinal epithelium and aspermatogenesis can be induced by the injection of autogenous or allogeneic testes with Freund's adjuvant. Sometimes a similar condition follows mumps orchitis. Lymphocytic infiltration of the testes and circulating antibodies to the sperms and germinal cells can be demonstrated in this condition.

4. Myasthenia Gravis

Myasthenia gravis is the prototype autoimmune disease mediated by blocking antibodies. Antibodies form against acetylcholine receptors (AChRs) present on the postsynaptic membrane of the neuromuscular junction. These autoantibodies that bind the acetylcholine receptors on the motor end-plates of muscles, blocking the normal binding of acetylcholine and also inducing complement mediated lysis of the cells. The result is a progressive weakening of the skeletal muscles. Ultimately, the antibodies destroy the cells bearing the receptors. Circulating antibodies to AChRs have been demonstrated in nearly all patients with myasthenia gravis, and the disease can be passively transferred to experimental animals by injecting serum from MG patient.

5. Autoimmune Diseases of the Eye

Two types of autoimmune diseases are seen in the eye.

i. Phacoanaphylaxis

Cataract surgery sometimes leads to intraocular inflammation caused by the autoimmune response to the lens protein. This is known as phacoanaphylaxis.

ii. Sympathetic Ophthalmia

Perforating injuries of the eye, particularly those involving the iris or ciliary bodies are often followed by sympathetic ophthalmia in the opposite eye. The disease can be produced in experimental animals by immunization with uveal or retinal tissue in Freund's adjuvant and can be passively transferred with the spleen or lymph node cells but not with serum.

6. Pernicious Anemia

Pernicious anemia is caused by autoantibodies to intrinsic factor, a membrane-bound intestinal protein on gastric parietal cells. Intrinsic factor facilitates uptake of vitamin B_{12} from the small intestine. Binding of the autoantibody to intrinsic factor blocks the intrinsic factor-mediated absorption of vitamin B_{12}. In the absence of sufficient vitamin B_{12}, which is necessary for proper hematopoiesis, the number of functional mature red blood cells decreases below normal. Pernicious anemia is treated with injections of vitamin B_{12}, thus circumventing the defect in its absorption.

7. Autoimmune Diseases of the Nervous System

i. Neuroparalytic Accidents

The 'neuroparalytic accidents' following rabies vaccination represent injury to the nervous system by the immune response against the sheep nervous tissue in the vaccine, which crossreacts with human nerve tissue leading to encephalitis. An essentially similar condition suh as experimental allergic encephalomyelitis (EAE), can be produced in animals by immunization with nervous tissue in Freund's adjuvant. The myelin basic protein (MBP) has been identified as the encephalogenic protein which shows no species specificity.

ii. Idiopathic Polyneuritis (Guillain-Barré Syndrome)

It is considered an autoimmune response against the peripheral nervous tissue. It can be reproduced in experimental animals by immunization with peripheral nervous tissue in an adjuvant.

8. Goodpasture's Syndrome

In Goodpasture's syndrome, autoantibodies specific for certain basement membrane antigens bind to the basement membranes of the kidney glomeruli and the alveoli of the lungs leading to complement activation and direct cellular damage and an ensuing inflammatory response. Damage to the glomerular and alveolar basement membranes leads to progressive kidney damage and pulmonary hemorrhage.

9. Insulin-Dependent Diabetes Mellitus

It is caused by immunological destruction of insulin-secreting cells of the pancreas. The autoimmune attack destroys beta cells, resulting in decreased production of insulin and consequently increased levels of blood glucose. Several factors are important in the destruction of beta cells.

10. Male Infertility

Yet another example of autoimmune disease is seen in rare cases of male infertility where antibodies to spermatozoa lead to clumping of spermatozoa, either by their heads or by their tails, in the semen.

11. Autoimmune Diseases of the Skin

Three serious diseases of the skin are considered to have an autoimmune basis. Pemphigus vulgaris may be caused by an antibody to the intercellular cement substance. In bullous pemphigoid, antibodies directed against the dermal epithelial junction have been demonstrated. Specific antibodies in dermatitis herpetiformis have not been identified.

12. Hemocytolytic Autoimmune Diseases

i. Autoimmune hemolytic anemia: An individual with autoimmune hemolytic anemia makes autoantibody to RBC antigens, triggering complement-mediated lysis or antibody-mediated opsonization and phagocytosis of the red blood cells. Serologically, two groups of autoimmune anemias can be distinguished, characterized by 'cold' and 'warm' antibodies, respectively.

a. Cold autoantibodies: The cold autoantibodies are, generally, complete agglutinating antibodies belonging to the IgM class and agglutinate erythrocytes at 4°C but not at 37°C. They are referred to as cold agglutinins because they agglutinate red cells at low temperatures. Cold agglutinins are seen in primary atypical pneumonia. trypanosomiasis and blackwater fever. The clinical symptoms result from an *in vivo* agglutination of red cells and fixation of complement in the parts such as tip of the nose, ears, fingers and toes where the temperature may drop to below 30ºC. Thus, there may be intravascular hemolysis.

b. Warm autoantibodies: Warm autoantibodies are generally incomplete, nonagglutinating antibodies usually belonging to the IgG class. They can be shown coating the erythrocytes in the direct Coombs test. Warm antierythrocyte antibodies are frequently seen in patients taking certain drugs such as sulphonamides, antibiotics, and alpha methyl dopa.

ii. Autoimmune Thrombocytopenia

Platelets are destroyed by the formation of antiplatelet antibodies idiopathic thrombocytopenic purpura result from the synthesis of autoantibodies to platelets. Sedormid purpura is an instance of immune response against drug-induced neoantigens on platelets. This condition is traditionally considered antibody-mediated hypersensitivity.

iii. Drug-induced Hemolytic Anemia

One form of autoimmune anemia is drug-induced: when certain drugs such as penicillin or the antihypertensive agent methyldopa interact with red blood cells, the cells become antigenic.

iv. Autoimmune Leukopenia

Nonagglutinating antileukocyte antibodies can be demonstrated in the serum of patients with systemic lupus erythematosus and rheumatoid arthritis.

B. Systemic (Nonorgan-Specific) Autoimmune Diseases (Table 21.1)

In systemic autoimmune diseases, the response is directed toward a broad range of target antigens and involves a number of organs and tissues. These diseases reflect a general defect in immune regulation that results in hyperactive T cells and B cells. Tissue damage is widespread, both from cell-mediated immune responses and from direct cellular damage caused by autoantibodies or by accumulation of immune complexes.

Classification: Klemperer (1942, classified a number of diseases of unknown origin with the common feature of connective tissue lesions as collagen diseases. Included in this category are systemic lupus erythematosus (SLE), rheumatoid arthritis, polyarteritis nodosa. Sjöogren's syndrome dermatomyositis and scleroderma. All these conditions are associated with generalized autoimmune processes.

1. Systemic Lupus Erythematosus (SLE)

Systemic lupus erythematosus is one of the best examples of a systemic autoimmune disease, which typically appears in women between 20 and 40 years of age. This is a chronic, multisystem disease with remissions and exacerbations terminating fatally. Affected individuals may produce autoantibodies to a vast array of tissue antigens, such as DNA, histones, RBCs, platelets, leukocytes, and clotting factors. Interactions of these autoantibodies with their specific antigens produces various symptoms. Biological false positive reaction is seen in standard tests for syphilis. The abundance and variety of autoantibodies suggest a breakdown in the central control of immunological homeostasis. SLE results from tissue damage caused by pathogenic subsets of autoantibodies and immune complexes.

LE Cell Phenomenon

The first immunological feature identified in SLE was the *LE cell phenomenon* in 1948. LE cell is a neutrophil containing a large, pale, homogeneous body *(LE body)* almost filling the cytoplasm. In tissues, nuclei of damaged cells react with ANAs, lose their chromatin pattern, and become homogenous, to produce so called LE bodies. Sometimes instead of being intracellular, the LE body can be seen free surrounded by a rosette of neutrophils.

The LE body is the immunologically damaged nucleus of a leukocyte. Related to this phenomenon is the LE cell, which is readily seen *in vitro*. Basically the LE cell is any phagocytic leukocyte (neutrophil or macrophage) that has engulfed the denatured nucleus of an injured cell. The fact that LE cell formation is due to an antibody *(LE factor)* present in SLE can be demonstrated by incubating normal blood with serum from an SLE patient. The LE cell test is positive in up to 70 percent of the patients with SLE.

Laboratory Diagnosis

Antinuclear antibodies (ANA) are directed against several nuclear antigens. ANAs are the best screening test. Laboratory diagnosis of SLE focuses on the characteristic antinuclear antibodies, which are directed against double-stranded or single-stranded DNA, nucleoprotein, histones and nucleolar RNA. Antinuclear antibodies (ANA) are the best screening test.

Indirect immunofluorescence staining with serum from SLE patients produces various characteristic nucleus-staining patterns. Four basic patterns are recognized. ANA tests are sensitive but not specific for SLE, as they may be positive in many other autoimmune conditions, viral infections, chronic inflammatory processes as well as in persons using certain medicines and in the aged.

Antinuclear antibodies can also be detected by agglutination of latex particles coated with deoxyribonucleoprotein. It is more specific and sensitive than LE cell test.

2. Rheumatoid Arthritis

Rheumatoid arthritis is a common autoimmune disorder, most often affecting women from 40 to 60 years old. This is a symmetric polyarthritis with muscle wasting and subcutaneous nodules, commonly associated with serositis, myocarditis, vasculitis and other disseminated lesions. The major symptom is chronic inflammation of the joints, although the hematologic, cardiovascular, and respiratory systems are also frequently affected. The synovial membranes of the affected joints are swollen and edematous, with dense infiltration of lymphocytes and plasma cells.

Many individuals with rheumatoid arthritis produce a circulating autoantibody called the 'Rheumatoid factor' (RF). Rheumatoid factor is an IgM antibody but they may be of other classes (lgG and IgA). Rheumatoid factor acts as an antibody against the Fc fragment of immunoglobulins. These factors are found in 70 percent of individuals suffering from rheumatoid arthritis. RF may be absent in some (seronegative) and may be found in some non-RA patients such as systemic lupus erythematosus, dermatomyositis and even in otherwise healthy people.

Such autoantibodies bind to normal circulating IgG, forming IgM-IgG complexes that are deposited in the joints. These immune complexes can activate the complement cascade, resulting in a type III hypersensitive reaction, which leads to chronic inflammation of the joints. The chronic inflammation caused by this deposition eventually leads to severe damage to the cartilage and bone of the joint.

Detection of RF—Agglutination Tests

RF is detected by agglutination tests using, as antigens, particles coated with globulins. In the Rose-Waaler test, the original technique for detection of RF, sheep erythrocytes coated with a subagglutinating dose of antierythrocyte antibody (amboceptor) are used as the antigen in an agglutination test. Latex and bentonite are used as the carrier particles for IgG in modifications of the test. Antinuclear antibodies are frequently found in rheumatoid arthritis.

3. Polyarteritis Nodosa

Polyarteritis nodosa (PAN) is a noninfective, acute necrotizing vasculitis involving the small and medium arteries leading to weakness of vessel wall, aneurysm formation and thrombus formation in some cases, ending fatally due to coronary thrombosis, cerebral hemorrhage or gastrointestinal bleeding. Though it has been suggested that PAN may be an autoimmune disease, the antibody responsible has not been identified. Hepatitis B is a risk factor. Immune complexes of hepatitis B virus antigen (HbsAg) in affected tissues, including the kidneys, have been demonstrated in 30 to 40 percent of patients.

4. Sjögren's Syndrome

Sjögren's syndrome is a clinicopathologic entity characterized by dry eyes (keratoconjunctivitis sicca) and dry mouth (xerostomia) resulting from immunologically-mediated destruction of the lacrimal and salivary glands. This syndrome may occur in association with other autoimmune diseases, e.g. rheumatoid arthritis and SLE. Antinuclear antibodies and rheumatoid factor commonly occur in the patient serum. The disease is associated with a 40-fold increased lifetime risk of lymphoma and can be viewed as being at the cross-roads of autoimmunity and malignancy.

Pathogenesis of Autoimmune Disease

Many diseases are considered to be of autoimmune origin, based on their association with cellular or humoral immune responses against self-antigens.

Humoral and cellular immune processes: Autoantibodies are more easily detected than cellular autosensitization. The relative importance of humoral and cellular immune processes in the etiology of autoimmune diseases is not known. Antibodies may cause damage by the cytolytic or cytotoxic (type 2) and toxic complex (type 3) reactions. They are obviously important in hemocytolytic autoimmune diseases.

A third mechanism of autoimmune tissue damage is by sensitized T lymphocytes (type 4 reaction). It is likely that humoral and cellular immune responses may act synergistically in the production of some autoimmune diseases. For example, experimental orchitis can be induced only when both types of immune responses are operative.

Once initiated, most autoimmune responses tend to be self-perpetuating. Autoimmune disease are usually treated with immunosuppressants, or drugs that interfere with T cell signaling, steroids and other anti-inflammatory drugs. Replacement therapy is necessary in some autoimmune diseases.

KNOW MORE

Pathogenesis of Autoimmune Disease

When autoantibodies are found in association with a particular disease there are three possible inferences:

1. The autoimmunity is responsible for producing the lesions of the disease.
2. There is a disease process which, through the production of tissue damage, leads to the development of autoantibodies.
3. There is a factor which produces both the lesions and the autoimmunity.

☞ KEY POINTS

- Autoimmunity is a condition in which structural or functional damage is produced by the action of immunologically competent cells or antibodies against the normal components of the body.
- A variety of mechanisms have been proposed for induction of autoimmunity, including release of sequestered antigens, molecular mimicry, inappropriate class II MHC expression on cells, and polyclonal B cell activation.
- Autoimmune diseases can be divided into organ-specific or widespread and systemic diseases.
- Autoimmune diseases are usually treated with drugs that suppress the immune and/or inflammatory responses.

IMPORTANT QUESTIONS

1. What is autoimmunity? What are the mechanisms of autoimmune diseases giving suitable examples?
2. Write short notes on:
   i. Classification of autoimmune diseases
   ii. Sequestrated antigens or hidden antigens.

FURTHER READING

Goldsby RA, et al. Immunology, 5th Edn. New York : WH Freeman and Company 2003.

King C, Sarvetnick N. Organ specific autoimmunity. Curr Opin Immunol 1997;9:863.

Levin Mc, et al. Autoimmunity due to molecular mimicry as a cause of neurological disease. Nat Med 2002;8:509.

Roitt I, et al. Immunology, 6 Edn. London: Mosby 2001.

Wucherpfennig KW. Mechanisms for the induction of autoimmunity by infectious agents. J Clin Invest 2001;108:1097.

CHAPTER 22 Immunology of Transplantation and Malignancy

Learning Objectives

After reading and studying this chapter, you should be able to:

- Discuss the following:
  - Types of transplants
  - Histocompatibility antigens
  - Graft-versus-host (GVH) reaction
  - Tumor antigens
  - Immunological surveillance.

DEFINITION

Transplantation refers to the act of transferring cells, tissues, or organs from one site to another. The tissue or organ transplanted is known as the **transplant or graft.** The individual from whom the transplant is obtained is known as the **donor** and the individual to whom it is applied, the **recipient or host.**

History

Transplantation is one of mankind's ancient dreams. Chimeras, fanciful creatures composed of parts from different species, figure in the mythology and pantheon of all ancient nations. The earliest application of transplantation appears to have been skin grafting for reconstruction of the severed nose, using the patient own skin flaps—a technique described in the *Sushruta Samhita* (circa 800 BC). **Alexis Carrel** reported the first systematic study of transplantation in 1908. The first human kidney transplant, attempted in 1935 by a Russian surgeon and failed. The first successful human kidney transplant, which was between identical twins, was accomplished in 1954. Only in 1940s the work of Medawar and his colleagues conclusively proved its immunological basis.

TYPES OF TRANSPLANTS (TABLE 22.1)

i. **Autograft:** Autograft is self-tissue transferred from one body site to another in the same individual.
   Examples
   - Transferring healthy skin to a burned area in burn patients
   - Use of healthy blood vessels to replace blocked coronary arteries.

ii. **Isograft:** Isograft is tissue transferred between genetically identical individuals. In humans, an isograft performed between genetically identical (monozygomatic) twins and in inbred strains of mice, an isograft performed from one mouse to another syngeneic mouse are examples of isografts.

iii. **Allograft:** Allograft is tissue transferred between genetically different members of the same species. In mice, an allograft is performed by transferring tissue or organ from one strain to another. In humans, organs grafts from one individual to another are allografts unless the donor and recipient are identical twins.

Table 22.1: Types of grafts

Donor	Name	Synonyms
Self	Autograft Isograft	Autogenous or autogenic graft.
Different individual, genetically identical with recipient. Identical twin or member of same inbred strain	Isograft	Isologous or syngeneic graft or syngraft
Genetically unrelated member of same species	Allograft	Allogeneic graft. Formerly called homograft
Different species	Xenograft	Xenogeneic. Formerly called heterograft

iv. **Xenograft:** Xenograft is tissue transferred between different species (e.g. the graft of a baboon heart into a human).

ALLOGRAFT REACTION

1. **First set response:** When a skin graft from an animal (such as a rabbit) is applied on a genetically unrelated animal of the same species, the graft appears to be accepted initially. The graft is vascularized and seems morphologically and functionally healthy during the first two or three days. By about the fourth day, inflammation becomes evident and the graft is invaded by lymphocytes and macrophages. The blood vessels within the graft are occluded by thrombi, the vascularity diminishes and the graft undergoes ischemic necrosis. The graft assumes a scab-like appearance with extending necrosis and sloughs off by the tenth day. This sequence of events resulting in the rejection of the allograft is known as the **first set response** (also known as the **'first set rejection or reaction'**).
2. **Second set response:** If in an animal which has rejected a graft by the first set response, another graft from the same donor is applied, it will be rejected in an accelerated fashion. Vascularization commences but is soon interrupted by the inflammatory response. Thrombosis of vessels is a feature. Necrosis sets in early and the graft sloughs off by the sixth day. The accelerated allograft rejection is known as the *second set response.*

Mechanism of Allograft Rejection

Graft rejection is an immunologic response displaying the attributes of specificity, memory, and self/nonself recognition. The immunological basis of graft rejection is evident from the specificity of the second-set rejection. Accelerated rejection is seen only, if the second graft is from the same donor as the first. Application of a skin graft from another donor will evoke only the first set rejection. Graft rejection is caused principally by a cell-mediated immune response to alloantigens (primarily, MHC molecules) expressed on cells of the graft.

Stages of Graft Rejection

The process of graft rejection can be divided into two stages:

1. **A sensitization phase**, in which antigen-reactive lymphocytes of the recipient proliferate in response to alloantigens on the graft, and
2. An **effector stage**, in which immune destruction of the graft takes place.

Classification of Allograft Rejection

Allograft rejection is classified as hyperacute, acute or chronic.

i. **Hyperacute rejection:** This occurs very rapidly (within minutes-hours) in patients who already have antibodies against a graft. Anti-HLA antibodies are induced by prior blood transfusion, multiple pregnancies or the rejection of a previous transplant. In addition, antibodies against the ABO blood group system can cause hyperacute rejection. The antigen-antibody complexes that form activate the complement system, resulting in an intense infiltration of neutrophils into the grafted tissue. The ensuing inflammatory reaction causes massive blood clots within the capillaries, preventing vascularization of the graft.

ii. **Acute rejection**

First-set rejection: This takes days or weeks to become manifest and is due to the primary activation of T cells and the consequent triggering of various effector mechanisms. Acute rejection is the typical **first-set rejection (a primary response)** and the recipient is not immune to the antigens of the donor. If a transplant is given to someone who has been presensitized to antigens on the graft, a secondary reactivation of T cells occurs, leading to an accelerated cell-mediated rejection response.

Second-set rejection: Accelerated or **'second-set' rejection** of skin grafts is particularly dramatic—so called **'white graft rejection'** in which the graft is rejected before it has time to heal. Histopathologic examination reveals a massive infiltration of macrophages and lymphocytes at the site of tissue destruction suggestive of Th cell activation and proliferation. The effector mechanisms involved in allograft rejection. The generation or activity of various effector cells depends directly on cytokines secreted by activated Th cells.

iii. **Chronic rejection:** Chronic rejection reactions develop months or years after acute rejection reactions have subsided. The mechanisms of chronic rejection include both humoral and cell-mediated responses by the recipient. The walls of the blood vessels in the graft thicken and eventually become blocked. This is called chronic rejection and may be due to several different causes, such as a low grade cell-mediated rejection or the deposition of antibodies or antigen-antibody complexes in the grafted tissue, which damage or activate the endothelial cells lining the vessel and trigger inappropriate repair responses.

Histocompatibility Antigens

Tissues that are antigenically similar are said to be **histocompatible;** such tissues do not induce an immunologic response that leads to tissue rejection. Tissues that display significant antigenic differences are **histoincompatible** and induce an immune response that leads to tissue rejection. The antigens primarily responsible for rejection of genetically different tissues are known as **histocompatibility (i.e. tissue compatibility) antigens** and the genes coding for these antigens are referred to as **histocompatibility genes**.

The various antigens that determine histocompatibility are encoded by more than 40 different loci, but the loci responsible for the most vigorous allograft-rejection reactions are located within the **major histocompatibility complex (MHC)**. In mice the organization of the MHC is called **H-2 complex**, while in humans it is known as **human leukocyte antigen (HLA) complex.** The products of allelic variants of other histocompatibility genes individually cause weaker rejection responses and are consequently known as **minor histocompatibility antigens**. These antigens are normal cellular constituents. None the less, combinations of several minor antigens can elicit strong rejection responses.

Immune response against transplants depends on the presence in the grafted tissue of antigens that are absent in the recipient and hence recognized as foreign. It follows, therefore, that if the recipient possesses all the antigens present in the graft, there will be no immune response, and consequently no graft rejection, even when the donor and the recipient are not syngeneic. Inbred mouse strains are **syngeneic** or identical at all genetic loci.

Eichwald-Silmser Effect

While transplants between members of a highly inbred strain of animals are successful, an exception is seen when the donor is a male and the recipient is a female. The grafted male tissue (XY) will have antigens determined by the Y chromosomes which will be absent in the female (XX) and such grafts are rejected. Grafts from the female to the male will succeed. This unilateral sex linked histocompatibility is known as *Eichwald-Silmser effect.*

HISTOCOMPATIBILITY TESTING

For matching of donor and recipient for transplantation following procedures are undertaken:

1. **ABO grouping:** ABO incompatibility is a major barrier to transplantation, and matching of donor and recipient for these antigens is of utmost importance because blood group antigens are strong histocompatibility antigens.
2. **Tissue typing (detection of MHC antigens):** In serological tissue typing, the laboratory uses **standardized antisera** or **monoclonal antibodies** that are specific for particular HLAs. Class I antigens are identified by means of antisera. Antisera for HLA typing were originally obtained from multiparous women (i.e. women who have had multiple pregnancies), placental fluid and from individuals who have received multiple transfusions, from individuals who have received and rejected grafts, and from volunteers who have been immunized with cells from another individual with a different HLA haplotype. These are being replaced by monoclonal antibodies. Following methods are used:
   i. **Microcytotoxicity test:** In this test, white blood cells from the potential donors and recipient are distributed into a series of wells on a microtiter plate, and then antibodies for various class I and class II MHC alleles are added to different wells. After incubation, complement and a dye, such as trypan blue, are then added. Cells carrying antigens corresponding to the HLA antiserum are killed by complement-mediated membrane damage. If antibodies in the antiserum have reacted specifically with the lymphocyte, the cell is damaged. The damaged cell will take up the dye (undamaged cells will not), thus indicating that the lymphocyte possesses a certain antigen. This method is simple and rapid.
   ii. **Molecular methods:** Recently, more discriminating molecular methods have been developed for tissue typing which include restriction fragment length polymorphism (RFLP) with southern blotting, and polymerase chain reaction (PCR) amplification using sequence specific primers to identify HLA genes in the DNA of donors and recipients.

Tissue Matching

Once a set of HLA compatible donors is available (commonly, siblings of the patient), the best donors among them can be chosen by tissue matching. In this situation, a one way **mixed lymphocyte reaction (MLR)** can be used to determine identity of class II HLA antigens between a potential donor and recipient.

Mixed Lymphocyte Reaction (MLR)

It depends on the fact that T lymphocytes in culture, when exposed to HLA incompatible antigen, will undergo blast transformation, the intensity of the reaction being a measure of the antigenic disparity between the donor and recipient lymphocytes.

Factors Favoring Allograft Survival

1. **Blood group and major histocompatibility antigens**
   Transplantation antigens are, by definition, those antigens that affect the survival of tissue or organ allografts. There are three classes of transplantation antigens: (1) ABO group antigens; (2) MHC antigens; (3) Minor histocompatibility antigens.
2. **Immunosuppression**
   a. **General immunosuppressive therapy:** The three nonspecific agents that are most widely used in current practice are steroids, cyclosporine and azathioprine.
   b. **Specific immunosuppressive therapy:** Specific immunosuppression reduces antigraft response without increasing susceptibility to infection. Specific immunosuppression to allografts has been achieved in animal experiments using antibodies or soluble ligands reactive with cell-surface molecule. Neonatal exposure to donor antigens can induce unresponsiveness to transplants in animals. In humans procedures such as total lymphoid irradiation (TLI) and use of antilymphocyte serum (ALS), widely used in heart transplant recipients to deplete circulating T cells.

3. **Immune tolerance to allografts:** Obviously, in the case of tissues that lack alloantigens, such as cartilage or heart valve, there is no immunological barrier to transplantation.
4. **Privileged sites:** An allograft can be placed without engendering a rejection reaction, in immunologically privileged sites. These sites include the **anterior chamber of the eye, the cornea, the uterus, the testes, and the brain.** Each of these sites is characterized by an absence of lymphatic vessels and in some cases by an absence of blood vessels as well. Immunologically privileged sites fail to induce an immune response because they are effectively sequestrated from the cells of the immune system.

Examples

i. **Cornea:** The privileged location of the **cornea** has allowed corneal transplants to be highly successful. Lack of vascularity at the site prevents graft rejection.
ii. **Brain: The brain** is an immunologically privileged site because the blood-brain barrier prevents the entry or exit of many molecules, including antibodies.
iii. **Pancreatic islet cells:** The successful transplantation of allogeneic **pancreatic islet cells** into the thymus in a rat model of diabetes suggests that the thymus may also be an immunologically privileged site.
iv. **The cheek pouch of the Syrian hamster** is a privileged site used in experimental situations.

FETUS AS ALLOGRAFT

Fetus is always a mixture of maternal and paternal genes, therefore, fetal MHC antigens are always different from those of the mother. The fetus can be considered an intrauterine allograft as it contains antigens which are foreign to the mother. In spite of this fetus is not treated as foreign transplanted tissue and rejected. The fetus does not reject the mother which may be due to the fact that the developing immune system of the fetus can not respond. The reason why the fetus is exempt from rejection is not clear, though many explanations have been offered.

1. **Immunological barrier:** The placenta acts as an immunological barrier by generating a hormone which is locally immunosuppressive.
2. **Mucoproteins:** Mucoproteins are produced by cells of the placenta which coat fetal cells, thus masking histocompatibility antigens and prevent recognition.
3. **Soluble inhibitory factor:** Soluble inhibitory factor can be produced by placental giant cells which suppresses T cell proliferation and antibody production, and induces a population of Ts cells. Hormones (human chorionic gonadotropin, produced by placenta, and the high levels of maternal progesterone produced during pregnancy) also cause immunosuppression.
4. **Blocking antibodies:** The mother produces specific blocking antibodies to fetal antigens of the fetal cells thus, blocking immune recognition and immune attack by maternal Tc cells.
5. **Major histocompatibility complex (MHC) antigens:** Major histocompatibility complex (MHC) antigens are present only in low density on trophoblastic cells and the cell membranes are relatively resistant to attack by T or K cells.

   An incomplete mucopolysaccharide barrier rich in sialic acid surrounds the trophoblastic cells, protecting them from cytotoxic lymphocytes.
6. **Alpha fetoprotein:** The high concentration of alphafetoprotein in fetal blood also may be a factor, as it has immunosuppressive properties, which may protect the fetus against immunological damage from any maternal leukocytes entering fetal circulation.

GRAFT-VERSUS-HOST REACTION

Graft rejection is due to the reaction of the host to the grafted tissue (host-versus-graft response). The contrary situation, in which the grafted tissue may react to and reject the host, is known as the graft-versus-host (GVH) reaction.

The GVH reaction occurs when the following conditions are present:

1. The graft (bone marrow, lymphoid tissue, splenic tissue, etc.) contains immunocompetent T cells.
2. The recipient possesses transplantation antigens that are absent in the graft.
3. The recipient must not reject the graft. The host's immunological responsiveness must be either destroyed or so impaired (following whole-body irradiation) that he cannot reject a graft (allograft or xenograft).

GVHD is a major complication of **bone marrow transplantation** and affects 50 to 70 percent of bone marrow transplant patients. It develops as donor T cells recognize alloantigens on the host cells. The activation and proliferation of these T cells and the subsequent production of cytokines generate inflammatory reactions in the skin, gastrointestinal tract and liver.

The major clinical features of the GVH in animals are retardation of growth, emaciation, diarrhea, hepatosplenomegaly, lymphoid atrophy and anemia, terminating fatally. The syndrome has been called *runt disease*.

IMMUNOLOGY OF MALIGNANCY

The development from a normal cell to a cancerous cell is usually a multistep process of clonal evolution driven by a series of somatic mutations that progressively convert the cell from normal growth to a precancerous state and finally a cancerous state. When a cell undergoes malignant transformation, it acquires new surface antigens. It may lose some normal antigens and this makes a tumor antigenically different from the normal tissues of the host. A tumor can, therefore, be considered an allograft and be expected to induce an immune response.

Clinical Evidence of Immune Response in Malignancy

Several clinical observations indicate the presence of an immune' response that prevents, arrests and occasionally cures malignancies.

1. **Spontaneous regression:** Instances of spontaneous regression of established tumors have been reported, especially with neuroblastoma and malignant melanoma. It is believed that recovery from malignancy also may represent an immune process on the analogy of the role played by the immune response in recovery from infections.
2. **Chemotherapy and cures:** Sometimes dramatic cures follow chemotherapy of choriocarcinoma and Burkitt's lymphoma. Even a single dose of cytotoxic drug may, on occasion, result in a complete cure. Again, removal of the primary tumor often leads to a regression of the metastases in some types of tumors, such as hypernephroma with pulmonary metastases. These observations suggest that once a large mass of tumor has been removed, mopping up operations can be effected by the immune process. The immune response appears to be effective only when the tumor is below a 'critical mass'.
3. **Overcome the defense mechanisms:** Postmortem data suggest that there may be more tumors than become clinically apparent. This indicates that the immune system is able to deal with malignant cells as they arise and that only some of them are able to overcome the defence mechanisms and develop into clinical cancer.
4. **Cellular response:** Histological evidence of immune response against malignancy is provided by the presence of lymphocytes, plasma cells and macrophages infiltrating tumors. The cellular response resembles that seen in the allograft reaction. Tumors showing such cellular infiltration have a better prognosis than those that do not.
5. **Immune deficiency states:** A high incidence of malignancy should be expected in immune deficiency states if the immune system plays a natural role in preventing tumor development,. This is indeed so. An increased incidence of cancer, particularly lymphoreticular malignancies, is found in congenital immunodeficiency states, in AIDS and in patients undergoing chronic immunosuppressive therapy.

TUMOR ANTIGENS

Tumor antigens are antigens that are present in malignant cells but absent in the corresponding normal cell of the host.

Tumor antigens were first demonstrated by transplantation tests. When a tumor was grafted onto an animal previously immunized with inactivated cells of the same tumor, resistance to the graft was seen. Two types of tumor antigens have been identified on tumor cells:

1. **Tumor-specific transplantation antigens (TSTAs)**
2. **Tumor-associated transplantation antigens (TATAs).**

1. **Tumor-specific transplantation antigens (TSTAs)** Tumor-specific antigens are unique to tumor cells and do not occur on normal cells in the body. Tumor-specific antigens have been identified on tumors induced with chemical or physical carcinogens and on some virally induced tumors. Different tumors possess different TSTA, even though induced by the same carcinogen. In contrast, TSTA of virus induced tumors is virus specific in that all tumors produced by one virus will possess the same antigen, even if the tumor occur in different animal strains or species.
2. **Tumor-associated transplantation antigens (TATAs):** These are present on tumor cells and also on some normal cells. Since they are also present on some normal cells, therefore, they do not evoke an immune response and are of little significance in tumor rejection. Tumor-associated antigens may also be proteins that are normally expressed at extremely low levels on normal cells but are expressed at much higher levels on tumor cells. Detection of these antigens is nevertheless of value in the diagnosis of certain tumors and antibodies raised against them can be useful for immunotherapy. The more common tumor-associated antigens are oncofetal antigens and increased levels of normal oncogene products. TATAs fall into three categories:
   - i. **Tumor-associated carbohydrate antigens (TACAs):** They represent abnormal forms of widely expressed glycoproteins and glycolipids such as mucin-associated antigen detected in pancreatic and breast cancers.
   - ii. **Oncofetal tumor antigens:** Oncofetal tumor antigens, as the name implies, are found not only on cancerous cells but also on normal fetal cells. These antigens appear early in embryonic development, before the immune system acquires immunocompetence. If these antigens appear later on cancer cells, they are recognized as nonself and induce an immunologic response. Two well-studied oncofetal antigens are **alpha fetoprotein (AFP)** in hepatoma and **carcinoembryonic** antigen **(CEA)** found in colonic cancer. There synthesis represents a differentiation of malignant cells into more primitive forms.
   - iii. **Differentiation antigens:** They are peculiar to the differentiation state at which cancer cell are arrested. For example, CD10, an antigen expressed in early B lymphocytes, is present in B cell leukemias. Similarly, prostate specific antigen (PSA) is expreseed on the normal as well as cancerous prostatic epithelium. Both serve as useful differentiation markers in the diagnosis of lymphoid and prostatic cancer.

IMMUNE RESPONSE IN MALIGNANCY

Tumors can induce potent immune responses. In experimental animals, tumor antigens can induce both humoral and cell-mediated immune responses that result in the destruction of the tumor cells. In general, the cell-mediated response appears to play the major role. The immune response to tumor includes CTL-mediated lysis, NK-cell activity, macrophage-mediated tumor destruction, and destruction mediated by ADCC. Several cytotoxic factors, including, TNF-α and TNF-β, help to mediate tumor-cell killing.

CTL-Mediated Lysis

T cell activation generates helper T (Th), delayed-type hypersensitivity T (Td) and cytotoxic T (Tc) cells. Of special interest is the role played by Td cells. These cells affect tumor killing by means of the lymphokines that they release.

NK Cell Activity

NK cells are lymphocytes that are capable of destroying tumor cells without prior sensitization. After activation with IL-2, NK cells can lyse a wide range of human tumors, including many that appear to be nonimmunogenic for T cells. So NK cells may provide the first line of defense against many tumors.

Macrophage-Mediated Tumor Destruction

Activated macrophages also exhibit some selective cytotoxicity against tumor cells *in vitro*. T cell derived cytokine (interferon-gamma) activates macrophages which acquire antitumor activity.

Destruction-Mediated by ADCC

Humoral mechanisms may also participate in tumor cell destruction by activation of complement and induction of ADCC by NK cells.

IMMUNOLOGICAL SURVEILLANCE

The immune surveillance theory was first conceptualized in the early 1900s (1906) by Paul Ehrlich. He suggested that cancer cells frequently arise in the body but are recognized as foreign and eliminated by the immune system. Some 50 years later, Lewis Thomas revived it in 1950s and was developed by Burnet. It postulates that the primary function of cell-mediated immunity is to 'seek and destroy' malignant cells that arise by somatic mutation. Such malignant mutations are believed to occur frequently and would develop into tumors but for the constant vigilance of the immune system. Inefficiency of the surveillance mechanism, either as a result of ageing or in congenital or acquired immunodeficiencies, leads to an increased incidence of cancer.

The development of tumors represents a lapse in surveillance.

It is evident that tumor cells must develop mechanisms to escape or evade the immune system in immunocompetent hosts. Several such mechanisms may be operative:

1. **Weak immunogenicity:** Some tumors are weakly immunogenic, so in small numbers they do not elicit an immune response. But when their numbers increase enough to provoke immune response the tumor load may be too great for the host's immune system to mount an effective response.
2. **Modulation of surface antigens:** Certain tumor-specific antigens disappear from the surface of tumor cells in the presence of serum antibody and then to appear after the antibody is no longer present.
3. **Masking tumor antigens:** Certain cancers produce copious amounts of a mucoprotein called sialomucin. It binds to the surface of the tumor cells. Immune system does not recognize these tumor cells as foreign since sialomucin is a normal component.
4. **Induction of immune tolerance:** Some tumor cells can synthesize various immunosuppressants. They may also activate specific Ts cells. Both these suppress the effector T and B cell clones.
5. **Production of blocking antibodies:** Antitumor antibody itself acts as a blocking factor. The antibody, binds to tumor-specific antigens and masks the antigens from cytotoxic T cells. Some tumor cells invoke immune system to produce blocking antibodies that cannot fix and activate complement, so lysis of tumor cell is not possible. Blocking antibodies also cover the surface of cancer cells, preventing Tc cells from binding to hidden receptors.
6. **Low levels of HLA class I molecules:** In some instances, tumor cells express reduced levels of HLA class I molecules. This impairs presentation of antigenic peptides to cytotoxic T cells.

IMMUNOTHERAPY OF CANCER

Different approaches have been attempted in the immunotherapy of cancer—active or passive, specific or nonspecific.

Active
- Nonspecific
- Specific

Passive-nonspecific
- Specific
- Combined

Nonspecific Active Immunotherapy

Biological response modifiers (BRMs) are used to enhance immune responses to tumors and fall into four major groups: **(i) Bacterial products; (ii) Synthetic molecules; (iii) Cytokines; (iv) Hormones.**

i. **Bacterial products:** Broadly speaking, bacterial products such as BCG and nonliving *Corynebacterium parvum* have adjuvant effects on macrophages. Most attempts at systemic therapy in man have not been conspicuously successful, but intralesional BCG can cause regression of melanoma and nonspecific local immunization with BCG is effective against bladder tumors.

ii. **Synthetic molecules:** A variety of synthetic polymers, nucleotides and polynucleotide induce IFN production and release. The cytokines administered directly act on macrophages and NK cells. Dinitrochlorobenzene has been tried in the treatment of squamous and basal cell carcinoma of the skin.

Glucan, a pyran polymer derived from microorganisms, and levamisole, originally introduced as an antihelminthic, have been tried for stimulating cell-mediated immunity and macrophage functions.

iii. **Cytokines:** Immunotherapy with cytokine can cause tumor regression. IFN-α can induce prolonged remission of the rare hairy-cell leukemia and IL-2 is effective in a proportion of melanomas and renal carcinomas. There are also encouraging results in the treatment of intraperitoneal ovarian tumors and IFN-γ and tumor necrosis factor-α (TNF-α).

iv. **Hormones:** A variety of hormones including the thymic hormones can be used to enhance T cell function.

Specific Active Immunotherapy

Specific active immunotherapy includes therapeutic vaccines of tumor cells, cell extracts, purified or recombinant antigens, peptides, heat shock proteins or DNA antigen-pulsed dendritic cells. Specific active immunotherapy by the injection of tumor cell 'vaccines' was tried early in this century but was given up unprofitable.

Passive Immunotherapy

Passive immunotherapy may be:

i. **Nonspecific** (lymphokine-activated killer (LAK) cells)

ii. Specific (antibodies alone or coupled to drugs, prodrugs, toxins or radioisotope, bispecific antibodies T cells)

iii. Combined (LAK cells and bispecific antibody).

STRATEGIES FOR VACCINATION AGAINST CANCER

Key elements in the design of strategies for vaccination against cancer are the identification of tumor antigens by genetic or biochemical approaches; the development of strategies for the effective presentation of tumor antigens; and the generation of activated populations of helper or cytotoxic T cells.

KNOW MORE

Histocompatibility Testing

ABO grouping: When tissue transplantation is anticipated, grouping and crossmatching of blood from donor and recipient are performed as a first step. If there exists any discrepancy in the ABO blood group, then the use of the prospective donor's tissue is absolutely contraindicated.

Tissue Matching

Mixed Lymphocyte Reaction (MLR)

Lymphocytes from the donor are irradiated or treated with mitomycin C to prevent cell division (serve as the stimulator cells) and then added to the cells from the recipient (serve as responder cells). If the class II antigens on the two cell populations are different, the recipient cells will divide rapidly and take up large quantities of radioactive nucleotides, i.e. [^{3}H] thymidine into the newly synthesized nuclear DNA. The amount of radioactive nucleotide uptake is roughly proportional to the MHC class II differences between the donor and recipient lymphocytes.

KEY POINTS

- Transplantation can be defined as the transfer of cells, tissues, or organs from one site in an individual to another, or between two individuals.
- There are four different basic types of transplants: autograft, isograft, allograft, and xenograft.
- There are three major types of rejection reactions: Hyperacute rejection, acute graft rejection and chronic rejection.
- The immune response to tissue antigens encoded within the major histocompatibility complex is the strongest force in rejection.
- **Graft-versus-host (GVH) reaction**—to occur requires three important components:
- The donor graft must contain immunocompetent T cells. The host must be immunocompromised. The recipient should express antigens such as MHC proteins, which will be identified as foreign to the donor.
- Tumor antigens can be classified as tumor specific antigens (TSAs) and tumor associated transplantation antigens (TATAs).
- Cancer should not occur if immunological surveillance is effective. Cancer immunoediting is the present concept being suggested for **immune surveillance**.

IMPORTANT QUESTIONS

1. What is a transplant or graft? Define various types of grafts. Describe allograft reaction.
2. What are histocompatibility antigens? Describe various procedures for histocompatibility testing.
3. Write short notes on:
   Mechanism of allograft rejection
   Graft-versus-host (GVH) reaction.
   Tumor antigens.
   Immunological surveillance.

FURTHER READING

Armstrong A. Cellular immunotherapy for cancer. BMJ 2001; 323:1289.

Barrett AJ. Graft-versus-host reaction- a review. J Royal Soc Med 1987;80:368.

Bishop JM. The molecular genetics of cancer. Science 1987; 235:305.

Boon T, Cerrottini JC, et al. Tumor antigens recognized by T lymphocytes. Annu Rev Immunol 1994;12:337-66.

Roitt I, et al. Immunology, 5th (Edn) London: Mosby 1998.

Rosenberg SA, et al. New approaches to immunotherapy of cancer. Annals Int Med 1988;108:853.

CHAPTER

23 Immunohematology

Learning Objectives

After reading and studying this chapter, you should be able to:

- Describe Rh blood groups.
- Discuss complications of blood transfusion.
- List various infectious agents transmitted via blood transfusion.

INTRODUCTION

Blood has held a mysterious fascination for us from the dawn of time. It was considered the essence of life and was believed to cure diverse diseases and restore youth and vitality to the aged. Blood transfusion has been attempted from very early times but such attempts were fruitless and often fraught with disastrous consequences. Blood transfusion became scientifically feasible only after the discovery of blood groups by Landsteiner.

HISTORY

The **ABO system** is the most important of all the blood group systems and its discovery made blood transfusion possible. No other blood group antigens were discovered for the next 25 years. Subsequently other blood groups (MN, P, Rh, Lutheran, Lewis, Kell, Duffy, Kidd, Diego, Yt, Kg, Dombroc and Colton) were reported.

ABO Blood Group System

The ABO blood group system was originally described by Landsteiner (1900) and now contains four blood groups. The blood group is determined by the presence or absence of two distinct antigens A and B on the surface of the erythrocytes. It is these antigens (also called agglutinogens because they often cause blood cell agglutination) that cause blood transfusion reactions. Red cells of group A carry antigen A, cells of group B antigen B and cells of group AB have both A and B antigens, while group O cells have neither A nor B antigen.

The serum contains the isoantibodies specific for the antigen that is absent on the red cell. The serum of a group A individual has anti-B antibody, group B has anti-A and group O both anti-A and anti-B, while in group AB both anti-A and anti-B are absent (Table 23.1).

The frequency of ABO distribution antigens differs in different people. Group O is the most common group and AB the rarest. The ABO distribution in Britain is approximately O—47 percent, A—42 percent, B—8 percent and AB—3 percent. In India, the distribution is approximately O—40 percent, A—22 percent, B—33 percent, AB—5 percent.

H Antigen

Red cells of all ABO groups possess a common antigen, the H antigen or H substance which is the precursor for the formation of A and B antigens. H antigen is not ordinarily important in grouping or blood transfusion due to its universal distribution. However, Bhende et al. (1952) from Bombay reported a very rare instance in which A and B antigens as well as H antigens were absent from the red cells. This is known as **Bombay or OH blood group**. Sera of these individuals have anti-A,

Table 23.1: Distribution of ABO antigens on the red blood cells and antibodies in the serum

Blood group	*Antigens on red blood cells*	*Antibodies in serum*	*Occurrence (%) in India*
A	A	Anti-B	22
B	B	Anti-A	33
AB	A and B	None	5
O	None	Anti-A and anti-B	40

anti-B and anti-H antibodies, therefore, they can accept the blood only from the same rare blood group.

A, B and H antigens are glycoproteins. They are also present in almost all the tissues and fluids of the body in addition to erythrocytes. They are found in secretions (saliva, gastric juice, sweat) of only about 75 percent of all persons while these antigens are always present in tissues. Such persons are called '**secretors**' and those who lack blood group antigens in secretions are called "**nonsecretors**".

Rh (Rhesus) Blood Group System

It was discovered by Landsteiner and Weiner in 1940. Their experiment was to produce an antibody to the red cells of the Rhesus monkey in rabbits and guinea pigs, but they discovered that not only did the antibody in the rodents' serum agglutinate the Rhesus monkey red cells, it also agglutinated the red cells of 85 percent of the human population. If an individual's red cells were clumped together by this antiserum, they were said to have the Rhesus factor on their red cells (i.e. Rh positive). If an individual's cells were not agglutinated by the antiserum, they were said to lack the Rhesus factor (i.e. Rh negative).

Rh Antigens—"Rh Positive" and "Rh Negative" People

There are six common types of Rh antigens, each of which is called an Rh factor. These types are designated as **C, D, E, c, d, and e**. A person who has a C antigen does not have the c antigen, but the person missing the C antigen always has the c antigen. The same is true for the D-d and E-e antigens. Also, because of the manner of the inheritance of these factors, each person has one of each of the three pairs of antigens.

The designations employed by the two system for the different Rh types are as follows:

The type D antigen (Rho) is widely present in the population and considerably more antigenic than the other Rh antigens. Rh positive or Rh negative blood depends on the presence or absence of D-antigen on the surface of red cells respectively. It can be accomplished by testing with anti-D (anti-Rh) serum. About 15 percent of the population have no RhD antigens and thus are "Rh negative". Among Indians, approximately 93 percent are Rh positive and about 7 percent negative. The Rh factor can be detected by testing the blood with anti-D (anti-Rh) serum.

A variant of D is known as Du red cells of Du subtype react with some but not all anti-D . Though Du cells may not be agglutinated by anti-D sera, they absorb the antibody on their surface. The Du subtype can therefore be detected by reacting red cell with anti-D serum and then doing a direct Coombs' test. For the purpose of blood donation, Du cells are considered Rh positive. But when a Du individual requires transfusion, advisable to use Rh negative blood because he is capable of being immunized by standard Rh positive blood.

There are no natural anti-Rh antibodies in the serum. They arise only as a result of Rh incompatible pregnancy or transfusion.

OTHER BLOOD GROUP SYSTEMS

Blood group systems other than ABO and Rh are of little clinical importance as they do not usually cause transfusion reactions or hemolytic disease. They have applications in genetics, anthropology, tissue typing and forensic medicine. As blood group antigens are inherited from the parents, they are often useful in settling cases of disputed paternity.

Lewis Blood Group System

It differs from other blood group systems in that the antigens are present primarily in the plasma and saliva and antigens do not form an integral part of the red cell membrane.

MN System

The antigens are M, N, S, and s. This system has expanded to include at least 28 antigens.

MEDICAL APPLICATIONS OF BLOOD GROUPS

1. **Blood transfusion:** Ideally, the donor and recipient should belong to the same ABO group. It used to be held that O group cells could be transfused to recipients of any group as they possessed neither A nor B antigen. Hence the O group was designated as the '**universal donor'.** The anti-A and anti-B antibodies in the transfused O blood group do not ordinarily cause any damage to the red cells of the A or B group recipients because they will be rendered ineffective by dilution in the recipient's plasma. But some O group plasma may contain isoantibodies in high titers (1:200 or above) so that damage to recipient cells may result. This is known as the '**dangerous O group**'. The AB group persons were designated '**universal recipients**' due to the absence of isoantibodies in plasma.
2. **Rh compatibility:** An Rh positive person may safely receive either Rh positive or negative blood. But an Rh negative individual receiving Rh positive blood may form antibodies against the Rh antigen because the donor's RBCs stimulate the production of anti-Rh antibodies in the recipient. A subsequent transfusion with Rh positive blood may then cause a rapid, serious hemolytic reaction.

Rh Typing of the Donor and Recipient

Cross matching: Besides ABO grouping and Rh typing of the donor and recipient, it is invariably necessary before transfusion to perform a **'cross matching'** to ensure that the donor's blood is compatible with the recipient' blood.

The routine procedure used in most blood banks is a rapid cross match by the tile or slide method. This is done in two parts—the major cross match where the donor red cells are tested against the recipient's serum, and the minor cross match where the recipient's cells are tested against the donor serum.

COMPLICATIONS OF BLOOD TRANSFUSION

The complications of blood transfusion may be divided into **immunological** and **nonimmunological.**

Immunological Complications

Some transfusion reactions may be due to immunological processes other than blood group incompatibility. Immunological complications may be caused by red cell, leukocyte or platelet incompatibility or allergic reaction to plasma components. Red cell incompatibility leads to acute intravascular hemolysis or the red cells may be coated by antibodies and engulfed by phagocytes, removed from the circulation and subjected to extravascular lysis. Hemolysis may also be due to transfusion of group O whole blood or plasma to group A or group B or group AB recipients.

Fever, pulmonary infiltrates, dyspnea, nonproductive cough and chest pain may be caused due to leukocyte incompatibility. The risk of these reactions can be reduced by using leukocyte—poor red cells. Fever is sometimes a complication of blood transfusion. Leukoagglutinins are probably the commonest cause of fever. Platelet incompatibility, allergy and infection may also lead to fever.

Nonimmunological Complications

Nonimmunological complications of blood transfusion include transmission of infectious agents and circulatory overload. Infectious agents which may be transmitted during blood transfusion may be viruses, bacteria and protozoa (Table 23.2). Massive transfusion may lead to circulatory overload.

Table 23.2: Nonimmunological complications of blood transfusion

A. Transmission of infectious agents

Viruses
- Hepatitis B virus*
- Hepatitis C virus**
- Hepatitis C virus**
- Human immunodeficiency virus 1 and 2*
- Human T cell lymphotrophic virus 1 and 2

Cytomegalovirus

Bacteria
- *Treponema pallidum**
- *Leptospira interrogans*
- *Borrelia burgdorferi*

Parasites

*Plasmodium spp.**
- *Babesia spp.*
- *Trypanosoma cruzi.*
- *Leishmania donovani*
- *Toxoplasma gondii*

B. Circulatory overload

*Mandatory tests in India.

**Mandatory tests in India since June 2001.

HEMOLYTIC DISEASE OF THE NEWBORN

When an Rh^- woman and an Rh^+ man produce a child, there is a 50 percent chance that the child will be Rh^+. If the child is Rh^+, the Rh^- mother can become sensitized to this antigen during birth, when the placental membranes tear and the fetal Rh^+ RBCs enter the maternal circulation, causing the mother's body to produce anti-Rh antibodies of the IgG type. During subsequent pregnancy, Rh antibodies of the IgG class pass from the mother to the fetus and damage its erythrocytes. The fetal body responds to this immune attack by producing large numbers of immature RBCs called erythroblasts. Thus, the term *erythroblastosis fetalis* was once used to describe what is now called **hemolytic disease of the newborn (HDNB).** It is an example of an antibody-mediated cytotoxicity disorder.

Factors Influencing the Incidence of Hemolytic Disease

The following factors influence the incidence of hemolytic disease due to Rh incompatibility:

1. **Immunological unresponsiveness to the Rh antigen:** Following antigenic stimulation not every Rh negative individual forms Rh antibodies. Some Rh negative individuals fail to do so even after repeated injection of Rh positive cells. They are called **nonresponders**. The reason for this immunological unresponsiveness is not known.
2. **Fetomaternal ABO incompatibility:** Rh immunization is more likely to result when the mother and fetus possess the same ABO group. Rh sensitization in the mother is rare when Rh and ABO incompatibility coexist. In this situation the fetal cells entering the maternal circulation are believed to be destroyed rapidly by the ABO antibodies before they can induce Rh antibodies.
3. **Number of pregnancies:** The risk to the infant increases with each successive pregnancy. The first child usually escapes disease because sensitization occurs only during its delivery.
4. **Zygosity of the father:** An individual may be homozygous or heterozygous with respect to D antigen. When the father is homozygous all his children will be Rh positive. When he is heterozygous, half his children will be Rh positive.

Detection of Rh Antibodies

Most Rh antibodies are of the IgG class, and they do not agglutinate Rh positive cells in saline being **'incomplete antibodies'**. IgG anti-D antibodies may be detected by the following techniques:

1. Using a colloid medium such as 20 percent bovine serum albumin.
2. Using red cells treated with enzymes such as trypsin, pepsin, ficin or bromelain.

3. By the indirect Coombs' test: This is the most sensitive method.

Prevention of Rh Isoimmunization

HDNB is usually prevented today by passive immunization of the Rh⁻mother at the time of delivery (within 24-48 hours) of any Rh⁺ infant with anti-Rh antibodies, which are available commercially (Rhogam). To be effective, this should be employed from the first delivery onwards. The Rh immune globulin for the purpose is prepared from human volunteers.

ABO Hemolytic Disease

The majority of cases (65%) of hemolytic disease of the newborn have minor consequences and are caused by ABO blood group incompatibility between the mother and fetus. Maternofetal ABO incompatibility is very common and in a proportion of these, hemolytic disease occurs in the newborn. Natural antibodies are IgM in nature in persons of blood group A or B, and so do not cross the placenta to harm the fetus. However, in persons of blood group O, the isoantibodies are predominantly IgG in nature either through natural exposure or through exposure to fetal blood group A or B antigens in successive pregnancies. Hence ABO hemolytic disease is seen largely in O group mothers, bearing A or B group fetus. As ABO hemolytic disease is due to naturally occurring maternal isoantibodies, it may occur even in the first born, without prior immunization.

ABO hemolytic disease is much milder than Rh disease, probably because erythrocytes of the newborn have fewer A or B antigenic sites as compared to adult erythrocytes. The direct Coombs' test is therefore often negative in this condition, while the indirect Coombs' test (neonatal serum with type-specific adult erythrocytes) is more commonly positive. Peripheral blood smear characteristically shows spherocytosis.

Blood Group and Diseases

It has been shown that some diseases may influence blood group antigens. Blood group antigens have been reported to become weak in **leukemia.**

Red cell suspensions contaminated with certain bacteria, such as *Pseudomonas aeruginosa,* become agglutinable by all blood group sera and even by normal human sera. This phenomenon is known as the **Thomsen-Friedenreich phenomenon**. It is due to the unmasking of a hidden antigen normally present on all human erythrocytes called the T antigen.

Several investigators have attempted to correlate blood group and susceptibility to certain diseases. It has been shown that **duodenal ulcer** is more frequent in persons of **blood group O** than in others. An association has also been established between **group A** and **cancer of the stomach**.

KEY POINTS

- Immunohematology is the study of blood group antigens and antibodies and their interactions in health and disease.
- *ABO blood group system:* The ABO blood group substances are glycopeptides with oligosaccharide side chains.
- *Rh antigens:* There are six common types of Rh antigens designated as C, D, E, c, d, and e. Of all the Rh antigens, antigen D (Rho) is most important.
- Transmission of infectious agents especially HIV I and II, HBV, and HCV through blood is the most important complication.
- Hemolytic disease of the newborn is an alloimmune condition. Rh antibodies of the IgG class pass from the mother to the fetus through the placenta and damage its erythrocytes. This disease usually occurs even in the second or successive child.

IMPORTANT QUESTIONS

1. Name various blood group systems and discuss Rh blood group system. Add a note on hazards of incompatible blood transfusion.
2. Write short notes on:
   - Complications of blood transfusion.
   - Hemolytic disease of the newborn (or) erythroblastosis fetalis.

FURTHER READING

Bloy C, Blanchard D, et al. Characterization of the D, C, E and G antigens of the Rh blood group system with human monoclonal antibodies. Mol Immunol 1988;25:926-30.

Mauro I, Colin Y, et al. Molecular genetic basis of the human Rhesus blood group system. Nature Genet 1993;5:62-5.

Race R, Sanger R. Blood Groups in Man, 6th edn, Oxford: Blackwell Scientific Publications 1975.

Schraber GB, et al. The risk of transfusion-transmitted viral infections. New Engl J Med 1996;334:1685.

SECTION THREE

Systemic Bacteriology

CHAPTER 24

Staphylococcus

Learning Objectives

After reading and studying this chapter, you should be able to:

- Describe species of *Staphylococcus*.
- Describe morphology and culture characteristics of *Staphylococcus aureus*.
- List characteristics of *Staph. aureus* strains.
- Explain coagulase test.
- List and describe toxins and enzymes of *Staphylococcus aureus*.
- Describe staphylococcal diseases.
- Discuss laboratory diagnosis of infections caused by *Staphylococcus aureus*.
- Explain methicillin-resistant staphylococci and its clinical problem.
- Describe the following: Coagulase-negative staphylococci (CNS); micrococci.
- Distinguish characteristics of *Staph. aureus*, *Staph. epidermidis* and *Staph. saprophyticus*.

INTRODUCTION

Staphylococci, micrococci and *Stomatococcus* belong to the **family Micrococcaceae.** Staphylococci are gram-positive cocci that occur in grape-like clusters. They are ubiquitous and are the most common cause of localized suppurative lesions in human beings. Their ability to develop resistance to penicillin and other antibiotics enhances their importance as a human pathogen, especially in the hospital environment.

Staphylococci were first observed in human pyogenic lesions by von Recklinghausen in 1871. Pasteur (1880) obtained liquid cultures of the cocci from pus and produced abscesses by inoculating them into rabbits. It was Sir Alexander Ogston, a Scottish surgeon, who established conclusively the causative role of the coccus in abscesses and other suppurative lesions (1880). He also gave it the name *Staphylococcus (Staphyle,* in Greek, meaning 'bunch of grapes'; *kokkos,* meaning a berry) due to the typical occurrence of the cocci "in grape-like clusters" in pus and in cultures. Ogston noticed that non-virulent staphylococci were also often present on skin surfaces.

Species

The genus *Staphylococcus* contains 33 defined species and 20 species found in man. Species of staphylococci are initially differentiated by the coagulase test and are classified into two groups: the **coagulase-positive** and **coagulase-negative staphylococci (CNS).**

Coagulase-positive Staphylococci

Of the 20 species found in man, *Staphylococcus aureus* (formerly also called *Staph. pyogenes)* is coagulase positive.

Coagulase-negative Staphylococci

There are 19 coagulase-negative staphylococci (CNS). *S. epidermidis* and *S. saprophyticus* are the most clinically significant species in this group. *S. epidermidis* has been known to cause various hospital-acquired infections, whereas S. *saprophyticus* is associated mainly with urinary tract infections, predominately in female adolescents and young women.

STAPHYLOCOCCUS AUREUS

Morphology

They are spherical cocci, approximately 1 µm in diameter, arranged characteristically in grapelike clusters (Fig. 24.1). Cluster formation is due to cell division occurring in three planes, with daughter cells tending to remain in close proximity. They may also be found singly, in pairs and in short chains of three or four cells, especially when examined from liquid culture. Long chains never occur.

They are nonsporing, nonmotile and usually noncapsulate with the exception of rare strains. They stain readily with aniline dyes and are uniformly gram-positive but old and phagocytosed organisms may be gram-negative.

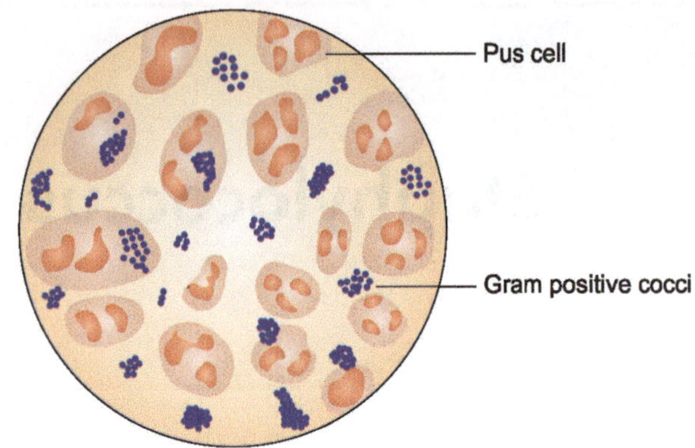

Fig. 24.1: *Staphylococcus* in a smear of pus

Cultural Characteristics

They are aerobes and facultative anaerobes. Optimum temperature for growth is 37°C (range being 12-44°C). Optimum pH is 7.5. They can grow readily on ordinary media.

1. Nutrient Agar

After aerobic incubation for 24 h at 37°C, colonies are 1-3 mm in diameter and have a smooth glistening surface, an entire edge, a soft butyrous consistency and an opaque, pigmented appearance. Most strains produce golden-yellow *(aureus)* pigment, though some strains may form white (nonpigmented) colonies. These white-colonied strains of *S. aureus* are fully virulent. Pigmentation is characteristic of this species when grown aerobically.

Pigmentation is enhanced on fatty media such as Tween agar, by prolonged incubation, and by leaving plates at room temperature. Nonpigmented strains are not uncommon. The pigment is believed to be lipoprotein allied to carotene. Grown anaerobically, colonies are often smaller and grayish in color.

2. Blood Agar

The colonies have the same appearance as on nutrient agar, but may be surrounded by a zone of **β-hemolysis.** Hemolysis is more likely to be present if sheep, human or rabbit blood is used instead of horse blood and if incubation is in air with 20 percent added carbon dioxide. Hemolysis is weak on horse blood agar.

3. MacConkey Agar

Colonies are smaller and are pink due to lactose fermentation.

4. Milk Agar

This medium is prepared by mixing 200 ml of sterile nutrient agar containing 2 percent agar and 100 ml of fresh or sterilized milk and poured into plates. On this medium, after overnight incubation, the colonies of *S. aureus* are larger than those on nutrient agar and pigmentation is well developed and easily recognized against the opaque white background.

5. Phenolphthalein Phosphate Agar

This is an indicator medium and assists in the identification of *S. aureus* in mixed cultures. The culture plate is inoculated with the test culture and incubated aerobically at 37°C for 18 hours or for 3 days at 30°C. Appearance of the colonies of *S. aureus* on this medium is similar to those on nutrient agar. All strains of *S. aureus* produce phosphatase which liberates phenolphthalein from sodium phenolphthalein diphosphate.

Detection of Phosphatase

To detect it, 0.1 ml of ammonia solution is placed in the lid of the plate, and the culture plate with the culture is inverted over the ammonia for a minute or so. Colonies of *S. aureus* become bright pink because phenolphthalein is pink in alkaline pH. Most other staphylococci form colonies that remain uncolored.

6. Selective Salt Media

Selective medium may be useful for the isolation and enumeration of staphylococci from materials, such as feces, food and dust, likely to contain a predominance of other kinds of bacteria. Therefore, **7 to 10 percent** of sodium chloride may be added to **nutrient agar (salt agar)** or **milk agar (salt milk agar); mannitol salt agar** containing 1 percent mannitol, 7.5 percent NaCl, and phenol red in nutrient agar; and **Ludlam's medium** containing lithium chloride and tellurite; and salt cooked meat broth (10% NaCl).

Biochemical Reactions

1. **Sugar fermentation:** *S. aureus* ferments a range of sugars (glucose, maltose, lactose, sucrose, including mannitol) producing acid but no gas. Sugar fermentation is of no diagnostic value except for mannitol, which is usually fermented anaerobically by *Staph. aureus* but not by other species.
2. **Catalase:** Catalase positive (unlike streptococci).
3. **Lipolytic:** When grown on media containing egg-yolk, produce a dense opacity because most strains are lipolytic.
4. **Phosphatase test:** They also produce phosphatase. This is a useful screening procedure for differentiating *Staph. aureus* from *Staph. epidermidis* in mixed cultures, as the former gives prompt phosphatase reaction, while the latter is usually negative or only weakly positive.
5. **Deoxyribonulease (DNAase) test:** It produces a deoxyribonulease (DNAase), and a heat-stable nuclease (thermonuclease, TNAase).
6. **Other biochemical tests:** Indole negative, MR positive, VP positive, urease positive, hydrolyzes gelatin and reduces nitrates to nitrites.

Table 24.1 shows characteristics of Staphylococcus aureus.

Table 24.1: Characteristics of *Staph. aureus* strains
1. Beta hemolysis—produce clear hemolysis on blood agar.
2. Golden yellow pigment.
3. Coagulase positive.
4. Greater biochemical activity, ferment mannite.
5. Liquefy gelatin.
6. Produce phosphatase.
7. Black colonies on potassium tellurite blood agar—produce black colonies in a medium containing potassium tellurite by reducing tellurite metallic tellurium.
8. Produce thermostable nucleases which can be demonstrated by the ability of boiled cultures to degrade DNA in an agar diffusion test.

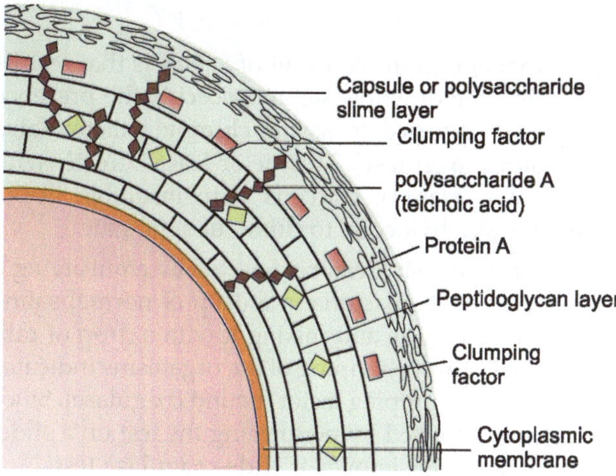

Fig. 24.2: Structure of staphylococcal cell wall

Resistance

S. aureus and the other micrococcaceae are among the hardiest of the nonsporing bacteria. Dried on threads, they retain their viability for 3-6 months. They have been isolated from dried pus after 2-3 months. It withstands moist heat at 60°C for 30 min but is killed after 60 min. Most strains grow in the presence of 10 percent NaCl and some even in 15 percent NaCl. These features are of significance in food preservation.

It is readily killed by phenolic and hypochlorite disinfectants at standard in-use concentrations, and by antiseptic preparations such as hexachlorophane, chlorhexidine and povidone-iodine. They are very sensitive to aniline dyes; thus they are inhibited on blood agar medium containing 1 in 500,000 crystal violet, which permits the growth of streptococci. Fatty acids inhibit the growth of staphylococci, the highly unsaturated acids having a more powerful action on coagulase positive than on coagulase negative strains. Staphylococci are uniformly resistant to lysozyme but some micrococci are sensitive to it. Staphylococci are generally sensitive to *lysostaphin*—a mixture of enzymes produced by a particular strain of *Staph. epidermidis.*

Antigenic Structure of *Staphylococcus* Aureus

The antigenic structure of *S. aureus* (Fig. 24.2) is complex. It is as follows:

A. Cell-Associated Polymers

1. Capsule

Capsular polysaccharide surrounding the cell wall inhibits opsonization.

2. Peptidoglycan

The cell wall polysaccharide peptidoglycan confers rigidity and structural integrity to the bacterial cell. It activates complement and induces release of inflammatory cytokines.

3. Teichoic Acids

Teichoic acid, an antigenic component of the cell wall, facilitates adhesion of the cocci to the host cell surface and protects them from complement-mediated opsonisation.

B. Cell Surface Proteins

1. Protein A

Protein A is a group-specific antigen unique to *S. aureus* strains. Ninety percent of protein A is found in the cell wall, covalently linked to the peptidoglycan. It is absent in both coagulase negative staphylococci (CNS) and micrococci.

Biologic Properties

Protein A has many biologic properties including chemotactic, anticomplementary, and antiphagocytic and elicits hypersensitivity reactions and platelet injury. It is mitogenic and potentiates natural killer activity of human lymphocytes.

The uniqueness of protein A is centered on its ability to interact with normal IgG of most mammalian species. Unlike a specific antigen-antibody reaction, binding involves not the Fab fragment but the Fc portion of the immunoglobulin. Protein A binds IgG molecules, nonspecifically, through Fc region leaving specific Fab sites free to combine with specific antigen. When suspension of such sensitized cells is treated with homologous (test) antigen, the antigen combines with free Fab sites of IgG attached to staphylococcal cells. This is known as coagglutination. Its ability to bind to the Fc region of IgG has led to numerous applications in immunochemical and cell-surface structural studies.

2. Cytoplasmic Membrane

The cytoplasmic membrane is made up of a complex of proteins, lipids, and a small amount of carbohydrates. It serves as an osmotic barrier for the cell and provides an anchorage for the cellular biosynthetic and respiratory enzymes.

3. Clumping Factor (Bound Coagulase)

The component on the cell wall of *S. aureus* that results in the clumping of whole staphylococci in the presence of plasma is referred to as the clumping factor (also called **bound coagulase**). This factor reacts directly with fibrinogen in plasma, converts it to insoluble fibrin, causing the staphylococci to clump or aggregate.

Slide coagulase test: It can be detected by emulsifying a few colonies of the bacteria in a drop of normal saline on a clean glass slide and mixing it with a drop of rabbit plasma. Prompt clumping of the organisms indicates the presence of clumping factor (bound coagulase). Since this factor is detected by performing the test on a slide, therefore, the test is known as **slide coagulase test.**

Pathogenicity and Virulence

Staph. aureus strains possess a large number of **cell-associated** and **extracellular factors**, which contribute to virulence. It is probable that virulence is multifactorial in the staphylococci.

Toxins and Enzymes

S. aureus produces a number of toxins, and enzymes. They are important virulence factors.

A. Toxins

1. Cytolytic Toxins

At least five cytolytic or membrane-damaging toxins (alpha, beta, delta, gamma, and Panton-Valentine [P-V] leukocidin are produced by *S. aureus.*

a. Alpha (α) Hemolysin

Alpha (α) lysin is the most important among them. It is a protein consisting of a single polypeptide chain, with a molecular weight of 28,000-30,000 daltons. It is inactivated at 60°C, however, its activity is regained paradoxically, if it is further heated to between 80-100°C. This is due to the fact that the toxin combines with a heat-labile inhibitor at 60°C but at higher temperature (80-100°C), the inhibitor is inactivated thus, toxin regains its activity. Above 100°C toxin itself gets inactivated.

The toxin lyses rabbit and sheep erythrocytes. It is leucocidal dermonecrotic and lethal. It is antigenic and treatment with formalin at 37°C leads to rapid disappearance of its hemolytic and toxic but not of its antigenic activity.

b. Beta (β) Hemolysin

Beta (β) toxin, also called sphingomyelinase C, is a 35,000 *Da* heat-labile protein produced by most strains of *S. aureus.* This enzyme has a specificity for sphingomyelin and lysophosphatidylcholine. Therefore, the susceptibility of red cells from different species depends upon their sphingomyelin content and is toxic to a variety of cells, including erythrocytes, leukocytes, macrophages, and fibroblasts.

β lysin is strongly active on sheep and weakly active on rabbit and human red blood cells. It does not lyse horse red blood cells. It exhibits a 'hot-cold phenomenon', the hemolysis being initiated at 37°C, but becoming evident only after chilling.

It is not dermonecrotic and kills experimental animals only when injected in large doses. It lyses platelets.

c. Gamma (γ) Hemolysin

Gamma (γ) toxin has pronounced hemolytic activity, but its precise mode of action is not known. It consists of two protein components that act synergistically, both being essential for hemolysis and toxicity. γ lysin acts on sheep, rabbit and human red blood cells but not on those of horse.

It is antigenic and the finding of elevated levels of specific neutralizing antibodies in human staphylococcal bone disease suggests a possible role of this toxin in the disease state.

d. Delta (δ) Hemolysin

The toxin has a wide spectrum of cytolytic activity. It lyses red blood cells of sheep, rabbit, horse and man. It also possesses lethal, dermonecrotic and leukocidal activity.

e. Leukocidin

Leukocidin (called the Panton-Valentine toxin after its discoverers) is also a two component (bicomponent) toxin, like the gamma lysin, S (Slow) and F (Fast). These components act synergistically to damage polymorphonuclear leukocytes and macrophages and to produce dermonecrosis. Cell lysis is mediated by pore formation with subsequent increased permeability to cations and osmotic instability.

2. Enterotoxins

This toxin is responsible for the manifestations of staphylococcal food poisoning nausea, vomiting and diarrhea, 2-6 hours after consuming contaminated food containing preformed toxin. Enterotoxins are commonly produced by about two-thirds of *Staph. aureus* strains, growing in carbohydrate and protein foods. The toxin is relatively heat stable, resisting 100°C for 10 to 40 minutes, depending on the concentration of the toxin and nature of the medium.

Eight serologically distinct staphylococcal enterotoxins (A-E, G-I) and three subtypes of enterotoxin C have been identified. Enterotoxin A is most commonly associated with disease. Enterotoxins C and D are found in contaminated milk products, and enterotoxin B causes staphylococcal pseudomembranous enterocolitis. Less is known about the prevalence of the other enterotoxins. The precise mechanism of toxin activity is not understood. The toxin is believed to act directly on the autonomic nervous system to cause the illness, rather than on the gastrointestinal mucosa. The toxin is antigenic and neutralized by the specific antitoxin. Type A toxin is responsible for most cases. For detection of the toxin sensitive serological tests such as latex agglutination and ELISA are available.

These toxins are superantigens, however, capable of inducing nonspecific activation of T cells and cytokine release. The toxin is potent, microgram amounts being capable of causing the illness. Some cases of postantibiotic-diarrhea are caused by enterotoxin-forming staphylococci. The toxin also exhibits pyrogenic, mitogenic, hypotensive, thrombocytopenic and cytotoxic effects.

3. Toxic Shock Syndrome Toxin-1 (TSST-1)

S. aureus is associated with toxic shock syndrome (TSS), a severe and often fatal disorder characterized by multiple organ dysfunction. TSST-1, formerly called pyrogenic exotoxin C and enterotoxin F, is heat and proteolysis resistant, chromosomally mediated exotoxin, Enterotoxin B and, rarely, enterotoxin C are responsible for approximately half the cases of nonmenstruation-associated TSS. It is antigenic and most persons over 30 years of age have circulating antibodies.

The enterotoxins and TSST-1 belong to a class of polypeptides known as **superantigens** which are potent activators of T lymphocytes leading to the release of cytokines such as interleukins and tumor necrosis factor. This results in clinical condition of TSS. The systemic effects of these diseases result from this process.

4. Epidermolytic Toxins (Exfoliative Toxins)

This toxin, also known as ET or 'exfoliatin' is responsible for the 'staphylococcal scalded skin syndrome' (SSSS). Two distinct forms of exfoliative toxin (ETA and ETB) have been identified, and either can produce disease. ETA is heat-stable and the gene is chromosomal, whereas ETB is heat-labile and plasmid-mediated. SSSS is seen mostly in young children and only rarely in older children and adults.

B. Extracellular Enzymes

Staph. aureus produces a number of enzymes such as coagulase, catalase, hyaluronidase, fibrinolysin, lipases, nucleases and penicillinase.

Coagulase

S. aureus produces an extracellular enzyme called **coagulase** which brings about clotting of human or rabbit plasma. It acts along with a **'coagulase reacting factor' (CRF)** present in plasma, binding to prothrombin and converting fibrinogen to fibrin. Coagulase does not clot plasma of guinea pigs and some other species because they lack CRF. Calcium or other clotting factors are not required for coagulase action. CRF is similar to prothrombin but is probably not identical with it. *Staph. aureus* strains usually secrete both coagulase and clumping factor. Coagulase test is the standard criterion for the identification of *S. aureus* isolates. The role of coagulase in the pathogenesis of disease is speculative, but coagulase may cause the formation of a fibrin layer around a staphylococcal abscess, thus localizing the infection and protecting the organisms from phagocytosis.

Eight antigenic types (A-H) have been described. Most human strains form coagulase type A.

$$\textit{Fibrinogen} \xrightarrow[\textit{CRF}]{\textit{Coagulase}} \textit{Fibrin (Clotting)}$$

Coagulase and clumping factor ('bound coagulase') differ in many respects (Table. 24.2).

Coagulase Test

Coagulase test is done by two methods—slide and tube coagulase test. The slide or tube coagulase test is performed to distinguish *Staph. aureus* from coagulase-negative species.

Slide Coagulase Test

The slide test detecting bound coagulase is much simpler and usually gives results parallel with the tube test. It can be detected by emulsifying a few colonies of the bacteria in a drop of normal saline on a clean glass slide and mixing it with a drop of rabbit plasma. Prompt clumping of the organisms indicates the presence of clumping factor (bound coagulase). Positive and negative controls also are set up.

Since this factor is detected by performing the test on a slide, therefore, the test is known as slide coagulase test. Almost all the clumping factor producing strains of *S. aureus* produce coagulase, whereas 12 percent of coagulase producing strains may not produce clumping factor.

Tube Coagulase Test

The tube coagulase test detects free coagulase. Place 1 ml volume in small test tubes of a 1-in-6 dilution of the human or rabbit plasma in saline. Emusify a colony of the staphylococcus under test in a tube of the diluted plasma or 0.1 ml of an overnight broth culture is added to 0.5 ml of undiluted plasma. The mixture is incubated in a water bath at 37°C for up to 4 hours. Examine at 1,2 and 4 hours for clot formation. If positive, the plasma clots and does not flow when the tube is tilted or inverted. If clot does not appear it is left overnight at room temperature and re-examined. On contin¬ued incubation, the clot may be lysed by fibrinoly¬sin produced by some strains. Controls with plasma alone, known coagulase-positive and coagulase negative cultures must be set up with each batch of tests.

False Positive Reaction

Citrated plasma should not be used because contaminating gram-negative bacilli (e.g. *Pseudomonas*) may utilize the citrate and produce false positive reaction. Oxalate, EDTA or heparin are suitable anticoagulants.

Bound coagulase (clumping factor)	Free coagulase
1. Heat stable.	1. Heat labile.
2. Constituent of cell wall.	2. Secreted into the medium.
3. It does not require the cooperation of CRF for its action.	3. Requires the cooperation of CRF for its action.
4. Clot plasma of guinea pigs and some other species.	4. Does not clot plasma of guinea pigs and some other species because they lack CRF.
5. Only one type of clumping factor has been identified.	5. Eight antigenic types (A-H) have been described. Most human strains form coagulase type A.
6. Detected by slide test.	6. Detected by tube test.

Table 24.2: Differences between bound and free coagulase

Epidemiology

Staphylococci are ubiquitous. *Staphylococcus* is a normal component of man's indigenous microflora and is carried asymptomatically in a number of body sites. The primary reservoir for staphylococci is the nares, with colonization also occurring in the axillae, vagina, pharynx, and other skin surfaces. About 10-30 percent healthy persons carry staphylococci in the nose and about 10 percent in the perineum and also on the hair. Vaginal carriage is about 5-10 percent, which rises greatly during menses, a factor relevant in the pathogenesis of TSS related to menstruation. During the first day or two of life, most babies become colonized in the nose and skin by staphylococci acquired from their mother, nurse or environment. The source of staphylococcal infection is a patient or a member of the hospital staff with a staphylococcal lesion. Staphylococcal disease may follow endogenous or exogenous infection. Cross-infection is an important method of spread of staphylococcal disease, particularly in hospitals. Staphylococci are a common cause of postoperative wound infection and other hospital infections. Most of these are due to certain strains of staphylococci the so-called 'hospital strains' that are present in the hospital environment. They belong to a limited number of phage types and are commonly resistant to penicillin and other antibiotics routinely used in hospitals. Epidemics of hospital cross infections are caused by some of them, the 'epidemic strains'. Phage type **80/81** was the first to be recognized for most of staphylococcal infections in hospitals throughout the world. Epidemics caused by antibiotic-resistant strains of the 80/81 complex during 1950s provided a betters understanding of the origin and epidemiology of staphylococese disease.

Staphylococcal Diseases

Staphylococcal infections are among the most common of bacterial infections and range from the trivial to the fatal. Staphylococcal infections are characteristically localized pyogenic lesions, in contrast to the spreading nature of streptococcal infections. *S. aureus* causes disease through the direct invasion and destruction of tissue or through the production of toxin.

A. Cutaneous Infections

These include: wound and burn infection, pustules (small cutaneous abscesses), furuncles or boils (large cutaneous abscesses), carbuncles, styes, impetigo and pemphigus neonatorum.

B. Deep Infections

These include: osteomyelitis, periostitis, tonsillitis, pharyngitis, sinusitis, bronchopneumonia, empyema, septicemia, meningitis, endocarditis, breast abscess, renal abscess and abscesses in other organs.

C. Toxin-Mediated Diseases

i. Food Poisoning

Staphylococcal food poisoning (nausea, vomiting and diarrhea) may follow 2-6 hours after the ingestion of food in which *S. aureus* has multiplied and formed enterotoxin. The types of food usually responsible are meat, fish, milk and milk products. The illness is usually self limited, with recovery in a day or so. The illness is rarely fatal.

ii. Toxic Shock Syndrome (TSS)

Toxin-producing strains of *S. aureus* have been implicated in most cases of TSS, a multisystem disease that primarily afflicts young women. Most cases occur in menstruating women who use tampons. However, nonmenstruating women, children, and men with boils or staphylococcal infections of wounds can also have TSS. The disease is initiated with the localized growth of toxin-producing strains of *S. aureus* in the vagina or a wound, followed by release of the toxin into the blood stream. Though tampon-related TSS is now rare, the syndrome occurs in other infections of the skin, mucosa and other sites and also in some surgical wounds.

iii. Exfoliative Diseases

These lesions are produced by the strains of *S. aureus* which produce epidermolytic toxins. This toxin is responsible for the **'staphylococcal scalded skin syndrome' (SSSS),** exfoliative skin diseases in which the outer layer of epidermis gets separated from the underlying tissues. SSSS is seen mostly in young children and only rarely in older children and adults. The severe form of SSSS is known as **Ritter's disease** in the newborn and **toxic epidermal necrolysis** in older patients. Milder forms are **pemphigus neonatorum** and **bullous impetigo**. Bullous impetigo is a localized form of SSSS.

Bacteriophage Typing

Staphylococci may be typed, based on their susceptibility to bacteriophages. An internationally recognized set of 23 standard typing phages is used for epidemiological studies and tracing the source of infection (Table 24.3). Staphylococci can seldom be characterized by lysis by a single phage, but many different patterns of lysis are obtained with a set of phages. The strength of the method is more to demonstrate differences between strains than to confirm relatedness.

Staphylococcal phage typing is done by a pattern method. The strain to be typed is inoculated on a plate of nutrient agar to form a lawn culture. After drying, the phages are applied over marked squares in a fixed dose (routine test dose). After overnight incubation, the culture will be observed to be lysed by some phages but not by others (Fig. 24.3).

Virulent phages cause lysis of staphylococci that they can infect, and thus produce a clearing in the lawn of growth. The phage type of a strain is expressed by designation of the phages that lyse it, and there is international agreement on the interpretation of results. Thus, if a strain is lysed only by phages 3C, 55 and 71, it is called phage type 3C/55/71.

The reference center for staphylococcal phage typing in India is located in the Department of Microbiology, Maulana Azad Medical College, New Delhi.

Laboratory Diagnosis

1. Specimens

The specimens to be collected depend on the type of lesion, for example:

Pus from suppurative lesions; sputum from respiratory infections; food remains and vomit from cases of food poisoning; nasal and perineal swabs from suspected carriers.

Swabs of the perineum, pieces of hair and umbilical stump may be necessary in special situations.

Table 24.3: International basic set of phages for typing *Staph. aureus* of human origin

Lytic group of phages	*Designation of phages*				
I	29	52	52A	79	80
II	3A	3C	55	71	95
III	6	42E	47	53	54
	75	77	83A	84	85
V	94	96			
Unclassified	81				

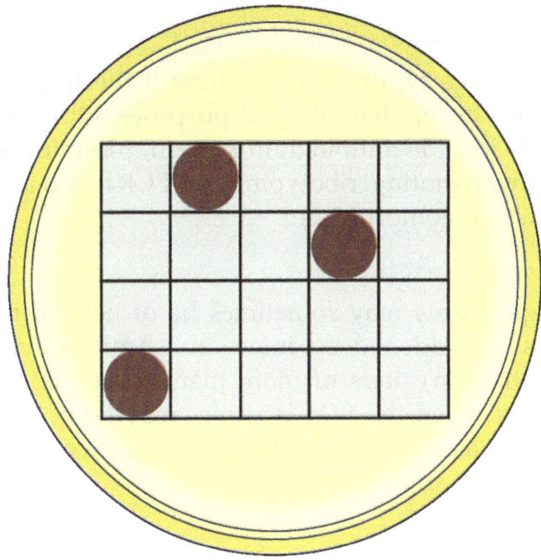

Fig. 24.3: Bacteriophage typing of staphylococci

2. Direct Microscopy

Direct microscopy with Gram stained smears is useful in the case of pus, where cocci in clusters may be seen. This is of no value for specimens like sputum where mixed bacterial flora are normally present.

3. Culture

The specimens are cultured on a **blood agar plate.** Staphylococcal colonies appear after overnight incubation. Specimens, where staphylococci are expected to be outnumbered by other bacteria (e.g. wound swab and feces), are inoculated on selective media like **Ludlam's or salt-milk agar or Robertson's cooked meat medium** containing 10 percent sodium. The inoculated media are incubated at 37°C for 18-24 hours. On blood agar plate, look for hemolysis around the colonies. The plates are inspected for golden-yellow or white colonies. Smears are examined from the culture and coagulase test done when staphylococci are isolated.

4. Identification

Relatively simple biochemical tests (e.g. positive reactions for coagulase [clumping factor], heat-stable nuclease, alkaline phosphatase, and mannitol fermentation) can be used to differentiate *S. aureus* and the other staphylococci.

5. Coagulase Test

Coagulase test is done by two methods—slide and tube coagulase test.

6. Antibiotic Sensitivity Tests

As a guide to treatment, antibiotic sensitivity tests should be performed appropriate to the clinical situation. This is important as staphylococci readily develop resistance to drugs.

7. Bacteriophage Typing

Bacteriophage typing may be done if the information is desired for epidemiological purposes. Other typing methods include antibiogram pattern, plasmid profile, DNA fingerprinting, ribotyping and PCR based analysis for genetic pleomorphism.

8. Serological Tests

Serological tests may sometimes be of help in the diagnosis of hidden deep infections. Antistaphylolysin (antialphalysin) titers of more than two units per ml, especially when the titer is rising, may be of value in the diagnosis of deep seated infections such as bone abscesses.

Treatment

Benzyl penicillin is the most effective antibiotic, if the strain is sensitive. Patients allergic to penicillins may be given erythromycin, vancomycin or first-generation cephalosporins. Cloxacillin, oxacillin, flucloxacillin and methicillin are penicillinase-resistant penicillins.

Methicillin-resistant strains (MRSA) are also resistant to other penicillins and cephalosporins. Glycopeptides (vancomycin or teicoplanin) are the agents of choice in the treatment of systemic infection, but these agents are expensive and may be toxic.

For cutaneous infections, oral therapy with a semisynthetic penicillin, such as cloxacillin or dicloxacillin, is usually efficacious. For **mild superficial lesions**, systemic antibiotics may not be necessary. Topical applications of drugs not used systemically, as bacitracin, chlorhexidine or mupirocin may be sufficient.

The treatment of carriers is by local application of suitable antibiotics such as bacitracin and antiseptics such as chlorhexidine. In resistant cases posing major problems, rifampicin along with another oral antibiotic may be effective in long-term suppression or elimination of the carrier state. Clindamycin is useful in cases of osteomyelitis.

Control

1. The focus of infection (e.g. abscess) must be identified and drained.
2. Treatment is symptomatic for patients with food poisoning.
3. Proper cleansing of wounds and use of disinfectant.
4. Thorough hand washing and covering of exposed skin helps medical personnel prevent infection or spread to other patients.
5. Asymptomatic nasopharyngeal carriage—the use of chemoprophylaxis consisting of vancomycin and rifampin to prevent spread of oxacillin-resistant organisms

OTHER COAGULASE-POSITIVE STAPHYLOCOCCI

Other staphylocoagulase producing (coagulase positive) staphylococci are *S. intermedius, S. delphini, S. lutrae*, and some strains of *S. hyicus*. These are often animal-associated species and are infrequently isolated from human samples.

Coagulase-negative Staphylococci

Coagulase-negative staphylococci (CNS) comprise a large group of related species, which are commonly found on the surface of healthy persons in whom they are rarely the cause of infection. They are opportunistic pathogens that cause infection in debilitated or compromised patients. *Staph. epidermidis* accounts for about 75 percent of all clinical isolates, probably reflecting its preponderance in the normal skin flora. Other species include *Staph. haemolyticus, Staph. hominis, Staph. capitis* and *Staph. saprophyticus.*

Staphylococcus epidermidis

Staph. epidermidis is invariably present on normal human skin. It is nonpathogenic ordinarily but can cause disease when the host defences are breached. *S. epidermidis* has a distinct predilection for foreign bodies, such as artificial heart valves, indwelling intravascular catheters, central nervous system shunts, and hip prostheses.Many strains of *S. epidermidis* are capable of producing large amounts of polysaccharide glycocalyx known as slime. Their etiological role is proved by repeated isolation.

Clinical Infection

1. Stitch abscesses—it is a common cause.
2. Endocarditis of native and prosthetic valves
3. Intravenous catheter infections
4. CSF shunt infections
5. Peritoneal dialysis, catheter-associated peritonitis
6. Bacteremia
7. Osteomyelitis
8. Wound infections
9. Vascular graft infections
10. Prosthetic joint infections
11. Mediastinitis
12. Urinary tract infections, especially in elderly hospitalized men
13. Natural valve endocarditis in intravenous drug abusers occasionally.

Staphylococcus saprophyticus

S. saprophyticus occurs on the normal skin and in the periurethral and urethral flora. It is a common cause of urinary tract infections in sexually active young women. The infection is symptomatic and may involve the upper urinary tract also. It may also cause urethritis in men and women, catheter-associated urinary tract infections, prostatitis in elderly men, and rarely bacteremia, sepsis and endocarditis.

This coagulase-negative staphylococcus can be distinguished from *S. epidermidis* by its resistance to novobiocin and by its failure to ferment glucose anaerobically. It is nonhemolytic and does not contain protein A. Table 24.4 lists the features useful for distinguishing the major species of staphylococci.

Characteristic	S. aureus	S. epidermis	S. saprophyticus
Anaerobic growth and fermentation of glucose	+	+	–
Mannitol			
Acid aerobically	+	V	V
Acid anaerobically	+	–	–
Coagulase	+	–	–
DNAase	+	–	–
Phosphatase	+	–/weak+	–
α-Toxin	+	–	–
Protein A in cell wall	+	–	–
Novobiosin sensitivity	Sensitive	Sensitive	Resistant

Table 24.4: Characteristics distinguishing three species of the genus *Staphylococcus*

Other Coagulase-negative Staphylococci

Other species of CoNS are found as normal flora in humans and animals. *S. haemolyticus* has been reported in wounds, bacteremia, endocarditis, and UTIs. Other species include *S. lugdunensis, S. warneri, S. capitis, S. simulans,* and *S. schleiferi.* A wide range of infections have been associated with these organisms, for example, endocarditis, septicemia, and wound infections.

Sensitivity to Antibiotics

Before the introduction of penicillin, most of the strains of *S. aureus* were sensitive to this antibiotic but from 1945 onwards penicillinase-producing (penicillin-resistant) strains were encountered. Staphylococci quickly developed drug resistance after penicillin was introduced, and today less than 10 percent of the strains are susceptible to this antibiotic. Similarly, they have also developed resistance against sulfonamides and other antibiotics. Penicillin resistance is of three types:

1. Production of Beta Lactamase (Penicillinase)

Production of beta lactamase (penicillinase) which inactivates penicillin by splitting the beta-lactam ring. Staphylococci produce four types of penicillinases, A to D. Hospital strains usually form type A penicillinase. Penicillinase is an inducible enzyme and its production is usually controlled by plasmids which are transmitted by' transduction or conjugation.

Penicillinase plasmid may also carry markers for resistance to heavy metals such as mercury, arsenic, cadmium, lead and bismuth and other antibiotics such as erythromycin, and fusidic acid. Penicillinase plasmids are transmitted to the sensitive staphylococci by transduction and also possibly by conjugation.

2. Changes in Bacterial Surface Receptors

Changes in bacterial surface receptors, reducing binding of beta-lactam antibiotics to cells. This change is normally chromosomal in nature and is expressed more at 30°C than at 37°C. This resistance also extends to cover beta lactamase resistant penicillins such as methicillin and cloxacillins. Some of these strains may show resistance to other antibiotics and heavy metals also and cause outbreaks of hospital infection. These strains have been called 'epidemic methicillin resistant *Staphylococcus aureus'* or EMRM (As methicillin is an unstable drug, cloxacillin is used for sensitivity testing instead).

Resistace to other antibioticsis achieved by a number of different mechanism depending on the class of antibiotic; these include membrane impermeability, alteration of the target site, and enzymatic degradation of the antibiotic.

3. Development of Tolerance

Development of tolerance to penicillin, by which the bacterium is only inhibited but not killed. Staphylococci also exhibit plasmid-borne resistance to erythromycins, tetracyclines, aminoglycosides and almost all clinically useful antibiotics except vancomycin. So far, all *S. aureus* remain sensitive to the glycopeptide antibiotics vancomycin and teichoplanin which are the mainstay of treatment of serious MRS infections.

Some strains of *S. aureus,* particularly those of phage type 11, carry plasmids which code for the production of bacteriocin (staphylococcin). It is thermostable and active against other gram-positive bacteria including staphylococci, streptococci, pneumococci, corynebacteria and aerobic spore-forming bacilli. It is ineffective against gram-negative bacteria.

Methicillin-resistant Staphylococci (MRSA)

Methicillin was the first compound developed to combat resistance due to penicillinase (beta lactamase) production by staphylococci. Due to the limitations in clinical use of methicillin, cloxacillins are used instead against penicillinase-producing strains. But strains of methicillin resistant *Staph. aureus* (MRSA) became common, which were resistant not merely to penicillin, but also to all other beta lactam antibiotics and many others besides. Isolates that are resistant have been traditionally termed **methicillin-resistant staphylococci**, with *S. aureus* being called MRSA and *S. epidermidis* referred to as **MRSE.** When any staphylococcus isolated is identified as being resistant to methicillin, this implies that it is also resistant to nafcillin and oxacillin and to all β-lactam antibiotics, including the cephalosporins. MRSA is also becoming more common in the community, especially in long-stay institutions.

Treatment: Glycopeptides (vancomycin or teicoplanin) are the agents of choice in the treatment of systemic infection, but these agents are expensive and may be toxic. Concerns with rising resistance to glycopeptides call for the restrictive use of these drugs.

Laboratory Diagnosis

1. For laboratory purposes, oxacillin is generally used for detection of methicillin resistance. The use of an oxacillin-salt agar plate, such as the oxacillin resistance screening agar can be used as a screening test for MRSA in clinical samples. A high salt concentration (5.5% NaCl) and polymyxin B make the medium selective for staphylococci.
2. The gold standard for MRSA detection is the detection of the ***mecA* gene** by using **nucleic acid probes or polymerase chain reaction (PCR) amplification.**

Control of MRSA

Control of MRSA requires strict adherence to infection-control practices such as barrier.

MICROCOCCI

Micrococci are catalase positive, gram-positive, coagulase-negative and usually oxidase-positive. They may occasionally colonize the skin or mucous membrane of humans, but they are only rarely associated with infections.

Only two species, *Micrococcus luteus* and *Micrococcus lylae,* remain in the genus. Micrococci, especially *M. luteus,* have a tendency to produce a yellow pigmented colony. Nine species of genus *Micrococcus* have been described.

Table 24.5 gives some differentiating features of Staphylococcus and Micrococcus.

STOMATOCOCCUS

Stomatococcus mucilaginosus, the only species in this genus, is a commensal organism that resides in the oropharynx and upper respiratory tract. The organism resembles staphylococci. However, a prominent mucoid capsule is present, which allows the organism to adhere to the body surfaces and foreign material (e.g. catheters, shunts, prosthetic valves and joints). In recent years, this organism has been reported to be the cause of an increasing number of opportunistic infections (endocarditis, septicemia, and catheter-related infections) in immunocompromised patients.

ALLOIOCOCCUS

Alloiococcus otitidis is the only species in this genus. It is an aerobic, gram-positive coccus that has been implicated in chronic middle ear infections in children. Because this organism grows slowly, its role in disease may be underappreciated.

KNOW MORE

Clumping Factor (Bound Coagulase)

Almost all the clumping factor producing strains of S. *aureus* produce coagulase, whereas 12 percent of coagulase producing strains may not produce clumping factor. This factor is also not detectable in capsulated strains of *S. aureus* presumably because the clumping factor is covered by extracellular polysaccharide.

Coagulase-negative staphylococci are morphologically similar to *Staph. aureus,* and the methods for isolation are the same. Colonies are usually nonpigmented (white), and they can be distinguished from *Staph. aureus* by their failure to coagulate plasma and by their lack of clumping factor and deoxyribonuclease. Because *Staph. epidermidis,* which accounts for most isolates, may contaminate clinical specimens, care has to be exercised in assessing its significance, especially from superficial sites.

☞ KEY POINTS

Staphylococcus

- *Staphylococcus* is gram-positive cocci arranged in clusters.
- Species of staphylococci are classified by the coagulase test into two groups: the **coagulase-positive** (*Staphylococcus aureus)* and **coagulase-negative** staphylococci (CNS). *S. epidermidis* and *S. saprophyticus* are the most clinically significant species in this group.
- They can grow readily on ordinary media.
- *Coagulase test:* The slide or tube coagulase test is performed to distinguish *Staph. aureus* from coagulase-negative species.

Staphylococcus aureus

- *Staph. aureus* strains usually exhibit following characteristics: (1) Beta hemolysis; (2) Golden yellow pigment; (3) Coagulase positive; (4) Greater biochemical activity, ferment mannite; (5) Liquefy gelatin; (6) Produce phosphatase; (7) Black colonies on potassium tellurite blood.
- *S. aureus* produces many virulence factors including: (1) Cytolytic or membrane-damaging toxins (alpha, beta, delta, gamma, and Panton-Valentine [P-V] leukocidin); (2) Exfoliative toxins: (3) Enterotoxins (A-E, G-I); (4) Toxic shock syndrome toxin-1 (TSST-1).

	Property	Staphylococcus	Micrococcus
1.	Gram staining	Gram-positive. Grape-like clusters. Uniform staining.	Gram-positive, darkly stained. In groups of four (tetrad) or eight. Often staining is not uniform
2.	Colony characters	Colonies are golden yellow. Size 1 mm.	Colonies are white in color generally. Size larger than staphylococcus.
3.	Anaerobic acid production from glucose	+	–
4.	Aerobic acid production from glycerol in the presence of erythromycin	+	–
5.	Modified Oxidase	–	+
6.	Bacitracin sensitivity (0.04 unit disc)	Resistant	Sensitive
7.	Lysostaphin sensitivity	Sensitive	Resistant
8.	Furazolidone susceptibility (100 µg) of furazolidone disc	Sensitive	Resistant

Table 24.5: Differentiation between staphylococci and micrococci

- *Diseases: S. aureus* causes cutaneous infections such as folliculitis, boils, carbuncles, impetigo, and purulent abscesses. These cutaneous infections can progress to **deeper abscesses** involving other organ systems and progress to septicemia and bacteremia. Toxin-induced diseases, such as food poisoning, scalded skin syndrome (SSS), and toxic shock syndrome (TSS), are also associated with this organism.

  Other systemic diseases (frequently associated with bacteremia) include pneumonia, empyema, septic arthritis, osteomyelitis, acute endocarditis, and catheter-related bacteremia.
- *Diagnosis:* It is done by microscopy, culture, antibiotic sensitivity tests and serological tests.
- Bacteriophage typing may be done if the information is desired for epidemiological purposes.
- *Treatment:* The antibiotics of choice are oxacillin (or other penicillinase-resistant penicillin) or vancomycin for oxacillin-resistant strains.
- Methicillin-resistant strains (MRSA) are also resistant to other penicillins and cephalosporins. Glycopeptides (vancomycin or teicoplanin) are the agents of choice in the treatment of systemic infection, but these agents are expensive and may be toxic.
- *Coagulase-Negative Staphylococci (CONS):* Coagulase-negative staphylococci are opportunistic pathogens that cause infection in debilitated or compromised patients. *Staph. epidermidis* accounts for about 75 percent of all clinical isolates. Other species include *Staph. haemolyticus, Staph. hominis, Staph. capitis* and *Staph. saprophyticus*.
- *Staph. saprophyticus* can be distinguished from *S. epidermidis* by its resistance to novobiocin and by its failure to ferment glucose anaerobically.
- Micrococci may occasionally colonize the skin or mucous membrane of humans, but they are only rarely associated with infections.

IMPORTANT QUESTIONS

1. Describe the morphology, cultural characteristics and antigenic structure of *Staphylococcus aureus*.
2. Name various virulence factors of *Staphylococcus aureus*.
3. Classify staphylococci. Discuss pathogenicity and laboratory diagnosis of *Staph. aureus*.
4. Write short notes on:
   - Coagulase or staphylocoagulase.
   - Clumping factor.
   - Toxins and enzymes produced by *Staphylococcus aureus*.
   - Staphylococcal food poisoning.
   - Toxic shock syndrome.
   - Epidermolytic toxins of *Staphylococcus aureus*.
   - Drug resistance in staphylococci.
   - Methicillin-resistant staphylococci.
   - Coagulase-negative staphylococci.
   - Micrococci.

FURTHER READING

Cafferkey MT. Methicillin-resistant Staphylococcus aureus. Clinical management and laboratory aspects. Marcel Dekker, New York 1992.

Dinges MM, et al. Exotoxins of Staphylococcus aureus, Clin-Microbiol Rev 2000;13:16-34.

Kloos WE, Bannerman TL. Update on clinical significance of coagulase-negative staphylococci. Clin Microbiol Rev 1994; 7:117-40.

Lowy FO. Staphylococcus aureus infections. N Engl J Med 1998;339:520.

Mulligan ME, et al. Methicillin-resistant Staphylococcus aureus: a consensus review of the microbiology, pathogenesis, and' epidemiology with implications for prevention and management, Am J Med 1993;94:313-323.

Rupp ME, Archer GL. Coagulase-negative staphylococci: pathogens associated with medical progress. Clin Infect Dis 1994;19:231-25.

Waldvogel FA. New resistance in Staphylococcus aureus, New Engl J Med 1999;340:556-7.

CHAPTER 25

Streptococcus and Enterococcus

Learning Objectives

After reading and studying this chapter, you should be able to:

- Classify streptococci.
- Describe antigenic structure of *Str. pyogenes.*
- List and describe toxins and enzymes of *Streptococcus pyogenes.*
- Discuss pathogenicity of streptococci.
- List and describe toxins and enzymes of *Streptococcus pyogenes.*
- Describe nonsuppurative complications of *Str. pyogenes* infections.
- Discuss laboratory diagnosis of streptococcal infections.
- Discuss group B streptococci, group D streptococci and viridans group.

INTRODUCTION

The genus *streptococcus* comprises a large and biologically diverse group of gram-positive cocci that grow in pairs or chains (Fig. 25.1). They are normal flora of humans and animals. They inhabit various sites, notably the upper respiratory tract, and live harmlessly as commensals. A number of species within the genus cause major human diseases. The most important of them is *Streptococcus pyogenes* causing pyogenic infections, with a characteristic tendency to spread, as opposed to staphylococcal lesions which are typically localized. In addition, it is responsible for the nonsuppurative lesions, acute rheumatic fever and glomerulonephritis.

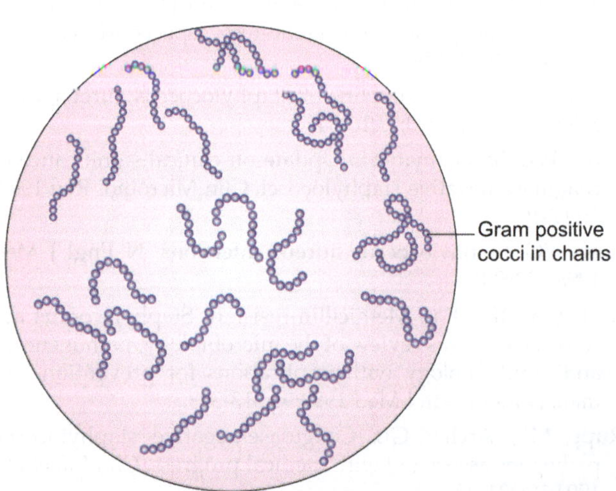

Fig. 25.1: Streptococci

Streptococci were first described by Billroth (1874) in exudates from erysipelas and wound infections, who called them streptococci (*streptos*, meaning twisted or coiled; *coccus*, a grain or berry). Pasteur (1879) found similar organisms in the blood of a patient with puerperal sepsis. Ogston (1881) isolated them in acute abscesses, distinguished them from staphylococci and by animal inoculation established their pathogenicity. Rosenbach (1884) gave the name streptococcuss pigeons to coeir isolated from human suppurative lesions.

CLASSIFICATION (FIG. 25.2)

Streptococci are first divided into obligate anaerobe and facultative anaerobes. Obligate anaerobe are designated as peptostreptococci.

Three different schemes are used to classify the organism as follows:

1. *Hemolytic patterns:* Complete (β-hemolysis), incom plete (α-hemolysis), and no hemolysis (γ-hemolysis).
2. *Serologic properties:* Lancefield groupings A to H, K to M, and O to V.
3. Biochemical (physiologic) properties.

1. Hemolytic Activity

The aerobic and facultative anaerobic streptococci are classified on the basis of their hemolytic properties. The type of hemolytic reaction displayed on blood agar has long been used to classify the streptococci. Brown (1919) categorized them into three varieties based on their growth in 5 percent horse blood agar pour plate culture.

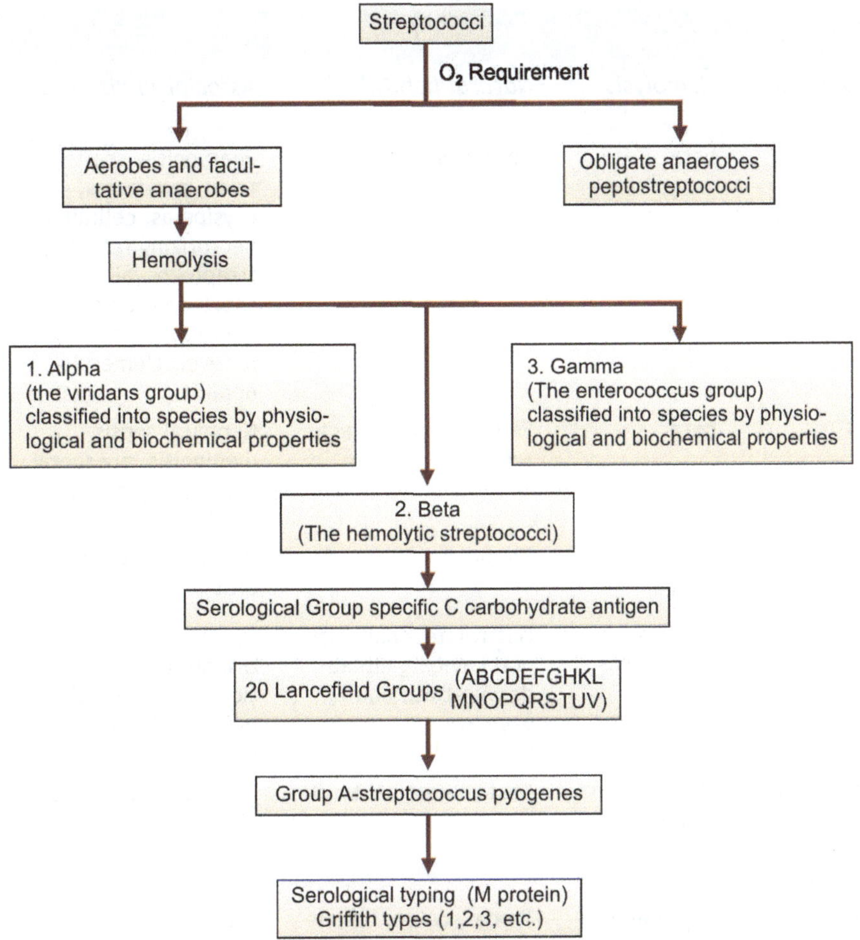

Fig. 25.2: Classification of streptococci

a. Alpha (α)-hemolytic Streptococci

They produce a zone of partial hemolysis with a greenish discoloration around the colonies on blood agar. The zone of lysis is small (1-2 mm wide) with indefinite margin, within this zone unlysed erythrocytes can be made out microscopically. The streptococci producing α-hemolysis are also known as viridans streptococci (from 'viridis' meaning green).

b. Beta (β)-hemolytic Streptococci

They produce a sharply defined, clear, colorless zone of hemolysis (2-4 mm wide) around the colony, caused by complete lysis of red blood cells in the agar medium induced by bacterial hemolysins. No red blood cell is visible on microscopic examination in clear zone of complete hemolysis.

The term '**hemolytic streptococci**' strictly applies only to beta lytic strains. β-hemolysis constitutes the principal marker for potentially pathogenic streptococci in cultures of throat swabs or other clinical samples.

c. Gamma (γ) or Nonhemolytic Streptococci

They produce no hemolysis on blood agar. Enterococcus faecalis is an important organism of this group.

2. Serological Properties

Lancefield Grouping

The work of Rebecca Lancefield in 1933 laid the groundwork for the serological classification of β-hemolytic streptococci. On the basis of group-specific carbohydrate (C) antigens in the cell wall, β-hemolytic streptococci are divided into 21 serological groups from A to W (without I and J). These are known as **Lancefield groups**. Groups A, B, C, D, and G are most commonly found associated with human infections.

Griffith Typing

Hemolytic streptococci of group A are known as *Str. pyogenes*. These are further divided into *types* based on the protein (M, T and R) antigens present on the cell surface (**Griffith typing**). About eighty types of *Str. pyogenes* have been recognized so far (types 1, 2, 3 and so on).

3. Biochemical (Physiologic) Properties

Biochemical and other criteria are also used in defining various species within a single serogroup, and some species contain strains of more than one serogroup.

Table 25.1 shows characteristics and clinical significance of important streptococci and enterococci.

Table 25.1: Characteristics and clinical significance of important streptococci and enterococci

Species	Lancefield group	Hemolysis	Natural habitat	Associated diseases	Laboratory tests
Str. pyogenes	A	beta	Throat, skin	Pharyngitis, scarlet fever, pyoderma, erysipelas, cellulitis, necrotizing fasciitis, streptococcal toxic shock syndrome, bacteremia, rheumatic fever, glomerulonephritis	Bacitracin sensitive; PYR test positive; Ribose not fermented
Str. agalactiae	B	beta	Female genital tract, rectum	Neonatal sepsis, meningitis, puerperal fever, pyogenic infections	Hippurate hydrolysis, CAMP test
S. equisimilis	C	beta	Throat	Pharyngitis, endocarditis	Ribose and trehalose fermentation
Enterococcus sp. *(Enterococcus faecalis* and other enterococci)	Group D	variable hemolysis	Gastrointestinal tract, oral cavity, gallbladder, urethra, and vagina	Urinary tract infections, endocarditis, bacteremia, abdominal infections	Growth in 6.5% NaCl; PYR positive
Nonenterococcal group D species (*Streptococcus bovis*)	Group D	alpha-hemolytic or nonhemolytic	Gastrointestinal tract	Neonatal meningitis	No growth in 6.5% NaCl
Str. anginosus group	A, C, F, G, untypable	Beta (alpha, gamma)	Throat, colon, female genital tract	Pyogenic infections	Group A strains, bacitracin resistant, PYR negative colony variants of other groups
Viridans streptococci (*Str. mitis, Str. mutans, Str. salivarius*, and many other species)	Not typed	Alpha (gamma)	Mouth, throat, colon, female genital tract	Dental caries; endocarditis	Optochin resistant, species classification on biochemical properties

STREPTOCOCCUS PYOGENES

Morphology

S. pyogenes are gram-positive, spherical to ovoid organisms 0.5 to 1.0 µm in diameter. The organism grows in short or moderately long chains, the chain length being dependent on the strain and culture medium. Chain formation is due to the cocci dividing in one plane only and the daughter cells failing to separate completely. There is often an appearance of pairing within the chains. Chains are longer in liquid media than in solid media.

Streptococci are nonmotile and nonsporing. Some strains of *S. pyogenes* and some group C strains produce a capsule of hyaluronic acid, while polysaccharide capsules are encountered in members of group B and D which may be demonstrable in very young cultures.

Cultural Characteristics

They are aerobe and facultative anaerobes, growing best at a temperature of 37ºC (range 22-42ºC). The optimal pH for growth is 7.4 to 7.6. It is exacting in nutritive requirements, growth occurring only in media containing fermentable carbohydrates or enriched with blood or serum.

On **blood agar**, *S. pyogenes* colonies are small (0.5-1 mm in diameter), circular, semitransparent, low convex disks surrounded by a wide zone of β-hemolysis, several times greater than the diameter of the colony after incubation for 24 hours. β-hemolysis serves as the marker for primary isolation. An enhancement of growth and hemolysis are promoted by 10 percent CO_2. The matt (finely granular) colonies contain M antigen, which is virulent strain, while avirulent strains form glossy colonies. **Mucoid colonies** may occur when a strain is heavily capsulate. Very rarely, nonhemolytic group A streptococci are encountered, which are typical of *S. pyogenes* in other respects.

Crystal violet blood agar and **PNF medium** (blood agar containing polymyxin B, neomycin and fusidic acid) are selective for beta hemolytic streptococci. **Pike's medium** is a **transport medium** for clinical specimens containing group A streptococci. Pike's medium is prepared by adding crystal violet (1 in 1,000,000) and sodium azide (1 in 16,000) to blood agar.

In **liquid media**, such as glucose or serum broth, growth occurs as a granular turbidity with a powdery deposit. No pellicle is formed.

Biochemical Reactions

i. *Str. pyogenes* is **catalase negative**.
ii. Insoluble in 10 percent bile unlike *S. pneumoniae.*
iii. It ferments several sugars producing acid and no gas. Fermentation of the sugars has been the basis of a physiological classification of the genus but in the case of hemolytic streptococci this has been replaced by serological grouping.
iv. Hydrolysis of pyrrolidonyl naphthylamide (PYR test) is positive and failure to ferment ribose distinguishes it from nongroup A hemolytic streptococci.

Resistance

Strep. pyogenes is a delicate organism, can be killed by heating at 54°C for 30 minutes. It can, however, survive in dust for several weeks, if protected from sunlight. Laboratory cultures should be stored at 3-5°C in blood broth or cooked-meat medium, or else freeze dried. It is sensitive to most antiseptics and is rapidly inactivated. It is more resistant to crystal violet than many other bacteria including *Staphylococcus aureus* hence it is used for preparation of selective media. Crystal violet (1 mg/L), nalidixic acid (15 mg/L) and colistin sulphate (10 mg/L) added to blood agar provide a good selective medium for the isolation of streptococci, including pneumococci. It is sensitive to benzylpenicillin and a wide range of antimicrobial drugs. It is susceptible to sulphonamides, but unlike *Staph. aureus* does not develop resistance to drugs. It is sensitive to bacitracin and this property is employed as a convenient method for differentiating *Str. pyogenes* from other hemolytic streptococci.

Antigenic Structure

Cell wall of *Strept. pyogenes* (Fig.) consists of:

1. Inner layer of peptidoglycan.
2. Middle layer of group specific carbohydrate.
3. Outer layer of protein and lipoteichoic acid, containing protein antigens (M, T, R).
4. The capsule when present is composed of hyaluronic acid.

1. Inner Layer of Peptidoglycan

Peptidoglycan (mucoprotein) is responsible for cell wall rigidity. It also has biological properties such as pyrogenic and thrombolytic activity.

2. Group-specific Polysaccharide Antigen

Serologic classification of β-hemolytic streptococci is based on their cell wall polysaccharide antigen. As this antigen is integral part of the cell wall, it has to be extracted for grouping by a precipitation test with group antisera.

Techniques for the extraction of the group antigens: A variety of techniques have been used for the extraction of the group antigens:

i. Lancefield's acid extraction method: For the test, streptococci are grown in Todd-Hewitt broth and extracted with hydrochloric acid (Lancefield's acid extraction method) or
ii. Formamide (Fuller's method);
iii. By enzyme produced by *Streptococcus albus* (Maxted's method);
iv. By autoclaving (Rantz and Randall's method).

Test Performance

The extracts thus obtained are then tested against group-specific antisera by capillary tube precipitation reactions. The extract and the specific antisera are allowed to react in capillary tubes. A white disk of precipitation may be seen within five minutes at the interface between extract and the homologous antiserum. Tests may be performed in the narrowing neck of small Pasteur pipette. A variety of agglutination techniques using extracts or whole cells may also be used.

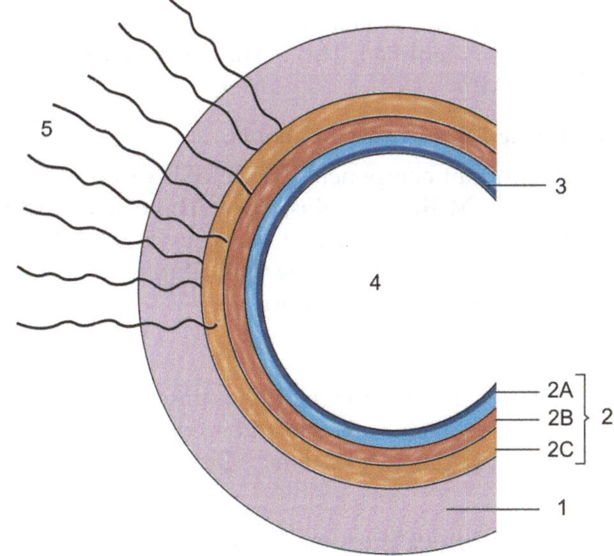

Fig. 25.3: Antigenic structure of *Str. pyogenes*: 1. Hyaluronic acid capsule; 2. Cell wall comprising; 2A, peptidoglycan; 2B. Group specific carbohydrate, and 2C. protein lipoteichoic acid fimbria; 3. Cytoplasmic membrane; 4. Cytoplasm; 5. Pili covered with lipoteichoic acid

3. Outer Layer

a. Type-specific Antigens

Several protein antigens have been identified in the outer layer of the cell wall. *Streptococcus pyogenes* can be typed, based on the surface proteins M, T and R.

i. M Protein

The M protein is the most important of these three and is a major type-specific protein associated with virulent streptococci. It acts as virulence factor by inhibiting phagocytosis. It is antigenic and specific anti-M antibody develops after infection, which enhances phagocytosis of the coccus and is protective. M protein is **resistant to heat and acid** but **susceptible to trypsin**. It can be extracted by the Lancefield acid extraction method and M typing is performed by capillary tube precipitation test using type-specific antisera and acid extract. About 100 M protein types have been recognized.

ii. T Protein

T (**Trypsin-resistant**) protein is heat and acid labile but resistant to trypsin present in many serotypes of *S. pyogenes*. It may be specific but many different M types possess the same antigen. T typing is performed by a slide agglutination test that uses trypsin-treated whole streptococci. T protein has no relation to virulence.

iii. R Protein

Some strains of *S. Pyogenes* (types 2, 3, 28 and 48) and some strains of groups B, C and G contain third antigen R protein. R protein has no relation to virulence.

iv. M Associated Protein (MAP)

A nontype-specific protein, associated with the M protein, has been identified. This is known as M associated protein (MAP).

v. Other Cell Surface Components

Other important components in the cell wall of *S. pyogenes* include **M-like proteins, lipoteichoic acid**, and **F protein**. Lipoteichoic acid, and F protein facilitate binding of host cells by complexing with fibronectin, which is present on the host cell surface.

4. Capsule

The outermost layer of the cell wall is the capsule, which is composed of hyaluronic acid and is not immunogenic. It has an antiphagocytic effect like other bacterial capsule.

Cross Reaction of Human Tissues with *S. pyogenes* Components

Various structural components of *S. pyogenes* exhibit antigenic cross reaction with different tissues of the human body (Table 25.2). It has been postulated that these antigenic cross reactions may account for some of the manifestations of rheumatic fever and other streptococcal diseases, the tissue damage being of an immunological nature.

Table 25.2: Structural components of *Streptococcus pyogenes* which cross-react with human tissues

Structural components of S. pyogenes	*Human tissue with which it cross-reacts*
Capsular hyaluronic acid	Synovial fluid
Cell wall protein	Myocardium
Cell wall carbohydrate	Cardiac valves
Cytoplasmic membrane	Vascular intima
Peptidoglycan	Skin antigens

Virulence Factors of *Str. pyogenes*

Str. pyogenes forms several exotoxins and enzymes which contribute to its virulence. Besides these, the M protein also acts as a virulence factor by inhibiting phagocytosis. The C polysaccharide has been shown to have a toxic effect on connective tissue in experimental animals.

Toxins and Enzymes

1. Hemolysins

Two hemolytic and cytolytic toxins—streptolysin O (SLO) and streptolysin S (SLS)—are produced by most strains of group A streptococci and many strains of groups C and G. They are responsible for the clear zones of β-hemolysis around the colonies in blood agar media.

i. Streptolysin O

Streptolysin O is so called because it is oxygen-labile hemolysin. It is inactivated in the oxidized form but may be reactivated by treatment with mild reducing agents. On blood agar, streptolysin O activity is seen only in pour plates and in anaerobic cultures and not in surface cultures. It may be obtained in the active state by growing streptococci in broth containing reducing agents such as sodium hydrosulphite. SLO is heat-labile protein, cytolytic and capable of lysing erythrocytes, leukocytes, platelets and cultured cells. It is lethal on intravenous injection into animals and has a specific cardiotoxic activity.

Streptolysin O resembles the oxygen labile hemolysins produced by *Streptococcus pneumoniae, Clostridium tetani, Clostridium perfrigens Bacillus cereus* and *Listeria monocytogenes*.

ASO Test

Streptolysin O is strongly antigenic and anti-streptolysin O (ASO) antibodies appears in sera following streptococcal infection. Estimation of this antibody are useful for documenting recent group A streptococcal infection (ASO test). ASO test is now done by the serological method of latex agglutination due to the complexity of the hemolysis inhibition test. An ASO titer in excess of 200 Todd units/ml is considered significant and suggests either recent or recurrent infection with streptococci.

ii. Streptolysin S (SLS)

Streptolysin S is **oxygen-stable** and causes the hemolysis around the colonies on aerobic blood culture. It is produced in the presence of serum (S indicates serum dependence). It is protein but nonantigenic which may be due to its small size. It is cell-bound hemolysin that can lyse erythrocytes, leukocytes, and platelets.

2. Pyrogenic Exotoxins (Erythrogenic, Dick, Scarlatinal Toxin)

The primary effect of the toxin is induction of fever and so it was renamed **Streptococcal pyrogenic exotoxin (SPE)**. This toxin was originally called '**erythrogenic**' toxins because its intradermal injection into susceptible individuals produced an erythematous reaction (Dick test, 1924). Historically, much attention has been paid to the scarlet fever rash caused by the *Str. pyogenes* strains producing this toxin. Antitoxin injected into the skin of a patient with scarlet fever causes localized blanching as a result of neutralization of erythrogenic toxin **(Schultz-Charlton reaction)**. The Dick test and Schultz Charlton reaction are now only of historical value as scarlet fever is no longer a common or serious disease.

Three immunologically distinct type heat-labile toxins (SPE A, B, and C) have been described in *S. pyogenes*. Type A and C are encoded by lysogenic phages; the gene for B is located on the bacterial chromosome. SPEs act as 'superantigens' (like staphylococcal enterotoxins and TSS toxin) interacting with both macrophages and helper T cells with the release of (1) interleukin-1 (IL-1), IL-2, and IL-6, (2) tumor necrosis factor-α (TNF-α) and TNF-β, and (3) interferon-α. These cytokines cause important effects, including fever, shock organ failure and the rash observed in patients with scarlet fever.

3. Enzymes

1. Deoxyribonucleases (Streptodornase DNase)

These enzymes depolymerize free DNA present in the pus. They are capable of depolymerizing the highly viscous DNA which accummlates in thick pus as a result of disintegration of polymorphonuclear leucocytes. Streptodornase helps to liquefy the thick pus and may be responsible for the thin serous character of streptococcal exudates. This property has been applied therapeutically for breaking down blood clots, thick pus and fibrinous exudates in closed spaces such as joints or pleural cavity.

There are four antigenically distinct nucleases (A, B, C, and D), of which B is the most antigenic in human beings. All strains of *S. pyogenes* produce at least one nuclease, usually the B enzyme. Nucleases A and C have only DNase activity, whereas B and D also possess RNase activity. Antibody titers to DNase B are of great value in the serodiagnosis of pharyngeal or skin infection, especially the latter, where the ASO titer may be low.

2. Streptokinase (Fibrinolysin)

Streptokinase, also known as **fibrinolysin**, is another spreading factor. Two different streptokinases (A and B) are produced by group A streptococci. This acts on **plasminogen**, a factor present in normal plasma, which is converted into plasmin, an active proteolytic enzyme that lyses fibrin. Fibrinolysin appears to play a biological role in streptococcal infections by breaking down the fibrin barrier around the lesions and facilitates the spread of infection.

Plasminogen $\xrightarrow{\text{Streptokinase}}$ plasmin (lyses fibrin)

It is an antigenic protein and neutralizing antibodies appear in convalescent sera. Antistreptokinase antibodies provide retrospective evidence of streptococcal infection. Streptokinase is given intravenously for the treatment of early myocardial infarction to prevent the formation of blood clot and other thromboembolism disorders.

3. Hyaluronidase

Hyaluronidase splits hyaluronic acid, an important component of connective tissue as well as the organism's own capsule. Thus, hyaluronidase aids in spreading infecting microorganisms (spreading factor). This might favor the spread of infection along the intercellular spaces. Hyaluronidases are antigenic and specific for each bacterial or tissue source. Specific antibodies appear in convalescent serum and may thus be of value in serodiagnosis.

4. Proteinase

Many strains of streptococci also produce a proteinase (especially as the environmental pH falls during growth). It destroys several proteins formed by the *Streptococcus* itself, including SLO, streptokinase, hyaluronidase and M protein. It is antigenic and antibodies can be demonstrated in convalescent sera.

5. Serum Opacity Factor (SOF)

Serum opacity factor (SOF) is an enzyme, lipoproteinase. It is called SOF, as it produces opacity when applied to agar gel containing horse or swine serum. It is produced by certain M types of group A streptococci. The exact biological significance is not known but there is strong correlation between the production of this enzyme and particular M types, and it is produced mainly by strains causing skin infections.

6. Nicotinamide Adenine Dinucleotidase (NADase)

This acts on the coenzyme NAD and liberates nicotinamide from the molecule. This is produced by all members of some types of *S. pyogenes* and some strains of groups C and G. The biologic significance of NADase is not known, though it is believed to be leucotoxic.

7. Other Enzymes

Many strains of streptococci also produce ATPase, phosphatase, esterases, amylase, N-acetylglucosaminidase, neuraminidase and other toxins or enzymes. It is not known whether, and to what extent, these contribute to pathogenesis.

Epidemiology

Group A streptococci commonly colonize the oropharynx of healthy children and young adults and there is asymptomatic colonization in upper respiratory tract and transient colonization of skin. The most common route of entry of *S. pyogenes* is the upper respiratory tract. Person-to-person spread by respiratory droplets (pharyngitis) or through breaks in skin after direct contact with infected person, fomite, or arthropod vector. Although the organism is ubiquitous, there are seasonal incidences of specific diseases: pharyngitis and associated rheumatic fever or glomerulonephritis (more common in cold months); pyoderma and associated glomerulonephritis (more common in warm months).

Pharyngitis due to *S. pyogenes* is primarily a disease of childhood between the age of 5 and 15 years, but infants and adults are also susceptible. The incidence of carriage is reported to be 15-20 percent. Symptomless infection is common and helps to maintain the organism in the community. Spread from person-to-person by respiratory droplets or through breaks in skin after direct contact with infected person, fomite, or arthropod vector.

Crowding, such as in classrooms and daycare facilities, increase the opportunity for the organism to spread, particularly during the winter months. No seasonal distribution has been identified in the tropics. Immunity is type specific and appears to be associated with antibody to the M protein. Reinfections occur because of the multiplicity of the serotypes.

Pathogenesis

Acute diseases associated with *Streptococcus pyogenes* occur chiefly in the **respiratory tract, bloodstream**, or the *skin*. Two post streptococcal sequelae (rheumatic fever following respiratory infection and glomerulonephritis following respiratory or skin infection), occur in 1-3 percent of untreated infections.

Suppurative Streptococcal Disease

Str. pyogenes produces pyogenic infections with a tendency to spread locally, along lymphatics and through the bloodstream.

1. Respiratory Infections

The most common route of entry of *S. pyogenes* is the upper respiratory tract.

i. **Sore throat** is the most common of streptococcal disease. It may be localized as **tonsillitis** or may involve the pharynx more diffusely (**pharyngitis**). **Tonsillitis** is more common in older children and adults than in younger children.

   Streptococcal pharyngitis ('strep throat') is usually associated with group A organisms, although sporadic cases and epidemics have been reported with groups C and G. Virulent group A streptococci adhere to the pharyngeal epithelium by means of lipoteichoic acid covering the surface pili. The glycoprotein fibronectin on the epithelial cells probably serve as the lipoteichoic acid ligand.

ii. **Scarlet fever:** Scarlet fever is a complication of streptococcal pharyngitis that occurs when the infecting strain is lysogenized by a temperate bacteriophage that stimulates production of a pyrogenic exotoxin. **Scarlet fever or scarlatina** is a diffuse erythematous rash of the skin and mucous membranes. Scarlet fever has become very uncommon in recent decades due to unknown reasons.

iii. **Suppurative complications:** The infection may spread to the surrounding tissues from the throat which may cause **suppurative complications** of streptococcal pharyngitis such as peritonsillar or retropharyngeal abscess, otitis media, mastoitidis, quinsy, Ludwig's angina (diffuse cellulites of the floor of the mouth), suppurative adenitis and disseminated infections to brain, heart, bone, and joints. It may rarely lead to meningitis. Streptococcal pneumonia may occur as a complication of influenza or other respiratory viral diseases.

2. Skin and Soft Tissue Infections

Str. pyogenes causes a variety of suppurative infections of the skin, including infection of wounds or burns, with a predilection to produce lymphangitis and cellulitis. Infection of minor abrasions may at times lead to fatal septicemia. The two typical streptococcal skin infections are erysipelas and impetigo.

i. Erysipelas

Erysipelas is an acute infection of the skin. It is a diffuse infection involving the superficial lymphatics. The involved skin area is red, swollen and indurated and typically raised and distinctly differentiated from the uninvolved skin. Erysipelas occurs most commonly in young children or older adults.

ii. Pyoderma (Impetigo)

Pyoderma (Impetigo) is seen primarily in young children (2 to 5 years old) with poor personal hygiene. Group A streptococci causing impetigo are frequently nephritogenic, that leads to acute glomerulonephritis. The main causes leading to acute glomerulonephritis in children in the tropics are impetigo and streptococcal infection of scabies lesions.

Antibody response to streptolysin O is not high in pyoderma and ASO estimation does not have as much clinical significance as in pharyngeal infections. For retrospective diagnosis of pyoderma antecedent to acute glomerulonephritis, antibody to DNAase B and hyaluronidase are more useful.

iii. Cellulitis

Unlike erysipelas, **cellulitis** typically involves the skin and deeper subcutaneous tissues, and the distinction between infected and noninfected skin is not clear. Local inflammation and systemic signs are observed, as in erysipelas.

iv. Necrotizing Fasciitis (Streptococcal Gangrene)

Necrotizing fasciitis, an infection that occurs deep in the subcutaneous tissue, spreads along the fascial planes and is characterized by an extensive destruction of muscle and fat. Bacteria other than group A streptococci can also cause necrotizing fasciitis. The group A streptococci that cause necrotizing fasciitis is referred to by the news media as "**flesh-eating bacteria**". It is introduced into the tissue through a break in the skin (e.g. minor cut or trauma, vesicular viral infection, burn, surgery). Systemic toxicity, multiorgan failure, and death (mortality exceeds 50%) are the hallmarks of this disease. Prompt medical intervention is necessary to prevent a poor prognosis.

Isolation of *Str. pyogenes* from the affected site and rising titers of antistreptolysin and anti-DNase B can be demonstrated.

v. Streptococcal Toxic Shock Syndrome

Patients with invasive and bactremic *Str. pyogenes* infections, and in particular necrotizing fasciitis, may develop **Streptococcal toxic shock syndrome** resembling staphylococcal TSS. It is a result of release of streptococcal toxins to the blood stream.

3. Other Suppurative Infections

S. pyogenes has been associated with a variety of other suppurative infections:

a. **Puerperal sepsis:** Historically, *Str. pyogenes* was an important cause of **puerperal sepsis** with the infection being exogenous. Streptococcal puerperal sepsis used to take a heavy toll of life before antibiotics became available. **Puerperal fever** is now much more commonly due to **endogenous infection** with anaerobic streptococci.
b. Abscesses in internal organs such as brain, lungs, liver and kidneys.
c. Septicemia and pyemia.

Non-suppurative Complications

Str. pyogenes infections lead to two important nonsuppurative sequelae: **acute rheumatic fever (ARF)** and **acute glomerulonephritis (AGN)**. Both are caused by immune reactions induced by the streptococcal infection. These complications ensue 1-5 weeks after the acute infection so that the organism may not be detectable when sequelae set in. They differ in their natural history in a number of respects (Table 25.3).

1. Acute rheumatic fever (ARF)—a potential sequela to pharyngitis (including scarlatina).
2. Acute glomerulonephritis (AGN)—primarily but not exclusively, associated with skin infections.

1. Acute Rheumatic Fever (ARF)

Rheumatic fever is a nonsuppurative inflammatory reaction that is epidemiologically and serologically related to antecedent group A streptococcal infection. Typically, rheumatic fever follows persistent or repeated streptococcal throat infection with a strong antibody response. The disease is caused by specific M types. Recovery from acute rheumatic fever occurs without residual injury to the joints but permanent damage to the heart may occur.

Pathogenesis

The pathogenesis of rheumatic fever is poorly understood. The disease is autoimmune in nature and is believed to result from the production of autoreactive (and polyspecific) antibodies and T lymphocytes induced by cross-reactive components of the bacteria and host tissues.

Various theories have been Proposed:

i. Antigenic cross-reactivity between streptococcal antigens and heart tissue.
ii. Direct toxicity due to streptococcal exotoxins.
iii. Actual invasion of the heart by streptococci.
iv. Localization of antigens within damaged muscle or valvular tissues.

2. Acute Poststreptococcal Glomerulonephritis (AGN)

In contrast to rheumatic fever, which occurs only after pharyngitis, AGN may be seen after either a pharyngeal

Table 25.3: Distinguishing features of rheumatic fever and glomerulonephritis

Feature	*Acute rheumatic fever*	*Acute glomerulonephritis*
Primary site of infection	Throat	Throat or skin
Latent period	Longer (2-5 weeks)	Shorter (1-3 weeks)
Prior sensitization	Essential	Not necessary
Repeated attacks	Common	Absent
Genetic susceptibility	Present	Not known
Serotype of *Str. pyogenes*	Any	Pyodermal types 49, 53-55, 59-61 and pharyngitis strains 1 and 12
Immune response	Marked	Moderate
Complement level	Unaffected	Lowered
Course	Progressive or static	Not indicated
Prognosis	Variable	Good
Penicillin prophylaxis	Essential	Not indicated

or a cutaneous infection. Specific nephritogenic strains of group A streptococci are associated with this disease. AGN is most often seen in children.

Pathogenesis

1. Immune complex deposition in the glomeruli.
2. Reaction of antibodies cross-reactive with streptococcal and glomerular antigen.
3. Alterations of glomerular tissues by streptococcal products such as streptokinase.
4. Direct complement activation by streptococcal components that have a direct affinity for glomerular tissues.
5. Circulating immune complexes have been found in the serum of patients with acute poststreptococcal glomerulonephritis.

LABORATORY DIAGNOSIS

In acute infections, diagnosis is established by the isolation and identification of β-streptococci from the patient, while in **nonsuppurative complications**, diagnosis is based mainly on the examination of the patient's serum for a rising titer of antibody to one or more streptococcal antigens.

A. Acute Suppurative infections

1. Specimens

Throat and nose swabs, high vaginal swabs, pus or pus swabs are the usual specimens collected. Serum is obtained for antibody demonstration.

2. Microscopy

Presumptive information may be obtained by an examination of Gram stained films from pus and CSF. The observation of typical gram-positive cocci in chains may indicate the likelihood of the presence of streptococcal infection. In contrast, smears are of no value in infections of throat or genitalia, where streptococci may form part of the resident flora and has a poor predictive value.

3. Culture

For culture, swabs should be collected under vision from the affected site and either plated immediately or if there is likely to be delay, swab should be sent to the laboratory in **Pike's medium** (blood agar containing 1 in 1,000,000 crystal violet and in 1 in 16,000 sodium azide). The specimen is plated on blood agar and incubated at 37°C anaerobically or under 5-10 percent CO_2, as hemolysis develops better under these conditions. Sheep blood agar is recommended for primary isolation because it is inhibitory for *Hemophilus haemolyticus*, colonies of which may be confused with those of hemolytic streptococci but their hemolysis is stronger on aerobic than on anaerobic plates.

Crystal violet blood agar and **PNF medium** are selective media that inhibit many throat commensal bacteria and may facilitate the detection of small numbers of *S. pyogenes* in throat swabs. They are rarely used for routine culture. *S. pyogenes* does not grow on MacConkey's bile-salt medium.

For isolating group A streptococci from throat swabs, the most common medium is blood agar supplemented with antibiotics trimethoprim/sulfamethoxazole to suppress the growth of normal flora.

4. Identification

Colonies of *S. pyogenes* on SBA are small, transparent, and smooth with a well-defined area of β-hemolysis. A **Gram stain** will reveal gram-positive cocci with some short chains. Hemolytic streptococci are grouped by the **Lancefield technique** using serologic methods, or biochemical tests can be performed. The **fluorescent antibody technique** has been employed for the rapid identification of group A streptococci. A key test that should be done is **bacitracin susceptibility** or **PYR hydrolysis.**

Bacitracin susceptibility: *S. pyogenes* is more sensitive to bacitracin than most other streptococci, though some strains of groups B, C and G are also sensitive (Maxted 1953, Coleman et al 1977). A disk containing 0.04 units of bacitracin is applied on the surface of an inoculated blood culture plate. *S. pyogenes* should show a large inhibition zone (e.g. > 15 mm diameter) and most other streptococci should show little or no inhibition. *S. pyogenes* is susceptible to bacitracin and hydrolyzes PYR, whereas the other β-hemolytic groups are resistant to bacitracin and are PYR negative.

Typing: Typing of *Str. pyogenes* is required only for epidemiological purposes in specialized reference laboratories. This may be done by precipitation or agglutination.

5. Antigen Detection

Numerous commercial kits are available for the detection of *Str. pyogenes* directly in throat swabs without cultivation. Enzyme immunoassay (EIA) or agglutination tests are used to demonstrate the presence of the antigen.

B. Nonsuppurative Complications

Serological Tests

Detection of antibodies against antigens of *Str. pyogenes* is an important means of establishing the diagnosis of poststreptococcal rheumatic fever and glomerulonephritis. Some immunologic tests used to detect past infection with *S. pyogenes* include ASO, anti-DNase, antistreptokinase, and anti-hyaluronidase titers.

Antistreptolysin O (ASO) test is used most frequently. ASO titers higher than 200 Todd units/ml are indicative of prior streptococcal infection. High levels are usually found in acute rheumatic fever but in glomerulonephritis, titers are often low.

Antideoxyribonuclease B (anti-DNase B) estimation is also commonly employed. Titers higher than 300 or 350 are taken as significant. **Anti-DNase B** and **anti-hyaluronidase (ASH) tests** are very useful for the retrospective diagnosis of streptococcal pyoderma, for which ASO is of much less value.

Commercial products are available for detection of **antistreptococcal antibodies**. **Streptozyme test** detects a mixture of antobodies. It is a passive slide hemagglutination test using erythrocytes sensitized with a crude preparation of extracellular antigens of streptococci, is a convenient, sensitive and specific screening test. It becomes positive after nearly all types of streptococcal infections, whether of the throat or the skin.

Treatment

S. pyogenes is very sensitive to penicillin and penicillin resistance has not yet been observed in *S. pyogenes*. Erythromycin or an oral cephalosporin can be used in patients with a history of penicillin allergy. Adequate treatment during acute infection prevents the complications of acute rheumatic fever.

Prophylaxis

Patients with a history of rheumatic fever require long-term antibiotic prophylaxis to prevent recurrence of the disease. Those persons who have recovered from ARF are given oral penicillin for many years to prevent recurrence. Because damage to the heart valve predisposes these patients to endocarditis, they also require antibiotic prophylaxis before they undergo procedures that can induce transient bacteremias (e.g. dental procedures).

Specific antibiotic therapy does not alter the course of acute glomerulonephritis, however, and prophylactic therapy is not indicated because recurrent disease is not observed in these patients. Antimicrobial drugs have no effect on established cases of AGN and ARF.

OTHER STREPTOCOCCI PATHOGENIC FOR HUMANS

Besides *Str. pyogenes*, streptococci belonging to groups B, C, D, F, G and rarely H, K, O and R may also cause human infections.

Data from Streptococcal Reference Laboratories in India (Lady Hardinge Medical College, New Delhi; Christian Medical College, Vellore) showed that while approximately 45 percent of hemolytic streptococcal isolates tested belong to group A, 10 to -15 percent belong to groups B and C each, about 25 percent to group G and 5 percent to group F.

Group B Streptococci: Streptococcus Agalactiae

Streptococcus agalactiae belongs to Lancefield group B and is the only species that carries the group B antigen. Human pathogenic group B strains possess a polysaccharide capsule which appears to confer virulence. Nine capsular serotypes have been identified, antibodies to which confer type specific protection.

Previously, *Streptococcus agalactiae* was recognized primarily as a cause of bovine mastitis (*agalactia*, want of milk) and their association with human diseases was recognized in 1930s. However, since 1960 it has become the leading cause of *neonatal infections* in industrialized countries and is also important cause of morbidity among peripartum women and nonpregnant adults with chronic medical conditions.

Strept. agalactiae is found in the vaginocervical tract of female carriers, and the urethral mucous membranes of male carriers, as well as in the gastrointestinal tract (GI tract), especially the rectum. Transmission occurs from an infected mother to her infant at birth, and venereally (propagated by sexual contact) among adults.

Clinical Diseases

1. Infection in the Neonate

Two different entities are recognized:

A. Early-onset Disease

Clinical symptoms of group B streptococcal disease acquired *in utero* or at birth develop during the first week of life. Infection is acquired from the maternal vagina during birth. Early-onset disease is characterized by bacteremia, pneumonia, or meningitis and is often fatal.

B. Late-onset Disease

This type of infection develops between second and twelfth weeks of life. The predominant manifestation is bacteremia with meningitis, but septic arthritis. Disease in older infants is acquired from exogenous source (e.g. mother, other infant).

Other group B infections in neonates: Osteomyelitis, conjunctivitis, sinusitis, otitis media, endocarditis and peritonitis may also occur.

2. Infections in the Adult

Puerperal sepsis and pneumonia. Other infections with group B streptococci are endometritis, urinary tract infection, wound infection, and bacteremia.

1. Diagnosis

Antigen tests—too insensitive.

2. Culture

Using a selective broth is the diagnostic test of choice.

3. Identification

Presumptive identification method is based on their ability to hydrolyse hippurate. They may be identified by the CAMP reaction (Christie, Atkins and Munch-Peterson), which can be demonstrated as an accentuated zone of hemolysis (arrowhead-shaped area of enhanced hemolysis) when *Str. agalactiae* is inoculated perpendicular to a streak of *Staph. aureus* grown on blood agar (Fig. 25.4). *S. agalactiae* produces a CAMP factor that enhances the lysis of sheep red cells by staphylococcal β-lysin. Group B streptococci are identified definitively by the demonstration of the **group-specific carbohydrate** or the use of commercially prepared **molecular probes** Occasional strains are bacitracin sensitive.

Group C Streptococci

Streptococci of this group are predominantly animal pathogens and comprise four species: *S. equi, S. equisimilis, S. dysgalactiae* and *S. zooepidemicus*. Group C strains

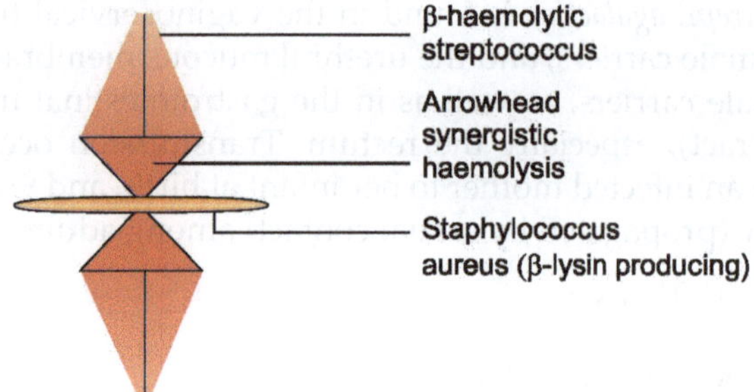

Fig. 25.4: CAMP reaction

isolated from human sources usually belong to *S. equisimilis*. It can cause upper respiratory infections, as well as deep infections such as endocarditis, osteomyelitis, brain abscess, pneumonia and puerperal sepsis. All species of group C are β-hemolytic, with the exception of *S. dysgalactiae*, which may be α-hemolytic or nonhemolytic. It resembles *Str. pyogenes* in fermenting trehalose but differs in fermenting ribose. It produces streptolysin O streptokinase (antigenically distinct from that produced by *Str. pyogenes*) and other extracellular substances. *S. equisimilis* serves as the source of streptokinase used in thrombolytic therapy in humans.

Group F

They have been called the '**minute streptococci**'. These grow poorly on blood agar unless incubated under CO_2. They are sometimes found in suppurative lesions. Streptococcus MG belongs to group F streptococci.

It has been isolated from the sputum of normal individuals and those suffering from primary atypical pneumonia. Demonstration of agglutinins sera of patients had been used as diagnostic test for primary atypical pneumonia.

Group G

These are commensals in the throats of human beings, monkeys or dogs. These may cause sore throat, pneumonia, septicemia, endocarditis, and bone, joint, skin and wound infections.

Group H and K

They sometimes cause **infective endocarditis.**

Group O

Group O is isolated mainly from the healthy human throat. It is suspected to cause **acute tonsillitis** and **endocarditis.**

Group R Streptococci

Group R strains are natural pathogens of pigs. *Str. suis* serotype 2 (group R streptococci) cause septicemia and meningitis in pigs. They occasionally infect people in contact with contaminated pork or infected pigs, and may cause septicemia, meningitis and respiratory tract infections. Abattoir workers, butchers and, to a lesser extent, housewives are at risk.

Group D Streptococci

Until the mid-1980s, the group D streptococci were divided into the two groups:

1. Enterococcus group (enterococci or fecal streptococci) which have been reclassified as a separate genus called Enterococcus.
2. Nonenterococcal group, for example, *Str. bovis, Str. equinus*.

ENTEROCOCCUS

The enterococci ("enteric cocci") were previously classified as group D streptococci. In 1984, the enterococci were reclassified into the new genus **Enterococcus.** Enterococcus can be differentiated from group D streptococci by a number of tests (Table 25.4). There are currently 16 species in this genus. This group consists of gram-positive cocci that are natural inhabitants of the intestinal tracts of humans and animals.

Species

i. Enterococcus faecalis ("pertaining to feces") is the enterococcus most often isolated from human sources.
ii. Enterococcus faecium ("of feces")
iii. Other species: E. durans, E. avium, E. casseliflavus, E. gallinarum, and *E. raffinosus* are observed occasionally.

All species produce the cell wall associated teichoic acid antigen, also referred to as the group D antigen in the Lancefield classification system.

It should be noted that enterococci sometimes exhibit a pseudocatalase reaction.

Table 25.4: Differences between group D streptococci and *Enterococcus* spp.

Characteristics	*Group D streptococci*	*Enterococcus spp.*
Hemolysis type	α, none	α, β, none
Growth in 6.5% sodium chloride	+	–
Growth in presence of 40% bile	+	–
Growth at 45°C	+	–
PYRase test	–	+
Susceptibility to penicillin	+	–

Characteristics of Enterococci

The enterococci are gram-positive cocci typically arranged in pairs and short chains, and is non-motile and non-capsulate. The cocci are facultatively anaerobic and grow optimally at 35°C, although most isolates can grow in the temperature range 10°C to 45°C. They grow readily on blood agar media, with large, white colonies appearing after 24 hours of incubation; the colonies are typically nonhemolytic but can be α-hemolytic or β-hemolytic. It grows readily on ordinary nutrient media and on MacConkey agar, on which it forms small (0.5-1 mm), usually magenta-colored colonies.

Distinctive Features of Enterococci

The Enterococci possess several distinctive features separating them from streptocooci: The enterococci grow in the presence of 6.5 percent NaCl, 40 percent bile, at pH 9.6, at 45°C and in 0.1 percent methylene blue. It survives heating at 60°C for 30 min, a feature distinguishing it from streptococci, and also grows within a wider range of temperatures (10-45°C). On MacConkey medium they produce deep pink colonies. Enterococci are PYRase test positive. They do not hydrolyze hippurate.

Identification

The identification of enterococcus species is made on biochemical characteristics. *E. faecalis* can be identified by its ability to ferment mannitol, sucrose, sorbitol and aesculin, and to grow on tellurite blood agar producing black colonies with gas production. It is VP positive.

Clinical Infections

The enterococci inhabit the gastrointestinal tract and the genitourinary tract in humans and other animals. Enterococci are frequent causes of nosocomial infections and may cause urinary tract infection, bacteremia, infective endocarditis, biliary tract infection, intraabdominal abscess complicating diverticulitis, peritonitis and wound infection.

Treatment

Most strains of enterococci are resistant to penicillin. They are also resistant to sulfonamides. Recently they have developed resistance to newer penicillins and cephalosporins, streptomycin and gentamicin.

Nonenterococcal Species of Group D

Nonenterococcal species of group D (*Str. bovis, Str. equinus*) are generally susceptible to penicillin and are inhibited by 6.5 percent sodium chloride or bile. They may cause **urinary infection** or **endocarditis**, rarely.

VIRIDANS STREPTOCOCCI

The viridans streptococci are commensals of mouth and upper respiratory tract infection. The viridans group of streptococci are a heterogeneous collection of α-hemolytic and nonhemolytic streptococci. The term *viridis* means "green" (Latin for "green") because many of these bacteria produce a green pigment (α-hemolysis) on blood agar media. Some of them may be nonlytic. Viridans streptococci are fastidious, with some strains requiring CO_2 for growth.

Classification

The current classification assigns streptococci species in the viridans group to one of the four groups:

1. **Anginosus group** (*S. anginosus, S. constellatus,* and *S. intermedius*): Organisms of the anginosus group may possess the Lancefield group A, C, F, G, or N antigen and in some instances may not be groupable.
2. **Mitis group** (*S. sanguis, S. parasanguis, S. gordonii, S. crista, S. infantis, S. mitis, S. oralis,* and *S. peroris).*
3. **Mutans group** *(S. criceti, S. downei, S. macacae, S. mutans, S. rattus,* and *S. sobrinus).*
4. **Salivarius groups** (*S. salivarius, S. thermophilus,* and *S, vestibularis*).

The organisms may also cross-react with other grouping sera. Thus identification using the Lancefield sera is of little value.

Clinical Infections

The viridans streptococci are ordinarily nonpathogenic but can on occasion cause disease. Although these organisms can cause a variety of infections, two clinically important phenomena are associated with viridans streptococci. **dental caries** and **subacute endocarditis**. *Streptococcus* anginosus is responsible for causing **pyogenic infections**, as was found with the groupable strains. They have also been implicated in meningitis, abscesses, osteomyelitis, and empyema.

Dental Caries

S. mutans is the principal cause of **dental caries** (tooth decay). *S. mutans* adheres to dental surfaces via extracellular carbohydrates (dextran) and erodes the teeth by converting sucrose to acetic acid and lactate. It breaks down dietary sucrose, producing acid and a tough adhesive dextran. The acid damages dentine and the dextrans bind together food debris, epithelial cells, mucus and bacteria to form **dental plaques**, which lead to caries. Their proportions in dental plaque are closely related to sugar consumption, and they are a major cause of dental caries because of their ability to produce large amounts of lactic acid even at pH values below pH 5.0. Experimental caries in monkeys has been prevented by a *Str. mutans* vaccine, but its extention to human use is fraught with problems.

Subacute Endocarditis

Viridans streptococci are the most common cause of subacute bacterial endocarditis. About two-thirds of the viridans-associated cases are due to *S. sanguis* and *S. mutans*. Transient bacteremia is associated with endocarditis. Most patients have underlying valvular heart disease, and the course of the endocarditis is generally subacute. Dental procedures, vigorous tooth brushing or chewing, and use of a water pick to clean teeth can cause a transient bacteremia that is enough to initiate valvular infection in a person with damaged valvular tissue. While viridans streptococci are generally penicillin sensitive, some strains may be resistant. It is, therefore, essential that in endocarditis, the causative strain is isolated and its antibiotic sensitivity determined so that appropriate antibiotics in adequate bactericidal concentration can be employed for treatment.

Nutritionally Variant Streptococci

Nutritionally variant streptococci (NVS) were first described in 1961. These bacteria grow as satellite colonies around other bacteria and require sulfhydryl compounds for growth.

They are usually α-hemolytic but may be nonhemolytic. They are part of normal flora and a occasionally cause bacteremia or endocardutus and can be faund in brain abscesses and other infections.

KNOW MORE

A. Alpha-hemolytic (α) streptococci:

The streptococci producing α-hemolysis are also known as viridans streptococci (from *'viridis'* meaning green). They are widely found as normal flora in upper respiratory tract of humans, but may cause opportunistic infections rarely. *Streptococcus salivarius* is the most commonly encountered species of this group.

Group B Streptococci: Streptococcus agalactiae

Streptococcus agalactiae belongs to Lancefield group B and is the only species that carries the group B antigen. It is gram-positive cocci arranged in long chains. It is facultative anaerobe, catalase-negative; positive CAMP and hippurate hydrolysis reactions (important identification tests). Group B organisms also produce DNases, hippuricase, neuraminidase, proteases, hyaluronidase and hemolysins. Group-specific carbohydrate (B antigen) in cell wall and type-specific antigens in capsule.

☞ KEY POINTS

- The organisms included in the family Streptococcaceae and the *Streptococcus-like* organisms are gram-positive cocci usually arranged in pairs or chains that are catalase negative.
- Streptococci are gram-positive cocci arranged in long chains.
- Three different schemes are used to classify the organism:
  1. **Hemolytic patterns**
  2. **Serologic properties:** Lancefield groupings A to H, K to M, and O to V; *Streptococcus pyogenes* belongs to Lancefield group A.
  3. **Biochemical (physiologic) properties.**

Streptococcus pyogenes

- *Strep. pyogenes* are gram-positive, spherical to ovoid organisms 0.5 to 1.0 µm in diameter.
- Catalase-negative; PYR-positive; bacitracin-susceptible are important identification tests.
- **Crystal violet blood agar** and **PNF medium** (blood agar containing polymyxin-B, neomycin and fusidic acid) are selective for beta hemolytic streptococci.
- **Antigenic structure**: Many streptococci can be categorized based on Lancefield group antigens. Group-specific carbohydrate (A antigen) and type-specific antigen (M protein) in cell wall.
- **Virulence factors** are cellular components, toxins and enzymes, streptolysin O, S and DNase B. ASO, anti-DNase B are clinically important.

Diseases:

A. Respiratory infection

1. Streptococcal pharyngitis
2. Scarlet fever (complication of pharyngitis).

B. Pyogenic cutaneous infections: Impetigo, erysipelas, cellulitis; necrotizing fasciitis involving deep subcutaneous tissues; streptococcal toxic shock syndrome.

Nonsuppurative sequelae: Rheumatic fever and acute glomerulonephritis (complications of pharyngitis or cutaneous infections).

Laboratory diagnosis is done by microscopy, culture, antigen detection and serological tests such as ASO, anti-DNase, antistreptokinase, and anti-hyaluronidase titers to detect past infection with *S. pyogenes*.

- *Streptococcus agalactiae* is a significant cause of invasive disease in newborns. Positive CAMP and hippurate hydrolysis reactions are important identification tests.
- *Diseases:* Two forms of neonatal disease: early-onset and late-onset. Other infections with group B streptococci are endometritis, urinary tract infection, wound infection, and bacteremia
- Enterococci: *Enterococcus faecalis* is the enterococcus most often isolated from human sources. Enterococci are frequent causes of nosocomial infections and may cause urinary tract infection, bacteremia, infective endocarditis, biliary tract infection, intraabdominal abscess complicating diverticulitis, peritonitis and wound infection.
- **Viridans Streptococci**: The viridans streptococci are commensals in mouth and upper respiratory tract that are regarded as opportunistic pathogens. Two clinically important phenomena are associated with viridans streptococci: **dental caries** (tooth decay) and **infective endocarditis**

- **Nutritionally variant streptococci:** Nutritionally variant streptococci are now classified within the genera *Granulicatella* and *Abiotrophia*.

IMPORTANT QUESTIONS

1. Classify streptococci. Describe the laboratory diagnosis of streptococcal sore throat.
2. Write short notes on:
   Antigenic structure of *Str. pyogenes*
   Lancefield grouping.
   Toxins and enzymes of *Streptococcus pyogenes*
   Pathogenicity of streptococci.
   Nonsuppurative complications of *Str. pyogenes* infections
3. Write briefly about:
   a. Group B streptococci.
   b. CAMP reaction.
   c. Group D streptococci.
   d. Enterococci (or) fecal streptococci.
   e. Viridans streptococci.
   f. Heat test.

FURTHER READING

Beighton D, et al. A scheme for identification of viridans streptococci. J Med Microbial 1991;3S:367.

Bisno AL. Group A streptococcal infections and acute rheumatic fever. New Eng J Med 1991;325:783.

Bisno AL, Stevens DL. Streptococcal infections of skin and soft tissues. New Engl J Med 1996;334:240.

Charles D, Larsen B. Streptococcal puerperal sepsis and obstetric infections: a historical perspective. Clin Microbiol Rev 1986;2:315.

Cunningham M. Pathogenesis of group A streptococcal infections, Clin Microbiol Rev 2000;13:470-511.

Moellering RC. Emergence of Enterococcus as a significant pathogen. Clin Infect Dis 1992;14:1173.

CHAPTER 26

Pneumococcus (Diplococcus pneumoniae: Str. pneumoniae)

Learning Objectives

After reading and studying this chapter, you should be able to:

- Describe morphology and cultural characters of pneumococci.
- Describe Quellung reaction.
- Explain C-reactive protein.
- Discuss laboratory diagnosis of pneumococcal infections.
- Differentiate between *Str. pneumoniae* and *Str. viridans*.

INTRODUCTION

Pneumococcus, a gram-positive lanceolate diplococcus, formerly classified as *Diplococcus pneumoniae*, has been reclassified as *Str. pneumoniae* because of its genetic relatedness to streptococcus. In contrast to other streptococci, *Str. pneumoniae* generally occurs as characteristic diplococci. Pneumococcus differs from other streptococci chiefly in its morphology, bile solubility, optochin sensitivity and possession of a specific polysaccharide capsule.

Pneumococci are normal inhabitants of the human upper respiratory tract. They are the single most prevalent bacterial agent in pneumonia and in otitis media in children. They can also cause sinusitis, bronchitis, infectious bacteremia, meningitis and other infections.

PNEUMOCOCCI (*DIPLOCOCCUS PNEUMONIAE, STREPTOCOCCUS PNEUMONIAE*)

Morphology

Pneumococci are gram-positive cocci in pairs (diplococci). The cocci are about 1 µm, slightly elongated cocci, with one end broad or rounded and the other pointed, presenting a **flame shaped or lanceolate appearance (Fig. 26.1)**. They may occur singly, in pairs, or in short chains but most often are seen as pairs (diplococci), with the broad ends in apposition, the long axis of the coccus parallel to the line joining the two cocci in a pair. They are nonmotile and nonsporing.

All freshly isolated strains are **capsulate.** The capsule encloses each pair. The capsule may be demonstrated as a clear halo in Indian ink preparations (Fig. 26.2) or may be stained directly by special techniques or by use of homologous type-specific antibody in the Quellung reaction.

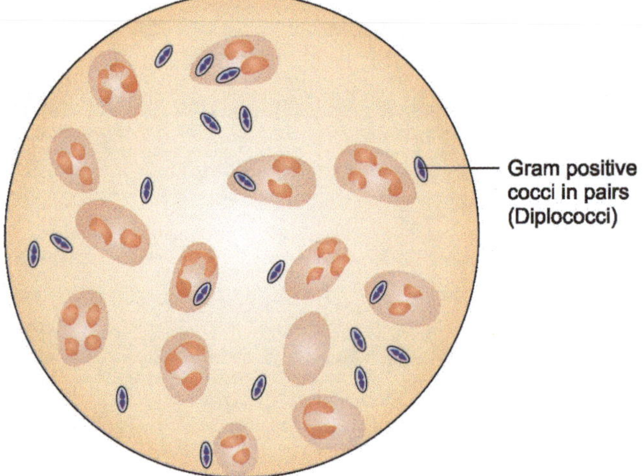

Fig. 26.1: *Str. pneumoniae* in pus

Cultural Characteristics

They are aerobes and facultative anaerobes. It grows best in air or hydrogen with 5-10 percent CO_2, for which some strains have a strict requirement. Optimum temperature being 37°C (range 25-40°C) and pH 7.8 (range 6.5-8.3). The pneumococcus has complex nutritional requirements and grow only in enriched media. It grows on ordinary media, but better on media with serum, blood or heated blood, which supplies nutrient, pH buffers and catalase.

On **blood agar**, after incubation for 18 hours, the colonies are small (0.5-1 mm), dome shaped and glistening, with an area of green discoloration (alpha

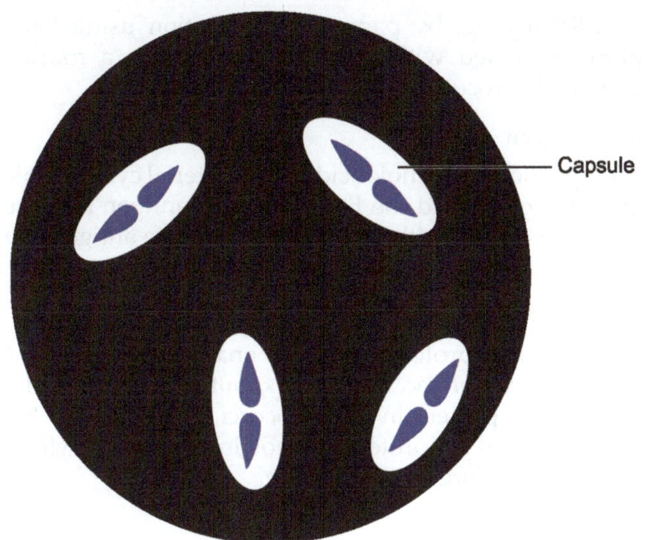

Fig. 26.2: Pneumococci. Indian ink preparation to show capsules

hemolysis) around them similar to the greenish discoloration observed with the viridans streptococci. On further incubation, the colonies become flat with raised edges and depressed centrally, so that concentric rings are seen on the surface when viewed from above **(draughtsman** or **carrom coin appearance)** which is due to autolysis of bacteria within the flat pneumococcal colonies. Some strains, e.g. of type 3, which form very large capsules, tend to form larger, mucoid colonies.

Under anaerobic conditions, however, a zone of beta hemolysis is produced around the colony by an oxygen labile pneumolysin O. In **liquid media** such as glucose broth, growth occurs as uniform turbidity. The cocci readily undergo autolysis in cultures due to the activity of intracellular enzymes. Autolysis is enhanced by bile salts, sodium lauryl sulphate and other surface active agents. Heat killed cultures do not undergo autolysis

Biochemical Reactions

1. Inulin Fermentation

Pneumococci ferment several sugars—glucose, lactose, sucrose and inulin with the production of acid and no gas. Fermentation is tested in Hiss's serum water or serum agar slopes. Fermentation of inulin by pneumococci is a useful test for differentiating them from streptococci as the latter do not ferment it.

2. Bile Solubility Test

Basis

Bile solubility test for identifying pneumococci is based on the presence in pneumococci, but not in the viridans streptococci, of an autolytic amidase that cleaves the bond between alanine and muramic acid in the peptidoglycan. The amidase is activated by surface-active agents such as bile or bile salts, resulting in lysis of the organisms.

Procedure

For bile solubility test, grow the isolate to be tested for 18 hours at 37°C in 5 ml serum, digest broth or infusion broth. While still warm, add 0.5 ml of 10 percent sodium deoxycholate solution and reincubate at 37°C. Pneumococci are lysed within 15 minutes and the initially turbid culture becomes clear and transparent.

Pneumococci are soluble in bile; viridans and other streptococci are not.

Alternatively, a rapid presumptive test may be made on the primary plate culture. Touch a suspected pneumococcal colony with a loopful of 2 percent sodium deoxycholate solution at pH 7.0. Incubate the plate for 30 minutes at 37°C. Colonies of pneumococcus disappear, leaving an area of α-hemolysis on the blood agar.

3. Pneumococci are Catalase and Oxidase negtive

Resistance

Pneumococci are delicate organisms and are killed by moist heat at 55°C in 10 min, and readily by most disinfectants. Strains may be maintained on semisolid blood agar or by lyophilization. Most strains are highly sensitive to benzylpenicillin, other penicillins, such as amoxycillin, cephalosporins, erythromycin and cotrimoxazole. A drug resistant *Strep. pneumoniae* (DRSP) strain originating in Spain has spread to most parts of the world posing problems in treament. The mode of resistance is not production of beta lactamase but alteration in the penicillin binding proteins on the bacterial surface. Such strains are also resistant to multiple drugs.

Optochin Sensitivity

Pneumococci are highly sensitive to killing by optochin (ethyl hydrocuprein hydrochloride), in a concentration of 1/500,00 and is useful in distinguishing them from viridans streptococci. For testing, place a paper disk containing 5 μg of optochin on an area of a blood agar plate inoculated with pneumococcus-like colonies from the primary diagnostic plate. A growth of pneumococcus will be inhibited in a zone extending radially for at least 5 mm from the margin of the disk on incubation. Viridans streptococci will grow right up to the disk.

Antigenic Structure

1. Capsular Antigens

The most important antigen of the pneumococcus is the type specific capsular polysaccharide. It is also called the **'specific soluble substance' (SSS)** as this polysaccharide diffuses into the culture medium or infective exudates and tissues. These polysaccharides are antigenic and form the basis for the separation of pneumococci into different serotypes. A total of 90 different capsular serotypes have been identified. The serotypes are designated by numbers, and those that are structurally related are grouped together (1, 2, 3, 4, 5, 6A, 6B, etc.).

The capsule is antiphagocytic, inhibiting complement deposition and phagocytosis where type-specific opsonic antibody is absent. Active or passive immunization against a specific polysaccharide produces a high level of resistance to infection with pneumococci of homologous type.

The antigenicity of the capsular polysaccharide varies in different species. It is antigenic in human beings and rabbits. But in mice, large doses (50 µg) induce no immunological response (immunological paralysis), while small doses (0.5 µg) are antigenic. The type of a pneumococcus is determined by its reactions with type-specific antisera, tested first in pools and then singly. The tests may be done by:

1. Agglutination of washed capsulate cocci.
2. Precipitation of SSS from culture supernates.
3. Quellung reaction or capsule swelling reaction.

It was described by Neufeld (1902). In the capsule swelling or 'quellung' reaction (quellung = swelling), a suspension of pneumococci is mixed on a slide with a drop of the type specific antiserum and a loopful of methylene blue solution and then examined using the oil-immersion objective. The capsule becomes apparently swollen, sharply delineated and refractile in the presence of the homologous antiserum. This reaction can be used to identify the organism directly from sputum, CSF, and other sources. Although currently seldom used, the quellung reaction identifies an isolate as *S. pneumoniae* and determines its capsular type. It used to be a routine bedside procedure in the past when the specific antiserum was used for the treatment of pneumonia.

2. Somatic Antigens

a. C polysaccharide

The cell wall of *S. pneumoniae* contains a species specific carbohydrate antigen, referred to as **C *substance***, which is similar to the C carbohydrate of the various Lancefield groups. A β-globulin in human serum, called the ***C-reactive protein***, reacts with C substance to form a precipitate. *C-reactive protein (CRP)* is an abnormal protein (beta globulin) that precipitates with the somatic 'C' antigen of pneumococci. It appears in acute phase sera of cases of pneumonia but disappears during convalescence. It is known as C-reactive protein because it precipitates with C antigen of pneumococci. CRP is present in low concentrations in healthy people but in elevated concentrations in patients with acute inflammatory diseases.

This is a chemical reaction and not an antigen-antibody combination. It is not an antibody but a protein that is present in low concentrations in normal blood. It is an 'acute phase' substance, produced in hepatocytes. Its production is stimulated by bacterial infections, inflammation, malignancies and tissue destruction. It disappears when the inflammatory reactions subside. It is used as an index of response to treatment in rheumatic fever and certain other conditions.

CRP testing, by passive agglutination using latex particles coated with anti-CRP antibody is a routine diagnostic procedure.

b. F antigen

The lipid bound teichoic acid in the bacterial cytoplasmic membrane is called the F or Forssman antigen because it can cross-react with the Forssman surface antigens on mammalian cells.

c. M protein

Type-specific protein antigens analogous to the M protein of *Streptococcus pyogenes*, but immunologically distinct, are present in pneumococci. Antibodies to the pneumococcal M protein do not inhibit phagocytosis and are, therefore, not protective.

Genetic Variation

1. Smooth to Rough (S-R) Variation

On repeated subculture, pneumococci undergo a **smooth-to-rough (S-R) variation**. In the R form, the colonies are rough and the cocci are noncapsulated, autoagglutinable and avirulent. R forms arise as spontaneous mutatants outgrow the parental S forms in artificial culture; in tissues, such R mutants are eliminated by phagocytosis.

2. Transformation

Rough pneumococci derived from capsulated cells of one serotype may be made to produce capsules of the same or different serotypes, on treatment: with DNA from the respective serotypes of pneumococci. This transformation, which may be demonstrated *in vivo* or *in vitro*, was first observed by Griffith (1928) and is of considerable historical interest as the first instance of genetic exchange of information in bacteria.

Virulence Factors

1. Capsule

The capsular polysaccharide is a crucial virulence factor. The capsule is antiphagocytic, inhibiting complement deposition and phagocytosis where type-specific opsonic antibody is absent. Noncapsulated strains are avirulent. The enhanced virulence of type 3 pneumococcus is due to the abundance of its capsular material. The antibody to the capsular polysaccharide affords protection against infection.

2. IgA1 Protease

Pneumococci produce an extracellular protease that specifically cleaves human IgA1 in the hinge region.

3. Pneumolysin

Pneumococci produce toxin known as pneumolysin, a cytotoxin similar to the streptolysin O in *S. pyogene.* It has cytotoxic and complement activating properties and so may be a virulence factor. Pneumolysin is immunogenic.

4. Autolysin

When activated, the pneumococcal autolysin breaks the peptide cross-linking of the cell wall peptidoglycan, leading to lysis of the bacteria.

Pathogenesis

Pneumococci colonize the human nasopharynx and may cause infection of the middle ear, paranasal sinuses and respiratory tract by direct spread. Infection of the meninges can also occur, by contiguity or through blood. Pneumococcal bacteremia may also lead to distant infections as in heart, peritoneum or joints. Infection is commonly endogenous, but exogenous infection may also occur, especially with highly virulent strains. Between 5-70 percent of normal human adults carry one or more serological types of *S. pneumoniae* in their throats, yet epidemics of pneumococcal pneumonia are rare and morbidity low.

1. Pneumonia

Pneumococci are one of the most common bacteria causing pneumonia, both **lobar** and **bronchopneumonia.** They also cause **acute tracheobronchitis** and **empyema. Bacteremia** may complicate pneumococcal pneumonia. This can result in metastatic involvement of the meninges, joints and, rarely, the endocardium.

i. Lobar Pneumonia

In adults, types 1-8 are responsible for about 75 percent of cases of pneumococcal pneumonia. In children, types 6, 14, 19 and 23 are frequent causes.

Pneumonia results from aspiration of pneumococci contained in upper airway secretions into the lower respiratory tract. When the normal defences are compromised by viral infection, anesthesia, chilling or other factors, pneumococci multiply, penetrate the bronchial mucosa and spread through the lung along peribronchial tissues and lymphatics. Contiguous spread commonly results in inflammatory involvement of the pleura. This may progress to **empyema. Pericarditis** is another uncommon but well recognized complication.

ii. Bronchopneumonia

Bronchopneumonia is almost always a secondary infection. This may be caused by any serotype of pneumococcus. Other causative agents responsible for bronchopneumonia include *Staph. aureus, K. pneumoniae, Str. pyogenes, H. influenzae, Fusobacterium* species and Bacteroides. Bronchopneumonia is frequently a terminal event in aged and debilitated patients.

2. Acute Exacerbations in Chronic Bronchitis

Pneumococci are commonly associated with the **acute exacerbations in chronic bronchitis.** The copious respiratory secretions in chronic bronchitis aid pneumococcal invasion. Another bacterium commonly associated with this condition is *Haemophilus influenzae.*

3. Meningitis

S. pneumoniae is among the leading causes of bacterial meningitis. It can spread into the central nervous system after bacteremia, infections of the ear or sinuses, or head trauma. Other bacterial agents of pyogenic meningitis include *N. meningitidis, H. influenzae, Str. agalactiae* (group-B) and *Listeria monocytogenes.* Bacterial meningitis can occur in patients of all ages but is primarily a pediatric disease. Untreated cases are almost invariably fatal. Even with antibiotic therapy, the case fatality rate is about 25 percent.

4. Bacteremia

Bacteremia occurs in 25 percent to 30 percent of patients with pneumococcal pneumonia and in more than 80 percent of patients with meningitis. Endocarditis can occur in patients with normal or previously damaged heart valves.

5. Other Infections

Pneumococci may also produce suppurative lesions in other parts of the body—empyema, pericarditis, otitis media, sinusitis, conjunctivitis, suppurative arthritis and peritonitis, usually as complications of pneumonia.

Epidemiology

S. pneumoniae is a common inhabitant of the throat and nasopharynx in healthy people. A 5 to 75 percent incidence of such carriage has been reported. Colonization is more common in children than in adults, and common in adults living in a household with children. Colonization with *S. pneumoniae* initially occurs at about 6 months of age. Subsequently, the child is transiently colonized with other serotypes of the organism. Although new serotypes are acquired throughout the year, the incidence of carriage and associated disease is highest during the cool months. The carriage rate for adults who have no contact with children is only 5 percent. In military installations where the incidence of pneumococcal infections also is very high, rates may be as high as 60 percent.

Infection usually leads only to pharyngeal carriage. Disease results only when the host resistance is lowered by contributory factors such as respiratory viral infections, pulmonary congestion, stress, malnutrition, immunodeficiency or alcoholism. Splenectomy and sickle cell disease are important predisposing conditions.

Pneumonia occurs when the endogenous oral organisms are aspirated into the lower airways. Although strains can spread on airborne droplets from one person to another in a closed population, epidemics are rare. Pneumococcal disease is most commonly associated with an antecedent viral respiratory disease. The case fatality rates of pneumonia may vary according to the virulence of the infecting serotype. Type 3 is the most virulent.

Lobar pneumonia is usually a sporadic disease but epidemics may occur among closed communities as in army camps. The incidence of bronchopneumonia increases when an epidemic of influenza or other viral infection of the respiratory tract occurs. Cases are more common in winter and affect the two extreme age groups more often.

Laboratory Diagnosis

1. Specimens

Sputum, lung aspirate, pleural fluid, cerebrospinal fluid, urine or blood are collected according to the site of lesion. Sputum specimens must be mucus expectorated from the lungs rather than samples of saliva. Blood culture is useful in pnemococcal septicemia.

2. Collection and Transport

All the specimens should be collected in sterile containers under all aseptic conditions. They should be processed immediately. CSF specimen should never be refrigerated in case of delay and should be kept at 37°C (*H. influenzae*, another causative agent of pyogenic meningitis may die at cold temperature).

3. Microscopy and Antigen Detection

Gram stain of sputum specimens is a rapid way to diagnose pneumococcal disease. If the smears are positive for gram-positive lancet-shaped diplococci, a presumptive diagnosis of pneumococcal pneumonia may be made. A centrifuged deposit of the CSF should be examined immediately in a Gram film in case of meningitis and presumptive diagnosis may be made by finding gram-positive diplococci both inside the polymorphs and extracellularly.

Pneumococcal antigen is often detectable by coagglutination (COA), latex agglutination (LA) or counterimmunoelectrophoresis (CIE) and ELISA. COA test for antigen gives positive result in larger proportion of specimens than either Gram film or culture. Moreover, by COA test, result is available within a sort time.

In addition to CSF, capsular polysaccharide can be demonstrated in the blood and urine by counterimmunoelectrophoresis

4. Capsule Swelling Tests

If typing sera are available, the most simple, rapid, and accurate method for the identification of pneumococci by direct examination is the quellung reaction. In this test, polyvalent anticapsular antibodies are mixed with the bacteria, and then the mixture is examined microscopically. A greater refractiveness around the bacteria is a positive reaction for *S. pneumoniae.*

5. Culture

Specimen is inoculated on plates of blood agar and heated blood agar incubated in air with 5-10 percent CO_2 for 18-24 hours. Typical colonies develop with α-hemolysis. The colonies are small (0.5-1 mm), dome shaped and glistening, with an area of green discoloration (α-hemolysis) around them. On further incubation, the colonies have **draughtsman** or **carrom coin appearance.**

6. Identification

Procedures commonly used to distinguish *S. pneumoniae* from the viridans streptococci are optochin susceptibility, bile solubility, and the quellung reaction. *S pneumoniae* is susceptible to optochin, whereas other α-hemolytic species are resistant (Table 26.1). Additional biochemical, serologic, or molecular diagnostic tests can be performed for a definitive identification.

7. Animal Pathogenicity Test

Isolation may be obtained by intraperitoneal inoculation in mice from specimens where pneumococci are expected to be scanty. Inoculated mice die in 1-3 days, and pneumococci may be demonstrated in the peritoneal exudate and heart blood. The test may be negative with occasional strains that are avirulent for mice (type 14 strains).

8. Blood Culture

In the acute stage of pneumonia, the organism may be obtained from blood culture in glucose broth. The finding of pneumococci in the blood is much better evidence of their pathogenic role in the lung than is their finding in sputum. Isolation of pneumococci from blood indicates bad prognosis.

9. Antibiotic Sensitivity Test

It is especially useful in strains which are resistant.

Prophylaxis

Immunity is type specific and associated with antibody to the capsular polysaccharide. The existence of some

Table 26.1: Differential characters of pneumococci and viridans streptococci

Character	*Pneumococcus*	*Viridans streptococci*
Morphology	Ovoid or lanceolate diplococci; some short chains	Short or long chains of rounded cocci
Capsule	Present	Usually absent
Colonies	Become flattened or draughtsman	Convex
Effect on blood agar	Narrow zones of α-hemolysis	Wide or narrow zone of α-hemolysis
Optochin sensitivity	Sensitive	Resistant
Bile solubility	+	–
Inulin fermentation	+	–
Virulence in mice	+	–

90 serotypes makes a complete polyvalent vaccine impracticable The current vaccine contains 23 different capsular polysaccharides which is stated to give 80-90 percent protection. The vaccine is immunogenic in normal adults, and the immunity is long-lived. It is meant for only in persons at enhanced risk of pneumococcal infection such as patients with absent or dysfunctional spleen, sickle cell disease, coeliac disease, chronic renal, lung, heart and liver diseases, diabetes mellitus and immunodeficiencies including human immunodeficiency virus (HIV) infection and renal transplant.

Vaccination is contraindicated in young children and elderly with lymphoreticular malignancies and immunosuppressive therapy.

Treatment

Penicillin is the drug of choice for susceptible strains, although resistance is increasingly common. Cephalosporins, erythromycin, chloramphenicol, or vancomycin are used for patients allergic to penicillin or for treatment of penicillin-resistant strains.

KNOW MORE

Str. pneumoniae has a long and fascinating history. It was isolated from human saliva in 1881 in independent studies by Sternberg and Pasteur. But the relationship between pneumococci and pneumonia was established only later by Fraenkel and Weichselbaum independently in 1886.

Some of the most important achievements in biology and medicine have resulted from studies on the pneumococcus.

Bacteremia may complicate pneumococcal pneumonia. This can result in metastatic involvement of the meninges, joints and, rarely, the endocardium.

KEY POINTS

- ***Streptococcus pneumoniae*** are elongated or "lancet-shaped," gram-positive cocci arranged in pairs (diplococci). They are **capsulated.**
- The pneumococcus has complex nutritional requirements and grow only in enriched media with serum, blood or heated blood. On **blood agar,** the colonies are small (0.5-1 mm), dome shaped and glistening, with an area of green discoloration (alpha hemolysis) around them. On further incubation, the colonies become flat with raised edges and depressed centrally, so that concentric rings are seen on the surface when viewed from above **(draughtsman or carrom coin appearance).**
- Pnemococci differs from viridans streptococci in their morphology (lanceolate diplococci), optochin sensitivity, bile solubility, inulin fermentation and virulence in mice

Diseases: Pneumonia, meningitis, sinusitis and otitis media. It can cause a variety of systemic infections, including bacteremia and endocarditis.

IMPORTANT QUESTIONS

1. Describe the laboratory diagnosis of pneumococcal infections
2. Differentiate between *Str. pneumoniae* and *Str. viridans* in a tabulated form.

FURTHER READING

Quie PG, et al. Symposium on the Pneumococus. Rev Infect Dis 1981;3:183.

Tomasz A. Streptococcus pneumoniae. Molecular Biology and Mechanisms of Disease. Larchmont, NY: Mary Ann Liebert, Inc., 2000.

Tuomanen EI, et al. Pathogenesis of pneumococcal infection. New Engl J Med 1995;332:1280.

CHAPTER 27

Neisseria and Moraxella

Learning Objectives

After reading and studying this chapter, you should be able to:

- Describe morphology, culture characteristics, biochemical reactions and antigenic structure of *Neisseria meningitidis*.
- Discuss pathogenicity and lab diagnosis of meningococcal meningitis.
- Describe morphology, culture characteristics, biochemical reactions of *Neisseria gonorrhoeae*.
- Discuss pathogenicity of *N. gonorrhoeae*.
- Discuss laboratory diagnosis of gonorrhoea.
- Explain nongonococcal urethritis (or) non-specific urethritis.
- Descrbe *Moraxella (Branhamella) catarrhalis*.

INTRODUCTION

The family **Neisseriaceae** includes five genera: *Neisseria, Kingella, Eikenella, Simonsiella* and *Alysiella.* Members of the genus *Neisseria* are aerobic, gram-negative cocci typically arranged in pairs (diplococci) with adjacent sides flattened together (resembling coffee beans). All species are oxidase-positive, and most produce catalase-properties that, combined with the Gram-stain morphology, allow for a rapid, presumptive identification of a clinical isolate.

Species

Important species of the genus *Neisseria* are: *N. meningitidis, N. gonorrhoeae, N. flavescens, N. subflava, N. sicca, N. mucosa, N. lactamica* and *N. polysacchareae* (Table 27.1). *N. gonorrhoeae* and *N. meningitidis* are the primary human pathogens of the genus. *N. gonorrhoeae* is always pathogenic, but *N. meningitidis* may be found as a commensal inhabitant of the upper respiratory tract of carriers. Other *Neisseria* species occur as commensals in the upper respiratory tract.

NEISSERIA MENINGITIDIS (MENINGOCOCCUS; DIPLOCOCCUS INTRACELLULARIS MENINGITIDIS)

Meningococcus was first described and isolated in 1887 by Weichselbaum from the spinal fluid of a patient.

MORPHOLOGY

Meningococci are **Gram-negative** oval or spherical **cocci** (0.6-0.8 μm in size), typically arranged in pairs, with the adjacent sides flattened or concave opposing edges and the long axes parallel. They are typically seen in large numbers inside polymorphonuclear leukocytes (Fig. 27.1). Considerable variations occur in size, shape and staining properties, especially in older cultures, due to autolysis.

Most fresh isolates are capsulated. They are nonsporing and nonmotile.

Cultural Characteristics

Meningococci have exacting growth requirements and do not grow on ordinary media. Growth occurs on media enriched with blood, serum or ascitic fluid, which promote growth by neutralizing certain inhibiting substances in culture media rather than by providing additional nutritional needs. Strains will grow on Mueller-Hinton medium without the addition of blood or serum but grow poorly if at all on most unenriched media.

They are strict aerobes, no growth occurring anaerobically. The optimum temperature for growth is 35-36°C. No growth takes place below 30°C. Optimum pH is

Table 27.1: *Neisseria* species and their diseases

Organism	*Diseases*
N. meningitidis	Meningitis, meningoencephalitis, bacteremia, pneumonia, arthritis, urethritis
N. gonorrhoeae	Urethritis, cervicitis, salpingitis, pelvic inflammatory disease, proctitis, bacteremia, arthritis, conjunctivitis, pharyngitis
Other *Neisseria* species	Opportunistic infections

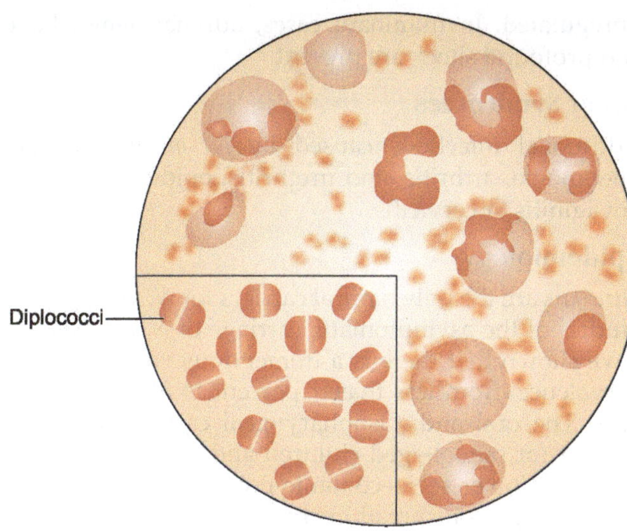

Fig. 27.1: *N. meningitidis* in cerebrospinal fluid. Inset—enlarged view showing flat adjacent side of cocci

7.0-7.4. Growth is facilitated by 5-10 percent CO_2 and high humidity. Blood agar, chocolate agar and Mueller-Hinton starch casein hydrolysate agar are the media commonly used for culturing meningococci. Modified Thayer-Martin (with vancomycin, colistin and nystatin) is a useful selective medium.

On **blood agar** after 24 hours incubation, colonies are 1-2 mm in diameter, round, convex, gray, translucent and nonhemolytic. After 48 hours, colonies are larger with an opaque raised center and thin transparent margins which may be crenated.

Heated blood (chocolate) agar colonies are slightly larger on heated blood (chocolate) agar than on ordinary blood agar. Growth is poor in liquid media, producing a granular turbidity with little or no surface growth.

Biochemical Reactions

They are **catalase** and **oxidase positive.**

Oxidase Test

The oxidase test is a key test for identifying them. When 1 percent solution of oxidase reagent (tetramethyl-paraphenylene-diamine-dihydrochloride) is poured on culture media, *Neisseria* colonies quickly turn deep-purple. This prompt oxidase reaction helps in the identification of meningococci and gonococci in mixed cultures. The test may also be performed by rubbing a little of the growth with a loop on a strip of filter paper moistened with the oxidase reagent (Kovacs' method). A deep purple color appears immediately. This test is used for screening species of *Neisseria* (both pathogenic and nonpathogenic), *Alcaligenes, Aeromonas, Vibrio, Campylobacter* and *Pseudomonas.*

Sugar Fermentation

Meningococci ferment glucose and maltose, with the production of acid but no gas, but not lactose or sucrose (gonococci ferment glucose but not maltose). Fermentation is usually tested on peptone serum agar slopes containing the sugar and indicator.

Indole and hydrogen sulfide are not produced and **nitrates** are not reduced.

Antigenic Classification

Meningococci are capsulated, unlike other neisseriae. Based on their capsular polysaccharide antigens, meningococci are classified into at least 13 serogroups: A, B, C, X, Y, Z, Z1 (29E) and W135. Further serogroups H, I, K and L have also been described. A group D was described but no capsular polysaccharide specific for this group has yet been demonstrated. Groups A, B and C are the most important. Group A is usually associated with epidemics and group C mostly with localised outbreaks, while group B causes both epidemics and outbreaks. Groups 29-E, W-135 and Y also frequently cause meningitis. Serogroups H, I, K and L have also been isolated from carriers and have not been associated with diseases.

Serogroups are further classified into **serotypes** and **subtypes.** The serotype classification of isolates is based on differences in the proteins in the outer membrane. However, molecular analysis (e.g. multilocus enzyme electrophoresis, DNA fingerprinting) has replaced serologic analysis in epidemiologic investigations.

Resistance

Meningococci are very delicate organisms, being highly susceptible to heat, dessication, alterations in pH and to disinfectants. They die within a few days at room temperature but cultures may be maintained on Dorset's egg medium or heated blood agar slopes in screw-capped bijou bottles for several weeks. Colonies emulsified in peptone water will survive at –70°C or in liquid nitrogen for years, but freeze-drying is preferable for long-term storage. They are killed by heating at 55°C in 5 minutes.

Pathogenicity

Meningococci are strict human parasites inhabiting the nasopharynx. Infection is usually asymptomatic. In some, local inflammation ensues, with rhinitis and pharyngitis. Dissemination occurs only in a small proportion.

Stages of Meningococcal Infections

If we follow the fate of the infecting organism, meningococcal infections can be categorized into three stages:

First Stage—Nasopharyngeal Infection

The organisms appear in nasopharynx leading to **nasopharyngeal infection**, which is usually asymptomatic but might result in a minor inflammation. This state can last for days to months and induces the formation of protective antibodies within a week, even though the infection remains asymptomatic. Dissemination occurs only in a small proportion.

Second Stage—Meningococcal Septicemia

In a small percentage of cases, the meningococci enter the bloodstream from the posterior nasopharynx, probably through the cervical lymph nodes where they rapidly multiply. This stage is called **meningococcemia.** The patient develops fever, malaise and petechial skin lesions due to foci of infection in the capillaries. Meningococci may be isolated from the petechial lesions. The organisms may also cause lesions in the joints and lungs and rarely cause massive bilateral hemorrhages in the adrenals **(Waterhouse-Friderichsen syndrome).** It is an overwhelming and usually fatal condition, characterized by shock, **disseminated intravascular coagulation (DIC)** and **multisystem failure**. Rarely chronic meningococcemia may be seen. Meningococcal disease is favored by deficiency of the terminal complement components (C5-C9). The hemorrhagic lesions occurring in both the skin and the internal organs, particularly the adrenals, are believed to result from the release of endotoxin.

Third Stage—Meningitis

In the third stage of meningococcal infection, the organisms can cross the blood-brain barrier and infect the meninges, causing the major symptoms of severe headache, stiff neck, and vomiting accompanied by delirium and confusion. The route of spread from the nasopharynx to the meninges is controversial. The spread may be directly along the perineural sheath of the olfactory nerve, through the cribriform plate to the subarachnoid space, or more probably, through the bloodstream. In certain cases, the site of entry of the meningococcus may be the conjunctiva. Cases of meningococcal purulent conjunctivitis occur.

On reaching the central nervous system, a suppurative lesion of the meninges is set up, involving the surface of the spinal cord as well as the base and cortex of the brain. The cocci are invariably found in the spinal fluid, both free and within the leucocytes. Case fatality is variable but in untreated cases may be as high as 80 percent. Survivors may have sequelae such as blindness and deafness. Some cases develop chronic or recurrent meningitis.

The pathogenic agent in meningococcal disease appears to be the endotoxin (LPS) released by autolysis. The vascular endothelium is particularly sensitive to the endotoxin. All major inflammatory cascade systems as well as cytokines and nitric oxide are triggered and upregulated. In fulminant cases, adrenal hemorrhage and profound shock are present.

Other Syndromes

Additional infections caused by *N. meningitidis* are pneumonia, arthritis, and urethritis, epididymitis, vulvovaginitis or cervicitis.

Epidemiology

Humans are the only natural carriers for *N. meningitidis.* Studies of the asymptomatic carriage of *N. meningitidis* have shown that there is a tremendous variation in its prevalence, from less than 1 percent to almost 40 percent. The oral and nasopharyngeal carriage rates are highest for school-aged children and young adults are higher in lower socioeconomic populations (caused by person-to-person spread in crowded areas), and do not vary with the seasons even though disease is most common during the dry, cold months of the year.

An increase in carrier rate heralds the onset of an epidemic. The carrier state may last for a few days to months. The carrier rate is higher in the members of the household of a patient with meningococcal disease. Household contacts of a case are 500-800 times more likely to develop meningococcal infection than the general population. During epidemics, the carrier rates in closed communities may go upto 90 percent. Endemic disease is most common in children younger than 5 years, particularly infants. People who are older and who live in closed populations (e.g, military barracks, prisons) are prone to infection during epidemics.

Natural infection is limited to human beings. Intraspinal inoculation of large numbers of cocci may produce a picture of meningitis in monkeys. Intraperitoneal inoculation of the cocci suspended in hog gastric mucin brings about a fatal infection in mice.

Laboratory Diagnosis

1. Specimens

i. Cerebrospinal fluid (CSF).
ii. Blood for culture (which may come from a patient with meningitis, a hemorrhagic rash or pyrexia of uncertain origin).
iii. Aspirate from skin lesions or pus from an infected joint.
iv. Throat or nasopharyngeal swabs from suspected cases.

Swabs should be transported in Stuart's transport medium. All specimens where meningococcal infection is suspected must be submitted to the laboratory immediately.

2. Examination of CSF

If meningitis is suspected, a lumbar puncture should be performed as soon as possible unless there are signs of raised intracranial pressure. In meningococcal meningitis, the fluid will be under pressure and turbid, with a large number of pus cells.

i. Cell Count

Perform a cell count. The exudate in meningococcal meningitis is typically polymorphonuclear.

ii. Microscopy

Centrifuge the remaining CSF. Make a smear of the centrifuged deposit and stain with Gram stain. CSF from a typical case of meningococcal meningitis will show gram-negative diplococci inside a limited proportion of the pus cells; many are extracellular.

Stain a second film with methylene blue to determine the cell type; occasionally, diplococci may be seen more easily with this stain.

If fluorescein isothiocyanate coupled antiserum is available, a smear of the deposit may be examined for the direct identification of the meningococcal serogroup responsible for infection.

iii. Antigen Detection

Divide the supernatant CSF into two aliquots: One to be kept if necessary for **biochemical examination**, the other to be examined for the presence of **meningococcal polysaccharide antigen** by counterimmunoelectrophoresis, latex agglutination or coagglutination using meningococcal antisera. Similar tests are also available for pneumococcus, *H. influenzae* type b and group B streptococcus antigens. Antigen detection is particularly useful in partially treated patients in whom smear and culture tests may be negative.

iv. Culture

Plate out the centrifuged deposit on both blood and heated blood agar (chocolate agar) and incubate at 37°C in 5-10 percent CO_2. Colonies appear after 18-24 hours which may be identified by morphology and biochemical reactions. It is important to remember that morphologically similar organisms such as *N. flavescens*, *N. flava* and *Acinetobacter* may also cause purulent meningitis occasionally.

Subculture

Add Robertson's cooked meat broth to the remaining deposit, incubate overnight and subculture in the same way. This method may sometimes succeed where direct plating fails. In the absence of visible meningococci, glucose broth may be added to the remaining sediment of the centrifuged deposit to facilitate the isolation.

v. Biochemical Reactions

Sugar utilization tests or commercial kits are used to identify any gram-negative diplococci. The oxidase test is performed on colonies on solid medium.

vi. Antibiotic Sensitivity Tests

Set up antibiotic sensitivity tests.

vii. Serogrouping

Serogrouping is performed by slide agglutination with hyperimmune sera. Serogrouping and serotyping procedures are done that provide important epidemiological information.

3. Blood Cultures

Blood culture is often positive in meningococcemia and in early cases of meningitis. Cultures should be incubated for 4-7 days, with daily subcultures. Subculture to blood agar and chocolate. Incubate cultures in 5-10 percent CO_2 for 24 hours and examine oxidase-positive colonies of gram-negative diplococci as above.

Meningococcal antigens can be found in the blood in active disease.

4. Pus, Aspirates and Swabs

Gram-stained films are examined. In addition to blood agar and chocolate agar, Thayer-Martin selective medium is used for the culture of materials expected to yield a mixture of organisms. Such as pus, aspirates, and throat, nasopharyngeal and genital swabs.

5. Petechial Lesions

Meningococci may sometimes be demonstrated in petechial lesions by microscopy and culture.

6. Serological Diagnosis

Paired sera may be tested for the presence of complement-fixing antibodies. This test is helpful in cases where no organisms have been isolated or in obscure pyrexias, which may be due to chronic meningococcal septicemia.

Specific antibodies to capsular polysaccharide may be demonstrated by a:

i. Hemagglutination test.
ii. ELISA tests are being standardized internationally.

7. Polymerase Chain Reaction (PCR)

Group specific diagnosis of infection can be made by detection of meningococcal DNA sequence in CSF or blood by PCR amplification.

Treatment

Sulfonamides, once the mainstay, are not used now due to widespread resistance. **Penicillin** is currently the antibiotic of choice, but resistance to it is also becoming more common. Intravenous penicillin G is the treatment of choice. **Chloramphenicol** is effective, but risks of blood dyscrasia have limited its use. Either chloramphenicol or a third generation cephalosporins such as cefotaxime or ceftriaxone is used in persons allergic to penicillin allergy.

At the end of a course of therapy with penicillin, it is important to give eradicative treatment with **rifampicin** or **ciprofloxacin** because penicillin does not eradicate meningococci from the nasopharynx and a patient returning home as a carrier may infect others. This probably does not apply to ceftriaxone or cefotaxime.

Prophylaxis

Chemoprophylaxis

Sulphonamides were used for prophylaxis, but are not effective due to resistance. In addition, penicillin is ineffective in eliminating the carrier state. **Minocycline** and **rifampin** have been used effectively for antibiotic mediated chemoprophylaxis. **Ciprofloxacin** is widely used as a prophylactic for adolescents and adults as a single, oral dose. All household and other intimate (e.g. mouth kissing) contacts of a case should be given chemoprophylaxis as a routine.

Immunoprophylaxis

A polyvalent vaccine effective against serogroups A, C, Y, and W135, which can be administered to children older than 2 years, has been developed. The vaccine cannot be administered to children in younger age groups because they do not respond to polysaccharide antigens. The immunity is group specific.

Conjugation of polysaccharide antigens to protein carriers has been used successfully with the *Haemophilus influenzae* vaccine and a similar conjugated *N. meningitidis* vaccine is under evaluation.

Unfortunately, the group B polysaccharide is a weak immunogen and cannot induce a protective antibody response. There is no group B vaccine available at present. Vaccination with a suspension containing serogroup A can be used for control of an outbreak of disease, for travelers to hyperendemic areas, or for people at increased risk for disease (e.g. patients with complement deficiency).

NEISSERIA GONORRHOEAE (GONOCOCCUS)

N. gonorrhoeae causes the venereal disease gonorrhea. The gonococcus was first described in gonorrheal pus by Neisser in 1879. Bumm in 1885 cultured the coccus and proved its pathogenicity by inoculating human volunteers. Gonococci resemble meningococci very closely in many properties.

Morphology

Morphology and staining of *N. gonorrhoeae* are identical to those of *N. meningitidis*. In smears from the urethral discharge in acute gonorrhea, the organism appears as a diplococcus with the adjacent sides concave, being typically kidney shaped. It is found predominantly within the polymorphs, some cells containing as many as a hundred cocci (Fig. 27.2).

Gonococci possess pili on their surface. Pili facilitate adhesion of the cocci to mucosal surfaces and promote virulence by inhibiting phagocytosis. Piliated gonococci agglutinate human red blood cells but not red cells from other mammals. The hemagglutination is not inhibited by mannose.

Cultural Characteristics

N. gonorrhoeae is a delicate organism with exacting nutritional and environmental requirements. Gonococci are more difficult to grow than meningococci. They are aerobic but may grow anaerobically also. Growth occurs best at pH 7.0-7.4 and at a temperature of 35-36°C. It is essential to provide 5-10 percent CO_2.

They grow well on **chocolate agar** and **Mueller-Hinton agar.** A popular selective medium is the **Thayer-Martin medium (chocolate agar containing vancomycin, colistin and nystatin)** which inhibits most contaminants including nonpathogenic *Neisseria*. Trimethoprim lactate may be added to Thayer-Martin medium to inhibit swarming *Proteus* species that are occasionally present in cervicovaginal and rectal specimens.

Colony Morphology

Colonies are small, round, translucent, convex or slightly umbonate, with finely granular surface and lobate margins after incubation for 24 hours in a moist aerobic environment enriched with 5-10 percent CO_2. They are soft and easily emulsifiable. After 48 hours, the colonies are larger (1.5-2.5 mm), sometimes with a crenated margin and an opaque raised center.

Considerable variation in size occurs with gonococcal colonies and on most culture media, the colony outline is irregular, unlike the circular colonies of *N. meningitidis*. On Thayer-Martin medium, growth is slower; although colonies are similar to those on MNYC medium they are usually smaller.

Types of gonococci: Kellogg divided gonococci into four types (T1-T4) on the basis of, colonial appearance, auto-agglutinability and virulence.

Types T1 and T2 form small brown colonies and bear numerous fimbriae (pilated types P1 and P2). They are auto-agglutinable and virulent.

Types T3 and T4 are nonpiliated (P–), form smooth suspensions and are avirulent.

Fresh isolates from acute cases of gonorrhea generally form T1 and T2 colonies. On serial subculture, they change to T3 or T4 colonial morphology. T1 and T2 types are also known as P+ and P++, respectively, while T3 and T4 are known as P^-.

Biochemical Reactions

The gonococcus is oxidase positive and resembles meningococci except in the fermentation of maltose. Gonococci ferment only glucose and not maltose and neither species ferments lactose or sucrose. This can be remembered by **G for gonococcus** and **M+G for meningococcus.**

Antigenic Structure

The structure of *N. gonorrhoeae* is typical of gram-negative bacteria, with the thin peptidoglycan layer sandwiched between the inner cytoplasmic membrane and the outer membrane. The outer surface is not covered with a true carbohydrate capsule, as is found in *N. meningitidis*.

Gonococci are antigenically heterogeneous and are capable of changing their surface structures *in vitro*. They probably do so *in vivo* as well, to avoid host defence. The surface structures include the following:

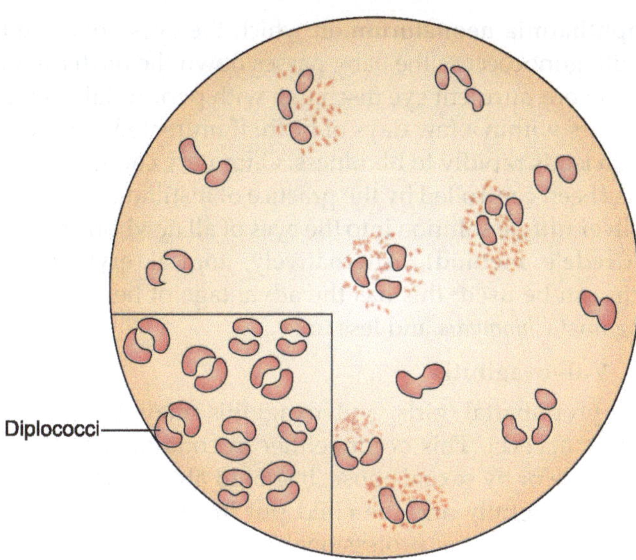

Fig. 27.2: *N. gonorrhoeae* in urethral pus. Inset— enlarged view showing diplococci with adjacent surfaces concave

1. Pili

Pili are hair-like appendages that extend up to several micrometers from the gonococcal surface. They act as virulence factors by promoting attachment to host cells and inhibiting phagocytosis. The pili are composed of repeating protein subunits (pilins), whose expression is controlled by the *pil* gene complex. Pilin proteins have a conserved region at the amino terminal end and a highly variable region at the exposed carboxyl terminus. Pili undergo antigenic and phase variation.

2. Por Proteins (Protein I)

The trilaminar *outer membrane* of gonococci contain many different proteins. The Por proteins (formerly protein I) are porin proteins that form pores or channels in the outer membrane. Protein I of a single strain is antigenically constant, though it shows considerable heterogeneity between different strains. Two classes of Por proteins (PorA and PorB), each with a variety of antigenic variations, have been identified. Each strain of gonococcus expresses only one type of Por, but the Por of different strains is antigenically different. Any one strain carries only either IA or IB but not both. Using monoclonal antibodies to protein I epitopes, gonococci can be classified into several serovars, AI to 24 and BI to 32.

3. Opa Proteins (Protein II)

Opa proteins (opacity proteins; formerly protein II) is related to the opacity of the gonococcal colonies and so is called the 'opacity associated' (OPA) outer membrane protein. Bacteria expressing the Opa proteins appear opaque (versus transparent) when grown in culture. A strain may express 0 to 3 serological varieties of the OPA protein at a time. These proteins facilitate bacterial adherence to each other and to eukaryotic cells and also for the clumping of cocci seen in urethral exudate smears.

4. Rmp (Protein III)

The third group of proteins in the outer membrane is the highly conserved Rmp proteins (reduction-modifiable proteins; formerly protein III).

These proteins stimulate antibodies that block serum bactericidal activity against *N. gonorrhoeae.*

5. Lipooligosaccharide (LOS)

This antigen is composed of lipid A and a core oligosaccharide similar to gram-negative lipopolysaccharide (LPS) and possess endotoxic activity.

6. Other Proteins

Other important gonococcal proteins are **IgA1 protease** which degrades secretory IgA and **β-lactamase**, which degrades penicillin. **Fbp (iron-binding protein),** similar in molecular weight to Por, is expressed when the available iron supply is limited, e.g. in human infection.

Resistance

The gonococcus is a very delicate organism, readily killed by drying, soap and water, and many other cleansing or antiseptic agents at their correct use and dilution. Organisms may remain viable for a day or so in pus contaminating linen or other fabrics. In cultures, the coccus dies in 3-4 days at room temperature. Freeze-drying is the most reliable method for long-term storage of gonococci but storage at –70°C or in liquid nitrogen may be more convenient for intermediate storage.

Pathogenesis

Gonorrhea

Gonorrhea is a venereal disease. The name 'gonorrhoea' derives from the Greek words *gonos* (seed) and *rhoia* (flow) and described a condition in which semen flowed from the male organ without erection. The name gonorrhea was first employed by Galen in 130 AD.

The disease is acquired by sexual contact. Gonorrhea is a disease essentially confined to the mucus-secreting epithelial cells of humans. Adhesion of gonococci to the urethra or other mucosal surfaces is the first step in infection. The presence of pili is important for the initial attachment. Adhesion is rapid and firm so that micturition after exposure offers no protection against infection. The cocci penetrate through the intercellular spaces and reach the subepithelial connective tissue by the third day after infection. Gonococci apparently penetrate between columnar epithelial cells. Stratified squamous epithelium is relatively resistant to infection.

The incubation period is 2-8 days. Gonococci attack mucous membranes of the genitourinary tract, eye, rectum, and throat, producing acute suppuration that may lead to tissue invasion; this is followed by chronic inflammation and fibrosis.

1. Diseases in Men

The commonest clinical presentation is **acute urethritis** in the male a few days after unprotected vaginal or anal sexual intercourse. **Dysuria** and a **purulent penile discharge** make most sufferers seek treatment rapidly. The infection extends along the urethra to the prostate, seminal vesicles and epididymis.

Although complications are rare, epididymitis, prostatitis, and periurethral abscesses can occur. **Chronic urethritis** may lead to stricture formation. The infection may spread to the periurethral tissues, causing abscesses and multiple discharging sinuses **(watercan perineum)**.

2. Diseases in Women

In women, the endocervix is the primary site of infection and extends to the urethra and vagina, giving rise to mucopurulent discharge. The vaginal mucosa is not usually affected in adults because the stratified squamous epithelium is resistant to infection by the cocci and also because of the acid pH of vaginal secretions, but severe vulvovaginitis can occur in prepubertal girls. Asymptomatic carriage in women is common, especially in the endocervical canal.

Symptomatic patients commonly experience vaginal discharge, dysuria, and abdominal pain. The infection may extend to Bartholin's glands, endometrium and fallopian tubes. At menstruation or after instrumentation, particularly termination of pregnancy, gonococci ascend to the fallopian tubes to give rise to **acute salpingitis**, which may be followed by **pelvic inflammatory disease** and a high probability of sterility if inadequately treated. Peritoneal spread occasionally occurs and may produce a perihepatic inflammation (**Fitz-Hugh-Curtis syndrome).** Clinical disease is as a rule less severe in women, many of whom may carry gonococci in the cervix without developing any clinical symptoms. Asymptomatic carriage of gonococci is rare in men.

Proctitis occurs in both sexes. In men this follows homosexual rectal intercourse. In women, it can follow rectal intercourse, but may also arise as a result of autoinoculation of rectal mucosa with infected vaginal discharge (direct contagious spread). Gonococcal pharyngitis may follow orogenital contact in either sex. Conjunctivitis may occur usually by autoinoculation with fingers.

3. Disseminated Gonococcal Disease

Blood invasion may occur from the primary site of infection and may lead to metastatic lesions such as arthritis, ulcerative endocarditis and very rarely meningitis. *N. gonorrhoeae* is a leading cause of purulent arthritis in adults. Occasional cases of pyemia have been reported.

4. Diseases in Childern

i. Ophthalmic Neonatorum

A nonvenereal infection is **ophthalmia neonatorum** in the newborn. Babies born to infected women may suffer **ophthalmia neonatorum**, in which the eyes are coated with gonococci as the baby passes down the birth canal. A severe purulent eye discharge with periorbital edema occurs within a few days of birth. If untreated, ophthalmia leads rapidly to blindness. Once very common, this has been controlled by the practice of instilling 1 percent silver nitrate solution into the eyes of all newborn babies **(Crede's method).** Alternatively, topical erythromycin can be used; this has the advantage of being active against *Chlamydia* and less toxic.

ii. Vulvovaginitis

In prepubertal girls, vulvovaginitis may be caused by gonococci. This occurs either in conditions of poor hygiene or by sexual abuse; it should always be investigated carefully and the child put in touch with social services and other professionals capable of dealing with this difficult condition.

Epidemiology

Gonorrhea occurs only in humans. It has no other known reservoir. It is never found as a normal commensal although a proportion of those infected, particularly women, may remain asymptomatic.

N. gonorrhoeae is transmitted primarily by sexual contact. Women have a 50 percent risk of acquiring the infection as the result of a single exposure to an infected man, whereas men have a risk of approximately 20 percent as the result of a single exposure to an infected woman.

The only nonvenereal infection is **ophthalmia neonatorum**. A severe purulent eye discharge with periorbital edema occurs within a few days of birth. The major reservoir for gonococci is the asymptomatically infected person. Asymptomatic carriage is more common in women than in men.

Acute gonorrhoea is usually easily diagnosed and treated, and was well controlled in much of the world until the 1960s. In 1970, the global incidence of new cases was estimated at 16 million, making it one of the most common infective diseases. With the AIDS scare in the 1980s, there was a noticeable decline in the incidence of gonorrhea, but this has not been kept up. A higher incidence of gonorrhea has been observed in persons belonging to blood group B. The basis for this is not known.

Laboratory Diagnosis

Diagnosis can be established readily in the acute stage but chronic cases sometimes present great difficulties.

1. Specimens

A. Specimens in Men

Urethra

In **acute gonorrhea,** the urethral discharge contains gonococci in large numbers. The meatus is cleaned with a gauze soaked in saline and a sample of the discharge collected with a platinum loop for culture, or directly on slide for smears. Purulent discharge may be expressed at the anterior urethra and collected with a swab.

In **chronic infections**, there may not be any urethral discharge. The morning drop of secretion may be examined or some exudate may be obtained after prostatic massage. It may also be possible to demonstrate gonococci in the centrifuged deposits of urine in cases where no urethral discharge is available.

Anal canal—rectal culture in homosexual males.

B. Specimens in Women

In women, urethral, cervical and rectal specimens should always be examined. A single well taken endocervical swab will detect approximately 90 percent of gonococcal infections in women. A high vaginal swab is not suitable. Throat infection also occurs and should be sought where appropriate.

C. Specimens in both Sexes

i. **Blood, swabs of skin lesions, or pus aspirated from a joint.**
ii. **Conjunctival swab,** particularly in neonatal ophthalmia.

2. Transport

For culture, specimens should be inoculated on prewarmed plates, immediately on collection. If this is not possible, specimens should be collected with charcoal impregnated swabs and sent to the laboratory in Stuart's transport medium.

3. Direct Microscopy

Do the **Gram staining** which shows characteristic kidney-shaped gram-negative diplococci lying within polymorphonuclear leucocytes with a few extracellular organisms which is typical of gonococcal infection and the smear is reported as positive. Approximately 95 percent of infected men will yield a positive smear. It has to be emphasized that diagnosis of gonorrhea by smear examination is unreliable in women as some of the normal genital flora have an essentially similar morphology.

The use of fluorescent antibody techniques for the identification of gonococci in smear is more sensitive and specific diagnosis by microscopy.

4. Culture

In acute gonorrhea, cultures can be obtained readily on chocolate agar or Mueller-Hinton agar incubated at 35-36°C under 5-10 percent CO_2. In chronic cases, where mixed infection is usual and in the examination of lesions such as proctitis, however, it is better to use a selective medium such as the Thayer-Martin medium. Examine plates after 24 hours incubation and the growth is identified by morphology and biochemical reactions. Incubation of primary isolation plates is continued for 48 hours and cultures are re-examined by the above procedures before any specimen can be reported negative.

Colonies are small, round, translucent, convex or slightly umbonate, with finely granular surface and lobate margins after incubation for 24 hours in a moist aerobic environment enriched with 5-10 percent CO_2. They are soft and easily emulsifiable. After 48 hours, the colonies are larger (1.5-2.5 mm), sometimes with a crenated margin and an opaque raised center.

Smear is made from the colony and Gram staining is done. Gonococci are gram-negative cocci arranged in pairs (diplococci) with adjacent sides concave (pear or bean shaped).

5. Identification

N. gonorrhoeae is identified preliminarily on the basis of the isolation of oxidase-positive, gram-negative diplococci that grow on chocolate blood agar or on media that are selective for pathogenic *Neisseria* species. *N. gonorrhoeae* is oxidase positive. It ferments glucose with acid only. It does not ferment maltose unlike meningococci.

6. Genetic Probes

Probes specific for the nucleic acids of *N. gonorrhoeae* have been developed for the direct detection of bacteria in clinical specimens. Tests using these probes are sensitive, specific, and rapid (results are available in 2 to 4 hours).

7. Serological Diagnosis

It may not be possible to obtain gonococci in culture from some chronic cases or from patients with metastatic lesions such as arthritis. Serological tests such as may be of value in such instances, such as

i. Complement fixation tests.
ii. Immunoblotting, radioimmunoassay, ELISA (enzyme-linked immunosorbent assay) tests.

However, no serological test has been found useful for routine diagnostic purposes. These tests are neither sensitive nor specific, and their use is not recommended.

Treatment

Penicillin is no longer the antibiotic of choice for treatment of gonorrhea since the development and widespread use of penicillin, gonococcal resistance to penicillin has gradually risen, owing to the selection of **chromosomal mutants,** so that many strains now require high concentrations of penicillin G for inhibition (MIC 2 μg/ml).

Penicillinase Producing Gonococci (PPNG)

In 1976, gonococci producing β-lactamase (penicillinase) have appeared, rendering penicillin treatment ineffective. These **Penicillinase producing gonococci (PPNG)** were isolated from widely separated areas in the United States and England and have spread widely. Penicillinase production, in gonococci, is plasmid-mediated.

Chromosomally Mediated Resistance (CMRNG)

Strains of penicillin-resistant *N. gonorrhoeae* that do not produce β-lactamase have also been isolated. This **chromosomally mediated resistance (CMRNG)** is not limited only to penicillin but extends to tetracyclines, erythromycin, and aminoglycosides and results from changes on the cell surface that keep the antibiotic from penetrating into the gonococcal cell. Resistance to fluoroquinolones such as ciprofloxacin has also become prevalent in Africa, Southeast Asia, Australia, and some US cities.

Empirical Therapy

Selection of effective empirical therapy is problematic because the incidence of antibiotic resistance in gonococci is growing. Currently, the Centers for Disease Control and Prevention (CDC) recommends that **ceftriaxone, cefixime, ciprofloxacin, or ofloxacin** be used as the initial therapy for cases of **uncomplicated gonorrhea. Doxycycline** or **azithromycin** should be added for infections complicated by dual infections with *Chlamydia.*

Prophylaxis

Control of gonorrhea consists of early detection of cases, contact tracing, health education and other general measures.

Barrier methods of contraception, condoms in particular, greatly reduce the rate of transmission. Chemoprophylaxis is of limited value because of the rise in antibiotic resistance of the gonococcus. As even clinical disease does not confer any immunity, vaccination has no place in prophylaxis.

NONGONOCOCCAL (NONSPECIFIC) URETHRITIS

Nongonococcal urethritis (NGU), also known as nonspecific urethritis (NSU), refers to chronic urethritis where gonococci cannot be demonstrated. In some of these, urethritis forms part of a syndrome consisting of conjunctivitis and arthritis in addition (Reiter's syndrome). Some of these cases may be due to gonococcal infection, the cocci persisting as L-forms and hence undetectable by routine tests. The majority of such cases are, however, the result of infections of diverse etiology.

Causative Agents

This condition is caused both by nonmicrobial factors such as catheters and drugs and by infectious microorganisms. The most important causative agents are:

A. Bacterial

- *Chlamydia trachomatis*
- *Ureaplasma urealyticum*
- *Mycoplasma hominis*
- *Gardnerella vaginalis,*
- *Acinetobacter wolfii,*
- *Acinetobacter calcoaceticus*

B. Viral

Herpes virus

Cytomegalovirus.

C. Fungi

Candida albicans

D. Protozoa

Trichomonas vaginalis

E. Mechanical or Chemical Irritation

Note: Most infections are acquired sexually, and of these, approximately 50 percent are *Chlamydia* infections. As etiological diagnosis is seldom achieved, the management of this syndrome is difficult.

Diagnosis

1. Demonstration of a leukocyte exudate.
2. Exclusion of urethral gonorrhea by Gram stain and culture.
3. Several rapid tests for detecting *Chlamydia* in urine specimens are also available.

Treatment

Treatment is with tetracycline, doxycycline, erythromycin, or sulfisoxazole.

COMMENSAL NEISSERIAE

These organisms occur on various mucous surfaces of the body. Several species of neisseriae inhabit the normal respiratory tract. They are regularly found in the throat, nose and mouth and, less frequently, on the genital mucosae. The characteristic features of some of the common species are listed in Table 27.2. Their pathogenic significance is uncertain though some of them (for example, *N. flavescens, N. catarrhalis)* have been reported occasionally as having caused meningitis.

Neisseria lactamica

Neisseria lactamica, frequently isolated from the nasopharynx is closely related to meningococci, though it is virtually avirulent. It grows readily on selective media and some strains cross-react with antisera raised against gonococci and meningococci. It differs from pathogenic neisseriae in being positive in the ONPG test for beta galactosidase. *Nasopharyngeal* colonization by *N. lactamica* in young children may be responsible for the presence in them of antibodies protective against meningococcal infection.

N. sicca, N. subflava, N. cinera, N. mucosa, and *N. flavescens* are also members of the normal flora of the respiratory tract, particularly the nasopharynx, and very rarely produce disease. *N. cinera* sometimes resembles *N. gonorrhoeae* because of its carbohydrate fermentation pattern.

Table 27.2: Differential characteristics of commonly isolated neisseriae

Species	Colonies	Growth		Fermentation			Serological classification
		On nutrient agar	At 22°C	Glucose	Maltose	Sucrose	
N. menigitidis	Round, smooth, shiny, creamy consistency	–	–	A	A	–	Thirteen antigenic groups
N. gonorrhoeae	Same as above, but smaller and more *opalescent*	–	–	A	–	–	Antigenically heterogenous
N. flavescens	Resemble meningococcus but pigmented yellow	+	+	–	–	–	Antigenically distinct homogeneous group
N. sicca	Small, dry, opaque, wrinkled, brittle	+	+	A	A	A	Autoagglutinable
N. catarrhalis (Branhamella catarrhalis)	Variable, smooth and translucent or adherent and opaque, not easily emulsifiable	+	+	–	–	–	Autoagglutinable

MORAXELLA

The genus *Moraxella* is a member of **the family Neisseriaceae**. The exact taxonomic delineation of the genus *Moraxella* is currently the subject of debate. It has been proposed that members of the genera *Moraxella* and *Acinetobacter* should be classified in the new family Moraxellaceae.

Species

The *Moraxella* spp. of medical importance are *M. lacunata, M. catarrhalis, M. osloensis, M. phenylpyruvica, M. atlantae* and *M. nonliquefaciens.*

MorAxella (Branhamella) Catarrhalis

Moraxella catarrhalis was previously named *Branhamella catarrhalis* and before that *Neisseria catarrhalis. Neisseria catarrhalis* is now classified as *Moraxella (Branhamella) catarrhalis.*

Morphology

They are oval gram-negative cocci. Sometimes organisms are single, but more often in pairs with adjacent sides flattened. They typically occur in pairs and may simulate gonococci. They are non-capsulate and nonmotile.

Cultural Characteristics

They are aerobes. Most strains grow on nutrient agar, blood agar or chocolate agar.

Biochemical Reactions

- It is oxidase positive and catalase positive
- It does not produce acid from glucose, maltose, sucrose, lactose or fructose
- It reduces nitrate to nitrite.

Pathogenesis

They are now recognized as an important respiratory pathogen. They form part of the normal pharyngeal flora but can cause respiratory infections, including otitis media, sinusitis, tracheobronchitis and pneumonia.

Treatment

As many strains produce beta lactamases, penicillins are not useful in treatment unless given in combination with clavulanate or sulbactam.

MORAXELLA LACUNATA (MORAX-AXENFELD BACILLUS)

Formerly this was included in the genus *Haemophilus* but it has been separated into the genus *Moraxella* as it does not require X or V factor. *M. lacunata* was first reported as the cause of angular conjunctivitis by Morax (1896) and Axenfeld. Hence it is also known as the Morax-Axenfeld bacillus.

Morphology

These are short, plump, gram-negative bacilli usually arranged in pairs. They are nonflagellated but have been reported to be sluggishly motile.

Culture

They are strictly aerobes, grow on ordinary media but require blood or serum for growth. On Löffler's serum slope, the colonies form pit or lacunae (hence the name lacunata).

Biochemical Reactions

They are nonfermentative and do not ferment sugars; oxidase and catalase positive, indole and H_2S negative.

Pathogenesis

The moraxellas occur as components of the normal flora of the upper respiratory tract, the conjunctiva, the skin and the genital tract.

- *Moraxella lacunata* causes a form of purulent conjunctivitis classically presenting as an **angular blepharoconjunctivitis.**

- **Endophthalmitis** may be caused by *M. nonliquefaciens.*
- From cases of sinusitis, conjunctivitis, bronchitis, endocarditis, meningitis and septicemia, *M. phenylpyruvica* and *M. osloensis* have been isolated.
- Systemic infections may occasionally be caused by *M. atlantae.*

Treatment

M. lacunata is very sensitive to zinc salts. They are sensitive to penicillin and most other antibiotics.

KINGELLA

The genus *Kingella,* comprising some species of oxidase positive, nonmotile, gram-negative rods, with a tendency to occur as coccobacillary and diplococcal forms was formerly grouped under the genus *Moraxella*. The genus contains three species *(K. kingae, K. indologenes* and *K. denitrificans).*

K. kingae is the most commonly isolated species. It is part of the normal oral flora and has been associated with endocarditis and infections of bones, joints and tendons.

HACEK Group of Oral Bacteria

Kingella species are included in the so-called **HACEK** group of oral bacteria *(Haemophilus* spp. *Actinobacillus actinomycetemcomitans, Cardiobacterium hominis, Eikenella corrodens* and *Kingella* spp.). These organisms are known to colonize endovascular tissue and produce vegetations on heart valves. In common with some other gram-negative rods, such as *Cardiobacterium hominis, Eikenella corrodens* and *Actinobacillus actinomycetemcomitans, Kingella* species (usually *K. kingae)* are sometimes found in endocarditis. They have also been implicated in joint infections.

KNOW MORE

N. meningitides

Eradication of the pool of healthy carriers of *N. meningitidis* is unlikely. For this reason, efforts have been concentrated on the prophylactic treatment of people exposed to diseased patients and on the enhancement of immunity to the serogroups most commonly associated with disease.

N. gonorrhoeae

The major reservoir for gonococci is the asymptomatically infected person. The site of infection also determines whether carriage occurs, with rectal and pharyngeal infections more commonly asymptomatic than genital infections.

- Although there is tremendous interest in developing a vaccine against *N. gonorrhoeae,* an effective vaccine is not available currently.

KEY POINTS

Neisseria

- The genus *Neisseria* are aerobic, gram-negative cocci typically arranged in pairs (diplococci). Oxidase and catalase-positive; acid produced from glucose oxidatively.

  Important species of the genus *Neisseria* are *N. meningitides* and *N. gonorrhoeae.*

Neisseria meningitidis

- Gram-negative diplococci with fastidious growth requirements.
- Growth is facilitated by 5-10 percent CO_2 and high humidity and grows best at 35°C to 37°C. Blood agar, chocolate agar and Mueller-Hinton starch casein hydrolysate agar are the media commonly used for culturing meningococci. Modified Thayer -Martin (with vancomycin, colistin and nystatin) is a useful selective medium.
- Oxidase and catalase positive; acid produced from glucose and maltose oxidatively.

Antigenic Classification

Based on their capsular polysaccharide antigens, meningococci are classified into at least 13 serogroups: A, B, C, X, Y, Z, Z1 (29E) and W135. Further serogroups H, I, K and L have also been described. Groups A, B and C are the most important. Groups 29-E, W-135 and Y also frequently cause meningitis.

- **Diseases:** Meningitis, meningoencephalitis, bacteremia, pneumonia, arthritis, urethritis.
- For immunoprophylaxis, vaccination is an adjunct to chemoprophylaxis; it is used only for serogroups A, C, Y, and W135; no effective vaccine is available for serogroup B. Polysaccharide vaccines conjugated with protein carriers offer protection for infants younger than 2 years.

Neisseria gonorrhoeae

- *N. gonorrhoeae* appears as a diplococcus with the adjacent sides concave, being typically kidney shaped. It is found predominantly within the polymorphs, some cells containing as many as a hundred cocci.
- It has fastidious growth requirements. Growth best at 35°C to 37°C in a humid atmosphere supplemented with CO_2. They grow well on **chocolate agar** and **Mueller-Hinton agar**. A popular selective medium is the **Thayer-Martin medium** (chocolate agar containing vancomycin, colistin and nystatin).
- **Diseases:** Humans are the only natural hosts. Asymptomatic carriage is the major reservoir. Transmission primarily by sexual contact. It causes urethritis, cervicitis, salpingitis, pelvic inflammatory disease, proctitis, bacteremia, arthritis, conjunctivitis, pharyngitis.

- **Treatment:** Gonococci producing β-lactamase (penicillinase) have appeared, called **penicillinaseproducing gonococci (PPNG)** and is plasmid mediated. **Chromosomally mediated resistance (CMRNG)** have also been isolated. Ceftriaxone, cefixime, ciprofloxacin, or ofloxacin can be administered in uncomplicated cases.
- **Nongonococcal urethritis (NGU)**, also known as nonspecific urethritis (NSU), refers to chronic urethritis where gonococci cannot be demonstrated.

Commensal neisseriae: They are regularly found in the throat, nose and mouth and, less frequently, on the genital mucosae. Their pathogenic significance is uncertain though some of them (for example, *N. flavescens, N. catarrhalis*) have been reported occasionally as having caused meningitis.

Moraxella (Branhamella) catarrhalis: They are now recognized as an important respiratory pathogen.

Kingella: *K. kingae* is part of the normal oral flora and has been associated with endocarditis and infections of bones, joints and tendons.

IMPORTANT QUESTIONS

1. Describe the morphology, pathogenicity and laboratory diagnosis of meningococcal meningitis.
2. Discuss laboratory diagnosis of gonorrhoea.
3. Write short notes on:
   - Antigenic structure of *Neisseria gonorrhoeae*
   - Nongonococcal urethritis (or) Nonspecific urethritis.
   - *Moraxella (Branhamella) catarrhalis.*

FURTHER READING

Briligan BE, Sparling PF. Gonococcal infection. New Engl J Med 1985;312:1683.

Carlwrighl K (Ed). Meningococcal Disease, Chichester: John Wiley 1995.

Tunkel AR, Scheld MW. Acute baclerial meningitis, Lancet 1995;346:1675.

Van Deuren M, et al. Update on meningococcal disease with emphasis on pathogenesis and clinical management, Clin Microbiol Rev 2000;13:144-66.

CHAPTER 28 Corynebacterium

LEARNING OBJECTIVES

After reading and studying this chapter, you should be able to:

- Describe morphology, cultural characteristics, biochemical characters, toxin production and pathogenesis of diphtheria.
- Differentiate among three different biotypes: gravis, intermedius and mitis of *C. diphtheriae*.
- Discuss diphtheria toxin.
- Discuss laboratory diagnosis of diphtheria.
- Toxigenicity tests/virulence tests of *C. diphtheriae*.
- Describe the following: Schick test; DPT vaccine or triple vaccine.
- Diphtheroids.
- Differentiate between of *C. diphtheriae* and diphtheroids.

INTRODUCTION

The *coryneform* group consists of *Corynebacterium* and related genera that are aerobic, nonsporing and irregularly shaped, nonspore-forming and gram-positive rods.

CORYNEBACTERIUM

The genus *Corynebacterium* comprises 66 species; 38 of them associated with human disease. Corynebacteria are gram-positive, nonacid fast, nonmotile rods with irregularly stained segments, and sometimes granules. They frequently show club shaped swellings and hence the name corynebacteria (from ***coryne,*** meaning club). The terms '**diphtheroid**' and '**coryneform**' have been used to describe other corynebacteria that do not commonly cause human infection and were considered less significant. Corynebacteria are closely related to mycobacteria and nocardiae. These three groups collectively may be referred as ***CMN group.***

CORYNEBACTERIUM DIPHTHERIAE

The diphtheria bacillus was first observed and described by Klebs (1883) but was first cultivated by Löffler (1884). It is hence known as the Klebs-Löffler bacillus. Löffler studied the effect of the bacillus in experimental animals and concluded that the disease was due to some diffusible product of the bacillus. Roux and Yersin (1888) discovered the diphtheria exotoxin and established its pathogenic effect. The antitoxin was described by von Behring (1890). *C. diphtheriae* owes its pathogenicity to the production of a potent exotoxin active on a range of tissues, including heart muscle and peripheral nerves.

Diseases

The major disease caused by *C. diphtheriae* is ***diphtheria*** (Greek, ***diphtheria,* "leathery skin",** referring to the **pseudomembrane** that initially forms on the pharynx). Bretonneau, a French army surgeon, described in 1821 the unique clinical characteristics of the disease and used the term '**diphtherie**' to signify the leathery membrane that occurs in the oropharynx, or sometimes the nasopharynx, and which is the hallmark of the disease. Occasionally, the organism is implicated in wound and chronic skin infections. Nontoxigenic strains have been associated with **endocarditis, meningitis, cerebral abscess** and **osteoarthritis** throughout the world.

Morphology

They are thin, slender **gram-positive bacilli** but are easily decolorized, particularly in old cultures, measuring approximately 3-6 µm × 0.6-0.8 µm. They have a tendency to **clubbing** at one or both ends. They are highly **pleomorphic.** Cells often show septa, and branching is infrequently observed. They are nonmotile, nonspore forming, and nonacid fast.

The bacilli are arranged in a characteristic fashion in smears. They are usually seen in pairs, palisades (resembling stakes of a fence) or small groups or as individual cells lying at sharp angles to another, resembling the letters V or L. This particular arrangement with *C. diphtheriae* has been called the **Chinese letter or cuneiform arrangement (Fig. 28.1).** This is due to the incomplete separation of the daughter cells after binary fission.

This organism has granular and uneven staining. When stained with methylene blue or toluidine blue, the

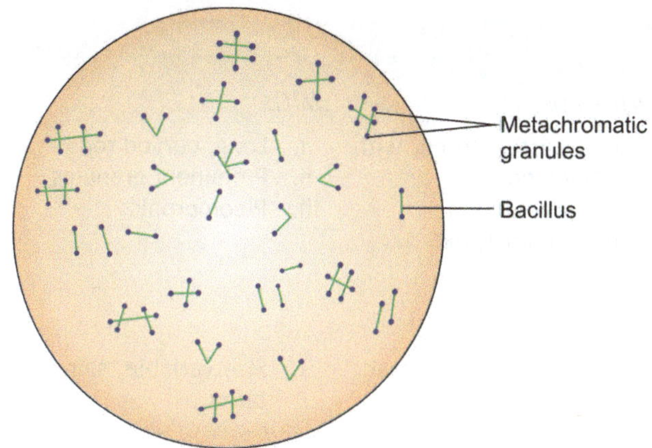

Fig. 28.1: Corynebacterium diphtheriae showing metachromatic granules and Chinese letter arrangement.

granules in the cell stain metachromatically reddish–purple. These granules are known as **metachromatic granules, volutin granules** or **Babes-Ernst granules (Fig. 28.1)**. They are often situated at the poles of the bacilli and are called **polar bodies**. Special stains, such as Albert's Neisser's and Ponder's have been devised for demonstrating the granules clearly. With Albert's stain, the granules stain **bluish black** and the protoplasm green. The granules represent accumulation of polymerized polyphosphates. The granules formation is best seen on Löffler's serum slope. It seems that they represent storage depots for materials needed to form high-energy phosphate bonds.

Cultural Characteristics

C. diphtheriae is an aerobe and facultative anaerobe; the optimum temperature for growth is 37°C (range 15-40°C) and optimum pH 7.2. Complex media are required for primary isolation and characterization. It can grow on ordinary nutrient agar, but its growth is improved by the presence of animal proteins such as blood or serum. Two media are useful for this purpose:

1. **Löffler's serum slope.**
2. **Blood agar containing fresh, lysed or heated blood.**

Löffler's Serum Slope

Diphtheria bacilli grow on Löffler's serum slope very rapidly and colonies can be seen in 6-8 hours, long before other bacteria grow. Colonies are at first small, circular white opaque disks but enlarge on continued incubation and may acquire a distinct yellow tint. Löffler's medium is also useful because it does not support growth of streptococci and pneumococci that may be present in the clinical specimen and inhibits the growth of most oral commensals and retards the growth of others such as *Candida albicans* and *Staphylococcus aureus,* acting as a selective agent but has little effect on diphtheria bacilli.

Tellurite Blood Agar

The addition of **potassium tellurite (0.03-0.04%)** makes the medium selective for corynebacteria by inhibiting most other pathogenic and commensal bacteria. On this medium, *C. diphtheriae* give gray/black, shiny or dull black colonies because the tellurite ion passes through the cell wall and membrane into the cytoplasm, where it is reduced to the metal tellurium and precipitated inside the cells. Some other corynebacteria, staphylococci, and yeasts reduce the tellurite salts producing characteristic gray to black colonies. The addition of cystine to a tellurite containing medium **(Tinsdale's medium)** has greatly helped the isolation of diphtheria bacilli. The growth of diphtheria bacilli may be delayed on the tellurite medium and colonies may take two days to appear.

Based on colonial morphology on the tellurite medium and other properties, McLeod and Anderson described three different biotypes: ***gravis, intermedius*** and ***mitis*** (Table 28.1). The names were originally proposed to relate to the clinical severity of the disease produced by the three types—**gravis,** causing the most serious, and **mitis** the mildest variety, with **intermedius** being responsible for disease of intermediate severity. However, this association is not constant. In general, biotype mitis is predominant in endemic areas, while intermedius and gravis tend to be epidemic.

Biochemical Reactions

C. diphtheriae ferments glucose and maltose with the production of acid (but no gas) but not lactose, mannitol, trehalose or sucrose. Starch and glycogen are used for biochemical differentiation of three biotypes of *C. diphtheriae* (Table 28.1). *Gravis* strains utilize glycogen and starch, while *mitis* and *intermedius* do not. Fermentation of sugars are usually done in Hiss's serum peptone water medium.

C. diphtheriae is **H_2S positive** and **reduces nitrate to nitrite.** It does not liquefy gelatin or hydrolyze urea or form phosphatase.

Pyrazinamidase (PYZ) Test

In pyrazinamidase (PYZ) test, pyrazinamide is converted into pyrazinoic acid by the organisms which produce pyrazinamidase (PYZ). This test is helpful to distinguish *C. diphtheriae* (PYZ-negative) from other corynebacterium species (mostly PYZ-positive) except *C. ulcerans* and *C. pseudotuberculosis* which are also PYZ-negative but they are urease test positive which differentiate them from *C. diphtheriae* (urease negative).

Toxin

Toxigenic strains of *C. diphtheriae* produce a a very powerful exotoxin. The toxicity observed in diphtheria is directly attributed to the toxin secreted by the bacteria at the site of infection.

Synthesis

Almost all strains of gravis, 95-99% of intermedius and 80-85% of mitis produce this toxin. Strains of all three types are invariably virulent when isolated from acute cases. Avirulent strains are common among

Table 28.1: Type differentiation of *Corynebacterium diphtheriae*			
	Gravis	*Intermedius*	*Mitis*
1. Morphology	i. Usually short rods, with uniform staining ii. Few or no granules iii. Pleomorphism (some degree), with irregularly barred, snow-shoe and teardrop forms.	i. Long barred forms with clubbed ends ii. Poor granulation iii. Very pleomorphic	i. Long, curved rods ii. Prominent granules iii. Pleomorphic.
2. Colony on tellurite blood agar	i. **In 18 hours.** Colony is 1-2 mm in size, grayish black center, paler, semitranslucent periphery and commencing crenation of edge ii. **In 2-3 days** 3-5 mm in size, flat colony with raised dark center and crenated edge with radial striation **'daisy head' colony.**	i. **In 18 hour:** Colony small, 1 mm in size, misty. ii. **In 48 hours:** Does not enlarge, dull granular center with smoother, more glistening periphery and a lighter ring near the edge **'frog's egg' colony.**	i. Size variable, shiny black. ii. **In 2-3 days:** Colonies become flat, with a central elevation **'poached egg' colony.**
3. Consistency of colonies	i. Brittle, moves as a whole on the plate like 'cold margarine' ii. Not easily picked out or emulsifiable.	Intermediate between gravis and mitis.	i. Soft, buttery. ii. Easily emulsifiable.
4. Hemolysis	Variable	Nonhemolytic	Usually hemolytic.
5. Growth in broth	Surface pellicle Deposit granular Turbidity little or no.	Turbidity in 24 hours, clearing in 48 hours, with fine granular sediment.	Turbidity diffuse with soft pellicle later.
6. Glycogen and starch fermentation	Positive	Negative	Negative.
7. Toxigenic strains	Almost 100%	95-99%	80-85%.
8. Virulence	Severe	Moderate	Mild.
9. Predominant strains in	Epidemic areas	Epidemic areas	Endemic areas.

convalescents, contacts and carriers, particularly in those with extrafaucial infection. There is considerable variation in the amount of toxin produced by the different strains, some strains producing it abundantly and others only poorly. However, toxin produced by different biotypes is antigenically similar. The strain almost universally used for toxin production is the 'Park Williams 8' strain, which has been variously described as a mitis (Toplev and Wilson) and an intermedius strain (Cruickshank). The classic Park-Williams strain (PW8) of *C. diphtheriae* isolated in 1896, is still used as a source of toxin for preparation of diphtheria toxoid (vaccine).

Lysogeny and Toxin Production

The toxigenicity of the diphtheria bacillus depends on the presence in it of **corynephages (tox+),** which act as the genetic determinant controlling toxin production. Nontoxigenic strains can be converted to **tox+** by infection with the appropriate bacteriophage. This is known as ***lysogenic or phage conversion.*** The bacillus loses the toxigenicity when it is cured of its phage, as by growing it in the presence of antiphage serum.

Iron for Toxin Production

Toxin production is also influenced by the concentration of iron in the medium. The optimum level of iron for toxin production is 0.1 mg per liter, while a concentration of 0.5 mg per liter inhibits the formation of toxin. The toxin is released in significant amounts only when the available iron in the culture medium is exhausted.

Properties of Toxin

Diphtheria toxin is an iron-free, crystalline, heat labile protein. It is extremely potent and is lethal for humans in amounts of 130 ng per kg of body weight. In highly susceptible species (guinea pig, rabbit), the lethal dose of diphtheria toxin is 0.1 μg/kg or less. The diphtheria toxin is a protein and has a molecular weight of about 62,000. Fragment A has all the enzymatic activity whereas fragment B is responsible for binding the toxin to the cells. The toxin is heat labile. It is extremely potent (0.0001 mg kills a guinea pig of 250 gm weight). The toxicity of the toxin is due to its ability to block protein synthesis in eukaryotic cells. The toxin consists of two fragments **A (active)** and **B (binding)** of MW 24,000 and 38,000, respectively. Both fragments are necessary for the toxic effect.

It has a special affinity for certain tissues such as the myocardium, adrenals and nerve endings. Prolonged storage, incubation at 37°C for 4-6 weeks, treatment with 0.2-0.4 percent formalin or acid pH converts it to toxoid. Toxoid is toxin that has lost its toxicity but not its antigenicity. It is capable of inducing antitoxin and reacting specifically with it.

Mode of Action

The toxin is secreted by the bacterial cell and is nontoxic until exposed to trypsin. The trypsinization results in two polypeptide fragments, A and B, which are linked together by a disulfide bridge. Fragment A is responsible for the cytotoxicity; fragment B binds to receptors on the eukaryotic cells and mediates the entry of fragment A into the cytoplasm. The antibody to fragment B is protective by preventing the binding of the toxin to the cells. The diphtheria toxin acts by inhibiting protein synthesis. Specifically, fragment A splits nicotinamide adenosine dinucleotide (NAD) to form **nicotinamide** and **adenosine diphosphoribose (ADPR)**. ADPR binds to and inactivates elongation factor 2 (EF-2), an enzyme required for elongation of polypeptide chains on ribosomes. Inhibition of protein synthesis is probably responsible for both the necrotic and neurotoxic effects of the toxin.

The reaction can be summarized as follows:

$$NAD^+ + \underset{\text{Active}}{EF\text{-}2} = \underset{\text{Inactive}}{ADPR\text{-}EF\text{-}2} + \text{nicotinamide} + H^+$$

Resistance

C. diphtheriae is more resistant to the action of light, desiccation, and freezing than are most nonsporeforming bacilli. On dried fragments of pseudomembranes, organisms survive for at least 14 weeks. They are readily killed, however, by a 1 minute exposure to 100°C or a 10 minute exposure to 58°C. They are susceptible to most of the routinely used disinfectants. It dies rapidly in 0.85% NaCl solution, but remains alive for weeks in dust and on fomites when dry and protected from sunlight. It is susceptible to penicillin, erythromycin and broad spectrum antibiotics.

Antigenic Structure

Diphtheria bacilli are antigenically heterogeneous. They possess three distinct antigens:

1. A deep-seated antigen found in all corynebacterial species as well as in *Mycobacterium tuberculosis*
2. A heat-labile protein (K antigen).
3. A heat-stable polysaccharide (O antigen).

Serotypes

On the basis of agglutination reaction, biotypes gravis, intermedius and mitis have been divided into 13, 4 and 40 serotypes, respectively. No connection has been established between type specificity and other characters.

Bacteriophage Typing

The susceptibility of *C. diphtheriae* to bacteriophage strains has been most comprehensively studied in Romania; 22 phages were used to type the *gravis* strains into 14 types, the *intermedius* into 3 and the *mitis* into 4. An additional set of 33 phages has also been used. The only other corynebacteria reported as susceptible to the diphtheria typing phages are *C. ulcerans* and *C. pseudotuberculosis.*

Bacteriocin Typing

With the help of nine indicator strains of *C. diphtheriae*, Gibson and Colman (1973) demonstrated 10 patterns of bacteriocin types amongst Australian strains. Other methods of typing include bacterial polypeptide analysis, DNA restriction patterns and hybridization with DNA probes.

Pathogenesis

In the upper respiratory tract, diphtheria bacilli elicit an inflammatory exudate and cause necrosis of the cells of the faucial mucosa. The diphtheria toxin possibly assists colonization of the throat or skin by killing epithelial cells or neutrophils.

Diphtheria is a toxemia. The organisms do not penetrate deeply into the mucosal tissue and bacteremia does not usually occur. The exotoxin is produced locally and is spread by the bloodstream to distant organs, with a special affinity for heart muscle, the peripheral nervous system and the adrenal glands.

Clinical Diseases

The organism is carried in the upper respiratory tract and spread by droplet infection or hand-to-mouth contact. The incubation period of diphtheria is 2-5 days, with a range of 1-10 days. Diphtheria, which occurs in two forms **(respiratory** and **cutaneous),** is found worldwide but is uncommon in North America and Western Europe.

A. Respiratory Diphtheria

The illness begins gradually and is characterized by low-grade fever, malaise, and a mild sore throat. The most common site of infection is the **tonsils** or **pharynx.** The organisms rapidly multiply on the epithelial cells, and the toxigenic strains of *C. diphtheriae* produce toxin locally, causing **tissue necrosis** and **exudate formation** triggering an inflammatory reaction. This combination of cell necrosis and exudate forms a tough gray to white pseudomembrane, which attaches to the tissues-commonly over the tonsils, pharynx, or larynx. Any attempt to remove the pseudomembrane results in bleeding. In nasopharyngeal infection, the pseudomembrane may involve nasal mucosa, the pharyngeal wall and the soft palate. In this form, edema involving the cervical lymph glands may occur in the anterior tissues of the neck, a condition known as **bullneck diphtheria.** Laryngeal involvement leads to obstruction of the larynx and lower airways.

B. Systemic Effects

The toxin also is absorbed and can produce a variety of systemic effects involving the **kidneys, heart,** and **nervous system,** although all tissues possess the receptor for the toxin and may be affected. Intoxication takes the form of **myocarditis** and **peripheral neuritis,** and may be associated with **thrombocytopenia.** Visual disturbance, difficulty in swallowing and paralysis of the arms and legs also occur but usually resolve spontaneously. Complete heart block may result from myocarditis. Death is most commonly due to congestive heart failure and cardiac arrhythmias.

C. Complications

The common complications are:

1. **Asphyxia** due to mechanical obstruction of the respiratory passage by the pseudomembrane for which an emergency tracheostomy may become necessary.
2. **Acute circulatory failure**, which may be peripheral or cardiac.
3. **Postdiphtheritic paralysis**, which typically occurs in the third or fourth week of the disease; palatine and ciliary but not pupillary paralysis is characteristic, and spontaneous recovery is the rule.
4. **Septic, such as pneumonia and otitis media.**

D. Cutaneous Diphtheria

In the **cutaneous diphtheria**, which is prevalent in the tropics, the toxin also is absorbed systemically, but systemic complications are less common than from upper respiratory infections with *C. diphtheriae.*

Laboratory Diagnosis

Diagnostic laboratory tests serve to confirm the clinical impression and are of epidemiologic significance but not for the treatment of individual cases. Specific treatment should be instituted *immediately* on suspicion of diphtheria without waiting for laboratory tests. Any delay may be fatal. Laboratory diagnosis consists of **isolation of the diphtheria bacillus** and **demonstration of its toxicity.**

1. Specimens

Swabs from the nose, throat, or other suspected lesions must be obtained before antimicrobial drugs are administered. In suspected cases, whether of faucial or nasal diphtheria, swabs should be taken both from the throat and from the nose, and preferably two swabs from the site most affected. Swabs should also be taken from skin lesions and wounds where diphtheritic infection is suspected, and both throat and nose swabs should be taken from suspected carriers.

2. Microscopy

Direct microscopy of a smear is unreliable since *C. diphtheriae* is morphologically similar to other coryneforms. Smears stained with alkaline methylene blue or Gram stain show beaded rods in typical arrangement. Hence smear examination alone is not sufficient for diagnosing diphtheria but is important in identifying **Vincent's angina.** For this, a Gram or Leishman stained smear is examined for Vincent's spirochetes and fusiform bacilli. Toxigenic diphtheria bacilli may be identified in smears by immunofluorescence.

3. Culture

The swab should be inoculated on Löffler's serum slope, tellurite blood agar, and blood agar. The cultures should be incubated aerobically at 37°C. Unless the swab can be inoculated promptly, it should be kept moistened with sterile horse serum so the bacilli will remain viable.

i. Löffler's Serum Slope

After incubation for 6 hours or overnight, make a smear of growth from all parts of the slope mixed in the condensation water, stain by the Albert-Laybourn method and look for the presence of slender green-stained bacilli containing the purple-black granules characteristic of *C. diphtheriae.* In 12-18 hours, the Löffler slant may yield organisms of typical "diphtheria-like" morphology.

ii. Tellurite Blood Agar

Blood tellurite agar is examined after 24 hours and after 48 hours, as growth may sometimes be delayed.

iii. Blood Agar

It is used for differentiating streptococcal or staphylococcal pharyngitis, which may simulate diphtheria.

4. Identification Tests

Identification is based on carbohydrate fermentation reactions and enzymatic activities. *C. diphtheriae* ferments glucose and maltose, producing acid but not gas, and is catalase positive. It reduces nitrate to nitrite and is nonmotile. Commercial kits such as the API Coryne strip provide a reliable identification.

5. Virulence Tests

Any diphtheria-like organism cultured must be submitted to a "virulence" test before the bacteriologic diagnosis of diphtheria is definite. Such tests are really tests for toxigenicity of an isolated diphtheria-like organism. Diagnosis of diphtheria depends on showing that the isolate produces diphtheria toxin. Virulence testing may be by *in vivo* or *in vitro* methods. *In vivo* testing is rarely done because the *in vitro* methods are reliable, less expensive, and free from the need to use animals.

A. *In Vivo* Tests

i. Subcutaneous test.
ii. Intracutaneous test.

B. *In Vitro* Test

i. Precipitation test.
ii. Tissue culture test.
iii. Enzyme-linked immunosorbent assays.
iv. Polymerase chain reaction (PCR).

A. *In Vivo* Tests

i. Subcutaneous Test

The growth from an overnight culture on Löffler's slope is emulsified in 2-4 ml broth and 1 ml of the emulsion injected **subcutaneously** into two guinea pigs or rabbits, one of which has been protected with the diphtheria antitoxin (500-1000 units) 18-24 hours previously and was used as control. If the strain is virulent, the unprotected animal will die within four days. Postmortem examination would show hemorrhage at the site of injection and injected blood vessels, with typically hemorrhagic adrenal necrosis.

Simple and reliable subcutaneous toxigenicity tests in rabbits or larger guinea pigs were used in the past, at a time when laboratories had many isolates each day. **The method is not usually employed as it is wasteful of animals.**

ii. Intracutaneous (Intradermal) Test

The broth emulsion of the culture is inoculated **intracutaneously** into two guinea pigs (or rabbits) so that each receives 0.1 ml in two different sites. One animal acts as the **control** and should receive antitoxin (500 units) the previous day. After four hours of the skin test, the other is given 50 units of antitoxin intraperitoneally in order to prevent death. In the test animal, toxigenicity is indicated by inflammatory reaction at the site of injection, progressing to necrosis in 48-72 hours and in **the control animal,** no change.

Advantages of Intracutaneous Test

a. The animals do not die.
b. As many as ten strains can be tested at a time on a rabbit.

B. *In Vitro* Test

i. Elek's Gel Precipitation Test

The *in vitro* diphtheria toxin detection procedure is an **immunodiffusion test** first described by **Elek.** In the Elek test, organisms (controls and unknowns) are streaked on media of low iron content to optimize toxin production.

Procedure: A rectangular strip of filter paper impregnated with diphtheria antitoxin **(1000 units/ml)** is placed on the surface of a 20% normal horse serum agar in a petridish while the medium is still fluid. Sheep or rabbit serum may be used, if horse serum is not available. When the agar has set, the surface is dried. The plate should be streaked with the **test strain** as well as the **control positive** and **negative strains** at right angles to the strip in a single straight line parallel to each other and 10 mm apart. The plate is incubated at 37°C and examined after 24 and 48 hours.

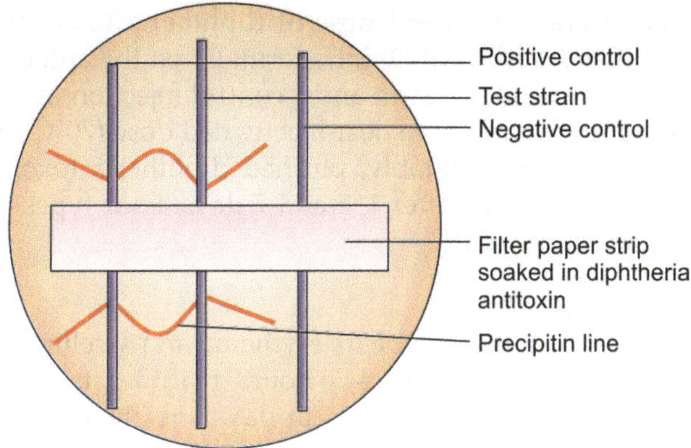

Fig. 28.2: Elek's test

Interpretation: Toxins produced by the bacterial growth will diffuse in the agar and where it meets the antitoxin at optimum concentration will produce a line of precipitation (Fig. 28.2). If an isolate is **positive** for toxin production, and it is placed next to the positive control, the toxin line of the positive control should join the toxin line of the positive unknown to form **an arch of identity**. A negative control should be free of any line. No precipitate will form in the case of nontoxigenic strains.

ii. Tissue Culture Test

The toxigenicity of diphtheria bacilli can be demonstrated by incorporating the strains in the agar overlay of cell culture monolayers. The toxin produced diffuses into the cells below and kills them.

iii. Enzyme-linked Immunosorbent Assays (ELISA)

Rapid, enzyme-linked immunosorbent assays and immunochromatographic strip assays are also available for the detection of diphtheria toxin.

iv. Polymerase Chain Reaction (PCR)

In addition, procedures for detecting the *C. diphtheriae tax* gene by the polymerase chain reaction (PCR) have been developed. The PCR assay can also be applied directly to clinical specimens.

Schick Test

Schick (1913) introduced an **intradermal test** (Schick test) for distinguishing between susceptible and immune persons.

Principle

This test depends upon the principle of toxin-antitoxin neutralization, *in vivo*, and the test is carried out by injecting one Schick test dose of diphtheria toxin (0.2 ml containing 1/50 MLD) intradermally on the anterior surface of the left forearm and a control injection in the right forearm contains a heat-inactivated dose (70°C for 30 minutes) or preferably, purified diphtheria toxoid. Readings are taken after 1, 4 and 7 days. Four types of reactions may occur:

i. Positive Reaction

In the test arm there appears erythema and swelling at the site of inoculation in 24-36 hours, reaching its maximum (1-5 cm) by the 4th to 7th days and then fading with superficial scaling and persistent brownish pigmentation. On the control arm there is no reaction.

A positive Schick test indicates that the individual is susceptible to diphtheria and little or no antitoxin (less than 0.01 unit/ml) is present. The subject is not immune and should be immunized.

ii. Negative Reaction

There is no reaction of any kind in either arm. This indicates that the toxin has been neutralized by the circulating antitoxin and the person is immune to diphtheria and does not need immunization. In quantitative terms, the test will be negative if the blood serum contains the antitoxin concentration of 0.0 I unit or more/ml and is not sensitive to any of the antigens in the toxin or toxoid and does not need immunization.

iii. Pseudoreaction

There is erythema occurring within 6-24 hours and disappearing within four days. The reaction is the same on both arms. This indicates that the individual is immune to diphtheria and also that he is hypersensitive to one or more antigens in the toxin preparation. This individual does not need immunization.

iv. Combined Reaction

Here the initial picture is that of pseudoreaction, but while the erythema in the control arm fades, within four days, it progresses in the test arm to a typical positive reaction. This indicates that the individual is susceptible to diphtheria and is sensitive to one or more antigens in the toxin preparation making immunization necessary but likely to induce reaction. Doses of the vaccine should be greatly induced, and the number of injections increased.

Note: The Schick test has largely been replaced by the measurement of serum antitoxin level by the **hemagglutination.**

Epidemiology

Diphtheria is a disease found worldwide, particularly in poor urban areas where there is crowding and the protective level of vaccine-induced immunity is low. *C. diphtheriae* is maintained in the population by asymptomatic carriage in the oropharynx or on the skin of immune people. Infection is confined to man and usually involves contact with a diphtheria case or a carrier. Humans are the only known reservoir, with carriage in oropharynx or on skin surface.

Diphtheria is primarily a pediatric disease, but the highest incidence has shifted toward older age groups in areas where there are active immunization programs for children. Acquired immunity to diphtheria is due primarily to toxin-neutralizing antibody (antitoxin) Passive immunity *in utero* is acquired transplacentally and can last for 1 or 2 years after birth. Skin infection with toxigenic *C. diphtheriae* (cutaneous diphtheria) also occurs.

Prophylaxis

The methods of immunization available are **active, passive** or **combined.** Of these only active immunization can provide herd immunity and lead to eradication of the disease. Passive and combined immunization can only provide emergency protection to susceptible individuals exposed to risk.

A. Active Immunization

The preparations used for active immunization are as follows:

1. Toxin-antitoxin mixture: It is not without hazards.
2. Single vaccines: It is less frequently used.
3. Combined preparations.

Combined Preparations

- DPT (diphtheria-pertussis-tetanus) vaccine
- DT (diphtheria-tetanus toxoid)
- DT (diphtheria-tetanus, adult type).

DPT Vaccine

Diphtheria toxoid is usually given in children as a trivalent preparation containing **tetanus toxoid** and **pertussis vaccine** is known also as the **DTP, DPT or triple vaccine.** A **quadruple vaccine** containing in addition the inactivated poliovaccine is also available. The WHO recommends that only **adjuvant DPT preparations** be utilized in immunization programs.

Schedule of primary immunization: The schedule of primary immunization of infants and children consists of **DPT given** at the age of 6 weeks, 10 weeks, 14 weeks and 16-24 months followed by booster dose **DT** at the age of **5-6 years (school entry)**.

Reactions: Fever and mild local reactions following DPT immunization are common. The most severe complications following DPT immunization are neurological (encephalitis/encephalopathy, prolonged convulsions, infantile spasms and Reye's syndrome).

B. Passive Immunization

This is an emergency measure to be employed where susceptibles (nonimmunized) are exposed to infection, as when a case of diphtheria is admitted to general pediatric wards. It consists of the subcutaneous administration of 500-1000 units of antitoxin (antidiphtheritic serum,

ADS). As this is a horse serum, precaution against hypersensitivity should be observed.

C. Combined Immunization

This consists of administration of the first dose of adsorbed toxoid, while ADS is given on the other arm, to be continued by the full course of active immunization since protection conferred by passive immununization is of short duration. Ideally, all cases that receive ADS prophylactically should receive combined immunization.

Treatment

Specific treatment of diphtheria consists of **antitoxic** and **antibiotic therapy**. Antitoxin should be given immediately as soon as clinical diagnosis is made to neutralize the toxin being produced, because antitoxin is ineffective if given after the toxin is bound to cell receptor sites. The dosage recommended is 20,000 units intramuscularly for moderate cases and 50,000 to 100,000 units for serious cases, half the dose being given intravenously.

C. diphtheriae is sensitive to most antibiotics, including **penicillin** and **erythromycin** and are used for the treatment of patients as well as carriers. The antibiotics do not neutralize circulating toxin. They prevent further toxin production by killing diphtheria bacilli. Penicillin-sensitive individuals can be given erythromycin. Erythromycin is more active than penicillin in the treatment of carriers.

OTHER MEDICALLY IMPORTANT CORYNEBACTERIA

The nondiphtheria corynebacteria are diverse; usually isolated from the environment and commensals of the skin and mucous membranes. The principal species involved and the main clinical syndromes associated with infection are shown in Table 28.2.

Corynebacterium ulcerans

A veterinary pathogen causing mastitis in cattle and other domestic and wild animals, *C. ulcerans* has been isolated from patients with diphtheria-like illness. It resembles gravis type of *C. diphtheriae* but it liquefies gelatin, ferments trehalose slowly and does not reduce nitrate to nitrite. It is PYZ negative and urease positive. It commonly produces diphtheria toxin as well as the separate toxin produced by *C. pseudotuberculosis*.

It is pathogenic to guinea pigs. The lesions produced resembling those caused by *C. diphtheriae*. It has been isolated from raw milk and can cause mastitis in cattle and man exposed to infected animals or milk may develop infection. Man to man transmission has not been reported. In man, infection usually takes the form of acute pharyngitis with pseudomembranes, and cardiac or neurological complications may occur.

Corynebacterium pseudotuberculosis (C. ovis)

Like *C. ulcerans, C. pseudotuberculosis* (Preisz-Nocard bacillus) is primarily an animal pathogen and rarely infects man. Human infections mainly occur in patients with animal (sheep) contact. It causes *caseous lymphadenitis in sheep and goats* and *abscesses* or *ulcerative lymphangitis in horses*. Rarely, it may cause *subacute* and *chronic lymphadenitis* involving axillary or cervical lymph nodes in those with prolonged exposure to sheep and horses.

Corynebacterium minutissimum

It is believed to be the causative agent of **erythrasma**. It is a localized infection of the stratum corneum which pro-

Table 28.2: Medically important nondiphtheria corynebacteria and disease associations of these corynebacteria

Organism	*Major habitat*	*Disease association*
C. ulcerans	Human throat and skin; animals; raw milk	Man: Diphtheria (toxigenic strains), pharyngitis and wound infection; cattle: Mastitis.
C. pseudotuberculosis	Sheep, horses, goats	Man: Lymphadenitis Animals: Abscesses and abortion
C. jeikeium	Skin	Bacteremia, endocarditis; infection of foreign bodies and CSF shunts.
C. urealyticum	Skin, urinary tract	Urinary tract infection, pyelonephritis, endocarditis.
C. amycolatum	Man and animals	Man: Bacteremia, endocarditis, peritonitis and wound infection; cattle: mastitis.
C. glucuranalyticum	Urinary tract of man and animals	Urogenital tract infection
C. minutissimum	Skin, urinary tract	Erythrasma, bacteremia.
C. striatum	Respiratory tract, skin	Respiratory tract infection, wound infection, bacteremia
C. pseudodiphtheriticum	Respiratory tract	Respiratory tract infection, endocarditis.
Arcanobacterium haemolyticum	Throat	Pharyngitis, skin ulcers, endocarditis.
Rhodococcus equi	Animals, soil	Pulmonary infection and soft tissue infection.

duces reddish-brown scaly patches in the intertriginous sites. Lesions usually involve the groin, toe, and axillae.

Corynebacterium jeikeium

Cornyebacterium jeikeium is named after Johnson and Kaye who first linked this organism with human infections. Infections have been limited to patients who are immune compromised, have undergone invasive procedures, or have a history of intravenous drug abuse. It is the most common cause of diphtheroid prosthetic valve endocarditis in adults. *C. jeikeium* has been reported to be resistant to a wide range of antimicrobials. Most strains are susceptible to vancomycin.

Corynebacterium xerosis

C. xerosis is commonly found on skin and mucocutaneous sites. Human infection with *C. xerosis* is rare, and affected patients are invariably immunosuppressed.

Corynebacterium bovis

C. bovis, commensal of cow's uddar, which may cause bovine mastitis. Many of them cause infections in immunocompromised patients.

DIPHTHEROIDS

Corynebacteria resembling *C. diphtheriae,* occur as normal commensals in the throat, skin and other areas. These may be mistaken for diphtheria bacilli and are known as diphtheroids. They stain more uniformly than diphtheria bacilli, are arranged in V forms or palisades rather than Chinese letter arrangement and possess few or no metachromatic granules. They can be differentiated from *C. diphtheriae* on the basis of biochemical characters and toxigenicity tests (Table 28.3). The common diphtheroids are *C. pseudodiphtheriticum* and *C. xerosis.*

OTHER CORYNEFORM GENERA

Other genera of irregularly shaped, gram-positive bacilli are **Arcanobacterium, Brevibacterium, Oerskovia** and **Turicella**. They have been found to colonize humans and cause disease.

Arcanobacterium

Arcanobacterium can cause pharyngitis with a "scarlet fever-like" rash, polymicrobic wound infections, and, less commonly, systemic infections such as septicemia and endocarditis. Infections can be treated with penicillin or erythromycin.

Brevibacterium

Brevibacterium colonize the skin surface and have been blamed for malodorous feet in some colonized people. It causes septicemia, osteomyelitis, and foreign body infections.

Oerskovia

Oerskovia is an environmental organism found in the soil and decaying organic matter. This has been associated with septicemia, endocarditis, meningitis, soft tissue infections, and infections in the presence of foreign bodies.

Turicella

Turicella (Turicella otitidis is the only species) has been isolated in the ears of healthy and infected individuals.

KNOW MORE

- Diphtheria does not occur naturally in animals but infection can be produced experimentally. Susceptibility varies in different species. Subcutaneous inoculation of a guinea pig with a culture of virulent diphtheria bacillus will cause death in 1-4 days. At autopsy, the following features can be observed:
  1. Gelatinous, hemorrhagic edema and, often, necrosis lit the site of inoculation,
  2. Swollen and congested draining lymph nodes,
  3. Peritoneal exudate which may be clear, cloudy or bloodstained,
  4. Congested abdominal viscera,
  5. Enlarged hemorrhagic adrenals, which is the pathognomonic feature,
  6. Clear, cloudy or bloodstained pleural exudate, and
  7. Sometimes, pericardial effusion.

☞ KEY POINTS

- Corynebacterium is gram-positive bacilli with an irregular shape, tendency to clubbing at one or

Table 28.3: Differences between *C. diphtheriae* and *diphtheroids*

Feature	*C. diphtheriae*	*Diphtheroids*
1. Morphology	i. Weakly gram-positive and thin bacilli	Strongly gram-positive, short and thick bacilli
	ii. Metachromatic granules present	Few or absent
	iii. Arranged in Chinese letter pattern	Pallisade arrangement
	iv. Pleomorphism present.	Very little pleomorphism present.
2. Culture	Grow on enriched media.	Can grow on ordinary media
3. Biochemical tests	Ferments glucose only and does not ferment sucrose.	Ferments both glucose and sucrose.
4. Toxin production	Toxic	Nontoxic.
5. Virulence tests	Positive	Negative.

both, highly pleomorphic with Chinese letter or cuneiform arrangement. The granules in the cell are known as metachromatic granules, volutin granules or Babes-Ernst granules. Special stains, such as Albert's, Neisser's and Ponder's have been devised for demonstrating the granules clearly. With Albert's stain, the granules stain **bluish black** and the protoplasm green.

- **Cultural characteristics:** Two media are useful. 1. Löffler's serum slope—Diphtheria bacilli grow very rapidly and colonies can be seen in 6-8 hours. 2. Tellurite blood agar—*C. diphtheriae* give gray/black, shiny or dull black colonies. Three different biotypes: gravis, intermedius and mitis are described.
- Fermentation of sugars are usually done in Hiss's serum peptone water medium. *C. diphtheriae* is H_2S positive and reduces nitrate to nitrite.
- **Toxin:** Toxigenic strains of *C. diphtheriae* produce a very powerful exotoxin. The toxigenicity of the diphtheria bacillus depends on the presence in it of corynephages (tox+). Diphtheria toxin is, heat labile protein, and consists of two fragments: A (active) and B (binding). Inhibition of protein synthesis is probably responsible for both the necrotic and neurotoxic effects of the toxin.
- **Clinical diseases:** Diphtheria occurs in two forms (**respiratory** and **cutaneous**). A tough gray to white pseudomembrane, may appear on the tonsils and then spread downward into the larynx and trachea. Systemic effects involve the kidneys, heart, and nervous system. In cutaneous diphtheria, systemic complications are less common.

  Complications: (1) Asphyxia (2) Acute circulatory failure (3) Postdiphtheritic paralysis (4) Septic, such as pneumonia and otitis media.
- **Laboratory diagnosis:** It depends upon microscopy, culture and virulence tests. Virulence testing may be by *in vivo* or *in vitro* methods. Virulence tests demonstrate the production of exotoxin by bacteria isolated on culture. *In vivo* tests are: (i) Subcutaneous test; (ii) Intracutaneous test. *In vitro* test include (i) Precipitation test; (ii) Tissue culture test; (iii) Enzyme-linked immunosorbent assays (ELISA); (iv) Polymerase chain reaction (PCR).
- **Epidemiology:** Diphtheria is primarily a pediatric disease and has worldwide distribution. Humans are the only known reservoir, with carriage in oropharynx or on skin surface. Spread from person to person by exposure to respiratory droplets or skin contact.
- **Prophylaxis:** Administration of diphtheria vaccine and booster shots are given. DPT given at the age of 6 weeks, 10 weeks, 14 weeks and 16-24 months followed by booster dose DT at the age of 5-6 years (school entry).
- **Diphtheroids:** Corynebacteria resembling *C. diphtheriae*, occur as normal commensals in the throat, skin and other areas. These may be mistaken for diphtheria bacilli and are known as diphtheroids. They stain more uniformly than diphtheria bacilli, are arranged in V forms or palisades rather than Chinese letter arrangement and possess few or no metachromatic granules. The common diphtheroids are *C. pseudodiphtheriticum* and *C. xerosis*.

IMPORTANT QUESTIONS

1. Discuss morphology, cultural characteristics and biochemical characters of *Corynebacterium diphtheriae*.
2. Name differrent species of genus *Corynebacterium*. Discuss in detail laboratory diagnosis of diphtheria.
3. Write short notes on:
   Diphtheria toxin.
   Pathogenicity of *C. diphtheriae.*
   Toxigenicity tests/virulence tests of *C. diphtheriae.*
   Schick test.
   Prophylaxis of diphtheria.
   Nondiphtheria corynebacteria.
   Diphtheroids.

FURTHER READING

Coyle MB, Lipsky BA. Coryneform bacteria in infectious diseases: clinical and laboratory aspects. Clin Microbiol Rev 1990;3:227-46.

Funke G, et al. Clinical microbiology of coryneform bacteria, Clin Microbiol Rev 1997;10:125-159,

Hofler W. Cutaneous diphtheria. Lnt J Dermatol 1991;30:845.

Lipsky BA, et al. Infections caused by nondiphtheria coryne. bacteria. Rev Infect Dis 1982;4:1220-1235.

CHAPTER 29

Bacillus

Learning Objectives

After reading and studying this chapter, you should be able to:

- Describe the morphology, cultural characters and pathogenicity of *Bacillus anthracis*.
- Discuss laboratory diagnosis of anthrax.
- Describe the following: Anthracoid bacilli; *Bacillus cereus* food poisoning.

INTRODUCTION

The family Bacillaceae consists of a diverse collection of bacteria comprising obligate aerobes and strict anaerobes, cocci and bacilli, and gram-positive and gram-negative organisms. The feature they all share is formation of endospores. The two clinically important genera are ***Bacillus*** (the aerobic and facultative anaerobic spore-formers) and ***Clostridium*** (the strict anaerobic spore-formers). In the past few years, *Bacillus* has been subdivided into **six genera**.

GENERAL CHARACTERISTICS OF BACILLUS

1. The genus *Bacillus* consists of aerobic bacilli forming heat resistant spores.
2. They are gram-positive but tend to be decolorized easily so as to appear Gram-variable, or even frankly gram-negative.
3. They are generally motile with peritrichate flagella, the anthrax bacillus being a notable exception.
4. Most form catalase and most produce acid but not gas from glucose.
5. The genus includes psychrophilic, mesophilic and thermophilic species, the maximum temperatures for vegetative growth ranging from about 25°C to above 75°C and the minimum from about 5°C to 45°C.
6. Their salt tolerance varies from less than 2 to 25 percent NaCl.
7. Their spores are ubiquitous, being found in soil, dust, water and air and constitute the commonest contaminants in bacteriological culture media.

SPECIES

Despite the subdivision of *Bacillus,* more than 50 species remain in the genus. Fortunately, the species that are of medical interest are relatively limited.

1. ***Bacillus anthracis:*** The organism responsible for anthrax, is the most important member of this genus.
2. ***Bacillus cereus:*** It is commonly implicated in episode of food poisoning and responsible for opportunistic infection.

BACILLUS ANTHRACIS

Historically, considerable attention was early focused on the genus *Bacillus* because of the economic importance of anthrax, the disease caused by *B. anthracis.*

1. It was **the first pathogenic bacterium to be observed** under the microscope (Pollender, 1849)
2. **The first communicable disease** shown to be transmitted by inoculation of infected blood (Davaine, 1850) was anthrax.
3. *B. anthracis* was first bacillus **to be isolated in pure culture** and shown **to possess spores (Koch, 1876).**
4. It was in studies on anthrax that Koch demonstrated for the first time a **set of criteria or postulates** that must be satisfied before an organism can be identified as the etiologic agent of a specific infection.
5. The first bacterium used for the preparation of **an attenuated vaccine** by Pasteur was *B. anthracis* (Pasteur, 1881).
6. Nobel Prize winner Metchnikoff studied virulent and attenuated strains of *B. anthracis,* in his pioneering work on **phagocytosis.**

Morphology

B. anthracis is one of the largest of pathogenic bacteria. 3 to 8 by 1 to 1.3 μm and is gram-positive nonacid fast, straight, sporing bacillus. It is rectangular in shape and arranged in filamentous chains in cultutre. In cultures, the bacilli are arranged end to end in long chains. The ends of the bacilli are truncated or often concave and somewhat swollen so that a chain of bacilli presents a **'bamboo stick' appearance.**

The spore is oval (ellipsoidal), refractile, central in position and of the same diameter as the bacillus and not swelling the mother cell (Fig. 29.1). Spores are formed in culture, in the soil, and in the tissue and exudates of dead animals but never in the blood or tissues of living animals. Sporulation occurs under unfavorable conditions for growth and is encouraged by distilled water, 2% NaCl or growth in oxalated agar. Anaerobic conditions and calcium chloride inhibit sporulation. Spores seen as unstained spaces in Gram-stained bacilli and, when free, faintly outlined with Gram counterstain. The anthrax bacillus is nonmotile, unlike most other members of this genus.

It is found singly, in pairs or in short chains **in tissues.** The entire chain being surrounded by a **capsule** which is polypeptide in nature, being composed of a **polymer of d(–) glutamic acid.** Capsules are formed in the animal body but in culture only if the media contain added bicarbonate or are incubated under 10 to 25 percent CO_2. Capsule formation may occur in the absence of CO_2 if grown in media containing serum, albumin, charcoal or starch.

When blood films containing anthrax bacilli are stained with **polychrome methylene blue** for a few seconds and examined under the microscope, an amorphous purplish material is noticed around the blue bacilli. This represents the capsular material and is characteristic of the anthrax bacillus. This is called the **McFadyean's reaction** and is employed for the presumptive diagnosis of anthrax in animals. Purple bacillus with red capsule is seen with **Giemsa's stain.** Fat globules may be made out within the bacilli when stained with **Sudan black B.** Spores seen as unstained spaces in Gram-stained bacilli and, when free, faintly outlined with Gram counterstain.

Cultural Characteristics

It is aerobe and facultative anaerobe. Temperature range for growth is 12-45°C (optimum 37°C). Sporulation requires aerobic conditions and is optimal at 25-30°C. Optimum pH for growth is 7.4. Germination of spores requires fresh nutrients and aerobic conditions. Good growth occurs on ordinary media.

1. **Nutrient agar:** On **nutrient agar,** colonies are irregularly round, 2-3 mm in diameter, raised, dull, opaque, grayish white, with a frosted glass appearance. The edge of the colony is composed of long, interlacing chains of bacilli; resembling locks of matted hair under the low power microscope. This is called the **'Medusa head appearance'** (Fig. 29.2).
2. **Blood agar:** Colonies on horse or sheep **blood agar** are virtually nonhemolytic, though occasional strains produce a narrow zone of hemolysis.
3. **In broth:** Growth develops as silky strands, a surface pellicle and a floccular deposit.
4. In a **gelatin stab**, there is growth down the stab line with lateral spikes, longer near the surface, giving

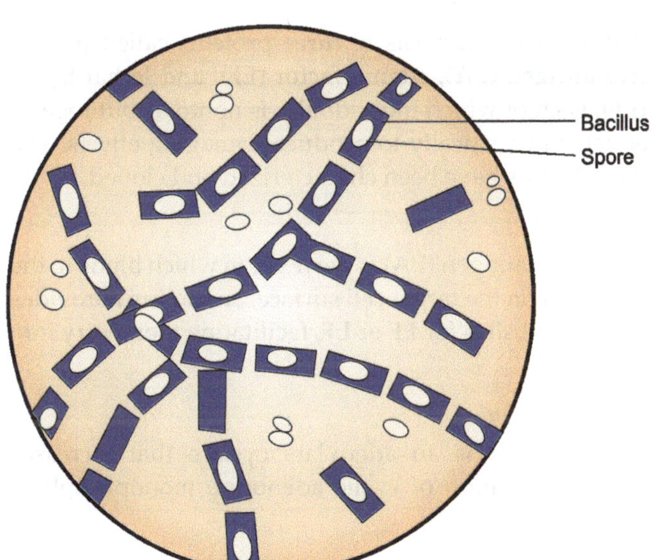

Fig. 29.1: Anthrax bacilli

Fig. 29.2: Medusa head appearance colony of anthrax bacilli

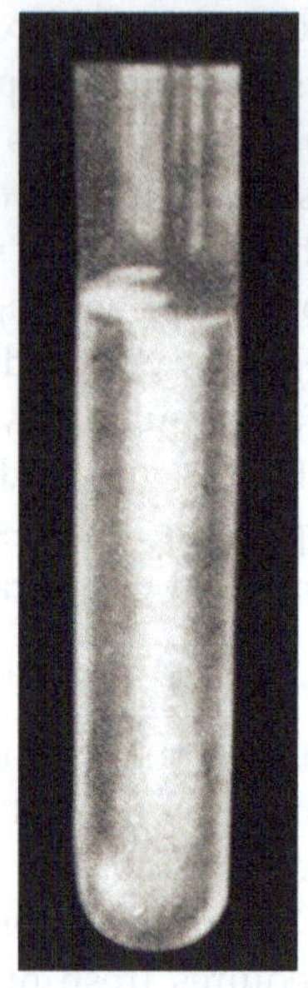

Fig. 29.3: Inverted fir tree appearance in gelatin stab culture of anthrax bacillus

an **'inverted fir tree' appearance** with slow liquefaction commencing from the top (Fig. 29.3).

5. **Selective medium: A selective medium (PLET medium),** consisting of polymyxin. lysozyme, ethylene diamine *tetra acetic* acid (EDTA) and thallous acetate added to heart infusion agar, has been devised to isolate *B. anthracis* from mixtures containing other spore-bearing bacilli.

Biochemical Reactions

B. anthracis ferments glucose, maltose, sucrose, trehalose and dextrin with the production of acid and no gas. Nitrates are reduced to nitrites. Catalase is formed. There is a weak lecithinase reaction on egg-yolk agar which gives a narrow zone of opalescence around the colonies.

Resistance

In the dry state and in certain soils the spores may survive for 50 years or more. In the carcasses of animals which have died of anthrax, the bacilli remain viable in the bone marrow for a week and in the skin for two weeks. Normal heat fixation of smears may not kill the bacilli in blood films.

Spore Form

The spores are resistant to chemical disinfectants and heat. With moist heat, the vegetative bacilli are killed at 60°C in 30 min and the spores at 100°C in 10 min. With dry heat the spores are killed at 150°C in 60 min. The spores are also killed by 4 percent formaldehyde or 4 percent potassium permanganate in a few minutes. Destruction of the spores in animal products imported into nonendemic countries is achieved by 'duckering' in which formaldehyde is used as 2 percent solution at 30-40°C for 20 minutes for disinfection of wool and as 0.25 percent at 60°C for six hours for animal hair and bristles.

The bacilli are sensitive to benzylpenicillin, streptomycin, tetracyclines, chloramphenicol, ciprofloxacin, the cephalosporins and sulfonamides. Occasional strains resistant to penicillin have been met with.

Antigenic Structure

Three main antigens have been characterized:

1. **Capsular polypeptide:** It is polypeptide consisting exclusively of D-glutamic acid.
2. **Somatic polysaccharide:** It is a component of the cell wall. Antibodies to this antigen is also not protective.
3. **Complex protein toxin (anthrax toxin):** Anthrax toxin, derived from the thoracic and peritoneal exudates of infected animals, is a complex exotoxin consisting of three protein components: **protective antigen (PA), lethal factor (LF), and edema factor (EF).** Each of the separate components is serologically active and distinct and is also immunogenic (see below).

Virulence Factors

The pathogenesis of *B. anthracis* depends on two important virulence factors: **a poly (D-glutamic acid) capsule** and a three-component protein exotoxin. Fully virulent organisms have two large plasmids that code for these products.

1. Capsule

The capsule interferes with phagocytosis and is especially important during the early stages of infection. Loss of the plasmid which controls capsule production leads to loss of virulence. This is how the live attenuated anthrax spore vaccine **(Sterne strain)** was obtained. Antibodies against the capsular antigen are produced, but they are not protective against the disease.

2. Anthrax Toxin

Anthrax toxin consists of three proteins called **protective antigen (PA), edema factor (EF), and lethal factor (LF),** each of which individually is nontoxic but together act synergistically to produce damaging effects. The three factors have been characterized and cloned.

Protective Antigen (PA)

Protective antigen (PA) is the fraction which binds to the receptors on the target cell surface, and in turn provides attachment sites for **EF or LF,** facilitating their entry into the cell.

Edema Factor

Edema factor is an adenylate cyclase that increases the concentration of cyclic adenosine monophosphate (cAMP) in host cells.

LF

LF is a protease that kills host cells by an unknown mechanism.

Epidemiology

Anthrax is primarily a disease of herbivores. Humans are infected through exposure to contaminated animals or animal products. Human anthrax may be **nonindustrial (agricultural)** or **industrial. Nonindustrial anthrax** is often an occupational disease in those who associate frequently with animals, such as veterinarians, butchers and farmers. It may also be found in the general population.

The exact incidence in humans and animals is unknown, but many countries in Africa and Asia regularly report cases to the World Health Organization. Large epidemics of anthrax were reported from Russia and Zimbabwe during 1978-1980.

Anthrax is enzootic in India, the number of animals infected running into tens of thousands annually. The disease is rare in some countries such as Britain where infection is imported through contaminated hides, bone meal fertilizer and other animal products The extent of anthrax in human beings is not clear but about 20,000 to 100,000 cases are believed to occur annually throughout the world, mostly in rural areas. An epizootic of anthrax in sheep has been active near the Andhra–Tamil Nadu border, causing many cutaneous and meningoencephalitic human infections, with high mortality rate. Anthrax infection in human beings provides permanent immunity and second attacks are extremely rare.

Industrial cases result from contact with contaminated animal products, such as hides, goat hair, wool, and bones imported from Africa, the Middle East, and Asia. A wide variety of finished products have been linked to human infection, including such items as shaving brushes, ivory piano keys, bongo drums, and wool products.

Although anthrax is rarely encountered in developed countries, the threat of biological warfare has renewed the concern about this disease. At least 17 nations and an unknown number of independent terrorist groups have biological warfare programs. There is significant concern that the spores will be used in bioterrorism.

Clinical Infections

Animal Infection

Anthrax is a **zoonosis.** Animals are infected by the ingestion of the spores present in the soil. Direct spread from animal to animal is rare. The disease is generally a fatal septicemia but may sometimes be localized, resembling the cutaneous disease in human beings. Infected animals shed in the discharges from the mouth, nose and rectum, large numbers of bacilli, which sporulate in soil and remain as the source of infection.

Human Anthrax

Based on the mode of infection, Human anthrax presents in one of three ways: (1) Cutaneous, (2) Pulmonary, or (3) Intestinal. All types leading to fatal septicemia or meningitis.

1. Cutaneous Anthrax

Cutaneous anthrax used to be caused by shaving brushes made with animal hair. *Stomoxys calcitrans* and other biting insects may occasionally transmit infection mechanically. Cutaneous anthrax accounts for more than 95 percent of human cases. It begins 2 to 5 days after infection as a small papule that develops within a few days into a vesicle filled with dark bluish black fluid. Rupture of the vesicle reveals a **black eschar** at the base, with a very prominent inflammatory ring of reaction around the eschar. (The name ***anthrax,*** which means coal, comes from the black color of the eschar). This is sometimes referred to as a **malignant pustule.**

The lesion is classically found on the hands, forearms, or head and is painless. It is rarely found on the trunk or lower extremities. The disease used to be common in dock workers carrying loads of hides and skins on their bare backs and hence was known as the **'hide porter's disease'**. Cutaneous anthrax generally resolves spontaneously, but 10-20 percent of untreated patients may develop fatal septicemia or meningitis

2. Pulmonary Anthrax

Pulmonary anthrax, known as **'wool-sorter's disease,** because it used to be common in workers in wool factories, due to inhalation of dust from infected wool. It occurs in patients who handle raw wool, hides, or horsehair and acquire the disease by the inhalation of spores.This is a **hemorrhagic pneumonia** with a high fatality rate. Hemorrhagic meningitis may occur as a complication.

3. Intestinal Anthrax

Intestinal anthrax, is rare and occurs mainly in **primitive communities** who eat the carcasses of animals dying of anthrax. An individual may suffer after a day or so from hemorrhagic diarrhea, and dies rapidly from septicemia. Often these episodes occur as small outbreaks in a family or village.

Complications

Approximately 5 percent of patients with anthrax (cutaneous, inhalation, gastrointestinal) develop **meningitis**. Recovery from infection appears to confer immunity.

Laboratory Diagnosis of Human Anthrax

In laboratories unfamiliar with the disease, additional precautions for staff safety need to be organized. All procedures connected with the handling of *B. anthracis* should be carried out with greatest care in a safety cabinet.

1. Specimens

Material from a malignant pustule, sputum from pulmonary anthrax, gastric aspirates, feces or food in intestinal anthrax and in the blood in the septicemic stage of all forms of the infection. Specimens should be taken before antibiotic therapy has been instituted.

2. Microscopy

Prepare smears of each specimen and stain with Gram's method and McFadyean's method or with Giemsa stain. Gram's stain may show typical large gram-positive bacilli. Capsule appears as a clear halo around the bacterium by India-ink staining.

Direct fluorescent antibody test (DFA) for capsule specific staining and for polysaccharide (cell wall) antigen confirms the identification.

3. Culture

Culture the exudate on **nutrient agar, blood agar, PLET medium** and **nutrient broth**. Incubate at 37°C for 18 hours. Examine plates for the medusa head colonies characteristic of *B. anthracis,* nonhemolytic on the blood agar plate. Prepare a smear, stain it by Gram's method and look for tangled chains of large gram-positive bacilli some of which have central, oval, nonbulging spores. In nutrient broth, look for a pellicle and a deposit.

4. Confirmatory Tests

i. Biochemical and Physiological Reactions

Demonstration of nonmotility, gelatin liquefaction, growth in straight chains and enhanced growth aerobically, as seen in the characteristic *inverted fir tree* appearance in a gelatin stab, will generally identify *B. anthracis* completely.

Toxin production can be demonstrated by immunological or gene probe methods in reference laboratories.

ii. Animal Inoculation

Inoculate intraperitoneally in mice 0.1 ml of 1 in 100 dilution of a 24 hours broth culture. The animal dies in 48-72 hours. Autopsy reveals gelatinous hemorrhagic edema and petechial hemorrhages in the peritoneum, black blood slowly clotting when shed, and an enlarged, dark red spleen. Make smears from heart blood and spleen, stain by Gram's and McFadyean's methods, and look for typical anthrax bacilli.

5. Serological diagnosis

Serological diagnosis by enzyme-linked immunosorb ent assay **(ELISA)** may be of value retrospectively, but is seldom used diagnostically

Laboratory Diagnosis in Domestic Animals

1. Specimens

When an animal is suspected to have died of anthrax, autopsy is not permissible, as the spilt blood will lead to contamination of the soil. An ear may be cut off from the carcass and sent to the laboratory. Alternatively, swabs soaked in blood or several blood smears may be sent.

2. Microscopy

The demonstration of gram-positive bacilli with the morphology of anthrax bacilli and a positive **McFadyean's reaction** will enable the presumptive diagnosis to be made. Immunofluorescent microscopy can confirm the identification.

3. Culture

Culture some of the blood on nutrient agar and blood agar plates and look for typical, nonhemolytic colonies. Isolation of the bacillus is easy if gross contamination has not occurred.

4. Animal Inoculation

The anthrax bacillus can often be isolated from contaminated tissues by applying them over the shaven skin of a guinea pig. It is able to penetrate through minute abrasions and produce fatal infection.

5. Serology

Ascoli's thermoprecipitin test: If the sample received is putrid so that viable bacilli are unlikely, diagnosis may be established by **Ascoli's thermoprecipitin test** by demonstration of the anthrax antigen in tissue extracts. The original thermoprecipitin test devised by Ascoli (1911) was a ring precipitation by letting the boiled tissue extract in a test tube react with the anthrax antiserum. The tissue is ground up in saline, boiled for 5 minutes, filtered and layered over antianthrax serum in a narrow tube. If tissue contains anthrax antigen, a ring of precipitation will appear at the junction of the two liquids within 5 minutes at room temperature.

With the availability of purified anthrax toxin antigen, Ascoli' s test has been replaced by highly sensitive and specific immunoassays. EIA can also detect antibody in the serum of animals surviving anthrax infection.

6. Polymerase Chain Reaction (PCR)

A sensitive and specific PCR technique has been developed for the detection of anthrax contamination of animal and agricultural products.

Treatment

Ciprofloxacin is the drug of choice. Penicillin, doxycycline, erythromycin, or chloramphenicol can be used (if susceptible), but the bacteria are resistant to sulfonamides and extended-spectrum cephalosporins.

Prophylaxis

Control of human anthrax ultimately depends on control of the disease in animals since anthrax is a zoonosis. Animals with known or suspected anthrax should be handled with care and carcasses of animals suspected to have died of anthrax are incinerated or buried deep in quicklime to prevent soil contamination and the spread of spores to new pastures. Wool, horsehair, and hides coming from areas where epidemic anthrax is present should be gas sterilized. A vaccine is available for control of outbreaks of human anthrax in an industrial setting.

Immunization

Live-attenuated bacilli were first used by Louis Pasteur in May 1881. Pasteur's vaccine was the anthrax bacillus attenuated by growth at 42-43°C . The original Pasteur's anthrax vaccine is of great historical importance. It was Pasteur's convincing demonstration of the protective effect of his anthrax vaccine in a public experiment on a farm in Pouilly-Ie-Fort that marked the beginning of scientific immunoprophylaxis.

Sterne Strain of Live Spore Vaccine

Subsequently, the **Sterne strain of live spore vaccine** has been used for animal immunization. The Sterne vaccine contained spores of a noncapsulated avirulent mutant strain. Live bacterial vaccines are not considered safe for man.

Alum-precipitated toxoid prepared from the protective antigen has been shown to be a safe and effective vaccine for human use. It has been used in persons occupationally exposed to anthrax infection. Three doses intramuscularly at intervals of six weeks between first and second, and six months between second and third doses induce good immunity, which can be reinforced if necessary with annual booster injections. Frequent booster doses are necessary. New recombinant protective antigen preparations may give better immunity and fewer adverse reactions.

ANTHRACOID BACILLI

A large number and variety of nonpathogenic aerobic spore bearing bacilli appearing as common contaminants in cultures and morphologically having a general resemblance to anthrax bacilli have been collectively called **pseudoanthrax or anthracoid bacilli**. The important species include *B. cereus, B. subtilis, B. licheniformis, B. pumilus*. Of them, the most important is *B. cereus* which from 1970 has been recognized as a frequent cause of foodborne gastroenteritis. It has also been associated with septicemia, meningitis, endocarditis, pneumonia, wound infections and other suppurative lesions, particularly as an opportunist pathogen. *B. subtilis, B. licheniformis, B. pumilus* and a few other species have also been occasionally implicated in causing food poisoning and other lesions, but they do not form toxins. Table 29.1 lists the main differentiating features between them.

Bacillus cereus

B. cereus has recently assumed importance as a cause of food poisoning. It is widely distributed in nature may be readily isolated from soil, vegetables and a wide variety of foods including milk, cereals, spices, meat and poultary.

Pathogenesis

Infections include emetic (vomiting) and diarrheal forms of gastroenteritis (Table 29.2); ocular infection following trauma to eye; and other opportunistic infections.

1. Emetic Form (Short Incubation Type)

The emetic form results from the consumption of contaminated rice. Most bacilli are killed during the initial cooking of the rice, but the heat-resistant spores survive. If the cooked rice is not refrigerated, the spores germinate, and the bacilli can multiply rapidly. The heat-stable enterotoxin that is released is not destroyed when the rice is reheated. After ingestion of the enterotoxin and a 1 to 6 hours incubation period, a disease of short duration (less than 24 hours) develops. Symptoms consist of vomiting, nausea, and abdominal cramps. Fever and diarrhea are generally absent (Table 29.1).

Two mechanisms of action have been described for the **enterotoxin** of *B. cereus*, one involving stimulation of CAMP system and the other independent of it. The short incubation type (emetic type) is caused by serotypes 1, 3 or 5 of *B. cereus* strains. Strains causing the emetic type of disease produce a toxin which causes vomiting when fed to Rhesus monkeys, resembling staphylococcal enterotoxin.

2. Diarrheal Form

The diarrheal form of *B. cereus* food poisoning results from the consumption of contaminated meat, vegetables or sauces. There is a longer incubation period occurring 8 to 24 hours after ingestion, during which the organism multiplies in the patient's intestinal tract and produces the **heat-labile enterotoxin**. Then the diarrhea, nausea, and abdominal cramps develop (Table 29.1).

The diarrheal disease is mostly caused by serotypes 2, 6, 8, 9, 10 or 12 of *B. cereus* strains. Isolates from the diarrheal type of disease produce enterotoxin which causes fluid accumulation in ligated rabbit ileal loop, resembling the heat labile enterotoxin of *Escherichia coli* and *Vibrio cholerae*.

3. Ocular Infection

B. cereus is also a common cause of **post-tramatic ophthalmitis.**

4. Other Opportunistic Infections

Intravenous catheter and central nervous system shunt infections and endocarditis (most common in drug abusers) as well as pneumonitis, bacteremia, and meningitis in severely immunosuppressed patients.

Laboratory Diagnosis

If food is available for testing, and this is often not the case in investigations of gastroenteritis, laboratory confirmation is easy. High numbers of *B. cereus*, often 10^8 or more per gram, in the absence of other food poisoning bacteria are sufficient to make the diagnosis.

Large facultatively anaerobic gram-positive bacilli that produce anthracoid colonies on **blood agar** after overnight incubation at 37°C are almost certain to be *B.cereus*. Because *B. cereus* is part of the normal fecal flora, the isolation of *B. cereus* from the patient's feces is not clinically relevant. The inability to recover organisms from fried rice does not necessarily rule out. *B. cereus* as cause of the emetic form of illness.

Table 29.1: Differentiating features between *B. anthracis* and Anthracoid bacilli		
Features	*B. anthracis*	*Anthracoid bacilli*
1. Motility	Nonmotile	Generally motile
2. Capsule	Capsulated	Noncapsulated
3. Chain formation	Grow in long chains	Grow in short chains
4. Colony on nutrient agar	Medusa head colony	Not present
5. Growth in penicillin agar (10 units/ml)	No growth	Grow usually
6. Hemolysis on blood agar	Hemolysis absent or weak	Usually well marked
7. Gelatin stab culture	Inverted fir tree growth and slow gelatin liquefaction	Rapid liquefaction
8. Turbidity in broth	No turbidity	Turbidity usual
9. Salicin fermentation	Negative	Usually positive
10. Growth at 45°C	No growth	Grows usually
11. Growth inhibition by chloral hydrate	Growth inhibited	Not inhibited
12. Susceptible to gamma phage	Susceptible	Not susceptible
13. Pathogenic to laboratory animals	Pathogenic	Not pathogenic
14. McFadyean's reaction	Positive	Negative
15. Ascoli's precepitin test	Positive	Negative
16. Fluorescent antibody test with antrax antiserum	Positive	Negative

Treatment

Both the emetic and diarrheal syndromes are shortlived and no specific treatment is needed.

Prevention

Prevention is best accomplished by the prompt refrigeration of boiled rice and other foods because it is nearly impossible to eliminate *B. cereus* spores from food.

OTHER BACILLUS SPECIES

Bacillus subtilis

Bacillus subtilis is one of the commonest saprophytes found as contaminants in foods, clinical specimens and laboratory cultures. It is usually a common laboratory contaminant but, like *B. cereus,* is capable of producing infection in the compromised host. The organism is sometimes found in opportunistic infections or food poisoning, also in overwhelming bacteremias and eye infections in heroin addicts.

Bacillus subtilis grows well on ordinary media, forming large colonies, that are circular or irregular, gray yellow, granular and difficult to emulsify.

Table 29.2: Bacillus cereus food poisoning		
Characteristics	*Emetic form*	*Diarrheal form*
1. Food implicated	Rice	Meat, vegetables
2. Incubation period (hours)	<6 (mean, 2)	> 6 (mean, 9)
3. Symptoms	Vomiting, nausea, abdominal cramps	Diarrhea, nausea, cramps
4. Duration of illness (hours)	8-10 (mean, 9)	20-36 (mean, 24)
5. Enterotoxin	Heat-stable	Heat-labile

Bacillus stearothermophilus

Bacillus stearothermophilus spores are used to evaluate the efficacy of autoclaving and other sterilization procedures.

KNOW MORE

Sterilization Test Bacilli

Filter paper strips impregnated with spores of *B. subtilis* subsp. *niger* have been used to test the efficacy of hot air oven. *B. stearothermophilus* has been used to test the autoclave and low temperature steam-formaldehyde sterilizer. Spores withstand 121°C for up to 12 min, and this has made the organism ideal for testing autoclaves that run on a time-temperature cycle designed to ensure the destruction of spores.

B. globigii, a red-pigmented variant of *B. subtilis,* has been used to test ethylene oxide sterilizers, and *B. pumilus* has been used to test the efficacy of ionizing radiation.

KEY POINTS

Bacillus

The genus *Bacillus* consists of, gram-positive, aerobic bacilli forming heat resistant spores. These spores are ubiquitous, being found in soil, dust, water and air and constitute the commonest contaminants in bacteriological culture media. Bacillus has two impotant species; *B. anthracis* (the causative agent of anthrax, is the most important pathogen of the bacillus and *B. cereus* (can cause food poisoning).

Bacillus anthracis

It is spore-forming gram-positive, a capsulated bacilli. *B. anthracis* is a McFadyean's reaction positive. It grows as 'Medusa head' appearance on nutrient agar medium.

- Ascoli's thermoprecipitin test has been used for rapid diagnosis of anthrax.
- *Epidemiology: B. anthracis* primarily infects herbivores with humans as accidental hosts. There is significant concern that the spores will be used in bioterrorism.
- *Diseases: B. anthracis* primarily infects herbivores with humans as accidental hosts. It cause cutaneous anthrax, inhalation anthrax and gastrointestinal anthrax.
- *Diagnosis:* Isolation of the organism from clinical specimens (e.g. papule or ulcer, blood). Ascoli's thermoprecipitin test has been used for demonstaion of the anthrax antigen in tissue.
- *Treatment, prevention, and control:* Ciprofloxacin is the drug of choice. Animal vaccination is effective, but human vaccines have limited usefulness. Alum precipitated toxoid has been used in persons occupationally exposed to anthrax infection. It is given in three doses intramuscularly at intervals of 6 weeks and 6 months. A booster dose may be given after one year.
- Aerobic spore bearing bacilli resembling *B. anthracis* are called anthracoid bacilli.
- *Bacillus cereus infections:* People at risk include those who consume food contaminated with the bacterium (e.g. rice, meat, vegetables, sauces), those with penetrating injuries (e.g. to eye), and those who receive intravenous injections.
- *Diseases:* Emetic (vomiting) and diarrheal forms of gastroenteritis; ocular infection; and other opportunistic infections. Gastrointestinal infections are treated symptomatically.

IMPORTANT QUESTIONS

1. Discuss laboratory diagnosis of anthrax.
2. Write short note on:
   Bacillus cereus food poisoning

FURTHER READING

Dixon TC, et al. Anthrax, N Engl J Med 1999;341:815-26.

Drobniewski FA. *Bacillus cereus* and related species. Clin Microbial Rev 1993;6:324-38.

Ingl TV, et al. Anthrax as a biological weapon-medical and public health management, JAMA 1999;281:1735-45.

Laforce FM. Anthrax. Clinlnfect Dis 1994;19:1009.

CHAPTER

30 Clostridium

LEARNING OBJECTIVES

After reading and studying this chapter, you should be able to:

- Describe classification of clostridia and diseases produced by different clostridia.
- Discuss morphology, cultural characteristics of *Cl. welchii*.
- Discuss Nagler's reaction.
- Discuss laboratory diagnosis and prophylaxis of gas gangrene.
- Discuss morphology, cultural characteristics of *Cl. tetani*.
- Describe toxins produced by *Cl. tetani*.
- Describe the following: Pathogenesis of tetanus; prophylaxis of tetanus.
- Discuss morphology and cultural characteristics of *Cl. botulinum*.
- Discuss laboratory diagnosis of botulinum.
- Describe the following: Gas gangrene; botulinum toxin; *Clostridium difficile*.

INTRODUCTION

The genus Clostridium includes all anaerobic, gram-positive bacilli capable of forming endospores. Spores of clostridia are usually wider than the diameter of the rods in which they are formed, giving the bacillus a swollen appearance resemling a spindle (*Clostridium* is Latin for **'little spindle'**). The name Clostridium is derived from the word **'Kloster'** (meaning a spindle).

Most of the species are saprophytes that normally occur in soil, water and decomposing plant and animal matter. They play an important part in natural processes of putrefaction. A few species are opportunistic pathogens. The genus contains bacteria responsible for three major diseases of human beings—gas gangrene, food poisoning and tetanus.

GENERAL FEATURES OF CLOSTRIDIA

1. Morphology

The clostridia are **gram-positive** typically large, straight or slightly curved rods, 3-8 × 0.6-1 µm with slightly rounded ends. Gram-positive, gram-negative and pleomorphism forms are common. Most species of clostridia are motile with peritrichous flagella except *Cl. perfringens* and *Cl. tetani* type VI which are nonmotile. All clostridia are noncapsulated with the exception of *Cl. perfringens* and *Cl. butyricum.*

All produce endospores. Spores of clostridia are usually wider than the diameter of the rods in which they are formed. In the various species, the spore is placed centrally, subterminally, or terminally. The position of the developing sporoes within the vegetative cell is useful in identifying the species (Table 30.1).

2. Culture

Most species are obligate anaerobes. A few species grow in the presence of trace amounts of air and some actually grow slowly under normal atmospheric conditions. Clostridia grow on enriched media in the presence of reducing agent such as cysteine or thioglycollate (to maintain a low oxidation-reduction potential), or in an O_2-free gaseous atmosphere provided by an air evacuated glove box, sealed jar, or other device.

Growth is relatively slow on solid media. Some organisms produce large raised colonies with entire margins (e.g. *C. perfringens*). Others produce smaller colonies that extend in a meshwork of the fine filaments (e.g. *C. tetani*). Many clostridia produce a zone of hemolysis on blood agar.

Liquid media like cooked meat broth (CMB) or thio glycollate media (containing reducing agent thioglycollste and 0.1% agar) are very useful for growing clostridia.

A very useful medium is Robertson's cooked meat broth. It contains unsaturated fatty acids which take up oxygen the reaction being catalyzed by hematin in the meat, and also sulfhydryl compounds which bring about a reduced OR potential. Clostridia grow in the medium, rendering the broth turbid. Most species produce gas.

Table 30.1: A morphological and biochemical classification of clostridia as human pathogens					
Position of spores	Both proteolytic and saccharolytic		Slightly proteolytic but not saccharolytic	Saccharolytic but not proteolytic	Neither proteolytic nor saccharolytic
	Proteolytic predominating	Saccharolytic predominating			
1. Central or subterminal	*Cl. bifermentans* *Cl. botulinum A.B.F.* *Cl. histolyticum* *Cl. sordellii* *Cl. sporogenes*	*Cl. perfringens* *Cl. septicum* *Cl. chauvoei* *Cl. novyi*		*Cl. fallax* *Cl. botulinum C.D.E*	
2. Oval and terminal	–	*Cl. difficile*	–	*Cl. tertium*	*Cl. cochlearium*
3. Spherical and terminal	–	–	*Cl. tetani*	*Cl. tetanomorphum* *Cl. sphenoides*	

3. Biochemical Reactions

These organisms are biochemically active. On the basis of biochemical reactions many clostridia can be divided into: **1. Predominantly saccharolytic clostridia; 2. Predominantly proteolytic clostridia; 3. Slightly proteolytic clostridia; (Table 30.1). Proteolytic clostridia turn the meat black and produce foul odor and saccharolytic species turn the meat pink.**

4. Resistance

The vegetative cells of clostridia do not differ from nonsporing bacilli in their resistance to physical and chemical agents. Spores of *Cl. botulinum* may withstand boiling after 3 to 4 hours and even at 105°C may not killed be completely in less than 100 minutes. Spores of most strains of *Cl. perfringens* are destroyed by boiling for less than five minutes, but those of some Type A strains that cause food poisoning survive for several hours. *Cl. tetani* spores persist for years in dried earth or dust. All species are killed by autoclaving at 121°C within 20 minutes. Among hospital disinfectants the greatest sporicidal activity is shown by alcoholic hypochlorite and glutaraldehyde.

In general, clostridia are susceptible to metronidazole, penicillin chloramphenicol and erythromycin; less so to tetracyclines, and resistant to aminoglycosides and quinolones.

5. Diseases Produced

Clostridia are more commonly associated with skin and soft tissue infections, food poisoning, and antibiotic-associated diarrhea and colitis (Table 30.1).

CLASSIFICATION

The traditional method for classifying an isolate in the genus *Clostridium* was based on a combination of diagnostic tests, including the demonstration of spores, optimal growth in anaerobic conditions, a complex pattern of biochemical reactivity such as saccharolytic and proteolytic capacities (Table 30.1) and the findings yielded by gas chromatography analysis of the metabolic by-products. With these methods, **more than 130 species** have been defined. Fortunately, most of the clinically important isolates fall within a few species.

Clostridia of medical importance may also be considered under the diseases they produce classification below).

CLOSTRIDIUM PERFRINGENS

(*Cl. welchii*, Bacillus aerogenes capsulatus, *B. phlegmonis* emphysematosae)

The bacillus was originally cultivated by Achalme (1891) but was first described in detail by Welch and Nuttall (1892) as *Bacillus aerogenes capsulatus*, who isolated it from the blood and organs of a cadaver. It has been commonly known as *C. welchii*, especially in the UK.

Cl. perfringens is a normal inhabitant of the large intestines of human beings and animals. The spores are commonly found in soil, dust and air.

Morphology

It is a relatively large gram-positive bacillus (about 4-6 × 1 µm) with straight, parallel sides and rounded or truncated ends, occurring singly or in chains or small bundles. It is pleomorphic, and filamentous and involution forms are common. It is capsulated and nonmotile.

Spores are typically oval, central or subterminal and not bulging but are rarely seen in artificial culture or in material from pathological lesions, and their absence is one of the characteristic morphological features of *Cl. perfringens*. Special media normally must be used to demonstrate sporulation.

Cultural Characteristics

It is an anaerobe but can also grow under microaerophilic conditions. It grows over a pH range of 5.5 to 8.0 and temperature range of 20°C to 50°C (optimum temperature range 37-45°C). **Robertson's cooked meat**

broth inoculated with mixtures of *Cl. perfringens* and other bacteria and incubated at 45°C for 4 to 6 hours serves as enrichment. Blood agar plates streaked after that time and incubated at 37°C will have proportionally higher numbers of *Cl. perfringens* (yield pure or predominant growth of *Cl. perfringens).* Good growth occurs in **Robertson's cooked meat medium**. The meat is turned pink but is not digested. The culture has an acid reaction and a sour odor.

It grows best on carbohydrate-containing media such as **glucose blood agar.** Surface colonies are large, smooth, regular, convex, slightly opaque disks. Colonies of most strains demonstrate a **'target hemolysis'** after overnight incubation on rabbit, sheep, ox, or human blood agar. It results from a narrow zone of complete hemolysis due to theta toxin and a much wider darker zone of incomplete hemolysis due to the α-toxin. On longer incubation this double zone pattern of hemolysis may fade.

C. perfringens also produces a characteristic pattern of synergistic β-hemolysis when streaked alongside *Streptococcus agalactiae* (the reverse CAMP test) (Fig. 30.1). Other types of colonies include one with a raised opaque center and a flat radially striate transparent border. Rough flat colonies with an irregular edge resembling a vine leaf may also occur. A variant occasionally produces very mucoid broth cultures and tenacious colonies on blood agar.

Biochemical Reactions

It is actively saccharolytic. Glucose, maltose, lactose and sucrose are fermented with the production of acid and gas. It is indole negative, MR positive and VP negative. Hydrogen sulfide is produced abundantly; sulfite is actively reduced; most strains reduce nitrates to nitrites.

In litmus milk medium, fermentation of lactose leads to formation of acid, which is indicated by the change in the color of litmus from blue to red. The acid clots the milk—casein (acid clot) and the clotted milk is disrupted due to the vigorous gas production. This is known as **'stormy fermentation** or **'stormy clot' reaction** that is produced by almost all strains of C. *perfringens* but is not specific for this organism.

Resistance

Spores are usually destroyed within five minutes by boiling but those of the 'food poisoning' strains of Type A and certain Type C strains resist boiling for 1 to 3 hours. Autoclaving of 121°C for 15 minutes is lethal. Spores generally resist routinely used antiseptics and disinfectants with the exception of formaldehyde and glutaraldehyde *C. perfringens* is sensitive to penicillin, erythromycin, many cephalosporins and metronidazole. It is generally sensitive to clindamycin and typically resistant to aminoglycosides.

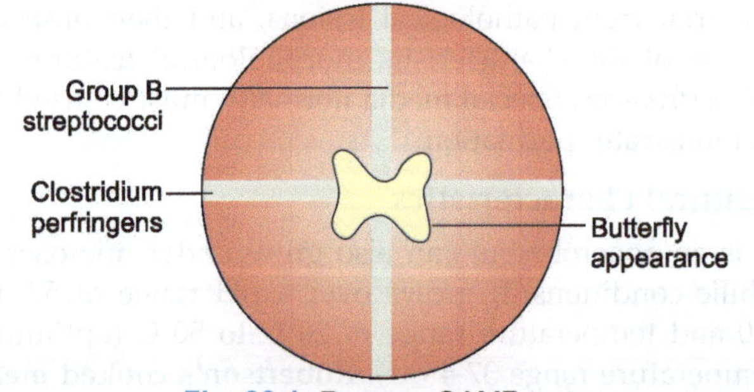

Fig. 30.1: Reverse CAMP test

Toxins

Cl. perfringens is one of the most prolific of toxin-producing bacteria, forming at least 12 distinct soluble substances or toxins, all of which are of protein in nature and antigenic. Major lethal toxins include: α (alpha), β (beta), ε (epsilon) and ι (iota) and minor lethal toxins include: γ (gamma), δ (delta), κ (kappa), λ (lambda), μ (mu), η (nu), θ (theta) and ν (eta). The four 'major toxins', alpha, beta, epsilon and iota, are predominantly responsible for pathogenicity (Table 30.2).

Classification

C. perfringens can be divided into five types, A to E on the basis of four major toxins (Table 30.3). Typing is done by neutralization tests with specific antitoxins by intracutaneous injection in guinea-pigs or intravenous injection in mice. Strains of *C. perfringens* type A that produce enterotoxin are associated with a mild form of food poisoning.

Alpha Toxin

The alpha *(α)* toxin is produced by all types of *Cl. perfringens* and most abundantly by Type A strains, is a lecithinase (phospholipase C) that lyses erythrocytes, platelets, leukocytes, and endothelial cells. It is lethal, dermonecrotic and hemolytic for the red cells of most species, except horse and goat. The lysis is of the **hot-cold variety,** being best seen after incubation at 37°C followed by chilling at 4°C. This toxin increases vascular permeability, resulting in massive hemolysis and bleeding, tissue destruction, hepatic toxicity, and myocardial

Table 30.2: Clostridia as human pathogens

A. The gas gangrene group:	
1. Established pathogens	*Cl. perfringens* *Cl. septicum* *Cl. novyi*
2. Less pathogenic	*Cl. histolyticum* *Cl. fallax*
3. Doubtful pathogens	*Cl. bifermentans* *Cl. sporogenes*
B. Tetanus:	*Cl. tetani*
C. Food poisoning:	
1. Gastroenteritis	*Cl. perfringens* (Type A)
2. Necrotizing enteritis	*Cl. perfringens* (Type C)
3. Botulism	*Cl. botulinum*
D. Acute colitis	*Cl. difficile*

Table 30.3: Toxins produced by *Cl. perfringens* types

Type	*Pathogenicity*	*Major toxins*				*Minor toxins*							
		α	β	ε	ι	γ	δ	η	θ	κ	λ	μ	ν
A	Gas gangrene: Wound infections, septicemia	+++	–	–	–	–	–				–		
	Food poisoning	+++	–	–	–	–	–	–			–		
B	Lamb dysentery	+++	+++	++	–	+		–	++	–	+++	+++	++
C	Enteritis in animals	+++	+++	–	–	–	–	–	+++	+++	–	–	
	Enteritis necroticans in human beings	+++	++	–	–	++	–	–	–	–	–	–	+++
D	Enterotoxemia of sheep	+++	–	+++	–	–	–	–	++	++	++		
E	Doubtful pathogen of sheep and cattle	+++	–	–	+++	–	–	–	++	++	++		

dysfunction. It is relatively heat stable and is only partially inactivated by boiling for five minutes.

Nagler's Reaction

Basis

The alpha (α) toxin is lecithinase C (or phospholipidase C) splits lecithin into phosphoryl choline and diglyceride, in the presence of Ca++ and Mg++ ions because the toxin is activated by Ca++ and Mg++ ions. This reaction is seen as an opalescence in serum or egg-yolk media and is specifically neutralized by the antitoxin. This is the basis of Nagler's reaction.

Procedure

For rapid detection of *C. perfringens,* a culture plate containing 6 percent agar, 5 percent peptic digest of sheep blood and 20 percent human serum or 5 percent egg-yolk is prepared. The incorporation of neomycin sulphate in the medium makes it more selective, inhibiting coliforms and aerobic spore bearers. Human serum may be replaced by 5 percent egg-yolk. Willis and Hobbs medium also incorporates lactose and neutral red to indicate lactose-fermenting organisms and milk to indicate proteolysis.

The plate is dried. On one half of the plate, 2 to 3 drops of *C. perfringens* antitoxin are spread and allowed to dry. The plate is then inoculated with the test organisms or the exudate under investigation and incubated anaerobically at 37°C for 18 hours.

Interpretation

On the section containing no antitoxin, *C. perfringens* colonies show surrounding **zone of opalescence, i.e. Nagler's reaction.** There will be **no opacity** around the colonies on the half of the plate with the antitoxin, due to the specific neutralization of the alpha toxin. (Fig. 30.2).

This reaction, however, is not totally specific for *C. perfringens* since the opalescence in the egg-yolk media may be produced by other lecithinase forming bacteria (*Cl. novyi, Cl. bifermentans,* some vibrios, some aerobic spore bearers). The reaction produced by *C. perfringens* is specifically neutralized by *C. perfringens* antitoxin, but serologically related phospholipases of *C. bifermentans* and *C. sordellii* and some other phospholipases are also

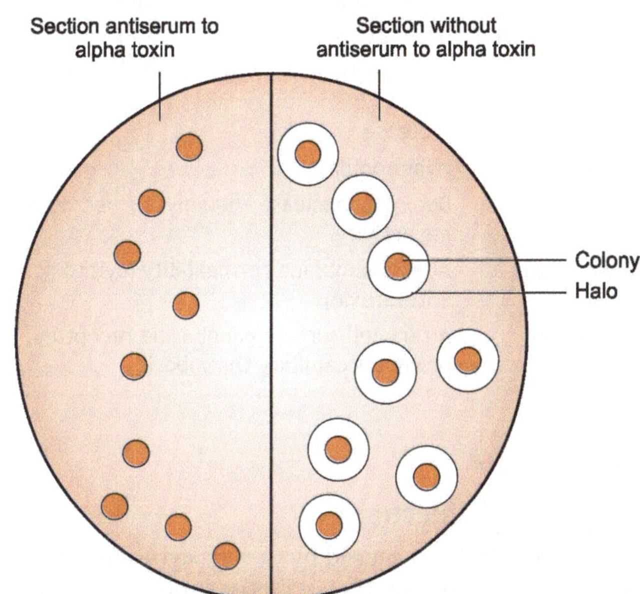

Fig. 30.2: Nagler's reaction. *Cl. perfringens* colonies on the right half of the plate are surrounded by haloes, while colonies on the left half (containing antiserum to alpha toxin) have no haloes around them

inhibited. These organisms can be separated by other tests.

Other Major Toxins (Table 30.4)

Beta (β), epsilon (ε) and iota (ι) toxins have lethal and necrotizing properties.

Minor Toxins

Gamma and eta toxins have minor lethal toxins. **Delta toxin** has a lethal effect and is hemolytic for the red cells of even goat, pigs and cattle. **Theta toxin** is oxygen-labile hemolysin antigenically related to streptolysin O. It is also lethal and a general cytolytic toxin.

Enterotoxin

C. perfringens type A strains produce a potent enterotoxin. Properties of the enterotoxin are erythema after intracutaneous injection, fluid accumulation in ligated rabbit ileal loop, lethality for mice, and diarrhea when fed orally to monkeys, rabbits and human volunteers.

Table 30.4: Virulence factors of *Clostridium perfringens*

Virulence factors	*Biologic activity*
α toxin	Lethal toxin; phospholipase C (lecithinase); increases vascular permeability; hemolysin; produces necrotizing activity
β toxin	Lethal toxin; necrotizing activity
ϵ toxin	Lethal toxin; permease
δ	Lethal binary toxin; necrotizing activity; adenosine diphosphate (ADP) ribosylating
i toxin	Hemolysin
θ toxin	Heat and oxygen-labile hemolysin; cytolytic
κ toxin	Collagenase; gelatinase, necrotizing activity
λ toxin	Protease
μ toxin	Hyaluronidase
ν toxin	Deoxyribonuclease; hemolysin; necrotizing activity
Enterotoxin	Alters membrane permeability (cytotoxic, enterotoxic)
Neuraminidase	Alters cell surface ganglioside receptors; promotes capillary thrombosis

Pathogenesis

1. Soft Tissue Infections

Soft tissue infections caused by *C. perfringens* are subdivided into: (1) cellulitis, (2) fasciitis or suppurative myositis, and (3) myonecrosis or gas gangrene.

i. Cellulitis

Anaerobic cellulitis is a more serious form of wound infection. Clostridial species can colonize wounds and skin with no clinical consequences.

ii. Fasciitis or Suppurative Myositis

Cellulitis process can progress to suppurative myositis characterized by an accumulation of pus in the muscle planes, but muscle necrosis and systemic symptoms are absent.

iii. Clostridial Myonecrosis or Gas Gangrene

The disease is characterized by rapidly spreading edema, myositis, necrosis of tissues, gas production and profound toxemia occurring as a complication of wound infection. The disease has been referred to in the past as 'malignant edema'. Other descriptive terms that have been used are **'anaerobic (clostridial) myositis'** and **'clostridial myonecrosis.**

Etiology

Generally, several species of clostridia are found in association with anaerobic streptococci and facultative anaerobes such as *E. coli* proteus and staphylococci. Amongst the pathogenic clostridia, *Cl. perfringens* is the most frequently encountered (approximately 60%), and *Cl. novyi* and *Cl. septicum* being the next common (20-40%), and *Cl. histolyticum* less often. Other clostridia usually found are *Cl. sporogenes, Cl. fallax, Cl. bifermentans, Cl. sordellii, Cl. aerofoetidum* and *Cl. tertium*. Since anaerobic infections are due to wound contamination, they are always polymicrobial.

Mechanism of Infection

Infection usually results from contamination of a wound with soil, particularly from manured and cultivated land. Clostridial spores are introduced into tissue, e.g. by contamination with dirt, or by endogenous transfer from the intestinal tract. After injury there is an incubation period may be as short as seven hours or as long as six weeks, usually of 12 to 48 hours, before symptoms suddenly appear.

The spores germinate and grow rapidly if the normal tissue oxidation-reduction potential is lowered, as occurs when there is considerable cell injury or compromise of circulation. These lesions almost always involve coinfection with several species of organisms, including **clostridia** and **other anaerobes,** and **facultative species** that use up available O_2, thus protecting the anaerobes from oxygen's toxic effects.

Germination and outgrowth of clostridia spores occurs. Alpha toxin and other exotoxins are secreted and extensive cell killing ensues. The production of enzymes that break down ground substance facilitates the spread of infection. Fermentation of tissue carbohydrates yields gas, and an accumulation of gas bubbles in the subcutaneous spaces produces a crinkling sensation on palpation **(crepitation),** hence the name gas gangrene. The exudates are copious and foul smelling. As the disease progresses, increased capillary permeability leads to the exotoxins being carried by the circulation from damaged tissue to other organs, resulting in systemic effects such as shock, renal failure, and intravascular hemolysis. Untreated clostridial myonecrosis is uniformly fatal within days of the initiation of gangrene.

2. Septicemia

The isolation of C. *perfringens* or other clostridial species in blood cultures can be alarming. Invasion of the bloodstream may occur in association with malignancy and may involve a localized myonecrosis in addition to a fulminating clostridial septicemia.

3. Food Poisoning

The organisms usually involved are strains of type A that produce heat resistant spores and minimal amounts of theta toxin. Meat, chicken, fish, and their by-products are the most common vehicles for clostridial food poisoning.

Clostridial food poisoning, a relatively common but underappreciated bacterial disease, is characterized by:

(1) a short incubation period (8-24 hours), (2) a clinical presentation that includes abdominal cramps and watery diarrhea but no fever, nausea, or vomiting, and (3) a clinical course lasting less than 24 hours. The illness is self-limited and recovery occurs in 24 to 48 hours. No specific treatment is indicated.

4. Enteritis Necroticans (Necrotizing Jejunitis, Necrotic Enteritis)

This is a severe and often fatal enteritis known by different names in different countries: Germany (Darmbrand), New Guinea (pigbel) East Africa, Thailand and Nepal. β-Toxin-producing *C. perfringens* type C is responsible for this disease.

5. *C. perfringens* Colitis

A sporadic diarrheal syndrome, usually occurring in elderly patients during treatment with antibiotics, has been described. An enterotoxin with a cytopathic effect can be detected in the patient's feces.

6. Clostridial Endometritis

This condition is a grave complication of incomplete abortion, or the use of inadequately sterilized instruments. Gangrenous infection of uterine tissue is followed by toxemia and bacteremia.

Laboratory Diagnosis

Gas Gangrene

Gas gangrene is a medical emergency. The diagnosis of gas gangrene must be made primarily on clinical grounds, and the function of the laboratory is only to provide confirmation of the **clinical diagnosis** as well as identification and enumeration of the infecting organisms.

1. Specimens

(1) Edge of the affected muscles; (2) Exudates from the wound; and (3) Necrotic tissue and muscle fragments.

2. Microscopy

Gram stained films give presumptive information about the species of clostridia present and their relative numbers. If gas gangrene is present, gram-positive rods may predominate. Thick, stubby, gram-positive rods suggest *C. perfringens* or *C. sordellii*, 'citron bodies', boat- or leaf-shaped pleomorphic bacilli with irregular staining, may indicate *C. septicum*; slender rods with round terminal spores suggest *C. tetani* and large rods with oval subterminal spores indicate *C. novyi*.

3. Culture

Fresh and **heated blood agar** are used for aerobic and anaerobic cultures. To prevent swarming by some species of clostridia, the use of plates containing increased agar (5-6%) are considered. A plate of **serum or egg-yolk agar,** with *Cl. perfringens* antitoxin spread on one half is used for the 'Nagler's reaction'.

Four tubes of **cooked meat broth** are inoculated and heated at 100°C for 5, 10, 15 and 20 minutes, incubated and subcultured on **blood agar plates** after 24 to 48 hours, to differentiate the organisms with heat resistant spores. **Blood cultures** are often positive especially in *Cl. perfringens* and *Cl. septicum* infections. However, *Cl. perfringens* bacteremia may occur without gas gangrene.

4. Identification

Examine plates for typical colonies. The isolates are identified based on their morphological, cultural, biochemical and toxigenic characters.

5. Animal Pathogenicity

Laboratory Diagnosis of Food Poisoning

A diagnosis can be made from isolation of C. *perfringens* in higher than normal numbers from the **feces** of infected patients and from samples of the ingested **food.** CMB is inoculated and heated at 100°C for 30 minutes. It is cooled and is incubated at 37°C for 18 hours and subcultured on selective medium which is then incubated anaerobically at 37°C for 18 hours. Identification of the bacterial isolates is done by morphology, cultural characteristics, biochemical reactions and Nagler's reaction. Isolation from feces, except in large numbers is not meaningful as *Cl. perfringens* may be present in normal intestines.

A more specific diagnosis is possible by use of an enzyme-linked immunosorbent assay (ELISA) to detect *C. perfringens* enterotoxin in the feces of affected persons.

Prophylaxis and Therapy

1. Surgery

All damaged tissues should be removed promptly and the wounds irrigated to remove blood clots, necrotic tissue and foreign materials. In established gas gangrene, uncompromising excision of all affected parts may be life-saving.

2. Antibiotics

Penicillin, metronidazole and an aminoglycoside may be given in combination. Alternatively, clindamycin plus an aminoglycoside or a broad-spectrum antibiotic such as meropenem or imipenem, may be considered.

3. Passive Immunization

A polyvalent antiserum used to be available but it has now been replaced by intensive antimicrobial therapy.

4. Hyperbaric Oxygen

Hyperbaric oxygen may be beneficial in treatment and is introduced in the depth of wound to reduce anaerobiosis.

5. Active Immunization

Toxoids induce antitoxic response experimentally but it has not been come into use practically.

CLOSTRIDIUM TETANI

Clostridium tetani is the causative agent of tetanus, a disease that is now relatively rare in well-developed countries. Tetanus has been known from very early times, having been described by Hippocrates and Aretaeus.

Morphology

It is a gram-positive, slender bacillus, 2 to 5 × 0.4-1 μm with rounded ends. The spores are spherical, terminal and twice the diameter of vegetative cells giving them typical **drumstick appearance** (Fig. 30.3). The spore does not stain with the Gram stain and appears as a colorless round structure. It tends to be pleomorphic and sometimes filamentous. It is noncapsulated and motile by peritrichate flagella (except *Cl. tetani* type VI) with peritrichate flagella. Young cultures of the organism usually stain gram-positive, but in older cultures and in smears made from wounds, they are Gram variable and even be gram-negative.

Cultural Characteristics

Cl. tetani is an obligate anaerobe. The optimal temperature for growth is 37°C, and the optimal pH is 7.4. It can grow well in cooked meat broth (CMB), thioglycollate broth, nutrient agar and blood agar. In **cooked meat broth (CMB),** growth occurs as turbidity and there is also some gas formation. The meat is not digested but becomes black on prolonged incubation. On **blood agar** the bacilli produce a swarming (thin spreading film) growth. On **horse blood agar**, the colonies of *Cl. tetani* are surrounded by a zone of α-hemolysis, which subsequently develops into β-hemolysis, due to the production of an oxygen-labile hemolysin known as tetanolysin. On **egg-yolk agar,** it does not produce opalescence or pearly layer.

In **deep agar shake cultures,** the colonies are spherical fluffy balls, 1-3 mm in diameter, made up of filaments with a radial arrangement. In **gelatin stab cultures** a fir tree type of growth occurs, with slow liquefaction.

Biochemical Reactions

Cl. tetani has feeble proteolytic but no saccharolytic property. It does not attack any sugar. Gelatin liquefaction occurs very slowly. Coagulated serum is rendered more transparent and softened but not liquefied. It is indole positive and MR, VP, H2S and nitrate reduction negative. A greenish fluorescence is produced on media containing neutral red (as on MacConkey's medium).

Resistance

The spores may be highly resistant to adverse conditions, but the degree of resistance varies with the strain. Most are killed by boiling for 10 to 15 minutes but some resist boiling for up to three hours. They can, however, be killed by autoclaving at 121°C for 15 minutes.

Spores are able to survive in soil for years, and are resistant to most antiseptics. They are killed by exposure

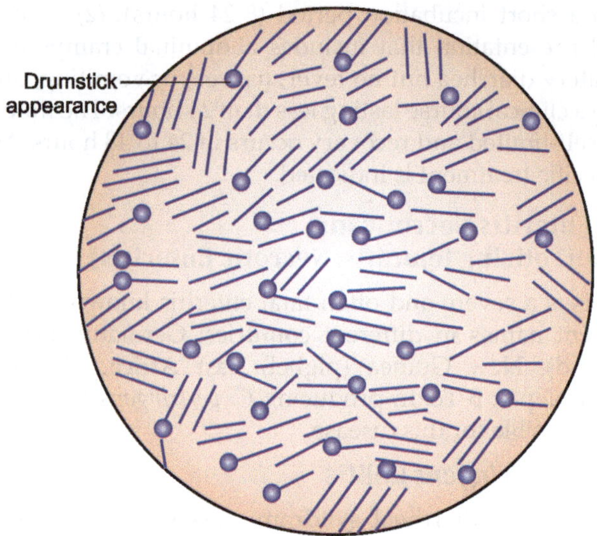

Fig. 30.3: *Cl. tetani*, some with spores and some without spores

to iodine (1% aqueous solution), hydrogen peroxide (10 volumes) and glutaraldehyde (2%) within a few hours.

Antigenic Structure

Flagella (H), somatic (O), and spore antigens have been demonstrated in *C. tetani.* The spore antigens are different from the H and O antigens of the somatic cell.

1. Flagella (H) Antigen

Ten serological types have been recognized based on agglutination (types I to X) of which type I and III are the commonest. This typing is on the basis of their flagellar (H) antigens. Type VI contains nonflagellated strains.

2. Somatic (O) Antigen

There is a single somatic agglutination group for all strains that permits identification of the organism by use of fluorescein-labeled antisera.

Tetanus Toxin

Cl. tetani produces at least two distinct toxins - an oxygen-labile **hemolysin (tetanolysin***)* and a powerful plasmid-encoded, heat-labile **neurotoxin (tetanospasmin).** The two are antigenically and pharmacologically distinct and their production is mutually independent. A third toxin, a **nonspasmogenic**, peripherally active neurotoxin, has been identified. It is not known whether this plays any role in the pathogenesis of tetanus.

1. Tetanospasmin

All of the symptoms in tetanus are attributable to an extremely toxic neurotoxin, tetanospasmin, which is an intracellular toxin released by cellular autolysis. The toxin is a heat-labile, oxygen stable, powerful plasmid-encoded **neurotoxin** that may be inactivated by heating for 20 minutes at 60°C. The toxin has been crystalized. It gets toxoided spontaneously or in the presence of low concentrations of formaldehyde. It is a good antigen and is specifically neutralized by the antitoxin.

Tetanus toxin is one of the most poisonous substances known. It is a simple protein composed of a single polypeptide chain. On release from the bacillus, it is autolyzed to form a heterodimer consisting of a heavy chain (93,000 MW) and a light chain (52,000 MW) joined by a disulfide bond. The purified toxin is active in extremely small amounts and has an MLD for mice of about **50-75 × 10^{-6} mg.** Its MLD for human beings is about 130 nanograms.

2. Tetanolysin

Tetanolysin is a heat labile, oxygen labile hemolysin, antigenically related to the oxygen labile hemolysins produced by *Cl. perfringens, Cl. novyi* and *Str. pyogens.* It is not relevant in the pathogenesis of tetanus.

Pathogenicity

Setting of the wound-**oxidation reduction potential** must be properly poised-**multiplication of the organism-toxigenesis.**

The spores of *Cl. tetani* are ubiquitous. They occur in the gastrointestinal tracts of man and animals. They are also present in the soil especially in manured soil.

Tetanus develops following the contamination of wound with C. *tetani* spores. The most typical focus of infection in tetanus is a puncture wound caused, for example, by a splinter. Introduced foreign bodies or small areas of cell killing create a nidus of devitalized material in which tetanus spores can germinate and grow. The lowering of the oxidation-reduction potential is associated with **tissue necrosis** after traumatic injuries or the injection of necrotizing substances. Germination of spores is dependent upon the reduced oxygen tension occurring in devitalized tissue. After germination of the spores, toxin is elaborated and gains entrance to the central nervous system.

Cl tetani has little invasive power.Infection strictly remains localized in the wound and the disease is due to the effect of a potent diffusible exotoxin (tetanospasmin). It is transported from an infected locus by retrograde neuronal flow or by blood. Tetanospasmin acts by blocking the release of neurotransmitters (e.g. gamma-aminobutyric acid [GABA], glycine) for inhibitory synapses, thus causing excitatory synaptic activity to be unregulated (spastic paralysis). The abolition of spinal inhibition causes uncontrolled spread of impulses initiated anywhere in the central nervous system. The toxin exerts its effects on the spinal cord, the brain stem, peripheral nerves, at neuromuscular junctions and directly on muscles.

When tetanus occurs naturally, the tetanus bacilli stay at the site of the initial infection and are not generally invasive. Toxin diffuses to affect the relevant level of the spinal cord **(local tetanus)** and then to affect the entire system **(generalized tetanus).** These stages, including the intermediate one of **'ascending tetanus'**, are demonstrable in experimental animals but the stages tend to merge in their clinical presentation in man.

Clinical Manifestations of Tetanus

The incubation period is variable—from two days to several weeks but is commoniy 6 to 12 days. The duration of the incubation period is directly related to the distance of the primary wound infection from the central nervous system.

The onset of signs and symptoms is gradual, usually starting with some stiffness and perhaps pain in or near a recent wound. **Generalized tetanus** is the most common form. The initial complaint may be of stiffness of the jaw **(lockjaw***)* in some cases. Pain and stiffness in the neck and back may follow. The stiffness spreads to involve all muscle groups; facial spasms produce the 'sardonic grin', and in severe cases spasm of the back muscles produces the opisthotonos (extreme arching of the back). A severe case with a relatively poor prognosis shows rapid progression from the first signs to the development of generalized spasms. The autonomic nervous system is involved in patients with more severe disease; the signs and symptoms include cardiac arrhythmias, fluctuations in blood pressure, profound sweating, and dehydration.

In cephalic tetanus the primary site of infection is the head and has high mortality. The most feared form of tetanus, **tetanus neonatorum**, is a significant cause of morbidity and mortality in developing nations. This form of tetanus usually results from cutting the umbilical cord with unsterile instruments or from improper care of the umbilical stump. The mortality in infants exceeds 90 percent.

Epidemiology

Tetanus is more common in the developing countries, where the climate is warm, and in rural areas where the soil is fertile and highly cultivated, where human and animal populations are substantial and live in close association and where unhygienic practices are common and medical facilities poor. In rural India, tetanus was a common cause of death, particularly in the newborn. But immunization of infants and expectant mothers has reduced the incidence to a large extent.

Tetanus was a serious disease with a high rate of mortality (80-90%), before specific treatment became available. The case fatality rate varies from 15 to 50 percent even with proper treatment. Tetanus neonatorum and uterine tetanus have very high fatality rates (70-100%), while otogenic tetanus is much less serious.

Laboratory Diagnosis

The diagnosis of tetanus is made on clinical grounds because isolation of the organism can occur in the absence of disease and also because it is possible to have the disease but be unable to isolate the organism. Laboratory tests only help in confirmation. Not infrequently, it may not be possible to establish a laboratory diagnosis at all.

1. Specimen

Wound exudate or tissue removed from the wound.

2. Microscopy

Microscopy is unreliable and the demonstration of the **typical 'drumstick' bacilli** in wounds itself is not diagnostic of tetanus. It may not also be possible to distinguish by microscopy between *Cl. tetani* and morphologically similar bacilli such as *Cl. tetanomorphum* and *Cl. Sphenoides.* Hence, microscopy is unreliable. Simple light microscopy is often unsuccessful. Immunofluorescence microscopy with a specific stain is possible but not generally available.

3. Culture

The material is inoculated on one half of a **blood agar plate.** *Cl. tetani* produces a swarming growth which may be detected on the opposite half of the plate after 1 to 2 days incubation anaerobically. The incorporation of polymyxin B, to which clostridia are resistant, makes the medium more selective.

The material is also inoculated into three tubes of cooked meat broth, one of which is heated to 80°C for 15 minutes, the second for five minutes, and the third left unheated. The purpose of heating for different periods is to kill vegetative bacteria, while leaving undamaged tetanus spores, which vary widely in heat resistance. The cooked meat tubes are incubated at 37°C and subcultured on one-half of blood agar plates daily for up to four days. For identification and toxigenicity testing,blood agar plates (with 4% agar to inhibit swarming), having tetanus antitoxin (1500 units per ml) spread over one-half of the plate are used.

4. Toxigenicity Test

Toxigenicity is best tested in animals. **Control mice** are protected with tetanus antitoxin. Two mice, one **unprotected** and other **protected** are used for each test. One animal is protected by giving 1,000 units of tetanus antitoxin intraperitoneally 1 hour before the test. Inject 0.1 ml of a 48 hour CMB culture supernate ot the organism intramuscularly into the hind limb of one mouse **(the test)** and the same amount in the another animal **(control animals).** The protected mouse remains well. Signs of ascending tetanus develop in the unprotected animal after several hours, they begin in the inoculated leg and extend to the tail, then the other hind limb is affected and then generalized signs appear. The animal dies within 2 days but may be killed earlier as the appearance of ascending tetanus is diagnostic.

Prophylaxis

The available methods of prophylaxis are:

1. Surgical attention
2. Antibiotics
3. Immunization—passive, active or combined.

1. Surgical Prophylaxis

This includes prompt and adequate wound toilet and proper surgical debridement of wound, removal of foreign material, necrotic tissue and blood clots to prevent an anaerobic environment for the growth of *C. tetani.*

2. Antibiotic Prophylaxis

Antibiotics destroy or inhibit *C. tetani* and pyogenic bacteria in the wound, so that the production of the toxin can be prevented. Long-acting penicillin injection is the drug of choice. An alternative is erythromycin. Antibiotic prophylaxis does not replace immunoprophylaxis but serves as a useful adjunct.

3. Immunoprophylaxis

It includes 3 types of immunization:

i. Active immunization
ii. Passive immunization
iii. Combined immunization

i. Active Immunization

Tetanus is best prevented by active immunization with tetanus toxoid. Two preparations are available for active immunization are:

a. Combined vaccine (DPT)
b. Monovalent vaccines
   i. Plain or fluid (formol) toxoid
   ii. Tetanus vaccine, adsorbed (PTAP, APT)

Tetanus toxoid (formol toxoid), which is available either as 'plain toxoid', or adsorbed on aluminum hydroxide or phosphate (APT), is commonly used for active immunization. Three doses of 0.5 ml tetanus toxoid (APT) each are given intramuscularly, with an interval of 4 to 6 weeks between first two doses and 6 to 12 months between the second and third dose. A full course of three doses confers immunity for a period of at least 10 years. A 'booster dose' of toxoid is recommended after 10 years.

Tetanus toxoid is given along with diphtheria toxoid and pertusis vaccine called **DPT** in children as **triple vaccine.** Pertusis vaccine acts as adjuvant. Three doses are given intramuscularly at interval of 4-6 weeks, starting at age as early as 6 weeks. Booster doses are given at age of 18 months and then at five years. Thereafter, booster doses of TT (tetanus toxoid) are given at the age of 10 and 16 years. Subsequently, immunity to tetanus can be maintained by booster doses of toxoid every 10 years.

ii. Passive Immunization

Passive immunity may be conferred by the administration of antitoxin. Temporary protection against tetanus can be provided by an injection of human tetanus hyperimmune globulin (TIG) or ATS.

Antitetanus Serum (ATS)

Antitetanus serum (ATS) from hyperimmune horses was the preparation originally used. The dose employed was 1500 IU given subcutaneously or intramuscularly in nonimmune persons soon after receiving any tetanus prone injury. ATS gives passive protection for about 7 to 10 days.

Disadvantages of ATS

Equine ATS carried two disadvantages implicit in the use of any heterologous serum—**'immune elimination** and **hypersensitivity reactions.** It is, therefore, mandatory to test for hypersensitivity before administration of ATS. A 'trial' dose given subcutaneously would be a better index of hypersensitivity. A dose of 0.5 ml of ATS is given subcutaneously and the patient observed for at least half an hour for general reactions. As even this dose may precipitate anaphylaxis in some cases, a *syringe loaded with adrenaline (1/1000) must be kept ready.* The trial dose should be 0.05 ml of a *1/10* dilution of ATS in persons with a history of any allergy.

Human Antitetanus Immunoglobulin (HTIG)

Passive immunity without risk of hypersensitivity can be obtained by the use of **human antitetanus immunoglobulin (HTIG).** This is effective in smaller doses. The prophylactic dose of HTIG is 250 units by intramuscular injection.

iii. Combined Immunization

It consists of administering to a nonimmune person ATS or HTIG at one site, along with the first dose of a course of active immunization with adsorbed toxoid at the same time at another site, followed by second and third doses of TT at appropriate intervals. The active immunization course must be subsequently completed. The purpose of antitoxin is for immediate temporary protection, and the purpose of toxoid is for long-lasting protection.

Treatment

1. The patient remains conscious and requires skilled sedation and constant nursing.
2. Treatment of tetanus requires debridement of the primary wound, use of metronidazole, passive immunization with human tetanus immunoglobulin, and vaccination with tetanus toxoid.
3. Full wound exploration and debridement is arranged, and the wound is cleansed and left open with a loose pack.
4. The patient is given 10,000 units of human tetanus immunoglobulin (RTIG) in saline by slow intravenous infusion.
5. Penicillin or metronidazole is given for as long as considered necessary to ensure that bacterial growth and toxin production are stopped. The antitoxin and antibiotics are given immediately, and preferably before surgical excision but delay must be avoided.
6. Vaccination with a series of three doses of tetanus toxoid followed by booster doses every 10 years is highly effective in preventing tetanus.

CLOSTRIDIUM BOTULINUM

C. *botulinum* (from the Latin ***botulus,* "sausage"**) causes **botulism. Botulism** is a severe, often fatal, form of food poisoning characterized by pronounced neurotoxic effects. The disease has been caused by a wide range of foods, usually preserved hams, large sausages of the German variety, home-preserved meats and vegetables, canned products such as fish, liver contact with the organism itself is not required; hence the disease can be a pure intoxication.

Morphology

C. botulinum is a strictly anaerobic gram-positive bacillus (about 5 x 1 µm). It is noncapsulated, motile with peritrichous flagella and produces spores which are oval, subterminal and bulging.

Cultural Characteristics

It is a strict anaerobe. Optimum temperature is 35°C but some strains may grow even at 1 to 5°C. Good growth occurs on ordinary media. Surface colonies are large, irregular, semitransparent, with fimbriate border. On **horse blood agar**, all strains except those of type G are β-hemolytic. On **egg-yolk agar (EYA)** all types except G produce opalescence and a pearly effect

Resistance: Spores are heat and radiation resistant, surviving several hours at 100°C and for up to 10 minutes at 120°C. Spores of nonproteolytic types of B, E and F are much less resistant to heat. The resistance of the spores to radiation is of special relevance to food processing.

Classification

The species *C. botulinum* includes a very heterogeneous group of strains that have been divided into eight serologically distinct types—A, B, C1, and C2, D, E, F, and G—on the basis of the type of toxin produced. The toxins produced by different types are antigenically distinct but pharmacologically similar. All types can cause human disease but types A, B and E are most common. The toxins produced by the different types of *Cl. botulinum* appear to be identical, except for immunological differences. Toxin production appears to be determined by the presence of bacteriophages, at least in types C and D.

Botulinum Toxin

A powerful exotoxin produced by *Cl. botulinum* is responsible for its pathogenicity. It differs, however, from a classic exotoxin in that it is not released during the life of the organism but appears in the medium only after death and autolysis of the organism. It is believed to be synthesised initially as a nontoxic protoxin or progenitor toxin. Trypsin and other proteolytic enzymes activate progenitor toxin to active toxin. The production of botulinum toxin is governed by specific bacteriophages.

Properties of Toxin

Botulinum toxin is one of the most potent toxins known. It is isolated as a pure crystalline protein with MW

70,000. It has a lethal dose for mice of 0.000,000,033 mg and lethal dose for human beings is probably 1 to 2 µg. It is a neurotoxin and acts slowly, taking several hours to kill.

The toxin is relatively stable being inactivated at 80°C for 30 to 40 minutes and at 100°C for 10 minutes. Food suspected to be contaminated with botulinum toxin can be rendered completely safe by pressure cooking or boiling for 20 minutes. It can be toxoided. It is a good antigen and is specifically neutralized by the antitoxin.

Botulinum toxin gains access to the peripheral nervous system, where it acts preferentially on cholinergic nerve endings to block the release of the neurotransmitter, acetylcholine, from the nerve terminals of neuromuscular junctions. A symmetric descending paralysis is the characteristic pattern, ending in death bv respiratory paralysis. In minute doses, botulinum toxin preparations have been used therapeutically to relieve spastic conditions such as poststroke spasticity and. strabismus.

Pathogenicity

It is noninvasive and its pathogenicity is entirely due to the toxin produced by it. The disease caused by this organism is known as **botulism.** It is of 3 types: food-borne botulism, wound botulism and infant botulism.

1. Food-Borne Botulism

It is due to the ingestion of preformed toxin. The causative organism, *C. botulinum,* multiplies in the food before it is consumed, and produces a powerful soluble toxin. The source of botulism is usually preserved food such as meat and meat products, fish, and vegetables. Food responsible for botulism is usually abnormal in appearance and odor. Bulging of tins and the presence of gas bubbles on opening suggest contamination with *C. botulinum.* However, at times food may look normal.

Symptoms usually begin 18 to 36 hours after ingestion of food and may include nausea, vomiting, thirst, constipation, double vision, difficulty in swallowing, speaking and breathing. This may be followed by muscular weakness, blurred vision, and death as a result of respiratory failure. Case fatality varies from 25 to 70 percent.

2. Wound Botulism

Wound botulism is a very rare condition resulting from wound infection with *Cl. botulinum.* Toxin is produced at the site of infection and is absorbed. The symptoms are those of food-borne botulism except for the gastrointestinal components which are absent. Type A has been responsible for most of the cases studied.

3. Infant Botulism

Infant botulism was first recognized in 1976 and is now the most common form of botulism in the United States. This is a toxico-infection. *Cl. botulinum* spores are ingested in food, get established in the gut and there produce the toxin. The disease typically affects infants younger than 1 year (most between 1 and 6 months). The most common food source in infant botulism is **honey** contaminated with botulinum spores.

Clinically, infant botulism is an acute flaccid paralysis. After a period of normal development, the infant develops constipation, listlessness, difficulty in sucking and swallowing, weak or altered cry, muscle weakness, ptosis and loss of head control. Eventually the baby appears 'floppy' **(floppy child syndrome)** and develops respiratory insufficiency or respiratory arrest. Fulminant forms may resemble the **sudden infant death syndrome (SIDS or crib death).** Patients excrete toxin and spores in their feces. Toxin is not generally demonstrable in blood.

Management consists of supportive care and assisted feeding. Antitoxins and antibiotics are not indicated. Degrees of severity vary from very mild illness to fatal disease.

Laboratory Diagnosis

Botulism confirmed by isolating the organism or detecting the toxin in food products or the patient's feces or serum.

1. Specimens

Feces, food, vomitus, gastric fluid, serum, environmental samples and occasionally wound exudate.

2. Culture

For the isolation of *C. botulinum,* the specimen is inoculated on egg yolk agar (EYA), blood agar and CMB. Culture of the heated specimen on nutritionally enriched anaerobic media allows the heat resistant *C. botulinum* spores to germinate. The strains of *C. botulinum* associated with human botulism are characterized by **lipase production** (appears as an iridescent film on colonies grown on egg-yolk agar) as well as the ability to digest milk proteins, hydrolyze gelatin and ferment glucose.

Presence of bacilli in food or feces in absence of toxin is of no significance. Hence, toxin in culture fluid must be demonstrated by toxigenicity test in mice.

3. Demonstration of Toxin

Demonstration of toxin production must be done with a **mouse bioassay.** This procedure consists of the preparation of two aliquots of the isolate, mixing of one aliquot with antitoxin, and intraperitoneal inoculation of each aliquot into mice. Toxin activity is confirmed if the antitoxin treatment protects the mice, control animals protected by polyvalent antitoxin remain healthy. Typing. is done by passive protection with type specific antitoxin. Samples of the implicated food, stool specimen and patient's serum should also be tested for toxin activity.

A retrospective diagnosis may be made by detection of antitoxin in the patient's serum but it may not be seen in all cases.

Diagnosis of Infant Botulism

The diagnosis of infant botulism is supported if (1) *C. botulinum* is isolated from feces or (2) toxin activity is detected in feces or serum. The organism can be isolated from stool cultures in virtually all patients, because carriage of the organism may persist for many months even after a baby has recovered.

Diagnosis of Wound Botulism

Wound botulism is confirmed by isolation of the organism from the wound or by detection of toxin activity in wound exudate or serum.

Prophylaxis

Control can be achieved by proper canning and preservation. Infant botulism has been associated with the consumption of honey contaminated with *C. botulinum* spores, so children younger than 1 year should not eat honey.

A prophylactic dose of polyvalent antitoxin should be given intramuscularly to all persons who have eaten food suspected of causing botulism.

Active immunization should be considered for laboratory staff who might have to handle the organism or specimens containing the organism or its toxin. Two injections of aluminum sulphate adsorbed toxoid may be given at an interval of ten weeks, followed by a booster a year later.

Treatment

Patients with botulism require the following treatment measures:

1. Elimination of the organism from the gastrointestinal tract through the judicious use of gastric lavage and metronidazole or penicillin therapy.
2. The use of polyvalent antitoxin to neutralize unfixed toxin
3. Adequate ventilatory support.

CLOSTRIDIUM DIFFICILE

Cl. difficile was first isolated in 1935 from the feces of newborn infants. It was so named because of the unusual difficulty in isolating It was not considered pathogenic till 1977 when it was found to be responsible for antibiotic associated colitis. This organism was infrequently isolated in fecal cultures and rarely associated with human disease.

Morphology

C. *difficile* is a motile gram-positive rod with oval subterminal spores. Spores are large, oval and terminal. It is nonhemolytic, saccharolytic and weakly' proteolytic.

Toxins

The organism produces an **enterotoxin (toxin A)** and a **cytotoxin (toxin B). 1. Toxin A** is an enterotoxin that is primarily responsible for diarrhea. It is capable of producing fluid accumulation in ligated rabbit ileal loop assay. **2. Toxin B** is a potent cytotoxin capable of producing cytopathogenic effects in several tissue culture cell lines.

Pathogenesis

It is a proven cause of **antibiotic associated diarrhea (AAD)**, and **pseudomembranous colitis (PMC)**—a life-threatening condition. The disease develops in people taking antibiotics because the agents alter the normal enteric flora, either permitting the overgrowth of these relatively resistant organisms or making the patient more susceptible to the exogenous acquisition of *C. difficile.* The disease occurs if the organisms proliferate in the colon and produce their toxins there. Virtually all antimicrobial drugs have been reported as predisposing to clostridial AAD and colitis. The three drugs most commonly implicated are **clindamycin, ampicillin and the cephalosporins.** The severity of disease varies widely from mild diarrhea through varying degrees of inflammation of the large intestine to a fulminant PMC. It is postulated that in the immature large intestine, exotoxin receptors may not yet be present or accessible.

Laboratory Diagnosis

1. Isolation of Bacilli

C. difficile can be isolated from the feces by enrichment and selective culture procedures.

2. Demonstration of Toxin

Toxin B can be demonstrated in the feces of patients by its characteristic effect on **Hep-2 and human diploid cell cultures** or both toxins may be demonstrated by immunological methods, e.g. enzyme-linked immunosorbent assay (ELISA).

Treatment and Prophylaxis

The disease is treated by discontinuing the antibiotic that is presumed to have precipitated the disease and to suppress the growth and toxin production of *C. difficile* by giving oral metronidazole or vancomycin.

☞ KEY POINTS

Clostridium

- The genus *Clostridium* includes all anaerobic, gram-positive bacilli capable of forming endospores.
- Clostridia are more commonly associated with skin and soft tissue infections, food poisoning and antibiotic-associated diarrhea and colitis.
- The shape and position of spores varies in different species and is useful in identification of clostridia, spores may be subterminal in *Cl. perfringens* and drumstick appearance in *Cl. tetani.*
- Clostridia grow on enriched media in the presence of reducing agent, or in an O_2- free gasseoues atmosphere. Clostridia grow well on blood agar medium under anaerobic conditions. Liquid media

like cooked meat broth (CMB) is very useful for growing clostridia.

Clostridium perfringens

- Toxins: *Cl. perfringens* is forming at least 12 distinct soluble substances or toxins. The four 'major toxins, alpha, beta, epsilon. and iota, are predominantly responsible for pathogenicity. It produces lecithinase (phospholipase C). Subdivided into 5 types (A-E) on the basis of toxin production.
- Diseases: 1. Soft tissue infections (cellulitis, suppurative myositis, myonecrosis); 2. Food poisoning. 3. Septicemia.
- Nagler reaction and reverse CAMP test are useful in identification *Cl. perfringens*, a causative agent of gas gangrene and food poisoning.
- Systemic infections require surgical debridement and high-dose penicillin therapy; antiserum against a toxin not used now.

Clostridium tetani

- Cl. tetani is the causative agent of tetanus.
- Tetanus toxin: Two distinct toxins—an oxygen-labile hemolysin (tetanolysin) and a neurotoxin (tetanospasmin). Tetanospasmin blocks release of neurotransmitters (i.e. gamma-aminobutyric acid, glycine) for inhibitory synapses, thus causing excitatory synaptic activity to be unregulated (spastic paralysis).
- Treatment requires debridement, antibiotic therapy (metronidazole), passive immunization with antitoxin globulin, and active immunization with tetanus toxoid.Prevention through use of vaccination, consisting of three doses of tetanus toxoid followed by boosters every 10 years.

Clostridium botulinum

- *Cl. botulinum* forms a powerful exotoxin which is responsible for the disease botulism. It can produce one of eight distinct botulinum toxins (A-G).
- *Diseases:* The disease caused by this organism is known as botulism. It is of 3 types: food-borne botulism, wound botulism and infant botulism.

Clostridium difficile

It is a proven cause of antibiotic associated diarrhea (AAD), and pseudomembranous colitis (PMC)—a life-threatening condition. The three drugs most commonly implicated are clindamycin, ampicillin and the cephalosporins. Treatment is done with metronidazole or vancomycin.

IMPORTANT QUESTIONS

1. Classify clostridia. Discuss the laboratory diagnosis of gas gangrene.
2. Discuss the pathogenicity and prophylaxis of gas gangrene.
3. Enumerate various pathogenic clostridia. Describe the pathogenesis and laboratory diagnosis of tetanus.
4. Write short notes on:
   - Alpha toxin
   - Stormy clot reaction
   - Nagler's reaction
   - Gas gangrene.
   - Tetanospasmin
   - Prophylaxis of tetanus.
   - *Clostridium botulinum*
   - Botulism.
   - Botulinum toxin.
   - Laboratory diagnosis of botulism
   - *Clostridium difficile*

FURTHER READING

Bartlett JG. *Clostridium difficile*. Clin Infect Dis 1994;18:S265.

Boone J, Carman R. *Clostridium penfringens*: Food poisoning and antibiotic-associated diarrhea; Clin Microbiol Newsletter 1997;19:65-67.

Brazier JS. Laboratory diagnosis of *Cl. difficile* associated disease. Rev Med Microbial 1995;l6:236

Bryant A, et al. Clostridial gas gangrene. I and II, J Infect Dis 2000;182:799-807,808-815.

Sakurai J. Toxins of *Cl. Perfringens*. Rev Med Microbiol 1995; 6:175.

Sanford JP. Tetanus - forgotten, but not gone. New Engl J Med 1995;332:812.

CHAPTER

31 Nonsporing Anaerobes

Learning Objectives

After reading and studying this chapter, you should be able to:
- Describe classification of nonsporing anaerobes.
- List infections caused by nonsporing anaerobes.
- Discuss laboratory diagnosis of infections caused by nonsporing anaerobes.

INTRODUCTION

The anaerobic bacteria are widespread in nature. A bewildering range of anaerobes is found in the mouth and oropharynx, gastrointestinal tract and female genital tract of healthy individuals as part of the commensal flora. They constitute the predominant part of our normal indigenous flora on skin and membrane surfaces and outnumber facultatively anaerobic bacteria in the gut by a factor of 1000:1. The numbers of anaerobes present have been estimated to be 10^4–10^5/ml in the small intestine, 10^8/ml in saliva and 10^{11}/g in the colon. On the skin, in the mouth, in the upper respiratory tract, and often in the female lower genital tract, they outnumber facultatively anaerobic bacteria by a factor of 5:1 to 10:1. Many of these anaerobic organisms, previously considered to be harmless commensals of our indigenous flora, are now recognized as opportunistic pathogens that may produce disease when the host's resistance is reduced.

CLASSIFICATION

Current classification is based on DNA base composition and analysis of the fatty acid end products of metabolism. Medically important anaerobes may be broadly classified as shown in Table 31.1.

ANAEROBIC COCCI

Strictly anaerobic gram-positive cocci have been assigned to the genera *Peptococcus, Peptostreptococcus, Coprococcus, Ruminococcus* and *Sarcina.* Strictly anaerobic gram-negative cocci are included in *Veillonella, Megasphera* and *Acidaminococcus.*

This brief account is restricted to *Peptococcus, Peptostreptococcus* and *Veillonella,* as these are the genera most commonly associated with clinical infections. Gram-positive anaerobic cocci comprise part of the normal microbial flora of the mouth, gastro-intestinal tract, genito-urinary tract and skin.

Peptococcus

Most of the peptococci have now been reclassified as peptostreptococci. *Peptococcus niger* is the only surviving member of the genus *Peptococcus.* They are gram-positive, nonsporing, anaerobic cocci, that occur singly

Table 31.1: Classification of nonsporing anaerobes

I. Cocci
   A. Gram-positive
      a. *Peptostreptococcus*
      b. *Peptococcus.*
   B. Gram-negative
      Veillonella.
II. Bacilli
   1. Endospore forming
      Clostridia.
   2. Nonsporing
   A. Gram-positive
      a. *Eubacterium*
      b. *Propionibacterium*
      c. *Lactobacillus*
      d. *Mobiluncus*
      e. *Bifidobacterium*
      f. *Actinomyces.*
   B. Gram-negative
      a. *Bacteroides*
      b. *Prevotellla*
      c. *Porphyromonas*
      d. *Fusobacterium*
      e. *Leptotrichia*
III. Spirochetes
   a. *Treponema*
   b. *Borrelias.*

or in pairs or in clusters. They produce black colonies on blood agar on prolonged incubation due to the production of H_2S. They occur as normal flora of skin, intestine and genitourinary tract. They may cause pyogenic infections of wounds, puerperal sepsis and urinary tract infections.

Peptostreptococcus

Anaerobic gram-positive cocci had been classified into the genera ***Peptostreptococcus*** and ***Peptococcus*** originally, based on morphology, chain forming and paired cocci placed in the former and cluster forming cocci in the latter. However, DNA base ratio studies have led to most of the species formerly considered as peptococci being reclassified as peptostreptococci. Most species known to be clinically significant are regarded as belonging to the genus *Peptostreptococcus*. They are cocci of small size (0.2-2.5 µm). Many of them are aerotolerant and grow well under 10 percent CO_2 in an aerobic atmosphere.

Peptostreptococcus anaerobius is most often responsible for puerperal sepsis and *Pst. magnus* for abscesses. *Pst. asaccharolyticus, Pst. tetradius* and *Pst. prevotii* are some other species commonly present in clinical specimens.

Other genera, including *Coprococcus, Gaffkya, Gemmiger, Ruminococcus* and *Sarcina,* are found as part of the flora of the bowel, but are not usually considered to be significant in infections.

GRAM-NEGATIVE ANAEROBIC COCCI

Veillonella

Veillonellae are gram-negative cocci of varying sizes, and measure 0.3–2.5 µm in diameter, occurring as diplococci, short chains or groups. They are obligate anaerobes, oxidase-negative, catalase-negative, nonmotile. They are normal inhabitants of the mouth, intestinal and genital tracts. *Veillonella parvula* has been reported from clinical specimens but its pathogenic role is uncertain.

ANAEROBIC, NONSPORE-FORMING, GRAM-POSITIVE BACILLI

The nonspore-forming, gram-positive bacilli are a diverse collection of facultatively anaerobic or strictly anaerobic bacteria that colonize the skin and mucosal surfaces. This group contains many genera, of which medically relevant are *Eubacterium, Propionibacterium, Lactobacillus, Mobiluncus, Bifidobacterium* and *Actinomyces* (Table 31.1). Association of anaerobic bacilli with human disease is shown in Table 31.2.

Eubacterium

Members of the genus *Eubacterium* are strictly anaerobic and grow very slowly. They are members of the normal mouth and intestinal flora. Some species *(E. brachy, E. timidum, E. nodatum)* are commonly seen in periodontitis. *E. lentum* is commonly isolated from nonoral clinical specimens.

Bifidobacterium

Bifidobacterium is gram-positive bacilli, nonmotile, nonsporing and pleomorphic, showing true and false branching. The name is derived from the frequent bifid Y-shaped cells. They occur as normal flora of mouth, gastrointestinal tract (GI) tract and genitourinary tract. They are usually nonpathogenic. *B. bifidium* is commonly found in the feces of infants and adults.

Lactobacillus

Lactobacilli are gram-positive bacilli, straight or slightly curved which frequently show bipolar and barred staining. They are nonsporing and most strains are nonmotile. They form considerable amount of lactic acid from carbohydrates and grow best at pH of 5 or less.

They are found as part of the normal flora of the mouth, stomach, intestines, and genitourinary tract. Lactobacilli are normally present in the mouth and have been incriminated in the pathogenesis of **dental caries.** It is believed that lactobacilli ferment sucrose to produce lactic acid, which dissolves the mineral components of enamel and dentine causing dental caries.

Several species of lactobacilli are present in the intestine, the commonest being *L. acidophilus.* Intestinal lactobacilli are beneficial in synthesizing vitamins such as biotin, vitamin B_{12} and vitamin K, which may be absorbed by the host.

Lactobacillus species are major members of the normal flora of the vagina and, typically, in the adult vagina and these are collectively known as **Doderlein's bacilli.** They ferment the glycogen deposited in the vaginal epithelial cell and form lactic acid, which accounts for the highly acidic pH of vaginal mucus epithelia. They protect adult vagina from infections. In prepubertal and postmenopausal vagina, lactobacilli are scanty.

Table 31.2: Anaerobic, nonsporing, gram-positive bacilli

Anaerobic bacilli	*Human disease*
1. *Eubacterium*	Opportunistic infections
2. *Bifidobacterium*	Opportunistic infections
3. *Lactobacillus*	Endocarditits, opportunistic infections
4. *Mobiluncus*	Bacterial vaginosis, opportunistic infections
5. *Propionibacterium*	Acne,lacrimal canaliculus, opportunistic infections
6. *Actinomyces*	Actinomycosis (cervicofacial, thoracic, abdominal, pelvic, central nervous system)

It is generally nonpathogenic though *L. catenaforme* has been associated with **bronchopulmonary infections**. Lactobacilli are acidophilic and grow best at acidic pH. Some species have strict growth requirements and are used for the microbiological assay of growth factors (vitamins).

Mobiluncus

Members of the genus *Mobiluncus* are obligate anaerobic, gram-variable or gram-negative, curved bacilli with tapered ends. Two species, *Mobiluncus curtisii* and *Mobiluncus mulieris,* have been identified in humans.

Mobiluncus curtisii and *Mobiluncus mulieris* have been isolated from the vagina in bacterial vaginosis, along with *Gardnerella vaginalis.* The organisms colonize the genital tract in low numbers but are abundant in women with **bacterial vaginosis**. Their microscopic appearance is a useful marker for this disease, but the precise role of these organisms in the pathogenesis of bacterial vaginosis is unclear.

Bacterial vaginosis is a polymicrobial infection characterized by a thin malodorous vaginal discharge. Its 'rotten fish' smell is accentuated by mixing it with a drop of KOH solution. The vaginal pH is less than 4.5. Clue cells (epithelial cells with surface covered by adherent bacilli) are seen in stained or unstained films.

Propionibacterium

Propionibacterium species are members of the normal flora of the skin and cause disease when they infect plastic shunts and appliances. Their metabolic products include propionic acid, from which the genus name derives.

Species

The two most commonly isolated species are *Propionibacterium acnes* and *Propionibacterium propionicus.*

P. acnes is responsible for two types of infections:

i. Acne (as the name implies) in teenagers and young adults;
ii. Opportunistic infections in patients with prosthetic devices (e.g. artificial heart valves or joints) or intravascular lines (e.g. catheters, cerebrospinal fluid shunts).

Propionibacterium Propionicus

When injected into experimental animals, *P. propionicus* causes **lacrimal canaliculitis** (inflammation of the tear duct) and **abscesses.**

Actinomyces (See Chapter 35)

ANAEROBIC GRAM-NEGATIVE BACILLI

Gram-negative, anaerobic, nonspore-forming, nonmotile rods were previously classified in the the **family** ***Bacteroidaceae*** within three genera, *Bacteroides, Fusobacterium* and *Leptotrichia.*

1. Bacteroides

The genus *Bacteroides* can be divided into:

a. Bile-tolerant

i. Members of the *B. fragilis* group. *B. fragilis, B. ovatus, B. distasonis, B. vulgatus, B. thetaiotaomicron,* and others.
ii. Two other species: *Bacteroides eggerthii* and *Bacteroides splanchnicus.*

b. Bile-sensitive Species

These are subdivided into:

I. Pigmented: Have been reclassified into the genera ***Porphyromonas*** and ***Prevotella.*** *(Prevotella melaninogenica:*

II. Nonpigmented species: Many nonpigmented species of *Bacteroides* also have been transferred to the genus ***Prevotella.***

2. *Fusobacterium (Bacilli with pointed ends)*

Fusobacterium necrophorum, Fusobacterium nucleatum, Fusobacterium necrogenes.

3. *Leptotrichia (Large bacilli)*

Leptotrichia buccali—the only species.

I. Bacteroides

Bacteroides species have been classified based on their saccharolytic effects. Bacteria once thought of as typical members of the genus *Bacteroides,* especially those isolated from humans, form three broad groups according to whether they are asaccharolytic, moderately saccharolytic or strongly saccharolytic .

a. Bile-Tolerant Species

Bile-tolerant *Bacteroides* spp. include members of the *B. fragilis* group and two other species, *Bacteroides eggerthii* and *Bacteroides splanchnicus. B. eggerthii* is considered by some authorities to be a member of the *B. fragilis* group. Members of the *B. fragilis* group, which contains about 10 related species, are especially pathogenic. *B. fragilis* is the most common species of anaerobic bacteria isolated from infectious processes of soft tissue and anaerobic bacteremia.

B. fragilis is the most frequent of the nonsporing anaerobes isolated from clinical specimens. It is often recovered from blood, pleural and peritoneal fluids, CSF, brain abscesses, wounds and urogenital infections.Classification is based on colonial and biochemical features and on characteristic short-chain fatty acid patterns in gas liquid chromatography. Their polysaccharide capsule is an important virulence factor, conveying resistance to phagocytosis. The bacterial cell wall lipopolysaccharide has little or no endotoxic activity, but several enzymes (for example, heparinase, collagenase, etc.) probably contribute to tissue destruction.

b. Bile-Sensitive Species

1. Pigmented Species

Prevotella

Prevotella spp. appear as gram-negative coccobacilli or bacilli, very similar to *Bacteroides* spp. ***Prevotella melaninogenica*** is part of the normal flora of the mouth and upper alimentary and respiratory tracts. *Prevotella* infections are usually associated with the upper respiratory tract, causing, for example, dental and sinus infections, pulmonary infections and abscesses, brain abscesses, and infections caused by a human bite. The source of these infections is generally the oral flora.

Laboratory identification is similar to that of *B. fragilis.* In addition, *P. melaninogenica* forms **black colonies on blood agar**—a characteristic from which its name was derived. The color is not due to the melanin pigment as was once thought but to produce protoporphyrin, a dark pigment that causes their colonies to become brown to black with age. Cultures of *P. melaninogenica* and even dressings from wounds infected with the bacillus give a characteristic red fluorescence when exposed to ultraviolet light.

Porphyromonas

The porphyromonas species also are gram-negative bacilli that are part of the normal oral flora and occur at other anatomic sites as well. Porphyromonas species can be cultured from gingival and periapical tooth infections and, more commonly, breast, axillary, perianal, and male genital infections.

II. Nonpigmented Species

Bile-sensitive nonpigmented *Bacteroides* spp. that pit the agar and those that are nonpitting. Recent reports have shown that the so-called pitting anaerobes are of the *Bacteroides ureolyticus* group. Nonpigmented *Prevotella* spp. include *Prevotella bivia, Prevotella buccae, Prevotella buccalis, Prevotella disiens, Prevotella oralis, Prevotella oris, Prevotella oulorum,* and *Prevotella veroralis.*

2. Fusobacterium

Fusobacteria are spindle-shaped gram-negative bacilli with pointed ends, a morphology characteristically referred to as *fusiform*. They are found as part of the normal flora of the mouth, female genital tract, and colon. They grow slowly *in vivo,* and are therefore of limited virulence.

Species

i. ***F. nucleatum,*** the most studied species, is frequently recovered from mixed infections of the head and neck region, including dental abscesses and the central nervous system, and is a1so quite commonly isolated from transtracheal aspirates and pleural fluid.
ii. ***F. necrophorum*** is an important animal pathogen. It is associated with human necrobacillosis and infections similar to those caused by *F. nucleatum* in man, but is isolated less often.

3. Leptotrichia

They are long, straight or slightly curved rods, often with pointed ends. This species was originally classified in the genus **Fusobacterium,** and was formerly known as **Vincent's fusiform bacillus** or **Fusobacterium fusiforme**. The genus *Leptotrichia* contains the single species, ***L. buccalis.*** The association of *L. buccalis* with disease is not clear-cut, although it has been reported in **acute necrotizing ulcerative gingivitis (Vincent's gingivitis)** or **Vincent's angina**, together with *Treponema, Porphyromonas* and *Fusobacterium* species. It is seen in patients with malnutrition, debility and poor oral hygiene. **Vincent's angina** may resemble diphtheria, with the inflamed pharyngeal mucosa showing a grayish membrane which peels easily. *L. buccalis* and *T. vincentii* are present in the exudate and pseudomembrane, and diagnosis is made by direct microscopy. Stained smears show large fusiform and spiral bacilli.

ANAEROBIC INFECTIONS

Most of the anaerobic bacteria that cause infection are members of our normal indigenous flora. Anaerobic infections are usually endogenous and are caused by tissue invasion by bacteria normally resident on the respective body surfaces. Anaerobic bacteria are normally present on the skin, mouth, nasopharynx and upper respiratory tract, intestines and vagina (Table 31.3).

Anaerobic infections generally follow some precipitating factor such as trauma, tissue necrosis, impaired circulation, hematoma formation or the presence of foreign bodies. Diabetes, malnutrition, malignancy or prolonged treatment with aminoglycoside antibiotics, corticosteroids and cytotoxic agents may act as predisposing factors. Anaerobic infections are typically polymicrobial. Anaerobic infections of the head, neck and respiratory tract are often associated with organisms found in the mouth, whilst infections in the abdominal and pelvic regions are more commonly associated with gut bacteria.

Table 31.3: Normal anaerobic flora of the human body

Anaerobe	*Skin*	*Mouth-nasopharynx*	*Intestine*	*Vagina*
Clostridium			++	
Actinomyces		+		
Bifidobacterium		+	++	+
Propionibacterium	++			
Bacteroides fragilis			++	
P. melaninogenica		++	+	++
Fusobacterium		++	+	
Gram-positive cocci		++	++	++
Gram-negative cocci		++	+	++
Spirochetes		+		

Table 31.4: Selected infections typically involving nonsporing anaerobes		
Site	*Type of infection*	*Bacteria commonly responsible*
A. Central nervous system	Brain abscess.	*B.fragilis; Peptostreptococcus.*
B. Ear, nose, throat	Chronic sinusitis, otitis media, mastoiditis, orbital cellulitis.	Fusobacteria (aerobes frequently responsible).
C. Mouth and jaw	Ulcerative gingivitis (Vincent's) dental abscess, cellulitis; abscess and sinus of jaw.	Fusobacteria spirochetes, mouth anaerobes, *Actinomyces*, other mouth anaerobes.
D. Respiratory	Aspiration pneumonia, lung abscess, bronchiectasis, empyema.	Fusobacteria, *P. melaninogenica*, anaerobic cocci; *B. fragilis* rarely.
E. Abdominal	Subphrenic, hepatic abscess; appendicitis, peritonitis; ischiorectal abscess; wound infection after colorectal surgery.	*B. fragilis.*
F. Female genitalia	Wound infection following: genital surgery; puerperal sepsis; tubo-ovarian abscess; Bartholin's abscess, septic abortion.	*P. melaninogenica*, anaerobic cocci; *B. fragilis* genital anaerobes and *Cl. perfringens.*
G. Skin and soft tissue	Infected sebaceous cyst. Breast abscess, axillary abscess Cellulitis, diabetic ulcer, gangrene.	Anaerobic cocci. Anaerobic cocci; *P. melaninogenica* (*Staph. aureus* commonest cause) *B. fragilis* and others.

Table 31.4 lists the common sites and type of anaerobic infections and the bacteria responsible.

Clinical Features of Anaerobic Infection

There are some clinical features which suggest the presence of anaerobic infection.

i. **Production of a foul or putrid odor:** A common, but not invariable, feature is the **production of a foul or putrid odor.** Foul-smelling pus or discharge should always alert the clinician to the likelihood that anaerobes are present, since no other organisms produce this effect.
ii. **Pronounced cellulitis** is a common feature of anaerobic wound infections.
iii. **Toxemia and fever** are not marked.

Laboratory Diagnosis

As anaerobes form part of the normal flora of the skin and mucous surfaces, their isolation from specimens has to be interpreted cautiously. The mere presence of an anaerobe does not prove its causal role.

1. Specimen Collection and Transport

Specimens should be collected in such a manner as to avoid resident flora. For example, from a suspected case of lung abscess, the sputum is unsatisfactory for culture and only material collected by aspiration would be acceptable. In general, material for anaerobic culture is best obtained by tissue biopsy or by aspiration using a needle and syringe. Aspirated or tissue specimens are preferable to swabs whenever feasible. Swabs are generally unsatisfactory but where they are to be used, they should be sent in Stuart's transport medium.

Ideally, specimens should be placed in an anaerobic transport device that consists of a tube or vial containing an anaerobic gas mixture substituted for air, which protects the organisms from O_2-exposure and drying during transport to the laboratory. Specimens should be delivered immediately (within 20 minutes) for culture. Recapping a syringe and transporting the needle and syringe to the laboratory is no longer acceptable because of safety concerns involving needle stick injuries. Therefore, even aspirates must be injected into some type of oxygen-free transport tube or vial. In some laboratories, **gas-liquid chromatography** is carried out directly on pus and other clinical specimens in order to detect metabolic products, such as butyric and propionic acids, that are characteristic of certain anaerobes.

2. Direct Microscopy

Examination of a Gram stained smear is very useful. Pus in anaerobic infection usually shows a large variety of different organisms and numerous pus cells.

Examination of the specimen under ultraviolet light may show the **bright red fluorescence *of P. melaninogenica.***

3. Culture

Several special media have been described for anaerobes but for routine diagnostic work, freshly prepared blood agar with neomycin, yeast extract, hemin and vitamin K is adequate. Plates are incubated at 37°C in an anaerobic jar, with 10 percent CO_2. **The Gas Pak system** provides a convenient method of routine anaerobic cultures. Plates are examined after 24 or 48 hours. Some anaerobes, such as fusobacteria, require longer periods of incubation. Since many anaerobes are relatively slow-growing, it is essential that cultures are incubated for several days before being discarded. In mixed infections, fast-growing aerobic or facultatively anaerobic organisms are often detected within 24 h, whereas some anaerobes may require incubation for 7-10 days before their colonies can be recognized.

Other anaerobic media, such as **cooked meat broth (CMB)** and **thioglycollate broth**, may also be used for inoculating the specimens. Parallel aerobic cultures (such as *Pseudomonas aeruginosa*) should always be set up. This is necessary as a control for the growth on anaerobic plates and also because in most anaerobic infections aerobic bacteria are also involved.

4. Identification

Colony morphology, pigmentation, and fluorescence are helpful in identifying anaerobes. Biochemical activities and production of short-chain fatty acids as measured by gas-liquid chromatography are used for laboratory confirmation. It takes time and is difficult, but it is possible to report on the following: (1) Whether the infection is solely aerobic, anaerobic or mixed. (2) The identification of the commoner anaerobes, particularly of *B. fragilis*. (3) An indication of antimicrobial agents likely to be used.

5. Antibiotic Sensitivity Tests

Antibiotic sensitivity tests can be done by disk diffusion or dilution methods.

6. Other Anaerobic Techniques

Gloved anaerobic chambers with continuous gas flow may be used for culture of specimens. Pre-reduced anaerobically sterilized media (PRAS) can also be employed but are not essential for routine diagnostic procedures.

TREATMENT OF ANAEROBIC INFECTIONS

Treatment of mixed anaerobic infections is by surgical drainage (under most circumstances) plus antimicrobial therapy.

The most active drugs for treatment of anaerobic infections are clindamycin and metronidazole. Clindamycin is preferred for infections above the diaphragm. Penicillin G remains the drug of choice for treatment of anaerobic infections that do not involve β-lactamase producing bacteroides and prevotella species.

KNOW MORE

Bacteroides Species

Bacteroides species are normal inhabitants of the bowel and other sites. Normal stools contain 10^{11} *B. fragilis* organisms per gram (compared with 10^8/g for facultative anaerobes). They are part of the normal flora, and only cause disease when they gain access to the blood during bowel penetration, for example, during surgery or trauma.

☞ KEY POINTS

The anaerobic bacteria are widespread in nature. Many anaerobic bacteria are pathogenic for human beings, and they outnumber aerobic bacteria in many habitats, including most sites of the human or animal body.

- *Anaerobic Gram-positive Cocci*: Most of important anaerobic cocci belong to the genus *Peptostreptococcus*. They can cause cellulites, infections of the soft tissue, intra-abdominal infection, pelvic abscess, salpingitis and endometritis, and infections in bones and visceral organs.
- *Anaerobic Gram-negative Cocci*: *Veillonella parvula* is the species frequently reported from clinical specimens.
- *Anaerobic Nonspore-forming Gram-positive Bacilli*: These include *Eubacterium, Propionibacterium, Lactobacillus, Bifidobacterium* and *Mobiluncus*.
- *Anaerobic Gram-negative Bacilli*: Medically important anaerobic gram-negative bacilli belong to the family Bacteroidaceae and are classified into the genera *Bacteroides, Prevotella, Porphyromonas, Fusobacterium* and *Leptotrichia*.

  Anaerobic infections are usually endogenous and are caused by tissue invasion by bacteria normally resident on the respective body surfaces.
- *Laboratory Diagnosis*: Specimens should be placed in an anaerobic transport device. Examination of a Gram stained smear is very useful. **The Gas Pak system** provides a convenient method of routine anaerobic cultures. Other anaerobic media, such as **cooked meat broth (CMB)** and **thioglycollate broth**, may also be used.

IMPORTANT QUESTIONS

1. Classify nonsporing anaerobes. Discuss the laboratory diagnosis of infections caused by nonsporing anaerobes.
2. Write short notes on:
   – Anaerobic cocci
   – *Lactobacillus*
   – *Bacteroides*
   – *Fusobacterium*
   – Anaerobic gram-positive bacilli.

FURTHER READING

Duerden BI, Drasar BS. (Eds) Anaerobes in Human Disease. London Edward Arnold 1991.

Finegold SM. Anaerobic bacteria: general concepts. In: Mandell GL, Bennett E, Oolin R (Editors) Mandell, Douglas and Bennetts Principles and Practice of Infectious Diseases. 5th edn. Churchill Livingstone, 2000.

Fineg SM. Overview of clinically important anaerobes, Clin Infect Dis 1995;20:S205.

Moncla B, Hillier SL. Peptostreptococcus, Propionibacterium, Lactobacillus, Actinomyces and other non-spore-forming anaerobic gram-positive bacteria. In: Murray PR et al (Eds). Manual of Clinical Microbiology. 8th edn. ASM Press. 2003.

CHAPTER 32

Mycobacterium tuberculosis

LEARNING OBJECTIVES

After reading and studying this chapter, you should be able to:

- Classify mycobacteria.
- Discuss morphology and culture characteristics and biochemical characteristics of *Mycobacterium tuberculosis*.
- Differentiate between *Mycobacterium tuberculosis* and *M. bovis*.
- Describe pathogenesis of *Mycobacterium tuberculosis*.
- Discuss the laboratory diagnosis of pulmonary tuberculosis.
- Describe the following: Koch's phenomenon; Tuberculin test; BCG vaccine.

INTRODUCTION

The genus *Mycobacterium* belongs to the family Mycobacteriaceae. There are over 80 named species of mycobacteria. The most familiar of the species are *Mycobacterium tuberculosis* (MTB) and *Mycobacterium leprae*, the causative agents of tuberculosis (TB) and Hansen's disease (leprosy), respectively. The name mycobacterium, meaning **'fungus like bacterium'** is derived from the mould-like appearance of *Mycobacterium tuberculosis* when growing in liquid media.

They are aerobic, nonmotile, noncapsulated and nonsporing. Growth is generally slow. The genus includes obligate parasites, opportunistic pathogens and saprophytes (Table 32.1). They do not stain readily but once stained with hot carbol fuchsin or other aryl methane dyes, they resist decolorization with dilute mineral acids (or alcohol). Mycobacteria are, therefore, known as acid-fast bacilli (AFB). Although other species of bacteria can be acid-fast (i.e. *Nocardia, Rhodococcus, Tsukamurella, Gordona)*, they stainless intensely (are partially acid-fast), their mycolic acid chains are shorter and their G + C content is lower.

The first member of this genus to be identified was the *Lepra bacillus* discovered by Armaller Hansen in 1868. Robert Koch (1882) isolated the mammalian tubercle bacillus and proved its causative role in tuberculosis by satisfying Koch's postulates. Tuberculosis in humans was subsequently shown to be caused by two types of the bacillus—the human and bovine types, designated *Mycobacterium tuberculosis* and *M. bovis* respectively. Saprophytic mycobacteria were isolated from a number of sources. These included *M. butyricum* from butter, *M. phlei* from grass, *M. stercoris* from dung and *M. smegmatis* from smegma.

M. TUBERCULOSIS COMPLEX (MTC)

M. tuberculosis is a member of the *Mycobacterium tuberculosis* complex, which also includes *M. bovis* (including the vaccination strain, bacillus Calmette Guerin),

Table 32.1: Classification of mycobacteria

Tubercle bacilli
1. Human—*M. tuberculosis*
2. Bovine—*M. bovis*
3. Murine—*M. microti*
4. Aviam—*M. avium*
5. Cold blooded—*M. marinum*

Lepra bacilli
Human—*M. leprae*
Murine—*M. lepraemurium*

Mycobacteria causing skin ulcers
1. *M. ulcerans*
2. *M. balnei*

Atypical mycobacteria
1. Photochromogens
2. Scotochromogens
3. Nonphotochromogens
4. Rapid growers

Johne's bacillus
M. paratuberculosis

Saprophytic mycobacteria
M. butyricum, M. phlei, M. stercoris.

M. africanum, M. canetti, and *M. microti. Mycobacterium* spp. produce a spectrum of infections in humans and animals in addition to tuberculosis and Hansen's disease. A large group of mycobacteria, excluding the *M. tuberculosis* complex and *M. leprae,* normally inhabit the environment and can cause disease that often resembles tuberculosis in humans. These organisms are sometimes referred to as *atypical mycobacteria* or *mycobacteria other than the Tubercle bacillus* (MOTT). They are opportunistic pathogens and can lead to many diseases. Classification of mycobacteria are shown in Table 32.1.

MYCOBACTERIUM TUBERCULOSIS

Morphology

M. tuberculosis is a slender, straight or slightly curved rod with rounded ends , about 3 μm × 0.3 μm, in pairs or as small clumps. The bacilli are nonmotile, nonsporing, noncapsulated and acid-fast. They are gram-positive but are difficult to stain.

When stained with carbol fuchsin by the Ziehl-Neelsen method or by fluorescent dyes (auramine O, rhodamine), they resist decolorization by 20 percent sulfuric acid and absolute alcohol for 10 minutes **(acid and alcohol fast).** With this stain, the *Tubercle bacilli* stain **bright red,** while the tissue cells and other organisms are stained blue (Fig. 32.1). Organisms in tissue and sputum smears often stain irregularly and have a beaded or barred appearance, presumably because of their vacuoles and polyphosphate content.

Acid fastness has been ascribed variously to the presence in the bacillus of an **unsaponifiable wax (mycoloic acid)** or to **a semipermeable membrane around the cell.** It is related to the **integrity of the cell** and appears to be a property of the lipid-rich waxy cell wall. Staining may be uniform or granular. In *M. tuberculosis* beaded or barred forms are frequently seen, but *M. bovis* stains more uniformly. *M. bovis* appear straighter, stouter and shorter with uniform staining.

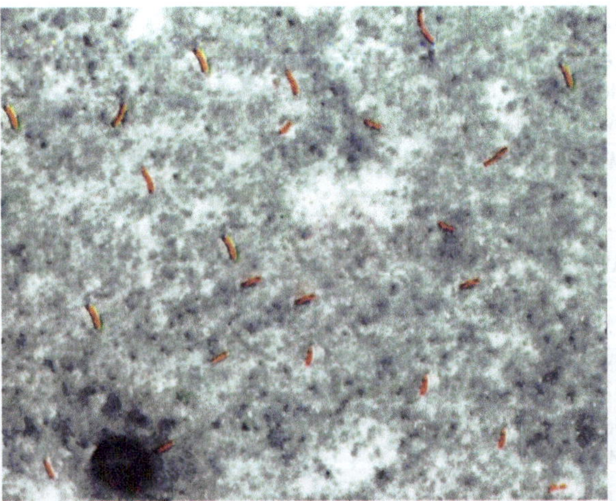

Fig. 32.1: *Mycobacterium tuberculosis* in Ziehl-Neelsen stained smear

Much (1907) demonstrated the presence of gram-positive granules in cold abscess pus where there was no evidence of acid-fast bacilli but which could produce tuberculosis when injected into susceptible animals. Much suggested that these granules (Much's granules) were nonacid-fast form of tubercle bacilli.

Cultural Characteristics

M. tuberculosis is an obligate aerobe while *M. bovis* is microaerophilic on primary isolation, becoming aerobic on subculture. The optimal growth temperature of *tubercle bacilli* is 35 to 37°C but they fail to grow at 25°C or 41°C. Most other mycobacteria grow at one or other, or both, of these temperatures. Optimum pH is 6.4 to 7.0. The bacilli grow slowly, the generation time *in vitro* being 14 to 15 hours. Colonies appear in about two weeks and may sometimes take up to eight weeks.

Solid Medium

Tubercle bacilli are able to grow on a wide range of enriched culture media and do not have exacting growth requirements but are highly susceptible even to traces of toxic substances like fatty acids in culture media. The toxicity is neutralized by serum albumin or charcoal. The solid media contain egg (Lowenstein Jensen, Petragnini, Dorset), blood (Tarshis), serum (Loeffler) or potato (Pawlowsky).

The solid medium most widely employed for routine culture is **Lowenstein-Jensen (LJ) medium** without starch, as recommended by the International Union Against Tuberculosis (IUAT). This consists of coagulated hens' egg, mineral salt solution, asparagine and malachite green, the last acting as a selective agent inhibiting other bacteria and to provide a contrasting color against which colonies of mycobacteria are easily seen. In this medium, egg acts as a solidifying agent. The addition of 0.5 percent glycerol improves the growth of *M. tuberculosis,* but has no effect on or may even impair the growth of *M. bovis.* Sodium pyruvate helps the growth of both types. Agar-based media or broths enriched with bovine serum albumin are also used, particularly in automated culture systems.

Human tubercle bacilli produce visible growth on LJ medium in about 2 weeks, although on primary isolation from clinical material colonies may take up to 8 weeks to appear. On solid media, *M. tuberculosis* forms dry, rough, raised, irregular colonies with a wrinkled surface. They are creamy white, becoming yellowish or buff colored on further incubation. They are tenacious and not easily emulsified. *Mycobacterium tuberculosis* has a luxuriant growth **(eugenic growth)** as compared to *Mycobacterium bovis* which grows poorly on LJ glycerol medium **(dysgonic growth)** and colonies, in comparison are flat, smooth, moist, white and break up easily when touched. The growth of *M. bovis* is much better on LJ pyruvate medium (media containing sodium pyruvate in place of glycerol).

Liquid Media

Among the several liquid media described, Dubos', Middlebrook's, Proskauer and Beck's, Sula's and Sauton's media are the more common. Dubos' medium and Middlebrook 7H9 are two commonly used liquid media. Liquid media are not generally employed for routine cultivation but are used for sensitivity testing, chemical analyzes and preparation of antigens and vaccines.

In liquid media without dispersing agents the growth begins at the bottom, creeps up the sides and forms a prominent surface pellicle which may extend along the sides above the medium. Ordinarily, mycobacteria grow in clumps or masses because of the hydrophobic character of the cell surface. Diffuse growth is obtained in Dubos' medium containing Tween-80 (sorbitan monooleate). In liquid cultures, they often grow as twisted rope-like colonies termed *serpentine cords*. Virulent strains tend to form long serpentine cords in liquid media, while avirulent strains grow in a more dispersed manner. The cord factor itself is not a virulence factor as it is present in some avirulent strains as well.

Resistance

Mycobacteria are rapidly killed by ultraviolet light (including the component in daylight and sunlight), even through glass, and by heat (60°C for 15 to 20 min, or by autoclaving). Cultures may be killed by exposure to direct sunlight for two hours but bacilli in sputum may remain alive for 20 to 30 hours. Bacilli in droplet nuclei may retain viability for 8 to 10 days under suitable conditions. Cultures remain viable at room temperature for 6 to 8 months and may be stored for up to two years at –20°C.

Tubercle bacilli are relatively resistant to chemical disinfectants, surviving exposure to 5 percent phenol, 15 percent sulfuric acid, 3 percent nitric acid, 5 percent oxalic acid and 4 percent sodium hydroxide. They are sensitive to formaldehyde and gluteraldehyde. They are destroyed by tincture of iodine in five minutes and by 80 percent ethanol in 2 to 10 minutes. Ethanol is a suitable disinfectant for skin, gloves and clinical thermometers.

Figs 32.2A and B: LJ media without growth and with growth

Biochemical Reactions

Several biochemical tests are used in identifying and differentiation of mycobacterial species.

1. Niacin Test

Most mycobacteria possess the enzyme that converts free niacin to niacin ribonucleotide. *M. tuberculosis* lacks this enzyme and accumulates niacin (nicotinic acid) in the culture medium. Accumulation of niacin, detected as nicotinic acid, is the most commonly used biochemical test for identification of MTB. When 10 percent cyanogen bromide and 4 percent aniline in 96 percent ethanol are added to a suspension of the culture, a canary yellow color indicates a positive reaction. The test is positive with human type and negative with bovine type of bacilli. *M. simiae* and *M. chelonae* also give positive niacin test.

2. Nitrate reduction test

M. tuberculosis, M. kansasii, M. fortuitum, M. chelonei, M. szulgai, M. flavescens, M. terrae and *M. triviale* produce enzyme nitroreductase. Therefore, all these reduce nitrate to nitrite. *M. bovis* and *M. avium* do not do so.

The test organism is suspended in a buffer solution containing nitrate and incubated at 37°C for 2 hours. Then sulphanilamide and N-(1-naphthyl) ethylenediamine dihydrochloride solution is added. Positive reaction is indicated by development of pink or red color within 30 to 60 seconds.

3. Catalase Activity

Catalase is an enzyme that can split hydrogen peroxide into water and oxygen. Mycobacteria are catalase positive. However, not all srains produce a positive reaction after the culture is heated to 68°C for 20 minutes. Isolates that are catalase positive after heating have heat stable catalase. Most *M. tuberculosis* complex organisms do not produce heat stable catalane except some isoniazid (INH) resistant strains, *M. gastri, M. haemophilum* and *M. marinum.* Semiquantitation of catalase production is done by using the addition of Tween 80 (a detergent) and hydrogen peroxide to a 2 week-old culture in an agar deep. The reaction is read after 5 minutes, and the resulting column of bubbles is measured. Based on the semiquantitative catalase test, mycobacteria are divided into two groups: Those producing less than 45 mm of bubbles and those producing more than 45 mm of bubbles.

4. Tween 80 Hydrolysis

Some mycobacteria possess a lipase that can split detergent Tween 80 into oleic acid and polyoxyethylated sorbitol which modifies the optical characteristics of the test solution from a straw yellow to pink whereas pathogenic species do not. This test is useful in distinguishing scotochromogenic and nonchromogenic mycobacteria. *M. kansasii, M. gordonae, M. flavescens* and *M. smegmatis are positive* while *M. bovis, M. africanum, M. ulcerans, M. scrofulaceum, M. xenopi* and *M. avium* complex are Tween 80 hydrolysis negative. *M tuberculosis* gives variable result.

5. Arylsulfatase Test

The enzyme arylsulfatase is formed by certain mycobacteria. The bacilli are grown in a medium containing 0.00 1M tripotassium phenolphthalein disulfate. If arylsulphatase is produced, it will split free phenolfthalein from tripotassium phenolphthalein disulphate. This can be detected by adding 2N NaOH dropwise to the culture. A pink color indicates a positive reaction.

6. Neutral Red Test

Virulent strains of tubercle bacilli are able to bind neutral red in alkaline buffer solution, while avirulent strains are unable to do so. *M. tuberculosis, M. bovis, M. avium* and *M. ulcerans* give positive tests.

7. Amidase Test

Atypical mycobacteria can be differentiated by their ability to split amides. A useful pattern is provided by testing five amides, viz., acetamide, benzamide, carbamide, nicotinamide and pyrazinamide.

A 0.00165 M solution of the amide is incubated with the bacillary suspension at 37°C and 0.1 ml of $MnSO_4$. 4 H_20, 1.0 ml of phenol solution and 0.5 ml of hypochlorite solution are added. The tubes are placed in boiling water for 20 minutes. Development of blue color in the tube indicates a positive test. *M. tuberculosis* produces nicotinamidase and pyrazinamidase which splits nicotinamide and pyrazinamide.

8. Susceptibility to Pyrazinamide

M. tuberculosis is sensitive to 50 µg/ml pyrazinamide, while *M. bovis* and other mycobacteria are resistant.

9. Susceptibility to Thiophen-2-Carboxylic Acid Hydrazide (TCH)

This test is used to distinguish *M. bovis* from M. tuberculosis, because only *M. bovis* is usually susceptible and is unable to grow in the presence of 10 g/ml of TCH, while *M. tuberculosis* is usually not susceptible to this chemical.

Antigenic Structure

Mycobactria contain many unique immunoreactive substances, most of which are components of the cell wall. Mycobacteria possess two types of antigens, cell wall (insoluble) and cytoplasmic (soluble) antigens.

1. Cell wall antigens

The basic structure of the cell wall is typical of gram-positive bacteria: an inner cytoplasmic membrane overlaid with a thick peptidoglycan layer and no outer membrane. The cell wall consists of lipids, proteins and polysaccharides. These lipids constitute 60% of the cell wall weight and contributes to several biological propreies. Lipids of the cell wall particularly mycolic acid fraction are responsible for acid-fastness of bacteria and the cellular reaction of the body.The cell wall is made up of four distinct layers (Fig. 32.3).

(i) Peptidoglycan (murein) layer

It is the innermost layer. The peptidoglycan skeleton is covalently linked with polysaccharides (arabinogalacta) whose terminal ends are esterified to high-molecular weight mycolic acids. It maintains the shape and rigidity of the cell.

(ii) Arabinogalactan layer

Arabinogalactan (polysaccharides) lies external to the peptidoglycan layer.

(iii) Mycolic acid layer

It is the principal constituent of cell wall and is a dense band on long chain α-alkyl and β-hydroxy fatty acids attached by ester bonds to the terminal arabinose units of arabinogalactan.

(iv) Mycosides (peptidoglycolipids or phenolic glycolipids)

These form the outermost layer of the cell wall. The outer surfce of the mycobacterium is formed by the intercaltion of medium chain (e.g. mycoserosates), short chain (e.g. acylglerols) lipids, glycolipids,and peptidogylolipids into the uneven hydrophobic layer of mycolic acids. Proteins (e.g. porins, transport proteins) are found throughout the various layers.

The cell wall antigens include arabinomanan, arabinogalactan and lipoarabinomanan.

2. Cytoplasmic Antigens

Cytoplasmic antigens are protein antigens which are employed to type the mycobacteria. These include antigen 5, antigen 6, antigen 14, antigen 19, antigen 32, antigen 38 and antigen 60. All are protein in nature except antigen 60 which is a lipopolysaccharide protein complex.

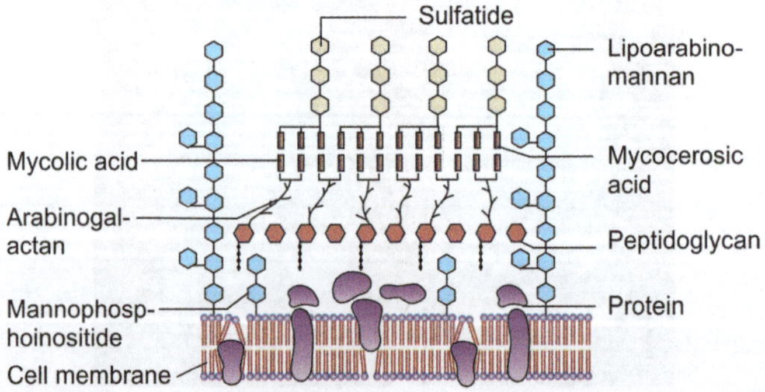

Fig. 32.3: Cell wall of *Mycobacterium tuberculosis*

Mycobacteriophages

Numerous phages with activities on many mycobacterial species, including *M. tuberculosis,* have been isolated from both clinical and environmental sources. Many mycobacteriophages have been isolated from soil, water and other environmental sources as well as from lysogenic strains. Many mycobacteria infected with phage are not truly lysogenic. Instead of being integrated with the bacterial chromosome, the phage genome appears free, like a plasmid. This is called *pseudolysogeny*.

M. tuberculosis has been classified into four phage types—A,B,C, and I (a type intermediate between A and B). Phage type A is the commonest and is present worldwide. Type B is common in Europe, the Middle East and North America. Type C is seen rarely. Type I is common in India and neighboring countries. Phage 33D can *M tuberculosis, M. bovis, M. africanum* and *M. microti.*

Pathogenesis

The source of infection is usually an open case of pulmonary tuberculosis. The mode of infection is by direct inhalation of aerosolized bacilli contained in droplet nuclei of expectorated sputum tubercle bacilli are acquired from persons with active disease who are excreting viable bacilli by means of coughing, sneezing or talking. Airborne droplet nuclei containing bacteria, 1 to 5 µm, enter the respiratory tract of an exposed individual and are deposited in the lung alveoli. Infection also occurs infrequently by ingestion, for example, through infected milk and rarely by inoculation.

The initial infection with *M. tuberculosis* is referred to as a **primary infection**. Subsequent disease in a previously sensitized person, either from an exogenous source or by reactivation of a primary infection, is known as **postprimary (secondary or reinfection) tuberculosis** with quite different pathological features.

1. Primary Tuberculosis

Primary tuberculosis is the initial infection by tubercle bacilli in a host. In endemic countries like India this usually occurs in young children. This begins with inhalation of the mycobacteria and ends with a T cell-mediated immune response that induces hypersensitivity to the organisms and controls 95 percent of infections.

The site of the initial infection is usually the lung, following the inhalation of bacilli. These bacilli engulfed by alveolar macrophages multiply and give rise to a subpleural focus of tuberculous pneumonia, commonly located in the lower lobe or the lower part of the upper lobe to form the **initial lesion or Ghon focus**. Some bacilli are carried in phagocytic cells to the hilar lymph nodes where additional foci of infection develop. The Ghon focus, together with the enlarged hilar lymph nodes, form the **primary complex.** This occurs about 3-8 weeks from the time of infection and is associated with the development of tuberculin hypersensitivity. In addition, bacilli are seeded by further lymphatic and hematogenous dissemination in many organs and tissues, including other parts of the lung.

After infection, *M. tuberculosis* cells are phagocytized by alveolar macrophages and are capable of intracellular multiplication. In a person with adequate cellular immunity, T cells arrive within 4 to 6 weeks with macrophage—activating polypeptides called *lymphokines.* This enables the macrophage in the area of infection to destroy the intracellular mycobacteria. There is then a regression and healing of the primary lesion and any disseminated foci of infection by the *M. tuberculosis* organisms.

In many exposed individuals, the immune system does not eliminate the bacteria. The pathologic features of TB are the result of a hypersensitivity reaction to mycobacterial antigen. If there is a little antigen and a great deal of hypersensitivity reaction, a **hard tubercle or granuloma** may be formed. When fully developed, this lesion, a **chronic granuloma**, consists of three zones:

1. **A central area** of large, multinucleated giant cells containing tubercle bacilli.
2. **A mid zone** of pale epithelioid cells, often arranged radially.
3. **A peripheral zone** of fibroblasts, lymphocytes, and monocytes.

Later, peripheral fibrous tissue develops, and the central area undergoes **caseation necrosis.** Such a lesion is called a **tubercle.** A caseous tubercle may break into a bronchus, empty its contents there, and form a cavity. It may subsequently heal by **fibrosis or calcification.**

The granuloma is an organization of lymphocytes, macrophages, giant cells, fibroblasts, and capillaries. With this granuloma formation, healing occurs, along with fibrosis, encapsulation and scar formation as a reminder of the past infection.

Naive macrophages are unable to kill the mycobacteria, the latter multiply, lyse the host cell, infect other macrophages and sometimes disseminate through the blood to other parts of the lung and elsewhere in the body. Within about 10 days of infection, clones of antigen-specific T lymphocytes (CD4+ helper T cells and CD 8+ suppressor T cells) are produced interacting with macrophages. These release cytokines, notably interferon-gamma, which activates macrophages to kill intracellular mycobacteria. These activated macrophages are termed epithelioid cells from their microscopical resemblance to epithelial cells. These form a compact cluster or epithelioid cell granuloma around the foci of infection. Some of them fuse to form multinucleated giant cells. The center of the granuloma contains a mixture of necrotic tissue and dead macrophages, which, from its cheese-like appearance and consistency, is referred to as *caseation.*

2. Postprimary (Secondary) Tuberculosis

The postprimary (secondary or adult) type of tuberculosis is due to reactivation of latent infection

Table 32.2: Differences between primary and post-primary tuberculosis in nonimmunocompromised patients		
Characteristics	*Primary*	*Postprimary*
Site	Any part of lung	Apical region
Local lesion	Small	Large
Cavity formation	Rare	Frequent
Lymphatic involvement	Yes	Minimal
Infectivity[a]	Uncommon	Usual
Tuberculin reactivity	Negative (initially	Positive
Local spread	Uncommon	Frequent

[a]Pulmonary cases.

(post-primary progression, endogenous reactivation) or exogenous reinfection and differs from the primary type in many respects (Table 32.2). Progression from infection to active disease varies with age and the intensity and duration of exposure. Reactivation TB occurs when there is an alteration or a suppression of the cellular immune system in the infected host that favors replication of the bacilli and progression to disease.

Reactivation tuberculosis is characterized by chronic tissue lesions, the formation of tubercles, caseation, and fibrosis. Regional lymph nodes are only slightly involved, and they do not caseate. The reactivation type almost always begins at the apex of the lung, where the oxygen tension (PO_2) is highest. These differences between primary infection and reinfection or reactivation are attributed to (1) resistance and (2) hypersensitivity induced by the first infection of the host with tubercle bacilli. If the disease has been chronic, fibrosis, loss of lung volume, and calcifications will be demonstrated. The PPD test may be negative in up to 25 percent of these cases; diagnosis is confirmed by smear and culture of sputum, gastric aspirates, or bronchoscopy specimens.

Two special features of secondary tuberculosis are the presence of **caseous necrosis** and of **cavities,** which may rupture into blood vessels, spreading mycobacteria throughout the body, and break into airways, releasing infectious mycobacteria in aerosols and sputum **(open tuberculosis).** In the immunodeficient, cavity formation is unusual. Instead there is widespread dissemination of lesions in the lungs and other organs.

In any case of pulmonary TB disease, there may be complications if diagnosis and treatment are delayed. These include **empyema, pleural fibrosis, massive hemoptysis, adrenal insufficiency** (rare), and **hypercalcemia** (up to 25 percent of cases).

Epidemiology

Tuberculosis is an ancient disease. Tuberculosis also remains the leading cause of death among notifiable infectious diseases. It has been called the 'white plague' and 'the captain of all the men of death'.

The most frequent source of infection is the human who excretes, particularly from the respiratory tract, large numbers of tubercle bacilli. Close contact (e.g., in the family) and massive exposure (e.g., in medical personnel) make transmission by droplet nuclei most likely. Infection occurs at an earlier age in urban than in rural populations.

According to WHO estimates one-third of the world's population has been infected. About 100 million individuals are infected annually and 8-10 million develop overt disease 4 to 5 million become open or infectious and around 3 million die.

The large majority of the cases and deaths are from the poor nations. India accounts for nearly one-third of global burden of tuberculosis. Half a million people die from the disease every year in India - one every minute. More than 40 percent of the population are infected and some 15 million suffer from tuberculosis in the country, of whom over three million are highly infectious open cases.

Poverty and tuberculosis go together. Tuberculosis had declined rapidly in the affluent nations with improvements in standards of living. But with the progress of the AIDS pandemic, tuberculosis became a problem for the rich nations also. A close relationship has emerged between tuberculosis and HIV. A third complication that has made the situation graver is the emergence and spread of multiple drug resistance among tubercle bacilli because patients who receive inadequate treatment may remain infectious for a longtime.

M. bovis infection in humans is zoonotic acquired through direct contact or by ingestion of raw milk from animals. Bovine tuberculosis is spread from animal to animal, and sometimes to human attendants, in moist cough spray.

Laboratory Diagnosis

The definitive diagnosis of tuberculosis is based on

microscopy, culture, by transmitting the infection to experimental animals, demonstration of hypersensitivity to tuberculoprotein and molecular diagnostic methods.

Pulmonary Tuberculosis

1. Specimens

i. Sputum

The most usual specimen for diagnosis of pulmonary tuberculosis is sputum which consists of pus and mucus secretions coughed up from the lung. Patient is instructed to cough up the sputum into a clean wide-mouthed container. *Disposable waxed cardboard containers are ideal.* If sputum is scanty, a 24-hour sample may be tested. Three or more consecutive samples should be examined, collected first thing in the morning if possible. Sputum sampling on three days increases chances of detection.

ii. Laryngeal Swabs or Bronchial Washings

Laryngeal swabs or bronchial washings may be collected where sputum is not available.

iii. Gastric Lavage

Gastric lavage can be examined in small children who tend to swallow the sputum.

2. Microscopy

Direct or concentration smears of sputum are examined. Smears should be prepared from the thick purulent part of the sputum. Smears are dried, heat fixed and stained by the **Ziehl-Neelsen technique(hot stain procedure).**

In the Ziehl-Neelsen (ZN) staining technique, heat-fixed smears of the specimens are flooded with a solution of carbol fuchsin (a mixture of basic fuchsin and phenol) and heated until steam rises but without boiling. Allow the preparation to stain for 5 minutes, applying heat at intervals to keep the stain hot. The slide is then washed with water and decolorized with 20 percent sulfuric acid till no more stain comes off and then with 95 percent alcohol (ethanol) for two minutes. Decolorization may be carried out as a single step with acid alcohol (3% HCl in 95% ethanol). After washing, the smear is counterstained with Loeffler's methylene blue, 1 percent picric acid or 0.2 percent malachite green for one minute.

Under the oil immersion objective, acid fast bacilli are seen as bright red rods while the background is blue, yellow or green depending on the counterstain used. At least **10,000** acid-fast bacilli should be present per ml of sputum for them to be readily demonstrable in direct smears.

Smears should be examined carefully by scanning at least 300 oil immersion fields (equivalent to three full horizontal sweeps of a smear that is 2 cm long and 1 cm wide) before reporting a smear as negative.

The classic carbolfuchsin (Ziehl-Neelsen) stain requires heating the slide for better penetration of stain into the mycobacterial cell wall. Hence, it is also known as the **hot stain procedure**. Kinyoun acid-fast stain is similar to the Ziehl-Neelsen stain but without heat; hence the term **cold stain.**

It is more convenient to use fluorescent microscopy when several smears are to be examined daily. Smears are stained **with auramine phenol or auramine rhodamine fluorescent dyes.** This stain is more sensitive than the conventional carbolfuchsin stains, because the fluorescent bacilli stand out brightly against the background. The smear can be initially examined at lower magnifications (250x to 400x), and therefore more fields can be visualized in a short period. In addition, a positive fluorescent smear may be restained by the conventional Ziehl-Neelsen or Kinyoun procedure, thereby saving the time needed to make a fresh smear. Screening of specimens with rhodamine or rhodamine-auramine will result in a higher yield of positive smears and will substantially reduce the amount of time needed for examining smears.

Table 32.3: Methods for reporting numbers of acid-fast bacilli observed in stained smears

No. of bacilli	*Observed in (oil immersion field)*	*Report*
0	300 Fields	Negative
1-2	100 Fields	±
1-9	100 Fields	1+
1-9	10 Fields	2+
1-9	1 Field	3+
More than 9	1 Field	4+

Fields-oil immersion field

The overall sensitivity of an acid-fast smear ranges from 20 to 80 percent. In general, specificity of acid-fast smear examination is very high. Microscopic demonstration of acid fast bacilli provides only presumptive evidence of tuberculous infection, as even saprophytic mycobacteria may present a similar appearance. However, most saprophytic species stain uniformly without appearing barred or beaded and are usually only acid fast without being alcohol fast.

3. Concentration Methods

Several methods have been described for the homogenization and concentration of sputum and other specimens. Concentration methods that do not kill the bacilli and so can be used for culture and animal inoculation. Several methods are in use:

i. Petroff's Method

This simple method is widely used. Equal volumes of sputum and 4 percent sodium hydroxide are mixed and incubated at 37°C with frequent shaking till it gets liquefied and becomes clear. On the average, it takes 20 to 30 minutes. It is then centrifuged at 3,000 r.p.m. for 30 minutes. The supernatant fluid is pipetted off and the deposit is neutralized by adding 8 percent hydrochloric acid in the presence of a drop of phenol red indicator. and used for smear, culture and animal inoculation. Excessive exposure to alkali is deleterious and should be avoided.

ii. Other Methods

Instead of alkali, homogenization can be achieved by treatment with dilute acids (6% sulphuric acid, 3% hydrochloric acid or 5% oxalic acid), N acetyl cysteine with NaOH, pancreatin, desogen, zephiran and cetrimide. Flocculation methods have also been described.

4. Culture

It is a very sensitive diagnostic technique for tubercle bacilli, detecting as few as 10 to 100 bacilli per ml. The concentrated material is inoculated onto at least two bottles of LJ medium. A direct drug sensitivity test may also be set up if the specimen is positive by microscopy.

Other media available include Middlebrook's 7HI0 and 7HII agar media and Middlebrook's 7H9 broth medium more often used for sensitivity testing than for isolation. Inoculated media are incubated at 35 to 37°C. Growth of most strains of *M. tuberculosis* may appear in 2 to 8 weeks. Longer incubation is necessary for strains originating from patients treated with antitubercular agents.

Cultures are examined for growth after incubation at 37°C for four days (for rapid growing mycobacteria, fungi and contaminant bacteria) and at least twice weekly thereafter. A negative report is given if no growth occurs after 8 to 12 weeks. Any bacterial growth is stained by the ZN method and, if acid-fast, it is subcultured for further identification. The first step in identification is to determine whether an isolate is a member of the *M. tuberculosis* complex (Fig. 32.4).

For routine purposes, a slow growing, nonpigmented, niacin positive acid fast bacillus is taken as *M. tuberculosis*. Confirmation is by detailed biochemical studies. When the isolate is niacin negative, a battery of tests may be needed for identification, including growth at 25°C and 45 °C, animal pathogenicity and biochemical tests (Tables 32.4 and Fig. 32.4).

In recent years, isolation of mycobacteria from clinical samples has been improved by newer techniques, notably radiometric respirometry. This is usually by means of the Bactec radiometric technique, using the Bactec 460 TB instrument. The procedure utilizes vials containing a 14 C-labeled palmitic acid substrate which during microbial growth releases $^{14}CO_2$ into the atmosphere above the medium. The $^{14}CO_2$ is detected by the instrument and converted proportionally to a quantitative growth index (GI) on a scale of 0-999. A vial giving a positive GI (> 50) is checked by AP and ZN smears to determine the presence of acid-fast bacilli. The sensitivity of radiometric method is slightly more than that of traditional culture method.This method allows a mean isolation time for all mycobacteria of about 12 days as opposed to 26 days or more.

5. Animal Inoculation

Sputum is decontaminated and concentrated by Petroff's method. 0.5 ml each of the neutralized and centrifuged deposit is inoculated intramuscularly into the thigh of two healthy, tuberculin-negative guinea-pigs about 12 weeks old. Subcutaneous inoculation is not recommended as it leads to a local ulcer which may be infectious. The animals are weighed before inoculation and at intervals thereafter at weekly intervals and tuberculin test is done after 3 to 4 weeks. Progressive loss of weight and positive tuberculin test are indications of development of tuberculosis. One animal is killed after four weeks and autopsied. If it shows no evidence of tuberculosis, the other is autopsied after eight weeks.

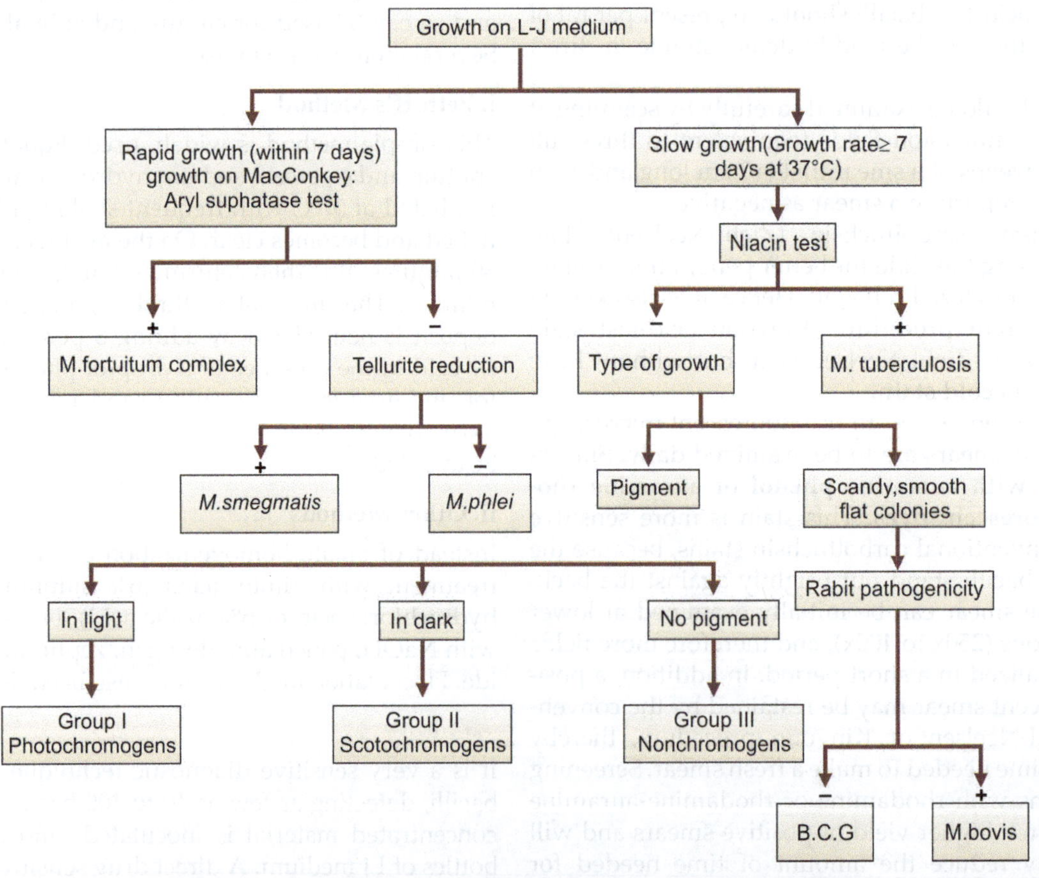

Fig. 32.4: Identification of tubercle bacilli and related mycobacteria

Table 32.4: Distinguishing features of Mycobacterium tuberculosis and *M. bovis*		
Features	*Mycobacterium tuberculosis*	*M. bovis*
1. Morphology	Slender, straight or slightly curved rods with barred or beaded appearance	Straight, stout, short uniformly
2. Growth on LJ medium	Growth is eugonic (luxuriant)	Growth is dysgonic
3. Effect of glycerol	In the concentration of 0.75% enhances the growth	In the concentration of 0.75 percent growth is inhibited.
4. Oxygen requirement	Obligate aerobe	Microaerophilic on primary isolation but becomes aerobic on subculture
5. Colony morphology	Dry, rough, raised, wrinkled, off-white to buff colored and not easily emulsifiable (rough, buff and tough)	Moist, smooth, flat, friable and white
6. Biochemical reactions		
• Nitrate reduction	+	–
• Niacin production	+	–
• Tween 80 hydrolysis	Variable	–
7. Susceptibility to pyrazinamide (50 µg/ml)	+	–
8. Susceptibility to thiophen 2-carboxylic acid hydrazide (10 µg/ml)		+
9. Animal pathogenicity Guineapig	+	+
• Rabbit	–	+

Necropsy shows:

i. Caseous lesion at the site of inoculation.
ii. The draining and internal lymph nodes are enlarged and caseous. The infection may spread to lumbar, portal, mediastinal and cervical lymph nodes.
iii. The spleen is enlarged with irregular necrotic areas.
iv. Tubercles 1 to 2 mm in diameter are seen in the peritoneum and sometimes in the lung, but the kidneys are unaffected.

The autopsy lesions have to be confirmed as tuberculous by acid fast staining of smears, to exclude *Y. pseudotuberculosis*, *Brucella*, *Salmonella* and several fungi which may produce infections and resemble the lesions of tuberculosis but will be smear negative. *M. tuberculosis* is highly pathogenic for guineapigs and virtually nonpathogenic for rabbits, while *M. bovis* is highly pathogenic for both guineapigs and rabbits. Guinea pig inoculation, once so commonly used, is now seldom resorted to because it is cumbersome, costly and less sensitive than culture, particularly with catalase negative, INH resistant strains isolated from south India. The animal inoculated with strains of low virulence may have to be observed for 12 weeks or longer and sometimes the only lesion demonstrable may be an enlarged lymph node.

Guineapig inoculation was formerly used extensively for the diagnosis of tuberculosis, but this has been largely superseded by more sensitive modem techniques and it is very rarely used nowadays.

6. Immunodiagnosis

i. Serology

Serological tests are not useful in diagnosis, though antibodies to many bacillary antigens have been demonstrated in the sera of patients. Detection of antibody to mycobacterial lipoarabinomannan has been reported to be of some value.

Various serological tests like enzyme-linked immunosorbent assay (ELISA), radioimmunoassay (RIA) and latex agglutination have been tried for the serodiagnosis of tuberculosis. ELISA is considered to be the most sensitive and specific. ELISA test has been attempted using several antigenic materials such as antigen 5, antigen 6, antigen 60, purified mycobacterial glycolipids, unheated sterile culture filtrate of *M. tuberculosis* and purified protein derivative derived from *M. tuberculosis*.

ii. Tuberculin Testing

Demonstration of hypersensitivity to tuberculoprotein (tuberculin testing) is a standard procedure. Its scope and limitations are discussed below.

7. Hybridization and Nucleic Acid Technology

i. **Nucleic acid probes** are available for the identification of the *M. tuberculosis* complex and specifically, *M. tuberculosis*, and also for certain other species are commercially available.
ii. **Polymerase chain reaction (PCR)** and **ligase chain reaction (LCR)** are used as diagnostic techniques.

iii. **Transcription mediated amplification, targeting ribosomal RNA** has been introduced as an improvement on PCR based DNA amplification.

iv. **DNA 'fingerprinting:** Most members of the *M. tuberculosis* complex contain 1 to 20 copies of the insertion sequence IS6110, which has been used to develop **DNA 'fingerprinting'** methods for epidemiological purposes. Alternatively, detection of spacer oligonucleotides, short DNA sequences found around the sites of the insertion sequences, is useful for typing isolates (spoligotyping). Demonstration of mutation in specific drug sensitivity genes is a useful indicator of drug resistance. Such tests for rifampicin resistance are already available.

8. Chromatography

The cell walls of *Mycobacterium* organisms contain long chain fatty acids called mycolic acids, which may be detected chromographically. The type and quality of mycolic acids are species specific. Earlier methods, such as **column chromatography** and thin-layer chromatography, have been replaced by **gas-liquid chromatography** and, most recently**, high-pressure liquid chromatography (HPLC).** Species identifications made with HPLC have been shown to agree well with biochemical and nucleic acid probe identifications. Chromatography is rapid and highly reproducible, but the initial cost of equipment is high.

Extrapulmonary Tuberculosis

For diagnosis of extrapulmonary tuberculosis, microscopy, culture and occasionally animal inoculation are also used though it is difficult to get conclusive results as the bacilli are present in far fewer numbers in these lesions than in pulmonary disease.

CSF from tuberculous meningitis often develops a spider web clot on standing, examination of which may be more successful than of the fluid. The use of PCR and DNA probes may help to detect the bacilli speedily and more often.

Bone marrow and liver biopsy specimens from miliary tuberculosis and blood from those with HIV coinfection are useful for culture. Pus from tuberculous abscesses often yields positive results in smear and culture.

Pleural effusion and other exudates may be collected with citrate to prevent coagulation. If free from other bacteria, they may be used for culture after centrifugation. If other bacteria are present, prior concentration is necessary.

Urinary excretion of bacilli in renal tuberculosis is intermittent. Hence, it is advisable to test 3 to 6 morning samples of urine. Each sample is centrifuged for 3000 rpm for 30 minutes and the sediment used for culture after concentration.

Sensitivity testing

As drug resistance is an important problem in tuberculosis, it is desirable to have sensitivity of isolates tested as an aid to treatment. Drug-resistant mutants continuously arise at a low rate in any mycobacterial population. The purpose of sensitivity testing is to determine whether the great majority of the bacilli in the culture are sensitive to the antitubercular drugs currently in use. Several methods have been described:

1. Absolute Concentration Method

For each drug tested, a standardized inoculum is inoculated to control (drug-free) media and media containing several appropriate graded drug concentrations. Resistance is expressed as the lowest concentration of drug that inhibits all or almost all of the growth, that is, the minimum inhibitor concentration (MIC). This method is inferior to resistance ratio method because known sensitive strain is not tested for MIC.

2. Resistance Ratio Method

The resistancce of the test organism is compared with that of a standard laboratory strain in which the susceptibilty patern is known. The resistance-ratio method in which test strains and susceptible controls i.e. a known sensitive strain of M. tuberculosis are inoculated on sets of LJ medium containing doubling dilutions of drug. Culture media are examined for growth after 3 weeks of incubation at 37° C. Resistance is expressed as the ratio of the minimum inhibitor concentration (MIC) of the test strain divided by the MIC for the standard strain for each drug. Susceptible strains have ratios of 1 or 2, while higher ratios indicate resistance

3. Proportion Method

This test is often referred as the proportion method because it allows one to predict the possibility for determining that the 1% is resistant or not.For each drug tested, several dilutions of standardized inoculum are inoculated onto control and drug-containing agar medium. The extent of growth in the absence or presence of drug is compared and expressed as a percentage. The isolate is considered clinically resistant if growth at the critical concentration of a drug is >1 %. The methods used drtrmining such resistance iclude agra dilution, disc dilution, disk elution, and the BACTEC system.

4. Radiometric Method

Employing the principles of the proportion method, this rapid method uses liquid medium containing ^{14}C-labeled growth substrate. Growth is indicated by the amount of ^{14}C-labeled-carbon dioxide (CO_2) released, as measured by the BACTEC 460TB instrument. For each drug tested, a standardized inoculum is inoculated into a drug-free and drug-containing vial. The rate and amount of CO_2 produced in the absence or presence of drug is then compared to determine susceptibilty or resistace.

5. Non-Radiometric method

Mycobacterial growth indicator tube (MGIT) can also be used for sensitivity testing of *M. tuberculosis.*

6. New Approaches

(i) **Luciferase reporter mycobacteriophage:** Luciferase reporter mycobacteriophage (containing the firefly luciferase gene) has been used for susceptibility testing of M. tuberculosis. Only viable mycobacteria can be infected with and replicate this mycobacteriophage while dead tubercle bacilli cannot. The isolate of M. tuberculosis to be tested is grown in the presence and absence of drug and the reporter mycobacteriophage is added. Then a substrate of luciferase, luciferin is added following infection. The luciferin is broken down and light is emitted and can be measured if bacteria are viable. This method is named as **chemiluminescence**. Light will be emitted if the isolate is resistant to drug while bacteria susceptible to the drug will not emit any light. The amount of light emitted is directly proportional to the number of viable bacteria.

(ii) **Epsilometer test (E-test):** It has also been applied for susceptibility testing of *M. tuberculosis*.

(iii) **Detection of restance gene:** Mutation in specific genes for different drugs can be dmontrated and is is a useful indicator of drug resistance. Kits able to detect about 95% of mutations to rifampicin resistance caused by mutations in the *rpoB* gene are commercially available. Nucleic acid technology is even more rapid.

Allergy and Immunity

Infections caused by *M. tuberculosis* produce a range of immunologic reactions, but since its first demonstration by Koch immunity to tuberculosis has remained an elusive concept. Two immunologic responses-antituberculous immunity and tuberculin hypersensitivity-develop simultaneously in the naturally infected host. Both these are mediated by T-cells sensitized to bacterial antigen. The resultant of these two processes determines the course of the infection. Although humoral antibodies are produced in response to naturally occurring tuberculous infection, they appear to play no beneficial role in host defense. Acquired antituberculous immunity is the prototype of cell-mediated immunity invoked by facultative intracellular bacteria. In the nonimmune host, the bacilli are able to multiply inside phagocytes and lyse the host cells, while in immune host CD 4+ helper T cells and CD 8+ suppressor T cells are produced. The former secrete interferon-gamma which activates macrophages to kill intracellular mycobacteria and the latter kill the macrophages that are infected with mycobacteria.

Koch Phenomenon

Robert Koch (1890, 1891) originally described the response of a tuberculous animal to reinfection. In a normal guinea pig subcutaneous injection of virulent tubercle bacillus produces no immediate response, but after 10 to 14 days a nodule develops at the site, which breaks down to form an ulcer that persists till the animal dies of progressive tuberculosis. The draining lymph nodes are enlarged and caseous. If on the contrary, virulent tubercle bacillus is injected in a guinea pig which had received a prior injection of the tubercle bacillus 4 to 6 weeks earlier, an indurated lesion appears at the site in a day or two. This undergoes necrosis in another day or so to form a shallow ulcer, followed by rapid healing and no lymph node involvement or other tissues. This is known as the *Koch phenomenon* and is a combination of hypersensitivity and immunity.

Components of Koch Phenomenon

Koch phenomenon has three components :

1. A local reaction of induration and necrosis.
2. A focal response in which there occurs acute congestion and even hemorrhage around tuberculous foci in tissues.
3. A systemic response of fever which may sometimes be fatal.

Tuberculin Test

Principle

The principle of this test is delayed (Type IV) hypersensitivity reaction. The test is based on the fact that persons infected with tubercle bacilli develop hypersensitivity to the proteins of the organism.

Reagents

i. Original or Old Tuberculin (OT)

Koch prepared a protein extract of tubercle bacillus bacillus by concentrating ten-fold by evaporation, a 6 to 8 week culture filtrate of the bacillus grown in 5 percent glycerol broth. This was called 'original' or 'old tuberculin' (OT). Initially Koch employed OT in the treatment of tuberculosis but it was soon given up as it was not only of no benefit but also caused serious reactions in some due to the focal and systemic components of the Koch phenomenon.

ii. Purified Protein Derivative (PPD)

OT was first used for allergic (tuberculin) testing by Von Pirquet (1906). It was replaced by a partially purified protein antigen introduced by Seibert as OT was a crude product and batches tended to vary in purity and potency. This is known as the purified protein derivative (PPD).

Dose of PPD

Purified protein derivative (PPD) is the skin test reagent that is primarily used to detect hypersensitivity in these persons. One large batch of PPD made by Seibert in 1939, PPD-S was recognized by the WHO as the international standard PPD-tuberculin and arbitrarily designated to contain 50,000 tuberculin units (TU) per mg, 1 TU equal to 0.01 ml of OT or 0.00002 mg of PPD-S.

Method

Many methods had been described for tuberculin testing.

i. Mantoux Test

The method used routinely is the technique of Mantoux (1910). In the Mantoux test, 0.1 ml of PPD containing 5 TU is injected intradermally on the flexor aspect of the forearm with a tuberculin syringe and when properly performed, will produce a discrete pale elevation of the skin (wheal).

Interpretation

Tuberculin tests should be read 48 to 72 hours after injection. The reading is based on the presence or absence of induration, which may be determined visually and by palpation. Erythema is not taken into account. Induration of diameter 10 mm or more is considered positive, 5 mm or less negative and 6 to 9 mm is of doubtful significance because it may be due to other mycobacterial infections. A PPD dose of I TU is used when extreme hypersensitivity is suspected and doses of 10 or 100 when 5 *TU* test is negative.

ii. Heaf Test

Multiple puncture testing as Heaf test is done with a spring-loaded gun which fires six prongs into the skin through a drop of PPD **(Heaf method**). They are not recommended for diagnostic use but are widely used for screening and survey purposes. Single-test disposable devices with PPD dried onto prongs **(tine tests)** are also available for individual testing.

Result

Positive Test

A positive tuberculin test indicates hypersensitivity to tuberculoprotein denoting infection with tubercle bacillus or BCG immunization, recent or past, with or without clinical disease. The test becomes positive 4-6 weeks after infection or immunization. Tuberculin allergy wanes gradually and disappears after 4-5 years in the absence of subsequent contact with the bacillus. In endemic areas, the allergy is maintained by repeated contacts with the bacillus.

False-Positive Reactions

False-positive reactions may be seen in infections with related mycobacteria ('atypical' mycobacteria). These are usually low-grade reactions and can be differentiated by testing with tuberculin prepared from these mycobacteria.

False-Negative Tests (Tuberculin Anergy)

The test may become negative in the following conditions:

1. Early tuberculosis; 2. Advanced tuberculosis; 3. Miliary tuberculosis; 4. In patients with measles and other exanthematous reactions; 5. Occasionally after chemotherapy and removal of lung lesion; 6. Advanced age; 7. Immunosuppressive therapy and defective cell mediated immunity (CMI); 8. Lymphoreticular malignancy; 9. Sarcoidosis; 10. Severe malnutrition; 11. False negative results may also be due to inactive PPD preparations and improper injection technique.

Uses of Tuberculin Test

1. To diagnose active infection in infants and young children.
2. To measure prevalence of infection in an area.
3. To select susceptibles for BCG vaccination.
4. Indication of successful BCG vaccination.

Tuberculin testing of cattle has been of great value in the control of bovine tuberculosis.

Prophylaxis

In the prevention of tuberculosis general measures such as adequate nutrition, good housing and health education are as important as specific antibacterial measures. The latter consists of early detection and treatment of cases, BCG vaccination and by chemoprophylaxis.

BCG Vaccination

Immunoprophylaxis is by intradermal injection of the live attenuated vaccine developed by Calmette and Guerin (1921), the Bacille Calmette Guerin or BCG. This is a strain of *M. bovis* attenuated by 239 serial subcultures in a glycerine-bile-potato medium over a period of 13 years between the years 1908 and 1920 which was avirulent for man while retaining its capacity to induce an immune response. This species was selected rather than *M. tuberculosis* on the dubious assumption that it was of limited virulence in man.

The first human was vaccinated by the intradermal technique in 1927. Recognition of the value of BCG came in 1948 when it was accepted by tuberculosis workers from all over the world as a safe preventive measure.

Aim

The aim of BCG vaccination is to induce a benign, artificial primary infection which will stimulate an acquired resistance to possible subsequent infection with virulent tubercle bacilli and thus reduce the morbidity and mortality from primary tuberculosis among those most at risk.

Dose and Administration

BCG vaccine is available in liquid form and freeze-dried (lyophilized) form. Freeze-dried (lyophilized) form is more stable preparation and commonly used. The lyophilized vaccine is reconstituted by sterile physiological saline to make a final concentration of 0.1 mg (moist weight) in 0.1 ml of the vaccine and it is supplied by BCG vaccine laboratory, Chennai. Vaccine should be utilized within 3 to 6 hours once reconstituted. The organisms grow to a limited extent in the tissues following injection of 0.1 ml of vaccine intradermally. BCG vaccine should be administered soon after birth failing which it may be given at any time during the first year of life.

Phenomena After Vaccination

Two to three weeks after injection a small nodule develops at the site of inoculation. It increases slowly in size and reaches a diameter of 4 to 8 mm by about 5 weeks. It then subsides or breaks into a shallow ulcer which heals spontaneously within 6 to 12 weeks leaving a permanent, tiny, round scar, typically 4 to 8 mm in diameter. Normally the individual becomes Mantoux-positive after a period of 8 weeks has elapsed, but sometimes about 14 weeks are needed. A few cases have been recorded where BCG has given rise to progressive tuberculosis.

Protective Efficacy

A tuberculin negative recipient is converted to a positive reactor following BCG vaccination. The immunity is similar to the immunity following natural infection except that it does not carry any risk of disease due to reactivation, as in the latter case. The duration of protection is from 15 to 20 years. Several field trials have been conducted to assess the efficacy of the BCG vaccine. Studies have shown that the range of protection offered by BCG varied from **0 to 80 percent in** different parts of the world (Table 32.6).

The consensus opinion is that BCG may not protect from the risk of tuberculosis infection but gives protection to infants and young children against the more serious types of the disease, such as meningitis and disseminated tuberculosis. The WHO recommends that the use of BCG should be continued as an antituberculosis measure in endemic countries such as India. BCG vaccine be administered to babies by intradermal injection on the deltoid immediately after birth, or as early as possible thereafter, before the age of 12 months. After the age of two years the vaccine need not be administered.

BCG induces a nonspecific stimulation of the immune system providing some protection against leprosy and leukemia. Multiple injection of BCG has been tried as adjunctive therapy in some malignancies. Some workers have reported that BCG is superior to PPD for tuberculin testing.

Complications of BCG Vaccine

1. **Local:** Abscess, indolent ulcer, keloid, tuberculides, confluent lesions, lupoid lesions, lupus vulgaris.
2. **Regional:** Enlargement and suppuration of draining lymph nodes.
3. **General:** Fever, mediastinal adenitis, erythema nodosum, tendency to keloid formation after wounding at other sites, and very rarely nonfatal meningitis. The very few cases of progressive tuberculosis reported are believed to have been in immunodeficient subjects.

Contraindications

BCG should not be given to patients suffering from generalized eczema, infective dermatosis, hypogammaglobulinemia and to those with a history of deficient immunity. The effect of BCG may be exaggerated in these patients.

Chemoprophylaxis

Chemoprophylaxis or preventive chemotherapy is the administration of antituberculous drugs (usually only isoniazid) to persons with latent tuberculosis (asymptomatic tuberculin positive) and a high-risk of developing active tuberculosis or to the uninfected exposed to high-risk of infection. It is particularly indicated in infants of mothers with active tuberculosis and in children living with a case of active tuberculosis in the house. Isoniazid 5 mg/kg daily for 6 to 12 months is the usual course. Trials have shown that this reduces the risk of developing active disease by 90 percent. HIV infected contacts of active tuberculosis also benefit from this prophylaxis.

Treatment

The **bactericidal drugs,** along with the **bacteriostatic drug** ethambutol (E) constitute the **first line drugs** in antituberculous therapy (Table 32.7). The old practice of daily administration of drugs for two years or so has been replaced by short course regimens of 6 to 7 months, which are effective and convenient. A typical example of such a schedule for a new smear positive case is a combination of four drugs (HRZE) given three times a week during an initial intensive phase of two months, followed by 4 to 5 months of continuing phase with only two drugs (HR) three times a week.

Multidrug-Resistant *Mycobacterium tuberculosis* (MDR-TB)

A very serious consequence of unchecked drug resistance has been the emergence and spread of **multidrug resistant tuberculosis (MDR-TB).** WHO defines a multidrug resistance (MDR) strain as one that is at least resistant to rifampicin (R) and isoniazid (H). This is because R and H form the sheet anchor of short-term chemotherapy and any strain resistant to both these drugs is unlikely to respond to treatment.

Risk factors for drug resistance may include previous treatment for TB, residence in an area endemic for drug resistance, or close contact with an individual who is infected with MDR-TB. Drug resistance is usually acquired by spontaneous mutations as a result of the inappropriate use of antimicrobial agents to treat *M. tuberculosis* and the lack of patient compliance. Another serious condition *extensively drug resistant-tuberculosis* (XDR-TB) has emerged recently. XDR-TB is due to *M. tuberculosis* strains which are resistant to any fluoroquinolone and at least one of three injectable second line drugs (capreomycin, kanamycin and amikacin) in addition to isoniazid and rifampicin.

Table 32.5: Some differential characteristics of tubercle bacilli causing human disease							
Species	*Oxygen preference*	*Glycerol enhanced*	*Niacin*	*Nitrate reduction*	*TCH*	*Pyrazinamide*	*Pathogenicity*
M. tuberculosis[a]	Aerobic	Yes	Positive	Positive	Resistant[b]	Sensitive	Pathogen
M. bovis	Microaerophilic	No	Negative	Negative	Sensitive	Resistant	Pathogen
M. bovis BCG	Aerobic	Yes	Negative	Negative	Sensitive	Resistant	Opportunistic Pathogen
M. africanum	Microaerophilic	No	Variable	Variable	Sensitive	Sensitive	Pathogen

[a]Includes the rare *M. canetti* variant.

[b]Strains from south India may be sensitive.

TCH, thiophen-2-carboxylic acid hydrazide.

Table 32.6: Protective efficacy of BCG vaccinations in nine major trials			
Immunization period	*Country or population*	*Age range of vaccinees*	*Efficacy (%)*
1935-38	North American Indian	0-20	80
1937-48	Chicago, USA	Neonates	75
1947	Georgia, USA	6-17	0
1949-51	Puerto Rico	1-18	31
1950	Georgia and Alabama, USA	>5	14
1950-52	UK	14-15	78
	Madanapalle, South India (Rural)	All ages	60
1968-71	Chingleput, South India (Rural)	All ages	30

Between 50 and 100 million people worldwide are thought to be infected with strains of drug resistant tuberculosis. MDR-TB requires an extended treatment period compared with drug-susceptible isolates. For cases of resistance to isoniazid or rifampin, second-line. antituberculosis drugs may include aminoglycosides (kanamycin, amikacin, capreomycin) and fluoroquinolones. With the numbers of cases of multidrug-resistant *M. tuberculosis* increasing, newer agents are being tested *in vitro* to determine their efficacy. If compliance is an issue, a single daily dose of all first line antitubercular drugs is preferred, and **'directly observed treatment strategy' (DOTS)** has been recommended by WHO. Otherwise resistance may be assumed and tested for *in vitro*.

Table 32.7: Antituberculosis drugs (in vivo)		
Sterilizing	*Bactericidal*	*Bacteristatic*
Rifampicin	Isoniazid	Ethionamide
Pyrazinamide	Streptomycin	Prothionamide
	Ethambutol	Thiacetazone
	Quinolones	p-Aminosalicylic acid
	Macrolides	Cycloserine

'In early stages of therapy'

KNOW MORE

Mycobacteria are rich in lipids. These include mycolic acids (long-chain fatty acids C78-C90), waxes, and phosphatides. Lipids are to some extent responsible for acidfastness. Their removal with hot acid destroys acid-fastness, which depends on both the integrity of the cell wall and the presence of certain lipids. Acid-fastness is also lost after sonication of mycobacterial cells.

Virulent strains of tubercle bacilli form microscopic **"serpentine cords"** in which acid-fast bacilli are arranged in parallel chains. Cord formation is correlated with virulence. A **"cord factor" (trehalose-6,6'-dimycolate)** has been extracted from virulent bacilli with petroleum ether. It inhibits migration of leukocytes, causes chronic granulomas and can serve as an immunologic "adjuvant."

Host Range

M. tuberculosis causes natural infection in humans, other primates, dogs and some other animals which have close contact with humans. Humans and guinea pigs are highly susceptible to *M. tuberculosis* infection, whereas fowl and cattle are resistant.

M. tuberculosis and *Mycobacterium bovis* are equally pathogenic for humans. *M. bovis* is more pathogenic for animals. Experimentally, it is highly pathogenic for guinea pigs and calves, moderately pathogenic for dogs, cats, horses and rats, and nonpathogenic for fowl. In man the portal of entry is usually the gastrointestinal tract. When inhaled, *M. bovis* can also cause pulmonary tuberculosis indistinguishable from that caused by *M. tuberculosis.* In developed countries, *M bovis* has become very rare.

Strains of tubercle bacilli isolated from parts of Africa, that show properties intermediate between human and bovine types have been called **'African strains' or *M. africanum*.** The name **'Asian type'** has been given to strains of tubercle bacilli originally isolated from south India, which are of low virulence for guinea pigs, susceptible to hydrogen peroxide, isoniazid sensitive and usually of phage type I from some other Asian countries.

KEY POINTS

- Mycobacteria are aerobic, nonmotile, noncapsulated and nonsporing. Growth is generally slow
- Mycobacteria do not stain readily, but once stained with hot carbol fuchsin or other aryl methane dyes, they resist decolorization with dilute mineral acids (or alcohol). Mycobacteria are, therefore, known as acid-fast bacilli (AFB). Acid fastness has been ascribed variously to the presence in the bacillus of an unsaponifiable wax (mycoloic acid) or to a semipermeable membrane around the cell. It is related to the integrity of the cell and appears to be a property of the lipid-rich waxy cell wall.

Mycobacterium tuberculosis

- *Mycobacterium tuberculosis* is weakly gram-positive, strongly acid-fast, aerobic bacilli.
- Ziehl-Neilsen acid-fast stain is useful in staining organisms either from cultures or from clinical material. With this stain, the tubercle bacilli stain bright red, while the tissue cells and other organisms are stained blue. Tubercle bacilli may also be stained with the fluorescent dyes (auramine O, rhodamine) and appear yellow luminous bacilli under the fluorescent microscope.
- They are aerobes, slow growers, produce luxuriant eugonic growth after 4-6 weeks.

  The solid medium most widely employed for routine culture is Lowenstein-Jensen (LJ) medium without starch. This consists of coagulated hens' egg, mineral salt solution, asparagine and malachite green, glycerol or sodium pyruvate
- On LJ media, *M. tuberculosis* forms dry, rough, raised, irregular colonies with a wrinkled surface. They are creamy white, becoming yellowish or buff colored on further incubation. They are tenacious and not easily emulsified. *Mycobacterium tuberculosis* has a luxuriant growth (eugenic growth) as compared to *Mycobacterium bovis* which grows poorly on LJ glycerol medium (dysgonic growth)
- Liquid media are Dubos', Middlebrook's, Proskauer and Beck's, Sula's and Sauton's media are the more common.
- They are weakly catalase positive, neutral-red positive, amidase positive, nitrate reduction test positive, niacin test negative, and aryl sulfatase negative.
- Mycobacteria protein (tuberculin) is responsible for development of delayed hypersensitivity in humans.
- **Pathogenesis and immunity:** The source of infection is usually an open case of pulmonary tuberculosis. The initial infection with *M. tuberculosis* is referred to as a primary infection. Subsequent disease in a previously sensitized person, either from an exogenous source or by reactivation of a primary infection, is known as postprimary (secondary or reinfection) tuberculosis with quite different pathological features. *M. bovis* infection is transmitted by ingestion of raw milk of the cows infected with M. bovis.
- **Diseases:** *M. tuberculosis* causes primarily pulmonary tuberculosis. Complications include miliary tuberculosis, disseminated tuberculosis, tubercular meningitis, tuberculosis of the skin, tuberculosis of the middle ear and ocular structures.
- **Diagnosis:** Bacteriological diagnosis of tuberculosis can be established by direct microscopy, culture examination or by animal inoculation test.

1. Sputum is the specimen of choice for pulmonary tuberculosis.
2. Microscopy—ZN staining and auramine-rhodamine staining for demonstration of AFB in stained smears is most useful and is the presumptive diagnosis of tuberculosis.
3. Culture is the definite method to detect and identify *M. tuberculosis* and is sensitive and specific.
4. Serology is of limited value in the diagnosis of pulmonary tuberculosis.
5. Direct detection by molecular probes is relatively insensitive.
6. The recent rapid and automated methods include automated radiometric culture methods (e.g. BACTEC) , SEPTICHEK, MGITs, etc.

- Resistance ratio method, absolute concentration method and proportion method are used to determine the sensitivity testing of *M. tuberculosis*. Other methods for sensitivity testing include BACTEC radiometric method, MGIT, chemiluminescence and Epsilometer test (E-test).
- **Koch phenomenon:** Koch phenomenon demonstrates acquired increased resistance in the infected animal but the level of immunity is inadequate to protect the animal against death from the initial infection.
- **Tuberculin test:** The test is based on the fact that persons infected with tubercle bacilli develop hypersensitivity to the proteins of the organism

and this test is delayed (Type IV) hypersensitivity reaction. Many methods such as 1. Mantoux test and 2. Heaf test had been described for tuberculin testing. Result may be positive test, false-positive reactions or false negative tests (tuberculin anergy).

Treatment, Prevention and Control

Chemotherapy forms the mainstay of treatment of tuberculosis. Drug resistance in *M. tuberculosis* is due to mutation. The emergence of multidrug resistance-tuberculosis (MDR-TB) is a very serious problem. The term multidrug resistance refers to resistance to rifampicin and isoniazid, with or without resistance to one or more other drugs. Another serious condition extensively drug resistant-tuberculosis (XDR-TB) has emerged recently. XDR-TB is due to *M. tuberculosis* strains which are resistant to any fluoroquinolone and at least one of three injectable second line drugs (capreomycin, kanamycin and amikacin) in addition to isoniazid and rifampicin.

Preventive measures against tuberculosis include chemoprophylaxis, vaccination, and general health measures. Immunoprophylaxis with BCG in endemic countries. The range of protection offered by BCG varied from 0 to 80 percent in different parts of the world

IMPORTANT QUESTIONS

1. Describe the morphology, cultural characteristics and pathogenicity of *M. tuberculosis.*
2. Classify mycobacteria. Describe the laboratory diagnosis of pulmonary tuberculosis
3. Differentiate between *Mycobacterium tuberculosis* and *Mycobacterium bovis* in a tabulated form.
4. Write short notes on:
   - Antigenic structure of *Mycobacterium tuberculosis*
   - Pulmonary tuberculosis.
   - Koch's phenomenon
   - Tuberculin test
   - Mantoux test.
   - BCG vaccine
   - Sensitivity testing of *Mycobacterium tuberculosis*
   - Multidrug-resistant *Mycobacterium tuberculosis* (MDR-TB).

FURTHER READING

American Thoracic Society Workshop. Rapid Diagnostic Tests for Tuberculosis. American J Respir. Crit. Care Med 1997; 155,1804.

Reichman LB, Hershfield, ES. Tuberculosis, New York: Marcel Dekker 2000.

Rom WN, Gray SM (Eds). Tuberculosis. Boston: Little Brown 1996.

Suri JC, Sen MK. Multidrug Resistant Tuberculosis. JAMA India 1998;1: 27.

Zumla A, Grange JM. Tuberculosis. BMJ 1998;316:1962.

CHAPTER 33

Mycobacterium leprae

Learning Objectives

After reading and studying this chapter, you should be able to:

- Describe morphology of *M. leprae*.
- Discuss cultivation of lepra bacilli.
- Explain animal models in leprosy.
- Describe pathogenesis of leprosy.
- Describe the following: Lepromatous leprosy; Tuberculoid leprosy; Lepromin test.
- Differentiate between lepromatous and tuberculoid leprosy.
- Discuss laboratory diagnosis of leprosy.

INTRODUCTION

Leprosy is probably the oldest disease known to mankind. Leprosy was described as 'kushta' in Sushruta Samhita written in India in 600 BC. Many accounts of leprosy may also be found in the ancient Hebrew writings, and although some of the skin lesions considered to be leprosy in the Old Testament of the Bible were probably not leprosy, many undoubtedly were.

It is caused by *Mycobacterium leprae* which was discovered by Hansen in 1874 in Norway. Although this was one of the first descriptions of a microorganism as the cause of human disease, the organism is still an enigma. This is because it has not been possible to grow the bacillus in culture media and due to lack of suitable animal model for experimental transmission of the disease.

MYCOBACTERIUM LEPRAE

Morphology

M. leprae is a straight or slightly curved rod, 1-8 µm × 0.2-0.5 µm in size, showing considerable morphological variations. The rods may stain uniformly or show granules and beads that are slightly larger than the average diameter of the cell. Polar bodies and other intracellular elements may be present. Clubbed forms, lateral buds or branching may be observed. They are nonmotile and nonsporing.

They are gram-positive and stain more readily than *M. tuberculosis*. With **Ziehl-Neelsen stain**, they are less acid-fast than tubercle bacilli, so 5 percent sulphuric acid is employed for decolorization after staining with carbol fuchsin. The bacilli are seen singly, and in groups, intracellularly or lying free outside the cells. Inside the cells they are present as bundles of organisms bound together by a lipid-like substance, the glia. These masses are known as *globi.* Large numbers of bacilli may be packed in the cells in an arrangement that suggests packets of cigars. The parallel rows of bacilli in the globi give appearance of a *cigar bundle.* In tissue sections, the bacilli are arranged in clumps resembling cigarette ends. The globi are present in Virchow's *lepra cells* or *foamy cells* which are large undifferentiated histiocytes.

Differentiation of Live and Dead Bacilli

It is possible to differentiate live and dead bacilli in smears stained with Ziehl-Neelsen method. Live bacilli in the smear appear solid and uniformly stained, while dead or dying forms appear fragmented, beaded and granular. The percentage of uniformly stained bacilli in the tissues is known as morphological index (MI). This provides a method of assessing the progress of patients on chemotherapy and is more meaningful than the old criterion of *bacteriological index* (the number of bacilli in tissues). Poorly stained bacilli are probably dead. A continuing fall in the MI is encouraging and a fall succeeded by a rise indicates development of drug resistance in the bacteria. Bacteriological index of a smear is the total number of acid-fast bacilli in an oil immersion field.

Generation Time of the Lepra Bacillus

The generation time of the lepra bacillus has been found to be exceptionally long in animal experiments, 12-13 days on the average but may vary from 8-42 days, in comparison with about 14 hours in the case of the tubercle bacillus and about 20 minutes in the case of coliform bacilli.

Cultivation

In spite of the efforts of many workers, it has not so far been possible to cultivate lepra bacilli either in bacteriological media or in tissue culture. There have been several reports of successful cultivation but none has been confirmed. One of the best known of such reports (1962) came from the Indian Cancer Research Centre (ICRC) Mumbai, where an acid fast bacillus was isolated from leprosy patients, employing human fetal spinal ganglion cell culture. This is known as ICRC bacillus. This ICRC bacillus has been adapted for growth on Lowenstein-Jensen medium. Its relation to the lepra bacillus is uncertain and taxonomical studies suggest that this organism is not *M. leprae* and belongs to *M. avium intracellulare* group.

There have been many attempts to transmit leprosy to experimental animals. However, the real breakthrough was in 1960, when Shepard discovered that *M. leprae* could multiply in the footpads of mice kept at a low temperature (20°C). Following animals have been used for experimental infection with *M. leprae:*

1. Mouse (foot-pad).
2. Nine-banded armadillos.
3. Chimpanzees.
4. Monkeys.
5. Slender loris.
6. Indian pangolin.
7. Chipmunks.
8. Golden hamsters.
9. European hedgehog.

Mouse

In the mouse, infections can be initiated with as few as 1 to 10 bacilli. A granuloma develops at the site in 1-6 months following intradermal inoculation into the **footpads of mice.** If cell mediated immunity is suppressed by thymectomy or the administration of antilymphocyte serum, a generalized infection is produced, simulating lepromatous leprosy.

Uses of Foot-pad of Mouse Model

i. Identification of *M. leprae.*
ii. To study the susceptibility of organisms to chemotherapeutic agents.
iii. To study the development of drug resistance in *M. leprae* in patients under treatment.
iv. To study the properties of bacilli isolated from different forms of the disease.
v. To study the efficacy of various vaccines to suppress multiplication of *M. leprae.*
vi. To determine viability of *M. leprae.*

Disadvantages of the Foot-pad of Mouse Model

i. There occurs only a limited multiplication of *M. leprae* (1×10^6 bacilli per foot-pad) following intradermal inoculation. Therefore, the yield of the bacilli is not sufficient for comprehensive research on *M. leprae.*
ii. It is not possible to study the pathogenesis of the disease due to short life span of mice and long incubation period and chronic course of leprosy.
iii. The lesion produced is not of lepromatous leprosy type.

Nine-Banded Armadillo

The **nine-banded armadillo *(Dasypus novemcinctus)*** is highly susceptible to infection with lepra bacilli. This is presumably due to low body temperature. Following inoculation into armadilloes, a generalized infection occurs with extensive multiplication of the bacilli and production of lesions typical of lepromatous leprosy. Natural infection by a mycobacterium resembling *M. leprae* has been observed in some wild armadillos in Texas and Mexico. Wild armadillo should, therefore, be screened for mycobacterial infection and kept in quarantine for 3 months before inoculation.

Advantages of Armadillo

i. Relatively long life-span (12-15 years).
ii. Relatively low body temperature (32-35°C).
iii. The yield of *M. leprae* from armadillo skin leproma is 100-1,000 times more than that from human leproma.
iv. Unique mode of reproduction by means of monozygous genetically identical quadruplets.
v. An abundant source of *M. leprae,* for research and the preparation of lepromin or of a vaccine, has now become available because *M. leprae* can be transmitted from one animal to another without any change in microbe.

Disadvantages of Using Armadillo as an Experimental Model

i. Only about 40 percent of the animals are susceptible to disseminated form of leprosy.
ii. For the development of lepromatous leprosy in armadillo, limits of incubation period are not known which may be many years long. Therefore, it might be hazardous to conclude after an arbitrarily chosen length of time, that an inoculated armadillo shall not develop disease in future.
iii. Some of the armadillos caught from wild life are naturally infected with mycobacteria resembling *M. leprae.*
iv. Their cost is very high because these animals are reported to be found only in southern parts of the USA and their nonavailability in other parts of the world.

Other Animals

'Natural disease' has also been identified in **chimpanzees** and **mangabey monkeys** from West Africa but it is not known whether they have any relevance to human infection.

Antigenic Structure

Mycobacterial antigens may be classified on the basis of:

1. Solubility as (a) soluble (cytoplasmic) and (b) insoluble (cell-wall-lipid—bound);
2. Chemical structure as (a) carbohydrate and (b) protein, and (3) distribution within the species. Up to 90 soluble antigens have been demonstrated by counterimmunoelectrophoresis in mycobacteria. They can be divided into 4 groups:
   Group I Common to all mycobacteria.
   Group II Occur in slowly growing species.
   Group III Occur in rapidly growing species.
   Group IV Unique to each individual species.
   M. leprae possesses antigens of groups I and IV.

The cell wall of *M. leprae* is made up of four layers like other mycobacteria. The innermost is a peptidoglycan layer which gives the cell its shape and rigidity. External to this is **lipoarabinomannan-B (LAM-B) layer,** attached to which is a dense palisade of characteristic long chain fatty acids known as mycolic acid. The outermost layer is composed of mycosides. A major component of this layer is phenolic glycolipid-1 (PGL-1).

A number of highly immunogenic substances have been identified in *M. leprae*. LAM-B is a dominant antigen of *M. leprae,* is highly immunogenic and is used in the serodiagnosis of leprosy. It is immunologically cross-reactive with a similar product from *M. tuberculosis.* It is distinct from lipoarabinomannan of *M. tuberculosis* from which it can be differentiated by monoclonal antibodies. PGL-1 protects pathogens against host cell enzymes and suppresses cell mediated immunity. Patient, infected with *M. leprae,* develops antibodies against polysaccharide constituent of PGL-1 which has been used for the serodiagnosis of leprosy.

In addition, *M. leprae* possesses a large number of **protein antigens,** namely 18 kDa, 28 kDa, 35 kDa, 36 kDa, 65 kDa and so on, according to molecular weight of antigens in kilodaltons (kDa) (Table 33.1).

Table 33.1: Protein antigens of *Mycobacterium leprae* and their functions

Antigen	Function
18 kDa	Stimulates CMI and AMI
28 kDa	Superoxide dismutase enzyme
35 kDa	Possesses epitopes for antibody and T cells
36 kDa, 65 kDa, 70 kDa } HSP	Elicits both CMI and AMI Stimulates CMI
30-32 kDa	Elicits an early protective immune response
85 complex	–
85 B	–
15 kDa	–
28 kDa	–
45 kDa	–

Resistance

In a warm humid environment, lepra bacilli have been found to remain viable for 9-16 days and in moist soil for 46 days. They survive exposure to direct sunlight for two hours and ultraviolet light for 30 minutes.

Classification

Madrid Classification

The disease may be classified into four types (Madrid, 1953):

1. Lepromatous.
2. Tuberculoid.
3. Dimorphous.
4. Indeterminate.

Classification System of Ridley and Jopling

The spectrum of disease activity in leprosy is very broad, characterized by pronounced variations in clinical, histopathologic, and immunologic findings. On the basis of these properties, Ridley and Jopling (1966) have established a classification scheme consisting of five forms of leprosy (Table 33.2):

1. Tuberculoid (TT).
2. Borderline tuberculoid (BT).
3. Borderline (BB).
4. Borderline lepromatous (BL).
5. Lepromatous (LL).

Hyper-reactive tuberculoid (TT) leprosy is at one pole and anergic lepromatous (LL) leprosy at the other. The type of disease is a reflection of the immune status of the host. It is therefore not permanent and varies with chemotherapy and alterations in host resistance. TT and LL are stable. The others are unstable, especially BT, which in the absence of treatment can regress to BB or BL.

Pathogenesis

Leprosy (Hansen's disease) is a chronic granulomatous disease of humans primarily involving the skin, peripheral nerves and nasal mucosa but capable of affecting any tissue or organ.

M. leprae is an obligate intracellular parasite that multiplies very slowly within the mononuclear phagocytes, especially the histiocytes of the skin and Schwann cells of the nerves. *M. leprae* has an especially strong predilection for nerves and is the only known human pathogen that preferentially attacks the peripheral nerves. The resulting nerve damage is responsible for the main clinical features of leprosy: anesthesia and muscle paralysis. Repeated injury to, and infections of, the anesthetic extremities leads to their gradual destruction. Infiltration of the skin and cutaneous nerves by bacilli leads to the formation of visible lesions, often with pigmentary changes. The two extreme or polar forms of the disease are the lepromatous and tuberculoid (TT) types (Table 33.3).

Table 33.2: Characteristics of the five forms of leprosy

	TT	*BT*	*BB*	*BL*	*LL*
Bacilli in skin	–	+/ –	+	++	+++
Bacilli in nasal secretions	–	–	–	+	+++
Granuloma formation	+++	++	+	–	–
Lepromin test	+++	+	+/ –	–	–
Antibodies to *M. leprae*	+/ –	+/ –	+	++	+++
Main phagocytic cell	Mature epithelioid	Immature epithelioid	Immature epithelioid	Macrophage	Macrophage
In vitro correlates of CMI	+++	++	+	+/–	–
Type 1 reactions	–	+	+	+	–
Type 2 reactions	–	–	–	+/–	++

CMI, cell-mediated immunity; see text for other abbreviations.

Table 33.3: Characteristics of lepromatous and tuberculoid leprosy

Feature	*Lepromatous leprosy*	*Tuberculoid (TT) leprosy*
1. Resistance	Seen in persons whose resistance is low.	Seen in persons whose resistance is high
2. Skin lesions	Many erythematous macules, papules, or nodules; extensive tissue destruction (e.g., nasal cartilage, bones, ears); diffuse nerve involvement with patchy sensory loss; lack of nerve enlargement.	Few erythematous or hypopigmented plaques with flat centers and raised, demarcated borders; peripheral nerve damage with complete sensory loss; visible enlargement of nerves.
3. Histopathology	Predominantly "foamy" macrophages with few lymphocytes; lack of Langhans' cells; numerous acid-fast bacilli in skin lesions and internal organs.	Infiltration of lymphocytes around center of epithelial cells; presence of Langhans' cells; few or no acid-fast bacilli observed.
4. Infectivity	High	Low
5. Bacilli in skin	+++	–
6. Bacilli in nasal secretions	+++	–
7. Granuloma formation	+++	–
8. Lepromin test	–	+++
9. Antibodies to *M. leprae*	Hypergammaglobulinemia	Normal
10. Erythema nodosum leprosum	Usually present	Erythema nodosum leprosum
11. Prognosis	Poor	Good

Lepromatous Leprosy

Lepromatous form is the generalized form of the disease and is found in individuals where the host resistance is low. Patient develops numerous nodular skin lesions **(lepromata)** on face, ear lobes, hands, feet and less commonly trunk. Skin lesions contain many macrophages, often seen as large foamy cells packed with AFB. Skin biopsy specimens may contain up to 10^9 bacilli per gram of tissue. This is known as **'multibacillary disease'**. Nodular skin lesions ulcerate due to repeated trauma as a result of loss of sensation. The ulcerated nodules become secondarily infected that leads to distortion and mutilation of extremities.

Bacilli invade the mucosa of the nose, mouth and upper respiratory tract and are shed in large numbers in nasal and oral secretions. **Bacillemia** is common. *M. leprae* tends to invade vascular channels, resulting in a continuous bacteremia in lepromatous patients and consistent involvement of the reticuloendothelial system. Eyes, testes, kidneys and bones are also involved.

Lepromatous leprosy is more infectious than other types and has a **poor prognosis. Cell mediated immunity is deficient** and the **lepromin test is negative** in lepromatous leprosy. Humoral antibodies against mycobacterial antigens are produced in high concentrations which play no protective role. Autoantibodies are also produced. Most cases show biological false positive reaction in standard serological tests for syphilis.

Tuberculoid (TT) Leprosy

At the other end of the spectrum is hyper-reactive *tuberculoid* (TT) leprosy, which is seen in patients with high degree of resistance where cell-mediated immunity is intact. The skin lesions are few and sharply demarcated, consisting of macular anesthetic patches. The lesions occur as large maculae (spots) in cooler body tissues such as skin (especially nose and outer ears, and testicles). Neuritis leads to patches of anesthesia. In tuberculoid leprosy, skin biopsies show mature granuloma formation in the dermis consisting of epithelioid cells, giant cells, and rather extensive infiltration of lymphocytes.

There are very few acid-fast bacilli (AFB) so that they are generally not seen microscopically (paucibacillary disease) and infectivity is minimal. The organisms invade the nerves and selectively colonize the Schwann cells. Neural involvement occurs early and may be pronounced, leading to deformities, particularly in the hands and feet. The local nerves are involved in the early stage and gradually the infection extends into the bigger nerve trunks which are thickened, hard and tender. This leads to deformities of hand and feet. The patient develops a strong cell mediated immune response, develops delayed hypersensitivity and the lepromin test is positive. Antimycobacterial and autoimmune antibodies are rare. The prognosis is good.

An intermediate form classified as borderline tuberculoid (BT), mid-borderline (BB) or borderline lepromatous (BL).

Borderline (BB) Leprosy

Borderline leprosy (BB), sometimes called dimorphous or intermediate leprosy, has features of both tuberculoid and lepromatous forms. This is an unstable form of the disease and may shift to the lepromatous or tuberculoid part of the spectrum depending on chemotherapy or alterations in host resistance.

Indeterminate Type

There is an early unstable tissue reaction with mild transient tissue lesion, often resembling maculo-anaesthetic patches which are not characteristic of either the lepromatous or the tuberculoid type. In many persons, the indeterminate lesions undergo healing spontaneously. In others, the lesions may progress to the tuberculoid or lepromatous types.

Epidemiology

Leprosy is an exclusively human disease and the only source of infection is the patient. Very large numbers of bacilli are shed in the nasal secretions and smaller numbers in discharges from superficial lesions of the lepromatous cases. Disease is spread by person-to-person contact. The mode of entry may be either through the respiratory tract or through the skin. Asymptomatic infection appears to be quite common in endemic areas. Bacilli may also be transmitted via breast milk from lepromatous mothers, by insect vectors, or by tattooing needles.

Leprosy has a long incubation period, an average of 3 to 5 years or more for lepromatous cases. It is rare in children aged less than 5 years. It has been estimated to vary from a few months to as long as 30 years. The tuberculoid leprosy is thought to have a shorter incubation period. It is generally held that intimate and prolonged contact is necessary for infection to take place. The disease is more likely if contact occurs during childhood.

Once worldwide in distribution, leprosy is now confined mainly, but not exclusively, to the underdeveloped areas of the tropics and the southern hemisphere. Leprosy is widely prevalent in India. Although the disease is present throughout the country, the distribution is uneven.

Immunity

Leprosy is a disease of low infectivity. A high degree of innate immunity against lepra bacilli seems to exist in human beings so that only a minority of those infected develop clinical disease.

Infection with lepra bacilli induces both humoral and cellular immune responses. Humoral antibodies do not have deleterious effect on the bacilli, while cellular immune mechanisms are able of destroying them. Also, in leprosy there is a close correlation between the various clinical forms and the cell-mediated immune response of the host. There is inverse relationship between the intensity of the delayed hypersensitivity response to *M. leprae* and the humoral response throughout the clinical spectrum of the disease. When it is adequate, the lesions are of the tuberculoid type.

Patients with **tuberculoid leprosy** exhibit a strong delayed-type hypersensitivity to lepromin, and the histology of lesions is that of hypersensitivity granulomas. The macrophages phagocytose the bacilli and destroy them. Specific humoral antibodies are not prominent. There is no increase in the immunogloblin level and the albumin:globulin ratio in the serum is not altered.

The **lepromatous type** of disease develops when cell mediated immunity is deficient. As the disease progresses across the leprosy spectrum, there is a progressive loss of hypersensitivity and development of an anergic state in the patient with lepromatous leprosy. A concomitant loss of cell-mediated immunity parallels the decline of delayed hypersensitivity to *M. leprae* antigens. Delayed hypersensitivity to the lepra bacillus protein is absent. The macrophages are able to phagocytose the bacilli but instead of being destroyed, the bacilli proliferate inside the cells. This is because of suppression of CMI in these patients. Conversely, a high serologic response characterizes lepromatous leprosy, and polyclonal hypergammaglobulinemia is a characteristic feature. In lepromatous leprosy, there is extensive polyclonal B cell activation with large amounts of antibodies being produced, both antimycobacterial and autoimmune.

Antibodies to *M. leprae* that cross-react with other mycobacteria may be detected in the sera of 75 percent to 95 percent of the patients with the lepromatous form of the disease. The albumin globulin ratio in the serum is reversed. Humoral antibodies play no protective role, however, in immune defense. On the other hand, they may have an enhancing effect. A number of abnormal serologic activities are also associated with lepromatous leprosy, including a biologic falsepositive reaction in routine serologic tests for syphilis.

HLA Genes and Leprosy

Although HLA genes do not determine susceptibility to leprosy, they do control the form of leprosy in susceptible individuals. Considerable information has been reported that suggest that HLA genes determine the type of leprosy that develops by controlling leprosy-specific immune responses. HLA-DR2 is seen preponderantly in persons with the tuberculoid type of reaction, while HLA-MTI and HLA-DQ I are associated with lepromatous disease.

Reactions

Though leprosy is a chronic disease, its course is sometimes interspersed with acute exacerbations due to immune reactions. These are of two types (Table 33.4):

1. Type 1 (Reversal reaction or the 'lepra reaction')
2. Type 2 (Erythema nodosum leprosum, ENL).

Type 1 (Reversal Reaction or the 'Lepra Reaction')

This occurs in borderline cases, occurring spontaneously or more often during chemotherapy. It is a cell mediated immune reaction, with an influx of lymphocytes into lesions, and a shift to tuberculoid morphology. The lesions are infiltrated with lymphocytes and epithelioid cells with reduction in number of bacilli. The lesions develop erythema and swelling, along with pain and tenderness. It may rapidly cause severe and permanent nerve damage. A similar clinical picture is seen in the 'downgrading reaction' which occurs usually in untreated or pregnant patients.

Type 2 (Erythema Nodosum Leprosum, ENL)

This is an immune-complex reaction seen only in lepromatous and borderline lepromatous cases, usually a few months after institution of chemotherapy. Clinically crops of red nodules appear in the skin, lasting for 1 or 2 days. Constitutional disturbances like fever, arthritis, iridocyclitis, orchitis and painful neuritis are common. The lesions show intense neutrophilic infiltration and the blood vessels in the dermis show acute fibrinoid necrosis. The histological picture is that of an Arthus reaction or immune complex disease. The immunological basis of type 2 is vasculitis associated with the deposition of antigen-antibody complexes.

Lepromin Test

Till recently, the only method for studying immunity in leprosy was a skin test for delayed hypersensitivity, the lepromin test first described by Mitsuda in 1919.

Lepromins

The lepromins used as antigen in lepromin test may be of human origin (lepromin-H) or of armadillo origin (lepromin-A) and are of two types.

Integral Lepromin (Mitsuda Lepromin)

The original antigen (lepromin) was boiled, emulsified, lepromatous tissue rich in lepra bacilli and this was developed by Mitsuda in 1919. The original crude Mitsuda antigens extracted from skin lesions of lepromatous patients (integral lepromins) were standardized on the basis of tissue content. Modern antigens are standardized according to their lepra bacillus content (4×10^7 lepra bacilli per ml) and has a shelf life of 2 years at 4°C. Standard lepromins are being prepared increasingly from armadillo derived lepra bacilli (Lepromin A), replacing human derived lepromin-H.

Bacillary Lepromin

This contains more of bacillary components and less of tissue. An important example of bacillary lepromin is Dharmendra antigen which is prepared by floating out the bacilli from finely ground lepromatous tissue with chloroform, evaporating it dry and removing the lipids by washing with ether. The antigen is made up in phenol saline for use.

Procedure

The test is performed by injecting intradermally 0.1 ml of lepromin into the inner aspect of the forearm of the individual. As a routine, the reaction is read at 48 hours and 21 days. The response to the intradermal injection of lepromin is typically biphasic.

Table 33.4: Main characteristics of the reactions in leprosy

Characteristics	*Type 1 (reversal reaction)*	*Type 2 (erythema nodosum leprosum)*
Immunological basis	Cell-mediated	Vasculitis with antigen-antibody complex deposition
Type of patient	BT, BB, BL	BL, LL
Systemic disturbance	No (or mild)	Yes
Hematological changes	No	Yes
Proteinuria	No	Frequently
Relation to therapy	Usually within first 6 months	Rare during first 6 months

Early or Fernandez Reaction

The early reaction is also known as **Fernandez reaction.** It consists of erythema and induration at the site of inoculation developing in 24-48 hours and usually remaining for 3-5 days. This reaction indicates that the patient has been infected at some time in the past and is a measure of pre-existing delayed hypersensitivity. This is analogous to the tuberculin reaction. This is usually poorly defined and carries little significance.

Late or Mitsuda Reaction

This is the classical Mitsuda reaction. The reaction develops late, becomes apparent in 7-10 days following the injection and reaching its maximum in 3 or 4 weeks. At the end of 21 days, if there is a *nodule* more than 5 mm in diameter at the site of inoculation, the reaction is said to be positive. The nodule may even ulcerate and heal with scarring if the antigen is crude. It takes several weeks to heal. Histologically, there is infiltration with lymphocytes, epithelioid cells and giant cells. Mitsuda late reaction does not indicate pre-existing DTH but is a measure of the CMI induced by the injected lepromin itself. It thus distinguishes between persons who can mount a CMI response against the lepra bacillus antigens and those who cannot. Therefore, this test is more meaningful to determine lepromin positivity.

Patients with tuberculoid leprosy usually exhibit both early and late lepromin reactions, but lepromatous patients never show these reactions because of complete anergy to the antigens of M. *leprae.* This anergy is very persistent in spite of long-term therapy.

The lepromin test is not used to diagnose leprosy, nor does it indicate prior contact with the lepra bacillus. Positive lepromin tests are elicited in patients with tuberculosis and also in normal, healthy children by vaccination with BCG. Healthy persons in nonendemic areas with no chance of contact with the bacillus may give a positive lepromin test.

Uses of Lepromin Reaction

The test is employed for the following purposes:

1. To Classify the Lesions of Leprosy Patients

The reaction is positive in tuberculoid and negative in lepromatous leprosy patients and variable in dimorphous and indeterminate types of disease.

2. To Assess the Prognosis and Response to Treatment

A positive reaction indicates good prognosis and a negative one bad prognosis. Conversion to lepromin positivity during treatment is evidence of improvement.

3. To Assess the Resistance of Individuals to Leprosy

It is desirable to recruit only lepromin positive persons for work in leprosaria as lepromin negative persons are more prone to develop the disease.

4. To Verify the Identity of Candidate Leprabacilli

Cultivable acid-fast bacilli, claimed to be lepra bacilli, should give matching results when tested in parallel with standard lepromin.

Laboratory Diagnosis

M. leprae cannot grow in cell-free cultures. Thus, laboratory confirmation of leprosy requires histopathologic findings consistent with the clinical disease and either skin test reactivity to lepromin or the presence of acid-fast bacilli in the lesions. The diagnosis consists of demonstration of acid fast bacilli in the lesions.

Specimens

For routine examination, specimens are collected from the nasal mucosa, skin lesions and ear lobules. Biopsy of the nodular lesions and thickened nerves, and lymph node puncture may be necessary in some cases. A brief account of method of skin smear and nasal smear examination is as follows:

Skin Smears

Slit and Scrape Method

Material from the skin is obtained from an active lesion, and also from both the ear lobes by the "slit and scrape" method. Samples from the skin should be obtained from the edges of the lesion rather than from the center. The skin is pinched up tight to minimize bleeding. A cut about 5 mm long is made with a scalpel, deep enough to get into the infiltrated layers. After wiping off blood or lymph that may have exuded, the blade of the scalpel is then turned at right angle to the cut (slit) and the bottom and the sides of the slit are scraped with the point of the blade, several times in the same direction so that tissue fluid and pulp (not blood) collects on one side of the blade which is smeared uniformly on a slide. When smear dries, it is fixed by passing the slide twice or thrice over a flame with the surface carrying the smear uppermost. About 5-6 different areas of the skin should be sampled, including the skin over the buttocks, forehead, chin, cheek and ears.

Nasal Scrapings

A blunt, narrow scalpel is introduced into the nose and the internal septum scraped sufficiently to remove a piece of mucous membrane, which is transferred to a slide and teased out into a uniform smear.

The skin or nasal smear is immediately fixed by lightly passing the underslide of the slide over the spirit lamp flame and transported to the laboratory for staining with Ziehl-Neelsen method.

Microscopy

Smears are stained by Ziehl-Neelsen method using 5 percent instead of 20 percent sulfuric acid for decolorization. Acid-fast bacilli (AFB) arranged in parallel bundles within macrophages *(Lepra-cell)* confirm the diagnosis of lepromatous leprosy. The viable bacilli stain uniformly and the dead bacilli are fragmented, irregular or granular.

The smears are graded, based on the number of bacilli as follows:

1-10 bacilli in 100 fields	= 1+
1-10 bacilli in 10 fields	= 2+
1-10 bacilli per field	= 3+
10-100 bacilli per field	= 4+
100-1,000 bacilli per field	= 5+
More than 1,000 bacilli, clumps and globi in every field	= 6+

Bacteriological Index (BI)

The bacteriological index is calculated by totalling the grades (number of pluses, +s scored in all the smears and divided by number of smears. Thus if 7 (seven) smears examined have a total of fourteen pluses (14+), BI will be 2. For calculating BI, a minimum of four skin lesions, a nasal swab and both the ear lobes are to be examined.

Morphological Index (MI)

It is defined as the percentage of uniformly stained bacilli out of the total number of bacilli counted.

Animal Inoculation

Injection of ground tissue from lepromatous nodules or nasal scrapings from leprosy patient into the **foot pad of mouse** produces typical granuloma at the site of inoculation within 6 months. **Nine-banded armadillo** is another animal used for inoculation of material. The lesions which develop in these animals can be identified by histological examination and Ziehl-Neelsen staining.

Lepromin Test

It is not a diagnostic test but is used to assess the resistance of patient to *M. leprae* infection. The test can be used to assess the prognosis and response to treatment.

Serological Test

Detection of antibody against *M. leprae* phenolic glycolipid antigen has been claimed to be a specific diagnostic test. Various serological tests like latex agglutination, *Mycobacterium leprae* particle agglutination (MLPA) and ELISA have been described. Anti PGL-1 antibody titers are higher in lepromatous patients but the tuberculoid patients show low titres. The antibody titers decrease following effective chemotherapy.

Molecular Diagnostic Methods

Attempts to develop molecular diagnostic methods are in progress. The PCR is increasingly used to detect *M. leprae* in clinical specimens. Meanwhile, microscopic demonstration of lepra bacilli and histology remain the most useful diagnostic procedures.

Treatment

Dapsone (4,4′-diaminodiphenyl sulfone; DDS) was the first effective chemotherapeutic agent against leprosy. Its use as a monotherapy for several years led to the development of resistant strains of lepra bacilli. Due to emergence of dapsone resistance, WHO recommended multiple drug therapy (MDT) for all leprosy cases based on dapsone, rifampicin and clofazimine. The regimen is determined by whether the patient has paucibacillary or multibacillary disease. Patients with paucibacillary (I, TT, BT) leprosy are given rifampicin 600 mg once a month (supervised) and dapsone 100 mg daily (unsupervised) for six months. For multibacillary (BB, BL, LL leprosy, rifampicin 600 mg once a month (supervised, dapsone 100 mg daily (unsupervised), clofazimine 300 mg once monthly supervised and 50 mg daily, unsupervised are given for two years or until skin smears are negative.

Where clofazimine is, totally unacceptable owing to the coloration of skin, its replacement by 250 to 375 mg self administered daily doses of ethionamide or protionamide has been suggested.

Clinical surveillance of cases after completion of treatment is an important part of the current recommendations for multidrug therapy. It is essential for the assurance of long-term success of treatment and for the early detection of any relapses.

An immunotherapeutic vaccine (Mycobacterium w) developed at the National Institute of Immunology, New Delhi is claimed to enhance the effect of MDT.

Immunotherapy

Since lepromatous leprosy patients do not possess CMI, efforts are being made to induce effective CMI to *M. leprae* in these patients. Procedures for trying to achieve this objective include:

1. Intravenous injection of peripheral blood lymphocytes obtained from patients with tuberculoid leprosy or from healthy donors possessing vigorous CMI and showing a strongly positive lepromin (Mitsuda) reaction.
2. Intravenous injection of transfer factor, an extract of lymphocytes, from patients suffering from tuberculoid leprosy. Sensitized lymphocytes secrete certain substances called lymphokines which stimulate the macrophages to ingest and kill the bacilli. If these substances are isolated and obtained from blood of patients with tuberculoid leprosy and injected into the lepromatous leprosy cases, CMI can possibly be induced.
3. Intradermal injection of vaccine (see below).

Prophylaxis

Case finding and adequate therapy have been the methods employed for prophylaxis. Long-term chemoprophylaxis has given encouraging results in child contacts of infectious cases in India and the Philippines.

Immunoprophylaxis

At present no effective vaccine against leprosy exists. A number of candidate vaccines (Table 33.5) have been tried and are still under trial. However, none of these have reached the stage for universal use. A candidate vaccine should be able to:

1. Induce upgrading in LL patients,
2. Bring about lepromin conversion both in the patients and healthy persons, and
3. Offer protection in animal models.

There is some degree of antigenic relationship between the lepra and tubercle bacilli. It is an old clinical observation that leprosy and tuberculosis do not usually coexist. BCG vaccine was observed to induce lepromin positivity and hence its use in the prevention of leprosy was suggested by Fernandez as early as in 1939. Shepard (1978) found that lepra bacilli did not multiply in the footpads of mice immunized with BCG. Controlled trials gave divergent results, from high to no protection.

BCG Vaccine

There is now considerable evidence that BCG vaccine can provide some protection against clinical leprosy. The results of controlled trials with BCG vaccine have demonstrated significant but varying levels of protection in four different countries—Uganda, Myanmar, Papua New Guinea and in India, the Chingleput (Chennai), ranging from 23-80 percent. Field trials in Venezuela with killed *M. leprae* and BCG provided no better protection than BCG alone.

Candidate Vaccines

In view of the variable protective effect of BCG vaccine against leprosy, several alternative vaccine preparations are under development (Table 33.5). These should more appropriately be called **"candidate vaccines".**

All the reported "candidate" vaccines have shown a similar degree of lepromin conversions in lepromatous patients (50-70%) and lepromin negative healthy individuals (90%). Field trials with different leprosy vaccines (BCG + killed lepra bacilli; ICRC bacillus) have not given conclusive results so far. Maximum work has been done with BCG + heat-killed *M. leprae.* However, none of the candidate vaccines have attained as yet "vaccinehood".

MYCOBACTERIUM LEPRAEMURIUM

M. lepraemurium, a causative agent of rat leprosy, was first described by Stefan sky in 1901 at Odessa. It was a chance finding when a large number of rats were being slaughtered during an attack of human plague, rat leprosy was found in 4-5 percent of them. It has been subsequently reported from several countries. The disease is probably transmitted naturally from rat to rat by fleas. Rat leprosy characterized by subcutaneous indurations, swelling of lymph nodes, emaciation, and sometimes ulceration and loss of hair.

Table 33.5: Candidate antileprosy vaccines

I. First generation vaccines
- BCG
- Armadillo-derived killed *Mycobacterium leprae*
- BCG and killed *M. leprae*
- ICRC bacillus.

II. Possible second generation vaccines; natural or recombinant form of 18, 31, 65 and 70 kDa proteins.

Acid-fast bacilli resembling *M. leprae* are found in large numbers in the mononuclear cells of the subcutaneous tissues, lymph nodes, and nodules in the liver and lungs. Because of the similarity of the disease to human leprosy, it was thought at one time that rats could be a potential source of the human disease. The geographic distribution of rat leprosy, however, does not correspond with the distribution of human leprosy. DNA studies have revealed that *M. lepraemurium* and *M. leprae* are not related species but that there is a relatedness between *M. lepraemurium* and *M. avium.*

M. lepraemurium can be maintained for months in tissue cultures of monocytes, where it has a generation time of about 7 days. It has also been cultured in rat fibroblasts, and *in vitro* in a cysteine-containing medium. *M. lepraemurium* has provided a useful model system for the study of host-parasite relationships of an intracellular parasite.

KNOW MORE

- The word *leper* comes from a Greek word meaning scaly. In India, leprosy is known since ancient times as *kustha roga* and attributed to punishment or curse from God. As early as 1400 BC, reference to it as an old disease in India may be found in the sacred Hindu writings of the Veda.
- Armadillos are naturally infected and represent an indigenous reservoir. Lepromatous form of disease, but not the tuberculoid form, is highly infectious. Person-to-person spread by direct contact or inhalation of infectious aerosols. People in close contact with patients who have lepromatous disease are at greatest risk. It has been suggested but not proved that insect vectors may have a role in transmission of leprosy. Mosquitoes which had fed on lepromatous patients were shown to be capable of infecting mice but it is not known whether this holds good for human beings.

KEY POINTS

- *Mycobacterium leprae* are weakly gram-positive, strongly acid-fast bacilli with parallel sides and rounded ends. With **Ziehl-Neelsen stain**, they are less acid-fast than tubercle bacilli, so 5 percent sulfuric acid is employed for decolorization after staining with carbol fuchsin. The bacilli are seen singly and in groups, intracellularly or lying free outside the cells. Inside the cells they are present as bundles of organisms bound together by a lipid-like substance, **the glia.** These masses are known as *globi.* The parallel rows of bacilli in the globi give appearance of a *cigar bundle.*

- They fail to grow in cell-free culture media. Animal models for culture include footpads of mice, thymectomized mice, the nine-banded armadillo *(Dasypus novemcinctus),* slender loris, Indian pangolin, and Korean chipmunks.
- **Lipid-rich cell wall.** Lipoarabinomannose-B (LAMB) is a major antigen of *M. leprae.* Phenolic glycolipid 1 (PGL-1) is another antigen which protects lepra bacilli against host cell enzymes. Diagnosis made with specific skin test (tuberculoid form of disease) or acid-fast stain (lepromatous form).

Pathogenesis and Immunity

The lepra bacilli are mainly virulent due to their capabilities for intracellular multiplication and growth, and host immune response that influences the clinical form of the disease. Disease primarily from host response to infection.

Lepromin test is a skin test for delayed hypersensitivity for studying immunity in leprosy. It is used to classify leprosy, to assess the prognosis and response to treatment, to assess the resistance of individuals to leprosy and to verify the identity of candidate leprabacilli

Epidemiology

Leprosy is an exclusively human disease. Person-to-person spread by direct contact or inhalation of infectious aerosols. Lepromatous form of disease, but not the tuberculoid form, is highly infectious. People in close contact with patients who have lepromatous disease are at greatest risk.

Diseases

Leprosy is classified into five types (tuberculoid leprosy, borderline tuberculoid leprosy, mid-borderline leprosy, borderline lepromatous leprosy, and lepromatous leprosy).

Diagnosis

Specimens are collected from the nasal mucosa, skin lesions and ear lobules. Biopsy of the nodular lesions and thickened nerves, and lymph node puncture may be necessary in some cases.

Material from the skin is obtained from an active lesion, and also from both the ear lobes by the "slit and scrape" method.

Microscopy is sensitive for the lepromatous form but not the tuberculoid form. ZN staining using 5 percent instead of 20 percent sulfuric acid is useful to demonstrate red-colored acid-fast bacilli in stained smear.

Culture cannot be used. Skin testing required to confirm tuberculoid leprosy. The serology is useful primarily in patients with untreated lepromatous leprosy. PCR is used to monitor treatment, diagnose relapses, or determine the need for chemotherapy.

Treatment, Prevention, and Control

Dapsone with or without rifampin is used to treat the tuberculoid form of disease; clofazimine is added for the treatment of the lepromatous form. Therapy is prolonged. Dapsone is recommended for long-term prophylaxis in treated patients. Disease is controlled through the prompt recognition and treatment of infected people. The vaccines *(Mycobacterium* ICRC vaccine, BCG vaccine; the *Mycobacterium* w vaccine; the BCG plus heat-killed *M. leprae, M. tufu,* and *M. habana* vaccine) have been evaluated against leprosy with limited success.

IMPORTANT QUESTIONS

1. Describe mycobacteria. Discuss the etiology, pathogenesis and laboratory diagnosis of leprosy.
2. Write short notes on:
   - Cultivation of leprae bacilli.
   - Animal models in leprosy.
   - Lepromatous leprosy.
   - Tuberculoid leprosy.
   - Differences between lepromatous and tuberculoid leprosy.
   - Lepromin test.
   - *Mycobacterium lepraemurium.*

FURTHER READING

Dharmendra. Leprosy, vol. I and 2. Bombay: Samant & Co 1985.

Fine PEM. Leprosy: the epidemiology of a slow bacterium. Epidemiol Rev 1982;4:161.

Hastings RC, et al. Leprosy Clin Microbiol Rev 1988;1:330-348.

CHAPTER 34

Nontuberculous Mycobacteria

Learning Objectives

After reading and studying this chapter, you should be able to:

- Classify nontuberculous mycobacteria and examples of different groups of nontuberculous mycobacteria.
- Name the diseases caused by nontuberculous mycobacteria.
- Discuss the clinical significance of nontuberculous mycobacteria.

INTRODUCTION

Mycobacteria other than human or bovine tubercle bacilli, may occasionally cause human disease resembling tuberculosis. This large group of mycobacteria have been known by several names; **atypical, anonymous, unclassified, paratubercle, tuberculoid, environmental or nontuberculous mycobacteria (NTM), opportunistic MOTT (mycobacteria other than tubercle bacilli).** The names **'environmental'** or **'opportunistic mycobacteria'** are better suited as their natural habitat appears to be soil or water and they cause opportunistic infections in human beings. The name **'nontuberculous mycobacteria (NTM)'** has gained wide acceptance in recent years.

NTM are present everywhere in the environment and sometimes colonize healthy individuals in the skin and respiratory and gastrointestinal tracts. In contrast to *M. tuberculosis* complex organisms, NTM are not usually transmitted from person-to-person nor does their isolation necessarily mean that they are associated with a disease process. They are distinct from the saprophytic mycobacteria such as *M. smegmatis* and *M. phlei* which are incapable of infecting human beings or animals. While human infection with them is common in some areas, disease is rare. They are unable to cause progressive disease when injected into guinea pigs.

Differentiating characterstics of *M. tuberculosis* and 'nontuberculous mycobacteria (NTM)' are shown in Table 34.1.

Table 34.1: Differentiating characteristics of *M. tuberculosis* and *nontuberculous mycobacteria (NTM)*

Characteristics	*M. tuberculosis*	*Nontuberculous mycobacteria (NTM)*
1. Culture		
Temperature	37°C	25-45°C
Rate of growth	Slow	Slow or rapid
Growth on LJ medium	Eugonic	Dysgonic
Colony characters	Dry, rough, tough, buff colored, difficult to emulsify	Dry, yellow, orange or creamy and easily emulsifiable
Growth in the presence of p-nitrobenzoic acid (PNB) 500 µg/ml	–	+
2. Biochemical reactions		
Niacin test	+	–
Nitrate reduction test	+	–
3. Animal pathogenicity	Pathogenic to guinea pig	Pathogenic to guinea pig
4. Transmission	Person-to-person	Soil or water

Properties of Nontuberculous Mycobacteria (Table 34.1)

Nontuberculous mycobacteria or atypical mycobacteria show the following features:

1. **Temperature:** They can grow at 25°C, 37°C, and even at 44°C.
2. **Rate of growth:** Some of them are rapid growers. They produce colonies within 1-2 weeks of incubation in the Lowenstein-Jensen (LJ) medium.
3. **Colony characters:** Some of them may produce bright yellow or orange pigments during their growth on the LJ medium.
4. **Staining:** They are acid fast as well as alcohol fast. They *may* differ or resemble in their morphology from those of tubercle bacilli.
5. **Biochemical reactions:** They are arylsulfatase test positive, but are niacin and neutral red reactions negative.
6. **Animal pathogenicity:** They are nonpathogenic for guinea pigs but pathogenic for mouse.
7. **Treatment:** They are usually resistant to antitubercular drugs such as streptomycin, INH, and para-aminosalicylic acid.
8. **Transmission:** Soil or water.
9. **Classification:** They are classified by Runyon into four groups on the basis of their rate of growth and their ability to produce pigments in the presence or absence of light.

CLASSIFICATION

Runyon (1959) classified NTM into four groups (Table 34.2) based on phenotypic characteristics of the various species, most notably pigment (yellow or orange) production and rate of growth. These include:

Table 34.2: Runyon classification scheme of nontuberculous mycobacteria

Runyon group	Name	Species
I	Photochromogens	*M. kansasii,* *M. marinum* *M. simiae,*
II	Scotochromogens	*M. scrofulaceum* *M. gordonae,* *M. szulgai*
III	Nonphotochromogens	*M. avium* *M. intracellulare* *M. xenopi,* *M. ulcerans,* *M. malmoense*
IV	Rapid growers	*M. chelonei* *M. fortuitum*

Group I: Photochromogens.
Group II: Scotochromogens.
Group III: Nonphotochromogens.
Group IV: Rapid growers.

Though other methods of classification have been described, Runyon's classification has found universal acceptance. Species identification depends on several additional characteristics (Table 34.3).

Group 1: Photochromogens

Photochromogens, which are colorless when incubated in the dark, but develop a bright yellow or orange coloration if young cultures are exposed to a light source for an hour or more and then re-incubated for 24-28 hours,. The caps of the culture bottles must be loosened during exposure to light, as oxygen is essential for pigment formation. They are slow growing, though growth is faster than that of tubercle bacilli. The important species in this group are *M. kansasii, M. marinum* and *M. simiae.*

Mycobacterium kansasii

Mycobacterium kansasii, which grows well at 37°C and is principally isolated from cases of pulmonary disease. Natural reservoir is tap water. Pulmonary disease is the most common clinical form of *M. kansasii* infection. It occurs primarily in middle-aged or elderly white men, most of whom have some pre-existing form of lung disease. Infections are more common in cities and in industrial areas such as Midwest United States and the mining areas in Wales.

M. kansasii may also occasionally cause infections of the cervical lymph nodes, penetrating wound infections, and granulomatous synovitis. It can produce generalized infection in HIV-positive patients.

Mycobacterium marinum

M. marinum (previously termed the fish tubercle bacillus) the cause of a warty skin infection known as **swimming pool granuloma**. Microscopically, it resembles *M. kansasii* but can be differentiated by its poor growth at 37°C, and cultures from skin lesions should be incubated at 33°C, negative nitratase, positive pyrazinamide hydrolase and L-fucosidase activity.

Mycobacterium simiae

M. simiae, which, like *M. kansasii,* grows at 37°C and is occasionally involved in pulmonary disease. Several photochromogenic mycobacteria were isolated in 1964 from monkeys exported from India. They have been classified into two species: niacin positive *M. simiae* and niacin negative *M. asiaticum.* They have subsequently been associated with pulmonary disease in human beings.

Group II: Scotochromogens

The scotochromogens are slow-growing NTM whose colonies are pigmented (yellow-orange-red) when grown in the dark or the light. They are widely distributed in the environment and sometimes contaminate cultures of tubercle bacilli.

Table 34.3: Differentiation between tubercle bacilli and some species of atypical mycobacteria

Test	*M. tuberculosis*	*M. bovis*	*M. microti*	*M. kansasii*	*M. marinum*	*M. scrofulaceum*	*M. avium intracellurae-complex*	*M. fortuitum*	*M. chelonei,*	*M. phlei*	*M. smegmatis*
Growth in 7 days	–	–	–	–	–	–	–	+	+	+	+
Growth at 25°C	–	–	–	+	+	+	±	+	+	+	
Growth at 37°C	+	+	+	+	±	+	+	+	+	+	+
Growth at 45°C	–	–	–	–	–	±	–	–	–	+	+
Pigment in light	–	–	–	+	+	+	–	–	–	+	–
Pigment in dark	–	–	–	–	–	+	–	–	–	+	–
Niacin	+	–	±		–	–	–	–	–	–	–
Nitrate reduction	+	–	–	+	–	–	–	+	–	–	+
Urease	+	+	+	–	+	+	–	+	+	+	+

Mycobacterium scrofulaceum

M. scrofulaceum, as its name suggests, is principally associated with **scrofula or cervical lymphadenitis** in children, but also causes pulmonary disease.

Mycobacterium gordonae

M. gordonae (formerly *M. aquae),* frequently found in water and a common contaminant of clinical material, is a rare cause of pulmonary disease. It differs from *M. scrofulaceum* in failing to hydrolyse urea, nicotinamide and pyrazinamide.

Mycobacterium szulgai

M. szulgai, an uncommon cause of pulmonary disease and bursitis. It is a scotochromogen when incubated at 37°C but a photochromogen at 25°C.

Group III: Nonphotochromogens

The nonphotochromogens are slow-growing NTM whose colonies produce no pigment whether they are grown in the dark or the light. Colonies may resemble those of tubercle bacilli.

Of the organisms classified in this group, those belonging to nonpathogenic for humans are *M. terrae* complex (*M. terrae, M. triviale,* and *M. nonchromogenicum)* and *M. gastri.* Medically important species are *M. avium. M. intracellulare, M. xenopi* and the skin pathogen *M. ulcerans.* The most prevalent and important opportunistic pathogens of man are: *M. avium* and *M. intracellulare.*

Mycobacterium avium

M. avium (the avian tubercle bacillus) which causes natural tuberculosis in birds and lymphadenopathy in pigs is one of the most common opportunist human pathogens. The closely related *M. intracellulare* is commonoly known as the **'Battey bacillus'** because it was first identified as a human pathogen at the Battey State Hospital for Tuberculosis, Georgia, USA. *M. avium* and *M. intracellulare* are so closely similar that these two species are usually grouped together as the ***M. avium* complex (MAC)**.

Organisms of the *M. avium* complex cause tuberculosis in birds and lymphadenitis in pigs as well as occasional disease in various other wild and domestic animals. **In man,** they are responsible for lymphadenopathy, pulmonary lesions and disseminated disease, notably in patients with the acquired immune deficiency syndrome (AIDS).

Mycobacterium xenopi

M. xenopi is probably the most easily recognized of potential mycobacterial pathogens. It was first isolated from a skin lesion in a South African toad *(Xenopus laevis).* It is unique among mycobacteria in that it grows poorly at 37°C, is a thermophile and grows well at 45°C. It has been isolated from water, from both hot and cold taps, and from granulomatous lesions in swine. *M. xenopi* produces a chronic slowly progressive pulmonary disease, which is clinically and radiologically similar to tuberculosis. Two phylogenetically similar species, *M. celatum* and *M. branderi,* have been described.

Mycobacterium malmoense

M. malmoense grows very slowly, often taking as long as 10 weeks to appear on primary culture. It causes pulmonary disease and lymphadenitis. It was first isolated from patients from Malmo in Sweden. For unknown reasons, isolation of this pathogen is increasing in several European countries. It is resistant to isoniazid and rifampicin and sometimes also to streptomycin and ethambutol.

Mycobacterium ulcerans

The bacillus grows on Lowenstein-Jensen medium slowly, in 4-8 weeks. The temperature of incubation is critical; growth occurs between 30°C and 33°C but not at 25°C or 37°C. Colonies are nonpigmented or a pale lemon-yellow color Inoculation into the foot pad of mice leads to edema of the limb though ulceration is infrequent. A toxin is produced by *M. ulcerans* that causes inflammation and necrosis when injected into the skin of guinea pigs. This is the only known instance of toxin produced by any mycobacterium species.

Other Nonchromogens

M. shinshuense, M. paratuberculosis, M. sylvaticum, M. lepraemurium, M. terrae, M. nonchromogenicum and *M. triviale* and *M. haemophilum are* other nonchromogens.

Group IV: Rapid Growers

This is a heterogeneous group of mycobacteria capable of rapid growth, colonies appearing within seven days of incubation at 37°C or 25°C. Within the group, photochromogenic, scotochromogenic, and non-chromogenic species occur. Most of these are purely environmental saprophytes. Only two of the rapidly growing species are well recognized human pathogens: *M. chelonae and M. fortuitum.*

Mycobacterium chelonae

M. chelonae (the turtle tubercle bacillus), some isolates of which are sometimes classified as *M. abscessus. M. chelonae* grows better at 25°C than at 37°C.

Mycobacterium fortuitum

M. fortuitum (the frog tubercle bacillus), some isolates of which are sometimes classified as *M. peregrinum. M. fortuitum* can further be differentiated from *M. chelonei* in reducing nitrate and assimilation of iron from ferric ammonium citrate.

Disease

Both species are nonchromogenic. They occasionally cause pulmonary or disseminated disease but are principally responsible for postinjection abscesses and wound infections, including corneal ulcers. Outbreaks of abscesses following injection of vaccines and other preparations contaminated by these mycobacteria have been reported on a number of occasions. The bacilli are found in the soil, and infection usually follows some injury.

M. fortuitum and *M. chelonei* do not produce any pigment. Pulmonary *M. fortuitum* infection cannot be distinguished by X-ray from typical tuberculosis. *M. fortuitum, M. chelonei* organisms are highly resistant *in vitro* to antituberculosis drugs. If drug therapy fails, surgical resection may be necessary.

Some noncultivable or poorly growing mycobacteria identified from the blood of AIDS patients have been characterised by their 16S RNA base sequences. They grow sparsely in some liquid media. Examples are *M. genavense, M. confluentis* and *M. intermedium.*

SAPROPHYTIC MYCOBACTERIA

All the chromogenic rapid growers are saprophytes (for example, *M. smegmatis, M. phlei). Mycobacterium gordonae* and *Mycobacterium terrae* are among the saprophytic species that have been associated with human disease on rare occasions.

Mycobacterium smegmatis

Since *M. smegmatis* is normally present in smegma, a whitish secretion around the orifice of urethra, it is a frequent contaminant of urine specimens. Some strains of *M. smegmatis* are acid-fast but not alcohol-fast. Therefore, they are not seen in a Ziehl-Neelsen smear if acid alcohol is used as decolorizer. Other strains are both acid and alcohol-fast. In such cases, rapid growth on LJ medium and guinea-pig inoculation distinguishes it from *M. tuberculosis.*

Mycobacterium phlei

M. phlei is rarely encountered and is nonpathogenic. It can be differentiated from *M. smegmatis* by its ability to grow at 52°C and survive heating at 60°C for 4 hours. Two species, *Mycobacterium smegmatis* and *Mycobacterium phlei,* do not cause disease.

PATHOGENESIS

Environmental mycobacteria are of low virulence compared with tubercle bacilli, although man is frequently infected, overt disease is very uncommon except in those who are profoundly immunosuppressed. Four main types of opportunist mycobacterial disease have been described in man (Table 34.4).

A. Localized lymphadenitis
B. Skin lesions following traumatic inoculation of bacteria.
C. Tuberculosis-like pulmonary lesions
D. Disseminated disease.

Lymphadenitis

In most cases a single node, usually tonsillar, is involved, and most patients are children aged less than 5 years. The *M. avium* complex is the predominant cause worldwide. Some reports claim a high incidence of *M. scrofulaceum,* but these strains were probably misidentified members of the *M. avium* complex.

Skin Lesions

Three main types have been described:

1. Postinjection (and posttraumatic) abscesses
2. Swimming pool granuloma
3. Buruli ulcer.

Postinjection Abscesses

These are usually caused by the rapidly growing pathogens *M. chelonae* and *M. fortuitum.* Abscesses occur sporadically, particularly in the tropics, or in small epidemics when batches of injectable materials are contaminated by these bacteria.

Table 34.4: Principal types of opportunist mycobacterial disease in man and the usual causative agents	
Disease	*Usual causative agent*
A. Lymphadenopathy	*M. avium complex*
	M. scrofulaceum
B. Skin lesions	
1. Post-trauma abscesses	*M. chelanae*
	M. fortuitum
	M. terrae
2. Swimming pool granuloma	*M. marinum*
3. Buruli ulcer	*M. ulcerans*
C. Pulmonary disease	*M. avium complex*
	M. kansasii
	M. xenopi
	M. malmoense
D. Disseminated disease	
1. AIDS-related	*M. avium complex*
	M. genevense
2. Non-AIDS-related	*M. avium complex*
	M. chelonae

Absesses are painful and last for many months. Infections by *M. terrae* have occurred in farmers and others who have been injured while working with soil.

Swimming Pool Granuloma

It is caused by *M. marinum* and is also known as *fish tank granuloma* and *fish fancier's finger*. *M. marinum* is a natural pathogen of cold-blooded animals, causing tuberculosis in fish and amphibia. It may also occur as a saprophyte in fresh or salt water. Human infection originates from contaminated swimming pools or fish tanks. The bacilli enter scratches and abrasions and cause warty lesions similar to those seen in skin tuberculosis. The lesion, beginning as a papule and breaking down to form an indolent ulear, usually follows abrasions and therefore occurs on the prominences—elbows, knees, ankles, nose, fingers or toes.

It was first described from Sweden under the name **'swimming pool granuloma',** and the bacillus was named *M. balnei* (from balneum, meaning bath). It has since been reported from other European countries and from North America. Its distribution is in temperate areas in contrast to *M. ulcerans,* which has a tropical prevalence. The disease is usually self-limiting although chemotherapy with minocycline, cotrimoxazole or rifampicin with ethambutol hastens its resolution.

Buruli Ulcer

This disease, caused by *M. ulcerans,* was first described from human skin lesions in Australia (1948). It has subsequently been recovered from similar lesions in several tropical countries, notably Uganda (Buruli ulcer), Congo, Nigeria, Ghana, Zaire, Mexico, Malaysia and Papua New Guinea. The name is derived from the Buruli district of Uganda where a large outbreak was extensively investigated.

Ulcers are usually seen on the legs or arms and are believed to follow infection through minor injuries. After an incubation period of a few weeks, indurated nodules appear, which break down forming indolent ulcers which slowly extend under the skin. Initially, smears from the edge of the ulcer show large clumps of bacilli which are acid-fast and alcohol fast. Later, the immunoreactive phase sets in and the bacilli disappear. The ulcers then heal with disfiguring scars.

PULMONARY DISEASE

These infections resemble tuberculosis. In most but not all cases, there is some predisposoing lung disese.This is most frequently seen in middle-aged or elderly men with lung damage. The disease may be caused by many species, but the most frequent are the *M. avium* complex and *M. kansasii.*

Diagnosis is made bacteriologically. Organism must be isolated repeatedly from the sputum to diffrentiate true disease from transient colonization.

Table 34.5 shows the range of human infections produced by different species of atypical mycobacteria.

DISSEMINATED DISEASE

Up to a half of all persons dying of AIDS in the USA had disseminated mycobacterial disease in the 1980s and early 1990s, almost always due to the *M. avium* complex.

LABORATORY DIAGNOSIS

1. Specimen

Sputum, pus or exudates.

2. Microscopy

Ziehel-Neelsen staining of smear shows acid fast bacilli. Repeated smear examination is necessary.

3. Culture

They grow well on LJ medium. Several LJ media should be inoculated with the specimen. These are incubated in the dark and in the light at different temperatures for distinguishing the species.

4. Identification

There is no universally recognized identification scheme, although reliance is usually placed on cultural characteristics (rate and temperature of growth and pigmentation), various biochemical reactions and resistance to antimicrobial agents (Table 34.3). The most discriminative methods are based on the detection of sequence differences in 16S ribosomal RNA.

NTM	Species	Natural habitat	Types of infection
Photochro-mogens	M. kansasii	Water, animals	Pulmonary, systemic, skin joints, lymph nodes
	M. simiae	primates, water	Pulmonary
	M. marinum	Aquarium water, fish	Cutaneous (swimming pool granuloma), joints
	M. asiaticum	Primates	Pulmonary
Scotochro-mogens	M. scrofulaceum	Soil, water, fomites	Lymphadenitis (usually cervical); pulmonary disseminated
	M. szulgai	Water and soil	Pulmonary, lymphadenitis, cutaneous, subcutaneous bursitis cutaneous
Non-photo-chrogens	M. avium inintracellulare	Soil, seawater animals	Pulmonary, systemic, gastrointestinal, lymphadenitis
	M. xenopi	Soil, water	Pulmonary, epididymitis
	M. ulcerans	Unknown	Cutaneous
	M. malmoense	Unknown	Pulmonary
	M. haemophilum	Unknown	Cutaneous, subcutaneous
	M. schimoidei	Unknown	Pulmonary
Rapidly growers	M. chelonae s.s. ehelonae	Soil, seawater, animals	Porcine heart valves, surgical wound, pulmonary
	M. chelonae s.s. ehelonae	Soil, seawater, animals	Cutaneous, surgical abscess,wound pulmonary-systemic
	M. fortuitum	Water, soil animals	Pulmonary, surgical wound cutaneous, systemic, bone and joint

Table 34.5: Atypical mycobacteria associated with human diseases

EPIDEMIOLOGY

Environmental mycobacteria are widely distributed in nature. Infection with them is quite common, from soil, water and air. Infection is mainly asymptomatic, though it may result in sensitization, causing weak positive Mantoux reaction, due to cross reaction with tubercle bacillus protein. In countries in which tuberculosis is 'uncommon, opportunist mycobacterial infections are relatively common. In addition, the absolute incidence is increasing as a result of the growing number of immunocompromised individuals, notably patients with AIDS. Some opportunist species may colonize tap water. When staining reagents were prepared from contaminated water, false-positive sputum smear examinations for acid-fast bacilli have occurred.

Treatment

Most environmental mycobacteria are resistant to the usual antituberculosis drugs although infections often respond to various combinations of these drugs. Pulmonary disease caused by *M. avium* complex or *M. kansasii* may respond to prolonged treatment with rifampicin, isoniazid and ethambutol.

KNOW MORE

Swimming polgramuloma

Bacilli are scanty in smears from ulcers. This organism will grow only at 30°C on primary culture but will grow at 37°C on subculture. It is a photochromogen and colonies are nonpigmented in the dark. However, they become intense orange yellow to red on exposure to light. *M. marinum* is not pathogenic for guinea pigs but intradermal inoculation in rabbits leads to a superficial granulomatous lesion. Footpad inoculation in mice leads to a more severe lesion than with *M. ulcerans,* local inflammation being followed by a purulent ulcer formation.

KEY POINTS

- Mycobacteria other than human or bovine tubercle bacilli, may occasionally cause human disease resembling tuberculosis have been known by several names; atypical, anonymous, unclassified, paratubercle, tuberculoid, environmental or nontuberculous mycobacteri (NTM), opportunistic MOTT (mycobacteria other than tubercle bacilli).

Their natural habitat appears to be soil or water and they cause opportunistic infections in human beings.

- *Classification:* Runyon (1959) classified NTM into four groups:

  Group I photochromogens: Photochromogens produce pigment when exposed to light, e.g. *M. kansasii.*

  Group II scotochromogens: Scotochromogens produce pigment in the dark. e.g. *M. scrofulaceum.*

  Group III nonphotochromogens: Nonphotochromogens do not form pigment even on exposure to light, e.g. *M. avium* and *M. intracellulare.*

  Group IV rapid growers: Rapid growers may be photo, scoto or nonphotochromogens. *M. chelonei* is an example of rapid growers which can grow rapidly (in four or five days) on culture medium.
- *Disease:* Four types of opportunist mycobacterial disease in man: A. Localized lymphadenitis; B. Skin lesions following traumatic inoculation of bacteria; C. Tuberculosis-like pulmonary lesions; D. Disseminated disease.
- *Laboratory diagnosis:* Ziehel-Neelsen staining of smear shows acid fast bacilli. They grow well on LJ medium. They also grow in the presence of p-nitrobenzoic acid (PNB) 500 µg/ml. This feature differentiates N1M from *M. tuberculosis* which do not grow in the presence of PNB.

IMPORTANT QUESTIONS

1. Discuss the classification of atypical mycobacteria and name the diseases caused by these bacteria.
2. Write short notes on:
   - Atypical mycobacteria.
   - Differentiation of typical mycobacteria from atypical mycobacteria
   - Photochromogenic atypical mycobacteria
   - Scotochromogenic atypical mycobacteria
   - *Mycobacterium ulcerans*
   - Mycobacteria causing skin ulcers
   - Swimming pool granuloma.

FURTHER READING

Falkinham JO. Epidemiology of infection by nontuberculous mycobacteria, Clin Microbial Rev 1996;9:173.

Wallace RJ Jr, O Brien R, et al. Diagnosis and treatment of disease caused by nontuberculous mycobacteria. Am Rev Respir Dis 1990;142:940-53.

Wansbrough-Jones MH. Non-tuberculous or atypical mycobacteria. In: James DG, Zumla A (Eds). The Granulomatous Disorders. Cambridge: Cambridge University Press 1999; 189-204.

CHAPTER 35

Actinomycetes, Nocardia

Learning Objectives

After reading and studying this chapter, you should be able to:

- Describe morphology of actinomyces.
- Explain clinical forms of actinomyces.
- Discuss laboratory diagnosis of actinomycosis.
- Differentiate between the genera *Actinomyces* and *Nocardia*.
- Describe the following: Nocardiosis; mycetoma.

INTRODUCTION

Gram-positive bacteria with branching filaments that sometimes develop into mycelia are included in the rather loosely defined order Actinomycetales. Actinomycetes are traditionally considered to be transitional forms between bacteria and fungi. They form a mycelial network of branching filaments like fungi but, like bacteria, they are thin, possess cell walls containing muramic acid, have prokaryotic nuclei and are susceptible to antibacterial antibiotics. They are therefore true bacteria, bearing a superficial resemblance to fungi. Actinomycetes are related to mycobacteria and corynebacteria.

They are gram-positive, nonmotile, nonsporing, noncapsulated filaments that break up into bacillary and coccoid elements. Although mostly soil saprophytes, occasionally cause chronic granulomatous infections in animals and man.

Important Genera

The family *Actinomycetes* contains three major medically important genera, *Actinomyces*, *Nocardia* and *Actinomadura*. Another genus, *Streptomyces* rarely causes disease in man, but its medical importance lies in the production of antibiotics by its several species. *Tropheryma whippelii* is a newly recognized species thought to be an actinomycete on the basis of nucleic acid studies. In addition, inhalation of some thermophilic actinomycetes such as *Micropolyspora faeni* and *Thermoactinomyces* sp. may cause allergic alveolitis (farmer's lung and bagassosis) in those who are occupationally exposed to mouldy vegetable matter.

Actinomyces is anaerobic or microaerophilic and non-acid-fast, while *Nocardia* is acid-fast and aerobic. *Streptomyces* and *Actinomadura* are non-acid-fast and aerobic.

ACTINOMYCES

A mould-like organism in the lesion of 'lumpy jaw' (actinomycosis) in cattle was found by Bollinger (1877). The name actinomyces was coined by Harz to refer to the raylike appearance of the organism in the granules that characterize the lesions (*actinomyces*, meaning ray fungus). Wolff and Israel (1891) isolated an anaerobic bacillus from human lesions and produced experimental infection in rabbits and guinea pigs. This was named *Actinomyces israelii*. It causes human actinomycosis. Actinomycosis in cattle is produced by *A. bovis*.

Morphology

Actinomyces are gram-positive, nonmotile, nonsporing, nonacid-fast. They often grow in mycelial forms and break up into coccoid and bacillary forms. Most show true branching.

Cultural Characteristics

They are facultative anaerobes. They grow best under anaerobic or microaerophilic conditions with the addition of 5-10 percent CO_2. The optimum temperature for growth is 35-37°C. They can be grown on brain-heart infusion agar, heart infusion agar supplemented with 5 percent defibrinated horse, rabbit or sheep blood. Suitable liquid media include brain-heart infusion broth and thioglycollate broth which may be supplemented with 0.1-0.2 percent sterile rabbit serum. Most species show good growth after 3 to 4 days incubation, however, *A. israelii* may take 7 to 14 days.

Pathogenesis

Actinomyces colonize the upper respiratory tract, gastrointestinal tract, and female genital tract. These bacteria are not normally present on the skin surface. The organisms have a low virulence potential and cause disease only when the normal mucosal barriers are disrupted by trauma, surgery, or infection.

Actinomycosis*:* The *Actinomyces* cause the disease known as actinomycosis. Actinomycosis is a chronic disease characterized by multiple abscesses and granulomata, tissue destruction, extensive fibrosis and the formation of sinuses. Within diseased tissues, the actinomycetes form large masses of mycelia embedded in an amorphous protein-polysaccharide matrix and surrounded by a zone of gram-negative, weakly acid-fast, club-like structures (Fig. 35.1). The mycelial masses may be visible to the naked eye and are called sulfur granules, as they are often light yellow in color. The sulfur granules may be dark brown and very hard in older lesions because of the deposition of calcium phosphate in the matrix.

In man, actinomycosis is usually caused by *Actinomyces israelii*. Less common causes include *A. gerencseriae, A. naeslundii, A. odontolyticus, A. viscosus, A. meyeri, Propionibacterium propionicum* and members of the genus *Bifidobacterium*. Concomitant bacteria, notably a small gram-negative rod, *Actinobacillus actinomycetemcomitans*, but also *Haemophilus* species, fusiforms and anaerobic streptococci, are sometimes found in actinomycotic lesions but their contribution to pathogenesis of the diseases, if any, is unknown.

Human Actinomycosis

Human actinomycosis may take several forms:

1. **Cervicofacial:** This is the commonest type and it occurs mainly in cheek and submaxillary regions. The disease is endogenous in origin. Dental caries is a predisposing factor, and infection may follow tooth extractions or other dental procedures. Men are affected more frequently than women, and in some regions, the disease is more common in rural agricultural workers than in town dwellers, probably owing to lower standards of dental care in the former.
2. **Thoracic:** Thoracic actinomycosis commences in the lung, probably as a result of aspiration of actinomyces from the mouth, that may involve the pleura and pericardium and spread outwards through the chest wall.
3. **Abdominal:** The lesion is usually around the cecum, with the involvement of the neighboring tissues and the abdominal wall. Sometimes the infection spreads to the liver via the portal vein.
4. **Pelvic:** Pelvic actinomycosis occasionally occurs in women fitted with plastic intrauterine contraceptive devices.
5. **Punch actinomycosis:** It is a rare infection of the hand acquired by injury of the knuckles on an adversary's teeth.

Laboratory Diagnosis

1. Specimens

Pus, sinus discharge, bronchial secretions, sputum or infected tissues are collected aseptically. These specimens may contain innumerable sulfur granules. The granules may also be present on dressings removed from a draining sinus tract.

2. Microscopy

'Sulfur granules' may be demonstrated in pus by shaking it up in a test tube with some saline. On standing, the granules sediment may be withdrawn with a capillary pipette. Granules may also be obtained by applying gauze pads over the discharging sinuses.

The granules are white or yellowish and range in size from minute specks to about 5 mm. Granules are crushed between two slides and stained with Gram and Ziehl-Neelsen staining using 1 percent sulfuric acid for decolorization. Gram staining shows a dense network of thin gram-positive filaments, surrounded by a peripheral zone of swollen radiating club shaped structures, presenting a *sun ray appearance* (Fig. 35.1). The 'clubs' are believed to be antigen-antibody complexes. Acid-fast staining shows central part as nonacid-fast surrounded by acid-fast 'clubs'. In absence of sulfur granules, Gram's staining of pus shows gram-positive branching filaments.

Sulfur granules and mycelia in tissue sections can also be identified by direct fluorescence microscopy.

3. Culture

Sulfur granules or pus containing actinomycetes are washed and inoculated into thioglycollate liquid medium or streaked on brain-heart infusion agar (BHI agar), blood agar and incubated anaerobically at 37°C. *A. bovis* produces general turbidity whereas *A. israelii* grows as fluffy balls at the bottom of the tube. On solid media,

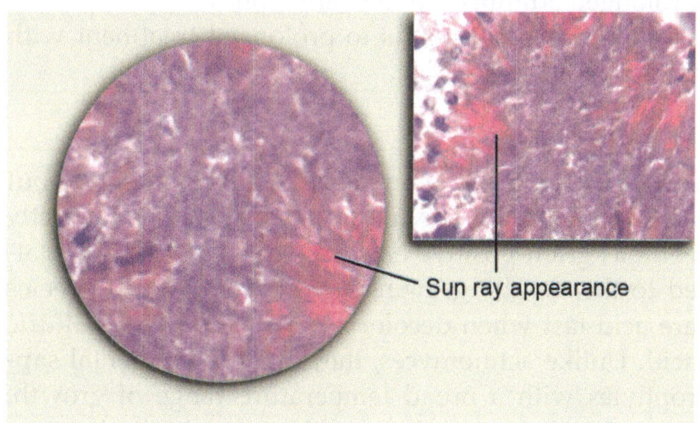

Fig. 35.1: Sulfur granule. Section of tissue showing an actinomycotic colony, the clubs at the periphery giving a 'sun ray' appearance

A. *israelii* may form so-called *spider colonies* that resemble molar teeth in 48-72 hours that become heaped up, white and irregular or smooth, large colonies in 10 days. Other species have different types of colonies.

4. Identification

The identity may be confirmed by direct fluorescence microscopy and biochemical tests or by gas chromatography of metabolic products of carbohydrate fermentation. Gel diffusion and immunofluorescence can differentiate *A. israelii* from other actinomycete species and from other filamentous anaerobes that may produce granules in tissues.

5. Biopsy

In hematoxylin and eosin stained sections, the sulfur granules are deeply stained with hematoxylin except in the periphery which is stained by eosin, which shows short, radiate, club-like structures. On Gram staining, the filaments are gram-positive and periphery gram-negative. The tissue reaction is a chronic suppurative, fibrosing, inflammatory process.

Epidemiology

Actinomycosis is an endogenous infection with no evidence of person-to-person spread or disease originating from an external source such as soil or water. The disease occurs throughout the world but its incidence in the advanced countries has been declining probably as a result of the widespread use of antibiotics. Actinomycosis is more common in rural areas and in agricultural workers. Young male persons (10-30 years old) are most commonly affected. The reason for this predisposition is not known. About 60 percent of the cases are cervicofacial and some 20 percent abdominal. Pelvic actinomyces is seen mainly in women using intrauterine devices.

Treatment

Treatment for actinomycosis involves the combination of surgical debridement of the involved tissues and the prolonged administration of antibiotics.

The disease responds to prolonged treatment with penicillin or tetracycline.

NOCARDIA

Nocardia resemble Actinomycetes morphologically but are strict aerobic. The nocardiae are branched, strictly aerobic, gram-positive bacteria, which are closely related to the rapidly growing mycobacteria. Most species are acid-fast when decolorized with 1 percent sulfuric acid. Unlike actinomyces, they are environmental saprophytes with a broad temperature range of growth. Nocardia are frequently found in soil and infection may be exogenous. Differentiating features of *Actinomyces* and *Nocardia* are shown in Table 35.1.

Table 35.1: Differences between the genera Actinomyces and Nocardia

Actinomyces spp.	*Nocardia* spp.
1. Facultative anaerobes	1. Strict aerobes
2. Grow at 35-37°C	2. Wide temperature range of growth
3. Oral commensals	3. Environmental saprophytes
4. Nonacid-fast mycelia	4. Usually weakly acid-fast
5. Endogenous cause of disease	5. Exogenous cause of disease

Species

The species most commonly associated with human diseases are: *N. asteroides, N. brasiliensis, N. farcinica. N. otitidiscaviarum, N. nova* and *N. transvalensis.* Many species of nocardiae are found in the environment, notably in soil, but opportunist disease in man is most always caused by *Nocardia asteroides,* so named because of its star-shaped colonies

Morphology

Nocardiae are gram-positive bacteria and form a mycelium, that fragments into rod shaped and coccoid elements. *Nocardia* resembles *Actinomyces,* but some species are acid-fast, and a few are nonacid-fast.

Cultural Characteristics

They are strict aerobes. Nocardiae readily grow in ordinary media. They are slow growing (require 5-14 days). Nocardiae readily grow on nutrient agar, Sabouraud dextrose agar, brain-heart infusion agar and yeast extract-malt extract agar. The inoculated plates should be incubated at 36°C for up to 3 weeks. They can grow at wide range of temperature. Selective growth is favored by incubation at 45°C. In addition, the technique of *paraffin baiting* may be used. A paraffin wax-coated glass rod is placed in inoculated carbon free broth. Nocardiae grow on the rod at the air-liquid interface and may be subcultured onto agar media.

Pathogenesis

Organisms in the *N. asteroides* complex cause approximately 90 percent of human *Nocardia* infections. They cause bronchopulmonary disease in immunocompromised patients, with a high predilection for hematogenous spread to the central nervous system (CNS) or skin.

Nocardiae produce opportunistic pulmonary disease known as *nocardiosis* in immunocompromised individuals including those with AIDS. Pre-existing lung disease, notably alveolar proteinosis, also predisposes to nocardial disease. Soil is known to be natural habitat of *Nocardia.* Man acquires infection by inhalation of the bacteria from environmental sources. The infection is exogenous, resulting from inhalation of the bacilli.

Bronchopulmonary Disease

Systemic nocardiosis, usually caused by *N. asteroides*, manifests primarily as pulmonary disease, pneumonia, lung abscess or other lesions resembling tuberculosis. Systemic nocardiosis occurs more often in immunodeficient persons.

Cutaneous Infection

Primary or secondary cutaneous infection may lead to mycetoma, lymphocutaneous infections, cellulitis, subcutaneous abscesses.

Laboratory Diagnosis

Diagnosis is by demonstration of branching filaments microscopically and by isolation in culture.

1. Specimens

Pus or purulent sputum.

2. Microscopy

The smears are stained with Gram staining and Ziehl-Neelsen (ZN) technique using decolorization with 1 percent sulfuric acid. Nocardia stain poorly with Gram stain and are typically partially acid-fast. Gram positive filamentous bacteria can be seen on Gram staining. Acid-fast bacilli are detected on ZN technique though some species are nonacid-fast.

Nocardiae are not so easily seen in tissue biopsies stained by the Gram or modified Ziehl-Neelsen methods but they may be seen in preparations stained by the Gomori methenamine silver method.

3. Culture

The specimens are inoculated on nutrient agar, Sabouraud's dextrose agar (SDA) and brain-heart infusion agar (BHI agar) and incubated at 36°C for 3 weeks. Colony morphology is seen and bacteria are identified by staining.

Nocardia can be isolated from sputum by *paraffin bait* technique. The specimen is homogenized with sterile glass beads and 2 ml of it is inoculated into carbon-free broth containing paraffin coated glass rod. The organisms grow on the rod at the air-liquid surface which may be subcultured onto agar media.

4. Identification

Colonies of nocardiae are cream, orange or pink colored. Their surfaces may develop a dry, chalky appearance and they adhere firmly to the medium. Identification of species is not easy and is usually undertaken in reference laboratories by sequence analysis of the 16S ribosomal RNA (ribotyping).

Treatment

Nocardia infections are treated with combination of antibiotics and appropriate surgical intervention. Sulphonamide are the antibiotics of choice. They are also susceptible to amikacin, imipenem, minocycline, tobramycin and vancomycin.

ACTINOMYCOTIC MYCETOMA

Mycetoma is a localized chronic, granulomatous involvement of the subcutaneous and deeper tissues, commonly affecting the foot and less often the hand and other parts. It presents as a tumour with multiple sinuses. This clinical syndrome was first described from Madura by Gill (1842) and came to be known as *Maduramycosis*. The disease is worldwide but common in tropical countries (Sudan, North Africa and West Coast of India). The disease occurs among bare footed persons who are prone to contamination by soil derived organisms.

Etiology

Mycetomas are usually caused by fungi but may be caused by bacteria as well. Even *Staph. aureus* and other pyogenic bacteria may occasionally cause a mycetoma-like lesion **(botryomycosis)**. Bacterial mycetomas are usually caused by actinomycetes—*Actinomyces (A. israelii, A. bovis), Nocardia (N. asteroides, N. brasiliensis, N. caviae), Actinomadura (A. madurae, A. pelletierii), Streptomyces (S. somaliensis).*

Diagnosis

Etiological diagnosis of mycetoma is important in choosing appropriate treatment. The color of the granules gives some indication. In actinomycotic mycetoma, the granules are white to yellow, while in eumycotic mycetomas, the granules are generally black. Examination of crushed smears of the granules helps to differentiate actinomycotic from mycotic mycetomas. In the former, the filaments are thin (about 1 µm), while in the latter they are stout (about 4-5 µm). Isolation of the agent in culture establishes the diagnosis.

Actinomycetes and Hypersensitivity Pneumonitis

Spores of some thermophilic actinomycetes such as *Faenia* and *Saccharomonospora* species, present in mouldy hay, when inhaled may induce allergic alveolitis leading to chronic obstructive pulmonary disease (COPD, farmer's lung).

KNOW MORE

The lymphatics are not usually involved in actinomycosis, but hematogenous spread to the liver, brain and other internal organs occasionally occurs. Involvement of bone is much less common in man than in animals and is usually the result of direct extension of adjacent soft tissue lesions.

KEY POINTS

- Actinomycetes are traditionally considered to be transitional forms between bacteria and fungi.

- Actinomycetes are gram-positive, nonmotile, nonsporing, noncapsulated filaments that break up into bacillary and coccoid elements. Most are free living, particularly in the soil.
- The family *Actinomycetes* contains important genera, *Actinomyces, Nocardia, Actinomadura, Streptomyces, Tropheryma whippelii* (a newly recognized species).
- *Actinomyces* is anaerobic or microaerophilic and nonacid-fast, while *Nocardia* is acid-fast and aerobic. *Streptomyces* and *Actinomadura* are nonacid-fast and aerobic.
- **Clinical forms of Actinomyces:** The *Actinomyces* causes the disease known as actinomycosis. The disease occurs in five clinical forms: cervicofacial, thoracic, abdominal, pelvic and punch actinomycosis.
- **Nocardia:** Nocardia resemble Actinomycetes morphologically but are aerobic. Most species (such as *N. asteroides* and *N. brasiliensis*) are acid-fast when decolorized with 1 percent sulfuric acid.
- **Pathogenesis:** *Nocardia* species cause primary cutaneous nocardiosis, bronchopulmonary infection, and secondary CNS infection.
- **Actinomycotic mycetoma:** Mycetoma is a localized chronic, granulomatous involvement of the subcutaneous and deeper tissues, commonly affecting the foot and less often the hand and other parts. Mycetomas are usually caused by fungi but may be caused by bacteria as well.

IMPORTANT QUESTIONS

I. Write short notes on:
   a. Actinomycosis
   b. Laboratory diagnosis of actinomycosis
   c. Nocardiosis
   d. Laboratory diagnosis of nocardiosis
   e. Mycetoma.

FURTHER READING

Collins CH, et al. Presumptive identification of nocardias. J Applied Bact 1988;65:55.

Curry WA. Human nocardosis—a clinical review with selected case reports. Archives of lnternal Medicine 1980;140:818-826.

Schaal KP, Lee HJ. Actinomycete infections in humans—a review. Gene 1992;115:201-11.

CHAPTER 36

Enterobacteriaceae: Escherichia, Klebsiella, Proteus and Other Genera

Learning Objectives

After reading and studying this chapter, you should be able to:

- Describe general characters of the family Enterobacteriaceae.
- Classify the family Enterobacteriaceae.
- Describe morphology, culture characteristics and biochemical reactions of *Escherichia coli*.
- Discuss pathogenicity of *Escherichia coli*.
- Discuss various groups of *Escherichia coli* producing diarrhea.
- Differentiate between heat labile toxin (LT) and heat stable toxin (ST) of *E. coli*.
- Discuss laboratory diagnosis of urinary tract infections caused by *Escherichia coli*.
- Discuss morphology, culture characteristics and biochemical reactions of Klebsiella sp.
- Differentiate between *E. coli* and *Klebsiella* sp.

INTRODUCTION

The family Enterobacteriaceae is the largest, most heterogenous collection of medically important gram-negative bacilli. It is composed of a large number of closely related species that are found in soil, water, decaying matter, and the large intestines of humans, animals, and insects. Because of their normal habitat in humans, these organisms are referred to as the **"enteric bacilli"** or **"enterics"**.

CHARACTERISTICS OF THE FAMILY ENTEROBACTERIACEAE

Members of the family Enterobacteriaceae have the following characteristics:

i. They are gram-negative bacilli.
ii. They are aerobes or facultative anaerobes and grow readily on ordinary laboratory media including MacConkey's lactose bile-salt agar.
iii. All species ferment glucose with the production of acid or acid and gas.
iv. They are either nonmotile or motile with peritrichous flagella.
v. They are catalase positive (except for *Shigella dysenteriae* type 1 which is catalase-negative)
vi. They are oxidase-negative.
vii. They reduce nitrate to nitrites (with the exception of certain biotypes of *Enterobacter agglomerans* and some strains of *Erwinia* and *Yersinia* which do not reduce nitrate to nitrites).
viii. They are typically intestinal parasites of humans and animals, though some species may occur in other parts of the body, on plants and in soil.

Taxonomy of the Enterobacteriaceae

In 1972, Edwards and Ewing described 11 genera and 26 species belonging to the *Enterobacteriaceae.* Currently a total of 32 genera and more than 130 species have been described. These genera have been classified based on biochemical properties, antigenic structure, and nucleic acid hybridization and sequencing.

CLASSIFICATION OF ENTEROBACTERIACEAE

The classification of *Enterobacteriaceae* has been controversial and there have been successive changes in their grouping and nomenclature.

Lactose Fermentation

The oldest method was to classify these bacteria into three groups based on their action on lactose. **Lactose fermenters (LF), Late lactose fermenters and nonlactose-fermenting (NLF).** The specimen is plated on a medium containing lactose and neutral red indicator. (MacConkey agar). The organisms fermenting the lactose form acid and in acidic pH, neutral red is red in color, therefore, the colonies of lactose-fermenting bacteria are red or pink and those of nonlactose-fermenting (NLF) bacteria are pale. All lactose-fermenting enterobacteria, e.g. *Escherichia, Klebsiella, Enterobacter* and *Citrobacter* are popularly known as **'coliform bacilli'** as the most common member of this group is the colon bacillus or *Escherichia coli.* The major intestinal pathogens, *Salmonella* and *Shigella* are nonlactose-fermenters **(NLF).** There remained a small group which showed delayed fermentation of lactose (2-8 days) and with the exception of *Shigella sonnei,* they were all commensals.

This heterogenous group of late lactose fermenters was called **paracolon bacilli.**

CLASSIFICATION OF ENTEROBACTERIACEAE BY TRIBES

The tribe concept provides with a convenient method of grouping together the major genera within the family that share similar biochemical reactions and are of similar diagnostic importance. The current practice is to group together bacteria that possess a number of common morphological and biochemical properties, and similar DNA base compositions. Three systems of nomenclature have been proposed (Bergey's manual, Kauffmann, Edwards-Ewing) and though they have certain differences, the general approach is the same.

The **family** is first classified into its major subdivision—**group or tribe.** Each tribe consists of one or more **genera** and each genus one or more **subgenera** and **species**. The species are classified into **types—biotypes, serotypes, bacteriophage types, colicin types**. **Table 36.1** lists the family *Enterobacteriaceae* and their respective tribes and genera. Table 36.2 shows the biochemical features that differentiate different genera of Enterobacteriaceae.

The genus *Yersinia,* including the plague bacillus, has been placed in the family *Enterobacteriaceae* but because of the special importance of plague, the major disease caused by yersinieae and its lack of similarity to enteric disease, it is dealt with separately.

ESCHERICHIA COLI

Introduction

This genus is named after the German pediatrician Theodar Escherich who first identified *Escherichia coli* under the name *Bacterium coli commune* (1885). The genus *Escherichia* consists of five species (*Escherichia coli, E. fergusonii, E. hermanii, E. vulneris and E. blattae*) of which *E. coli* is the most common and clinically most important.

Table 36.1: Classification of the family Enterobacteriaceae

Tribe	*Genus*
i. Escherichieae	1. Escherichia 2. Shigella
ii. Edwardsielleae	Edwardsiella
iii. Salmonelleae	Salmonella
iv. Citrobactereae	Citrobacter
v. Klebsielleae	1. Klebsiella 2. Enterobacter 3. Serratia 4. Hafnia
vi. Proteeae	1. Proteus 2. Morganella 3. Providencia
vii. Yersinieae	Yersinia

Unlike other coliforms, *E. coli* is an obligate parasite living only in the human or animal intestine that cannot live free in nature. Voided in feces, it remains viable in the environment only for some days. Its presence in water supplies, therefore, is evidence of recent fecal contamination with human or animal feces.

Morphology

E. coli is a gram-negative, straight rod measuring 1-3 × 0.4-0.7 µm arranged singly or in pairs. It is motile by peritrichate flagella, though some strains may be nonmotile. Fimbriae and capsules are found in some strains. It is nonsporing and noncapsulated.

Cultural Characteristics

It is an aerobe and a facultative anaerobe. The temperature range is 10-40°C (optimum 37°C). It can grow on ordinary media like **nutrient agar.** Colonies are large, thick, grayish white, moist, smooth opaque or partially translucent disks. This description applies to the smooth (S) form seen on fresh isolation, which is easily emulsifiable in saline. The rough (R) forms give rise to colonies with an irregular dull surface and are often autoagglutinable in saline. The S-R variation occurs as a result of repeated subcultures and is associated with the loss of surface antigens and usually of virulence. Many pathogenic isolates have polysaccharide capsules. Some strains may occur in the 'mucoid' form.

On **blood agar,** many strains, especially those isolated from pathogenic conditions, are hemolytic. On **MacConkey's medium**, colonies are red or pink due to lactose fermentation. On selective media such as **DCA** or **SS agar,** growth is largely inhibited and is used for the isolation of salmonellae and shigellae. In broth, growth occurs as general turbidity and a heavy deposit, which disperses completely on shaking.

Biochemical Reactions

i. *E. coli* ferments glucose, lactose, mannitol, maltose and many other sugars with the production of acid and gas. Typical strains do not ferment sucrose.
ii. Indole and MR positive, and VP and citrate negative (IMViC + + – –).

    The four biochemical tests widely employed in the classification of enterobacteria are the indole, methyl red (MR), Voges-Proskauer (VP) and citrate utilization tests, generally referred to by the mnemonic 'IMViC'.
iii. It is negative for phenylalanine deaminase test, urease test, H_2S production, gelatin liquefication, growth in the presence of KCN, and malonate utilization.

Antigenic Structure

Serotyping of *E. coli* is based on three antigens—the flagellar antigen H, somatic antigen O and the capsular antigen K as detected in agglutination assays with specific rabbit antibodies.

Table 36.2: Important distinguishing features of the different genera of Enterobacteriaceae

Test	Escherichia	Shigella-1	Edwardsiella	Klebsiella	Enterobacter	Serratia	Hafnia	Citrobacter	Salmonella–2	Proteus	Morganella	Providentia
Motility	+	–	+	–	+	+	+	+	+	+	+	+
Gas from glucose	+	–	+	+	+	d	+	+	+	d	+	+
Acid from lactose	+	–	–	+	+	–	–	+	–	–	–	–
Acid from sucrose	d	–	–	+	+	+	–	d	–	d	–	d
Growth in KCN	–	–	–	+	+	+	+	+	d	+	+	+
Indole	+	d	+	–	–	–	–	d	–	d	+	+
MR	+	+	+	–	–	–	–	+	+	+	+	+
VP	–	–	–	+	+	+	+	–	–	–	–	–
Citrate	–	–	–	+	+	+	+	+	+	d	d	d
H_2S	–	–	+	+	–	–	–	+	+	+	–	–
Urease	–	–	–	+	d	–	–	–	–	+	+	d
Phenylalanine deaminase (PPA)	–	–	–	–	–	–	–	–	–	+	+	+
Arginine dehydrolase	d	–	–	–	d	–	–	d	+	–	–	–
Lysine decarboxylase	+	–	+	d	d	+	+	–	+	–	–	–
Ornithine decarboxylase	d	d	+	–	+	+	+	d	+	d	+	

(d = results different in different species or strains).
Important exceptions:
1. *Sh. sonnei* ferments lactose and sucrose late.
2. *S. Typhi* does not produce gas from sugars.

1. H Antigens

These are thermolabile. Most are monophasic but rare diphasic strains have been reported. So far 75 antigens have been identified. There are only a few significant cross-reactions between them and with the H antigens of other members of the *Enterobacteriaceae.* Strains may need to be grown in semisolid agar to induce flagella expression because certain strains of *Esch. coli* cease to express flagella during growth *in vitro.*

2. Somatic Antigen (O Antigen)

These are heat-stable, lipopolysaccharide (LPS) antigens of cell walls. Over 170 different O antigens have been described. Serotyping may detect cross-reactions because of shared epitopes on the LPS expressed by strains of *Esch. coli* and organisms belonging to the genera *Brucella, Citrobacter, Providencia, Salmonella, Shigella* and *Yersinia.* In some instances, the antigens appearing in the different genera are identical.

Several different serotypes of *E. coli* are found in the normal intestine. Most of them do not have K antigens. The normal colon strains belong to the 'early' O groups (1, 2, 3, 4, etc.), while the enteropathogenic strains belong to the 'later' O groups (26, 55, 86, 111, etc).

3. Capsular Antigen (K Antigen)

K antigen refers to the acidic polysaccharide antigen located in the 'envelope' or microcapsule. (K for *Kapsel,* German for capsule). It encloses the O antigen and renders the strain inagglutinable by the O antiserum. It may also contribute to virulence by inhibiting phagocytosis.

In the past, these antigens were divided into three classes—L, A and B (the thermolabile L antigens, the thermostable A and B antigens) according to the effect of heat on the agglutinability, antigenicity and antibody binding power of bacterial strains that express them. Later it was shown that the B antigen was not a separate entity. K antigens are therefore currently classified into two groups, I and II, generally corresponding to the former A and L antigens (Table 36.3).

4. Fimbrial Antigen (F Antigen)

These are thermolabile proteins. Heating the organisms at 100°C leads to detachment of fimbriae. The F antigen has no role in antigenic classification of *Esch. coli.*

Virulence Factors

Two types of virulence factors have been recognized in *E. coli*—surface antigens and toxins.

A. Surface Antigens

1. Somatic Antigen (O Antigen)

The somatic lipopolysaccharide surface O antigen, besides exerting endotoxic activity, also protects the bacillus from phagocytosis and the bactericidal effects of complement.

2. K Antigen

The envelope or K antigens also afford protection against phagocytosis and antibacterial factors in normal serum, though it is not effective in the presence of antibody to O or K antigen. Most strains of *E. coli* responsible for **neonatal meningitis** and **septicemia** carry the KI envelope antigen which is a virulence factor resembling the group B antigen of meningococcci.

3. Fimbriae

Like many other members of the Enterobacteriaceae, strains of *Esch. coli* exhibit common fimbriae which are chromosomally determined, present in large numbers and causing mannose sensitive hemagglutination and probably not relevant in pathogenesis. Filamentous protein structures resembling fimbriae cause mannose-resistant hemagglutination and play an important part in the pathogenesis of **diarrheal disease** and in **urinary tract infection.** They include the **K88 antigen** found in strains causing enteritis of pigs, the **K99 antigen** found in strains causing enteritis of calves and lambs, and the **colonization factor antigens** (CFAs) (CFAI, CFAII, CFA/III expressed by enterotoxigenic *Esch. coli* (ETEC) causing diarrheal disease in humans.

Fimbriae that are of importance in urinary tract infection and cause mannose-resistant hemagglutination are distinguished according to their receptor specificities. These include the *P. fimbriae* that bind specifically to receptors present on the P blood group antigens of human erythrocytes and uroepithelial cells.

B. Toxins

1. Exotoxins

E. coli produce two kinds of exotoxins: *hemolysins* and *enterotoxins.*

Table 36.3: K antigens (group I and group II) of Esch. coli

Properties	Group I	Group II
1. Molecular weight	> 100,000	<50,000
2. O groups	08, 09	Many
3. Acidic component	Hexuronic acid	Glucuronic acid, phosphate, KDO, Neu-NAc
4. Electrophoretic mobility	Low	High
5. Expressed at 17-20°C	Yes	No
6. Chromosome site	His	Ser A

KDO, ketodeoxyoctonate; NeuNAc, N-acetylneuraminic acid.

a. Hemolysins

Hemolysins do not appear to be relevant in pathogenesis though they are produced more commonly by virulent strains than by avirulent strains.

b. Enterotoxins

Enterotoxins are important in the pathogenesis of diarrhea. Three distinct types of *E. coli* enterotoxins have been identified:

i. Heat labile toxin (LT)
ii. Heat stable toxin (ST)
iii. Verotoxin (VT) also known as Shiga-like toxin (SLT).

i. Heat-Labile Toxin (LT)

The *E. coli* enterotoxin LT was discovered in 1956 by De and colleagues in isolates from adult diarrhea cases in Calcutta, by the rabbit ileal loop method which they had earlier used for identifying the cholera enterotoxin (CT), viz. injection of *E. coli* culture filtrates into closed ligated loops of rabbit ileum induced outpouring of fluid and ballooning of the loops.

Mechanism of Action of LT

E. coli LT is closely related to the toxin produced by strains of *Vibrio cholerae* in its structure, antigenic properties and mode of action. There are two main forms, termed LT-I and LT-II. Different forms of LT-I associated with human, porcine and chicken infection have been described. Similarly, two forms of LT-II (LT-IIa and LT- IIb) have been detected.

LT is a complex of polypeptide subunits–each unit of the toxin consisting of one subunit A (A for *active)* and five subunits B (B for *binding).* The toxin binds to the GMl ganglioside receptor on intestinal epithelial cells by means of subunit B, following which the subunit A is activated to yield two fragments—Al and A2. The A1 fragment activates adenyl cyclase in the enterocyte to form cyclic adenosine 5′ monophosphate (cAMP), leading to increased outflow of water and electrolytes into the gut lumen, with consequent diarrhea.

Though the mechanism of action of LT and CT is the same, the latter is about a hundred times more potent than the former. LT is a powerful antigen and can therefore, be detected by a number of serological as well as biological tests (Table 36.4).

ii. Heat-Stable Toxin (ST)

The heat stable toxins of *E. coli* (ST), first identified in 1970, have a low molecular weight which is probably responsible for their heat stability and poor antigenicity in contrast to LTs. There are two major classes, designated ST-I (or ST_a) and ST-II (or ST_b). Variants of ST-I have been associated with porcine and human infections. ST_a but not ST_b is associated with human disease

a. **ST-I (or ST_a):** ST-I (or ST_a) was originally detected by an infant mouse test in which secretion occurs in the intestine within 4 hours after intragastric administration. It can now be detected by immunoassay. STa is a small, monomeric toxin that binds to guanylate cyclase, leading to an increase in the level of cyclic guanosine monophosphate and subsequent hypersecretion of fluids. STI is methanol soluble, is plasmid encoded and these plasmids may also encode the genes for LT, adhesive factors and antibiotic resistance.

b. **ST-II (ST_b):** ST-II (ST_b) is distinguished from ST-I (ST_a) by its biological activity and by its insolubility in methanol. It stimulates fluid accumulation in ligated intestinal loops of young piglets (upto nine weeks) but not in the infant mouse test. The mechanism of action is not known but it appears not to act via cAMP or cGMP. ST genes are carried on plasmids which may also carry other genes, such as for LT and drug resistance. However, ST_a and ST_b genes are not seen to be carried on the same plasmid.

Table 36.4: Differential properties of heat-stable toxins of Esch. coli

	ST-I (ST_a)	*ST-II (ST_b)*
Molecular weight (kDa)	2	5
Infant mouse test	+	–
Methanol	Soluble	Insoluble
Pig intestinal loop	+	+
Rabbit ileal loop	+	–
Rat gut loop	+	–
Action	Activates cyclic guanosine monophosphate	Unknown (cyclic nucleotides)

iii. Verocytotoxin or Verotoxin (VT)

E. coli verocytotoxin or verotoxin (VT) was so named because it was first detected (1977) by its cytotoxic effect on Vero cells, a cell line derived from African green monkey kidney cells. The biological properties, physical characteristics and antigenicity of VT are very similar to those of Shiga toxin (Stx), produced by strains of *Sh. dysenteriae* type 1 so it is also known as 'Shiga-like toxin' (SLT).

VT1 and **VT2 form:** Serological tests have revealed two antigenically distinct forms, termed **VT1** and **VT2.** Antibodies prepared to VT1 neutralize Shiga toxin, while antibodies specific for VT2 do not. Variant forms of VT2 ($VT2_v$) have been described in strains of human and porcine origin. The genes controlling production of these variant toxins are not phage-encoded, and the toxin receptor also differs from that used by VT1 and VT2 (Table 36.5).

Mechanism of Action of VT1 and VT2

Like Shiga toxin, VT1 and VT2 comprise one A and five B subunits. For both toxins, the A subunit possesses the biological activities of the toxin while the B subunits mediate specific binding and receptor-mediated uptake of the toxin. VT1 and VT2 bind to globotriosylceramide

Table 36.5: Differentiating properties of verocytotoxins by *Esch.coli*

	VT1	*VT2*	*VT2v**
Synonyms	SLT1	SLT2	SLT2v
Cytotoxicity			
Vero cells	+	+	+
HeLa cells	+	+	–
Genes phage-encoded	+	+	–

*Human and porcine variants.

(Gb3) molecules present on the surface of certain eukaryotic cells. In contrast VT2 variant toxins bind to globotetraosylceramide (Gb4). After the A subunit is internalized, it is cleaved into two molecules: A_1 and A_2. A_1 fragments binds to 28S ribosomal ribonucleic acid and disrupts protein synthesis. Destruction of the intestinal villus results in decreased absorption with a relative increase in fluid secretion.

VT1 and VT2, like Shiga toxin, are cytotoxic for Vero and HeLa cells, although VT2 variant toxins do not bind to HeLa cells since this particular cell line does not express Gb3 receptors. Enterotoxicity in ligated rabbit gut loops and mouse paralytic lethality can also be used to detect VTs.

Clinical Infections

Four main types of clinical syndromes are caused by *E. coli:*

1. Diarrhea
2. Urinary tract infection
3. Pyogenic infections
4. Septicemia.

1. Diarrhea

Although *E. coli* is normally carried in the gut as a harmless commensal, it may cause gastrointestinal disease ranging in severity from mild, self-limiting diarrhea to hemorrhagic colitis. At least five different types of diarrhegenic *E. coli* are now recognized, each associated with specific serotypes and with different pathogenic mechanisms.

A. Enteropathogenic *E. coli*

These have been associated mainly with diarrhea in infants and children, usually occurring as institutional outbreaks but they can also cause sporadic diarrhea in children and, less often, in adults. Enteropathogenic *E. coli* (EPEC) diarrhea was common worldwide from the late 1940s to the 1960s. Afterwards, it has become less common. EPEC enteritis is still common in communities with poor hygiene.

EPEC strains were originally identified epidemiologically as a cause of diarrheal disease in infants. Certain strains belonging to characteristically EPEC serogroups, such as O26 and O111, were later shown to express verocytotoxins. EPEC were identified by serotyping, initially by their O and B antigens (for example, O26:B6, O55:B5, O111:B4 and so on). After the existence of the B antigens became suspect, only O typing is practiced.

Pathogenesis of EPEC Diarrhea

The pathogenesis of EPEC **diarrhea** is not fully understood. EPEC do not ordinarily produce enterotoxins, nor are they invasive. In infantile enteritis, the bacilli are seen to be adherent to the mucosa of the upper small intestine, intimately attached to cup-like projections ('pedestals') of the enterocyte membrane, causing disruption of the brush border microvilli. The name *enteroadherent E. coli* has been proposed for these strains, which can be identified by their adhesion to HEp-2 cells.

Laboratory Diagnosis of EPEC

The diagnosis of EPEC diarrhea is relatively easy during outbreaks but very difficult in sporadic cases. Fresh diarrheal feces is plated on blood agar and MacConkey media. After overnight incubation, *E. coli* colonies are emulsified in saline on a slide and are examined by slide agglutination with polyvalent rabbit antisera designed to detect the somatic antigens of strains belonging to EPEC-associated serogroups. Bacteria agglutinated by these sera are then identified with monovalent antisera to individual serogroups. At least ten colonies per plate should be tested, as many serotypes are present in a single culture. If isolated colonies are negative, the confluent growth is emulsified and tested.

During outbreaks, if the causative serotype is known, cultures need be tested only with the particular antiserum. EPEC antisera are now difficult to obtain and so specific diagnosis is available only in few laboratories.

When the outbreak is caused by a strain with some readily demonstrable feature such as failure to ferment sorbitol, rapid identification is possible by using appropriate culture media.

B. Enterotoxigenic *E. coli*

Diarrhea caused by Enterotoxigenic *E. coli* (ETEC) is endemic in the developing countries in the tropics, among all age groups in the local population. In developing countries, ETEC are a major cause of mortality in children under the age of 5 years. Persons from developed countries visiting endemic areas, often suffer from ETEC diarrhea—a condition known as **'traveller's diarrhea'.** ETEC diarrhea came into prominence from the late 1960s. Infection is usually of brief duration, often beginning with the rapid onset of loose stools and accompanied by variable symptoms, including nausea, vomiting and abdominal cramps.

ETEC produce a heat-stable enterotoxin (ST) or a heat-labile (LT) cholera toxin, like enterotoxin or both **(See under virulence factors above).** Toxin production alone may not lead to illness. The organism must initially be able to adhere to the mucosal surface of the epithelial cells of the small intestine. This adhesion is usually mediated by fimbriae that bind to specific receptors in the intestinal cell membrane. These adhesins have been termed **colonization factor antigens (CFAs)** of which a number of have been identified (CFAI, II, III, IV) and, no doubt, others remain to be discovered. Plasmids that simultaneously carry genes for both CFAs and enterotoxin production have been described.

Though plasmids with enterotoxin genes may be present in any strain of *E. coli*, in practice only a small number of serotypes become enterotoxigenic (for example, O6, O8, O15, O25, O27, O167).

Laboratory Diagnosis of ETEC

Diagnosis of ETEC diarrhea depends on the demonstration of enterotoxin in *E. coli* isolates by any of the methods listed in Table 36.6 strain of ETEC may produce either LT or ST or both.

Detection of LT

In vitro methods, such as tissue culture tests (rounding of YI mouse adrenal cells and elongation of CHO cells due to intracellular increase of cAMP concentration), and serological tests (ELISA, passive agglutination and immunolysin tests) are available (Table 36.6).

A precipitin test (the **Biken test**) performed directly on bacterial cell lines growing on a special agar medium may be suitable for use in field laboratories. Rabbit antibodies specific for LT are incorporated into the agar culture medium. As the bacteria grow and secrete LT, the toxin binds to the anti-LT antibodies, forming a precipitin line. In vivo tests such as rabbit loop or intradermal tests may be used when in vitro tests are not available.

Detection of ST

The detection of ST is more difficult. The infant mouse test is still widely employed. The poor antigenicity of ST has prevented the development of serological tests, though ST ELISA using monoclonal antibody has been introduced. **Genetic probes** are available for detection of ST and LT in *E. coli* cultures, or directly in feces, food or water (Table 36.6).

C. Enteroinvasive *E. coli* (EIEC)

These are closely related by phenotypic and pathogenic properties to *Shigella*. Many of these strains are nonmotile, do not ferment lactose or ferment it late with acid, but without producing any gas and do not form lysine decarboxylase. Many of these show O antigen cross reaction with shigellae. These **'atypical'** ***E. coli*** **strains** had earlier been grouped under the **'Alkalescens-Dispar Group'** and given names such as *'Shigella alkalescens'* (resembling *Sh. flexneri* except in fermenting dulcitol and forming alkali in litmus milk) and *'Sh. dispar'* (late lactose fermenter like *Sh. sonnei* but indole positive). Besides these **'atypical strains'** many typical *E. coli* strains can also cause clinical illness resembling shigellosis. These have been termed *enteroinvasive E. coli* because they have the capacity to invade interstitial epithelial cells *in vivo* and penetrate HeLa cells in tissue culture. EIEC strains usually belong to serogroups O28ac, O112ac, O124, O136, O143, O114, O152, O154. The most common serogroup is O124.

Pathogenesis

EIEC, like those of Shigella species, can penetrate the epithelial cells of the large intestine and multiply intra-

Table 36.6: Methods for detection of ETEC enterotoxins

Assay	LT	ST
1. *In vivo* tests		
Ligated rabbit ileal loop		
Read at 6 hours	±	+
Read at 18 hours	+	–
Infant rabbit bowel	+	+
Infant mouse intragastric (4 hours)	–	+
Adult rabbit skin (vascular permeability factor)	+	–
2. *In vitro* tests		
i. Tissue culture tests		
ii. Rounding of Y1 mouse adrenal cells		
iii. Elongation of Chinese hamster ovary (CHO) cells	+	–
iv Serological tests		
ELISA	+	(ST-ELISA with monocolonial antibody)
Passive agglutination tests, passive immune hemolysis, precipitin (Eiken's) test	+	–
v. *Genetic tests*		
DNA probes	+	+

Species	Test/substrate[a]						
	ind	mal	H_2S	KCN	adon	arbtl	mel
C. freundii	–	–	±	+	–	–	±
C. koseri	+	+	–	–	+	+	–
C. amalonaticus	+	–	–	+	–	–	–

Table 36.7: Distinguishing reactions of *Citrobacter*

[a] Ind, indole production; mal, utilization of malonate; H_2S, H_2S produced in TSI agar; KCN, growth in KCN medium; adon, arbtl, mel, fermentation of adonitol, D-arabitol, melibiose.

cellularly, giving rise to blood and mucus in the stool. Infection is by ingestion. Spread to neighbouring cells leads to tissue destruction and consequent inflammation which is the underlying cause of the symptoms of bacillary dysentery. Pathogenicity in shigellae and EIEC depends on both chromosomal and plasmid genes.

Infections are usually food-borne but there is also evidence of cross-infection. Clinically EIEC infection resembles shigellosis, ranging from mild diarrhea to frank dysentery, and occurs, in children as well as adults.

Laboratory Diagnosis of EIEC

i. **Sereny test:** For laboratory diagnosis of EIEC, the **Sereny test** used to be employed (that is, instillation of a suspension of freshly isolated EIEC or shigella into the eyes of guinea pigs leads to mucopurulent conjunctivitis and severe keratitis). Mice may be used instead of guinea pigs.
ii. **Tissue culture and DNA hybridization methods-** The original *Sereny test* has been superseded by **tissue culture** and **DNA hybridization methods.** Cell penetration of HeLa or HEP-2 cells in culture is a more humane diagnostic test. This ability to penetrate cells is determined by a large plasmid, detection of which can also be a diagnostic test.
iii. **ELISA (VMA ELISA) test:** The plasmid codes for outer membrane antigens called the **'virulence marker antigens' (VMA)** which can be detected by the ELISA (VMA ELISA) test.

D. *E. coli* Verocytotoxin or Verotoxin (VT)

E. coli verocytotoxin or verotoxin (VT) was so named because it was first detected (1977) by its cytotoxic effect on Vero cells, a cell line derived from African green monkey kidney cells and the link with two diseases of previously unknown etiology: **hemorrhagic colitis** and **hemolytic uremic sydrome**, was established. Outbreaks were first recognized in the USA in 1982 and strains of VTEC belonging to serogroup O157 emerged as the major cause. Since then, outbreaks and sporadic cases have been reported in several other countries and VTEC belonging to many other serotypes have been described.

The ability to cause hemorrhagic colitis has led some workers to refer to these strains as enterohemorrhagic *Esch.coli* or EHEC. Since 1995, well publicized major episodes in the USA, Canada, Japan and Scotland have heightened general awareness of the importance of this disease.

The disease may occur sporadically or as outbreaks of food poisoning. Outbreaks of infection with VTEC have occurred in the community, in nursing homes for the elderly and in day care centers for young children. The most severe clinical manifestations are usually seen in the young and the elderly.

The source of infection is contamination by human or animal feces, directly or indirectly. Changing lifestyles and eating habits, with growing popularity of fast foods have led to a remarkable increase in EHEC food poisoning.

Pathogenesis

The biological properties, physical characteristics and antigenicity of VT are very similar to those of Shiga toxin (Stx), produced by strains of *Sh. dysenteriae* type 1 (See under heading virulence factors). Strains of O157 VTEC express an attaching and effacing (AE) phenotype and, in common with strains of EPEC, the genes involved are located on a pathogenicity island located on the *Esch. coli* chromosome.

Laboratory Diagnosis of VTEC

Demonstration of the bacilli or VT in feces

Laboratory diagnosis of VTEC diarrhea can be made by demonstration of the bacilli or VT in feces directly or in culture. The proportion of VTEC in, the faecal flora may be low, often less than 1 percent, so that testing of individual colonies from culture plates may not always detect the presence of VTEC. Most strains of O157 VTEC produce colorless colonies after overnight incubation and these can be tested with an O157 LPS-specific antiserum in a simple agglutination assay.

Sorbitol MacConkey Medium

Most VTEC strains belong to the serotype O157:H7 which does not ferment sorbitol, unlike the majority of *E. coli.* So the use of sorbitol MacConkey medium helps in screening for OI57 VTEC.

Confirmation

Toxigenicity is confirmed by **gene probes**, by **PCR,** by testing strains for a cytotoxic effect on Vero cells or by a specific **ELISA.**

Serology

Demonstration of VT neutralizing antibodies in convalescent sera may help in retrospective diagnosis. Evidence for VTEC O157 infection has also been obtained by detecting patients serum and salivary antibodies directed against the O157 LPS antigens.

E. Enteroaggregative *E. Coli* (EAEC)

These strains are so named because they are characterized by their ability to adhere to particular laboratory-cultured cells, such as HEp-2, in an aggregative or 'stacked brick' pattern. They have been associated with persistent diarrhea, especially in developing countries. Strains of EAggEC were first reported in 1987, as a cause of chronic diarrhea in malnourished young children living in Chile, and were reported subsequently in Brazil, India, Mexico and Zaire.

Pathogenesis

EAEC stimulate secretion of mucus, which traps the bacteria in a biofilm overlying the epithelium of the small intestine. Shortening of microvilli, mononuclear infiltration, and hemorrhage are then observed. Strains have been reported to produce an ST-like toxin, a low molecular weight heat stable enterotoxin called EAST ('enteroaggregative heat stable enterotoxin-1), the mechanisms by which these strains cause a diarrheal illness are only poorly understood. Some isolates express hemolysins.

Diagnosis

The only methods currently available for detecting these bacteria are the—HEp-2 cell test for determining the aggregative phenotype, and DNA probes. The HEp-2 cell test involves allowing strains of *Esch. coli* to adhere to cell monolayers *in vitro* and observing the pattern of adhesion by microscopy. Although tissue culture tests are laborious, the pattern of adhesion remains the key assay for detecting EAggEC.

2. Urinary tract infection

E. coli and coliforms account for the large majority of naturally acquired urinary tract infections. Those acquired in the hospital, following instrumentation, are more often caused by other bacteria such as *Pseudomonas* and *Proteus*.

Most frequently encountered O serotypes of *E. coli* in UTI include O1, O2, O4, O6, O7, O18 and O75. These are also known as nephritogenic strains. Special nephropathogenic potential of these strains appears to be due to:

a. **The polysaccharides of the O and K antigens** protect the organism from the bactericidal effect of complement and phagocytes in the absence of specific antibodies. Strains possessing K1 or K5 antigen appear to be more virulent than those with other K antigens.
b. **Fimbriae** present on the surface of *E. coli* mediate the adherence of the organism to the uroepithelial cells. The receptor is part of the P blood group antigen and therefore the fimbriae have been termed P fimbriae.

E. coli that cause UTI often originate in the gut of the patient. The bacteria may gain access to the urinary tract by the ascending or the hematogenous route. The ascending route of infection is believed to be usual one. The bacteria from the fecal flora spread to the perineum and from there they ascend into the bladder.

UTI occurs more often in females than in males. This is due to short urethra, pregnancy, infrequent voiding and sexual intercourse which may lead to **'honeymoon' cystitis**. Shorter, and wider female urethra appears to be less effective in preventing access of the bacteria to the bladder. The high incidence of UTI in pregnant women can be attributed to impairment of urine flow due to pressure on the urinary tract and due to hormonal changes. About 5 to 7 percent of pregnant women have been reported to have urinary infection without any symptoms. Such ***asymptomatic bacteriuria,*** undetected and untreated may lead to symptomatic infection later in pregnancy, pyelonephritis and hypertension in the pregnant women, as well as to prematurity and perinatal death of the fetus.

Relative infrequency of UTI in men may be due to longer male urethra and the bactericidal activity of the prostatic fluid. Other causes of urinary stagnation that may predispose to UTI include enlarged prostate, urinary calculi, congenital malformations and neurological disorders. Catheterization and cystoscopy may introduce endogenous or exogenous bacteria into tbe bladder leading to infection.

Etiology of UTI

1. Other members of family Enterobacteriaceae that usually cause UTI are *Kiebsiella, Proteus, Citrobacter,* and those which rarely produce UTI are salmonellae, edwardsiellae and *Enterobacter.*
2. The gram-positive organisms which can cause UTI are *Staphylococcus aureus,* coagulase-negative staphylococci, *Streptococcus faecalis, S. pyogenes, S. agalactiae, S. milleri,* other streptococci and anaerobic streptococci.
3. Rarely, *Gardnerella vaginalis* may cause UTI.
4. *Candida albicans* may cause UTI in diabetic and immunocompromised patients.
5. The hospital-associated infection following instrumentation and catheterization is mostly caused by *Pseudomonas* and *Proteus.*

Laboratory Diagnosis of UTI

a. Collection of Specimen

i. Catheter specimen
ii. Midstream urine specimen
iii. Suprapubic stab.

i. Catheter Specimen

Bacteriological diagnosis of urinary tract infection has undergone a marked change following the development by Kass of the concept of 'significant bacteriuria'. Normal urine is sterile, but during voiding may become contaminated with genital commensals. In order to avoid such contamination, urine used to be collected by catheterization for culture. Any bacterial growth from catheterized urine was considered to denote infection. Catheterization for this purpose is no longer considered justifiable because even under ideal conditions, catheterization leads to urinary infection in at least two percent and when precautions are inadequate, the risk is much higher. Instead, clean-voided midstream samples of urine are employed for culture.

ii. Midstream Urine Specimen

In **male patients**, retract the prepuce and clean the glans penis with wet cotton. **In case of female**, anogenital toilet is more important and should consist of careful cleaning with soap and water. Patient should be asked to remove underclothing and should sit comfortably on the seat and swing one leg to the side as far as possible. Periurethral and perineal region is washed with soap and water followed by with nonirritant antiseptic solution such as chlorhexidine. Separate labia majora with fingers of one hand and collect midstream urine in a sterile wide-mouthed container. The first portion of urine that flushes out commensal bacteria from the anterior urethra is discarded. The next portion of the urine (midstream sample) is collected directly into a sterile wide mouthed container and transported to the laboratory.

In most patients with bacterial cystitis, quantitative examination of a mid-stream specimen will accurately show the presence of significant bactriuria. In the investigation of urethritis and prostatitis, the initial flow of urine, rather than a midstream specimen, should be examined. Urethritis can be detected by examining the discharge or the first 5 to 10 ml of urine passed.

For the diagnosis of bacterial prostatitis, patient is asked to pass few ml of urine, to wash anterior urethra, and withhold it thereafter. Prostatic massage is then carried out and **expressed prostatic massage (EPS)** is collected. Patient is asked to pass a few ml of urine in a sterile tube to wash out EPS from the urethra, if no EPS is available. The volume of urine specimen should not exceed 5 ml to prevent overdilution of small number of bacteria from the prostate.

iii. Suprapubic Stab

In childen and young infants, urine may be aspirated from the bladder into a syringe with a needle introduced aseptically through the skin and abdominal wall just above the pubis (suprapubic stab).

b. Transport

Urine is a good medium for the growth of coliforms and other urinary pathogens, and hence delay in processing will vitiate the results of quantitative culture because contaminating bacteria, from anterior urethra, can readily multiply to reach significant number. If delay of more than 1 to 2 hours is unavoidable, the specimen should be refrigerated at 4°C, or by transport in some form of refrigerated container, or by collection and transport in a container with boric acid at a final, bacteriostatic concentration of 1.8 percent

c. Microscopy of Urine

Microscopic examination of urine is done principally to detect the presence of increased number of polymorphs as an indication of the UTI. In the past, the microscopical examination was commonly done on a wet film or Gram stained film of deposit centrifuged from the urine. The deposit of the centrifuged urine can be examined under microscope to find out the presence of pus cells, red blood cells and bacteria in it. Presence of more than 3 pus cells per high power field is suggestive of infection. Red blood cells indicate damage to urinary tract. **Note:** Nowadays centrifugation is not recommended.

d. Semiquantitative Culture

For quantitative culture, serial tenfold dilutions of urine are tested by the pour plate or surface culture methods. This, however, is too complicated for routine diagnostic work, for which semiquantitative techniques are more convenient.

Standard Loop Method

The most widely used technique employs a standard loop. Measured quantity of urine (0.004 ml or 1/250 ml) with the help of standardized loop (internal diameter 3.26 mm) of nichrome or platinum wire of SWG 28 is inoculated on blood agar and another loopful on MacConkey agar and incubated overnight at 37°C. Blood agar medium gives a quantitative measurement of bacteriuria, while MacConkey agar enables a presumptive diagnosis of the bacterium. The number of colonies is counted or estimated and multiplied by 250 to get the bacterial cell count per ml of urine. Depending on this result, it can be reported whether the patient has a significant bacteriuria or not.

Other methods of semiquantitative estimation of bacterial counts are: **filter paper method, dip spoon, dip slide methods**, etc.

Interpretation of Results

Kass (1956) gave a criterion for active bacterial infection of urinary tract as follows:

Significant bacteriuria: When bacterial count is more than 10^5/ml of a single species.

Doubtful significance: Between 10^4 to 10^5 bacteria per ml. Specimen should be repeated for culture.

No significant growth: $<10^3$ bacteria per ml and are regarded as contaminant.

e. Identification

The organisms are identified by colony characters, Gram's staining, motility, biochemical reactions and slide agglutination test.

f. Antibiotic Sensitivity Test

E. coli and other common urinary pathogens develop drug resistance so frequently that no antibacterial therapy can be instituted meaningfully without testing individual strains. Resistance is often to multiple drugs and is of the transferable variety.

g. Localization of the Site of Urinary Infection

For the localization of the site of urinary infection, the antibody coated bacteria test has been employed. This is based on the assumption that bacteria coated with specific antibodies are present in the urine only when the kidneys are infected and not when the infection is confined to the bladder. Antibody coated bacteria are detected by immunofluorescence using fluorescent tagged antihuman globulin or by staphylococcal coagglutination.

3. Pyogenic Infections

E. coli form the most common cause of intra-abdominal infections, such as peritonitis and abscesses resulting from spillage of bowel contents. They also cause pyogenic infections in the perianal area. They are an important cause of neonatal meningitis, but is much less so in older patients.

Laboratory Diagnosis

i. Specimens

The specimens are usually pus and wound swab.

ii. Culture

Cultures are made on McConkey's agar.

iii. Identification

The isolate is identified by colony morphology, staining, motility and biochemical reactions.

4. Septicemia

Blood stream invasion by *E. coli* may lead to fatal conditions like septic shock and 'systemic inflammatory response syndrome' (SIDS). As *E. coli* commonly show multiple drug resistance, antibiotic sensitivity testing of strains is important in treatment.

Laboratory Diagnosis

Diagnosis depends on the isolation of the organism by blood culture and its identification by colony morphology, staining, motility and biochemical reactions.

EDWARDSIELLA

This bacterium resembles *Citrobacter* species or the salmonellae in its production of H_2S in TSI and its failure to utilize lactose. Genus *Edwardsiella* is separated from *Escherichia* by its ability to produce hydrogen sulphide in triple sugar iron agar. The genus contains the species *Edwardsiella tarda.* The name *tarda* refers to slow or weak fermentation of sugars by the organism. It is the only recognized human pathogen.

It is a gram-negative bacillus, motile, noncapsulated.

It ferments only glucose and maltose with weak fermentative powers. It is indole and MR positive, and VP and citrate negative (IMViC + + – –). It produces H_2S, decarboxylates lysine and ornithine.

Clinical Infection

E. tarda is a normal intestinal inhabitant of snakes and other cold-blooded animals. The occasional human infections probably originate from contact with cold-blooded animals. It has been cultured from normal and diarrheic human feces. However, its role in the causation of diarrhea has yet to be established. It mainly causes **wound infection**, but **meningitis** and **septicemia** have also been reported.

CITROBACTER

Members of the genus *Citrobacter* are motile enterobacteria confused with both *Escherichia* and *Salmonella.* They are motile, indole positive or negative, MR positive, VP negative, citrate positive (IMViC – + – +), urease weak positive and may or may not ferment lactose but they nearly always produce β-galactosidase (ONPG positive). They do not decarboxylate lysine but most strains decarboxylate ornithine.

Species

Three species are recognized, *Citro. freundii* which gives typical reactions and *Citro. koseri* (formerly *Citro. diversus)* and *Citro. amalonaticus* which do not form H_2S (Table 36.7).

Ballerup-Bethesda Group

The genus *Citrobacter* was first proposed for a group of lactose-negative or late lactose-fermenting coliform bacteria that share certain somatic antigens with salmonellae and were also known as the **Ballerup-Bethesda group**. These organisms are now known as *Citrobacter freundii.* They exhibit extensive antigenic sharing with salmonellae and may cause confusion in the diagnostic laboratory. Some strains (for exmaple, the Bhatnagar strain) have a Vi antigen serologically identical to the antigen of *S. Typhi* and *S. Paratyphi C.* These may be used for the estimation of Vi antibodies or for raising Vi antisera. However, they can be distinguished by their negative lysine decarboxylase and positive KCN reactions.

Clinical infection: *Citrobacter* spp. are often found in human feces and may be isolated from a variety of clinical specimens. It may cause infections of the **urinary tract, gall bladder, middle ear and meninges.** *C. koseri* occasionally causes **neonatal meningitis.**

KLEBSIELLA

Introduction

Members of the genus *Klebsiella* are gram-negative, nonsporing, nonmotile bacilli that grow well on ordinary media, produce pink mucoid colonies on MacConkey's agar. They can be differentiated by simple biochemical tests and often have a pronounced capsule. In these conditions the growth on agar is luxuriant, grayish white and extremely mucoid. They are usually found in the intestinal tract of humans and animals or free-living in soil, water, and on plants.

Classification

Their classification has undergone various modifications. DNA re-association studies have shown that the previouly named species *K. pneumoniae, K. ozaenae K. rhinoscleromatis* and *K. aerogenes* belong to a single species. The name *K. pneumoniae* is used for the species as a whole. It is further divided into 4 subspecies. The most frequently encountered, biochemically typical form of it is known as *K. pneumoniae* subsp. *aerogenes.*

K. pneumoniae subsp. *ozaenae, K. pneumoniae* subsp. *pneumoniae, K. pneumoniae* subsp. *rhinoscleromatis* (Table 36.8). Indole-producing strains that resemble *K. pneumoniae subsp. aerogenes* biochemically are classified in a separate species, *K. oxytoca.*

Morphology

They are short, plump, gram-negative, nonsporing, capsulated, nonmotile bacilli, 1-2 µm long and 0.5-0.8 µm wide with parallel or bulging sides and slightly pointed or rounded ends.

Cultural Characteristics

Klebsiellae grow well on ordinary media at temperatures between 12 and 43°C (optimum, 37°C) in 18-24 hours. On MacConkey agar, the colonies typically appear large, mucoid and red in color. Mucoid nature of colonies is due to capsular material produced by the organism. However, some strains are not mucoid. Capsular material is produced in greater amounts on media rich in carbohydrate

Biochemical Reactions

They ferment sugars (glucose, lactose, sucrose, mannitol) with production of acid and gas. They are urease positive, indole negative, MR negative, VP positive and citrate positive (IMViC – – ++). These reactions are typical of *K. pneumoniae* subsp. *aerogenes.*

Glucose	Lactose	Sucrose	Mannitol	
+(AG)	+	+	+	
Urease	Indole	MR	VP	Citrate
+	–	–	+	+

Biochemical reactions of different subspecies of *K. pneumoniae* and *K. oxytoca* are given in Table 36.8.

Antigenic Structure

Klebsiella possess capsular (K) and somatic (O) antigens.

1. Capsular (K) Antigen

On the basis of capsular (K) antigens, the klebsiellae have been differentiated into 80 serotypes. Members of capsular types 1-6 occur most frequently in the human respiratory tract.

Capsular antigens are usually detected by means of the capsular 'swelling' reaction, countercurrent immuno-electrophoresis and enzyme-linked immunosorbentassay (ELISA).

2. Somatic (O) Antigen

Five different somatic or O antigens (01-05) occur in various combinations with the capsular antigens. Four of the five *Klebsiella* O antigens are identical to or related to *Esch. coli* O antigens. In capsulated strains, O antigens are masked by K antigens and because the latter are heat-stable at 100°C for 2.5 hours, therefore, O antigens are identifiable only in non-capsulated mutants.

Typing Methods

1. Bacteriocin (klebocin or pneumocin) typing; 2. Phage typing; 3. Biotyping; and 4. Resistotyping; 5. Molecular typing methods: Plasmid analysis, DNA profiling by random amplified polymorphic DNA (RAPD) and pulsed-field gel electrophoresis.

KLEBSIELLA PNEUMONIAE

(Friedlander's *bacillus, Bacillus mucosus capsulatus)*

This bacillus was first isolated by Friedlander (1883) from fatal cases of pneumonia. It ferments sugars (glucose, lactose, sucrose, mannitol) with the production of acid and abundant gas. It is indole and MR negative and VP and citrate positive (IMViC – – + +). Biochemically variant strains are common. It forms urease. Strains, formerly labeled as nonmotile *Aerobacter aerogenes (K. aerogenes),* are now considered to be *K. pneumoniae* subspecies aerogenes.

It is the second most populous member of the aerobic bacterial flora of the human intestine. It has become a very important cause of nosocomial infections, even replacing *E. coli* in some centers.

Pathogenicity

Klebsiella pneumoniae can cause a primary communty-acquired pnuemonia, nosocomial infections, urinary tract infections, wound infections, bacteremia and meningitis and rarely diarrhea. In some hospitals, *K. pneumoniae* has replaced *E. coli* as the leading blood culture isolate. As most strains are resistant to antibiotics, treatment poses serious problems.

Pneumonia

***Klebsiella* pneumonia** is a serious disease with high case fatality. The typical patient is a middle- or older-aged man who have medical problems such as alcoholism, chronic bronchopulmonary disease or diabetes mellitus. The disease is characterized by massive mucoid inflammatory exudate of lobar or lobular distribution, involving one or more lobes of the lung. Necrosis and abscess formation are more frequent than in pneumococcal pneumonia. Serotypes 1, 2 and 3 are usually responsible

Tests	K. pneumoniae subspecies				K. oxytoca
	aerogenes	pneumoniae	ozaenae	rhinoscleromatis	
Gas from glucose	+	+	V	–	+
Acid from lactose	+	+	V	–	+
Urease	+	+	V	–	+
Citrate	+	+	V	–	+
Malonate	+	+	–	+	+
MR	–	+	+	+	V
VP	+	–	–	–	V
Lysine decarboxylase	+	+	V	–	+
KCN	+	+	+	±	+

Table 36.8: Distinguishing reactions of *Klebsiella species*

V-variable

for pneumonia. Positive blood cultures can be obtained in about 25 percent of the cases.

Diarrhea

Some strains of *K. pneumoniae* isolated from cases of **diarrhea** have been shown to produce an enterotoxin very similar to the heat stable toxin of *E. coli*. The production of this toxin is determined by the presence of a plasmid.

Laboratory Diagnosis

Diagnosis is made by culturing appropriate specimens on blood agar and MacConkey agar and identifying the isolate by biochemical reactions. Antibiotic sensitivity should invariably be done. Many strains carry plasmids determining multiple drug resistance.

K. ozaenae

K. pneumoniae spp. *ozaenae* is a bacillus associated with ozena, an uncommon, chronic disease in which there is atrophy of the nasal mucosa characterized by foul smelling nasal discharge. Identification is difficult due to wide variations in the biochemical reactions of individual strains. *K. ozaenae* belongs to capsular types 3-6.

K. rhinoscleromatis

K. rhinoscleromatis causes rhinoscleroma, a chronic granulomatous hypertrophy of the nose prevalent in south-eastern Europe, India and Central America, where it is associated with prolonged exposure to crowded and unhygienic conditions. The lesions occur in the nose, larynx, throat and, to a lesser extent, in the trachea and consist of granulomatous infiltrations of the submucosa. The bacilli are seen intracellularly in lesions. It can be identified by biochemical reactions and belongs to capsular type 3.

K. oxytoca

K. oxytoca may be rarely isolated from clinical specimens.

Treatment

Clinical isolates of *Klebsiella* are resistant to ampicillin, amoxicillin and other penicillins, but combinations of these drugs with β-lactamase inhibitors such as clavulanic acid are usually effective. They are normally susceptible to cephalosporins, especially β-lactamase stable derivatives such as cefuroxime and cefotaxime, and to fluoroquinolones. They are often sensitive to gentamicin and other aminoglycosides.

Klebsiella infection of the urine often responds to trimethoprim, nitrofurantoin, co-amoxiclav or oral cephalosporins. Pneumonia and other serious infections require vigorous treatment with aminoglycoside or a cephalosporin such as cefotaxime.

ENTEROBACTER

Enterobacter is a motile, capsulated, lactose fermenting bacillus which is indole and MR negative and VP and citrate positive (IMViC – – + +). These characteristics are similar to those of *Klebsiella* but can be differentiated from *Klebsiella* because it is motile and ornithine positive. Two clinically relevant species are *E. cloacae* and *E. aerogenes* (Table 36.9).

Clinical Infections

They are normally found in feces, sewage, soil and water and rarely in urine, pus and other pathological materials. They may cause urinary tract infections and hospital infections. They are occasionally associated with meningitis and septicemia.

Treatment

Aminoglycosides are often effective in the treatment of *Enterobacter* infections.

HAFNIA

These organisms are probably best regarded as non-lactose fermenter (NLF) member of genus *Enterobacter*.

Table 36.9: Biochemical reactions of *Enterobacter aerogenes, E. cloacae* and *Hafnia alvei*			
Biochemical test	*E. aerogenes*	*E. cloacae*	*H. alvei*
Indole	–	–	–
Methyl red test	–	–	–
Voges-Proskauer	+	+	+
Citrate	+	+	+
Decarboxylation of:			
• Osnithine	+	+	+
• Lysine	+	–	+
• Arginine	–	–	+
Urease	–	v	–
Growth in KCN	+	+	+
Acid production from:			
• Dulcitol	–	v	–
• Adonitol	+	v	–
• Raffinose	+	+	–
• Inositol	+	v	–
• Melibiose	+	+	–

v—variable.

This is a motile, nonlactose-fermenting bacillus which is indole and MR negative and VP and citrate positive (Table 36.9). Biochemical reactions are evident best at 22°C but at 37 °C they may be negative or irregular. *Hafnia alvei is* the only species.

Strains are isolated from feces of man and other animals and are also found in sewage, soil, water and dairy products. They are occasionally encountered as opportunistic pathogens that have been recovered from infected wounds, abscesses, sputum, urine, blood and other sites.

SERRATIA

Although numerous species of the genus *Serratia* have been described, *S. marcescens is* the one most commonly encountered in clinical specimens. Several others, including *S. liquefaciens* (formerly known as *Ent. liquefaclens)* and *S. odorifera,* are sometimes isolated.

It is small, motile, gram-negative bacillus. This differs from *Hafnia* in forming a pink, red or magenta, nondiffusible pigment called **prodigiosin** which is formed optimally at room temperature, is soluble in absolute alcohol and other organic solvents but is insoluble in water. Pigment is formed only in the presence of oxygen and at a suitable temperature. Prodigiosin is also formed by certain organisms unrelated to *S. marcescens* including an actinomycete, and certain gram-negative rods isolated from sea water.

Pathogenesis

It is a saprophyte found in water, soil and food. It may grow in sputum after collection and may suggest hemoptysis because of the pigment formed ('pseudohemoptysis'). Nosocomial infections due to *S. marcescens* are being reported with increasing frequency. The bacillus has been associated with infections of the urinary and respiratory tracts, meningitis, wound infections, septicemia and endocarditis. Multiple drug resistance is common in hospital strains.

KNOW MORE

Bacteriocin (Klebocin or Pneumocin) Typing

Many *Klebsiella* strains produce bacteriocins known as klebocins or pneumocins which show a narrow range of activity on other *Klebsiella* strains. Kelbocin typing can be done by the help of liquid preparations of bacteriocins. For epidemiological studies, Klebocin typing and capsular serotyping together may be very useful.

KEY POINTS

- Members of the family Enterobacteriaceae are gram-negative bacilli. They are aerobes or/and facultative anaerobes and grow readily on ordinary laboratory media, ferment glucose with the production of acid or acid and gas, are either non-motile or motile with peritrichous flagella.
  They are catalase positive (except for *Shigella dysenteriae* type 1 which is catalase-negative), oxidase-negative, reduce nitrate to nitrites and are typically intestinal parasites of humans and animals.
- All lactose-fermenting enterobacteria, e.g. *Escherichia, Klebsiella, Enterobacter* and *Citrobacter* are popularly known as **'coliform bacilli'**, as the most common member of this group is the colon bacillus or *Escherichia coli.* The major intestinal pathogens, *Salmonella* and *Shigella* are nonlactose-fermenters **(NLF)**.

Escherichia coli

- Gram-negative bacilli, aerobe and a facultative anaerobe, can grow on ordinary media like nutrient agar. On MacConkey's medium, colonies are red or pink due (lactose fermenter). Selective media such as DCA or SS agar growth.
- **Antigenic structure:** Serotyping of *E. coli* is based on three antigens—the flagellar antigen H, somatic antigen O and the capsular antigen K.
- **Virulence factors:**
  1. **Somatic antigen (O antigen):** Endotoxin;
  2. **K antigen;**
  3. Fimbriae;

Toxins

1. **Exotoxins:** *Hemolysins* and *enterotoxins.*
   Enterotoxins: (e.g., heat-stable and heat-labile enterotoxins, and verotoxin (VT) also known as Shiga-like toxin (SLT).

Clinical infections: **1. Urinary tract infection** (most common cause of bacterial UTIs; limited to bladder (cystitis) or can spread to kidneys (pyelonephritis) or prostate (prostatitis).

2. **Diarrhea:** At least six different pathogenic groups cause gastroenteritis (ETEC, EPEC, EIEC, EHEC, EAEC, DAEC).
3. **Pyogenic infections:** Neonatal meningitis (usually with strains carrying the K1 capsular antigen).
4. **Bacteremia** (most commonly isolated gram-negative bacillus).

Edwardsiella: The genus contains the species *Edwardsiella tarda.* It mainly causes wound infection, but meningitis and septicemia have also been reported.

Citrobacter: Three species, *Citro. freundii, Citro. koseri* (formerly *Citro. diversus)* and *Citro. amalonaticus.* It may cause infections of the urinary tract, gall bladder, middle ear and meninges. *C. koseri* occasionally causes neonatal meningitis.

- **Klebsiella:** Members of the genus *Klebsiella* are gram-negative, nonsporing, nonmotile bacilli that grow well on ordinary media, produce pink mucoid colonies on MacConkey's agar.
- The name *K. pneumoniae* is used for the species as a whole. It is further divided into 4 subspecies known as *K. pneumoniae* subsp. *aerogenes,* subsp. *ozaenae,* subsp. *pneumoniae,* subsp. *rhinoscleromatis* strains that resemble *K. pneumoniae* subsp. *aerogenes* biochemically are classified in a separate species, *K. oxytoca.*
- **Klebsiella pneumoniae**—can cause a primary communiy-acquired pnuemonia, nosocomial infections, urinary tract infections, wound infections, bacteremia and meningitis and rarely diarrhea.
  - *K. ozaenae*—is associated with ozena.
  - **K. rhinoscleromatis**—causes rhinoscleroma. A chronic granulomatous hypertrophy of the nose
  - **K. oxytoca**—may be rarely isolated from clinical specimens.
  - **Enterobacter:** Two species are *E. cloacae* and *E. aerogenes.* They may cause urinary tract infections and hospital infections, occasionally associated with meningitis and septicemia.
- **Hafnia:** They have been recovered from infected wounds, abscesses, sputum, urine, blood and other sites.
- **Serratia:** *S. marcescens is* the one most commonly encountered in clinical specimens. It is associated with infections of the urinary and respiratory tracts, meningitis, wound infections, septicemia and endocarditis.

IMPORTANT QUESTIONS

1. Discuss various mechanisms by which *Escherichia coli* produces diarrhea. Describe the laboratory diagnosis of bacterial diarrheas.
2. Discuss the pathogenicity of *Escherichia coli.*
3. Discuss the laboratory diagnosis of various infections caused by *Escherichia coli.*
4. Discuss the pathogenesis and laboratory diagnosis of urinary tract infections caused by *Escherichia coli.*
5. Write short notes on:
   a. Antigenic structure of *Escherichia coli*
   b. Enterotoxins of *Escherichia coli*
   c. Verotoxin (VT) or Shiga-like toxin (SLT).
6. Write briefly about:
   a. *Citrobacter*
   b. *Klebsiella pneumoniae*
   c. *Enterobacter*
   d. *Serratia.*

FURTHER READING

Abbott S. Klebsiella, Enterobacter, Citrobacter, and Serratia. In Murray PR et al, (Eds).: Manual of clinical microbiology, 7th Edition, Washington, DC, 1999, American Society of Microbiology.

Doran TI. The role of citrobacter in clinical disease of children: review. Clin Infect Dis 1999;28:384-94.

Gross RJ. The pathogenesis of Escherichia coli diarrhea. Rev Med Microbiol 1991;2:37.

Nataro JP, Kaper JB. Diarrheagenic Escherichia coli. Clin Microbiol Rev 1998;11:142-201.

Sanders WE, Sanders CC. Enterobacter spp. pathogens poised to flourish at the turn of the century. Clin Microbiol Rev 1997;10:220-241.

Chapter 37 Tribe Proteae: Proteus, Morganella and Providentia

Learning Objectives

After reading and studying this chapter, you should be able to:

- Describe morphology, cultural characteristics and biochemical reactions of Proteus sp.
- List various differences among Proteus, Morganella and Providentia.

CLASSIFICATION

The tribe *Proteeae* is classified into three genera *Proteus, Morganella* and *Providencia.* Most of them except for some *Providencia* strains, produce a powerful urease which rapidly hydrolyzes urea to ammonia and carbon dioxide. A characteristic feature which distinguishes tribe *Proteeae* from other enterobacteria is the presence, in all members of the tribe, of the enzyme phenylalanine deaminase which converts phenylalanine to phenyl pyruvic acid (PPA reaction). Virtually no other member of the Enterobactericaea synthesize the required enzyme phenylalanine deaminase. All members of this tribe fail to ferment lactose.

The major differentiating features of medically important species of Proteus bacilli are shown in Table 37.1.

PROTEUS

Proteus Bacilli

Proteus bacilli are normal intestinal commensals and opportunistic pathogens like coliforms. The name 'Proteus' refers to their pleomorphism, after the Greek god Proteus who could assume any shape.

Genus *Proteus* has four species: *P. mirabilis, P. vulgaris, P. myxofaciens and P. penneri. P. mirabilis, P. vulgaris* are widely recognized as human pathogens. These are motile, gram-negative bacilli, characterized by *swarming* growth on agar.

Morphology

They are gram-negative coccobacilli, 1-3 um long and 0.6 um wide. Pleomorphism is frequent–short coccobacilli to long filaments. In young swarming cultures, many of the bacteria are long, curved and filamentous, sometimes reaching up to 80 um in length. They may be arranged singly, in pairs or in short chains. They are actively motile with peritrichous flagella. However, non-flagellate and nonmotile variants are also encountered. They also have more type of fimbriae and are noncapsulated.

Cultural Characteristics

They are aerobe and facultative anaerobes. All grow well on laboratory nutrient media. Proteus organisms

Table 37.1: Biochemical features of species of Proteus, Morganella and *Providentia*

Test	*Pr. vulgaris*	*Pr. mirabilis*	*Morg. morganii*	*Prov. alcalifaciens*	*Prov. stuarti*	*Prov. rettgeri*
Swarming	+	+	–	–	–	–
Gas from Glucose	+	+	+	+	–	–
Indole	+	–	+	+	+	+
Phenyl pyruvic acid (PPA) test	+	+	+	+	+	+
Urease	+	+	+	–	±	+
H_2S production	+	+	–	–	–	–
Ornithine decarboxylase	–	+	+	–	–	–
Fermentation of adonitol	–	–	–	+	±	±
Fermentation of trehalose	±	+	±	–	+	–

are usually first recognized by their characteristic putrefactive odor described as **'fishy' or 'seminal'** and swarming appearance on noninhibitory solid media such as nutrient agar and blood agar. Swarming appears as a thin, colorless, transparent film extending from the margin of a young colony and spreding in several waves demarcated by a raised margin until most or all of the surface of culture plate is covered. Swarming is a striking feature of *Pr. mirabilis* and *Pr. vulgaris* and does not occur with other species at 37ºC though it may be induced by growing on soft agar at 20 to 28ºC. Swarming of *Proteus* appears to be due to vigorous motility of the organisim although, the exact cause is not yet established.

Swarming growth is a problem in the laboratory when mixed growth is obtained in which proteus bacilli are present with other bacteria. The swarming growth may cover most or all of the agar surface as well as colonies of other organisms. A number of methods have been devised to inhibit swarming, mainly to avoid interference with the isolation of clinically more important orgamsms. Swarming of *Proteus* can be inhibited by (1) increasing concentration of agar (6%) and by (2) incorporation of chloral hydrale (1:500), sodium azide (1:500), alcohol (5-6%), sulfonamide, surface. active agents or boric acid (1:1000).

Swarming does not occur on MacConkey's medium, on which smooth colorless (NLF) formed. Nonmotile variant do not swarm. *Proteus* produces uniform turbidity with a slight powdery deposit and an ammonical odor in liquid medium (peptone water).

Dienes Phenomenon

When two identical strains of *Proteus* are inoculated at different places of the same culture plate, the resulting swarms of growth coalesce without signs of demarcation. No line is formed between swarming culture of the same strain. When, however, two different strains of *Proteus* species are inoculated, the spreading films of growth fail to coalesce and remain separated by a narrow but easily visible furrow. This is known as **Dienes phenomenon**.

It has been used to determine the identity or nonidentity of various strains of *Proteus*.

Biochemical Reactions

The distinctive characters of this genus are:

i. **PPA test**—Deamination of phenylalanine to phenyl pyruvic acid (PPA test) is always positive.
ii. **Urea hydrolysis**—Urea hydrolysis by enzyme urease is another characteristic of Proteus but is negative in some Providencia strains.
iii. All species of Proteus produce acid from glucose.
iv. Lactose is not fermented.
v. They are malonate utilization negative.
vi. Indole is formed by *Pr. vulgaris* but is negative in *Pr. mirabilis*.
vii. They are MR positive and VP negative.
viii. H_2S is produced by *Pr. vulgaris* and *P. mirabilis*.
ix. Nitrate reduction positive.

Other biochemical characters of four species of Proteus are given in Table 37.1.

Antigenic Structure

Proteus bacilli possess somatic O and flagellar H antigens, which are of considerable historical interest. Weil and Felix (1916) studying Proteus bacilli observed that flagellated strains growing on agar formed a thin surface film resembling the mist produced by breathing on glass and named this variety the 'Hauch' form (from *Hauch,* meaning film of breath). Nonflagellated variants grew as isolated colonies without the surface film and were called 'Ohne Hauch'. (meaning without film of breath). These names came to be abbreviated as the H and O forms. Subsequently, the H and O were extended to refer to the flagellar and somatic antigens of other bacilli as well.

O Antigens

There are 32 O antigens specific for *P. mirabilis* and 22 for *P. vulgaris* and 5 are found in both species. *P. mirabilis* and *P. vulgaris* have been divided into 54 O groups on the basis of their O antigens.

H Antigens

In P. *mirabilis* and *P. vulgaris* nineteen H antigens have been detected. There are numerous cross-reactions between H antigens. These groups are further subdivided according to different H antigens into a large number of serotypes.

Clinical Use of the Antigen Typing

The major clinical use of the antigen typing of *Proteus* species has been the diagnosis rickettsial diseases. Weil and Felix also observed that certain nonmotile strains of *Pr. vulgaris,* called the 'X strains', were agglutinated by sera from typhus fever patients. This heterophilic agglutination due to the sharing of an alkali stable carbohydrate antigen by certain strains of Proteus (OX2, OX19 and OXK) and rickettsiae forms the basis of the Weil-Felix reaction for the diagnosis of some rickettsial infections. Three nonmotile Proteus strains OX2, OX19 and OXK are used in the agglutination test. OX19, OX2 are the strains of *P. vulgaris* serotype 01 and serotype 02 and OXK is the strain of *P. mirabilis* serotype 03.

Typing Methods

Phage typing, bacteriocin (proticin) typing and serotyping schemes have been developed for *Proteus* and *Providencia* species. Swarming *Proteus* strains exhibit the Dienes phenomenon and this forms the basis for a precise method of differentiation among such strains.

Pathogenesis

Proteus bacilli are widely distributed in nature as saprophytes, being found in decomposing animal matter, in sewage, in manured soil and in human and animal feces. They are frequently present on the moist areas of the skin. They are opportunistic pathogens, commonly responsible for urinary and septic infections, often nosocomial.

P. mirabilis accounts for the majority of human infections seen with this group of organisms. All members of the tribe can cause **urinary tract infections (UTI), wound infections, pneumonia, infection of the ear, respiratory tract infection, septicemia** and **nosocomial infections**. Strains of Pr. *mirabilis* are a prominent cause of urinary tract infection in children and in domiciliary practice. Indole-producing strains of *Proteus and Providencia* are usually isolated from hospital patients, especially in elderly men following surgery or instrumentation.

UTI caused by *Proteus* tends to be more serious than that caused by *E. coli* and other coliforms, because these organisms are usually confined to the bladder whereas *Proteus* has a predilection for the upper urinary tract. It produces urease which splits urea into carbon dioxide and ammonia. Ammonia inactivates complement, damages renal epithelium and makes the urine alkaline. This increase in pH causes precipitation of calcium and magnesium salts from the urine and results in the formation of urinary calculi. It may also lead to hyperammonaemic encephalopathy and coma. Patients with long-term indwelling urinary catheters are prone to developing bladder colonization with *P. mirabilis* and *P. stuartii.*

Laboratory Diagnosis

Culture

Laboratory diagnosis of the infections caused by species *Proteus* can be carried out by culture of the specimen on MacConkey agar or DCA.

Identification

The isolate is identified by its morphological, biochemical and agglutination reactions.

Treatment

Proteus bacilli are resistant to many of the common antibiotics. An exception is *P. mirabilis* which is sensitive to ampicillin and cephalosporıns.

MORGANELLA

Morganella morganii is the only species in the genus *Morganella* (formerly *Pr. morganii*). It is motile and lactose-nonfermenting. Unlike *Proteus*, they do not swarm on solid media.

M. morganii has been isolated from sputum, infected wounds and occasionally from urine. It is commonly found in human and animal feces and causes urinary infection infrequently. Nosocomial wound infections also occur.

Providencia

Five *Providencia* species include *P. alcalifaciens, P rettgeri, P. rustigianii and P. stuartii* and *P. heimbachae*. All species may be recovered from feces. However, only *P. alcalifaciens* may be associated with diarrhea. *P rettgeri* and *P. stuartii* have been associated with hospital-acquired urinary tract, wound and other infections, particularly in immunocompromised patients.

P. stuartii may occasionally cause hospital-associated urinary tract infection, infection of burns and pneumonia and septicemia in elderly and immunodeficient patients.

P. rettgeri is part of the normal fecal flora of reptiles and amphibians and sometimes causes nosocomial infection of the urinary tract, wounds, burns and blood.

Laboratory Diagnosis

Culture

Laboratory diagnosis of the infections caused by species *Proteus, Morganella* and *Providencia* can be carried out by culture of the specimen on MacConkey agar or DCA.

Identification

The isolate is identified by its morphological, biochemical and agglutination reactions.

Treatment

Proteus bacilli are resistant to many of the common antibiotics except *Pr. mirabilis which is* moderately sensitive to benzylpenicillin and fully sensitive to ampicillin and most other β-lactam antibiotics. Providencia are the most resistant, particularly *Providencia stuartii* which is also resistant to disinfectants such as chlorhexidine, cetrimide, benzalkonium chloride and heavy metal compounds such as silver sulfonamide. It is sensitive to phenol and glutaraldehyde. Amikacin and ciprofloxocin are generally effective in treatment.

ERWINIA

These are anaerogenic bacilli forming a yellowish pigment, usually found in soil and causing plant infections. *E. herbicola* has occasionally been isolated from respiratory and urinary infections in predisposed or hospitalized patients.

KNOW MORE

Most strains of *Proteus, Morganella and Providen*ia are motile, methyI red positive, voges-Proskauer negative, can degrade tyrosine and grow in the presence of KCN. They fail to acidify lactose, dulcitol or malonate; do not form arginine or lysine decarboxylase or beta galactosidase.

All of them with few exceptions show the following features:

Gram negative, pleomorphic, motile rods, noncapsulated, resistant to KCN; degrade tyrosine.

Fail to acidify lactose, dulcitol or malonate; Do not form arginine or lysine decarboxylase or beta galactosidase; MR positive, VP negative.

KEY POINTS

- The tribe *Proteeae* is classified into three genera *Proteus, Morganella* and *Providencia.*
- Genus *Proteus* has four species: *P. mirabilis, P. vulgaris, P. myxofaciens and P. penneri.*
- **Cultural characteristics**—Proteus organisms are usually first recognized by their characteristic putrefactive odor described as 'fishy' or 'seminal' and swarming appearance on noninhibitory solid media such as nutrient agar and blood agar. Swarming does not occur on MacConkey's medium.
- **Biochemical reactions**—PPA test and urea hydrolysis positive. Indole is formed by *Pr. vulgaris* but is negative in *Pr. mirabilis.* H_2S is produced by *Pr. vulgaris* and *Pr. mirabilis.*
- **Pathogenesis**—*P.mirabilis* accounts for the majority of human infections seen with this group of organisms. All members of the tribe can cause urinary tract infections (UTI), wound infections, pneumonia.
- *Morganella morganii* causes urinary infection infrequently. Nosocomial wound infections also occur.
- Five *Providencia* species include *P. alcalifaciens, P. rettgeri, P. rustigianii and P. stuartii* and *P. heimbachae.*
- **Erwinia**—*E. herbicola* has occasionally been isolated from respiratory and urinary infections in predisposed or hospitalized patients.

IMPORTANT QUESTIONS

1. Write short notes on:
   - Classification of tribe *Proteeae*
   - Dienes phenomenon
   - Genus *Morganella*
   - Genus *Providencia*

FURTHER READING

Alison C, Coleman N, et al. Ability of *Proteus mirabilis* to invade human urothelial cells is coupled to motility and swarming diffefertiation. Infect Immune 1992;60:4740-6.

Johnson DE, Ryssel Rg, et al. Contribution of *Proteus mirabilis* urease to persistence, urolithiasis and acue pyelonephritis in a mouse model of ascending urinary tract infection. Infect Immune 1993;61:2748-54.

Yakubu DE, Old DC, Senior BW. The haemagglutinins and fimbriae of *Proteus penneri.* J Med Microbiol 1989;30:279-84.

CHAPTER 38

Shigella

LEARNING OBJECTIVES

After reading and studying this chapter, you should be able to:

- Describe morphology and various culture media for the isolation of Shigella.
- Discuss the antigenic structure and toxins of Shigella.
- Discuss laboratory diagnosis of bacillary dysentery.

INTRODUCTION

The genus *Shigella* belongs to the tribe Escherichieae and is closely related to the genus *Escherichia. Shigella* species, however, are not members of the normal gastrointestinal flora. The genus *Shigella* is named after the Japanese microbiologist Kiyoshi Shiga, who first isolated the organism in 1896. It was then called *S. shigae* and is now known as *S. dysenteriae* serotype 1. *S. flexneri* by Flexner (1900), *S. sonnei* by Sonne (1915) and *S. boydii* Boyd (1931) were subsequently described.

Dysentery is a clinical condition of multiple etiology, characterized by the presence of blood, mucus, and pus in the stool. The two common types of dysentery are bacillary and amebic. All *Shigella* species can cause **bacillary dysentery**. Some other bacilli, such as enteroinvasive *E. coli, Vibrio parahaemolyticus* and *Campylobacter,* can also cause the clinical picture of dysentery.

SHIGELLA

Morphology

Shigellae are nonsporing, noncapsulate, Gram-negative rods, 2-4 × 0.6 µm, nonmotile and nonflagellate. Fimbriae (Type 37.1) occur only in *Shigella flexneri,* though not in serotype 6 and some strains in other flexneri serotypes. Other species of *Shigella* do not possess fimbriae.

Cultural Characteristics

They are aerobes and facultative anaerobes, with a growth temperature range of 10-40°C and optima of 37°C and pH 7.4. They grow well on conventional media.

1. Nutrient Agar and Blood Agar

Colonies are smooth, greyish or colorless, translucent, often 2-3 mm in diameter, circular, convex, resembling those of salmonellae.

2. MacConkey Agar

Colonies are colorless, i.e. pale and yellowish (non-lactose-fermenting) due to the absence of lactose fermentation. An exception is *S. sonnei,* a late fermenter of lactose, become pink when incubation is prolonged beyond 24 hours.

3. Deoxycholate Citrate Agar (DCA)

Deoxycholate citrate agar (DCA) is an excellent selective plating medium. Colonies are pale and similar to, though usually slightly smaller, e.g. 1-1.5 mm in diameter, and more translucent than those of salmonellae. They do not form a black center.

4. Wilson and Blair's Brilliant Green Bismuth Sulphite Medium

Growth is inhibited on this medium.

5. Xylose Lysine Deoxycholate (XLD)

XLD is probably the best selective medium for shigellae. Colonies are red and, unlike those of most salmonellae, without black centers.

6. Salmonella-Shigella (SS) Agar

Salmonella-Shigella (SS) agar is a highly selective medium for the isolation of Salmonella and Shigella. Colonies of Shigella on this medium are colorless with no blackening, while those of Salmonella are colorless with black centers.

7. Hektoen Enteric (HE) Agar

Hektoen enteric (HE) agar is useful as a direct plating medium for fecal specimens *for the isolation of Shigella and Salmonella. Colonies of Shigella on this medium are green while those of Salmonella are blue green with black centers due to H_2S production.*

8. Peptone Water and Nutrient Broth

Good growth with uniform turbidity on incubation overnight at 37°C. Some strains, especially fimbriate ones, form a surface pellicle on longer incubation.

9. Enrichment Broths

a. Selenite F broth

Selenite F broth will grow and enrich *S. sonnei* and *S. flexneri* serotype 6, but is inhibitory to other shigellae. Sodium selenite in this enrichment medium inhibits coliform bacilli while permitting salmonellae and many shigellae to grow. Therefore, it is recommended for the isolation of these organisms from feces.

Tetrathionate broth and brilliant green media are inhibitory and unsuitable for enrichment cultures.

b. Gram-Negative (GN) Broth

Enrichment is best done in uninhibitory nutrient broth or weakly inhibitory GN broth. Most strains of Shigella grow well.

Resistance

Shigellae are not specially resistant. All types are killed by moist heat at 55°C in 1 hour and fairly readily by strong disinfectants, e.g. by 1 percent phenol in 15 min. They mostly die within a few hours if dried. *S. sonnei* is more resistant to adverse environmental conditions as compared to other species. Cultures retain their viability well, e.g. for many years on Dorset's egg.

Biochemical Reactions

The shigellae are divided into four **groups**, or **species,** by their biochemical reactions and antigenic structure (Table 38.1). The groups A, B, C and D correspond to the species *S. dysenteriae, S. flexneri, S. boydii* and *S. sonnei.*

Glucose is fermented with the production of acid, without gas, except for two serotypes, *Sh. flexneri* serotype. 6 (Newcastle and Manchester varieties) and *S. boydii* serotype 14, which form gas. Fermentation of mannitol is of importance in classification and shigellae have traditionally been divided into mannitol fermenting and nonfermenting species. *Sh. flexneri, Sh. boydii and Sh. sonnei* ferment mannitol, while *Sh. dysenteriae* does not. Exceptions are not infrequent. Lactose and sucrose are not fermented, except by *Sh. sonnei* which ferments them late. Dulcitol is not fermented by the majority of shigellae. Adonitol, inositol and salicin are not fermented.

S. dysenteriae serotype 1, *S. flexneri* serotype 6 and *S. sonnei* are always indole negative. Strains of other serotypes differ in their reactions. Shigellae are MR positive and negative for VogesProskauer reaction, citrate and malonate utilization. They reduce nitrates to nitrites, do not form H_2S and are inhibited by KCN. Catalase is produced, except by *Sh. dysenteriae* type 1 which are catalase negative.

Members of groups A, B and C fail to decarboxylate lysine and' ornithine. *S. sonnei* decarboxylates ornithine but not lysine.

Antigenic Structure

The shigellae are divided into four 'major' **O antigenic groups**, designated A, B, C, and D. In addition to the major O antigens, groups A, B, and C contain **'minor' O antigens**, which allow for subgrouping. Some strains possess **K or envelope antigens,** which are not important for serologic typing but, when present, may interfere with the serologic reactions of the O antigens. There is considerable antigenic sharing between some members of the genus as well as between shigellae and *E. coli.* Common fimbrial antigens may also occur, particularly in *Sh. flexneri.* It is, therefore, important that the identification of shigellae should be made by a combination of antigenic and biochemical properties and not by slide agglutination alone. Since all *Shigella* are nonmotile, there are no H antigens.

Classification

Shigellae are classified into four species or subgroups (A, B, C, D) based on a combination of biochemical and serological characteristics. Serotypes are distinguished within the species. *Sh. sonnei* is serologically homogeneous and is classified by colicin typing.

Fermentation of mannitol is of importance in classification and shigellae have traditionally been divided into mannitol fermenting species. (*Sh. flexneri, Sh. boydii and Sh. Sonnei)* ferment mannitol, and mannitol nonfermenting species (*Sh. dysenteriae) which* does not ferment mannitol.

Table 38.1: Distinguishing features of Shigella species

Subgroup *Species*	*A* *Sh. dysenteriae*	*B* *Sh. flexneri*	*C* *Sh. boydii*	*D* *Sh. sonnei*
Mannitol	–	A	A	A
Lactose	–	–	–	ALate
Sucrose	–	–	–	ALate
Dulcitol	–	–	d	
Indole	d	d	d	
Ornithine decarboxylase	–	–	–	+
Serotypes	10	6 + variants	15	Only one

A = Acid

Group A (S. Dysenteriae)

This species of mannitol nonfermenting bacilli consists of 12 serotypes, each characterized by a different type antigen. **Serotypes I and 2** are the organisms formerly called ***S. shigae*** and ***S. schmitzii***. ***S. shigae*** is indole negative and is the only member of the family that is always catalase negative (*Sh. schmitzi* and *Sh. sonnei* are invariably catalase positive, while among other shigella species, some strains may be catalase negative). Serotype 2 shares a minor antigen with serotype 10.

***Sh. dysenteriae* type 2 (*Sh. schmitzi*)** forms indole and ferments sorbitol and rhamnose. **Serotypes 3-7** were described by Large and Sachs in India and hence used to be known as the **Large-Sachs group**. Three further serotypes have been described making a total of ten.

Exotoxins

Sh. dysenteriae type 1 produces a powerful exotoxin (**Shiga toxin),** the earliest example of an exotoxin produced by a gram-negative bacillus. It acts as **enterotoxin** as well as **neurotoxin.** As enterotoxin, it acts on the intestinal mucosa causing transudation of fluid in the lumen and can be demonstrated with induction of fluid accumulation in ligated rabbit ileal loop. As neurotoxin, it damages endothelial cells of small blood vessels of the central nervous system which results in neurological complications like polyneuritis, coma and meningism.

Cytotoxin

S. dysenteriae produces also an cytotoxin which is active on vero cells snd is known as **Verotoxin.** This appears to be the same as **Verotoxin 1 (or Shiga-like toxin)** produced by certain strains of *E. coli* (VTEC). Shiga toxin is a subunit toxin comprising an **A (active)** portion and **five B (binding)** subunits. The A subunit possesses the biological activity of the toxin while the B subunits mediate specific binding and receptor-mediated uptake of the toxin. Within the host cell the A subunit is divided into two fragments A1 and A2. Fragment A I appears to inactivate host cell 60 S ribosome, interfering with protein synthesis. The primary manifestation of toxin activity is damage to the intestinal epithelium; however, in a small subset of patients, the Shiga toxin can mediate damage to the glomerular endothelial cells, resulting in renal failure (HUS).

Group B (Sh. flexneri)

This group is named after Flexner, who described the first of the mannitol fermenting shigellae from Philippines (1900). This structure was elucidated by Boyd in the decade 1930-40.

S. flexneri was known to have a complex antigenic structure for more than 50 years. This group is biochemically heterogeneous and antigenically the most complex among shigellae. Based on type specific and group specific antigens, they have been classified into six serotypes (1-6) and several subtypes (la; Ib; 2a, 2b; 3a, 3b, 3c, 4a, 4b, 5a, 5b). In addition, two antigenic 'variants' called X and Y are recognized, which have lost their type antigens and are distinguished by their different group antigens (Table 38.2).

Table 38.2: Antigens of *Shigella flexneri* serotypes

Serotype	*Subserotype*	*Type antigen*	*Group antigens.*
1	1a	I	4
	1b	I	6
2	2a	II	3, 4
	2b	II	7, 8
3	3a	III	6, 7, 8
	3b	III	3, 4, 6
4	4a	IV	3, 4
	4b	IV	6
5	5a	V	3, 4
	5b	V	7, 8
6	...	VI	–
X variant	...	–	7, 8
Y variant	...	–	3, 4

Not all group antigens are listed.

Serotype 6 is always indole negative and occurs in three biotypes: Boyd 88, Manchester and Newcastle, some of which form gas from sugars (Table 38.3).

Group C (Sh. Boydii)

The group is named after Boyd, who first described these strains from India (1931). This group consists of dysentery bacilli that resemble *Sh. flexneri* biochemically but not antigenically. Eighteen different serotypes of *S. boydii* are recognized. *Sh. boydii* are isolated least frequently from cases of bacillary dysentery.

Group D (S. Sonnei)

This bacillus, first described by Sonne (1915) in Denmark, ferments lactose and sucrose late. It is indole negative. It is antigenically homogeneous. It may, however undergo an antigenic variation and may occur in two forms, **phase 1 and phase II. Phase 1** gives rise to smooth colonies (S) while the **phase II** forms colonies that are larger, flatter and more irregular (R). On subculture, phase I produces both types of colonies but phase

Table 38.3: Biotypes of *Sh. flexneri* Type 6

	Fermentation of	
	Glucose	*Mannitol*
Boyd 88	A	A
Manchester	AG	AG
Newcastle	A or AG	–

A=Acid
AG=Acid and Gas

II is considered to be a loss variation. Cultures often contain a mixture of both forms, but the colonies of one form can be selected by addition of antiserum of the other form to the culture medium because form II variant is antigenically different from form I variant. Organisms in phase II may be isolated from patients but are more common in convalescents and carriers.

Sh. sonnei causes the mildest form of bacillary dysentery. In many cases the disease may only be a mild diarrhea. However, *Sh. sonnei* infection persists as the most common shigellosis in the advanced countries. For epidemiological purposes, *Sh. sonnei* has been classified into 26 colicin types.

Pathogenic Mechanisms

1. Surface Properties

The ability to survive the passage through the host defences may be due to the O antigens. Lipopolysaccharide (LPS) has been implicated in causing localized cytokine release, and the resultant inflammatory response and cellular disruption enables these bacteria to enter intestinal cells.

2. Invasiveness

Sh. dysenteriae type 1 forms an exotoxin which appears to be much less important in pathogenesis than the ability of the bacillus to penetrate and multiply in colonic mucosa. Invasive property is related to the presence in the bacillus of large plasmids (M.W. 140 × 106) coding for the outer membrane protein responsible for cell penetration. These proteins are called **'virulence marker antigens' (VMA).** Detection of VMA by ELISA serves as a virulence test for Shigellae, as for enteroinvasive *E. coli.* The invasive property of the bacillus can be demonstrated by its ability to penetrate cultured HeLa or Hep-2 cells or by the Congo red binding test.

3. Toxins

Sh. dysenteriae type 1 forms a produces a powerful exotoxin **(Shiga toxin**) (details see under heading Group A (*S. dysenteriae).*

Pathogenicity

Shigellae cause bacillary dysentery. Humans are the only known reservoir of *Shigella* organisms infection occurs by ingestion. The infection is highly communicable because of the low infective dose required to produce the disease. The minimum infective dose is low, as few as 10-100 bacilli being capable of initiating the disease, probably because they survive gastric acidity better than other enterobacteria. *Shigella* spp. are pathogens of man and other primates, and the pathogenesis of infection with these bacteria and entero-invasive *E. coli* (EIEC) is very similar.

Shigella cause disease by invading and replicating in cells lining the colonic mucosa. After reaching the large intestine, the shigellae multiply in the gut lumen. The shigellae multiply within the epithelial cells and spread laterally into adjacent cells, where cell-to-cell passage occurs, and deep into the lamina propria. The infected epithelial cells are killed and the lamina propria and submucosa develop an inflammatory reaction with capillary thrombosis. Patches of nerotic epithelium are sloughed and ulcer form. The cellular response is mainly by polymorphonuclear leukocytes which can be seen on microscopic examination of stool, together with red cells and sloughed epithelium.

The ulcers of the bacillary dysentery are much shallower than amoebic ulcers. The intervening mucosa is inflamed and edematous. Bacteremia may occur in severe infections, particularly in malnorished children and in acquired immunodeficiency syndrome (AIDS). There is no macroscopic lesion in small intestine.

Bacillary dysentery has a short incubation period (1-7 days, usually 48 hours). The onset and clinical course are variable and are largely determined by the virulence of the infecting strain. The clinical manifestations of shigellosis vary from asymptomatic to severe forms of the disease. The main clinical features are frequent passage of loose, scanty feces containing blood and mucus, along with abdominal cramps and tenesmus. Fever and vomiting may be present. Infection is usually self-limited, although antibiotic treatment is recommended to reduce the risk of secondary spread to family members and other contacts.

In dysentery caused by *S. dysenteria* type 1, patients experience more severe symptoms. Bloody diarrhea that progresses to dysentery may appear within a few hours to a few days. Patients suffer from extremely painful bowel movements, which contain predominantly mucus and blood. In young children, abdominal pain is quite intense, and rectal prolapse may result from excessive straining. Severe ilness may also be caused by members of *Sh. flexneri* and *Sh. boydii* groups. In contrast dysentery associated with *Sh. sonnei* (*Sonne dysentery)* in an otherwise healthy person may be confined to the passage of a few loose stools with vague abdominal discomfort and the patient often continues at school or work.

Complications

Complications are most often seen in patients with *S. dysenteriae* serotype 1 infection. These include arthritis, toxic neuritis, conjunctivitis, parotitis, and, in children, intussusception. Hemolytic uremic syndrome (HUS) may occur as a complication in severe cases. Death from bacillary dysentery is uncommon in the developed countries. It occurs mostly at the extremes of life or in individuals who are suffering from some other disease or debilitating condition.

The severity of the disease may vary from acute fulminating dysentery to mild diarrhea. As the term bacillary dysentery refers only to the more severe cases, the term **'shigellosis'** has been employed to include the whole spectrum of disease caused by shigellae.

Although the effects of shigella toxin have been implicated as the mechanism responsible for the signs of the disease, the connection between the toxin hypothesis and the symptoms remains unclear. Nontoxigenic mutants can still cause dysentery but not noninvasive ones. However, it has been reported that the detectable toxin levels produced by *S. dysenteriae* type 1 are higher than those produced by other *Shigella* spp.

Epidemiology

Bacillary dysentery has a global distribution. It is mostly associated with the overcrowding and bad hygienic conditions encountered in times of war and other disasters, and in jails and mental institutions. In several campaigns, more men have died of dysentery than were killed in battle. A recent instance was the major epidemic affecting many thousands, with high case fatality, which occurred during the Rwandan civil war in 1994. Epidemics in civilian communities are associated with poverty and lack of sanitation.

Humans are the only known reservoir of *Shigella* organisms. Transmission may occur by direct person-to person contact, and spread may take place via the fecal. oral route, with carriers as the source. From the carries the organisms can be spread by **flies, fingers, food** and **feces.** Personal hygiene plays a major role in the transmission of *Shigella* organisms. Young children in daycare centers, people living in crowded and less-than-adequate housing, and young male homosexuals as part of the **gay bowel syndrome** who participate in anal-oral sex are most likely affected.

Sh. sonnei is the predominant infecting agent. *Sh. sonnei* is the main type in the north in the USA, while *Sh. flexneri* is more common in the south. In India, *Sh. flexneri* has been the predominant species, having formed 50-85 percent of isolates in different series. *Sh. dysenteriae* (8-25 percent) and *Sh. sonnei* (2-24 percent) are the next common species. *Sh. boydii* (0-8 percent) has been isolated least frequently.

Human beings are the only natural hosts for shigellae. Captive monkeys have been found infected but such infections may have been of human origin. Experimentally, dysentery can be produced only in monkeys. Human volunteer studies have clarified the spectrum of shigellosis.

Laboratory Diagnosis

Diagnosis depeds on isolating the bacillus from feces.

A. Specimens

i. Feces: A specimen of feces is always preferable to a rectal swab
ii. Rectal swabs: A direct swab may be taken from the ulcer by sigmoidoscopic examination.

B. Transport

Fresh feces should be inoculated without delay or transported in a suitable medium such as Sachs' buffered glycerol saline. pH 7.0-7.4, which prevents the dysentery bacilli from being destroyed by the acid produced during the growth of other organisms.

C. Microscopy

Make a wet film of a suspension of the feces in saline. This will show numerous erythrocytes and polymorphs and some macrophages.

D. Culture

For inoculation it is best to use mucus flakes if they are present in the sample. The feces are inoculated on **MacConkey agar** and **DCA. SS agar** and **XLD medium** can also be used. After overnight incubation at 37°C, the plates are inspected for pale (non-lactose-fermenting) colonies on MacConkey agar and DCA, and red and colorless colonies with no blackening on XLD and SS agar, respectively.

E. Identification

i. **Biochemical reactions:** These are tested for motility and biochemical reactions. Any nonmotile bacillus that is urease, citrate, H_2S and KCN negative should be further investigated by biochemical tests (Table 38.1).
ii. **Slide agglutination:** Identification is confirmed by slide agglutination with polyvalent and monovalent sera and then with type-specific sera (belonging to subgroups A, B, C) unless the strain is S. *sonnei.*

F. Colicine Typing

Plasmid pattern analysis and **colicine typing** may help elucidate patterns of spread of *S.sonnei.*

G. Serology

Demonstration of antibodies in sera is not useful have no place in diagnosis of the disease.

Treatment

Most of the cases of bacillary dysentery especially those due to *S. sonnei,* are mild and self-limiting and do not require antibiotic therapy. Replacement of fluids and electrolytes by oral rehydration salt solution is all that is required.

Routine antibacterial treatment is not indicated in dysentery. As with salmonella infections, drugs that impair gut motility should be avoided. Treatment with a suitable antibiotic is necessary in the very young, the aged or the debilitated, and in severe infections. Ampicillin, co-trimoxazole, tetracycline, the quinolone antibiotics such as nalidixic acid and ciprofloxacin are appropriate choices.

Multiple drug resistance plasmids are widely prevalent in shigellae. Indiscriminate antibiotic treatment will only worsen the problem of drug resistance in intestinal bacteria. The choice of antibiotic should be based on the sensitivity of the prevailing strain.

Control

Control consists essentially improving personal an environmental sanitation. Antibiotics have no place in prophylaxis. No effective vaccine is available.

KNOW MORE

Group A (S. Dysenteriae)

Three types of toxic activity have been demonstrated in shigella culture filtrates—neurotoxic, cytotoxic, and enterotoxic

- **Neurotoxicity**, demonstrable by paralysis and death on injection into mice or rabbits. Though known as 'neurotoxin', the primary site of its action appears to be not the nervous tissue but the blood vessels, mainly of the central nervous system.
- Two new shigella enterotoxins have been identified, designated as **Sh. ET-1 and 2. Sh.ET-1 is** confined to ***Sh. flexneri* 2a** and **Sh. ET- 2** is more widespread.

KEY POINTS

- The genus *Shigella* belongs to the tribe Escherichieae and are gram-negative bacilli, non-motile. They are facultative anaerobe and grow well on conventional media.
- **Classification:** The shigellae are divided into four **groups**, or **species.** The groups A, B, C and D correspond to the species *S. dysenteriae, S. flexneri, S. boydii* and *S. sonnei.*
- **Pathogenesis:** It causes gastroenteritis (shigellosis). These bacteria are highly infectious, as 10-100 organisms can initiate disease. Bacteria are acquired by fecal-oral spread. Humans are the only reservoir. Most common form is an initial watery diarrhea progressing within 1 to 2 days to abdominal cramps and tenesmus (with or without bloody stools). A severe form of disease is caused by *S. dysenteriae* (bacterial dysentery). Exotoxin (Shiga toxin) is produced by *S. dysenteriae;* disrupts protein synthesis and produces endothelial damage. Hemolytic colitis and hemolytic uremic syndrome (HUS) associated with *Shigella.*
- Diagnosis: Culture of feces reveals nonmotile bacillus that is urease, citrate, H_2S and KCN negative and should be further investigated by biochemical tests.

IMPORTANT QUESTIONS

1. Enumerate the various causes of bacillary dysentery. Discuss laboratory diagnosis of *Shigella dysentery.*
2. Write short notes on:
   - Classification of shigella
   - Pathogenesis of shigella.

FURTHER READING

Admoni O, Yagupsky P, et al. Epidemiological, clinical and microbiological features of shigellosis among hospitalized children in northern Israel. Scand Infect Dis 1995;27:139-44.

Brenner DJ. Recommendations on recent proposals for the classification of shigellae. International Journal of Systematic Bacteriology 1984;34:87-88.

Duguid JP. Shigella. Chapter in Mackie & McCartney. Practical Medical Microbiology. Edinburg, Churchill-Livingstone 14(Edn):405.

Hale TH. Bacillary dysentery. In: Topley and Wilsons Microbiology and Microbial Infections. London: Arnold 1998; 9(Edn)3.

Paton JC, Paton AW. Pathogenesis and diagnosis of Shiga toxin-producing Escherichia coli infections, Clin Microbiol Rev 1998;11:450-79.

CHAPTER 39

Enterobacteriaceae III: Salmonella

Learning Objectives

After reading and studying this chapter, you should be able to:

- Describe morphology, culture charactrestics and biochemical reactions of Salmonella sp.
- Describe the following: antigenic structure of Salmonella; antigenic variation of Salmonellae; Kauffmann-White scheme.
- Discuss laboratory diagnosis of enteric fever.
- Describe vaccination against enteric fever.
- Discuss salmonella gastroenteritis.

INTRODUCTION

Genus *Salmonella* has been named after American microbiologist, DE Salmon and it consists of bacilli that parasitizes the intestines of a large number of vertebrate species and infect human beings, leading to enteric fever, gastroenteritis, septicemia with or without focal suppuration and the carrier state.

The most important member of the genus is Salmonella Typhi, the causative agent of typhoid fever. Eberth (1880) first observed the typhoid bacillus in the mesenteric nodes and spleen of fatal cases of typhoid fever and was isolated by Gaffky (1884). It came to be known as the Eberth-Gaffky bacillus or *Eberthella typhi.* Salmon and Smith (1885) described a bacillus which was believed to cause hog cholera (mistakenly, as it is a virus disease). This bacillus (later called S. Choleraesuis) was the first of a series of similar organisms to be isolated from animals and human beings-the genus *Salmonella.* It was redesignated S. Typhi, the genus *Eberthella* having been abolished.

SALMONELLA

Of all the Enterobacteriacae, the *Salmonella* is the most complex, currently comprise above 2463 serotypes or species, all of them potentially pathogenic. In addtion to humans, they infect many animals and are capable of invading extraintesinal tissues. For practical purposes, they may be divided into two groups:

1. Enteric Fever Group

It consists of the typhoid and paratyphoid bacilli that are exclusively or primarily human parasites;

2. Food Poisoning Group

These are essentially animal parasites but which can also infect human beings, producing gastroenteritis, septicemia or localized infections.

Morphology

Salmonellae are gram-negative bacilli, 2-4 × 0.6 µm in size. They are motile with peritrichous flagella except S. Gallinarum-Pullorum which is non-motile. Non-motile variants are occasionally found in other serotypes. They are nonacid-fast, noncapsulate and nonsporing.

Cultural Characteristics

Salmonellae are aerobes and facultatively anaerobes, growing readily over a range of pH 6 to 8 and temperature 15 to 45°C (optimum 37°C). They can grow on simple laboratory media. Various media are as follows:

A. Nutrient Agar and Blood Agar

Colonies of most strains are moderately large (e.g. 2-3 mm in diameter), grey-white, moist, circular disks with a smooth convex surface and entire edge after 24 hours at 37°C.

B. Peptone Water and Nutrient Broth

In liquid media most strains give abundant growth with uniform turbidity. A thin surface pellicle usually forms on prolonged incubation.

C. Differential and Selective Solid Media

These media are valuable for the isolation of salmonellae from feces and other materials contaminated with many bacteria of other kinds. They include:

1. MacConkey Agar

The colonies are 1 to 3 mm in diameter pale yellow or nearly colorless due to the absence of lactose fermentation.

2. Brilliant Green Maconkey Agar

Salmonellae appear as low convex, pale-green translucent colonies 1 to 3 mm in diameter.

3. Deoxycholate-Citrate Agar (DCA)

The colonies of salmonellae on DCA are similar to or slightly smaller in size than those on MacConkey agar. They are pale, nearly colorless, smooth, shiny and translucent. A useful modification of D CA contains both lactose and sucrose (DCLS).

4. Wilson and Blair's Brilliant-Green Bismuth Sulphite Agar (BBSA)

Jet black colonies with a metallic sheen are formed due to production of H_2S. This is due to the reduction of sulphite to sulphide. This medium is particularly valuable for the isolation of Typhi. S. Paratyphi A and other species that do not form H_2S produce green colonies.

5. Xylose Lysine Deoxycholate (XLD) Agar

Most salmonellae produce hydrogen sulphide which reacts with ferric ammonium citrate in the medium to produce black centers in their red colonies. The H_2S-negative salmonellae (e.g. Paratyphi A) form red colonies without black centers.

Enrichment Media

Tetrathionate broth and Selenite F broth are commonly used enrichment media for inoculation of specimens escepially feces.

Biochemical Reactions

1. Salmonellae ferment glucose, mannitol, arabinose, maltose, dulcitol and sorbitol, forming acid and gas except S. Typhi, Gallinarum and rare anaerogenic variants in other serotypes form only acid and no gas.
2. Lactose, sucrose, salicin or adonitol are not fermented.
3. Indole is not produced. They are MR positive, VP negative and citrate positive (IMViC –+ – +) except by S. Typhi and S. Paratyphi A which are citrate negative as they need tryptophan as the growth factor.
4. Hydrogen sulphide is produced except by S. Paratyphi A, S. Choleraesuis, S. Typhisuis and S. Sendai.
5. Urease is not hydrolysed.
6. Salmonellae decarboxylate the amino acids lysine, ornithine and arginine, but not glutamic acid. S. typhi is exceptional in lacking ornithine decarboxylase and paratyphi in lacking lysine decarboxylase.

The enteric fever group may be separated biochemically (Table 39.1).

Table 39.1: Biochemical characters of typhoid and paratyphoid bacilli

	Glucose	*Xylose*	*d-Tartrate*	*Mucate*
S. Typhi	A	d	A	d
S. Paratyphi A	AG	–	–	–
S. Paratyphi B	AG	AG	–	AG
S. Paratyphi C	AG	AG	AG	–

Resistance

Salmonellae are readily killed by moist heat, at 55°C in 1 hour or at 60°C in 15 minutes and most strong disinfectants. Boiling or chlorination of water and pasteurization of milk destroy the bacilli. In polluted water and soil, they survive for weeks and in ice for months. They are killed within five minutes by mercuric chloride (I: 500) or 5 percent phenol. Cultures on slopes of Dorset's egg kept tightly capped to prevent drying and stored in the dark at room temperature usually remain viable for at least 10 to 20 years. They die more quickly when dried. Many salmonellae, however, survive for fairly long periods when dried in foodstuffs, e.g. in dried egg, milk or coconut, and infections have been carried in such products from one country to another.

Antigenic Structure

Antigenic structure of salmonellae is shown in Fig. 39.1. Salmonellae possess the following antigens based on which they are classified and identified:

1. Flagellar antigen H
2. Somatic antigen O
3. Surface antigen Vi, found in some species.

Several strains carry fimbriae. Fimbrial antigens are not important in identification but may cause confusion due to their nonspecific nature and widespread sharing among enterobacteria. Salmonella antigens are also

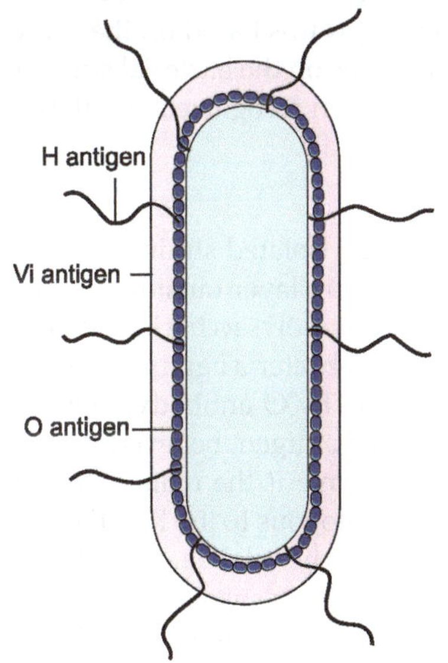

Fig. 39.1: Anigenic structure of salmonellae

found in some members of other genera such as *Escherichia, Shigella, Citrobacter* and *Proteus.*

1. H Antigen

These antigens represent determinant groups on the flagellar protein. They are heat-labile and alcohol-labile, but are well preserved in 0.04 to 0.2 percent formaldehyde. In most salmonellae, H antigen exists in two 2 alternative phases called *phase* 1 and *phase* 2.

When mixed with antisera, H suspensions agglutinate rapidly, producing large, loose, fluffy clumps. The H antigen is strongly immunogenic and induces antibody formation rapidly and in high titres following infection or immunization.

2. O Antigens

The somatic O antigen is a phospholipidprotein-polysaccharide complex which forms an integral part of the cell wall. O polysaccharide have a core structure that is common to all enterobacteria and side chains of sugars attached to the core determine core specificity. It is identical with endotoxin.

The O antigen is unaffected by boiling (**heat-stable**), alcohol (**alcohol-stable**) or weak acids. The O antigens are unaffected by suspension of the bacteria in 0.2 percent formaldehyde but if flagella are present, their fixation by the formaldehyde renders the bacteria inagglutinable by O antibodies. They are hydrophilic and enable the bacteria to form stable, homogeneous suspensions in saline (0.85% NaCl) solution.

The O antigen is less immunogenic than the H antigen and the titre of the O antibody induced after infection or immunization is generally lower than that of the H antibody. O agglutination takes place more slowly and at a higher temperature optimum (50-55°C) than H agglutination (37°C).

The O antigen is not a single factor but a mosaic of two or more antigenic factors. Salmonellae are classified into a number of groups based on the presence of characteristic O antigens on the bacterial surface. Sixty seven O antigens have been recognized and they are designated by arabic numerals.

3. Vi Antigen

Almost all recently isolated strains of S. Typhi form Vi antigen as a covering layer outside their cell wall. This antigen is an acidic polysaccharide. When fully developed it renders the bacteria agglutinable by Vi antibody and inagglutinable by O antibody. Felix and Pitt, who first described this antigen, believed that it was related to virulence and gave it the name 'Vi antigen' (Vi for virulence). It is analogous to the K antigens of coliforms.

The Vi antigen is heat-labile. It can be removed from the bacteria by heating a suspension for 1 hour at 100°C and centrifuging the bacteria from the Vi-containing fluid. It is also destroyed by N HCI and 0.5 N NaOH. It is unaffected by alcohol or 0.2 percent formol.

Originally observed in S. Typhi, the Vi antigen with similar antigenic specificity is present in S. Paratyphi C and S. Dublin, as well as in certain strains of *Citrobacter* (the Ballerup-Bethesda group). The Vi antigen tends to be lost on serial subculture. The Vi polysaccharide acts as a virulence factor by inhibiting phagocytosis, resisting complement activation and bacterial lysis by the alternative pathway and peroxidase mediated killing. Strains possessing the Vi antigen were found to cause clinical disease more consistently than those lacking it in human volunteer experiments.

The Vi antigen is poorly immunogenic and following infection only low titres of antibody are produced. Vi antigen is not employed in the Widal test because detection of the Vi antibody is not helpful for the diagnosis of cases. The total absence of the Vi antibody in a proven case of typhoid fever indicates poor prognosis. The antibody disappears early in convalescence. Its persistence indicates the development of the carrier state. The Vi antigen affords a method of epidemiological typing of the S. Typhi strains based on specific Vi bacteriophages.

Phenolized vaccine induces no Vi antibody though low titres are produced by the alcoholized vaccine. The protective efficacy of the Vi antigen is demonstrated by the success of the purified Vi vaccine for typhoid now in routine use.

Antigenic Variations

The antigens of salmonellae undergo phenotypic and genotypic variations.

1. H-O Variation

This variation is associated with the loss of flagella. When salmonellae are grown on agar containing phenol (1:800), flagella are inhibited. This change is phenotypic and temporary. Flagella reappear when the strain is subcultured on media without phenol. Salmonellae may rarely lose flagella by mutation resulting in the development of stable or irreversible mutant. For example, a stable nonmotile mutant of S. Typhi is the 901-O strain which is widely employed for the preparation of O-agglutinable bacterial suspensions.

Craigie's tube-enerally, the loss of flagella is not total and there occurs only a diminution in the number of flagella and the quantity of the H antigen. Flagellated cells are found in small numbers in such cultures. To obtain a population of motile cells, rich in H antigen, from such cultures, selection may be carried out by using Craigie's tube. This consists of a wide tube containing soft agar (0.2%) at the center of which is embedded a short, narrow tube open at both ends, in such a way that it projects above the agar. The strain is inoculated carefully into the inner tube. After shortest period (8-16 hours) incubation, required for swarming, subcultures withdrawn from the top of the agar outside the central tube will yield a population of motile cells (Fig. 39.2).

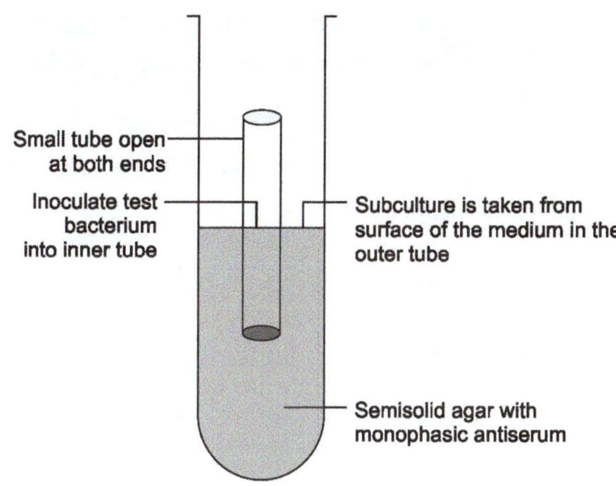

Fig. 39.2: Craigie tube

A U-tube of soft agar may be employed instead of Craigie tube. Inoculation is made into one limb and subculture taken from the other.

2. Phase Variation

The flagellar antigens of most salmonellae occur in one of two phases, that is, the flagella may exhibit one or the other of two alternative sets of antigens, defined by two separate sets of genes in the bacterial genome. Phase I antigens are either specific for a species or shared by a few species only. Hence, phase I is called the '**specific**' phase. Phase 2 antigens are widely shared and hence phase 2 is called the '**nonspecific' or 'group' phase**. The presumptive identification of serotypes, therefore, mainly depends on the identification of the H antigens in phase I, which are relatively 'specific'.

Phase I antigens are designated a, b, c, d, etc. and after z, as z I, z2, etc. Phase 2 antigens are far fewer and are termed I, 2, etc. In some species, antigens belonging to phase I may occur as the phase 2 antigens (for example, e, n, x, z 15). The strains that possess both phases are known as **'diphasic'** and the strains that possess only one phase are known as **monophasic** (e.g. S. Typhi, S. Paratyphi A, S. Enteritidis etc.).

A culture will contain cells with the flagellar antigens of both phases but generally one or the other phase will predominate so that the culture is agglutinated only by one of the phase antisera. It may be necessary to identify the flagellar antigens of both phases for serotyping of salmonella isolates. A culture in phase I can be converted to phase 2 by passing it through a Craigie's tube containing specific phase I antiserum, and the reverse conversion achieved by using phase 2 antiserum (Fig. 39.2).

3. V-W Variation

Almost all recently isolated strains of S. Typhi form Vi antigen as a covering layer outside their cell wall. When fully developed it renders the bacteria agglutinable by Vi antibody and inagglutinable by O antibody. This is called the 'V form'. The Vi antigen is either partially or completely lost after a number of subcultures. Such cultures are inagglutinable with the Vi antiserum but readily agglutinable with the O antiserum. This is called the 'W'form. With partial loss of Vi antigen, the organisms are agglutinable with both O and Vi antisera and these intermediate forms are known as VW forms.

Other Vi-containing bacilli such as S. Paratyphi C and S. Dublin do not completely mask the O antigen.

4. S-R Variation

The smooth(S) to rough(R) variation is associated with the change in colony morphology and loss of the O antigen and of virulence. Conversion into R forms occurs by mutation. Suspensions in saline are autoagglutinable. The colonies become large, rough and irregular. Rough variation occurs rarely in nature but is common in laboratory strains maintained by serial subcultivation. S-R variation may be prevented to some extent by maintaining cultures on Dorset's egg media in the cold or ideally by lyophilization.

5. Variation in O Antigen

Changes in the structural formulae of the O antigen may be induced by lysogenization with some converting phages, resulting in the alteration of the serotypes. Thus, S. Anatum (serotype 3,10:e, h:1,6) is converted by phage 15 into S. Newington (serotype 3,15:e, h:1,6) and S. Newington into S. Minneapolis (serotype 3,15, 34:e, h:1,6) by phage 34. It is likely that such changes occurring in nature contribute to the abundance of Salmonella serofypes.

Classification and Nomenclature

The classification and nomenclature of salmonellae have undergone several modifications over the years. Inclusion in the genus is based on common **biochemical properties**. Classification within the genus is on antigenic characterization based on the **Kauffmann-White scheme**.

A. Classification Based on Biochemical Reactions

On the basis of biochemical reactions, the genus salmonellae is classified into four subgenera (Table 39.2):

Subgenus I

It is the largest and medically the most important group contains all the species which commonly cause human and animal infections.

Subgenus II

It contains mostly species isolated from reptiles.

Subgenus III

It contains bacilli, formerly designated Arizona, originally isolated from lizards but subsequently found in reptiles, birds, domestic animals and human beings. Many of them are prompt lactose fermenters.

Table 39.2: Biochemical reactions of Salmonella subgenera

Test	*Subgenera*			
	I	*II*	*III*	*IV*
Dulcitol fermentation	+	+	–	–
Lactose fermentation	–	–	+	–
Malonate utilization	–	+	+	–
d-Tartrate	+	–	–	–
Salicin fermentation	–	–	–	+
Culture in KCN	–	–	–	+

Subgenus IV

Subgenus IV strains are rarely encountered and may be considered atypical members of subgenus II.

Ewing proposed that only three species should be recognized in the genus *Salmonella*-S Choleraesuis, S. Typhi and S. Enteritidis-all other species being considered serotypes of S. Enteritidis. This proposal is now not followed.

Modern taxonomical techniques, especially DNA studies, have shown that all the members of the genus *Salmonellae* and of the former genus *Arizona* are so closely related that they should all be considered as belonging to a single species, in a genetic, phylogenetic and evolutionary sense. Variations in properties such as antigenic structure, biochemical reactions and host preferences exhibited by different strains can be considered intraspecies divergences. Different antigenic types were originally classified as separate species but it now generally accepted that they represent serotypes of a single species, S. enterica. Thus, a new species name S. enterica has been coined to include all salmonellae based on DNA-reassociation tests S. enterica is classified into seven subspecies named *enterica, salamae, arizonae, diarizonae, houtenae, bongori* and *indica* (Table 39.3). Most of the serotypes that infect mammals are found in a subspecies also designated *enterica*. Subspecies *enterica* corresponds to the former subgenus 1.

Such classification and nomenclature would be too complicated for use in clinical bacteriology, while being taxonomically correct. For example, the taxonomically correct name for the typhoid bacillus would be '*Salmonella enterica*, subspecies enterica, serotype Typhi', but this is abbreviated to S.serotype Typhi or simply S. Typhi. Serotype name should be given in Roman and not in italics. Therefore, the old practice of referring to clinically important salmonellae serotypes by the species name continues in clinical bacteriology.

B. Kauffmann-White Scheme of Classification

Classification within the genus is on antigenic characterization based on the **Kauffmann-White scheme**. This scheme depends on the identification, by agglutination, of the structural formulae of the O and H antigens of the strains (Table 39.4).

Table 39.3: Biochemical reactions of subspecies *Salmonella enterica*

	Subspecies[a]						
	I	*II*	*IIIa*	*IIIb*	*IV*	*V*	*VI*
1. Dulcitol fermentation	+	+	–	–	–	+	±
2. Lactose fermentation	–	–	+	±	–	–	±
3. Sorbitol fermentation	+	+	+	+	+	+	–
4. 4. Malonate utilization	–	+	+	+	–	–	–
5. Salicin fermentation	–	–	–	–	+	–	–
6. Culture in KCN	–	–	–	–	+	+	–
7. Gelatinase	–	+	+	+	+	–	+
8. ONPG hydrolysis	–	–	+	+	–	+	±

[a] Subspecies are: I, *.enterica*; II, *salamae*; IIIa, *arizonae*; IIIb, *diarizonae*; IV, *houtenae*; V, *bongori*; VI, *indica*.

This scheme, first developed in 1934, classifies the salmonellae into different O *groups*, or O *serogroups*, each of which contains a number of serotypes possessing a common O antigen not found in other O groups. The O groups first defined were designated by capital letters A to Z and those discovered later by the number (51-67) of the' characteristic O antigen. It would be more logical to name the serogroups according to their characteristic O antigen factor number rather than by letters, i.e. to abandon the letters A-Z used to designate early O groups. Hence, O groups become: O2 (A), O4 (B), O7 (Cl), O8 (C2-C3), O9, 12 (Dl), 09, 46 (D2), 03, 10 (EI) etc. Some serogroups were subdivideed into subserogroups (C1-C4; E1-E4). Serogroups A-Z and 51-67 represented O antigens 2 to 50 and 51 to 67 respectively. Groups 02 to 03, 10 (A-EI) contain nearly all the salmonellae that are important pathogens in man and animals.

Within each O group the different serotypes are distinguished by their particular H antigen or combination of H antigens (Table 39.2). Kauffmann-White classification gave species status to each serotype. The species were named according to the disease caused (S. Typhi), the animal source (S. Gallinarium), the discoverer (S. Schottmulleri), the name of the patient from whom the first strain was isolated (S. Thompson), or the place of isolation (S. Poona). This was satisfactory so long as the serotypes were not too many but now with some more than 2400 serotypes of salmonellae, giving individual names is not realistic.

The antigenic structure of a *Salmonella* serotype is expressed as an antigenic formula which has three parts describing the O antigens, the phase 1 H antigens and

Table 39.4: Kauffmann-White classification-Antigenic formulae of some representative serotypes of Salmonella

Serogroup	O-antigen group.	Serotype name	O antigens and Vi	H antigens	
				Phase 1	Phase 2
2	A	Paratyphi A	1,2,12	a	[1,5]
4	B	Paratyphi B	1,4,[5],12	b	1,2
		Stanley	1.4,[5],12,27	d	1,2
		Typhimurium	1.4,[5],12	i	1,2
		Heidelberg	1.4,[5],12	r	1,2
7	C1	Choleraesuis	6,7	c	1,5
		Paratyphi C	6,7[Vi]	c	1,5
		Typhisuis	6,7	c	1,5
		Virchow	6,7	r	1,2
8	C2-C3	Muenchen	6,8	d	1,2
		Newport	6,8,20	–e,h	1,2
		Hadar	6,8	z10	e,n,x
		Miami	1,9,12	a	1,5
	-	Sendai	1,9,12	a	1,5
9	D1	Typhi	9,12[Vi]	d	–
		Enteritidis	1,9,12	g,m	[1,7]
		Dublin	1,9,12,[Vi]	g,p	–
		Panama	1,9,12	l,v	1,5
		Gallinarum	1,9,12	–	–
3,10	E1	Anatum	3,10,[12],[1Q,;H)	e,h	1,6

Antigens in brackets [x] are not always present.

phase 2 H antigens in that order. The three parts are separated by colons and the component antigens in each part by commas. More than 2400 serotypes have so far been described in Kauffmann-White scheme.

Differentiation of Antigenically Similar Strains

Serotypes that share the same antigenic formula may be distinguished from one another by biochemical tests, e.g. Paratyphi C has the same O and H antigens as Choleraesuis, and Typhisuis. Sendai has the same antigens as the less virulent Miami. Thus, Paratyphi C ferments d-tartrate and trehalose within 2 days, but not Stern's glycerol. Choleraesuis ferments d-tartrate, but not trehalose or Stern's glycerol, and Typhisuis only trehalose. Sendai ferments arabinose but not Stern's glycerol, whereas Miami ferments Stern's glycerol but not arabinose.

Bacteriophage Typing

Strains within a particular serotype may be differentiated into a number of *phage types* by their patterns of susceptibility to lysis by members of a series of phages with different specificities. Intraspecies classification of S. Typhi for epidemiolocal purpose was made possible by bacteriophage typing, first developed by Craige and Yen (1937). They observed that a bacteriophage acting on the Vi antigen of the typhoid bacillus(Vi phage II) is highly adaptable. The parent phage is called phage A. It could be made specific for a particular strain of typhoid bacillus by serial propagation in the strain. Such adaptation was obtained by phenotypic or genotypic variation. Phage type E_1, O and A are the most common in India.

Serotypes that are subdivided in existing systems of phage typing include Typhi, Paratyphi A, Paratyphi B, Enteritidis, Hadar, Thompson, Typhimurium and Virchow. Thus, over 100 different phage types of Typhi and over 230 phage types of Typhimurium are distinguished. Among the Paratyphi A isolated from India, phage types 1 and 2 are the most common.

Phage typing is carried out at the National Phage Typing Center and is coordinated by the International Reference center. The National Salmonella Phage Typing Center for India is located at the Lady Hardinge Medical College, New Delhi. Phage types A and E1 are the most common and present throughout India. However, the relative prevalence in different regions is subject to change from time to time.

Biotyping

Biotyping is a useful adjunct to phage typing for it can subdivide a large group of untypale strains or members of common phage types. Similarily, strains of the same biotype may be subdivided into different phage types.

Pathogenesis

Salmonellae are strict parasites of animals or humans. All vertbrates appear capable of harboring these bacteria in their gut and salmonellae can colonize virtually all animals

Host-Adpted Serotypes

Some serotypes exhibit host specificity. S. Typhi, S. Paratyphi A and usually, but not invariably S. Paratyphi B are confined to human beings. Other salmonellae are parasitic in various animals-domestic animals, rodents, reptiles-and birds. These salmonellae adapted to particular animal hosts include Cholerae-suis. (pigs), Dublin (cattle), Gallinarum-pullorum (poultry), Abortus-equi (horses) and Abortus-ovis (sheep). On the other hand, many serotypes such as S. Typhimurium have a wide host range affecting animals, birds and man.

Clinical Syndromes

Salmonellae cause the following four major syndromes:

1. Enteric fever
2. Septicemia with or without metastatc infection
3. Gastro-enteritis or food poisoning
4. Asymptomatic carrier state.

1. Enteric Fever

The term enteric fever includes typhoid fever caused by S. Typhi and paratyphoid fever caused by Paratyphi A, B or C. The clinical features tend to be more severe' with S. Typhi *(typhoid fever)* Other salmonellae have on occasion been reported to cause enteric fever. These have included S. Bublin, S. Barielly, S. Sendai, S. Enteritidis, S. Typhimurium, S. Eastboume, S. Saintpaul, S. Oranienburg and S. Panama. Infection with *Alkaligenes fecalis* also may sometimes cause a similar clinical picture.

a. Typhoid Fever

The infection is acquired by ingestion. In human volunteer experiments, the ID50 was found to be about 10^3 to 10^6 bacilli. On reaching the gut, the bacilli attach themselves to the microvilli of the ileal mucosa by means of adhesins on the bacterial surface, which adhere specifically to mannose-containing receptors on the epithelium. They then penetrate to the lamina propria and submucosa, where they are phagocytized by neutrophils and macrophages. They resist intracellular killing and multiply within these cells. The ability to resist intracellular killing and multiply within these cells is a measure of virulence. They enter the mesenteric lymph nodes, where after a period of multiplication they invade the bloodstream via the thoracic duct.During this period the bacilli are seeded in the liver, gall bladder, spleen, bone marrow, lymph nodes, lungs and kidneys, where further multiplication takes place.After multiplication in these organs, bacilli pass into the blood, causing a second and heavier bacteraemia, the onset of which approximately coincides with that of fever and other signs of clinical illness.

The salmonellae multiply abundantly in the gall bladder as bile is a good culture medium for the bacillus and are discharged continuously into the intestine where Peyer's patches and other gut lymphoid tissues of ileum become involved. These become inflamed and infiltration with mononuclear cells, followed by necrosis, sloughing and the formation of characteristic typhoid ulcers occurs. Ulceration of the bowel leads to two major complications of the disease-intestinal perforation and hemorrhage.

Clinical Course

The incubation period is usually 7 to 14 days and appears to be related to the dose of infection. The clinical course may vary from a mild undifferentiated pyrexia to a rapidly fatal fulminating disease. The onset is usually gradual and early symptoms are often vague: Headache, malaise, anorexia, a coated tongue and abdominal discomfort with either constipation or diarrhea. Diarrhea is uncommon and early in the illness many patients complain of constipation.

In the untreated case the temperature shows a step" ladder rise over the first week of the illness, remains high for 7 to 10 days and then falls by lysis during the third or fourth week. **Physical signs** include a relative bradycardia at the height of the fever, hepatomegaly, splenomegaly and often a rash of *rose spots,* found on the front of the chest during the second or third week and fade on pressure. They are characteristic of, but not specific for, enteric fever and seldom noticeable in dark skinned patients. The whit blood cell count is normal or low.

Complications

The most important complications are **intestinal perforation, hemorrhage** and **circulatory collapse.** Some degree of **bronchitis or bronchopneumonia** is always found. Some develop **psychoses, deafness or meningitis**. Other complications include **cholecystitis, arthritis, periosteitis, nephritis, hemolytic anemia, venous thromboses, peripheral neuritis**. **Osteomyelitis** is rare sequel

Relapse

Apparent recovery can be followed by relapse in 5 to 10 percent of untreated cases. The relapse rate is higher in patients treated early with chloramphenicol (15-20%).

Morbidity and Mortality

Classic typhoid fever is a serious infection which, when untreated, has a mortality approaching 20 percent. In endemic areas, and particularly where it co-exists with schistosomiasis, chronic infection can present with fever of many months duration, accompanied by chronic bacteraemia.

b. Paratyphoid Fever

S. Paratyphi A and B cause paratyphoid fever which resembles typhoid fever but is generally milder. S Paratyphi C may also cause paratyphoid fever but more often it leads to a frank septicemia with suppurative complications.

2. Bacteremia with Focal Lesion

This is associated with S. Choleraesuis but may be caused by any salmonella serotype. Infection occurs by oral route and there is early invasion of the bloodsream with poassible focal lesions and in particular may cause septicaemic disease(with focal lesions, in lungs, bones, meninges etc.) but intestinal manifectations are often absent. Blood culture are positive.

3. Gasteroenteritis or Food Poisoning

This is the most common manifestation of salmonella infection. In the USA, Salmonella Typhimurium and Salmonella enteritis are prominent, but enreocoloitis can be caused by any of the more than 1400 group 1 serotype of salmonellae. (For detail see under the heading, "**Salmonella gasroentritis**".)

4. Asymptomatic Carrier State

Most people infected with salmonella continue to excrete the organism in their stools for days or weeks after complete clinical recovery but eventual clearance of the bacteria from the body is usual. A few patients continue to excrete the salmonellae for prolonged periods. The term *chronic carrier* is reserved for those who excrete salmonellae for a year or more. Up to 5 percent of convalescents from typhoid and a smaller number of those who have recovered from paratyphoid fever become chronic carriers, many for a lifetime. The development of carrier state is more common in women and in older age groups. After enteric fever, less than 1 percent of patients under 20 years old become carriers but this proportion rises to more than 10 percent in patients over 50 years of age.At all ages women become carriers twice as often as men.

Epidemiology

The typhoid and paratyphoid bacilli are essentially human parasites. Man is the reservoir host and most infections can be traced to a human source, or at least to a source of human sewage. All other salmonellae have animal hosts.

S. Paratyphi A is prevalent in India and other Asian countries, Eastern Europe and South America, S. Paratyphi B in Western Europe, Britain and North America; and S. Paratyphi C in eastern Europe and Guyana. Enteric fever is endemic in all parts of India.Typhoid fever has been estimated to affect 150 to 300 per 100,000 population, though there is no reliable information about its incidence in the country. The proportion of typhoid to paratyphoid A is about 10:1. Paratyphoid B is rare and C very rare. The disease occurs at all ages but is probably most common in the 5 to 20 year age group.

The source of infection is a patient, or far more frequently, a carrier. Patients who continue to shed typhoid bacilli in feces for three weeks to three months after clinical cure are called **'convalescent carriers'**. Those who shed the bacilli for more than three months but less than a year are called **'temporary carriers'** and those who shed the bacilli for a year or more are called **chronic carrier.**

Food handlers or cooks who become carriers are particularly dangerous. The best known of such typhoid carriers was Mary Mallon ('Typhoid Mary'), a New York cook who caused at least seven outbreaks affecting over 200 persons over a period of 15 years.

Typhoid fever occurs in two epidemiological types: endemic or residual and epidemic typhoid. The endemic or residual typhoid that occurs throghout the year though seasonal variations may sometimes be apparent. The epidemic typhoid, which may occur in endemic or nonendemic areas. Typhoid epidemics are water, milk or food borne.

The rate of clinical typhoid is about twenty five higher in the HIV infected than in others.

Laboratory Diagnosis

Bacteriological dignosis of enteric fever consists of:

A. Isolation and identification of the bacilli
B. Domonstration of circulating antigen.
C. Demonstration of antibodies in patient's serum.
D. Other laboratory teasts.

A. Isolation and Identification of the Bacilli

It may be done by culture of specimens such as patient's blood, feces, urine, bone marrow, duodenal drainage rose spots etc. For the levorotatory diagnosis of enteric fever, selection of relevant specimens depends upon duration of illness, e.g. blood for culture must be taken repeatedly. Urine culture may be positive after second week and stool second or third week.

Blood Culture

The organisms may be recovered from the blood stream at any stage of illness, but most commonly found during the first 7 to 10 days and during relapse. Blood cultures are positive in approximately 90 percent cases in the first week of fever in approximately 75 percent of cases in the second week, 60 percent in the third week and 25 percent thereafter till the subsidence of pyrexia (Fig. 39.3, Table 39.5). The popular belief that blood culture for diagnosis of typhoid fever is useful only in the first week is erroneous.

With all aseptic precautions, about 5 to 10 ml of blood is collected by venipuncture and inoculated into a culture bottle containing 50 to 100 ml of 0.5 percent bile broth. Large quantity (5-10 ml) of blood is required because the number of organisms are not numerous and may be

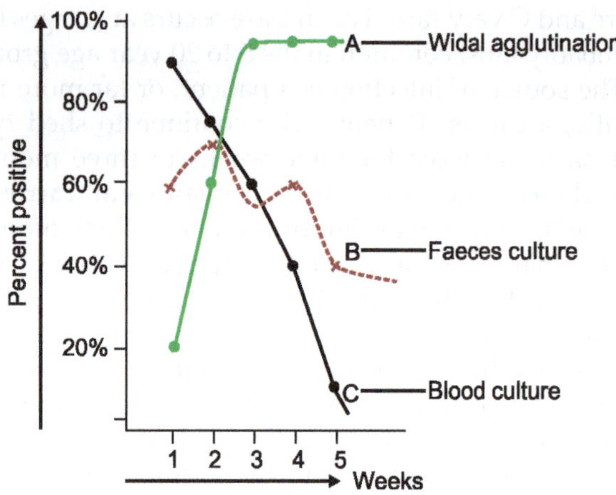

Fig. 39.3: Laboratory diagnosis of typhoid fever. The approximate percentages of tests found positive during different stages of the disease (from 1st to 5th week). A. Widal agglutination. B. Feces culture. C. Blood culture

Table 39.5: Positivity of various specimens at different phases of enteric fever

Duration	*Specimen*	*% positivity*
1st Week	Blood culture	90
	Feces culture	–
	Widal test	–
2nd Week	Blood culture	75
	Feces culture	50
	Widal test	Low titre
3rd Week	Blood culture	60
	Feces culture	80
	Widal test	80-100

quite small. Blood contains substances that inhibit the growth of the bacilii and hence it is essential that the broth be taken in sufficient quantity so that this effect may be annulled by diluting of blood, i.e. 5 to 10 ml of blood in 50 to 100 ml of the medium (10-fold dilution). The blood is diluted between 1 in 5 and 1 in 10 in culture medium to reduce the concentration of natural antimicrobilal constituents to a subeffective level.The addition of liquoid(sodium polyanethol sulphonate) counteracts the bactericidal action of blood.Every attempt should be made to collect samples before antibiotic therapy. Supplements are available to reduce an antibiotic effect.

After overnight incubation at 37ºC, the bile broth is subcultured on MacConkey agar and blood agar media. These plates are incubated at 37ºC for 24 hours. Pale nonlactose fermenting (NLF)colonies that may appear on MacConkey agar medium are picked out for biochemical tests and motility. Salmonellae will be gram-negative and motile bacilli.

Subcultures Repetition

If salmenellae are not obtained from the first subculture from bile broth, then subcultures should be repeated every other day till growth is obtained.It is usual to continue incubation and inspection for up to 5 to 7 days with a final subculture then. If no growth is obtained after 5 to 7 days, then the culture is declared negative.

Castaneda's Method of Culture

Castaneda's method of culture may be adopted to eliminate the risk of introducing contamination during repeated subculture and also for economy and safety. In this method, a double medium is used. The bottle of bile broth has an agar slant on one side. After inoculation of blood, the bottle is incubated in the upright position. For subculture, the bottle is merely tilted so that the broth runs over the surface of the agar. It is reincubated in the upright position. If salmonellae are present, colonies will appear on the slant.

Clot culture

An alternative to blood culture is the clot culture. Here, with strict aseptic precautions, 5 ml of blood is withdrawn from the patient into a sterile test tube and allowed to clot. The serum is pipetted off and used for Widal test. The clot is broken up with a sterile glass rod and added to a bottle of bile broth incorportating streotokinase(100 units/ml).Streptokinase in the broth facilitates lysis of the clot with release of bacteria trapped in the clot.

Advantage of Clot Culture

i. Clot cultures with streptokinase yields a higher rate of isolation than blood cultures as the bactericidal action of the serum is obviated.
ii. Another advantage is that a sample of serum also becomes available for Widal test. Even though agglutinins may be absent in the early stages of the disease, a Widal test provides a baseline titre against which the results of tests performed later may be evaluated.

Feces Culture

Salmonellae are shed in the feces throughout the course of the disease and even in convalescence, with varying frequency. Hence, fecal cultures are almost as valuable as blood cultures in diagnosis. Isolation from the feces is of less certain significance. A positive fecal culture may occur in carriers as well as in patients. The use of enrichment and selective media and repeated sampling increase the rate of isolation. Fecal culture is particularly valuable in patients on antibiotics as the drug does not eliminate the bacilli from the gut as rapidly as it does from the blood.

Fecal samples are plated diretly on MacConkey agar, DCA and Wilson and Blair's brilliant-green bismuth sulphite agar media. The last is highly selective and should be plated heavily. On MacConkey and DCA media, salmonellae appear as pale colonies. On Wilson and Blair's brilliant-green bismuth sulphite agar media, S. Typhi forms large black colonies, with a metallic sheen. S. Paratyphi A produces green colonies due to the absense of H_2S production.

For enrichment,one tube each of selenite F and tetrathionate broth are also inoculated. These are also incubated at 37°C for 8 to 12 hours with subsequent subculture on MacConkey agar and DCA media.

Urine Culture

Urine culture is less useful than the culture of blood and feces because salmonellae are shed in the urine irregularly and infrequently. Generally cultures are positive only in the second and third weeks and then only in about 25 percent of cases. The rate of isolation is improved by repeated sampling clean voided urine samples are centrifuged and the deposit inoculated into enrichment and selective media as for fecal culture.

Other Materials for Culture

Salmonellae may be isolated from several other sources but they are not usually employed. **Bone marrow culture** is valuable and it is positive in most cases even when blood culture is negative. **Culture of bile** obtained by duodenal aspiration is usually positive and this may be employed for the detection of carrires. Other matrials such as **rose spots, pus from suppurative lesions, CSF and sputum** may yield isolation at times. At autopsy, cultures may be obtained from the gall bladder, liver, spleen and mesenteric lymph nodes.

Colony Morphology

Pale nonlactose fermenting (NLF)colonies that may appear on MacConkey agar or DCA medium after incubating at 37°C for 24 hours are picked out for biochemical tests and motility. Salmonellae will be gram-negative and motile bacilli.

Biochemical Reactions

Salmonellae will be indole and urease negative, catalase positive, oxidase negative, **nitrate reduction positive** and ferment glucose, mannitol and maltose but not lactose or sucrose. S. Typhi will be anaerogenic and ferments glucose and mannitol with production of acid only, while paratyphoid bacilli (S. Paratyphi A, B and C) will form acid and gas from sugars.

Identification

Identification of the isolate is by slide agglutination.

Slide Agglutination Test

A loopful of the growth from an agar slope is emulsified in two drops of saline on a slide. One emulsion acts as a control to show that the strain is not autoagglutinable. If S Typhi is suspected (that is, when no gas is formed from glucose), a loopful of typhoid O antiserum (factor 9/group D) is added to one drop of bacterial emulsion on the slide and agglutination looked for after rocking the slide gently. Prompt agglutination indicates that the isolate belongs to Salmonella group D. Its identity as S. Typhi is established by agglutination with the flagellar antiserum (anti-d serum). Quite often, fresh isolates of S. Typhi are in the V form and do not agglutinate with the O antiserum. Such strains may be tested for agglutination against anti-Vi serum. Alternatively, the growth is scraped off in a small amount of saline, boiled for 20 minutes and tested for agglutination with the O antiserum.

Where the isolate is a nontyphoid Salmonella (producing gas from sugars), it is tested for agglutination with O and H antisera for groups A, B and C. For identification of unusual serotypes, the help of the National Salmonella Reference Center' should be sought. The National Salmonella Reference Center in India is located at the Central Research Institute, Kasauli. The reference center for salmonellae of animal origin is at the Indian Veterinary Research Institute, Izatnagar.

Colony Morphology and Staining

On MacConkey's agar or DCA, salmonellae grow as pale yellow, nonlactose fermenting (NLF) colonies. Gram staining from these colonies show gram negative bacilli and on hanging drop preparation, these are motile bacilli.

Biochemical Reactions

Salmonellae are catalase positive, oxidase negative, nitrate reduction positive and ferment glucose, mannitol but not lactose or sucrose. S. Typhi ferments glucose and mannitol with production of acid only but paratyphoid bacilli (S. Paratyphi A, B and C) form acid and gas.

B. Demonstration of Circulating Antigen

In the early phase of the disease, typhoid bacillus antigens are consistently present in the blood and also in the urine of patients. The antigen can be demonsrated by sensitized staphylococcal coagglutination. *Staphylococcus aureus* containing protein A (Cowan 1 strain) is stabilised with formaldehyde and then coated with S. Typhi anibody. When a 1 percent suspension of such sensitized cells is mixed on a slide with serum from patients in the first week of typhoid fever, the typhoid antigen present in the serum combines specifically with the antibody attached to the staphylococcal cells producing visible agglutination within two minutes. The test is rapid, sensitive and specific but is not positive after first week of the disease. ELISA test has also been employed to detect typhoid antigen in blood and urine.

C. Demonstration of Antibodies in patient's serum

i. Widal Test

Testing the patient's serum for salmonella antibodies is useful only in the diagnosis of enteric fever (Widal reaction) and even for this condition the significance of the results of the test is often doubtful. The Widal test is not recommended on a routine basis in areas of low endemicity. For the diagnosis of pyrexial illnesses, physicians should be advised to submit blood cultures and feces to the laboratory and not to rely on the fallible serological test.

Procedure

Tests for the presence of salmonella antibodies in the patient's serum may be of value in the diagnosis of enteric fever. This is a test for measurement of H and O agglutinins for typhoid and paratyphoid bacillli in the patient's serum.The patient's serum is tested by tube agglutination for its titres of antibodies against H, O and Vi suspensions of the enteric fever bacteria likely to be encountered, e.g. S. Typhi and S. Paratyphi A in India Two types of tubes are generally used for the test:

1. Dreyer's agglutination tube (narrow tubes with a conical bottom) for the H agglutination.
2. Felix tube (a short round bottomed tube) for the O agglutination.

Equal volumes (0.4 ml) of serial dilutions of the serum (from 1/10 to 1/640) and the H antigens of S. Typhi (TH) and S. Paratyphi A(AH) and O antigen of S. Typhi (TO) are mixed in Dreyer's and Felix's agglutination tubes respectively. Incubate H agglutinatiions for 2 to 4 hours in water bath at 37ºC and read after standing on the bench for half an hour. Incubate O agglutinations for 4 to 6 hours at 37ºC and read after overnight refrigeration at 4ºC. Controls tubes containing the antigen and normal saline are set to check for autoagglutination. The agglutination titers of the serum are read.

H agglutination leads to the formation of loose, cotton woolly clumps, while O agglutination is seen as a disk like pattern at the bottom of the tube. Control (Felix) tube shows a compact deposit. In both, the supernantant fluid is rendered clear. The highest dilution of the serum showing agglutination indicates the titre of the antibody.

The antigens used in the test are the H and O antigens of S. Typhi and S. Paratyphi A and B. The paratyphoid O antigens are not employed as they cross react with the typhoid O antigen due to their sharing of factor 12. Although suspensions may be prepared from suitable stock laboratory cultures, there is little need for that nowadays because ready made Widal kits of stained antigens available commertially are now widely used.

Interpretation of Widal Test

The results of the Widal test should be interpreted taking into account the following:

1. Stage of the Disease

The agglutinins titre will depend on their stage of the disease.Usually agglutinins appear by the end of first week (seventh to tenth day) of the illness, so that blood taken earlier may give a negative result and is inconclusive. The titre then increases steadily till the third or the fourth week, after which it declines gradually.

2. Rising Titre

Demonstration of a rise in titre of antibodies by testing two or more samples is more meaningful than a single test. However, if the first sample is taken late in the disease, a rise may not be demonstrable. Insead, a fall in titre may be seen in some cases.

3. Single Widal Test

The results of a single Widal test should be interpretated with caution. Levels of significance is difficult to lay down though it is generally stated that titres of 1/100 or more for O agglutinins and 1/200 or more for H agglutinins are significant. It is necessray to obtain information on the distribution of agglutinin levels in 'normal sera' in different areas.

4. Immunization

On account of prior disease, inapparent infection or immunization, agglutinuns may be present. Serum from an individual immunized with TAB vaccine will generally have antibodies to S. Typhi, S. Paratyphi A and B. However, in case of infection antibodies will be seen only against the infecting species.

5. O and H Agglutinis

H agglutinis persist longer (for many months) than O agglutinins which tend to disappear sooner, i.e. within 6 months. Therefore, rise in O agglutinins indicate recent infection.

6. Anamnestic Response

Persons who have prior infection or immunization may develop anamnestic response during an unrelated fever such as malaria, influenza etc. This may be differentiated by repetition of the test after a week. The anamnestic shows only a transient rise, while in enteric fever, the rise is sustained.

7. Fimbrial Antigen

Bacterial suspension used as antigens should be free from fimbria which may produce false positive results.

8. Effect of Treatment

Patients treated early with chloramphenicol may show a poor agglutinin response.

ii. Other Serological Tests

Enzyme-linked immunosorbent assay (ELISA) is a sensitive method of measuring antibody against the lipopolysaccharide of salmonellae. Indirect hemagglutination test and CIEP are other serological methods of diagnosis.

Detection of porins, the outer membrane proteins of S. Typhi, by ELISA method is useful for early serodiagnosis of typhoid fever.

D. Other Laboratory Tests

i. **Total leukocyte count (TLC):** A white cell count is useful. Leukopenia with a relative lymphocytosis is seen.
ii. **Diazo test in urine:** This test becomes positive generally between 5th and 14th day of fever and remains positive till the fever subsides.

Procedure

Equal volumes of patient's urine and the diazo reagent are mixed and a few drops of 30 percent ammonium hydroxide are added. if the test is positive, a red or pink froth develops on shaking the mixture.

Diazo Reagent

Solution A-Sulphanilic acid
Conc. H_2SO4
Distilled water
Solution B- Sodium nitrite
Distilled Water
For use, 40 parts of solution A are added to one part of solution B.

DIAGNOSIS OF CARRIERS

The detection of carriers is important for epidemilogical and public health purposes. It is also useful in screening food handlers and cooks.

1. Isolation of the Bacillus

The identification of the fecal carriers is by isolation of the bacillus from feces or from bile. The frequency and intensity of bacillary shedding vary widely and it is essential, therefore, to test repeated samples. Chance of isolation is increased by cholagogue purgatives. For the detection of urinary carriers,repeated urine culture should be carried out.

2. Widal Test

Widal test is of no value in the detection of carriers in endemic countries like India.The demonstration of Vi agglutinins (1:10 or more) has been claimed to indicate the carrier state. Whlie this is useful as a screening test but confirmation should be made culture.

3. Sewer-Swab Technique

The tracing of carriers in cities may be accomplished by the 'sewer-swab' technique. Gauze pads left in sewers and drains are cultured and by tracing positive swabs, one may be led to the house harboring a carrier.

4. Filtration Through Millipore Membrane

Another technique of isolating salmonellae from sewage is filtration through millipore membrane and culturing the membranes on highly selective media such as Wison and Blair media.

PROPHYLAXIS

Control depends essentially on safe water supply, proper sewage disposal, handling of food hygienically, and periodic examination of food handlers to ascertain that they are not carriers.

1. Vaccines Against Typhoid Fever

I. Killed whole cell vaccine:TAB vaccine

Heat-killed, phenol-preserved whole-cell vaccines containing a mixture of cultures of Typhi, Paratyphi A and Paratyphi B **(TAB)** have been used for many years in countries with a high endemic level of typhoid fever. The TAB vaccine which came into general use contained S. Typhi 1,000 million and S. Paratyphi A and B, 750 million each per ml killed by heating at 50 to 60°C and preserved in 0.5 percent phenol. **Acetone-killed vaccine** preserved in the dry state also provides similar protection.

In nonendemic areas, vaccination is recommended for troops. medical and paramedical personnel. In endemic areas vaccination is recommended for all children, in whom a single dose might give adequate protection. which may be maintained for several years by the booster effect of repeated natural subclinical infections.

Dose Schedule

The vaccine is given in 2 doses of 0.5 ml subcutaneusly at an interval of 4 to 6 weeks followed by a booster dose every 3 years.

Protection

Field trials have shown that such preparations confer protection against typhoid fever with an efficacy of 70 to 90 percent for 3 to 7 years.

Side Effects

Local and general reactions lasting for one or two days are quite frequent. There has always been doubt about the value of the paratyphoid components, as well as concern about the unpleasant side effects associated with the large antigenic load of the triple vaccine.

Note: In civilian practice, protection is mainly required against typhoid fever. Since only one of the paratyphoid A or B is found in an area, either A or B may be added, but not both. Hence in India, a divalent vaccine containing S. Typhi and S. Paratyphi A are now in use instead of the TAB vaccine, eliminating S. Paratyphi B which is rare in the country or the monovalent typhoid vaccine is preferred. In Europe and USA, monovalent vaccine containing S. Typhi is employed as paratyphoid A and B infections are rare in these countries.

2. Oral Vaccine-Live Oral (Ty21) Typhoid Vaccine

The live oral vaccine **(typhoral)** is a stable mutant of S. Typhi strain (Ty21a), lacking the enzyme UDP-galactose-4-epimerase (Gal E mutant). On ingestion, it initiates infection but 'self destructs' after four or five cell divisions, and therefore, it cannot induce any illness. This oral vaccine is available in enteric coated capsule containing 10^9 viable lyophilized mutant bacilli.

Dose Schedule

Three doses of the vaccine are given on alternate days to children. The course consists of one capsule orally, taken an hour before food, with a glass of water or milk, on days 1, 3, and 5. No antibiotic should be taken during this period.

Protection

It is safe and confers 65 to 96 percent protection for 3 to 5 years. Because immunity to typhoid fever appears to be primarily cell-mediated, a live vaccine will undoubtedly be more effective than killed vaccines. Therefore, several live avirulent vaccines are undergoing experimental trials.

3. Vaccine of Purified Vi Antigen (Typhim-Vi)

Dose Schedule

This injectable vaccine (*typhim-Vi*) contains purified Vi polysaccharide antigen (25 µg per dose) from S. Typhi strain Ty2.

Protection

It is given as a single subcutaneous or intramuscular injection, which causes only minimal reaction. This vaccine is not recommended for children younger than 2 years. Booster doses are recommended every 2 years.

In the trials conducted in black South African children and in young adults in Nepal the protection conferred by the vaccine was 64 percent and 72 percent respectively. The conjugation of purified Vi polysaccharide with a protein increases its immunogenicity.

TREATMENT

Specific antibacterial therapy for enteric fever became available only in 1948 with the introduction of chloramphenicol, which continued as the sheet anchor against the disease till the 1970s when resistance became common. Ampicillin, amoxicillin, and trimethoprim-sulfamethoxazole have been used successfully.

In the past decade, third-generation cephalosporins, particularly ceftriaxone and cefoperazone and fluoroquinolones including norfloxacin, ciprofloxacin, ofloxacin, and pefloxacin all are at least as effective as chloramphenicol in treating typhoid fever. Ciprofloxacin has emerged as the drug of choice for the treatment of adult typhoid, and is proving equally effective and free from side effects in children.

DRUG RESISTANCE

S. Typhi resistant to chloramphenicol was first reported in England in 1950, 2 years after this antibiotic was first successfully used in the treatment of typhoid fever. Resistance to chloramphenicol did not pose any problem in typhoid fever till 1972, when resistant strains emerged in Mexico and Kerala in India. Chloramphenicol resistant typhoid fever appeared in epidemic form first in Calicut (Kerala) in early 1972. It became endemic and was confined to Kerala till 1978. Subsequently such strains carrying drug resistance plasmids appeared in many other parts of India. Resistance was originally confined to phage type D1-N, but later to types C5, A and O. This multidrug resistance was due to a transmissible plasmid carrying resistance determinants to chloramphenicol, streptomycin, sulphadiazine and tetracycline (CSSuT).

By late 1980s, typhoid bacillus strains resistant to many or all of these drugs began to spread in most parts of India. However, many isolates of typhoid bacilli are still sensitive to chloramphenicol. A ciprofloxacin-resistant S. Typhi isolate was first reported from the United Kingdom in 1992 from a 1-year-old who acquired the infection in India and did not respond to therapy. Chromosomally mediated resistance to fluoroquinolones, associated with Vi phage type E1, was subsequently identified in other MDRTF isolates from patients returning from the Indian subcontinent.

At present, the drugs useful in treatment of such multiresistant typhoid cases are the later fluoroquinolones (such as ciprofloxacin, pefloxacin, ofloxacin) and the third generation celphalosporins (such as ceftazidime, ceftrioxone, cefotaxime).

SALMONELLA GASTROENTERITIS

Salmonella gastroenteritis (more appropriately **enterocolitis) or *food* poisoning** is generally a zoonotic disease. The source of infection being animal products. It may be caused by any salmonella except S. Typhi. In most parts of the world, S. Typhimurium is the commonest (30-40%) species. Some other common species have been S. Enteritidis, S. Haldar, S. Heidelberg , S. Agona, S. Virchow, S. Seftenberg, S. Indiana, S. Newport and S. Aanatum. S. Dublin.

The first instance of salmonella food poisoning was identified in 1888, when Gaerbler in Gennany isolated a bacillus (S.Enteritidis) from the meat of an emergency-slaughtered cow and from the cadaver of a fatal case of food poisoning caused by the meat. Durham in England and de Nobele in Belgium isolated S. Typhimurium from meat and from food poisoning cases in 1898. A very large number of salmonellae have since been identified from cases of gastroenteritis and food poisoning but a few species account for the majority of cases.

Source of Infection

The most frequent sources of salmonella food poisoning are poultry, meat, milk and milk products. Poultry (particularly hens), ducks and turkeys are the most significant reservoirs of food poisoning salmonellae in the UK. Of great concern are eggs and egg products. Hen eggs acquire their shells higher in the oviduct can be contaminated on the outside if laid on soil contaminated by infected hen feces or contaminated chicken feed.

Food contamination may also result from the droppings of rats, lizards or other small animals. Human carriers do occur but their role is minimal. Even salads and other uncooked vegetables may cause infection if contaminated through manure or by handling. Gastroenteritis may occur without food poisoning as in cross infection in hospitals.

Clinical Features

Clinically, the disease develops after a short incubation period of 24 hours or less, with diarrhea, vomiting, abdominal pain and fever. It usually subsides in 2 to 4 days but in some cases a more prolonged enteritis develops. In a few, typhoidal or septicemic type of fever may develop. Dehydration and eletrolyte imbalance constitute the major threats in very severe cases and in very young and old.

Laboratory Diagnosis

The laboratory diagnosis depends on the isolation of the causal organism from samples of feces or suspected foodstuffs. Isolation of salmonellae from the article of food confirms the diagnosis.

Treatment

Treatment of uncomplicated, noninvasive salmonellosis is symptomatic. Antibiotics should not be used. Not only do they not hasten recovery but they may actually increase the period of fecal shedding of the bacilli. But for the serious invasive cases, antibiotic treatment is needed.

Control

Control of salmonella food poisoning requires the prevention of food contamination which may be prevented by proper cooking of food and control of natural infection in animals.

SALMONELLA SEPTICEMIA

S. Choleraesuis, in particular may cause septicemic disease with focal suppurative lesions, such as osteomyelitis, deep abscesses, endocarditis, pneumonia and meningitis. Focal lesions may develop in any tissue.

Salmonella septicemia is prolonged and characterized by fever, chills, anorexia, and anemia. Gastroenteritis is minor or even absent, and the organism is rarely cultured from feces. Infection occurs by oral route and the incubation period is short. The case fatality may be as high as 25 percent.

Salmonelle may be isolated from the blood or from the pus from the suppurative lesions. Feces culture may sometimes may be positive. Septicaemic samonellosis should be treated with chroramphenicol or other appropriate antibiotics as determined by sensitivity tests.

A chronic bacteremia has also been described in patients with schistosomiasis. The *Schistosoma* carry the bacterium, so cure of the bacterimia is not achieved until the underlying parasitic infection is cured.

MULTIRESISTANT SALMONELLAE

R factors conferring multiple drug resistance have become widely disseminated among salmonellae. The clinical significance of this phenomenon was first observed during the studies of human and veterinary infections with drug resistant S. Typhimurium phage type 29 in England in the 1960s. Human infections were initially gastroenteritis due to spread from infected animals through food. Subsequently, some salmonellae appear to have changed their ecology in some ways. From being responsible for zoonotic infections only, as in the past, some multiresistant salmonellae have now become important agents of hospital cross infections. Such nosocomial salmonellosis manifests particularly in neonates as septicemia, meningitis and suppurative lesions. Diarrhea may not always be present.

In India, several hospital outbreaks of neonatal septicemia caused by multiresistant salmonellae have occurred in recent years. Mortality in neonates is very high unless early treatment is started with antibiotics to which the infecting strJn is sensitive.

KNOW MORE

Enrichment Media

1. **Tetrathionate broth:** It enriches salmonellae, including typhi, and sometimes shigellae, but permits the growth of *Proteus* species, which may reduce the tetrathionate and thus impair the selectivity for salmonellae.
2. **Kauffmann: Muller tetrathionate broth with brilliant green**-The addition of brilliant green (0.001%) inhibits the growth of *Proteus* and so improves the selectivity of the tetrathionate broth, but also makes it rather too inhibitory to Typhi and shigellae. As an alternative to brilliant green, the addition of 40 mg novobiocin/liter of medium before the addition of iodine serves to overcome interference by proteus.
3. **Selenite F broth:** It is excellent enrichment medium for Typhi and Dublin, but some salmonellae, e.g. Paratyphi A and Choleraesuis, and some shigellae may fail to multiply.

EPIDEMIOLOGY

Typhoid fever has been virtually eliminated from the advanced countries during the last several decades mainly as a result of improvements in water supply and sanitation but it continues to be endemic in the poor nations of the world. The control of paratyphoid fever has not been so successful.

In developing countries, S. Typhi is common, but data are incomplete. It has been estimated that between 10 and 500 cases of typhoid per 100 000 of the population occur annually throughout the developing world.

Some persons may become carriers following inapparent infection (symptomless excretor). The shedding of the bacilli is usually intermittent. The bacilli persist in the gallbladder or kidney and are eliminated in the feces (fecal carrier) or urine (urinary carrier). Urinary carriage is less often frequent and is generally associated with some urinary lesions such as calculi or schistisomiasis.

KEY POINTS

Salmonella

- Gram-negative bacilli, nonsporing, and noncapsulated, motile bacilli (Salmonella Gallinarum and Salmonella Pullorum are nonmotile).
- Aerobic and facultative anaerobic grow easily on a variety of nonselective media (e.g. Mueller Hinton agar, nutrient agar) and selective (e.g. Wilson and Blair's bismuth sulfite medium, xylose, lysine deoxycholate agar, etc). Selenite F and tetrathionate broth are enrichment media.

- Salmonellae possess: (1) Flagellar antigen H, (2) Somatic antigen O (3) Surface antigen Vi, found in some species.
- More than 2400 O serotypes (commonly referred to as individual *Salmonella* species).
  Salmonella are classified by Kauffman-White scheme based on structural formulae of the 0 and H antigens of the strains.
- Epidemiology- S. Typhi and S. Paratyphi are strict human pathogens (no alternative reservoir). Other salmonellae are parasitic in various domestic animals, rodents, reptiles, and birds. Infected patient and more frequently carriers are important reservoirs of infections for enteric fever. Food, vegetables, and water contaminated by human feces-by S. Typhi- are the common sources of infection. Animal-to-human zoonotic transmission is common in nontyphoidal salmonellae.
- Bacteriophage typing, biotyping, and recently, molecular methods are used for intraspecies classification of S. Typhi.
- **Diseases-**Asymptomatic colonization.
  Enteric fever.
  Enteritis.
  Bacteremia.
- Diagnosis-Isolation of the S. Typhi or S. Paratyphi from blood, feces, urine, bone marrow, duodenal drainage rose spots etc. Fecal culture is particularly valuable in patients on antibiotics. Urine culture are positive only in the second and third weeks and then only in about 25 percent of cases.
  In the early phase of the disease, typhoid bacillus antigens are consistently present in the blood and also in the urine of patients. Bile culture is useful for detection of carriers.
  Widal test is the traditional serologic test used for the diagnosis of typhoid fever.
  Indirect hemagglutination, counter-current immunoelectrophoresis, indirect fluorescent Vi antibody, and indirect ELISA for IgM and IgG antibodies to S. Typhi polysaccharide are other serological tests.
- DNA probes have been evaluated for identifying S. Typhi from bacterial culture isolates and directly from blood.
- Treatment-Infections with S. Typhi and S. Paratyphi or disseminated infections with other organisms should be treated with an effective antibiotic such as fluoroquinolones (e.g. ciprofloxacin), chloramphenicol, trimethoprim/sulfamethoxazole or a broad-spectrum cephalosporin can be used. Antibiotic treatment not recommended for enteritis because the duration of disease may be prolonged.
- Prevention and Control-Safe drinking water, proper food hygiene, and sanitary disposal of excreta are the most cost-effective strategies for reducing the incidence of typhoid fever. Carriers of S. Typhi and S. Paratyphi should be identified and treated.
  TAB vaccine, Vi capsular polysaccharide antigen vaccine and acetone-inactivated parenteral vaccine are the killed vaccines and Ty21 a vaccine is the oral vaccine used for immunization against typhoid fever.
- **Salmonella gastroenteritis** is generally a zoonotic disease. It may be caused by any salmonella except S. Typhi (S. Typhimurium, S. Enteritidis, S. Haldar, S. Heidelberg, S. Agona, S. Virchow, S. Seftenberg, S. Indiana, S. Newport and S. A anatum. S. dublin)
- The most frequent sources of salmonella food poisoning are poultry, meat, milk and milk products.
- **Salmonella septicemia:** S. Choleraesuis, in particular may cause septicemic disease with focal suppurative lesions, such as osteomyelitis, deep abscesses, endocarditis, pneumonia and meningitis.
- **Multiresistant salmonellae-**R factors conferring multiple drug resistance have become widely disseminated among salmonellae.

IMPORTANT QUESTIONS

1. Name the salmonellae causing enteric fever. Describe in detail the laboratory diagnosis of enteric fever.
2. Describe the pathogenesis and laboratory diagnosis of enteric fever.
3. Write short notes on :
   Antigenic structure of Salmonella
   Vi-antigen or surface antigen
   Kauffmann-White scheme
   Clot culture
   Widal test
   Vaccination against enteric fever
   Salmonella gastroenteritis
   Salmonella septicaemia.

FURTHER READING

Anand AC, Kataria VK, et al. Epidemic multiresistant enteric fever in eastern India. Lancet 1990;335:52.

Bhutta ZA, Naqui SH, et at. Multidrug-resistant 'typhoid in children: Presentation and clinical features. Rev Infect Dis 1991;13:832-6.

Brenner F, et al. Salmonella nomenclature, J Clin Microbiol, 2000;38:2465-67.

Chart H, Cheesbrough J S, Waghorn D J. The serodiagnosis of infection with Salmonella typhi. Journal of Clinical Pathology 2000;53:851-53.

Kumar S, Kumari S, Saigal SR. Non-typhoidal salmonellosis. Indian J Clin Practice 2005;15(8):43-51.

Kumar S, Sabherawal U, Arora DR. T and B lymphocytes in patients with enteric fever. Indian J Pathol Microbiol 1991;34:88-91.

Kumar S, Rizvi M, Berry N. Rising prevalence of enteric fever due to multidrug resistant Salmonella: an epidemiological study. J Med Micro 2008;57:1247-50.

Mandel BK. Salmonella typhi and other salmonellas. Gut 1994; 35:726.

Mirza SH et al. Multidrug resistant Salmonella typhi: A global problem. J Med Micriobiol 1996;44:317.

WHO. Control of Salmonella infections in animals and prevention of human foodborne Salmonella infections. Bulletin of the World Health Organization 1994;72:831-33.

CHAPTER

40 Vibrio, Aeromonas and Pleisomonas

Learning Objectives

After reading and studying this chapter, you should be able to:

- Describe morphology, cuture characteristics and biochemical reactions of *Vibrio cholerae*.
- Discuss antigenic structure of *Vibrio cholerae*.
- Describe the following: Pathogenesis of cholera; mechanism of action of cholera toxin.
- Differentiate between classical and El Tor vibrios.
- Discuss laboratory diagnosis of cholera.
- Describe the following: Cholera vaccine; non-agglutinating vibrios; halophilic vibrios.

INTRODUCTION

The second major group of gram-negative, facultatively anaerobic, fermentative bacilli are the genera *Vibrio, Aeromonas,* and *Pleisomonas*. These organisms were at one time classified together in the family **Vibrionaceae.** Molecular biology techniques have established, however, that these genera are only distantly related and belong in three separate families. *Vibrio* and *Aeromonas* are now classified in the families **Vibrionaceae** and **Aeromonadaceae,** respectively. *Pleisomonas* are closely related to proteus and have now been placed in the **Enterobacteriaceae** family, notwithstanding the differences noted previously. It is appropriate to consider these bacteria together because their epidemiology and range of diseases are similar, despite this taxonomic reorganization.

VIBRIO

Of the 35 *Vibrio* species recognized, 12 have been implicated in gastrointestinal and extra-intestinal infections in man. The genus can be divided into nonhalophilic vibrios, including *V. cholerae* and other species that are able to grow in media without added salt, and halophilic species that do not grow in these media. The most important member of the genus is *Vibrio cholerae,* the causative agent of cholera. It was first isolated by Koch (1883) from cholera patients in Egypt, though it had been observed earlier by Pacini (1854) and others. The species most frequently isolated from clinical specimens are strains of *V. cholerae, V. parahaemolyticus, V. vulnificus, V. minicus* and *V. alginolyticus.*

VIBRIO CHOLERAE

Morphology

These are gram-negative, short, curved, cylindrical rods, about 1.5 µm × 0.2-0.4 µm in size, with rounded or slightly pointed ends. The cell is typically comma shaped (hence the old name *V. comma)* but the curvature is often lost on subculture. Upon serial transfers, the organisms revert to straight forms. S shaped or spiral forms may be seen due to two or more cells lying end to end. In old cultures, they are frequently highly pleomorphic. The vibrios are seen arranged in parallel rows, described by Koch as the **'fish in stream' appearance** in stained films of mucous flakes from acute cholera cases.

It is actively motile, by means of a single, polar sheathed flagellum. The motility is of the **darting type,** and when acute cholera stool or a young culture is examined under the microscope, the actively motile vibrios suggest a **'swarm of gnats'.**

They are nonsporing, noncapsulated and nonacid-fast (Fig. 40.1).

Cultural Characteristics

The cholera vibrio is strongly aerobic, growth being scanty and slow anaerobically. It grows within a temperature range of 16-40°C (optimum 37°C). Growth is better in an alkaline medium and it occurs freely between pH 7.4 and 9.6 (optimum pH 8.2). *V. cholerae* is a nonhalophilic vibrio. It grows well on ordinary media.

A. Ordinary Media

i. Nutrient Agar

On **nutrient agar,** after overnight growth, colonies are moist, translucent, round disks, about 1-2 mm in

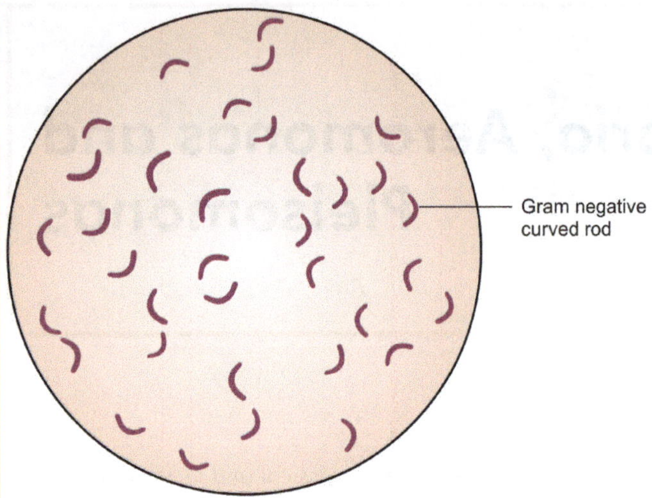

Fig. 40.1: Cholera vibrios

diameter, with a bluish tinge in transmitted light. The growth has a distinctive odor.

ii. MacConkey Agar

On **MacConkey agar**, the colonies are smaller than those on nutrient agar and are colorless, but become reddish on prolonged incubation due to the late fermentation of lactose. Growth is poor on most other enteric selective media such as DCA and XLD.

iii. Blood Agar

On **blood agar**, colonies are initially surrounded by a zone of greening, which later becomes clear due to hemodigestion.

iv. Gelatin Stab Culture

In **gelatin stab culture**, at first a white line of growth appears along the track of the inoculating wire. Liquefaction of gelatin begins at the top which spreads downwards in infundibuliform (funnel shaped) or napiform (turnip shaped) in 3 days at 22°C.

v. Peptone Water

When incubated at 37°C in liquid media, such as **peptone water**, it forms a fine surface pellicle because of its affinity for oxygen, which on shaking breaks up into membranous pieces. Turbidity and a powdery deposit develop on continued incubation.

B. Special Media

A number of special media have been employed for the cultivation of cholera vibrios. They may be classified as follows:

a. Holding or Transport Media

V. cholerae is quite sensitive to drying, exposure to sunlight and extreme changes in pH. It is also inhibited by normal intestinal flora. The stool samples should be transported in transport (holding) media if the cultures cannot be put up immediately.

i. Venkatraman-Ramakrishnan (VR) Medium

A simple modified form of this medium is prepared by dissolving 20 g crude sea salt and 5 g peptone in one liter of distilled water and adjusting the pH to 8.6-8.8. It is dispensed in screw capped bottles in 10-15 ml amounts. About 1-30 m1 stool is to be added to each bottle. Vibrios do not multiply in this medium but remain viable for several weeks. Moreover, it prevents overgrowth by other organisms.

ii. Cary-Blair Medium

This is a buffered solution of disodium hydrogen phosphate (1.1 g), sodium thioglycollate (1.5 g), sodium chloride (5.0 g) and agar (5.0 g) to 1 liter of distilled water and pH 8.4. It is a suitable transport medium for Salmonella and Shigella as well as for vibrios

iii. Autoclaved Sea Water

Autoclaved sea water also serves as a holding medium.

b. Enrichment Media

The rapid growth and tolerance of vibrios for alkaline conditions is exploited in the formulation of media used for their isolation.

i. Alkaline Peptone Water

Alkaline peptone water at pH 8.6 is useful for preliminary enrichment of vibrios from feces or other contaminated materials.

ii. Monsur's Taurocholate Tellurite Peptone Water at pH 9.2

Both these are good transport as well as enrichment media.

c. Plating Media

i. Alkaline Bile Salt Agar (BSA); pH 8.2

This simple medium has stood the test of time and is still widely used. This is modified nutrient agar medium containing 0.5 percent sodium taurocholate. The colonies on BSA are similar to those on nutrient agar medium.

ii. Monsur's Gelatin Taurocholate Trypticase Tellurite Agar (GTTA) Medium; pH 8.5

This medium is useful for the isolation of cholera and other vibrios from feces. High (8.5) pH and potassium tellurite, in this medium, are inhibitory to most enterobacteria (except *Proteus)* and gram-positive bacteria.

After 24 hours incubation, vibrios produce small (1-2 mm) translucent colonies with greyish black center and a turbid halo, due to hydrolysis and denaturation of gelatin. After 48 hours incubation, colonies increase in size to 3-4 mm.

iii. Thiosulphate-Citrate-Bile-Sucrose (TCBS) Agar; pH 8.6

This is most used selective plating medium for vibrios. Constituents of this medium are sodium thiosulphate, sodium citrate, ox bile, sucrose, yeast extract, peptone, sodium chloride, ferric citrate, thymol blue, bromothymol blue (indicator) and water. This medium resembles DCA except that it has the high pH value of 8.6 and contains sucrose instead of lactose. On this differential medium, the colonies of sucrose-fermenting vibrios, e.g. *V. cholerae*, are yellow, those of sucrose-non-fermenting vibrios, e.g. *V. parahaemolyticus*, are green.

Biochemical Reactions

1. **Sugar fermentation:** It ferments glucose, mannitol, maltose, mannose and sucrose and ferments lactose only after several days (late lactose-fermenter). Arabinose and dulcitol are not fermented.
2. **Cholera red reaction:** *V. cholerae* is strongly indole positive and reduces nitrates to nitrites. These two properties contribute to the **'cholera red reaction'** which is tested by adding a few drops of concentrated sulfuric acid to a 24-hour peptone water culture at 37°C. A reddish pink color is developed due to the formation of nitroso-indole with cholera vibrios.
3. It is **catalase and oxidase-positive, methyl red and urease negative.**
4. It decarboxylates lysine and ornithine but does not utilize arginine.
5. Gelatin is liquefied.
6. Voges Proskauer reaction and hemolysis of sheep RBCs are positive in El Tor biotype and both these reactions are negative in classical biotype.
7. **String test:** Vibrio colonies may be identified by the 'string test'. A loopful of the growth is mixed with a drop of 0.5 percent sodium deoxycholate in saline on a slide. If the test is positive, the suspension loses its turbidity, becomes mucoid and forms a 'string' when the loop is drawn slowly away from the suspension.

Glucose	Mannitol	Maltose	Mannose	Sucrose	Lactose
A	+	+	+	+	–
Indole	NO_3 reduction	Catalase	Oxidase	MR	VP*
+	+	+	+	–	–
Lysine	Ornithine	Arginine	Gelatin	Sheep RBCs hemolysis*	
+	+	–	+	+	
*In case of El Tor biotypes VP and sheep RBCs haemolysis are positive and all biochemical reactions are similar.					

RESISTANCE

Cholera vibrios are susceptible to heat, drying and acids, but resist high alkalinity. Vibrios are killed by heating at 56°C for 30 minutes or within a few seconds by boiling. They die quickly on dry fomites and in sewage-polluted water, but survive for 1-2 weeks in clean, nonacid fresh or sea water. In general, the El Tor vibrio survives longer than the classical cholera vibrio.

In the laboratory, vibrios survive for months in sterile sea water, and this has been suggested as a method for the survival of vibrios in nature. In grossly contaminated water, such as the Ganges water of India, the vibrios do not survive for any length of time, due to the apparently large amounts of vibriophages present. They survive in clean tap water for 30 days. On fruits, they survive for 1-5 days at room temperature and for a week in the refrigerator. They may survive up to a few days on moist fruit, vegetables, fish and cooked foods.

They are easily killed by disinfectants and short exposure to normal gastric juices for a few minutes but they may survive for 24 hours in achlorhydric gastric juice. Most strains are sensitive to tetracyclines, chloramphenicol, ampicillin, aminoglycosides and trimethoprim-sulfamethoxazole.

Antigenic Structure

1. **Flageller antigen (H antigen):** Many vibrios share a single heat-labile flagellar antigen. Antibodies to the H antigen are not involved in the protection of susceptible host.
2. **O antigen (LPS, endotoxin):** *V. cholerae* has O lipopolysaccharides (LPS, endotoxin) that confer serologic specificity. This apparently plays no role in the pathogenesis of cholera but is responsible for the immunity induced by killed vaccines. It may cause the fatal illness produced experimentally by peritoneal inoculation in mice. Antibodies to the O antigens tend to protect laboratory animals against infections with *V cholerae.*

There are at least more than 200 O antigen groups. *V. cholerae* strains of O group 1 and O group 139 cause classic cholera. Occasionally, *non-Ol/non0139 V. cholerae* causes cholera-like disease.

Classification

In the past, many oxidase positive, motile, curved rods were rather loosely grouped as vibrios. Precise criteria have been laid down for differentiating vibrios from related genera (Table 40.1).

Vibrios were classified by Heiberg (1934) into six groups based on the fermentation of mannose, sucrose and arabinose. Two more groups were added later. Cholera vibrios belong to Group I (Table 40.2).

Serological Classification

A serological classification was introduced by Gardner and Venkatraman (1935). Cholera vibrios and biochemically similar vibrios, possessing a **common flagellar (H) antigen** were classified as **Group A vibrios**, and the rest as **Group B vibrios** comprising a heterogeneous collection.

Table 40.1: Differentiation of vibrios from allied genera

Genus	Oxidation-Fermentation (High-Liefson Test)		Utilization of amino acids			String test
	Oxidation	Fermentation	Lysine	Arginine	Ornithine	
Vibrio	+	+1	+	–	+	+
Aeromonas	+	+2	–	+	–	V
Pseudomonas	+	–	V	V	V	–
Pleisomonas	+	+	+	+	+	–

Table 40.2: Heiberg grouping of vibrios

Group	Fermentation of mannose	Sucrose	Arabinose
I	A	A	–
II	–	A	–
III	A	A	A
IV	–	A	A
V	A	–	–
VI	–	–	–
VII	A	–	A
VIII	–	–	A

The **somatic (O) antigen** structure is of fundamental importance in the identification of this organism. Based on the major somatic (O) antigen, Group A vibrios were classified into **'subgroups'** (now called **O serogroups or serovars),** more than 200 of which are currently known. (Fig. 40.2). All isolates from epidemic cholera (till 1992) belonged to serogroup 0-1. Therefore in the diagnostic laboratory group 0-1 antiserum (commonly called **'cholera nondifferential serum'**) came to be used for identifying pathogenic cholera vibrios (which are referred to as **'agglutinable vibrios'**). Other vibrio isolates which were not agglutinated by the O-1 antiserum came to be called **nonagglutinable or NAG vibrios**. They were considered nonpathogenic and hence also called **noncholera vibrios (NCV).**

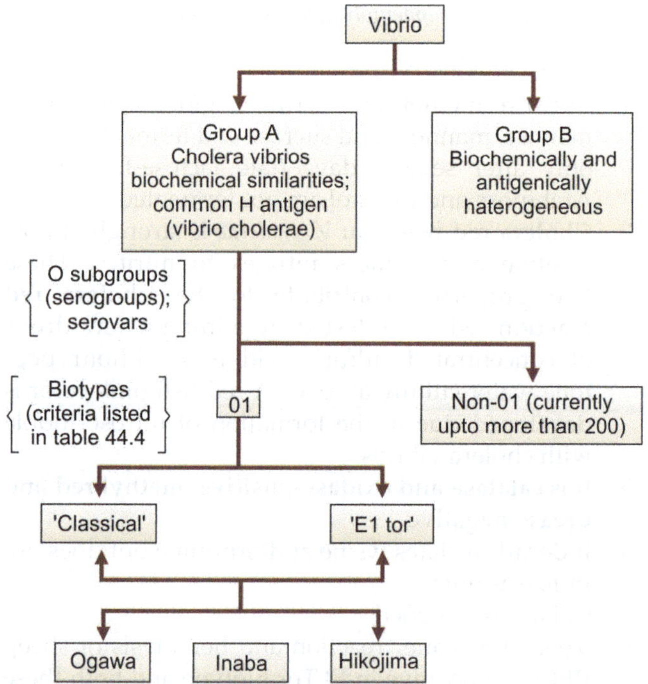

Fig. 40.2: Antigen classification of vibrios (Gardner and Venkatraman, updated)

Biotypes of *V. cholerae* 01

There are two biotypes of V. *cholerae* 01: *classical* and *El Tor* biotypes. The differences between classical and El Tor biotypes *V. cholerae* are shown in Table 40.3. El Tor biotype produces acetoin in the Voges-Proskauer test, agglutinates fowl erythrocytes, lyses sheep erythrocytes in a heart infusion broth with glycerol, grows in the presence of polymyxin (50 unit disk), is resistant to Mukerjee's group IV phage and sensitive to the group V phage of Basu and Mukerjee. The Classical biotype has the opposite properties.

1. **Chick Red Cell Agglutination Test:** A loopful of the organisms from an agar cultures is emulsified in a drop of saline on a slide and a drop of 2.5 percent chick erythrocyte suspension is added. Clumping of erythrocytes within .a minute indicates a positive test. The test is positive with all EI Tor strains but classical cholera vibrios are negative.
2. **Sensitivity to Polymyxin B:** The organism is tested by the disk diffusion method using disks containing 50 units of polymyxin B. All strains of classical cholera vibrio are sensitive and all strains of EI Tor vibrio are resistant.
3. **Sensitivity to Cholera Phage IV:** All strains of classical cholera vibrio are lysed by Mukherjee's group IV phage routine test dilution (RTD), while all EI Tor strains are not lysed. This is considered to be the most dependable test for differentiating between EI Tor and classical strains.

Subtypes of *V. cholerae* 01

Strains of *V. cholerae* 01 may be further subdivided on the basis of their O antigens into three subtypes, Ogawa, Inaba and Hikojima. This is on the basis of differences

Table 40.3: Differeces between classical cholera and El Tor vibrios

Test	Classical cholera	El. Tor
Hemolysis	–	+*
Voges-Proskauer	–	+*
Chick erythrocyte agglutination	–	+
Polymyxin B sensitivity†	+	–
Group IV phage susceptibility	+	–
El Tor phage 5 susceptibility	–	+

* Strain isolated after 1961 give variable results;
† 50 i. u. disk.

Table 40.4: Serotypes of cholera vibrios

Serotype	O antigens
Ogawa	AB
Inaba	AC
Hikojima	ABC

in minor O antigens (A, B and C). Antigen A is present in all the three subtypes. O antigens present in Ogawa, Inaba and Hikojima are A and B, A and C, and A, Band C respectively (Table 40.4). Ogawa and Inaba strains agglutinate with their specific absorbed antisera, while Hikojima strains which are rare are agglutinated by both Ogawa and Inaba antisera. Thus, the pandemic strain of 01 that started in South-east Asia in 1961 was of the Ogawa serotype, whereas the epidemic strain that recently spread through South America is of the Inaba serotype.

Non-Agglutinating (NAG) Vibrios

Though NAG vibrios are not agglutinable by the 0-1 antiserum, they are readily agglutinated by their own antisera. The term **noncholera vibrio** is not correct as some of them can cause a disease clinically indistinguishable from cholera. However, by and large, NAG vibrios were nonpathogenic and commonly isolated from environmental sources and healthy human intestines.

The non-Ol vibrios (the so called NAG vibrios) have been classified into many serogroups, currently upto more than 200. The serogroup 0-139, identified in 1992 causes epidemics of cholera, emphasising that they can no longer be considered as noncholera vibrios. *V. cholerae* 01 and 0139 are responsible for causing classic cholera, which can occur in epidemics or worldwide pandemics. Occasionally, non-01/non0139 *V. cholerae* causes cholera-like disease.

V. cholerae 0139

In 1992 cases of cholera indistinguishable from that caused by *V. cholerae* 01 were reported in Madras (Chennai), India. Similar outbreaks soon followed in different parts of India. By mid-January 1993 similar isolates were found in neighboring Bangladesh, and these rapidly spread north, following the course of the major rivers and raising fears of a new pandemic. The new strain did not agglutinate with antisera to any of the O serogroups and was assigned to a new serogroup, 0139 Bengal. However, it closely resembles *V. cholerae* 01 El Tor biochemically and physiologically, and it may eventually attain the status of a subtype *(Bengal)* of that organism alongside the subtypes *lnaba* and *Ogawa* already recognized.

V. cholerae 0139 may have evolved from *V. cholerae* 01, but with modified lipopolysaccharide structure. *V. cholerae* 0139 makes a polysaccharide capsule like other non-O 1 *V. cholerae* strains, while *V. cholerae* 01 does not make a capsule. In the affected areas, this strain replaced the El Tor vibrios as the epidemic and environmental serovar. It also showed a tendency to be more invasive, causing bacteremic illness in some. As it possessed novel surface antigens, the 0-1 strain vaccines could not protect against 0-139 infection. It was therefore considered likely that the 0-139 strain may initiate the next pandemic of cholera. Both 0-1 El Tor and 0-139 strains began to coexist in endemic areas.

Modern taxonomical criteria, particularly DNA studies, have led to the recognition that all the cholera vibrios that belong to Gardner and Venkatraman's group A and share similar biochemical properties and a common H antigen are so closely related that they constitute a single species *Vibrio cholerae,* which can be classified into serogroups (or serovars), biotypes and serotypes. Accordingly the present nomenclature will be indicative of all these features, as for example, *V. cholerae* serovar 01, biotype El Tor, serotype Ogawa.

Phage Typing

Further classification can be made by phage typing. Phage typing schemes have been standardized for classical and El Tor biotypes. New molecular methods like ribotyping have added further refinements to strain typing.

Strains of the classical biotype of V. cholerae 01 can be divided into 5 types by means of 3 phages (I-III) and a fourth phage (IV) lyses all classical but not El Tor strains (Table 40.5). On the basis of lysis by 4, phages, El Tor strains can be divided into 6 types. All these strains are lysed by a fifth phage (V) (Table 40.6).

Pathogenesis

In human infection, the vibrios enter orally through contaminated water or food. Vibrios are highly susceptible to acids, and gastric acidity provides an effective barrier against small doses of cholera vibrios. Any medication or condition that decreases stomach acidity makes a person more susceptible to infection with *V. cholerae.* Achlorhydria predisposes to cholera in the field.

Table 40.5: Phase types of strains of classical biotype of *Vibrio cholerae* 01 (Mukerjee, 1968)

Phage type	*Sensitivity to phase group*			
	I	*II*	*III*	*IV*
1	+	+	+	+
2	−	+	+	+
3	+	−	+	+
4	−	−	+	+
5	+	+	−	+

Table 40.6: Phase types of strains of El Tor biotype of *Vibrio cholerae* 01 (Basu and Mukerjee, 1968)

Phage type	*Sensitivity to phase group*				
	I	*II*	*III*	*IV*	*V*
1	+	+	+	+	+
2	+	+	+	−	+
3	+	+	−	+	+
4	+	+	−	−	+
5	+	−	−	−	+
6	−	+	−	−	+

The sequence of events leading to cholera is confined to the gut. The cholera vibrios are ingested in drink or food and, in natural infections, the dose must often be small. After passing the acid barrier of the stomach the organisms begin to multiply in the alkaline environment of the small intestine.

Once in the small intestine, strains of *V. cholerae* 0I migrate towards epithelial cells, facilitated by active motility and the production of mucinase and other proteolytic enzymes. A hemagglutinin-protease (formerly known as 'cholera lectin') cleaves mucus and fibronectin. It also helps in releasing vibrios bound to bowel mucosa, facilitating their spread to other parts of the intestine and also their fecal shedding. Adhesion to the epithelial surface and colonization may be facilitated by special fimbria such as the 'toxin coregulated pilus' (TCP). Throughout the course of infection, the vibrios remain attached to the epithelium but do not damage or invade the cells. The changes induced are biochemical rather than histological.

Cholera Toxin (CT)

Vibrios multiplying on the intestinal epithelium produce a toxin (choleragen, cholera enterotoxin, cholera toxin, CT, or CTX) which is very similar to the heat labile toxin (LT) of *E. coli* in structural, chemical, biological and antigenic properties, though CT is far more potent than LT in biological activity. CT production is determined by a filamentous phage integrated with the bacterial chromosome. It can also replicate as a plasmid which can be transmitted to nontoxigenic strains, rendering them toxigenic. Multiple chromosomal genes involved in the virulence of *V. cholerae* 01 have been characterized. These include genes for the two subunits of cholera toxin (*ctxA* and ctxB), toxin coregulated pilus (*tcp*) gene complex, accessory colonization factor (acf) genes, the hemagglutination-protease (*hap*) gene, and neuraminidase. Regulatory genes (e.g., *ToxR* regulator) control the expression of these genes.

Mechanism of Action

The toxin molecule, of approximately 84,000 MW consists of one A and 5 B subunits. The B (binding) subunit of cholera toxin binds to the ganglioside GM_1 receptors on the intestinal epithelial cells, which promotes entry of subunit A into the cell. The A (active) subunit, on being transported into the enterocyte dissociates into two fragments A_1 and A_2. The A_2 fragment only links the biologically active A_1 to the B subunit. The active portion (A_1) of the A subunit enters the cell and activates adenyl cyclase. The cholera enterotoxin causes the transfer of adenosine diphosphoribose (ADP ribose) from nicotinamide adenine dinucleotide (NAD) to a regulatory protein, which is part of the adenylate cyclase enzyme responsible for the generation of intracellular cyclic adenosine monophosphate (cAMP). The result is irreversible activation of adenylate cyclase and overproduction of cAMP. This in turn causes inhibition of uptake of Na^+ and Cl^- ions by cells lining the villi, together with hypersecretion of Cl^- and HCO_3 ions. This blocks the uptake of water which normally accompanies Na^+ and Cl^- absorption, and there is a passive net outflow of water across mucosal cells, leading to serious loss of water and electrolytes.

Other Biological Effects of CT

Other biological effects are also exhibited by CT which can be used for its detection and estimation. These include

i. Activation of lipolysis in rat testicular tissue
ii. Elongation of Chinese hamster ovary (CHO) cells in culture.
iii. Histological changes in adrenal tumor (YI) cell culture and vero cells.
iv. It also increases skin capillary permeability, and so has been called the **'permeability factor'** (PF). It can be demonstrated by the 'skin blueing test'—when CT is injected intradermally in rabbits or guinea pigs and pontamine sky blue injected intravenously afterwards, the site of toxin injection becomes blue.

Note: Cholera enterotoxin is antigenically related to LT of *Escherichia coli* and can stimulate the production of neutralizing antibodies. However, the precise role of antitoxic and antibacterial antibodies in protection against cholera is not clear. CT can be toxoided.

Cholera

Cholera is an acute diarrheal disease caused by *V. cholerae*. The incubation period varies from less than 24 hours to about five days. The clinical illness may begin slowly with mild diarrhea and vomiting in 1-3 days or abruptly with sudden massive diarrhea.

About 60 percent of infections with classic *V. cholerae* are asymptomatic, as are about 75 percent of infections with the El Tor biotype. In its most severe form, cholera is a dramatic and terrifying illness in which profuse painless watery diarrhea and copious effortless vomiting may lead to hypovolemic shock and death in less than 24 hours. In treated cases, the disease may last 4-6 days, during which period the patient may pass a total volume of liquid stool equal to twice his body weight. All the clinical features of severe cholera result from this massive loss of fluid and electrolytes.

The cholera stool is typically a colorless watery fluid with flecks of mucus, said to resemble water in which rice has been washed (hence called **'rice water stools'**). It has a characteristic inoffensive sweetish odor. In composition it is a bicarbonate-rich isotonic electrolyte solution, with little protein. Its outpouring leads to diminution of extracellular fluid volume, hemoconcentration, hypokalemia, base-deficit acidosis and shock. There is rapid loss of fluid and electrolytes, which leads to profound dehydration, circulatory collapse, and anuria.

The common complications are muscular cramps, renal failure, pulmonary edema, cardiac arrhythmias and paralytic ileus. The mortality rate without treatment is between 25 percent and 50 percent. The El Tor biotype tends to cause milder disease than the classic biotype.

Epidemiology

Cholera is both an epidemic and endemic disease.Seven major pandemics of cholera have occurred since 1817, resulting in thousands of deaths and major socioeconomic changes.

Cholera is an exclusively human disease. Infection is generally spread by contaminated water or foods such as uncooked seafood or vegetables. The source of the contamination is usually the feces of carriers or patients with cholera.

The seasonal incidence is fairly consistent in different endemic regions but the climatic conditions during epidemic waves may be distinctive for each region. For example, in Bangladesh the cholera season (November to February) follows the monsoon rains and ends with the onset of the hot dry months. In Calcutta the main epidemic wave (May to July) rises to its peak in the hot dry season and ends with the onset of the monsoon but extends inland to neighboring states during the rainy season.

India, more specifically the large deltaic area of the Ganges and Brahmaputra in Bengal, is its homeland, where it has been known from very ancient times. Till early in the nineteenth century, cholera was virtually confined to India, periodically causing large epidemics in different parts of the country. From 1817 to 1923 cholera vibrios had spread from Bengal, in six separate pandemic waves, involving most parts of the world. After the end of the 6th pandemic in 1923, till 1961 the disease remained confined to its endemic areas, except for an isolated epidemic in Egypt in 1947. It was largely due to the threat of pandemic cholera that international health organizations came into being.

The seventh pandemic occurred in 1961 and was first to be caused by the El Tor biotype. It originated from Sulawesi (Celebes), Indonesia, *V. cholerae* 0 I biotype El Tor was first isolated by Gotschlich at the El Tor Quarantine Station in Egypt. El Tor vibrio is hardier and more capable of surviving in the environment. Furthermore, it leads to a larger proportion of mild cases, higher incidence of carriers and greater chances of endemicity as compared to classical vibrio. After spreading to Hongkong and the Philippines, it spread steadily westwards, invading India in 1964. By 1966, it had spread throughout the Indian subcontinent and West Asia. In the 1970s the pandemic extended to Africa and parts of Southern Europe.

The vibrios had invaded affluent countries also during the course of the pandemic. In the I970s small outbreaks had occurred in Queensland, Australia and the Gulf Coast in the USA from special environmental foci in the coastal waters.

In January, 1991, the pandemic reached Peru, thus encircling the globe in thirty years time. The epidemic had involved all Latin American countries except Uruguay. By 1994 most parts of Central and South America had been involved and rendered endemic.

In 1992 cases of cholera indistinguishable from that caused by *V. cholerae* 01 were reported in Madras (Chennai), India. Similar outbreaks soon followed in different parts of India. By mid-January 1993 similar isolates were found in neighboring Bangladesh, and these rapidly spread north, following the course of the major rivers and raising fears of a new pandemic. The new strain was assigned to a new serogroup, 0139 Bengal. The new strain continued spreading, eastwards to the South East Asian countries, and westwards to Pakistan, China and some parts of Europe. But surprisingly, by 1994 the El Tor strain regained its dominance and the threat of an 0139 pandemic diminished.

Laboratory Diagnosis

The diagnosis of cholera can never be made with certainty on clinical grounds. Laboratory methods of diagnosis are required to confirm the diagnosis:

A. Specimen

Watery stool
Rectal swabs

Collection of Specimen

Stool

A fresh specimen of stool should be collected for laboratory examination. Sample should be collected before the person is treated with antibiotics. Collection may be made generally in one of the following ways:

i. **Rubber catheter:** Fecal specimens from early acute cases should be collected into a sterile container, e.g. universal container, preferably through a soft sterile rubber catheter (e.g. No. 24-26) inserted into the rectum. The specimen is best collected by introducing into the rectum a lubricated catheter and letting the liquid stool flow directly into a screw capped container.
ii. **Rectal swab:** Rectal swabs may be used, provided they are made with good quality cotton wool, absorbing about 0.1-0.2 ml of fluid. They are useful in collecting specimens from convalescents who no longer have watery diarrhea. In such cases, the swabs should be moistened with transport medium before sampling. If no transport medium is available, a cotton-tipped rectal swab should be soaked in the liquid stool, placed in a sterile plastic bag, tightly sealed and sent to the testing laboratory.

Collection from a bedpan should be avoided because of the risk of contamination or the presence of disinfectant.

Vomitus

Vomitus is not useful.

B. Transportation

It is necessary to preserve the specimen at 4 °C or in some appropriate holding medium as cholera vibrios may die in a few hours at tropical temperatures. If possible, specimens should be processed without delay but, if there is to be a delay of more than 6 hours in their reaching the laboratory, feces or rectal swabs should be placed in a liquid alkaline transport medium suitable for preventing the overgrowth of vibrios by other organisms1-3 ml feces may be added to 10-20 ml transport medium. Stool samples may be preserved in VR fluid or CaryBlair medium for long periods.

If the specimen can reach the laboratory in a few hours, it may be transported in enrichment media such as alkaline peptone water or Monsur's medium, thus saving the time required for isolation.VR medium can be used if larger stool specimens can be collectted. The specimen should be transported in alkaline peptone water or Cary-Blair medium if it is collected by a rectal swab.

Strips of blotting paper may be soaked in the watery stool and sent to the laboratory packed in plastic envelopes if transport media are not available.

Whenever possible, specimens should be plated at the bedside and the inoculated plates sent to the laboratory.

C. Microscopy

Direct microscopic examination of feces is not recommended as the results are not reliable. Gram-stained films of stools may prove equally difficult to interpret. For rapid diagnosis, the characteristic motility of the vibrio and its inhibition by antiserum can be demonstrated under the dark field or phase contrast microscope.

In the dark field, the vibrios evoke the image of many shooting stars in a dark sky. If motility ceases on mixing with polyvalent anti -cholera diagnostic serum, the organisms are presumed to be cholera vibrios. A presumptive diagnosis of cholera can thus be established. Demonstration of vibrios in stools by direct immunofluorescence has been attempted but nonspecific fluorescence is common and the technique is too complicated for use in the field.

D. Culture

In the laboratory, the sample should be plated both directly and after enrichment culture, on to suitable solid media. The specimens sent in enrichment media should be incubated for 6-8 hours including transit time. The specimens sent in holding media should be inoculated into enrichment media, to be incubated for 6-8 hours before being streaked on a selective and a nonselective medium.

Selective Media

The plating media used vary in different laboratories but the media employed usually are bile salt agar (BSA), MacConkey agar for nonselective and GTTA and TCBS agar for selective plates. It is possible to identify vibrio colonies on nonselective media after incubation for 4-5 hours by examination under a stereoscope with oblique illumination. Generally, the plates are examined after overnight incubation at 37°C.

Colony Morphology

On **MacConkey medium**, they form translucent colonies, on GTTA medium they form translucent colonies with greyish-black center and a turbid halo and on TCBS, they form yellow colonies.

E. Identification

Do the Gram staining from the suspected colonies and look for gram negative curved or comma-shaped rods. Perform motility and oxidase tests. Cholera vibrios show characteristic motility and are oxidase positive.

Slide Agglutination

Pick up oxidase-positive colonies with a straight wire and test by slide agglutination with *V. cholerae* 01 antiserum. If positive, agglutination may be repeated using specific Ogawa and Inaba antisera. Hikojima strains will agglutinate well with both Ogawa and Inaba antisera. If agglutination is negative with one colony, repeat the test with at least five more colonies as 01 and non-01 vibrios may co-exist in the same specimen.

If slide agglutination is positive, the isolate is tested for chick red cell agglutination. This is employed for presumptive differentiation between El Tor and classical cholera vibrios. A report can be sent at this stage, usually the day after the specimen is received. If no vibrios are isolated, a second cycle of enrichment and plating may succeed in some cases.

Biochemical Reactions

The identity of the organism should be confirmed biochemically in a set of conventional tests. The use of a Hugh & Leifson O/F test, Moller's arginine dihydrolase, lysine and ornithine decarboxylases, arabinose, inositol, mannose and sucrose peptone water sugars and a test for growth in the absence of NaCI by inoculating either a 1 percent tryptone water with no added NaCl or a CLED plate. Alternatively, laboratories that isolate vibrios infrequently may find it easier to use commercial kits for identification.

For further characteriration of the biotype of the *V. cholerae* 01 isolate, do VP test, agglutination of fowl RBCs, hemolysis of sheep RBCs, and sensitivity to polymyxin B, Mukerjee phages IV and V (Table 40.4). The strain may be sent to the International Reference Center for vibrio phage typing at the National Institute of Cholera and Enteric Disease (NICED) at Calcutta for confirmation of identity and for O-serotyping.

When the isolated strain is not agglutinated by *V. cholerae* 01 antiserum, it should be tested for agglutination with *V. cholerae* H antiserum. Any vibrio which is agglutinated by H antiserum and not by 01 antiserum is considered to be to be non-Ol cholera vibrio. Specific antiserum against 0-139 is available. In the fully equipped laboratory, diagnostic tests in cholera and other diarrheal diseases should consist of a battery of tests designed to isolate other known pathogens also.

Serological Diagnosis

In patients who have not been immunized, a retrospective diagnosis of infection can be made by the demonstration of a rising titer of agglutinins or vibriocidal antibodies in paired sera, one taken in the first 3 days of illness and the other after 7-10 days or antibody against cholera toxin (CT) may be detected in an ELISA test.

Detection of Carriers

For isolation of vibrios from carriers, essentially the same techniques are to be followed, except that more than one cycle of enrichment may be necessary. As vibrio excretion is intermittent, repeated stool examination will yield better results. Examination of stools after a purgative (magnesium sulfate 15-30 g or Mannitol 30 g), or of bile after duodenal intubation is of special value.

Serological examination is of little use in the diagnosis of cases though it may be helpful in assessing the prevalence of cholera in an area. The tests available are agglutination using live or killed vibrio suspensions, indirect hemagglutination, vibriocidal test and antitoxin assay. Of these, the complement dependent vibriocidal antibody test is the most useful.

Examination of Water Samples

For examination of water samples for vibrios, enrichment or filtration methods may be employed.

Enrichment Method

In this method, 900 ml of water are added to 100 ml tenfold concentrated peptone water at pH 9.2, incubated at 37°C for 6-8 hours and a second enrichment done before plating on selective media.

Filtration Method

For the filtration technique the water to be tested should be filtered through the Millipore membrane filter, which is then placed directly on the surface of a selective medium and incubated. Colonies appear after overnight incubation. Sewage should be diluted in saline, filtered through gauze and treated as for water.

Treatment

1. Oral Rehydration Therapy (ORT)

In cholera absolute priority must be given to the lifesaving replacement of fluid and electrolytes. Oral rehydration therapy (ORT) is often sufficient, but severe cases may require intravenous rehydration.

2. Antibacterial Therapy

Antibacterial therapy is of secondary importance. Oral tetracycline was recommended for reducing the period of vibrio excretion and the need for parenteral fluids. This dramatically reduces infectivity. Alternatively, the long-acting tetracycline (doxycycline) may be used for chemoprophylaxis, if the prevailing strains are not resistant.

Initially cholera vibrios were uniformly susceptible to all antibiotics active against gram-negative bacilli, but since 1979, multiple drug resistant strains have become increasingly common.

Immunity

Gastric acid provides some protection against cholera vibrios. In cholera, the vibrios remain confined to the intestine, where they multiply and elaborate the enterotoxin which is responsible for the disease. Immunity, therefore, may be directed against the bacterium or against the toxin-antibacterial or antitoxic. Natural infection confers some amount of immunity but it does not seem to last for more than 6-12 months and reinfections are known after this period.

An attack of cholera is followed by immunity to reinfection, but the duration and degree of immunity are not known. In experimental animals, specific IgA antibodies occur in the lumen of the intestine. Similar antibodies in serum develop after infection but last only a few months. Vibriocidal antibodies in serum (titer~1:20) have been associated with protection against colonization and disease. The presence of antitoxin antibodies has not been associated with protection.

Immunity may be local, in the intestine, or systemic. The appearance of local antibodies in the intestine has been known for a long time. These are known as 'coproantibodies' as they appear in the feces. They consist of IgG, IgM and IgA.

Prophylaxis

1. General Measures

Most important are the provision of safe drinking water supplies and the proper disposal of human feces. Control rests on education and on improvement of sanitation, particularly of food and water.

2. Specific Measures-Vaccines

Following vaccines are available:

a. Killed Whole Organism Vaccine

It is killed suspension containing 8000 million *V. cholerae* per ml, composed of equal numbers of Ogawa and Inaba serotypes. The concentration of the vaccine has been increased to 12,000 million per ml, in order to improve the antigenic stimulus. Primary immunization consists of 2 equal doses, injected subcutaneously, at an interval of 4 to 6 weeks.

The vaccine trials in Calcutta, Bangladesh and the Phillippines showed that this vaccine offers about 60 percent protection for 3-6 months. A single dose of vaccine is ineffective in children below five years of age while two doses at 1-4 week intervals are protective. However, it confers protection in adults due to its action as a booster because of prior natural infection.

Cell free somatic antigen preparations are as effective as whole cell vaccine. Aluminium hydroxide and phosphate adjuvant vaccines produced better immunity, particularly in young children. Toxoid vaccines have not been successful.

In recent years, doubts have been raised about the usefulness of cholera vaccine as a preventive measure. They are of no value in controlling epidemics. Traditional whole-cell vaccines are not very effective and are no longer recommended for travellers.

b. Oral Vaccine

Two types of oral cholera vaccines are available in some countries:

i. Non-Living Oral B Subunit-Whole Cell (BS-WC) Vaccine

This vaccine consists of killed whole-cell *V. cholerae* 01 in combination with a recombinant B-subunit of cholera toxin (WC/rBS) is available since early 1990s. This vaccine contains CT subunit B, 2.5×10^{10} heat killed vibrios each of Ogawa and Inaba serotypes of classical biotype and equal number of formalin killed vibrios each of Ogawa and Inaba serotypes of El Tor biotype.

It is given orally in two dose schedule, 10-14 days apart. On an average, the vaccine confers 50-60 percent protection for at least 3 years.

ii. Live Oral Vaccine

Recombinant DNA vaccine with expression of *V. cholerae* 01 in attenuated strain of S. Typhi Ty21 as a carrier bacterium has been developed. The live salmonellae colonize the Peyer's patches of the small intestine and induce IgA response by local immune system of the gut.

It is a single dose vaccine. The overall protective effect against El Tor cholera of any severity is 80 percent. There is no vaccine for the 139 strains.

The World Health Assembly in May 1973 abolished the requirement of a cholera vaccination certificate for international travel.

HALOPHILIC VIBRIOS

Vibrios that have a high requirement of sodium chloride are known as halophilic vibrios. Their natural habitat is sea water and marine life. Some halophilic vibrios have been shown to cause human disease - *V. parahaemolyticus, V. alginolyticus* and *V. vulnificus.*

Vibrio parahaemolyticus

Morphologically, it resembles *V. cholerae* except that it is capsulated, shows bipolar staining and has a tendency to pleomorphism, especially when grown on 3 percent salt agar and in old cultures. However, being a halophilic species, it does not grow in peptone water without sodium chloride or on cysteine-lactose electrolyte-deficient (CLED) agar but grows well in peptone water with 8 percent (but not 10%) sodium chloride.

It grows well on blood agar. On MacConkey agar it forms pale, non-lactosefermenting colonies and on sheep blood agar it produces β-hemolysis. On TCBS agar, the colonies are green (non-sucrose-fermenting). Unlike other vibrios, it produces peritrichous flagella when grown on solid media. Polar flagella are formed in liquid cultures.

Biochemical Reactions

It is oxidase, catalase, indole and citrate positive. It reduces nitrate to nitrite. It ferments glucose, maltose, mannitol, mannose and arabinose with the production of acid only. Lactose, sucrose salicin, dulcitol or inositol are not fermented. It is VP positive and decarboxylates lysine and ornithine but not arginine.

Pathogenesis

Not all strains of *V. parahaemolyticus* are pathogenic for human beings. Strains isolated from environmental sources (such as water, fish, crabs or oysters) are nearly always nonhemolytic when grown on a special high salt blood agar **(Wagatsuma's agar),** while strains from human patients are almost always hemolytic. This hemolysis is known as the **Kanagawa phenomenon** and is due to a heat stable hemolysin.

The significance of this hemolysis is not known but it is used as a laboratory test for pathogenicity. Kanagawa positive strains being considered pathogenic for human beings and negative strains nonpathogenic. No enterotoxin has yet been isolated from this organism. The vibrio is believed to cause enteritis by invasion of the intestinal epithelium.

Vibrio parahaemolyticus causes acute gastroenteritis following ingestion of contaminated seafood such as raw fish or shellfish. After an incubation period of 12-24 hours, nausea and vomiting, abdominal cramps, fever, and watery to bloody diarrhea occur. Fecal leukocytes are often observed. The enteritis tends to subside spontaneously in 1-4 days with no treatment other than restoration of water and electrolyte balance. A few extra-intestinal infections have been reported, particularly from wounds.

Epidemiology

The disease occurs worldwide, with highest incidence in areas where people eat raw seafood. Cases are more common in summer, and in adults than in children. In Calcutta, *V. parahaemolyticus* could be isolated from 5-10 percent of diarrhea cases admitted to the Infectious Diseases Hospital. *V. parahaemolyticus* is common in sea fish in some other parts of India but human cases are much less frequent.

Laboratory Diagnosis

The feces of patients with a history of recent consumption of seafood may be examined by the methods used for *V. cholerae.* In the examination of seafood and sea or estuarine waters for halophilic species, including *V. parahaemolyticus,* enrichment culture in alkaline peptone water containing 1 percent sodium chloride is used.

Vibrio Vulnificus

V. vulnificus, previously known as L+ vibrio or *Beneckea vulnifica,* is a marine vibrio of medical importance. After cholera, the second most serious type of *Vibrio*-associated infections are those caused by *V. vulnificus.*

This halophilic vibrio resembles *V. parahaemolyticus* in forming green, non-sucrose-fermenting colonies on TCBS medium. It is VP negative and ferments lactose but not sucrose. It has a salt tolerance of less than eight percent. The ability of *V. vulnificus* to ferment lactose ('lactose positive vibrios') a key identifying characteristic because the other *Vibrio* species most commonly encountered in clinical laboratories are lactose negative. It can grow in the presence of a 300 unit disk of polymyxin.

Vibrio vulnificus can cause severe wound infections, bacteremia, and probably gastroenteritis.

It is responsible for rapidly progressive wound infections after exposure to contaminated seawater. Following ingestion of the vibrio, usually in oysters, it penetrates the gut mucosa without causing gastrointestinal manifestations and enters the bloodstream, rapidly leading to septicemia with high mortality. Infections are most severe in patients with hepatic disease, hematopoietic disease, or chronic renal failure and in those receiving immunosuppressive drugs.Acute diarrhoea follows the consumption of shellfish. This is less common.

Diagnosis is by culturing the organism on standard laboratory media; TCBS is the preferred medium for stool cultures, where most strains produce blue-green (sucrose-negative) colonies.

Tetracycline appears to be the drug of choice for *V vulnificus* infection; ciprofloxacin may be effective also based on in vitro activity.

Vibrio Alginolyticus

Vibrio alginolyticus is a halophilic organism formerly regarded as biotype 2 of *V. parahaemolyticus.* It resembles *V. parahaemolyticus* in many respects. It has a higher salt tolerance, is VP positive and ferments sucrose (Table 40.7). It forms large, yellow (sucrose fermenting) colonies on TCBS. There is pronounced swarming on non-selective solid media. It has been associated with infections of eyes, ears and wounds in human beings exposed to sea water.

Other Vibrio Species

V. hollisae, V. mimicus, V. fluvialis, V. furnissii, and *V. damsela* are responsible for causing gastroenteritis, wound infections, and bacteremia.

Single cases of infections with *V. metschnikovii* (bacteremia), *V. cincinnatiensis* (meningitis), and *V. carchariae* (wound infection) have been documented.

AEROMONAS

The taxonomy of the genus aeromonas is in transition. Until recently, Aeromonas *was classified in the family Vibrionaceae.* The genus has been placed in the new **family Aeromonadaceae** from the family Vibrionaceae.

Aeromonas is a gram-negative, facultative anaerobic bacillus that morphologically resembles members of the Enterobacteriaceae. A total of 16 species of Aeromonas has been described, including 11 associated with human disease. The most important pathogens are *Aeromonas hydrophila, Aeromonas caviae,* and *Aeromonas veran;*

Table 40.7: Some characteristics of *V. parahaemolyticus* and *V. alginolyticus*

	V. parahaemolyticus	*V. alginolyticus*
Indole	+	+
V.P.	–	+
Nitrate reduction	+	+
Urease	–	–
Sucrose fermentation	–	+
Swarming	–	+
Growth in 0% NaCl	–	–
7% NaCl	+	+
10% NaCl	–	+

biovar soma. The organisms are ubiquitous in fresh and brackish water. They form exoenzymes, such as amylase, deoxyribonuclease, esterases and peptidases, and many other hydrolytic enzymes.

The two major diseases associated with *Aeromonas* are **gastroenteritis** and **wound infections (with or without bacteremia).**

Gastroenteritis typically occurs after the ingestion of contaminated water or food, whereas wound infections result from exposure to contaminated water. They are opportunistic pathogens.

Aeromonas strains are susceptible to tetracyclines, aminoglycosides, and cephalosporins.

PLEISOMONAS

Pleisomonas are closely related to Proteus and have now been placed in the **Enterobacteriaceae family.** It has only one species, *P. shigelloides* which is taxonomically related to Proteus species and serologically related to *Shigella sonnei.*

They are oxidase positive, motile, have multiple polar flagella gram-negative bacilli and may be mistaken for vibrios. The organism is found in fresh water and estuarine waters and is acquired through contact with fresh water, the consumption of seafood, or exposure to amphibians or reptiles.

P. shigelloides is ubiquitous in surface waters and in soil. It has also been isolated from a variety of mammals, including dogs, cats, goats, sheep and monkey. It is rarely recovered from human feces. It commonly infects various cold-blooded animals like frogs, snakes, turtles and lizards. Man is infected primarily by ingesting contaminated water or food. It causes:

1. **Gastroenteritis:** Which manifests as a mild,watery diarrhea in which stools are free of blood and mucin.
2. **Extraintestinal lesions:** Including septicoemia, endophthalamitis, septic arthritis, meningitis, cellulitis, and acute cholecystitis.

KNOW MORE

Animal Models for V. cholerae

Under natural conditions, *V. cholerae* is pathogenic only for humans and not in animals. A number of animal models have been developed which have helped in understanding the pathogenic mechanisms in cholera. Rabbit ileal loop model of De and Chatterjee (I 953) was the first of these models. Injection of cholera culture or culture filtrate into the ligated ileal loop caused fluid accumulation and ballooning. Intestinal loops of many other animals and also of chickens have been shown to behave in a similar manner.

KEY POINTS

- *Vibrio* and *Aeromonas* are classified in the families Vibrionaceae and Aeromonadaceae, respectively. *Pleisomonas* are closely related to Proteus and have now been placed in the Enterobacteriaceae family.
- *Vibrio cholerae is* the causative agent of cholera and is short, typically comma-shaped, show typical darting type motility. It is strongly aerobic.
- It grows well on a wide variety of media including ordinary and special media.
  Transport media are VR medium and Cary-Blair medium. Enrichment media are alkaline peptone water and Monsur's taurocholate tellurite peptone. Selective media are TCBS medium, Monsur's GTTA medium, and alkaline BSA.
- Two biotypes of *V. cholerae* 01 strains-EI tor and classical. Non-01 *V. cholerae,* which do not agglutinate with 01 group antisera are designated as noncholera vibrios or nonaggluti nating vibrios.
- Classical and El Tor biotype vibrios can be differentiated by acetoin in the Voges-Proskauer test, agglutination of fowl erythrocytes, lysis of sheep erythrocytes, polymyxin B senisitivity, susceptibility to Mukerjee's group IV phage and sensitivity to the group V phage.
- **Disease:** Cholera is a toxin-mediated disease. Cholera toxin is primarily responsible for the watery diarrhea characteristic of this species. Spread is by consumption of contaminated food or water.
- **Laboratory diagnosis:** Fresh stool specimen is used for dark-field microscopy and direct immunofluorescence. The specimen inoculated on a selective and nonselective media. Suspected *V. cholerae* are tested by slide agglutination using specific *V. cholerae* 01 antisera. If the colony is identified as *V. cholerae* 01 then it is tested by various tests to determine whether isolated *V. cholerae* 01 is classical or Eltor.
- **Treatment, Prevention, and Control:** Treatment of cholera is based on prompt and adequate replacement of fluid and electrolytes. Antibiotic therapy reduces the bacterial burden and exotoxin production, as well as duration of diarrhea. Doxycycline (adults), trimethoprim-sulfamethoxazole (children), or furazolidone (pregnant women) is administered.
  Improved hygiene is critical for control. The killed parenteral vaccine is of no value, but the newer oral vaccine has some protective value.
- **Halophilic vibrios:** *Vibrio parahaemolyticus, Vibrio alginolyticus,* and *Vibrio vulnificus* are three important halophilic vibrios species known to cause infection in humans. *V. parahaemolyticus* in humans causes gastroenteritis.
- **Aeromonas:** Aeromonas species in humans cause gastroenteritis and wound infections.

- **Pleisomonas:** *P. shigelloides* (the only species) in humans cause gastroenteritis, celullitis, septic arthritis, septicemia, neonatal meningitis, etc.

IMPORTANT QUESTIONS

1. Discuss laboratory diagnosis of cholera.
2. Write short notes on:
   Classification on vibrios
   Non-cholera vibrios
   Differences between classical and El Tor vibrios
   Pathogenesis of cholera
   Cholera toxin
   Prophylaxis against cholera
   Halophilic vibrios
   Kanagawa phenomenon.
   Aeromonas
   Pleisomonas

FURTHER READING

Albert MJ. Vibrio cholerae 0139. J Clin Microbiol 1994;32:2345-9.

Altwegg M, Martinetti Luchini G, et al. Aeromonas associated gastroenteritis after consumption of contaminated shrimp. Eur J Clin Microbiol Infect Dis 1991;10:44-5.

Calia KE, Murtagh M, et al. Comparison of Vibrio cholerae 0139 with V. cholerae 0I classical and El Tor biotypes. Infect Immune 1994;62:1504-6.

Calia KE. Vibrio cholerne 0139: An emerging pathogen, Clin Microbiol Newsletter 1996;18:17-22.

Holmberg SD, Farmer JJ ID. Aeromonas hydrophila and Pleisomonas shigelloides as causes of intestinal infections, Rev Infect Dis 1984;6:633-39.

Janda JM, et al. Aeromonas species in septicemia: Laboratory characteristics and clinical observations, Clin Infect Dis1994; 19:77-83.

CHAPTER 41

Campylobacter and Helicobacter

Learning Objectives

After reading and studying this chapter, you should be able to:

- Describe morphology, culture characterstics and biochemical reactions of *C ampylobacter.*
- Discuss morphology, culture characterstics and biochemical reactions of Nelicobacter.
- Discuss laboratory diagnosis of *Helieobaeter pylori* infections.

CAMPYLOBACTER

Introduction

Campylobacter and *Arcobacter* are grouped into the family **Campylobacteriaceae;** *Helicobacter, Wolinella,* and *Flexispira* are grouped into **an unnamed family**. Because *Campylobacter* and *Helicobacter* species are the most commonly isolated and clinically most important members of this superfamily.

Campylobacters were first isolated in 1906 from aborting sheep in the UK. Originally thought to be vibrios, they were later placed in their own genus with C. *fetus* as the type species and is a major cause of abortion in sheep and cattle worldwide. The discovery that C. *jejuni* and C. *coli* commonly cause acute enteritis in man was not made until the late 1970s. They can, on occasion, also cause systemic infections. They are important veterinary pathogens.

Species

A total of 18 species and subspecies are now recognized; 13 have been associated with human disease. **C. *jejuni*** is the prototype organism in the group and is very common cause of diarrhea in humans.

CAMPYLOBACTER JEJUNI AND CAMPYLOBACTER COLI

Campylabacter jejuni and *Campylabacter coli* have emerged as common human pathogens, causing mainly enteritis and occasionally systemic infection. C. *jejuni* and C. *coli* are closely related, but C. *jejuni* hydrolyses hippurate the only species of campylobacter to do so. C. *jejuni* accounts for 90 to 95 percent of campylobacter infections in most parts of the world.

Morphology

The genus *Campylobacter* (Greek, meaning **curved rod**) consists of small, comma-shaped, gram-negative bacilli that microscopically resemble vibrios. They are motile by means of a single polar flagellum. Old cultures are coccoid and pleomorphic. They are nonsporing.

Cultural Characteristics

Most species are microaerobic, requiring an atmosphere with decreased oxygen (5% oxygen) and increased hydrogen and carbon dioxide (CO_2) level for aerobic growth. Many pathogenic species are thermophilic, growing well at 42°C.

Selective media for isolation of C. *jejuni* are Butzler's selective medium, Skirrow's *Campylobacter* selective medium, Preston *Campylobacter* selective medium and Blaser's medium (Campy-BAP). Plates are incubated for 48 hours. Colonies are circular and convex but those of thermophilic group, particularly C. *jejuni,* are flat and tend to swarm on moist agar.

Biochemical Reactions

Campylobacters do not ferment carbohydrates and utilize a respiratory (oxidative) pathway. They are strongly oxidase positive. They are catalase positive and reduce nitrates to nitrites. C. *jejuni* has the ability to hydrolyze sodium hippurate.

Pathogenesis

Infection is acquired by ingestion. The jejunum and ileum are the first sites to become colonized, but the infection extends distally to affect the terminal ileum and usually the colon and rectum. The organisms are invasive and may involve mesenteric lymph nodes and cause bacteremia. **Heat-labile enterotoxin** along with the **invasive property** of this organism may contribute to the production of the damage.

Mechanism of Producing Diarrhea

It can produce diarrhea by following mechanisms:

1. It produces a **heat-labile enterotoxin** resembling CT that raises intracellular levels of cAMP leading to watery diarrhea.
2. Like *Shigella* and *Salmonella,* it **penetrates** gut epithelium leading to edematous exudative enteritis of jejunum, ileum and colon, with infiltration by polymorphonuclear leukocytes and ulceration of the mucosa.

Clinical Manifestations

Clinical manifestations are acute onset of crampy abdominal pain, profuse diarrhea that may be grossly bloody, headache, malaise, and fever. Usually the illness is self-limited to a period of 5 to 8 days, but occasionally it continues longer. Prolonged carriage occurs only in patients with immunodeficiency.

Complications are reactive (aseptic) arthritis and Guillain-Barre syndrome, a form of peripheral polyneuropathy.

Epidemiology

Campylobacters were first isolated in 1906 from aborting sheep in the UK. Originally thought to be *vibrios,* they were later placed in their own genus with *C. fetus* as the type species. *C. fetus* is a major cause of abortion in sheep and cattle worldwide. Campylobacters first gained prominence in the 1970s as a common cause of human diarrheal disease, affecting children and adults. They can, on occasion, also cause systemic infections. They are important veterinary pathogens.

Campylobacter infections are **zoonotic**, with a variety of animals serving as reservoirs (see Table 42.1). Humans acquire the infections with *C. jejuni* and *C. coli* after consumption of contaminated food, milk, or water. Contaminated poultry are responsible for more than half of the *Campylobacter* infections in developed countries. It is a part of the normal intestinal flora of domestic animals and birds and is shed in their feces. It can be isolated frequendy from surface waters. **Campylobacter** enteritis is the commonest form of acute infective diarrhea in most developed countries.

Laboratory Diagnosis

Laboratory diagnosis depends on isolation of the campylobacter from feces.

A. Specimens

Diarrheal stool is the usual specimen.

B. Microscopy

Gram-stained smears of stool may show the typical "gull wing" -shaped rods. **Dark-field or phase contrast microscopy** may show the typical darting or tumbling motility of the spiral rods.

C. Culture

Feces or rectal swabs are plated on **selective media.** A transport medium has to be employed in case of delay in culturing. Campylobacters survive for 1 to 2 weeks at 4°C in Cary-Blair transport medium but glycerol-saline is not satisfactory. **Selective media** for isolation of C. *jejuni* are Butzler's selective medium, Skirrow's *Campylobacter* selective medium, Preston *Campylobacter* selective medium and Blaser's medium (Campy-BAP). Skirrow's medium contains vancomycin, polymyxin B, and trimethoprim to inhibit growth of other bacteria.

Inoculated plates are incubated at 42°C to favour growth of the thermophilic campylobacters *(C.jejuni, C. coli, C. lari* and *C. hyointestinalis)* over that of other fecal bacteria. If, however, the presence of *C. fetus* (nonthermophile) is suspected, additional plates should be incubated at 37°C to allow growth of this nonthermophile. Incubation must be done in an atmosphere of 5 percent O_2, 10 percent CO_2 and 85 percent N_2 plates are incubated for 48 hours.

Colonies are typically flat and effuse, with a tendency to spread on moist agar. They are nonhemolytic, grey or colorless, moist, and flat or convex D.

D. Identification

Suggestive colonies are screened by gram staining, motility and oxidase tests. Confirmation is by further biochemical tests, including positive catalase and nitrate reduction tests.

E. Serology

Serology can be useful in patients presenting with aseptic arthritis or the Guillain-Barre syndrome after a bout of diarrhea that was not investigated. Complement fixation test and **enzyme-linked immunosorbent assay (ELISA)** are group-specific tests that can detect recent infection with C. *jejuni* or C. *coli.*

Treatment, Prevention, and Control

For gastroenteritis, infection is self-limited and is managed by fluid and electrolyte replacement. Severe gastroenteritis and septicemia are treated with erythromycin (drug of choice), tetracyclines, quinolones. Gastroenteritis is prevented by proper preparation of food and consumption of pasteurized milk; prevention of contaminated water supplies also controls infection.

Treatment

Campylobacter gastroenteritis is typically a self-limited infection managed by the replacement of lost fluids and electrolytes. Antimicrobial treatment should be reserved for patients with severe or complicated infections. Severe gastroenteritis and septicemia are treated with erythromycin (drug of choice), tetracyclines, quinolones.

Control

Gastroenteritis is prevented by proper preparation of food (particularly poultty), and consumption of pasteurized milk; prevention of contaminated water supplies also controls infection.

Other Campylobacters

Campylobacter species other than *C jejuni* are encountered infrequently.

1. Campylobacter fetus

Campylobacter fetus subspecies *fetus* is a very important veterinary pathogen. It causes infective abortion in cattle and sheep. It is an opportunistic pathogen that causes **systemic infections** in man in immunocompromised patients. It may occasionally cause **diarrhea.** The gastrointestinal tract may be the portal of entry when *C fetus* causes bacteremia and systemic infection. Unlike other species, *Campylobacter fetus* is most commonly responsible for causing systemic infections such as **bacteremia, septic thrombophlebitis, arthritis, septic abortion, and meningitis.**

2. Campylobacters concasus

It has been isolated from cases of gingivitis and periodontal disease. It has also been isolated from feces.

3. Campylobacter fetus subsp. venerealis

It causes enzootic sterility (infectious infertility) of cattle but has not been associated with human infection.

4. C. jejuni subsp. doylei

This organism can be distinguished from other campylobacters because it does not reduce nitrate to nitrite and hydrolyzes hippurate. The pathogenicity of this organism is unknown. It has been isolated from human gastric epithelium biopsy and from feces of children with diarrhea.

5. C. coli

It causes an infection clinically indistinguishable from that of *C. jejuni C. coli* is believed to account for 3 to 5 percent of campylobacter diarrheas. It is commonly found in healthy pigs. It can be differentiated from *C. jejuni* by hippurate hydrolysis test which is positive in *C. jejuni* and negative in *C. coli* .

6. C. lari

C. lari (formerly known as *C. laridis)* is regularly found in birds and a wide variety of other animals, notably dogs. It causes enteritis simulating *C. jejuni* infections in humans. *C. lari* septicemia, in an immunocompromised patient, has also been reported.

7. C. hyointestinalis

It was initially found only in animals, principally as a cause of ileitis in swine. A few isolations have been reported from human patients with diarrhea and homosexual men with proctitis.

8. Campylobacter upsaliensis

It is *from* dogs occasionally causes diarrhea in humans.

9. C. sputorum subsp

Sputorum: It constitutes a part of normal flora of respiratory tract and gingival crevices of man. It has also been isolated from feces of 2 percent of healthy people. It may occasionally cause diarrhea, abscess and septicemia.

HELICOBACTER

These are strict microaerophiles with a spiral or helical morphology. They possess sheathed flagella.

Species

Various species included in this genus are: *H pylori, H. cinaedi, H fennelliae, H. canis, H pullorum H. rappininae* and *H. canadensis.* Of these, first three are medically important.

Helicobacter pylori

Warren and Marshall in Australia in 1983 observed spiral, campylobacter-like bacteria in close apposition to the gastric mucosa in several cases of gastritis and peptic ulcer. They were originally named ***Campylobacter pyloridis*** then ***C. pylori*** and now redesignated as ***Helicobacter pylori.*** Features that distinguish this organism from campylobacters are its multiple-sheathed flagella, its strong hydrolysis of urea and its unique fatty acid profile.

H. pylori, colonizes the stomachs of roughly one-half of the world's population. It is associated with antral gastritis, duodenal (peptic) ulcer disease, gastric ulcers, and gastric carcinoma. Other helicobacter species that infect the gastric mucosa exist but are rare.

Morphology

H. pylori is a gram-negative spirally-shaped bacterium, 0.5 to 0.9 µm wide by 2 to 4 µm long. It is motile by means of a tuft of sheathed unipolar flagella, unlike the unsheathed flagella of campylobacters. It is nonsporing. Helicobacters have a spiral shape in young cultures but can assume coccoid forms in older cultures.

Cultural Characteristics

Like campylobacters, *H pylori* is microaerophic. The optimum temperature for its growth is 35 to 37°C, some grow poorly at 42°C but none grows at 25°C. It can grow in an atmosphere of 5 percent O_2, 10 percent CO_2 and 85 percent N_2. It does not grow anaerobically or in air. It can be grown on moist freshly prepared **chocolate agar** and **Skirrow's *Campylobacter* selective medium**. *H pylori* produces circular, convex and translucent colonies after incubation at 35 to 37°C in a microaerophilic atmospherefor 3 to 5 days.

Biochemical Reactions

H pylori gives the following reactions.

i. **Abundant urease production-** A distinctive feature is the production of **abundant urease,** and this property has been used as a rapid diagnostic test in gastric biopsy samples. The urease enzyme produced by *H. pylori* is almost 100 times more active than that of *Proteus vulgaris.*

ii. It produces oxidase, catalase, phosphatase and H_2S.

iii. It does not metabolize carbohydrates or reduce nitrate.

Virulence Factors

1. *H. pylori* produces a cytotoxin causing vacuolation of gastric mucosa. Its production is determined by Cag A Gene. Virulence has been associated with certain alleles in genes, such as ***cag* (cytotoxin associated gene)** and ***vac* (vacuolating cytotoxin gene).** The genes controlling the production of VacA and CagA, together with over 40 other genes, are located on a large region of DNA called the **cag pathogenicity island.**
2. Urease released by *H. pylori* produces ammonia ions that neutralize stomach acid in the vicinity of the organism, thus favoring bacterial multiplicatefun.
   Ammonia may also both cause injury and potentiate the effects of a cytotoxin produce and by *H. pylori.*
3. The bacterial antigens cross-react with antran gastric antigens stimulating an autoimmune response.
4. Protease produced by the organism degrades gastric mucosa.
5. *H. pylori* infected patients show hypergastrinemia that upsets gastrin-HCI homeostasis.

Pathogenesis

H. pylori colonizes the surface of the gastric mucosa, especially of the antrum but any part of the stomach may be colonized. The bacteria are present in large numbers in the mucus overlying mucosa where the pH is about 7.0. Colonization often extends into gastric glands, but the mucosa is not invaded by the bacteria. It has been isolated from gastric biopsy specimens and occasionally from gastric juices, saliva and bile.

The exact pathogenic mechanisms are not clearly understood. Bacterial protease, toxins or ammonia released by urease activity or autoimmune responses to gastric antigens may all contribute.

Although gastric acid is potentially destructive to *H. pylori,* protection is provided by its powerful urease. *H. pylori* produces potent urease activity, which yields production of ammonia and further buffering of acid and neutralizes acid around the bacteria. Thus, the enzyme urease is important colonization of the stomach by this organism. Where they are numerous, the underlying mucosa usually shows a superficial gastritis of the type known as **chronic active or type B gastritis.**

H. pylori is associated with **antral gastritis, duodenal (peptic) ulcer disease, gastric ulcers.** It is also recognised as a risk factor for **gastric malignancies**, namely, **'adenocarcinoma'** and **'mucosa associated lymphoid tissue' (MALT) lymphomas.**

Laboratory Diagnosis

Diagnostic tests are of two kinds:

A. Non-Invasive Tests

1. Serology
2. Urea breath test
3. Fecal antigen test
4. Polymerase chain reaction (PCR).

B. Invasive Tests

1. Microscopy
2. Culture
3. Biopsy urease test.

A. Non-invasive Tests

In practice, noninvasive tests are used for initial screening.

1. **Serology:** Antibodies to *H. pylori* or its products can be detected in the patient serum by ELISA test. The titre of antibodies falls after several months if infection is eradicated.
2. **Urea breath test:** This test detects bacterial urease activity in the stomach by measuring the output of CO_2 resulting from the splitting of urea into CO_2 and ammonia. Urea tagged with an isotope of carbon (carbon-14 or -13) is fed to the patient. If the patient's stomach is colonized with *H. pylori,* urea is converted into ammonia and tagged CO_2. The latter appears in the breath where it can be measured. Patients infected with *H. pylori* give high readings of the isotope.
3. **Fecal antigen test:** In this polyclonal antibodies are used to detect *H. pylori* antigens in feces. It has the potential to supplant serology as a routine screening test.
4. **Polymerase chain reaction (PCR):** Various DNA probes have been developed for the direct detection of *H. pylori* by PCR in gastric juice, feces, dental plaque and water supplies.

B. Invasive Tests

1. **Specimens-Endoscopic biopsy of gastric mucosa** for examination by microscopy, culture and urease tests.
2. **Microscopy:** The biopsy specimen can be examined by microscopic examination of gram staining, silver staining, hematoxylin and eosin (H and E)staining, Giemsa staining or immunofluorescence for the presence of bacteria. Warthin-Starry silver stain is the most sensitive.
3. **Culture-** Culture is done on nonselective medium such as **chocolate agar** and a selective medium. **Skirrow's *Campylobacter* selective medium.** Plates are incubated for 2 to 7 days in a moist, microaerophilic atmosphere at 35 to 37°C in the presence of

5 to 10 percent CO_2. High humidity is essential. The organism is identified on the basis of its colonial morphology, gram staining, biochemical properties and positive urease tests.

Biopsy Urease Test

A specimen is placed into a small quantity of urea solution with an indicator that detects alkalinity resulting from the formation of ammonia by urease The biopsy urease test can be performed by crushing biopsy tissue in 0.5 ml urea solution with an indicator and incubated at 37°C. If *H. pylori* is present, the pH changes within a few minutes to 2 hours due to the production of ammonia. The abundance of urease produced by the organism permits detection of the alkaline byproduct in less than 2 hours.

Treatment

The standard treatment is a combination of bismuth subsalicylate, tetracycline (or amoxycillin) and metronidazole for two weeks. An alternative schedule employs a proton pump inhibitor like omeprazole and clarithromycin.

Prevention and Control

Prophylactic treatment of colonized individuals has not been useful and potentially has adverse effects, such as predisposing patients to adenocarcinomas of the lower esophagus. Human vaccines are not currently available.

HELICOBACTER CINAEDI

H. cinaedi (formerly known as *Campylobacter cinaedi)* has been associated with proctitis in homosexual men. It has also been described as a cause of bacteremia in homosexual men with concurrent tuberculosis, and in HIV -positive individuals and AIDS cases.

HELICOBACTER FENNELLIAE

H. fenllelliae (formerly known as *Campylobaeter fennelliae)* like *H. cinaedi* has been associated with proctitis in homosexual men.

KNOW MORE

- Campylobacters cause both diarrheal and systemic diseases and are among the most widespread causes of infection in the world.
- *H. pylori* is present on the gastric mucosa of less than 20 percent of persons under age 30 but increases in prevalence to 40 to 60 percent of persons age 60, including persons who are asymptomatic. In developing countries, the prevalence of infection may be 80 percent or higher in adults.

KEY POINTS

Campylobacter Infections

- Campylobacters are thin, curved gram-negative bacilli. They are are microaerobic and strongly oxidase positive.
- *Campylabacter jejuni* and *Campylabacter coli* have emerged as common human pathogens.
- Diseases-Zoonotic infection; improperly prepared poultry is a common source of human infections. *C. jejuni* is associated with gastroenteritis, septicemia, meningitis, spontaneous abortion, proctitis, Guillain-Barre syndrome. *C. fetus* is associated with septicemia and is disseminated to multiple organs.
- **Helicobacter pylori-**Curved gram-negative bacilli. Urease production at very high levels is typical of gastric helicobacters (e.g. *H. pylori).*
- Diseases-H. *pylori causes* gastritis, peptic ulcers, gastric adenocarcinoma.
- Diagnosis-A.Noninvasive tests-1. Serology 2. Urea breath test. 3. Fecal antigen test 4. Polymerase chain reaction (PCR). B. Invasive Tests:
  1. Microscopy
  2. Culture
  3. Biopsy urease test.

IMPORTANT QUESTIONS

1. Discuss pathogenesis and laboratory diagnosis of diarrhea caused by *Campylobacter.*
2. Write short notes on:
   Helicobacter pylori
   Laboratory diagnosis of *Helicobacter pylori* infections
   Urea breath test.

FURTHER READING

Altekruse SF, Stern NJ, Fields PI, Swerdlow Dl. Campylabacter jejuni-an emerging foodborne pathogen. Emerging Infectious Diseases 1999;5:28-35.

Bateson MC. Helicabacter pylori. Postgraduate Medical Journal 2000;76:141-44.

Dunn B, et al. Helicobacter pylori, Clin Microbiol Rev 1997; 10:720-41.

On SL W. Identification methods for campylobacters, helicobacters, and related organisms, Clin Microbiol Rev 1996;9: 405-22,

Skirrow MB, Blaser MJ. Campylobacter jejuni. In: Blaser MJ, Smith PD, Ravdin JI, et al. (Eds) Infections of the Gastrointestinal Tract, 2nd edn. Raven Press, New York 2002.

Van Enk R. Serologic diagnosis of Helicobacter pylori infection, Clin Microbiol Newsletter 1996;18:89-92.

CHAPTER 42 Pseudomonas, Stenotrophomonas, Burkholderia

Learning Objectives

After reading and studying this chapter, you should be able to:

- Describe morphology, cultural characteristics, biochemical reactions and laboratory diagnosis of Pseudomonas aeruginosa.
- Describe the following: Pigments of Pseudomonas; pathogenicity of Pseudomonas aeruginosa; Burkholderia mallei.

INTRODUCTION

The term pseudomonads describes a large group of aerobic, nonfermentative, nonsporing Gram-negative bacilli, motile by polar flagella. They belong to over 100 species that were originally contained within the genus *Pseudomonas.* Most are saprophytes found widely in soil, water and other moist environments. Some pseudomonads are pathogenic for plants, insects and animals. A few cause human infection, typically opportunistic.

Molecular analyses of the group have led to revised taxonomic classifications, and many species have been allocated to new genera which include *Burkholderia, Comammona, Stenotrophomonas, Ralstonia* and *Brevundimonas.*

Pseudomonas aeruginosa is the most common pseudomonad and this is the species most commonly associated with human disease. *Burkholderia* (previously *Pseudomonas) pseudomallei* is an important pathogen in tropical areas of the world. *Burkholderia* (previously *Pseudomonas) cepacia,* which is now encompassed within the *B. cepacia* complex, has emerged as an important pathogen in immunocompromised patients, particularly in individuals with cystic fibrosis or chronic granulomatous disease. *Stenotrophomonas, maltophilia* also infects immunocompromised patients, with a mortality reaching 60 percent in patients with hematological malignancies.

PSEUDOMONAS AERUGINOSA

Morphology

It is a slender gram-negative bacillus, 1.5-3 µm *x* 0.5 µm, actively motile usually with a single polar flagellum. Occasional strains have two or three flagella. Fimbriae may be present and are usually polar and non-hemagglutining. It is nonsporing, noncapsulated but many strains have a mucoid slime layer. Mucoid strains, particularly isolates from cystic fibrosis patients have an abundance of extracellular polysaccharides composed of alginate polymers, known as glycocalyx capsule.

Cultural Characteristics

It is a strict aerobe but can grow anaerobically if nitrate is available. Growth occurs at a wide range of temperatures, 6-42°C, the optimum being 37°C. Optimum pH 7.4-7.6. It grows well on ordinary media and in the laboratory, it can be isolated on virtually any medium.

1. **Nutrient agar:** After aerobic incubation on **nutrient agar** at 37°C for 24 hours, the colonies are large, 2-3 mm in diameter, smooth, translucent, irregularly round and emit a characteristic fruity odor.

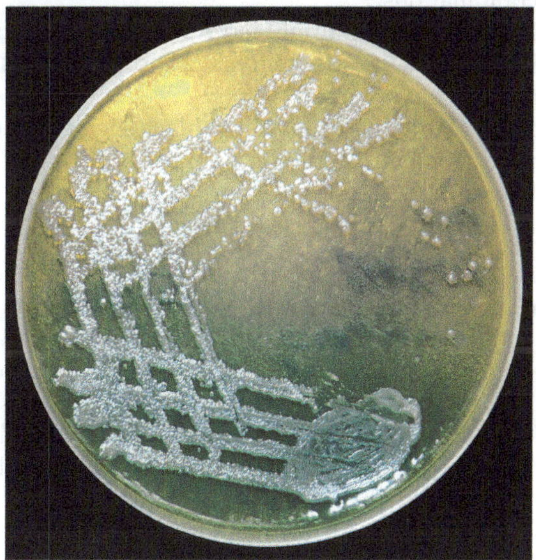

Fig. 42.1: Pseudomonas aeruginosa on nutrient agar

This grape-like smell is due to the production of aminoacetophenone from tryptophan. Many *P. aeruginosa* strains exhibit a moth-eaten type of colonial lysis with a metallic sheen known as iridescence are seen in cultures on nutrient agar. Mucoid strains often produce copious amounts of an extracellular polysaccharide on agar culture. These strains are particularly common in the sputum of patients with cystic fibrosis.

2. It grows on **MacConkey and DCA media,** forming nonlactose-fermenting colonies. On MacConkey agar, the characteristic pigments are often poorly observed.
3. **Blood agar:** Colonies on **blood agar** may be surrounded by a zone of hemolysis.
4. In broth, it forms a dense turbidity with a surface pellicle.
5. **Cetrimide agar** is selective medium for *Ps. aeruginosa.*

Pigment Production

P. aeruginosa produces at least 4 distinct pigments:

i. **Pyocyanin:** It is a bluish-green phenazine pigment soluble in chloroform and water. It diffuses into the surrounding medium. This pigment is not produced by other species of this genus. Demonstration of the presence of the blue phenazine pigment pyocyanin is absolute confirmation of a strain as *P. aeruginosa* and thus the major diagnostic test.

ii. **Pyoverdin (fluorescein):** The yellow/green pigment pyoverdin (fluorescein) is also produced by most strains, giving the characteristic blue-green appearance of infected pus or cultures. It is insoluble in chloroform but soluble in water. It imparts a yellowish tinge to cultures but this is sometimes not easy to detect unless cultures are examined under ultraviolet light. Other pigments produced are pyorubin (red) and pyomelanin (brown) in various combinations.

iii. **Pyorubrin:** It is a bright red water soluble pigment. It is a phenazine pigment that is insoluble in chloroform.

iv. **Pyomelanin:** It is a brown to black pigment and its production is uncommon. It is chemically unrelated to animal melanin.

Biochemical Reactions

P. aeruginosa differs from members of the Enterobacteriaceae by deriving energy from carbohydrates by an oxidative rather than a fermentative metabolism. Peptone water sugars are unsuitable for detecting acid production, since this is weak and gets neutralised by alkali produced from peptone. An ammonium salts medium in which the sugar is the only carbon source is the best. Special media such as the O-F medium of Hugh and Leifson must be used for diagnostic tests because the amount of acid produced by oxidative pathways is less than that produced by fermentation.

It utilizes glucose oxidatively with the production of acid only. Lactose and maltose are not utilized. Indole, MR, VP and H2S tests are negative.

However, all strains give a rapid positive oxidase reaction (within 30 seconds) and utilize citrate as sole source of carbon.

It reduces nitrates to nitrites and further to gaseous nitrogen.

It is catalase, arginine dihydrolase and gelatinase positive and lysine decarboxylase and aesculin hydrolysis negative.

O-F medium of Hugh and Leifson-Oxidative	Glucose A	*Lactose* –	*Maltose* –	*Mannitol* –	Sucrose –
Citarte +	Indole, MR, VP - - -	H2S -	NO_3 reduction -		

ANTIGENIC CHARACTERISTICS

O Antigens

P. aeruginosa possesses 19 distinct, group-specific O antigens. O antigens are heat-stable and can be extracted with acid or formamide.

H antigens

On the basis of slide agglutination, at least two heat-labile H antigens have been recognized. Serological characterization is primarily used as an epidemiological tool rather than for diagnostic confirmation of species identity.

Epidemiological Typing Methods

1. Serotyping
2. Bacteriocin (pyocin) typing
3. Phage typing
4. Molecular methods.

Serotyping

Identification of group-specific heatstable lipopolysaccharide antigens by agglutination forms the basis of O serotyping. Typically, nine serotypes account for over 90 percent of isolates. Serotype discrimination can be improved by including identification of H antigens.

Serogroups 06 and 011 predominate in clinical material and 011 has been found to be responsible for most of the hospital-associated infections.

Bacteriocin (Pyocin) Typing

Three types of bacteriocins (pyocins) are produced by *P. aeruginosa* which are known as R, F and S. Pyocin-producing strains are resistant to their own pyocins though they may be sensitive to those produced by other strains. Pyocin produced by the test strain is employed to assess the growth inhibition of 13 indicator strains of *P. aeruginosa* The pattern of inhibition of the indicator strains determines the type of the strain. Depending upon the growth inhibition of these 13 indicator strains, 105 types are recognized. Pyocin typing is easy to perform, results are available by the third day and has a reasonable reproducibility and good discrimination.

Phage Typing

In bacteriophage typing considerable difficulties have been encounterd.

Molecular Methods

Restriction endonuclease typing wih pulsed field gel electrophoresis (PAGE) is the most reliable typing method and discriminatory of the present **DNA-based** typing methods and is considered to be the gold standard.

Virulence Factors

Most strains produce two exotoxins, exotoxin A and exoenzyme S, and a variety of cytotoxic substances including proteases, phospholipases; rhamnolipids and the blue-green pigment pyocyanin; an alginate-like exopolysaccharide. The importance of these putative virulence factors depends upon the site and nature of infection:

1. **Capsule:** *P. aeruginosa* produces a polysaccharide capsule (also known as mucoid exopolysaccharide, alginate coat, or glycocalyx) that has multiple functions and is responsible for the mucoid phenotype.
2. **Pili:** Adherence of *P. aeruginosa* to host cells is mediated by pili and nonpilus adhesins.
3. **Lipopolysaccharide (LPS):** It has Endotoxin activity
4. **Pyocyanin:** It mediates tissue damage through production of toxic oxygen radicals (i.e., hydrogen peroxide, superoxide, hydroxyl radicals). This pigment also stimulates the inflammatory response and impairs ciliary function.
5. **Exotoxin A and Exotoxin S:** Exotoxin A is believed to be one of the most important virulence factors produced by pathogenic strains of *P. aeruginosa.* Mechanism of action of exotoxin A is identical to that of diphtheria toxin. Exotoxin S Inhibits protein synthesis and is immunosuppressive
6. **Extracellular enzymes and hemolysins:** *P. aeruginosa* produces proteases (general protease, alkaline protease and elastase), hemolysins (phospholipase C and heat-stable rhamnolipid) and lipase. These play a key role in the formation of local lesions.
7. **Antibiotic Resistance:** *P. aeruginosa* is inherently resistant to many antibiotics and can mutate to even more resistant strains during therapy.

Resistance

The bacillus is not particularly heat resistant, being killed at 55°C in one hour but exhibits a high degree of resistance to chemical agents. It is resistant to the common antiseptics and disinfectants such as quaternary ammonium compounds, chloroxylenol and hexachlorophane and may even grow profusely in bottles of such antiseptic lotions kept for use in hospitals. Therefore, dettol and cetrimide can be incorporated in selective media for isolation of *Pseudomonas.* It is sensitive to a 2 percent aqueous alkaline solution of glutaraldehyde (cidex), acids, silver salts and strong phenolic disinfectants.

P. aeruginosa possesses a considerable degree of natural resistance to antibiotics. Examples of clinically effective antibiotics are aminoglycosides (gentamicin, amikacin), cephalosporins (cefotaxime, ceftazidime, cefoperazone), fluoroquinolones (ciprofloxacin, ofloxacin, pefloxacin), penicillins (piperacillin, ticarcillin, azlocillin). topical colistin, polymyxin B or 1 percent acetic acid may be useful for localized infections.

Epidemiology

Opportunistic pathogens: Pseudomonads are opportunistic pathogens present in a variety of environments. Its ability to persist and multiply, particularly in moist environments and on moist equipment (e.g. humidifiers) in hospital wards, bathrooms and kitchens, is of particular importance in cross-infection control. Consumption of salad vegetables contaminated with pseudomonas is a potential risk for immunocompromised patients in intensive care units.

Resistance to antibiotics and disinfectants: Pseudomonads have minimal nutritional requirements, can tolerate a wide range of temperatures (4°C to 42°C), and are resistant to many antibiotics and disinfectants. It can transiently colonize the respiratory and gastrointestinal tracts of hospitalized patients, particularly those treated with broad-spectrum antibiotics, exposed to respiratory therapy equipment, or hospitalized for extended periods.

The recovery of *Pseudomonas,* particularly species other than *P. aeruginosa,* from a clinical specimen may represent simple colonization of the patient or environmental contamination of the specimen during collection or laboratory processing.

Pathogenesis

P. aeruginosa has many virulence factors, including structural components, toxins, and enzymes **(See under Virulence factors).** *P. aeruginosa* disease begins with attachment to and colonization of host tissue. Pili on the bacteria mediate adherence, and a glycocalyx capsule reduces the effectiveness of normal clearance mechanisms. Host tissue damage facilitates adherence and colonization. *P. aeruginosa* produces numerous toxins and extracellular products that promote local invasion and dissemination of the organism.

'Blue pus' was known as a surgical entity long before Gessard (1882) isolated *Ps. aeruginosa* from such cases. Both the specific names of the bacillus refer to its capacity to cause 'blue pus', the term aeruginosa, meaning verdigris which is bluish green in color and *pyocyanea,* being a literal translation of 'blue pus'.

P. aeruginosa can infect almost any external site or organ. Most community infections are mild and superficial, but in hospital patients, infections are more common, more severe and more varied. Individuals most at risk include those with impaired immune defenses. A prior antibiotic therapy that eliminate normal flora can also provide *P. aeruginosa* with increased access for colonizing tissue. *P. aeruginosa* is regularly a cause of **nosocomial pneumonia, nosocomial urinary tract infections, surgical site infections, infections of severe burns and infections of patients undergoing either chemotherapy for neoplastic diseases or antibiotic therapy.**

A. Community Infections

1. **Otitis externa** and **varicose ulcers.**
2. **Corneal infections** resulting from contaminated contact lenses or other sources can be rapidly destructive and painful.
3. **Jacuzzi rash or whirlpool rash:** Recreational and occupational conditions associated with pseudomonas infections include **Jacuzzi rash or whirlpool rash** (an acute self-limiting folliculitis)
4. Industrial eye injuries, which may lead to **panophthalmitis.**

B. Hospital Infections

In the hospital, it may cause localized or generalized infections.

1. **Localized lesions:** Localized lesions are commonly **infections of wounds and bed sores, eye infections and urinary infections following catheterization.** *P. aeruginosa* is the most common and most serious cause of **infection in burns.** It is also one of the agents responsible for **iatrogenic meningitis** following lumbar puncture. It frequently causes **post-tracheostomy pulmonary infection.**
2. **Septicemia and endocarditis:** Septicemia and endocarditis may occur in patients who are debilitated due to concomittant infection, malignancy or immunosuppressive therapy.
3. **Ecthyma gangrenosum and many other types of skin lesions:** They occur either alone or as part of generalized infection, mainly in patients with leukemia and other types of malignancy.
4. **Infection of the nail bed:** Infection of the nail bed is not uncommon following excessive exposure or hands to detergents and water.
5. **Infantile diarrhea and sepsis:** *P. aeruginosa* has been described as one of the agents responsible for infantile diarrhea and sepsis.
6. **Shanghai fever:** *P. aeruginosa* has been reported to cause a self-limited febrile illness (Shanghai fever) resembling typhoid fever in some tropical areas.
7. **Other Infections:** *P. aeruginosa* is also the cause of a variety of other infections, including those localized in the **gastrointestinal tract, central nervous system, and musculoskeletal system.**

Laboratory Diagnosis

1. **Specimens:** Pus, wound swab, urine, sputum, CSF or blood.
2. **Microscopy:** Gram-negative rods are often seen in smears.
3. **Culture:** They grow easily on common isolation media such as **blood agar** and **MacConkey agar.** It may be necessary to use selective media such as **cetrimide agar** for isolation of *P. aeruginosa* from feces or other samples with mixed flora such as wound swab. As *P aeruginosa* is a frequent contaminant, isolation of the bacillus from a specimen should not always be taken as proof of its etiological role. Repeated isolations help to confirm the diagnosis.
4. **Identification:** The isolates are identified by their **colonial morphology** and **biochemical characters.** The colonial morphology (e.g., colony size, hemolytic activity, pigmentation, odor) combined with the results of selected rapid biochemical tests (e.g., positive oxidase reaction) is sufficient for the preliminary identification of these isolates.
5. **Typing method:** Biochemical profiles, antibiotic susceptibility patterns, phage typing, production of pyocins, serologic typing, and the molecular characterization of DNA or ribosomal RNA are used for the specific classification of isolates for epidemiologic purposes.
6. **Antibiotic sensitivity tests:** It is useful to select out proper antibiotic as multiple reistance to antibiotics is quite common in *P. aeruginosa.*

Treatment

P. aeruginosa is intrinsically resistant to most commonly employed antimicrobial agents. Many strains are, however, susceptible to carbenicillin, azlocillin, ticarcillin, cefotaxime, ceftazidime, gentamicin and tobramycin. Ciprofloxacin exhibits good activity against *P. aeruginosa* and penetrates well into most tissues. Since this can be given orally, therefore, it may be used if therapy has to be prolonged. However, resistance to ciprofloxacin may sometimes develop during therapy.

Control

Prevention of *P. aeruginosa* cross infection in hospitals requires constant vigilance and strict attention to asepsis. The inappropriate use of broad-spectrum antibiotics should also be avoided. Immunotherapy in human burns cases with antiserum to *P. aeruginosa* may be useful. Vaccine from appropriate types administered to high-risk patients provides protection against pseudomonas sepsis.

STENOTROPHOMONAS MALTOPHILIA (FORMERLY PSEUDOMONAS MALTOPHILA)

Stenotrophomonas maltophilia is one of the most commonly isolated nonfermentative, gram-negative bacilli. It is usually oxidase negative and acidifies maltose in addition to glucose, lactose and sucrose.

It is responsible for infections in debilitated patients with impaired host defense mechanisms. The spectrum of nosocomial infections with *S. maltophilia* includes **bacteremia, pneumonia, meningitis, wound infections**, and **urinary tract infections.**

Trimethoprim-sulfamethoxazole is the agent most active against the organism. Good activity is also seen with chloramphenicol and ceftazidime.

BURKHOLDERIA CEPACIA (FORMERLY PSEUDOMONAS CEPACIA)

B. cepacia was previously best known as the causative agent of soft rot of onions (*cepacia*, Latin= onion). However, in the past two decades it has emerged as a human pathogen causing life-threatening respiratory infection in patients with chronic granulomatous disease or cystic fibrosis in whom it causes fatal necrotizing pneumonia.

Morphology

B. cepacia is a slender, motile, gram-negative rod. The bacillus accumulates poly-β-hydroxybutyrate as granules, so stains irregularly.

Culture

It is nutritionally very versatile. It can grow many common disinfectants and can even use penicillin G as a sole source of carbon. It is aerobic and grows well on nutrient agar optimally at 25-30°C. On prolonged incubation, colonies become reddish. purple due to the formation of a non-diffusible phenazine.

Biochemical Reactions

lt is weak oxidase-positive. It utilizes glucose, maltose, lactose, and mannitol and is lysine decarboxylase positive, ornithine decarboxylase positive, and arginine dihydrolase negative. Isolates are motile by means of polar tuft of flagella.

Pathogenicity

Like *P. aeruginosa, B. cepacia* can colonize a variety of moist environmental surfaces and is commonly associated with nosocomial infections. With the exception of pulmonary infections, *B. cepacia* has a relatively low level of virulence, and infections with the organism do not commonly result in death.

Infections caused by this organism include the following:

1. **Respiratory tract infections** in patients with cystic fibrosis or chronic granulomatous disease.
2. **Urinary tract infections** in catheterized patients.
3. **Septicemia**, particularly in patients with contaminated intravascular catheters.
4. **Endocarditis**, especially in drug addicts, pneumonitis, osteomyelitis, dermatitis and wound infections.

Treatment

B. cepacia is susceptible to trimethoprim-sulfamethoxazole.

BURKHOLDERIA MALLEI (FORMELY PSEUDOMONAS MALLEI).

The bacillus had also been classified variously as *Loefferella, Pfeifferella, Malleomyces, Actinobacillus* and *Acinetobacter.* The bacillus was discovered by Loeffler and Schutz (1882) from a horse dying of glanders. It is the causative agent of glanders (*malleus, in Latin*), a disease primarily of equine animals-horses, mules and asses-but capable of being transmitted to other animals and to human beings.

Morphology

P. mallei is a slender, nonmotile, gram-negative bacillus, 2-5 μm × 0.5 μm staining irregularly and often giving a beaded appearance.

Culture

It is an aerobe and facultative anaerobe, growing on ordinary media under a wide range of temperature. It does not grow on MacConkey medium. Colonies which are small and transluscent initially become yellowish and opaque on ageing. On potato, a characteristic amber, honey-like growth apperars, becoming greenish yellow resembling *P. aeruginosa.*

Biochemical Reactions

It is quite inactive biochemically, attacking only glucose. In its morphology and biochemical reactions *P. mallei* closely resembles *P. pseudomallei* (Table 36.1) but shows a number of negative differences. It is the only nonmotile species in the genus *P. seudomonas,*

Animal pathogenicity

The natural disease in equines occurs in two forms - glanders and farcvaria.

1. **Glanders:** In glanders, respiratory system is affected. The infected animal develops profuse catarrhal discharge from the nose and the nasal septum shows nodule formation. Later the nodules break down with the production of irregular ulcers.
2. **Farcy:** This follows infection through skin with involvement of superficial lymph vessels and lymph nodes. The lymph vessels are thickened and stand out as hard cords under the skin which are called 'farcy pipes'.

Strauss Reaction

Guinea-pigs are susceptible and intraperitoneal injection into male guinea pigs induces the *Straus reaction.* This consists of swelling of the testes, inflammation of tunica vaginal is and ulceration of the scrotal skin. The Straus reaction is not diagnostic of glanders, as it may also be produced by inoculation of other bacteria such as *Brucella* species, Preisz-Nocard bacillus, *Actinobacillus ligniersi* and *P. pseudomallei.*

Human Pathogenicity

Human infection is usually occupational, found in ostlers, grooms and veterinarians. Humans may become infected via skin abrasions or wounds which come into contact with the discharges of a sick animal. Glanders is uncommon and almost totally restricted to persons handling horses in those countries from which the disease has not been eradicated, principally in Asia and South America.

Human disease may take the form of an acute fulminant febrile illness or a chronic indolent infection producing abscesses in the respiratory tract or skin. The fatality rate is high. While human infection is acquired only rarely from infected animals, laboratory cultures are highly infectious and *P. mallei* is one of the most dangerous bacteria to work with.

Laboratory Diagnosis

The diagnosis is based on:

1. Culture of the organism from local lesions of humans or horses.
2. Rising agglutin titers
3. Mallein test

Animals suffering from glanders develop a delayed hypersensitivity to the bacterial protein. This.is the basis of the *mallein* test used for diagnosing glanders. This is analogous to the tuberculin test and may be performed by the subcutaneous, intracutaneous or conjunctival methods.

Treatment

Human case can be treated effectively with tetracycline plus an aminoglycoside.

BURKHOLDERIA PSEUDOMALLEI

Burkholderia pseudomallei (formerly *Pseudomonas pseudomallei,* also known as *Whitmore's bacillus, Actinobacillus whitmori, Malleomyces pseudomallei, Loefflerella pseudomallei).*

B. pseudomallei is a saprophyte found in soil, water, and vegetation. It is endemic in Southeast Asia, India, Africa, and Australia. This is the causative agent of melioidosis, a glanders-like disease, (The name is derived from 'melis', a disease of asses-glanders, and eidos, meaning resemblance). This organism was isolated by Whitrnore and Krishnaswami (1912) from a glanders-like disease of man in Rangoon. It was designated *Bacillus pseudomallei* by Whitmore (1913). Stanton and Fletcher (1921, 1925) gave the name of *B. whitmori* to the causative agent and melioidosis (glanders-like) to the disease.

Morphology

It resembles *P. mallei* but differs in being motile.

Culture

B. pseudomallei is easily cultured and produces characteristic wrinkled colonies after several days of growth on nutrient agar. Fresh cultures emit a characteristic pungent odour of putrefaction.

Biochemical Reactions

It resembles *Ps. mallei* but differs in liquefying gelatin and forming acid from several sugars. The organism is oxidase-positive and motile but, unlike some other pseudomonads, does not produce diffusible pigments (Table 42.1).

Toxins

Two thermolabile exotoxins, one lethal and the other necrotising have been identified in culture filtrates.

Table 42.1: Distinguishing characters of clinically relevant Pseudomonas species and S. maltophila

	Pyocyanin	Fluorescein	Growth at 42°C	Anginine dihydrolase	Lysine decaboxylase	Gelatinase	Aesculin hydrolysis	Lactose (oxidative)	Maltose (oxidative)	Poly-β-hydroxybutyrate	NO_2 reduction to N_2
P. aeruginosa	+	+	+	+	−	+	−	−	−	−	+
P. fluorescens	−	+	−	+	−	+	−	−	−	−	−
P. putida	−	+	−	+	−	−	−	−	−	−	−
S. maltophilia	−	−	±	−	+	+	+	−	+	−	−
P. stutzeri	−	−	±	−	−	−	−	−	+	−	+
P. cepacia	−	−	+	−	+	+	+	+	+	+	−
P. pickettii	−	−	+	−	−	−	−	+	+	+	+
P. pseudomallei	−	−	+	+	−	+	+	+	+	+	+
P. mallei	−	−	±	+	−	+	−	−	±	+	±

Organisms	Oxidase test	Habitat	Pathogenicity
Acinetobacter spp.	–	Saprophytes found in soil, water and sewage, and occasionally as commensals of moist areas of human skin.	Opportunist pathogens, Serious infections include meningitis, pneumonia and septicemia
Alcaligenes spp.	+	Human feces	UTI, wound infection
Achromobacter spp.	+	–	CSOM, Postoperative meningitis
Flavobacterium spp.	+	Saprophyte of soil and moist environments	Opportunistic nosocomial infections, particularly in infants and associated with meningitis
Eiknella spp.	+	Commensal of mucosal surfaces	Endocarditis, meningitis, pneumonia, and infections of wounds and various soft tissues.

Table 42.2: Some charactristics of Non-Fermenters

Pathogenesis

Human infection is mainly acquired cutaneously through skin abrasions or by inhalation of contaminated particles. It may also get transmitted from the animals by the bite of hematophagous insects. Agriculture workers, especially those who work in moist soil (e.g. paddy fields), are particularly prone to infection.

The disease may take one of four forms: acute, subacute, chronic, or latent. It may be an acute septicemia with involvement of many organs, a subacute typhoid like disease, or pneumonia and hemoptysis resembling tuberculosis. In chronic form, there may be multiple caseous or suppurative foci, with abscess formation in the skin and subcutaneous tissues, bones and internal organs. Acute melioidosis has a high case fatality rate.

Serological evidence indicates that inapparent infection is common in endemic areas. The organism can survive intracellularly within the reticulo-endothelial system, and this may account for latency and the emergence of symptoms many years after exposure. Suppurative parotitis is a characteristic presentation of melioidosis in children.

In India, cases of melioidosis have been reported from Maharashtra, Kerala, Tamil Nadu, Orissa, West Bengal and Tripura.

Laboratory Diagnosis

1. **Microscopy:** A **Gram stain** of an appropriate specimen will show small gram-negative bacilli; **bipolar staining (safety pin appearance)** is seen with Wright' stain or methylene blue stain.
2. **Culture:** The organism may be observed may be cultured from sputum, urine, pus or blood on selective media. Isolation of *B. pseudomal1ei* for diagnostic purposes should be approached carefully because the organism is highly infectious.
3. **Serology-Enzyme-linked immunosorbent assay (ELlS A)** for detection of bacterial antigen, specific IgG and IgM antibody to *B. pseudomallei,* as well as an **indirect hemagglutination test**, are useful screening tests in subclinical melioidosis.
4. **Polymerase chain reaction (PCR)** methods are also available.

Treatment

Ceftazidime is the drug of choice, along with cotrimoxazole, tetracycline, amoxycillin clavulanate, or chloramphenicol. Prolonged treatment, for many months may be necessary.

GLUCOSE NONFERMENTERS

The heterogeneous group of aerobic Gram-negative bacilli commonly referred to as *glucose nonfermenters* is taxonomically distinct from the carbohydrate fermenting Enterobacteriaceae and the oxidative pseudomonads. They grow easily on common culture media, but unequivocal identification may be difficult as most species are relatively inert in the biochemical tests used in identification of Gram-negative bacteria. Their clinical relevance is based on their role as opportunistic pathogens in hospital-acquired infections and their intrinsic resistance to many antimicrobial agents. Susceptibility to antibiotics of glucose non-fermenters is very variable, and treatment should be based on the results of laboratory tests. These are *Acinetobacter, Alcaligenes* and *Achromobacter, Eikenella corrodens,, Flavobacterium meningosepticum* (See chapter 48). Some characteristics of these organisms are shown in Table 42.2.

KNOW MORE

Burkholderia Cepacia (Formerly Pseudomonas Cepacia)

In ironic contrast to its role as a human pathogen, the organism has been developed as a **biopesticide** for protecting crops against fungal diseases and as a bioremediation agent for breaking down recalcitrant herbicides and pesticides in contaminated soils. Some isolates demonstrate an unusual form of antibiotic resistance by an ability to use penicillin as a sole carbon and energy source.

☞ KEY POINTS

The term pseudomonads describes a large group of aerobic, non-fermentative, nonsporing Gram-negative bacilli, motile by polar flagella. They were originally contained within the genus *P. seudomonas.*

Pseudomonas aeruginosa

- It is small gram-negative bacilli.
- It is a strict aerobe. It grows well on ordinary media in the laboratory. Cetrimide agar is selective medium for *P. aeruginosa.*
- *P. aeruginosa* produces at least 4 distinct pigments: (i) Pyocyanin; (ii) Pyoverdin (fluorescein); (iii) Pyorubrin; (iv) Pyomelanin
- **Diseases**
  A. Community infections-1. Otitis externa and varicose ulcers; 2. Corneal infections; 3. Jacuzzi rash or whirlpool rash; 4.Indusrial eye injuries.
  B. Hospital infections-1. Localized lesions (infections of wounds and bedsores, eye infections and urinary infections following catheterisation), infection in burns, iatrogenic meningitis, post-tracheostomy pulmonary infection; 2. Septicemia and endocarditis; 3. Skin lesions; 4. Infection of the nail bed; 5. Infantile diarrhoea and sepsis; 6. Shanghai fever; 7. Other Infections (gastrointestinal tract, central nervous system, and musculoskeletal system).
- Diagnosis: Readily grow on common laboratory media. *P. aeruginosa* is identified by colonial characteristics and simple biochemical tests (e.g., positive oxidase reaction).
- Treatment, Prevention, and Control-Combined use of effective antibiotics (e.g., aminoglycoside and β-lactam antibiotics) frequently required.
- *Stenotrophomonas maltophia* is usually oxidase negative. The spectrum of nosocomial infections with *S. maltophilia* includes **bacteremia, pneumonia, meningitis, wound infections,** and **urinary tract infections.**
- *Burkholderia cepacia:* Diseases caused are respiratory tract infections, particularly in patients with cystic fibrosis; urinary tract infections; septic arthritis; peritonitis; septicemia; opportunistic infections".
- *Burkholderia pseudomallei: causes a*symptomatic colonization; "cutaneous infection with regional lymphadenitis, fever, and malaise; pulmonary disease ranging from bronchitis to necrotizing pneumonia
- *Burkholderia mallei:* Diseases caused is glanders in livestock.

IMPORTANT QUESTIONS

1. Describe the morphology, cultural characteristics, pathogenicity and laboratory diagnosis of *Pseudomonas aeruginosa.*
2. Write short notes on:
   Pigments of *Pseudomonas.*
   Pathogenicity of *Pseudomonas aeruginosa*
   Burkholderia mallei
   Pyocin typing
   Burkholderia pseudomallei
   Stenotrophomonas maltophila.

FURTHER READING

Baltch AL and Smith RP (eds) 1994. Pseudomonas aeruginosa infections, New York: Marcel Dekker.

Dance DAB. Melioidosis: The tip of the iceberg, Clin Microbiol Rev 1991;4:52-60.

Denton M, Kerr K. Microbiological and clinical aspects of infection associated with Stenotrophomonas matophila, Clin Microbiol Rev 1998;11:57-80.

Robin T, Janda JM. Pseudo-, Xantho-, Stenotrhomonas moltophilia: An emerging pathogen in search of a genus, Clin Microbiol Newsletter 1996;18:9-13.

CHAPTER 43

Legionella

Learning Objectives

After reading and studying this chapter, you should be able to:

- Describe morphology, culture characterstics and biochemical reactions of *Legionella pneumophila*
- Describe the following: *Legionella pneumophilia*; diseases caused by *Legionella*.

INTRODUCTION

In the summer of 1976, public attention was focused on an outbreak of severe pneumonia that caused many deaths in members of the American Legion convention in Philadelphia. The disease was characterized by fever, cough and chest pain, leading on to pneumonia and often ending fatally. The causative agent has been called *Legionella pneumophila*. Subsequent studies found this organism to be the cause of multiple epidemic and sporadic infections. It is now recognized to be a ubiquitous aquatic saprophyte.

Species

Taxonomic studies have shown that the family Legionellaceae consists of one genus, *Legionella,* with 40 species and more than 60 serogroups. Approximately half of these species and serogroups have been implicated in human disease, with the others found in environmental sources. The original isolate in this genus is designated *L. pneumophila* serogroup 1 (SG1), which accounts for nearly all severe infections. Examples of other species that cause human infection less often are *L. micdadei, L. bozemanii, L. dumoffii* and *L. gormanii.*

LEGIONELLA PNEUMOPHILA

Morphology

Legionellae are thin, noncapsulated bacilli, 25 m x 0.3-0.1 11 m, coccobacillary in clinical material and assuming longer forms in culture. Most are motile with polar or subpolar flagella. They are gram-negative but stain poorly, particularly in smears from clinical specimens. They stain better by silver impregnation but are best visualized by direct fluorescent antibody (DFA) staining with monoclonal or polyclonal sera.

Culture

Legionellae are nutritionally fastidious. Their growth is enhanced with iron salts and depends on the supplementation of media the L-cysteine. They have fastidious requirements and grow on complex media such as buffered charcoal, yeast extract (BCYE) agar, with L-cysteine and antibiotic supplements, with 5 percent CO_2, at pH 6.9, 35°C and 90 percent humidity. Growth is slow and colonies take 3 to 6 days to appear.

Biochemical Reactions

The organisms are nonfermentative and derive energy from the metabolism of amino acids. Most species are motile and catalase-positive, liquefy gelatin and do not reduce nitrate or hydrolyze urea.

Epidemiology

Sporadic and epidemic legionellosis has a worldwide distribution. Legionellae are widely distributed in natural water sources, such as stagnant waters, mud and hot springs, where the nutritional and growth requirements for these fastidious bacteria are provided by some types of algae. The bacteria are commonly present in natural bodies of water, such as lakes and streams, as well as in air conditioning cooling towers and condensers and in water systems (e.g. showers, hot tubs). Legionellae survive and multiply inside free-living amoebae and other protozoa. They also multiply in some artificial aquatic environments, which serve as amplifiers.

Human infection is typically by inhalation of aerosols produced by cooling towers, air conditioners and shower heads which act as disseminators. Aerosolized legionellae can survive for long and can be carried over long distances. No animal reservoir exists and infection is limited to human beings. No carrier state is established. Man-to-man transmission does not occur.

The outcome of inhalation of legionellae depends on the size of the infecting dose, virulence of the strain and resistance of the host. Known risk factors are smoking, alcohol, advanced age, intercurrent illness, hospitalization and immunodeficiency. Men are more often affected than women. In the developed countries, legionellosis accounts for 1 to 3 percent of community acquired, and 10 to 30 percent of hospital acquired pneumonias. Its prevalence in the developing countries is not adequately known.

Pathogenesis

Respiratory tract disease caused by *Legionella* species develops in susceptible people who inhale infectious aerosols.Legionellae are facultative intracellular parasites that can multiply in alveolar macrophages and monocytes following entry into the alveoli through aerosols. Dissemination occurs by endobronchial, hematogenous, lymphatic and contiguous spread. Because of their intracellular location, humoral antibodies are ineffective. Cellular immunity is responsible for recovery.

Clinical Diseases

Asymptomatic *Legionella* infections are relatively common. Symptomatic infections primarily affect the lungs and present in one of two forms

1. Pontiac fever- An influenza-like illness.
2. Legionnaires' disease- A severe form of pneumonia.

Pontiac Fever

Pontiac fever is a milder, nonfatal 'influenza like' illness with fever, chills, myalgia malaise, and headache but no clinical evidence of pneumonia. Outbreaks with high attack rates may occur.

Legionnaire's Disease

The incubation period is 2 to 10 days. The disease presents with fever, nonproductive cough and dyspnea, rapidly progressing, if untreated, to pneumonia. The primary manifestation is pneumonia. Multiorgan disease involving the gastrointestinal tract, central nervous system, liver and kidneys is common.

Case fatality may be 15 to 20 percent but can be much higher in patients with severely depressed cell-mediated immunity, the cause of death being progressive respiratory failure and shock. All age groups are susceptible, though more cases have occurred in the elderly. Legionnaires' disease (legionellosis) is characteristically more severe and causes considerable morbidity, leading to death unless therapy is initiated promptly. Legionnaire's disease may be either epidemic or sporadic.

Laboratory Diagnosis

1. Microscopy

Legionellae in clinical specimens stain poorly with gram stain. Nonspecific staining methods, such as those using Dieterle's silver or Gimenez's stain, can be used to visualize the organisms. The most sensitive way of detecting legionellae microscopically in clinical specimens is to use the direct fluorescent antibody (DFA) test, in which fluorescein-labeled monoclonal or polyclonal antibodies directed against *Legionella* species are used . The test is specific, with false-positive reactions observed only rarely if monoclonal antibody preparations are used.

2. Culture

Although legionellae were difficult to grow initially, commercially available media now make growth easy. Legionellae require L-cysteine and their growth is enhanced with iron (supplied in hemoglobin or ferric pyrophosphate). The medium most commonly used for the isolation of legionellae is buffered charcoal-yeast extract (BCYE) agar, although other supplemented media have also been used. Antibiotics can be added to suppress the growth of rapidly growing contaminating bacteria. Legionellae grow in air or 3 to 5 percent carbon dioxide at 35°C after 3 to 5 days. Their small (1- to 3 mm) colonies have a groundglass appearance

3. Antigen Detection

Enzyme-linked immunoassays, radioimmunoassays, the agglutination of antibody-coated latex particles, and nucleic acid analysis studies have all been used to detect legionellae in respiratory specimens and urine. If an antigen test is used, culture should always be performed. Nucleic acid analyzes have been disappointing to date.

4. Serology

Detection of serum antibody is done by ELISA or indirect immunofluorescent assay.

Treatment

For treatment, the newer macrolides,ciprofloxacin, and tetracyclines are effective. Rifampicin is employed in severe cases. Beta lactamase antibiotics and aminoglycosides are ineffective.

Prevention

Prevention of legionellosis requires identification of the environmental source of the organism and reduction of the microbial burden. Hyperchlorination of the water supply and the maintenance of elevated water temperatures have proved moderately successful. However, complete elimination of *Legionella* organisms from a water supply is often difficult or impossible. Because the organism has a low potential for causing disease, reducing the number of organisms in the water supply is frequently an adequate control measure. Hospitals with patients at high risk for disease should monitor their water supply on a regular basis for the presence of *Legionella* and their hospital population for disease. If hyperchlorination or superheating of the water does not eliminate disease (complete elimination of the organisms in the water supply is probably not possible), continuous copper-silver ionization of the water supply may be necessary.

☞ KEY POINTS

- *Leginella pneumophila* serogroup 1 is most common human pathogen.
- *L. pneumophila* are small, slender, pleomorphic, gram-negative bacilli.
- Legionellae are widely distributed in natural water sources, such as stagnant waters, mud and hot springs. Human infection is typically by inhalation of aerosols produced by cooling towers, air conditioners and shower heads .
- *L. pneumophila* causes Legionnaire's disease, a life-threatening multifocal pneumonia and Pontiac fever, a self-limited, febrile, flu-like illness.
- Laboratory diagnosis depends on direct fluorescent antibody (DFA) test, culture, detection of antigen and demonstration of serum antibodies.
- Erythromycin or tetracycline is useful.
- Hyperchlorination of the water and superheating of water is of value in preventing the disease.

IMPORTANT QUESTIONS

1. Discuss pathogenicity and laboratory diagnosis of Legionnaire's disease.
2. **Write short notes on :**
   Legionella . pneumophila
   Legionnaire's disease
   Pontiac fever

FURTHER READING

Fang GD, et al. Disease due to the Legionallaeceae, Medicine (Baltimore) 1989;68:116.

Stout JE and VE Yo. Legionella. New Eng] Med 1997; 337:682.

Stout J, Yu V: Legionellosis, N Engl J Med 1997;337:681-687

CHAPTER 44

Yersinia, Pasteurella, Francisella

Learning Objectives

After reading and studying this chapter, you should be able to:

- Describe morphology, culture characteristics and biochemical reactions of *Yersinia. Pestis*.
- Describe laboratory diagnosis of plague.
- Descrbe the following: Prophylaxis against plague; *Yersinia enteroeolitiea; Yersinia pseudotuberculosis; Pasteurella multocida; Francisella tularensis.*

INTRODUCTION

The organisms within these three genera(**Yersinia, Pasteurella, Francisella)** are animal pathogens that, under certain conditions, are transmissible to man either directly, or indirectly through food and water or via insect vectors. They are gram-negative coccobacilli, formerly contained within one genus, **Pasteurella.** Molecular genetics has indicated a completely separate identity for the **three genera**-*Yersinia. Pasteurella* and *Francisella.*

Genus Yersinia-was so named after Alexandre Yersin, who had isolated the plague bacillus in 1894. *Yersinia* belongs to the family Enterobacteriaceae and the tribe Yersinia **The genus Yersinia** currently consists of 11 named species.

Medically important species are:

Y. pestis (the causative agent of plague);

Y. pseudotuberculosis (a primary pathogen of rodents);

Y. enterocolitica (which causes enteric and systemic disease in animals and human beings).

Genus Pasteurella

The genus Pasteurella is now restricted to a number of animal pathogens and contains several related bacteria causing hemorrhagic septicemia in different species of animals and occasionally producing local and systemic infections in human beings, grouped under a common species named *P. multocida*. One of these, *P. aviseptica* is the chicken cholera bacillus used by Pasteur for the development of the first attenuated bacterial vaccine. Hence, the name Pasteurella.

Genus Francisella

The third new **genus *Francisella*,** consisting of ***F. tularensis***, is named after Francis for his pioneering studies on tularemia caused by this bacillus.

YERSINIA PESTIS (FORMERLY PASTEURELLA PESTIS)

The plague bacillus was discovered independently and simultaneously by Yersin and Kitasato (1894) in Hong Kong at the beginning of the last pandemic of the disease.

Morphology

Y. pestis is a gram-negative, short, oval coccobacilli with rounded ends and convex sides, about 1.5 × 0.7 µm, occurring singly, in short chains or in small groups. In smears stained with Giemsa or methylene blue, it shows **bipolar staining (safety pin appearance)** with the two ends densely stained and the central area clear (Fig. 44.1).

Pleomorphism is marked *in* culture, especially in old cultures, involution forms are seen-coccoid, club saped, filamentous and giant forms. This involution in culture can be hastened by the addition of 3 percent NaCI, and the phenomenon may be used as a means of identification. Bipolar staining is less obvious in culture than in smears from the tissues and the involution forms stain only faintly.

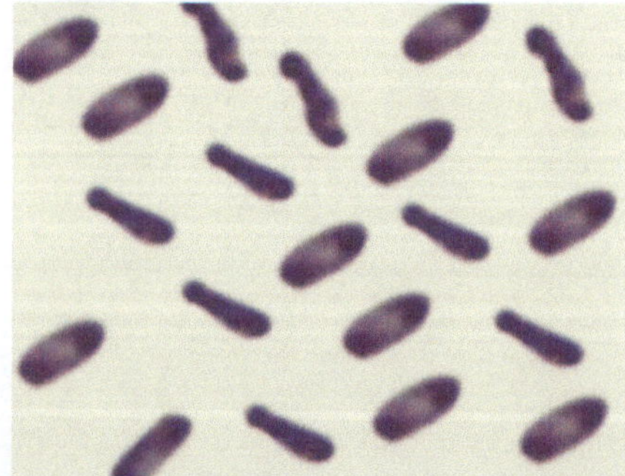

Fig. 44.1: Smear from gland puncture in a case of plague showing *Y. pestis* with bipolar staining (safety pin appearance)

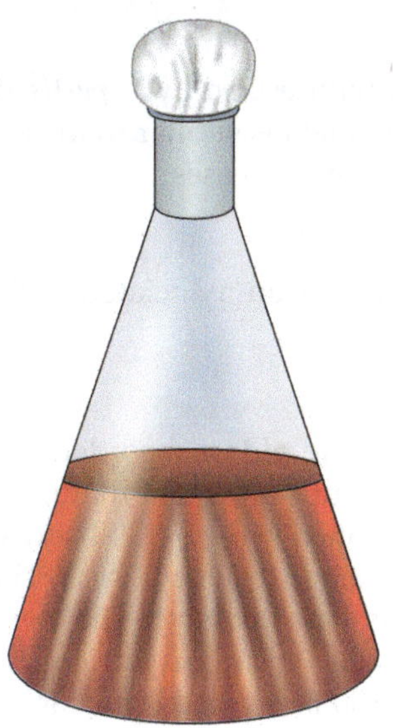

Fig. 44.2: Y. pestis in ghee broth culture. Stalactite growth

The bacillus is surrounded by a **slime layer (envelope or capsule).** It is nonmotile, nonsporing and non-acid fast.

Cultural Characteristics

The plague bacillus is aerobe and facultative anaerobe. Growth occurs over a wide range of pH 5-9.6 (optimum pH 7.2). Optimum temperature for primary culture is 27°C (range 14-37°C). Several phenotypic characteristics are best expressed at room temperature but the envelope develops at 37°C.

It can grow on ordinary laboratory media and is not nutritionally exacting.

1. Nutrient Agar

On nutrient agar, colonies are small (as small as 0.1-0.2 mm), delicate, transparent disks and become opaque on continued incubation.

2. Blood agar

On **blood agar** colonies are dark brown due to absorption of the haemin pigment.

3. DCA and MacConkey Agar

On **DCA** and **MacConkey agar** it grows poorly, producing pin-point, reddish colonies after 24 hours incubation.

Broth

4. In **broth,** a flocculent growth occurs at the bottom and along the sides of the tube, with little or no turbidity. A delicate pellicle may form later.

5. Ghee Broth

If grown in a flask of **broth with sterile oil or ghee** (clarified butter) floated on top (ghee broth) a characteristic growth occurs which hangs down into the broth from the surface, resembling stalactites **(stalactite growth)** (Fig. 44.2).

Biochemical Reactions

It ferments glucose, mannitol and maltose with the production of acid but no gas.Lactose and sucrose or rhamnose are not fermented.

It is catalase positive, indole negative, MR positive, VP and citrate negative (IMViC - + - -), nitrate reduction positive, aesculin positive and oxidase and urease negative. Gelatin is not liquefied.

Physiological Varieties of Y. Pestis

Devignat has distinguished three physiological varieties of *Y. pestis* based on the fermentation of glycerol and reduction of nitrate. These biotypes have been designated orientalis, mediaeval and antigua, and are characterized by differences in their geographic distribution. This typing appears to be of epidemiological significance because of the different geographical distribution of the types (Table 44.1).

Resistance

The plague bacillus is easily destroyed by exposure to heat, sunlight, drying and chemical disinfectants. It is killed by moist heat at 55°C in 5 min and by 0.5 percent phenol in 15 minutes. It is very susceptible to drying. It remains viable for long periods in cold, moist environments.It can survive for several months, and even multiply, in the soil of rodent burrows. All strains are lysed by a specific antiplague bacteriophage at 22°C.

Antigens, Toxins and other Virulence Factors

At least 20 different antigens have been detected in *Y. pestis* by gel diffusion and biochemical analysis, several of the antigens are virulence determinants, some of which are plasmid encoded.

Table 44.1: Various diseses caused by species of Yersinia, Pasteurella and Francisella

Yersinia,	*Y. pestis*	Causative agent of plague
	Y. pseudotuberculosis	Pseudotuberculosis in animals
	Y.enterocolitica	Enteric and systemic disease in animals and human beings
Pasteurella	*Pasteurella multocida*	Hemorrhagic septicemia in animals and occasionally producing local and systemic infections in human beings.
Francisella	*Francisella tularensis*	Tularemia

1. F-L Antigen or Envelope Antigen

The heat-labile Fraction I (FI) protein capsular antigen is a soluble antigen contained within the bacterial envelope. Maximal production occurs at 37°C. It helps the organism to resist phagocytosis. This plastid encoded antigen has been considered a virulence determinant but occasional strains deficient in "Fraction I antigen have been isolated from fatal human cases. The antigen is highly immunogenic and the antibodies to it appear to be protective in both humans and experimental animals.

2. V and W Antigens

These antigens are always produced together. These antigens have been considered to be the virulence factors as they inhibit phagocytosis and intracellular killing of the bacillus. Production of V and W antigens is plasmid mediated.

3. Pigment Binding and Iron-Regulated Surface Proteins

In yersiniae, avirulent and low-pathogenicity strains require iron overload to produce septicemia in humans or lethality in mice. Highly pathogenic species do not depend on an exogenous iron supply to produce the same effects.

4. Pesticin I, Coagulase and Plasminogen Activator

The production of these components is genetically linked and mediated by a plasmid. **Pesticin I** is a bacteriocin produced by *Y. pestis* that inhibits the growth of *Y. pseudotuberculosis* as well as some strains of *Escherichia coli* and *Y. enterocolitica*. It is thought that these enzymes, especially the plasminogen activator, may contribute to the highly invasive fulminant character of the disease

5. Other Virulence—Associated Factors

A number of additional factors have been proposed as virulence factors-a endotoxin, murine toxin and a protein antigen, the pH 6 Ag.

a. Endotoxin

The role of *Y. pestis* endotoxin in the disease process is ill-defined.

b. Murine Toxins

The second class of toxins is protein in nature, possessing some properties of both exotoxins and endotoxins. They are thermolabile and may be toxoided but do not diffuse freely into the medium and are released only by the lysis of the cell. They are called ***'murine toxins'*** as they are active in rats and mice but not in guinea pigs, rabbits and primates. On injection into experimental animals, plague toxins produce local edema and necrosis with systemic effects on the peripheral vascular system and liver. The role of plague toxins in natural disease in human beings is not known.

c. pH 6 Ag

The pH 6 Ag is synthesized by *Y. pestis* at a temperature of 37°C and at pH levels similar to those in macrophage phagolysosomes or abscesses.

6. Purine Synthesis

Virulence has also been associated with the ability for purine synthesis.

Plague

Plague is one of the oldest recorded infectious diseases and is an ancient scourge of mankind. Epidemics of plague are mentioned in the Bible. The disease was familiar to the ancient civilizations of Asia. The association of plague with rats was known to the ancients. It is mentioned in Bhagwat Puran (an ancient Indian text written about 1500-800 BC) that as soon as the dead rats are seen, the residence should be immediately abandoned. Down the ages, plague was known as "Mahamari", the great death.

Central Asia is believed to have been the original home of plague, from where it has, in wave after wave, spread far and wide, causing epidemics and pandemics, exacting a toll of human life surpassing any other disease. The identity of the Biblical plague of the Philistines (1320 BC) is in doubt. Since the dawn of Christian era, there have been three great pandemics: The first began in the year 542 (Justinian plague) and is estimated to have caused 100 million deaths; the second began in the year 1346, lasted for three centuries and claimed 25 million lives; and the last began in 1894 and continued until 1930s. More than 150 epidemics, most of them associated with three main pandemics, have been responed.

The second pandemic, known as the Black Death. started in the fourteenth century. Historians of plague identify 41 epidemics before the birth of Christ and 109 epidemics in the next 15 centuries. There are records of 45 pandemics between 1500 and 1720 AD. The disease was quiescent in the 18th and 19th centuries and confined to endemic foci.

The third pandemic originated in Burma, spread to China in 1894, and from Hong Kong was carried to other continents, induding North America, via rat-infested ships. India was one of the countries worst hit by this pandemic. Plague reached Bombay in 1896 and spread all over the country during the next few years, causing more than 10 million deaths by 1918. It gradually receded thereafter, though occasional cases continued to occur in endemic foci till 1967. No further plague cases were seen in India till 1994, when in August a nonfatal outbreak of bubonic plague was reported from Maharashtra (Beed district). In September pneumonic plague was reported in Surat and adjoining areas of Gujarat and Maharashtra, causing much panic and consternation. A few cases were reported from different parts of North India also, probably caused by the exodus from affected areas. 4780 suspected cases were reported, out of which 167 tested positive for plague and 53 deaths were reported. More recently as of 19th Feb. 2002, the Ministry of Health reported a total of 16 cases of pneumonic plague (including 4 deaths) in Hat Koti village, Shimla Dist. Himachal Pradesh since the onset of the outbreak on 4th Feb. 2002. The disease was confirmed by NICD (National Institute of Communicable Diseases).

Pathogenesis

Plague is a **zoonotic** disease. The plague bacillus is naturally parasitic in rodents. Infection is transmitted among them by **rat fleas.** The fleas acquire the infection by feeding on infected rodents. The most efficient involves ingestion of the organism by the flea during a blood meal from a bacteremic host. When a rat flea, commonly *X cheopsis,* bites a diseased rat, it sucks blood. In the flea, the bacilli multiply in the stomach to such an extent that they block the proventriculus. When such a **'blocked flea'** bites another rodent, it cannot suck in blood because the bacterial mass blocks the passage mechanically. The blood, mixed with the bacteria is regurgitated into the bite, transmitting the infection. Infection may also be transferred by contamination of the bite wound with the feces of infected fleas.When a diseased rat dies (rat fall), the fleas leave the carcass and in the absence of another rat, may bite human beings, causing bubonic plague.

Forms of Human Plague

Traditionally, three severe forms of human plague are described. All of these may occur at different stages in the same patient.

1. Bubonic plague
2. Pneumonic plague
3. Septicemic plague.

1. Bubonic Plague

After an incubation period of 2 to 5 days, the lymph nodes draining the site of entry of bacillus become infected. As the plague bacillus usually enters through the bite of infected flea on the legs, the inguinal lymph nodes are involved, hence the name bubonic plague (**bubo** means enlarged gland in groin). The glands become enlarged and suppurate. The bubo may be preceded by. prodromata of chills, fever, malaise, confusion, nausea and pains in the limbs and back. The bacilli enter the bloodstream and produce septicemia. Sometimes, there are hemorrhages into the skin and mucosa. Disseminated intravascular coagulation is common and may lead to gangrene of the skin, fingers and penis. The case fatality in untreated cases may be 30 to 90 percent.

Pestis Minor

The transfer of *Y. pestis* from rats to man through the bites of infected fleas may occasionally result in a localized infection, known as ***pestis minor,*** with mild constitutional symptoms.

2. Pneumonic Plague

This can develop in patients presenting with bubonic or septicemic plague. It may also be acquired as a primary infection by inhalation of droplets infected with *Y. pestis,* usually from an individual with pneumonic disease or as a result of exposure to aerosols generated from cultures. Pneumonic plague may be seen sometimes during epidemics of bubonic plague.The bacilli spread through the lungs producing hemorrhagic pneumonia. The sputum becomes thin and blood-stained. It contains numerous plague bacilli that are demonstrable in stained films or on culture of the sputum.

This type of plague is highly contagious and is almost invariably fatal unless treated very early almost 100 percent and with treatment it is 5 to 30 percent.

3. Septicemic Plague

This may occur as a primary infection or as a complication of bubonic or pneumonic plague. Purpura may develop in the skin, giving the skin a blackish coloration which, in the past, led to the name **"black death". Disseminated intravascular coagulation** is usually present. **Meningitic involvement** may occur rarely.

Epidemiology

Several species of fleas may act as vectors, the most important being *Xenopsylla cheopis.* X. *astia* and *Ceratophyllus fasciatus.* However, the Oriental rat flea, *Xenopsylla cheopis,* is highly efficient and has been the classic vector in urban rat-borne epidemics. X. *cheopis,* the predominant species in north India is a more efficient vector than the south Indian species X. *astia.* This has contributed to the more extensive nature of plague outbreaks in the north as compared tos south India.

Plague is perpetuated by three cycles: (1) natural foci among commensal rodents with transmission by fleas (**sylvatic plague, wild plague**), (2) urban rat plague, which is transmitted by the rat flea (**domestic plague, urban plague)**, and (3) **human plague**, which may be acquired by contact with either of the former cycles and which may be transmitted by pneumonic spread or rarely, by the bite of a human flea.

In India, the gerbil *(Tatem indica)* and the bandicoot are infected. Human infection may occur during skinning and handling of carcasses of infected wild animals. Carnivores, including cats and dogs can get infected by eating infected rodents or through their fleas. Clinical plague is seldom seen in dogs but may develop in cats. Human infection from inhalation of respiratory droplets from infected cats has been reported.

The studies of the various governmental Plague Commissions in Bombay during the early years of this century, helped to clarify the epidemiology of plague. It was found that plague produced epizootics first in *Rattus norvegicus* (sewer rat). When their number dwindled, the disease passed to the domestic rat, *R. rattus.* It was from the domestic rat that the infection spread to human beings.

In the 1990s, there has been a re-emergence of plague in countries where it had ceased to be noticed for many years. This has happened in the developing and the developed countries - India and China in Asia, Malawi and Zimbabwe in Africa, the erstwhile USSR in Europe and in the USA. Plague bacillus strains carrying plasmid borne resistance to multiple antibiotics were reported from Madagascar in 1995. These have the potential to spread and pose a great threat.

Y. pestis has been employed as a biological warfare agent. Its potential application in bio-terrorism is of major concern.

Laboratory Diagnosis

Pneumonic plague is easily acquired in the laboratory by inhalation of aerosols generated from *Y. pestis* cultures. These clinical specimens suspected of containing the organism should only be handled under containment conditions appropriate for Class 3 pathogens. Laboratory animals used for diagnostic tests must be housed under insect-free containment conditions.

Plague is confirmed by demonstrating the bacilli in fluid from buboes or local skin lesions, in the sputum and in blood films. Blood culture may be intermittently positive in all forms of the disease. Postmortem, the bacilli can usually be isolated from a wide range of tissues, especially spleen, lung and lymph nodes.

1. Specimens

i. Bubonic plague-Pus or fluid aspirated.
ii. Pneumonic plague-sputum and blood.
iii. Septicemic plague-blood.
iv. Meningeal plague-cerebrospinal fluid (CSF)
v. On postmortem-splenic tissue

2. Microscopy

Smears of exudate or sputum are stained with methylene blue or Giemsa stain. Characteristic gram-negative coccobacilli and bacilli showing bipolar staining with methylene blue suggest plague bacilli. The smears are also stained by gram's method and observed for gram-negative, ovoid coccobacilli with bipolar bodies. The fluorescent antibody technique may be of use in identifying plague bacilli.

3. Culture

Culture the samples on **blood agar plates, MacConkey agar, nutrient agar** and **ghee broth** and incubated at 27°C. Colonies on blood agar are dark brown due to absorption of hemin pigment. Colonies on MacConkey agar are colorless. In ghee broth, a characteristic stalactite growth is produced. The growth is identified by biochemical tests and slide agglutination tests. Demonstration of the FI capsular antigen by immunospecific staining will confirm the presence of *Y. pestis*.

4. Animal Inoculation

If exudate is inoculated subcutaneously into guinea-pigs or white rats, or on to their nasal mucosa, infection follows and the animals die within 2 to 5 days. The bacilli may then be isolated from the blood or from smears of spleen tissue taken postmortem.

Postmortem examination shows a marked local inflammatory condition at the site of inoculation with necrosis and edema. The regional lymph nodes are enlarged, the spleen is enlarged and congested and may show greyish-white patches in the tissue. Prepare films from the local lesions, lymph nodes, spleen pulp and heart blood; stain and examine for characteristic plague bacilli.

5. Antigen Detection

Demonstration of the FI capsular antigen by immunospecific staining and ELISA test will confirm the presence of *Y. pestis.*

Dipstick Test

FI glycoprotein can be detected by dipstick test using monoclonal antibodies. It is rapid test and produces reliable result within 15 minutes.

6. Serology

The antibodies to F-I antigen can be demonstrated in patient's serum by **complement fixation test** (CFT), **hemagglutination test** and enzyme linked immunosorbent assay (ELISA)

7. Polymerase Chain Reaction (PCR)

A polymerase chain reaction (PCR), with primers based on FI gene sequences, offers a rapid and less hazardous means of diagnosis than culture.

Diagnosis of Plague in Wild Rats

Before examining rats that may have died as a result of plague, immerse them in disinfectant to kill any infected fleas. It is important to differentiate plague from infection due to *Y. pseudotuberculosis.* Necropsy has the following features in plague:

i. Enlargement of the lymphatic nodes with periglandular inflammation and edema that is most frequent in cervical lymph nodes due to the tendency for the flea to attack the rat's neck region
ii. Pleural effusion
iii. Enlargement of the spleen which may show small white areas in the pulp
iv. Liver congested and mottled;
v. Congestion and hemorrhages under the skin and in internal organs.

Make films and cultures of heart blood, lymph nodes and spleen.

Inoculate material from lesions on to the nasal mucosa of guinea pigs or white rats as with sputum (see above). Examine stained films and pure cultures for characteristic morphology and biochemical reactions of *Y. pestis.*

Prophylaxis

Plague is one of the internationally quarantinable diseases and reporting of cases is mandatory. Public health authorities may institute enforced quarantine and disinfection of persons, ships and aircraft arriving with known or suspected infected persons or animals.

Prophylaxis can be carried out by:

A. General Measures

1. Control of fleas and rodents.
2. Spray insecticide(DDT) inside the rodent, burrows and houses to kill the fleas. If they are resistant to DDT, other insecticides should be tried.
3. After the fleas have been killed, kill the rat with rat-poison.

Other control measures include the construction of rat-proof dwelling houses and buildings such as warehouses in dockland areas.

B. Vaccination

Two types of vaccine have been in use—killed and live attenuated vaccines.

1. Killed Vaccine

The killed vaccine used in India (prepared at the Haffkine Institute, Bombay) is a whole culture antigen. A virulent strain of the plague bacillus is grown in case in hydrolysate broth for 2 to 4 weeks at 32°C and killed by 0.05 percent formaldehyde and preserved with phenyl mercuric nitrate (Sokhey's modification of Haffkine's vaccine). In Asia, Haffkine Institute, Mumbai is the only vaccine producing center.

Vaccine Dose Schedule

The vaccine is given **subcutaneously,** two doses at an interval of 1 to 3 months, followed by a third six months later. Vaccination can confer significant protection against bubonic but not pneumonic plague. The protection does not last for more than six months.

Indications for Vaccine

The vaccine is recommended only in those exposed occupationally or otherwise to infection, such as plague laboratory or hospital personnel and troops deployed in known plague areas. It is of no value in plague outbreaks and mass vaccination is not advised.

Side Effects

Fever, headache. malaise. lymphadenopathy, and erythema and induration at the site of inoculation. It may cause fetal damage and abortion in pregnant women. Vaccination is not very effective, therefore, even vaccinated individuals should also be given chemoprophylaxis when exposed to plague.

2. Live Attenuated Vaccines

Live vaccines are prepared from two avirulent strains of *Y. pestis, Otten's Tjiwidej* strain from Jawa and *Girard's EV 76 strain* from Malagasey. Killed vaccine is recommended for general use because live vaccines are difficult to prepare and may provoke unacceptable reactions. Live vaccines are not in use now.

Chemoprophylaxis

A person exposed to definite risk of infection, whether vaccinated or not, should be given chemoprophylaxis-cotrimoxazole or tetracycline orally for at least five days. Close contacts of patients with plague should be given a course of tetracycline (500 mg 6 hourly for one week).

Treatment

Y. pestis infections are treated with streptomycin; tetracyclines, chloramphenicol, or trimethoprim/sulfamethoxazole can be administered as alternative therapy. Streptomycin is the drug of choice. Chloramphenicol is recommended in patients with meningitic symptoms. Tetracycline may be adequate in uncomplicated bubonic plague. Gentamicin and ciprofloxacin are also effective. Because of the potential for overwhelming septicemia, rapid institution of antibiotic therapy is crucial. Supportive therapy is essential for patients with signs of shock.

YERSINIOSIS

The term *yersiniosis* denotes infection with *Yersinia* species other than *Y. pestis,* namely, *Y. pseudotuberculosis* and *Y. enterocolitica*. They are found in the intestinal tract of a variety of animals, in which they cause diseases, and are transmissible to humans, in which they produce a variety of clinical syndromes. These are **zoonotic** diseases.

These are non-lactose fermenting gram-negative rods that are urease-positive and oxidase-negative. They resemble *Y. pestis* because they are small, gram-negative rods with bipolar staining and rodents, wild and domestic animals are reservoirs of infection. They differ from *Y. pestis* by motility when grown at 22°C (non-motile at 37°C), noncapsulated, urease positive, oxidase negative and insusceptible to *Y. pestis* bacteriophage (Table 44.2)

Yersinia pseudotuberculosis

Y. pseudotuberculosis is a small, oval, gram-negative, bipolar-stained bacillus. It is nonsporing, noncapsulated and slightly acid-fast. Genetically it is very similar to *Y. pestis.*

The organisms may be differentiated from *Y. pestis* by (Table 44.2).

- its relatively poor growth on MacConkey agar
- motility when grown at 22°C(but not at 37°C)
- ability to produce urease
- fermentation of rhamnose and melibiose
- failure to be lysed by the anti plague bacteriophage at 22°C
- lack of the FI antigen as shown by immunospecific staining or PCR.

There are eight major O serotypes, several of which can be separated into subtypes based on thermostable LPS somatic antigens. Thermolabile flagellar antigens are present in cultures grown at 18 to 26°C. The majority of human cases of *Y. pseudotuberculosis* infection are due to serotype 1.

Table 44.2: Biotypes of Yersinia pestis			
Variety	*Glycerol fermentation*	*Nitrate reduction*	*Geographical distribution*
Y. pestis var. orientalis	–	+	Primary foci in India, Myanmar, and China. Causative agent of 1894 pandemic. Responsible for wild plague in Western USA, South: America, South Africa.
Y. pestis var. antiqua	+	+	Transbaikalia, Mongolia, Manchuria, perhaps responsible for Justinian plague.
Y. pestis var. medievalis	+	–	Southeast Russia

Pathogenesis

Animal infection

It causes pseudotuberculosis which is a **zoonotic diseses. In animals,** the natural mode of infection is by gastrointestinal tract. In infected guinea pigs, the liver, spleen and lungs show multiple nodules resembling tuberculosis lesions (hence the name **pseudotuberculosis**).

Human Infection

Human infection probably results from ingestion of materials contaminated with animal feces usually through contaminated food or water. Infection may be subclinical but occasionally results in a **severe typhoid-like illness** with fever, purpura and enlargement of the liver and spleen, which is usually fatal. More frequently it causes **mesenteric lymphadenitis** and **terminal ileitis** simulating acute or subacute appendicitis. It has also been reported to cause immunological sequelae such as erythema nodosum or reactive arthritis in some patients.

Laboratory Diagnosis

Laboratory diagnosis may be made by isolation of the organism in culture from blood, local lesions or mesenteric nodes, or demonstration of antibodies in patient serum during the acute phase of the illness by **tube agglutination tests Hemagglutination of red cells** sensitized with LPS, or **ELISA** can also be used.

Treatment

Ileitis and mesenteric lymphadenitis are usually self-limiting but septicemia may be treated with parenteral ampicillin, chloramphenicol, gentamicin or tetracycline.

Yesinia Enterocolitica

It is a gram negative coccobacillus showing pleomorphism in older cultures. This bacillus resembles *Y. pseudotuberculosis* in being motile at 22°C but differs from it in fermenting sucrose and cellobiose and decarboxylating ornithine. It does not ferment rhamnose or melibiose. Many strains are indole and VP positive (Table 44.2).

Cultural Characters

It is aerobe and facultative anaerobe. Optimum temperature for growth is 22 to 29°C;. On **blood agar**, it forms nonhemolytic, smooth, translucent colonies, 2 to 3 mm in diameter in 48 hours. On **MacConkey medium**, it forms pinpoint colonies.

Antigenic Structure

Based on O and H antigens, more than 60 different O antigens and 19 H factors have been identified. Serotypes O3, O8 and O9 account for most human infections. Serological cross reactions between serotype 9 and brucella strains occur.

Pathogenesis

Y. enterocolitica has been isolated from a wide range of domestic and wild animals and, in recent years, is increasingly being reported from human clinical material. Human disease usually results from ingestion of contaminated food or from contact with the environment. Blood transfusion is a significant hazard.

Types of Disease in Humans

1. **Gastroenteritis or enterocolitis**, which is self limited and occurs in **young children.** It may be either inflammatory or noninflammatory.
2. **Mesenteric lymphadenitis and terminal ileitis** in **older children** that may mimic appendicitis, and
3. **Septicemia**, which is often fatal, is most common in the elderly or in patients with predisposing conditions.
4. **Pneumonia and meningitis** are rare presentations.
5. **Postinfectious complications** include **erythema nodosum, polyarthritis, Reiter's syndrome and thyroiditis.**

Treatment

Y. enterocolitica is sensitive to many antibiotics, including aminoglycosides, chloramphenicol, cotrimoxazole, quinolones and tetracyclines. Sensitivity to other β-lactam antibiotics is variable.

Uncomplicated gastrointestinal infection is usually self-limiting and treatment is indicated only in severe cases. Tetracycline is probably the drug of choice. Invasive infections such as septicemia require intensive parenteral antibiotic treatment.

PASTEURELLA MULTOCIDA (FORMERLY PASTEURELLA SEPTICA)

A group of related bacteria isolated from hemorrhagic septicemia in a variety of animals and birds had, in the past, been named according to their species of origin –

Pasterella boviseptica. lepiseptica, aviseptica, etc. Though they show some degree of host specificity, they are so alike in other respects that they are now considered strains of a single species designated *P. multocida.*

Morphology

They are gram-negative, nonmotile, nonsporing coccobacilli which show bipolar staining. Some strains are encapsulated.

Cultural Characteristics

Pasteurellae are aerobes and facultative anaerobes. They grow on blood agar and chocolate agar. It resembles *Yersinia* but differs in being oxidase positive, producing indole and failing to grow on MacConkey agar.

Epidemiology

The bacillus is usually normal inhabitant of the upper respiratory tract of a variety of animals such as dogs, cats, cattle and sheep. *Pasteurella multocida* occurs worldwide in the respiratory and gastrointestinal tracts of many domestic and wild animals. It is perhaps the most common organism in human wounds inflicted by bites from cats and dogs. Thus, Pasteurella infections are considered **zoonoses.**

Pathogenesis

The most common presentation is a history of animal bite. **Human infections** usually present as:

1. **A local abscess** at the site of a cat or dog bite, with cellulitis, adenitis and, sometimes, osteomyelitis.
2. **Infections of the respiratory system** such as pleurisy, pneumonia, empyema, bronchitis, bronchiectasis and nasal sinusitis.
3. **Meningitis or cerebral abscess** (usually following head injury), endocarditis, pericarditis or septicemia and infections of the eye, liver, kidney, intestine and genital tract are rare manifestations of disease.

Animals and Birds

P. multocida can be extremely virulent to many species of animals and birds, causing **fowl cholera** and **hemorrhagic septicemia**, which are usually fatal. It also causes respiratory infections and contributes to the pathogenesis of atrophic rhinitis in pigs.

Laboratory Diagnosis

1. **Culture-**Swabs from bite wounds from blood, from CSF in cases of meningitis and from secretions in suppurative conditions of the respiratory tract are cultured on **blood agar plates** incubated at 37°C for 24 hours. The organisms are identified by various cultural and biochemical tests.
2. **Serology-**Serology is of no value in diagnosis of acute human infection.
3. **Polymerase chain reaction (PCR)-PCR** is potentially useful but rarely available.

Treatment

Penicillin G is considered the drug of choice for *P multocida* infections resulting from animal bites. Tetracyclines and fluoroquinolones are alternative drugs.

FRANCISELLA TULARENSIS (PASTEURELLA TULARENSIS, BRUCELLA TULARENSIS)

Francisella tularensis produces *tularaemia* in man and certain small mammals, notably rabbits, hares, beavers and various rodent species. Tularaemia *was* originally described in Tulare county, California. It can be transmitted by direct contact, by biting flies, mosquitoes and ticks, by contaminated water or meat or aerosols.

Morphology

It is a very small, nonmotile, nonsporing, capsulate, gram-negative coccobacillus, about 0.3 to 0.7 µm × 0.2 µm in size. In culture it tends to be pleomorphic to and larger, even filamentous, forms are present. It stains poorly with methylene blue but dilute carbol fuchsin (10%) produces characteristic bipolar staining.

Cultural Characters

F. tularensis is strictly aerobic. It will not grow on ordinary nutrient media but grows well on **blood agar** containing 2.5 percent glucose and 0.1 percent cysteine hydrochloride. Minute droplet-like colonies develop in 72 hours.

Biochemical Reactions

Under suitable conditions acid is formed from glucose and maltose. Indole and urease tests are negative. For other reactions see Table 44.2.

Two biovars are recognized. Strains of *F. tularensis* have been subdivided into biotypes based on their virulence and epidemiological behaviour. Highly virulent strains are found only in N. America, while strains of low virulence are seen in Europe and Asia also.

Pathogenesis

The infection, which is a typical **zoonosis**, is mainly spread by insects or ticks among lagomorphs and rodents. It is transmitted to man through handling of infected animals, e.g. rabbits or hares tick, mosquito or fly bites, inhalation of contaminated dust,ingestion of contaminated water or meat. Laboratory workers are especially at risk through handling infected laboratory animals or cultures of the organism. Man-to man transmission of infection apparently does not occur.

In human beings, **tularemia** may present as a local ulceration with lymphadenitis, a typhoid like fever with glandular enlargement or an influenza like respiratory infection. The severity of disease is much greater with type A strains and case fatality rates may exceed

5 percent. Disease caused by type B strains is much less severe, with very low mortality.

Laboratory Diagnosis

F. tularensis is extremely dangerous to handle in the laboratory and Category 3 containment is required for all manipulations and animal work.

Diagnosis may be made by **culture** or **by inoculation into guinea pigs or mice.** A **PCR** has been described, but is not widely available. **Serology** is most likely to be positive after 3 weeks. Rising titres of agglutinins to *F. tularensis* or individual titres of 160 are diagnostic. Serum from cases of brucellosis may cross-react with *F. tularensis* and vice versa, usually to relatively low titre An **intradermal delayed hypersensitivity test** has been used in the past but the antigen is not readily available.

Treatment

Streptomycin or gentamicin are the antibiotics of choice in tularaemia and are usually curative.

Prophylaxis

A vaccine based on the live-attenuated LVS strain confers some protection. It can be administerd by scarification to persons who are subject to high risk of infection.

F. tularensis has been developed as a biological warfare agent and has potential application in bioterrorism.

KNOW MORE

Plague epidemics generally occur in the cool, humid seasons that favor the multiplication of fleas, leading to a high 'flea index' (mean number of fleas per rat). In the hot, dry weather, fleas do not thrive and the transmission of infection is interrupted. Some infective wild rodent fleas survive in burrows for long periods of time even after the rodent hosts have died. The development of dichlorodiphenyltrichloroethane (DDT)-resistant fleas in some areas may influence the epidemic potential or plague. Dog and cat fleas are very poor vectors and have been associated with individual cases of human plague but not with outbreaks.

☞ KEY POINTS

- Yersinia are gram-negative bacilli that are facultative anaerobes, positive by the catalase test and negative by the oxidase test.
- The genus *Yersinia* consists of 10 species, with *Y. pestis, Yersinia enteroeolitiea* and *Yersinia pseudotuberculosis* the well-known human pathogens.
- *Y. pestis* is a gram-negative, shows **bipolar staining (safety pin appearance),pleomorphic,** nonmotile, and is capsulated. It grows at 27°C but the envelope develops at 37°C. It can grow on ordinary laboratory media.
- **F-l antigen or envelope antigen**-This **antigen** has been considered a virulence determinant.
- **Epidemiology-**Plague is a zoonotic infection in humans, who become accidental hosts when they come into contact with infected rodents or their fleas. Disease is spread by flea bites or direct contact with infected tissues or person-to-person by inhalation of infectious aerosols from a patient with pulmonary disease.
- **Diseases-***Yersinia pestis* causes *plague,* which manifests in one of three forms: Bubonic plague; pneumonic plague; septicemic plague.
- **Diagnosis-**For diagnosis, material is aspirated from the bubo, for demonstration of bipolar staining short rods (coccobacilli) using Wright's or gram stain for demonstration of the bacteria by immunofluorescence. Culture of the bacteria on blood agar. Serology can be used to detect antibody to capsular antigen.
- **Yersinia *pseudotuberculosis*-**is found in domestic animals and birds. It causes pseudotuberculosis, a zoonotic disease.
- Human infection may be subclinical or a severe typhoid-like illness,mesenteric lymphadenitis and terminal ileitis.

Yesinia enterocolitica

- *Y. enterocolitica* has been isolated from a wide range of domestic and wild animals.
- It produces in humans.-gastroenteritis or enterocolitis, mesenteric lymphadenitis and terminal ileitis in older children, septicaemia, pneumonia and meningitis, erythema nodosum, polyarthritis, Reiter's syndrome and thyroiditis.

Pasteurella multocida

- Pasteurella infections are considered **zoonoses.** The most common presentation is a history of animal bite. **Human infections** usually present as :
  1. **A local abscess;**
  2. **Infections of the respiratory system;**
  3. **Meningitis or cerebral abscess.**
- *P. multocida **fowl cholera*** and hemorrhagic septicemia in many species of animals and birds.

Francisella tularensis

- *Francisella tularensis* produces *tularaemia* in man and certain small mammals. The infection is a typical **zoonosis.**

IMPORTANT QUESTIONS

1. Describe the pathogenesis and laboratory diagnosis of plague.
2. Write short notes on:
   Prophylaxis against plague

Yersinia pseudouberculosis
Yersinia enterocoliica
Pasteurellosis
Pasteurella multocida
Francisella tularensis

FURTHER READING

Abdel-Haq NM, Asmar BI, Abrahamson WM, Brown WJ Yersinia enterocolitica infection in children. The Pediotric Infectious Disease Journal 2000;19:954-958.

Bottone EJ Yersinia enterocolitica; overview and epidemiologic correlates. Microbes and Infection 1999;1:323-33

Butler, T. Yersinia infections, Clin Infect Dis 1994;19,655.

Campbell and Hughes 1 M. Plague in India. Ann Int Med, 1995;122:55.

Nakajima H, Inoue M, et at. Detection and identification of Yesinia pseudo/uberculosls and pathogenic Yersinia enterocolitica by an improved polymerase chain reaction method. J Clin Microbiol 1992;30:2484-6.

Solomon T. Alexandre Yersin and the plague bacillus. J Trop Med Hyg 1995;98:209-12.

CHAPTER 45

Haemophilus

Learning Objectives

After reading and studying this chapter, you should be able to:

- Describe morphology and culture characteristics of *Haemophilus influenzae*.
- Describe the following: X and V factors; Satellitism; Antigenic structure of *Haemophilus influenzae*.
- Describe pathogenicity of *H. influenzae*.
- Discuss laboratory diagnosis of infections caused by *Haemophilus influenzae*.
- Discus *Haemophilus influenzae* biogroup aegyptius.
- Describe morphology and culture characteristics of *Haemophilus ducreyi*.
- Discuss *Haemophilus ducreyi* or Chancroid or Soft sore and its lab diagnosis.

INTRODUCTION

The genus *Haemophilus* comprises a group of small, nonmotile, nonsporing, non-acid-fast, gram-negative coccobacilli or rods, often markedly pleomorphic and sometimes filamentous that are parasitic on human beings or animals. They are characterized by their requirement of one or both of two accessory growth factors (X and V) present in blood. The name of the genus comes from the requirement by these organisms for accessory growth factors found in blood, that is, *haemo* (Greek for blood) and *philos* (Greek for loving).

In **1883** Robert Koch first documented observation of hemophili and is known as *H. aegyptius*. *Pfeiffer* (1892) noted the constant presence of large numbers of small bacilli in the sputum of patients from the influenza. pandemic of 1889-92 and proposed this as the causative agent of human influenza. This came to be known as the **'Influenza bacillus' (Pfeiffer's bacillus)**, later renamed ***Haemophilus influenzae***. Smith, Andrewes and Laidlaw (1933) confirmed that the true etiological agent was a virus.

SPECIES

Haemophilus influenzae is the species most commonly associated with disease such as meningitis, epiglottitis, pneumonia and septic arthritis, and localized disease of the respiratory tract including bronchitis and otitis media.

Haemophilus ducreyi is well-recognized as the etiologic agent of the sexually transmitted disease soft chancre, or chancroid.

Haemophilus aphrophilus is an uncommon but important cause of endocarditis. The other members of the genus are commonly isolated in clinical specimens but are rarely pathogenic, being responsible primarily for **opportunistic infections.** (Table 45.1) *H. influenzae* is the first free living organism whose complete genome has been sequenced.

HAEMOPHILUS INFLUENZAE

Morphology

H. *influenzae* is a small, Gram-negative rods or coccobacilli (0.3-0.5×1-2 μm size), exhibiting considerable pleomorphism (Fig. 45.1). In sputum, it usually occurs as clusters of coccobacillary forms while in CSF from meningitis cases, usually appears coccobacilli but bacillary and filamentous forms predominate. Cells from young cultures (18-24 hours) are usually coccobacillary, while older cultures are distinctly pleomorphic. Virulent strains are often capsulated, which can be demonstrated by negative staining or quellung reaction. The bacilli are relatively difficult to stain. Staining for 5-15 minutes with Loeffler's methylene blue or dilute carbol fuchsin gives good results. It is nonmotile, nonsporing and non-acid-fast.

Cultural Characteristics

H.influenzae is an aerobe and facultative anaerobe. Growth is enhanced by a moist atmosphere supplemented with 5-10% CO_2. The optimum temperature is 37°C. *H. influenzae* requires two accessory growth factors: X factor (haemin, hematin) and /or Vfactor (nicotinamide -adenine dinucleotide [NAD]). The accessory growth factors present in blood are essential for growth.

X factor: X factor (haemin, hematin) is a heat-stable protoporphyrin IX, haemin or some other iron-containing

Table 45.1: Growth characteristics of Haemophilus species				
Species	*Growth requirements*			*Hemolysis on horse blood agar*
	X	*V*	*CO_2*	
H. influenzae	+	+	–	–
H. aegyptius	+	+	–	–
H. ducreyi	+	–	Variable	Variable
H. parainfluenzae	–	+	–	–
H. haemolyticus	+	+	–	–
H. parahaemolyticus	–	+	–	–
H. aphrophilus	+	–	–	–
H. paraphrophilus	–	+	–	–

porphyrin. It is required for the synthesis of the iron-containing respiratory enzymes cytochrome c, cytochrome oxidase, peroxidase and catalase.The X factor is not required for anaerobic growth.

V factor: V factor is coenzyme, nicotinamide adenine dinucleotide (NAD, coenzyme 1), NAD phosphate (NADP, coenzyme II) which acts as a hydrogen acceptor in the metabolism of the cell. The V factor was so named because it was originally thought to be a bacterial vitamin. It is **heat labile** being destroyed at 120°C in a few minutes. It is present in red blood cells and in many other animal and plant cells. It is synthesized by some fungi and bacteria such as Staph. aureus. The differential requirement of X and V factors helps to differentiate various species of Hemophillus (Table 45.1). Species of Hemophillus with the prefix para only require V factor for their growthsuch as *H. parainfluenzae, H. paarhemolyticus* and *H. parphrophilus*)

Growth is scanty on **blood agar** because only X factor is available in this medium and V factor is present inside the RBC. **Chocolate agar (heated blood agar** at 75-1000C) is superior to plain blood agar for the growth of H. injluenzae, because V factor is released from within the erythrocytes into the medium and inactivates serum NADase. Clear transparent media may be prepared by boiling and filtering a mixture of blood and nutrient broth **(Levinthal's medium)** or by adding a peptic digest of blood to nutrient agar **(Fildes agar).**

Colonies: Blood agar with Staph. aureus streak and chocolate agar are routinely used for the identification of *H. influenzae* The colonies are small, translucent and non haemolytic on blood agar. Capsulated strains produce larger, distinctive iridescent colonies. Fildes agar is best for primary isolation of *H. influenzae* and gives a copious growth. Capsulated strains produce transluscent colonies with a distinctive iridescence on Levinthal's agar.

Satellitism: The V factor is intracellular and is present inside the RBC. It is synthesized by some fungi and bacteria such as *Staph. aureus* in excess of their requirements and released into the surrounding medium. When *Staph. aureus* is streaked across a plate of blood agar on which a specimen containing *H. influenzae* has been inoculated, the colonies of *H. influenzae* will be large and well developed alongside the streak of staphylococcus, and smaller farther away after overnight incubation. This phenomenon that helps in the recognition of Hemophillus species that require V factor is called **satellitism**. It demonstrates the dependence of *H. influenzae* on the V factor, which is available in high conoentrations near the staphylococcal growth and only in smaller quantities away from it (Fig 45.1).

This is a routine test in clinical bacteriology for the identification of *H. influenzae* It is, however, not very specific as it will also be positive with other V factors requiring hemophilus species as well as with occasional strains of ohter bacterial species of unrelated genera such as some srains of streptococci, neisseriae and diphtheroids.

Biochemical Reactions

H. influenzae ferments glucose and galactose but does not ferment sucrose, lactose and mannitol. Catalhase and oxidase reactions are positive. It reduces nitrates to nitrites. *H. influenzae* can be divided into eight biotypes on the basis of indole production, urease activity and ornithine decarboxylase reactions (Table 45.2). Biotypes I-III are the most common, and Biotype I is most frequently responsible for meningitis.

Resistance

H. influenzae is a delicate organism. It is readily killed by moist heat (at 55°C in 30 minutes), refrigeration (4°C), drying and disinfectants. It also dies in I-48 hours in dried secretions and airborne droplet-nuclei. In culture, the cells die within two or three days due to autolysis.

Antigenic Structure

There are three major surface antigens:

1. Capsular polysaccharide,
2. Outer membrane proteins (OMP)
3. Lipooligosaccharide (LOS).

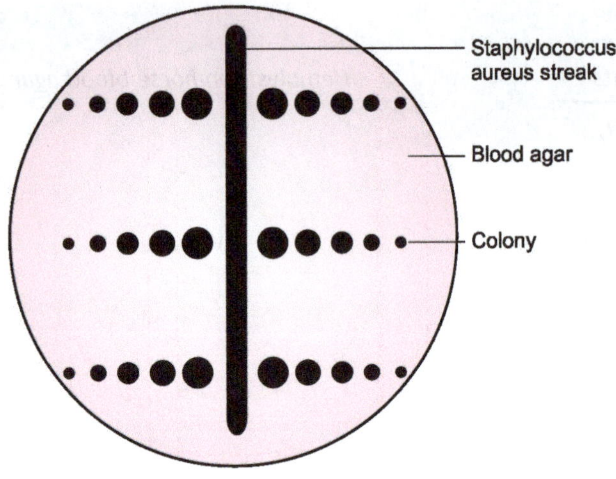

Fig. 45.1: Satellitism. *H. influenzae* colonies are large near growth of *staphylococcus*, and smaller away from it

Table 45.2: Characters of six biotypes of Haemophilus influenzae

	Biotype							
Character	*I*	*II*	*III*	*IV*	*V*	*VI*	*VII*	*VIII*
Indole production	+	+	–	–	+	–	+	–
Urease activity	+	+	+	+	–	–	–	–
Ornithine decarboxylase	+	–	–	+	+	+	–	–

1. Capsular Antigens

The major antigenic determinant of capsulated strains is the capsular polysaccharide based on which *H. influenzae* strains have been classified by Pittman into six capsular types - types a to f. Typing was originally done by agglutination but other methods such as Quellung reaction (swelling of the capsule) with type-specific antisera, precipitation, coagglutination, counterimmunoelectrophoresis (CIE) and enzyme immunosorbent assay (ELISA) may also be used. Diagnostic kits for the identification of *H. influenzae* type b (Hib) are commercially available.

The type b capsular polysaccharide has a unique chemical structure, containing the pentose sugars ribose and ribitol instead of the hexoses and hexosamines as found in the other five serotypes. The casular polyribosylribitol phosphate (PRP) antigen of Hib induces IgG, IgM and IgA antibodies which are bactericidal, opsonic and protective. Antibodies to this polyribosylribitol phosphate (PRP) capsular antigen plays a key role in protection from *H. influenzae* type b (Hib) infection. Hib PRP is therefore employed for immunization. *H. influenzae* strains lacking a capsule can not be typed and are called 'nontypable strains'.

2. Somatic Antigen

The cell envelope of *H. inflenzae* consists of an outer and inner membrane containing protein and lipooligosaccharide (LOS) antigens. Outer membrane protein (OMP) antigens of Hib have been classified into at least 13 subtypes. OMP and LOS subtyping may be of epidemiological value.

Variation

Colonies of *H. influenzae* show a smooth to rough (S-R) variation associated with loss of capsule and virulence. As in case of *Streptococcus pneumoniae,* genetic transformation has been demonstrated in *H. influenzae* also. The characters transformed are the capsular antigens and antibiotic resistance.

Virulence Factors

The virulence of *H. influenzae* is determined by the following factors:

1. Capsular Polysaccharide

This confers on the bacterium ability to resist phagocytosis. Loss of capsule is associated with the loss of virulence.

2. Adherence

Most noncapsulate strains are adherent to human epithelial cells and may explain the tendency of these strains to cause more localized infections. Most serotype b strains are not adherent and explain the tendency for type b strains to cause systemic infections.

3. Outer Membrane Proteins

They also contribute in adhesion and invasion of host tissues.

4. IgAl Protease

Production of IgA1 protease allows pathogens to inactivate the primary antibody found on mucosal surfaces and thereby eliminate protection of the host by the antibody.

Pathogenesis

H. influenzae is exclusively a human parasite, which resides principally in the upper respiratory tract. Noncapsulate organisms are present in the nasopharynx or throat of 25 to 80 percent of healthy people. Capsulate strains (about half of which are capsular type b) are present in 5 to 10 percent. It is not naturally pathogenic for animals but intraperitoneal inoculation of large doses is fatal in mice, guinea pigs and rabbits.

H. influenzae causes invasive and **non-invasive infections.** The former are caused by capsulate strains, type b accounting for the most cases, and the latter by noncapsulate strains which are quite distinct in their epidemiological profiles.

A. Invasive Infections

1. Meningitis

The most serious of the diseases produced by *H. influenzae* is an acute bacterial meningitis. The bacilli reach the meninges from the nasopharynx, apparently through the blood stream. The disease is more common in children between two 'months and three years of age. Case fatality rates is up to 90 percent in the untreated . Majority of the cases are due to type b strains.

2. Epiglottitis

This is an acute inflammation of the epiglottis, with obstructive laryngitis, seen in children above two years. This condition is always associated with bacteremia and blood cultures are usually positive in more than 97 percent of cases.

3. Pneumonia

Haemophilus pneumonia typically occurs in infants and is accompanied by empyema and sometimes meningitis as well. The picture is of lobar pneumonia in older children and adults. While these are primary infections due to capsulated strains, bronchopneumonia may occur as a secondary infection with the noncapsulated strains.

4. Suppurative Lesions

Suppurative lesions such as arthritis, endocarditis and pericarditis may result from hematogenous dissemination.

5. Cellulitis

Cellulitis particularly in the buccal and periorbital areas is seen in young children.

B. Non-invasive Disease

The commonest are:

1. Acute Sinusitis and Otitis Media

Acute sinusitis and otitis media are usually initiated by viral infections.

2. Acute Exacerbations of Chronic Obstructive Airway Disease

Acute exacerbations of chronic obstructive airway disease are similarly initiated by acute viral infections. *H. influenzae* is an important pathogen associated with pneumococci in the acute exacerbations of chronic bronchitis and bronchiectasis.

C. Conjuctivitis

Epidemiology

There is considerable similarity between the epidemiology of *H. influenzae* and pneumococci. Both are indigenous to human beings primarily parasitic in the upper respiratory tract. Infection is transmitted by the respiratory route. Carriage in the upper respiratory tract is common particularly in young children but such strains are usually noncapsulated and not responsible for acute invasive infection. The frequency of invasive infections is inversely related to age; only a small percentage occurs in adults and older children. Infections in the first 2 months of life are rare, probably because of transplacental tranfer of maternal antibody. *H. influenzae* diseases are worldwide in their occurrence and for the most part endemic in nature.

Laboratory Diagnosis

1. Specimens

The following specimens may be collected depending upon the type of lesion:

i. Blood culture
ii. Cerebrospinal fluid,
iii. Throat swabs,
iv. Sputum,
v. Pus,
vi. Aspirates from joints, middle ears or sinuses, etc.

2. Collection and Transport

For optimal yield, specimens should be transported to the laboratory and seeded on to appropriate culture media without delay. As hemophili are poorly viable in clinical specimens particularly at 4°C, therefore, the specimens should never be refrigerated.

3. Direct Microscopy

i. Gram-Stained Smear

Gram-stained smear of clinical material showing poorly stained gram-negative coccobacilli and occasionally slender filamentous forms should arouse the suspicion of *H. influenzae* infection and provides a rapid, presumptive identification. Hemophili tend to stain poorly and dilute carbol fuchsin is a better counterstain than neutral red or safranin.

ii. Immunofluorescence and Quellung Reaction

Immunofluorescence and Quellung reaction can be employed for direct demonstration of *H. influenzae* after mixing with specific rabbit antiserum type b.

iii. Antigen Detection

Type b capsular antigen can be detected in patient serum, urine, CSF or pus by several methods. These include latex agglutination, coagglutination, countercurrent immunoelectrophoresis, RIA and ELISA.

a. **Latex agglutination:** Demonstration of agglutination of latex particles (LA) coated with rabbit antibody to type b antigen.
b. **Coagglutination (COA):** *S. aureus is* coated with antibody to type b antigen and mixed with specimen. Agglutination occurs if positive.
c. **Counterimmunoelectrophoresis (CIE):** In this test specific antiserum is put in one well of agarose gel while specimen is put in other well. Current is passed and it will give precipitation line between two wells in positive test.

4. Culture

i. CSF Culture

CSF should be plated promptly on blood agar or chocolate agar. A strain of *S. aureus* should be streaked across the blood agar plate on which the specimen has already been inoculated. Plate is then incubated at 37°C with 5-10 percent CO_2 and high humidity overnight. It may also be cultured on Levinthal and Fildes blood-digest agar. The growth is identified by colony morphology, Gram staining, satellite phenomenon, biochemical reactions and serotyping.

ii. Blood Culture

Blood cultures are often positive in cases of epiglottitis and pneumonia.

iii. Sputum Culture

Sputum should be homogenized by treatment with pancreatin or by haking with sterile water and glass beads for 15-30 minutes. The rate of isolation is increased by culturing several samples of sputum from the patient.

5. Identification

H. influenzae colonies have a characteristic seminal odor. Confirmation of the identity depends on demonstrating a requirement for one or both of the growth factors, X and V *H. influenzae* requires both.

6. Molecular Techniques

Polymerase chain reaction (PCR) is used to identify *Haemophilus* species in clinical specimens and as confirmatory tests on isolates.

7. Antibiotic Sensitivity Tests

Accurate determination of the antibiotic susceptibility of *H. influenzae* requires careful standardization of the methodology because of the fastidious nature of the organism.

Treatment

H. influenzae is susceptible to sulphonamides, trimethoprim, ampicillin, chloramphenicol, tetracycline, coamoxiclav, ciprofloxacin, cefuroxime, cefotaxime and ceftazidime. Ceftriaxone and related cephalosporins such as cefotaxime are the antibiotics of first choice for the treatment of meningitis. Cefotaxime, cefuroxime and ceftazidime are highly effective for the treatment of *H. influenzae* infection Ampicillin resistance is due to the production of β-lactamase. Amoxycillin-clavulanate or clarithromycin is more effective.

Prophylaxis

Active Immunization

Active immunization against Hib is effective in preventing invasive disease, and also reduces respiratory carnage.

i. A Purified Type b Capsular Polysaccharide Vaccine

It is used in children of 18-24 months. Vaccine is administered in two doses at an interval of two months.

ii. Conjugate Vaccines

Hib PRP vaccine in which the polysaccharide is covalently coupled to various proteins (e.g. tetanus toxoid, *Neisseria meningitidis* outer membrane protein, and diphtheria toxoid) produce a lasting anamnestic response. Such Hib PRP are available for use in young children. The widespread use of conjugate vaccine may soon render chemoprophylaxis unnecessary.

iii. Household Contacts of Patients

Rifampicin given for four days prevents secondary infection in contacts and also eradicates carrier state.

HAEMOPHILI OTHER THAN H. INFLUENZAE

Haemophilus Influenzae Biogroup Aegyptius

(Koch-Weeks Bacillus, Formerly H. Aegyptius)

Even before Pfeiffer described the 'influenza bacillus' , Koch (1883) had observed, a small bacillus in conjunctivitis cases in Egypt. It was first cultivated by Weeks (1887) in New York. This organism, formerly known as the **Koch-Weeks bacillus** and ***H. aegyptius***, is now thought to be a subgroup of *H. influenzae*.

It is indistinguishable from *H. influenzae* biotype III in routine tests, but can be identified by a PCR method.It is worldwide in distribution. It is especially common in the tropics and subtropics and may occur in epidemic forms.

Diseases

It causes **purulent conjunctivitis** and ***Brazilian purpuric feve*** **(BPF).**

1. **Pink eye:** It is acute, contagious, purulent conjunctivitis commonly referred to as pink eye. It is worldwide in distribution It responds to local sulphonamides or gentamicin.
2. Brazilian purpuric fever (BPF): Brazilian purpuric fever (BPF), a clinical syndrome which is characterized by conjunctivitis, high fever, vomiting, petechiae, purpura, septicemia, and shock. It occurs in infants and children with high fatality.Mortality rate of BPF may reach as high as 70 percent. First recognized in Brazil in 1984, BPF is now endemic in South America.

Treatment

Ampicillin in combination with chloramphenicol has been successful when treatment has been started sufficiently early.

Haemophilus ducreyi

Ducreyi. (1890) demonstrated this bacillus in chancroid lesions and by inoculation into the skin on the forearm, was able to transmit the lesion through several generations. Bezancon et al. (1900) succeeded in isolating the organisms on 30 percent rabbit blood agar.

Morphology

H. ducreyi is a Gram negative short, ovoid bacillus (1-1.5 um × 0.6 µm) with a tendency to occur in end to end pairs or short chains and frequently shows bipolar staining. The bacilli may be arranged in small groups or whorls or in parallel chains giving a 'school-of fish; or **'rail road track' appearance**. The organims are pleomorphic and may occur both extracellularly and intracellularly.

Cultural Characteristics

Primary isolation is difficult. It requires X factor but not V factor for its growth. It can be grown on **rabbit-blood agar, fresh clotted rabbit blood** or **chocolate agar enriched with 1 percent Iso Vitalex, and** containing vancomycin as a selective agent. It requires 10 percent CO_2 and high humidity for primary isolation. Cultures should be incubated for up to 5 days at 35-37^0 C in a humid atmosphere with addtional CO_2. It may also be grown on chorioallantoic membrane of the chick embryo.

The colonies of *H. ducreyi* are small, pin-point to 0.5 mm in diameter, nonmucoid, grey, yellow or tan, translucent or semiopaque after 24 hours incubation. After 48-72 hours, the colonies are 1-2 mm in diameter and semiopaque.

Identification

The species is antigenically homogeneous and cultures may be identified by agglutination with the antiserum. Intradermal inoculation of the culture into rabbits produces a local ulcerative lesion.

Biochemical Reactions

H. ducreyi is biochemically inert with the exception of positive nitrate reduction test.

Pathogenicity

H ducreyi is the etiologic agent of a highly communicable sexually transmitted disease (STD) **chancroid or soft sore** characterized by tender nonindurated irregular ulcers on the genitalia. Genital lesions caused by *H. dureyi* are also known as **soft chancres or soft sores** because they are characterized by nonindurated irregular ulcers whereas primary lesion of syphilis **(chancre)** is sharply demarcated and indurated. The lesions are generally on the penis in male and they may be present on the labia and within the vagina in females.The infection remains localized, spreading only to the inguinal lymph nodes which are enlaged and painful.

There is no immunity following infection but a hypersensitivity develops, which can be demonstrated by intradermal inoculation of killed bacilli.

Laboratory Diagnosis

1. Specimens

Take specimens from the base of an ulcer or aspirate material from bubo.

2. Direct Microscopy

Typical small Gram-negative bacilli can be seen in material from the ulcers or in pus from lymph node aspirates.

3. Culture

Inoculate heated blood agar with I percent Iso Vitalex. Vancomycin 3 mg/L may be added to make the medium selective. Cultures should be incubated for up to 5 days at 35-37°C in a humid atmosphere with additional CO_2 and look for characteristic colonies.

4. Agglutination

H. ducreyi is antigenically homogeneous and cultures are identified by agglutination with the antiserum.

Treatment

Chancroid may be treated with azithromycin. Erythromycin, ciprofloxacin or ceftriaxone. Resistance to sulphonamides, trimethoprim and tetracyclines has reduced the usefulness of these agents. Strains with intermediate resistance to ciprofloxacin or erythromycin have been reported.

Haemophilus parainfluenzae

It can be distinguished from *H. influenzae* by its non requirement of X factor, ability to ferment sucrose but not D-xylose, and low pathogenicity.

H. parainfluenzae is normally present as a commensal in the mouth and throat. Clinical infection is the result of local or bloodstream invasion from these sites, usually after **dental disease, dental procedures or other oral trauma.**

It may occasionally cause conjunctivitis, acute pharyngitis, infective endocarditis, uretritis and bronchopulmonary infections in patients with cystic fibrosis.

Haemophilus Aphrophilus

It requires the X factor but not the V factor. Its name refers to its high CO_2 requirement for optimal growth. Its distinguishing characters are its production of gas in the fermentation of glucose and its dependence on the addition of 10 percent CO_2 to air for growth on heated blood agar.

It occurs as a commensal in the mouth and dental plaque.

It may cause bacterial endocarditis, brain abscess, pneumonia, sinusitis and abscesses elsewhere.

Haemophilus Paraphrophilus

H. paraphrophilus closely resembles *H. aphrophlus* in its production of gas from glucose, requirement for extra CO_2 in occurrence as commensal and pathogenicity. But it differs from *H. aphrophilus* in requiring V factor and not X factor for primary isolation.

Haemophilus haemolyticus

This haemophilus has also been regarded as a variety of *H. influenzae*, from which it differs in forming a zone of β-haemolysis around its colonies on blood agar. The hemolysis is strongest on sheep or ox blood, but is also seen on horse or human blood. Colonies on blood agar may be mistaken for those of hemolytic streptococci. It requires both X and V factors.

It is found as a commensal in the throat, but not mouth, and appears to be nonpathogenic.

Strains that do not require the X factor have been designated *H. parahaemolyticus.*

Hacek Group Bacteria

HACEK is an acronym consisting of the first initial of each genus represented in the group:

1. Haemophilus species (parainfluenzae, aphrophilus, pamphrophilus)
2. Actinobacillus actinomycetemcomitans
3. Cardiobacterium hominis
4. Eikenella corrodens
5. Kingella kingae

The acronym HACEK refers to a group of fastidious slow growing bacteria, normally resident in the mouth, which can sometimes cause severe infections, particularly **endocarditis.** Members of the HACEK group include both fermentative and nonfermentative, gram-negative bacilli. All of the members can be normal flora of the oral cavity, allowing for their introduction in the bloodstream and resultant infections. All members are opportunists and generally require a compromised host. Risk factors for infective (bacterial) endocarditis include tooth extraction, history of endocarditis, gingival surgery, heart valve surgery, and mitral valve prolapse.

Members of this group of gram-negative bacilli have in common the need for an environment with increased CO_2 (capnophilic). Their predilection for attachment to heart valves, usually damaged or prosthetic, makes many of them an important cause of endocarditis. Additional organisms that make up the majority of cases of endocarditis are the viridans group of streptococci (most common after 1 year of age), S. *aureus,* S. *pneumoniae,* the coagulase-negative staphylococci, the so-called "nutritionally variant streptococci" *(Abiotrophia* spp.), and enterococci.

Blood cultures from HACEK patients take 7 to 30 days to become positive. Antibiotic sensitivity tests are essential for effective therapy as drug resistance is very common.

KNOW MORE

Invasive Infections

In the invasive group, the bacillus acts as a primary pathogen, causing acute invasive infections. The bacilli spread through blood, being protected from phagocytes by their capsule. Haemophilus meningitis is the most common manifestation, but *H. influenzae* also causes epiglottitis, septic arthritis, osteomyelitis, pneumonia and cellulitis. In some cases the patient develops a bacteremia without a clearly defined focus of infection. Most bacteremic infections are caused by Hib. The polysaccharide capsule is the major virulence factor for Hib.

KEY POINTS

- The genus *Haemophilus* comprises a group of small, pleomorphic, gram-negative bacilli or coccobacilli. Most species require X and/or V factor for growth.
- *Hae mophilus influenzae* and *Haemophilus ducreylt* are important pahogens.
- Satellitism: This phenomenon demonstrates the dependence of *H. influenzae* on the V factor.
- *H. influenzae* subdivided serologically (types a to f), biochemically (biotypes I to VIII).
- **Diseases:** *H. influenzae* is responsible for meningitis, epiglottitis, cellulitis, arthritis, otitis, sinusitis, lower respiratory tract disease, conjunctivitis.
- **Diagnosis:** Microscopy is a sensitive test for detecting *H. influenzae* in CSF, synovial fluid, and lower respiratory specimens. Culture is performed using chocolate agar.

  Antigen detection: Type b capsular antigen can be detected in patient serum, urine, CSF or pus by several methods such as latex agglutination, coagglutination, countercurrent immunoelectrophoresis, RIA and ELISA.
- **Haemophilus influenzae biogroup aegyptius-** causes purulent conjunctivitis and Brazilian purpuric feve (BPF).
- **Haemophillus ducreyi:** It is the etiologic agent of a highly communicable sexually transmitted disease (STD *chancroid or soft sore.* characterized by tender nonindurated irregular ulcers on the genitalia.
- **Haemophilus parainfluenzae:** It may occasionally cause conjunctivitis, acute pharyngitis, infective endocarditis, uretritis and bronchopulmonary infections in patients with cystic fibrosis.

IMPORTANT QUESTIONS

1. Discuss pathogenesis and laboratory diagnosis of infections caused by *Haemophilus influenzae.*
2. Write short notes on:

   X and V factors

   Satellitism.

   Haemophilus influenzae biogroup aegyptius

   Haemophilus ducreyi or Chancroid or Soft sore.

FURTHER READING

Albritton WL. Infections due to Haemophilus species other than H. influenzae, Ann Rev Microbial 1982;36:199-216.

Daum R. et al. Epidemiology, pathogenesis, and prevention of Haemophilus influenzoe disease, J Infect Dis 1992;165 (suppl):I :206.

Jordan, JZ and Slack, MPE Haemophilus influenzae then and now, Eur J Clin Microbial 1995;14:935.

Swaminathan B, Mayer LW, Bibb WF, et al. The Brazilian Purpuric Fever Study Group 1989 Microbiology of Brazilian Purpuric fever and diagnostic tests. Journal of Clinical Microbiology 17:605-608.

Trees D, Morse S. Chancroid and Haemophilus ducreyi: An update, Clin Microbial Rev 1995;8:357-75.

CHAPTER 46

Bordetella

LEARNING OBJECTIVES

After reading and studying this chapter, you should be able to:

- Describe various factors which determine the virulence of Bordetella pertussis.
- Describe culture media for Bordetella pertussis.
- Discuss pathogenesis of pertussis.
- Discuss laboratory diagnosis of pertussis.
- Describe the following:cough plate method; vaccination against pertussis.

INTRODUCTION

The genus *Bordetella* constitutes a group of minute, gram-negative, non-acidfast, non-sporing, coccobacilli, often described as **parvobacteria.** Bordetellae used to be classified in the genus *Haemophilus.* However, growth is not dependent on either of the nutritional factors X and V, and *B. parapertussis* and *B. bronchiseptica* do not require blood for their growth. The three species resemble each other in being small Gram-negative bacilli, in causing infection of the respiratory tract, and in sharing some surface antigens.

SPECIES

The genus contains four species.

1. ***Bordetella pertussis*** (pertussis, meaning intense cough)., a highly communicable and important pathogen of humans, causing whooping cough (pertussis). It was first observed by Bordet and Gengou (1900).
2. ***Bordetella para pertussis* can cause a similar disease and was isolated from mild cases of whooping cough** in United States **by Bradford and Slavin (1937).**
3. ***Bord. Bronchiseptica:*** Originally isolated from dogs with bronchopneumonia (1911) may occasionally infect human beings, producing a condition resembling pertussis.
4. ***Bard. avium*** causes respiratory disease in turkeys was first isolated by Flion and coworkers (1967). It causes rhinotracheitis in them.

BORDETELLA PERTUSSIS (BORDET-GENGOU BACILLUS; FORMERLY *HAEMOPHILUS PERTUSSIS)*

Morphology

The bacteria are small, gram-negative coccobacilli with slight pleomorphism measuring 0.2-0.3 μm by 0.5 to 1.0 μm. They appear singly, in pairs, and in small clusters. On primary isolation, cells are uniform in size, but in subcultures they become quite pleomorphic and filamentous, and thick bacillary forms are common. It is nonmotile and nonsporing. **Bipolar metachromatic staining** may be demonstrated with toluidine blue. **Capsules** may be demonstrable in young, freshly isolated cultures only by special stains. In culture films, the bacilli tend to be arranged in loose clumps, with clear spaces in between giving a **'thumb print' appearance.** Freshly isolated strains of *Bord. pertussis* have fimbriae.

Cultural Characteristics

It is an obligate aerobe. The optimum temperature for growth is 35-36°C. It does not require X and V factors for its growth. Complex media are necessary for primary isolation. The medium in common use is the **Bordet-Gengou medium (potato-blood-glycerol agar).** Primary isolation of *B. pertussis* requires the addition of charcoal, ion exchange resins, or 15-20 percent blood to neutralize growth-inhibiting effects. **Potatoes** impart a high starch content to the medium that neutralizes toxic materials. **Glycerol** acts as a stabilizing agent. **Charcoal blood agar** is a useful medium.

The plates are incubated at 35-36°C in a moist environment (*e.g.*, a sealed plastic bag). After incubation for 48-72 hours, colonies on Bordet-Gengou medium are small, dome shaped, smooth, opaque, viscid, greyish white, refractile and glistening, resembling **'bisected pearls' or 'mercury drops'**. Colonies are surrounded by a hazy zone of hemolysis. Confluent growth presents an **'aluminium paint'** appearance. Subcultures of *B. pertussis* may be obtained on less exacting media, e.g. **nutrient agar** to which charcoal or starch has been added.

Biochemical Reactions

It is biochemically inactive. It does not ferment sugars, form indole, reduce nitrates, utilize citrate or split urea (Table 46.1). It produces oxidase and usually catalase also.

Resistance

B. pertussis is a delicate organism. It can be killed by heating at 55°C for 30 minutes, drying and disinfectants. Outside the body it can survive for five days on glass, three days on cloth and a few hours on paper. The organism is usually sensitive to ampicillin and erythromycin and these drugs have a reasonable therapeutic record.

Antigenic Constituents and Virulence Factors

B. pertussis produces a number of factors that are involved in the pathogenesis of disease.

A. Adhesins

1. Agglutinogens

Bordetellae possess genus specific and species specific surface 14 agglutinogens associated with the capsular K antigens or fimbriae. Factors 1 to 6 are found only in strains of *Bord. pertussis,* all of which carry Factor I and one or more of the other factors. Factor 7 is common to all three mammalian species of bordetellae. Factor 12 is specific' for *Bard. bronchiseptica* and Factor 14 for *Bord. parapertussis.* Agglutinogens promote virulence by helping bacteria to attach to respiratory epithelial cells. They are useful in serotyping strains and in epidemiological studies. Bordetellae are classified into various types based on the agglutinogens they carry.

2. Pertussis toxin (PT)

PT, also known as lymphocytosis-promoting factor, pertussigen, histaimine-sensitizing factor and islet-activating factor, has a wide spectrum of biologic activity. Pertussis toxin promotes lymphocytosis, sensitization to histamine, and enhanced insulin secretion. PT is expressed on the surface of the bacillus and secreted into the surrounding medium.

PT has a molecular weight of 117,000 and is a classic A-B toxin(dissociated into A and B subunits) made up of 6 polypeptide chains consisting of a A subunit, **toxic subunit (SI)** and five **binding subunits** (S2 to S5; two S4 subunits are present in each toxin molecule). B unit consists of the remaining 5 polypeptide chains binds the toxin to the target cells and helps A unit to cross the membrane. It can be toxoided. Pertussis toxoid is the major component of acellular pertussis vaccines. It is apparently responsible for many of the clinical signs and symptoms of pertussis, as well as the relative and absolute lymphocytosis observed during the clinical illness.

Antibody against PT is also protective in animal models of pertussis. It can be toxoided. PT toxoid is the major component of acellular pertussis vaccines. Anti-

Table 46.1: Differentiatial characterstics of Bordetella species

Characterstics	*Bord. pertussis*	*Bord. parapertussis*	*Bord. bronchiseptica*	*Bord. avium*
1. Motility	–	–	+	+
2. Growth on:				
Nutrient agar	–	+	+	+
Growth on Bordet-Gengou medium (days)	3-6	1-2	1	1
Mac Conkey agar	–	+	+	+
3. Urease	–	–	+	+
4. Citrate utilization	–	V	+	+
5. Nitrate reduction	–	–	+	–
6. Toxins				
Pertussis toxin	+	–	–	–
Adenylate cyclase toxin	+	+	+	–
Heat labile toxin	+	+	+	+
Tracheal cytotoxin	+	+	+	+
Lipopolysaccharide	+	+	+	+
7. Agglutinogens	1-7,13	7-10,14	7-13	Not known

body to PT can protect mice against intranasal, intraperitoneal or intracerebral challenge. The filamentous hemagglutinin and pertussis toxin are secreted proteins and are found outside the *B. pertussis* cells.

3. Filamentous Hemagglutinin (FHA)

The filamentous hemagglutinin and pertussis toxin are secreted proteins and are found outside of the *B. pertussis* cells. It is a protein that acquired its name through its ability to agglutinate erythrocytes. It mediates adhesion to ciliated epithelial cells. FHA is used in acellular pertussis vaccines along with PT toxoid. Antibodies against filamentous hemagglutinin are protective.

FHA and PT hemagglutinins also promote secondary infection by coating other bacteria such as *Haemophilus influenzae* and or *Streptococcus pneumoniae* and assisting their binding to respiratory epithelium besides facilitating adhesion of *Bord. pertussis* to respiratory epithelium,. This potential "piracy of adhesins" by other organisms may contribute to secondary bacterial invasion in pertussis .

4. Adenylate Cyclase (AC)

All mammalian bordetellae but not *Bord. avium* produce adenylate cyclase . At least two types of AC are known, only one of which has the ability to enter target cells and act as a toxin. This is known as **AC toxin (ACT).** It has the ability to enter target cells (leukocytes) and act as a toxin. It can be activated by eukaryotic calcium-dependent regulatory protein, calmodulin. Inside the cell, after activation by calmodulin, this enzyme synthesizes cAMP(as pertussis toxin does). which is responsible for the biological effects such as interfering with leukocyte functions (inhibition of phagocytosis and chemotaxis).

5. Heat Labile Toxin (HLT) or Dermonecrotic Toxin

The heat-labile toxin (HLT) produced by all species of *Bordetella* appears to be a cytoplasmic protein it is a heat-labile toxin it is dermonecrotic and at high doses, this toxin causes fatal reactions in mice. Role in disease is unknown.

6. Tracheal Cytotoxin (TCT)

Tracheal cytotoxin is a low-molecular-weight cell wall peptidoglycan monomer that has a specific affinity for ciliated epithelial cells. It induces ciliary damage in hamster tracheal ring cultures and inhibition of DNA synthesis in the ciliated respiratory epithelial cells, resulting in accumulation in the lungs of mucus, bacteria and inflammatory debris leading to severe cough. The disruption of ciliary function may also contribute to the secondary bacterial infections. The toxin also stimulates the release of the cytokine interleukin-l, which leads to fever.

7. Lipopolysaccharide (Heat-Stable Toxin)

It is present in all bordetellae and exhibits features of gram-negative bacterial endotoxins. Their role in the disease process is unknown.

Pathogenesis

Whooping cough is predominantly a pediatric disease. Whooping cough in 95 percent of cases is caused by *Bord. pertussis. Bord. parapertussis* causes about 5 percent of the cases and by *Bard.bronchiseptica* very infrequently (0.1%).

Stages of Disease

In **human beings**, after an incubation period of about 1-2 weeks, the disease takes a protracted course comprising three stages-the catarrhal, paroxysmal and convalescent -each lasting approximately two weeks.

1. Prodromal or Catarrhal Stage

The first stage, the **catarrhal stage,** resembles a common cold, with serous rhinorrhea, sneezing, malaise, anorexia, and low grade fever. Clinical diagnosis in the catarrhal stage is difficult. During this stage, large numbers of organisms are sprayed in droplets, and the patient is highly infectious but not very ill. This is unfortunate as this is the stage at which the disease can be arrested by antibiotic treatment.

2. Paroxysmal Stage

After 1 to 2 weeks, the **paroxysmal stage** begins. As the catarrhal stage advances to the paroxysmal stage, the cough increases in intensity and comes on in distinctive bouts. During the paroxysm, the patient is subjected to violent spasms of continuous coughing, followed by a long inrush of air into the almost empty lungs, with a characteristic **whoop** (hence the name). The paroxysms of coughing may be so severe that cyanosis, vomiting and convulsions follow, completely exhausting the patient.

3. Convalescent Stage

After 2 to 4 weeks, the paroxysmal stage is followed by **convalescence stage**, during which the frequency and severity of coughing gradually decrease but secondary complications can occur.

The disease usually lasts 6-8 weeks though in some it may be very protracted.

Complications

1. Subconjunctival hemorrhage, subcutaneous emphysema, inguinal hernia or rectal prolapse due to pressure effects during the violent bouts of coughing.
2. Respiratory (bronchopneumonia, lung collapse)-Respiratory complications are self limited, the atelectasis resolving spontaneously.
3. Neurological (convulsions, coma.)-neurological complications may result in permanent sequelae such as epilepsy, paralysis, retardation, blindness or deafness.

Blood Changes

Blood changes in the disease are distinctive and helpful in diagnosis. Pertussis typically causes an elevated white cell count, sometimes in excess of 50,000 cells/µl (normal range=4500-11000 white blood cells/µl during the latter part of the catarrhal or early paroxysmal phase. A marked leukocytosis occurs, with relative lymphocytosis (total leukocytic counts 20,000-30,000 per/lm with 60-80 percent lymphocytes). The erythrocyte sedimentation rate is not increased, except when secondary infection is present.

Epidemiology

Whooping cough is predominantly a pediatric disease, the incidence and mortality being highest in the first year of life. Maternal antibodies do not seem to give protection against the disease. Immunization should, therefore, be started early. The disease is commoner in the female than in the male at all ages. It is worldwide in distribution. It occurs in epidemic form periodically but the disease is never absent from any community.

The source of infection is the patient in the early stage of the disease. Infection is transmitted by droplets and fomites contaminated with oropharyngeal secretions. Whooping cough is one of the most infectious of bacterial diseases and nonimmune contacts seldom escape the disease.

The secondary attack rates are highest in close household contacts. The disease is often atypical. In adolescents and adults and may present as bronchitis. They may serve as a source of infection in infants and children. Natural infection confers protection though it may not be permanent, and second attacks have been reported.

Laboratory Diagnosis

Three methods are available:

1. Isolation of *B. pertussis* by culture from a pernasal swab;
2. Identification of the organism in a smear from a pernasal swab by immunofluorescence microscopy;
3. Serological demonstration of specific antibodies in the patient's serum.

1. Microscopy

Microscopic diagnosis depends on demonstration of the bacilli in respiratory secretions by the fluorescent antibody technique.

2. Specimen Collection and Transport

Though the disease is mainly in the lower respiratory tract, the organism can be recovered readily from the nasopharynx. 'Cough plates' and postnasal swabs are unsatisfactory because of overgrowth by commensal bacteria. The optimal diagnostic specimen is a nasopharyngeal aspirate.

i. The Cough Plate Method

Here a culture plate is held about 10-15 cm in front of the patient's mouth during about of spontaneous or induced coughing so that droplets of respiratory exudates impinge directly on the medium. This has the advantage that specimen is directly inoculated at the bedside.

ii. The Postnasal (Peroral) Swab

Secretions from the posterior pharyngeal wall are collected with a cotton swab on a bent wire passed through the mouth. Salivary contamination should be avoided. A West's postnasal swab may be conveniently employed. Cotton swabs should not be used because they contain fatty acids that are toxic to *B. pertussis* so it is preferable to use dacron or calcium alginate swabs for specimen collection.

iii. The Pernasal Swab

A sterile swab on a flexible wire is passed gently along the floor of the nose until it meets resistance. The swab, which will collect mucopus, is withdrawn and either plated immediately on charcoal blood agar, or placed in transport medium. The use of transport medium reduces the isolation rate. A single swab may yield a negative culture, but isolation rates of up to 80 percent may be achieved by taking specimens on several successive days. The pernasal swab has generally replaced the cough plates or postnasal swabs that were used in the past

3. Culture

The swab is inoculated immediately on charcoalhorse blood agar and Bordet-Gengou medium both with and without methicillin or çephalexin and incubated for at least seven days before being discarded as negative. The specimen may be transported in Regan-Lowe semi-solid medium if delay in transport is unavoidable.

Plates are incubated in high humidity at 35-36°C and colonies appear in 48-72 hours. Typical 'bisected pearl' colonies appearing after 3-5 days must be investigated further.

4. Identification

Identification is confirmed by microscopy and slide agglutination with specific antisera. lmmunofluorescence is useful in identifying the bacillus in direct smears of clinical specimens and of cultures. The differentiating features of bordetellae are listed in Table 47.1.

5. Detection of Bacterial Antigens

Bordetella antigens may be detected in serum and urine in tests with specific antiserum. Alternatively, bacteria in nasopharyngeal secretions are labelled with fluorescein-conjugated antiserum and examined by ultraviolet microscopy.

This method has the theoretical advantage, compared with culture, of detecting dead bordetellae.

6. Polymerase Chain Reaction (PCR)

Polymerase chain reaction (PCR) is used for the detection of bordetella DNA in nasopharyngeal specimens, by the use of various primers with a sensitivity of 80 percent to 100 percent.

7. Serology

Rise in antibody titer may be demonstrated in paired serum samples by ELISA, agglutination. Complement fixation, immunoblotting, indirect hemagglutination, and toxin neutralization. Demonstration of specific secretory IgA antibody in nasopharyngeal secretions by ELISA has been proposed as a diagnostic method in culture negative cases.

Treatment

Bord. pertussis is susceptible to several antibiotics (except penicillin). The drug of choice is erythromycin (or one of the newer macrolides such as clarithromycin), which may reduce the severity of the illness if given before the paroxysmal stage. Chloramphenicol and cotrimoxazole are also useful.

Treatment with pertussis immunoglobulin has been tried, but with limited success.

Prophylaxis

Preventing the spread of infection by isolation of cases is seldom practicable as infectivity is highest in the earliest stage of the disease when clinical diagnosis is not easy.

Treatment and Quarantine

Antibiotics and immunoglobulins currently available are not very effective for the treatment of patients or the protection of contacts. Control of the disease by quarantine is unrealistic.

Vaccination

Specific immunization with killed *Bord. pertussis* vaccine has been found very effective. It is of utmost importance to use a smooth **phase I strain** for vaccine production. The vaccines in general use are suspensions of whole bacterial cells, killed by heat or chemicals. Adsorption of the bacteria on to an adjuvant, such as aluminium hydroxide, enhances the immune response (particularly important with factor3) and also causes fewer adverse reactions.

In India, National policy is to immunize against diphtheria, whooping cough and tetanus (DPT) simultaneously, by administering 3 doses (each dose 0.5 ml) of DPT vaccine intramuscularly, at 1-2 months interval, starting when the infant is about 6 weeks old . A booster dose of DPT is indicated at the age of 18-24 months. *Bord. pertussis* acts as an adiuvant for the toxoids (diphtheria and tetanus toxoid) producing better antibody response.

Infants and young children should be kept away from cases. Those known to have been in contact with whooping cough may be given prophylactic antibiotic (erythromycin or ampicillin) treatment for 10 days to prevent the infecting bacteria to become established. The best protection that can be given to an infant is to administer a booster dose of DPT/DT to his siblings before he is born.

Adverse Reactions

Pertussis vaccination may induce reactions ranging from local soreness and fever to shock and neurological complications like convulsions and encephalopathy. Provocation poliomyelitis is a rare complication.

Contraindications

If severe complications such as encephalopathy, seizures. 'shock or hyperpyrexia develop following the vaccine, subsequent doses of the vaccine are contraindicated. Routine pertussis vaccination is not advisable after the age of seven years as adverse reactions are likely and the risk of severe disease is low.

Acellular Pertussis Vaccine

Acellular vaccines containing the protective components of the pertussis bacillus (PT. FHA. agglutinogens 1, 2. 3) first developed in Japan, cause far fewer reactions, particularly in older children. Both whole cell and acellular vaccines have a protetion rate of about 90 percent.

Bordetella parapertussis

This organism is readily distinguished from *B. pertussis* by its ability to grow on nutrient agar, with the production of a brown diffusible pigment after 2 days (Table 47.1). It also grows more rapidly than *B. pertussis* on charcoalblood agar and is agglutinated more strongly by parapertussis than by pertussis antiserum.

This is an infrequent cause of whooping cough (5% cases) and disease is mild. The pertussis vaccine does not protect against *Bord. parapertussis* infection. It usually causes less severe illness than *B. pertussis*, and is uncommon in most countries.

Bordetella Bronchiseptica (Bord. bronchicanis)

It differs from the other species by also being motile by peritrichate flagella and by producing an obvious alkaline reaction in the Hugh and Leifson medium that is used to differentiate oxidative from fermentative action on sugars. It is therefore placed by some taxonomists in the genus *Alcaligenes.* However, it is readily distinguished from the intestinal commensal *Alcaligenes faecalis* by its rapid hydrolysis of urea.

It can grow on nutrient agar and is antigenically related to *Bord. pertussis* and *Brucella abortus.* It occurs naturally in the respiratory tract of several species of animals. It has been found to cause a very small proportion **(0.1 percent)** of cases of whooping cough.

Bordetella avium

It was first isolated by Flion and coworkers (1967). It produces HT,TCT and does not produce ACT and PT. It causes respiratory disease in turkeys (rhinotracheitis).

KNOW MORE

Pathogenesis

Infection with *B. pertussis* and the development of whooping cough require exposure to the organism, bacterial attachment to the ciliated epithelial cells of the respiratory tract, proliferation of the bacteria, and production of localized tissue damage and systemic toxicity. The source of infection is the patient in the early stage.

KEY POINTS

- The genus *Bordetella* constitutes a group of minute, gram-negative, non-acidfast, nonsporing, coccobacilli, often described as **parvobacteria.**
- Three important species of *Bordetella* include *Bordetella pertussis, B. parapertussis* and *B. bronchiseptica.*
- *B. pertussis* are extremely small, ovoid, Gram-negative coccobacilli showing pleomorphism; nonmotile and nonsporing.
- *Bordetella pertussis* are strict aerobes and nutritionally fastidious, grow on specialized media such as Bordet-Gengou agar and charcoal agar with 10 percent blood.
- *B. pertussis* produces a number of factors that are involved in the pathogenesis of disease such as **pertussis toxin (PT).**
- **Diseases:** Pertussis and is characterized by three stages: catarrhal, paroxysmal, and convalescent stages. Children younger than 1 year at greatest risk for infection
- **Laboratory diagnosis-depends on** microscopy, culture and polymerase chain reaction (PCR). Detection of IgG or IgA can confirm the clinical diagnosis.
- **Treatment, Prevention, and Control:** Treatment with macrolide (i.e., erythromycin, azithromycin) is effective in. Erythromycin has been used for prophylaxis.

  Vaccination with whole-cell vaccines is effective but associated with side effects. Acellular vaccines are effective and associated with fewer adverse effects.
- ***Bordetella parapertussis*** is responsible for only about 5 percent of cases of whooping cough and relatively causes a mild form.
- ***Bordetella bronchiseptica:*** It is responsible for causing cases of whooping cough in very small proportion (0.1%) of.
- ***Bord. Avium:*** It causes respiratory disease in turkeys (rhinotracheitis).

IMPORTANT QUESTIONS

1. Discuss laboratory diagnosis of whooping cough
2. Write short notes on:
   Culture media for *Bordetella pertussis*
   Pathogenesis of pertussis
   Laboratory diagnosis of pertussis
   Cough plate method
   Vaccination against pertussis
   Acellular pertussis vaccine

FURTHER READING

Editorial Pertussis. Adults, infant and herds. Lancet 1992; 339:526-527.

Friedman RL. Pertussis. The disease and new diagnostic methods. Clin Microbiol Rev 1988;1:365-76.

Hodder SL, Mortimer EA. Epidemiology of pertussis and reactions to pertussis vaccine. Epidemiol Rev 1992;14:243-68.

Preston NW Pertussis (whooping-cough). The road to eradication is well sign-posted but erratically trodden. Infectious Diseases Review 2000;2:5-11.

Preston NW, Matthews RC. Acellular pertussis vaccines: progress but deja vu. Lancet 1998;351:1811-12.

CHAPTER

47 Brucella

Learning Objectives

After reading and studying this chapter, you should be able to:

- Discuss classification of Brucella.
- Describe culture characteristics and biochemical reactions of Brucella sp.
- Discuss pathogenesis of brucellosis.
- Discuss laboratory diagnosis of brucellosis.
- Describe Castaneda method of blood culture.

INTRODUCTION

The genus *Brucella* consists of very small, nonmotile, aerobic, gram-negative coccobacilli that grow poorly on ordinary media and have little or no fermentative powers. Brucellae are facultatively intracellular pathogens that are essentially pathogens of goats, sheep, cattle and pigs. Man acquires infection by direct or indirect contact with infected animals. Human infection is a **zoonosis** that is acquired from animals or animal products.

The first human cases of brucellosis were described in **1861** by JA Marston, a physician with the British army stationed at Malta. A British army doctor, David Bruce (1886) cultivated causative agent of brucellosis from spleen tissues of victims of Malta fever. This was named *Brucella melitensis* (Brucella after *Bruce, melitensis* after *Melita,* the Roman name for Malta). B Bang, a Danish veterinarian (1897), described *Br. abortus,* the cause of contagious abortion in cattle. Meyer and Shaw proposed the new genus *Brucella* in honor of Sir David Bruce to include both the agents of Malta fever and Bang's disease as B. *melitensis* and *B. abortus,* respectively. Traum (1914) isolated the third major species in the genus, *Br. suis* from pigs in the USA. Other species causing animal infections include *Br. canis,* isolated from cases of canine abortion, *Br. ovis* from abortion in sheep and *Br. neotomae* from desert wood rats. *Br. canis* may occasionally cause a mild human disease, but the other two are not pathogenic for humans.

Species

Six species are currently recognized: *B. melitensis, B. abortus, B. suis, B. ovis, B. neotomae* (isolated from desert wood rats), and *B. canis.*

BRUCELLA

Morphology

Brucellae are coccobacilli or short rods 0.5 to 0.7 μm × 0.6-1.5 μm. In gram-stained preparations of pure cultures they are arranged singly and less frequently, in pairs, short chains, or small groups. In older cultures, irregular forms appear. They are gram-negative non-acid fast, nonmotile, noncapsulated and nonsporing. Bipolar staining is usually not observed.

Cultural Characteristics

Brucellae are strict aerobes and do not grow anaerobically. Many strains of *Br. abortus* and nearly all of *B. ovis* are capnophilic, requiring 5-10 percent CO_2 for growth. The optimum temperature is 37°C (range 20-40°C) and pH 6.6 to 7.4. They may grow on simple media, though growth tends to be slow and scanty. Growth is improved by serum or blood. The media employed currently are **serum dextrose agar, serum potato infusion agar, trypticase soy agar or tryptose agar.** The addition of bacitracin, polymyxin and cycloheximide to the above media makes them selective. On solid culture media, colonies may take 2 to 3 days to develop. They are small, smooth, transparent, low convex with an entire margin. Mucoid, smooth and rough types of colonies appear, associated with changes in antigenic structure and virulence. *B. canis* and *B. ovis* characteristically produce nonsmooth colonies. In **liquid media**, growth is uniform, and a powdery or viscous deposit is formed in old cultures.

Biochemical Reactions

The metabolism is oxidative and not fennentative. They are catalase-positive, urease positive (variable in

B. melitensis) and usually oxidase-positive (except *B. neotomae, B.* ovis and some strains of *B. abortus* which are oxidase-negative).

Indole is not produced and MR and VP tests are negative. Citrate is not utilized.

Most *Brucella* strains reduce nitrates to nitrites(except *B. ovis*).

H_2S is produced from sulphur-containing amino acids by *B. neotomae, B. abortus*(many strains) and *B. suis* (some strains).

Resistance

Brucellae are destroyed by heat at 60°C in 10 minutes; therefore they are killed in milk by pasteurization. They are very sensitive to direct sunlight and moderately sensitive to acid, so that they tend to die out in sour milk and in cheese that has undergone lactic acid fermentation. The organisms can survive in soil, manure and dust for weeks or months. They remain viable for 10 days in refrigerated milk, one month in ice cream, four months in butter and for varying periods in cheese depending on its pH. They are susceptible to common disinfectants if used at appropriate concentration and temperature. They can be killed by 1.0 percent phenol in 15 minutes.

They are sensitive to many disinfectants and antibiotics, including ampicillin, co-amoxiclav, cephalosporins, aminoglycosides, tetracyclines, chloramphenicol, ciprofloxacin, sulphonamides and cotrimoxazole; relatively resistant to vancomycin, nalidixic acid and polymyxins.

Antigenic Structure

There are at least two antigenic determinants-**A (abortus)** and **M (melitensis)**-present on the lipopolysaccharide (LPS) protein complex, which constitutes the major agglutinogen of fresh or smooth isolates. They are present in different amounts in the three major species. *Br. abortus* contains about 20 times A as M , *Br. melitensis* about 20 times M as A. *Br. suis* has an intermediate antigenic pattern. Absorption of the minor antigenic component from an antiserum will leave most of the major antibody component and such absorbed A and M monospecific sera are useful for species identification by the agglutination test. The species identification of brucella strains is not, however, so straightforward and strains are often seen that behave biochemically as abortus and serologically as melitensis and vice versa. Species and biotype identification depends on a variety of other factors besides antigenic structure.

Antigenic cross reactions exist between brucellae and *V. cholerae,* with *E. coli* 0: 116; 0: 157, Salmonella serotypes group N (0:30 antigen*), Ps. maltophila, Y. enterocolitica* and *F. tularensis.*

Brucella Bacteriophage

The Tblisi (Tb) phage has been designated as the reference phage and at RTD (Routine Test Dose). It lyses *Br. Abortus* at both RTD and 10,000 TTD. *Br. suis* is lysed at 10,000 RTD, while *Br. melitensis* is not lysed at all. They have been classified into six groups based on their host specificity.

Classification of Brucellae

Brucellae may be categorized into **species** and **biovars** by the following tests (Table 47.1):

CO_2 requirement
H_2S production
Sensitivity to dyes (basic fuchsin and thionin),
Agglutination with monospecific antisera
Lysis by specific bacteriophage

Biotypes

The three major species are *Br. melitensis, Br. abortus, B. suis* infecting primarily goats or sheep, cattle and swine, respectively. Many biotypes have been recognized in these species (Table 47.1).

i. *B. melitensis*- three biotypes,
ii. *B. abortus*- 7 biotypes (1-6 and 9, biotypes 7 and 8 have been discarded as invalid).
iii. *B. suis*- 5 biotypes

Br. suis strains that produce H_2S are known as **'American' strains** and those that do not as **'Danish' strains.**

Pathogenesis

Although each species of brucella has a preferred host, all can infect a wide range of animals, including humans. All three major species of brucellae are pathogenic to human beings. *Br. melitensis* is the most pathogenic, *Br. abortus* and *Br. suis* of intermediate pathogenicity.

Mode of Infection

The modes of infection are by ingestion, contact, inhalation or accidental inoculation. Person to person spread does not ordinarily occur, but very rarely transmission has been reported through the placenta, breastfeeding and sex.

Ingestion

The most important vehicle of infection is raw milk. Milk products, meat from infected animals and raw vegetables or water supplies contaminated by the feces or urine of infected animals may also be responsible.

Contact

Infection by contact occurs when brucellae in vaginal discharges, fetuses, placenta, urine-, manure or carcasses enter through the skin, mucosa or conjunctiva. Contact infection is especially important as an occupational hazard in agricultural workers, veterinarians, butchers, animal handlers, and others in occupations that involve handling of animals or uncooked animal tissues are at higher risk for direct inoculation and is particularly common during the calving season.

Inhalation

Infection is transmitted by inhalation of dried material of animal origin such as dust from wool. Infection by

Table 47.1: Differential characteristics of the species and biotypes of brucella

species	biotypes	Lysis by phage		Growth on dye media				Agglutination by mono specific serum			Most common host
		RTD	RTD x 10^4	CO_2 requirement	H_2S production	Basic Fuchsin 1:50,000	Thionin 1:25,000	Thionin 1:50,000	A	M	R	
Br. meliten-sis	1	–	–	–	–	+	–	+	–	+		sheep, goats
	2	–	–	–	–	+	–	+	+	–	–	
	3	–	–	–	–	+	–	+	+	+	–	
Br.abortus	1	+	+	±	+	+	–	–	+	–	–	Cattle
	2	+	+	+	+	–	–	–	+	–	–	
	3	+	+	±	+	+	+	+	+	–	–	
	4	+	+	±	+	+	–	–	–	+	–	
	5	+	+	–	–	+	+	+	–	+	–	
	6	+	+	–	–	+	+	+	+	–	–	
	9	+	+	±	±	+	+	+	–	+	–	
Bt.suis	1	–	+	–	+	–	–	+	+	–	–	Pigs
	2	–	+	–	–	–	–	+	+	–	–	Pigs, here
	3	–	+	–	–	+	+	+	+	–	–	Pigs
	4	–	+	–	–	+	+	+	+	+	–	Reindeer
Br. neotoma'e		–	+	–	+	–	–	–	+	–	–	Wood rat
Br. ovis		–	–	+	–	+	+	+	–	–	+	Sheep
Br. canis		–	–	–	–	–	+	+	–	–	+	Dogs

RTD: Routine test dilution
A—abortus; M—melitensis; R—rough

inhalation is a serious risk in laboratory workers handling brucellae. Infection by accidental inoculation is not infrequent among veterinarians and laboratory workers.

Course of Disease

Brucellosis is primarily an intracellular pathogen affecting reticuloendothelial system. Brucellae have a special predilection for intracellular growth and may be demonstrated inside phagocytic cells. The brucellae spread from the initial site of infection through lymphatic channels to the local lymph glands, in the cells of which they multiply. They then spill over into the bloodstream and are disseminated throughout the body. Once liberated into the bloodstream, they may infect a variety of organs but are most often localized in the reticuloendothelial system, where they reside within phagocytic cells. Granulomas or abscesses most often develop in the bone marrow, liver, spleen, lymph nodes, or lungs. Other sites of infection may include subcutaneous tissue, testes, epididymis, ovary, gallbladder, kidneys, and brain. Meningitis and endocarditis are commonly reported complications.

The brucellae that infect humans have apparent differences in pathogenicity. *B abortus* usually causes mild disease without suppurative complications; noncaseating granulomas of the reticuloendothelial system are found. *B canis* also causes mild disease. *B suis* infection tends to be chronic with suppurative lesions; caseating granulomas may be present. *B melitensis* infection is more acute and severe.

Placentas and fetal membranes of cattle, swine, sheep, and goats contain erythritol, a growth factor for brucellae. The proliferation of organisms in pregnant animals leads to placentitis and abortion in these species. There is no erythritol in human placentas and abortion is not part of brucella infection of humans.

Types of Human Infection

The incubation period is usually about 10 to 30 days but. infection may persist for several months without causing any symptoms. Human infection may be of three types:

1. Latent or subclinical infection
2. Acute or subacute brucellosis
3. Chronic brucellosis.

1. Latent or Subclinical Infection

There is no clinical evidence of disease but is detectable only by serological tests.

2. Acute Brucellosis

Acute brucellosis is mostly due to *Br. melitensis.* It is associated with prolonged bacteremia and irregular fever. The onset is insidious, with malaise, fever, weakness, aches, and sweats. The fever usually rises in the afternoon; its fall during the night is accompanied by drenching sweat. It is also known as **undulant fever or Malta fever** because of the periodic noctural fever that may occur over weeks, months or years particularly in untreated cases.

There may be gastrointestinal and nervous symptoms. Lymph nodes enlarge and the spleen becomes palpable. Hepatitis may be accompanied by jaundice. Deep pain and disturbances of motion, particularly in vertebral bodies, suggest osteomyelitis.

These symptoms of generalized brucella infection generally subside in weeks or months, although localized lesions and symptoms may continue. In addition, localized foci of infection may occur in the liver, reticuloendothelial system, bones and joints, genitourinary tract, central nervous system, eyes, skin, lungs and heart (endocarditis).

Complications

Debilitating neuropsychiatric disorders.

Infection of a bone or joint (including the vertebral column).

Endocarditis in unusual cases.

Other viscera, such as spleen and liver and bone marrow may have evidence of infection for a significant period of time.

3. Chronic Brucellosis

Chronic brucellosis is a low grade infection with periodic exacerbations. It is usually nonbacteremic. The symptoms are generally related to a state of hypersensitivity in the patient. The common clinical manifestations being sweating, lassitude and joint pains with minimal or no pyrexia. Brucellae cannot be isolated from the patient at this stage but the agglutinin titer may be high. The illness lasts for years.

Epidemiology

Brucella organisms are distributed throughout the world. Human brucellosis is acquired from animals, directly or indirectly. The animals that commonly act as sources of human infection are goat, sheep, cattle, buffaloles, and swine. In some parts of the world, infection may also come from dogs, reindeer, caribuou, camels and yaks. Other farm animals, such as horses and poultry, may become infected with brucellae in very unusual situations but they do not constitute a large reservoir or significant source of human infections.

The common routes of infection in humans are the intestinal tract (ingestion of infected milk), mucous membranes (droplets), and skin (contact with infected tissues of animals). Cheese made from unpasteurized goats' milk is a particularly common vehicle.

Brucellae have a wide host range hut exhibit a degree of host preference in natural infections- *Br. melitensis* predominantly in goats and sheep, *Br. abortus* in cattle and *Br. suis* in swine. Brucellae exist as naturally occurring parasites in a wide variety of animal species. Infection is transmitted among animals directly or through blood-sucking arthropods, particularly ticks.

Brucellosis is a **zoonosis** of worldwide importance, particularly in developing countries. This disease is highly endemic in the mediterranean basin, the Middle East, western Asia, Africa, and Latin America. Brucellosis may also be potentially transmitted in association with biowarfare, bioterrorism, or biocriminal activities.

Almost all human infections in various parts of India are due to *B. melitensis* acquired from goats and sheep. Animal brucellosis is reported from practically every state in India. However, no stastical information is available about the extent of infection in man in various parts of the country.

Laboratory Diagnosis

The clinical manifestations of human brucellosis are variable, so clinical diagnosis is almost impossible and laboratory aid is therefore essential. Cultures must be handled under containment conditions appropriate to Class 3 pathogens. Laboratory methods inclue culture of brucellae, serology, polymerse chain reaction and hypersensitivity type skin tests.

1. Specimens

Blood culture is most impotant. Material from bone marrow or liver biopsy, is also cultured, lymph nodes, cerebrospinal fluid, urine and abscesses and on occasion, also from sputum, breast milk, vaginal discharges and seminal fluid.

In animals culture may be attempted from abortion material, placenta, milk, semen or from samples of lymphoid tissue, mammary gland, uterus or testis collected postmortem. Laboratory methods for diagnosis include:

2. Culture

i. Blood Culture

Blood culture is the most definitive method for the diagnosis of brucellosis. When brucellosis is suspected, blood culture should be attempted repeatedly, not only during the febrile phase. Because the organisms may be scanty, at least 10 ml of blood should be withdrawn on each occasion, 5 ml being added to each of two blood culture bottles containing serum dextrose (SD) broth. One of these bottles should be incubated in an atmosphere containing 10 percent carbon dioxide at 37°C. Subcultures are made on solid media (serum dextrose agar) every 3 to 5 days, beginning on the fourth day. Growth may often be delayed and blood cultures should be retained for 6 to 8 weeks before being discarded as negative. Preliminary lysis and centrifugation of the blood improves the isolation rate. Automated blood culture systems may also be used.

ii. Castaneda's Method of Blood Culture

Advantages

The **Castaneda method** of blood culture has several advantages and is recommended.

i. A two-phase Castaneda culture system, in which the broth is periodically allowed to flow over agar contained within the blood culture bottle, may be used. Here, both liquid and solid media are available in the same bottle. The blood is inoculated into the broth and the bottle incubated in the upright position. For subculture, it is sufficient, if a bottle is tilted so that the broth flows over the surface of the agar slant. It is again incubated in an upright position. Colonies appear on the slant.
ii. This method **minimizes materials and manipulation.**
iii. Reduces chances of contamination during the period of incubation and risk of infection to laboratory workers.

Blood cultures are positive only in about 30 to 50 per cent cases, even when repeated samples are tested. *B. melitensis* and *B. suis* are more frequently isolated from blood than are *B. abortus* or *B. canis*.

iii. Bone Marrow or Liver Biopsy

Isolation rates can be markedly improved if material from bone marrow or liver biopsy is also cultured. Bone marrow cultures have been positive more often than blood cultures, especially when patients have taken antibiotics. The intracellular localization of *Brucella* within reticuloendothelial cells may account for the posititve cultures from bone marrow aspirates at a time when blood cultures from the same patient are negative.

iv. Other Material

Cultures may also be obtained from lymph nodes, cerebrospinal fluid, urine and abscesses, if present and on occasion also from sputum, breast milk, vaginal discharges and seminal fluid.

Identification-depends upon biochemical tests and may be classified into different **species** and **biovars,** based on CO_2 requirements, H_2S production, sensitivity to dyes (basic fuchsin and thionin), agglutination with monospecific sera, lysis by specific phage and oxidative metabolic tests with amino acids and carbohydrates.

3. Serological Tests

In the absence of positive cultures, the diagnosis of brucellosis usually depends on serological tests, the results of which tend to vary with the stage of the infection.

Antibodies appear within 7 to 10 days of onset of the disease. IgM antibodies appear first which are rapidly followed and superseded by IgG and to a lesser extent IgA antibodies. These antibodies reach their maximum titres in the **third or fourth week** of disease and then slowly decline but they usually persist throughout the active phase of the disease and in some cases long thereafter. As the disease progresses, IgM antibodies decline, while the IgG antibodies persist or increase in titre.

In chronic infections, IgM may often be absent and only IgG can be demonstrated.

The agglutination test identifies mainly the IgM antibody, while both IgM and IgG can fix complement. However, as IgG and IgA antibodies are formed during the course of the infection some of them bind with antigen, thus preventing its agglutination by larger IgM molecule. These IgG and IgA antibodies are known as **blocking or non-agglutinating antibodies** which may prevent agglutination. It is thus evident that the agglutination test is usually positive in acute infection but may be only weakly positive or even negative in chronic cases.

Brucella antibodies can be detected by a variety of serological tests. The most useful are the standard tube agglutination test (SAT), 2-mercaptoethanol (2ME) agglutination test, complement fixation test, anti-human globulin (Coombs') test, enzymelinked immunosorbent assay (ELISA) and radioimmunoassay (RIA).

i. Standard Agglutination Test (SAT)

This is a tube agglutination test in which equal volumes of serial dilutions of the patient's serum and the standardized antigen (a killed suspension of a standard strain of Br. abortus) are mixed and incubated at 37°C for 24 hours or 50°C for 18 hours. It detects antibodies against *B. abortus*, *B. suis*, and *B. melitensis* but not *B. canis*. A titre of 160 or more is considered significant.

Prozone Phenomenon

Several sources of error have to be guarded against. Sera often contain 'blocking' or 'nonagglutinating' antibodies. A blocking factor may interfere with agglutination at low serum dilutions **(the prozone phenomenon)** although positive in higher dilutions. It is essential that several serum dilutions be tested in brucellosis.

The blocking effect may sometimes be removed by prior heating of the serum at 55°C for 30 minutes or by using 4 percent saline as the diluent for the test. The most reliable method for obviating the blocking effect and detecting the 'incomplete' antibodies is the **antiglobulin (Coombs) test.**

Cross-Reactions

Cross-reactions may be observed with antibodies directed against *Francisella tularensis Vibrio cholerae,* or *Yersinia enterocolitica or* immunization. Cholera induced agglutinins may be differentiated by the agglutinin absorption test and also as they are removed by treatment with 2-mercapto-ethanol. In order that results from, different laboratories are comparable, it is the practice to express

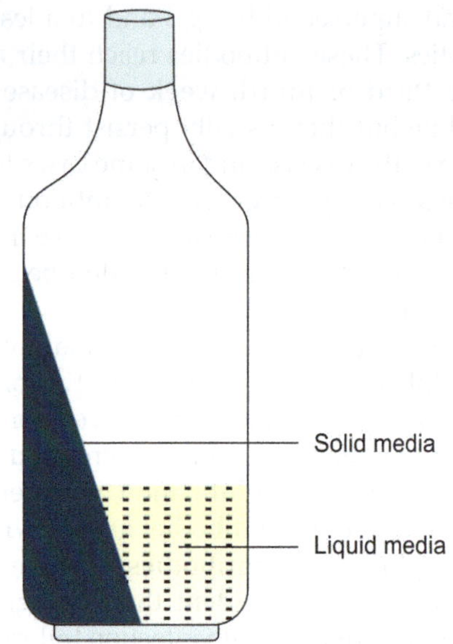

Fig. 47.1: Castaneda's method of blood culture

agglutinin titres in International Units. This is done by using a standard reference serum for comparison.

ii. 2-Mercaptoethanol (2ME) Agglutination Test

The mercaptoethanol test is carried out simultaneously and in the same manner as the standard agglutination test except that the saline diluent contains 0.05 M 2-mercaptoethanol. The agglutinating ability of IgM and sometimes IgA, is destroyed by 2-mercaptoethanol, and therefore, agglutination in this test is indicative of the continuing presence of IgG and the likelihood of persisting infection.

iii. Complement Fixation Test

The complement fixation test is more useful in chronic cases as it detects IgG antibody also. In latent or chronic infection, the complement fixation test is likely to be positive whereas in cases of past infection it is negative.

iv. Enzyme-Linkedimmunosorbentassay (ELISA) and RIA

These tests are very sensitive useful for differentiation between the acute and chronic phases of brucellosis. ELISA is sensitive, specific and can detect IgM and IgG antibody separately.

v. Rose Bengal Plate Test

This is a rapid slide agglutination test with a buffered stained antigen. It is widely used as a screening test in farm animals. It also gives good results in human brucellosis. It is not affected by prozones or immunoglobulin switching.

vi. Rapid Dipstick Assay

The rapid dipstick assay, a relatively simple screening procedure for *Brucella* specific IgM, shows promise as a field test in areas without direct access to a reference laboratory.

4. Polymerase Chain Reaction (PCR)

The PCR with primers specific for the *omp2*, *omp25* and *rrs-rrl* genes can detect *Brucella* specifically and also give an indication of species and biovar. Promising results have been obtained in clinical studies.

5. Hypersensitivity Test (Brucellin Skin Test)

Delayed hypersensitivity type skin tests with brucella antigens ('brucellins') are not useful in diagnosing acute brucellosis. This test, similar to the tuberculin test, is no longer recommended as it does not differentiate active from past or subclinical infection. They parallel the tuberculin test in indicating only prior sensitization with the antigens, and may remain positive for years. Brucellin testing may lead to a rise in titre of antibodies.Moreover, intradermal antigen preparations are not readily available and some interfere with the serological response to infection.

6. Detection of Animal Infection

The methods used for the laboratory diagnosis of human brucellosis may also be employed for the diagnosis of animal infections. In addition, brucellae may be demonstrated microscopically in pathological specimens by suitable staining or by immunofluorescence.

Several rapid methods have been employed for the detection of brucellosis in herds of cattle. These include the *rapid plate agglutination test* and the *Rose Bengal card test*. For the detection of infected animals in dairies, pooled milk samples may be tested for bacilli by culture and for antibodies by several techniques such as **milk ring test.**

i. Milk Ring Test

1. In the *milk ring test* a sample of whole milk is mixed well with a drop of the stained brucella antigen (a concentrated suspension of killed *Br. abortus* stained with hematoxylin).
2. It is incubated in a water bath at 70ºC for 40 to 50 minutes.

Result

Positive test (If antibodies are present in the milk)- the bacilli are agglutinated and rise with the cream to form a blue ring at the top, leaving the milk unstained.

Negative(If antibodies are absent)- no colored ring is formed and the milk remains uniformly blue.

ii. Whey Agglutination Test

The whey agglutination test is another useful method for detecting the antibodies in milk.

Prophylaxis

1. As the majority of human infections are acquired by consumption of contaminated milk, prevention consists of checking brucellosis in dairy animals.
2. Control of this disease in cattle has been achieved by serologic surveillance, vaccination *(B. abortus* strain 19), and elimination of reactor cattle. In many advanced countries, this is achieved by the

detection of infected animals, their elimination by slaughter and the development of certified brucella free herds.

3. Pasteurization eliminates the risk of brucellosis from the consumption of infected milk or milk products.
4. Vaccines have been developed for use in animals. The live-attenuated *B. abortus* strain S19 vaccine has been is now being replaced by the rough strain *B. abortus* RB51, which gives comparable protection but does not induce interfering antibodies and is less hazardous to man.

   The live-attenuated smooth strain *B. melitensis* Rev I is used to protect sheep and goats from *B. melitensis* infection. Vaccination of pigs is not widely practised but the attenuated *B. suis* strain 2 has been used in China.
5. Human vaccination is not recommended because effective and nonreactogenic vaccines are not currently available.

Treatment

Brucella infections respond to a combination of streptomycin or gentamicin and tetracycline or to rifampicin and doxycycline. Tetracycline alone is often adequate in mild cases. Treatment should be continued for at least 6 weeks.

Co-trimoxazole and rifampicin can be used in children. In children younger than 8 years aminoglycoside has resulted in successful treatment without the side effects of tetracyclines in this age group.

KNOW MORE

EPIDEMIOLOGY

Brucellosis may also be potentially transmitted in association with biowarfare, bioterrorism or biocriminal activities. B. *suis* was one of the agents included in the U.S. offensive biological weapons research program, prior to its termination in 1969. Brucellosis is associated with a high incidence of spontaneous abortion in pregnant women in areas of hyperendemic infection, such as Saudi Arabia.

KEY POINTS

- The genus *Brucella* consists of very small, nonmotile, aerobic, gram-negative coccobacilli. They are essentially pathogens of goats, sheep, cattle and pigs.
- **Six species** are currently recognized and three major species of the genus Brucella are *B. melitensis, B. abortus, B. suis.*
- Brucellae are strict aerobes. *Br. abortus* and nearly all of *B. ovis* are capnophilic, requiring 5 to 10 percent CO_2 for growth.
- They grow best on media employed currently are **serum dextrose agar, serum potato infusion agar, trypticase soy agar, or tryptose agar.** They are catalase and oxidase positive.
- They are catalase-positive, urease positive (variable in *B. melitensis)* and usually oxidase-positive
- *Brucella abortus* consists of seven biotypes, *Brucella melitensis* three biotypes and *Brucella suis* five biotypes.
- **Epidemiology**-Animal reservoirs are goats and sheep *(Brucella melitensis),* cattle *(Brucella abortus),* swine *(Brucella suis),* and dogs *(Brucella canis).* Individuals at greatest risk for disease are people who consume unpasteurized dairy products, people in direct contact with infected animals, and laboratory workers.
- **Disease**-Human brucellosis is primarily a zoonotic bacterial infection (acute brucellosis, chronic brucellosis and localized infection).
- **Laboratory diagnosis**-depends on the culture of brucellae and serology. **Castaneda's method of blood culture** is a useful method to culture blood. **Serology** can be used to confirm the clinical diagnosis and include standard tube agglutination tests, indirect immunofluorescent tests, ELISA, etc. PCR is available for rapid detection of *Brucella* species. **Milk ring test** is a screening test used for demonstration of antibodies in the milk of animals.

IMPORTANT QUESTIONS

1. Discuss laboratory diagnosis of brucellosis.
2. Write short notes on:

   Castaneda method of blood culture.

   Serodiagnosis of brucellosis.

   Diagnosis of brucellosis in animals.

FURTHER READING

Adams LG (Ed). Advances in Brucellosis Research,Texas A & M University Press, Austin 1990.

Corbel, MJ. Brucellosis, an overview, Emerging Infect Dis 1997;3:2.

Corbel, MJ. Vaccines against bacterial zoonoses, J Med Microbiol 1997;46:267.

Corbel MJ. Brucella. In: Balows A, Duerden B I (eds) Principles of Bacteriolagy, Virology and Immunity, 9th edn, Vol 2. Edward Arnold, london 1998;829-952.

Expert Committee on Brucellosis. 1986. Sixth Report, Technical Rep. Series No. 740. W.H.O., Geneva.

Young EJ. An overview of human brucellosis, Clin Infect Dis. 1995;21:283.

CHAPTER 48

Spirochetes

Learning Objectives

After reading and studying this chapter, you should be able to:

- Describe diseses caused by different spirochetes.
- Name various pathogenic treponemes.
- Describe morphology of *Treponema pallidum*
- Discuss pathogenesis of syphilis
- Describe diseses caused by *Treponema pallidum* their laboratory diagnosis
- Explain serological tests for syphilis.
- Discuss standard tests for syphilis (STS)
- Describe the following:fluorescent treponemal antibody-absorption (FTA-ABS) test; TPHA (or) *T. pallidum* hemagglutination test.
- Discuss BFP (Biological false positive) reactions.
- Describe the following i. Endemic syphilis (or) Bejel; ii. Yaws; iii. *Treponema pertenue*; iv. Pinta; v. *Borrelia recurrentis*; vi. *Borrelia vincentii*; vii. *Borrelia burgdorferi* or Lyme disease; viii. Leptospirosis; ix. Weil's disease

INTRODUCTION

The spirochetes (from *Speira,* meaning coil and *chaite,* meaning hair) are elongated, motile, slender, helically coiled, flexible organisms with one or more complete turns in the helix. Multiplication is by transverse fission. Many are free-living saprophytes, while some are obligate parasites. They may be aerobic, anaerobic or facultative. Reproduction is by transverse fission.

DESCRIPTION

Spirochetes are slender unicellular helical or spiral rods with a number of distinctive ultrastructural features used in the differentiation of the genera. They are gram-negative and are 0.1 to 3.0 μm wide and 5 to 250 μm in length. They are structurally more complex than other bacteria. The spirochetes have gram-negative type cell wall composed of an outer membrane, a peptidoglycan layer and a cytoplasmic membrane. They are structurally more complex than other bacteria.

They possess a varying number of fine fibrils which are attached subterminally at each pole of the cell and extend, towards the opposite pole between outer membrane and peptidoglycan layer.

They differ from other bacteria that they have flexible cell wall around which several fibrils are wound. These fibrils termed as *endoflagella.* Their most distinctive morphologic property is the presence of varying numbers of *endoflagella.* These *endoflagella* are polar flagella wound along the helical protoplasmic cylinder, and situated between the outer membrane and cell wall (Fig. 48.1). The spiral shape and serpentine motility of the spirochetes depend upon the integrity of these endoflagella. Motility is of three types: (1) flexion and extension of cells, (2) corkscrew-like rotatory movement around the long axis and (3) translatory motion, i.e. from one site to another. Some are very actively motile'while others are sluggish.

Larger spirochetes like *Borrelia* are gram-negative but other spirochetes cannot be stained by routine methods. However, the spirochetes can be seen by dark ground microscopy, silver impregnation method and immunofluorescence.

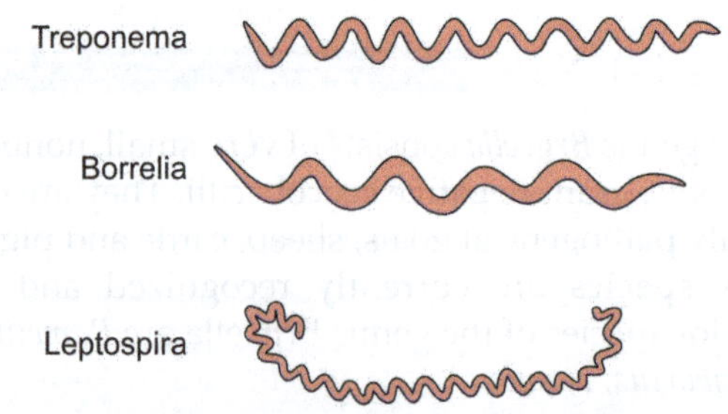

Fig. 48.1: Species designation of spirichetes based on morphology

CLASSIFICATION

Spirochetes belong to the order **Spirochetales**. It has two families:

i. Family Spirochetaceae

Spirochetes are anaerobic, facultative anaerobic or microaerophilic. They are not hooked.

Family *Spirochetaceae* has genera: *Borrelia, Cristispira, Serpulina, Spirocheta and Treponema;*

ii. Family *Leptospiraceae*

These spirochetes are obligate aerobes and are hooked. It contains the genera *Borrelia, Cristispira, Serpulina, Spirochaeta and Treponema.*

Note: Out of eight genera, three ***(Treponema, Borrelia,* and *Leptospira)*** are responsible for human disease. Members of the genus *Spirochaeta* are saprophytes found in water and sewage, while *Cristispira* are found in molluscs. The spirochetes also fall into genera based loosely on their morphology (Fig. 48.1). Treponema are slender with tight coils; *Borrelia* are somewhat thicker with fewer and looser coils; and *Leptospira* resemble *Borrelia* except for their hooked ends.

Table 48.1 shows diseases caused by spirochetes.

TREPONEMA

The name *Treponena* is derived from the Greek words *trepo: to turn* and *nema,* meaning *thread)* are relatively short slender spirochetes with fine spirals and pointed or rounded ends. Some of them are pathogenic, while others occur as commensals in the mouth, intestines, and genitalia. Morphological and antigenic similarities between pathogenic and commensal spirochetes may cause problems in the clinical and serological diagnosis.

The two treponemal species that cause human disease are *Treponema pallidum* (with three subspecies) and *Treponema carateum.* The species *T. pallidum* is now considered to include **three subspecies**—subspecies *pallidum, endemicum* and *pertenue.*

Treponemes Cause the following Diseases in Humans:

1. *T. pallidum* subspecies *pallidum causes* **Venereal syphilis**
2. *T. pallidum,* subspecies *endemicum (T. endemicum)* causese **endemic syphilis.**
3. *T. pallidum* subspecies *pertenue* causes **yaws**
4. *T. carateum* causes **pinta.**

Treponema Pallidum Subspecies *Pallidum*

Treponema pallidum, the causative agent of syphilis, was discovered by Schaudinn and Hoffmann (1905) in the chancres and inguinal lymph nodes of syphilitic patients. The name *pallidum* refers to its pale staining.

Morphology

It is a very delicate, spiral filament 6-14 µm (average 10 µm) by 0.2 µm, with 6-12 coils which are comparatively small, sharp and regular. The length of the coils is about 1 µm and the depth 1-1.5 µm. The ends are pointed and tapering. Spirochetes show rotary corkscrew-like motility and also movements of flexion; angulation, with the organism bending almost to 90° near its center, is highly characteristic of *T. pallidum.* Its progression is relatively slow compared to that of many motile bacteria. During motion, secondary curves appear and disappear in succession but the primary spirals are unchanged (Fig. 48.2).

T. pallidum cannot be seen under the light microscope in wet films but can be made out by negative staining with **Indian ink.** It is feebly refractile. Its morphology and motility can be seen under the dark ground or phase contrast microscope (Fig. 48.2). It does not take ordinary bacterial stains but stains light rose red with prolonged Giemsa staining. It can be stained by **silver impregnation methods. Fontana's method**

Table 48.1: Diseases caused by spirochetes

Genus	*Species*	*Diseases*	*Transmission*
Treponema	*T. pallidum*	Syphilis	Sexual contact or congenital
	T. pertenue	Yaws	Traumatized skin comes in contact with an infected lesion
	T. caratem	Pinta	Traumatized skin comes in contact with an infected lesion
	T. endemicum	Endemic syphilis	Mouth to mouth by utensils
Borrelia	*B. recurrentis*	Epidemic Relapsing fever Endemic Relapsing fever	Body louse Soft-shelled tick
	B. vincentii	Vincent's angina	
	B. *burgdorferi*	Lyme disease	Tick bites
Leptospira	*L. interrogans* *L. biflexa*	Leptospirosis Saprophyts	

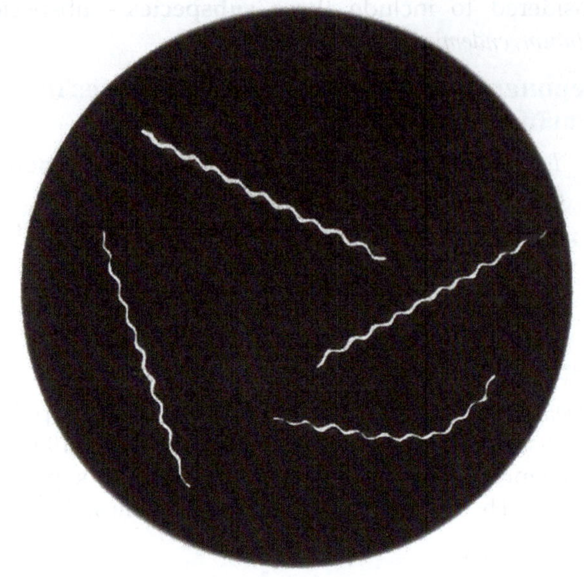

Fig. 48.2: Treponema pallidum-dark ground illumination

is useful for staining films and **Levaditi's method** for tissue sections. **Immunofluorescence methods** can now be used to detect treponemes in tissues and body fluids.

Ultrastructurally, the cytoplasm of *T. pallidum* is surrounded by a **trilaminar cytoplasmic membrane,** enclosed by a cell wall containing **peptidoglycan** which gives the cell rigidity and shape. In *Treponema* species, **fine cytoplasmic filaments** are visible in the bacterial cytoplas but these are absent in *Borrelia* species. External to this is the rigid rich outer membrane layer. Usually three, but occasionally four, **endoflagella** lie inside the outer membrane and are inserted at the tapering portion at each end of the cell. The endoflagella are more than half the length of the organism and run along the axial aspect of the spiral body. The endoflagella are presumed to be responsible for motility although there is no direct evidence for this. In contrast to other motile bacteria, these flagella do not protrude into the surrounding medium but are enclosed within the bacterial outer membrane. A capsular or slime layer has been observed occasionally on the surface of *T. pallidum* and may explain the lack of serological reactivity of organisms freshly isolated from animal tissues (Figure 49.1).

Saprophytic spirochetes are generally coarser in appearance, lack the uniform spirals with regular spacing, and show lashing motility.

Cultivation

There have been many claims of cultivation of *T. pallidum* in cultures but none has been substantiated. It is possible to maintain *T. pallidum* in motile and virulent form for 10-12 days in complex media under anaerobic conditions. Virulent *T. pallidum* strains have been maintained by serial testicular passage in rabbits for many decades. One such strain **(Nichol's strain)** was isolated in 1912 from the CSF of a patient with neurosyphilis from the brain of a fatal case of general paralysis of the insane is still virulent for man and is being propagated and used for diagnostic and research purposes. Cultivable treponemes such as ***T. phagedenis*** **(Reiter's treponeme)** and ***T. refringens*** **(Noguchi strain)** are nonpathogenic. They can be grown under strict anaerobic conditions.

Resistance

T. pallidum is very delicate, being readily inactivated by drying or by heat (41-42°C in one hour). Susceptibility of *T. pallidum* to heat was the basis of the 'fever therapy' for syphilis. It is inactivated by contact with oxygen, distilled water, soap, arsenicals, mercurials, bismuth, common antiseptic agents and antibiotics. The organism dies rapidly in water and is very sensitive to drying. It is killed in 1-3 days at 0-4°C, so that transfusion syphilis can be prevented by storing blood for at least four days in the refrigerator before transfusion. Stored frozen at –70°C in 10 percent glycerol, or in liquid nitrogen (–130°C), it remains viable for 10 to 15 years. Freezing followed by desiccation (freeze-drying) kills the organism.

Antigenic Structure

The antigenic structure of *T. pallidum* is complex. Treponemal infection induces at least three types of antibodies. On the basis of these antibodies, the treponemal antigens may be divided into non-specific and specific antigens.

A. Non-specific Antigen

The first is the *reagin* antibody in which a hapten extracted from the beef heart is used as the antigen. This lipid hapten is known as *cardioliom* and is chemically a diphosphatidyl glycerol. This lipid bas been detected in *T. pallidum* but it is not known whether the reagin antibody is induced by cardiolipin that is present in the spirochete or released from damaged host tissues.

B. Specific Antigens

1. Group-Specific Antigen

It is a protein antigen present in *T. pallidum* as well as in non-pathogenic treponemes, such as Reiter treponema.

2. Species-Specific Treponemal Antigen

It appears to be polysaccharide in nature. *T. pallidum* is used as antigen for detection of species-specific antibody.

Animal Pathogenicity

Natural infection with *T. pallidum* occurs only in human beings. Intratesticular injection leads to a syphilitic orchitis in rabbits. Intradermal inoculation also produces lesions. At present the only source of *T. pallidum* for preparing antigens and for experimental work is from the testes of infected rabbits. Monkeys and anthropoid apes can also be infected experimentally. A disease resembling syphilis can be produced experimentally in chimpanzees, with typical lesions of primary and secondary syphilis. Hamsters are also susceptible.

Syphilis: Syphilis is a disease of blood vessels and of the perivascular areas. *T. pallidum* is a strict parasite and its life outside the animal body is short.

Syphilis was first recognized in Europe at the end of the fifteenth century, when the disease first appeared in the Mediterranean areas and rapidly reached epidemic proportions at that time. The name 'syphilis' was derived from a poem written by Fracastorius of Verona in 1530 describing the legend of a shepherd named Syphilus, who had been struck with the disease. The natural history of the disease has undergone alterations since then but syphilis continues to be one of the most important and widespread of human infections.

Syphilis initially was called the *Italian disease,* the *French disease,* and the *great pox* as distinguished from smallpox. Its venereal transmission was not recognized until the eighteenth century. Delineation of the characteristics of syphilis was hindered by confusion of its symptoms with those of gonorrhea. In 1767, John Hunter, a great English experimental biologist and physician, inoculated, himself with uretheral exudate from a patient with gonorrhea. Unfortunately, the patient also had syphilis, and the subsequent symptoms experienced by Hunter convinced two generations of physicians of the unity of gonorrhea and syphilis. The separate nature of gonorrhea and syphilis was demonstrated in 1838 by Ricord. Recognition of the stages of syphilis followed, and in 1905 Schaudinn and Hoffman discovered the causative agent. The following year Wassermann introduced the diagnostic serologic test that bears his name.

Pathogenesis

Treponema pallidum subsp. *pallidum* causes syphilis. Venereal syphilis is acquired by sexual contact. Syphilis can also be acquired by nongenital contact with a lesion (e.g. on the lip) or transplacental transmission to a fetus, resulting in congenital syphilis. *T. pallidum* enters tissues by penetration of intact mucosae or through abraded skin. Clinical disease sets in after an incubation period of about a month (range 10-90 days). The natural course of syphilis can be divided into **primary, secondary, and tertiary stages** based on the clinical manifestations. Co-infection with human immunodeficiency virus (HIV) can result in variation of the natural course of the disease. Furthermore, ulcers caused by syphilis may contribute to the efficiency of HIV transmission in populations with high rates of both infections.

i. Primary Disease

After inoculation, the spirochetes multiply rapidly and disseminate to local lymph nodes and other organs via the bloodstream. The bacteria multiply at the initial entry site and a the primary lesion in syphilis is the **chancre.** The chancre is a painless, relatively avascular, circumscribed, indurated, superficially ulcerated lesion. It is covered by a thick, glairy exudate very rich in spirochetes. It is known as **'hard chancre'** to distinguish it from the nonindurated lesions of '**soft sore**' caused by *H. ducreyi,* and **Hunterian chancre**. The chancre is most frequently on the external genitalia, but it may occur on the cervix, perianal area, in the mouth or anal canal. In some cases the chancre may not be visible, as when it occurs on the uterine cervix.

The regional lymph nodes are swollen, discrete, rubbery and non tender. Even before the chancre appears, the spirochetes spread from the site of entry into the lymph and bloodstream, so the patient may be infectious during the late incubation period. The chancre invariably heals in **about 10-40 days,** even without treatment, leaving a thin scar. Chancres usually occur singly but multiple or persistent chancres may develop in immunocompromised individuals, such as those infected with the human immunodeficiency virus (HIV).

ii. Secondary Syphilis

Secondary syphilis sets in 1-3 months after healing of primary lesion. The secondary lesions are due to widespread multiplication of the spirochetes and their dissemination through the blood. In this stage, patients typically experience a **"flu-like" syndrome, lymphadenopathy,** and a **generalized mucocutaneous rash.** Characteristic lesions are **roseolar or papular skin rashes, mucous patches in the oropharynx and condylomata at the mucocutaneous junctions.** Condylomata lata occur around moist areas, such as the anus and vagina. Spirochetes are abundant in the lesions and consequently during the secondary stage the patient is most infectiousas with the primary chancre. There may also be **ophthalmic, osseous and meningeal involvement.**

Secondary lesions are highly variable in distribution, intensity and duration but they usually undergo spontaneous healing, in some instances taking as long as four or five years. The rash and symptoms gradually resolve spontaneously, and the patient enters the latent or clinically inactive stage of disease.

iii. Latent Syphilis

After the secondary lesions disappear there is a period of quiescence known as 'latent syphilis'. No clinical manifestations are evident and diagnosis during this period is possible only by serological tests. Individuals with late latent syphilis are not generally considered infectious, but may still transmit infection to the fetus during pregnancy and their blood may remain infectious. In many cases, this is followed by natural cure but in others, after several years, manifestations of **tertiary syphilis** appear.

iv. Tertiary Syphilis or Late Syphilis

Tertiary or late syphilis, which may develop decades after the primary infection, is a slowly progressive, destructive inflammatory disease that may affect any organ. Isolation of *T. pallidum* from patients with late syphilis is usually impossible, and much of the observed pathology may be due to auto-immune phenomena. The three most common forms of late syphilis are cardiovascular syphilis, gummatous syphilis and neurosyphilis.

a. Cardiovascular Syphilis

These consist of cardiovascular lesions including aneurysms, chronic granulomata (gummata) and meningovascular manifestations.

b. Gummatous Syphilis

It is a rare granulomatous lesion of the skeleton, skin or mucocutaneous tissues.

c. Neurosyphilis

Neurological manifestations such as tabes dorsalis or general paralysis of the insane develop several decades after the initial infection in a few cases. These are known as **late tertiary or quaternary syphilis.**

Congenital Syphilis

In conngenital syphilis infection is transmitted from mother to fetus transplacentally. Transplacental transmission can take place at any stage of pregnancy and Congenital syphilis affects many body systems and is therefore severe and mutilating. The lesions of congenital syphilis usually develop only after the fourth month of gestation, the time when fetal immune competence starts appearing. Congenital syphilis can be prevented if the mother is given adequate treatment before the fourth month of pregnancy. The obstetric history in an untreated syphilitic woman is typically one of abortions and stillbirths followed by live births of infants with stigmata of syphilis and finally of healthy infants. Abortion because of congenital syphilis usually occurs during the second trimester of pregnancy.

Syphilis Acquired Nonvenereally

In syphilis acquired nonvenereally (as occupationally in doctors or nurses), the natural evolution is as in venereal syphilis except that the promary chancre is extragenital, usually on the fingers. In the rare instances where syphilis is transmitted by blood transfusion, the primary chancre does not occur.

Laboratory Diagnosis

Laboratory diagnosis consists of **demonstration of the spirochetes** under the microscope and of **antibodies in serum or CSF.**

A. Demonstration of Treponemes in the Exudates

1. Dark-ground Microscopy

Specimens should be collected with care as the lesions are highly infectious. The lesion is cleaned with a gauze soaked in warm saline and the margins gently scraped so that the superficial epithelium is abraded. Gentle pressure is applied to the base of the lesion and the serum that exudes is collected preventing admixture with blood. If it is blood stained, it should be wiped away and process repeated until a clear fluid is obtained. Wet films are prepared with the exudate and after applying thin coverslips, examined under the dark ground microscope. Oral lesions should not be examined; nonpathogenic spirochetes are numerous in these specimens and will lead to misinterpretation. Culture methods are not available, so if dark-field microscopy is inconvenient, a serologic-based diagnosis is pursued.

Diagnosis by microscopy is applicable in primary and secondary stages and in cases of congenital syphilis with superficial lesions. This bacterium is too thin to be visualized with a standard gram stain so two techniques to visualize it with a light microscope are **dark field microscopy** and **immunofluorescence**. *T. pallidum* is recognized by its slender structure, regularity of its spirals and slightly pointed ends. Numerous commensal spirochetes that are found as the normal flora of the genital and rectal mucosal surfaces must be carefully distinguished.

If the initial test is negative, the procedure should be repeated daily for at least 3 days. Dark ground examination is useful but negative results do not exclude the diagnosis of syphilis, because of its low sensitivity. A treponemal concentration of 10^4 per ml in the exudates is required for the test to be positive.

2. Direct fluorescent-antibody staining for Treponema pallidum (DFA- Tp)

For direct fluorescent-antibody staining for *T. pallidum* (DFA-TP) test whereby a smear of exudate is made on a slide, fixed in acetone, and sent to the laboratory, where the DFA-TP test is done using fluorescent tagged anti *T. pallidum* antiserum. The use of specific monoclonal antibody has made the test more reliable. The treponemes appear distinct, sharply outlined and exhibit an apple green fluorescence. It is a better and safer method for microcpic examination.

3. Silver Impregnation Method (Levaditi's Method)

Treponemes in the tissues can be demonstrated by immunofluorescence staining.or silver impregnation method of staining *(Levaditi's stain).*

4. Enzyme Immuneassay and Plynarese Chain Reation (PCR)

T. pallidum antigen in the lesion can be detected by enzyme immunoassay and polymerase chain reaction (PCR) and appears to hold some promise.

B. Serological Tests

These tests form the mainstay of laboratory diagnosis. (Table 48.2). Two major types of serologic tests exist: **nontreponemal tests** and **treponemal tests. In nontreponemal tests or Standard tests for syphilis (STS),** cardiolipin or lipoidal antigen is used. While in **treponemal tests** treponemes are used as the antigen.

Nontreponemal Tests or Standard Tests for Syphilis (STS)

Reagin antibodies are detected by cardiolipin antigen in standard tests for syphilis (STS). The antigen used in these tests is an alcoholic extract of beef heart tissue (cardiolipin) to which lecithin and cholesterol are added (cardiolipin-lecithin complex made from bovine hearts). The STS includes Wassermann, Kahn, Venereal Diseases Research Laboratory (VDRL) and the rapid plasma reagin (RPR) tests. All these tests are *flocculation tests* except Wassermann reaction which is a complement fixation test (CFT). The Wassermann reaction is no longer in use. Similarly Kahn test is rarely done these days.

The two nontreponemal tests widely used today are the **Venereal Disease Research Laboratory (VDRL)** and **RPR tests.** These tests are inexpensive to perform, demonstrate rising and falling reagin titers, and correlate with the clinical status of the patient.

i. VDRL (Veneral Disease Research Laboratory) Test

The Kahn test has been replaced by the simpler and more rapid VDRL test. which gives more quantitative results (VDRL, for Venereal Disease Research Laboratory, USPHS, New York, where the test was developed). These tests provide similar clinical information and have similar advantages. These tests are cheaper, more rapid and simpler to perform and control.

VDRL is the most widely used simple and rapid test which requires only a small quantity of serum. The VDRL test uses a cardiolipin antigen that is mixed with the patient's serum or CSF. Flocculation occurs in a positive reaction and. is observed microscopically.

In its original form the VDRL test is performed by mixing heat-inactivated patient's serum with a freshly prepared suspension of cardiolipin-lecithin-cholesterol antigen on a glass slide. The mixture is rotated, usually mechanically, for 4 minutes after which the flocculation (aggregation of antigen-antibody complexes in suspension) can be detected under a low power objective of a microscope. By testing serial dilutions, the antibody titer can be determined.

Method

1. The test is done in a specially prepared slide, with depressions of 14 mm diameter each. Serum is inactivate heated to it prior to the test, whereas CSF need not be heated.
2. Inactivated patient serum (0.05 ml) is pipetted into the paraffin ring on the glass slide. Each (0.05 ml) of positive and negative control sera are pipetted into other paraffin rings.
3. One drop of working antigen suspension is added to each of these paraffin rings from a syringe delivering 60 drops in 1 ml.
4. Mix with wooden sticks and rotate slide at 180 revolutions per minute for four minutes on a mechanical VDRL or manually. Flocculation occurs in a positive reaction and is observed microscopically.

Interpretation

The results of qualitative test are reported as **'reactive', 'weak reactive' or 'non reactive'. Reactive**' means positive (large clumps of antigen with marked background clearing are obtained.) while **'non-reactive'** is negative (The antigen particles are seen as small fusiform needles which remain more or less evenly dispersed). The reciprocal of the end point is given as the titer for reporting of quantitative test, e.g. reactive in 1:4 dilution is reported as 'reactive 4 dilution' or R4.

Sometimes VDRL test may give *false negative* reaction due to high titers of antibody in patient's serum (prozone phenomenon). The test is performed with diluted serum in such cases and it becomes positive.

VDRL-ELISA Test

An automated VDRL-ELISA test has been developed which can measure IgG and IgM antibodies separately and is suitable for large scale testing of sera.

ii. Rapid Plasma Reagin (RPR) Test

RPR is the more common test used; it employs carbon particles and is read macroscopically. The black carbon particles are bound to cardiolipin; when mixed with a positive serum on a disposable card, the particles clump together. Agglutination is easily observed without a microscope. Reactive or weakly reactive sera should undergo titration and be tested with treponemal tests.

Rapid Plasma Reagin (RPR) test is the most popular modifications of the VDRL test. RPR test employs a stabilized VDRL carbon antigen which make the result more clear cut and is read macroscopically.

Advantages of RPR Test

i. It enables the result to be read by eye instead of microscopically.
ii. Useful in field studies in developing countries.
iii. RPR test can be done with unheated serum or plasma.
iv. A fingerprick sample of blood is sufficient.

Disadvantage

It is not suitable for testing cerebrospinal fluid (CSF) while VDRL remains the standard test for use with cerebrospinal fluid.

Toluidine Red Unheated Serum Test (TRUST)

A modification of the RPR test has been used. Instead of carbon particles it uses paint pigment toner toludine red particles.

Automated RPR Test (ART)

Automated RPR test (ART) is available for large scale tests.

Disadvantages of STS

Since cardiolipin antigen tests detect antibodies against a non-specific antigen shared by treponemes and mammalian tissues, a positive result is sometimes obtained with sera from healthy individuals or patients without clinical evidence of syphilis. These reactions are termed **Biological False Positives (BFP)** rections. **Biological False Positives (BFP)** reactions are defined as positive reactions obtained in cardiolipin tests, with negative results in specific treponemal tests, in the absence of past or present treponemal infections-and not caused by technical faults. They represent nontreponemal cardiolipin antibody responses.

BFP reactions may occur in about one percent of normal sera. BFP antibody is usually IgM, while reagin antibody in syphilis is mainly IgG. Clinically, BFP reactions may be classified as acute or chronic.

Acute BFP Reactions

Acute BFP reactions last only for a few weeks or months and are usually associated with acute infections, injuries or inflammatory conditions and occur in patients with other acute illnesses, especially pneumonia, hepatitis, vaccinations, and viral exanthematous disease.

Chronic BFP Reactions

Chronic BFP reactions persist for longer than six months. These may occur in a wide variety of infectious and non-infectious conditions associated with tissue damage and are typically seen in:

1. SLE and other collagen diseases; 2. Leprosy; 3. Malaria; 4. Relapsing fever; 5. Infectious mononucleosis 7. Tropical eosinophilia.

Reagin antibody becomes detectable 7-10 days after the appearance of primary chancre (or 3-5 weeks after acquiring the infection). The sensitivity in the primary stage is 60-75 percent with the titers being low, upto eight. The titre diminishes and the test tends to become negative after treatment. In the secondary stage, the sensitivity is 100 percent and titers range from 16 to 128 or more. Prozone phenomenon may be a problem in high titre sera and it is therefore essential to test sera in dilutions. A positive reaction is usually obtained on diluting the serum.

Another stage of syphilis in which such high titers are seen is congenital syphilis. After the secondary stage, titers diminish and about a third of patients with late syphilis are seronegative. The titers may rise in patients developing cardiovascular, neurological or gummatous lesions. In some cases of neurosyphilis, reagin tests may be negative with serum but positive with the CSF.

b. Treponemal tests (Table 48.2)

Treponemal tests in which treponemes are used as the antigen. These are of two types (Table 48.2):

1. Group specific tests using cultivable treponemes, such as Reiter treponemes *(T. phagedenis)* as the antigen-Reiter protein complement fixation (RPCF) test
2. **Specific tests using pathogenic *T. pallidum* (Nichol's strain)**

1. Group-specific Tests using Reiter Treponeme

Reiter Protein Complement Fixation (RPCF) Test

The principle of this test is the same as that of Wassermann test.

These employed the cutivable Reiter treponemes for prepation of antigen. Its sensitivity and specificity were lower than those of tests using *T. pallidum*. RPCF and other Reiter treponeme tests are not now in generaral use.

2. Specific Tests Using Pathogenic *T. Pallidum* (Nichol's Strain)

These tests use the virulent Nichol's strain of *T. pallidum* maintained by serial inoculation in rabbit testis.

I. Tests using live *T. pallidum*

Treponema Pallidum Immobilization (TPI) Test

The first in this group is the *Treponema pallidum* immobilization (TPI) test introduced in 1949 (Nelson and Mayer 1949). The test serum is incubated with complement and *T. pallidum* maintained in a complex medium anaerobically. If antibodies are present, the treponemes are immobilized, that is, rendered nonmotile, when examined under dark ground illumination. The test is considered positive if the percentage of treponemes immobilized is 50 or more, negative if 20 or less, and inconclusive if in between. Sera from patients on penicillin or other antitreponemal drugs may show false positive results.

In its time, TPI was the most specific test available for diagnosis of syphilis and was considered the gold standard in syphilis serology. Because the TPI test employs live treponemes it is time-consuming, expensive and technically demanding. A few reference laboratories still perform the TPI test on selected sera for research purposes.

II. Tests Using Killed *T. pallidum*

The *Treponema pallidum* Immune Adherence (TPIA) test and *Treponema pallidum* Agglutination (TPA) test were used for some time but have now been given up.

a. Treponema Pallidum Immune Adherence (TPIA) Test

A suspension of *T. pallidum,* inactivated by formalin, is mixed with the test serum and examined under dark ground microscopy. The treponemes are found agglutinated in the presence of antibodies. The test is not very specific and false positive reactions are common.

b. Treponema Pallidum Agglutination (TPA) Test

A suspension of inactivated *T. pallidum* is mixed with the test serum, complement and fresh heparinized whole blood from a normal individual and incubated. The

Table 48.2: Diagnostic tests for syphilis

A. Demonstration of treponemes in the exudate	1. Dark-ground microscopy 2. Direct fluorescent-antibody staining for *Treponema pallidum* (DFA- Tp) 3. Silver impregnation method (Levaditi's stain) 4. Enzyme immunoassay, Polymerase chain reaction (PCR).
B. Serological tests	**a. Nontreponemal tests** **Nonspecific *(reagin antibody)* tests using the cardiolipin antigen *(standard tests for syphilis or STS).*** 1. Wassermann complement fixation test 2. Kahn flocculation test 3. Venereal Disease Research Laboratory (VDRL) test 4. Rapid Plasma Reagin (RPR) test 5. Toluidine red unheated serum test (TRUST) **b. Treponemal tests** **a. Group specific test using cultivable treponemal (Reiter strain) antigen** I. Reiter Protein CF (RPCF) test (1953) **b. Specific tests using pathogenic treponemes *(T. palidum)*** I. Test using live *T. pallidum* *Treponema pallidum* Immobilization (TPI) test **II. Tests using killed *T. pallidum*** a. Treponema pallidum agglutination (TPA) test b. Treponema pallidum immune adherence (TPIA) test c. Fluorescent Treponemal Antibody Absorption (FTA-ABS) test **III. Tests using *T. pallidum* extract** a. *Treponema pallidum* Hemagglutination Assay (TPHA) Microhemagglutination test for *Treponema pallidum* (MHA-TP) b. *Treponema pallidum* Enzyme Immunoassays (TP-EIA):

treponemes will be found to adhere to the erythrocytes in the presence of antibodies. *Immune* adherence will not occur in absence of antibodies (negative test). Both TPA and TPIA are not used in diagnostic laboratories.

c. Fluorescent Treponemal Antibody-Absorption (FTA-ABS) Test

It is an indirect immunofluorescence test. Smears are prepared on slides with Nichol's strain of *T. pallidum* as antigen. The slides can be stored for several months in deep freeze. The currently used modification of the test is the FTA-Absorption (FTA-ABS) test in which the test serum is preabsorbed with a sonicate of the Reiter treponemes *(sorbent)* to eliminate group specific reactions. After specific antibody from the patient is allowed to react with the organisms, unbound antibodies in the serum are removed by washing. The presence of anti-*T. pallidum* antibody is then detected by application of fluorescein-labeled, anti-human globulin and examination of the slide with an ultraviolet (UV) microscope. Positive results are indicated by fluorescence of the *T. pallidum* organisms. An FTA-ABS 19S-lgM test has also been developed for evaluation of congenital syphilis; however, the test is still considered provisional.

FTA-ABS is as specific as the TPI test and is now accepted as a standard reference test and is highly specific and sensitive at all stages of syphilitic infection although a small percentage of false positive reactions occurs. The FTA-Abs test is positive in approximately 80, 100 and 95 percent of primary, secondary and late syphilitics, respectively, and, unlike the VDRL test, remains positive following successful therapy. However, as it can be done only in suitably equipped laboratories, it is not available for routine testing.

III. Tests using an extract of *T. pallidum*

a. T. pallidum Hemagglutination Assay (TPHA)

The *Treponema pallidum* hemagglutination assay (TPHA) uses tanned erythrocytes sensitized with a sonicated extract of *T. pallidum* as antigen.When these sensitized ertyrocytes are mixed with patiet's serum (test sera) containing antitreponemal antibidies it causes hemagglutination. As in the FTA-Abs assay, sera are preabsorbed with a non-pathogenic treponeme to remove antibody against commensal spirochetes.

The test sera for TPHA are absorbed with a diluent containing components of the Reiter treponeme, rabbit testis and sheep erythrocytes. Sera are screened in an initial dilution of 1 : 80 but titers of 5120 or more are common in the secondary stage.

Advantages

i. TPHA is just as specific as FTA-ABS and almost as sensitive, except in the primary stage.

Table 48.3: Frequency of reactive serological tests in untreated syphilis (percentage) in common use			
Stage	*VDR/RPR*	*FTA-ABS*	*TPHA*
Primary	70-80	85-100	65-85
Secondary	100	100	100
Latent/late	60-70	95-100	95-100

ii. This assay can be used to detect localized production of anti-treponemal antibodies in cerebrospinal fluid, a marker of neurosyphilis.
iii. It is also much simpler and more economical.
iv. No special equipment is needed. Kits are available commercially. These advantages have made TPHA a standard confirmatory test.

Microhemagglutination Test (MHA-TP)

TPHA may be performed in microtiter plates and is called microhemagglutination test (MHA- TP). Hemagglutination treponemal test for syphilis (HATTS) is an automated conversion of TPHA test. The only disadvantage of MHA-TP and HATTS is that they lack sensitivity in primary syphilis.

Treponema Pallidum Particle Agglutination (TP-PA)

The TP-PA test has largely replaced the microhemagglutination assay for antibodies to *T. pallidum* (MHA-TP) test. The TP-PA test uses gelatin particles sensitized with *T. pallidum* antigens instead of the sensitized red blood cells used in the MHA-TP assay. The TP-PA test is simpler to perform than the FTA-ABS test, does not require an absorption step or expensive UV microscope, and is more specific than the MHA-TP test.

Table 48.3 shows the relative sensitivities of the serological tests in common use.

b. Enzyme Immunoassays (EIA)

Enzyme immunoassays (EIA) have been developed using *T. pallidum* antigens and are available commercially (Bio-Enza Bead test; Captia Syphilis-G test). Assays that detect either IgM or IgG are available and are increasingly replacing the TPHA and VDRL tests for routine screening.

Interptretation of Various Serological tests

1. Serological Screening

Because of their simplicity and accuracy, the cardiolipin antigen tests are used as **screening or first line procedures** for both routine diagnosis and mass screening programes. When used together, the VDRL and TPHA tests provide a highly efficient screen for the detection or exclusion of treponemal infection.

The practice for serological screening for syphilis varies in different countries. In the UK, a **combination of VDRL and TPHA tests** is used. This is an efficient combination for the detection or exclusion of syphilis at all stages, except the early primary stage. A repeat test 1-3 months later will bring even this to light. In the USA, screening is by VDRL or RPR test alone. This may fail to detect about one percent of secondary syphilis due to the prozone effect and about 30 percent of latent or late syphilis.

2. Response to Treatment

Reagin tests usually become negative 6-18 months after effective treatment of syphilis, depending on the stage at which treatment is given. Reagin tests are preferred because they usually become negative following treatment. Serial quantitative VDRL testing provides the best means of measuring response to treatment in most stages of treponemal infection. Treatment in the primary stage leads to seroreversal in about four months; in the secondary and early latent stages, it takes 12-18 months; in later stages, it may take five years or more. In some cases low titer reactivity may persist indefinitely, in spite of effective treatment. Specific treponemal tests tend to remain positive in spite of treatment so they are of little value as indicators of clinical cure. TPHA titers may fall rapidly following treatment in secondary syphilis but remain positive for life in low titers.

3. Biological False Positive (BFP) Reactions

TPHA and FTA-ABS are helpful in excluding or confirming the diagnosis of syphilis and for identifying BFP reactions. Both TPHA and FTA-ABS can give false positive results, though very rarely. All serological tests for syphilis may be positive in nonvenereal treponematoses, and some in a few other spirocheatal infections as well. VDRL test is negative but FTA-ABS may be positive in Lyme disease.

A negative TPHA virtually excludes the diagnosis of syphilis, past or present, except in the very early stages. A negative CSF VDRL test may not be conclusive in neurosyphilis but a negative TPHA test eliminates the possibility of neurosyphilis.

4. Diagnosis of Congenital Syphilis

In the diagnosis of congenital syphilis it is necessary to differentiate between passive transplacental transfer of maternal antibody to the fetus and production by the fetus of endogenous antitreponemal antibody. Being the initial type of antibody to appear, IgM is detectable by the second week of infection. As IgM does not cross the placenta, its presence in neonatal serum confirms congenital syphilis and helps differentiate it from seropositivity due to passively transferred maternal antibody (svphilotoxemia). For the selection of IgM antibodies many techniques have been developed which include modifications of the FTA ABS, TPHA, ElA and VDRL tests, using whole sera or separated IgM fractions. If IgM-FTA-ABS test gives a reactive test result with infant blood then it is strong evidence of active congenital disease. Serial testing is also useful because the titer of passively transferred antibody decreases rapidly, the VDRL test becoming negative by three months.

Syphilis Associated with Human Immunodeficiency Virus (HIV) Infections

The many interactions between syphilis and HIV demand that every patient with positive syphilis serology should be tested for HIV (with the patient's informed consent) and that every HIV-positive patient should be tested for syphilis. HIV-infected patients who acquire syphilis may fail to produce anti-treponemal antibodies. This has extremely serious implications for the diagnosis and control of syphilis.

Epidemiology

Syphilis is found worldwide in distribution. Natural syphilis is exclusive to humans and has no other known natural hosts. The most common route of spread is by direct sexual contact. The disease can also be acquired congenitally or by transfusion with contaminated blood. Syphilis is not highly contagious.

The incidence of disease has steadily decreased since the advent of penicillin therapy in the early 1940s. With the discovery of the dramatic therapeutic response to penicillin, it was hoped that it may even be possible to eradicate syphilis, as the disease has no extra human reservoir. However, not only has it not been possible to eliminate the disease but an increase has occurred in its incidence, due to the changing customs, habits and values in society.

The advent of the AIDS pandemic has had an impact on syphilis. The advent of HIV and the acquired immune deficiency syndrome (AIDS) in the 1980s reduced the incidence among this group owing to changes in sexual practices, but this trend did not continue everywhere. Concurrent infection with syphilis HIV is common and may lead to earlier evolution of neurosyphilis. High-risk sexual behavior and co-infection with HIV continue to complicate syphilis control efforts.

Immunity

The immune mechanisms in syphilis are not adequately understood. Humoral immune response against the treponeme does not appear to be effective but **Cell mediated immunity** may be more relevant. In early syphilitic lesions, T lymphocytes and macrophages are predominant. Specifically sensitized Th1 cells secrete cytokines favoring clearance of spirochetes by activated macrophages. The tissue destruction and lesions observed in syphilis are primarily the consequence of the patient's immune response to infection.

In a person already having active infection reinfections do not appear to occur. It was believed that **premunition or infection immunity**, as seen in some parasitic infections, holds good in syphilis also and that a patient becomes susceptible to reinfection only when his original infection is cured.

Prophylaxis

1. As transmission is by direct contact, it is possible to protect against syphilis by avoidance of sexual contact with an infected individual. An infected individual may serve as a source of infection for 3-5 years during early syphilis.
2. When this is not complied with, the use of physical barriers (such as condoms), antiseptics (potassium permanganate) or antibiotics may minimize the risk.
3. Educating people about sexually transmitted diseases, including the proper use of barrier contraceptives, reporting each case of syphilis to the health authorities for contact investigation, and treating all sexual contacts of persons infected with syphilis are cornerstones of syphilis control efforts.
4. Protective vaccines are not available. The control of syphilis and other venereal diseases has been complicated by an increase in prostitution among drug abusers.

Treatment

Penicillin is the drug of choice for treating infections with *T. pallidum*. So far there have been no reports of penicillin-resistance. **In early cases**, a single injection of 2.4 million units of benzathine penicillin is adequate. **For late syphilis**, this may be repeated weekly for three weeks. In patients allergic to penicillin, **erythromycin or tetracycline** may be used. Only penicillin can be used for the treatment of neurosyphilis. Thus, penicillin-allergic patients must undergo desensitization. This is also true for pregnant women, who should not be treated with the tetracyclines.

Jarisch-Herxheimer Reaction

Antibiotic therapy of syphilitics, particularly with penicillin, characteristically induces a systemic response called the **Jarisch-Herxheimer reaction.** This is characterized by the rapid onset (within 2 hours fever, chills, myalgia, tachycardia, hyperventilation, vasodilatation and hypotension. It is frequent but harmless in primary and secondary syphilis and is rare in late syphilis but can be dangerous in some cases of gummatous, cardiovascular or neurosyphilis. It is believed to be due to the liberation of toxic products like tumor necrosis factors from the massive destruction of treponemes or due to hypersensitivity.

NONVENEREAL TREPONEMATOSES

Nonvenereal treponemal diseases occur in endemic foci in several parts of the world. These treponematoses are found in developing countries where hygiene is poor, little clothing is worn, and direct skin contact is common because of overcrowding.

Infection is usually transmitted by direct body to body contact. These infections are rarely transmitted by sexual contact, and congenital infections do not occur. All three diseases have primary and secondary stages, but tertiary manifestations are uncommon. All diseases respond well to penicillin or tetracycline. The three non-venereal treponemal diseases are:

1. Endemic syphilis
2. Yaws
3. Pinta

1. Endemic Syphilis (Bejel)

E*ndemic syphilis* or *bejel* sphilis, transmitted nonvenereally, was endemic in several foci. Under the name of *sibbens*, it was present in Scotland in the last century. It is known *bezel* in the Middle East, *njovera* in Zimbabwe, *dichuchwa* in Bechua, *Skerlievo* in Eastern Europe and *siti* in Gambia. It has also been reported from India. Bejel is endemic in Africa, western Asia and Australia, and mainly affects children in rural populations where living conditions and personal hygiene are poor.

The etiologic agent is *T. pallidum* subspecies *endemicum* and closely resembles yaws in clinical manifestations. **Transmission** is by direct person-to-person contact and by sharing of contaminated eating or drinking utensils.

Clinical Manifetations

The primary chancre is not usually seen, except sometimes on the nipples of mothers infected by their children. The disease is usually seen with manifestations of secondary syphilis, such as mucous patches in the mouth and skin eruptions. The disease progresses to tertiary lesions particularly syphilitic gummata on the skin, bone and nasopharynx. As in yaws, the cardiovascular and central nervous systems are not involved, and congenital infection is rare because of the early age of infection. **The laboratory diagnosis and treatment** are as for venereal syphilis.

2. Yaws (Frambesia)

Yaws is a disease that is endemic among rural populations in tropical and subtropical countries such as Africa, South America, Southeast Asia and Oceania. Yaws, also known as **frambesia, Rian, parangi** and by many other synonyms. Yaws eradication campaigns by mass penicillin injections in endemic areas led to the virtual eradication of the disease. Termination of these programes has led to a resurgence of pockets of disease, particularly in West Africa. In India, cases of yaws have been identified in Andhra Pradesh, Orissa and Madhya Pradesh.

Causative agent: *T. pallidum* subspecies *pertenue* (still known as *T. pertenue)* which is morphologically and antigenically indistinguishable from *T. pallidum.*

Clinical Manifestations: The primary lesion (mother yaw) is an extragenital papule. It enlarges and breaks down to form an ulcerating granuloma. As in syphilis, secondary and tertiary manifestations follow but neurological and cardiovascular involvement is rare. In the late stages of yaws, ulcerative skin lesions and bone lesions develop.

Transmission: Infection is acquired by direct contact with open ulcers. Flies may act as mechanical vectors. The small fly, *Hippolates pallipes* been found feeding on open sores but its epidemiological importance is not known.

Laboratory diagnosis and treatment are similar to those of venereal syphilis. There appears to be some cross immunity in between yaws and syphilis, in that venereal syphilis is rare in communities where yaws is endemic.

3. Pinta

Pinta (carate, mal del pinto) is a contagious inoculable disease endemic in Central and South America and the neighboring islands. It is caused by *T. carateum* which is morphologically and antigenically indistinguishable from *T. pallidum* so cross immunity between pinta and syphilis is only partial.

Spread of infection is by direct contact with infectious lesions. Transmission appears to require direct person to person contact. The skin bears the brunt of the disease. Incubation ranges from 7-21 days. **The primary lesion** is an extragenital papule, which does not ulcerate but develops into a lichenoid or psoriaform. **Secondary skin lesions** are characterized by hyperpigmentation or hypopigmentation. Tissues other than skin are seldom affected. **The laboratory diagnosis and treatment** are similar to those of venereal syphilis.

NONPATHOGENIC TREPONEME

Several commensal treponemes occur on the buccal and genital mucosa and may cause confusion in the diagnosis of syphilis by dark field examination. Best known among them is the oral spirochete, *T. dentium,* which can be readily cultivated. Treponemes also occur on the surface of gastric and colonic epithelium in human beings and animals.

During early attempts to grow the syphilis spirochete in cultures, several treponemes had been grown and mistakenly called *T. pallidum*-for example, the Reiter and Kazan strains which have been identified *as P. phagedenis* and !be aviru.knt Nichols and Noguchi strains which have been recognized as *T. refringens. T. refringens and T. gracilis* may be found as normal commensal in genitalia. In experimental work on *T. pallidum* in rabbits, *T. paraluiscuniculi* (formerly, *T. cuniculi),* which has a very similar appearance and causes natural venereal infection in rabbits, may pose problems.

BORRELIA

Borreliae are large, motile, refractile spirochetes with irregular, wide, open coils. They are usually 10-30 μm long and 0.3-0.7 μm wide. They are readily stained by ordinary methods and are gram-negative. Members of the genus *Borrelia* can be easily distinguished from spirochetes of the genera *Leptospira* and *Treponema* by their much larger size (10-30 in length by 0.3-0.7 in width), by their irregular, wide-open, loosely wound primary coils and by the ease with which they can be stained with the usual laboratory aniline dyes.

Species of Borrelia

Several species of Borrelia occur as commensals on the buccal and genital mucosa. Although pathogens for mammals and birds, borrelias are the causative agents of tick borne and louse borne relapsing fever and tick-borne Lyme disease in humans.

Borreliae of Medical Importance

1. *B. recurrentis* causing relapsing fever
2. *B. vincentii* sometimes causes fusospirochetosis
3. *B. burgdorferi*- causative agent of Lyme disease.

fevers are characterized clinically by recurrent periods of fever and spirochetemia. Relapsing fever.

Morphology

Borrelia are helical organisms 0.2 to 0.5 µm wide and 8 to 20 µm in length. They are gram-negative, actively motile and possess 5-8 irregular spirals at intervals of 2 µm with pointed ends (Fig. 48.3). Spirals are coarser and more irregular than those of the treponemes or leptospires and usually can be seen with light microscopy in preparations stained with aniline dyes, such as Wright's or Giemsa stains.

Cultural Characteristics

Borrelia are microaerophilic. Optimum temperature for growth is 28-30°C. Cultivation is difficult but has been successful in complex media containing serous fluids. The organism can be grown on Noguchi's medium (asctic fluid containing rabbit kidney), chorioallantoic membrane (CAM) of chick embryos and in mice or rats intraperitoneally. For primary isolation, the best method is to inoculate mice or rats intraperitoneally. Great care has to be taken to ensure that the animals are free from pre-existing brreliosis when using experimental animals.

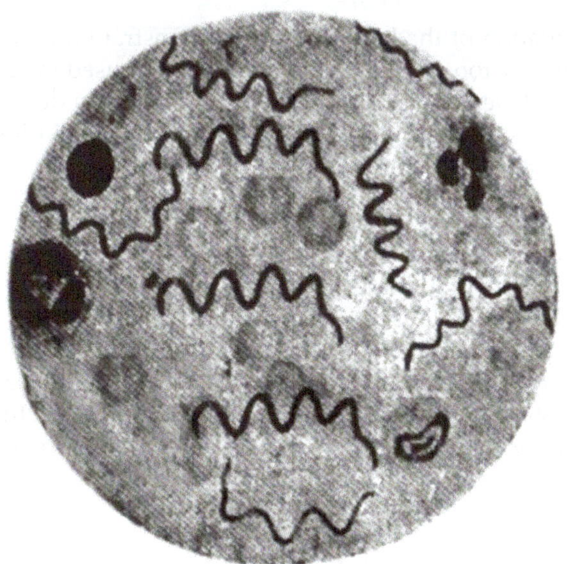

Fig. 48.3: Borrelia recurrentis in peripheral blood smear

Antigenic Properties

The most striking property of relapsing fever is the capacity of *Borrelia* to undergo several antigenically distinct variations within a given host during the course of a single infection. This is believed to be the reason for the occurrence of relapses in the disease. **Antigenic variations** have been shown to be caused by DNA rearrangements in linear plasmids present in borrelia. Ultimate recovery after a number of relapses may be due to the development of immunity to all the antigenic variants.

Agglutinating, complement fixing and lytic antibodies develop during infection but their demonstration is not possible as a routine diagnostic test due to the difficulty in preparing satisfactory antigens.

Relapsing Fever

Relapsing Fever (RF) has been known since the time of Hippocrates and has occurred in epidemic, endemic or sporadic form throughout the world. RF is an **arthropod-borne infection**, and two types of which occur **louse-borne** and **tick-borne**. The borreliae causing them are indistinguishable in morphology and many other features but differ in their arthropod hosts.

Epidemic or Louse-Borne Relapsing Fever

The causative agent of louse borne or epidemic RF is *B. recurrenti.* It is an exclusive human pathogen, being transmitted from person to person through body lice ***(Pediculus humanus corporis)*** and disease is found worldwide. Humans are the only reservoir for *B. recurrentis* No extra-human reservoir is known. It is an obligate human pathogen first observed by Obermeier (1873) in the blood of patients during an epidemic in Berlin.

Endemic or Tick-Borne Relapsing Fever

The second form of relapsing fever is endemic and **tick-borne**. It is caused by as many as 15 species of borreliae and cause RF *(B. duttonni, B. hermsii, B. parkeri B. turicatae.* etc.) and is spread by infected soft ticks of the genus *Ornithodoros,* that vary according to the country where the infection occurs. The natural hosts for these organisms include rodents and other small mammals on which the ticks normally feed. The disease occurs worldwide, reflecting the distribution of the tick vector. Human infection is only an accidental event. Borrelliae have been assigned to various species based on the ticks that carry them.

Pathogenicity

After an incubation period of 2-10 days relapsing fever sets in as fever of sudden onset. During this period, borreliae are abundant in the patient's blood (up to 10^5 spirochetes per cubic millimeter of blood). The fever subsides in 3-5 days. Another bout of fever sets in after an afebrile period of 4-10 days during which borreliae are not demonstrable in blood. The borreliae reappear in blood during the relapses of fever. The disease ultimately subsides after 3-10 relapses. Subsequent relapses are usually milder and of shorter duration.

Generally, there are more relapses associated with cases of untreated tick borne relapsing fever, but louse borne relapsing fevers tend to be more severe. The periodic febrile and afebrile cycles of relapsing fever stem from the ability of the borreliae to undergo antigenic variation.

Experimentally, rodents such as rats, mice and, less readily, guinea pigs may be infected by intraperitoneal injection. Borreliae may survive in the brains of infected animals after they have disappeared from the blood.

Epidemiology

The etiologic agent of louse-borne epidemic relapsing fever is *B. recurrentis*. The vector is the human body-louse, and humans are the only reservoir. Lice become infected after feeding on an infected person. The organisms are ingested, pass through the wall of the gut, and multiply in hemolymph and is not shed in excreta. Disseminated disease is not believed to occur in lice. So the infection is transmitted not by the bite of lice but by their being crushed and rubbed into abraded skin. *B. recurrentis* is not transmitted transovarially in lice.

Louse-borne relapsing fever tends to occur as epidemics whenever poverty, overcrowding and lack of personal hygiene encourage louse infestation. Epidemic relapsing fever used to be very common during wars and in jails of former days but with improvements in hygiene and the discovery of insecticides, it has now become rare. Although epidemics of louse-borne relapsing fever swept from Eastern to Western Europe in the past century, disease now appears to be restricted to Ethiopia, Rwanda, and the Andean foothills. It survives in some areas, as in parts of Africa and appears as outbreaks whenever civil strife and famine encourage large scale louse infestation. The louse borne disease presents a more severe clinical picture than the tickborne variety and is associated with jaundice, hemorrhages and, in some outbreaks, a high rate of fatality.

Tick-borne endemic relapsing fever is a **zoonotic disease**, with rodents, small mammals, and soft ticks (*Ornithodoros* species) the main reservoirs and many species of *Borrelia* responsible for the disease. Tick-borne relapsing fever occurs as sporadic cases in endemic areas. It is a 'place disease' and is frequently associated with certain dwellings or other locations that are inhabited by infected ticks. The disease is milder but relapses are more frequent than in louse-borne fever. The borrelia invades all parts of the body of the tick and is shed in its saliva and feces. So the infection is transmitted to humans through the bite of ticks or their discharges.

Several species of soft ticks belonging to the genus Ornithodoros act as vectors, different species being responsible in different regions. In India, the vector species are *O. tholozoni, O. crossi, O. lahorensis* and the fowl tick, *Argas versicus*. These soft ticks can live for ten years or more with only an occasional blood meal. They feed usually while the host is sleeping, and painlessly so that the feed goes unnoticed. In some areas human beings are the only mammal infected but in other areas, rodents and other animals act as the reservoir of infection.

Very rarely relapsing fever may be acquired congenitally by transplacental transfer. Laboratory infection may occur through contact with the blood of patients or experimental animals.

Laboratory Diagnosis

Because of the difficulty of culturing borrelias and the unreliability of serological tests due to antigenic variation, routine laboratory diagnosis of relapsing fever depends on demonstrating the spirochetes in peripheral blood samples, either in the living state or after staining.

1. Dark Ground Microscopy

During the pyrexial phase of the illness a drop of blood may be examined as a wet film under the dark ground or phase contrast microscope and borreliae detected by their lashing movements.

2. Giemsa or Leishman Stain

Blood smears are stained with Giemsa or Leishman stain and examined for borreliae

3. Animal Inoculation

Inoculate intraperitoneally 1-2 ml blood into six young white mice. The borreliae multiply in the animals and appear in large numbers in peripheral blood within two days. Examine drops of blood obtained by clipping the tip of the tail 48 hours later and daily thereafter for up to 1 week. Examine by darkfield microscopy for living spirochetes or after staining by Giemsa.

4. Culture and Serology

Cultivation of the borreliae and demonstration of antibodies are too difficult and unreliable to be used in diagnosis. Patients with relapsing fever sometimes develop false positive serological tests for syphilis. Agglutinins for Proteus OXK are sometimes seen in high titers in louse-borne relapsing fever.

Prophylaxis

Prevention of louse-borne relapsing fever consists of prevention of louse infestation along with the use of insecticides whenever necessary.

Prevention of tick-borne disease is less easy and consists of identification of tick infested places and their avoidance, or eradication of the vectors. No vaccine is available.

Treatment

Tetracyclines, chloramphenicol, penicillin, and erythromycin are effective.

Borrelia vincenti (*Treponema vincentii*)

T. vincentii (old name *Borrelia vincentii)* is a motile spirochete, about 5-20 µm long and 0.2-0.6 µm wide, with 3-8 coils of variable size. It is easily stained with dilute carbol fuchsin and is gram-negative.

T. vincentii is a normal commensal of mouth. It may give rise to **ulcerative gingivostomatitis or oropharyngitis (Vincent's angina)** under predisposing conditions such as malnutrition or viral infections, In Vincent's angina, *T. vincentii* is often associated with anaerobic gram negative 'fusiform bacillus known as *Fusobacterium fusiforme* now (known as *Leptotrichia buccalis).* This symbiotic infection is known as **fusospirochetosis.** Large numbers of spirochetes and fusiform bacilli may also be demonstrated in some cases of **lung abscess, phagedenous skin ulcers and gangrenous balanitis.** Their significance is uncertain. They are not primary pathogens but may cause opportunistic disease in devitalized tissues. Fusospirochetal infection of the intestine has been reported to cause choleraic diarrhea or dysentery but this needs further confirmation.

Diagnosis

Microscopic Examination

Diagnosis may be made by demonstrating spirochetes and fusiform bacilli in stained smears of exudates from the lesions.

Culture

B. vincenti may be cultivated with difficulty in enriched media anaerobically.

Fusiform bacilli also grow in the culture and it is very difficult to obtain a pure growth.

Treatment

Penicillin and metronidazole are effective in treatment.

Lyme Disease: Borrelia Burgdorferi

Lyme disease was identified in 1975. *Lyme disease* or *Lyme borreliosis* (originally *Lyme arthritis),* as it was first observed in Lyme, Connecticut, USA. The disease is widespread in the USA, where it is the most common vector-borne infection. Lyme disease has also been reported in Scandinavia, eastern Europe, China, Japan and Australia. It is caused *by Borrelia burgdorferi* transmitted by the bite of Ixodid ticks.

Morphology

It measures 4-30 µm × 0.2-0.25 µm. It is flexible, helical and gram-negative.

Culture

It is a microaerophilic spirochete. It is fastidious bacterium and can be grown in a modified Kelley's (BSK) medium, after incubation for two weeks or more. Optimum temperature for growth is 33°C.

Pathogenesis

The natural hosts for *B. burgdorferi* are wild and domesticated animals, including mice and other rodents, deer, sheep, cattle, horses and dogs. *B. burgdorferi* is transmitted to man by ixodid ticks that become infected while feeding on infected animals. **The principal vectors in the USA** are *Ixodes dammini* and *I. pacificus,* and in Europe, *I. ricinus.* The bacterium grows primarily in the midgut of the tick, and transmission to man occurs during regurgitation of the gut contents during the blood meal.

A complex of at least 10 *Borrelia* species is responsible for Lyme disease in animals and humans. Three species (i.e. *B. burgdorferi, Borrelia garinii, Borrelia afzelii)* cause human disease, each of which is prevalent in different geographical regions, causing regional variations in clinical features.

Stages of Lyme disease

Lyme disease may be a progressive illness, and is divided into three stages:

Stage 1: The first stage of 'localized infection' appears as a small red macule or papule at the site of bite *(erythema migrans* or EM) after an incubation period of 3-30 days.

Stage 2: It develops in some patients after several weeks or months. The second stage of 'disseminated infection' develops with headache, fever, myalgia and lymphadenopathy. Some develop meningeal or cardiac involvement.

Stage 3: The third stage of 'persistent infection' sets in months or years later with chronic arthritis, polyneuropathy, encephalopathy and acrodermatitis.

Congenital infection may occur with serious, potentially fatal, consequences for the fetus.

Laboratory Diagnosis

1. Isolation of the Borrelia

The borrelia has been isolated from ticks as well as from skin lesions, CSF and the blood of patients, but culture is too slow and difficult to be of use in diagnosis.

2. Serology

Serological tests such as ELISA and immunofluorescence (IF) have been described and immunoblotting recommended for confirmation. Antibodies take 1-2 months to appear.

False positive syphilis serology may be seen, with FTA-ABS being positive and VDRL test negative.

Treatment

Penicillins, the newer macrolides, cephalosporins and tetracyclines have all been used successfully in Lyme disease. Jarisch-Herxheimer reactions occur in 15 percent of patients following antibiotic therapy.

LEPTOSPIRA

Introduction

Leptospires are actively motile, delicate spirochetes, possessing a large number of closely wound spirals and characteristic hooked ends. They are too thin to be seen under the light microscope (*leptos*, meaning fine or thin). They may be visualized under dark ground illumination. They do not stain readily.

Several leptospires are saprophytic, while many are parasitic in rodents and other animals. Infection in natural hosts is generally asymptomatic but when other animals or human beings are infected, clinical disease may result.

Leptospira: The first recognized leptospiral disease of human beings was the spirochetal jaundice described by Adolf Weil of the University of Heidelber (1886). Stimson (1907) observed slender spirochetes in silver stained sections of kidneys from a fatal case of jaundice. He named the organism *Spirochaeta interrogans* from a fancied resemblance of its shape to an interrogation (question) mark. The causative agent of Weil's disease was isolated in 1915 by Inada and named *L.icterohaemorrhagiae.* In 1917, another Japanese scientist, Hideyo Noguchi proposed the genus name *Leptospira,* meaning a 'slender coil'. Subsequently, large numbers of leptospires have been isolated from human patients and animals from different parts of the world. These were given different names and considered different species of leptospires. Several saprophytic leptospires were also isolated from water, sewage and other sources.

Classification

The **family Leptospiraceae** belongs to the **order Spirochetales** and can be subdivided into three morphologically indistinguishable genera:

Leptospira, Leptonema and Tumeria. Only *Leptospira* spp. are considered to be pathogenic for animals or man.

General Characteristics of Leptospira

The genus *Leptospira* is now classified into two species *L. interrogans*, and *L. biflexa*. Pathogenic species are called *Leptospira interrogans,* and most saprophytic leptospires are called L. *bifiexa.* Within each species are **serogroups,** which are further classified into **serotypes (serovars).** *L. interrogans* included all the pathogenic strains and *L. biflexa* contained the saprophytic leptospiras. More than 200 different serovars (serotypes) of *L. interrogans* have been reported.

Leptospira interrogans

It comprises the parasitic and pathogenic leptospires. *L. interrogans* is classified into 23 serogroups **(Icterohaemorrhagiae, Canicola, Pyrogenes, Autumnalis, Australis, Pomona, Hebdomadis, Grippotyphosa, etc).** Within each serogroup over 200 serovars are recognised, for example, the serogroup Icterohaemorrhagiae contains the serovars *icterohaemorrhagiae, copenhageni, smithi,* etc).

L. biflexa

It contains saprophytic leptospires found predominantly in surface waters but cause no disease. *L. biflexa* can be subdivided into 38 serogroups containing more than 60 serovars.

Morphology

Leptospires are delicate spirochetes about 6-20 µm long and 0.1 µm thick. They have numerous closely wound primary coils, so closely set together that they are difficult to demonstrate in stained preparations although they are quite obvious in the living state by darkfield microscopy or by electron microscopy. Their ends are hooked and resemble umbrella handles (Fig. 48.4). They are actively motile, Gram-negative, but take up conventional stains poorly. They can be visualized by Giemsa or silver deposition methods or by use of fluorescent antibody.

Cultural Characteristics

They are aerobic and microaerophilic. Optimum temperature is 25-30°C and optimum pH 7.2-7.5. The generation time in laboratory media is 12-16 hours and 4-8 hours in inoculated animals. Leprospires can be grown in media enriched with rabbit serum.

Liquid medium consists of a buffered salts base with or without peptone, to which is added either rabbit serum, as in Stuart's or Korthofs medium. **Semisynthetic media,** such as EMJH (Ellinghausen, McCullough, Johnson, Harris) medium are now commonly used. A simple **semisolid medium** is Fletcher's medium which consists of nutrient agar and rabbit serum. In semisolid media, growth occurs characteristically a few millimeters below the surface. They may be difficult to see without oblique light against a dark background.

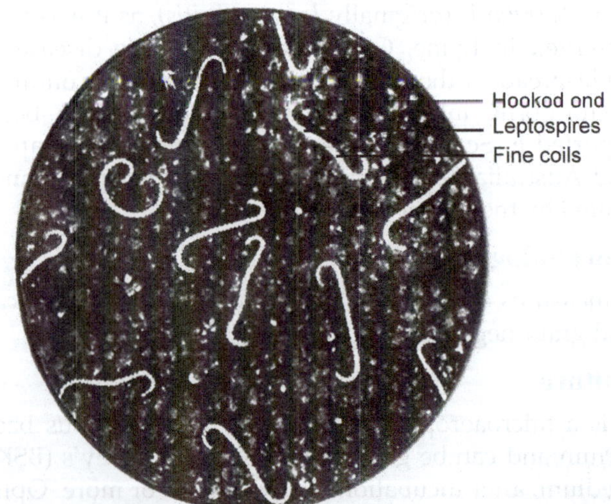

Fig. 48.4: Darkground microscopy shows the appearance of living leptospires

Leptospires may be grown on the **chorioallantoic membrane (CAM)** of chick embryos. The use of 5-fluorouracil has been recommended for the inhibition of contaminating bacteria in cultures. A simple method for obtaining cultures free of contaminants is to inoculate the material intraperitoneally in guinea pigs and culture the heart blood collected ten minutes later.

Resistance

Leptospires are susceptible to heat; 10 min at 50°C or 10 seconds at 60°C kills them. They are also sensitive to acid and are destroyed by gastric juice in 30 minutes. They are rapidly killed by bile or trypsin.

They are also readily destroyed by chlorine and most other antiseptics and disinfectants. Their survival in water or soil depends on temperature, acidity, salinity and nature and amount of pollution, dying rapidly in acid urine, nonaerated sewage, saltish or brackish water. They can survive for days in moist conditions at pH 6.8-8.

Antigenic Properties

Leptospires exhibit considerable antigenic cross reaction. A genus specific **somatic antigen** is present in all members of the genus. Classification into **serogroups** and **serotypes** (now referred to as **serovarieties** or **serovars**) is based on surface antigens.

Pathogenesis

The pathogenesis of leptospirosis is incompletely understood, but a vasculitis resulting in damage to the endothelial cells of small blood vessels is probably the main underlying pathology. In natural reservoir hosts, leptospiral infection is asymptomatic. It is transmitted to humans when the leptospires in water contaminated by the urine of carrier animals enters the body through cuts or abrasions on the skin or through intact mucosa of mouth, nose or conjuctiva. During the acute phase of the disease, leptospires are seen in the blood but can seldom be demonstrated after 8-10 days. They persist in the internal organs, and most abundantly in the kidneys, so that they may be demonstrated in the urine in the later stages of the disease.

Diseases (Table 48.4)

Mild virus-like syndrome: The incubation period is usually about 10 days (range 2-26 days). The onset of clinical illness is usually abrupt, with nonspecific, influenza-like constitutional symptoms such as fever, chills, headache, severe myalgia, and malaise.

Systemic leptospirosis with aseptic meningitis: The patient may develop a more advanced disease-including **aseptic meningitis.**

Severe systemic disease (Well's disease) includes renal failure, hepatic failure, and intravascular disease, and may result in death.

Duration of the illness varies from less than 1 week to 3 weeks. Late manifestations may be caused by the host immunologic response to the infection.

Serious cases of leptospirosis are caused most often by serotype **icterohaemorrhagiae,** though they may also be due to **colenhagen** and less often **batavitae, gripotyphosahosa, pyrogenes** and some others. **Aseptic meningitis** is common in canicola infection and abdominal symptoms in grippotyphosa infections. However, clinical syndromes are not serotype specific and any type of illness can be produced by any serotype.

Table 48.4: Important leptospiral diseases

Serotype	*Disease in humans*	*Clinical picture*	*Source of infection*	*Distribution*
1. Icterohemorrhagiae	Weil's disease	Fever, jaundice, hemorrhages aseptic meningitis	Rat urine water	Worldwide
2. Canicola	Canicola fever	Influenza like illness, aseptic-meningitis	Dog urine	Worldwide
3. Bataviae	Indonesian Weil's disease	Fever	Rat	South-east Asia, Africa, Europe
4. Grippotyphosa	Swamp or marsh fever	Fever, prostration, aseptic-meningitis	Field mice	Europe, Africa, South-east Asia, USA
5. Pyrogenes	Febrile spirochetosis	Fever	Pig	South-east Asia, Europe, USA
6. Pomona	Swineherd's disease	Fever, prostration, aseptic meningitis	Pig, cattle	America, Europe, Middle East, Indonesia, Australia
7. Hebdomadis	Seven day fever	Fever, lvrnphadenopathy	Field mice	Japan, Europe, USA
8. Fortbragg	Pretibial fever, Fort Bragg fever	Fever, rash over tibia	Not known	Japan, South-east Asia, USA
9. Hardjo	Dairy farm fever	Fever	Cattle	UK, USA, New Zealand

Laboratory Diagnosis

Diagnosis may be made by:

1. Demonstration of the leptospira microscopically in blood or urine;
2. Isolation in culture;
3. Animal inoculation;
4. Serological tests.

1. Demonstration of Leptospiras in Blood or Urine

1. Microscopy

As leptospires disappear from the blood after the first week, blood examination is helpful only in the early stages of the disease, preferably before antibiotics are given. Leptospires may be demonstrated by examination of the blood under the dark field microscope or by immunofluorescence but this is of little practical value.

Leptospires may be found in the urine during the second week and intermittently for 4-6 weeks or even longer. Examine the urine immediately after it has been voided as leptospires are sensitive to acid urine and are lysed. Centrifuged deposit of the urine may be examined under dark ground illumination.

2. Culture

Daily culture of the blood may result in isolation of the infecting strain. Three or four drops of blood are inoculated into each of several bijou bottles containing EMJH or similar medium. The bottles are incubated at 37°C for two days and left thereafter at room temperature in the dark for two weeks. Samples from the cultures are examined every third day for the presence of leptospires under dark ground illumination. Primary isolation may be delayed and may take many weeks to months. Chances of isolation are increased by culturing blood daily at the early stage of the disease. Leptospires may sometimes be isolated from the CSF also.

Direct culture of urine is seldom successful because of contamination but isolation is usually possible by inoculation into guinea pigs.

3. Animal Inoculation

The blood or urine from the patient is inoculated intraperitoneally into young guinea pigs. The animals develop fever and die within 8-12 days with jaundice and hemorrhage into the lungs and serous cavities with virulent serotypes like icterohaemorrhagiae.

With other serotypes such as canicola and pomona the animal may not become ill and infection will have to be identified by demonstration of the leptospires in the peritoneal fluid, by blood culture or by serology.

The identification of the isolates of leptospires is made by agglutination with type specific sera. In recent years techniques for identifying strains based on their genetic similarities and differences have been developed. Due to the large number of serotypes and the high degree of antigenic cross reactions between them, identification of isolates is a complicated procedure and is generally confirmed by one of the WHO/FAO Reference Laboratories.

4. Serological Diagnosis

Antibodies are usually detectable in the serum towards the end of the first week after onset. The level of antibody increases until the end of the third week and then begins to decline. Tests for detection of leptospiral antibodies in sera are of two kinds: A. Genus-specific tests; B. Sero-group specific tests.

i. Genus-Specific Tests

The broadly reactive or genus specific tests identify leptospiral infection without indicating the exact infecting serovar. The antigens for these tests are prepared from the nonpathogenic *L. biflexa* Patoc 1 strain. The tests employed include **sensitized erythrocyte lysis (SEL), complement fixation, agglutination** and **indirect immunofluorescence**. **ELISA** has been used to detect IgM and IgG antibodies separately, in order to indicate the stage of infection. A simple and rapid **dip-stick assay** has been developed for the assay of leptospira-specific IgM antibody in human sera.

ii. Serogroup-Specific Tests

Serogroup-specific tests are reactive mainly with strains of the same serogroup as the infecting strain. They comprise agglutination tests that are either **macroscopic,** or **microscopic**. In **macroscopic agglutination tests,** formalinized suspensions of prevalent leptospira serovars are tested for macroscopic agglutination with serial dilutions of the test serum.

The microscopic agglutination test (MAT) generally uses live cultures of different serotypes and agglutination is observed under the low power dark field microscope. This test is more specific and is generally accepted as **"gold standard"**. Due to the presence of some degree of cross reaction between different serovars, agglutinin absorption tests may sometimes become necessary for accurate diagnosis.

Diagnosis of Leptospiresis in Animal

Infection in rodents and other animals may be diagnosed by serological tests or by culturing pieces of kidneys.

Examination of Water for Pathogenic Leptospires

Infection takes place through the abrasions, if a shaved and scarified area of the skin of a young guinea pig is immersed in water for an hour.

Epidemiology

Leptospirosis is most common **zoonotic bacterial disease** throughout the world. Rodents are most important reservoirs. Pathogenic leptospires survive for long periods in the convoluted tubules of the kidneys in natural hosts, multiply and are shed in the urine. Humans: accidental end-stage host. People at risk are those exposed to urine-contaminated streams, rivers, and standing water; occupational exposure to infected animals for farmers, meat handlers, veterinarians.

When human beings come into contact with such water, the leptospires enter the body through abraded skin or mucosa and initiate infection. Certain occupational groups such as agricultural workers in rice or cane fields, miners and sewer cleaners are more often exposed to infection, and so leptospirosis is more common in them.

Several animals act as carriers. **Rats** are particularly important as they are ubiquitous and carry the most pathogenic serotype icterohaemorrhagiae in the convoluted tubules of the kidney long-term, resulting in chronic excretion of viable leptospires in their urine. **Field mice** carry grippotyphosa, **pigs pomona and dogs canicola serotypes**. Infection among animals is also transmitted by urinary contamination of water and fodder. Human beings are an aberrant or. 'end' host. Person-to-person spread has not been documented.

From being predominantly a rural disease of agricultural workers, leptospirosis has, in recent times also become an urban problem in the developing countries. This is perhaps due to overcrowding, insanitation, increasing rat population and the habit of walking barefooted.

Prophylaxis

General measures of prevention such as:

i. Rodent control
ii. Disinfection of water
iii. Wearing of protective clothing.

Vaccination

Mass immunization of domestic livestock will prevent clinical disease in the animals and reduce the risk of human acquisition of infection. Vaccination has been attempted with some success in dogs, cattle and pigs. Immunity following vaccination or infection is serotype specific. Vaccination has also been tried in persons at high risk such as agricultural workers.

TREATMENT

Penicillin is given intravenously (IV), 1-2 million units 6 hourly for 7 days in serious cases. For milder infections a 7-10-day course of oral amoxicillin is appropriate. Patients allergic to penicillins can be treated with erythromycin. Doxycycline 200 mg orally once a week is effective in prophylaxis.

☞ KEY POINTS

- Spirochetes are slender unicellular, motile, helical or spiral rods. They are motile because of endoflagella.
- The members of genera *Treponema, Borrelia* and *Leptospira* are pathogenic to man.

Treponema pallidum

- *T. pallidum is* thin, coiled spirochete;cannot be seen with Gram or Giemsa stains observed by darkfield or phase contrast microscopy.
- **Diseases:** Venereal syphilis *(Treponema pallidum* subspecies *pallidum).* Syphilis is a sexually transmitted disease found worldwide.

  Lab Diagnosis: Dark field microscopy is useful for demonstration of treponemes in the clinical specimen. Direct fluorescent antibody *T. pallidum* (DFA-TP) is a sensitive method for direct detection of treponemal antigen in the exudates for diagnosis of syphilis.

  Seroloical tests: Two major types of serologic tests exist: **nontreponemal tests** and **treponemal tests. In nontreponemal tests or Standard tests for syphilis (STS) cardiolipin or lipoidal antigen** is used. Nontreponemal tests are Venereal Diseases Research Laboratory (VDRL) test and rapid plasma reagin (RPR). These are *flocculation tests.*
- VDRL or RPR tests are used for screening or for diagnostic purposes of large number of sera. These tests are also more useful for the assessment of cure following treatment and are of prognostic value.
- *Biological false positive (BFP)* reactions may occur in certain conditions because of use of nonspecific antigen (cardiolipin) in STS. BFP are not caused by technical faults.
- **Treponemal tests** in which treponemes are used as the antigen. These tests use live *T. pallidum* strains (e.g., *T. pallidum* .immobilization test), killed *T. pallidum* (e.g. *T. pallidum* agglutination test, *T. pallidum* immune adherence test, and fluorescent treponema antibody test), or *T. pallidum* extracts as antigens (e.g. *Treponema pallidium* hemagglutination (TPHA test) and enzyme immunoassay (EIA)].
- *IgM-FTA-ABS* test, a modification of FTA-ABS, can detect IgM antibodies in *congenital syphilis* and helps to differentiate it from seropositivity due to passively transferred maternal antibodies.
- **Endemic syphilis or bejel** is caused by *T. pallidum* subspecies *endemicum;* **Yaws by** *T. pallidum* subspecies *pertenue and* **Pinta** caused by *Treponema carateum.*

Borrelia

- The important pathogenic borreliae of medical importance include B. *recurrentis,* B. *vincentii* and B. *burgdorferi.* B. *recurrentis* is the causative agent of *relapsing fever* while B. *vincentii* causes *vincent's angina. Lyme disease* is caused by B. *burgdorferi.*
- **Diseases: Epidemic relapsing fever:** transmitted person to person, reservoir-humans, vector-human body louse. **Endemic relapsing fever:** transmitted rodents to humans, reservoirs-rodents, small mammals, and soft ticks, vector-soft ticks.

Lyme disease: Transmitted by hard ticks from mice to humans, reservoir-mice, deer, ticks, vectors

Leptospira

- They stain poorly with aniline dyes (e.g. Gram stain) but stain well with silver impregnation methods (e.g. Levaditi's and Fontana's staining). They are obligate aerobes.
- **Diseases:** Leptospirosis is most common zoonotic bacterial disease throughout the world. Rodents are most important reservoirs. It causes-mild virus-like syndrome, systemic leptospirosis with aseptic meningitis, overwhelming disease (Weil's disease) with vascular collapse, thrombocytopenia, hemorrhage, and hepatic and renal dysfunction.

IMPORTANT QUESTIONS

1. Classify spirochetes. Name different spirochetes. and diseses caused by them. Discuss the laboratory diagnosis of syphilis.
2. Discuss pathogenesis of syphilis.
3. Write short notes on:
   Standard tests for syphilis (STS)
   VDRL test
   Repid plasma reagin (RPR) test
   Treponemal tests for serodiagnosis of syphilis
   BFP (Biological false positive) reactions.
   Fluorescent treponemal antibody-absorption (FTA-ABS) test.
   TPHA (or) T. *pallidum* hemagglutination test.
   Nonvenereal treponematoses
   Borrelia recurrentis
   Lyme disease
   Write short notes on:
   Weil's disease

FURTHER READING

Anderson JF, Magnarelli LA. Epizootiology of Lyme disease and methods of cultivating Borrelia burgdorferi. Ann NY Acad Sci 1992;653: 52-63.

Barbour AG, Hayes SF. Biology of Borrelia species, Microbiol-Rev 986;50:381-400, 1.

Cutler S, et al: Borrelia recurrentis characterization and comparison with relapsing fever, Lyme-associated, and other Borrelia spp., Int J Syst Bacteriol 1997;47:958-968.

Faine S. Leptospira and Leptospirosis, CRC Press, Boca Raton, Florida. USA 1994.

Larsen S, Steiner B, Rudolph A. Laboratory diagnosis and interpretation of tests for syphilis, Clin Microbiol Rev 8:1-21, 1995.

Norris S1. Syphilis. Immunology of Sexually Transmitted Diseases, ed. Wright DJM, Kluver Academic Publishers, Dordrecht 1988;1-31.

CHAPTER 49

Mycoplasma and Ureaplasma

Learning Objectives

After reading and studying this chapter, you should be able to:

- Describe the general characteristics of the *Mycoplasma* spp. and how they differ from other bacterial species.
- Describe morphology and characteristics of *Mycoplasma* spp.
- Compare the clinical diseases caused by *Mycoplasma pneumoniae*, *Mycoplasma hominis*, and *Ureaplasma urealyticum*.
- Analyze the diagnostic methods appropriate for detection of mycoplasmal and ureaplasmal infections.
- Discuss the use of serologic tests for diagnosing *M. pneumoniae* infections.
- Compare mycoplasmas and L forms
- Describe *Ureaplasma urealyticum*.

INTRODUCTION

Mycoplasmas are the smallest prokaryotes capable of self replication. They are found in man, animals, plants, insects, soil and sewage. The first to be recognized, *Mycoplasma mycoides* ssp. *mycoides* was isolated by Nocard and Roux in 1898 from cattle with pleuropneumonia. A similar organism was found to cause contagious agalactia in sheep. As other pathogenic and saprophytic isolates accumulated from veterinary and human sources they became known as *pleuropneumonia-like organisms* (PPLO) because of their similarity to the original isolate. This unsatisfactory name has been replaced by the term Mycoplasma. 'Mycoplasma' (Greek: *mykes* = fungus; *plasma* = something molded) refers to the filamentous (fungal-like) nature of the organisms of some species and to the plasticity of the outer membrane resulting in pleomorphism.

Mycoplasmas are distributed widely in nature, and various species cause economically important infections in cattle, goats, sheep, swine, other mammals birds and cold-blooded animals (alligators, crocodiles, tortoises) as well as man. In addition, Mycoplasmas are of concern to those who use cell cultures because of the problems of contamination.

CLASSIFICATION

Mycoplasmas are members of the class Mollicutes *(mollis,* soft; *kutis,* skin), and the order Mycoplasmatales. The term *mycoplasmas* is used rather loosely to denote any species in this class. The class Mollicutes contains five families (Fig. 49.1).

Family Mycoplasmataceae

Parasitic mycoplasmas belong to this family requiring cholesterol or other sterols as an essential growth factor. It is subdivided into two genera:

a. **Genus *Mycoplasma***—utilize glucose or arginine but do not split urea. It has more than 110 named species.
b. **Genus *Ureaplasma***—hydrolyze urea. It has six species.

Family Acholeplasmataceae

It does not depend on sterol for growth and is mostly saprophytic mycoplasmas. It contains a single genus,

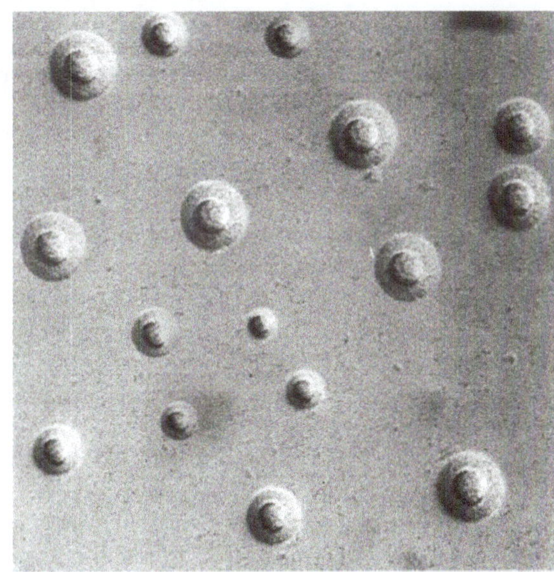

Fig. 49.1: Colonies of *M. hominis* with typical 'fried egg' appearance.

Table 49.1: Mycoplasmas of humans

Organism	Site	Disease
M. pneumoniae	Respiratory tract	Upper respiratory tract disease, tracheobronchitis, atypical pneumonia
Mycoplasma hominis	Genitourinary tract	Pyelonephritis, pelvic inflammatory disease, post partum fever
Mycoplasma genitalium	Genitourinary tract	Urethritis
Mycoplasma alivarium, Mycoplasma buccale, Mycoplasma faucium, Mycoplasma pophilum	Respiratory tract	Unknown
Mycoplasma primatum	Respiratory and genitourinary tracts	Unknown
Mycoplasma fermentans	Genitourinary tract	Unknown
Mycoplasma fermentans var. incognitus	Blood	Fulminant disseminated disease in patients with acquired immunodeficiency syndrome (AIDS)
Mycoplasma pirum	Blood	Septicemia in patients with AIDS
Ureaplasma urealyticum	Genitourinary tract	Urethritis

Acholeplasma, comprising at least 14 species, of which *Acholeplasma laidlawii* was the first to be named.

Family Spiroplasmataceae

They are sterol dependent. It contains the genus *Spiroplasma,* which parasitize arthropods and plants.

Family Anaeroplasmataceae

Genus *Anaeroplasma,* which contains strict anaerobes found in the rumen of cattle and sheep.

Family Entomoplasmataceae

It contains the genera *Entomoplasma* and *Mesoplasma* found in insects and plants.

More than 60 species of mycoplasma are known to cause disease in a variety of mammalian, insect and plant hosts. About sixteen species, belonging to three families are found in human beings (Table 49.1).

General Characteristics

The unique property that characterizes these simple forms and separates them from the true bacteria are:

1. *Mycoplasma* and *Ureaplasma.* organisms are the smallest free-living bacteria.
2. Mycoplasmas differ from other bacteria in that they lack a rigid cell wall and due to absence of a cell wall are highly pleomorphic, with no fixed shape or size. Because of the absence of a cell wall, they were originally grouped under the general term **cell wall deficient** bacteria.
3. These organisms are bounded by a soft trilaminar unit membrane containing sterols.
4. They lack even cell wall precursors like muramic acid or diaminopimelic acid.
5. The mycoplasmas form pleomorphic filaments with an average diameter of 0.1 to 0.3 µm, and many can pass through the 0.45 µm, filters used to remove bacteria from solutions.
6. The organisms divide by binary fission (typical of all bacteria), grow on artificial cell-free media, and contain both RNA and DNA.
7. Mycoplasmas are generally slow-growing, highly fastidious, facultative anaerobes requiring complex media containing cholesterol and fatty acids for growth.
8. Mycoplasma species grow embedded beneath the surface of solid media, therefore transferring colonies with a loop is ineffective. On solid media, some species (e.g., *M. hominis)* form colonies with slightly raised centers giving the classic **"fried egg" appearance".**
9. The mycoplasmas adhere to the epithelium of mucosal surfaces in the respiratory and urogenital tracts and are not eliminated by mucous secretions or urine flow.
10. The human mycoplasmas are susceptible to adverse environmental conditions such as heat and drying.
11. Transmission of mycoplasmas and ureaplasmas in humans may occur via direct sexual contact, from mother to child during delivery or *in utero,* and by respiratory secretions or fomites in cases of *M. pneumoniae* infections.

Morphology

Mycoplasmas are the smallest free living microorganisms. They can pass through bacterial filters. They lack cell wall but are bounded by a trilaminar membrane, 8 to 10 nm thick, which is rich in cholesterol and other lipids. They are very small pleomorphic cells and range in size from 0.2 to 0.8 µm in diameter which may range from spherical through coccoid, coccobacillary, ring and dumb-bell forms, to short and long branching, beaded or segmented filaments. The filaments are slender, of varying leng ths and show true branching. Mycoplasmas are gram-negative but are better stained by Giemsa stain.

Replication is basically by binary fission, but genome replication is not necessarily synchronized with cell division and accounts for the budding forms and chains of beads often observed. A distinctive feature seen in some species is a bulbous enlargement, with a differentiated tip structure, by means of which the organisms get attached to suitable host cells carrying neuraminic acid receptor. They may be responsible for the hemadsorption shown by some species.

Mycoplasmas do not possess spores, flagella or fimbria. Some mycoplasmas, including *M. pneumoniae,* exhibit a gliding motility on liquid-covered surfaces.

Cultural Characteristics

Most mycoplasmas are facultatively anaerobic but, since organisms from primary tissue specimens frequently grow only under anaerobic conditions, an atmosphere of 95 percent nitrogen and 5 percent carbon dioxide is preferred for primary isolation. They grow within a temperature range of 22-41°C, the parasitic species growing optimally at 35-37 °C and the saprophytes at lower temperatures.

Media for cultivating mycoplasma are enriched with 20 percent horse or human serum and yeast extract. The high concentration of serum is necessary as a source of cholesterol and other lipids. Mycoplasmas may be cultivated in liquid or solid media. A medium widely used for the isolation of mycoplasmas consists of bovine heart infusion broth pleuropneumonia like organisms broth (PPLO broth) to which are added 20 percent horse serum and 10 percent fresh yeast extract along with glucose and phenol red as a pH indicator. This medium can be solidified by the addition of agar. Penicillin, ampicillin and polymyxin B may be added in the medium to inhibit contaminating bacteria and amphotericin B to inhibit fungi. A diphasic medium in screw-capped bottle containing an agar phase that is overlain with broth medium of similar composition may also be used. Colonies appear after incubation for 2-6 days and are about 10-600 µm in size. On agar, colonies are typically biphasic that have a 'fried egg' appearance, with an opaque central zone of growth within the agar and a translucent peripheral zone on the surface (Fig. 49.1).

Colonies may be seen with a hand lens but are best studied after staining by Dienes method. For this, a block of agar containing the colony is cut and placed on a slide. It is covered with a cover slip on which has been dried an alcoholic solution of methylene blue and azure. Colonies cannot be picked with platinum loops. Subculture is done by cutting out an agar block with colonies and rubbing it on fresh plates. Most mycoplasma colonies are hemolytic.

Biochemical Reactions

1. Mycoplasmas are mainly fermentative. Most mycoplasmas use glucose (or other carbohydrates) or arginine as a major source of energy.
2. Urea is not hydrolyzed, except by ureaplasmas.
3. They are generally not proteolytic.

Antigenic Structure

The surface antigens presented to the infected host are components of the cell membrane and consist of glycolipids and proteins.

Glycolipid Antigen

The glycolipid antigens of *M. pneumoniae* are major membrane antigens found in the lipid fraction. The purified glycolipids act as antigens in the complement-fixation test and other *in vitro* antigen-antibody reactions but are poor immunogens *in vivo* unless attached to proteins. Glycolipid antigens are identified by complement fixation.

Protein Antigens

During the course of infection, a number of protein antigens are recognized. Among them are two major surface antigens, one of which is the P1 protein that mediates attachment. Antibody to these surface proteins is found consistently in convalescent human sera and in respiratory secretions by radioimmunoprecipitation, gel electrophoresis, and Western blot analysis.

The cold agglutinins that are induced in patients with *M. pneumoniae* infection are directed against the I antigen determinant of erythrocytes, which is also present on the surface of other blood cells. Shared idiotypic antigens have been demonstrated on cold agglutinins from patients with *M. pneumoniae* infection.

Resistance

Mycoplasmas are killed by heating at 56°C for 30 minutes. For long term preservation, lyophilization or freezing broth cultures at–70°C (or in liquid nitrogen) are the preferred storage methods.

Mycoplasmas are believed to be resistant to ultraviolet light and the photodynamic action of methylene blue. *M. pneumoniae* can grow in the presence of 0.002 percent methylene blue in agar, while many other species are inhibited. Antiseptic solutions such as chlorhexidine and cetrimide inhibit the growth of mycoplasmas. They are relatively resistant to lysis by osmotic shock but are very sensitive to lysis by surface active agents and lipolytic agents such as taurocholate and digitonin.

They are resistant to penicillin and cephalosporin as well as to lysozymes that act on the bacterial cell walls but are sensitive to tetracycline and many other antibiotics that inhibit protein synthesis.

Pathogenicity

Parasitic mycoplasmas exhibit host specificity. They generally produce surface infections by adhering to the mucosa of the respiratory, gastroin testinal and genitourinary tracts. Mycoplasma cause two types of diseases in humans—pneumonia and genital infections.

Mycoplasma pneumoniae

Infection with *M. pneumoniae* typically produces mild **upper respiratory tract disease.** More severe disease with **lower respiratory tract symptoms** occurs in less than 10 percent of patients. **Tracheobronchitis** can occur. **Pneumonia (referred to as primary atypical pneumonia or walking pneumonia)** can also develop. School-age children and young adults are especially susceptible to infection. Clinical disease is uncommon in very young children and older adults. Other groups at risk include closed-in populations such as prisoners, college students and military personnel. Epidemics are known to occur in these populations. Infection is not considered seasonal, but many cases occur in autumn and early winter. Transmission is probably through aerosol droplet spray produced while coughing.

Pneumonias not attributable to any of the common bacterial causes were labelled historically as **primary atypical pneumonias.** Eaton (944) was the first to isolate the causative agent of the disease and because it was filterable, it was considered to be a virus **(Eaton agent).** The most important species is *Mycoplasma pneumoniae* (also called Eaton's agent after the investigator who originally isolated it).

The disease has an incubation period of 1-3 weeks, and early symptoms are nonspecific, consisting of headache, low-grade fever, malaise, and anorexia. Sore throat, dry cough, and earache are accompanying symptoms. However, patients usually do not appear seriously ill and few warrant admission to hospital.

The disease is characterized by paucity of respiratory signs on physical examination but radiological evidence of consolidation, which is usually patchy involving one of the lower lobes, starting at the hilum, and fanning out to the periphery. About 20 percent of patients suffer bilateral pneumonia, but pleurisy and pleural effusions are unusual. The disease is usually self-limited, recovery occurring in 1 2 weeks, but can be prolonged.

Extrapulmonary complications—including cardiovascular, central nervous system, dermatologic, and gastrointestinal problems, are rare occurrences. Secondary complications include otitis media, erythema multiforme (Stevens-Johnson syndrome), hemolytic anemia, myocarditis, pericarditis, and neurologic abnormalities. The disease resolves slowly. Secondary infections can occur because immunity is incomplete.

Ureaplasma urealyticum

Some strains of mycoplasma frequently isolated from the urogenital tract of human beings and animals form very tiny colonies, generally 15-50 μm in size. They were called **T strain or T form mycoplasmas (T for tiny)** because of the small colonies they produce. They are peculiar in their ability to **hydrolyze urea,** with the production of ammonia, which is an essential growth factor in addition to cholesterol. Human T strain mycoplasmas have been reclassified as ***Ureaplasma urealyticum.***

Clinical Manifestations

1. *U. urealyticum* may cause nonchlamydial, non-gonococcal urethritis (NGU), epididymitis, vaginitis and cervicitis.
2. They may cause chorioamnionitis, prematurity, postpartum endometritis, chronic lung disease of the premature infant and infection of wounds and soft tissues.
3. Ureaplasmas are the commonest organisms isolated from the CNS or lower respiratory tract of sick premature or newborn infants.
4. Ureaplasmas have also been blamed to cause male and female infertility and low birth weight but there are conflicting reports.

Mycoplasma hominis

M. hominis is found in the lower genitourinary tracts of approximately 50 percent of healthy adults and has not been reported as a cause of **nongonococcal urethritis (NGU).** The organism may, however, invade the upper genitourinary tract and cause salpingitis, pyelonephritis, pelvic inflammatory disease (PID), or postpartum fevers.

Mycoplasma genitalium

M. genitalium has been associated with some cases of non-gonococcal urethritis and pelvic inflammatory disease.

Urogenital Infections

M. fermentans, M. genitalium, M. hominis, M. penetrans, M. pneumoniae, M. primatum, M. salivarium, M. spermatophilum and ureaplasmas have been isolated from the urogenital tract. *M. hominis, M. genitalium* and ureaplasmas occur most frequently.

Genital infections are caused by *M. hominis* and *U. urealyticum.* They are transmitted by sexual contact, and may cause urethritis, proctitis, balanoposthitis and Reiter's syndrome in men, and acute salpingitis, pelvic inflammatory disease, cervicitis and vaginitis in women. They have also been associated with infertility, abortion, postpartum fever, chorioamnionitis and low birth weight of infants.

Epidemiology

M. pneumoniae is worldwide in its distribution and is found at all ages. Transmission is by droplets of nasopharyngeal secretion. The infection is endemic in the population throughout the year, with epidemics occurring at 4 to 6-year intervals. Disease is most common in schoolaged children and young adults (5 to 15 years). Close contact is necessary for transmission, and the infection usually occurs among classmates or family members. Spread is favoured by close contact, as in families and most typically among military recruits. The mycoplasma may remain in the throat for two or more months after recovery from the disease.

Laboratory Diagnosis

Laboratory diagnosis of mycoplasma and ureaplasma infections may be established either by isolation of the mycoplasma or by serological methods.

1. Specimens

M. pneumoniae may be recovered from throat swabs, nasopharyngeal swabs, sputum, throat washings, bronchoalveolar lavage, tracheal aspirate and lung tissue specimens. **Genital mycoplasmas** may be isolated from urethral, vaginal and cervical swabs, semen, prostatic secretions, urine, blood, CSF, amniotic fluid, respiratory tract secretions, synovial fluid, pericardial fluid and biopsy specimens from endometrium, fallopian tubes, placenta and aborted fetus.

2. Culture

In the laboratory, if inoculation is not possible immediately, then the specimen may be held up to 24 hours at 4°C. If delay more than 24 hours is expected, then the specimen should be frozen at –70°C. A widely used isolation medium contains bovine heart infusion (PPLO broth) with fresh yeast extract and horse serum supplemented with penicillin (to inhibit other bacteria), thallium acetate, glucose and with phenol red as a pH indicator because mycoplasmas do not produce turbidity in broth media. For genital mycoplasma, polymymin B and amphotericin B and licomycin are also added to mycoplasmal broth.

Another medium is **SP-4 broth** (sucrose phosphate buffer) and SP4 agar containing glucose for isolation of spiroplasmas, *M. pneumoniae* and other human mycoplasmas (e.g. *M. genitalium)*. The *Mycotrim RS* is the only commercially available systems that have been used extensively for the cultivation of *M. pneumoniae*.

3. Isolation and Identification

In general, the growth of *M. pneumoniae* from clinical specimens is detected by the ability of these organisms to produce acid from glucose. Most broth media are diphasic such as methylene blue-glucose diphasic medium. Broth cultures are incubated at 35°C with the caps tightened. Tubes are inspected daily for color changes (from salmon to yellow) in the medium. A slight, gradual shift in the pH indicator over an 8 to 15 day period without gross turbidity suggests a true-positive culture. The broth must be subcultured to appropriate agar medium as soon as color changes in the medium are apparent. Inspection of the agar surface under the low power of the microscope will reveal small colonies of the organisms. In the absence of obvious color change in diphasic media, a blind subculture to agar media should be performed after 1 and 3 weeks of incubation.

i. Colonies

M. pneumoniae may take 21 days or more, while *M. hominis* and *Ureaplasma* spp. colonies may appear within 2 to 4 days. Colonies of *M. pneumoniae* are small, betahemolytic and have a homogeneous granular appearance **("mulberry shaped")**, (Fig. 49.2) unlike the fried-egg morphology of other mycoplasmas. Because the organisms do not stain well with Gram or acridine orange stains, mycoplasma-like colonies are stained with the Dienes or methylene blue stains. *M. hominis* colonies have a typical large **"fried egg" appearance.** *Ureaplasma* spp. form extremely small colonies (15-60 µm in diameter) that are difficult to see with the naked eye; hence, mycoplasmal cultures on solid media should always be examined with a stereoscopic microscope.

ii. Species Identification

Procedures for species identification include hemadsorption with guinea pig erythrocytes, reduction of tetrazolium, agar growth inhibition with homologous antisera, epi-immunofluorescence or immunoperoxidase staining, immunoblotting with monoclonal antibodies, metabolism inhibition tests, and PCR assays. These colonies may be identified by:

a. Hemadsorption Test

The characteristic of guinea pig red blood cells (0.4% in saline) adhering to colonies of *M. pneumoniae* and not *M. hominis* is another standard assay that helps distinguish the two species.

b. Tetrazolium Reduction Test

Colonies of *M. pneumoniae* appear red when these are flooded with solution of tetrazolium compound which is colorless. *M. pneumoniae* reduces tetrazolium (colorless) to red colored compound.

c. Serological Techniques

Identification of isolates can be confirmed by inhibition of their growth with specific antisera. This includes inhibition of colony development around disks impregnated with specific antiserum or fluorescent-labeled anti-*M. pneumoniae* antibody is flooded on colonies on the plate; the plate is, then washed and examined for immunofluorescence.

d. Polymerase Chain Reaction (PCR) Amplification

Several groups have shown that it is possible to detect *M. pneumoniae* in respiratory exudates or secretions by PCR amplification of a chosen sequence in its genome. DNA primers specific for *M. fermentans. M. genitalium* and *U urealyticum* have been developed and used for amplification by the PCR.

4. Antigen Detection Techniques

Detection of antigen in respiratory exudates by direct immunofluorescence and counterimmunoelectropho-

Fig. 49.2: Typical large *Mycoplasma pneumoniae* colony showing **'mullberry shaped'.** (Dr Surinder Kumar, Director Professor, Mycoplasma Research Laboratory, Department of Microbiology, Maulana Azad Medical College, New Delhi-110002, INDIA)

resis techniques, immunoblotting with monoclonal antibodies and antigen capture enzyme immunoassay (EIA).

5. DNA Probes

Detection of RNA nucleotide sequences by probe hybridization in respiratory-exudate. *M pnemoniae* DNA probe hybridizes with the 16S rRNA of the organism and uses an ^{125}I radioactive label to generate a detection signal. Probe has a sensitivity and specificity of more than 90 percent and results are available in about 2 hours while culture requires several weeks.

6. Serological Tests

Serologic tests are available only for *M. pneumoniae*. Serological diagnosis may be made by:

A. Specific tests using mycoplasmal antigens
B. Nonspecific methods.

A. Specific Tests Using Mycoplasmal Antigens

The development of antibody to *M. pneumoniae* by infected subjects may be measured by a range of techniques of widely differing sensitivity, viz. complement fixation, metabolic inhibition, inhibition of tetrazolium reduction, immunofluorescence on sections of chick embryo lung, direct or antibody capture enzyme-immunosorbent assays (EIA), or agglutination of antigen-coated erythrocytes, latex or gelatin particles.

i. Complement Fixation Test

Detection of antibodies directed against *M. pneumoniae* by complement fixation is a useful but technically cumbersome test. Titers peak in 4 weeks and persist for 6-12 months; diagnostic titer is ≥ 1: 32, or a fourfold increase; has good sensitivity but can be nonspecific. Because the antibodies are directed against outer membrane glycolipids common to other organisms and tissues, false-positive reactions are observed.

ii. Enzyme-linked Immunosorbent Assays (ELISA)

Alternative antibody-directed tests, such as enzyme-linked immunosorbent assays and immunofluorescence, have been developed and are now used more commonly than the complement fixation test. The tests are easier to perform and are highly sensitive if both immunoglobulin M (IgM) and IgG are measured. Nevertheless, problems with test specificity persist.

B. Nonspecific Serological Tests

The nonspecific serological tests are *Streptococcus MG* and cold agglutination tests. About half of the patients infected with *M. pneumoniae* develop cold hemagglutinins to their own or group 'O' erythrocytes and a smaller proportion develops agglutinins to *Streptococcus* MG. These heterogenetic reactions probably depend on fortuitous similarities between glycolipid haptens in the mycoplasma membrane and carbohydrate determinants of the streptococcal cell or in the 'I' antigen of the erythrocyte. As cold hemagglutinins develop rapidly, it may be of value to estimate them as a bedside test, as well as subsequently testing for complement fixing or other antibody to *M. pneumoniae.*

i. Streptococcus MG Test

It is done by mixing serial dilutions of the patient's unheated serum and a heat killed suspension of *Streptococcus MG,* and observing agglutination after overnight incubation at 37°C. A titer of 1:20 or over is considered suggestive.

ii. Cold Agglutination Test

The cold agglutination test is based on the appearance in a high proportion of cases with primary atypical pneumonia, of macroglobulin antibodies that agglutinate human group O cells at low temperature. Cold agglutinins appear about one week after infection with a peak at 4-5 weeks. Thereafter, titer declines rapidly and the test becomes negative after about 5 months.

This test is easily performed by mixing serial dilutions of the patient's serum with an equal volume of a 0.2 percent washed human O group erythrocytes and clumping observed after incubating the mixture overnight. The test is based on the development of hemagglutination, which can be reversed by placing the tubes at 37ºC. A titer of 1:32 or over is suggestive but demonstration of rise in titer in paired serum samples is more reliable. Because paired sera are not always available, a single antibody titer of 64-128 or more with either test, in a suggestive clinical setting, should be sufficient to institute therapy.

Cold agglutinins are occasionally induced in other diseases such as infectious mononucleosis, rubella, adenovirus infections, psittacosis, tropical eosinophilia, trypanosomiasis, cirrhosis of liver, paroxysmal hemoglobinuria and hemolytic anemia.

Treatment

Mycoplasmas and ureaplasmas are resistant to penicillins and cephalosporins which act on the cell wall and are sensitive to tetracyclines and erythromycin which inhibit protein synthesis. Therefore, tetracyclines and erythromycin are drugs of choice for the treatment of *Mycoplasma* and *Ureaplasma* infections.

The genital mycoplasmas are more variable in their antimicrobial susceptibility patterns and their resistance to both these antibiotics is fairly common. Patients with NGU should receive one of the tetracyclines and ureaplasmas resistant to tetracyclines, patients should then be treated with erythromycin, to which most tetracycline-resistant ureaplasmas are sensitive.

Prophylaxis

The best prevention is to avoid contact with person suspected of having the disease. Tetracyclines or erythromycin may be administered prophylactically to prevent the spread of infection to the contacts. At present, no vaccine against *Mycoplasma* and *Ureaplasma* is available.

Infections in Immunocompromised Patients

M. pneumoniae may cause severe pneumonia in immunodeficient patients and may persist for many months in the respiratory tract of hypogammaglobulinemic patients, despite apparently adequate treatment. A few of these patients develop suppurative arthritis, and mycoplasmas are responsible for at least two-fifths of the cases. The mycoplasmas involved have been *M. pneumoniae, M. salivarium* (usually regarded as nonpathogenic), *M. hominis* and, particularly, ureaplasmas. In some cases involving ureaplasmas, the arthritis has been associated with subcutaneous abscesses, persistent urethritis and chronic cystitis.

Mycoplasma and HIV (Human Immunodeficiency virus) Infection

Mycoplasmas tend to cause more severe and prolonged infections in the HIV infected and other immunodeficient subjects. A synergetic effect had been proposed between HIV and mycoplasma (particularly *M. fermentans).*

MYCOPLASMA AS CELL CULTURE CONTAMINANTS

Few primary cell cultures become infected with mycoplasmas, but continuous cell lines do so frequently. The contamination may originate from the worker or from animal sera or trypsin used in cell culture. The mycoplasmas most responsible are *M. arginini. M. fermentans, M. hyorhinis, M. orale, M. salivarium* and *A. laidlawii.*

Contamination generally does not produce cytopathic effects but may interfere with the growth of viruses in such cell cultures and may also produce misleading results in serological tests. Mycoplasmas growing in cell cultures have often been mistaken for viruses. Eradication of mycoplasmas from infected cells is difficult.

Detection of Mycoplasma Contamination

Culture and an indicator cell system with staining (e.g. Hoechst DNA dye) are used to detect contamination, but these may be superseded by PCR methods.

Procedures to Eliminate Mycoplasmas From Cell Cultures

Numerous procedures have been described to eliminate mycoplasmas from cell cultures, but none is consistently successful.

1. Discard the Cultures

Whenever possible it is easier to discard the cultures, replace them with mycoplasma free cells and adhere to simple guidelines to prevent contamination.

2. Treatment with an Antibiotic

If it is imperative to save cells, treatment with an antibiotic that is likely to have mycoplasmacidal activity, such as a fluoroquinolone, identification of the contaminant and use of a specific antiserum, or a combination of these methods is most likely to be successful.

MYCOPLASMAS AND L FORMS OF BACTERIA

Kleineberger (1935) found pleuropneumonia-like forms in a culture *Streptobacillus moniliformis* and termed them L forms, after Lister Institute, London, where the observation was made. It was subsequently shown that many bacteria, either spontaneously or induced by certain substances like penicillin, lost part or all of their cell wall and develop into L forms. Such L forms may be 'unstable', when they revert to their normal morphology, or 'stable' when they continue in the cell wall deficient state permanently.

Cell wall deficient forms (L forms, protoplasts, spheroplasts) may not initiate disease but may be important in bacterial persistence during antibiotic therapy and subsequent recurrence of the infection.

Differences between L-forms and Mycoplasma

It has been suggested that mycoplasmas may represent stable L forms of bacteria but genetic, antigenic and biochemical evidence are against the possibility. L-forms also produce 'fried egg' colonies like mycoplasma, but they differ in the following respects:

1. L-forms are not filterable.
2. Do not require sterol for growth.
3. The remnants of cell wall components can be demonstrated in L-forms though they lack cell walls.
4. The stable L-forms continue in the cell wall deficient state permanently but resemble parent bacteria both biochemically and antigenically. Despite

some colonial similarities, mycoplasmas are quite distinct from L-phase variants of bacteria and do not revert to bacteria when cultured in media free of inhibitors of bacterial cell wall synthesis or other L-phase inducers. Unstable L-forms revert to their normal morphology.
5. L-forms play an important role in persistence of chronic infection during antibiotic therapy and subsequent recurrence of the infection but may not initiate disease.
6. Nonpathogenic to laboratory animals.

As agglutinins to *Streptococcus* MG are frequently detected following infection with *M. pneumoniae*, it is thought that the latter is an L-form for the former, but several investigations of postulated L-phase-bacterial relationships (e.g. *Streptococcus* MG and *M. pneumoniae)* have failed to show the same G+C ratio or sequence homology between the genomes of the pairs of organisms.

ATYPICAL PNEUMONIA

Atypical pneumonia was defined in the 1930s as lower respiratory infection that did not resemble classic lesions that had been described. Around the turn of century, any patient who presented with a sudden onset of fever with shaking chills, pleuritic pain and the production of rust-coloured sputum was thought to have typical pneumonia attributable to *Streptococcus pneumoniae.* All other patients who did not show this characteristic picture were referred to as patients with **atypical pneumonia** or **walking pneumonia.** The major difference is that sputum production is minimal in atypical pneumonia.Infection is often milder than in classic pneumonia, but not necessarily so. The primary pathogen responsible for atypical pneumonia are *Mycoplasma pneumoniae, Chlamydophila pneumoniae* and *Legionella pneumophila.*

Atypical pathogens do not respond to β-lactam antibiotics and their presence is often overlooked until patients fail to respond to standard penicillin or cephalosporin therapy. β-lactam antibiotics act by inhibiting cell wall synthesis. This makes them ineffective against *Mycoplasma*. Neither penicillin nor cephalosporins have good intracellular penetration. Hence they are also ineffective against *Chlamydia* and *Legionella.*

KNOW MORE

The term 'mycoplasma(s)' is used often, as here, in a trivial fashion to refer to any member of the class Mollicutes, irrespective of whether they belong to the genus *Mycoplasma,* although the term 'mollicute(s)' is also used.

☞ KEY POINTS

- Mycoplasmas are the smallest free-living bacterium; able to pass through 0.45 µm pore filters. They are gram-negative, but are better stained by Giemsa stain. They were previously named as *pleuropneumonia like organisms* or PPLO.
- Absence of cell wall and a cell membrane containing sterols are unique among bacteria. Due to lack of rigid cell wall, they are extremely pleomorphic. Three-layer (trilaminar) cell membrane contains sterols, strict aerobe.
- The members of the genus *Ureaplasma* can hydrolyze urea while other genera cannot do so. All mycoplasmas except *Acholeplasma,* require sterol for their growth.
- A medium widely used for the isolation of mycoplasmas is PPLO broth. This medium can be made solid by the addition of agar. On agar, colonies are typically biphasic that have a 'fried egg' appearance.
- ***Mycoplasma pneumoniae*** Primarily infects children aged 5 to 15 years, but all populations susceptible to disease. Transmitted by inhalation of aerosolized droplet. Strict human pathogen.
- **Diseases:** Upper respiratory infections; lower respiratory infections, including tracheobronchitis and **pneumonia (referred to as primary atypical pneumonia or walking pneumonia).**
- ***Mycoplasma hominis*** causes *non-gonococcal urethritis (NGU).* This has also been incriminated in postpartum sepsis, proctitis, acute salpingitis, pelvic inflammatory disease, cervicitis and vaginitis. It is transmitted by *sexual contact.*

Ureaplasma urealyticum

- They were called **T strain** or **T form mycoplasmas (T for tiny)** because of the small colonies, peculiar in their ability to **hydrolyze urea.** Human T strain mycoplasmas have been reclassified as ***Ureaplasma urealyticum.***
- *U. urealyticum* may cause non-gonococcal urethritis (NGU), epididymitis, vaginitis and cervicitis, chorioamnionitis, prematurity, postpartum endometritis, chronic lung disease of the premature infant and infection of wounds and soft tissues.
- Ureaplasmas have also been blamed to cause male and female infertility and low birth weight but there are conflicting reports.
- **Laboratory diagnosis:** Laboratory diagnosis of mycoplasmal infections may be carried out by isolation of the organism or by serological tests.

  Culture: Test is slow (2-6 weeks before positive) and insensitive.

  Serology: Non-specific test—*Streptococcus* MG and *cold agglutination tests* are non-specific tests used for detection of antibody in serum. *Complement fixation test* and *enzyme immunoassay (EIA)* are specific tests to detect antibody against *M. pneumoniae.*
- **Treatment, Control, and Prevention:** Drug of choice is erythromycin or tetracycline (not used in young children). Immunity to reinfection is not life-

long, and vaccines have proved ineffective. Mycoplasmas tend to cause more severe and prolonged infections in the HIV infected and other immunodeficient subjects.

IMPORTANT QUESTIONS

1. Classify mycoplasmas. Describe laboratory diagnosis of mycoplasma infections.
2. Write short notes on:
   a. Morphology and general characters of *Mycoplasma* and *Ureaplasma.*
   b. *Mycoplasma pneumoniae* or PPLO or Eaton's agent.
   c. Streptococcus MG agglutination test.
   d. Cold agglutinin test for *Mycoplasma* infection.
   e. *Ureaplasma urealyticum.*
   f. Atypical pneumonia.

FURTHER READING

Kumar S, Maria A, Saigal SR, Maheshwari M. *Mycoplasma pneumoniae* as a cause of non-resolving pneumonia in a neonate. J Med Microbiol 2010;59:731-732.

Kumar S, Saigal SR, Sethi GR. Rapid diagnosis of *Mycoplasma pneumoniae* by polymerase chain reaction in community-acquired lower respiratory tract infections. Tropical Doctor 2011;41:160-162.

Kashyap B, Kumar S, Sethi GS, Das BC and Saigal SR. Comparison of PCR, culture and serological tests for the diagnosis of *Mycoplasma pneumoniae* in community-acquired lower respiratory tract infections in children. Indian J of Med Research 2008:128;134-139.

Maheshwari M, Kumar S, Sethi GR, Bhalla P. Detection of *Mycoplasma pneumoniae* in children with lower respiratory tract infections. Tropical Doctor 2011;41:40-42.

Taylor-Robinson D Mycoplasmas and their role in human respiratory tract disease. In: Myint S, Taylor-Robinson D (eds). Viral ond Other Infections of the Human Respiratory Tract. Chapman and Hall, London 1996;319.

Taylor-Robinson D. Infection due to species of Mycoplasma and Ureaplasma: an update. Clinical Infectious Diseases1996;23:671-684.

Taylor-Robinson D and Bradbury J. Mycoplasma diseases, In. Topley and Wilson's Microbiology and Microbial Infections, 9th edn. London: Arnold 1998.

Waites KB and Talkington DF. Mycoplasma pneumoniae and its role as a human pathogen. Clin Microbiol Reviews 2004;17(4);697-728.

Waites KB. Taylor-Robinson D. Mycoplasma and Ureaplasma. In: Murray PR, Baron EJ 1999.

CHAPTER 50

Miscellaneous Bacteria

LEARNING OBJECTIVES

After reading and studying this chapter, you should be able to:

- Describe morphology and culture characrestics of Listeria.
- Describe infections caused by Listeria monocytogenes and their laboratory diagnosis.
- Discuss rat bite fever.
- Describe Klebsiella granulomatis and disease caused by it.
- Acinetobacter spp.

LISTERIA MONOCYTOGENES

Organisms of the genus *Listeria* are nonsporing gram-positive bacilli. The genus contains eight species but almost all cases of human listeriosis are caused by *L. monocytogenes.* The disease chiefly affects pregnant women, unborn or newly delivered infants, the immunosuppressed and elderly. It is predominantly transmitted by the consumption of contaminated food.

L. ivanovii is associated with about 10 percent of infections in livestock animals. This species and *L. seeligeri* have been associated with a very small number of human infections.

Morphology

Listeria monocytogenes is a small, coccoid, gram-positive bacillus measuring approximately 0.5 × 2-3 µm. They occur singly or in pairs which are often angled at the point of contact and may resemble diphtheroids or diplococci. It exhibits a characteristic, slow, tumbling motility when grown at 25°C but at 37°C is nonmotile. This is because peritrichous flagella are produced by the bacillus optimally at 20 to 30°C but only scantily or not at all at 37°C. They are noncapsulate, nonsporing and nonacid-fast.

Cultural Characters

Listeriae are aerobes and facultative anaerobes.They can grow over a temperature range of 2 to 43°C, the optimum temperature for the growth is 35 to 37°C. They can grow on ordinary media containing fermentable carbohydrate but growth is better on blood agar or tryptose phosphate agar. After 24 hours incubation at 37°C, colonies are 0.5 to 1.5 mm in diameter, smooth, translucent and emulsifiable and nonpigmented.

On blood agar, *L. monocytogenes* develops zones of slightly hazy β-hemolysis. The colonial appearances may be indistinguishable from those of group B streptococci, although listeriae are never pigmented.

Biochemical Reactions

L monocytogenes ferments glucose, maltose, L rhamnose and alpha methyl D-mannoside, producing acid without gas. It is catalase positive. It grows in the presence of 0.1 percent potassium tellurite, 10 percent salt and at pH 9.6.

Various Species

On the basis of a few tests, the genus Listeria can be divided into 8 species (Table 50.1). Many serovars have been recognized. Two major tests of differentiation of various species include D-xylose fermentation and CAMP test with *Staph. aureus* and *Rhodococcus equi. L. monocytogenes* gives a positive CAMP test with *Staph. aureus.*

Epidemiology

L monocytogenes is widely distributed in nature. It has been isolated from a wide range of mammals, birds, fish, ticks and crustacea. It occurs as a saprophyte in soil, water and sewage. Asymptomatic carriage as well as disease is well-documented in humans and other mammals. Disease is associated with consumption of contaminated food products or transplancental spread from mother to neonate. Sporadic cases and epidemics occur throughout the year but peak in warmer months. The young and elderly as well as patients with defects in cellular immunity are at risk for disease.

L. monocytogenes has been isolated from uncooked meat, milk and cheese: It can grow in refrigerated food and can tolerate preserving agents. Several large outbreaks of listeriosis caused by foods contaminated with listeriae were reported in North America and Europe during the 1980s. Nosocomial outbreaks have also been reported.

Pathogenicity

The genus contains eight species but almost all cases of human listeriosis are caused by *L. monocytogenes.Listeria monocytogenes* is commonly ingested in food and is usually a harmless transient in the intestinal tract. Carriage rates of 5 to 10 percent have been reported, yet listeriosis remains a comparatively rare disease. The disease chiefly affects pregnant women, unborn or newly delivered infants, the immunosuppressed and elderly. It is predominantly transmitted by the consumption of contaminated food. *L. ivanovii* is associated with about 10 percent of infections in livestock animals. This species and *L. seeligeri* have been associated with a very small number of human infections.

Most human infections are caused by serovar 4. Experimental inoculation in rabbits causes a marked monocytosis (hence the name *monocytogenes*). Monocytosis is a feature of human listeriosis also. Instillation into the eyes of rabbits produces keratoconjunctivitis *(Anton test)*.

Human infection is believed to result from contact with infected animals, inhalation of contaminated dust or ingestion of contaminated milk or food. Outbreaks of food borne listeriosis have been known. Hospital acquired infections have also been reported. *L.monocytogenes* produces a hemolysin known as listeriolysin-O, which is a virulence factor antigenically related to streptolysin-O and pneumolysin. Non-hemolytic strains of *L. monocytogenes* produced by genetic deletion are avirulent.

Clinical Features

1. Intrauterine and Neonatal Infection

Intrauterine infection of the fetus may result in abortion, stillbirth, premature delivery, or acute-onset disseminated infection in the newborn infant (including the form known as granulomatosis infantisepticum). Asymptomatic infection of the female genital tract may cause infertility. Meningitis or septicemia may occur in neonates.

2. Adult and Juvenile Infection

It may cause meningitis or meningoencephalitis, particularly in neonates and in the elderly.

3. Disease in Healthy Adults

Most listeria infections in healthy adults are asymptomatic or occur in the form of a mild influenza-like illness. Several food-borne outbreaks of acute gastroenteritis with fever have been described.

4. Other Infections

Listeriosis may also present as abscesses, conjunctivitis, pharyngitis, urethritis, pneumonia, infectious mononucleosislike syndrome, endocarditis or septicemia.

Laboratory Diagnosis

1. Specimens

Blood, CSF, amniotic fluid, placenta, pus and biopsy material from the organs involved may be collected. Specimens may also be collected from neonate, stillbirth or products of conception.

2. Microcopy

If the gram stain shows organisms, they are intracellular and extracellular gram-positive coccobacilli. Care must be used to distinguish them from other bacteria, such as S. *pneumoniae, Enterococcus, Corynebacterium* and occasionally, *Hemophilus.*

3. Culture

Specimens should be inoculated on blood agar, chocolate agar and tryptose phosphate agar and incubated at 35 to 37°C for 1 to 3 days. Uncentrifuged CSF and blood may be added to nutrient broth and incubated at 35 to 37°C for 5 days followed by subculture on solid media. Greater success in isolation is achieved if the materials are stored in tryptose phosphate or thioglycollate broth at 4°C and subcultures are done at weekly intervals for 1 to 6 months *(cold enrichment* Isolates are likely to be missed as nonpathogenic diphtheroids unless properly investigated.

In case of listerial brain abscess, pus culture may be done.

Blood agar shows small colonies surrounded by a narrow zone of β-hemolysis. The bacteria are actively motile when grown at 25°C. The isolate is identified by its morphology and biochemical tests (Table 50.1).

Treatment, Prevention and Control

Currently, penicillin or ampicillin, either alone or with gentamicin, is the treatment of choice for infections with *L. monocytogenes.* Erythromycin can be used in patients allergic to penicillin but resistance to trimethoprim and the tetracyclines has been observed.

Because listeria are ubiquitous and most infections are sporadic, prevention and control are difficult. People at high risk of infection, however, should avoid eating raw or partially cooked foods of animal origin, soft cheeses and unwashed raw vegetables.

A vaccine is not available and prophylactic antibiotic therapy for high-risk patients has not been evaluated.

ERYSIPELOTHRIX RHUSIOPATHIAE

The genus *Erysipelothrix* contains two species, of which *Erysipelothrix rhusiopathiae* is responsible for human disease.

Species	Hemolysis	CAMP test with		Acid producition from				Nitrate reduction
		Staph. aureus	*Rhodococcus equi*	*D-Mannitol*	*L-Rhamnose*	*D-Xylose*	*α-Methyl D-mannoside*	
L. monocytogenes	+	+	–[a]	–	+	–	+	–
L. innocua	–	–	–	–	+/–	–	+	–
L. ivanovii	++	–	+	–	–	+	–	–
L. seeligeri	±	+	–	–	–	+	+/–	–
L. welshimeri	–	–	–	–	+/–	+	+	–
L. grayi	–	–	–	+	+/–	–	NK	+/–
L. murayi	–	–	–	+	V	–	NK	+
L. denitrificans	–	–	–	–	–	+	NK	+

Table 50.1: Distinguishing characters of *Listeria spp*

[a] Regarded as ±; +, positive reaction; -, negative reaction; V, Variable, NK, Not known

Morphology

E. rhusiopathiae is a slender, nonmotile, nonsporing, noncapsulated, straight or slightly curved, gram-positive rod, measuring 1-2 × 0.2-0.4 μm with tendency towards formation of long filaments.

Cultural Characteristics

It is microaerophilic on primary isolation but on subculture, grows as an aerobe or facultative anaerobe and growth is improved by 5 to 10 percent CO_2. It can grow on nutrient agar but the growth is improved by added glucose, serum or blood. Black colonies are developed in tellurite media.

Biochemical Reactions

E. rhusiopathiae ferments glucose without gas production and forms acid from fructose, galactose and lactose. No acid from mannitol, rhamnose, D-xylose or α-methyl-D-mannoside (cf. listeriae).

It is catalase negative with negative reactions for oxidase, indole production, urease, nitrate reduction, Voges-Proskauer and methyl red tests. H_2S is produced in triple sugar-iron agar (TSI).

Pathogenicity

Erysipelothrix is a a natural parasite of many animals. Disease is common in swine but rare in humans. It causes various diseases and syndromes in animals, notably a form of erysipelas in pigs and arthritis in pigs and sheep.

Human Infection

Human infection is an occupational hazard and usually occurs on the hand or fingers of persons handling animals fish or animal products.

Three forms of human infection with *E. rhusiopathiae* have been described:

1. Localized Skin Infection (Erysipeloid)

Inoculation of the organism through the skin causes *erysipeloid,* a painful, swollen, purplish lesion usually affecting the fingers or hands. It may be accompanied by local arthritis, lymphangitis or lymphadenitis.

2. Generalized Cutaneous Form

3. Septicemia

Septiaemia and/or endocarditis may occur, either as a rare complication of erysipeloid or from ingestion of the organism.

Treatment

Penicillin G is the drug of choice. *E. rhusiopathiae* is also highly susceptible to ampicillin, methicillin, piperacillin, cephalothin, cefotaxime, ciprofloxacin and clindamycin.

ALCALIGENES FAECALIS

Alcaligenes faecalis refers to gram-negative, short, nonsporing bacilli, which are strict aerobes and do not ferment sugars. They are motile by means of peritrichous flagella. They are usually oxidase positive. Nitrate reduction is variable. It does not ferment any of the sugars in peptone water. It produces alkaline reaction in litmus milk and sugar media. The name *Bacterium faecalis alkaligenes* was originally applied to an ill-defined group of gram negative bacilli isolated from human feces, which did not ferment sugars but produced an alkaline reaction in litmus milk.

Alc. faecalis is a saprophyte found in water and soil contaminated with decaying organic matter. They are also commensals in the intestines of man and animals. They have been isolated from a variety of clinical specimens such as urine pus and blood. They have been considered responsible for a typhoid-like fever, urinary infections, infantile gastroenteritis and suppuration in various parts of the body.

CHROMOBACTERIUM VIOLACEUM

Chromobacrium violaceum is a gram-negative, nonsporing bacillus, motile by means of a single polar and scanty lateral flagella.

It is a facultative anaerobe. It grows readily on simple nutrient agar at 35 to 37ºC including MacConkey agar producing violet pigment soluble in ethanol and insoluble in water and chloroform. It is catalase- and oxidase-positive and indole and urease negative.

It occurs as a saprophyte in soil and water in the tropical and subtropical regions. It causes rare but dangerous infection and consist of skin lesions with pyemia and multiple abscesses. It begins with cellulitis or lymphadenitismay and can rapidly progress to systeminc infection with abscess formation in various organs and septic shock.

It is sensitive to aminoglycosides chloramphenicol and tetracycline.

FLAVOBACTERIUM MENINGOSEPTICUM

Flavobacterium meningosepticum is a gram-negative nonmotile rod. It can grow on nutrient agar. It forms a yellow nondiffusible pigment after incubation at room temperature for 48 hours. It is catalase and oxidase-positive, proteolytic and weakly fermentative.

F. meningosepticum is a saprophyte whose natural habitat is soil and moist environments, including nebulizers. It may cause opportunistic nososcomial infections, particularly in infants. It has been responsible for outbreaks of meningitis in newbon infants. Infection in adults leads to a mild febrile illness.

It is usually sensitive to cotrimoxazole, novobiocin, rifampicin, clindamycin and cefoxitin.

DONOVANIA GRANULOMATIS (CALYMMATOBACTERIUM GRANULOMATIS) OR *KLEBSIELLA GRANULOMATIS*

The etiologic agent of granuloma inguinale, a granulomatous disease affecting the genitalia and inguinal area, has been called historically *Calymmatobacterium (Donovania) granulomatis)*. The organism was discovered by Donovan (1905) who described the presence of characteristic intracellular bodies in smears from ulcerated lesions of a disease now known as Donovanosis. Donovan's intracellular bodies have since been identified as bacteria and named *Donovania granulomatis*.

Recently, this organism was transferred into the genus Klebsiella based on genomic criteria and the fact this organism produces clinical and pathologic changes similar to two other species of *Klebsiella-K rhinoscleromatis* (causes a granulomatous disease of the nose) and *K ozaenae* (causes chronic atrophic rhinitis).

Morphology

K. granulomatis is a small, capsulate, gram-negative, coccobacillus. *Diagnosis* can be made by demonstration of **Donovan bodies** in Wright-Giemsa stained impression smears from the lesions. They appear as rounded cocobacilli, 1 to 2 µm, within cystic spaces in large mononuclear cells. They show bipolar condensation of chromatin, giving a closed **safety pin appearance** in stained smears. Capsules are usually seen as dense acidophilic areas around the bacilli. They are gram-negative, nonmotile, nonsporing and nonacid-fast.

Culture

It can be cultured readily in the egg yolk medium and on a modified Levinthal agar.

Pathogenicity

Donovanosis is a venereal disease, first described by McLeod in India in 1882 and seen mainly in the tropics. It can be transmitted after repeated exposure through sexual intercourse or nonsexual trauma to the genitalia. The incubation period ranges from 1 to 12 weeks. It begins as a painless papule on the genitalia, which leads to a slowly progressive, autoinoculable ulcers. The disease runs a chronic course. It causes a chronic granulomatous disease known as *granuloma inguinale, granulomatous venereum* or *donovanosis*.

Laboratory Diagnosis

Laboratory confirmation of granuloma inguinale is made by scraping the border of the lesion, by spreading the collected tissue on a slide and by staining it with Giemsa or Wright's stain. Pathognomonic **Donovan bodies** are observed within mononuclear phagocytes.

Treatment

Cases of donovanosis respond well to tetracycline, chloramphenicol, erythromycin, clindamycin, co-trimoxazole, streptomycin and other aminoglycosides.

ACINETOBACTER(MIMA POLYMORPHA; BACTERIUM ANITRATUM

The genus *Acinetobacter* contains strictly aerobic short, stout, often capsulate, non-motile Gram-negative bacilli or coccobacilli that grow well on simple media. They are oxidase negative.

The genus contains only one species, *Acinetobacter calcoaceticus*, which embraces two variants: *A. calcoaceticus* var. *anitratus* produces acid oxidatively from glucose whereas *A. calcoaceticus* var. *lwoffii* does not. This terminology supersedes earlier terms such as *Herellea vaginicola* and *A. anitratum* which correspond to the *anitratus* variant, and *Mima polymorpha* and *Moraxella lwoffii* which correspond to the *lwoffii* variant. In the past, *Achromobacter* species names were also assigned to these variants.

The genus *Acinetobacter* has also undergone taxonomic reorganization. At the present time, 7 species are recognized, but 21 DNA groups ("**genomospecies**") have been recognized. The most important species currently are *A. baumannii, Acinetobacter lwoffii*, and *Acinetobacter haemolyticus.* Strains commonly isolated in clinical laboratories are called the *Acinetobacter calcoaceticus-baumannii complex* subdivided as follows: glucose oxidising, nonhemolytic clinical strains as *A. baumannii* (corresponding to the former *A.antitratus); the glucose negative nonhemolytic strain as *A. lwoffi* (corresponding to the form *Mima polymorpha);* and the hemolytic strain as *A. hemolyticus..*

A. baumannii

These form pinkish colonies on MacConkey medium. Acid without gas is formed in glucose, arabinose, xylose, and occasionally in rhamnose. A characteristic reaction is the formation of acid in 10 percent, but not 1 percent lactose. Several serotypes have been identified by capsule swelling and immunofluorescence.

A. lowffi

This forms yellow colonies on MacConkey medium and does not acidify sugars. Some strains are oxidase positive.

Pathogenesis

Acinetobacter species are saprophytes found in soil, water and sewage and occasionally as commensals of moist areas of human skin. The organisms survive well in the hospital environment and are increasing in importance as opportunist pathogens. Serious infections, including meningitis, pneumonia and septicemia, are most commonly associated with *Acinetobacter baumannii.* Patients in intensive care units are at particular risk.

Treatment

Most strains are resistant to sulphonamides, penicillins including ampicillin, the cephalosporins, erythromycin, the tetracyclines and chloramphenicol. Specific therapy must be guided by *in vitro* susceptibility tests but empirical therapy for serious infections should consist of aβ-lactam antibiotic (e.g. ceftazidime, imipenem) and an aminoglycoside.

RAT BITE FEVER *(STREPTOBACILLUS MONILIFORMIS* AND *SPIRILLUM MINUS)*

Streptobacillus moniliformis and Spirillum minus are the causative agents of two distinct diseases referred to collectively as rat-bite fever (Table 50.2). Rat bite fever (RBF) is characterized by relapsing fever, rash and arthralgia occurring days or weeks after a rat bite. Two different bacteria can cause this condition - Streptobacillus moniliformis and Spirillum minus, both of which are natural parasites of rodents.

Streptobacillus Moniliformis

Morphology

S. moniliformis is a long, thin (0.1 to 0.5 × 1 to 5 µm), gram negative bacillus that tends to stain poorly and to be more pleomorphic in older cultures. Granules and bulbous swellings resembling a string of beads may be seen hence the species name 'moniliformis'. It may lose its cell wall and exists as L-form. In fact, L-forms were originally discovered during the study of this bacillus.

Culture

S. *moniliformis* is a nutritionally exacting aerobe and facultative anaerobe. Growth requires the presence of blood or other body fluids. It grows well on moist Loeffler's serum slope or moist plate of nutrient agar containing 20 percent horse serum and in 20 percent serum broth. Colonies appear on the surface after 48 hour incubation and are discrete granular and greyish yellow. L phase variants show minute colonies (0.1-0.2 mm diameter) with a **fried egg appearance.** They consist mainly of small coccoid bodies

Biochemical Reactions

It attacks sugars fermentatively and produces acid without gas from glucose, galactose, dextrin, raffinose and starch. It is catalase, oxidase, indole production, urease and nitrate reduction negative.

Pathogenesis

Both *Streptobacillus* and *Spirillum* organisms are found in the nasopharynx of rats and other small rodents as well as transiently in animals that feed on rodents (e.g. dogs, cats). Turkeys exposed to rats and mice, as well as contaminated water and milk, have also been implicated in *Streptobacillus* infections.

In humans, it causes streptobacillary rat-bite fever. Another type of clinically indistinguishable, rat-bite fever is caused by *Spirillum minus.*

In man, the organisms enter the body through the wound caused by the bite of rat or other animals. Streptobacillary RBF develops 2 to 10 days after exposure. Relapses are common in untreated cases. The disease can also occur as outbreaks, in the absence of rat bite. This condition, first observed in Haverhill, USA, is called **Haverhill fever** or *erythema arthriticum epidemicum.* It is believed to be caused also by consumption of raw milk or water contaminated by rat excrement.

Laboratory Diagnosis

1. Specimens

i. Blood—during acute phase of disease.
ii. Joint fluid—when arthritis develops.

2. Culture

Specimen is inoculated on blood agar or Loeffier's serum slope.

3. Animal Pathogenicity Test

Mice are highly susceptible to intraperitoneal inoculation of infected material. The animals develop a rapidly fatal generalized condition or a more progressive disease with swelling of feet and legs.

4. Serology

Diagnosis may also be established by serological tests including agglutination, complement fixation and fluorescent antibody tests.

Treatment

Penicillin is the antibiotic of choice for treating rat bite fever. S. *moniliformis* is also sensitive to cephalosporins, erythromycin, clindamycin, tetracycline and aminoglycosides. The L phase variant is penicillin-resistant but sensitive to tetracycline.

Spirillum Minus

The organism commonly known as *Spirillum minus*, one of the causes of rat-bite fever in man, is of uncertain taxonomic position. It was once regarded as a spirochete but was later placed in the genus *Spirillum.*

Morphology

S. *minus* is a short, spiral, gram-negative organism about 3 to 5 µm × 0.2-0.5 µm in size, with two or three regular spirals and 1 to 7 amphitrichous flagella. It is very actively motile, showing darting movements like those of a vibrio.The organisms can be demonstrated with the dark ground microscopy or by staining with Leishman or Giemsa stain.

Culture

The organism has not been cultivated on artificial media and many of its properties are therefore unknown.*S. minus* is best isolated by intraperitoneal inoculation of specimens (infected tissue or blood) into mice and guinea pigs.

Pathogenesis

It was first observed in a rat by Carter (1888) in India. Japanese workers identified it as the causative agent of one type of RBF, called *Sodoku.*

S. minus is a natural parasite of wild rats and other rodents. The organism is inoculated into humans through the bite of a rat. Local lymphadenopathy and lymphangitis develop with the onset of fever and systemic disease. A generalized rash with large brown to purple macules is usually observed but some patients present with urticarial lesions. A roseolar rash may spread from the area of the original bite (Table 50.2).

Rare complications are endocarditis, meningitis, hepatitis, nephritis and myocarditis.

Laboratory Diagnosis

Microscopic Examination

Darkfield microscopy examination of blood, ulcer exudates, or lymph node aspirates is used to detect S. *minus.* Blood smears can be stained with Giemsa's or Wright's stain.

Animal Inoculation

S. minus can be isolated by intraperitoneal inoculation of blood or material from lymph node into mice and guinea pigs. Spirilla may be demonstrated by dark field microscopy in blood and peritoneal fluid of animal 1 to 3 weeks after the intraperitoneal inoculation.

Treatment

Infections with S. *minus* respond to treatment with penicillin and tetracyclines. In the rare case of endocarditis, the addition of an amino glycoside may be of value.

EIKENELLA CORRODENS

Morphology

E. corrodens is a fastidious , small, nonmotile, nonsporeforming, facultatively anaerobic gram-negative bacillus. It lacks flagella and shows *twitching* motility which is due to contractile fimbria-like filamentous appendages.

Culture

It is an aerobe and facultative anaerobe. A slow-growing, fastidious organism, *E. corrodens* requires 5 percent to 10 percent carbon dioxide to grow. It can grow on blood agar or chocolate agar. After 48 hours incubation on blood agar or chocolate agar, the colonies are small (0.5-1 mm in diameter) with characteristic pitting or corroding of blood agar, hence the species name *corrodens.* The organism also produces a characteristic bleachlike odor.

Biochemical Characters

It is oxidase positive and catalase negative. It does not produce acid from carbohydrates. It is indole and urease negative but lysine and ornithine decarboxylase positive.

Pathogenesis

Eikenella corrodens is normally present in the mouth, upper respiratory tract and gastrointestinal tract of human beings. It is an opportunistic pathogen that causes infections in patients who are immunocompromised or have diseases or trauma of the oral cavity. *E. corrodens* is most commonly isolated in the settings of a **human bite wound or fistfight injury.** Other infections are endocarditis, sinusitis, meningitis, brain abscesses, pneumonia, and lung abscesses.

Treatment

E. corrodens is susceptible to penicillin, ampicillin, extended-spectrum cephalosporins, tetracyclines and fluoroquinolone.

CARDIOBACTERIUM HOMINIS

This gram-negative bacilli are nonmotile, characteristically small (1 X 1 to 2 µm) but sometimes pleomorphic. The organism grows slowly in culture. On blood agar the colonies are small, 1 to 2 mm in diameter, smooth, glistening, circular and opaque after 48 hours incubation at 35°C. The organism does not grow on MacConkey agar or other selective media commonly used for gram-negative bacilli. The bacteria are fermentative, indole- and oxidase-positive, and catalase and nitrate negative.

C. hominis occurs commonly as a commensal in the human nose and throat may cause endocarditis, particularly in those with pre-existing cardiovascular disease.

It can be diagnosed by isolation of the causative agent in blood culture.

C. hominis is susceptible to multiple antibiotics and most infections are successfully treated with penicillin or ampicillin for 2 to 6 weeks.

CAPNOCYTOPHAGA

The *Capnocytophaga* species are gram-negative,slow-growing, capnophilic, fusiform or filamentous bacilli.

They are fermentative and facultative anaerobes that require CO_2 for aerobic growth. They may show **gliding motility** which can be seen as outgrowths of colonies.

C.ochracea, C. gingivalis and *C. sputigena* are members of the normal oral flora of humans. They have been associated with severe periodontal disease in juveniles. They occasionally cause bacteremia and severe systemic disease in immunocompromised patients, especially granulocytopenic patients with oral ulcerations. C. ochracea is the most common clinical isolate.

Most *Capnocytophaga* infections can be treated with broad-spectrum cephalosporins, fluoroquinolones or penicillin.

GARDNERELLA VAGINALIS

Gardnerella vaginalis is a small, gram-negative, nonmotile, pleomorphic rod which shows metachromatic granules. It was formerly known as *Haemophilus vaginalis* or *Corynebacterium vaginale.* It grows on blood or chocolate agar aerobically under 5 percent CO_2, Minute colonies appear in 24 to 48 hours and are hemolytic on human or rabbit blood agar. It is catalase, oxidase, indole and urease negative.

G. vaginalis is considered responsible for bacterial vaginosis, a mild but common condition characterized by raised vaginal pH > 4.5, foul smelling discharge and the presence of 'clue cells', which are vaginal epithelial cells with their surface studded with numerous small bacteria. Bacterial vaginosis is also associated with anaerobic bacteria, particularly *Mobiluncus.* Metronidazole is effective in treatment.

KEY POINTS

- ***Listeria***-Organisms of the genus *Listeria* contains eight species but almost all cases of human listeriosis are caused by *L. monocytogenes.*
- *Listeria monocytogenes* is a small, coccoid, gram positive bacillus and exhibits a characteristic, slow, tumbling motility when grown at 25 °C but at 37 °C is nonmotile.
- The disease chiefly affects pregnant women, unborn or newly delivered infants, the immunosuppressed and elderly. It is predominantly transmitted by the consumption of contaminated food.
- Erysipelothrix rhusiopathiae—Disease is common in swine but rare in humans. Three forms of human infection-erysipeloid,generalized cutaneous form and septicemia.
- ***Alcaligenes faecalis***-*A. faecalis* causes urinary tract infection, infantile gastroenteritis and typhoid-like fever in humans.
- ***Chromobacterium violaceum***—It is associated with intestinal and genitourinary infections and septicemic illnesses with pneumonia.
- ***Flavobacterium meningosepticum***—causes opportunistic infections, neonatal meningitis in premature infants and pneumonia in immunocompromised hosts.
- ***Calymmatobacterium (Donovania) granulomatis*** is the causative agent of granuloma inguinale, a sexually transmitted disease.
- **Acinetobacter species** resemble Neisseriae, differing in being oxidase negative. Occasionally, serious infections are caused in immunocompromised patients; Acinetobacter species may cause nosocomial infections.
- Rat bite fever is caused by two different bacilli: *Streptobacillus moniliformis* and *Spirillum minus.* S moniliformis causes one type of rat bite fever, and is a commensal of the oral cavity of the rat.
- Eikenella corrodens—It is most commonly isolated in the settings of a human bite wound or fistfight injury.
- **Cardiobacterium hominis**—*C. hominis* may cause endocarditis, particularly in those with pre-existing cardiovascular disease.
- *Capnocytophaga* species have been associated with severe periodontal disease in juveniles, bacteremia and severe systemic disease in immunocompromised patients.
- *Gardnerella vaginalis* is considered responsible for bacterial vaginosis, a mild but common condition characterized by raised vaginal pH > 4.5, foul smelling discharge and the presence of 'clue cells'. Metronidazole is effective in treatment.

IMPORTANT QUESTIONS

1. Write short notes on:
   a. *Listeria monocytogenes*
   b. *Erysipelothrix rhlusiopathiae*
   c. Donovan bodies
   d. Rat-bite fever
   e. *Eikenella corrodens*
   f. Acinetbacter spp.
   g. *Alkaligenes faecalis*

FURTHER READING

Bergone- Berezin E and KJ Towner. Acinetobacter spp. As nosocomial pathogens. Clin Microbiol Rev 1996;9:148.

Catlin BW. Gardnerella vaginalis, Clin Microbiol Rev1 992; 5,213.

Levett PN. Bacterial vaginosis. Rev Med Microbio 1992;13:15.

Morrison VA and KF Wagner Kingella kingae infections. Rev Infect Dis 1989;11:776.

Parenti DM, DR Syndinan. Capnocytophaga species infections. J Infect Dis 1985;151:140.

Schuchat A et al. Epidemiology of human listeriosis. Clin Microbiol Rev 1991;4:169.

Sherertz RJ, MI Sullivan Infections with Acinetobacter calcoaceticus. J Infect Dis 1985;151:252.

Verghese A and SL Berk Moraxella catarrhalis. Infect Dis Clin N America 1991;5:523.

Wormser GP and EJ Bottone Cardiobacterium hominis. Rev Infect Dis 1983;5:680.

CHAPTER 51

Rickettsiaceae, Bartonellaceae and Coxiella

Learning Objectives

After reading and studying this chapter, you should be able to:

- Name various pathogenic treponemes.
- Describe diseases caused by different rickettsiae.
- Describe Brill-Zinsser disease.
- Discuss the laboratory diagnosis of rickettsial infections.
- Describe Weil-Felix reaction.
- Discuss Q fever, Trench fever and Oroya fever.

INTRODUCTION

Rickettsiae are small gram-negative bacilli adapted to obligate intracellular parasitism and transmitted by arthropod vectors. They are primary parasites of arthropods such as lice, fleas, ticks and mites, in which they are found in the alimentary canal. In vertebrates, including humans, they infect the vascular endothelial and reticuloendothelial cells. Many rickettsiae are transmitted transovarially in the arthropod, which serves as both vector and reservoir.

The family Rickettsiaceae is named after Howard Taylor Ricketts who discovered the spotted fever rickettsia (1906) and died of typhus fever contracted during his studies. These bacteria were originally thought to be viruses because they are small (0.3 × 1 to 2 µm), stain poorly with the gram stain, and grow only in the cytoplasm of eukaryotic cells. Nevertheless, these organisms have the characteristics of bacteria:

Characteristics Similar to Bacteria

1. They are structurally similar to gram-negative bacilli.
2. Their cell wall contains muramic acid.
3. They contain both DNA and RNA.
4. They contain enzymes for the Krebs cycle and ribosomes for protein synthesis.
5. They multiply by binary fission.
6. They are inhibited by antibiotics (e.g. tetracycline, chloramphenicol).

CLASSIFICATION

The family Rickettsiacae currently comprises three genera **Rickettsia**. **Orientia** and **Ehrlichia**, which appear to have descended from a common ancestor (Table 51.1).

Former members of the family, *Coxiella burnetii,* which causes Q fever and *Rochalimaea quintana* causing trench fever have been excluded because the former is not primarily arthropod-borne and the latter not an obligate intracellular parasite, being capable of growing in cell-free media, besides being different in genetic properties.

GENUS RICKETTSIA

The genus Rickettsia consists of the causative agents of two groups of diseases-typhus fevers and spotted fevers.

Morphology

In smears from infected tissues, rickettsiae appear as pleomorphic coccobacilli, 0.3-0.6 *µm* × 0.8-2*µm* in size. They are nonmotile and noncapsulated. They are gram negative, though they do not take the stain well. They stain bluish purple with giemsa and castaneda stains and deep red with Machiavello and Gimenez stains.

Table 51.1: The family Rickettesiacae causing human diseases

Genus	*Species*
A. Ricketesia	*R. prowazekii*
	R. typhi
	R. rickettsii
	R. conorii
	R. australis
	R. sibirica
	R. akari
B. Orientia	*O. tsutsugamushi*
C. Ehrlichia	*E. sennestu*
	E. chaffeensis
	E. phagocytophila

Under the electron microscope, rickettsiae are seen to have a three layered cell wall, a trilaminar plasma membrane and an outer slime layer.

Cultivation

Rickettsiae are unable to grow in cell free media. All the bacteria are strict intracellular parasites, growth generally occurs in the cytoplasm of infected cells but in the case of the spotted fever rickettsiae, growth may take place in the nucleus as well. Rickettsiae grow best in cells that are not metabolizing actively. The optimum temperature for growth is 32 to 35°C.

1. **Yolk sac**-They are readily cultivated in the **yolk sac** of developing chick embryos, as first shown by Cox. Pure preparations of rickettsiae for use in laboratory testing can be obtained by differential centrifugation of yolk sac suspensions.
2. **Cell culture**-Many strains of rickettsiae also grow in **cell culture**. They grow on mouse fibroblast, HeLa, HEp-2, Detroit 6 and other continuous cell lines but tissue cultures are not satisfactory for primary isolation. In cell culture, the generation time is 8 to 10 hours at 34°C.
3. **Laboratory animals** such as guinea pigs and mice are useful for the isolation of rickettsiae from patients. They may also be propagated in arthropods.

Resistance

Most rickettsiae survive only for short times outside of the vector or host. Rickettsiae are readily inactivated by physical and chemical agents. They are rapidly destroyed at 56°C and at room temperature when separated from host components, unless preserved in skimmed milk or a suspending medium containing sucrose, potassium phosphate and glutamate (SPG medium). Dried feces of infected lice may contain infectious *Rickettsia prowazekii* for months at room temperature.

Rickettsiae are susceptible to tetracycline, chloramphenicol and ciprofloxacin. Penicillin and sulphonamides are ineffective but para-aminobenzoic acid has an inhibitory action on rickettsiae. Rickettsial growth is enhanced in the presence of sulfonamides, and rickettsial diseases are made more severe by these drugs.

Antigenic Structure

Rickettsiae possess at least three types of antigens:

1. **Group specific soluble antigen**-Rickettsiae have group specific antigens. It is present on the surface of rickettsiae.
2. **Species specific antigen-** It is associated with the bodies of rickettsiae. In case of scrub typhus, it is strain specific. The immunodominant surface protein antigens (SPA) of *R. prowazekii* and *R. typhi* have both species specific and cross reactive epitopes.
3. **Alkali stable polysaccharide-** It is an alkali stable polysaccharide found in some rickettsiae and in some strains of Proteus bacilli. This sharing of antigens between rickettsiae and proteus is the basis for the Weil-Felix reaction used for the diagnosis of rickettsial infections by the demonstration of agglutinins to Proteus strains OX 19, OK 2 and OK.

Pathogenesis

Rickettsiae normally enter the body through the bite or feces of an infected arthropod vector. On entry into the human body, the rickettsiae multiply locally and enter the blood. They become localized chiefly in the vascular endothelial cells, which enlarge, degenerate and cause thrombus formation with partial or complete occlusion of the vascular lumen. The overall pathological features of the rickettsial diseases are similar and can be explained by the damage to the vascular endothelium. Observations in cell culture systems suggest that spotted fever and typhus group rickettsiae destroy the host cell by different mechanisms.

The long survival of rickettsiae in various organs and lymphatic tissues of infected men and animals is a distinctive feature in pathogenesis and is of importance in the epidemiology of some rickettsial diseases.

A. Typhus Fever Group

Typhus group rickettsiae cause:

1. Epidemic typhus
2. Brill-Zinsser disease
3. Endemic typhus (Murine typhus).

1. Epidemic Typhus (Louse-borne Typhus, Classical Typhus, Gaol fever)

Epidemic typhus is a louse-borne disease and had a tremendous impact on the history of man. It has been one of the great scourges of mankind, occurring in devastating epidemics during times of war and famine. Epidemic typhus occurs among people living in crowded, unsanitary conditions that favor the spread of body liceconditions such as those that arise during wars, famines and natural disasters. The disease has been reported from all parts of the world but has been particularly common in Russia and Eastern Europe. In recent times, the main foci have been Eastern Europe, Africa, South America and Asia. In India, the endemic spot is Kashmir.

Etiologic agent of epidemic typhus is ***R. prowazekii***.

The principal vector is the human body louse *Pediculus humanus corporis*. **The head louse** may also transmit the infection but not the pubic louse. Unlike with most other rickettsial diseases, humans are the **primary reservoir** of typhus.

The lice become infected by feeding on rickettsiaemic patients. The rickettsiae multiply in the gut of the lice and appear in the feces in 3 to 5 days. Lice succumb to the infection within 2 to 4 weeks, remaining infective till they die. They can transmit the infection after about a week of being infected. Lice defecate while feeding. Scratching produces minor excoriations that in turn become portals of entry for rickettsiae in the louse's feces. Infection is transmitted when the contaminated louse feces is rubbed through the minute abrasions caused by scratching. Occasionally, infection may also be transmitted by aerosols of dried louse feces through inhalation or through the conjuctiva. Being sensitive to temperature changes in the host, they leave the febrile patient or the cooling carcass and parasitise other persons.

The incubation period typically ranges from 10 to 14 days. Usually the onset is abrupt, with generalized myalgias, chills, fever and headache. Other symptoms such as gastrointestinal complaints, weakness, cough and meningismus may also be present and lead to diagnostic confusion. One of the hallmarks of epidemic typhus, **skin rash,** usually appears from 4 to 7 days after onset characteristically first on the trunk and later spreads to the extremities. The patient becomes stuporous and delirious towards the second week and develops the cloudy state of consciousness in the disease (from *typhos,* meaning cloud or smoke). The case fatality may reach 40 percent and increases with age.

2. Brill-Zinsser Disease (Recrudescent Typhus)

The rickettsiae may remain latent in the lymphoid tissues, or organs for years in some who recover from epidemic typhus. Such latent infection may, at times, be reactivated leading to **recrudescent typhus (Brill-Zinsser disease).** Brill (1898) first recognized and Zinsser (1934) isolated *R. prowazekii* from such cases and proved that they were recrudescences of infections acquired many years previously.

Brill-Zinsser disease explains the manner in which the rickettsia is able to survive without extrahuman reservoirs. In itself, the disease is not important but such cases. occurring in louse ridden communities may initiate epidemics of typhus fever. Brill-Zinsser disease is a milder illness than that of epidemic typhus and the duration of the disease is shorter. Case fatality is lower.

3. Endemic Typhus (Murine Typhus, Flea-borne Typhus, Rat Typhus)

Endemic or murine typhus is a milder disease than epidemic typhus.

It *is* caused by ***Rickettsia typhi(R. mooseri).***

The primary reservoir are rodents and the **principal vector** is the **rat flea *(Xenopsylla cheopis).*** A related species ***R. felis,*** maintained in a cycle involving opossums and the cat flea *(Ctenocephalus felis),* has been found to cause endemic typhus in the USA. Most cases occur during the warm months.

The rickettsia multiplies in the gut of the flea and is shed in its feces. The flea is unaffected but remains infectious for the rest of its natural span of life. Man is infected by the contamination of abraded skin, respiratory tract or conjunctiva with infective flea feces. Ingestion of food recently contaminated with infected rat urine or flea feces may also cause infection. Human infection is a dead end. Man to man transmission does not occur. The disease is an occupational hazard of working in rat-infested areas such as markets or ports. Human disease is known to occur in the USA, Mexico, northern South America, Israel, Pakistan, India, SouthEast Asia, China and Australia and Spain. It is an important cause of fever in Khmer refugees in Thailand. In Kashmir and China, lice have been known to transmit endemic typhus in human beings, producing smouldering outbreaks. Endemic typhus is worldwide in prevalence but is not of much public health importance as the disease is mild and sporadic and can be easily controlled now.

Differences between R. typhi (R. mooseri) and R. prowazekii

R.typhi and *R. prowazekii* are closely similar but may be differentiated by biological and immunological tests.

i. **Neill-Mooser or the tunica reaction.** When male guinea pigs are inoculated intraperitoneally with blood from a case of endemic typhus or with a culture of *R. typhi,* they develop fever and a characteristic scrotal inflammation. The scrotum becomes enlarged and the testes cannot be pushed back into the abdomen because of inflammatory adhesions between the layers of the tunica vaginalis. This is known as the *Neill-Mooser* or the tunica reaction. The Neill-Mooser reaction is negative with *R. prowazekii.*
ii. **Other methods** used include IFA, ELISA and PCR based DNA tests.

B. Spotted Fever Group

They are all transmitted by ticks, **except *R. akari,*** which is miteborne. Rickettsiae of this group possess a common soluble antigen and multiply in the nucleus as well as in the cytoplasm of host cells. The basic pathologic process in the spotted fever group is widespread vasculitis involving the skin (with production of a rash) and internal organs (producing dysfunctions of the brain, heart, lungs, and kidneys).

Tick Typhus

Many species have been recognized in this group.

1. *R. rickettsii*—the causative agent of Rocky Mountain spotted fever, discovered by Ricketts in 1906 was the first insect transmitted bacterial pathogen to be recognized.
2. *R. siberica*—causes Siberian tick typhus.
3. *R. conori*—causes Indian, Mediterranean, Kenyan and South African tick typhus.
4. *R. australis*—causes Queensland tick typhus.
5. *R. japonica*—causes the Oriental spotted fever.
6. *R. africae*—reported recently from sub Saharan Africa as the cause **of tick bite fever** has been observed in the Carribean islands also.
7. *R. akari*—causes mite-borne rickettsial pox.

The rickettsiae are shed in tick feces but transmission to human beings is primarily by bite, as the rickettsiae also invade the salivary glands of the ticks. Man becomes infected following the bite of infected ticks or through contamination of abraded skin or mucous membranes. People place themselves at risk when they enter areas infested with infected ticks. Individuals may also become infected if they are bitten by ticks of domestic dogs or if partially fed ticks rupture during manual deticking of dogs.

Rocky Mountain Spotted Fever

Rocky Mountain spotted fever (RMSF) is a potentially life-threatening infection. *R. rickettsi is* the etiologic agent of Rocky Mountain spotted fever.

Vector-ixodid (hard) ticks are the vectors. It is transmitted by *Dermacentor andersoni* and related species of ticks.The ecology and epidemiology of spotted fever are directly related to the life cycle of four species of **ixodid (hard) ticks** vectors. Humans are only accidentally infected

Clinical diseases—The incubation period is about one week. Clinical picture is similar to that of typhus fever but the rash appears earlier and is more pronounced. It is prevalent in many parts of North and South America.

Other Spotted Fever Rickettsiae (Table 51.2)

R. siberica causes the **Siberian tick typhus** which is mild rickettsial disease. **Boutonneuse fever** is caused by *R. conori*. *R. conori* strains are isolated from the Mediterranean littoral, Kenya, South Africa and India are indistinguishable. **Australian tick typhus** is caused by *R. australis*. All these three rickettsiae are maintained in nature in **ixodid ticks.** Humans accidentally enter the natural cycle. Diseases produced by these rickettsiae resemble RMSF but are of milder form. A black spot having a necrotic center (eschar) frequently occurs at the site of the tick bite in all spotted fever group infections except Rocky Mountain spotted fever. The tick Rhipicephalus sanguineus is the most important vector. **Hemaphysalis leachi, Amblyomma** and **Hyalomma ticks** can also transmit the infection.

Rickettsial Pox

Rickettsial pox is a relatively mild infection transmitted by mites. It is a self-limited, nonfatal, vesicular exantham first observed in New York (1946). Rickettsial pox is characterized by a local eschar, a papulovesicular rash, and a benign clinical course. The name is derived from the resemblance of the disease to chickenpox. It is also called **vesicular or varicelliform rickettsiosis.** Rickettsial pox is primarily an urban disease associated with mice-infested buildings.

The causative agent is.*R. akari* (from *akari,* meaning mite). The **reservoir of infection** is the domestic mouse, *Mus musculus*.**The vector is the mite, Liponyssoides sanguineus,** in which transovarial transmission occurs. *R. akari* has also been isolated from wild rodents in Korea. Foci of this infection are known to exist in the Eastern USA. *R. akari* is serologically more akin to *R. australis* than to *R. conorii* and *R. rickettsii.*

C. Scrub Typhus Group

Scrub typhus (chiggar-borne typhus, tsutsugamushi disease) is caused by ***Orientia tsutsugamushi*** (formerly *R. tsutsugamushi, R. orientalis*). It was first observed in

Table 51.2: Human Disease caused by *Rickettsia* and *Orientia* species

	Group	Species	Diseases	Vector	Vertebate reservoir	Geogrphical distribution
A.	Typhus group	*R.prowazekii*	Epidemic typhus	Louse	Human beings	Worldwide group
			Brill-Zinsser disease	"	Human beings	America,Europe
		R. typhi	Endemic typhus	Rat flea	Rat	Worldwide
		R felis	Endemic typhus	Cat flea	Oppossum	USA
B.	Spotted fever group	*R.rickettsii*	Rocky Mountain spotted fever	Tick	Rabbit, dog	N. America
		R. siberica	Siberian tick typhus	"	Wild Animals cattle	Russia, Mongolia
		R. conori	Boutonneuse Fever	"	Dog, rodents	Mediterranean
			S. African tick typhus	"	"	S. Africa
			Kenyan tick typhus	"	Rodents	Kenya
			Indian tick typhus	"	? Rodents	India
		R. australis	Qeensland tick typhus	"	Bush rodents	N. Australia
		R. japonica	Oriental spotted fever	"	?	Japan
		R. akari	Rickettsial pox	Gamasid mite	Mouse	USA, Russia
C.	Scrub typhus group	*O. tsutsugamushi*	Scrub typhus	Trombiculid mite	Small rodents,	East Asia, Pacific

Japan where it was found to be transmitted by **mites.** The disease was therefore called ***tsutsugamushi*** (from *tsutsuga,* meaning dangerous, and Mushi meaning insect or mite). It occurs all along East Asia, from Korea to Indonesia, and in the Pacific Islands including Australia.

The vectors are **tromiculid mites** belonging to the genus *Leptotrombidium* . *O. tsutsugamushi* is transmitted to man by the larval stages of mites of the genus ***Leptotrombidium***-*L. akamushi* in Japan and *L. deliensis* in India. The nymphal and adult stages of the mites do not feed on mammals. The parasitic larvae (chiggers) occur in habitats that have been disturbed by the loss or removal of the natural vegetation. The area becomes covered with scrub vegetation, which is the preferred habitat for chiggers and their mammalian hosts, and gives the disease its name. The disease is often localized because of the restricted habitat of the chiggers. Persons entering infected areas are at risk.

Zoonotic Tetrad

Four factors are essential for the establishment of a microfocus of infection, viz. coexistence and intimate relationship among *O. tsutsugamushi.* chiggers. rats and secondary or transitional forms of vegetation (known as the **zoonotic tetrad***).*

Clinical Diseases

The incubation period is 1 to 3 weeks. Patients typically develop a characteristic **eschar** at the site of the mite bite, with regional lymphadenopathy and maculopapular rash. The disease sets in with fever, headahe and coniunctival iniection. Death may result from encephalitis, respiratory failure and circulatory failure.

Scrub typhus confers only transient immunity and reinfection may occur with heterologous or homologous strains. No vaccine is available, so the disease is prevented by avoidance of exposure to chiggers (i.e. the wearing of protective clothing, the use of insect repellents).

Laboratory Diagnosis

Rickettsial diseases may be diagnosed in the laboratory either

1. Isolation of rickettsiae
2. Direct detection of the organisms and their antigens
3. Serology.

1. Isolation of Rickettsiae

Isolation of the organism provides conclusive proof of rickettsial infection. As rickettsiae are highly infectious and have caused several serious and fatal infections among laboratory workers, their isolation should be attempted with utmost care and only in laboratories equipped with appropriate safety provisions. Rickettsiae can be isolated in **laboratory animals** such as **mice or guinea-pigs, in embryonated chicken eggs** and in **cell culture.**

i. Laboratory Animals

Rickettsiae may be isolated in male guinea pigs or mice from patients in the early phase of the disease. Blood clot ground in skimmed milk or any suitable suspending medium is inoculated intraperitoneally. The animals have to be observed for 3 to 4 weeks and their temperature recorded daily. Their response to rickettsial infection varies.

1. **In Rocky Mountain spot fever(RMSF)**—guinea-pigs develop fever, scrotal necrosis and may even die due to overwhelming infection of *R. rickettsii.*
2. **In *R. typhi, R. conorii* and *R. akari* infection**—guinea-pigs develop fever and tunica reaction.
3. **In *R. prowazekii* infection**—the animals develop fever without any testicular inflammation.

For isolation of *O. tsutsugamushi,* mice are preferred over guinea-pigs. The infected animals become ill and develop ascites.

Smears from peritoneum, tunica and spleen of infected animals may be stained by Giemsa or Gimenez methods to demonstrate the rickettsiae.

ii. Embryonated Chicken Egg

Rickettsiae can also be grown in the yolk sac of chick embryo.

iii. Cell Culture

Cell culture is the most widely used method for isolation of rickettsiae from clinical samples. Rickettsiae grow well in 3 to 5 days on vero cell MRC 5 cell cover slip cultures and can be identified by immunofluroscence using group and strain specific monoclonal antibodies.

2. Direct Detection of the Organisms and their Antigens

a. Detection of Rickettsiae in Tissue

Skin biopsies from the center of petechial lesions can be examined for rickettsiae by immunofluorescence or immuno-enzyme methods. Biopsy specimens may be stained with Giemsa, Macchiavello or Gimenez stains and with direct and indirect immunofluorescence techniques. In tissue smears, rickettsiae are usually seen as bipolar rods occurring near cells or free in the cytoplasm. *R. rickettsii* may also be seen within the nuclei of infected cells.

b. Polymerase Chain Reaction (PCR)

Detection of rickettsial DNA by PCR is more rapid than isolation.

3. Serology

Serological diagnosis may be by the heterophile Weil-Felix reaction or by specific tests using rickettsial antigens.

i. Weil-Felix (WF) Reaction

The Weil-Felix reaction is an agglutination test in which sera are tested for agglutinins to the antigens of certain nonmotile Proteus strains OX 19, OX 2 and OX K.

The test was developed from the chance observation of Weil and Felix (1916) that a Proteus strain isolated from the urine of a patient of epidemic typhus was agglutinated by the patient's serum as well as by the sera of other typhus patients.

Basis of The Test

The basis of the test is the sharing of an alkali-stable carbohydrate antigen by some rickettsiae and by certain strains of Proteus, *P. vulgaris* OX 19, and OX 2 and *P. mirabilis* OX K.

Procedure

The test may be performed as a micro-agglutination reaction in microtitre plates with round bottomed wells with hematoxylin-stained antigen or as a tube agglutination test though rapid slide agglutination methods have been employed for screening.

Interpretation

Sera from epidemic and endemic typhus agglutinate OX 19 and sometimes OX 2 also. The test is negative or only weakly positive in Brill-Zinsser disease. In tickborne spotted fever both OX 19 and OX 2 are agglutinated. OX K agglutinins are found only in scrub typhus. (Table 52.3).

The Weil-Felix reaction is a simple and useful test for the diagnosis of some rickettsial diseases. The antibody appears rapidly during the course of the disease, reaches peak titres of up to 1: 1000 or 1:5000 by the second week and declines rapidly during convalescence. False positive reaction may occur in some cases of urinary or other infections by Proteus and at times in typhoid fever and liver diseases. Hence it is desirable to demonstrate a rise in titre of antibodies for the diagnosis of rickettsial infection.

Note: Although the Weil-Felix test (which involves the differential agglutination of *Proteus* antigens) has been used historically for the diagnosis of rickettsial infections, it is not recommended now because it is insensitive and nonspecific.

ii. Specific tests using rickettsial antigens

Serological methods using rickettsial antigens are specific, which include complement fixation test, latex agglutination test and enzyme immunoassay.

a. Complement Fixation Test (CFT)

It is the most frequently employed serological method using rickettsial antigens. The test is done by using the group specific soluble antigen or the type specific washed rickettsial antigen. The CFT using the group specific soluble antigen is in routine use. The CFT using type specific antigen helps to differentiate between epidemic and endemic typhus.

b. Latex Agglutination Test

Latex agglutination test is available for Rocky Mountain spotted fever. It is positive only during an acute infection, so a single positive test is diagnostic.

Table 51.3: Weil-Felix reaction in rickettsial diseases

Disease	*Agglutination pattern with*		
	OX 19	OX2	OXK
1. Epidemic typhus	+++	+	–
2. Brill-Zinsser disease	Usually negative	or	weak positive
3. Endemic typhus	+++	+/–	–
4. Tickborne spotted fever	++	++	–
5. Scrub typhus	–	–	+++

c. Enzyme Immunoassay

Enzyme immunoassay with particulate or extracted antigens has been used.

Treatment

Rickettsial infections may be treated with tetracyclines or chloramphenicol. Both these drugs are rickettsiostatic.

Prophylaxis

It is virtually impossible to eradicate rickettsial infections because of their enzootic nature.

1. General Measures

General measures such as control of vectors and animal reservoirs are useful to prevent rickettsial diseases.

2. Vaccination

Immunization is useful in special situations. There is no safe, effective vaccine for any of the rickettsial diseases.

i. Weigl's Vaccine

In earlier days, phenolised intestinal contents of lice infected per rectum with R. *prowazekii* (Weigl's vaccine) was developed. It was too complicated for mass production.

ii. A Live Vaccine

Containing attenuated *E strain* of *R. prowazekii* grown in yolk sac have been found to be effective but a proportion of vaccinees develop mild disease.

iii. Formalin Inactivated R. rickettsii

It has been used but does not prevent the disease completely.

iv. Recombinant Vaccines

It may be more successful.

GENUS EHRLICHIA

Ehrlichiae are small gram-negative, obligately intracellular bacteria which have an affinity towards blood cells that parasitize mononuclear and granulocytic phagocytes but not erythrocytes. The intracellular location of the organisms protects them from the host's antibody response. They multiply within membrane-bound cytoplasmic vacuoles, usually in various phagocytes and form characteristic microcolonies resembling mulberries, termed morulae (Latin *morum* = mulberry).

Speceis

Three human infections caused by this group of organisms have been identified (Table 51.4).

1. Ehrlichia sennestu

Ehrlichia *sennestu,* an intraleukocytic bacterium closely related to the Rickettsiae, was first isolated in Japan in 1954 from a patient resembling glandular fever who showed serological response against the agent of canine ehrlichiosis and named *accordingly-sennetsu* being the Japanese name for glandular fever. Similar cases have been reported in Malaysia and Philippines.

It causes **lymphoid hyperplasia** and **atypical lymphocytosis.** No insect vector has yet been implicated in its transmission. It is associated with ingestion of raw fish infested with ehrlichia-infected flukes.

2. Ehrlichia chaffeensis

Tick-borne ehrlichiosis was first recognized in the United States in 1986. The disease, human monocytic ehrlichiosis was believed initially to be caused by Ehrlichia *canis.* However, a serologically distinct species, *Ehrlichia chaffeensis,* within the *E. canis* group was recognized to be the etiologic agent. This organism was first isolated from an army reservist at Fort Chaffee, Arkansas (hence the species name).

It is transmitted by Amblyomma ticks. Deer and rodents are believed to be reservoir hosts. Human disease is associated with leukopenia, thrombocytopenia and elevated liver enzymes. Multisystem involvement and fatality may occur.

3. Ehrlichia ewingii and Ehrlichia phagocytophila

Human granulocytic ehrlichiosis is caused by two species of *Ehrlichia—Ehrlichia ewingii* and *Ehrlichia phagocytophila. E. ewingii* is a newly identified species in the *E. canis* group that has a geographic distribution similar to *E. chaffeensis.* It is transmitted by Ixodes ticks. Deer, cattle and sheep are the suspected reservoir. Leukopenia and thrombocytopenia are seen in patients. Giemsa stained blood films may show morula form of the ehrlichia. Doxycycline is recommended for treatment of ehrlichioses.

Laboratory Diagnosis

1. **Identification of characteristic morulae:** Human granulocytic ehrlichiosis can often be diagnosed by identification of characteristic morulae in **Giemsa stained peripheral blood neutrophils**. *E. chaffeensis* is rarely detected in monocytes in blood smears.
2. **Specific antibodies** can be demonstrated by indirect immunofluorescence methods.
3. **Polymerase chain reaction (PCR)** with specific primers to amplify the ehrlichial DNA.

GENUS COXIELLA: Q FEVER

The name *Q fever* (Q for "query") was first used when Derrick (1935) was investgating an outbreak of typhoid-like fever in abattoir workers in Australia. Q fever is caused by *Coxiella burnetit* (after the names of *Cox* and *Burnet* who identified the same agent independently). *Coxiella burnetii,* is an obligately intracellular prokaryote but genetic analysis suggests that it is more closely allied to *Legionella* species.

Morphology

Cox. burnetii is pleomorphic, occurring as small rods 0.2 to 0.4 µm × 0.4-1.0 µm or as spheres 0.3 to 0.4 µm in diameter. It is filterable. Generally regarded as gram-negative. It is better stained with Giminez and other rickettetsiae stains.

Resistance

C. burnetii may be the most infectious of all bacteria.In dried feces or wool it survives for a year or more at 4°C and in meat at least for one month. It is not completely inactivated at 60°C or by 1 percent phenol in one hour. In milk it may survive pasteurization by the holding method but the flash method is effective. It can survive in dust and aerosols, therefore, can be transmitted as an airborne infection. It can be inactivated by 2 percent formaldehyde, 5 percent hydrogen peroxide and 1 percent lysol. It grows well in the yolk sac of chick embryos and in various cell cultures.

Antigenic Variation

An important characteristic of *Coxiella* infections is the ability to undergo **antigenic variation** in expression of the cell wall LPS antigen. *Cox. bumetii* shows phase variation. Fresh isolates are in **Phase I**. It becomes **Phase II**

Table 51.4: Ehrlichia species responsible for human disease

Species	*Disesase*	*Vector or Source*	*Reservoir*
E. sennetu	Sennestu fever	Ingestion of raw fish infested with infected flukes	Unknown
E. chaffeensis	Monocyclic ehrlichiosis	*Amblyomma americanum* (Lone Star tick) *Amblyomma*	White-tailed deer, domestic dogs
E. ewingii	Granulocytic ehrlichiosis	*Amblyomma americanum*	Canines
E. phagocytophila	Granulocytic ehrlichiosis	*Ixodes* (including *1. scnpu/aris,1. pacificus, 1. ricinus)*	Small mammals (e.g., white footed mouse; chipmunks, voles)

on repeated passage in yolk sac but reverts to Phase I by passaging in guinea pigs.

Phase I cells are autoagglutinable and are phagocytosed in the absence of antibody. Phase I activity is due to a periodate-sensitive trichloracetic acid-soluble surface carbohydrate. Phase I is a more powerful immunogen than Phase II and elicits good antibody response to both I and I1 antigens. Acute disease is characterized by antibodies against the exposed phase II antigen, whereas high antibody titers against the phase I and II antigens are detected in patients with chronic infections. Phase II antigen is more suitable for complement fixation tests. Q fever sera do not cross react with rickettsial or proteus bacillus antigens.

Pathogenesis

Coxiella *burnetii* causes Q fever which has a worldwide distribution, as a **zoonosis** solidly established in **domestic livestock**.

Cycles of Infection

In nature, there are two cycles of infection of C. *burnetii.* One involves arthropods (especially ticks) and a variety of vertebrates. The other cycle is maintained among domestic animals (cattle, sheep and goats).

Reservoirs of the Disease

The primary reservoirs of the disease are wild and domestic ungulates, including cattle, sheep, goats, rabbits, cats and dogs. It is transmitted among them and to cattle, sheep and poultry by ixodid ticks. Transovarial transmission occurs in ticks. Coxiella are abundant in tick feces and survive in dried feces for long periods. They are shed in the milk of infected animals.

Animal Infection

Domestic animals have inapparent infections but may shed large quantities of infectious organisms in their urine, milk, feces and especially, their placental products.

Human Infection

Human infection may occur occupationally through consumption of infected milk, handling of contaminated wool or hides, soil contaminated by infected animal feces, infected straw and even to dusty clothing. *C. burnetii* may enter the body through the skin (e.g. a contaminated minor abrasion), lungs (e.g., inhalation of infectious aerosols), mucous membranes (e.g. conjunctival contact with infectious materials) or gastrointestinal tract (e.g. ingestion of contaminated raw milk). *Coxiella* proliferate in the respiratory tract and then disseminate to other organs. Person-to-person transmission is a rarity. Ticks do not seem to be important in human infection.

Incubation period is 2 to 4 weeks. Acute diseases include influenza-like syndrome, atypical pneumonia, hepatitis, pericarditis, myocarditis, meningoencephalitis. Chronic diseases include endocarditis, hepatitis, pulmonary disease and infection of pregnant

Laboratory Diagnosis

1. Isolation

Isolation of *C. burnetii* from patient specimens is a specialized procedure and is not generally recommended because of the extremely infectious nature of the organism.

2. Serology

Using complement fixation test (CFT) or indirect immunofluorescence assay.

Treatment

Tetracyclines are the drugs of choice for acute infections. Rifampin combined with either doxycycline or trimethoprim-sulfamethoxazole is used to treat chronic infections.

Prophylaxis

Measures for control of Q fever include:

i. Construction of separate facilities for animal parturition;
ii. Destruction of suspected placentas
iii. Heat treatment of milk at 74°C for 15 seconds;
iv. Abattoir workers should take care like wearing of gloves, glasses, mask, etc. while handling carcasses and animal hides.

Vaccine

For the prevention of Q fever, vaccines have been developed from formalin killed whole cells, attenuated *C. burnetii* and trichloroacetic acid extracted antigens. Vaccines derived from phase I organisms generate a much greater protective response than similar ones prepared from phase II organisms.

BARTONELLA

Family Bartonellaceae contains two genera:**Bartonella and Grahamella**

The genus *Rochalimaea* was transferred into the genus *Bartonella.* Members of genus *Bartonella* are very small gram-negative bacilli. They grow in close association with the surfaces of vertebrate erythrocytes, including those of humans. They are mainly arthropod-borne. They cause feverish illness in humans involving red blood cells. The genus *Bartonella,* which now consists of 11 species, including 5 that cause human disease (Table 52.5). Members of the genus *Grahamella* preferentially grow within the erythrocytes of vertebrates but do not infect humans.

Bartonella Bacilliformis

It is a pleomorphoic gram-negative rod, which is motile by a tuft of polar flagella. It can be cultivated in semisolid agar with rabbit or human blood. Growth is slow and becomes visible after 10 days incubation.

Species	Diseases
1. B. bacilliformis	Oroya fever (bartonellosis, Carrion's disease)
2. B. qintana	Trench fever; cutaneous, subcutaneous, and osseous manifestations of bacillary angiomatosis; endocarditis
3. B. henselae	Cat-scratch disease; bacteremia; cutaneous, lymphatic, and hepatosplenic (peliosis hepatis) manifestions of bacillary angiomatosis; endocarditis
4. B. clarridgeiae	Endocarditis, cat-scratch disease (rare)
5. B. elizabethae	Endocarditis (rare)

Table 51.5: *Bartonella* species associated with human infections

Pathogenesis

Oroya fever-*Bartonella bacilliformis* is responsible for bartonellosis, an acute febrile illness consisting of severe anemia **(Oroya fever)** followed by a chronic cutaneous form **(verruga).** The name Oroya fever was given after an epidemic of the disease in 1870 during the building of a railway between Lima and Oroya in Peru, when 7000 labourers died within a few weeks. The disease was called *Oroya fever,* which was seen in the mountainous parts of Peru, Columbia and Ecuador in South America. Oroya fever presents as fever and progressive anemia due to bacterial invasion of erythrocytes. Mortality is high in untreated cases.

Verruga Peruana

Some of the survivors developed nodular ulcerating skin lesions, called **verruga peruana**. It is a late sequel in survivors or in those with asymptomatic infection.

Etiology

It is spread by the **sandfly vector *Phlebotomus*.**

Carrion's Disease

The common etiology of these two conditions was established tragically in 1885 by the Peruvian medical student Daniel Carrion. He inoculated himself with material from verruga and developed oroya fever from which he died. Oroya fever is therefore also known as **Carrion's disease.**

Laboratory Diagnosis

i. *B. bacilliformis* can be demonstrated in blood smears stained by Giemsa. Organisms are seen in the cytoplasm as well as adhering to the cell surfaces.
ii. Organisms may be isolated from the blood, in semisolid medium containing rabbit serum and hemoglobin. Visible growth may take up to 10 days. Identification can be achieved by PCR or by cultural and serological tests.
iii. Guinea pig inoculation leads to verruga peruana but not oroya fever.
iv. Other tests
    a. **Polymears cahain reaction(PCR)** with primers for 16S ribosomal RNA sequences provides a rapid means of diagnosis.
    b. **Serology**-Antibodies to *Bart. bacilliformis* can be detected by indirect hemagglutination, indirect immunofluorescence and ELISA.

Treatment

B. bacillifornds is susceptible to penicillin, streptomycin, tetracycline and chloramphenicol. These may be used for the treatment of oroya fever and verruga peruana.

Prophylaxis

Insecticides such as DDT should be used to eliminate the sandfly inside and outside the houses.

Bartonella Quintana

This organism was formerly' classified among the rickettsiae as *Rochalimaea quintana* (from *quintana,* meaning fifth, referring to 'five-day fever', a synonym for trench fever). As it was found to differ from rickettsiae in a number of respects, including its ability to grow in cell-free culture media such as blood agar, it was separated into a new genus Rochalimaea (after da Rocha Lima, an early investigator of rickettsial diseases). In a subsequent taxonomical shift, it has been reclassified as Bartonella and named *B. quintna.*

B. quintana is a small, gram-negative bacillus. It does not possess flagella, although it may exhibit twitching movement caused by fimbriae. It grows slowly on rabbit or sheep blood agar at 35°C in 5 percent CO_2 in air. Colonies appear after two weeks in primary culture and 3 to 5 days in subsequent passages.

Pathogenesis

Trench fever or five-day fever- *B. quintana* causes **trench fever or five-day fever** occurred among soldiers fighting in the trenches in Europe during the First World War. The disease was not fatal but because of its slow course and prolonged convalescence, it caused very considerable loss of manpower.

Trench fever is an **exclusively human disease** and no animal reservoir is known. It is transmitted by the **body louse**. The feces of lice becomes infectious 5 to 10 days after an infectious meal. Infection is acquired when infected lice feces are scratched into the skin. The lice are unharmed and remain infectious for life. Vertical transmission does not occur in lice.

Clinical Diseases

Clinically, after an incubation period of 14 to 30 days patient develops headache, malaise, fever, chills, severe pain in the back and legs and a roseolar rash on the chest, abdomen or back. Recovery is frequently followed by relapses occurring at five day intervals (hence the species name, quintana). The infection is mild, recovery is slow and fatalities are rare.

More recently, *B. quintana* has been associated with **bacillary angiomatosis**, a vascular proliferative disorder seen primarily in immunocompromised patients e.g. patients with the human immunodeficiency virus (HIV) as well as with endocarditis in immunocompetent patients.

The disease frequently leads to a chronic or latent infection. Recrudescence may occur as in Brill-Zinsser disease and relapses have been reported as long as 20 years after the primary disease. The chronic infection and late relapses help to maintain the bartonella in the absence of animal reservoirs.

Trench fever was thought to have vanished with the world wars. But isolation of *B. quintana* from Tunisia and Mexico recently suggests that the disease may be more widely distributed than is realized. Trench fever cases' have been identified in some homeless persons living in unsanitary conditions in the USA.

Laboratory Diagnosis

i. *B. quintana* does not grow in mice or guinea-pigs. It can be isolated by allowing healthy lice to feed upon the patient and the bartonellae may be detected in the gut of these lice.
   It can also be cultivated from patient's blood on rabbit or sheep blood agar.
ii. **Polymearse chain reaction (PCR)—***B. quintana* has been detected in the tissues by PCR.
iii. **Weil-Felix reaction** in trench fever, is negative.

Bartonella Henselae

Bart. henselae is responsible for cat scratch disease. It is small, slightly curved gram-negative bacillus. It shows twitching motility. It can be grown on chocolate agar, Columbia agar with 5 percent blood and trypticase agar. The optimum temperature for growth is 35 to 37°C in 5 percent CO_2, Growth is slow and takes 5 to 15 days. Colonies are white, dry, cauliflower-like and embedded in the agar.

Pathogenesis

1. Cat scratch disease

The disease is acquired after exposure to cats (e.g. scratches, bites and contact with cat fleas). Typically, cat-scratch disease is a benign infection in children, characterized by chronic regional adenopathy of the lymph nodes draining the site of contact. Cat scratch disease is a severe condition of regional lymphadenopathy and fever resulting from the scratch or bite of an infected cat. *Bart. clarridgeiae,* can cause an identical syndrome. An organism known as *Afipia felis* has also been implicated in a small proportion of cases of cat scratch disease.

2. Bacillary angiomatosis

B. henselae is also responsible for bacillary angiomatosis, but primarily involving the skin, lymph nodes or liver and spleen (peliosis hepatis). *Bart. henselae* and less frequently, *Bart. quintana* and other species have been identified in the blood and tissues of individuals suffering from two severe clinical syndromes associated with human immunodeficiency virus (HIV) or other immunosuppressant conditions. These are *Bacillary angiomatosis,* which produces proliferative vascular lesions in the skin, regional lymph nodes and various internal organs, and *Bacillary peliosis,* which affects the liver and spleen.

3. Subacute bacterial endocarditis

B. henselae can cause subacute bacterial endocarditis, like *B. quintana.*

Laboratory Diagnosis

1. It can be demonstrated in lymph node biopsy smears and sections by Warthin-Starry staining show clusters of bacilli.
2. *B. henselae* has been isolated from the blood of patients in blood media after prolonged incubation and is now considered as its etiological agent.

Treatment

Cat-scratch disease is usually self limiting disease and does not appear to respond to antimicrobial therapy.

KNOW MORE

Spotted Fever Group

Tick-borne rickettsiae of the spotted fever group are maintained in enzootic cycles involving ticks and their wild animal hosts. Ticks are the primary reservoirs of the rickettsiae, and maintain the organisms by both transstadial transmission (larvae to nymph to adult tick) and transovarial or vertical transmission, which therefore act as both vectors and reservoirs. Ticks are not harmed by the rickettsiae and remain infected for life.

☞ KEY POINTS

- The family rickettsiaceae comprises three genera -Rickettsia. Orientia and Ehrlichia. The important species in the family include *Rickettsia prowazekii, Rickettsia typhi, Rickettsia rickettsii, Orientia tsutsugamushi* and the Ehrlichia species.
- **Genus Rickettsia**- Rickettsiae are gram-negative, small, intracellular bacteria. Stain poorly with gram stain; best with Giemsa or Gimenez stains. Replication occurs in cytoplasm of infected cells. They grow in various tissue cultures and in yolk sac of *embryo*nated egg.
- **Diseases**
  1. **Epidemic typhus (louse-borne typhus)- caused by** ***R. prowazekii.*** The principal vector is the human body louse.
  2. Brill-zinsser disease (Recrudescent typhus)- caused by *R. prowazekii.*
  3. **Endemic typhus** is caused by R. *mooseri* (R. *typhi).* Vector is rat flea *(Xenopsyella cheopis).*

- **R.** ***mooseri*** can be differentiated from R. *prowazekii* by *Neil-Mooser* or *tunica reaction.*
- **R.** ***rickettsii*** causes **Rocky Mountain spotted fever** and ticks being the vectors of transmission.
- **R.** ***conori*** *is* responsible for **boutonneuse fever** and transmitted by ixodid ticks.
- ***Rickettsia akari*** causes rickettsial pox. Infection is transmitted by the bite of mites.
- **O** ***tsutsugamushi*** causes scrub typhus. The disease is transmitted by trombiculid mite.
- **Laboratory diagnosis** of rickettsial diseases may be carried out by *isolation of rickettsiae* and *serology.*
  a. **Serology-**By the heterophile Weil-Felix reaction or by specific tests using rickettsial antigens. **The Weil-Felix reaction** is an agglutination test which detects anti-rickettsial antibodies in which sera are tested for agglutinins to the antigens of certain nonmotile Proteus strains OX 19, OX 2 and OX K.
  b. **Specific tests using rickettsial antigens-**include complement fixation test, latex agglutination test and enzyme immunoassay.

Ehrlichia

- These tick-borne bacteria cause three human infections: *E sennetsu* causes a type of glandular fever; *E chaffeensis* causes human monocytic ehrlichiosis; *E phagocytophi/a* causes human granulocytic ehrlichiosis

Coxiella burnetii

- It is small, pleomorphic coccobacillary bacterium with a gram-negative cell wall, intracellular bacteria.
- Disease-Q fever is a worldwide **zoonosis** affecting wild and domestic animals. Most disease acquired through inhalation; possible disease from consumption of contaminated milk.
  In humans, Q fever manifests as **acute diseases** such as influenza-like syndrome, atypical pneumonia, hepatitis, pericarditis, myocarditis, meningoencephalitis and **chronic diseases** such as endocarditis, hepatitis, pulmonary disease, and infection of pregnant women.
- Diagnosis-The diagnosis of Q fever is based mainly on serological tests. **PCR** and a real-time PCR assay can also be used.

Bartonella

The genus Bartonella contains *B. bacilliformis, B. quintana* and *B. henselae* which cause Oroya fever, trench fever and cat-scratch disease in man respectively.

IMPORTANT QUESTIONS

1. Write short notes on:
   Typhus fevers
   Brill-zinsser disease (Recrudescent typhus)
   Rocky Mountain spotted fever
   Scrub typhus
   Trench fever
   Coxiella burnetti or Q fever
   Weil-Felix reaction
   Cat-scratch disease
   Neil-Mooser reaction or Tunica reaction
   Ehrlichia
   Oroya fever
2. Discuss laboratory diagnosis of rickettsial infections.

FURTHER READING

Adal K A Bartonella: New species and new diseases. Reviews in Medical Microbiology 1995;6:155-64.

Dumler JS: Laboratory diagnosis of human rickettsial and ehrlichial infections, Clin Microbiol Newsletter 1996;18:57-61.

English CJ, Wear DJ et al. Cat-scratch disease. Isolation and culture of the bacterial agent. JAMA 1988;259:1347-52.

La Scola B, Raoult D: Minireview: Laboratory diagnosis of rickettsioses: current approaches to diagnosis of old and new rickettsial diseases, J Clin Microbiol 1997;35:2715-27.

Maurin M, Raoult D. Q fever, Clin Microbiol Rev 1999;12:518-53.

Raoult D, Roux V: Rickettsioses as paradigms of new fremerging infectious diseases, Clin Microbiol Rev 1997;10:694-719.

CHAPTER 52

Chlamydia and Chlamydophila

Learning Objectives

After reading and studying this chapter, you should be able to:

- Describe differences between chlamydiae and viruses.
- Describe serotypes of various chlamydiae and diseases produced by them.
- Associate each of the major diseases with the three most important human species of *Chlamydia* and *Chlamydophila*
- Discuss morphology and the unique growth cycle of *Chlamydia*, describing elementary (EB) and reticulate (RB) bodies.
- Discuss laboratory diagnosis of chlamydial infections.
- Describe the following: TRIC agents, inclusion conjunctivitis lymphogranuloma venereum (LGV), Frei's test

INTRODUCTION

Chlamydiaceae is a family of obligate intracellular bacterial parasites, small, non-motile and gram-negative with a tropism for columnar epithelial cells lining the mucous membranes. They are widely distributed in nature and are responsible for a variety of ocular, genitourinary and respiratory diseases in man. There is some evidence that they may be involved in atherosclerosis and, possibly, other chronic diseases.

They cause psittacosis, lymphogranuloma venerum (LGV) and trachoma in man and diverse diseases in birds and mammals. Based on the human diseases they were then known to cause, they were called **psittacosis-lymphogranuloma-trachoma (PLT) viruses**, or noncommittally as **'PLT agents'** or TRIC (trachoma-inclusion conjunctivitis) organisms. In recognition of the pioneering studies of Sir Samuel Bedson on psittacosis, the name ***Bedsonia*** was proposed for this group. However, the official term for this group now is *Chlamydia*. The name *Chlamydia* is derived from the characteristic appearance of inclusion body by these agents which enclose the nuclei of infected cells as a cloak or mantle (*chlamys*, meaning mantle).

Differences between Chlamydiae and Viruses

The Chlamydiaceae were once considered viruses because they are small enough to pass through 0.45 µm filters and are obligate intracellular parasites. However, they differ from viruses in many respects and the organisms have the following properties of bacteria:

1. They possess inner and outer membranes similar to those of gram-negative bacteria.
2. They contain both DNA and RNA.
3. They possess prokaryotic ribosomes.
4. They synthesize their own proteins, nucleic acids, and lipids.
5. They multiply by binary fission.
6. They do not have 'eclipse phase' following cellular infection.
7. They are susceptible to numerous antibacterial antibiotics.

CLASSIFICATION

Genus *Chlamydia* is in the order Chlamydiales and the family Chlamydiaceae. The family Chlamydiaceae was formerly classified into one genus, *Chlamydia*, with four species (*Chlamydia trachomatis, C. psittaci, C. pneumoniae* and *C. pecorum*). All except *C. pecorum* have been associated with human disease.In 1999, a new taxonomic classification of the family Chlamydiaceae has been proposed on the basis of genomic studies of these organisms. The proposed taxonomic classification for the family Chlamydiaceae consists of two genera: (1) *Chlamydia* to include *C. trachomatis* and (2) *Chlamydophila* to include *C. pneumoniae, C. psittaci,* and *C. pecorum.* Other species have been placed into the two genera, but they are uncommon human pathogens and are not discussed in this chapter (Table 52.1).

CHLAMYDIA SPECIES

The family Chlamydiaceae was formerly classified into four species, belonging to the single genus *Chlamydia:*

Table 52.1: Revised Classification of the Family Chlamydiaceae	
Genus Chlamydia	*Genus Chlamydophila*
C. *trachomatis*	C. *pneumoniae*
C. *muritlarum*	C. *psittaci*
C. *S'llis*	C. *pecorum*
	C. *abortus*
	C. *caviae*
	C. *felis*

Chlamydia trachomatis

C. *trachomatis* strains form compact inclusions with the glycogen matrix (iodine-staining cytoplasmic inclusions), are sensitive to sulphonamides. *C. trachomatis* is divided into two biovars—those causing *trachoma, inclusion conjunctivitis* (the so-called TRIC agents) and oculogenital infection.

Chlamydia psittaci

C. p*sittaci* strains form diffuse vacuolated inclusions without the glycogen matrix (do not produce iodine-staining inclusions), are resistant to sulphonamides.

Ch. psittaci is a natural parasite of birds (particularly in psittacine and ornithine bird) and animals. It may lead to severe and sometimes fatal pneumonia in man *(psittacosis* or *ornithosis).*

Chlamydophila pneumoniae

Ch. pneumoniae is the former TWAR (Taiwan Acute Respiratory) agent. It is an exclusive human pathogen with no animal or avian host. It is a common cause of acute respiratory disease worldwide. (Table 52.2).

Morphology

Chlamydiae are small, non-motile bacteria. Although they stain poorly with gram's stain they have the typical LPS of Gram-negative bacteria. There are two morphologically distinct forms of chlamydiae: elementary body (EB) and reticulate body (RB) (Table 52.3).

Elementary body (EB)

It is a spherical particle, 200-300 nm in diameter, with a rigid trilaminar cell wall similar to the cell walls of gram-negative bacteria, and an electron dense nucleoid. The elementary body is the extracellular, infective form, responsible for attaching to the target host cell and promoting its entry.

Reticulate Body (RB)

The reticulate body is the intracellular growing and replicative form, 500-1000 nm in size (larger than the EB) and is osmotically fragile. Its cell wall is fragile and pliable, leading to pleomorphism. They are metabolically active, so their cytoplasm is rich in ribosomes, which are required for protein synthesis.

Table 52.2: Differential characteristics of the family Chlamydiaceae that cause human disease

Property	*Chlamydia trachomatis*	*Chlamydophila psittaci*	*Chlamydophila pneumoniae*
1. Host range	Humans	Birds	Humans
2. Elementary body morphology	Round	Round	Pear-shaped
3. Inclusion morphology	Round, vacuolar	Variable, dense	Round, dense
4. Glycogen in inclusions	Yes	No	No
5. Plasmid DNA	Yes	Yes	No
6. Susceptibility to sulphonamides	Yes	No	No
7. Major diseases	Lymphogranuloma venereum (LGV) trachoma: ocular trachoma, oculogenital disease, infant pneumonia	Pneumonia (psittacosis)	Bronchitis, pneumonia, sinusitis, pharyngitis, coronary artery disease (?)

Table 52.3: Comparison of chlamydial elementary bodies and reticulate bodies

Characteristic	*Elementary body*	*Reticulate body*
1. Size	0.2-0.3 µm	1 µm
2. Morphology	Electron-dense core; rigid	Fragile, pleomorphic
3. Infectivity to host	Infectious	Non-infectious
4. RNA:DNA ratio	1:1 (condensed DNA core)	3:1 (increased ribosomes)
5. Metabolic activity	Relatively inactive	Active. Replicating stage
5. Trypsin digestion	Resistant	Sensitive
6. Projections and rosettes	Few	More

The chlamydiae are nonmotile, lacking flagella, and nonpiliated.

Growth Cycle

The chlamydia growth cycle is initiated by the attachment of an infectious elementary body to the surface of a susceptible epithelial cell, followed by its endocytosis (Fig. 52.1). Typically, chlamydiae attach to the host cell near the base of microvilli, from where they are actively enclosed in tight endocytic vesicles. Inside the host cell, the elementary body lies within the endosome, being separated from the host cell cytoplasm by the endosomal membrane throughout its active growth cycle.

By about eight hours, the elementary body within the endosome undergoes spheroplast-like transformation to the large reticulate body, which begins to divide by binary fission by 12 hours. By 20-24 hours, the pleomorphic progeny show central condensation and are converted to elementary bodies. Binary fission continues till about 40 hours. The developing chlamydial microcolony within the host cell is called the ***inclusion body.*** The mature inclusion body contains 100-500 elementary bodies which are ultimately released from the host cell.The host cell is severely damaged and release of the elementary bodies occurs within 48 hours by host cell lysis in *C. psittaci* infections. With *C. trachomatis,* the mature inclusion appears to be exocytosed in 72-96 hours, the host cell being left with a scar (Fig. 52.1).

The *Chlamydia*-specific lipopolysaccharides accumulate on the host cell surface during the active intracellular growth of the organism. This highly antigenic material induces inflammatory and immunological responses which contribute to the pathogenesis of chlamydial diseases.

Laboratory Propagation

Chlamydiae can be propagated in the mouse, chick embryo or in cell culture though they show individual variations in susceptibility. The presence of chlamydial inclusions is determined by microscopy in conjunction with a suitable staining method, preferably fluorescence microscopy with labelled monoclonal antibody.

Antigenic Structure

Chlamydiae possess three major groups of antigens: genus-specific antigens, species-specific antigens, and serotype-specific antigens.

1. Genus-specific Antigens

The heat-stable genus-specific antigen common to all *Chlamydia* is LPS, with an acidic polysaccharide as the antigenic determinant. The antigen, which is present throughout the developmental cycle, can be detected by the complement fixation test, SDS polyacrylamide gel electrophoresis, and by polyclonal or epitope specific monoclonal antibodies.

2. Species-specific Antigens

Several species-specific antigens have been detected on or near the envelope surface. These protein antigens are shared by all members of a chlamydial species. They help in classifying chlamydiae into the species—*trachomatis, psittaci, pneumoniae* and *pecorum.*

3. Serotype-specific Antigens

Serotype-specific determinants are common only to certain isolates within a species and help in intraspecies typing. They are located on the major outer membrane proteins (MOMP) and can be demonstrated by microim-

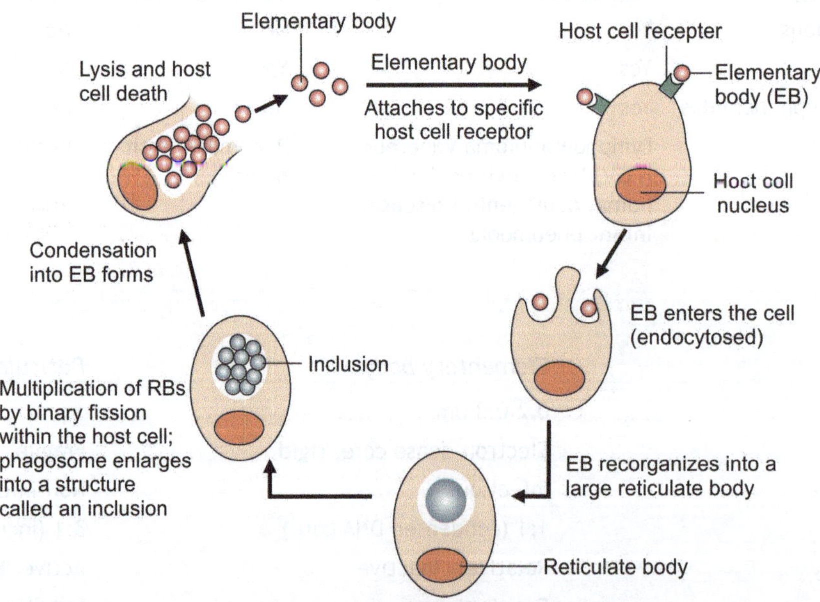

Fig. 52.1: The chlamydia growth cycle

munofluorescence. By micro-IF, chlamydiae have been classified into many serological variants (**serovars, serotypes**).

Chlamydia trachomatis has been divided into three biovars: **trachoma, lymphogranuloma venereum (LGV),** and **mouse pneumonitis** (renamed *C. muridarum*). *C. trachomatis* is a leading cause of ocular and genital infections worldwide.

Chlam. Trachomatis

Chlamydia trachomatis has been divided into three biovars (biological variants) which cause trachoma, inclusion conjunctivitis (the so-called TRIC agents) and lymphogranuloma venereum (LGV), **and mouse pneumonitis** (renamed *C. muridarum*). Both the trachoma and the LGV biovars can be divided into serotypes (serovars) on the basis of epitopes carried on their major outer membrane protein (MOMP).

Trachoma Biovar

There are 15 serovars in the trachoma biovar (with variants within them) which are given the letters serovars A through K (A, B, Ba, C, D, Da, E, F, G, Ga, H, I, Ia, J, and K). Serovars A, B, Ba and C cause blinding trachoma in endemic areas, and serovars D to K associated with the less serious ocular infection, inclusion conjunctivitis and with various genital infections.

LGV Biovar

There are four serovars Ll, L2, L2a and L3, in the LGV biovar. Serovars L_1, L_2, L_{2a}, and L_3 are associated with LGV, an invasive urogenital tract disease and hemorrhagic proctitis.

Chlaniydia psittaci

The serological classification of C. *psittaci* is complex, many serotypes having been identified.

Chlanuydaphila pneumoniae

C. pneumoniae has not been subclassified yet (Table 52.4).

Chlamydia trachomatis

C. trachomatis is a leading cause of ocular and genital infectios worldwide.

1. Ocular Infections

1. Trachoma

Trachoma is a chronic keratoconjunctivitis. It is characterized by follicular hypertrophy, papillary hyperplasia, pannus formation and in the late stages, cicatrisation caused by *C. trachomatis* serotypes A, B, or C. The name *trachoma,* derived from the Greek word *trakhus* (rough), refers to the pebbled appearance of the infected conjunctiva.

Trachoma is transmitted eye-to-eye by droplet, hands, contaminated clothing, and eye-seeking flies. The pathogen may also be transmitted by respiratory droplet or through fecal contamination. Trachoma generally is endemic in communities where the living conditions are crowded, sanitation is poor, and the personal hygiene of the people is poor—all risk factors that promote the transmission of infections.

According to the internationally accepted MacCallan classification of trachoma, there are four major stages of disease. Established trachoma progresses through stages I-IV. Infectivity is maximum in the early cases. Stage IV is noninfectious.

Though Halberstaedter and Prowazek in 1907 transmitted the infection to orangutans and demonstrated in conjunctival smears the characteristic inclusion body that bears their names, cultivation of the chlamydia became possible only half a century later, when Tang and colleagues (1957) grew it in the yolk sac of eggs.

Laboratory Diagnosis

1. Direct Cytopathologic Examination

The characteristic inclusion *(Halberstaedter Prowazek* or *HP bodies)* may be demonstrated in conjunctival scrapings, after staining by Giemsa, Castaneda or Machiavello methods. They may be stained with iodine solution also because they possess a glycogen matrix. The fluorescent antibody method enhances the sensitivity of smear diagnosis.

2. Culture

i. Embryonated Egg

The chlamydia may be grown in the yolk sac of 6-8 days old eggs. The material is treated with streptomycin or

Table 52.4: Human diseases caused by Chlamydiae

Species	*Serotype**	*Reticulate body*
C. trachomatis	A, B, Ba, C	Endemic blinding trachoma
C. trachomatis	D, E, F, G, H, I, J, K	Inclusion conjunctivitis (neonatal and adult) Genital chlamydiasis Infant pneumonia
C. trachomatis	L1, L2, L3	Lymphogranuloma venereum
C. psittaci	Many serotypes	Psittacosis
C. pneumoniae	Only one serotype	Acute respiratory disease

* Predominant types associated with the disease

polymyxin B before inoculation. The eggs are incubated at 35°C in a humid atmosphere. Blind passages may be necessary for isolation. This method is seldom used now as it is time consuming, cumbersome and relatively insensitive.

ii. Cell Culture

Tissue culture using stationary phase cells (non-replicating cells) is the method of choice for isolation. The bacteria infect a restricted range of cell lines *in vitro* (e.g., HeLa-229, McCoy BHK-21, Buffalo green monkey kidney cells), similar to the narrow range of cells they infect *in vivo*. McCoy cells rendered nonreplicating by irradiation or antimetabolites are used. HeLa or HL cells treated with DEAE dextran may also be used. The inoculum has to be driven into the cells by centrifugation upto 15,000 g to get a good growth.

Epidemiology

C. trachomatis is found worldwide and causes trachoma (chronic keratoconjunctivitis). An estimated 500 million people worldwide are infected with the serovar trachoma, 7 to 9 million of whom are blinded as a result. Trachoma is endemic in the Middle East, North Africa, and India. Infections occur predominantly in children, who are the chief reservoir of *C. trachomatis* in endemic areas.

Treatment and Control

Local application and oral administration of erythromycin and tetracycline or other suitable antibiotics should be continued for several weeks. Single-dose azithromycin treatment has been used with good results.

Vaccines that are efficacious and safe are not available. Basic, however, to the ultimate control of trachoma, are good standards of hygiene that accompany improvement in standards of living.

iii. Inclusion conjunctivitis

The epidemiology of this condition, first recognized by Halberstaedter and Prowazek in 1910 had to be reestablished in recent years. The natural habitat of *C. trahomatis* types D to K is the genital tract in both sexes.

a. Inclusion Blennorrhoea

'Inclusion blenorrhea', is the neonatal form of inclusion conjunctivitis. The disease in the newborn usually becomes clinically apparent 5 to 12 days after birth. The organisms are acquired from the mother during birth. About half of infants born through a Chlamydia infected cervix develop ocular infection. The disease can be prevented by local application of antibiotics.

b. Adult Inclusion Conjunctivitis

Inclusion conjunctivitis (paratrachoma) is most prevalent in sexually active young people, being spread from genitalia to the eye and occurs worldwide. It was known as **'swimming pool conjunctivitis'** as infection was associated with bathing in community swimming pools which presumably get contaminated with chlamydia from the genital secretions of bathers. Contamination of the eye with the patient's own genital secretion may be the cause more often. The disease is much milder, is usually self-limiting and rarely causes visual loss, presumably because, unlike trachoma, repeated infection is less common.

2. Infant Pneumonia

Neonatal pneumonia, *C. trahomatis* is a prevalent cause of pneumonitis in infants. The children characteristically become ill at 4 to 16 weeks of age, have prominent respiratory symptoms of wheezing and cough, and lack systemic findings of fever or toxicity. They may be eosinophilic and have elevated serum IgG and IgM, with a very pronounced titer to the infecting serovar. Immune response is believed to have a role in pathogenesis of this condition. Chlamydial neonatal conjunctivitis often precedes the onset of the pneumonia. Radiographic signs of infection can persist for months.

3. Genital Infections

C. trachomatis infections of the genital tract are of two types and are sexually transmitted:

A. Those caused by the **oculogenital serotypes D through K** collectively referred to as 'genital chlamydiasis'
B. LGV caused by serotypes Ll, L2, and L3.

A. Genital Chlamydiasis

i. Infection in Men

In men, they cause urethritis ('nongonococcal urethritis'), epididymitis, proctitis, conjunctivitis and Reiter's syndrome. (Reiter's syndrome is a triad of recurrent conjunctivitis, polyarthritis and urethritis or cervicitis, associated with many infections but most commonly with *C. trachomatis).* Anal intercourse may cause chlamydial proctitis in either sex. Associated symptoms include rectal pain and bleeding, mucopurulent discharge and diarrhea.

ii. Infection in Women

Most genital tract infections in women are asymptomatic (as many as 80%) but can nevertheless become symptomatic. **The clinical manifestations** include acute urethral syndrome, bartholinitis, mucopurulent cervicitis, endometritis, salpingitis, pelvic inflammatory disease (PID), conjunctivitis, perihepatitis (Fitz-Hugh-Curtis syndrome) and Reiter's syndrome. Genital chlamydiasis may cause infertility, ectopic pregnanc, premature deliveries, perinatal morbidity and postpartum fever.

B. Lymphogranuloma venereum

Lymphogranuloma inguinale, climatic bubo, tropical bubo, and esthiomene are synonyms of this sexually transmissible disease. It is characterized by suppurative inguinal adenitis. Humans are the sole natural hosts of this infection caused by the LGV biovar of *C. trachomatis,* Ll, L2 and L3 most commonly L2 (Table 52.4).

Clinical Manifestations.

The usual incubation period of LGV is I to 4 weeks.

1. Primary Lesion

The primary lesion is painless, small, inconspicuous, and vesicular and often escapes notice. Characteristically the presenting complaint concerns the enlarged matted inguinal and femoral lymph nodes which are moderately painful and firm and may become fluctuant.

2. Secondary Stage

The secondary stage results from lymphatic spread to the draining lymph nodes. In men, the inguinal lymph nodes are involved most often, and in women, the intrapelvic and pararectal nodes. Women and homosexual men may develop hemorrhagic proctitis with regional lymphadenitis. The nodes enlarge, suppurate, become adherent to the skin and break down to form sinuses discharging pus. Metastatic complications may sometimes occur, with involvement of joints, eyes, and meninges.

3. Tertiary Stage

The tertiary stage is chronic, lasting for several years, representing the sequelae of scarring and lymphatic blockage. Rectal stricture and rectal perforation are recognized late sequelae to LGV proctitis. The course of the disease is variable. Late sequelae are more distressing in women and lymphatic obstruction in women can lead to elephantiasis of the vulva, called *esthiomene.*

Laboratory Diagnosis

The primary lesion usually goes unnoticed and the disease is seen commonly first in the stage of inguinal adenitis (bubo).

A. Microscopy

Smears of material aspirated from the bubos may show the elementary bodies **(Miyagawa's granulocorpuscles).** The sensitivity of microscopic diagnosis is very low.

B. Culture

Isolation of the chlamydia by intracerebral inoculation into mice and into yolk sac of eggs has been replaced by cell cultures.

C. Serology

LGV patients develop high titers of circulating antibodies, with titers of 1:64 or more in CF test and 1:512 or more in micro-IF.

D. Intradermal Test (Frei's Test)

An intradermal test (Frei's test) originally described by Frei in 1925 was commonly used formerly. The crude chlamydial antigen originally obtained from the bubo pus, and later from mouse brain or yolk sac cultures (lygranum), was inoculated intradermally in the forearm, with a control on the other arm. A positive reaction is indicated by induration of 7 mm or more in 2-5 days. This test becomes positive 2-6 weeks after infection and remains positive for several years.

Frei's test is now not in use due to the frequent occurrence of false positive reactions,. At present, the test has few diagnostic indications.

Treatment

Treatment is with tetracycline, which should be given for at least three weeks.

Chlamydophila Psittaci

Chlamydophila psittaci is the cause of psittacosis *(psittacos* means parrot) among psittacine birds, also known as *ornithosis* (derived from the Greek word *ornithos* for "bird") or *parrot fever.* The organism is present in the blood, tissues, feces, and feathers of infected birds that may appear either ill or healthy. Human infections are mostly occupational. The respiratory tract is the main portal of entry, and infection usually is acquired by inhalation of organisms from infected birds and their droppings. The incubation period is about 10 days. The illness ranges from an 'influenza-like' syndrome, to a severe illness with delirium and pneumonia. Psittacosis is a septicemia and there may be meningo-encephalitis, arthritis, pericarditis or myocarditis, or a predominantly typhoidal state with enlarged liver and spleen, and even a rash resembling that of enteric fever. Endocarditis resembling that complicating Q fever has been described.

Laboratory Diagnosis

Culture

The chlamydia can be isolated from blood during the early stages of the disease and from sputum later on. Infected cells, including alveolar macrophages from patients, and mouse brain, yolk sac and cell cultures show inclusion bodies **(Levinthal-Cole-Lillie or LCL bodies).** These differ from *C. trachomatis* inclusion in being more diffuse and irregular, not stained by iodine and not inhibited by sulphadiazine or cycloserine.

It is generally difficult to recover the chlamydia from patients treated with antibiotics. Isolation should be attempted only in laboratories where special containment facilities are available, as laboratory infection is a serious hazard.

Antigen Detection

Antigen detection is done by direct fluorescent antibody staining or by immunoassay or molecular diagnosis by polymerase chain reaction.

Serology

Psittacosis is usually diagnosed on the basis of serologic findings. A fourfold increase in titer, shown by the group specific CF testing of paired acute and convalescent phase sera, is suggestive of *C. psittaci* infection, but the species-specific MIF (micro-IF) test must be performed to confirm the diagnosis.

Treatment and Prevention

Infections can be treated successfully with tetracyclines or macrolides.

Psittacosis can be prevented only through the control of infections in domestic and imported pet birds

Chlamydophila pneumoniae

Chlamydophila pneumoniae was formerly known as *Chlamydia* sp., strain TWAR, and it was originally identified in 1965 from a conjunctival culture of a child (TW) enrolled in a Taiwan trachoma vaccine study. In 1983, at the University of Washington, a similar organism was isolated in HeLa cells from a pharyngeal specimen of a college student (AR). Grayston and colleagues (1986) isolated a chlamydial strain from acute respiratory disease in adults in Taiwan and designated it as ***C. psittaci*** **strain TWAR** from Taiwan Acute Respiratory). It possessed the group-specific antigen in common with *C. psittaci* and *C. trachomatis* but could be distinguished from both of them by species-specific antigens, DNA hybridization and restriction endonuclease analysis. This new organism was initially called TWAR, then classified as *Chlamydia pneumoniae,* and finally placed in the new genus *Chlamydophila.* Infection is transmitted by respiratory secretions; no animal reservoir has been identified.

Spectrum of Disease

1. *C. pneumoniae* has been associated with pneumonia, bronchitis, pharyngitis, sinusitis, and a flulike illness. It also has been isolated from patients with otitis media with effusion, pneumonia with pleural effusion, and aseptic pharyngitis.
2. It has also been linked to chronic illnesses such as atherosclerosis, coronary heart disease, asthma and stroke.
3. Infection with *C. pneumoniae* has been established as a risk factor for Guillain-Barre syndrome (GBS) and also appears to be a relationship between sarcoidosis and *C. pneumoniae.*

Ch. pneumoniae has been suspected of playing a role in other **chronic human diseases**, including Alzheimer's disease, multiple sclerosis, stroke and sarcoidosis, but the evidence is even more slender than for heart disease.

Laboratory Diagnosis of Chlamydia Infections

The laboratory diagnosis of *Chlamydia* infections can be accomplished by various approaches: (1) microscopic demonstration of inclusion or elementary bodies, (2) isolation of the organisms, 3) demonstration of chlamydial antigen and (4) detection of specific antibodies (5) Skin test—Hypersensitivity against these bacteria.

The specimens collected are conjunctival scrapings, sputum, throat swab, bubo pus, genital swabs and blood.

1. Demonstration of Inclusion or Elementary Bodies

Smears may be prepared from conjunctival scrapings or bubo pus and stained by Giemsa, Castaneda, Machiavello or Giminez stains to demonstrate characteristic inclusion bodies under light microscope. Chlamydia are gram-negative but are stained better by Castaneda, Machiavello or Giminea stains. Chlamydial elementary bodies and inclusions are large enough to be seen under the light microscope.

Glycogen-containing inclusions of *C. trachomatis* can be stained with Lugol's iodine. Inclusions of *C. psittaci* are not stained by the iodine technique since this organism produces no glycogen during its development. The inclusion bodies of trachoma and inclusion conjunctivitis are named *Halberstaedter Prowazek* or HP bodies whereas the elementary bodies of *C. psittaci* are called *Levinthal Cole Lillie* or LCL bodies.

Immunofluorescence (IF) can identify not only inclusions but also extracellular elementary bodies. Besides ocular infections, IF is useful also in examination of cervical or urethral specimens, which may contain elementary bodies but few intact intracellular inclusions. It is also more sensitive than iodine staining for detection of inclusions in infected cell cultures.

2. Isolation of the Organisms

Traditionally the isolation of *Chlamydia* has been accomplished by the inoculation of infected material into either **embryonated eggs, experimental animals**, or selected **tissue culture cell lines**.

a. Yolk Sac Inoculation

All known strains of *Chlamydia* will infect the chick embryo, and group-specific antigen and characteristic inclusions can be found in yolk sac material from infected 6-to 8-day embryos. However, this procedure is impractical for use by clinical laboratories since it is tedious, time-consuming, and less sensitive than tissue culture.

b. Animal Inoculation

Chlamydiae can be propagated in experimental animals such as mice. Chlamydiae differ in their infectivity to mice. *C. psittaci* strains infect mice by intracerebral, intranasal, intraperitoneal and subcutaneous routes. Among *C. trachomatis* strains, only the LGV serovars (Ll, L2, L3) infect mice when injected intracerebrally, but other *C. trachomalis* strains(The TRIC serovars) will not infect mice by any route of infection. They will, however, cause a rapid toxic death in mice if injected intravenously. Isolation by mouse inoculation is only of historical importance.

c. Tissue Culture

The most widely used method of cultivation is by the use of tissue cultures. Monkey kidney cell lines, McCoy and HeLa cell lines and, more recently, the heteroploid HL cell line are commonly used.

Cell cultures used for isolation are pretreated by irradiation or chemicals such as 5-iodo-2-deoxyuridine, cytochalasin B, or cycloheximide to enhance chlamydial replication and to allow easier recognition of inclusions. Pretreatment of the cells with diethylaminoethyldextran (DEAE-dextran) or centrifugation of the chlamydiae onto the host cells increases contact between the infectious chlamydial particles and the host cell monolayer, with a subsequent increase in infectivity. The viral growth is detected by demonstration of inclusion bodies.

Both genital and ocular strains of ***C. trachomatis*** tend to vary widely in their infectivity for cell cultures. LGV strains grow well, while TRIC strains are less infective. ***C. psittaci*** can be isolated from infected material by methods similar to those used for other *Chlamydia*. However, suitable isolation or containment facilities should be available for this extremely infectious human pathogen; laboratory-acquired infections are common.

Propagation of *C. pneumoniae* in cell culture has proven to be difficult. HeLa 229 cells and, more recently, the heteroploid HL cell line have been described as being more sensitive than McCoy cells for the isolation of this organism.

3. Demonstration of Chlamydial Antigen

a. Immunofluorescence (IF)

For diagnosis by demonstration of chlamydial antigens, the method commonly used is **micro-IF**. The infected ocular or genital samples are smeared on a slide, stained with fluorescent conjugated antibody and examined under the UV microscope. This test approaches cultures in sensitivity.

b. ELISA

The ELISA method is preferred for screening as it enables rapid testing for LPS antigen in large numbers of specimens. The greatest application of FA and EIA methods is in the direct detection of *C. trachomatis* from infected cervical or urethral specimens and conjunctival scrapings of adults. They have also been successfully used to identify *C. trachomatis* in nasopharyngeal and conjunctival specimens from infants with chlamydial conjunctivitis and pneumonitis.

c. Molecular Methods

Molecular methods like **DNA probes** and amplification techniques (**polymerase chain reaction, ligase chain reaction**) have greatly increased the sensitivity and specificity of antigen detection.

i. DNA Probes

In certain chlamydial infections, it may not be possible to isolate viable organisms. In such cases either an antigen detection method or a DNA probe has to be used for diagnosis. DNA probes are highly valuable in trachoma and chlamydial infections of the rectum.

ii. Polymerase Chain Reaction (PCR)

DNA is amplified and detected by PCR. This method is more sensitive than culture.

4. Detection of Specific Antibodies or Hypersensitivity Response

Complement-Fixation (CF) Test

The CF test uses the genus-specific antigen (LPS) and is applicable for the diagnosis of infections caused by *C. psittaci, C. pneumoniae,* and the LGV strains of *C. trachomatis.* In the majority of patients for whom the identity of the infecting serovar of *C. trachomatis* is known, there is a type-specific antibody response with a titer of 1:8 or greater. CF test is of little value in TRIC infections, in which micro-IF is more useful.

ii. Microimmunofluorescence (Micro-IF)

Micro-IF can test IgG and IgM antibody separately. Type-specific antibodies are demonstrated by this method.

iii. ELISA Test

EIAs are genus-specific, and measurements of IgG and IgM can be accomplished like the CF test. The advantage of these tests is they are less technically cumbersome. The assay has been used successfully to detect chlamydial IgM in infants with pneumonia and can be readily applied to large epidemiologic studies.

5. Skin Test

The **Frei test** is an intradermal skin test that detects a delayed hypersensitivity response to chlamydial antigen and was widely used formerly for diagnosis of LGV but has been given up because false positive results are very frequent.

KNOW MORE

Cheamydiphila pneumoniae

C. pneumoniae is a human pathogen. No bird or animal reservoirs have been identified. The mode of transmission from person to person is by aerosolized droplets via the respiratory route. The laboratory diagnosis of *C. pneumoniae* infections is accomplished by:

1. Cultivation
2. Serology
3. Amplification of *C. pneumoniae-specific* rDNA with subsequent RFLP analysis.

KEY POINTS

- Chlamydiae are obligate, intracellular parasites which are small, non-motile and gram-negative.
- The family Chlamydiaceae has been divided into two genera, ***Chlamydia*** and ***Chlamydophila:***

1. *Chlamydia* to include C *trachomatis* and
2. *Chlamydophila* to include *C. pneumoniae, C. psittaci,* and *C. pecorum.*

- **Chlamydia trachomatis infections:** Two distinct forms: infectious elementary bodies and noninfectious reticulate bodies. Two human biovars: trachoma (with 15 serovars) and lymphogranuloma venereum (LGV; 4 serovars).
- **Diseases: Trachoma biovar** responsible for ocular trachoma, adult inclusion conjunctivitis, neonatal conjunctivitis, infant pneumonia, and urogenital infections. **LGV biovar** responsible for LGV and ocular LGV.
- ***C. pneumoniae:*** *C. pneumoniae,* recognized as a significant community acquired respiratory pathogen, has been implicated in other chronic afflictions such as asthma and cardiovascular diseases. It has also been linked to chronic illnesses such as atherosclerosis, coronary heart disease, and stroke.
- ***C. psittaci:*** *C. psittaci* is the cause of psittacosis, also known as parrot fever or ornithosis.
- **Laboratory diagnosis of chlamydial infections** depends on direct detection of antigens, isolation of organism and serology for antibody detection. *Frei's test* is a skin test previously employed for diagnosis of LGV.

IMPORTANT QUESTIONS

1. Discuss laboratory diagnosis of chlamydial infections.
2. Write short notes on:
   a. Developmental cycle of chlamydiae
   b. TRIC agents
   c. Inclusion conjunctivitis
   d. Frei's test
   e. Lymphogranuloma venereum (LGV)
   f. *Chlamydophila pneumoniae*

FURTHER READING

Beatty, WL et al. Persistent Chlamydiae, Microbiol Rev. 1994; 58,686.

Black C. Current methods of laboratory diagnosis of Chlamydia trachomatis infections, Clin Microbiol Rev 1997;10:160-184.

Boman J, Gaydos C, Quinn T. Minireview: molecular diagnosis of Chlamydia pneumoniae infection, J Clin Microbial 1999;37:3791-3799.

Gibbs R, Careey N, Davies A. Review: Chlamydia pneumoniae and vascular disease, Br J Surg 1998;85:1191-1197.

Grayston J, et al. Evidence that Chlamydia pneumoniae causes pneumonia and bronchitis, J Infect Dis 1993;168:1231-1235.

Kumar S, Saigal SR, Sethi GR. Detection of IgM and IgG antibodies to Chlamydophila pnemoniae in pediatric community-acquired lower respiratory tract infections. Indian J Pathol Microbiol 2011;54:782-785.

Morre S, et al. Urogenital Chlamydia trachomatis serovars in men and women with a symptomatic or asymptomatic infection: an association with clinical manifestations? J Clin Microbiol 2000;38:2292-2296.

SECTION FOUR

VIROLOGY

C H A P T E R 53

General Properties of Viruses

Learning Objectives

After reading and studying this chapter, you should be able to:
- Describe size, shape and symmetry of viruses.
- Describe cultivation of viruses.
- List various cell cultures.
- Discuss detection of virus growth in cell cultures.
- List DNA and RNA viruses.
- Describe prions and viroids.

INTRODUCTION

Viruses are the smallest known infective agents and are perhaps the simplest form of life known. Viruses do not possess a cellular organization and they do not fall strictly into the category of unicellular microorganisms. Even the simplest of microorganisms are cells enclosed within a cell wall, containing both types of nucleic acid (DNA and RNA), synthesizing their own macromolecular constituents and multiplying by binary fission.

MAIN PROPERTIES OF VIRUSES

1. Viruses do not have a cellular organization.
2. They contain only one type of nucleic acid, either DNA or RNA but never both.
3. They are obligate intracellular parasites.
4. They lack the enzymes necessary for protein and nucleic acid synthesis and are dependent for replication on the synthetic machinery of host cells.
5. They multiply by a complex process and not by binary fission.
6. They are unaffected by antibacterial antibiotics.

The major differences between viruses and microorganisms are shown in Table 53.1. Viruses are generally considered microorganisms in medical microbiology in spite of these basic differences.

MORPHOLOGY OF VIRUSES

Size

Viruses are much smaller than bacteria. The extracellular infectious virus particle is called the **virion.** It was their small size and 'filterability' (ability to pass through filters that can hold back bacteria) that led to their recognition as a separate class of infectious agents. Hence they were for a time known as '**filterable viruses**'. They were called '**ultramicroscopic**' as they were too small to be seen under the light microscope. Some of the larger viruses, such as poxviruses can be seen under the light microscope when suitably stained. The virus particles seen in this manner are known as **'elementary bodies'.**

The unit for measurement of virion size is nanometers (nm). Viruses vary widely in size from 20 nm to 300 nm. The largest among them is pox virus (300 nm) and is as large as the smallest bacteria (mycoplasma). The smallest viruses are the parvovirus (about 20 nm) and are nearly as small as the largest protein molecules such as hemocyanin.

Table 53.1: Properties of prokaryotes and viruses

	Cellular organization	*Growth on inanimate media*	*Binary fission*	*Both DNA and RNA*	*Ribosomes*	*Sensitivity to antibacterial antibiotics*	*Sensitivity to interferon*
1. Bacteria	+	+	+	+	+	+	–
2. Mycoplasmas	+	+	+	+	+	+	–
3. Rickettsiae	+	+	+	+	+	+	–
4. Chlamydiae	+	–	+	+	+	+	+
5. Viruses	–	–	–	–	–	–	+

Measuring the Size of Viruses

1. Passing them through Collodion Membrane

The earliest method of estimating the size of virus particles was by passing them through collodion membrane filters of graded porosity (gradocol membranes).

2. Electron Microscopy

Electron microscopy is the most widely used method for estimating particle size.

3. Sedimentation in the Ultracentrifuge

The virus size could be calculated from the rate of sedimentation of virus in the ultracentrifuge with the development of ultracentrifuge.

4. Comparatative Measurements

It can be done with reference to: *Staphylococcus*, bacterial viruses (bacteriophages) and representative protein molecules.

Shape of the Virus

The overall shape of the virus particle varies in different groups of viruses. Most of the animal viruses are roughly spherical, some are irregular and pleomorphic. Poxviruses are brick-shaped, rabies virus is bullet-shaped, tobacco mosaic virus is rod-shaped. Bacteriophages have a complex morphology. The extracellular infectious virus particle is known as **virion.**

STRUCTURE AND CHEMICAL COMPOSITION OF THE VIRUSES

A. Viral Capsid
B. Virus Symmetry
C. Viral Envelope
D. Viral Nucleic Acids

A. Viral Capsid

Viruses consist of nucleic acid core surrounded by a protein coat called **capsid.** The capsid with the enclosed nucleic acid is known as **nucleocapsid.** The capsid is composed of a large number of **capsomers** which form its morphological units. The chemical units of the capsid are polypeptide molecules which are arranged symmetrically to form molecules to form an impenetrable shell around the nucleic acid core (Fig. 53.1).

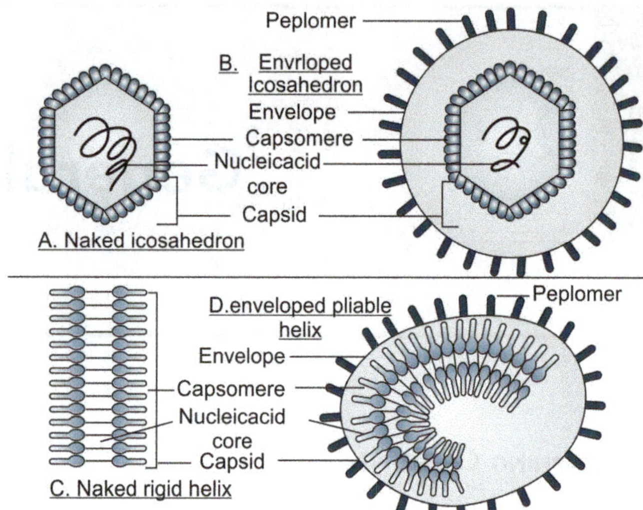

Fig. 53.1: Schematic diagram illustrating the components of the complete virus particle (the virion)

Functions of Capsid

i. **Protection:** It protects the viral genome from physical destruction and enzymatic inactivation by nucleases in biological material.
ii. **Binding sites:** It provides binding sites which enable the virus to attach to specific receptor sites on the host cell.
iii. It facilitates the assembly and packaging of viral genetic information.
iv. **Vehicle of transmission:** It serves as a vehicle of transmission from one host to another.
v. **Antigenic:** It is antigenic and specific for each virus type.
vi. **Host's defence:** It is of paramount importance in the host's defence to virus infection.
vii. It provides the **structural symmetry to the virus particle.**

B. Virus Symmetry

Viral architecture can be grouped into three types based on the arrangement of morphologic subunits: (1) Icosahedral symmetry (2) Helical symmetry (3) Complex structures.

1. Icosahedral Symmetry

An icosahedral (*icosa,* meaning 20 in Greek) is a polygon with 12 vertices or corners and 20 facets or sides. Each facet is in the shape of an equilateral triangle. Two types of capsomers constitute the icosahedral capsid. They are the pentagonal capsomers at the vertices (pentons) and the hexagonal capsomers making up the facets (hexons). There are always 12 pentons but the number of hexons varies with the virus group, e.g., adenoviruses (Fig. 53.2).

2. Helical Symmetry

The nucleic acid and the capsomers are wound together in the form of a helix or spiral.

Examples: Single-stranded RNA viruses such as influenza (Fig. 53.2), the parainfluenza viruses, and rabies.

3. Complex Symmetry

Viruses (e.g. poxviruses) which do not show either icosahedral or helical symmetry due to complexity of their structure are referred to have complex symmetry.

C. Viral Envelope

Virions may be enveloped or nonenveloped (naked).

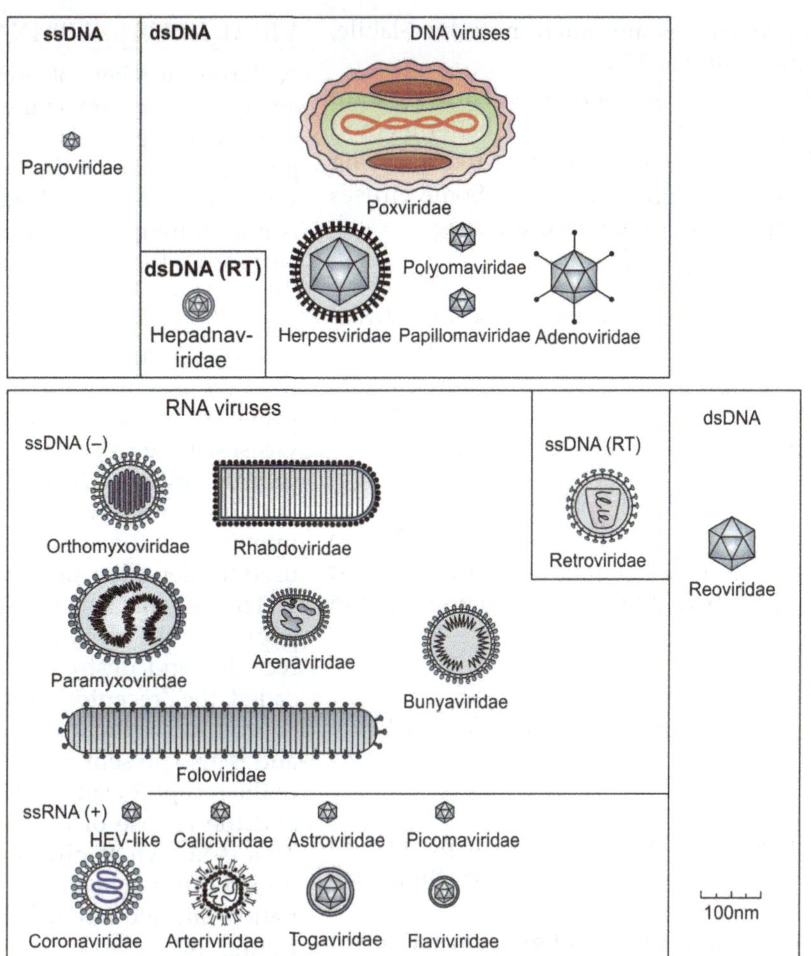

Fig. 53.2: Shapes and relative sizes of animal viruses of families that infect vertebrates

Enveloped Virus

The **envelope** or outer covering of virus containing lipid is derived from the plasma membrane of the host cell during their release by budding from the cell surface. The envelope is glycoprotein in nature. The lipid is largely of host cell origin while the protein is virus-encoded. Enveloped viruses are susceptible to the action of lipid solvents such as ether, chloroform and detergents, whereas most viruses existing as naked capsids are more likely to be resistant to them.

Peplomers

In mature virus particle, the glycoproteins often appear as projecting spikes on the outer surface of the envelope. These are known as peplomers (from *peplos*, meaning envelope). A virus may have more than one type of peplomers, e.g., the influenza virus carries two kinds of peplomers, the **hemagglutinin** which is a triangular spike and the **neuraminidase** which is a mushroom-shaped structure. Envelope confer chemical, antigenic and biological properties on viruses.

Functions of Peplomers

i. **Mediate attachment:** Many peplomers mediate attachment of the virus to the host-cell receptors to initiate the entrance of the virion into the cell.
ii. **Attach to receptors:** Some viral glycoproteins also attach to receptors on red blood cells, causing these cells to agglutinate (hemagglutination).
iii. **Enzymatic activity:** Other glycoproteins possess enzymatic activity like neuraminidase which cleave neuraminic acid from host cell glycoproteins.
iv. **Major antigens:** Glycoproteins are also major antigens for protective immunity.

D. Viral Nucleic Acids

Viruses contain a single kind of nucleic acid—either DNA or RNA—that encodes the genetic information necessary for replication of the virus. The genome may be single-stranded or double-stranded, circular or linear, and segmented or nonsegmented. The type of nucleic acid, its strandedness, and its size are major characteristics used for classifying viruses into families.

SUSCEPTIBILITY TO PHYSICAL AND CHEMICAL AGENTS

1. Heat and Cold

Heat Labile

With few exceptions, viruses are very heat labile. There are individual variations but in general, they are inactivated within seconds at 56°C, minutes at 37°C and days at 4°C. Enveloped viruses are much more heat-labile, rapidly dropping in titer at 37°C.

They are stable at low temperatures. For long term storage, they are kept frozen at –70°C. A better method for prolonged storage is lyophilisation or freeze drying (drying the frozen virus under vacuum). Some viruses (such as poliovirus) do not stand freeze drying.

2. pH

Viruses vary greatly in their resistance to acidity. Viruses are usually stable between pH values of 5.0 and 9.0. Some viruses (eg, enteroviruses) are resistant to acidic conditions while rhinoviruses are very susceptible. All viruses are destroyed by alkaline conditions.

3. Stabilization of Viruses by Salts

Molar concentrations of certain salts ($MgCI_2$ Na_2SO_4) also protect some viruses (for example poliovirus) against heat inactivation. The stability of viruses is important in the preparation of vaccines.

4. Radiation

Ultraviolet, X-ray, and high-energy particles inactivate viruses.

5. Disinfectants

They are, in general, more resistant than bacteria to chemical disinfectants, probably because they lack enzymes.

i. Bacteria are killed in **50 percent glycerol saline** but this acts as a preservative for many viruses (for example, vaccinia, rabies).
ii. **Phenolic disinfectants:** are only weakly virucidal.
iii. **Oxidising agents:** The most active antiviral disinfectants are **oxidizing agents** such as hydrogen peroxide, potassium permanganate and hypochlorites. **Organic iodine compounds** are actively virucidal. Chlorination of drinking water kills most viruses but its efficacy is greatly influenced by the presence of organic matter. Some viruses (such as **hepatitis virus, polio viruses)** are relatively resistant to chlorination. **Formaldehyde** and **beta propiolactone** are actively virucidal and are commonly employed for the preparation of killed viral vaccines.

6. Lipid Solvents

Enveloped viruses which possess lipid-containing envelope are sensitive to lipid solvents such as ether, chloroform and bile salts and the naked viruses are resistant to them. The selective action is useful in the identification and classification of viruses.

7. Antibiotics

Antibiotics active against bacteria are completely ineffective against viruses. This property is made use of in eliminating bacteria from clinical specimens by antibiotic treatment before virus isolation.

VIRAL HEMAGGLUTINATION

A large number of viruses contain hemagglutinin spikes (peplomers) on the capsid or envelope which can agglutinate red cells of different species. Viral hemagglutination was originally observed with the influenza virus by Hirst (1941). **Hemagglutinin** of influenza virus is due to the presence **of hemagglutinin spikes** on the surface of the virus. When RBCs are added to serial dilutions of viral suspension, the RBCs and the viruses collide in the suspension and adhere to each other resulting in **haemagglutination.** The hemagglutination reaction is important in laboratory work because it provides a simple and rapid method for detection of viruses in egg or tissue culture fluid. **Hemagglutination inhibition (HI)** provides a convenient test for the antiviral antibody, as hemagglutination is specifically inhibited by the antibody to the virus. This test can be used for detection and quantitation of antibody to virus.

The influenza virus also carries on its surface another peplomer, the enzyme **neuraminidase** which acts on the receptor and destroys it. Neuraminidase is, therefore, called the **'receptor destroying enzyme' (RDE).** RDE is produced by many viruses including cholera vibrios, and is also present in many vertebrate cells. Destruction of the receptor leads to the reversal of hemagglutination and the release of the virus from the red cell surface. This is known as **elution.** Elution is found only in the myxoviruses that possess neuraminidase. Hemagglutination and elution also help in purifying and concentrating the virus

Viral Hemagglutination Test

The **hemagglutination test** can be carried out in test tubes or special plastic trays. When RBCs are added to serial dilutions of viral suspension, the highest dilution that produces hemagglutination provides the **hemagglutination titer which is defined in the form of HA units.** Red cells which are not agglutinated settle at the bottom in the form of a **'button'**, while the agglutinated cells are seen spread into a **shield-like pattern** (Fig. 53.3). The test serves to titrate killed influenza vaccines as the inactivated virus can also hemagglutinate. Hemagglutination is a convenient method of detection and assay of the influenza virus.

Hemagglutination is stable with other viruses. Hemagglutination appears to be a reversible state of equilibrium between the virus and erythrocytes in the case of arboviruses, being influenced by slight variations in pH and temperature. Table 53.2 shows the characteristics of hemagglutination by different viruses.

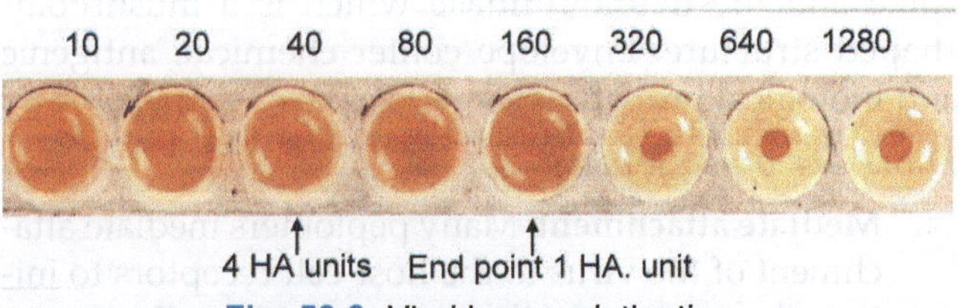

Fig. 53.3: Viral hemagglutination

Table 53.2: Characteristics of hemagglutination by different viruses	
Virus	*Erythrocyte species and other conditions*
Influenza virus	Fowl, human, guinea pig, others; Elution at 37°C
Parainfluenza, mumps, NDV	Fowl, human, guinea pig, others; Elution at 37°C Hemolysin present
Measles	Monkey, 37°C
Togavirus—several groups of Arbovirus	Goose, pigeon, one day old chick; pH and temperature critical
Rubella	Goose, pigeon, one day old chick; 4°C
Enterovirus, some Coxsackie and ECHO	Human; 4°C and 37°C
Rhinovirus, some serotypes	sheep; 4°C
Rabies	Goose; 4°C, pH 6.2
Reovirus	Human; 37°C

VIRAL REPLICATION

The genetic information necessary for viral replication is contained in the viral nucleic acid but lacking biosynthetic enzymes, the virus depends on the synthetic machinery of the host cell for replication. The viral multiplication cycle can be divided into six sequential phases, though the phases may sometimes be overlapping: 1. Adsorption or attachment, 2. Penetration, 3. Uncoating, 4. Biosynthesis, 5. Maturation, and 6. Release.

1. Adsorption or Attachment

Virions come in contact with cells by random collision but adsorption or attachment is specific and is mediated by the binding of virion surface structures, known as ligands, to receptors on cell surface. In case of **influenza virus**, a surface glycoprotein (the hemagglutinin) binds specifically to sialic acid residue of glycoprotein receptor sites on the surface of respiratory epithelium. In case of **human immunodeficiency virus-1 (HIV1),** surface glycoprotein gp 120 acts as a ligand. It binds to the CD4 60kDa glycoprotein on the surface of mature T lymphocytes. Similarly, rabies virus binds to the acetylcholine receptor found on neural cells.

2. Penetration

After binding, the virus particle is taken up inside the cell. In some systems, this is accomplished by receptor-mediated endocytosis (viropexis), with uptake of the ingested virus particles within endosomes. **Most non-enveloped** viruses enter the cell by receptor-mediated endocytosis or by viropexis. **Enveloped viruses** fuse their membranes with cellular membranes to deliver the nucleocapsid or genome directly into the cytoplasm.

3. Uncoating

This is the process of stripping the virus of its outer layers and capsid so that the nucleic acid is released into the cell. With most viruses, uncoating is effected by the action of lysosomal enzymes of the host cell. The genome of DNA viruses, except for poxviruses, must be delivered to the nucleus, whereas most RNA viruses remain in the cytoplasm.

4. Biosynthesis

This phase includes synthesis not merely of the viral nucleic acid and capsid protein but also of enzymes necessary in the various stages of viral synthesis, assembly and release. In addition, certain 'regulator proteins' are also synthesized which serve to shut down the normal cellular metabolism and direct the sequential production of viral components. The site of viral synthesize depends on the type of virus. In general, most DNA viruses synthesize their nucleic acid in the host cell nucleus. The exceptions are the poxviruses, which synthesize all their components in the host cell cytoplasm. Most RNA viruses synthesize all their components in the cytoplasm, except for orthomyxoviruses, some paramyxoviruses and retroviruzes which are synthesized partly in the nucleus. Viral protein is synthesised only in the cytoplasm.

Steps: of Biosynthesis

Transcription, translation, and replication of the genome are probably the most important steps in viral multiplication.

i. **Transcription** of messenger RNA (mRNA) from the viral nucleic acid.
ii. **Translation** of the mRNA into **'early proteins'**. These are enzymes which initiate and maintain synthesis of virus components. They may also induce shutdown of host protein and nucleic acid synthesis.
iii. **Replication** of viral nucleic acid.
iv. **Synthesis of 'late' or structural proteins,** which are the components of daughter virion capsids.

Viruses have been categorized into six classes by Baltimore (1970) based on their replication mechanisms (Table 53.3).

5. Maturation

Assembly of the various viral components into virions occurs shortly after the replication of the viral nucleic acid and may take place in either the nucleus (herpes and adenoviruses) or cytoplasm (picorna and poxviruses). In case of enveloped viruses, the envelopes are derived

Table 53.3: Baltimore classification of viruses based on replication mechanisms	
Class	
1.	**Single stranded DNA viruses:** With **single stranded DNA viruses**, the DNA molecule moves into the host cell nucleus and is converted into the duplex form. Transcription is achieved by host enzymes. for example parvovirus
2.	**Double stranded DNA viruses:** In the case of fully **double stranded DNA viruses**, the DNA enters the host cell nucleus and uses the host cell enzymes for transcription. e.g. **i. Hepadnaviruses; ii. Poxviruses.**
3.	**Single stranded RNA viruses:** Depending on the method of mRNA transcription, single stranded RNA viruses are classified into two categories. **i. Positive strand (plus strand, positive sense):** The viral RNA itself acts as the mRNA. Viral RNA is infectious by itself and is translated directly into viral proteins in the host cell cytoplasm, e.g. picorna, togaviruses. **ii. The negative strand (minus sense) RNA viruses:** The RNA is 'antisense', with polarity opposite to mRNA. They possess their own RNA polymerases for mRNA transcription, e.g. rhabdo-, orthomyxo-, paramyxoviridae.
4.	Double stranded RNA viruses: The double stranded RNA is transcribed to mRNA by viral polymerases, e.g. reoviruses.
5.	**Retroviridae** exhibit a unique replicative strategy. Their single stranded RNA genome is converted into an RNA:DNA hybrid by the viral reverse transcriptase (RNA directed DNA polymerase) enzyme. Double stranded DNA is then synthesized from the RNA:DNA hybrid. The double stranded DNA form of the virus (provirus) is integrated into the host cell chromosome. This integration may lead to transformation of the cell and development of neoplasia.

from the host cell nuclear membrane (herpes virus) and from plasma membrane when the assembly occurs in the cytoplasm of the host cell (orthomyxoviruses and paramyxoviruses).

6. Release

Viruses can be released from cells **after lysis of the cell**, by **exocytosis, or by budding** from the plasma membrane. Viruses that exist as naked nucleocapsids may be released by the lysis of the host cell **(polioviruses)** or they may be extruded by a process which may be called **reverse phagocytosis**. Release of many enveloped viruses occurs after budding from the plasma membrane without killing the cell.

ECLIPSE PHASE

The unique feature of viral multiplication is that the virus cannot be demonstrated inside the host cell from the stage of penetration till the appearance of mature daughter virions. This period during which the virus seems to disappear or go 'underground' is known as the **'eclipse phase'**. The time taken for a single cycle of replication is about 15-30 minutes for bacteriophages and about 15-30 hours for animal viruses. A single infected cell may release a large number of progeny virions. While this can be determined readily in bacteriophages **(burst size),** it is difficult to assess in the case of animal viruses that are released over a prolonged period.

ABNORMAL REPLICATIVE CYCLES

1. Incomplete Viruses

A proportion of daughter virions that are produced may not be infective. This is due to defective assembly. Such 'incomplete viruses' are seen in large proportions when cells are infected with a high dose of the influenza virus. The virus yield will have a high hemagglutinin titer but low infectivity. This is known as the ***von Magnus phenomenon.***

2. Abortive Infections

Abortive infections fail to produce infectious progeny, either because the cell may be nonpermissive and unable to support the expression of all viral genes or because the infecting virus may be defective, lacking some functional viral gene.

3. Latent Infection

A **latent** infection may ensue, with the persistence of viral genomes, the expression of none or a few viral genes, and the survival of the infected cell.

Defective Viruses

Viruses which are genetically deficient and therefore incapable of producing infectious daughter virions without the helper activity of another virus are known as' **defective viruses'**. Yield of progeny virions occurs only if the cells are simultaneously infected with a **helper virus,** which can supplement the genetic deficiency.

Example

i. **Hepatitis D virus** and **adeno-associated satellite viruses** which replicate only in the presence of their helper viruses—**hepatitis B** and **adenoviruses** respectively.
ii. **Rous sarcoma virus (RSV) is** other examples of defective viruses which cannot code for the synthesis of the viral envelope. When RSV infects a cell that harbors a helper virus (for example avian leukosis virus), infectious progeny results, the helper virus contributing to the synthesis of the envelope.

CULTIVATION OF VIRUSES

Because viruses are obligate intracellular parasites, their growth requires susceptible host cells capable of replicating them. They cannot be grown on any inanimate culture medium. Three methods are employed for the cultivation of viruses:

A. Animal inoculation
B. Embryonated eggs
C. Cell culture.

A. Animal Inoculation

Uses of Animal Inoculation

i. Primary isolation of certain viruses
ii. For the study of pathogenesis, immune response and epidemiology of viral diseases
iii. For the study of oncogenesis.

1. Monkeys

Monkeys were used for the isolation of the poliovirus but find only limited application in virology due to their cost and risk to handlers.

2. Mice

The use of white mice, pioneered by Theiler (1903) extended the scope of animal inoculation greatly. **Infant (suckling) mice** are very susceptible to coxsackie and arboviruses, many of which do not grow in any other system. Mice may be inoculated by several routes—intracerebral, subcutaneous, intraperitoneal or intranasal. The growth of the virus in inoculated animals may be indicated by death, disease or visible lesions. The viruses are identified by testing for neutralization of their pathogenicity for animals, by standard antiviral sera.

B. EmbrYonated Eggs

The embryonated hen's egg was first used for the cultivation of viruses by Goodpasture (1931) and the method was further developed by Burnet. The embryonated egg (8-11 day old) are inoculated by several routes for the cultivation of viruses such as chorioallantoic membrane (CAM), allantoic cavity, amniotic cavity and yolk sac (Fig. 53.4). After being inoculated, eggs are incubated for 2-9 days.

1. Chorioallantoic Membrane (CAM)

Inoculation on to chorioallantoic membrane (CAM) can be used to cultivate herpesvirus, smallpox virus (variola), vaccinia, myxoma virus, Rous sarcoma virus and eastern equine encephalitis virus.

Inoculation on the chorioallantoic membrane (CAM) produces visible lesions **(pocks).** Each infectious virus particle can form one pock under optimal conditions. Pock counting, therefore, can be used for the assay of pock-forming viruses, such as variola or vaccinia. Different viruses have different pock morphology.

2. Allantoic Cavity

Allantoic inoculation is employed for growing the influenza virus for vaccine production. Other chick embryo vaccines are yellow fever (17D strain) and rabies (Flury strain) vaccines. Duck eggs are bigger and have a longer incubation period than hen's eggs. Therefore, they provide a better yield of rabies virus and were used for the preparation of the inactivated, non-neural rabies vaccine.

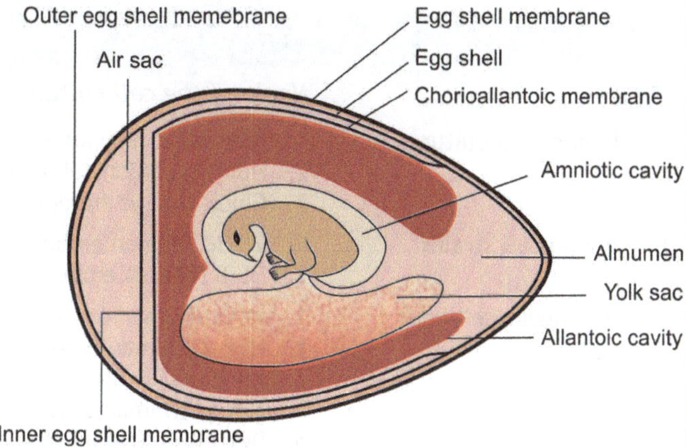

Fig. 53.4: Cross-section of an embryonated hen's egg.

3. Amniotic Sac

Inoculation into the amniotic sac is employed for the primary isolation of the influenza virus.

4. Yolk Sac

Yolk sac inoculation is used for the cultivation of some viruses, ***chlamydiae, Coxiella burnetti and rickettsiae.***

C. Tissue Culture

Three types of tissue cultures are available

1. Organ Culture

Small bits of organs can be maintained *in vitro* for days and weeks, preserving their original architecture and function. Organ cultures are useful for the isolation of some viruses which appear to be highly specialized parasites of certain organs. For example, the tracheal ring organ culture is employed for the isolation of **coronavirus,** a respiratory pathogen.

2. Explant Culture

Fragments of minced tissue can be grown as 'explant' embedded in plasma clots and was originally known as 'tissue culture'. This method is now seldom employed in virology. **Adenoid tissue explant cultures** were used for the isolation of adenoviruses.

3. Cell Cultures

This is the type of culture routinely employed for growing viruses. Tissues are dissociated into the component cells by the action of proteolytic enzymes such as trypsin and mechanical shaking. The cells are washed, counted and suspended in a growth medium. Such media will enable most cell types to multiply with a division time of 24-48 hours. The cell suspension is dispensed to the glass surface and on incubation, divide to form a confluent monolayer sheet of cells covering the surface within about a week. Cell culture tubes may be incubated in a sloped horizontal position, either as 'stationary culture' or may be rolled in special 'roller drums' to provide better aeration. Some fastidious viruses grow only in such roller cultures.

Type	Name of the cell culture
A. Primary cell cultures	1. Rhesus monkey kidney cell culture 2. Human amnion cell culture 3. Chick embryo fibroblast cell culture
B. Diploid cell strains	1. WI-38 (Human embryonic lung cell strain) 2. HL-8 (Rhesus embryo cell strain)
C. Continuous cell lines	1. HeLa (Human carcinoma of cervix cell line) 2. HEP-2 (Human epithelioma of larynx cell line) 3. KB (Human carcinoma of nasopharynx cell line) 4. McCoy (Human synovial carcinoma cell line) 5. Detroit-6 (Sternal marrow cell line) 6. Chang C/I/L/K (Human conjunctiva (C) Intestine (I), Liver (L) and Kidney (K) cell lines) 7. Vero (Vervet monkey kidney cell line) 8. BHK-21(Baby hamster kidney cell line)

Table 53.4: Some cell cultures in common use

Classification of Cell Cultures

Cell cultures are classified into three types based on their origin, chromosomal characters and the number of generations through which they can be maintained (Table 53.4).

1. Primary Cell Cultures

These are normal cells freshly taken from the body and cultured. They are capable of only limited growth in culture (5-10 passages) and cannot be maintained in serial culture. Primary cell cultures are useful for the isolation of viruses and their cultivation for vaccine production. Important examples of primary cell cultures are monkey kidney; human embryonic kidney, human amnion and chick embryo cell cultures.

2. Diploid (Semi-continuous) Cell Strains

These are cells of a single type that retain the original diploid chromosome number and karyotype during serial subcultivation for a limited number of times. They undergo 'senescence' after about fifty serial passages. Diploid cells developed from human fibroblasts are susceptible to a wide range of human viruses.

They are useful for isolation of some fastidious pathogens and for the production of viral vaccines, e.g. rabies vaccine is produced by cultivation of the fixed rabies virus in WI-38 human embryonic lung cell strain.

3. Continuous Cell Lines

These are the most widely used for diagnostic work. These are cells of a single type, usually derived from cancer cells, that are capable of continuous serial cultivation indefinitely. They have been derived from diploid cell lines or from malignant tissues. They invariably have altered and irregular number of chromosomes. Such cells often produce tumors if injected into susceptible animals.

Standard cell lines derived from human cancers, such as HeLa, Hep-2 and KB cell lines have been used in laboratories throughout the world for many years. These cell lines may be maintained by serial subcultivation or stored in the cold (–70°C) for use when necessary. Some cell lines are now permitted to be used for vaccine manufacture, for example, Verocell for rabies vaccine. Other cell lines (and their sources) are in common use (Table 53.4). The type of cell culture used for viral cultivation depends on the sensitivity of the cells to a particular virus.

DETECTION OF VIRUS GROWTH IN CELL CULTURE

Virus growth in cell cultures can be detected by the following methods:

1. Cytopathic Effect

Many viruses cause morphological changes in cultured cells in which they grow. These changes can be readily observed by microscopic examination of the cultures and are known as **'cytopathic effects' (CPE)** and the viruses causing CPE are called **'cytopathogenic viruses'**. Most viruses produce some obvious cytopathic effect in infected cells that is generally characteristic of the viral group (Figs 53.5A to D).

Main Types of CPE

i. **Rounding of cells:** Viral replication may lead to nuclear pyknosis, rounding, refractility and degeneration. This is seen in picornaviruses.
ii. **Cell necrosis and lysis:** Enteroviruses produce rapid CPE with crenation of cells and degeneration of the entire cell sheet.
iii. **Syncytium formation:** Some viruses (measles, respiratory syncytial virus, human immunodeficiency virus (HIV) lead to syncytium formation in which infected cells fuse with neighbouring infected or uninfected cells to form giant cells containing several (up to 100) nuclei.
iv. **Discrete focal degeneration:** Herpes virus causes discrete focal degeneration.
v. **Rounding and aggregation:** Adenovirus produces large granular clumps resembling bunches of grapes.
vi. **Cytoplasmic vacuolation:** SV 40 produces prominent cytoplasmic vacuolation

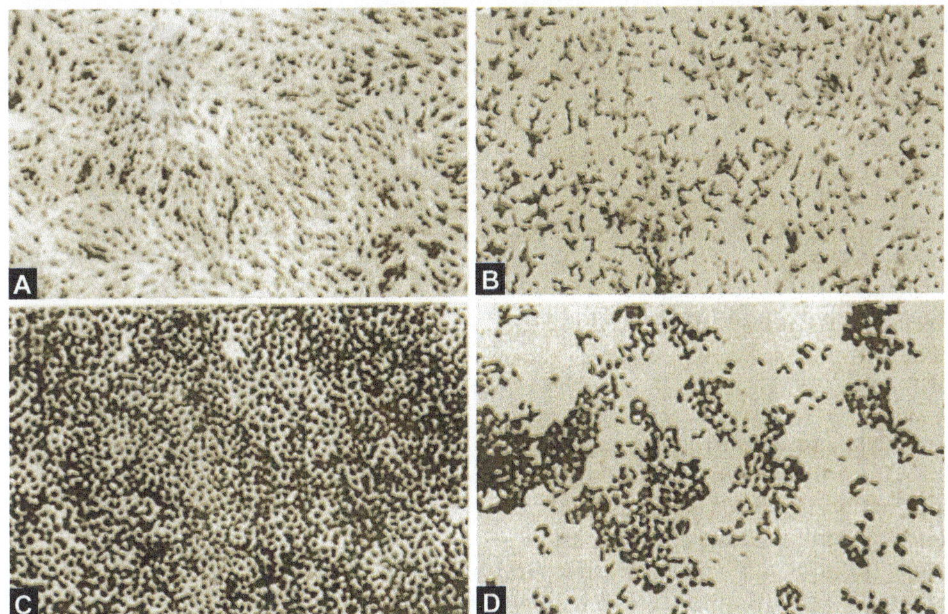

Figs 53.5A to D: A. Normal Vero cell monolayer. B. Vero cell monolayer infected with Coxsackie B virus C. HeLa cell monolayer. D. HeLa cell monolayer infected with Coxsackie virus B3, stained after 48 hours

2. Metabolic Inhibition

The medium turns acid due to cellular metabolism in normal cell cultures. When viruses grow in cell cultures, cell metabolism is inhibited and there is no acid production. This can be made out by the color of the indicator (phenol red) incorporated in the medium.

3. Hemadsorption

When hemagglutinating viruses (such as influenza and parainfluenza viruses) grow in cell cultures, their presence can be indicated by the addition of guinea pig erythrocytes to the cultures. The erythrocytes will adsorb onto the surface of cells if the viruses are multiplying in the culture. This is known as 'hemadsorption'. This reaction becomes positive before cytopathic changes are visible and in some cases occurs in the absence of cytopathic effects.

4. Interference

The multiplication of one virus in a cell usually inhibits the multiplication of a second virus, called the challenge virus, when it is added to the culture. The growth of a non-cytopathogenic virus in cell culture can be tested by the subsequent challenge with a known cytopathogenic virus. The growth of the first will inhibit infection by the second virus by interference.

Example

Rubella virus which do not produce cytopathic changes although they multiply within the cell. A known cytopathogenic challenge virus is then introduced into the cells. No CPE will be seen in the cell culture as replication of challenge virus will be prevented because of interference by rubella virus.

5. Transformation

Tumor forming (oncogenic) viruses induce cell 'transformation' and loss of contact inhibition, so that growth appears in a piled-up fashion producing microtumors. Some herpesviruses, adenoviruses, hepadnaviruses, papovaviruses and retroviruses (human T cell lymphotropic virus type I) can transform cells.

6. Immunofluorescence

Cells from virus infected cultures can be stained by fluorescent conjugated antiserum and examined under the UV microscope for the presence of virus antigen. This gives positive results earlier than other methods and, therefore, finds wide application in diagnostic virology.

7. Detection of Virus-specific Nucleic Acid

Molecular-based assays such as polymerase chain reaction provide rapid, sensitive, and specific methods of detection.

8. Detection of Enzymes

The virus isolate can be identified by detection of viral enzymes, such as reverse transcriptase in retroviruses, in the culture fluid.

9. Electron Microscopy

Viruses have distinctive appearances and can be detected by electron microscopy of ultra thin sections of infected cells.

VIRAL ASSAY

The virus content of a specimen can be assayed in two ways: either with reference to the **total virus particles** or with reference to **infectious virion** only.

A. Total Virus Particles Count

Two methods employed for total enumeration are electron microscopy and hemagglutination.

1. Electron Microscopy

By simple negative staining, the virus particles in a suspension can be counted directly under the electron microscope. The virus suspension can be mixed with a known latex particles. The ratio between particles under the electron microscope gives an indication of the virus count.

2. Hemagglutination

A convenient method of quntitation is the determination of hemagglutination titers with hemagglutinating viruses. Hemagglutination is not a very sensitive indicator of the presence of small amount of virus particles. Thus, approximately 10^7 influenza virions are required to produce macroscopic agglutination of convenient quantity of chicken erythrocytes (0.5 percent suspension). Hemagglutination is method of virus assay because of its simplicity.

B. Infectious Virion Assay

Two types of infectivity assays can be carried out—assays quantal and quantitative assay.

1. Quantal Assays

Quantal assays only indicate the presence or absence of infectious viruses. Using serial dilutions of virus suspensions and with the aid of statistical methods, reasonably accurate estimates of infectivity can be obtained in quantal assays.

Quantal assays of infectivity can be carried out in **animals, eggs** or **tissue culture.** Endpoints used for infectivity titration are the death of the animal, production of hemagglutinin in allantoic fluid or the appearance of CPE in cell cultures. This type of measurement, called a *quantal* assay, does not actually measure the number of infectious particles in a suspension but determines the extent to which a virus suspension can be diluted and still contain infectious viruses. The virus titer is usually expressed as the 50 percent infectious dose' (ID_{50}) per ml, which indicates the highest dilution of the inoculum that would produce an effect in 50 percent of animals, eggs or cell cultures inoculated.

2. Quantitative Infectivity Assay

Quantitative assays measure the actual number of infectious particles in the inoculum. Two methods are available—**plaque assay** in monolayer cell culture and **pock assay** on chick embryo CAM.

i. Plaque Assay

A viral suspension is added to a monolayer of cultured cells in a bottle or Petri dish, and after allowing time for absorption, the medium is removed and replaced with

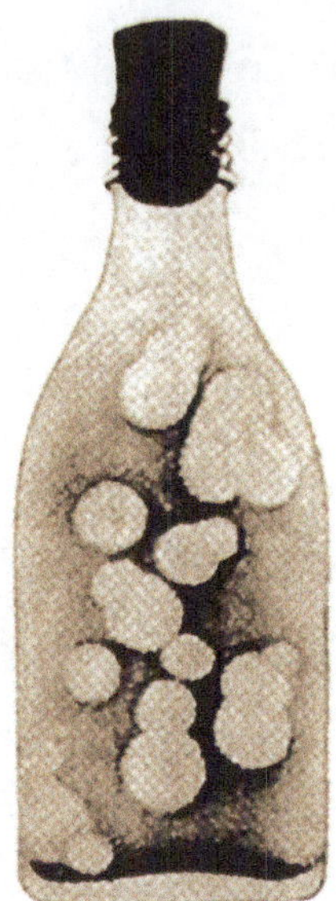

Fig. 53.6: Plaque formation in monkey kidney cells by poliovirus

a solid agar gel to prevent virus spreading throughout the culture, to ensure that the spread of progeny virions is confined to the immediate vicinity of infected cells. In this system, each infectious viral particle gives rise to a localised focus of infected cells that can be seen with the naked eye. Such foci are known as **'plaques'** and each plaque indicates an infectious virus (Fig. 53.6). Under controlled conditions, a single plaque can arise from a single infectious virus particle, termed a **plaque-forming unit (PFU).**

ii. Pock Assay

Certain viruses, e.g. herpes and vaccinia, form pocks when inoculated onto the chorioallantoic membrane of an embryonated egg. Such viruses can be assayed by counting the number of pocks formed on CAM by appropriate inocula of virus. This is known as pock assay.

VIRAL GENETICS

Viruses obey the laws of genetics like all other 'living beings'. Several properties of viruses, such as virulence and antigenicity are under genetic control. Genetic studies, therefore, have contributed to a better understanding of virus-host interactions and the development of better viral vaccines.

Mechanisms for Genetic Modification in Viruses

The two main mechanisms for genetic modification in viruses are:

A. Mutation.
B. Recombination.

A. Mutation

It is a random, undirected and heritable variation. The frequency of mutation in viruses is about 10^{-4} to

10^{-8}, approximately the same as in bacteria. Mutations, therefore, occur during every viral infection. Many mutations are lethal, because the mutated virus is unable to replicate. A mutant becomes evident only if the mutation confers some readily observable property or affords the mutant virus some selection or survival advantage. Mutation may occur spontaneously or may be induced by mutagens, physical agents such as UV light or irradiation or chemical agents such as 5-fluorouracil.

Conditional Lethal Mutant

Mutations in essential genes are termed **lethal mutations.** A class of mutants that are of great importance in laboratory studies is the **conditional lethal mutant.** These are mutants which are able to grow under certain conditions (called **permissive conditions**), but are lethal, that cannot grow under certain other specified conditions (called **nonpermissive or restrictive conditions)**. There are different types of conditional lethal mutants but the types most widely employed in genetic studies are the **'temperature sensitive' (ts) mutants.** These can grow at low (permissive) temperature (28-31°C), but not at a higher (restrictive) temperature (37°C). Because of their low virulence, the mutants have been used extensively in attempts to produce attenuated live virus vaccine.

B. Recombination

When two different, but related, viruses infect a cell simultaneously, genetic recombination may take place. The two viruses exchange segments of nucleic acid between them so that a hybrid is formed possessing genes from both parents. Recombinants may occur between 1) two active (infectious) viruses, 2) one active and one inactive virus, and (3) two inactive viruses.

i. **Recombinants between two active (infectious) viruses:** When two inactive viruses markers are grown together, recombinants may be derived that possess the distinctive properties of both parents. Thus, if a human and an avian strain of influenza virus (whose hemagglutinin and neuraminidase antigens are different and easily identifiable) are grown together, a hybrid may be obtained with the hemagglutinin of one parent and the neuraminidase of the other. This may be one of the ways by which the pandemic strains of the influenza virus originate in nature.

ii. **Recombinants between one active and one inactive virus—Cross-reactivation or marker rescue:** When a cell is 'infected' with an active virus and a different but related inactivated virus, progeny possessing one or more genetic traits of the inactivated virus may be produced. This phenomenon is known as **cross reactivation** or **marker rescue**.

   This finds application in the manufacture of the influenza virus vaccines.

iii. **Recombinants between two inactive viruses—Multiplicity reactivation:** When a cell is 'infected' with a large dose (high multiplicity) of a single virus inactivated by UV irradiation, live virus may be produced. The different virions that cause multiple infection of a cell may have suffered damage to different genes. Thus from the total genetic pool it may be possible to obtain a full complement of undamaged genes. This phenomenon is called **multiplicity reactivation.**

Pseudovirion

As a general rule, virus capsids enclose viral nucleic acids. Sometimes segments of host nucleic acid become encapsidated instead. This is known as pseudovirion.

NONGENETIC INTERACTIONS

i. Phenotypic Mixing

When two different viruses infect the same cell, some 'mix up' may take place during assembly so that progeny genome of one virus may be surrounded by the capsid belonging entirely or partly to the other virus. This is known as **phenotypic mixing**. This is not a stable variation. On subsequent passage, the capsid will be found to be of the original type only. If the genome is encased in a completely heterologous protein coat, this extreme example of phenotypic mixing may be called **"phenotypic masking"** or **"transcapsidation."**

ii. Genotypic Mixing or Heterozygosis

It results from the incorporation of more than one complete genome into a single virus particle. There is no recombination between the different genomes so that the two kinds of viral progeny are formed on passage.

iii. Complementation

This refers to the interaction of viral gene products in cells infected with two viruses, one or both of which may be defective. It results in the replication of one or both under conditions in which replication would not ordinarily occur. The basis for complementation is that one virus provides a gene product in which the second is defective, allowing the second virus to grow. The genotypes of the two viruses remain unchanged. If both mutants are defective in the same gene product, they will not be able to complement each other's growth.

iv. Interference

Infection of either cell cultures or whole animals with two viruses often leads to an inhibition of multiplication of one of the viruses, an effect called **interference.**

v. Enhancement

Mixed infection of cells may sometimes lead to increased virus yield or greater CPE. This is known as **'enhancement'.**

CLASSIFICATION OF VIRUSES

Viruses began to be classified into groups based on their physicochemical and structural features from the early

1950s. Nomenclature and classification are now the official responsibility of the International Committee on Taxonomy of Viruses.

Main Criteria Used for the Classification of Viruses (Figs 53.7, 53.8)

1. **Type of nucleic acid:** Viruses are classified into two main divisions depending on the type of nucleic acid they possess: **riboviruses** are those containing RNA and **deoxyriboviruses** are those containing DNA.
2. **Number of strands of nucleic acid:** Single- or double-stranded, linear, circular, circular with breaks, segmented.
3. **Polarity of the viral genome:** RNA viruses in which the viral genome can be used directly as messenger RNA are by convention termed 'positive-stranded' and those for which a transcript has first to be made are termed 'negative-stranded.
4. The symmetry of the nucleocapsid.
5. The presence or absence of a lipid envelope.

Universal System of Virus Taxonomy

A system has been established in which viruses are separated into major groupings called **families**. Virus family names have the suffix **-viridae**. Within each family subdivisions called **genera**. Genus names carry the suffix ***-virus***. In four families (Poxviridae, Herpesviridae, Parvoviridae, Paramyxoviridae), a larger grouping called **subfamilies** has been defined. Virus **orders** may be used to group virus families that share common characteristics. Currently, 24 families contain viruses that infect humans and animals.

Short Descriptions of the Major Groups of Viruses

DNA Viruses (Fig. 53.7)

1. Parvoviridae Family

Three genera have been described: *Parvovirus, Adenosatellovirus* and *Densovirus.*

2. Herpesviridae Family

These are medium sized viruses containing linear double stranded DNA. Only one genus, *Herpesvirus*, has been characterized, but several members of the family await classification.

Human herpesviruses include herpes simplex types 1 and 2 (oral and genital lesions), varicella-zoster virus (chickenpox and shingles), cytomegalovirus, Epstein-Barr virus (infectious mononucleosis and association with human neoplasms), human herpesviruses 6 and 7 (T lymphotropic), and human herpesvirus 8 (associated with Kaposi's sarcoma). Other herpesviruses occur in many animals.

3. Hepadnaviridae Family

This consists of the human hepatitis type B virus and related viruses of animals and birds. (The name comes from *hepa* = liver, and *dna* for DNA core). Three viral types are known that infect mammals (humans, woodchucks, and ground squirrels) and another that infects ducks.

4. Papovaviridae Family

Two genera have been recognized—*Papillomavirus* and *Polyomavirus.*

Papillomavirus are also a former member of the Papovaviridae family. Polyomaviruses were formerly a part of the Papovaviridae family before it was split into two families.

5. Adenoviridae Family

Members have been classified into two genera: Mastadenovirus (mammalian adenoviruses) and Aviadenovirus (adenoviruses of birds)

6. Poxviridae Family

The family is divided into several genera. All poxviruses tend to produce skin lesions. Some are pathogenic for humans (smallpox, vaccinia, molluscum contagiosum); others that are pathogenic for animals can infect humans (cowpox, monkeypox).

RNA Viruses (Fig. 53.8)

i. Orthomyxoviridae Family

Only one genus *Influenzavirus* has been recognised. Influenzavirus type C possesses several distinctive features and may have to be separated into a new genus.

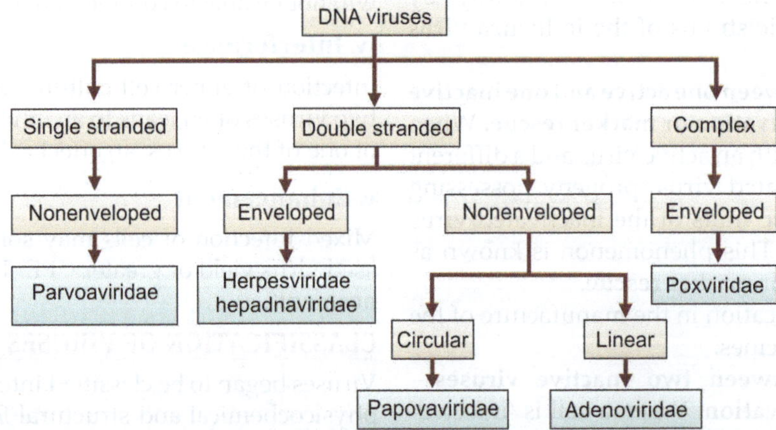

Fig. 53.7: Summary of the classification of DNA viruses

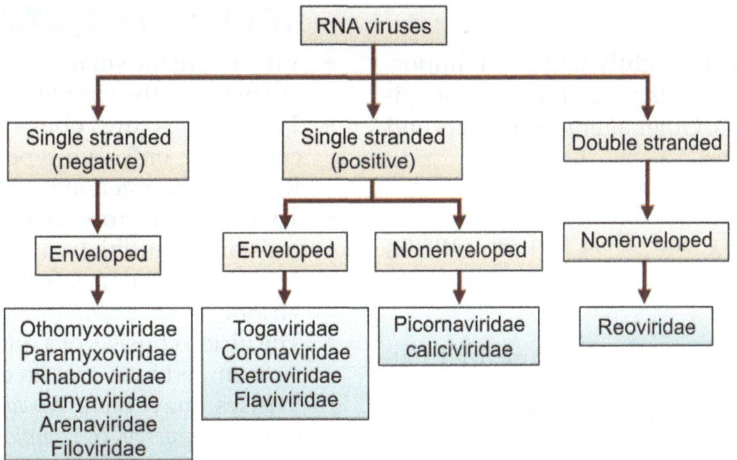

Fig. 53.8: Summary of the classification of RNA viruses

ii. Picornaviridae Family

Three genera are of medical importance:

1. *Enterovirus,* including polio, coxsackie, echo and several other related viruses.
2. *Rhinovirus,* including human, bovine and equine rhinoviruses.
3. *Hepatovirus:* Hepatitis A virus.

2. Paramyxoviridae Family

Three genera have been recognised:

1. *Paramyxovirus* which consists of the Newcastle disease virus, mumps virus and parainfluenza viruses of humans, other mammals and birds.
2. *Morbillivirus,* containing measles, canine distemper, rinderpest and related viruses.
3. *Pneumovirus,* containing respiratory syncytial virus of humans and related viruses.

4. Bunyaviridae Family

The majority of these viruses are transmitted to vertebrates by arthropods (arboviruses). Five genera are established—the large genus *Bunyavirus* containing about 150 species and four other genera-*Hantavirus, Nairovirus, Phlebovirus, Ukuvirus*-and many unassigned viruses.

5. Arenaviridae Family

Only one genus *Arenavirus* has been recognized. Species include, lymphocytic choriomeningitis virus, Lassa and members of the Tacaribe complex.

Rhabdoviridae Family

Two genera have been recognized:

1. *Vesiculovirus,* containing vesicular stomatitis virus, Chandipura virus (isolated from humans in India) and related species.
2. *Lyssavirus,* containing the rabies virus and related viruses such as Lagos bat, Mokola, Duvenhage and others.

Other genera have been suggested to include rhabdoviruses of insects and plants.

7. Togaviridae Family

Three genera have been described:

1. *Alpha virus,* consisting of viruses formerly classified as Group A arboviruses.
2. *Rubivirus,* consisting of the rubella virus and has no arthropod vector.
3. *Pestivirus,* consisting of the mucosal disease virus, hog cholera virus and related viruses.

8. Coronaviridae Family

Only one genus *Coronavirus* has been recognized. Members include human corona virus causing upper respiratory disease, avian infectious bronchitis virus, calf neonatal diarrhea corona virus, murine hepatitis virus and related viruses. Most human coronaviruses cause mild acute upper respiratory tract illnesses (colds) but a new coronavirus identified in 2003 causes a severe acute respiratory syndrome (SARS). Toroviruses, which cause gastroenteritis, form a distinct genus.

Retroviridae Family: (Re=Reverse, tr = Transcriptase)

These are RNA tumor viruses and related agents. The characteristic biochemical feature is the presence of RNA dependent DNA polymerase (reverse transcriptase) within the virus that produces a DNA copy of the RNA genome. This DNA becomes circularized and integrated into host chromosomal DNA. The virus is then replicated from the integrated "provirus" DNA copy. Three subfamilies are recognized:

1. *Oncovirinae,* the RNA tumor virus group.
2. *Spumivirinae,* the foamy virus group (Spuma =foam)
3. *Lentivirinae, (Lenti* = slow) visna and maedi viruses of sheep belonging to the slow virus group.

10. Flaviviridae Family

Flaviviruses, formerly grouped under togaviridae, as Group B arboviruses, have been classified as a separate family.

This group of arboviruses includes yellow fever virus and dengue viruses. Most members are transmitted by blood-sucking arthropods. Hepatitis C virus has no known vector.

12. Caliciviridae Family

Similar to picornaviruses but slightly larger. An important human pathogen is Norwalk virus, the cause of epidemic acute gastroenteritis. Other agents infect cats and sea lions as well as primates.

Filoviridae Family

This contains the *Marburg* and *Ebola* viruses causing human hemorrhagic fevers.

13. Reoviridae Family

Icosahedral, nonenveloped viruses, medium-sized (60-80 nm), with double layered capsids. Genome consists of double stranded RNA in 10-12 pieces. Three genera have been recognised.

1. *Reovirus,* containing reoviruses from humans, other mammals and birds.
2. *Orbivirus,* containing several species of arboviruses such as blue tongue virus, African horse sickness virus.
3. *Coltivirus* includes Colorado tick fever virus of humans.
4. *Rotavirus* including human rotaviruses, calf diarrhea virus and related agents. Other genera may have to be defined to include plant and insect viruses belonging to this family.

VIROIDS

Viroids are small infectious agents which are circular single stranded RNAs without a protein coat. They cause disease of plants. Viroids are agents that do not fit the definition of classic viruses.

PRIONS

Prions (proteinaceous infectious particles) are infectious particles composed solely of protein with no detectable nucleic acid. Unlike viruses, the agents are resistant to a wide range of chemical and physical treatments. They are highly resistant to inactivation by heat, formaldehyde, and ultraviolet light that inactivate viruses. The prion protein is encoded by a single cellular gene.

Prion Diseases

Prion diseases, called "transmissible spongiform encephalopathies," include scrapie in sheep, mad cow disease in cattle, and kuru and Creutzfeldt-Jakob disease in humans. Prions do not appear to be viruses. It has been suggested that they may also be responsible for some other chronic neurological degenerative diseases of humans (See Chapter 69).

KNOW MORE

Cultivation of viruses: The earliest method for the cultivation of viruses causing human diseases was inoculation into human volunteers. Reed and colleagues (1900) used human volunteers for their pioneering work on yellow fever. Human volunteers are used only when no other method is available and when the virus is relatively harmless due to the serious risk involved.

KEY POINTS

- Viruses are the smallest known infective agents and are perhaps the simplest form of life known.
- The viruses are obligate intracellular parasites containing only one type of nucleic acid (DNA or RNA) as their genome.
- They do not grow in inanimate media. They are resistant to antibiotics.
- The extracellular infectious virus particle is the virion.
- The virion consists of a nucleic acid core, the *genome,* surrounded by a protein coat, the capsid.
- Viruses may have *icosahedral (cubical) symmetry, helical symmetry,* or *complex symmetry.*
- The capsid together with the enclosed nucleic acid is known as the nucleocapsid. Some viruses are surrounded by envelopes.
- Three methods are employed for cultivation of viruses, namely *animal inoculation, embryonated egg inoculation* and *tissue culture.*
- Embryonated egg inoculation is done by one of the several routes such as chorioallantoic membrane (CAM), allantoic cavity, amniotic sac and yolk sac.
- Tissue culture are three types: organ culture; explant culture and cell culture.
- Depending on the type of nucleic acids viruses possess, they are classified into two groups: deoxyriboviruses that contain DNA (DNA virus) and riboviruses that contain RNA (RNA virus).
- **Prions (proteinaceous infectious particles)** are infectious particles composed solely of protein with no detectable nucleic acid.
- Prion diseases, called "transmissible spongiform encephalopathies, include scrapie in sheep, mad cow disease in cattle, and kuru and Creutzfeldt-Jakob disease in humans.

IMPORTANT QUESTIONS

1. Describe the various methods of isolation of viruses in the laboratory.
2. Write short notes on:
   Morphology of viruses.
   Determination of size of viruses.
   Viral hemagglutination.
   Recombination in viruses.
   Cultivation of viruses.
   Detection of virus growth in cell cultures.
   Prions.
   Viroids.

FURTHER READING

Belshe RB, et al. Textbook of Human Virology, 2nd edition, Chapters 1-6. St. Louis: Mosby Year Book 1991.

Diener TO. The viroid: big punch in a small package. Trends Microbiol 1993;1:289-94.

Prusiner SB Prions. Adv Virus Res 1984;29:I.

White DO, Fenner FJ, eds. Structure and composition of viruses. Medical Virology 4th edn, Academic Press, San Diego, New York, Boston, London, Sydney, Tokyo, Toronto 1994:1-73.

CHAPTER 54

Virus-Host Interactions: Viral Infections

LEARNING OBJECTIVES

After reading and studying this chapter, you should be able to:

- Describe inclusion bodies
- Describe routes of transmission of human virus infections
- Descrbe the following: Immunity in virus infections; interferons

INTERACTIONS BETWEEN VIRUSES AND HOST CELLS

Virus-host interactions may be considered at different levels—the cell, the individual and the community. The cellular response to viral infection may range from no apparent effect to cytopathology with accompanying cell death to hyperplasia or cancer.

Some viruses, like poliovirus cause cell death (cytocidal effect) or even lysis (cytolysis). Others may cause cellular proliferation (as molluscum contagiosum) or malignant transformation (as oncogenic viruses). In some instances the virus and host cell enter into a peaceful coexistence, both replicating independently without any cellular injury, a condition known as 'steady state infection'

1. Cellular factors—The presence of appropriate receptors on the surface of the cell determines whether virus can adsorb to it and the virus gets into the cell. It must replicate in order to establish infection, which may take several forms.

2. Cytopathic effects—Many viruses kill the cells in which they replicate, sometimes with characteristic appearances or cytopathic effects (CPEs) a property that is useful in the diagnostic laboratory. Furthermore, the type of CPE may give clues as to the sort of immune response to be expected.

a. **Cell lysis**—The 'early' virus-coded proteins may shut down synthesis of macromolecules, particularly polypeptides, by the host cell. Death of the cell is followed by lysis and release of large numbers of virions. We can think of these viruses as 'bursters'.
b. **Cell fusion**—Many viruses produce alterations in the cytoplasmic membrane of infected cells. Some cause fusion of adjacent cell membranes, leading to syncytium formation e.g. paramyxoviruses such as respiratory syncytial virus, parainfluenza viruses, and measles. Herpesviruses and some retroviruses also give rise to syncytia.
c. **Inclusion bodies**—Inclusion bodies are structures with distinct size, shape, location and staining properties that can be demonstrated in virus infected cells under the light microscope. The appearance of inclusion bodies is the most characteristic histological feature in virus-infected cells. They may be acidophilic (stained by eosin) or basophilic (stained by hematoxylin), single or multiple, large or small and round or irregular. The presence of inclusion bodies may be of considerable diagnostic aid, e.g. intracytoplasmic inclusion in nerve cells—the Negri body is pathognomonic for rabies. They may be situated in the nucleus, in the cytoplasm or in both (Fig. 54.1).

Intracytoplasmic inclusion bodies—Those viruses that have cytoplasmic assembly (mainly RNA viruses) yield cytoplasmic inclusions and are found in cells infected with rabies virus (Negri bodies), vaccinia (Guarnieri bodies), fowlpox (Bollinger bodies), molluscum contagiosum (molluscum bodies), paramyxoviruses and reoviruses.

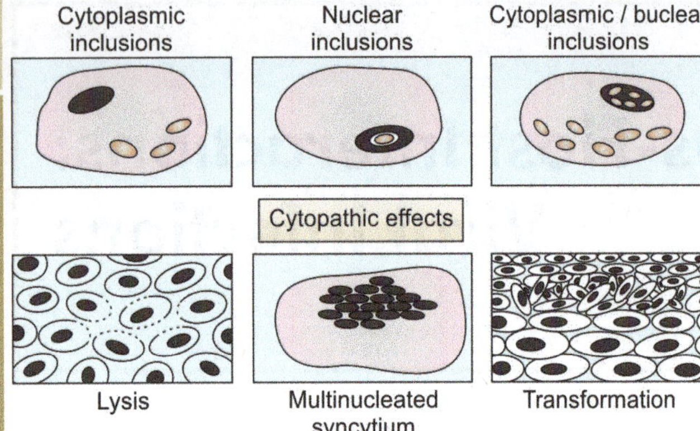

Fig. 54.1: Cytopathic effects of viral infection. (top) inclusion bodies: intracytoplasmic, e.g. rabies; intranuclear, e.g. herpesviruses; or both, e.g. measles virus. (bottom) The three main types of cytopathic effect: lysis, syncytium formation, and transformation

Intranuc lear inclusion bodies—In general, those viruses that are assembled in the nucleus (usually DNA viruses) produce intranuclear inclusions. They are found in cells infected with herpesviruses, adenoviruses and parvoviruses. Intranuclear inclusion bodies were classified into two types by Cowdry (1934). Cowdry type A inclusions are of variable size and granular appearance (as with herpesvirus, yellow fever virus), while type B inclusions are more circumscribed and often multiple (as with adenovirus, poliovirus).

Intranuclear and intracytoplasmic inclusion bodies—Some viruses such as measles virus and cytomegalovirus may produce both intranuclear and intracytoplasmic.

Nature of the Inclusion Bodies

The nature of the inclusions varies with the virus concerned. In many viral infections, the inclusion bodies are the site of development of the virions (the viral factories) or made up of virus antigens present at the site of virus synthesis, or simply degenerative changes produced by viral infection which confer altered staining properties on the cell. Variations in the appearance of inclusion material depend largely upon the tissue fixative used.

3. New cell-surface antigens—Another very important consequence of many virus infections is the induction of new antigens on the cell surface. These antigens may confer new properties on the cells.This is particularly important in the case of enveloped viruses that bud from the cell surface (e.g. herpes-, myxo-, paramyxo-, and retroviruses). Virus coded antigens also appear on the surface of cells transformed by oncogenic viruses.

4. Damage to the chromosomes of host cells— Certain viruses such as measles, mumps, adenoviruses, cytomegalovirus and varicella virus cause damage to the chromosomes of host cells. Chromatid gaps and breaks in chromosome 17 occur frequently in cultured cells infected with adenovirus types 12 and 31.

5. Latent and persistent infections—Viruses have evolved mechanisms to continue to survive in the face of a strong host immune response.

Examples

I. **Herpes simplex types 1 and 2 and varicella-zoster viruses**, following primary infection, travel up the peripheral nerves to the sensory ganglia. The viruses remain latent in the ganglia, to be reactivated periodically in some individuals causing recurrent lesions.

II. **Cytomegalovirus (CMV) and Epstein-Barr virus (EBV)**—Similarly, they are also known to establish latent infection.

III. **Hepatitis B virus (HBV)**—Some viruses like hepatitis B virus (HBV) may cause chronic infections which may remain clinically inapparent for many years. However, it may lead to more serious consequences, such as cirrhosis or hepatocellular carcinoma.

IV. **Subacute sclerosing panencephalitis (SSPE)**—Some viruses produce slow progressive infections. An important example of this type of persistent infection is subacute sclerosing panencephalitis (SSPE). This disease develops between 1 to 10 years after recovery from measles virus infection.

V. **Human immunodeficiency virus (HIV)**—Other viruses that produce slowly progressive fatal diseases in man are human immunodeficiency virus (HIV) which causes acquired immunodeficiency syndrome, and kuru agent and Creutzfeldt-Jakob agent which cause slowly progressive encephalopathy.

PATHOGENESIS OF VIRAL DISEASES

To produce disease, viruses must enter a host, come in contact with susceptible cells, replicate, and produce cell injury. Understanding of viral pathogenesis at the molecular level is necessary to design effective and specific antiviral strategies. Much of our knowledge of viral pathogenesis is based on animal models, because such systems can be readily manipulated and studied.

Steps in Viral Pathogenesis

Specific steps involved in viral pathogenesis are the following: viral entry into the host, primary viral replication, viral spread, cellular injury, host immune response, viral clearance or establishment of persistent infection, and viral shedding.

TRANSMISSION OF HUMAN VIRUS INFECTIONS

Viruses enter the body through the following routes (Table 54.1):

Table 54.1: Transmission of human virus infection

Route of transmission	*Viruses*
1. Respiratory tract	**Respiratory viruses:** Influenza A, B and C, parainfluenza types 1-4, RSV, rhinovirus, coronavirus, adenovirus and coxsackievirus A **Systemic viruses:** *Measles, mumps, rubella, varicella-zoster, CMV and EBV*
2. Alimentary tract	Many enteroviruses including poliovirus types 1-3, HAV, HEV, adenoviruses, rotaviruses, Norwalk and related viruses, coronavirus, astrovirus, coxsackievirus A and B and echovirus
3. Skin	**Minor abrasions:** Papillomaviruses, molluscum contagiosum, cowpox, orf, milker's nodes viruses, herpes simplex viruses and HB*V*. **Insect bite:** Arboviruses **Animal bite:** Rabies virus, herpes B virus **Injection:** *HBV, HCV, HIV-1, HIV-2,* HTL*V, CMV,* EBV and ebola virus
4. Genital tract	Papillomaviruses, herpes simplex viruses, *HIV, HTLV, HBV, HCV*
5. Conjunctiva	Some adenoviruses and few enteroviruses

RSV—Respiratory syncytial virus; CMV—Cytomegalovirus; EBV—Epstein-Barr virus; HAV— Hepatitis A virus; HEV—Hepatitis E virus; HIV—Human immunodeficiency virus; HTLV—Human T cell leukemia virus

1. Respiratory Tract

Many viruses enter the body through inhaled droplets expelled from the nose or mouth of infected persons during talking, coughing or sneezing.This is the most frequent means of viral entry into the host. All viruses that infect the host via the respiratory tract probably do so by attaching to specific receptors on epithelial cells. Some viruses, such as influenza A, B and C, parainfluenza types 1 to 4, respiratory syncytial virus (RSV), rhinovirus, coronavirus, adenovirus and coxsackievirus are restricted to respiratory tract where they multiply and produce local disease. Other viruses, multiply locally to initiate a silent local infection which is followed by lymphatic or hematogenous spread to other parts of the body where more extensive multiplication takes place before producing generalized disease. Measles, mumps, rubella, varicella-zoster, CMV and EBV are examples of such systemic diseases in which the portal of entry is the respiratory tract.

2. Alimentary Tract

The alimentary tract is the next most important route of entry for viruses. The viruses that initiate infection of humans via the alimentary tract are many enteroviruses including poliovirus types 1 to 3, hepatitis A virus (HAV), hepatitis E virus (HEV), adenoviruses, rotaviruses, Norwalk and related viruses, coronavirus, astrovirus, coxsackie A and B and echovirus. These are acid- and bile-resistant. Polioviruses, HAV and HEV do not produce any intestinal symptoms, but cause generalized infection.

3. Skin

The skin is a tough and impermeable barrier to the entry of viruses. However, a few viruses are able to breach this barrier and initiate infection of the host (Table 54.1). Of the viruses that enter through the skin, only a few produce local lesions. Some obtain entry through small abrasions of the skin (papillomaviruses, molluscum contagiosum, cowpox, orf, milker's nodes viruses, herpes simplex and HBV), insect bite (arboviruses), animal bite (rabies virus, herpes B virus) or injection (HBV, HCV, HIV, HTLV, CMV, EBV and ebola virus).

Viruses that are introduced deeper into the dermis have access to blood vessels, lymphatics, dendritic cells, and macro-phages and usually spread and cause systemic infections. Rabies virus travels along the nerves to the spinal cord or brain.

4. Conjunctiva

The conjunctiva also may act as a portal of entry for viruses. This may lead to local disease (adenovirus) or to systemic spread (measles).

5. Genital Tract

Papillomaviruses and herpes simplex viruses are sexually transmitted and produce lesions on the genitalia and perineum. HIV, HTLV, HBV and HCV are also sexually transmitted but do not produce local lesions.

6. Congenital Viral Infections

Congenital infection may occur at any stage from the development of the ovum up to birth. If the virus crosses the placenta and infection occurs in utero, serious damage may be done to the fetus. Rubella virus and cytomegalovirus are presently the primary agents responsible for congenital defects in humans and produce maldevelopment or severe neonatal disease. Congenital infections can also occur with herpes simplex, varicella-zoster, hepatitis B, measles, and mumps virus and with HIV, parvovirus, and some enteroviruses.

SPREAD OF VIRUS IN THE BODY

Generalized Infections

The manner in which the infecting virus spreads from the point of entry, multiplies in sites of election and causes lesions in target tissues was first studied by F. Fenner, in Australia, (1948) using mousepox as the experimental model.

Sequence of event s—The sequence of events, summarized as follows:

1. Entry—Virions enter through an epithelial surface, where they undergo limited replication.
2. Migration—They migrate to the regional lymph nodes where some are taken up by macrophages and inactivated, but others enter the bloodstream.
3. Primary viremia—Virions which enter the bloodstream. This is the primary viremia.
4. Secondary viremia—From the blood, the virus gains access to the large reticuloendothelial organs—liver, spleen, and bone marrow-in which it again multiplies, and a large amount of virus is produced which again spills over into the bloodstream, causing a secondary viraemia. This heralds the onset of clinical symptoms (the prodromal phase in eruptive fevers).
5. Target organ—The virus reaches the target organ (skin in eruptive fevers) through the bloodstream. Multiplication in the target sites produces the distinctive lesions. With minor modifications, this model holds good for most systemic virus diseases.

SIGNIFICANCE OF THE INCUBATION PERIOD

The incubation period represents the time taken for the virus to spread from the site of entry to the organs of viral multiplication and thence to the target organs for the production of lesions. Its duration is therefore influenced by the relation between the sites of entry, multiplication and lesion. They will be given in more detail in the chapters devoted to individual viruses, but for the moment, and as an aid to memory, we shall classify them into four main groups: short, medium, long, and very long .

1. Short means less than a week and primarily applies to viruses causing localized infections that spread rapidly on mucous surfaces. Some viruses injected directly (e.g. arboviruses transmitted by the bite of an arthropod) also, as a rule, have short incubation periods.
2. Medium incubation periods range from about 7 to 21 days.
3. Long refers to periods measured in weeks or months (e.g. 2-6 weeks for hepatitis A and 6-20 weeks for hepatitis B). Rabies may also have incubation periods extending for many months.
4. Very long incubation periods are measured in years, which is why the agents involved were originally termed 'slow' viruses. This group comprises the prions and a few'conventional' viruses, such as papovaviruses and measles, which very occasionally cause delayed disease of the central nervous system.

HOST RESPONSE TO VIRUS INFECTIONS

The outcome of a virus infection is influenced by the virulence of the infecting strain and the resistance offered by the host. Mechanisms of host resistance may be immunological or nonspecific.

A. Immunological Response

Virions in general are good antigens and induce both antibody- and cell-mediated immunity (CMI). For some diseases such as poxviruses, measles, herpes simplex virus and CMV infections, CMI appears to be the main immunospecific defence, for others, such as enterovirus and arbovirus infections, antibody production appears to be the main immunospecific defence.

1. Antibody-mediated Immunity

In mediating humoral antiviral immunity, the important classes of antibodies are IgG, IgM and IgA. IgG and IgM play a major role in blood and tissue spaces, while secretory IgA antibody is important in protecting against infection by viruses through the respiratory or gastrointestinal tracts. Humoral immunity protects the host against reinfection by the same virus.

Mechanisms of Antibodies Effecting Virus Neutralization

Antibodies effect virus neutralization by several mechanisms.

i. They may prevent adsorption of the virus to cell receptors, cause enhanced virus degradation or prevent release of the progeny virus from infected cells.
ii. Opsonization of virions for phagocytosis and killing by macrophages.
iii. Complement acts in conjunction with antibodies in causing surface damage to enveloped virions and in producing cytolysis of virus infected cells.

2. Cell-mediated Immunity

Cell mediated immunity prevents infection of target organs and promotes recovery from disease by destroying virus and virus-infected cells by any of the following four different processes:

1. Cytolysis by cytotoxic t cells (Tc) cells.
2. Cytolysis by natural killer (NK) cells.
3. Antibody-complement-mediated cytotoxicity.
4. Antibody-dependent cell-mediated cytotoxicity (ADCC).

In general, viral infections are followed by solid immunity to reinfection, which may often be lifelong. Apparent exceptions like the common cold and influenza are not due to lack of immunity but to reinfection being caused by antigenically different viruses. Live virus vaccines also induce more durable protection than bacterial vaccines.

B. Nonimmunological Responses

1. **Phagocytosis**—Polymorphonuclear leukocytes do not play any significant role in the defence against viral infections. On the other hand, macrophages phagocytose viruses and are important in clearing viruses from the bloodstream.
2. **Fever**—Fever may act as a natural defence mechanism against viral infections as most viruses are

inhibited by temperatures above 39°C. An exception is herpes simplex which is usually reactivated by fever to produce 'fever blisters'

3. **Hormones**—Corticosteroid administration enhances most viral infections. Treatment with steroids exacerbates the severity of herpes simplex and varicella-zoster infections. Normally mild infections such as varicella and vaccinia may be lethal in patients on cortisone. Injudicious use of steroids in the treatment of herpetic keratoconjunctivitis may cause blindness. Pregnancy significantly increases the likelihood of severe disease following infection with certain viruses which may be related to the hormonal changes associated with pregnancy. The deleterious effect of cortisone may be due to its depression of the immune response and inhibition of interferon synthesis.
4. **Malnutrition**—Malnutrition can exacerbate viral infections, for example, measles produces a much higher case fatality in malnourished than in well-fed children.
5. **Age—M**ost viral infections are commoner and more dangerous at the two extremes of age. Coxsackieviruses are able to infect suckling but not adult mice. Similarly in man, rotaviruses and respiratory syncytial viruses cause severe disease only in infants in the first year. Virus infections in elderly people can be very severe, herpes zoster (shingles) being a case in point.
6. **Interferons—**Interferons are a family of host coded proteins produced by cells on induction by viral or nonviral inducers. Isaacs and Lindenmann in 1957 discovered that virus infected cells produce a soluble factor that protects other cells from infection and they gave the name interferon to this antiviral substance.

Mode of action of interferon—On exposure to interferon, cells produce a protein (translation inhibiting protein, TIP) which selectively inhibits translation of viral mRNA, without affecting cellular mRNA. What has been called TIP is actually a mixture of at least three different enzymes (a protein kinase, an oligonucletide synthetase and an RNAase) which together block translation of viral mRNA into viral proteins. It has also been suggested that inhibition of viral transcription may also be responsible for the antiviral activity of interferon.

Synthesis of Interferons

Interferons are produced by all vertebrate species. Normal cells do not generally synthesize interferon until they are induced to do so. Viruses also vary in their capacity to induce interferon, cytocidal and virulent viruses being poor inducers and avirulent viruses being good inducers. Infection with viruses is a potent insult leading to induction; RNA viruses are stronger inducers of interferon than DNA viruses.

Types of Interferon

There are multiple species of interferons that fall into three general groups, designated IFN-α, IFN-β, and IFN-γ.

1. Alpha interferon (IFN-α)—Formerly known as leucocyte interferon, is produced by leukocytes following induction by suitable viruses. It is a non-glycosylated protein. At least 16 antigenic subtypes have been identified.
2. Beta interferon (IFN-β)—Formerly known as 'fibroblast interferon, is produced by fibroblasts and epithelial cells following stimulation by viruses or polynucleotides. It is a glycoprotein.
3. Gamma interferon (IFN-γ)—Formerly known as immune interferon, is produced by T lymphocytes, on stimulation by antigens or mitogens. It is a glycoprotein. It is more concerned with immunomodulatory and antiproliferative functions than with antiviral defence.

Main Biological Effects of Interferons

1. **Antiviral effects**—Induction of resistance to infections.
2. **Antimicrobial effects**—Resistance to intracellular infections, for example toxoplasma, chlamydia, malaria.
3. **Cellular effects**—Inhibition of cell growth and proliferation; and of DNA and protein synthesis; increased expression of MHC antigens on cell surfaces.
4. **Immunoregulatory effects**—Enhanced cytotoxic activity of NK, K and T cells; activation of macrophage cytocidal activity; modulation of antibody formation; activation of suppressor T cells; suppression of DTH.

KNOW MORE

Examples of potent inducers are togaviruses, vesicular stomatitis virus, Sendai virus and NDV. Nucleic acids (for example double stranded RNA and some synthetic polymers (for example Poly I:C) are particularly efficient inducers. Interferon production is increased by increasing the temperature up to about 40°C and is inhibited by steroids and increased oxygen tension. Interferon synthesis begins within about an hour of induction and reaches high levels in 6 to 12 hours. The promptness of interferon induction, much quicker than the antibody response, suggests that interferons may play a primary role inhost defence against viral infections. Cellular transcription and protein synthesis are necessary for interferon production.

☞ KEY POINTS

- Virus-host interactions may be considered at the level of the cell, individual and community.
- The cellular response to viral infection may range

from no apparent effect to cytopathology with accompanying cell death to hyperplasia or cancer.

- Some viruses, like poliovirus cause cell death (cytocidal effect) or even lysis (cytolysis). Others may cause cellular proliferation (as molluscum contagiosum) or malignant transformation (as oncogenic viruses). In some instances the virus and host cell enter into a peaceful coexistence a condition known as 'steady state infection'
- Inclusion bodies are virus-specific intracellular globular masses which are produced during replication of virus in host cells.
- Viruses enter the body through the respiratory tract, the alimentary tract, the skin, conjunctiva and the genital tract.
- The immune response is the best and most important means of controlling virus infections. Cell-mediated immune responses confer protection against viral diseases.
- Interferons are a family of host coded proteins produced by cells on induction by viral or nonviral inducers. Interferons fall into three general groups, designated IFN-α, IFN-β, and IFN-γ.

IMPORTANT QUESTIONS

1. Discuss various methods by which viruses can be transmitted to human beings.
2. Write short notes on:
   a. Inclusion bodies
   b. Interferons

FURTHER READING

Collier L, Oxford J. Human Virology. London: Oxford University Press, 2000.

Tyler KL, Nathanson N. Pathogenesis of viral infections. In: Knipe OM, et al (eds). Fields Virology, 4th ed. Lippincott Williams & Wilkins 2001.

CHAPTER 55

Laboratory Diagnosis, Prophylaxis and Chemotherapy of Viral Diseases

Learning Objectives

After reading and studying this chapter, you should be able to:

- Describe laboratory diagnosis of viral infections.
- Describe immunoprophylaxis of viral diseases.
- Describe the following: Live viral vaccines; killed viral vaccines.

LABORATORY DIAGNOSIS OF VIRAL INFECTIONS

Indications for laboratory diagnosis of viral infections:

1. **Diagnosis of diseases caused by viruses** for which antiviral chemotherapy is already available (herpesviruses, human immunodeficiency virus (HIV).
2. **For proper management of the patient.**
   Examples:
   i. If rubella is diagnosed in the first trimester of pregnancy, abortion is recommended.
   ii. If a person is bitten by an animal, an early establishment of the diagnosis of rabies in the animal and postexposure immunization of the patient prevents the development of rabies.
   iii. If a woman has primary genital herpes at the time of delivery, cesarean section is indicated.
   iv. If a baby is born to an HBsAg positive mother, immunization at birth is mandatory.
3. **Screening of blood donors:** It is an important procedure in the prevention of some diseases (such as screening for HBV and HIV in blood donors).
4. **Early detection of dangerous epidemics:** Such as yellow fever, encephalitis, influenza, poliomyelitis, etc is of vital importance to initiate appropriate control measures.
5. **To discover new viruses:** HIV was discovered in 1983.

Methods for Laboratory Diagnosis of Viral Infections

Laboratory diagnosis of viral infections can be carried out by the following methods:

A. Direct detection of virus
B. Virus isolation and growth
C. Detection of viral proteins
D. Detection of viral genetic material
E. Serology.

A. Direct Detection of Virus

1. Electron Microscopy (EM)

Viruses that are difficult to culture can be recognized by electron microscopy. The clinical material can be negatively stained with potassium phosphotungstate or uranyl acetate and scanned by EM. Clinical applications of electron microscopy include detection of rotavirus and hepatitis A virus in fecal specimens, poxviruses in vesicle fluid and herpesvirus in brain biopsy tissue. This method can also be used for the identification of virus isolates in cell culture.

2. Immunoelectron Microscopy

EM as a diagnostic tool has low sensitivity. The addition of virus-specific antibody to a sample can cause viral particles to clump, thereby facilitating the detection and simultaneous identification of the virus (immunoelectron microscopy). This method is useful for the detection of enteric viruses, such as rotavirus.

3. Fluorescence Microscopy

Direct or indirect fluorescent antibody technique can be used to detect virions or viral antigens in frozen tissue sections, acetone-fixed cell smears, cells from virus infected cultures or vesicles fluid.

Fluorescence microscopy of brain biopsy can be used for the diagnosis of herpes simplex encephalitis and subacute sclerosing panencephalitis (a late sequelae of measles) and for the verification of rabies in the brain of animals suspected to be rabid. This method is also useful for the rapid diagnosis of respiratory infections caused by paramyxoviruses, orthomyxoviruses, adenoviruses and herpesviruses.

4. Light Microscopy

Demonstration of the inclusion body is a routine diagnostic method. Rabies may be detected through the finding of Negri bodies (rabies virus inclusions) in brain cells of animals.

B. Virus Isolation

For virus isolation it is imperative that the specimen be collected properly and transported with least delay to the laboratory. The reasons are that many viruses are labile and that the samples are susceptible to bacterial and fungal overgrowth. Viruses are best transported and stored on ice and in special media that contain antibiotics and proteins such as serum albumin or gelatin. Specimens should be collected early in the acute phase of infection, before the virus ceases to be shed.

In general the methods used for isolation consist of inoculation into **animals, eggs** or **tissue culture,** after the specimen is processed to remove bacterial contaminants.

The isolates are identified by neutralization or other suitable serological procedures. Many viruses (for example, adenoviruses, enteroviruses) are frequently found in normal individuals. The results of isolation should always be interpreted in the light of the clinical data.

C. Detection of Viral Proteins

The viral proteins can be assayed by the following methods:

i. Protein patterns (by **electrophoresis)**
ii. Enzyme activities (e.g. reverse transcriptase)
iii. Hemagglutination and hemadsorption
iv. **Antigen detection:** Viral antigens can be detected by immunofluorescence, enzyme-linked imunosorbent assay (ELISA), radioimmunoassay (RIA), and latex agglutination (LA).

D. Detection of Viral Genetic Material

i. Electrophoretic patterns of RNA (influenza, reovirus).
ii. Different strains of HSV-l and HSV-2 can be distinguished in this way by restriction endomclease cleavage patterns from DNA—HSVI and HSV2.
iii. **DNA probes**: DNA probes with sequences complementary to specific regions of a viral genome can be used, like antibodies, as sensitive and specific tools for detecting a virus. Enzyme-labelled or radiolabeled nucleic acid (DNA or RNA) sequences complementary to unique regions in nucleic acid sequences of most viruses are now manufactured commercially. These labeled complementary sequences are known as nucleic acid probes. Two strands of the target DNA molecule in the clinical specimens are first separated, then, allowed to hybridize with a labeled single-stranded DNA or RNA probe present in excess. Depending on the type of label attached to the probe, hybridized-labelled probe can be detected by radiography, gamma-counting or a simple colorimetric evaluation (dot-blot hybridization). The DNA probes are detected with autoradiography or with fluorescent or EIA-like methods. By use of nucleic acid probes cytomegalovirus, papillomavirus and Epstein-Barr virus have been identified.
iv. **Southern, Northern and dot blot analysis:** Viral genomes can also be detected in clinical samples with the use of dot blot or Southern blot analysis. Electrophoretically separated viral RNA **(Northern blot:** RNA:DNA probe hybridization) blotted onto a nitrocellulose filter can be detected in a similar manner.
v. **Polymerase chain reaction (PCR):** The polymerase chain reaction (PCR) for DNA, reverse transcriptase polymerase chain reaction (RT-PCR) for RNA, and branched-chain DNA (DNA, RNA) assays are becoming very important for viral detection.

   This technique is especially useful for detecting latent and integrated sequences of viruses such as retroviruses, herpesviruses, papillomaviruses, and other papovaviruses as well as the sequences of viruses present in low concentrations.
vi. **Reverse transcriptase polymerase chain reaction (RT-PCR) for RNA:** This approach was very useful for identifying and distinguishing the Hantaviruses.
vii. **Branched-chain DNA assays:** Quantitate viral DNA or RNA much like ELISA.

F. Serological Diagnosis

The demonstration of a rise in titer of antibodies to a virus during the course of a disease is strong evidence that it is the etiological agent. For this, it is essential to examine paired sera, the 'acute' sample collected early in the course of the disease and the 'convalescent' sample collected 10-14 days later. Examination of a single sample of serum for antibodies may not be meaningful except when IgM specific tests are done which is present during the first 2 or 3 weeks of a primary infection, generally indicates a recent primary infection.

The serological techniques employed would depend on the virus but those in general use are neutralization, complement fixation, ELISA, hemagglutination inhibition tests, indirect fluorescent antibody test, latex agglutination test.

IMMUNOPROPHYLAXIS OF VIRAL DISEASES

A. Active immunization
B. Passive immunization

Active Immunization

Viral vaccines are of three types (Tables 55.1 and 55.2):

I. Live-virus Vaccines

Currently available live-virus vaccines are oral polio, measles, mumps, rubella, (MMR), yellow fever and varicella vaccines. Polyvalent measles-mumps-rubella (MMR) vaccine is also available.

Table 55.1: Viral vaccines in common use

Type of vaccine		Disease	Mode of preparation
A. Live viral vaccine	i.	Measles	Attenuated virus grown in tissue culture
	ii.	Mumps	Attenuated virus grown in human diploid cell culture
	iii.	Rubella	Attenuated virus grown in tissue culture
	iv.	Poliomyelitis (Sabin)	Avirulent strains grown in monkey kidney cell culture
	v.	Influenza	
		a. Attenuated	Virus attenuated by serial passage in eggs
		b. Mutant	*ts* mutants which are avirulent
		c. Recombinant	Recombinants with surface antigens of new strains and growth characters of established strains
	vi.	Yellow fever (17D)	Attenuated virus grown in chick embryos and lyophilized
B. Killed viral vaccine	i.	Polio (Salk)	Virulent strains grown in monkey kidney cell culture, formalin-killed
	ii.	Rabies	Fixed virus grown in sheep brain and inactivated by phenol or beta propiolactone
	iii.	Influenza (subunit)	Virus disintegrated with sodium deoxycholate
	iv.	Japanese B encephalitis	Virus grown in mouse brain and inactivated by formalin
	v.	Varicella	Attenuated virus grown in chick embryo fibroblastculture
			Virus grown in cell culture and inactivated with beta propiolactone
C. Subunit		Hepatitis B	HBsAg cloned in yeast

These are prepared from:

- Attenuated strains, e.g. yellow fever
- Temperature sensitive (*ts*) and cold-adapted (ca) mutants, e.g. influenza
- Recombinant live viral or bacterial vectors, e.g. influenza.

Advantages of Live-virus Vaccines

1. **Single dose:** A single dose is usually sufficient because they multiply in the human host and provide continuous antigenic stimulation over a period of time resulting in durable immunity.
2. **Local immunity:** They can be administered by the route of natural infection so that local immunity is induced.
3. **Wide spectrum of immunoglobulins:** They induce a wide spectrum of immunoglobulins including secretory IgA to whole range of viral antigens.
4. **Cell-mediated immunity:** They also induce cell-mediated immunity.
5. **Long-lasting immunity:** They provide more effective long-lasting immunity than killed vaccines.
6. **Primary vaccine failures are uncommon.**
7. **More economical:** They can, in general, be prepared more economically and more conveniently for mass immunization.

Disadvantages of Live-virus Vaccines

1. **Reversion of virulence:** There is risk, however, remote, of reversion of virulence.
2. **Heat-labile:** Live-virus vaccines are heat-labile and have to be preserved under strict refrigeration.
3. **Interference:** Interference by coinfection with a naturally occurring, wild-type virus may inhibit replication of the vaccine virus and decrease its effectiveness, e.g. with the vaccine strains of poliovirus inhibited by concurrent infections by various enteroviruses.
4. **Contamination:** The vaccine may be contaminated with potentially dangerous viruses such as oncogenic viruses or other infectious agents.
5. **Local remote complications:** Some live-virus vaccines may cause local remote complications.
6. **Virus may spread from the vaccinees to contacts:** While this is a serious danger in some situations, as when spread occurs to immunodeficient or other high risk contacts (e.g. in rubella, if the vaccine strain is teratogenic). In other cases, it may even be an advantage (as in poliomyelitis where the range of vaccination is extended by the natural spread of the vaccine virus among children and adults).

II. Killed Viral Vaccines

Killed vaccines have been prepared by inactivating viruses with heat, phenol, formalin or beta propiolactone. Adverse reactions may also be reduced by the use of 'subunit vaccines' in which the virus is split by detergents or other chemicals and only the relevant antigens incorporated in the vaccine. Vaccine production by cloning the desired antigen in bacteria or yeast is becoming increasingly common, as in hepatitis B vaccine.

Some of the currently available killed viral vaccines include polio (inatirated polio vaccine, rabies, Japanese B encephalitis, influenza and hepatitis B).

Table 55.2: Antiviral agents used for treatment of viral infections

Mode of Action	Drug	Target Virus
A. VIRAL POLYMERASE INHIBITORS	1. Acyclovir	Herpes simplex, varicella-zoster
	2. Ganciclovir	Cytomegalovirus
	3. Cidofovir	Cytomegalovirus, herpes simplex
	4. Foscarnet	Herperviruses, HIV-1, HBV
	5. Valacyclovir	Herpesviruses
	6. Vidarabine	Herpesviruses, vaccinia, HBV
B. BLOCKING OF VIRAL UNCOATING	1. Amantadine	Influenza virus A
C. HIV PROTEASE INHIBITORS	1. Saquinavir	HIV-1, HIV-2
	2. Indinavir	HIV-1, HIV-2
	3. Ritonavir	HIV-1, HIV-2
D. REVERSE TRANSCRIPTASE INHIBITORS	1. Zidovudine (azidothymidine, AZT)	HIV-1, HIV-2, HBV
	2. Didanosine (dideoxyinosine, ddl)	HIV-1, HIV-2
	3. Zalcitabine (dideoxycytidine, ddC)	HIV-1, HIV-2, HBV
	4. Stavudine (didehydrodeoxythymidine)	HIV-1, HIV-2
	5. Lamivudine (dideoxythiacytidine)	HIV-1, HIV-2, HBV
	6. Nevirapine	HIV-1
E. BLOCKING OF CAPPING OF mRNA	1. Ribavirin	Respiratory syncytial virus, Influ enza viruses A and B, Lassa fever
F. INHIBITORS OF PROTEIN SYNTHESIS	1. Interferons	HBV, HCV, CMV

HIV—Human immunodeficiency virus; HBV—Hepatitis B virus; HCV—Hepatitis C virus; CMV—*Cytomegalovirus*

Advantages

1. They can be given in combination as polyvalent vaccines.
2. There is also no danger of the spread of the virus from the vaccinee.
3. Stability and safety.

Disadvantages

1. Extreme care is required in their manufacture to make certain that no residual live virulent virus is present in the vaccine.
2. The immunity conferred is often brief and must be boosted.
3. Parenteral administration of killed-virus vaccine, is does not induce local (IgA) immunoglobulions.
4. Cell-mediated response to inactivated vaccines is generally poor.
5. Some killed-virus vaccines have induced hypersensitivity to subsequent infection.

III. Passive Immunization

Passive immunization with human gammaglobulin, convalescent serum or specific immune globulin gives temporary protection against many viral diseases such as measles, mumps and infectious hepatitis. These are indicated only when nonimmune individuals who are at special risk are exposed to infection. Combined active and passive immunization is an established method for the prevention of rabies.

CHEMOPROPHYLAXIS AND CHEMOTHERAPY OF VIRUS DISEASES

Because viruses are obligate intracellular parasites, antiviral agents must be capable of selectively inhibiting viral functions without damaging the host, making the development of such drugs very difficult. Viral infection may be checked at the level of attachment, penetration transcription of viral nucleic acid, translation of viral mRNA, replication of viral nucleic acid and inhibition of viral DNA polymerase and transcriptase (Table 55.2).

Available antiviral agents can be considered under various categories (Table 55.2).

Acyclovir (Acylguanosine)

Acyclovir (acylguanosine) is an analog of guanine, acting against herpesviruses through thymidine kinase. Herpes viruses that code for then own thymidine kinase (HSV, V-Z) are far more susceptible than those which do not (CMV-EBV). The related drug *Ganciclovir* is more active against CMV.

Amantadine and Rimantadine

These synthetic amines specifically inhibit influenza A viruses by blocking viral uncoating. *Amantadine (Adamantanamine hydrochloride, Symmetrol)* blocks host cell penetration by influenza A virus. A derivative *rimantadine* is less toxic and equally effective. They must be administered prophylactically to have a significant protective effect.

Azidothymidine (Zidovudine, AZT)

The widely publicized drug *Azidothymidine* (Zidovudine, AZT). AZT is used widely in HIV infection, but is toxic and costly.

A series of dideoxynucleosides *(Didanosine, Zalcitabine, Stavudine, Lamivudine)* have been synthesized and found to possess anti-HIV activity by blocking reverse transcriptase.

Interferons

Beneficial effect of interferons has been obtained in persistent infections such as hepatitis Band C, laryngeal papilloma and against CMV infections in transplant recipients.

KNOW MORE

Virus Isolation

The "gold standard" for proving a viral etiology of a syndrome is the recovery and growth of the infecting agent.

KEY POINTS

- Laboratory diagnosis of viral infections can be carried out by the methods such as I. Direct detection of virus and its components; II. Virus isolation; III. Detection of specific antibodies; IV. Cytological examination.
- Virus isolation: The methods used for isolation consist of inoculation into animals, eggs or tissue culture.
- **The viral proteins** can be assayed by protein patterns (by **electrophoresis),** enzyme activities (e.g. reverse transcriptase), hemagglutination and hemadsorption and antigen detection (e.g. indirect fluorescence, enzyme linked immunosorbent assay, Western blot).
- Detection of viral genome is done by DNA probes, dot blot or Southern blot analysis, Northern blot or RNA:DNA probe hybridization, PCR, and RT PCR.
- The serological techniques employed are neutralization, complement fixation, ELISA, hemagglutination inhibition tests, indirect fluorescent antibody test, latex agglutination test.
- Immunoprophylaxis of viral diseases is by active and passive immunization.
- Viral vaccines are used for active immunization which may be *live* or *killed*.
- Passive immunization with human gammaglobulin, convalescent serum or specific immune globulin gives temporary protection against many viral diseases.

IMPORTANT QUESTIONS

1. Discuss laboratory diagnosis of viral infections.
2. Discuss viral vaccines.
3. Write short notes on:
   Live viral vaccines.
   Acquired immunity in viral infections.
   Antiviral therapy.

FURTHER READING

Ada G. Vaccines and vaccination. N Engl J Med 2001;345:1042.

Collier L and J Oxford 2000. Human Virology. London: Oxford University Press.

Karupiah G, Xie QW, et al. Inhibition of viral replication by interferon Y-induced nitric oxide synthase. Science 1993;261: 1445-8.

CHAPTER 56

Bacteriophages

Learning Objectives

After reading and studying this chapter, you should be able to:

- Describe morphology of bacteriophage.
- Discuss lytic and lysogenic cycle of bacteriophage.
- Describe phage typing.

INTRODUCTION

Viruses that infect bacteria are called ***bacteriophages***, or simply phages. Twort (1915) described a degenerative change in staphylococcal colonies isolated from calf lymph, which could be transmitted serially by application of culture filtrates from the original growth. d'Herelle (1917) observed that the filtrates of feces cultures from dysentery patients induced transmissible lysis of a broth culture of a dysentery bacillus. He suggested that the lytic agent was a virus and gave it the name bacteriophage (*phage: to eat*). Phages occur widely in nature in close association with bacteria. Phages can readily be isolated from a wide range of environments such as feces, sewage and other natural sources of mixed bacterial growth.

ROLE OF BACTERIOPHAGES

1. They play an important role in the transmission of genetic information from one bacterium to another by the process of transduction.
2. They also play a role in the evolution of bacterial types and in the transmission of some virulence characters.
3. Phages may indeed be effective in treating bacterial infections, including those caused by antibiotic-resistant bacteria.
4. Phages have been used as cloning vectors in genetic manipulations.
5. They may have a role in the control of bacterial populations in natural waters.

MORPHOLOGY

Certain bacteriophages that infect E. coli, called the T even phages (T2, T4, T6), have been studied in great detail and traditionally serve as the prototypes in describing the properties of bacteriophages.

T even phages have a complex and characteristic morphology. Most phages are tadpole-shaped, possess a hexagonal head and a cylindrical tail.

Head

The head consists of a tightly packed core of nucleic acid (double stranded DNA) surrounded by a **protein coat or capsid.** The size of th e head varies in different phages from 28 nm to 100 nm. The size of the head varies in different phages from 28 nm to 100 nm. The head of phage T4 has a diameter of 65 nm and is 100 nm long.

Tail

The tail is cylindrical and is composed of a central hollow core or tube, a contractile sheath surrounding the core and a terminal base plate which has attached to it prongs or tail fibers (usually six in number) or both (Fig. 56.1) that bind to specific receptor sites on the bacterial surface.

The type of nucleic acid present in the phage particle varies, but no phage has more than one type. Although double-stranded DNA is the most common, singlestranded RNA or DNA is present in some phages. Singlestranded phage DNA often is circular. The nucleic acids range in size from about 3000 bases, sufficient to encode three or four proteins, up to 150,000 base pairs in the case of coliphage T4, which encodes more than 150 proteins.

LIFE CYCLE

Phages exhibit two different types of life cycle. In the *virulent* or *lytic cycle,* intracellular multiplication of the phage culminates in the lysis of the host bacterium and the release of progeny virions. In the *temprate* or *lysogenic cycle* the phage DNA becomes integrated with the bacterial genome, replicating synchronously with it, causing no harm to the host cell (Fig 56.2).

Lytic Cycle

Replication of a virulent phage can be considered in the following stages adsorption, penetration, synthesis of phage components, assembly, maturation and release of progeny phage particles.

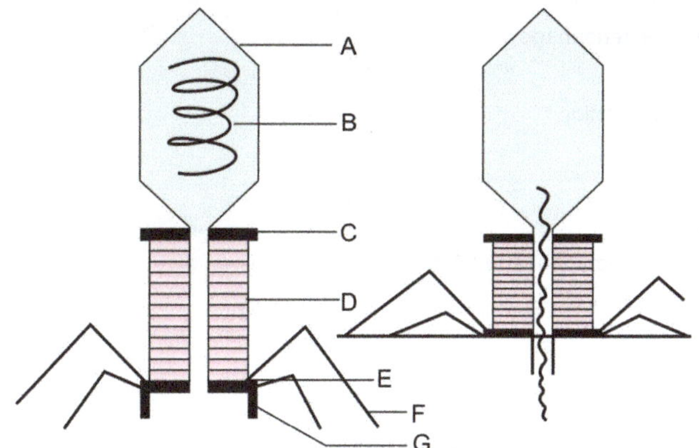

Fig. 56.1: Morphology of bacteriophage. A. hexagonal head, B. DNA core, C. demarcation between head and tail, D. tail, E. base plate, F. tail fibers, G. prongs; right, process of injection of phage DNA into host cell

i. Adsorption

By random collision, phage particles attach to virus-specific receptors on the host cell by its tail. Adsorption is a specific process and depends on the presence of complementary chemical groups on the receptor sites on the bacterial surface and on the terminal base plate of the phage. Adsorption is a very rapid process under optimal conditions, being complete within minutes. Any component on the bacterial surface can serve as receptor for some phage. Host specificity of phages is determined at the level of adsorption. Experimental infection by direct injection of phage DNA can be achieved even in bacterial strains that are insusceptible to infection by the whole phage. The infection of a bacterium by the naked phage nucleic acid is known as ***transfection.***

ii. Penetration

After adsorption, most phages inject their nucleic acid into the bacterial cytoplasm and leave their protein capsid outside. The process of penetration resembles injection through a syringe. Following adsorption, six tail pins make contact with the host cell surface and firmly attach the phage plate to it. The contractile tail sheath then contracts forcing the hollow interior tail tube into the bacterial cell wall. The phage DNA then passes through the hollow interior tail tube. The empty head and tail of the phage remain outside the bacterium as the shell or 'ghost' after penetration. Penetration may be facilitated by the presence on the phage tail of lysozyme which produces a hole on the bacterial wall for the entry of the phage core.

When bacteria are mixed with phage particles at high multiplicity (that is very large number of phages per bacterial cell), multiple holes are produced on the cell with the consequent leakage of cell contents. Bacterial lysis occurs without viral multiplication. This is known as **'lysis from without'**.

iii. Synthesis of Phage Nucleic Acid and Proteins

The synthesis of the phage components is initiated immediately after penetration of the phage nucleic acid. The first products to be synthesized (called ***early proteins)*** are the enzymes necessary for the building of the complex molecules peculiar to the phage. Subsequently, ***late proteins*** appear, which include the protein subunits of the phage head and tail. During this period, the synthesis of bacterial protein, DNA and RNA ceases and the cell is forced to make viral constituents.

Late genes are expressed only after the phage DNA has replicated. Late gene products include the structural components for the new phage particles, including heads, tails, and fibers. Late products also include the phage lysozyme, which will degrade the bacterial cell wall to liberate the mature phage particles.

iv. Assembly and Maturation

The phage structural proteins and nucleic acid are assembled in definite pathways to form the mature progeny phage particle. Phage DNA, head protein and tail protein are synthesized separately in the bacterial cell. The DNA is condensed into a compact polyhedron and 'packaged' into the head and, finally, the tail structures are added. This assembly of the phage components into the mature infective phage particle is known as **maturation.**

v. Release

Many phages lyse their host cells at the end of the intracellular phase. Release of the mature progeny phages typically occurs by lysis of the bacterial cell. The bacterial cell wall is weakened and it assumes a spherical shape during the replication of the phage. Phage enzymes act on the weakened cell wall causing it to burst or lyse resulting in the release of mature daughter phages.

Eclipse Phase

The interval between the entry of the phage nucleic acid into the bacterial cell and the appearance of the first infectious intracellular phage particle is known as the *eclipse phase*. It represents the time required for the synthesis of the phage components and their assembly into mature phage particles. The interval between the infection of a bacterial cell and the first release of infectious phage particles is known as the *latent period* (20 to 40 minutes). Immediately following the latent period, the number of phage particles released increases for a few minutes till the maximum number of daughter phages is attained. This period, during which the number of infectious phages released rises, is known as the *rise period*. The average yield of progeny phages per infected bacterial cell is known as the *burst size* (100 to 300 phages). This is estimated by experiments in which infection is established with one phage per bacterium and the release of infected phage particles is estimated serially over a period of time. The results of such an experiment plotted on a graph is known as the one-step *growth curve* (Fig. 56.3).

Lysogenic Cycle

Unlike virulent phages which produce lysis of the host cell, temperate phages enter into a symbiotic relationship with their host cells without destroying them. Following

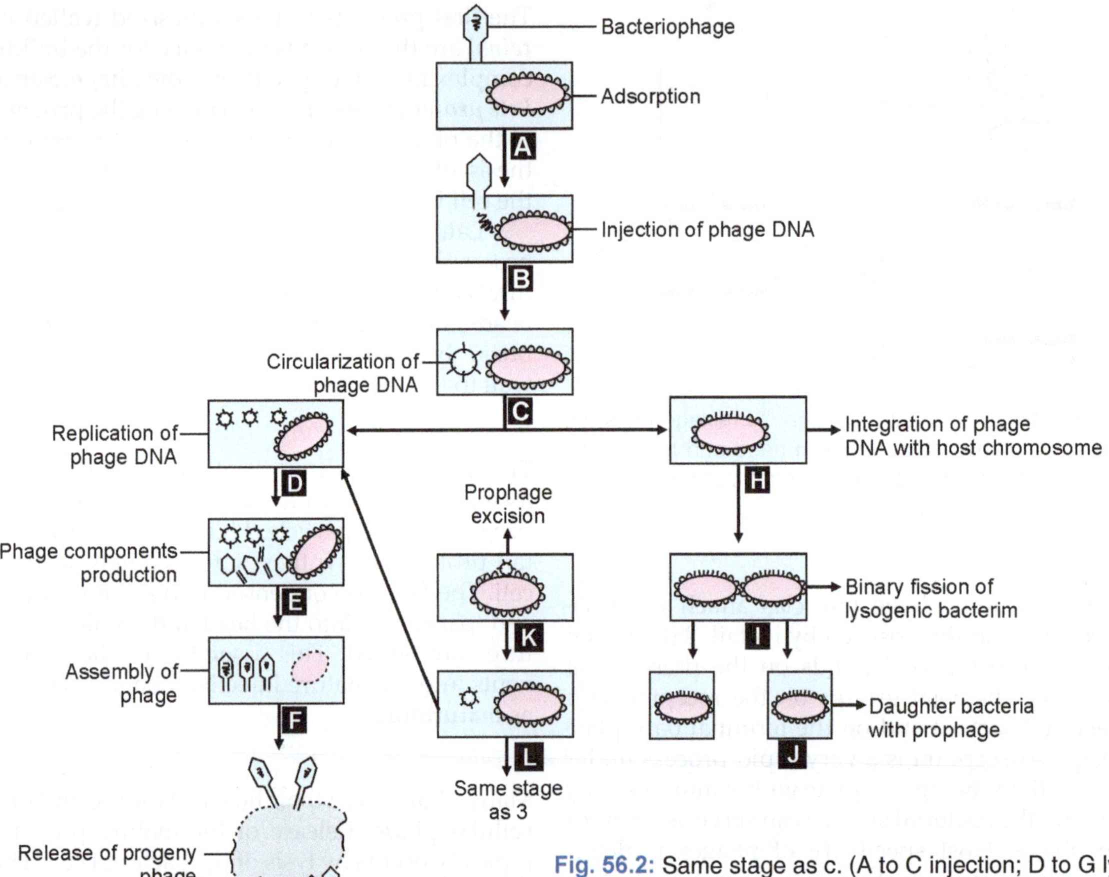

Fig. 56.2: Same stage as c. (A to C injection; D to G lytic; H to J lysogenic cycle; K to L induction)

entry into the host cell, the temperate phage nucleic acid becomes integrated with the bacterial chromosome. The integrated phage nucleic acid is known as the ***prophage.*** The prophage behaves like a segment of the host chromosome and replicates synchronously with it. This phenomenon is called ***lysogeny*** and a bacterium that carries a prophage within its genome is called a ***lysogenic bacterium*** or *lysogen* because the prophage retains the potential to lyse its host bacterium, and phages able to enter into this relationship are temperate phages. The prophage usually is integrated into the bacterial genome but sometimes exists independently. Lysogenization does not upset the bacterial metabolism. Temperate phages commonly are found in clinical isolates of gram-positive and gram-negative bacteria and, in some cases, contribute to the pathogenicity of the organism.

The prophage confers certain new properties on the lysogenic bacterium. This is known as **lysogenic or phage conversion.**

i. Toxin production in *Corynebacterium diphtheriae* is determined by the presence in it of the prophage β. The elimination of this prophage abolishes the toxigenicity of the bacillus and nontoxigenic strains can be made toxigenic by lysogenization.
ii. *Clostridium botulinum* types C and D produce toxin only if these are infected with phage CE β and DE β respectively.
iii. A wide variety of temperate phages of *Salmonella* can modify the antigenic properties of somatic O antigen. When *Salmonella* is infected by an epsilon phage, the structure of its outer lipopolysaccharide layer may be modified. The antigenic formula of S. Anatum is 3, 10: e, h: 1,6 but when it is lysogenised by a temperate phage its antigenic formula becomes 3, 15: e, h: 1,6.

The prophage may become 'excised' from occasional cells during the multiplication of lysogenic bacteria. The excised prophage initiates lytic replication and the daughter phage particles are released, which infect other bacterial cells and render them lysogenic. This is known as **spontaneous induction of prophage.** While this is a rare event, all lysogenic bacteria in a population can be induced to shift to the lytic cycle by exposure to certain physical and chemical agents. Such inducing agents include UV rays, hydrogen peroxide and nitrogen mustard. A lysogenic bacterium is resistant to reinfection by the same or related phages. This is known as ***superinfection immunity.***

SIGNIFICANCE OF PHAGES

1. Virulent Phage

i. Phage Assay

When a phage is applied on the lawn culture of a susceptible bacterium, areas of clearing occur after incubation. These zones of lysis are called ***plaques.*** The size, shape and nature of plaques are characteristic for different

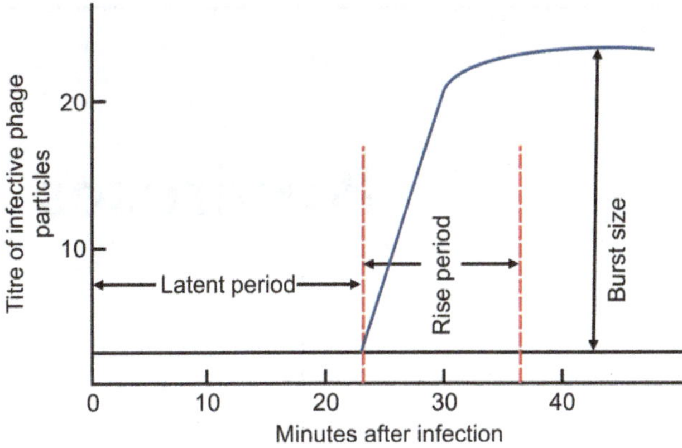

Fig. 56.3: One step growth curve of bacteriophage

phages. Plaque assay can be employed for titrating the number of viable phages in a preparation since under-optimum conditions a single phage particle is capable of producing one plaque. Plaquing is also useful for the purification of phages as plaques are analogous to bacterial colonies.

ii. Phage Typing

Bacteriophage (phage) typing is based on the specificity of phage surface receptors for cell surface receptors. The limited host range of many phages enables them to be used as an **epidemiological marker** to discriminate between bacterial strains that are biochemically or serologically indistinguishable. This method has been used to trace outbreaks of infection caused by ***Staphyolococcus aureus, S. Typhi* and *Vibrio cholerae.***

The strain to be typed is inoculated on a plate of nutrient agar to produce a lawn culture. After drying, the phages are applied overmarked squares in a fixed dose **(routine test dose).** The highest dilution of the phage preparation that just produces confluent lysis known as the **routine test dose (RTD).** After overnight incubation, the culture will be observed to be lysed by some phages but not by others. The phage type of the strain is expressed by the designation of the phage/phages that lyse it. Area of lysis caused by phage is known as **plaque.** Since a single phage particle is capable of producing one plaque, plaque assay can be employed for titrating the number of viable phages in a preparation.

The most important application of phage typing is for intraspecies typing of bacteria, as in the phage typing of ***S.typhi* and staphylococi.** Phages are available that lyse all members of a bacterial genus (for example, **genus-specific bacteriophage for *Salmonella*),** all members of a species (for example, **specific bacteriophage for *B. anthracis*),** and all members of a biotype or subspecies (for example, Mukerjee's phage IV which lyses—all strains of classical *V. cholerae* but not *V. cholerae* biotype EI Tor).

2. Temperate Phage

i. Transduction

Bacteriophages may act as carriers of genes from one bacterium to another. This is known as *transduction.* Two types of transduction are recognized: **generalized transduction**, in which any portion of the donor DNA can be transferred, and **specialized transduction**, in which only a specific set of genes can be carried to a recipient cell.

ii. Toxin Production

Toxin production in *Corynebacterium diphtheriae* and *Clostridium botulinum* types C and D are determined by genes carried in prohage DNA.

iii. Antigenic Property

A wide variety of temperate phages of *Salmonella* can modify the antigenic properties of somatic O antigen. Such acquisition of new properties by bacterial cells following phage infection is called *"phage conversion"*.

iv. Cloning Vector

Bacteriophages have been used as cloning vectors in genetic manipulations.

KEY POINTS

- Bacteriophages are bacterial viruses. They are associated with transmission of genetic material from one bacterium to another.
- **Morphology:** The best-studied phages are the T-even phages (T2, T4, T6, etc.). T- even phages have a complex and characteristic morphology. Most phages are tadpole-shaped, possess a hexagonal head and a cylindrical tail. Most of the phages usually consist of single, linear, and double stranded DNA molecule genome. The phages show high host specificity.
- **Life cycle:** Phages exhibit two different types of life cycle: lytic cycle and lysogenic cycle. In the lytic (virulent) cycle, intracellular multiplication of the phage ends in lysis of the host bacterium and release of progeny virions.
- In the lysogenic (temperate) cycle, phage DNA becomes incorporated within the bacterial genome and then replicates synchronous with it, causing no harm to the host cell.
- **Phage typing** has been used for *Staphylococcus aureus, Vibrio cholerae, Salmonella,* and *many* other bacteria.

IMPORTANT QUESTIONS

1. Draw a labelled diagram of a T-even phage.
2. Discuss the life cycle of bacteriophages.
3. Write short notes on:
   a. Lysogenic cycle of phages
   b. Lysogenic conversion
   c. Phage typing
   d. Significance of phages.

FURTHER READING

Black LW. DNA packaging in dsDNA bacteriophages. Annu Rev Microbiol 1989;43:267-92.

Day, M. Bacteriocins and bacteriophages, in: Topley, Wilson's (Eds). Microbiology and Microbial Infections. London, Armold 1998;01(2):185.

Young RY. Bacteriophage lysis: mechanism and regulation. Microbiol Rev 1992;56:430-81.

CHAPTER 57

Poxviruses

Learning Objectives

After reading and studying this chapter, you should be able to:

- Describe structure of vaccinia virus.
- Describe molluscum contagiosum.

INTRODUCTION

Poxviruses are the largest and most complex of viruses that infect vertebrates, large enough to be seen under the light microscope. This group contains several viruses that infect human beings, animals, birds and insects.

Smallpox, the great success story in the fight against infectious disease and one that provides many valuable lessons. This epic provides at least three 'firsts': the first vaccine, the first disease to be totally eradicated by immunization, and the first virus infection against which chemotherapy was clinically effective.

CLASSIFICATION

The family Poxviridae contains two subfamilies: the Chordopoxvirinae, the poxviruses of vertebrates, and Entomopoxvirinae, the poxviruses of insects. Chordopoxvirinae are placed in eight genera- *Orthopoxvirus, Parapoxvirus, Molluscipoxvirus, Yatapoxvirus, Capripoxvirus Leporipoxvirus, Avipoxvirus* and *Sui poxvirus*. Most of the poxviruses that can cause disease in humans are contained in the genera Orthopoxvirus and *Parapoxvirus;* tyere are also several that are classified in the genera *Yatapoxvirus and Molluscipoxvirus.* Poxviruses causing diseases are enumerated in Table 57.1.

MORPHOLOGY

These are the largest viruses of all. The orthopoxviruses are brick-shaped. They measure about 230 × 270 nm and when suitably stained can just be seen with an ordinary light microscope. The poxviruses are neither icosahedral nor helical: their structure is referred to as complex. Inside there is a core structure shaped like a dumbbell, and two accompanying lateral bodies, so named after their location in the virion (Fig. 57.1).

Variola virus was first demonstrated microscopically by Buist in 1887. Paschen in 1906 developed a staining technique for the virus particles and demonstrated the elementary bodies (Paschen bodies) in smears from smallpox lesions.

Table 57.1: Poxviruses that infect humans

Genus	*Virus*	*Primary host(s)*	*Clinical features in humans*
Orthopoxvirus	i. Variola	Man	Smallpox
	ii. Vaccinia	Man	Vesicular vaccination lesion
	iii. Cowpox	Cattle, cats, rodents	Lesions on hands
	iv. Monkeypox	Monkeys, squirrels	Resembles smallpox
	v. Buffalopox	Buffaloes	Human infections rare; localized disease
Parapoxvirus	Pseudocowpox	Cattle	Localized nodular lesions ('milkers' nodes)
	Orf	Sheep, goats	Localized vesiculo. granulomatous lesions
Yatapoxvirus	Tanapox	Monkeys	Vesicular skin lesions and febrile illness
Molluscipoxvirus	Molluscum contagiosum	Man	Multiple small skin nodules

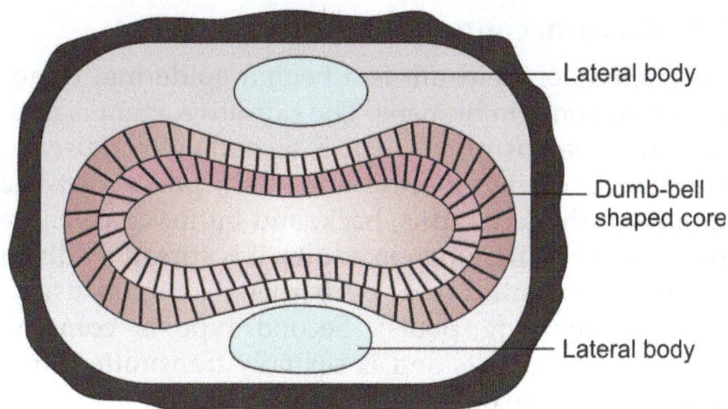

Fig. 57.1: Structure of vaccinia virus. The nucleic acid is contained within a dumb-bell shaped core (C). Fitting into the concavities of the core are two lateral bodies (LB). The virion is enclosed within a protein shell which has an irregular surface

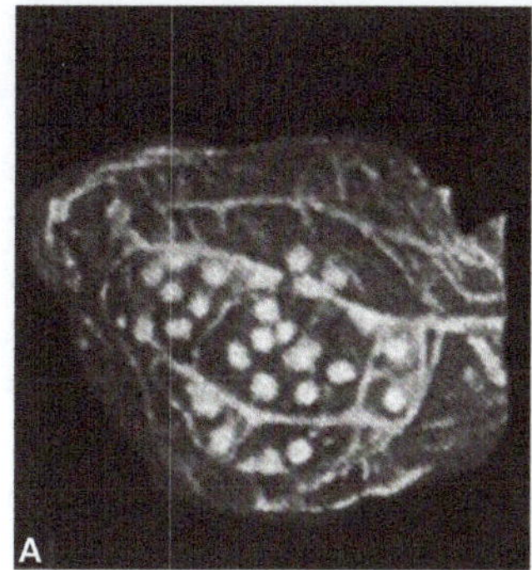

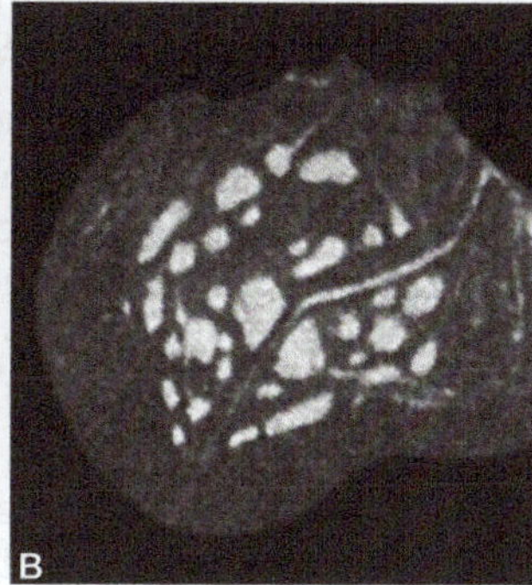

Fig. 57.2: Variola and vaccinia pocks on CAM. Left—variola, showing small, uniform pocks; Right—vaccinia, showing large, irregular pocks

PHYSICAL AND CHEMICAL PROPERTIES

Poxviruses are stable and if protected from sunlight may remain viable for months at room temperature. They survive for years in the cold or when freeze dried. They are susceptible to ultraviolet light and other irradiations. They are resistant to 50 percent glycerol and 1 percent phenol but are readily inactivated by formalin and oxidiing disinfectants.

ANTIGENIC STRUCTURE

All poxviruses share a common nucleoprotein (NP) antigen. Some twenty different antigens have been identified by immunodiffusion. These include the LS antigens (a complex of two antigens, the heat labile L and the heat stable S antigens), agglutinogen, and hemagglutinin, which is responsible for the agglutination of erythrocytes of those fowls which are also agglutinated nonspecifically by tissue lipids.

CULTIVATION AND HOST RANGE

The variola and vaccinia viruses can be differentiated by their growth characteristics and host range.

Chick Embryo

Virus isolation is carried out by inoculation of vesicular fluid onto the chorioallantoic membrane (CAM) of chick embryos. Both vaccinia and variola viruses grow on the CAM of 11-13 day old chick embryo producing **pocks** in 48-72 hours. **Variola pocks** are small, shiny, white, convex, non-necrotic, non-hemorrhagic lesions. **Vaccinia pocks** are larger, irregular, flat, grayish, necrotic lesions, some of which are hemorrhagic (Fig. 57.2).

Tissue Culture

Variola and vaccinia viruses can be grown in tissue cultures of monkey kidney, HeLa and chick embryo cells. Cytopathic effects are produced by vaccinia in 24-48 hours and more slowly by variola. Eosinophilic inclusion bodies—**Guarnieri bodies**—can be demonstrated in stained preparations. The inclusion bodies consist of aggregations of virus particles in a matrix. Vaccinia virus produces plaques in chick embryo tissue cultures but not variola virus.

Animals

Monkeys, calves, sheep and rabbits can be infected by scarification leading to vesicular lesions.

VARIOLA AND VACCINIA VIRUSES

Variola Viruses

The variola virus is the causative agent of smallpox. Variola has a narrow host range (only humans and monkeys). Smallpox used to occur in two distinct clinical varieties.

a. **Variola major or classical smallpox:** The florid, highly fatal disease typically seen in Asia.
b. **Variola minor or alastrim:** The mild, nonfatal disease (alastrim) typically seen in Latin America.

Variola major and minor were antigenically identical but they differed in certain biological characteristics. They were stable variants as the disease produced by each always bred true. Alastrim did not lead to smallpox and vice versa.

Vaccinia Virus

Vaccinia virus, the agent used for smallpox vaccination, is a distinct species of *Orthopoxvirus*. Vaccinia virus is unique in that it is an 'artificial virus' and does not occur in nature as such. It may be the product of genetic recombination, a new species derived from cowpox virus or variola virus by serial passage, or the descendant of a now extinct viral genus.

Variola has a narrow host range, whereas vaccinia has a broad host range that includes rabbits and mice. It is safer to work with vaccinia virus so vaccinia virus has been studied in greater detail than variola. For the development of recombinant vaccines, vaccinia virus is being employed as a vector. The vaccinia genome can accommodate about 25,000 foreign base pairs sufficient for introducing several genes. Many genes have been inserted, including those coding for the antigens of hepatitis B virus, HIV, rabies and for pharmacologically important products such as neuropeptides.

CONTROL OF SMALLPOX

The story of Edward Jenner's discovery in 1796 that inoculation with cowpox would prevent smallpox is well known. To Jenner, however, goes the credit for showing that, following the inoculation of young James Phipps with cowpox virus on 14 May 1796, deliberate inoculation with smallpox material failed to induce the disease.

For thousands of years, smallpox raged as a scourge of mankind, causing death and disfigurement. The global eradication of smallpox, achieved after 10 years of concerted campaigns under the auspices of the WHO, has been a most impressive medical achievement. Naturally occurring smallpox came to an end in 1977. On 8 May, 1980, the WHO formally announced the global eradication of smallpox.

OTHER POXVIRUS DISEASES (TABLE 57.1)

1. Cowpox

Cowpox is, as its name implies, a zoonosis. Cowpox lesions are seen on the udder and teats of cows and may be transmitted to humans during milking. The lesions in humans usually appear on the hands or fingers. The natural reservoir of cowpox seems to be a rodent.

2. Monkeypox

This orthopoxvirus zoonosis causes an illness in humans very similar to smallpox, Squirrels seem to be the main reservoir of infection, and the infection is seen mainly in children who may acquire it from playing with captive animals.

3. Buffalopox

Buffalopox was identified in cattle in India in 1934 and was considered an outbreak of vaccinia in them. Epizootics had occurred in buffaloes and lesions had been observed on the hands of persons in contact with infected animals.

4. Milker's Node

Milker's node (paravaccinia) is a trivial occupational disease that humans get by milking infected cows. The lesions are small ulcerating nodules.

5. Orf (Contagious Pustular Dermatitis or Sore Mouth)

Orf is a disease of sheep and goats transmitted to human beings by contact. It is an occupational disease of sheep handlers. In humans, the disease occurs as a single papulovesicular lesion with a central ulcer usually on the hand, forearm or face.

6. Tanapox

This virus takes its name from the Tana River in Kenya, where it was first diagnosed. It is prevalent in monkeys and appears to be spread by insect bites. Monkeys are the only animals susceptible. There is usually only one vesicular lesion but its appearance is preceded by fever and quite severe malaise. Recovery is uneventful.

5. Molluscum contagiosum

Molluscum contagiosum is a benign epidermal tumor that occurs only in humans. The causative agent is molluscum contagiosum virus. It is a contangious disease. The lesions of this disease are small, pink, wart-like tumors on the face, arms, back, and buttocks are more frequent in children than in adults. It is spread by direct and indirect contact (e.g. by barbers, common use of towels, swimming pools). Second type is common among young adults and is sexually transmitted. It is also seen in some patients with AIDS.

The diagnosis of molluscum contagiosum can usually be made clinically. Sections of skin lesions show large, eosinophilic intracytoplasmic inclusions, which displace the nuclei to the margin, known as **molluscum bodies** under light microscope. It can be detected under **electron microscope** and PCR can detect viral DNA sequences. The virus cannot be grown in animals or tissue cultures.

KNOW MORE

Control of Smallpox

The observation that smallpox could be prevented by inoculation of healthy people with material from the lesions seems to have originated in China. From there an account of the practice (variolation) was sent by Joseph Lister to The Royal Society in 1700 and was followed by others from Turkey.

KEY POINTS

- The family Poxviridae is a large family of ovoid or brick-shaped viruses, and large enough to be just visible under a light microscope.
- The genus Orthopoxvirus includes the viruses causing cowpox, vaccinia and variola.
- They can grow in chorioallantoic membrane (CAM) of chick embryos and tissue culture.
- Smallpox was the first viral disease to be eradicated from the world.
- **Molluscum contagiosum**-It is a benign epidermal tumour that occurs in humans. The virus is spread by close contact, often through sexual contact. It causes small, pink, papular, pearl-like benign tumors of the skin or mucous membranes.

IMPORTANT QUESTIONS

1. Write short notes on:
   a. Variola virus.
   b. Vaccinia virus.
   c. Molluscum contagiosum.

FURTHER READING

Behbehani, AM. The Smallpox Story: Life and death of an old disease. Microbiol Rev 1983;47:455.

Esposito JJ, Baxby D, et al. Poxviridae. Arch Virol, Suppl 10, 1995;79-91.

Jezek Z, et al. Human Monkeypox. J Infect Dis 1986;154:551.

CHAPTER 58

Herpesviruses

Learning Objectives

After reading and studying this chapter, you should be able to:

- Describe morphology of herpesvirus.
- Classify human herpesviruses.
- Describe infections caused by Herpes simplex virus type 1 and Herpes simplex virus type 2.
- Describe the following: Varicella-zoster virus; Cytomegalovirus.
- Discuss clinical manifestations of Epstein-Barr virus (EBV).
- Describe Paul-Bunnell test.

INTRODUCTION

The herpesvirus family contains several of the most important human pathogens and contains over a hundred species of enveloped DNA viruses that affect humans and animals. Some have a wide host-cell range, whereas others have a narrow host-cell range. The outstanding property of herpesviruses is their ability to establish latent infections, lifelong persistent infections in their hosts and to undergo periodic reactivation.

STRUCTURE

The herpesvirus capsid is icosahedral, composed of 162 capsomers, and enclosing the core containing the linear double stranded DNA genome. The nucleo-capsid is surrounded by the **lipid envelope** derived from the modified host cell nuclear membrane through which the naked virions bud during replication. The envelope carries surface spikes, about 8 nm long. Between the envelope and the capsid is an amorphous structure called the **tegument**, containing several proteins (Fig. 58.1). The enveloped virion measures about 200 nm and the naked virion about 100 nm in diameter.

Herpesviruses replicate in the host cell nucleus. They form **Cowdry type A intranuclear (Lipschutz) inclusion bodies**. Like other enveloped viruses, herpesviruses are susceptible to fat solvents like alcohol, ether, chloroform and bile salts. They are heat labile and have to be stored at 70°C.

CLASSIFICATION

The family Herpesviridae is divided into three subfamilies based on biological, physical and genetic properties (Table 58.1).

Eight different types of herpesviruses are known whose primary hosts are human. They have been officially designated 'Human herpesvirus types 1-8' but their common names continue to be in general use, except for types 6, 7 and 8 (Table 58.1).

HERPES SIMPLEX VIRUS (HSV)

Herpes simplex virus (HSV) was the first human herpesvirus to be recognized. The name herpes is derived from a Greek word meaning "to creep." "Cold sores" were described in antiquity, and their viral etiology was established in 1919.

Properties of the Viruses

There are two distinct herpes simplex viruses: type 1 and type 2 (HSV-1, HSV-2) (Table 58.1). They differ in their mode of transmission. HSV type 1 (Human herpes virus type 1 or HHV type 1) is usually isolated from lesions in and around the mouth and is transmitted by direct contact or droplet spread from cases or carriers. HSV type 2 (HHV type 2) is transmitted sexually or from a maternal genital infection to a newborn. This results in different clinical features of human infections.

Intracerebral inoculation in rabbits and mice leads to encephalitis, and corneal scarification produces keratoconjunctivitis in rabbits. The virus grows in a variety of primary and continuous cell cultures (monkey or rabbit kidney, human amnion, HeLa) producing cytopathic changes, well defined foci with heaped up cells and syncytial or giant cell formation. On chick embryo CAM, small (diameter less than 0.5 mm) white shiny non-necrotic pocks are produced. The two types of the virus cross-react serologically. They can be differentiated by the following features:

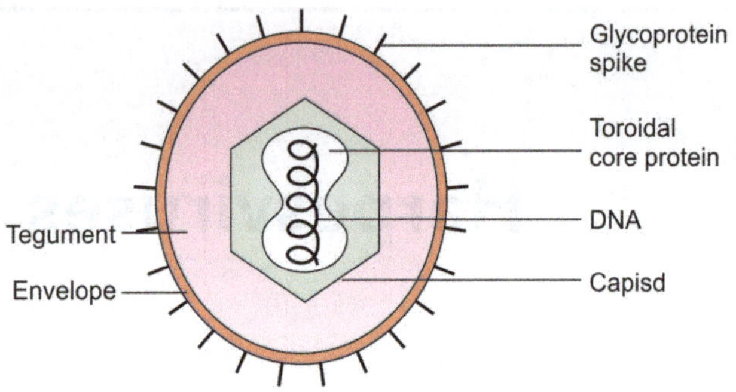

Fig. 58.1: Typical structure of the herpesviruses

1. Antigenic differences can be made out using type specific monoclonal antibodies.
2. On chick embryo CAM type 2 strains form larger pocks resembling variola.
3. Types 2 strains replicate well in chick embryo fibroblast cells, while type 1 strains do so poorly.
4. The infectivity of type 2 is more temperature sensitive than that of type 1.
5. Type 2 strains are more neurovirulent in laboratory animals than type I.
6. Type 2 strains are more resistant to antiviral agents like IUDR and cytarabine in culture.
7. Restriction endonuclease analysis of viral DNA enables differentiation between the two types as well as between strains within the same type.

Pathogenesis

The mechanisms involved in the pathogenesis of HSV-1 and HSV-2 are very similar. Both viruses initially infect and replicate in mucoepithelial cells and then establish latent infection of the innervating neurons. **Skin and mucous membranes** are the portals of entry in which the virus also multiplies, causing lysis of cells and formation of **vesicles**.

Soon after replication is underway in the skin or a mucous membrane, virions travel to the **root ganglia** via the sensory nerves supplying the area. Primary infections of the orofacial and genital areas involve respectively the **trigeminal** and **lumbosacral** dorsal root ganglia. The virus then becomes **latent** in the ganglia.

Once HSV has established itself in a ganglion *in vivo*, **reactivations** are liable to take place at intervals of weeks or months thereafter. Such attacks may be triggered by a variety of stimuli including axonal injury, fever, physical or emotional stress, and exposure to ultraviolet light. The virus follows axons back to the peripheral site, and replication proceeds at the skin or mucous membranes.

HSV generally causes disease at the site of infection. HSV-1 is usually associated with infections **above the waist**, and HSV-2 with infections **below the waist**, consistent with the means of spread for these viruses. Other differences between HSV-1 and HSV-2 are those in growth characteristics and antigenicity; also, HSV-2 has a greater potential to cause viremia with the associated systemic "flu"-like symptoms.

Clinical Features

1. **Cutaneous infections**—(i) The most common site is the face - on the cheeks, chin, around the mouth or on the forepead; (ii) **Napkin rash**- Lesions may also appear on the buttocks in infants as napkin rash; (iii) **'Fever blister' or herpis febrilis**- The typical lesion is the 'fever blister' or herpis febrilis, caused by viral reactivation in febrile patients. In some sensitive persons, very minor stimuli, like common cold,

Table 58.1: Classification of human herpesviruses

Subfamily	*Virus*	*Primary target cell*	*Site of latency*	*Mode of spread*
A. Alphaherpesvirinae				
Human herpesvirus 1	Herpes simplex type 1	Mucoepithelial cells	Neuron	Close contact
Human herpesvirus 2	Herpes simples type 2	Mucoepithelial cells	Neuron	Close contact (sexually transmitted disease)
Human herpesvirus 3	Varicella-zoster virus	Mucoepithelial cells	Neuron	Respiratory and close contact
B. Betaherpesvirinae				
Human herpesvirus 5	Cytomegalovirus	Monocyte, lymphocyte, and epithelial cells	Monocyte, lymphocyte, and ?	Close contact, transfusions, tissue transplant, and congenital
Human herpesvirus 6	Herpes lymphotropic virus	T cells and ?	T cells and ?	Respiratory and close contact?
Human herpesvirus 7	Human herpesvirus 7	T cells and ?	T cells and ?	?
C. Gammaherpesvirinae				
Human herpesvirus 4	Epstein-Barr virus	B cells and epithelial cells	B cell	Saliva (kissing disease)
Human herpesvirus 8	Kaposi's sarcoma-related virus	Lymphocyte and other cells	?	Close contactt (sexual), saliva?

exposure to sun or even mental strain or menses may bring on such reactivation; (iii) **Herpetic whitlow- Herpetic whitlow**-is an infection of the finger; (iv) **Herpes gladiatorum** -Herpes gladiatorum is an infection of the body. It is often acquired during wrestling or rugby; (v) **Eczema herpeticum**- A severe form of cutaneous herpes may occur in children with atopic eczema eczema herpeticum or Kaposi's varicelliform eruption.

2. **Oral infection**—Acute gingivostomatitis, herpetic stomatitis, pharyngitis, tonsillitis and localized lymphadenopathy may occur.
3. **Ophthalmic**—Severe **keratoconjunctivitis, follicular** conjunctivitis with vesicle formation on the lids, **dendritic keratitis** or **corneal ulcers** or as vesicles on the eyelids, corneal scarring and impairment of vision. Recently, acute retinal necrosis associated with HSV-2 has been recognized.
4. **Nervous system**—(i) HSV encephalitis; (ii) Sporadic encephalitis; (iii) HSV meningitis; (iv) Sacral autonomic dysfunction; (v) rarely transverse myelitis or the Guillain-Barre syndrome and Bell's palsy.
5. **Visceral**—(i) HSV esophagitis; (ii) Tracheobronchitis and pneumonitis; (iii) Hepatitis; (iv) Erythema multiforme; (v) Disseminated HSV infection.
6. **Genital infections**—Genital disease is usually caused by HSV-2, although HSV-l can also cause clinical episodes of genital herpes. Genital herpetic ulcers are known to increase the risk of transmission of infection with human immunodeiciecy virus(HIV).

   (i) **In male patients**—The lesions typically develop on the glans or shaft of the penis and occasionally in the urethra; (ii) **In female patients**- The lesions may be seen on the vulva, vagina, cervix, perianal area, or inner thigh; (iii) **Inguinal lymphadenopathy**- In patients of both sexes; (iv) **Herpetic proctitis**-Rectal and perineal lesions occur in homosexuals; (v) **Association between HSV-2 and carcinoma of the cervix uteri**- There have been several reports of an association between HSV -2 and carcinoma of the cervix uteri but a causal relationship has not been established.
7. **Neonatal Herpes**-*Congenital malformations*- Transplacental infection with HSV1 or 2 can lead to congenital malformations, but this is rare.
8. **Infections in immunocompromised hosts**-These include patients immunosuppressed by disease or therapy.

Laboratory Diagnosis

The diagnosis of herpes virus infection may be made by microscopy, antigen or DNA detection, virus isolation or serology.

1. Specimens

Specimens include vesicle fluid, skin swab, saliva, corneal scraping, brain biopsy and CSF according to site of involvement.

2. Microscopy

i. **Tzanck smear:** Characteristic cytopathologic effects (CPEs) can be identified in a Tzanck smear (a scraping of the base of a lesion), Papanicolaou smear, or biopsy specimen. CPEs include syncytia, "ballooning" cytoplasm, and **Cowdry type A intranuclear inclusions.**

   The Tzanck smear is a rapid, fairly sensitive and inexpensive diagnostic method. Smears are prepared from the lesions, preferably from the base of vesicles and stained with 1 percent aqueous solution of toluidine blue '0' for 15 seconds. Multinucleated giant cells with faceted nuclei and homogeneously stained 'ground glass' chromatin (Tzanck cells) constitute a positive smear.

ii. **Electron microscopy:** The virus particle may also be demonstrated under the electron microscope.

iii. **Fluorescent antibody technique:** The herpesvirus antigen may be demonstrated in smears or sections from lesions by the fluorescent antibody technique. The fluorescent antibody test on brain biopsy specimens provides reliable and speedy diagnosis in encephalitis.

3. Virus Isolation

Primary human embryonic kidney, human amnion and many other cells are susceptible, but human diploid fibroblasts are preferred. Typical cytopathic changes may appear as early as in 24–48 hours but cultures should be observed for two weeks before being declared negative.

HSV isolates can be typed by biochemical, biologic, nucleic acid, or immunologic methods. The restriction endonuclease cleavage patterns of the DNA of HSV-1 and HSV-2 and allow unequivocal typing of the isolates. HSV type-specific DNA probes, specific DNA primers for PCR and antibodies are also available for detecting and differentiating HSV-1 and HSV-2.

4. Serology

Antibodies develop within a few days of infection and rise in titer of antibodies may be demonstrated by ELISA, neutralization or complement fixation tests.

5. Polymerase Chain Reaction

Polymerase chain reaction is useful for detecting viral DNA in cerebrospinal fluid when herpetic infection of the CNS is suspected.

Treatment

Idoxuridine used topically in eye and skin infections was one of the first clinically successful antiviral agents. Several antiviral drugs have proved effective against HSV infections, including acyclovir, valacyclovir, and vidarabine. Acyclovir is currently the standard therapy. Acyclovir and vidarabine enabled the effective management of deep and systemic infections. Intravenous, oral, and topical preparations are available. Drug-resistant virus strains may emerge. Valaciclovir and famciclovir are more effective oral agents. When resistance to these drugs develop, drugs like foscarnet which are independent of viral thymidine kinase action may be useful.

HERPESVIRUS SIMIAE: B VIRUS

Herpes B virus of Old World monkeys is highly pathogenic for humans. This virus was isolated by Sabin and Wright (1934) from the brain of a laboratory worker who developed fatal ascending myelitis after being bitten by an apparently healthy monkey. It came to be known as the 'B' virus from the initials of this patient. Many similar cases have been reported since then.

The virus is transmitted to humans by monkey bites or saliva or even by tissues and cells.

Virus isolation or serologic tests can be used to establish the diagnosis of B-virus infections.

VARICELLA-ZOSTER VIRUS (VZV)

Varicella-zoster virus (VZV) causes chickenpox (varicella) and, with recurrence, causes **herpes zoster**, or **shingles**. Varicella (chickenpox) and herpes zoster are different manifestations of the same virus infection. The virus is therefore called varicella-zoster virus (VZV). Thus, chickenpox is 'caught' but not zoster.

Properties of the Virus

Varicella-zoster virus is morphologically identical to herpes simplex virus. It does not grow in experimental animals or chick embryos. It can be grown in cultures of human fibroblasts, human amnion or HeLa cells and produces typical intranuclear inclusion bodies. Cytopathic changes are more focal and spread much more slowly than those induced by HSV. By using highly specific antisera, it is possible to distinguish between herpes virus types 1, 2 and varicella-zoster viruses. Only one antigenic type of VZV is known.

Varicella (Chickenpox)

Varicella (chickenpox) is one of the mildest highly communicable and most common of childhood infections. The disease may, however, occur at any age.

It is usually a mild disease of childhood and is normally symptomatic, although asymptomatic infection may occur. The portal of entry of the virus is the respiratory tract or conjunctiva. After an incubation period of about two weeks (7-23 days) the lesions begin to appear. The patient is infectious for 2 days before and up to 5 days after onset, while new vesicles are appearing.

In children, there is little prodromal illness and the disease is first noticed when skin lesions appear. Initially macular, the rash rapidly evolves through papules to the characteristic clear vesicles. The evolution of the rash is so rapid that the various stages—macule, papule, vesicle, pustule and scab—cannot be readily followed in individual lesions. The rash is characteristically centripetal in distribution, affecting mainly the trunk and sparing the distal parts of the limbs and is very superficial without involving the. deeper layers of the skin, resembling a dew drop lying on the skin. The rash appears in successive crops, so that all stages of the eruption can be seen at the same time. It matures very quickly, beginning to crust within 48 hours. Recovery is usually uneventful.

Complications: Primary infection is usually more severe in adults than in children. The rash may become **hemorrhagic** and occasionally bullous lesions appear. Pitted scars on the skin may remain after recovery. **Varicella pneumonia** is more common in adults, and often fatal in the elderly. A variety of organs may be affected with complications like **myocarditis, nephritis, acute cerebellar ataxia, meningitis** and **encephalitis. Secondary bacterial infections,** usually due to staphylococci or streptococci, may occur. **Reyes' syndrome** may follow varicella in some cases with a history of administration of salicylates.

In most cases, chickenpox is an uneventful disease and recovery is the rule. One attack confers lasting immunity.

HERPES ZOSTER (SHINGLES, ZONA)

The name is derived from Herpes, meaning to creep; **Zoster**, meaning girdle).

While varicella is typically a disease of childhood, herpes zoster is one of old age.

Pathogenesis: Herpes zoster usually occurs in persons who had chickenpox several years earlier. The virus remaining latent in the sensory ganglia, may leak out at times but is usually held in check by the residual immunity. It is believed that years after the initial infection, when the immunity has waned, the virus may be reactivated, and triggered by some precipitating stimulus, travel along the sensory nerve to produce zoster lesions on the area of the skin or mucosa supplied by it. It usually starts with severe pain in the area of skin or mucosa supplied by one or more groups of sensory nerves and ganglia. The most common sites are the areas innervated by spinal cord segments D3 to L2 and the trigeminal nerve, particularly, its ophthalmic branch. Within a few days after onset, a crop of vesicles appears over the skin supplied by the affected nerves.

Complications

1. **Post-herpetic pain**—in the affected area is frequent, particularly in the elderly.
2. **Ophthalmic zoster**—It is a potentially serious complication, when, as often happens, the ophthalmic branch of the trigeminal nerve is affected.
3. **Generalized zoster**—is similar to generalized varicella. **Encephalitis** is a rare complication of herpes zoster.
4. **Ramsay Hunt syndrome**—It is a rare form of zoster affecting the facial nerve, with eruption on the tympanic membrane and external auditory canal, and often a facial palsy.

Immunity

Previous infection with varicella is believed to confer lifelong immunity to varicella. Zoster occurs in the presence of neutralizing antibody to varicella. The development of varicella-zoster virus-specific cell-mediated immunity is important in recovery from both varicella and zoster. Appearance of local interferon may also contribute to recovery.

Laboratory Diagnosis

Diagnosis is usually clinical. Laboratory diagnosis includes:

1. *Microscopy:* Multinucleated giant cells and **type A intranuclear inclusion bodies** may be seen in smears prepared by scraping the base of the early vesicles (Tzanck smear) and stained with toluidine blue, Giemsa or Papanicolaou stain. Direct examination by electron microscopy will reveal herpes particles.
2. *Virus isolation:* Virus isolation can be attempted from the buccal or cutaneous lesions in the early stages by inoculating human amnion, human fibroblast, HeLa or Vero cells.
3. *Virus antigen:* The virus antigen can be detected in scrapings from skin lesions by **immunofluorescence**, and in vesicle fluid by **counterimmunoelectrophoresis** with zoster immune serum. **ELISA** and PCR techniques are also in use.
4. *Serological diagnosis:* A rise in specific antibody titer can be detected in the patient's serum by various tests, including **fluorescent antibody, latex agglutination**, and **enzyme immunoassay**.

Prophylaxis and Treatment

1. Active Immunization

A live-attenuated varicella vaccine was developed by Takahashi in Japan in 1974 by attenuating a strain of varicella virus (*Oka* strain, so named after the patient) by serial passage in tissue culture. This vaccine, given by subcutaneous injection, has been found to be immunogenic with good antibody response but it was very labile and had to be stored frozen. A modified lyophilized form of the vaccine is now available, which can be stored between 2°C and 8°C. It is recommended as a single subcutaneuos dose for childern 1-12 years old, and for those older, as 2 doses, 6-10 weeks apart. It is safe and effective. It is not considered safe in pregnancy.

2. Passive Immunization

Varicella-zoster immunoglobulin (VZIG) prepared from the patients convalescing from herpes zoster seems to be of some use in preventing or modifying severe disease in immunodeficient patients who come into contact with chickenpox or herpes zoster. It is of no value for treating established infection.

Treatment: Acyclovir and vidarsabine are effective in the treatment of severe varicella and zoster.

CYTOMEGALOVIRUS (CMV)

The name Cytomegaloviruses (CMV) means 'large cell virus' and derives from the swollen cells containing large intranuclear inclusions that characterize these infections.Cytomegaloviruses (CMV), formerly known as salivary gland viruses, are a group of ubiquitous herpesviruses of humans and animals.

Properties of the CMV: The CMV s have the same general structure as other members of the herpes group. Cytomegalovirus has the largest viruses in the human herpesviruses, being 150-200 nm in size. The virus exhibits strict host specificity. Human cytomegaloviruses can be grown in human fibroblast cultures. Infection is spread primarily cell-to-cell.Cultures have to be incubated for prolonged periods, upto 50 days, as the cytopathic effects are slow in appearance.

CMV infection is particularly serious in AIDS patients because of their lack of CD4+ cells.

Pathogenesis

Primary infection with CMV may be acquired at any time, possibly from conception onwards. Transmission requires close person-to-person contact. The congenital, oral, and sexual routes, blood transfusion, and tissue transplantation are the major means by which CMV is transmitted. CMV persists in the host for life. Reactivation is common, and virus is shed in body secretions such as urine, saliva, semen, breast milk and cervical fluid.

A. Normal hosts—Primary cytomegalovirus infection of older children and adults is usually asymptomatic but occasionally causes a spontaneous infectious mononucleosis syndrome. An association has been observed between the presence of cytomegalovirus and restenosis following coronary angioplasty.

B. Immunocompromised hosts—CMV can cause severe and even fatal infections in the immunocompromised host. This occurs in transplant recipients, cancer patients on chemotherapy, and more particularly in the human immunodeiciecy virus(HIV) infected.

C. Congenital and perinatal infections—Intrauterine infection leads to fetal death or cytomegalic inclusion disease of the newborn which is often fatal. Clinical features include intrauterine growth retardation, jaundice, hepatosplenomegaly, thrombocytopenia, microcephaly, chorioretinitis and cerebral calcification resembling congenital toxoplasmosis. Survivors may show mental retardation.

Cytomegalic inclusion disease is seen almost exclusively in infants born to mothers developing primary CMV infection during pregnancy. Perinatal infection may be acquired from the infected mother through genital secretions or breast milk.

D. Postnatal infection—This can be acquired in many ways-such as saliva, sexual transmission, blood transfusion and donated organs

Laboratory Diagnosis

A. Specimens—CMV can be isolated from the urine, saliva, breast milk, semen, cervical secretions and blood leucocytes.

B. Demonsration of cytomegalic cell—The histologic hallmark of CMV infection is the cytomegalic cell, which is an enlarged cell that contains a dense, central, "owl's-eye," basophilic intranuclear inclusion body.

C. Isolation of virus—Human cytomegaloviruses can be grown in human fibroblast cultures. Cultures have to be incubated for prolonged periods and the cytopathic effects(swollen refractile cells with cytoplasmic granules) may take 2-3 weeks..

D. Antigen detection—A rapid, sensitive diagnosis can be obtained by histologic means through the use of antibodies (especially monoclonal), DNA probes to directly detect the CMV antigens in tissues or fluids.

E. Polymerase chain reaction(PCR)—A rapid, sensitive diagnosis can also be obtained by PCR to directly detect the genome in tissues or fluids.

F. Serology—IgM antibodies suggests a current infection and can bedetectedin errum by ELISA.

Treatment and Prevention

Ganciclovir (dihydroxypropoxymethyl guanine) and foscarnet (phosphonoformic acid) have been approved for the treatment of CMV infections and are especially useful for immunosuppressed patients.

Prevention is indicated only in high risk cases such as organ transplants, immunodeficient persons and in premature infants. Screening of blood and organ donors and administration of CMV immunoglobulins have been employed in prevention.

No vaccine is available. Both live and recombinant cyromegalovirus vaccines are under development.

EPSTEIN BARR VIRUS (EBV)

In 1958, Burkitt described an unusual lymphoma among children in certain parts of Africa and suggested that the tumour may be caused by a mosquito borne virus. This led to several attempts at isolating viruses from such tumours. The mosquito theory was wrong, but 6 years later in 1964, one virus was observed in the cultured lymphoma cells by Epstein, Barr and Achong which was a new type of herpesvirus, named the Epstein-Barr virus (EB) virus, specifically affecting cells of the B lymphocyte lineage.Epstein-Barr virus is in some respects the most sinister herpesvirus, for its association with malignant disease is now well established.

Only human and some subhuman primate B cells have receptors (CD 21 molecules) for the virus. EBV infected B cells are transformed so that they become capable of continuous growth in vitro.

Pathogenesis

EBV is commonly transmitted by infected saliva and initiates infection in the oropharynx. The virus enters the pharyngeal epithelial cells through CR2 (or CD21) receptors, which are the same as for the C3d component of complement. Viral replication occurs in epithelial cells (or surface B lymphocytes) of the pharynx and salivary glands. The diseases of EBV result from either an overactive immune response (infectious mononucleosis) or the lack of an effective immune re-sponse (lymphoma).

It multiplies locally, invades the bloodstream and infects B lymphocytes in which two types of changes are produced. In most cases, the virus becomes latent inside the lymphocytes, which become transformed or 'immortalised: so that they become capable of indefinite growth in vitro. They are polyclonally activated to produce many kinds of immunoglobulins. Lytic infection is a second type of effect, shown by a few infected B cells with cell death and release of mature progency virions.

The mononucleosis represents a polyclonal transformation of infected B lymphocytes. Infectious mononucleosis results from a "civil war" between the EBV-infected B cells and the protective T cells. The classic lymphocytosis (increase in mononuclear cells) associated with infectious mononucleosis results mainly from the activation and proliferation of T cells. These appear as atypical lymphocytes also called Downey cells).

Intermittent reactivation of the latent EB virus leads to clonal proliferation of infected B cells. In immunocompetent subjects, this is kept in check by activated T cells. In the immunodeficient, B cell clones may replicate unchecked, resulting in lymphomas. Hyperendemic malaria prevalent in Africa is believed to be responsible for the immune impairment in children with Burkitt' lymphoma. The frequency of lymphomas seen in many types of immunodeficiencies, most typically in AIDS, may have a similar pathogenesis..

Nasopharyngeal carcinoma seen in men of Chinese origin. Genetic and environmental factors are said to be important in the and EB virus DNA is regularly found in the tumour cells.

Epidemiology

The Epstein-Barr (EB) virus is ubiquitous in all human populations. Infection with EBV is transmitted by saliva, and requires intimate oral contact. As with other herpesviruses, infection with the EB virus leads to latency, periodic reactivation and lifelong persistence

Clinical conditions—Most primary infections in children are asymptomatic. In adolescents and young adults, the classic syndrome associated with primary infection is infectious mononucleosis (about 50% of infections). EBV is also associated with several types of cancer. The source of infection is usually the saliva of infected persons who shed the virus in oropharyngeal secretions for months following primary infection and intermittently thereafter. The EB virus is not highly contagious and droplets and aerosols are not efficient in transmitting infection. Saliva sharing between adolescents and young adults often occurs during kissing; thus, the nickname "kissing disease" for infectious mononucleosis.

EB virus infection may lead to the following clinical conditions:

1. Infectious mononucleosis.
2. EBV associated malignancies:
   a. Burkitt's lymphoma.
   b. Lymphomas in immunodeficient persons such as AIDS patients and transplant recipients.
   c. Nasopharyngeal carcinoma in persons of Chinese origin.

A. Infectious mononudeosis(Glandular fever): This is an acute self-limited illness usually seen in nonimmune young adults following primary infection with the EB virus. The incubation period is 4-8 weeks.Infectious mononucleosis is characterized by high fever, malaise, pharyngitis, lymphade-nopathy (swollen glands), and, often, hepatosplenomegaly. A mild transient rash may be present. Some patients treated with ampicillin may develop a maculopapular rash due to immune complex reaction to the drug. In most patients the spleen is palpable and there is some liver dysfunction, occasionally with frank jaundice. The typical illness is self-limited and lasts 2-4 weeks.

Complications—The disease is rarely fatal in healthy people.Complications are rare but some are serious such as: (i) Acute airway obstruction may occur as a result of the lymphoid enlargement and oedema. (ii) Splenic rupture. (iii) Neurological complications include meningitis, encephalitis and the Guillain-Barre syndrome.

B. Oral hairy leukoplakia: This lesion is a wart-like growth that develops on the tongue in some HIV-infected persons and transplant patients. It is an epithelial focus of EBV replication.

C. Chronic Disease: EBV can cause cyclic recurrent disease in some people. These patients experience chronic tiredness and may also have low-grade fever, headaches, and sore throat. This disorder is different from chronic fatigue syndrome, which has another etiology.

D. Burkitt's lymphoma: EBV is associated with the development of Burkitt's lymphoma (a tumor of the jaw in African children and young adults). Most African tumors (> 90%) contain EBV DNA and express EBNA1 antigen. Malaria, a recognized cofactor, may foster enlargement of the pool of EBV-infected cells.

E. Nasopharyngeal carcinoma (NPC): This cancer of epithelial cells is common in males of Chinese origin. It mainly affects people aged 20-50 years, males preponderating. This neoplasm is also asso-ciated with EBV, the evidence being similar to that for Burkitt's lymphoma. EBV DNA is regularly found in nasopharyngeal carcinoma cells, and patients have high levels of antibody to EBV.Genetic and environmental factors are believed to be important in the development of nasopharyngeal carcinoma.These tumours are relatively inaccessible to surgery or chemotherapy. Even after irradiation, the prognosis is poor.

F. Lymphoproliferative diseases in immunodeficient hosts: Immunodeficient patients are susceptible to EBV induced Iymphoproliferative diseases that may be fatal. On infection with EBV, people lacking T-cell immu-nity are likely to suffer life-threatening polyclonal leukemia-like B-cell proliferative disease and lymphoma instead of infectious mononucleosis.

AIDS patients are susceptible to EBV-associated lymphomas and hairy oral leukoplakia of the tongue. Virtually all central nervous system non-Hodgkin's lymphomas are associated with EBV, whereas less than 50% of systemic lymphomas are EBV-positive. In addition, EBV appears to be associated with classic Hodgkin's disease, with the viral genome detected in the malignant Reed-Sternberg cells in up to 50% of cases.

Laboratory Diagnosis

Diagnosis is based on the haematological findings and on serological tests. Virus isolation is impracticable.

1. Atypical Lymphocytes

The feature that gives infectious mononucleosis its name is the raised leucocyte count. Atypical lymphocytes are probably the earliest detectable indication of an EBV infection. These cells appear with the onset of symptoms and disappear with resolution of the disease. Atypical lymphocytes, accounting for 20% of the lymphocytosis common in this condition, are seen in blood films. Blood examination during the initial phase may show leucopenia due to a drop in the number of polymorphs. Later there is a prominent leucocytosis with the appearance of abnormal or atypical lymphocytes. These atypical cells are lymphoblasts derived from T cells reactive to the virus infection.

2. Paul Bunnell Test

Infectious mononucleosis is accompanied by production of heterophile agglutinins. These antibodies are IgM heterophile antibodies elicited by EBV infection and appear in 85-90% of patients sera during the acute phase of illness. The Paul-Bunnel antibody develops early during the course of infectious mononucleosis, and disappears within about two months.These can be detected by the Paul-Bunnell test or a rapid slide agglutination test . Agglutination of horse or sheep red cells by serum absorbed to exclude a natural antibody is the basis of this test.

Procedure

Inactivated serum.(56°C for 30 minutes) in doubling dilutions is mixed with equal volumes of a 1% suspension of sheep erythrocytes. After incubation at 37 °C for four hours the tubes are examined for agglutination. An agglutination titre of 100 or above is suggestive of infectious mononucleosis.

Confirmation

Such antibodies may also occur after injections of sera and sometimes even in normal individuals. In the course of infectious mononucleosis, most patients develop transient heterophil antibodies that agglutinate sheep cells. For confirmation, differential absorption of agglutinins

Table 58.2: Differential absorption test for Paul-Bunnell antibody. Result of absorption on		
	Guinea pig kidney	*Ox red*
Normal serum	Absorbed	Absorbed
Antibody after serum therapy	Absorbed	Not absorbed
Infectious mononucleosis	Not absorbed	Absorbed

with guinea pig kidney and ox red cells is necessary. Forssman antibody induced by injection of horse. serum is removed by treatment with guineapig kidney and ox red cells. Infectious mononucleosis antibody is removed by ox red cells but not guineapig kidney (Table 58.2).

This differential agglutination test has largely been replaced by a simple slide aggutination test ('Monospot') employing sensitised horse erythrocytes, with the same sensitivity and specificity.

3. EBV-Specific Antibodies

Tests are also available for the demonstration of specific EB virus antibodies.. The IgM antibody to VCA (virus capsid antigen) appears soon after primary infection and disappears in 1-2 weeks and is indicative of current infection.The IgG antiVCA antibody indicates past or recent infection and persists throughout life. Immunofluorescence and ELISA are commonly employed for their demonstration.

Early antigen (EA) antibodies are generally evidence of current viral infection and are often found in patients with Burkitt's lymphoma or nasopharyngeal carcinoma. Antibodies to the EB nuclear antigen(EBNA) reveal past infection with EBV, though detection of a rise in antiEBNA antibody would suggest a primary infection.

4. Antigen Detection

EBV antjgen can be detected by immunofluorescence using monoclonal antibodies.

5. Nucleic Acid Hybridization

It is the most sensitive means of detecting EBV in patient materials. Viral antigens can be demonstrated directly in lymphoid tissues and in nasopharyngeal carcinomas.

6. Virus Isolation

EBV can be isolated from saliva, peripheral blood, or lymphoid tissue by immortalization of normal human lymphocytes, usually obtained from umbilical cord blood. This assay is laborious and time-consuming (6-8 weeks), requires specialized facilities, and is seldom performed.

6. Polymerase chai reaction(PCR)-EBV DNA can be detected by PCR.

HUMAN HERPESVIRUSES 6 (HHV6)

HHV6 was first isolated from the blood of patients with AIDS and grown in T-cell cultures. Like EBV and CMV, HHV6 is lymphotropic and ubiquitous. It is present in the saliva of most adults and is spread by oral secretions. Two variants are recognized, A and B. Variant B is the cause of the mild but common childhood illness 'exanthem subitum' (roseola infantum or 'sixth disease'). In older age groups, it has been associated with infectious mononucleosis syndrome, focal encephalitis and, in the immunodeficient, with pneumonia and disseminated disease.

Laboratory dignosis: HHV6 can be isolated from peripheral blood mononuclear cells in early febrile stage of the illness by co-cultivation with lymphocytes.

Virus antigen-can be detected by immunofluorescence using monoclonal antibodies. ELISA is used for detecting both antigen and antibodies in patient serum.

HUMAN HERPESVIRUS 7 (HHV7)

Like HHV6, HHV7 also appears to be widely distributed and transmitted through saliva. However, HHV7 remains an orphan virus with no disease association. It shares with HIV the same CD4 receptor on T cells and could therefore contribute to a further depletion of CD4 T cells in HIV infected persons.

It, may cause an illness resembling infectious, mononucleosis, some cases of exanthem subitum, and a possible association with pityriasis rosea, a transient inflammatory rash, has also been reported.

HUMAN HERPESVIRUS 8 (HHV8)

A new herpesvirus, also called Kaposi's sarcoma-associated herpesvirus (KSHV), was first detected in 1994 in Kaposi's sarcoma specimens. KSHV is the cause of Kaposi's sarcomas, vascular tumors of mixed cellular composition, and is involved in the pathogenesis of body cavity-based lymphomas occurring in AIDS patients and of multicentric Castleman's disease.

HHV8 is limited to certain geographic areas (Italy, Greece, Africa) and to patients with AIDS. It appears to be sexually transmitted among men who have sex with men, who have a higher seroprevalence (30-60%). Infections are common in Africa (>50%), with infections acquired early in life by nonsexual routes, possibly through contact with oral secretions. The virus can be transmitted through organ transplants and places the recipients at risk of KSHV-related diseases.

Diagnosis depends mainly on detection of viral DNA by PCR. Direct virus culture is difficult and impractical. Serologic assays are available to measure persistent antibody to KSHV, using indirect immunofluorescence, Western blot, and ELISA formats.

Foscarnet, ganciclovir, and cidofovir have activity against KSHV replication.

VARICELLA IN PREGNANCY

Varicella virus can cross the placenta following viremia in the pregnant woman, and infect the fetus. The infection may be more serious for the mother herself in pregnancy, with pneumonia, the major problem. The baby may develop two types of complications, depending on the period of gestation when the woman develops chickenpox. Some infants may develop **fetal varicella syndrome** manifesting as cicatrizing skin lesions, hypoplasia of limbs, chorioretinitis and CNS defects. Some babies may not exhibit any defects, but may carry latent VZV infection.

When maternal varicella occurs near delivery, babies may develop **congenital (neonatal) varicella**, within two weeks of birth.

☞ KEY POINTS

- Herpesviruses are DNA viruses. The outstanding property of herpesviruses is their ability to establish latent infections, lifelong persistent infections in their hosts and to undergo periodic reactivation.
- HSVs are large, icosadeltahedral viruses containing a double stranded DNA genome and is surrounded by a lipid envelope containing peplomers.
- Human herpesviruses include human herpesvirus 1 (HV-1) to human herpesvirus 8 (HV-8). HHV 3, HHV 4 and HHV 5 are varicella-zoster virus, Epstein-Barr (EB) and cytomegalovirus (CMV) respectively. Herpes simplex virus type 1 and 2 are designated as HHV 1 and HHV 2.
- HSV-1 causes acute herpetic gingivostomatitis, acute herpetic pharyngotonsillitis, herpes labialis, herpes encephalitis, eczema herpeticum, and herpetic whitlow. HSV-2 causes genital herpes, neonatal infection, and aseptic meningitis.
- Varicella Zoster Virus (VZV) causes chickenpox (varicella) and herpes zoster or shingles, two distinct clinical entities in humans.
- Virus causes lifelong infection.

Cytomegalovirus (CMV)

- CMV or HSV-5 is the causative agent of mononucleosis syndrome in immunocompetent hosts. CMV causes latent infection; hence reactivation may result in disease in patients who are immunocompromised.
- Transmission—Virus is transmitted orally and sexually, in blood transfusions, in tissue transplants, *in utero*, at birth, and by nursing.
- **Clinical syndromes:** Congenital CMV infection, acquired CMV infection, CMV infection in immunocompromised patients, and CMV infection in immunocompetent adult hosts. CMV generally causes subclinical infection
- **Laboratory diagnosis:** CMV can be isolated from the urine, saliva, breast milk, semen, cervical secretions and blood leucocytes.

  Cytology and histology: Owl's-eye": inclusion body and basophilic intranuclear inclusion body is the diagnostic feature of the cell infected by CMV.

  PCR, cell culture in human diploid fibroblast cell lines, and serological techniques are also used.

Epstein-Barr Virus (EBV)

- EBV causes heterophile antibody-positive infectious mononucleosis and has been causally associated with Burkitt's lymphoma, Hodgkin's disease, and nasopharyngeal carcinoma. EBV has also been associated with B-cell lymphomas in patients with acquired or congenital immunodeficiencies. EBV is a mitogen for B cells and immortalizes B cells in tissue culture.
- Transmission occurs via saliva, close oral contact ("kissing disease"), or sharing of items. Malaria has been suggested as an important cofactor in the progression of chronic or latent EBV infection to acute Burkitt's lymphoma in Africa.
- **Laboratory diagnosis:** Atypical lymphocytes are probably the earliest detectable indication of an EBV infection.
- **Paul-Bunnell test** is most frequently used test to detect heterophile antibodies in infectious mononucleosis patients.
- Monospot test, IFA, ELISA, and Western blot, and DNA probe, PCR and virus isolation are used for diagnosis.
- Human herpesvirus 8 (HH8) is the cause of Kaposi's sarcomas, vascular tumors, lymphomas in AIDS and multicentric Castuman's disease.

IMPORTANT QUESTIONS

1. Name various viruses of the family Herpesviridae. Discuss the various infections caused by herpes simplex virus types 1 and 2.
2. Classify human herpesviruses. Discuss briefly their pathogenesis and laboratory diagnosis.
3. Describe the lesions caused by herpes simplex virus and their laboratory diagnosis.
4. Write short notes on:
   Varicella-zoster virus.
   Varicella or chickenpox.
   Cytomegalovirus.
   Epstein-Barr virus (or) EB virus.
   Infectious mononucleosis.
   Paul-Bunnel test .
   Human herpesvirus 6 (HHV6).

FURTHER READING

Ablashi DV, et al. Human herpesvirus-6 (HHV6) (short review), In Vivo 1991;5:193-200.

Arvin AM, Moffat JF, Redman R. Varicella-zoster virus: aspects of pathogenesis and the host response to natural infection and varicella vaccine, Adv Virus Res 1996;46:263-309.

Faulkner GC, Krajewski AS, Crawford DH. The ins and outs of EBV infection. Trends Microbiol 2000;8:185-189.

Jaffe, HW and Pellett, PE. Human herpesvirus 8 and Kaposi's sarcoma New Engl J Med, 1999. (Editorial, June 17),340, No.24.

Whitley RJ, Kimberlin DW, Roizman B. Herpes simplex virus: state of the art clinical article. Clin Infect Dis 1998;26:541-555.

CHAPTER 59

Adenoviruses

Learning Objectives

After reading and studying this chapter, you should be able to:

- Describe morphology of adenovirus.
- Describe diseases associated with adenovirus.
- Describe adenovirus-associated viruses (AAV).

INTRODUCTION

Adenoviruses are a group of medium sized, nonenveloped, double stranded DNA viruses that share a common complement fixing antigen. In fact, the name "**adenovirus**" (*adeno*, from adenoid) reflects the recovery of the initial isolate from explants of human adenoids and Rowe and associates (1953) first isolated it. Since then, approximately 100 serotypes, at least 49 of which infect humans, have been recognized. All human serotypes are included in a single genus within the **family Adenoviridae.** They infect humans, animals and birds, showing strict host specificity.

ADENOVIRUSES

Classification

The family Adenoviridae comprises two genera. The two genera are completely distinct antigenically.

A. Mastadenovirus—infects mammals. Human adenoviruses are further subdivided into six species A-F (also called subgroups or subgenera) based on properties such as hemagglutination, fiber length, DNA fragment analysis and oncogenic potential (Table 59.1).

In addition to at least 49 serotypes of human origin, mastadenoviruses include simian, bovine, equine, ovine, canine, murine, porcine and cetacean serotypes.

B. Aviadenovirus- infects birds.

Morphology

Adenoviruses are 70-90 nm in diameter and display icosahedral symmetry. There is no envelope. The capsid is composed of 252 capsomers arranged as an icosahedron with 20 triangular facets and 12 vertices. Of the 252 capsomers, 240 have six neighbours and are called hexons, while the 12 capsomers at the vertices have five neighbors and are called pentons.

Each penton unit consists of a penton base anchored in the capsid and a projection or fiber consisting of a rod like portion with a knob attached at the distal end. Thus, the virion has the appearance of a space vehicle (Fig. 59.1). The DNA is linear and double stranded.

Resistance

Adenoviruses are relatively stable, remaining viable for about a week at 37°C. They are readily inactivated at 50°C. They resist ether and bile salts.

Pathogenesis

Adenoviruses infect and replicate in epithelial cells of the respiratory tract, eye, gastrointestinal tract, urinary

Table 59.1: Classification of human adenoviruses

Group (Sub-genus)	*Serotype (Species)*	*Total*	*Hemagglutination pattern with red cells of*	*Oncogenicity in newborn hamsters*
A	12,18,21	3	Rat (Partial)	High
B	3,7,11,14,16,21,34,35	8	Monkey (Complete)	Low
C	1,2,5,6	4	Rat (Partial)	None
D	8-10,13,15,17,19,20,22-30 32,33,36-39,42-47	29	Rat (Complete)	None
E	4	1	Rat (Partial)	None
F	40,41	2	Rat (Partial)	None
	Total	**47**		

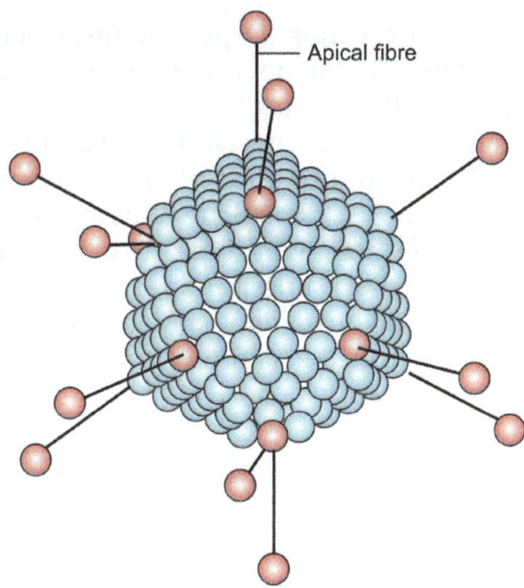

Fig. 59.1: Morphology of adenovirus

bladder, and liver. Several distinct clinical syndromes are associated with adenovirus infection (Table 59.2).

A. Respiratory Diseases

1. **Pharyngitis**—Adenoviruses are the major cause of nonbacterial pharyngitis and tonsillitis, presenting as febrile common cold. Types 1-7 are commonly responsible.
2. **Pneumonia**—Adenovirus types 3 and 7 are associated with pneumonia in adults and types 3, 7 and 21 are thought to be responsible for about 10-20 percent of pneumonias in childhood. In infants and young children types 7 may lead to more serious and even fatal pneumonia.
3. **Acute respiratory diseases (ARD)**—Adenoviruses are the cause of an acute respiratory disease syndrome among military recruits. Serotypes 4, 7 and 21 are the agents commonly isolated.

B. Eye Infections

1. **Pharyngoconjunctival fever**—Pharyngoconjunctival fever tends to occur in outbreaks, such as at children's summer camps ("swimming pool conjunctivitis"), and is associated with serotypes 3, 7 and 14.
2. **Epidemic keratoconjunctivitis (EKC)**—This is usually caused by type 8 and less often by types 19 and 37. This disease occurs mainly in adults and is highly contagious.
3. **Acute follicular conjunctivitis**—Types 3, 4 and 11 are commonly responsible. Adenoviral and chlamydial conjunctivitis are clinically similar.

C. Gastrointestinal Disease

Diarrhea—Some fastidious adenoviruses can cause diarrheal disease in children (for example types 40, 41).

D. Other diseases

Adenoviruses have been isolated from mesenteric lymph nodes in cases of mesenteric adenitis and intussusception in children. Adenovirus has also been associated with a pertussislike illness, acute hemorrhagic cystitis with dysuria and hematuria in young boys, musculoskeletal disorders, genital and skin infections.

E. Systemic Infection in Immunocompromised Patients

Adenoviral disease in immunocompromised patients include pneumonia and hepatitis.

Laboratory Diagnosis

A. Specimens

Depending on the clinical disease, virus may be recovered from stool or urine or from a throat, conjunctival, or rectal swab.

B. Direct Demonstration of Virus

i. **Electron microscopy:** Virus particles may be seen directly in stool extracts by electron microscopy.
ii. **Virus antigen:** The presence of viral antigen in the nasopharynx may be identified by immunofluorescence with group-specific antibodies (polyclonal or

Table 59.2: Diseases associated with adenovirus serotypes

Disease	*Those at risk*	*Associated serotypes*
1. Acute febrile pharyngitis		
Endemic	Infants, young children	1,2,5,6
Epidemic	Infants, young children	3,4,7
2. Pneumonia	Infants	1,2,3,7
3. Acute respiratory disease	Military recruits	4,7,14,21
4. Pharyngoconjunctival fever	Older school-age children	3,7
5. Epidemic keratoconjunctivitis (shipyard eye)	Adults	8,19,37
6. Follicular (swimming pool) conjunctivitis	Any age	3,4,11
7. Diarrhea and vomiting	Infants, young children	40,41
8. Intussusception	Infants	1,2,5
9. Hemorrhagic cystitis	Infants, young children	11,21
10. Disseminated infection	Immunocompromised, e.g. AIDS, renal, bone marrow and heart-lung transplant recipients	5,11,34,35,43-47

monoclonal) directly on aspirates or by enzyme immunoassays.

iii. **Viral DNA:** It is also possible to detect viral DNA directly from feces by polyacrylamide gel electrophoresis.

iv. **Latex agglutination method:** Enteric adenoviruses may be detected by latex agglutination method.

C. Virus Isolation

The clinical specimens are inoculated in tissue culture such as HeLa, Hep, KB and human embryo kidney cells. The development of characteristic cytopathic effects—rounding and clustering of swollen cells—indicates the presence of adenovirus in inoculated cultures.

Isolates can be identified as adenoviruses by immunofluorescence tests and hemagglutination inhibition and neutralization tests.

D. Polymerase Chain Reaction (PCR)

Polymerase chain reaction (PCR) assays can be used for diagnosis of adenovirus infections in tissue samples or body fluids.

E. Serology

For serological diagnosis, rise in titer of antibodies should be demonstrated in paired sera by complement fixation and neutralization tests used by reference laboratories or in research.

Treatment, Prevention, and Control

There is no known treatment for adenovirus infection. Live oral vaccines have been used to prevent infections with adenovirus types 4 and 7 in military recruits but are not used in civilian populations. However, genetically engineered subunit vaccines could be prepared and used in the future.

Gene Therapy

There is growing interest in the potential use of adenoviruses as gene delivery vehicles for gene therapy or DNA vaccination. Adenoviruses are attractive because replication-defective virus is able to lyse the endosome after internalization and release DNA into the cytoplasm. Efficient delivery of foreign DNA has been achieved by chemically coupling the DNA of interest with adenovirus particles.

Adenoviruses have been used and are being considered for more applications of gene delivery for correction of several human diseases, including immune deficiencies (e.g. adenosine deaminase deficiency), cystic fibrosis, lysosomal storage diseases, and even cancer.

Adeno-associated Viruses (AAV)

The adenovirus-associated viruses (AAVs) are members of the parvoviridae. They are about 22 nm in diameter, more hexagonal than circular in outline and contain insufficient single-stranded DNA to replicate on their own. They form a genus, Dependoviruses, indicating their dependence on adenoviruses (or herpes simplex virus) to provide the missing functions.

They can be detected by electron microscopy and complement fixation or immunofluorescence with specific antisera. Types 1, 2 and 3 are of human origin and cause natural infection, while type 4 is of simian origin. Their pathogenic role is uncertain.

True AAV (also known as adenovirus satellite virus) has not been implicated in clinical disease so this pathogenic role is uncertain.

KNOW MORE

Growth and Host Range

Adenoviruses are host specific and so laboratory animals are not susceptible to adenoviruses infecting humans. Human adenoviruses grow only in tissue cultures of human origin, such as human embryonic kidney, HeLa or HEP-2. Cytopathic changes may take several days (1 to 4 weeks) to develop and consist of cell rounding and aggregation into grape like clusters.

KEY POINTS

- Adenoviruses are non-enveloped, icosahedral, DNA viruses. The virion has the appearance of a space vehicle.
- Adenoviruses infect and replicate in epithelial cells of the respiratory tract, eye, gastrointestinal tract, urinary bladder, and liver. They usually do not spread beyond the regional lymph nodes.
- Adenovirus-associated viruses (AAVs) are members of the parvoviridae. They form a genus, *DEpendoviruses*, indicating their dependence on adenoviruses (or herpes simplex virus) to provide the missing functions. Their pathogenic role is uncertain.

IMPORTANT QUESTIONS

1. Discuss laboratory diagnosis of infections caused by adenoviruses.
2. Write short notes on:
   i. Adenoviruses
   ii. Adenovirus-associated viruses (AAV)

FURTHER READING

Fields BN, Knipe DM, Howley PM, editors. Virology, ed 3, New York, 1996, Lippincott-Raven.

Liu C. Adenoviruses. In, Textbook of Human Virology, 2nd Edn. Belshe, RB : Mosby Year Books 1991.

Mandell GL, Bennett JE, Dolin R. Principles and practice of infectious diseases, ed 5, New York, 2000, Churchill Livingstone.

CHAPTER 60

Papovaviruses

LEARNING OBJECTIVES

After reading and studying this chapter, you should be able to:
- Describe the following: *Papillomaviruses; Polyomaviruses.*

INTRODUCTION

The term 'Papova' is a sigla indicating the names of viruses included in this group: (*Pa,* papilloma; po, polyoma; va, vacuolating virus) belong to the family Papovaviridae and has two genera—*Papillomavirus* containing human and animal papilloma viruses and Polyoma virus which contains the simian vacuolating virus (SV 40) and polyomavirus.

They are small, nonenveloped (naked), have icosahedral nucleocapsids, and contain supercoiled, double-stranded, circular DNA. Papovaviruses induce both lytic infections, and either benign or malignant tumors, depending on the infected cell type.

PAPILLOMAVIRUSES

The Papillomaviridae family is a former genus in the Papovaviridae family. The papillomaviruses are slightly larger in diameter (55 nm) than the polyomaviruses (45 nm) and contain a larger genome (8 kbp versus 5 kbp).

Papilloma viruses are widely distributed in nature and are species-specific DNA viruses that infect the squamous epithelia and mucous membranes of vertebrates, including man. All papillomaviruses induce hyperplastic epithelial lesions in their host species.

Over seventy types of human papillomaviruses (HPVs) are now recognized. The HPVs exhibit great tissue, and cell specificity, infecting only surface epithelia of skin and mucous membranes. There is correlation between the virus type and the type of lesion produced.

Pathogenesis

Papillomaviruses cause several different kinds of warts in humans, including cutaneous warts, genital warts, respiratory papillomatosis, oral papillomas and cancer.

1. Cutaneous Warts

Cutaneous warts commonly infect the keratinized. epithelium of the hands and feet, producing typical warts frequently seen in young children and adolescents. The viruses associated with such lesions are HPV types: 1 and 4 (plantar warts); 3 and 10 (flat warts); 2, 4 and 7 (common warts). They usually disappear spontaneously but occasionally may be resistant to treatment. Regrowth of the lesions after treatment is thought to be due to persistence of the virus in the skin surrounding the original wart.

Epidermodysplasia Verruciformis

Another category of cutaneous lesion occurs in patients with what appears to be an inherited predisposition for multiple warts that do not regress, but instead spread to many body sites-a disease called epidermodysplasia verruciformis. These lesions give rise with high frequency to squamous cell carcinomas several years after initial appearance of the original warts, especially in areas of skin exposed to sunlight.

2. Anogenital Warts

These lesions (also known as *condylomata acuminata)* are commonest in sexually active adults. In women they are found:on the vulva,within the vagina and on the cervix. In men the most common sites for lesions are the shaft of the penis,peri-anal skin and the anal canal.

HPV types 6 and 11 are commonly found in benign vulval or penile warts. Cervical and anogenital warts may be due to HPV types 16 or I8, both of which are associated with malignant and premalignant lesions of the cervix and anogenital tract. Other types frequently found in genital cancers and precursor lesions include types 31, 33 and 45.

3. Recurrent Respiratory Papillomatosis.

This is a rare condition characterized by the presence of benign squamous papillomata on the mucosa of the respiratory tract, most commonly on the larynx. It has peaks of incidence in children under 5 years of age and adults after the age of I5 years. Children acquire the disease by passage through an infected birth canal, while adults acquire the disease from orogenital contact with an infected sexual partner. It is caused by infection of the respiratory mucosa with HPV types 6 and II.

4. Oral Papillomatosis

A variety of papillomata and benign lesions associated with HPV occur on the oral mucosa and tongue. Multiple lesions may develop on the buccal mucosa, a condition known as **oral florid papillomatosis**. The virus types here are those more commonly found in the genital tract, and infection is acquired during orogenital contact with an infected sexual partner. Several HPV types, including types 2, 7, 13 and 32, have been found.

5. Cancer

Malignant disease of the cervix is preceded by neoplastic change in the surface epithelium, a condition known as *cervical intra-epithelial neoplasia* (CIN). A similar pattern of events takes place in other sites in the genital tract of both men and women.

HPV DNA can be detected in all grades of the premalignant lesions of the female and male genital tract. HPV types 6 and II are most commonly found in low-grade disease whereas. HPV types 16 and 18 are more commonly associated with lesions of greater severity and invasive cancer. The association of wart viruses with invasive cancers of the skin, larynx and genital tract is well documented.

Laboratory Diagnosis

Morphological Identification

HPV infection may be readily diagnosed when there are typical clinical lesions.

Subclinical infection requires laboratory confirmation using

1. Cytological and histological detection
2. Immunocytochemical detection-HPV capsid antigen in sections of tissues or in cell smears can be detected by immunoperoxidase test using antiserum (commercially available). It detects all the genital HPV types
3. **Molecular methods:** Amplification of HPV DNA by polymerase chain reaction (PCR) using consensus primers to detect a wide range of HPV types has also been widely used.

Serology

Such assays are more useful for showing evidence of past exposure rather than current infection and' are appropriate for epidemiological studies.

POLYOMAVIRUSES

The Polyomaviridae family is a former genus in the Papovaviridae family (which no longer exists). These are small viruses (diameter 45 nm) that possess a circular genome of double-stranded DNA enclosed within a non enveloped capsid exhibiting icosahedral symmetry. Cellular histones are used to condense viral DNA inside virus particles.

The name is derived from 'poly' (many) and 'oma' (tumour). The viruses are species-specific, and although tumour induction is well described in experimental animals, there is to date no documented association with any naturally occurring tumour of man. Recognized members of this group include:

1. Mouse polyomavirus
2. SV 40 of monkeys
3. JC virus (JCV)
4. BK virus

Two viruses of man-JC virus (JCV) and BK virus (BKV), both named after the initials of the people from whom they were first isolated.

1. Mouse Polyomavirus

It causes harmless infections in mice by natural routes. However, it induces different types of malignant tumours when injected into infant rodents.

2. Simian Vacuolating Virus (SV 40)

The simian vacuolating virus (SV 40) was isolated from uninoculated rhesus and cynomolgus monkey kidney tissue cultures. The role SV 40 is playing in formation of human tumors is under investigation. SV40 is oncogenic in newborn hamsters. Its only medical importance is that because of its oncogenic potential, live viral vaccines should be manufactured only in monkey kidney tissue cultures tested and found free from SV 40 infection.

3. JC Virus (JCV)

The JC virus was isolated in 1971 from the brain of a patient with Hodgkin's disease and progressive multifocal leukoencephalopathy (PML). It was named after the initials of the person from whom it was first isolated. JC virus is the cause of progressive multi focal leukoencephalopathy (PML), so called because the lesions are restricted to the white matter that occurs in some immunocompromised persons. In recent years, PML has been seen especially in patients with AIDS. Following intracerebral inoculation in newborn hamsters, it is oncogenic producing malignant gliomas. JC virus has been associated with human brain tumors.

4. BK Polyomaviruses

The human polyomaviruses (BK and JC) have been isolated from immunocompromised patients. It was named after the initials of the person from whom it was first isolated. BK virus was isolated from the urine of a patient with kidney transplant. BK virus causes cystitis, nephropathy, and severe renal allograft dysfunction. It is believed to be the cause of polyomavirus-associated nephropathy in renal transplant recipients, which results in graft failure in up to 50 percent of infected patients.

Laboratory DiaGnosis

1. Electron Microscopy

Human polyomaviruses can be detected by electron microscopy from brain tissue in a case of PML (JC virus) and from the urine of a renal transplant case (BK virus).

2. Virus Isolation

JC polyomavirus can be isolated from the urine or the brain and BK polyomavirus from the urine. Human fetal glial cell culture and human diploid fibroblasts are used for the isolaton of JC polyomavirus BK virus respectively. Hemagglutination inhibition is used to differentiate these two viruses.

3. Viral Antigen Detection

The brain biopsy or autopsy material can be examined directly for JCV antigen by immunofluorescence or immunoperoxidase staining.

4. Viral Nucleic Acid Detection

Viral nucleic acid can be detected by nucleic acid hybridization and polymerase reaction (PCR).

5. Cytopathology

The cytology of urine in human polyomavirus infection is quite characteristic. Exfoliated urinary epithelial cells show the presence of enlarged deeply stained bacophilic nuclei with a single inclusion.

KEY POINTS

- Papovaviruses (from *pa*, papilloma, *po*, polyoma, *va*, vacuolating agent) are double stranded DNA viruses and has two genera, *Papillomavirus* and *Polyomavirus.*
- Papillomaviruses are species-specific DNA viruses that infect the squamous epithelia and mucous membranes of vertebrates, including man.
- Over seventy types of human papillomaviruses (HPVs) are now recognized.
- Papillomaviruses cause several different kinds of warts in humans, including cutaneous warts, genital warts, respiratory papillomatosis, oral papillomas and cancer.
- *Polyomaviruses* include. mouse polyomavirus, *simian virus* 40 (SV 40) of monkeys, JC virus (JCV) and BK polyomavirus

IMPORTANT QUESTIONS

1. Describe the pathogenesis and laboratory diagnoses of infections caused by papillomaviruses.
2. **Write short notes on:**
   SV40
   Papovavirus
   JC polyomaviruses
   BK polyomaviruses

FURTHER READING

Fields BN, et a1. 1996. Virology, 3rd edn. Lippincott-Raven, Philadelphia.

CHAPTER

61 Parvovirus

Learning Objectives

After reading and studying this chapter, you should be able to:

- Describe parvoviruses.

INTRODUCTION

The parvoviruses are the smallest of the DNA viruses (about 20 nm) and have been isolated from a wide range of organisms, from arthropods to humans.

The family Parvoviridae is divided into two subfamilies: the Parvovirinae and the Densivirinae. The latter group infects only invertebrates. The Parvovirinae contains three genera: *Parvovirus, Dependovirus* and *Erythrovirus.*

1. PARVOVIRUS

The **Parvovirus genus** contains the *autonomous* parvoviruses, which are widespread in nature and capable of autonomous replication. They cause a wide variety of diseases in different organs of their natural hosts. The group includes feline and canine parvoviruses (FPV and CPV) which are so important in veterinary medicine that immunization against them is a routine practice in developed countries. Minute virus of mice (MVM) was first discovered in 1966 and, although not very pathogenic, it has served as a model for understanding the function and molecular biology of other members of the genus. It is possible that some of the small round viruses seen in human feces may be shown to be parvoviruses although none have been recognized as pathogenic in humans.

2. DEPENDOVIRUS

In contrast to parvoviruses, dependoviruses require helper virus functions for replication. They infect a number of species, but the most studied are the human adeno associated viruses (AAV), but occasionally a herpesvirus, to assist in replication. Several serotypes have been noted but none have yet been associated with human disease. (See Chapter 59 for more detail).

3. ERYTHROVIRUS

There is only one parvovirus (B19) known to cause disease in humans and, together with closely related simian viruses, B19 has been, placed in a new genus, *Erythrovirus.*

PARVOVIRUS (B19)

The cell receptor for parvovirus B19 is the P antigen. This is present on red blood cells, erythroid progenitors, vascular endothelium and fetal myocytes. Thus the distribution of the cell receptor is linked to the clinical manifestations of disease. Human parvovirus B19 is found worldwide. It is usually endemic and infections can occur throughout the year. Infection is acquired in childhood and is often asymptomatic. Transmission of parvoviruses appears to be by the respiratory route. Parvovirus 19 is highly contagious.

Clinical Diseases

i. Minor Illness

In children, in whom B19 infection is most common, asymptomatic infection accounts for about half of all infections. Non-specific respiratory tract illness is the next most common illness, at least in boys. This can mimic influenza and coincides with the viremic phase of the infection.

ii. Erythema Infectiosum (Fifth Disease)

B19 virus causes an erythematous maculopapular rash, which in its most clinically distinct form is called *erythema infectiosum.* It is common in children aged 4-11 years, and is sometimes called *fifth disease* since it was the fifth of six erythematous rash illnesses of childhood in an old classification. Classically, it starts with an intense erythema of the cheeks, hence another of its names—

'slapped-cheek disease'. The rash then proceeds to involve the trunk and limbs. There may be associated lymphadenopathy and joint symptoms.

iii. Joint Disease

In addition to a rash, a complication accompanying B19 infection is an acute arthritis that usually involves joints symmetrically. This is considerably more frequent in adults than in children, and usually resolves within several weeks.

iv. Aplastic Crisis

Parvovirus B19 induces in childern with aplastic crisis with chronic hemolytic anemia (e.g. sickle cell anemia or thalassemia). With very low hemoglobin and disappearance of circulating reticulocytes.

v. Infection During Pregnancy

Parvovirus B19 infection during the second or third trimester of pregnancy may result in nonimmmune fetal hydrops.

vi. Infection in Immunosuppressed

Persistent infections have been described in patients with underlying immunodeficiency states, including Nezelof's syndrome, acute lymphatic leukemia and human immunodeficiency virus (HIV) infection. They have also been noted post-transplant. The illness is characterized by either persistent anemia or a remitting and relapsing anemia.

Laboratory Diagnosis

Diagnosis may be made by detection of virus in the blood in early cases, and of antibody later.

1. Virus Isolation

Parvovirus B19 may be cultured in cells from human bone marrow or fetal liver. It can be detected in patient's blood by electron microscopy.

2. Detection of Viral Nucleic Acid

Nucleic acid can be detected by nucleic' acid dot blot hybridization or by polymerase chain reaction-based assays.

3. Antigen Detection

Detection of viral antigen is done by counterimmunoelectrophoresis, ELISA, RIA or indirect immunofluorescence.

4. Antibody Detection

IgM antibodies or a significant rise in IgG antibodies can be detectd by ELISA or RIA. It is the most successful technique.

KEY POINTS

- Parvoviruses are the smallest of the DNA viruses (about 20 nm).
- The family Parvoviridae is divided into two subfamilies: the Parvovirinae and the Densivirinae.
- The Parvovirinae contains three genera: *Parvovirus, Dependovirus* and *Erythrovirus.*
- **The Parvovirus genus** the group includes feline and canine parvoviruses (FPV and CPV).
- **Dependovirus** require helper virus functions for replication but the most studied are the human adeno associated viruses (AAV), but occasionally a herpesvirus,
- **Erythrovirus** is only one **parvovirus (B19)** known to cause disease in humans and, together with closely related simian viruses.
- Human parvovirus B19 may cause respiratory infection with an erythematous maculopapular rash (erythema infectiosum—slapped cheek disease), joint disease, aplastic crisis in children with chronic hemolytic anemia (sickle cell disease), nonimmune fetal hydrops following infection during pregnancy and persistent anemia in immunodeficient individuals.

IMPORTANT QUESTIONS

Write short notes on:
Parvovirus.
Dependovirus.
Erythrovirus or Parvovirus (B19).

FURTHER READING

Schwartz TF, et al. Human parvovirus B 19 infection: Lancet 1987;2,738.

CHAPTER 62

Picornaviruses

Learning Objectives

After reading and studying this chapter, you should be able to:
- Describe enteroviruses.
- Discuss prophylaxis against poliomyelitis.
- Differentiate live and killed polio vaccines.
- Differentiate coxsackie A and coxsackie B viruses.
- Describe diseases caused by coxsackie viruses.
- Describe the following: acute hemorrhagic conjunctivitis; Rhinoviruses.

INTRODUCTION

The family Picornaviridae comprises a large number of very small RNA (*pico*, meaning small, rna:*RNA*) viruses with a diameter of 27-30 nm and containing single-stranded RNA. They are nonenveloped viruses, resistant to ether and other lipid solvents. It is one of the largest families of viruses and includes some of the most important human and animal viruses.

CLASSIFICATION

The Picornaviridae family has more than 230 members that are divided into six genera: Enterovirus, Rhinovirus, Cardiovirus, Aphthovirus, Hepatovirus and Parechovirus (Table 62.1).

Table 62.1: Picormaviridae

1. Enterovirus
Poliovirus types 1, 2, and 3
Coxsackie A virus types 1 to 22 and 24
Coxsackie B virus types 1 to 6
Echovirus (ECHO virus) types 1 to 9, 11 to 27, and 29 to 34
Enterovirus 68 to 72
2. Rhinovirus types 1 to 100+ –most important cause of the common cold.
3. Cardiovirus–of mice, including the encephalomyocarditis virus
4. Aphthovirus–causing the foot and mouth disease of cattle.
5. Heparnavirus–Hepatitis A virus.
6. Parechovirus–Echovirus 22 and 23.

Important properties of picornaviruses are summarized in Table 62.2.

IMPORTANT PROPERTIES OF PICORNAVIRUSES

Size: 28-30 nm in diameter

Virion: Icosahedral, contains 60 subunits.

Genome: Single-stranded RNA, linear, positive sense.

Proteins: Four major polypeptides cleaved from a large precursor polyprotein. Surface proteins VP1 and VP3 are major antibody-binding sites. Internal protein VP4 is associated with viral RNA.

Envelope: None

Replication: Cytoplasm

Culture: Many enteroviruses (polioviruses, echoviruses, some coxsackieviruses) can be grown at 37°C in human and monkey cells. Most rhinovirus strains can be recovered only in human cells at 33°C. Coxsackieviruses are pathogenic for newborn mice.

ENTEROVIRUSES

Enteroviruses of medical importance include polioviruses, coxsackieviruses and echoviruses because they are all found in the intestines and are excreted in the feces. At least 72 serotypes of human enteroviruses exist, including the Polioviruses types 1-3, CoxsackievirusesA types 1-24, Coxsackievirus B types 1-6, Echovirus types 1-34 and Enterovirus types 68-72.

Since 1969, new enterovirus types have been assigned enterovirus type numbers rather than being subclassified as coxsackie viruses or echoviruses. The vernacular

Table 62.2: Some properties of picornaviruses		
Property	*Enteroviruses*	*Rhinoviruses*
Size (nm)	22-30	30
1. Capsid form	Icosahedral	Icosahedral
2. Polypeptides	VP1, VP2, VP3, VP4	VP1, VP2, VP3,
3. RNA type	Single-stranded, positive-sense	VP4 Single-stranded, positive-sense
4.Optimal temperature for growth	37°C	33-34°C
5. Acid	Stable (pH 3-9)	Labile (pH 3-5)

names of the previously identified enteroviruses have been retained. Enterovirus 72 is the virus causing infectious hepatitis (Hepatitis type A), which has been reclassified as a separate genus *Hepatovirus*. Because of its special status, it is considered in the chapter on Hepatitis Viruses.

POLIOVIRUS

Poliomyelitis is an acute infectious disease that in its serious form affects the central nervous system. The destruction of motor neurons in the spinal cord results in flaccid paralysis. However, most poliovirus infections are subclinical.

Poliovirus has served as a model picornavirus in many laboratory studies of the molecular biology of picornavirus replication.

Morphology

Size: The virion is a spherical particle, about 27 nm in diameter

Capsid: It consists of a capsid shell of 60 subunits, each consisting of four viral proteins (VP1-VP4), arranged in icosahedral symmetry. VP1 which faces outside, carries the major antigenic site for combination with type specific neutralizing antibodies.

Genome: The genome is a single strand of positive sense RNA.

The virus can be crystallised, and arrays of virus crystals can be seen in the cytoplasm of infected cells.

Resistance

1. Enteroviruses are among the most stable viruses. Poliovirus is resistant to ether, chloroform, bile, proteolytic enzymes of the intestinal contents and detergents.
2. It is stable at pH 3.
3. In feces, virus can survive for months at 4°C, for years at –20 or –70°C and at room temperature for several weeks, depending on the amount of virus present, the amount of moisture present and other environmental conditions.
4. They are inactivated when heated at 55°C for 30 minutes, but molar $MgCl_2$ prevents this inactivation. Milk or ice cream also provides such protection.
5. Drying rapidly inactivates enteroviruses by ultraviolet light, and usually by drying.
6. Formaldehyde and oxidizing disinfectants destroy the virus.
7. Chlorination destroys the virus in water but organic matter delays inactivation. Phenolic disinfects are not effective.
8. Poliovirus does not survive lyophilization well

Antigenic Properties

Two antigens C and D (C = coreless or capsid; D = dense) can be recognized by complement fixation, enzyme linked immunosorbent assay (ELISA) or precipitation tests.

C Antigen

The C antigen, also called the Heated or H antigen associated with the 'empty' noninfectious virus, is less specific and reacts with heterotypic sera. Anti-C antibody does not neutralize virus infectivity.

D Antigen

The D antigen, also called the Naive or N antigen, is associated with the whole virion and is type-specific. The D antigen is converted into the C antigen by heating the virus at 56°C. Anti-D antibody is protective and therefore the potency of injectable poliovaccine can be measured in terms of D antigen units.

Types

There are three types (1, 2 and 3) of poliovirus, identified by neutralization tests. The prototype strains are:

- Type 1, the Brunhilde and Mahoney strains—Type 1 is the common epidemic type
- Type 2, which includes the rodent-adapted strains, the Lansing and MEFI strains is usually associated with endemic infections
- Type 3, the Leon and Saukett strains has caused recent epidemics.

Immunity is type specific.

Host Range and Cultivation

The virus grows readily in tissue cultures of primate origin. Primary monkey kidney cultures are used for diagnostic cultures and vaccine production. The infected cells round up and become refractile and pyknotic. Eosinophilic intranuclear inclusion bodies may be demonstrated in stained preparations. Well-formed plaques develop in infected monolayers with agar overlay.

Pathogenesis

The virus is transmitted by the fecal-oral route through ingestion. Inhalation or entry through conjunctiva of droplets of respiratory secretions may also be possible modes of entry in close contacts of patients in the early stage of the disease. All enterovirus infections follow a similar pattern, with differences in the target organs, e.g. central nervous system, skin, heart or muscle.

Poliovirus has one of the narrowest tissue tropisms, recognizing a receptor expressed on anterior horn cells of the spinal cord, dorsal root ganglia, motor neurons, skeletal muscle cells, lymphoid cells, and few other cells.

Virus is ingested and multiplies initially in the lymphoid tissue of the tonsil or Peyer's patches in the small intestine. It then spreads to the regional lymph nodes and enters the blood stream (minor or primary viremia). In fatal cases, the Peyer's patches and the mesenteric lymph nodes are found to be greatly swollen and inflamed and to contain large amounts of virus. Primary viremia spreads the virus to receptor-bearing target tissues, where a second phase of viral replication may occur, resulting in symptoms and a secondary viremia.

In the case of poliovirus, the virus crosses the blood-brain barrier or gains access to the brain by infecting skeletal muscle and traveling up the innervating nerves to the brain, like the rabies virus.

The paralytic effect of poliovirus results from infection of motor neurone cells in the anterior horns of the spinal cord or bulbar regions.

In addition to pathologic changes in the nervous system, there may be myocarditis, lymphatic hyperplasia, and ulceration of Peyer's patches.

Clinical Features

The incubation period is usually 7-14 days, but it may range from 3 days to 35 days. Following exposure to poliovirus, 90-95 percent of susceptible individuals develop only **inapparent infection**, which causes seroconversion alone. It is only in 5-10 percent that any sort of clinical illness results.

1. Asymptomatic Illness

Asymptomatic illness results if the viral infection is limited to the oropharynx and the gut. At least 90 percent of poliovirus infections are asymptomatic.

2. Abortive Poliomyelitis: The Minor Illness

It is a nonspecific febrile illness occurring in approximately 5 percent of infected people. Recovery occurs in a few days.

3. Nonparalytic Poliomyelitis or Aseptic Meningitis

Nonparalytic poliomyelitis or aseptic meningitis occurs in 1 percent to 2 percent of patients with poliovirus infections. In this disease, the virus progresses into the central nervous system and the meninges, causing back pain and muscle spasms in addition to the symptoms of the minor illness. The disease lasts 2-10 days, and recovery is rapid and complete. In a small percentage of cases, the disease advances to paralysis.

4. Paralytic Poliomyelitis: The Major Illness

Paralytic polio, the major illness, occurs in 0.1 percent to 2.0 percent of persons with poliovirus infections and is the most severe outcome. The predominating complaint is flaccid paralysis resulting from lower motor neuron damage. Paralysis is focal in distribution initially but spreads over the next 3-4 days. Depending on the distribution of paralysis, cases are classified as *spinal, bulbar* or *bulbospinal*.

Mortality ranges from 5-10 percent and is mainly due to respiratory failure. Recovery of the paralysed muscles takes place in the next 4-8 weeks and is usually complete after six months, leaving behind varying degrees of residual paralysis.

5. Progressive Postpoliomyelitis Muscle Atrophy

Post polio syndrome is a sequela of poliomyelitis that may occur much later in life (30 to 40 years later) in 20 percent to 80 percent of the original victims. Affected people suffer a deterioration of the originally affected muscles.

Precipitating Factors

Several provocative or risk factors have been found to precipitate an attack of paralytic polio in individuals already infected with polio viruses. They include fatigue, trauma, intramuscular injections, operative procedures such as tonsillectomy undertaken especially during epidemics of polio and administration of immunizing agents particularly alum containing DPT (diphtheria, pertussis, tetanus).

Laboratory Diagnosis

1. Specimens

Many specimens can be used, including blood, CSF, throat swabs and feces. Polioviruses may be isolated from the patient's pharynx during the first few days of illness, from the feces for as long as 30 days, but from the CSF only rarely but can be obtained from the spinal cord and brain, postmortem unlike other enteroviruses.

2. Culture

Primary monkey kidney cells are usually employed. The virus growth is indicated by typical cytopathic effects in 2-3 days. An isolated virus is identified and typed by neutralization with specific antiserum.

It must be remembered that the mere isolation of poliovirus from feces does not constitute a diagnosis of poliomyelitis. Virus isolation must be interpreted along with clinical and serological evidence.

3. Serological Tests

Serodiagnosis is less often employed. Antibody rise can be demonstrated in paired sera by **neutralization or complement fixation tests**.

Immunity

Immunity is permanent to the type causing the infection. Humoral immunity provided by circulating and secretory antibody is responsible for protection against poliomyelitis.

The virus also induces cell mediated immunity but its importance appears to be uncertain.

Prophylaxis

Immunization

Both killed and live attenuated vaccines are available. Two types of vaccines are used throughout the world; they are:

1. Inactivated polio vaccine (IPV)—Salk killed polio vaccine.
2. Oral polio vaccine (OPV)—Sabin live polio vaccine.

1. Inactivated polio vaccine (IPV)—Salk's Killed Polio Vaccine

By 1953, Salk had developed a killed vaccine. Salk's killed polio vaccine is a formalin inactivated preparation of the three types of poliovirus grown in monkey kidney tissue culture. Standard virulent strains are used. The three types of polioviruses are grown separately in monkey kidney cells. Viral pools of adequate titer are filtered to remove cell debris and clumps, and inactivated with formalin (1:4000) at 37°C for 12-15 days. Stringent tests are carried out to ensure complete inactivation and freedom from extraneous agents. The three types are then pooled and after further tests for safety and potency, issued for use.

Killed vaccine is given by injection and is therefore called inactivated or injectable poliovaccine (IPV). The primary or initial course of immunization consists of 4 inoculations. The first 3 doses are given at intervals of 1-2 months and 4th dose 6-12 months after the third dose. First dose is usually given when the infant is 6 weeks old to ensure that immune response is not impaired by residual maternal antibodies. Additional doses are recommended prior to school entry and then every 5 years until the age of 18. Inactivated vaccine is recommended for immunocompromised individuals and their contacts and others for whom a live vaccine is contraindicated.

IPV induces, humoral antibodies (IgM, IgG and IgA serum antibodies) but does not induce intestinal or local immunity. The circulating antibodies protect the individual against paralytic polio, but do not prevent reinfection of the gut by wild viruses. In the case of an epidemic IPV is unsuitable.

2. Oral Polio Vaccine (OPV)—Sabin Vaccine

Oral polio vaccine (OPV) was described by Sabin in 1957. It contains live attenuated virus (Types 1, 2 and 3) grown in primary monkey kidney or human diploid cell cultures. The vaccine is issued either in the monovalent or trivalent form, in pleasantly flavoured syrup after tests for neurovirulence, genetic stability and potency. The use of molar $MgCl_2$ or sucrose stabilizes the vaccine against heat inactivation, particularly under tropical conditions. It can be given to young infants, as the maternal antibody has little effect on intestinal infection. Theoretically, a single dose should be sufficient to establish infection and immunity.

OPV used in India is stated to contain Type 1 virus 10 lakh, Type 2 virus 2 lakh and Type 3 virus 3 lakh TCID50 per dose (0.5 ml). The liquid vaccine is thermostabilized with $MgCl_2$ which acts only at a pH below 7.0. The vaccine has to be kept in airtight containers to maintain the pH. The shelf life of the vaccine at 4-8°C is four months and at –20°C is two years. Improper storage conditions and 'cold chain' failure may be partly responsible for the apparent failure of OPV to control poliomyelitis in the developing countries.

Criteria of Attenuated Strains for Live Vaccine

1. They should not be neurovirulent as tested by intraspinal inoculation in monkeys;
2. They should be able to set up intestinal infection following feeding and should induce an immune response;
3. They should not acquire neurovirulence after serial enteric passage;
4. They should possess stable genetic characteristics (markers) by which they can be differentiated from the wild virulent strains.

Markers for Differentiating the Wild from the Attenuated Strains

Several markers have been described for differentiating the wild from the attenuated strains. Molecular epidemiological methods give better results.

1. **d marker:** Wild strains will grow well in low levels of bicarbonate but avirulent strains will not;
2. **rct 40:** Wild strains grow well at 40°C, while avirulent strains grow poorly;
3. **MS:** Wild strains grow well in a stable cell line of monkey kidney, while avirulent strains grow poorly;
4. **McBride's intratypic antigenic marker** shown by the rate of inactivation by specific antiserum.

Immunization Schedule

The WHO Programme on Immunization (EPI) and the National Immunization Programme in India recommend a primary course of 3 doses of OPV at one-month intervals, commencing the first dose when infant is 6 weeks old. One booster dose of OPV is recommended 12 to 18 months later. It has been recommended that in the tropics the number of doses of vaccine be increased to five, in order to enhance seroconversion in the vaccinees. It is very important to complete vaccination of all infants before 6 months of age. This is because most polio cases occur between the ages of 6 months and 3 years.

There has been much controversy about the relative merits of killed and live vaccines. The differences between IPV and OPV are given in Table 62.3.

Global Eradication

By global immunisation with OPV it is possible to eradicate the disease. A major campaign by the World Health Organization is under way to eradicate poliovirus from the world as was done for smallpox virus. The World Health Organisation Assembly in 1988 had proposed global eradication of poliomyelitis by the year 2000. The Americas were certified as free from wild poliovirus in 1994, the Western Pacific Region in 2000, and Europe in

Table 62.3: Differences between killed (IPV) and live polio vaccines (OPV)

	Killed polio vaccine (Salk type)	*Live polio vaccine (Sabin type)*
1. Virus	Killed formolized virus	Live attenuated virus
2. Route of Adminisration	Given subcutaneously or IM	Given orally
3. Nature of immunity	Induces circulating antibody, but no local (intestinal) immunity	Immunity is both humoral and intestinal. Induces antibody quickly
	Prevents paralysis, but does not prevent reinfection by wild polio viruses	Prevents not only paralysis, but also intestinal reinfection
4. Useful in controlling epidemics	Not useful in controlling epidemics	Can be effectively used in controlling epidemics. Even a single dose elicits substantial immunity (except in tropical countries)
5. Manufacture	More difficult to manufacture	Easy to manufacture
6. Duration of immunity	Lifelong	Booster vaccine needed for lifelong immunity
7. Cost	The virus content is 10,000 times more than OPV. Hence costlier	Cheaper
8. Storage	Does not require stringent conditions during storage and transportation. Has a longer shelf-life.	Requires to be stored and transported at sub-zero temperatures, unless stabilized
9. Contraindicated in immunodeficiency states or pregnancy	No	Yes

2002. Poor progress in immunization in many countries has been a setback to this objective. Progress is being made globally, but several thousand cases of polio still occur each year, principally in Africa and the Indian subcontinent

Epidemiology

Poliomyelitis is an exclusively human disease and the only source of virus is humans, the patient or much more commonly the symptomless carrier. Most infections are subclinical. The cases are most infectious 7 to 10 days before and after onset of symptoms. There are no chronic carriers. However, the virus may persist in the environment (sewage) for up to six months. Virus shed in throat secretions during the early part of the disease may also be a source of infection for the contacts of patients

The disease occurs in all age groups, but children are usually more susceptible than adults because of the acquired immunity of the adult population. In India, polio is essentially a disease of infancy and childhood. About 50 percent of cases are reported in infancy. In developed countries, before the advent of vaccination, the age distribution shifted so that most patients were over age 5 and 25 percent were over age 15 years. The case fatality rate is variable. It is highest in the oldest patients and may reach 5-10 percent.

Most infections are subclinical. It is estimated that for every clinical case, there may be 1000 subclinical cases in children and 75 in adults.

Poliovirus type 1 is responsible for most epidemics of paralytic poliomyelitis. Type 3 also causes epidemics to a lesser extent. Type 2 usually causes inapparent infections in the western countries but in India paralysis due to type 2 is quite common. Immunity is type-specific but there is a significant amount of cross protection between various types.

In temperate climates, infection with enteroviruses, including poliovirus, occurs mainly during the summer. Virus is present in sewage during periods of high prevalence and can serve as a source of contamination of water used for drinking, bathing, or irrigation. There is a direct correlation between poor hygiene, sanitation, and crowding and the acquisition of infection and antibodies at an early age.

COXSACKIEVIRUS

The prototype strain was isolated by Dalldorf and Sickles (1948) from the village of Coxsackie in New York. Several related viruses have been isolated since then from different parts of the world. The characteristic feature of this group is its ability to infect suckling but not adult mice. Coxsackieviruses are classified into two groups, A and B based on the pathological changes produced in suckling mice (Table 62.4).

Properties of the Virus

Coxsackieviruses are highly infective for newborn mice. Following inoculation in suckling mice, Group A viruses produce widespread myositis in the skeletal muscles of newborn mice, resulting in flaccid paralysis without other observable lesions leading to death within a week. Group B viruses may produce a patchy focal myositis, spastic paralysis, necrosis of the brown fat and, often, pancreatitis, hepatitis, myocarditis and encephalitis (Table 63.4). The genetic makeup of inbred strains of mice determines their susceptibility to coxsackie B viruses.

Mouse inoculation is no longer used as a diagnostic test. It has been replaced by RNA detection using the reverse transcription polymerase chain: reaction (RTP-CR).

Table 62.4: Features of coxsackievirus A and B infection in the laboratory

Features	Coxsackievirus A	Coxsackievirus B
1. Growth in monkey kidney	+	+
2. Effect in suckling mice	Generalized myositis Flaccid paralysis Death within a week	Patchy focal myositis Spastic paralysis Necrosis of the brown fat and, often, pancreatitis, hepatitis, myocarditis and encephalitis
3. Types	23 (1-24[a])	6 (1-6)

[a] Coxsackievirus A23 now classified as echovirus 9.

Antigenic Characters

Thirty antigenic types have been defined by cross-neutralization tests in mice or cell culture, and cross-complement fixation reactions. Twenty-three have the features of group A and six have those of group B. Coxsackie A 23 is the same as echo 9 (Coxsackievirus A23 has now been reclassified as echovirus 9) and Coxsackie A24 the same as ECHO 34.

Clinical Features

Like other enteroviruses, coxsackieviruses inhabit the alimentary canal primarily and are spread by the fecal-oral route. The incubation period of coxsackievirus infection ranges from 2 days to 9 days. The clinical manifestations of infection with various coxsackieviruses are diverse and may present as distinct disease entities (Table 62.5).

A. Group A Viruses

These viruses give rise to:

1. Herpangina (Vesicular Pharyngitis)

Herpangina is a severe febrile pharyngitis and is a common clinical manifestation of coxsackie group A infection in children. Fever, sore throat, pain on swallowing, anorexia, and vomiting characterize this disorder. The classic finding is vesicular ulcerated lesions around the soft palate and uvula. Less typically, the lesions affect the hard palate.

2. Aseptic Meningitis

Aseptic meningitis is caused by all types of group B coxsackieviruses and by many group A coxsackieviruses, most commonly A7 and A9. Type A7 had caused outbreaks of paralytic disease in Russia, Scotland and elsewhere.

3. Hand-Foot-and-Mouth Disease

Hand-foot-and-mouth disease is a vesicular exanthem and is characterized by oral and pharyngeal ulcerations and a vesicular rash of the palms and soles that may spread to the arms and legs. This disease has been associated particularly with coxsackievirus A16, but A5 and A10 have also been implicated.

4. Respiratory Infections

A number of the enteroviruses have been associated with common colds; among these are coxsackieviruses A21, A24, B1, and B3-5.

B. Group B Viruses

1. Epidemic Myalgia or Bornholm Disease

Epidemic myalgia or Bornholm disease, so called because it was first described on the Danish island of Bornholm. It is an acute illness in which patients have a sudden onset of fever and unilateral low thoracic, pleuritic chest pain that may be excruciating and muscles on the involved side may be extremely tender. Coxsackie B virus is the causative agent.

2. Myocardial and pericardial infections

Myocardial and pericardial infections caused by coxsackie B virus occur sporadically in older children and adults but are most threatening in newborns.

3. Aseptic meningitis

Group B viruses may cause aseptic meningitis with paralyses

4. Juvenile Diabetes

Juvenile diabetes has been claimed to be associated with coxsackie B4 infection.

5. Neonatal Infections

Transplacental and neonatal transmission has been demonstrated with coxsackie B viruses resulting in a serious disseminated disease that may include hepatitis, meningoencephalitis and adrenocortical involvement.

6. Chronic fatigue syndrome

There is some evidence of a possible association between chronic fatigue syndrome and infection with enteroviruses, particularly coxsackie B viruses.

Laboratory Diagnosis

A. **Virus isolation:** The virus is isolated readily from throat washings, conjunctival swabs, throat swabs, and feces.
   i. *Inoculation into suckling mice:* Specimens are inoculated into suckling mice. In suckling mice, signs of illness appear usually within 3-8 days with group A strains and 5-14 days with group B strains. Identification is by studying the histopathology in infected mice and by

neutralisation tests. Because of the difficulty of the technique, virus isolation in suckling mice is rarely attempted.

ii. *Tissue culture:* Specimens are inoculated into tissue cultures and a cytopathic effect appears within 5-14 days. Of group A coxsackieviruses, only A7 and A9 grow in monkey kidney cells, and coxsackievirus A21 can be grown in HeLa, HEp2 or human embryonic kidney cell cultures. All group B coxsackieviruses grow readily in monkey kidney cell cultures.

B. **Serology:** Serologic tests are difficult to evaluate (because of the multiplicity of types).

C. **Nucleic acid detection:** Reverse transcription-polymerase chain reaction tests can be broadly reactive (detect many serotypes) or more specific.

Prevention: Vaccination is not practicable as there are several serotypes and immunity is type specific.

ECHOVIRUSES

Echoviruses (Enteric cytopathogenic human orphan viruses) are grouped together because they infect the human enteric tract. They were called orphans as they could not be associated with any particular clinical disease then. They have been given the descriptive designation: **enteric cytopathogenic human orphan viruses'** and are generally known by the sigla '**echoviruses**'.

Antigenic Properties

Echoviruses have been classified into 34 serotypes by neutralisation tests. Types 10 and 28 have been removed from the group, the former becoming a reovirus and the latter a rhinovirus.

Epidemiology

The epidemiology of echoviruses is similar to that of other enteroviruses. Like other enteroviruses, echoviruses inhabit the alimentary tract primarily and are spread by the fecal-oral route.

Clinical Features

Most echovirus infections are asymptomatic but some have been associated with clinical syndromes.

1. Aseptic meningitis, paralysis, rash and fever.
2. Respiratory illnesses in children.
3. Infantile diarrhea.
4. Pericarditis and myocarditis conjunctivitis, muscle weakness, and spasm.

Laboratory Diagnosis

Specimen

The procedure of choice is isolation of virus from throat swabs, stools, rectal swabs, and, in aseptic meningitis, cerebrospinal fluid.

A. *Virus isolation:* Specimens may be inoculated into monkey kidney tissue cultures and virus growth detected by cytopathic changes. The large number of serotypes makes identification by neutralisation tests laborious.

B. *Serology:* Serological diagnosis is impractical due to large number of serotypes.

C. *Polymerase chain reaction(PCR): Polymerase* chain reaction, is more rapid than virus isolation for diagnosis.

Prevention: Vaccination has not been attempted.

OTHER ENTEROVIRUS TYPES

Four enteroviruses (types 68-71) grow in monkey kidney cultures, and three of them cause human disease (Table 62.6).

ACUTE HEMORRHAGIC CONJUNCTIVITIS

A pandemic of acute hemorrhagic conjunctivitis, apparently arising in West Africa in 1969 spread widely involving several parts of Africa, the Middle East, India, South East Asia, Japan, England and Europe. Acute hemorrhagic conjunctivitis has a sudden onset of subconjunctival hemorrhage ranging from small petechiae to large blotches covering the bulbar conjunctiva. The disease is most common in adults, with an incubation period of 1 day and a duration of 8-10 days. The symptoms are sudden swelling, congestion, watering and pain in the eyes. Subconjunctival hemorrhage is a characteristic feature. Complete recovery is the rule and is

Table 62.5: Summary of Clinical Syndromes caused by Major Enterovirus Groups

Syndrome	*Polioviruses*	*Coxsackie A virus*	*Coxsackie B virus*	*Echoviruses*	*Enterovirus types 68-71*
Aseptic meningitis	1-3	Many	1-6	Many	71
Paralysis	1-3	7, 9	2-5	2, 4, 6, 9, 11, 30	70, 71
Encephalitis	1-3	2, 5-7, 9	1-5	2, 6, 9, 19	70, 71
Fever with rash	-	9, 16, 23	-	4, 6, 9, 16	-
Herpangina	-	1-6, 8, 10	-	-	-
Hand, foot and mouth disease	-	5, 10, 16	-	-	71
Upper respiratory infection	-	21	-	11, 20	-
Pneumonitis, bronchiolitis	-	-	-	-	68
Bornholm disease	-	-	1, 5	-	-
Myocarditis, pericarditis	-	-	1, 5	-	-
Acute hemorrhagic conjunctivitis	-	24	-	-	70

usually complete in 3-7 days. Radiculomyelopathy has been reported as a complication from India. Sometimes it leads to paralysis resembling poliomyelitis.

Enterovirus 70 is the chief cause of acute hemorrhagic conjunctivitis. It grows only on cultured human cells (human embryonic kidney or HeLa) on primary isolation, but can be adapted to grow on monkey kidney cells. Coxsackievirus type A24 also produces the same disease.

Enterovirus 71: Enterovirus 71 has been isolated from patients with meningitis, encephalitis, and paralysis resembling poliomyelitis. It is one of the main causes of central nervous system disease, sometimes fatal, around the world.

In some areas—particularly in Japan, Taiwan, and Sweden—the virus has caused outbreaks of hand-foot-and-mouth disease and herpangina. In an epidemic in Taiwan in 1998, the chief neurologic complication was brain stem encephalitis, and most fatalities were due to pulmonary edema and hemorrhage.

Various clinical syndrome associated with enteroviruses are given in Table 62.5.

RHINOVIRUSES

Rhinoviruses are the most important cause of the common cold and upper respiratory tract infections. These viruses as well as coronaviruses, adenoviruses, enteroviruses, parainfluenza viruses, and influenza viruses cause upper respiratory tract infections, including the common cold syndrome. Recently the rhinoviruses have been associated with acute exacerbations of asthma. These viruses are of major economic significance because they cause the loss of many million human-hours of work.

Properties of the Virus

Rhinoviruses resemble other picornaviruses in size and structure. They differ from enteroviruses in being more acid labile, but more heat stable. Inactivation of rhinoviruses occurs below pH 6.0 and is more rapid at lower pH. Complete inactivation occurs at pH 3.0. They are relatively stable in the range from 20 to 37°C (Table 62.2).

By neutralization tests, they have been classified into over 100 serotypes. Immunity is type-specific.

Host Range and Growth

Rhinoviruses can be grown in tissue cultures of human or simian origin with cytopathic changes, if good oxygenation (achieved by rolling), low pH (around 7) and low temperature (33°C) are provided. Rhinoviruses were classified into three groups, H, M and O depending upon growth in tissue culture. H strains grew only in human cells, while M strains grew equally well in human and monkey cells. O strains could be grown only in nasal or tracheal ciliated epithelium. This classification is no longer in use as the growth characteristics are not stable and can be changed by adaptation.

Table 62.6: Illness associated with recently identified enteroviruses

Enterovirus type	*Clin ical illness*
68	Pneumonia and bronchiolitis
69	Isolated from an ill person in Mexico
70	Acute haemorrhagic conjuctivitis
70, 71	Paralysis, meningo-encephalitis
71	Hand, foot and mouth disease
72[a]	Hepatitis A

[a]Reclassified as Hepatavirus

Pathogenesis

The virus enters via the upper respiratory tract. Infection can be initiated by as little as one infectious viral particle. The virus attaches to receptors on nasal ciliated epthelial cells, enters and replicates within them, spreading to other cells. Local inflammation and cytokines may be responsible for the symptoms of common cold. Immunity to rhinoviruses is transient and is unlikely to prevent subsequent infection because of the numerous serotypes of the virus.

Clinical Syndromes

The incubation period is brief-from 2 to 4 days. The acute illness usually lasts for 7 days although a nonproductive cough may persist for 2-3 weeks. The average adult has 1-2 attacks each year. Usual symptoms in adults include sneezing, nasal obstruction, nasal discharge, and sore throat; other symptoms may include headache, mild cough, malaise, and a chilly sensation. There is little or no fever. The nasal and nasopharyngeal mucosa become red and swollen, and the sense of smell becomes less keen. Secondary bacterial infection may produce acute otitis media, sinusitis, bronchitis, or pneumonitis, especially in children. There are no distinctive clinical findings that permit an etiologic diagnosis of colds caused by rhinoviruses versus colds caused by other viruses.

Laboratory Diagnosis

The clinical syndrome of the common cold is usually so characteristic that laboratory diagnosis is unnecessary.

a. *Specimens:* Nose and throat swabs in virus transport medium are the specimens of choice for the recovery of virus from all age groups. Nasopharyngeal aspirates are excellent specimens from children.
b. *Culture:* Cell cultures of human origin such as MRC5 or WI38 are preferred for the isolation of rhinoviruses. Cultures are incubated at 33°C and observed microscopically for a CPE.
c. *Serology:* Serology is not feasible because of the multiplicity of serotypes and the lack of a common antigen.
d. *Nucleic acid detection:* Nucleic acid detection is likely to become a significant tool in diagnosis.

Treatment and Prophylaxis

Antiviral drugs are thought to be a more likely control measure for rhinoviruses because of the problems with vaccine development. Pleconaril is one such drug showing activity against rhino viruses and enteroviruses.

Hand washing and the disinfection of contaminated objects are the best means of preventing the spread of the virus.

KNOW MORE

- Most enteroviruses are host-specific, infecting only one or a few related species. Two outstanding characteristics of the viruses are their affinity for nervous tissue and the narrow host range, with only humans and primates susceptible.

☞ KEY POINTS

- The family picornaviridae comprises a large number of very small, nonenveloped, icosahedral viruses, which contain a single-stranded, nonsegmented, positive-sense RNA genome and four structural proteins.
- The Picornaviridae family has divided into six genera: *Enterovirus, Rhinovirus, Cardiovirus, Aphthovirus, Hepatovirus and Parechovirus*.
- *Enteroviruses* include the polioviruses, echoviruses, coxsackieviruses.

Poliovirus

- Poliovirus is the causative agent of poliomyelitis.
- Virus is ingested and multiplies initially in the lymphoid tissue of the tonsil or Peyer's patches in the small intestine. It then spreads to the regional lymph nodes and enters the blood stream (minor or primary viremia). Primary viremia spreads the virus to receptor-bearing target tissues, where a second phase of viral replication may occur, resulting in symptoms and a secondary viremia.
- Prevention of poliomyelitis is accomplished by an inactivated (killed) injectable vaccine (Salk vaccine) or an attenuated live, orally administered vaccine (Sabin vaccine), which confers immunity by raising neutralizing antibody. The simultaneous administration of oral polio vaccine to all children in a region on the same day *(pulse immunization)* has been found to be useful to interrupt the transmission of wild poliovirus by displacing it from intestine, where the wild poliovirus multiply.

Coxsackieviruses

- Based on the pathological changes produced in suckling mice, they are classified into two groups A and B. Group A viruses produce a generalized myositis and flaccid paralysis leading to death of suckling mice. Group B produces a patchy focal myositis, spastic paralysis, necrosis of brown fat, pancreatitis, hepatitis and myocarditis.
- The clinical manifestations of infection with various coxsackieviruses are diverse and may present as distinct disease entities such as herpangina (vesicular pharyngitis), aseptic meningitis, hand-foot-and-mouth disease, respiratory infections, pleurodynia, myocardial and pericardial infections, juvenile diabetes, orchitis, hepatitis, meningoencephalitis and adrenocortical involvement chronic fatigue syndrome.

Echoviruses

- Echoviruses are found in the intestinal tract of the infected humans.
- Most echovirus infections are asymptomatic but some have been associated with clinical syndromes such as aseptic meningitis, paralysis, rash and fever, respiratory illnesses in children, infantile diarrhea, pericarditis and myocarditis, conjunctivitis, muscle weakness, and spasm.
- Enterovirus type 68 causes pneumonia and bronchitis, type 70 causes acute hemorrhagic conjunctivitis (AHC), and type 71 causes meningoencephalitis and paralysis; and type 69 does not cause any human diseases.

Rhinoviruses

- Rhinoviruses are the most important causative agents of the common cold and upper respiratory tract infections.
  They differ from other picornavirus in being more acid labile and more heat stable. Rhinovirus causes common cold after an incubation period of 2-4 days.
- The rhinovirus is identified by their typical cytopathic effect and demonstration of acid liability.

IMPORTANT QUESTIONS

1. Classify the picornaviruses. Discuss pathogenesis and laboratory diagnosis of poliomyelitis.
2. Write short notes on:
   Prophylaxis against poliomyelitis
   Coxsackieviruses
   Echoviruses
   Enterovirus 70
   Acute hemorrhagic conjunctivitis
   Rhinoviruses.

FURTHER READING

Pallansch MA, Roos RP. Enteroviruses: Polioviruses, coxsackieviruses, echoviruses and newer enteroviruses. In: Fields Virology. 4th ed. Knipe OM et al [edirors]. Lippincott Williams & Wilkins 2001.

Rotbart, HA. Human enterovirus infections, Washington, DC: ASM Press 1995.

Tracy S, Chapman NM, Mahy BW. Coxsackie B viruses. Current topics in microbiology and immunology, Springer-Verlag. Vol 223, Berlin, 1997.

Yin-Murphy M. Acute hemorrhagic conjunctivitis Progr Med Virol 1984;29:23.

CHAPTER 63

Orthomyxovirus

Learning Objectives

After reading and studying this chapter, you should be able to:

- Differentiate between orthomyxoviruses and paramyxoviruses.
- Describe morphology of influenza virus.
- Discuss types and subtypes of orthomyxoviruses.
- Describe the following: Hemagglutinin (H) and neuraminidase (NA); antigenic variation in Influenza virus; antigenic shift and antigenic drift.
- Discuss laboratory diagnosis of influenza.
- Describe the following: Influenza pandemics; prophylaxis against influenza; influenza vaccines.

INTRODUCTION

The name myxovirus was used originally for a group of enveloped RNA viruses characterised by their ability to adsorb onto mucoprotein receptors on erythrocytes, causing hemagglutination. The name referred to the affinity of the viruses to mucins (from myxa, meaning mucus). Despite having certain similarities, the orthomyxoviruses and the paramyxoviruses are separated into two distinct groups because of fundamental differences in their structures and their patterns of replication. Table 63.1 lists the important differences between orthomyxovirus and paramyxovirus.

Properties of the Orthomyxoviruses

The family orthomyxoviridae comprises four genera: influenza A, B and C viruses and thogotoviruses. Antigenic changes continually occur within the type A group of influenza viruses and to a lesser degree in the type B group, whereas type C appears to be antigenically stable.

Influenza A—Influenza A strains are also known for aquatic birds, pigs, horses, and seals. Influenza A viruses can infect a variety of different host species.

Influenza B—Only infects humans.

Influenza C—Although assumed to be primarily a human infection, has been isolated from pigs in China.

Thogotoviruses—Form a newly discovered fourth genus of the orthomyxovirus family and are found in mosquitoes, ticks and the banded mongoose.

INFLUENZA VIRUSES

Morphology

Virus—The influenza virus is typically spherical, with a diameter of 80 to 120 nm but pleomorphism is common. Filamentous forms, upto several micrometers in length and readily visible under the dark ground microscope, are frequent in freshly isolated strains.

Genome—The virus core consists of ribonucleoprotein in helical symmetry. The negative sense single stranded RNA genome is segmented and exists as eight pieces. Also present within the virion is the viral RNA-dependent RNA polymerase: this is essential for infectivity as the virion RNA is of negative sense.

Envelope—The nucleocapsid is surrounded by an M1 protein shell, immediately exterior to which is a lipid envelope derived from the host cell. The M2 protein projects through the envelope to form ion channels, which allow pH changes in the endosome. The protein part of the envelope is virus coded but the lipid layer is derived from the modified host cell membrane, during the process of replication by budding.

Peplomers—Projecting from the envelope are two types of spikes (peplomers): hemagglutinin (HA) spikes which are triangular in cross section and the mushroom shaped neuraminidase (NA) peplomers which are less numerous (Fig. 63.1). These two surface glycoproteins are the important antigens that determine antigenic variation of influenza viruses and host immunity. The HA represents about 25 percent of viral protein, and the NA about 5 percent.

Resistance

The influenza virus withstands slow drying at room temperature on articles such as blankets and glass. It can be preserved for long periods at –70°C, and remains viable indefinitely when freeze-dried.

Influenza viruses may be stored at 0 to 4°C for weeks without loss of viability. Exposure to heat for 30 min at

Table 63.1: Differences between orthomyxoviruses and paramyxoviruses		
Property	*Orthomyxovirus*	*Paramyxovirus*
Size of virion	80-120 nm	100-300 nm
Shape	Spherical; filaments in fresh isolates	Pleomorphic
Genome	Segmented; eight pieces of RNA	Single linear molecule of RNA
Diameter of nucleocapsid	9 nm	18 nm
Site of synthesis of ribonucleoprotein	Nucleus	Cytoplasm
DNA-dependent RNA synthesis	Required for multiplication	Not required
Effect of actionomycin D	Inhibits multiplication	Does not inhibit
Antigenic stability	Variable	Stable
Hemolysin	Absent	Present

56ºC is sufficient to inactivate most strains. The viruses are inactivated by a variety of substances, such as 20 percent ether in the cold, phenol, formaldehyde, salts of heavy metals, detergents, soaps, halogens and many others. Iodine is particularly effective.

Antigenic Structure

The antigens of the influenza virus can be classified as the internal antigens and the surface antigens.

A. Internal Antigens

1. Ribonucleoprotein (RNP) antigen—The internal antigen is the ribonucleoprotein and is hence called the RNP antigen. It was also called the 'soluble' (S) antigens because it is found free in infected tissues and occurs in the supernatant when the virus containing fluid is centrifuged.

It is type-specific and based on its nature, influenza viruses are classified into types A, B and C. The RNP antigens of types A, B and C are distinct but all strains of anyone type possess the same antigen. The RNP antigen is stable and does not exhibit any significant antigenic variation. Anti-RNP antibody develops after infection but not following killed vaccines.

2. Matrix (M) protein—M protein antigen is also type-specific like the RNP antigen and distinct for A, B and C types of influenza viruses. The envelope lipid antigen is host-specific and is determined by the species in which virus replication takes place. The M1 proteins line the inside of the virion and promote assembly. The M2 protein forms a proton channel in membranes and promotes uncoating and viral release. The M2 of influenza A is a target for the antiviral drugs amantadine and rimantadine.

B. Surface Antigens

The term 'viral' or V antigen was formerly used to describe the surface antigen of the influenza virus. The V antigen is actually composed of at least two virus coded proteins, the hemagglutinin and the neuraminidase. These two surface glycoproteins are the important antigens that determine antigenic variation of influenza viruses and host immunity.

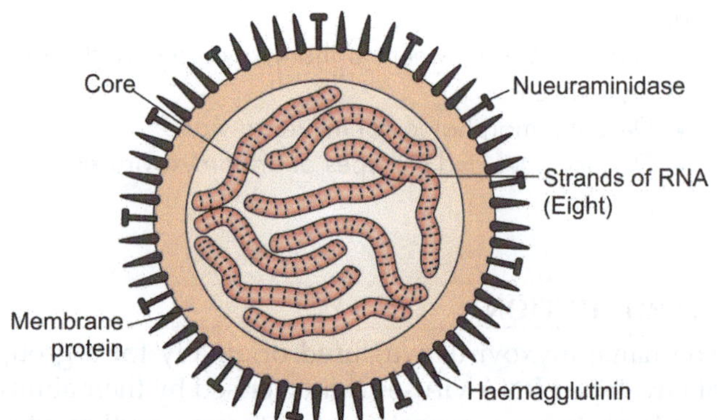

Fig. 63.1: Diagrammatic representation of influenza virus

1. Hemagglutinin (HA)—Hemagglutinin (HA1) is a glycoprotein composed of two polypeptides - HA1 and HA2. The HA has several functions. It is the viral attachment protein responsible for hemagglutination and hemadsorption. It enables the virus to adsorb to mucoprotein receptors on red cells as well as on respiratory epithelial cells. Antihemagglutinin antibodies are produced following infection and immunization.This antibody is protective by preventing adsorption of the virus to cells.The hemagglutinin is a strain specific antigen and is capable of great variation. Mutation-derived changes in HA are responsible for the minor ("drift") and major ("shift") changes in antigenicity. Fifteen distinct HA subtypes, named H1 to H15 have been identified in avian influenza viruses, but only four of them have been found in human isolates so far. *Shifts occur only with influenza A virus, and the different HAs are designated Hl, H2, and so on.*

2. Neuraminidase (NA)—Neuraminidase is a glycoprotein enzyme which destroys cell receptors by hydrolytic cleavage. Neuraminidase activity is also thought to be important in the final stages of release of new virus particles from infected cells. The antineuraminidaze antibody is formed following infection and immunization. It is not as effective in protection as the antihemagglutinin antibody. It does not prevent the adsorption of virus onto cells but can inhibit the release and spread of progeny virions and may thus contribute to limiting the infection. It is a strain-specific antigen and exhibits variation. Major differences acquire the designations N1, N2, and so on. Nine different subtypes have been identified (NI-N9).

Antigenic Variation

Influenza viruses are remarkable because of the frequent 'antigenic, changes that occur in HA and NA.This is of great importance in the epidemiology of the disease. Antigenic variability is highest in influenza virus type A and less in type B, while it has not been demonstrated in type C.

The internal RNP antigen and M protein antigen are stable but both the surface antigens, hemagglutinin and neuraminidase, undergo independent antigenic variations, which may be of two types antgenic drift (minor antigenic changes) antigenic shift (major antigenic changes) in HA or NA result in the appearance of a new subtype. Antigenic shift is most likely to result in an epidemic.

Antigenic drift—Antigenic drift refers to minor antigenic changes either in hemagglutinin or neuraminidase or both. It is the gradual sequential change in antigenic structure occurring regularly at frequent intervals. Here, the new antigens, though different from the previous antigens, are yet related to them, so that they react with antisera to the predecessor virus strains, to varying degrees. Antigenic drift is due to mutation and selection, the process being influenced by the presence of antibodies to the predecessor strains in the host population.

Antigenic drift accounts for the periodical epidemics of influenza.

Antigenic shift—Antigenic shift, is an abrupt, drastic, discontinuous variation in the antigenic structure, resulting in a novel virus strain unrelated antigenically to predecessor strains. Such changes may involve hemagglutinin, neuraminidase or both.

The mechanism for shift is genetic reassortment between human and avian influenza viruses. Antibodies to predecessor viruses do not neutralise the new variants and can, therefore, spread widely in the population causing major epidemics or pandemic. Influenza B and C viruses do not exhibit antigenic shift because few related viruses exist in animals.

Antigenic Classification

Antigenic differences exhibited by two of the internal structural proteins, the nucleocapsid (NP) and matrix (M) proteins, are used to divide influenza viruses into types A, B, and C. These proteins possess no cross reactivity among the three types. Antigenic variations in the surface glycoprotens, hemagglutinin (HA)and neuraminidase (NA), are used to subtype the viruses. Only type A has designated subtypes. Influenza virus type A strains can be classified into subtypes based on variations in their surface antigens.

Nomenclature

There is standard nomenclature system for influenza virus isolates. The complete designation of a strain will include: the type, host of origin, place of origin, serial number and year of isolation followed by the antigenic subtypes of the hemagglutinin and neuraminidase in parenthesis e.g. A/swine/Iowa/3/70 (H1N1). If isolated from a human host the origin is not given, e.g. A/Scotland/42/89 (H3N 2).So far, 15 subtypes of HA (HI-H 15) and nine subtypes of NA (NI-N9), in many different combinations, have been recovered from birds, animals, or humans. Four HA (HI-H3, H5) and two NA (NI, N2) subtypes have been recovered from humans.

Influenza virus type B also exhibits antigenic variation but the changes have not been marked enough for the subtypes to be delineated. The type C virus does not undergo any significant antigenic variation.

Host Range

1. Animals—The human influenza virus can cause experimental infection in a number of animal species. Intranasal inoculation in ferrets produces an acute respiratory disease. The virus can be 'adapted' by serial intranasal passage in mice to produce fatal pulmonary infection.

2. Egg inoculation—The virus grows well in the amniotic cavity of chick embryos. After a few egg passages, the virus grows well in the allantoic cavity also, except for the type C virus which does not generally grow in the allantoic cavity. The influenza virus does not damage chick embryos, which may hatch out normally. Virus growth is detected by the appearance of hemagglutinin in the allantoic and amniotic fluids.

3. Cell culture—The virus grows in primary monkey kidney cell cultures, as well as in some continuous cell lines. Cytopathic effects are not prominent and virus growth is detected by hemadsorption or demonstration of hemagglutinin in the culture fluid.

Von Magnus phenomenon—When passaged serially in eggs, using as inocula undiluted infected allantoic fluid, the progeny virus will show high hemagglutinin titers, but low infectivity. This has been called the Von Magnus phenomenon and is due to the formation of incomplete virus particles lacking nucleic acid.

Pathogenesis

Influenza virus spreads from person-to-person by airborne droplets. The viral neuraminidase facilitates infection by reducing the viscosity of the mucus film lining the respiratory tract and exposing the cell surface receptors for virus adsorption. The ciliated cells of the respiratory tract are the main sites of viral infection. Within a short time, many cells in the respiratory tract are infected and eventually killed. Influenza infections cause cellular destruction and desquamation of superficial mucosa of the respiratory tract. This renders the respiratory tract highly vulnerable to bacterial invasion, especially staphylococci, streptococci, and Hemophilus influenzae. Viral pneumonia, seen only in the more severe cases.

Clinical Features

A. Uncomplicated Influenza

The incubation period is 1 to 3 days. The disease varies in severity from a mild coryza to fulminating and rapidly fatal pneumonia. Most infections are subclinical. Symptoms of classic influenza usually appear abruptly and include chills, headache, and dry cough, followed closely by high fever, generalized muscular aches, malaise, and anorexia. The fever usually lasts 3 to 5 days, as do the systemic symptoms.

In children—Clinical symptoms of influenza in children are similar to those in adults, although children may have higher fever and a higher incidence of gastrointestinal manifestations (abdominal pain and vomiting). Finally, otitis media may develop.

The uncomplicated disease resolves within about seven days. The similarity to influenza of the prodromal stages of several infections has led to the use of the term 'flu-like' to describe these features.

Complications—Complications of influenza include primary viral pneumonia, secondary bacterial pneumonia, myositis and cardiac complications, such as congestive failure or myocarditis and, neurological involvement, such as Guillain-Barre syndrome, encephalopathy, encephalitis and Reye's syndrome may occur.

Reye's Syndrome

Influenza, particularly infection with type B, has been associated with Reye's syndrome. It is an acute encephalopathy of children and adolescents, usually between 2 and 16 years of age and is characterized by acute degenerative changes in the brain, liver and kidneys.

Type B infections may sometimes cause gastrointestinal symptoms (gastric flu).

Laboratory Diagnosis

Diagnosis of influenza relies on isolation of the virus, identification of viral antigens or viral nucleic acid in the patient's cells, or demonstration of a specific immunologic response by the patient.

1. Demonstration of the Virus Antigen

Rapid diagnosis of influenza may be made by demonstration of the virus antigen on the surface of the nasopharyngeal cells by immunofluorescence.This test is rapid but is not as sensitive as viral isolation.

Detection of influenza RNA by reverse transcriptase polymerase chain reaction may be more sensitive than antigen detection but is not widely available in diagnostic laboratories.

2. Isolation of the Virus

Nasal washings, gargles, and throat swabs are the best specimens for viral isolation and should be obtained within 3 days after the onset of symptoms but less often in later stages. Throat garglings are collected using broth saline or other suitable buffered salt solution.The sample should be held at 4°C until inoculation into cell culture, or if the delay is long, at –70°C. The specimen should be treated with antibiotics to destroy bacteria. Isolation may be made in eggs or in monkey kidney cell culture.

The material is inoculated into the amniotic cavity of 11 to 13 day old eggs, using at least six eggs per specimen. After incubation at 35°C for three days, the eggs are chilled and the amniotic and allantoic fluids harvested separately. The fluids are tested for hemagglutination using guineapig and fowl cells in parallel, at room temperature and at 4°C. Some strains of the influenza virus type A agglutinate only guineapig cells on initial isolation. The type B virus agglutinates both cells, while type C strains agglutinate only fowl cells at 4°C. Subtype identification is made by hemagglutination inhibition test. Some of the recent type A strains can be isolated by direct allantoic inoculation of the clinical specimen into 9 to 11 day old eggs. However, type B and C viruses will be missed if only allantoic inoculation is used.

For primary isolation the most suitable cells are primary monkey kidney or human embryo kidney cells, but since these tissues are scarce most laboratories now use secondary baboon kidney cells or Madin-Darby canine kidney cells. Incubation at 33°C in roller drum is recommended. The presence of virus may be detected by hemadsorption with human O group, fowl or guineapig red blood cells. Rapid results can be obtained by demonstrating virus antigen in infected cell cultures by immunofluorescence.

3. Serology

Complement fixation tests (CFTs) and hemagglutination inhibition (HI) tests are employed for the serological diagnosis of influenza. Paired acute and convalescent sera are necessary, because normal individuals usually have influenza antibodies. A fourfold or greater increase in titer must occur to indicate influenza infection.

i. *Complement fixation tests* with the RNP antigen of influenza virus types A, B and C are very useful as the antibodies are formed during infection only, and not following immunization with inactivated vaccines. Because of its complexity, CF tests are now used only rarely.
ii. *Hemagglutination inhibition (HI)* is a convenient and sensitive test for the serological diagnosis of influenza. The major drawback is the frequent presence in the sera of nonspecific inhibitors of hemagglutination. The sera suitably treated for the

removal of nonspecific inhibitors, are diluted serially in hemagglutination plates and the influenza virus suspension containing 4 HA units added to each cup. Fowl red cells are then added. The highest dilution of serum that inhibits hemagglutination is its HI titer.

iii. *Neutralization tests* are the most specific and the best predictor of susceptibility to infection.
iv. ELISA test is more sensitive than other assays.
v. Neuraminidase antibody by *enzyme neutralization tests*.
vi. Radial immunodiffusion tests in agarose gel have been described for the identification of antibodies to the RNP antigen, hemagglutinin and neuraminidase.

Immunity

Immunity to influenza is long-lived and subtype-specific. Antibodies against HA and NA are important in immunity to influenza, whereas antibodies against the other virus-encoded proteins are not protective.

An attack of influenza confers protection effective for about one or two years. The apparent short duration of immunity is due to the antigenic variation that the virus undergoes frequently. Following infection and immunization, circulating antibodies are formed against the various antigen of the virus. However, the local concentration of antihemagglutinin and, to a smaller extent, of antineuraminadase antibodies (mainly IgA) in the respiratory tract that is more relevant in protection.

When an individual experiences repeated infections with different antigenic variants of influenza virus type A, he responds by forming antibodies not only against each infecting strain but also against the strain that he first comes into contact with. The dominant antibody response will be against the strain that caused the earliest infection. This phenomenon called the doctrine of "original antigenic sin".

Influenza virus infection induces cell-mediated immunity also, but its role in protection has not been clarified.

Epidemiology

Influenza viruses occur worldwide: epidemics are local; pandemics are worldwide. Influenza infection is spread readily via small airborne droplets expelled during talking, breathing, and coughing.The most susceptible population is children, and school-aged children are most likely to spread the infection. Children, immunosuppressed people (including pregnant women), the el-derly, and people with heart and lung ailments (including smokers) are at highest risk for more serious disease, pneumonia, or other complications of infection.

The three types of influenza vary markedly in their epidemiologic patterns. Influenza C is least significant; Influenza B sometimes causes epidemics, but influenza type A can sweep across continents and around the world in massive epidemics called pandemics.

New influenza A strains are generated through mutation and reassortment. An exchange of the HA glycoproteins may generate a new virus that can infect an immunologically naive human population. For example, an HINI duck virus and an H3N2 human virus infected pigs, reassortants were isolated from the pig, and the resulting virus was able to infect humans.

The mere appearance of a new or hybrid strain may not lead to a pandemic. For this, the new strain.should be capable of spreading rapidly among people. The swine flu virus HI N1 caused a localised outbreak in a military camp in New Jersey, USA in 1976 but it did not spread. In 1997 in Hong Kong, the first documented infection of humans by avian influenza A virus (H5Nl) was isolated form at least I8 infected humans, six of whom died.

The virus resembled a chicken virus, AlChickenI-Hong Kong/258/97 (H5Nl), leading to the destruction of all 1.6 million chickens in Hong Kong to destroy the potential source of the virus.

Influenza pandemics have been recorded at irregular intervals from 1173. Pandemics of modern times date from 1889 probably caused by an H2N8 subtype and the epidemic of 1900 by an H3N8 virus. The most severe pandemic in recorded history occurred in 1918-1919 ('Spanish flu'), caused by the abrupt appearance of the HI N 1 subtype, the swinelike influenza during which over 200 million people were affected and more than 20 million perished. India suffered the most, with some 10 million deaths. An unusual feature of this pandemic was the very high rate of mortality among young adults. Mild epidemics occurred around 1933 and 1946 associated with minor variations in the H antigen (from Hsw to HO in 1933, HO to HI in 1946). Subsequent antigenic shifts have been documented by viral isolations; H2N2 (Asian flu) appeared in 1957 originated in China and spread throughout the world within a short period. It was replaced in 1968 by the H3N2 subtype (Hong Kong flu) also caused a pandemic but it was much less severe. The HINlstrain reappeared in 1977 (Russian flu). From 1977, both H3N2 and HINI viruses have been circulating together. Fig. 63.2 lists the sequence of appearance of these various subtypes.

A unique feature of influenza epidemiology was that once an antigenic variant emerged, it displaced completely the preexisting strain. However, this rule has not been observed in recent years.Until 1977, when HINI reappeared, it was the rule that when a 'new' virus appeared the 'old' one disappeared, but since that time two subtypes have been circulating concurrently, namely H3N2 and H1.N1.

Cases of influenza have appeared in Mexico and then in USA in April 2009. These were first thought to be due to swine influenza A virus (H1N1) but it was identified as a new H1N1 virus different from swine influenza virus. It can be transmitted from human to human. It spread to more than 100 countries across the world and WHO declared the situation as pandemic.

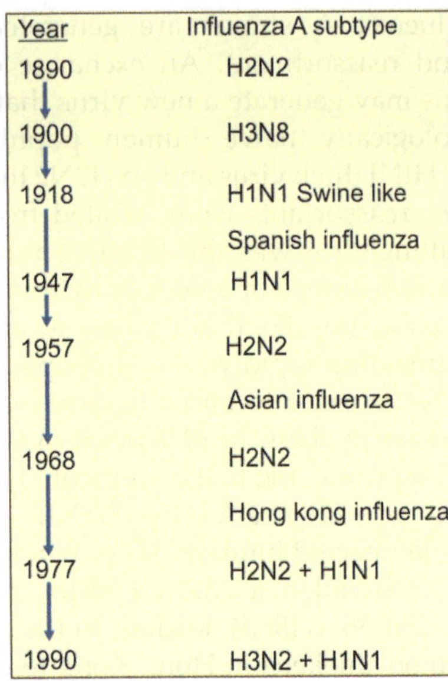

Fig. 63.2: Sequential changes in influenza A antigens associated with antigenic shift. Antigenic drift occurs after the appearance of a new subtype

Till 6th July 2009, 94,512 laboratory confirmed human cases including 429 deaths have been officially reported from 123 countries. Among these cases, 33,902 cases including 170 deaths have been reported only from United states. Mexico reported 10,262 cases including 119 deaths,. Canada 7,983 cases with 25 death. There are 7,376 reported cases and 14 deaths from Chile. Rest of the cases have been reported from remaining countries. In India , 129 laboratory confirmed cases were reported without any death. The pandemic is still continuing. This pandemic has occured 41 years after the last pandemic in 1968. Ostelamivir (Tamiflu) is the drug used for treatment. However, H1N1 strains resistant to ostelamivir have been reported from Denmark, Japan, Hongkong and China.

Prophylaxis

A. Chemoprophylaxis

Chemoprophylaxis has been reported to be successful with the antiviral drugs amantadine and rimantadine which block the viral M2 protein which functions as an ion channel. These act only with type A virus and not with type B, which lacks the M2 components.

B. Influenza Vaccines

Influenza vaccines have been in use for many decades. The aim of immunization is to produce hemagglutination inhibiting or neutralizing antibody in all vaccines. However, certain characteristics of influenza viruses make prevention and control of the disease by immunization especially difficult. Existing vaccines are continually being rendered obsolete as the viruses undergo antigenic drift and shift. Vaccines are of two types.

a. Inactivated Viral Vaccines

Vaccines are either whole virus (WV), subvirion (SV), or surface antigen preparations.

Whole virus (WV) vaccine—Inactivated vaccines are prepared from appropriate strains of influenza A and B grown in the chick embryo allantoic cavity. The infected fluids are harvested, purified by ultracentrifugation, and inactivated with formalin or β-propiolactone. Whole virus vaccine should not be given to those who are allergic to egg protein.

'Subunit' vaccines—The H and N antigens (purified HA and NA) may be separated from the whole virus by treatment with ether and detergent, and these subunit or split-virus vaccines are better tolerated, especially in young children.

Indications

1. **Annual influenza vaccination is recommended for high-risk groups**—These include individuals at increased risk of complications associated with influenza infection (those with either chronic heart or lung disease, including children with asthma, or metabolic or renal disorders; residents of nursing homes; and those 65 years of age and older) and persons who might transmit influenza to high-risk groups (medical personnel, employees in chronic care facilities, household members).
2. **Pandemic threat**—The most important indication for immunoprophylaxis is when a pandemic is threatened by a new virus. Here, the time taken for the manufacture of the vaccine with the new variant is crucial, as the virus is likely to spread fast and infect whole populations before the vaccine becomes available. To overcome these hurdles, recombinant vaccine has been introduced.

Contraindication—The only contraindication to vaccination is a history of allergy to egg protein.

b. Live-attenuated Vaccines

1. The earliest live vaccine was the virus attenuated by repeated egg passage. It was administered by intranasal instillation. However, it sometimes gave rise to clinical disease, especially in childern. These vaccines have been generally effective in provoking a good local (IgA) antibody response but are not at present widely used.
2. Use of temperature sensitive mutants—Another approach is to live vaccine is the use of temperature sensitive mutants. A cold-adapted donor virus, able to grow at 25°C but not at 37°C the temperature of the lower respiratory tract-should replicate in the nasopharynx, which has a cooler temperature (33°C). A live attenuated, cold-adapted, trivalent influenza virus vaccine administered by nasal spray has proved effective in clinical trials in children.

Treatment

The antiviral drug amantadine and its analog rimantadine inhibit an uncoating step of the influenza A virus but do not affect the influenza B or C virus. The target for their action is the M2 protein. Resistance to these drugs develops rapidly.

Zanamivir and oseltamivir (Tamiflu) inhibit both influenza A and B as enzyme inhibitors of the neuraminidase.Zanamivir is inhaled, whereas oseltamivir is taken orally as a pill. These drugs are effective for prophylaxis and for treatment during the first 24 to 48 hours after the onset of influenza A illness. Treatment cannot prevent the later host-induced immunopathogenic stages of the disease.

KNOW MORE

Hemagglutination

Hemagglutination is an important characteristic of influenza viruses. The virus is adsorbed onto the mucoprotein receptors on the cell surface when mixed with a suspension of fowl erythrocytes. The virus links together adjacent cells producing hemagglutination. The hemagglutinin peplomers on the viral surface are responsible for this activity. Hemagglutination is followed after a time by the detachment of the virus from the cell surface, reversing the hemagglutination. This process is known as elution and is caused by the enzyme neuraminidase (sialidase) present on the viral surface. The enzyme acts on the cell receptor, destroying it by splitting off N-acetylneuraminic acid from it.

Virus particles which have eluted from red cells are still capable of agglutinating fresh red cells but red cells that have been acted on by the virus are not susceptible to agglutination by the same strain of the virus. Such red cells may, however, be agglutinated by other myxoviruses. The inability of these red cells to be reagglutinated by the same virus is due to the destruction of the specific cell receptors by the initial treatment with the virus. Myxoviruses can be arranged in a series in which the treatment of red cells with anyone virus removes the receptors for that virus and the preceding viruses but not for the viruses later in the series. This is called the 'receptor gradient'. For myxoviruses in general, the gradient is mumps, Newcastle disease virus and influenza, in that order.

Hemagglutinin is more resistant to physical and chemical agents than infectivity. Therefore, hemagglutination can be used for the titration of the inactivated influenza virus also, as, for example, in the standardization of killed influenza virus vaccines.

Hemagglutination inhibition (HI) offers a convenient method for the detection and quantitation of the antibody to the virus. A disadvantage of this serological technique is the frequent presence in sera of certain substances that cause nonspecific inhibition of hemagglutination. Different kinds of nonspecific inhibitors have been identified in sera and have been given names such as alpha (Francis), beta (Chu) and gamma (Shimojo) inhibitors. They are mostly glycoproteins. A variety of techniques has been introduced for inactivating them without affecting the antibody content of sera. These include treatment with RDE, trypsin, potassium periodate, kaolin and CO_2. No single method has been found effective in completely destroying inhibitors to all types of viruses from all kinds of sera.

Hemagglutination and elution can be used for purifying and concentrating influenza viruses.

The plasma membranes of tissue culture cells in which the virus is multiplying contain the hemagglutinin. Therefore, red cells are adsorbed onto the surface of such cells. This is the basis of hemadsorption, a technique by which the growth of the influenza virus in cell cultures can be identified.

P-Q-R Variation

Influenza virus strains belonging to the same subtype even strains isolated during the course of a single outbreak - may behave differently in neutralization tests with antisera. Van der Veen and Mulder called this the P-Q-R variation. Strains in the P phase were neutralized by the homologous antiserum in high titers and by heterologous antiserum in low titers. Strains in the Q phase were neutralized poorly by either homologous or heterologous sera. R phase strains were neutralized by both homologous and heterologous sera in high titers.

O-D Variation

Burnet and Bull (1943) observed that influenza virus type A underwent certain changes when serially passaged in eggs. They called this the O-D variation. The O-D variation was considered a result of mutation.The fresh isolate was said to be in the 'Original' (O) phase and the passaged virus in the 'Derived' (D) phase.

☞ KEY POINTS

- The family orthomyxoviridae comprises four genera: influenza A, B and C viruses and thogotoviruses.
- Influenza viruses are spherical or filamentous, enveloped particles. The virion contains an RNA-dependent RNA polymerase. The viral genome is a single-stranded antisense RNA, characteristically; it is segmented and and exists as eight pieces. Virion has a lipoprotein envelope containing hemagglutinin (HA) triangular in cross section and mushroom shaped neuraminidase (NA) peplomers.
- The influenza viruses are of three types: A, B, and C.

- Hemagglutination is an important characteristic of influenza viruses.
- The antigens of the influenza virus can be classified as the internal antigens and the surface antigens(1. Hemagglutinin (HA) and neuraminidase (NA).
- HA is composed of two polypeptides HA1 and HA2 responsible for hemadsorption and hemagglutination. Fifteen subtypes of hemagglutinins (H1-H15) and nine of neuraminidases (N1-N9) are known in birds, some of which have also been found in various combinations in humans.
- Antigenic drift is the gradual sequential change in antigenic structure occurring regularly at frequent intervals and is due to mutation and selection. Antigenic drift accounts for the periodical epidemics of influenza.

Antigenic shift is an abrupt, drastic, discontinuous variation in the antigenic structure, resulting in a novel virus strain unrelated antigenically to predecessor strains. Such changes may involve hemagglutinin, neuraminidase or both. The mechanism for shift is genetic reassortment between human and avian influenza viruses. Antibodies to predecessor viruses do not neutralize the new variants and can, therefore, spread widely in the population causing major epidemics or pandemic. Influenza B and C viruses do not exhibit antigenic shift because few related viruses exist in animals.

- Influenza virus is transmitted from person to person primarily by airborne respiratory droplets. Adults: classic "flu" syndrome.
  Children: asymptomatic to severe respiratory tract infections.
- *Laboratory diagnosis* depends on demonstration of virus antigens, isolation of the virus and serology.
  i. Demonstration of the virus antigen may be made on the surface of the nasopharyngeal cells by immunofluorescence. RT-PCR technique is used for the detection of viral nucleic acid in the nasopharyngeal cells.
  ii. Isolation of the virus can be done by egg inoculation (amniotic cavity) or into certain cell cultures (e.g. primary monkey kidney or human embryo kidney cells).
  iii. Serological tests—include complement fixation test (CFT), hemagglutination inhibition test (HI), neutralization test, radial immunodiffusion test and ELISA.
- Influenza pandemics of modern times date from 1889. The most severe pandemic in recorded history occurred in 1918-1919 ('Spanish flu').
- Influenza virus vaccines have been used to prevent influenza, primarily influenza A and B. Vaccines are either inactivated viral vaccines or live virus vaccines. Killed vaccine contains predicted yearly strains of influenza A and B viruses.
- Amantadine and rimantadine are the antiviral agents effective for treatment of influenza A virus only.
- Zanamivir and oseltamivir (Tamiflu) are effective for treatment of both influenza A and B viruses

IMPORTANT QUESTIONS

1. Tabulate the differences between orthomyxoviruses and paramyxoviruses
2. Discuss the morphology and pathogenesis of influenza virus infection.
3. Write short notes on:
   Antigenic shift
   Antigenic drift
   Laboratory diagnosis of inflenza
   Prophylaxis against influenza

FURTHER READING

Helenius A. Unpacking the incoming influenza virus, Cell1 992;69:577-578.

Influenza. NIAID fact sheet [on line]. Available at http:// www.niaid.nih.gov/spotlightlfluOO/default.htm.

Lamb RA, Krug RM. Orthomyxoviridae: The viruses and their replication. In: (Eds) Knipe OM et al . Fields Virology, 4th ed, (editors). Lippincott Williams & Wilkins, 2001.

Prevention and control of influenza. Recommendations of the Advisory Committee on Immunization Practices. MMWR Morb Mortal Wkly Rep 2002;51(RR-3):1.

Webster RG. et al, Evolution and ecology of influenza viruses, Microbiol Rev1992; 56:152-79.

Wright PF, Webster RG: Orthomyxoviruses. In: Fields Virology, 4th ed. Knipe OM, et al. Lippincott Williams & Wilkins 2001.

CHAPTER

64 Paramyxoviruses

Learning Objectives

After reading and studying this chapter, you should be able to:

- Classify paramyxoviruses.
- Describe morphology of paramyxoviruses.
- Describe the following: Parainfluenza viruses; Measles virus; Respiratory syncytial virus (RSV).
- Discuss pathogenicity and complications of measles.
- Discuss Measles, Mumps and Rubella(MMR) vaccine.

INTRODUCTION

The family Paramyxoyiridae cause a variety of diseases, predominantly involving the respiratory tract, in humans, birds, and other animals. In humans, they include measles, respiratory infections caused by respiratory syncytial virus (RSV) and parainfluenza viruses, and the more innocuous salivary gland infection of mumps. These viruses, particularly RSV, cause fusion of infected cells with formation of multinucleated giant cells (syncytia). Though much less common, infections may also occur in adults.

MORPHOLOGY AND STRUCTURAL PROTEINS OF PARAMYXOVIRUSES

Paramyxoviruses resemble orthomyxoviruses in morphology but are larger and more pleomorphic. (Fig. 64.1). They are roughly spherical in shape and range in size from 100 to 300 nm, sometimes with long filaments and giant forms of upto 800 nm.

The helical nucleocapsid is much wider than in orthomyxoviruses, with a diameter of 18 nm (except in Pneumovirus. where it is 13 nm). The nucleocapsid consists of the negative-sense, single-stranded RNA associated with the nucleoprotein (NP), polymerase phosphoprotein (P), and large (L) protein. Genome of the Paramyxoviridae is an ssRNA molecule of negative polarity. Unlike the orthomyxoviruses, in which the segmented nature of the genome facilitates genomic reassortments and antigenic variation so typical of influenza viruses, the paramyxoviruses with their unsegmented genome do not undergo genetic recombinations or antigenic variations. Hence all paramyxoviruses are antigenically stable.

The nucleocapsid associates with the matrix (M) protein at the base of the lipid envelope. The virion envelope contains two glycoproteins, a fusion (F) protein, which promotes fusion of the viral and host cell membranes, and a viral attachment protein (hemagglutinin-neuraminidase [HN], hemagglutinin [H], or G protein). The F or fusion protein, formed by proteolytic cleavage of a larger precursor polypeptide, is important, since it mediates the fusion of infected cells which then form the syncytia so characteristic of infections with this group of viruses.

CLASSIFICATION

Within the family Paramyxoviridae two subfamilies, Paramyxovirinae and Pneumovirinae are recognized (Table 64.1).

Subfamily Paramyxovirinae

1. Respirovirus (*para-influenza* viruses 1, 3)
2. Rubulavirus (mumps virus, *para-influenza* viruses 2, 4a, 4b)

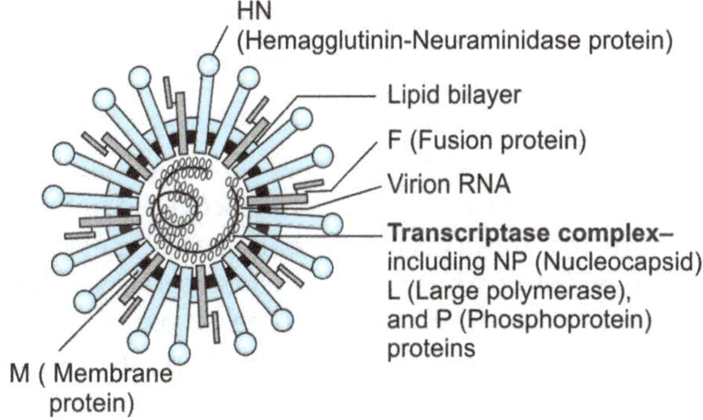

Fig. 64.1: Schematic diagram of a paramyxovirus showing major components

Table 64.1: Characteristics of genera in the subfamilies of the family paramyxoviridae

	Paramyxovirinae				Pneumovirinae	
Property	Respirovirus	Rubulavirus	Morbillivirus	Henipavirus[1]	Pneumovirus	Metapneumovirus
Human viruses	Parainfluenza 1,3	Mumps, parainfluenza 2, 4a, 4b	Measles	Hendra, Nipah	Respiratory syncytial virus	Human metapneumovirus
Serotypes	1 each	1 each	1	?	2	?
Diameter of nucleocapsid (nm)	18	18	18	?	13	13
Membrane fusion (F protein)	+	+	+	+	+	+
Hemolysin[2]	+	+	+	?	0	0
Hemaglutinin	+[3]	+[3]	+[4]	0	0	0
Hemadsorption	+	+	+	0	0	0
Neuraminidase	+[3]	+[3]	0	0	0	0
Inclusions[5]	C	C	N, C	?	C	?

[1]Zoonotic paramyxoviruses.
[2]Hemolysin activity carried by F glycoprotein.
[3]Hemagglutination and neuraminidase activities carried by HN glycoprotein.
[4]Hemagglutination of monkey erythrocytes only, by H glycoprotein that lacks neuraminidase activity.
[5]C, cytoplasm; N, nucleus.

3. Morbilivirus (Measles)
4. Henipavirus—Nipah virus and Hendra virus

Subfamily Peumovirinae

1. Pneumovirus (respiratory syncytial virus (RSV))
2. Metapneumovirus—Human metapneumovirus

Differentiation of Genera

Parainfluenza viruses and mumps virus have a surface hemagglutinin and neuraminidase, while measles virus has a hemagglutinin but no neuraminidase, and pneumovirus has neither. In addition, measles virus has a hemolysin not possessed by the others, while respiratory syncytial virus (RS) virus has a large surface glycoprotein, G, which has a similar cell-attaching function as a hemagglutinin (Table 64.1). Parainfluenza, mumps, measles and RS viruses are the common human patho gens. Unlike the orthomyxoviruses, the paramyxoviruses with their unsegmented genome cannot exchange genetic information by recombination and variation depends on mutational change.

PARAINFLUENZA VIRUSES

Parainfluenza viruses, which were discovered in the late 1950s, are respiratory viruses that usually cause mild coldlike symptoms but can also cause serious respiratory tract disease. There are four types of para-influenza viruses (1-4) with antigenically distinct epitopes.

Pathogenesis

Paramyxoviruses are acquired by droplets and contact with respiratory secretions. Incubation period varies from 2–6 days.

Parainfluenza viruses 1, 2, and 3 may cause respiratory tract syndromes ranging from a mild coldlike upper respiratory tract infection,

Primary infections in young children usually result in rhinitis and pharyngitis, often with fever and some bronchitis. However, children with primary infections caused by parainfluenza virus type 1, 2, or 3 may have serious illness, ranging from laryngotracheitis and croup (particularly with types 1 and 2) to bronchiolitis and pneumonia (particularly with type 3). The severe illness associated with type 3 occurs mainly in infants under the age of 6 months. Croup or laryngotracheobronchitis is more likely to occur in older children.

Parainfluenza virus type 4 causes only mild upper respiratory tract infections in children and adults.

Parainfluenza viral infection is confined to the respiratory tract, unlike mumps which is a systemic disease.

Epidemiology

Parainfluenza viruses are ubiquitous, and infection is common. The virus is transmitted by person-to-person contact and respiratory droplets.

Parainfluenza viruses are troublesome causes of nosocomial infection in pediatric wards in hospitals.

Laboratory Diagnosis

Definitive diagnosis relies on viral isolation from appropriate specimens. Isolation of the viruses is the best method of diagnosis.

1. Direct Identification

Immunofluorescence or ELISA—Direct identification of viral antigens in specimens is most commonly done.

Antigens may be detected in exfoliated nasopharyngeal cells by immunofluorescence or ELISA.

2. Virus Isolation

Throat and nasal swabs are inoculated in primary monkey kidney cell cultures or continuous monkey kidney cell lines (LLC-MK2) with trypsin. Parainfluenza viruses grow slowly and produce very little cytopathic effect. Virus growth is detected by hemadsorption. Typing is by immunofluorescence, hemadsorption inhibition, or hemagglutination inhibition.

3. Serology

Serodiagnosis should be based on paired sera and can be tested by neutralization, enzyme linked immunosorbent assay (ELISA) or hemagglutination inhibition test (HI) or complement fixation test for rise in titer of antibodies.

Treatment

No specific antiviral agents are available.

GENUS RUBULAVIRUS

Mumps Virus

Mumps is an acute contagious disease commonly affecting children characterized by nonsuppurative enlargement of one or both parotid glands. Other organs that may also be involved include the pancreas, testes, and ovaries as well as the central nervous system. The name probably originates from an old word meaning 'to mope', an apt description for the miserable child afflicted by this common illness.

Mumps was one of the first infections to be recognized and was described by Hippocrates as early as the fifth century BC. The viral etiology of mumps was demonstrated by Johnson and Goodpasture (1934) by its experimental transmission to monkeys. Habel in 1945 cultivated the virus in embryonated eggs. In 1955, Henle and Deinhardt grew it in tissue culture.

Morphology

Mumps virus is a typical paramyxovirus, possess both hemagglutinin and neuraminidase (HN) or a fusion (F) protein. The envelope also contains a matrix (M) protein. It agglutinates the erythrocytes of fowl, guinea pig, humans and many other species. Hemagglutination is followed by hemolysis and elution at 37°C.

Pathogenesis

Humans are the only natural hosts for mumps virus. Transmission is from person to person by large droplets. Primary replication occurs in nasal or upper respiratory tract epithelial cells. Viremia then disseminates the virus to the salivary glands and other major organ systems and infects the parotid gland. The virus is spread by the viremia throughout the body to the testes, ovary, pancreas, thyroid, and other organs such as infection of the central nervous system, especially the meninges. Immunity is lifelong.

Clinical Features

Infection is acquired by inhalation, and probably also through the conjunctiva. The incubation period may range from 7 days to 25 days but is typically about 16-18 days and is followed by a generalized illness with localization in the salivary glands, usually the parotids. The generalized phase is the usual 'flu-like' illness with fever and malaise, followed by developing pain in the parotid glands, which then swell rapidly. Parotitis is nonsuppurative and usually resolves within a week. However, involvement of the extraparotid sites may be more serious.

Complications: Various complications are epididymo orchitis, aseptic meningitis, postinfection encephalitis, mumps meningitis. Other less common complications are arthritis, oophoritis, nephritis, pancreatitis, thyroiditis and myocarditis.

Epididymoorchitis is a complication seen in about a third of postpubertal male patients and rarely causes sterility.

Epidemiology

Mumps occurs endemically worldwide, with humans the only known reservoir. Mumps is primarily an infection of children. The disease reaches its highest incidence in children aged 5-15 years, but epidemics may occur in army camps.

Mumps is quite contagious. The virus is transmitted by direct contact, airborne droplets, or fomites contaminated with saliva or urine. The period of communicability is from about 6 days before to about 1 week after the onset of symptoms. Infection appears to confer life-long immunity, and second infections do not occur. Most infants below the age of six months are immune because of maternal antibodies. The incidence of mumps and associated complications have declined markedly since introduction of the live-virus vaccine.

Immunity

There is only one antigenic type of mumps virus, and it does not exhibit significant antigenic variation. Immunity is permanent after a single infection.

Laboratory Diagnosis

Laboratory studies are not usually required to establish the diagnosis of typical cases.

The diagnosis may be established by virus isolation and serological tests.

1. Direct Demonstration

Direct demonstration by immunofluorescence on secretions is very rarely successful.

2. Virus Isolation and Identification

Virus can be recovered from the saliva, urine and CSF. Monkey kidney cells are preferred for viral isolation. For rapid diagnosis, immunofluorescence using mumps-specific antiserum can detect mumps virus antigens as early as 2-3 days after the inoculation of cell cultures in shell vials.

Isolation can also be made by inoculation into six to eight day old chick embryos by the amniotic route and testing the amniotic fluid after 5-6 days for hemagglutinins. The virus can be identified by hemagglutination inhibition using specific antisera.

3. Serology

The diagnosis can be made serologically by showing a rise in antibody titer in paired sera. Enzyme linked immunosorbent assay (ELISA) or hemagglutination inhibition test is commonly used. Mumps specific IgM antibodies can be detected in serum by enzyme linked immunosorbent assay (ELISA) for rapid diagnosis.

Prophylaxis

Vaccination

An effective attenuated live mumps virus vaccine based on the Jeryl Lynn or Urabe strains, has been available as a monovalent vaccine. Mumps vaccine is available in monovalent form (mumps only) or as combined vaccine, viz. combined mumps-rubella (MR) or into a triple vaccine against measles, mumps and rubella (MMR) live-virus vaccines. The vaccine is recommended for children over age 1 year as maternal antibodies may interfere with the multiplication of the vaccine virus if given earlier. The vaccine is given as a single dose (0.5 ml) intramuscularly. It provides effective protection for at least ten years. Contraindications are pregnancy, immunodeficiency, severely ill and hypersensitivity to neomycin or egg protein.

Immunoglobulin

A specific immunoglobulin (mumps immunoglobulin) is available and is of no value either for postexposure prophylaxis or treatment.

GENUS MORBILLIVIRUS

Measles (Rubeola)

Measles derives from an Anglo-Saxon word, measles. Its Latin name, morbilli, is a diminutive of morbus, a disease and thus signifies a minor illness. Measles is one of the five classic childhood exanthems, along with rubella, roseola, fifth disease, and chickenpox. Historically, measles was one of the most common and unpleasant viral infections with potential sequelae.

Morphology

The virus has the general morphology of paramyxoviruses (Fig. 64.2). It is a roughly spherical but often pleomorphic particle, 120-250 nm in diameter. The tightly coiled helical nucleocapsid is surrounded by the lipoprotein envelope carrying on its surface hemagglutinin (H) spikes. The envelope also has the F protein which mediates cell fusion and hemolytic activities. Matrix M protein is located below the lipoprotein envelope. The

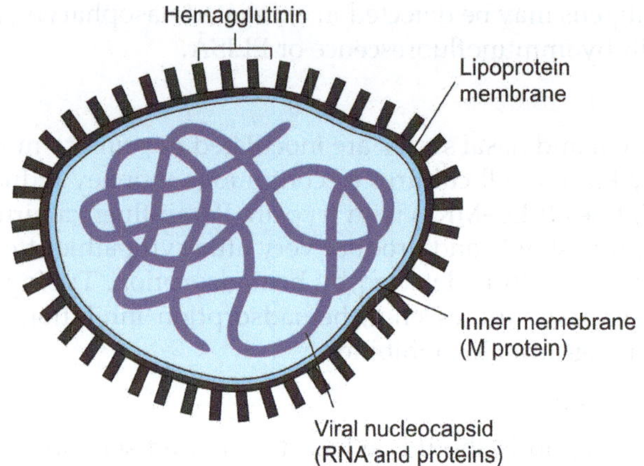

Fig. 64.2: Schematic diagram of measles virus

measles virus agglutinates monkey erythrocytes but there is no elution as the virus does not possess neuraminidase activity.

Pathogenesis

The virus gains access to the human body via the respiratory tract, or the conjunctiva where it multiplies locally. The infection then spreads to the regional lymphoid tissue, where further multiplication occurs. Primary viremia disseminates the virus, which then replicates in the reticuloendothelial system. Finally, a secondary viremia seeds the epithelial surfaces of the body, including the skin, respiratory tract, and conjunctiva, where focal replication occurs.

Clinical Features

Measles is a serious febrile illness. Incubation period is 9-11 from the time of exposure to infection for the first signs of clinical disease to appear. After 2 days of illness, the typical mucous membrane lesions, known as Koplik's spots, appear most commonly on the buccal mucosa across from the molars, and are pathognomic of measles.

Before the rash there is a prodromal stage lasting 2 or 3 days, with high fever and CCC and P-cough, coryza, and conjunctivitis, in addition to photophobia. The disease is most infectious at this time. The prodromal illness subsides within a day or two of the appearance of the rash. The red maculopapular rash of measles typically appears on the forehead first and spreads downwards, to disappear in the same sequence 3-6 days later, leaving behind a brownish discoloration and finely granular desquamation.

Complications

1. Complications may be due to the virus (croup, bronchitis) or to secondary bacterial infection (pneumonia, otitis media), giant cell pneumonia, particularly in children with immunodeficiencies or severe malnutrition.

2. Post-measles encephalitis: The most serious complication is meningoencephalitis. Many survivors have neurological sequelae.
3. Subacute sclerosing panencephalitis (SSPE)
   A rare late complication is subacute sclerosing panencephalitis (SSPE) which is an extremely serious, very late neurologic sequel of measles that afflicts about 7 in every 1 million patients. It occurs in children or early adolesents who have measles early in life. This disease occurs when a defective measles virus persists in the brain and acts as a slow virus. Within infected cells is a defective form of measles virus, which is unable to induce the production of a functional M protein and is not released as complete virus from the cell. Unusually high levels of measles antibodies are found in the blood and cerebrospinal fluid of patients with SSPE.
4. Protracted diarrhea is often seen as a complication in children in the poor nations.

Atypical Measles

Atypical measles occurred in people who received the older inactivated measles vaccine and were subsequently exposed to the wild measles virus. It may also rarely occur in those vaccinated with the attenuated virus vaccine. Prior sensitization with insufficient protection enhances the immunopathologic response to the challenge by wild measles virus. The illness begins abruptly and is a more intense presentation of measles.

Laboratory Diagnosis

Typical measles is reliably diagnosed on clinical grounds, usually in the patient's home. Laboratory diagnosis may be necessary in cases of modified or atypical measles.

1. Demonstration of Virus Antigen

A simple diagnostic test, which can be used even before the rash appears, is the demonstration of multinucleated giant cells in Giemsa stained smears of nasal secretions. The measles virus antigen can be detected in these cells by immunofluorescence.

2. Virus Isolation

The virus can be isolated from the nose, throat, conjunctiva and blood during the prodromal phase and upto about two days after the appearance of the rash. The virus may be obtained from the urine for a few more days. Primary human or monkey kidney and amnion cells are most useful. Cytopathic changes may take upto 7-10 days to develop but earlier diagnosis of viral growth is possible by immunofluorescence. Typical cytopathic effects (multinucleated giant cells containing both intranuclear and intracytoplasmic inclusion bodies) suggests the presence of measles virus.

3. Serological Diagnosis

Demonstration of measles-specific IgM in a single specimen of serum drawn between one and two weeks after the onset of the rash is confirmatory. Enzyme linked immunosorbent assay (ELISA), hemagglutination inhibition (HI), and neutralisation (Nt) tests all may be used to measure measles antibodies. A four-fold rise in titer is diagnostic.

High titer measles antibody in the CSF is diagnostic of SSPE.

Epidemiology

The key epidemiologic features of measles are as follows: the virus is highly contagious, there is a single serotype, there is no animal reservoir, inapparent infections are rare, and infection confers lifelong immunity. Transmission is person-to-person, probably by respiratory droplets, but the associated conjunctivitis may also be a source. In general, epidemics recur regularly every 2-3 years. The disease is ubiquitous throughout the world and, although a candidate for eradication, this may be difficult to achieve.

Malnutrition is one of the main underlying causes of this excess mortality. The measles virus has only one serotype and infects only humans, and infection usually manifests as symptoms. These properties facilitated the development of an effective vaccine program.

Prophylaxis

Passive Immunization

Passive protection with normal immunoglobulin (NHIG) given within six days of exposure can prevent or modify the disease, depending on the dose. Passive immunization is recommended in children with immunodeficiency, pregnant women and others at risk.

Active Immunization

A highly effective, safe, attenuated live measles virus vaccine is available. The original live vaccines used the Edmonston strain developed by multiple passage through human kidney, amnion and chick embryo cultures. Due to its high risk of causing febrile rash (vaccination measles), further attenuation became necessary. The Schwartz and Moraten strains so developed were safe but effective only in children older than 15 months. In the tropics, measles is common and serious in children below 12 months. Recent observations have suggested that the Edmonston-Zagreb strain, attenuated by passage in human diploid cells, may protect children from 4 to 6 months of age. This will be a boon to all countries. The recommended age for measles vaccination in the developing countries is 9 months, while in the advanced countries it remains 15 months.

The vaccine is given either by itself, or in combination as the MMR vaccine. A single subcutaneous injection of the measles vaccine provides protection beginning in about 12 days and lasting for over 20 years, probably for life. Contraindications are pregnancy, acute illnesses, immunodeficiency and untreated tuberculosis

A live attenuated vaccine has been developed which can be given by intranasal aerosol in young babies and gives good protection irrespective of the presence of maternal antibodies.

NIPAH AND HENDRA VIRUSES

During 1998-99, an outbreak of respiratory disease in pigs associated with encephalitis in humans occurred in Malaysia. There were over 200 human cases, with 105 deaths; a few survivors had persistent neurologic deficits. The causative agent was found to be a paramyxovirus given the name Nipah. It is distinct genetically from all the other paramyxoviruses and is most closely related to Hendra, another recently (1994) discovered paramyxovirus (a new equine morbillivirus) causing epidemic fatal respiratory disease in horses and which can be transmitted to man, resulting in at least one fatal infection in humans in Australia.

The taxonomic position of these new viruses has yet to be established, but they will most probably be put in a separate genus within the paramyxoviridae. Fruit bats appear to be the natural reservoir of both viruses, with transmission to mammals (including man) an exceptional event.

GENUS PNEUMOVIRUS

Respiratory Syncytial Virus (RSV)

Respiratory syncytial virus (RSV), first isolated from a chimpanzee in 1956, with coryza and was called the 'chimpanzee coryza agent' (CCA). A year later, the virus was obtained from children with lower respiratory tract infection. It was named respiratory syncytial virus (RSV) because it caused cell fusion and the formation of multinucleated syncytia in cell cultures.

Description

RSV is pleomorphic and ranges in size from 150-300 nm. The viral envelope has two glycoproteins—the G protein by which the virus attaches to cell surfaces, and the fusion (F) protein which brings about fusion between viral and host cell membranes. The F protein is also responsible for cell-to-cell fusion, which leads to the characteristic syncytial cytopathic changes in RSV infection.

It is placed in a separate genus—Pneumovirus because of these minor physical differences. RSV differs from other paramyxoviruses in not possessing hemagglutinin activity, neuraminidase or hemolytic properties. Another difference is that its nucleocapsid diameter (13 nm) is less than that of other paramyxoviruses (18 nm).

RSV does not grow in eggs but can be propagated on heteroploid human cell cultures, such as HeLa and Hep-2. It is highly labile and is inactivated rapidly at room temperature. It can be preserved by lyophilisation. It is antigenically stable and for most purposes there is only one serotype.

Clinical Features

The spectrum of respiratory illness ranges from the common cold in adults, through febrile bronchitis in infants and older children and pneumonia in infants, to bronchiolitis in very young babies.

1. Common cold: In adults, RSV infection may present as a febrile common cold. It can cause pneumonia in the elderly.
2. Febrile rhinitis and pharyngitis.
3. Bronchiolitis, pneumonia, or both—The most serious illness caused by RS virus is bronchiolitis in young babies. In infants, the disease may begin as febrile rhinorrhea, with cough and wheezing, progressing in 25-40 percent to lower respiratory involvement, including tracheobronchitis, bronch olitis and pneumonia.
4. Sudden infant death syndrome (SIDS): A relation between RSV and the sudden death syndrome (SIDS) in infants has been proposed but not proven. Respiratory syncytial virus is an important etiologic agent of otitis media in young children.

Epidemiology

Respiratory syncytial virus is distributed worldwide and is recognized as the major pediatric respiratory tract pathogen. It is the most common cause of viral pneumonia in children under age 5 years but may also cause pneumonia in the elderly or in immunocompromised persons. Respiratory syncytial virus infection of older infants and children results in milder respiratory tract infection than in those under age 6 months. The newborns are believed to be protected by high levels of maternal antibody. RSV is highly contagious.

RSV infections almost always occur in the winter. It causes nosocomial infections in nurseries and on pediatric hospital wards. The virus is transmitted byclose contact, and through contaminated fingers and fomites.

Laboratory Diagnosis

1. Demonstration of Virus Antigen

Immunofluorescence and enzyme immunoassay tests are available for direct detection of the viral antigen in infected cells and nasal washings.

2. Virus Isolation

Human heteroploid cell lines HeLa and HEp-2 are the most sensitive for viral isolation. The presence of respiratory syncytial virus can usually be recognized by development of giant cells and syncytia in inoculated cultures but cytopathic effects may take as long as 10 days for to appear. Definitive diagnosis can be established by detecting viral antigen in infected cells using a defined antiserum and the immunofluorescence test.

3. Serology

Serological diagnosis is by demonstration of rising antibody titers in paired serum samples by Enzyme linked immunosorbent assay (ELISA), complement fixation (CF), neutralization or immunofluorescence tests.

Prophylaxis

No vaccine is currently available for RSV prophylaxis. The use of recombinant DNA technology for making RSV vaccine is now being studied.

Treatment

In otherwise healthy infants, treatment is supportive care. Ribavirin is administered by inhalation (nebulization) and has been found beneficial in hospitalized patients, decreasing the duration of illness and of virus shedding.

METAPNEUMOVIRUS

Human metapneumovirus is a respiratory pathogen and is an important cause of respiratory tract infection in children. It can also cause disease in adults. It causes a disease similar to that of human respiratory syncytial virus.

It is a single stranded RNA virus like other paramyoviruses. Respiratory secretions are clinical specimen for test. The virus is difficult to grow. Polymerase chain reaction (PCR) can be used for diagnosis. No specific antiviral treatment or vaccine is available.

NEW CASTLE DISEASE VIRUS (NDV)

The Newcastle disease virus is an avian paramyxovirus and is a natural pathogen of poultry in which it causes pneumoencephalitis in young chickens and 'influenza' in older birds with high mortality. In India it is known as the Ranikhet virus. Control measures consist of vaccination, and slaughter of infected birds.

Human infection with NDV is confined to a selflimited conjunctivitis in poultry workers and others in contact with infected birds. Recovery is complete in 10-14 days

Other types of avian paramyxoviruses cause inapparent infection in many species of birds.

KNOW MORE

Mumps Virus

Properties

The virus can be grown in chick embryos—in the amniotic cavity for primary isolation and the allantoic cavity after adaptation. Eggs are inoculated at 6-8 days and incubated at 35°C for five days before harvesting.

Cell cultures are better suited for isolation—primary monkey kidney being the preferred cell. The cytopathic effect is slow and consists of syncytium formation and the presence of acidophilic cytoplasmic inclusions.

Growth is best identified by hemadsorption.

The mumps virus is labile, being rapidly inactivated at room temperature or by exposure to formaldehyde, ether or ultraviolet light. It can be preserved at –70°C or by lyophilisation.

The mumps virus is antigenicalfy stable and there is only one serotype, although monoclonal antibodies have shown minor variations in the various surface antigenic epitopes. Two complement fixing antigens can cognised, as in influenza viruses - the soluble (S) antigen and the 'viral' (V) antigen.

KEY POINTS

- The family Paramyxoviridae is of enveloped, helical, RNA viruses.
- Unlike the orthomyxoviruses, the paramyxoviruses with their unsegmented genome do not undergo genetic recombinations or antigenic variations.
- Within the family Paramyxoviridae two subfamilies—Paramyxovirinae and Pneumovirinae are recognized.
- Subfamily Paramyxovirinae contain respirovirus (para-influenza viruses1, 3), rubulavirus (mumps virus, para-influenza viruses 2, 4a, 4b), morbilivirus (measles) and henipavirus—Nipah virus and Hendra virus
- Subfamily Pneumovirinae contain pneumovirus (respiratory syncytial virus (RSV) and metapneumovirus (Human metapneumovirus).
- Parainfluenza viruses 1, 2, and 3 may cause respiratory tract syndromes ranging from a mild coldlike upper respiratory tract infection (coryza, pharyngitis, mild bronchitis, wheezing, and fever) to bronchiolitis and pneumonia. Parainfluenza virus type 4 does not cause serious disease, even on first infection.
- Mumps virus causes mumps.
- Measles virus causes measles, atypical measles, and subacute sclerosing panencephalitis (SSPE). Complications include otitis media, croup, bronchopneumonia, and encephalitis.
- Measles vaccine is a live attenuated vaccine which now uses Schwartz and Moraten attenuated strain of the original Edmonston B strain.
- Measles vaccine along with mumps and rubella (MMR) vaccine is currently used for universal immunization of the children. A single subcutaneous injection of the measles vaccine at the age of nine months.
- Respiratory syncytial virus (RSV) infection is confined more to the upper respiratory tract than the lower respiratory tract. RSV infection is highly contagious. RSV primarily causes infection of the respiratory tract, ranging from the common cold in adults, through febrile bronchitis in infants and

older children and pneumonia in infants, to bronchiolitis in very young babies. It may be the cause of some cases of sudden infant death syndrome.

IMPORTANT QUESTIONS

1. Classify and discuss the morphology of paramyxoviruses.
2. Write short notes on:
   i. Parainfluenza viruses
   ii. Mumps virus
   iii. Measles virus
   iv. Respiratory syncytial virus (RSV)
   v. Measles, Mumps and Rubella (MMR) vaccine.
   vi. Subacute sclerosing panencephalitis (SSPE).

FURTHER READING

Bellini WJ, et al. Paramyxoviruses. In: Collins, L, et al (eds). Topley & Wilson~Microbiology and Microbial Infections.9th edn, Vol.l. London: Arnold 1998.

Heilman CA. Respiratory syncytial and parainfluenza viruses. J Infect Dis 1990;161:402.

Katz SL, et al. Krugman's infectious diseases of children, St Louis, Mosby 1998;10 (Edn).

White DO, Fenner F. Medical virology, ed 4, San Diego, Academic 1994.

CHAPTER 65

Arboviruses

Learning Objectives

After reading and studying this chapter, you should be able to:

- Classify arboviruses.
- Describe laboratory diagnosis of arboviruses.
- List mosquito-borne and tick-borne arboviruses.
- List the arboviruses prevalent in India.
- Describe the following: Chikungunya; Japanese B encephalitis; dengue fever (or) break bone fever; Kyasanur forest disease (KFD)

INTRODUCTION

Arboviruses (Arthropod-borne viruses) are defined as viruses of vertebrates biologically transmitted by hematophagous insect vectors. They multiply in the tissues of arthropods, and are passed on to new vertebrates by the bites of arthropod after a period of extrinsic incubation.

Insect viruses and viruses of vertebrates that are sometimes mechanically transmitted by insects do not come into this category. Inclusion in this group is based on ecological and epidemiological and hence it contains members that are dissimilar in other properties. They are reassigned to better defined taxonomical groups with better understanding of the physical and chemical properties of individual viruses. The name 'arbovirus' is a useful biological concept though taxonomically unacceptable.

Certain viruses within the six families containing arboviruses are not transmitted by arthropods, but are maintained in nature within rodent reservoirs that may transmit infection directly to humans. These include the Hantavirus genus of the family Bunyaviridae

CLASSIFICATION

Arboviruses are classified within five families (Table 65.1). Most are members of the families Togaviridae, Flaviviridae and Bunyaviridae. Some are assigned to the families Reoviridae (the genera Coltivirus, e.g. Colorado tick fever virus, and Orbivirus, e.g. Bluetongue viruses), Orthomyxoviridae (e.g. Thogotovirus) and the Rhabdoviridae (members of the genera Vesiculovirus, e.g. vesicular stomatitis. Many arboviruses show close relationships with other arboviruses.Within each family, they are classified into genera, and antigenic groups, based on serological relationships.

PROPERTIES

Arboviruses share common biological attributes (Table 65.2).

1. Intracerebral inoculation in suckling mice is the most sensitive method for their isolation in the laboratory.
2. Most arboviruses agglutinate the red cells of goose or day-old chicks. Spontaneous elution does not occur.
3. They can be grown in the yolk sac or chorioal antoic membrane of chick embryo, in tissue cultures of primary cells like chick embryo fibroblasts or continuous cell lines like vero or HeLa, and in cultures of appropriate insect tissues.
4. Many arboviruses multiply in continuous tissue cultures of mosquito cells when incubated at 34°C or lower temperatures.
5. Mosquito-borne arboviruses multiply after oral feeding or intrathoracic injection of several Aedes and Culex mosquito species.
6. Tick-borne arboviruses multiply after oral feeding to larval or nymphal ixodid ticks.
7. In general, arboviruses are labile, being readily inactivated at room temperature and by bile salts, ether and other lipid solvents.

LABORATORY DIAGNOSIS

Diagnosis of arbovirus infections depends on virus isolation, detection of arbovirus-specific RNA and serology.

A. Specimen

As all arbovirus infections are viremic, blood collected during the acute phase of the disease may yield the

Table 65.1 Taxonomy of some important arboviruses		
Family	*Genus*	*Important species*
1. Togaviridae	Alphavirus	Eastern, Western and Venezuelan equine encephalitis viruses, Chikungunya, O'nyong-nyong, Mayaro, Semliki Forest, Sindbis, Ross River.
2. Flaviviridae	Flavivirus	St. Louis encephalitis, Ilheus, West Nile, Murray Valley encephalitis, Japanese encephalitis, Yellow Fever, Dengue types 1, 2, 3, 4, Kyasanur Forest Disease Russian Spring Summer encephalitis complex, Louping ill, Powassan, Omsk hemorrhagic fever
3. Bunyaviridae	Bunyavirus	California encephalitis, Oropouche, Turlock
	Phlebovirus	Sandfly fever viruses, Rift Valley fever virus
	Nairovirus	Crimean-Congo hemorrhagic fever viruses, Nairobi sheep disease virus, Ganjam virus
	Hantavirus	Hantan, Seoul, Puumala, Prospect Hill, Sin Nombre viruses
4. Reoviridae	Orbivirus	African horse sickness, Blue tongue viruses
	Coltivirus	Colorado tick fever
5. Rhabdoviridae	Vesiculovirus	Vesicular stomatitis virus, Chandipura virus

Table 65.2: Properties of Arboviruses					
Properties	*Alphavirus*	*Flavivirus*	*Bunyavirus*	*Rhabdovirus*	*Reovirus*
1. Symmetry	Cubic	Cubic	Helical	Bullet-shaped	Cubic
2. Size (Diameter in nm)	60-65	40-60	80-100	180 × 85	60-80
3. Nucleic acid	Single stranded positive sense RNA	Single stranded positive sense RNA	Single stranded negative sense RNA	Single stranded negative sense RNA	Double stranded RNA
4. Inactivation by diethyl ether/sodium deoxycholate	+	+	+	+	-

virus. Isolation may also be made from the CSF in some encephalitic cases but the best specimen for virus isolation is the brain.

B. Virus Isolation

i. Suckling Mice

Specimens are inoculated intracerebrally into suckling mice. The animals develop fatal encephalitis, though serial blind passages may be necessary in some cases. This is the most sensitive method for isolation of viruses

ii. Tissue Cultures

Some viruses may also be isolated in tissue cultures or, less readily, in eggs. Specimens are inoculated into Vero, BHK-21 and mosquito cell lines. Isolates are identified by hemagglutination inhibition, complement fixation, gel precipitation, immunofluorescence, immunochromatography, ELISA or neutralization with appropriate antisera.

iii. Virus Isolation from Insect Vectors and Reservoir Animal

Virus isolation from insect vectors and reservoir animal or avian species also aids in the identification of arboviruses activity in the area.

C. Arbovirus-specific RNA Detection

Viral RNA is extracted from serum or suspensions of tissues from patients, or from tissue culture cells or mosquito homogenates.This is amplified by reverse transcriptase polymerase chain reaction (RTPCR).

D. Serology

Diagnosis may also be made serologically by demonstrating rise in antibody titer in paired serum samples by hemagglutination inhibition, complement fixation or neutralization tests. Virus-specific IgM antibody may be detected within 1 day of onset of clinical symptoms using an IgM capture ELISA test.

PATHOGENESIS

The virus enters the body through the bite of the insect vector. After multiplication in the reticuloendothelial system, viremia of varying duration ensues and, in some cases, the virus is transported to the target organs, such as the central nervous system in encephalitides, the liver in yellow fever and the capillary endothelium in hemorrhagic fevers. Arboviruses cause the following clinical syndromes (Tables 65.3).

- Fever with or without rash and arthralgia
- Encephalitis
- Hemorrhagic fever
- The characteristic systemic disease, yellow fever

Arboviruses are maintained in natural transmission cycles involving reservoir hosts and arthropod vectors, typically: ticks, mosquitoes and other biting flies.

FAMILIES OF ARBOVIRUSES

Family Togaviridae

Morphology

Togaviruses are spherical enveloped viruses with a diameter of 50 to 70 nm. The genome is a molecule of single stranded RNA. The name Togavirus is derived from 'toga', meaning the Roman mantle or cloak, and refers to the viral envelope.

Table 65.3: Arboviruses associated with different clinical syndromes

Family	*Genus*	*Virus*	*Vector*	*Geographic distribution*	**Reservoir**
		Fever with or without rash and arthralgia			
Togaviridae	Alphavirus	Chikungunya	Mosquito	Africa, Asia	Not known (? Monkeys)
	Alphavirus	0' nyong-nyong	Mosquito	Africa	Not known
	Alphavirus	Ross River	Mosquito	Australia	Small animals
	Alphavirus	Sindbis	Mosquito	Africa, Asia	Birds, mammals
	Alphavirus	Mayaro	Mosquito	South America	Monkeys, marsupials
Flaviviridae	Flavivirus	Dengue, types 1-4	Mosquito	Widespread, especially Asia Pacific, Caribbean	Not known (? Monkeys)
	Flavivirus	West Nile	Mosquito	Asia, Africa, USA	Birds
Bunyaviridae	Bunyavirus	Sandfly fever	Sandfly	Mediterranean, Asia, Tropical America	not known (? Small mammals)
	Bunyavirus	Rift Valley fever	Mosquito	Africa	Sheep, cattle
	Bunyavirus	Oropouche	Mosquito	South America	Not known
Reoviridae	Orbivirus	Colorado tick fever	Tick	USA	Rodents
		Encephalitis			
Togaviridae	Alphavirus	Eastern equine encephalitis	Mosquito	Americas	Birds
	Alphavirus	Western equine encephalitis	Mosquito	America	Repties (? Birds)
	Alphavirus	Venezuelan equine	Mosquito	America	Rodents
Flaviviridae	Flavivirus	St. Louis encephalitis	Mosquito	America	Birds
	Flavivirus	West Nile	Mosquito	Africa, Europe, USA, West Asia	Birds
	Flavivirus	Japanese encephalitis	Mosquito	East and South East Asia	Birds

Contd...

Contd...

	Flavivirus	Murray Valley encephalitis	Mosquito	Australia	Birds
	Flavivirus	RSSE complex	Tick	East Europe, USSR	Rodents, other mammals, birds, ticks
	Flavivirus	Louping ill	Tick	Britain	Sheep
	Flavivirus	Powassan	Tick	North America	Rodents
Bunyaviridae	Bunyavirus	California	Mosquito	North America	Rodents
			Hemorrhagic fever		
Togaviridae	Alphavirus	Chikungunya	Mosquito	Africa, Asia	Not known (? Monkeys)
Flaviviridae	Flavivirus	Dengue types 1-4	Mosquito	Tropics	Not known (? Monkeys)
	Flavivirus	Yellow fever	Mosquito	Africa, South America	Monkeys, man
	Flavivirus	Kyasanur Forest Disease	Tick	Southwest India	Rodents (? Ticks)
	Flavivirus	Omsk hemorrhagic fever	Tick	USSR	Small mammals
	Flavivirus	Crimean-Congo hemorrhagic fever	Tick	USSR, Central Asia, Africa	Small. mammals

Classification

The family Togaviridae contains two genera: Alphavirus and rubivirus.

Alphavirus

The genus Alphavirus was formerly classified as 'Group A arboviruses' which explains the name Alphavirus (from Alpha, the first letter of the Greek alphabet, corresponding to the letter A). In the Togaviridae family, the Alphavirus genus consists of about 32 viruses of which at least 13 are known to infect humans. All of them are mosquito borne.

They have a worldwide distribution and produce epidemics of encephalitis in America and dengue-like fever in the tropics.

Rubivirus

Rubivirus, which is not arthropod-borne and which causes rubella (Chapter 64).

Viruses of Togaviridae

A. Alphavirus (Mosquito-borne)

1. Encephalitis Viruses

i. Eastern equine encephalitis (EEE)
ii. Western equine encephalitis (WEE)
iii. Venezuelan equine encephalitis (VEE)

2. Viruses Causing Febrile Illness

i. Chikungunya virus (CHIKV)
ii. O'nyong-nyong virus (ONNV)
iii. Semliki forest virus
iv. Sindbis virus
v. Ross River virus

B. Rubivirus,

Rubella virus

A. Alphavirus

1. Encephalitis Viruses

Three members of this group, Eastern, Western and Venezuelan equine encephalitis viruses, cause encephalitis in horses and humans. Equines, like humans, are unessential hosts for the maintenance of the virus.

i. Eastern equine encephalitis (EEE)—occurs along eastern Canada, USA and the Caribbean, causing sporadic cases and small epidemics. In addition, Eastern equine encephalitis produces severe epizootics in certain domestic game birds.
ii. Western equine encephalitis (WEE)—is more widely distributed in America and causes large epidemics.
iii. Venezuelan equine encephalitis (VEE)—prevalent in Central and South America, usually causes an influenza" like illness, with encephalitis in a small proportion of cases.

Reservoirs: Several species of Culex and Anopheles mosquitoes are the vectors, and wild birds the reservoirs.

Vaccine: Formalinized vaccines have been developed for EEE and WEE and a live attenuated vaccine for VEE.

2. Viruses Causing Febrile Illness

i. Chikungunya Virus (CHIKV)

Chikungunya virus (CHIKV) was first isolated during a 1952 epidemic in Tanzania from human patients and

Aedes aegypti mosquitoes. In Swahili, 'chikungunya' means 'that which bends up', and refers to the posture assumed by patients suffering from severe joint pains.

The incubation period is 1 to 12 days with an average of 2 to 3 days. The disease presents as a sudden onset of fever, crippling joint pains, lymphadenopathy and conjunctivitis. A maculopapular rash is common and some show hemorrhagic manifestations. The fever is typically biphasic with a period of remission after 1 to 6 days of fever. Clinical picture resembles that of dengue fever, with which it is often confused.

Vector—The vector is *Aedes aegypti*.

In India the virus first appeared in 1963, when along with dengue, it caused very extensive epidemics in Calcutta, Madras and other areas. Chikungunya outbreaks occurred at irregular intervals along the east coast of India and in Maharashtra till 1973. Since then the virus has been quiescent.

No animal reservoir has been identified. No vaccine is available.

ii. O'nyong-nyong Virus (ONNV)

O'nyong-nyong virus (ONNV), derived from the description by the Acholi tribe, meaning 'joint breaker', was first isolated during a 1959-1962 epidemic affecting 2 million people. This is confined to Africa, is closely related to the chikungunya virus antigenically and causes a similar disease.

This is transmitted by the Anopheles species (*Anopheles funestus* and *Anopheles gambiae*).

iii. Semliki Forest Virus

This virus was first isolated in 1942 in Uganda from Aedes mosquitoes has not been associated with clinical illness in humans though neutralizing antibodies to the virus have been demonstrated in Aricans.

iv. Sindbis Virus

The Sindbis virus, 'originally' isolated from Culex mosquitoes in the Sindbis district of Egypt in 1952, has subsequently been recovered from other parts of Africa, India, Philippines and Australia. In Africa, it is known to be associated with febrile illness in human beings. In India, antibodies have been detected in human sera but no association has been established with human disease.

v. Ross River Virus

The closely related Ross River virus has been associated with epidemic polyarthritis in Australia.

Family Flaviviridae

The family Flaviviridae (formerly classified with the Togaviridae) contains three genera: 1. Flavivirus; 2. Pestivirus; 3. Hepacivirus

Flavivirus

Only the Flavivirus genus contains arthropod-borne viruses. The name Flavivirus refers to the type species, the yellow fever virus (Flavus, L = yellow). The Flaviviridae family consists of about 70 viruses. Representative members of this group are distributed in all parts of the world, covering all the zoogeographic regions.

Morphology

They are somewhat smaller than alphaviruses, being 40 nm in diameter. They contain a single stranded positive sense RNA. Inner viral core is surrounded by a lipid envelope which is covered with glycoprotein peplomers and matrix or membrane protein.

They may be considered under two sections, the mosquito-borne and the tick-borne viruses (Table 65.3).

A. Mosquito-borne Group

a. Encephalitis viruses-St. Louis encephalitis, Ilheus, West Nile, Murray Valley encephalitis, Japanese encephalitis.
b. Yellow Fever
c. Dengue types 1, 2, 3, 4,

B. Tick-borne Group

These viruses produce-two clinical syndromes, encephalitis and hemorrhagic fevers

A. Mosquito-borne group

a. Encephalitis Viruses

Five members of this group cause encephalitis, each of them limited to a geographic zone:

1. St. Louis Encephalitis Virus (SLEV)

This is prevalent in North and Central America and is the most important mosquito-borne disease in the USA. The clinical picture ranging from mild febrile illness to frank encephalitis and the case fatality ranging from 2 to 20 percent.

Wild birds act as the reservoir and *Culex tarsalis* as the vector. No vaccine is yet available.

2. Ilheus Virus

This occurs in South and Central America, maintained in forests by a cycle involving mosquitoes, wild birds and monkeys. Human infection is largely subclinical or leads to febrile illness. Encephalitis is rare.

3. West Nile Virus (WNV)

WNV was first isolated from a febrile human in the West Nile district of Uganda in 1937, has since been reported from many African countries, Israel, Cyprus, France and India. Although WNV infects a wide variety of animals, including horses, cattle and humans, the major vertebrate hosts are wild birds.

Vector—Both mosquitoes and ticks have been reported as vectors. The principal vectors are considered to be mosquitoes of the Culex genus.

4. Murray Valley Encephalitis Virus (MVEV)

This is confined to Australia and New Guinea. MVEV caused epidemics of encephalitis in irrigated farming regions of the Murray-Darling River basin of southeastern Australia during the summer months (January to March) of 1951 and 1974.

Vector—Natural cycles of transmission of MVEV involve *culex annulirostris* as the principal mosquito vector and water birds as reservoirs.

5. Japanese Encephalitis

Japanese encephalitis (JE) is a mosquito-borne encephalitis caused by a group B arbovirus (Flavivirus) and transmitted by culicine mosquitoes. It is a zoonotic disease, i.e. infecting mainly animals and incidentally man. *Culex tritaeniorhynchus*, a rural mosquito that breeds in rice fields, is the principal vector.

Geographical Distribution

The disease has been recognized in Japan since 1871 and was named Japanese 'B' encephalitis to distinguish it from encephalitis A: (encephalitis lethargica, von Economo's disease) which was then prevalent. The virus was first isolated in Japan during an epidemic in 1935. Several large epidemics have occurred since then. Epidemics show a seasonal incidence (summer-autumn) in the temperate regions, though this is not evident in the tropical areas. More recently, JE has spread to Pakistan and India in the west, and to the Torres Strait and has been detected once in far north Queensland in Australia.

Problem in India

Recognition of JE, based on serological surveys, was first made in 1955 when the virus was isolated from mosquitoes of the *Culex vishnui* complex from Vellore during an outbreak of encephalitis in Tamil Nadu. Japanese encephalitis remained confined to the southeastern parts of India till 1973, when it caused a large outbreak of encephalitis in West Bengal. From 1976, there have been periodical outbreaks of the disease in various parts of India. Japanese encephalitis has become a major public health problem of national importance in India. In the year 2003, there has been a major upsurge of JE in Assam, Andhra Pradesh, Bihar, Karnataka, Haryana, Maharashtra, Tamil Nadu and Uttar Pradesh.

Clinical Features

The incubation period in man probably varies from 5 to 15 days, following mosquito bite. The course of the disease in man may be divided into three stages:

a. *Prodromal stage:* The onset of illness is usually acute and is heralded by fever, headache, and malaise.
b. *Acute encephalitic stage:* The prominent features are fever, nuchal rigidity, focal central nervous system (CNS) signs, convulsions and altered sensorium progressing in many cases to coma.
c. *Late stage and sequelae:* Convalescence may be prolonged and residual neurological deficits may not be uncommon. The case fatality rate varies between 20-40 percent, but it may reach 58 percent and over.

Residual neurological damage may persist in upto 50 percent of survivors. The large majority of infections are, however, asymptomatic. It has been estimated that 500-1000 inapparent infections occur for every case of clinical disease.

Epidemiology

The disease is transmitted to man by the bite of infected mosquitoes. Man is an incidental "dead-end" host. Man to man transmission has not so far been recorded. Available evidence indicates that the basic cycles of transmission are:

a. Pig → Mosquito → Pig
b. The Ardeid bird → Mosquito → Ardeid bird

a. Animal Host

Pigs have been incriminated as the major vertebrate hosts for JE virus and are considered as "amplifiers" of the virus.

b. Birds

Natural infection has been demonstrated in Ardeid birds (herons and egrets), as well as bird-to-bird transmission through *Culex tritaeniorhynchus*. Other birds such as ducks, pigeons and sparrows may also be involved.

Vectors of JE: Culicine mosquitoes, notably *C. tritaeniorhynchus*, *C. vishnui* and *C. gelidus* along with some anophelines have been incriminated as the vectors of JE. *Culex tritaeniorhynchus*, a rice-field breeding mosquito, is the main vector in north Asia and Japan.

Control of Japanese Encephalitis

Preventive measures include mosquito control and locating piggeries away from human dwellings.

Vaccination

a. Formalin Inactivated Mouse Brain Vaccine

A formalin inactivated mouse brain vaccine using the Nakayama strain has been employed successfully for human immunization in Japan and, in a small scale, in India also. For primary immunization, 2 doses of 1 ml each (0.5 ml for children under the age of 3 years) should be administered subcutaneously at an interval of 7 to 14 days. A booster injection of 1 ml should be given after a few months (before one year) in order to develop full protection. Protective immunity develops in about a month's time after the second dose. Revaccinations may be given after 3 years.

b. Live Attenuated Vaccine

A live attenuated vaccine has been developed in China from JE strain SA 14-14-2, passed through weanling mice. The vaccine is produced in primary baby hamster kidney cells. Administered in two doses, one year apart, the vaccine has been reportedly effective in preventing clinical disease.

Vaccination of pigs: Vaccination of pigs has been proposed in view of their importance as amplifier hosts.

During major epidemics, slaughter of pigs have been employed as a measure of containment.

c. Yellow Fever

Yellow fever virus is the prototype member of the Flaviviridae family. It causes yellow fever, an acute, febrile, mosquito-borne illness that occurs only in Africa and Central and South America. It does not exist in India.

Pathogenesis and Pathology

The virus is introduced by a mosquito through the skin and spreads to the local lymph nodes where it multiplies. From the lymph nodes, it enters the circulating blood and becomes localized in the liver, spleen, kidney, bone marrow, and lymph glands, where it may persist for days.The lesions of yellow fever are due to the localization and propagation of the virus in a particular organ. Infections may result in necrotic lesions in the liver and kidney.

Histologically, the liver shows cloudy and fatty degeneration and necrosis which is typically midzonal. The necrosed cells coalesce and become hyalinized leading to the formation of characteristic eosinophilic masses known as Councilman bodies. Acidophilic intranuclear inclusion bodies (Torres bodies) may be seen in the infected liver cells in the early stages.

Clinical Findings

The incubation period is 3 to 6 days. The disease starts as a fever of acute onset with chills, headache, nausea and vomiting. The pulse is usually slow despite a high temperature. Jaundice, albuminuria, and hemorrhagic manifestations develop and the patient may die of hepatic or renal failure.

Epidemiology

Two major epidemiologic cycles of transmission of yellow fever are recognized:

1. Urban yellow fever—In the urban cycle, humans act both as the natural reservoir and as the definitive case, the virus being transmitted by the domestic *Aedes aegypti* mosquito.
2. Forest or sylvatic cycle -in the forest or sylvatic cycle, wild monkeys act as the reservoirs and forest mosquitoes (*Haemagogus spegazzini* in South America and *Aedes africanus* and *A. simpsoni* in Africa) as the vectors. Human cases occur only when humans trespass into the forest or when the monkeys raid villages near the forest.

Yellow fever is largely confined to Central and South America and Africa. Yellow fever has never invaded Asia, even though the vector, *A aegypti*, is widely distributed there. It is likely that stray virus introduced may have been kept out, due to the prevalence in the local *Aedes aegypti* of Dengue virus, and of antibodies to a wide range of arboviruses in the local population. Another reason could have been that in Africa, yellow fever was mainly in the west, and in India, Aedes mosquitoes were along the east coast, so that even stray importations of virus by sea may not have found suitable vectors. This is no longer valid as yelow fever has in If ever yellow fever gets established in India, the consequences could be catastrophic.

Control

1. Vector Control

Vigorous mosquito abatement programs have virtually eliminated urban yellow fever but this is obviously impracticable with the sylvatic disease.

2. Vaccination

Two very effective vaccines have been developed for human use.

i. French Neurotropic Vaccine (Dakar)

French neurotropic reactive (Dakar) produced from the infected mouse brain was used as a vaccine the vaccine carries a high risk of producing encephalitis in the vaccinees, especially in children. It was later replaced by a nonneurotropic (17D) vaccine.

ii. 17D Vaccine

A safe and equally effective vaccine, the 17D vaccine was developed by Theiler in 1937 by passaging the Asibi strain serially in mouse embryo and whole chick embryo tissues and then in chick embryo tissue from which the central nervous tissue has been removed and has been used as a vaccine for over 40 years.

The 17D vaccine is thermolabile and is administered by subcutaneous inoculation. Immunity develops within 10 days of vaccination. Vaccination which is mandatory for travel to or from endemic areas is valid for 10 years beginning 10 days after vaccination. In India, the 17D vaccine is manufactured at the Central Research Institute, Kasauli.

c. Dengue

Dengue virus is widely distributed throughout the tropics and subtropics. The name 'dengue' is derived from the Swahili Ki denga pepo, meaning a sudden seizure by a demon. The term 'break bone fever' was coined during the Philadelphia epidemic in 1780.

Four types of dengue virus exist: DEN 1 first isolated from Hawai in 1944, DEN 2 from New Guinea in 1944 and DEN 3 and 4 from the Philippines in 1956. Immunity is type specific so that it is possible for a person to have four separate episodes of dengue fever. However, the febrile clinical symptoms associated with dengue are similar to those of other arboviruses from other families.

Clinical Findings

a. Classic Dengue Fever

Classic dengue usually affects older children and adults. After incubation period of 2 to 7 days, patient develops fever of sudden onset and often biphasic with severe headache, chills, retrobulbar pain, conjunctivitis and severe pain in the back, muscles and joints (break bone fever). A maculopapular rash generally appears on the trunk in 3 to 5 days of illness and spreads later to the face and extremities. Lymph nodes are frequently enlarged. Leukopenia with a relative lymphocytosis is a regular occurrence.

Classic dengue fever is a self-limited disease. Convalescence may take weeks, although complications and death are rare. Especially in young children, dengue may be a mild febrile illness lasting a short time.

b. Other Manifestations

Dengue may also occur in more serious forms, with hemorrhagic manifestations (*dengue hemorrhagic fever*) or with shock (*dengue shock syndrome*). They are more common in previously healthy children in the indigenous populations of endemic areas. It appears to be hyperimmune response.

Pathogenesis–The pathogenesis of the severe syndrome involves preexisting dengue antibody. It is postulated that virus-antibody complexes are formed within a few days of the second dengue infection and that the non-neutralizing enhancing antibodies promote infection of higher numbers of mononuclear cells, followed by the release of cytokines, vasoactive mediators, and procoagulants, leading to the disseminated intravascular coagulation seen in the hemorrhagic fever syndrome.

Dengue hemorrhagic fever may occur in individuals (usually children) with passively acquired (as maternal antibody) or pre-existing nonneutralizing heterologous dengue antibody due to a previous infection with a different serotype of virus. Dengue shock syndrome (DSS), a more severe form of the disease characterized by shock and hemoconcentration, may ensue. Circumstantial evidence suggests that secondary infection with dengue type 2 following a type 1 infection is a particular risk factor for severe disease.

These complications, first recognized in Thailand, have since occurred in many countries in Southeast Asia and the Western Pacific.

Laboratory Diagnosis

1. Specimens: For antibody detection—Serum

For isolation of virus and PCR—Serum, plasma, whole blood (washed buffy coat), autopsy tissues and, mosquitoes collected in nature.

2. Virus Detection

Isolation of the virus is difficult. Virus isolation can be done by inoculating clinical specimen into mosquitoes, mosquitoes cell lines (C6/36 or AP-61 cells), or suckling mice. Further identification is done by using fluorescent antibody test. Live mosquito inoculation is the most sensitive technique for isolation of virus.

3. Polymerase Chain Reaction (PCR)

PCR based methods are available for rapid identification and subtyping. Viral RNA can be detected in clinical specimens by reverse transcriptase polymerase chain reaction (RTPCR). Viral genomic sequences can also be done.

4. Serology

Demonstration of circulating IgM antibody provides early diagnosis, as it appears within two to five days of the onset of illness and persists for one to three months. IgM ELISA test offers reliable diagnosis. A strip immunochromatographic test for IgM is available for rapid diagnosis.

IgG antibody appears later than IgM antibody. ELISA is used for detection of IgG antibody. Detection of four fold rise in IgG titer in paired sera taken at an interval of ten days or more is confirmatory.

Haematological diagnosis: Thrombocytopenia (100,000 cells or less per mm^3); Haemoconcentration (720% in haematocrit).

Epidemiology

Dengue virus is transmitted from person-to-person by *Aedes aegypti* mosquitoes. No vertebrate hosts other than humans have been identified. Most subtropical and tropical regions around the world where Aedes vectors exist are endemic areas. Up to 100 million infections are thought to occur each year, with the majority in children.

Dengue was initially confined to the east coast of India and has caused epidemics, sometimes along with the chikungunya virus, as in 1963 when extensive outbreaks affected Calcutta and Madras. Subsequently it has spread westwards and in the 1990s Surat and Delhi had major epidemics with deaths due to DHF and DSS. All four types of dengue virus are present in this country. Occasionally, more than one type of the virus has been isolated from the same patient.

Control

Control depends upon antimosquito measures, e.g. elimination of breeding places and the use of insecticides as no vaccine is currently available.

A live attenuated vaccine containing all four dengue serotypes is under clinical trial in order to avoid DHF/DSS in immunized persons.

B. Tick-borne Group

These viruses produce two clinical syndromes, (i) Encephalitis, (ii) hemorrhagic fevers.

1. Tick-borne Encephalitis (TBE) Viruses

Tick-borne encephalitis is used to describe a serocomplex of related viruses that are transmitted by ticks and cause similar diseases.

Russian spring summer encephalitis (RSSE) complex: A number of viruses belonging to the Russian Spring Summer Encephalitis (RSSE) complex cause encephalitis along a wide area of the northern landmass from Scotland to Siberia. The names given to the disease vary from one area to another depending on the variations in the prominent clinical features. Thus, in Scotland, it is called 'louping ill' as the disease occurs primarily in sheep in which it causes a curious 'leaping' gait. It is called Central European Encephalitis in Central Europe, biphasic meningoencephalitis in Eastern Europe and RSSE in USSR.

Human infections by TBEV may range in severity from mild biphasic meningoencephalitis, to a severe form of polioencephalomyelitis. Biphasic meningoencephalitis may be transmitted to human beings by drinking the milk of infected goats and less commonly by entry through injured skin or mucosa. Rare aerosol transmission may occur.

i. Russian spring-summer encephalitis (RSSE): It is the most serious form, with high rates of fatality and permanent paralytic sequelae in some survivors. Infection is transmitted by the bite of lxodid ticks. The virus is transmitted transovarially in ticks so that they can act as vectors as well as reservoir hosts. Wild rodents and migrating birds are other reservoirs.

Control

The control of infection depends on the avoidance of tick bites. A formalin inactivated vaccine is commercially available for the Western subtype and for the Eastern subtype.

ii. Powassan virus—Powassan virus causes encephalitis in Canada and Northern USA.

2. Tickborne Hemorrhagic Fevers

a. Kyasanur Forest Disease (KFD)

Kyasanur Forest Disease (KFD) is a febrile disease associated with hemorrhages caused by an arbovirus flavivirus and transmitted to man by bite of infective ticks. A new arbovirus antigenically related to the RSSE complex, was isolated by investigators from the National Institute of Virology (then Virus Research Center), Pune, from the patients and dead monkeys. The disease was later named after the locality Kyasanur Forest—from where the virus was first isolated — Kyasanur Forest Disease (KFD).

History

KFD was first recognized in 1957 in Shimoga district of Karnataka State in South India. Earlier the disease was found to be limited mainly to an area around the original focus (Shimoga district). Between 1972 and 1975, a few other smouldering foci, developed in the adjacent areas in North Kanara. The situation changed suddenly in 1982 with the appearance of an epizootic and epidemic in Belthangadi taluk in South Kanara. This followed the clear felling of part of an evergreen reserve forest in the area in September 1982. The outbreak, known locally as monkey fever, started with dead monkeys being observed in October. The first human case was seen late in December. During the next five months, 1142 human cases were recorded with 104 deaths. The outbreak subsided with the onset of the monsoon in June but reappeared the following December. The ecological disturbance caused by clear felling of the virgin forest is believed to have activated a silent enzootic focus of the virus.

Epidemiology

a. *Agent:* Flavivirus—The agent KFD virus is a member of group B togaviruses (flaviviruses).
b. *Natural hosts and reservoirs:* Main reservoirs of the virus are small mammals particularly rats and squirrels and forest birds. Monkeys are only amplifer hosts.
c. *Vectors:* Infection is transmitted by the bite of ticks, the principal vector being *Hemaphysalis spinigera.*

Clinical Features

Incubation period is between 3 and 8 days. The disease appears with a sudden onset of fever, headache and severe myalgia, with prostration in some patients. Gastrointestinal disturbances and hemorrhages from nose, gums, stomach and intestine may occur in severe cases.

In a number of cases, there is a second phase characterized by mild meningoencephalitis after an afebrile period of 7 to 21 days. The case fatality rate has been estimated to be 5 to 10 percent.

Laboratory Diagnosis

1. Virus Isolation

Virus can be isolated from the blood up until the 12th day in suckling mice, hamster or monkey kidney cells or HeLa cells with cytopathic effect.

2. Serology

Serological diagnosis can be made with rising antibody titers in acute and convalescent sera, as well as by enzyme immunoassay (EIA) tests.

Control

Control is essentially a breaking of the tick—human contact. Alteration of the environment and keeping cattle out of the forest are important. Personal protection involves regular (daily) de-ticking of the body and the use of repellents and protective clothing.

Vaccine—A killed KFD virus vaccine was used in a small field, trial and appeared to provide some degree of protection.

b. Omsk Hemorrhagic Fever

This occurs in Russia and Romania. It is clinically similar to KFD and is caused by a related virus. Dermacentor ticks are the vectors.

Family Bunyaviridae

The Bunyaviridae family contains more than 300 viruses is the largest group of arboviruses, mostly athropod-transmitted.

Morphology

The virus is about 100 nm in diameter and has a complex structure, with a triple segmented genome of single stranded RNA.

Classification

The family Bunyaviridae contains four genera of medical importance

A. Bunyavirus—Mosquito-borne
B. Phlebovirus—Sandfly
C. Nairovirus—Tick-borne
D. Hantavirus—Nonarthropord-borne.
A number of viruses are yet ungrouped.

A. Bunyavirus

The genus contains over 150 species, of which only a few cause human infections. This genus includes California encephalitis virus, La Crosse virus, and Chittor virus. California encephalitis virus and La Crosse virus are isolated from the United States and Chittor virus from India.The clinical disease caused is encephalitis, aseptic meningitis and fever. All are mosquito-borne infections.

B. Phlebovirus

1. Sandfly fever (Phlebotomus fever): Sandfly fever (Phlebotomus fever) also known as Pappataci fever and three-day fever, is a self, nonfatal fever transmitted by the bite of sandfly *Phlebotomus papatasi. It occurs* along the mediterranean Coast and Central Asia, extending as far east as Pakistan and North West India. Cases have also been reported from South and Central America.

Natural vectors—are *Phlebotomus papatasi* and other phlebotomine sandflies.

Twenty antigenic types of the virus exist, of which only five cause human disease—Naples, Sicilian, Punta Toro, Chagres, Candiru. Its occurrence in India was thought to be doubtful.

2. Rift valley fever: The agent of this disease, a bunyavirus of the Phlebovirus genus, is a mosquito-borne zoonotic virus pathogenic primarily for sheep, cattle, and goats. Humans are secondarily infected during the course of epizootics in domesticated animals. Infection among laboratory workers is common. It is named after Rift Valley, Kenya, where it was first recognized.

C. Nairovirus

Members of the Crimean-Congo hemorrhagic group are the major human pathogens in this genus.

Crimean-Congo Hemorrhagic Fever (CCHF) virus: It is distributed widely throughout tropical Africa, from Mauritania to Uganda and Kenya, the Middle East and West Pakistan, and southwards to South Africa. It is also found in parts of Asia, including parts of China.

Cattle, sheep, goats and other domesticated animals act as natural reservoirs. It is transmitted by Hyalomma ticks. During the acute phase of the disease, the blood of the patients is highly infectious and direct transmission may occur through contact.

Hazara virus: A related virus, Hazara, has been isolated in Pakistan. It is also widespread in Iran, Iraq and the UAE. Antibodies to the CCHF group of viruses have been detected in human and animal sera from India.

Nairobi sheep disease: It is an acute, hemorrhagic gastroenteritis caused by a Nairovirus in sheep and goats in East Africa. It is transmitted by Rhipicephalus ticks. The virus produces a mild febrile illness in shepherds tending infected flocks.

Ganjam virus: Ganjam virus, isolated from ticks collected from sheep and goats in Orissa, India, is closely related to the Nairobi sheep disease virus. The Ganjam virus has also been isolated from human sources. Accidental infection in laboratory workers has caused mild febrile illness.

D. Hantavirus

Hantaviruses are classified in the Hantavirus genus of the Bunyaviridae family. The viruses are found worldwide and cause hemorrhagic fever with renal syndrome (HFRS) also known as endemic or epidemic nephrosonephritis, Manchurian epidemic hemorrhagic fever, nephropathia epidemic, a rodent-borne nephropathy and other names. Principal vertebrate reservoirs comprise *Apodemus agrarius* rodents in Asia and *Clethrionomys glareolus* (bank vole) in Europe.

Species: The genus contains at least four' species: 1. Hantaan virus; 2. Seoul virus; 3. Puumala virus; 4. Prospect Hill virus. Hantavirus species are natural pathogens of rodents-field mice *(Apodemus agrarius)* being the major host for Hantaan, rats *(Rattus rattus* and *R. norvegicus*) for Seoul, and voles for Puumala and Prospect Hill viruses. Transmission to humans occurs by inhaling aerosols of rodent excreta (urine, feces, saliva). The disease occurs in two forms—the milder epidemic nephritis (EN) common in Scandinavia and the more serious epidemic hemorrhagic fever (EHF) in the far east. The clinical picture resembles typhoid, leptospirosis and scrub typhus.

Domestic rats appear to be the source of infection in urban cases of HFRS. HFRS should be considered a robovirus and not stritly an arbovirus infection in the absence of proved arthropod transmission.

Sin Nombre Virus (SNV)

In 1993 an outbreak of severe respiratory illness in the United States, now designated the hantavirus pulmonary syndrome (HPS), was found to be caused by a novel hantavirus.

The disease is caused by a newly identified hantavirus, the *Sin Nombre* (meaning nameless) virus, which is associated with the deer mouse and other rodents of the sigmodontine subfamily. The principal rodent reservoir is *Peromyscus maniculatus* (deer mouse). No arthropod has been linked with transmission of the virus.

Infection appears to be caused by inhalation of the virus aerosols in dried rodent feces. Person-to-person transmission of hantaviruses seldom occurs.

Laboratory Diagnosis

Laboratory diagnosis depends on detection of viral nucleic acid by reverse transcription-polymerase chain reaction or detection of specific antibodies using recombinant proteins of different hantaviruses. Hantaviruses can be isolated in cultured cells, but those methods require the use of containment facilities.

Family Reoviridae

Family Reoviridae contains four genera—Orbivirus, Coltivirus, Orthoreovirus, and Rotavirus.

The genus Orbivirus contains arthropod-borne viruses. Rotaviruses and reoviruses have no arthropod vectors.

A. Genus Orbivirus

The genus Orbivirus of the family Reoviridae contains arthropod-borne viruses which infect animals and humans that differ from other arboviruses in having double stranded RNA genomes. African horse sickness and bluetongue viruses are in the genus Orbivirus.

i. *African horse sickness virus:* African horse sickness virus, transmitted by Culicoides, has for long been known to cause disease among equines in Africa. It caused extensive disease among army horses and mules in India.
ii. *Palyam, Kasba and Vellore viruses:* Palyam, Kasba and Vellore viruses belonging to the orbivirus group have been isolated from mosquitoes in India but their pathogenic significance is not known.

B. Genus Coltivirus

Colorado tick fever is classified in the genus Coltivirus.

Colorado Tick Fever

Colorado tick fever, also called mountain fever or tick fever, is transmitted by a tick. The disease in humans is self-limited. It is spread by the wood tick *Dermacentor andersoni* and the distribution of the disease in Western USA is limited to the habitat of the tick, which acts both as the vector and reservoir. Natural infection occurs in rodents.

C. Orthoreovirus

Cause aymptomatic infections in humans.

D. Rotavirus

Causes human infantile gastroenteritis.

Family Rhabdoviridae

Chandipura Virus

Chandipura virus, belonging to the vesiculovirus genus of Rhabdoviridae, was isolated in 1967 from the blood of patients during an epidemic of dengue chikungunya fever in Nagpur. The virus appears to multiply in sandflies and Aedes mosquitoes. The pathogenic significance of this virus has not been established.

UNGROUPED ARBOVIRUSES

Examples of ungrouped arboviruses isolated from India are the following:

1. Wanowrie Virus

This was isolated from *Hyalomma ticks* in India. It had also been isolated from the brain of a young girl who died after a two-day fever in Sri Lanka. The virus is also present in Iran and Egypt.

2. Bhanja Virus

This was isolated from *haemaphysalis ticks* from goats in Ganjam, Orissa. This virus is present in goats in West Africa and South East Europe. Human infections with disease and death have been reported from Yugoslavia. Laboratory infections also have been recorded.

ARBOVIRUS KNOWN TO BE PREVALENT IN INDIA

Some of the arboviruses known to be prevalent in India are as shown in Table 65.5.

KNOW MORE

Epidemiology of Yellow Fever

Yellow fever does not exist in India and it is important to us for this paradoxical reason. India offers a receptive area with a large population of *Aedes aegypti* and nonimmune humans. Strict vigilance is enforced on vaccination and quarantine for travel from endemic areas. This, no doubt, has checked the entry of the virus into India through legitimate passengers. It is estimated that annually, yellow fever strikes 200,000 persons, of whom about 30,000 die.

KEY POINTS

- Arboviruses are viruses which are maintained in nature principally, or to an important extent

Table 65.5: Some of the arboviruses known to be prevalent in India

A. Group A (Alphaviruses)	C. Others
Sindbis	Umbre
Cikungunya	Sathuperi
	Chandipura
B. Group B (Flaviruses)	Chittor
Dengue	Ganjam
KFD	Minnal Venkatapuram
JE	Dhori
West Nile	Kaisodi
	Sandfly fever
	African horse sickness
	Vellore

through biological transmission between susceptible vertebrate hosts by hematophagous arthropods.

- Arboviruses are classified within five families. Most are members of the families Togaviridae, Flaviviridae and Bunyaviridae. Some are Reoviridae and rhabdoviridae.
- Arboviruses cause the clinical syndromes such as fever with or without rash and arthralgia encephalitis, hemorrhagic fever, the characteristic systemic disease, yellow fever.
- The family Togaviridae contains two genera: Alphavirus and rubivirus.
  1. Alphavirus (mosquito-borne); A. Encephalitis viruses; B. Viruses causing febrile illness; i. Chikungunya virus (CHIKV); ii. O'nyong-nyong virus (ONNV); iii. Semliki Forest virus; iv. Sindbis virus; v. Ross River virus;
  2. Rubivirus; Rubella virus
- The Chikungunya virus has been implicated in epidemics in India. Humans are the host, and *Aedes aegypti* mosquitoes the vectors.
- Japanese encephalitis (JE) is a mosquito-borne encephalitis caused by a group B arbovirus (Flavivirus) and transmitted by culicine mosquitoes. *Culex tritaeniorhynchus*, a rural mosquito that breeds in rice fields, is the principal vector. Natural infections of Japanese B encephalitis occur in Adreid birds which act as reservoirs.
- Yellow fever virus causes yellow fever, an acute, febrile, mosquito-borne illness that occurs only in Africa and Central and South America. The disease yellow fever does not occur in India.
- Dengue virus is distributed worldwide. Four types of dengue virus exist: DEN1, DEN2, DEN3 and 4. The dengue virus causes classic dengue or break-down fever, dengue hemorrhagic fever(DHF), and dengue shock syndrome (DSS). Both dengue virus and chikungunya virus are transmitted by *Aedes aegypti* mosquito.
- Kyasanur Forest Disease (KFD) is a febrile disease associated with hemorrhages caused by an arbovirus flavivirus and transmitted to man by bite of infective ticks.
- Hantaviruses are classified in the Hantavirus genus of the Bunyaviridae family. The viruses cause hemorrhagic fever with renal syndrome (HFRS). The genus contains at least four species:
  Principal vertebrate reservoirs comprise *Apodemus agrariu* srodents in Asia and *Clethrionomys glareolus* (bank vole) in Europe. Wild rodents (species of mice, rats, voles) are the hosts; vectors have not been detected. Transmission to humans occurs by inhaling aerosols of rodent excreta (urine, feces, saliva).

IMPORTANT QUESTIONS

1. Classify arboviruses. Discuss various methods used for laboratory diagnosis of arboviruses.
2. Name the arboviruses which cause encephalitis. Decribe briefly Japanese B encephalitis.
3. List the arboviruses prevalent in India. Describe the pathogenicity and laboratory diagnosis of dengue virus.
4. Write short notes on:
   Chikungunya
   Yellow fever
   Dengue fever (or) Break bone fever
   Kyasanur Forest Disease (KFD)
   Bunyaviruses
   Sandfly fever (phlebotomus fever)
   Hantaviruses

FURTHER READING

Banerjee K. Emerging viral infections with special reference to India. Ind J Med Res 1996;103:177.

Burke OS, Monath TP. Flaviviruses. In: Knipe OM, et al (Eds) Fields Virology, 4th ed. Lippincott Williams & Wilkins, 2001.

Gibbons RV, Vaughan DW. Dengue: an escalating problem. Br Med J 2002;324:1563.

Griffin DE. OM, et al. (Eds) Alphaviruses. In: Fields Virology, 4th ed. Knipe. Lippincott Williams & Wilkins, 2001.

Khan AS, et al. Hantavirus pulmonary syndrome. Lancet 1996; 347:739.

CHAPTER 66

Rhabdoviruses

LEARNING OBJECTIVES

After reading and studying this chapter, you should be able to:

- Describe morphology of rabies virus.
- Describe the following: Street virus vs. fixed virus; pathogenesis and clinical features of rabies; Negri bodies.
- Discuss laboratory diagnosis of rabies.
- Discuss prophylaxis against rabies.
- Describe neural and nonneural vaccines against rabies.

INTRODUCTION

Bullet shaped, enveloped viruses with single stranded RNA genome are classified as rhabdoviruses (from *rhabdos,* meaning rod). The family Rhabdoviridae contains viruses that infect mammals, reptiles, birds, fishes, insects and plants. Some members multiply in vertebrates and arthropods.

Classification

Rhabdoviruses infecting mammals belong to two genera (Table 66.1):

A. ***Lyssavirus*** **(Greek,** ***lyssa*****—meaning madness or rage, a synonym for rabies)**
   It contains rabies virus and related viruses (Lagos bat virus, Mokola, Duvenhage).
B. ***Vesiculovirus***
   It contains vesicular stomatitis virus (VSV) and related viruses and affects horses and cattle. Other genera in the family affect only insects or plants.

Table 66.1: Some members of the family *Rhabdoviridae*

A. **Genus Lyssavirus Members**
   Rabies
   Lagos bat virus
   Mokola
   Duvenhage.
B. **Vesiculovirus**
   Vesicular stomatitis virus and other viruses infecting vertebrates and invertebrates.

RABIES VIRUS

Morphology

1. **Virion:** The rabies **virion** consists of a helical nucleocapsid contained in a **bullet-shaped lipoprotein envelope** 180 × 75 nm, with one end rounded or conical and the other plane or concave (Fig. 66.1).
2. **Proteins:** Protruding from the lipid envelope are approximately 200 glycoprotein (G) spikes of the virus, responsible for viral attachment to cellular receptors and subsequent fusion activity. G also has **hemagglutinin activity** and has important antigenic sites which are neutralized by specific antibody. Spikes do not cover the planar end of the virion.
3. **Membrane or matrix (M) protein**: Beneath the envelope is the **membrane or matrix (M) protein** layer and is the major structural protein of the virus which may be invaginated at the planar end. The membrane may project outwards from the planar end of some virions forming a bleb (Fig. 66.1).
4. **Genome:** The core of the virion consists of helically arranged ribonucleoprotein. The genome is single-stranded RNA, linear, nonsegmented, negative-sense. RNA-dependent RNA polymerase enzyme which is essential for the initiation of replication of the virus, is enclosed within the virion in association with the ribonucleoprotein core.

Resistance

Rabies virus is sensitive to ethanol, iodine preparations, quaternary ammonium compounds, soap, detergents and lipid solvents such as ether, chloroform and acetone. It is inactivated by phenol, formalin, beta propiolactone, ultraviolet irradiation, sunlight and heat at 50°C for 1 hour or 60°C for 5 minutes. Rabies virus survives storage at 4°C for weeks and at –70°C for years or by lyophilization. It is inactivated by CO_2 so on dry ice it must be stored in glass-sealed vials.

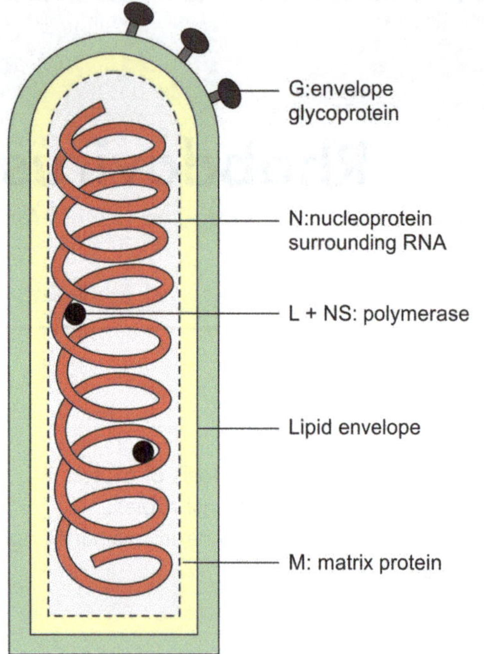

Fig. 66.1: Rabies virus: Bullet shaped virion, showing tightly wound helix of ribonucleoprotein in the core, and bilayered membranous envelope carrying glycoprotein spikes.

Antigenic Properties

There is a single serotype of rabies virus. However, there are strain differences among viruses isolated from different species (raccoons, foxes, skunks, canines, bats) in different geographic areas. These viral strains can be distinguished by epitopes in the nucleoprotein and glycoprotein recognized by monoclonal antibodies as well as by specific nucleotide sequences. There are at least five antigenic variants found in terrestrial animals and eight other variants found in bats in the United States.

A. Glycoprotein G

The surface spikes are composed of glycoprotein G, which is important in pathogenesis, virulence and immunity. It mediates the binding of the virus to acetyl choline receptors in neural tissues, induces hemagglutination inhibiting (HI) and neutralizing (protective) antibodies and stimulates cytotoxic T cell immunity. It is a serotype specific antigen. Purified spikes containing the viral glycoprotein elicit neutralizing antibody in animals. The purified glycoprotein may therefore provide a safe and effective subunit vaccine.

Hemagglutinating Activity

Rabies virus possesses hemagglutinating activity, optimally seen with goose erythrocytes at 0-4°C and pH 6.2. Hemagglutination is a property of the glycoprotein spikes. It is inactivated by heat (56°C for 30-60 minutes), ether, trypsin, pronase, deoxycholate or Tween 80 but not by beta propiolactone. HI antibodies develop following infection or immunization and parallel neutralizing antibodies. HI tests would therefore provide a useful method of assaying immunity to rabies but the low sensitivity of the test and the presence of nonspecific inhibitors in all sera limit its value. Nonspecific inhibitors can be destroyed by treatment with acetone or kaolin. The hemagglutinin antigen is species specific and distinct from the antigens on rabies related viruses.

B. Nucleocapsid Protein

Complement fixing antibodies are induced by the **nucleocapsid protein** and are not protective. This antigen is group specific and cross-reactions are seen with some rabies related viruses. Antiserum prepared against the purified nucleocapsid is used in diagnostic immunofluorescence for rabies.

C. Other Antigens

Other antigens identified include two **membrane proteins, glycolipid** and **RNA dependent RNA polymerase.**

Host range and Growth Characteristics

A. Animals

Rabies virus has a wide host range. All warm-blooded animals, including humans, can be infected. All mammals are susceptible to rabies infection, though differences in susceptibility exist between species. Susceptibility varies among mammalian species, ranging from very high (foxes, coyotes, wolves) to low (opossums); those with intermediate susceptibility include skunks, raccoons, and bats. Humans and dogs occupy an intermediate position. Pups are more susceptible than adult dogs. Experimental infection can be produced in any laboratory animal but mice are the animals of choice. They can be infected by any route. After intracerebral inoculation, they develop encephalitis and die within 5-30 days.

Street Virus

The rabies virus isolated from natural human or animal infection is termed the ***street virus***. Following inoculation by any route, it can cause fatal encephalitis in laboratory animals after a long and variable incubation period of about 1-12 weeks (usually 21-60 days in dogs). Intracytoplasmic inclusion bodies (**Negri bodies**) can be demonstrated in the brain of animals dying of street virus infection.

Fixed Virus

After several serial intracerebral passages in rabbits, the virus undergoes certain changes and becomes what is called the ***fixed virus*** that no longer multiplies in extraneural tissues. The **fixed (or mutant) virus** is more neurotropic, though it is much less infective by other routes. After intracerebral inoculation, it produces fatal encephalitis after a short and fixed incubation period of 6-7 days. Negri bodies are usually not demonstrable in the brain of animals dying of fixed virus infection. The fixed virus is used for vaccine production.

B. Chick Embryos

The rabies virus can be grown in chick embryos. The usual mode of inoculation is into the yolk sac. Serial propagation in chick embryos has led to the development of attenuated vaccine strains like Flury and Kelev. Strains adapted to duck eggs which give high yields of virus have been used for the preparation of inactivated vaccines.

C. Tissue Culture

The rabies virus can grow in several primary and continuous cell cultures such as chick embryo fibroblast, porcine or hamster kidney but cytopathic effects are not apparent and the yield of virus is low. The fixed virus strains adapted for growth in human diploid cell, chick embryo and vero cell cultures are used for the production of vaccines.

Pathogenesis

Rabies infection usually results from the bite of rabid dogs or other animals. The virus can also be transmitted following non-bite exposures through the inhalation of aerosolized virus (as may be found in bat caves), in transplanted infected tissue (e.g., cornea), and by inoculation through intact mucosal membranes. The virus present in the saliva of the animal is deposited in the wound. Rabies infection of the animal causes secretion of the virus in the animal's saliva and promotes aggressive behavior ("mad dog"), which in turn promotes transmission of the virus. If untreated, about half of such cases may develop rabies.

The virus appears to multiply in the muscles, connective tissue or nerves at the site of deposition for 48-72 hours. Virus may directly infect nerve endings by binding to nicotinic acetylcholine or ganglioside receptors of neurons or muscle at the site of inoculation. The virus remains at the site for days to months (Fig. 66.2) before progressing to the central nervous system (CNS). Rabies virus travels by retrograde axoplasmic transport to the dorsal root ganglia and to the spinal cord. Once the virus gains access to the spinal cord, the brain becomes rapidly infected. The affected areas are the hippocampus, brain stem, ganglionic cells of the pontine nuclei, and Purkinje cells of the cerebellum. The virus then disseminates from the CNS via afferent neurons to highly innervated sites, such as the skin of the head and neck, salivary glands, retina, cornea, nasal mucosa, adrenal medulla, renal parenchyma, and pancreatic acinar cells.

After the virus invades the brain and spinal cord, an encephalitis develops, and neurons degenerate. The presence of the virus in the saliva and the irritability and aggression brought on by the encephalitis ensure the transmission and survival of the virus in nature. The virus ultimately reaches virtually every tissue in the body, though the centrifugal dissemination may be interrupted at any stage by death. The virus is almost invariably present in the cornea and the facial skin of patients because of their proximity to the brain. The virus may also be shed in milk and urine. Rabies virus has not been isolated from the blood of infected persons. Despite the extensive CNS involvement and impairment of CNS function, little histopathologic change can be observed in the affected tissue other than the presence of Negri bodies.

With rare exception (three known cases), rabies is fatal once clinical disease is apparent. The length of the incubation period is determined by: (1) the concentration of the virus in the inoculum, (2) the proximity of the wound to the brain, (3) the severity of the wound, (4) the host's age, and (5) the host's immune status. There is a higher attack rate and shorter incubation period in persons bitten on the face or head. The lowest mortality occurs in those bitten on the legs.

Clinical Features

A. Humans

Rabies is primarily a disease of lower animals and is spread to humans by bites of rabid animals or by contact with saliva from rabid animals. Rabies is usually acquired from the bite of an infected animal, but simple licking of abraded skin may also transmit the virus. The infection has also been acquired from aerosols in bats' caves.

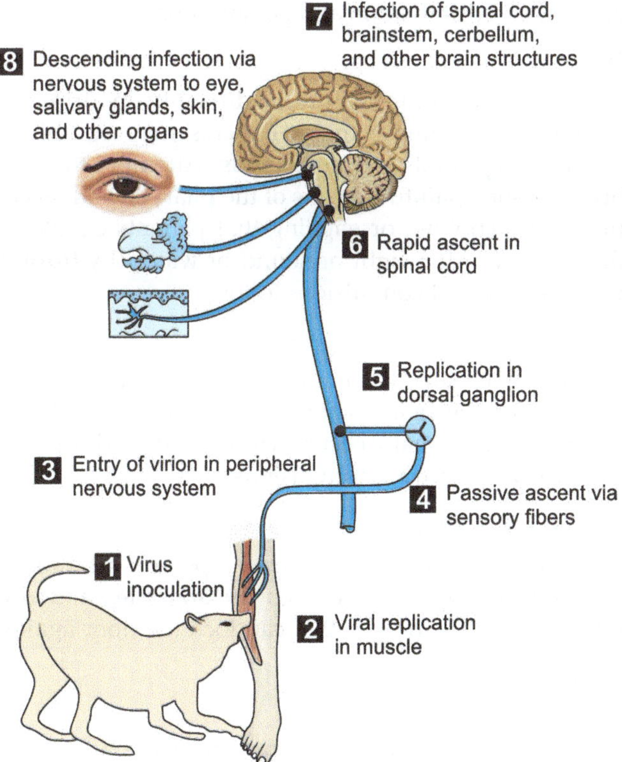

Fig. 66.2: Pathogenesis of rabies virus infection

The incubation period in humans is typically 1-2 months but may be as short as 1 week or as long as many years (up to 19 years). It is usually shorter in children than in adults. The incubation period depends on the site of the bite, severity of the bite, number of wounds, amount of virus injected, species of the biting animal, protection provided by the clothing and treatment undertaken, if any. In general, incubation period tends to be shorter in severe exposures and bites on face, head, neck and upper extremities and bites by wild animals. In no other specific communicable disease is the incubation period so variable and dependent on so many factors as in rabies.

Phases of Clinical Spectrum

The clinical spectrum can be divided into three phases: a short prodromal phase, an acute neurologic phase, and coma.

a. Prodromal Phase

The prodrome, lasting 2-10 days, may show any of the following nonspecific symptoms: malaise, anorexia, headache, photophobia, nausea and vomiting, sore throat, and fever. Excessive libido, priapism and spontaneous ejaculation may occur rarely.

b. Acute Neurologic Phase

During the acute neurologic phase, which lasts 2-7 days, patients show signs of nervous system dysfunction such as nervousness, apprehension, hallucinations, and bizarre behavior. General sympathetic overactivity is observed, including lacrimation, pupillary dilatation, and increased salivation and perspiration.

Hydrophobia

The pathognomonic feature is difficulty in drinking, together with intense thirst. Patients may be able to swallow dry solids but not liquids. Attempts to drink bring on such painful spasms of the pharynx and larynx producing choking or gagging that patients develop a dread of even the sight or sound of water (**hydrophobia**). Generalised convulsions follow.

c. Coma

Patients who survive the stage of acute neurological involvement lapse into coma, which may last for hours or days. Death usually occurs within 1-6 days due to respiratory arrest during convulsions.

B. Rabies in Dogs

In developing countries, over 90 percent of human deaths from rabies are caused by dog bites and dog rabies control is the key that can lock the door against human rabies.

Clinical Picture

The incubation period in dogs ranges from 3-8 weeks, but it may be as short as 10 days or as long as a year or more. Rabies in dogs may manifest itself in two forms – Furious rabies and Dumb rabies.

a. Furious Rabies

This is the typical "**mad dog syndrome**", characterized by:

i. **A change in behavior:** The animal loses its fear of people, becomes very aggressive, bites without provocation and bites unusual objects like sticks, straw and mud.
ii. **Running amuck**
iii. **Change in voice:** The dog barks or growls in a hoarse voice or often unable to bark because of paralysis of laryngeal muscles.
iv. **Excessive salivation and foaming at the angle of the mouth.**
v. **Paralytic stage:** Paralysis, convulsions and death follow.

b. Dumb Rabies

In this type, the excitative or irritative stage is lacking. The disease is predominantly paralytic form in which the animal lies huddled, unable to feed. The dog withdraws itself from being seen or disturbed. The dog may not bite but attempts to feed it are dangerous. The dumb form is as infectious as the furious type. Rabid dogs usually die in 3-5 days. Once the symptoms of rabies develop in an animal, it rarely survives more than a week.

Laboratory Diagnosis

Tests are performed on samples of saliva, serum, spinal fluid, and skin biopsies of hair follicles at the nape of the neck. Saliva can be tested by virus isolation or reverse transcription followed by polymerase chain reaction (RT-PCR). Serum and spinal fluid are tested for antibodies to rabies virus. Skin biopsy specimens are examined for rabies antigen in the cutaneous nerves at the base of hair follicles.

1. Rabies Antigens by Immunofluorescence

The method most commonly used for diagnosis is the demonstration of rabies virus antigens by immunofluorescence. The specimens tested are corneal smears and skin biopsy (from face or neck) or saliva antemortem, and brain postmorterm. Direct immunofluorescence is done using antirabies serum tagged with fluorescein isothiocyanate. The use of monoclonal antibody instead of crude antiserum makes the test more specific.

2. Virus Isolation

i. Mouse Inoculation

Samples of brain tissue, saliva, CSF, or urine may be injected intracerebrally into newborn mice for isolation of the virus. Infection in mice results in encephalitis and death. The inoculated mice are examined for signs of illness and their brains are examined at death or at 28 days postinoculation for Negri bodies, or by immunofluorescence rabies antigen.

ii. Isolation in Cell Culture

A more rapid and sensitive method is isolation of the virus in tissue culture cell lines (WI38, BHK21, CER). CPE is minimal and so virus isolations are identified by immunofluorescence. A positive IF test can be obtained as early as 2-4 days after inoculation. The identity of the isolate can be established by the neutralization test with specific antirabies antibody.

3. Serology

Rabies antibodies can be detected in the serum and CSF of the patient by ELISA. High titer antibodies are present in the CSF in rabies but not after immunization. Their demonstration can therefore be used for diagnosis.

4. Detection of Nucleic Acid

Reverse transcription-polymerase chain reaction (RTPCR) testing can be used to amplify parts of a rabies virus genome from fixed or unfixed brain tissue for detection of rabies virus RNA. This technique can confirm dFA results and can detect rabies virus in saliva and skin biopsy samples.

Animal Rabies

Laboratory diagnosis of rabies in dogs and other biting animals is of great importance in assessing the risk of infection and deciding postexposure treatment. The head of the animal is cut off and sent to the nearest testing laboratory, duly packed in ice in an air-tight container. Alternatively, the brain may be removed with antiseptic precautions and sent in 50 percent glycerol-saline for examination and the other in Zenker's fixative, sent for biological test and microscopy, respectively. The portion of brain sent should include the hippocampus and cerebellum as Negri bodies are most abundant there. The following tests are done in the laboratory:

1. Immunofluorescence Test

This is a highly reliable and the best single test currently available for the rapid diagnosis of rabies viral antigen in infected specimens. This test can establish a highly specific diagnosis within a few hours. Examination of salivary glands by immunofluorescence is useful. Further, fluorescent antibody titers in clinical rabies have been well in excess of 1:10,000, a feature which helps to distinguish between rabies and vaccine reaction.

2. Demonstration of Inclusion Bodies (Negri Bodies)

A definitive pathologic diagnosis of rabies can be based on the finding of **Negri bodies** in the brain or the spinal cord. Negri bodies, named after the Italian physician who first discovered them. This is still the method most commonly used as facilities for immunofluorescence and biological tests are not available in many laboratories.

Impression smears of the brain are stained by Seller's technique (basic fuchsin and methylene blue in methanol), which has the advantage that fixation and staining are done simultaneously. **Negri bodies** are seen as intracytoplasmic, round or oval, purplish pink structures with characteristic basophilic inner granules. Negri bodies vary in size from 3-27 µm. Other types of inclusion bodies may sometimes be seen in the brain in diseases such as canine distemper but the presence of inner structures in the Negri bodies makes differentiation easy. Failure to find Negri bodies does not exclude the diagnosis of rabies. The microscopic examination for Negri bodies identifies 75-90 percent of cases of rabies in dogs. Failure to find Negri bodies does not exclude the diagnosis of rabies.

Negri bodies contain rabies virus antigens and can be, demonstrated by immunofluorescence. Both Negri bodies and rabies antigen can usually be found in animals or humans infected with rabies, but they are rarely found in bats.

If impression smears are negative, the tissue should be sectioned and stained by Giemsa or Mann's method.

3. Isolation of the Rabies Virus (Biological Test)

This is done as described above, for human rabies diagnosis.

4. Corneal Test

Rabies virus antigen can be detected in live animals in corneal impressions or in frozen sections of skin biopsies by the fluorescent antibody test. A positive result is indicative of rabies, but a negative result does not rule out the possibility of infection.

Prophylaxis

This may be considered under:

A. Post-exposure prophylaxis
B. Pre-exposure prophylaxis.

A. Post-Exposure Prophylaxis

Specific prophylaxis is generally employed after exposure to infection and is therefore called antirabic treatment. This consists of:

a. Local treatment
b. Antirabic vaccines
c. Hyperimmune serum.

a. Local Treatment of Wound

Prompt and adequate local treatment of all bite wounds and scratches is the first requisite and is of utmost importance. The purpose of local treatment is to remove as much virus as possible from the site of inoculation before it can be absorbed on nerve endings. The local treatment comprises the following measures:

i. Cleansing

Immediate flushing and washing the wound(s), scratches and the adjoining areas with plenty of soap and water, preferably under a running tap, for at least 5 minutes is of paramount importance in the prevention of human rabies. If soap is not available, simple flushing of the wounds with plenty of water should be done as first-aid.

ii. Chemical Treatment

Whatever residual virus remains in the wound(s), after cleansing, should be inactivated by irrigation with *virucidal* agents—either alcohol (400-700 ml/liter), tincture or 0.01 percent aqueous solution of iodine or povidone iodine.

iii. Suturing

Bite wounds should not be immediately sutured to prevent additional trauma which may help spread the virus into deeper tissues.

iv. Antirabies Serum

The local application of antirabies serum or its infiltration around the wound has been shown to be highly effective in preventing rabies.

v. Antibiotics and Antitetanus Measure

The application of antibiotics and antitetanus procedures when indicated should follow the local treatment recommended above.

b. Antirabic Vaccines

Pasteur (1883) performed the first successful human antirabies vaccination. Antirabic vaccines fall into two main categories: neural and non-neural (Table 66.2). The former are associated with serious risk of neurological complications and have been replaced by the latter.

1. Neural Vaccines

These are suspensions of nervous tissues of animals infected with the fixed rabies virus. The earliest was Pasteur's cord vaccine prepared by drying over caustic potash, for varying periods, pieces of infected rabbit spinal cord. This was replaced by infected brain vaccines, of which there have been several preparations.

Table 66.2: Rabies vaccines

1. **Neural vaccines**
   - I. Semple vaccine
   - II. Beta-propiolactone (BPL) vaccine
   - III. Suckling mouse brain vaccine.
2. **Non-neural vaccines**
   - A. Duck egg vaccine
   - B. Cell culture vaccines
     - a. First-generation cell culture vaccine
       - Human diploid cell vaccine (HDCV)
     - b. Second-generation cell culture vaccines
       - Purified chick embryo cell vaccine (PCEC)
       - Purified vero cell rabies vacine (PVRV)
3. **Third-generation rabies vaccine**
   Poxvirus-rabies glycoprotein recombinant vaccine (undergoing clinical trials in humans)

Nervous tissue vaccines are crude products capable of causing severe and even fatal reactions. Although apparently effective, they are generally of low or variable potency and are usually administered to exposed subjects in a large number of doses.

i. Semple Vaccine

Semple (1911) developed this vaccine at the Central Research Institute, Kasauli (India), had been the most widely used vaccine for over half a century. It is a 5 percent suspension of sheep brain infected with fixed virus and inactivated with phenol at 37°C, leaving no residual live virus.

ii. Beta Propiolactone (BPL) Vaccine

This is a modification of the Semple vaccine, in which beta propiolactone is used as the inactivating agent instead of phenol. It is prepared from fixed virus grown in the brains of adult sheep (Semple type) or other animals. It is believed to be more antigenic.

iii. Infant Brain Vaccines

The encephalitogenic factor in brain tissue is a basic protein associated with myelin. It is scanty or absent in the nonmyelinated neural tissue of newborn animals. So vaccines were developed using infant mouse, rat or rabbit brain. To reduce the hazard from neuroparalytic factors, vaccines have been prepared on a large scale from the brains of suckling mice (less than 9 days old). This vaccine is considered to be devoid of neuroparalytic effect because of the absence or low content of myelin in the neonatal animal. Occasional cases of neurological reactions have occurred following infant brain vaccines also. Infant brain vaccine is impractical in India due to the very large quantities required.

Disadvantages of Neural Vaccines

Neural vaccines are unsatisfactory for many reasons:

i. **Poor immunogens:** They are poor immunogens as they contain mostly nucleocapsid antigen, with only small quantities of glycoprotein G, which is the sole protective antigen.
ii. **Contains infectious agents:** They may contain infectious agents which may not be inactivated during vaccine preparation and storage.
iii. **Encephalitogenic:** They are encephalitogenic. It causes sensitization to nerve tissue and results in postvaccinal encephalitis (an allergic disease) with substantial frequency (0.05%).

Note: In the developed countries, neural vaccines have been abandoned. The only reason for their continued production and use in a few developing countries is that they are cheap.

2. Non-neural Vaccines

i. Egg Vaccines

a. Duck Embryo Vaccine (DEV)

The rabies virus is grown in embryonated duck eggs and inactivated with beta propiolactone, but was discontinued because of its poor immunogenicity. Recently a highly purified DEV has been developed which contains reduced amounts of host tissue and therefore an improvement over currently used vaccines derived from adult animal nervous tissue. DEV is not available in India.

b. Live Attenuated Chick Embryo Vaccines

These vaccines were used for vaccination of animals. Two types of vaccines were developed with the Flury strain.

i. **Low Egg Passage (LEP) vaccine**—at 40-50 egg passage level for immunization of dogs.
ii. **High Egg Passage (HEP) vaccine**—at 180 passage level for cattle and cats. These are not in use now. Rabies viruses grown in various animal cell cultures have also been used as vaccines for domestic animals.

ii. Cell Culture Vaccines

They are more potent and much safer than the conventional brain tissue vaccines. Further, immunization requires fewer injections of smaller volume with relatively few side-effects. The cell culture vaccines are of two types:

a. Human Origin

First Generation Cell Culture Vaccine

Human Diploid Cell Vaccine (HDCV)

The first cell culture vaccine was the human diploid cell (HDC) vaccine developed by Koprowsky, Wiktor and Plotkin. It is a purified and concentrated preparation of fixed rabies virus (Pitman-Moore strain) grown on human diploid cells (WI 38 or MRC 5) and inactivated with beta propiolactone or tri-n-butyl phosphate. It is highly antigenic and free from serious side effects. Its only disadvantage is its high cost. HDC vaccine is now licensed for use in a number of countries including India for both pre- and post-exposure immunization.

b. Non-Human Origin

Second Generation Cell Culture Vaccines (Table 66.2)

e.g. purified chick embryo vaccine and purified vero cell retires vaccine.

Because of their potency and low cost, "Second generation" vaccines are being preferred. Second generation tissue culture vaccines are cheaper than HDC vaccines. The WHO has recommended that cultures of the human diploid cell line should be replaced by cultures of animal cell lines susceptible to rabies infection. They are derived from **"non-human" sources**. These include: **primary cell culture** vaccines grown on chick embryo, hamster kidney and dog kidney cells and **continuous cell culture** vaccines grown on *vero cell line* derived from the kidneys of vervet monkey or African green monkey (*Cercopithecus aethiops*).

Cell culture vaccines in India: In India, the following cell culture vaccines are available:

i. Human diploid cell (HDC) vaccine
ii. Purified chick embryo cell (PCEC) vaccine
iii. Purified vero cell (PVC) vaccine.
All three of them are equally safe and effective

Note: Until cell culture vaccines become more generally available, time-honored nervous tissue vaccines will have to be used in many Third World Countries.

c. Subunit Vaccine

The glycoprotein subunit on the virus surface, which is the protective antigen, has been cloned and recombinant vaccines produced. They are still in the experimental stage. This vaccine may prove valuable in the immunization of both wildlife reservoir species and domestic animals.

Indications for Antirabies Treatment

Antirabies treatment should be started immediately:

i. If the animal shows signs of rabies or dies within 10 days of observation.
ii. If the biting animal cannot be traced or identified.
iii. Unprovoked bites.
iv. Laboratory tests (e.g. fluorescent rabies antibody test or test for Negri bodies) of the brain of the biting animal are positive for rabies.
v. All bites by wild animals.

Vaccination Schedules

Neural Vaccines

The dosage of the vaccine depends on the degree of risk to which the patient has been exposed. Accordingly, patients are classified as follows:

Classification of Exposures (Table 66.3)

One of the factors determining the dose of anti-rabies vaccine is the degree of risk of rabies to which the person is exposed. Accordingly, patients are classified as follows:

Class I (Slight risk)
Class II (Moderate risk)
Class III (Severe risk).

Dosage Schedules

Two dosage schedules (one recommended by the Central Research Institute, Kasauli and the other recommended by the Pasteur Institute of Southern India, Coonoor) are followed in India. Vaccines should be given according to the schedule and dose recommended by the manufacturers. A full schedule consists of 7-10 daily inoculations followed by 1-2 boosters. Booster doses are imperative when combined serum and vaccine treatment is employed regardless of the vaccine or schedule used.

The recommended schedule of vaccination for the different classes is as follows:

	Semple vaccine	BPL vaccine
Class I	2 ml × 7 days	2 ml × 7 days
Class II	5 ml × 14 days	3 ml × 10 days
Class III	10 ml × 14 days	5 ml × 10 days

Table 66.3: Guidelines for post-exposure prophylaxis of rabies		
Category of risk	*Type of exposure*	*Recommendations*
I	Touching or feeding of animals; licks on intact skin	None if case history is reliable
II	Nibbling of uncovered skin; Minor scratches or abrasions without bleeding; Licks on broken skin	Administer vaccine immediately. Stop treatment if animal remains healthy throughout an observation period of 10 days, or if animal is killed humanely and found to be negative for rabies by appropriate laboratory techniques
III	Single or multiple transdermal bites or scratches. Contamination of mucous membrane with saliva (i.e. licks)	Administer rabies immunoglobulin and vaccine immediately. Stop treatment if animal remains healthy throughout an observation period of 10 days, or if animal is killed humanely and found to be negative for rabies, by appropriate laboratory techniques

The above schedule for the BPL vaccine is recommended by the Pasteur Institute, Coonoor. The Central Research Institute, Kasauli, recommends a slightly different dosage for its vaccine *(The manufacturer's instructions should be followed in every case).*

Site for Vaccination

The ideal site for vaccination is the anterior abdominal wall, for this area offers enough space to accommodate the large quantity of vaccine to be injected. The injections are given deep subcutaneously. The development of immunity after anti-rabies immunization is rather slow. The immunity following vaccination with neural vaccines is expected to last for six months only and any exposure later should receive fresh treatment.

Adverse Reactions

Nervous tissue vaccines contain neuroparalytic factors such as myelin. Despite prolonged treatment and the relatively large doses of the vaccine involved, the majority of patients suffer no inconvenience throughout the period of treatment. However, in some persons, the following complications are encountered:

1. *General:* Headache, insomnia, giddiness, palpitation, diarrhea.
2. *Local:* Itching irritation, pain, tenderness, redness and swelling at the site of injection.
3. *Allergic:* Urticaria, syncope, angioneurotic edema, anaphylactic reaction.
4. *Neuroparalysis*: Post-vaccinial paralysis due to sensitization. It may be of the neuritic type, the dorsilumbar type, the Landry type of ascending paralysis or encephalomyelitis.

The etiology of neurological complication is believed to be immune response to the injected brain tissue resulting in organ specific immunological damage as in experimental allergic encephalomyelitis.

When such complications are noticed during the course of vaccination, further vaccination should be withheld and the patient started on corticosteroids. If further vaccination is considered imperative, non-neural vaccine should be used. Severe exertion and the use of alcohol during vaccination have been said to increase the risk of neurological reactions.

Cell Culture Vaccines

The use of modern, inactivated, purified cell-culture and purified duck embryo vaccine should, where economically and technically feasible, replace those produced on brain tissue in both developing and developed countries. All three cell culture vaccines available in India (HDC, PCEC and PVC) have the same dosage schedule, which is the same for both adults and children.

Pre-exposure Prophylaxis

Pre-exposure prophylaxis requires three doses of the vaccine injected on day 0, 7, 21 or 0, 28 and 56. A booster dose is recommended after one year and then one every five years

Postexposure Prophylaxis

Postexposure prophylaxis requires five or six doses, on days 0, 3, 7, 14, 30 and optionally 90. The vaccine is to be given IM or SC in the deltoid region, or in children on the anterolateral aspect of the thigh. Gluteal injections are to be avoided as they are found to be less immunogenic. This course is expected to give protection for at least five years, during which period any further exposure may need only one or two booster doses (on days 0,3) depending on the degree of risk. After five years, it is advisable to give a full five injection course if exposed to infection.

It has been shown that a dose of 0.1 ml administered intradermally is as effective as a 0.5-1.0 ml dose SC or IM and that immunization may thus be made more economical. However, this is not recommended as a routine practice, as intradermal injection is technically difficult, and it will be ineffective if this dose is given subcutaneously by mistake.

Advantages of Cell-culture Vaccines Efficacy and Safety

The major advantages of cell-culture vaccines over conventional vaccines are their efficacy and safety. Results indicated that 3 to 4 injections of cell-culture vaccines produce antibody levels comparable with those induced by 10 injections of a nervous tissue vaccine .

Disadvantage

The disadvantage of cell-culture vaccines is their cost

Passive immunization

Passive immunization is an important adjunct to vaccination and should be invariably employed whenever the exposure is considered of high risk. Two preparations of anti-rabies serum are available for passive immunization:

i. Horse Anti-rabies Serum

Potent antirabies serum has been produced in horse and other animals (mules, donkeys, rabbits). It should be given on day 0 in a single dose of 40 International Unit per kg of body weight subject to a maximum of 3000 Units. Half of the serum is infiltrated around the bite wound and the rest is given intramuscularly. This should be followed by a course of vaccine.

ii. Human Rabies Immune Globulin (HRIG)

Human rabies Ig (HRIG) has now replaced equine antirabies serum in many countries. It is now commercially available. The dose recommended is a single administration of 20 IU per kg of body weight. The recommended procedure is to inject part of the dose around the wound and to administer the rest by IM in the gluteal region. In persons receiving the serum and vaccine, a booster dose of cell culture vaccine on day 90 may be given. It does not require any prior sensitivity testing.

Recommendations for postexposure prophylaxis, as endorsed by the WHO in 1988, are shown in *'Vaccine failures'* (persons developing rabies even after course of immunization) are not uncommon with neural vaccines, while they are extremely rare when immediate local treatment has been followed by rabies immunoglobulin and a full course of a cell culture vaccine. In view of the safety of the cell culture vaccine, it would be advisable to recommend the vaccine even when there is the slightest risk of exposure to rabies.

Vaccine for Animals

Antirabies immunization in animals is to be done as pre-exposure prophylaxis. Postexposure treatment is not generally of much use. Neural vaccines are not satisfactory as they are not adequately immunogenic, need multiple doses and have to be repeated every six months. Concentrated cell culture vaccines containing inactivated virus are now available, which give good protection after a single IM injection. Injections are given at 12 weeks of age and repeated at 1-3 year intervals. Rabies vaccines may be given separately or as combined vaccine for immunization against other common veterinary infections also.

Preexposure Prophylaxis

This is indicated for persons at high risk of contact with rabies virus (research and diagnostic laboratory workers, spelunkers) or with rabid animals (veterinarians, animal control and wildlife workers). The goal is to attain an antibody level presumed to be protective by means of vaccine administration prior to any exposure. It is recommended that antibody titers of vaccinated individuals be monitored periodically and that boosters be given when required.

Epidemiology

Rabies is the classic **zoonotic** infection, spread from animals to humans. Rabies is believed to be the tenth most common cause of death in humans due to infections. Rabies virus is present in terrestrial animals in all parts of the world except Australasia and Antartica, and some islands like Britain.

All warm blooded animals including man are susceptible to rabies. Rabies in man is a dead-end infection, and has no survival value for the virus. Major source of virus is saliva in bite of rabid animal. Direct person-to-person transmission of rabies has not been recorded. An unusual mode of transmission of rabies has occurred in some recipients of corneal grafts. Minor source is aerosols in bat caves containing rabid bats.

Epidemiological Forms of Rabies

Rabies exists in two epidemiological forms:

a. Urban Rabies

Transmitted by domestic animals like dogs and cats.

Most cases of human rabies follow dog bites but in endemic areas almost any animal can transmit rabies. In India, antirabic treatment is to be considered following the bite of any animal except rats. Where urban or domestic rabies has been controlled, as in the USA, the majority of infections are due to bites by wild animals.

b. Wild-life or Sylvatic Rabies

Involving animals in the wild, such as jackals, wolves, foxes, mongooses, skunks and bats. In South Africa, the disease is enzootic in the mongoose.

In certain Latin American countries and parts of USA, the vampire bat is an important host and vector of rabies. Bats present a special problem because they may carry rabies virus while they appear to be healthy, excrete it in saliva, and transmit it to other animals and to humans. These bats feed exclusively on the blood of animals and man. They can transmit rabies to animals and humans. Rabies transmitted by the vampire bat is thought to kill hundreds of thousands of cattle annually. Vampire bats have not been reported in India .

Reservoir of Rabies

Reservoir of rabies are wild animals. The primary source of the rabies virus in nature seems to be in the mustelids and viverrids, the ermine in the northern

Table 66.4: Lyssavirus sera/genotypes

Genotype/Serotype Virus	*Virus*	*Isolated from*	*Disribution*
1	Rabies	Warm blooded animals	Worldwide with few exceptions
2	Lagos bat/Natal bat	Bat/cat	Nigeria/Central and South Africa
3	Mokola	Shrew/cat/dog/human	Nigeria/other African countries
4	Duvenhage	Human/bat	South Africa
5	European bat lyssavirus: Type I	Bat/human	Europe
6	European bat lyssavirus: Type II	Bat/human	Europe
7	Australian bat lyssavirus:	Bat/human	Australia

coniferous forests, the skunk, mink and weasel in North America, the mottled pole cat in the USSR, the civet and pole cat in Africa and the mongoose in Asia. From the reservoir species, wild vectors such as foxes, wolves and jackals acquire the infection and occasionally epizootics occur in these species. Carnivorous animals may acquire the infection by eating carcasses containing the virus. From these species, the disease spreads to dogs and other domestic animals.

Rabies in India

Rabies is endemic in India and occurs in all the parts of the country with exception of Lakshadweep, and Andman and Nicobar islands. Rabies is not a notifiable disease and the 30,000 deaths reported by national authorities may not be a complete picture since these represent only the deaths reported from hospitals. It is estimated that the number of deaths due to rabies may be 10 times more than those reported. Every year approximately 1.1 to 1.5 million people receive post-exposure treatment with either nerve tissue or cell culture rabies vaccine. More than 95 percent of these cases are bitten by dogs. The dog population in India is estimated to be around 25 million, and most of them are not protected against rabies.

Control

Human rabies can be checked by control of rabies in domestic animals, by registration, licensing and vaccination of pets and destruction of stray animals. In countries where wildlife rabies exists and where contact between domestic animals, pets, and wildlife is inevitable, all domestic animals and pets should be vaccinated. With the dog population in India estimated to be around 25 million, the problem is immense. Vaccine baits (chicken head or other meat containing live attenuated rabies virus) have been used to immunize the red fox in an attempt to check the epizootic in the forests of Europe. The technique holds much promise for the future control of rabies not only in foxes but also in other wild-life species.

RABIES RELATED VIRUSES

The genus Lyssavirus consists of the rabies virus and other serologically related viruses. The relevance of rabies related viruses in human disease is not clear, though some of them have caused illness and death in humans. They are considered to represent a biological bridge between the rabies virus and other rhabdoviruses. Lyssaviruses have been classified into seven serotypes (Table 66.4):

1. **Lyssavirus serotype 1:** Rabies virus.
2. **Lagos bat virus:** The Lagos bat virus classified as Lyssavirus serotype 2, was isolated in 1956 from the pooled brains of frugivorous bats from Lagos Island, Nigeria. It causes a rabies-like illness following intracerebral inoculation.
3. **Mokola virus:** The Mokola virus, first isolated in 1968 from shrews captured near Ibadan, Nigeria, has later been found in many wild and domestic animals in Africa. It is classified as Lyssavirus serotype 3.
4. **Duvenhage virus:** The Duvenhage virus was reported in 1971 from the brain of a man who died in South Africa of clinical rabies after being bitten by a bat. It is classified as Lyssavirus serotype 4.
5. **Serotype 6 and Serotype 7 Rabies-like viruses:** Rabies-like viruses isolated from European bats have been classified into two groups: European bat lyssavirus types 1 and 2. They can infect humans, as was found in the UK in 2002, when a wildlife worker fell ill with 'rabies' and died. This was the first 'rabies' death in the UK in a century.
6. **Australian bat lyssavirus:** Australia was considered free of rabies and related viruses till 1996, when a lyssavirus was isolated from a frugivorous bat. Since then a number of similar isolates have been obtained from different types of bats in Australia. Fatal infections have occurred in persons having contact with bats. The virus antibody is widely prevalent among Australian bats which appear to be carriers. The virus, named Australian bat lyssavirus is closely related to, but distinct from the rabies virus. Antirabic vaccine and serum appear to protect against experimental infection.

KNOW MORE

Observation

If the animal is *apparently healthy* at the time of inflicting the bite, treatment appropriate to the degree of exposure should be started at once. The animal should be confined and kept under observation, preferably by a veterinarian for 10 days. The observation period of ten days is recommended because the virus may be present in the saliva 3-4 days before onset of symptoms and the animal usually dies within 5-6 days of developing the disease. If the animal remains healthy after this period, there is no risk of rabies and vaccination, if already started, may be discontinued.

☞ KEY POINTS

- Rhabdoviruses are bullet or rod shaped, enveloped viruses with single stranded RNA genome.
- The rabies virus causes rabies in humans and a wide variety of animals.
- The rabies virus isolated from natural human or animal infection is termed the ***street virus.***
- After several serial intracerebral passages in rabbits, the virus undergoes certain changes and becomes what is called the ***fixed virus*** **(or mutant) virus,**
- Rabies virus has a broad host range. The virus can infect all mammals but dogs, foxes, wolves, and bats are important for transmission of infection.
- Rabies is primarily a disease of lower animals. The virus is transmitted to humans by primarily a bite of a rabid dog or by other infected animals. The infection has also been acquired from aerosols in bats' caves.
- **Laboratory tests for human rabies:** Viral antigens can be demonstrated in the corneal smear, skin biopsy collected from the face or neck, and saliva (antemortem) or in the brain tissue (postmortem) directly by direct fluorescent antibody (DFA) test.
- Isolation of rabies virus can be done by Mouse inoculation and isolation in cell culture. Reverse transcription-polymerase chain reaction (RTPCR) testing can be used for detection of rabies virus RNA.
- Laboratory diagnosis of animal rabies in dogs and other biting animals can be done by demonstration of rabies virus antigen by immunofluorescence, demonstration of inclusion bodies (Negri bodies), isolation of the rabies virus (biological test) and by corneal test.
- Postexposure prophylaxis is started in the persons immediately after exposure to infection and consists of (a) Local treatment of wound (b) Antirabic vaccines and (c) Hyperimmune serum.
- **Cell culture vaccines** such as human diploid cell strain (HDCS), purified chick embryo cell (PCEC) vaccine, and purified vero cell (PVC) vaccines are now increasingly used. These are non-neural vaccines.
- All three cell culture vaccines available in India (HDC, PCEC and PVC) have the same dosage schedule, which is the same for both adults and children. **Postexposure prophylaxis** requires five or six doses, on days 0, 3, 7, 14, 30 and optionally 90. The vaccine is to be given IM or SC in the deltoid region, or in children on the anterolateral aspect of the thigh. This course gives protection for at least five years.

IMPORTANT QUESTIONS

1. Describe the morphology and draw a labeled diagram of rabies virus. Discuss the laboratory diagnosis of rabies.
2. Describe the prophylaxis of rabies.
3. Write short notes on:
   Street virus vs. fixed virus or Differences between street virus and fixed virus.
   Negri bodies.
   Neural vaccines against rabies.
   Nonneural vaccines used against rabies.
   Cell culture vaccines.
   Rabies related viruses.

FURTHER READING

Bogel K, Motschwiller E. Incidence of rabies and postexposure treatment in developing countries. Bull WHO 1986;64:883-7.

Charlton KM. The pathogenesis of rabies and other lyssaviral infections. Lyssaviruses, eds Rupprecht CE, DietzSchold B, Koprowski H. SpringerVerlag, Berlin 1994:95-119.

Fishbein BD and LE Robinson. Rabies, New Engl J Med 1993; 329:1632.

Rupprecht CE, et al. Lyssaviruses. Current Topics in Microbiol Immunol 1994;187.

Smith JS. New aspects of rabies. Clin Microbiol Rev 1996;9:166.

WHO. Expert Committee on Rabies, 8th Report. Technical Series No. 824, Geneva 1992.

Warrell DA, Warrell MJ. Human rabies and its prevention: an overview. Rev Infect Dis 1988;10 (supp14):S726-S731.

CHAPTER 67

Hepatitis Viruses

Learning Objectives

After reading and studying this chapter, you should be able to:

- Classify various hepatitis viruses.
- Tabulate differences between various hepatitis viruses.
- Compare various features of hepatitis A virus (HAV) and hepatitis B virus (HBV).
- Describe the following: Morphology of hepatitis B virus; antigenic structure of hepatitis B virus; modes of transmission of hepatitis B virus; hepatits B carriers.
- Describe laboratory diagnosis of hepatitis B virus.
- Discuss prophylaxis of hepatitis B infections or hepatitis B vaccine.
- Describe the following: Hepatitis C virus or Type C hepatitis; Hepatitis D virus or Delta agent; Hepatitis E virus; Hepatitis G virus

INTRODUCTION

Viral hepatitis is a systemic disease primarily involving the liver. At least six viruses, A through E and a newly discovered G, are considered hepatitis viruses. (The designation 'type F' had been proposed for a putative virus believed to cause transfusion-associated hepatitis, distinct from type A to E. But it proved to be a mutant of type B virus and not a separate entity. Type F was therefore, deleted from the list of hepatitis viruses). Although the target organ for each of these viruses is the liver, they differ greatly in their structure, mode of replication, and mode of transmission and in the course of the disease they cause. Hepatitis A virus (HAV) and Hepatitis B virus (HBV) are the best known, but three non-A, non B hepatitis (NANBH) viruses (C, G, and E) have been described, as has hepatitis D virus (HDV), the delta agent (Table 67.1). Other viruses are associated with hepatitis that cannot be ascribed to known agents, and the associated disease is designated non-A to E hepatitis.

HEPATITIS A VIRUS (HAV) INFECTIOUS HEPATITIS

Hepatitis A virus (HAV) causes infectious hepatitis and is spread by the fecal-oral route. It is a subacute disease of global distribution, affecting mainly children and young adults.

Properties of Hepatitis Viruses

Feinstone and coworkers in 1973, using immunoelectron microscopy (IEM) demonstrated this virus in the feces of experimentally infected human volunteers. Sensitive serologic assays and polymerase chain reaction (PCR) methods have made it possible to detect HAV in stools and other samples and to measure specific antibody in serum.Chimpanzees and marmosets can be infected experimentally. HAV can be grown in some human and simian cell cultures and is the only human hepatitis virus which can be cultivated *in vitro*. It has also been cloned.

Morphology

HAV is a 27 nm nonenveloped RNA virus belonging to the picornavirus family (Fig. 67.1). Although, it was first provisionally classified as enterovirus 72, but it has been placed into a new genus, Hepadnavirus, on the basis of its unique genome. Only one serotype is known.

Resistance

HAV is stable to treatment with 20 percent ether, acid (pH 1.0 for 2 hours), and heat (60°C for 1 hour). The virus is destroyed by autoclaving (121°C for 20 minutes), by boiling in water for 5 minutes, by dry heat (180°C for 1 hour), by ultraviolet irradiation (1 minute at 1.1 watts), by treatment with formalin (1:4000, or by treatment with chlorine (10-15 ppm for 30 minutes). It survives prolonged storage at a temperature of 40°C or below.

Pathogenesis

HAV is ingested and probably enters the blood stream through the oropharynx or the epithelial lining of the intestines to reach its target, the parenchymal cells of the liver. The virus can be localized by immunofluorescence in hepatocytes and Kupffer's cells. Virus is produced in these cells and is released into the bile and from there into the stool. Virus is shed in large quantity into the

Table 67.1: Comparative feature of hepatitis viruses					
Feature	*Hepatitis A*	*Hepatitis B*	*Hepatitis C*	*Hepatitis D*	*Hepatitis E*
1. Virus structure	HAV, 27 nm RNA, Picornavirus (Hepatovirus)	HBV, 47 nm DNA (Hepadnavirus)	HCV, 30-60 nm RNA, Flavivirus (hepacivirus)	HDV, 35-37 nm Defective RNA Deltavirus	HEV,32-34 nm RNA Herpesvirus
2. Modes of infection	Fecal-oral	Parenteral Vertical, Sexual	Parenteral	Parenteral	Fecal-Oral
3. Age Affected	Children	Any age	Adults	Any age	Young adults
4. Incubation Period (days)	15-45	30-180	15-160	30-180	15-60
5. Onset	Acute	Insidious	Insidious	Insidious	Acute
6. Illness	Mild	Occasionally severe	Moderate	Occasionally severe	Mild, except in pregnancy
7. Carrier state	Nil	Common	Present	Nil (only with HBV)	Nil
8. Oncogenicity	Nil	Present specially after neonatal infection	Present	Nil	Nil
9. Prevalence	Worldwide	Worldwide	Probably worldwide	Endemic areas (Mediterranean, N, Europe, Central and N. America)	Only developing countriest India, Asia, Africa, Central, America)
10. Laboratory diagnosis	Symptoms and anti-HAV IgM	Symptoms and serum levels of HBs Ag, HBe Ag, and anti-HBc IgM	Symptoms and anti-HCV ELISA	Anti-HDV ELISA	—
11. Specific prophylaxis	Ig and vaccine	Ig and vaccine	Nil	HBV vaccine	Nil

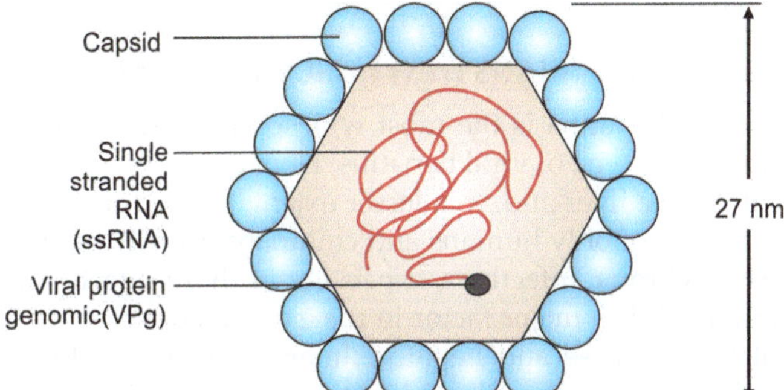

Fig. 67.1: The picornavirus structure of hepatitis A virus. The icosahedral capsid is made up of four viral polypeptides (VP1 to VP4). Inside the capsid is a single-stranded, positive-sense RNA (ssRNA) that has a genomic viral protein (VPg) on the 5' end

stool approximately 10 days before symptoms of jaundice appear or antibody can be detected.

HAV replicates slowly in the liver without producing apparent cytopathic effects. A brief viremia occurs during the preicteric phase, but ceases with the onset of jaundice. Chronic viremia does not occur.

Epidemiology

HAV transmission is by the fecal-oral route in contaminated water, in food, and by dirty hands. Shellfish, especially clams, oysters, and mussels, are important sources of the virus. HAV outbreaks usually originate from a common source (e.g. water supply, restaurant, daycare center). Under crowded conditions and poor sanitation, HAV infections occur at an early age.

The epidemiology of type A hepatitis resembles that of poliomyelitis. In the developing countries, infection is acquired in childhood and by the age of ten, 90 percent of the population possess antibody to the virus and are immune. In India, type A hepatitis is the most common cause of acute hepatitis in children, but is much less frequent in adults.

Natural infection with HAV is seen only in humans. Though primates such as chimpanzees have been shown to acquire the infection from humans and transmit it to human contacts, there is no evidence of any extrahuman source of the virus in nature.

Clinical Features

The incubation period is 2 to 6 weeks. Disease in children is generally milder than that in adults and is usually asymptomatic. The clinical disease consists of two stages: the *prodromal or preicteric* and the *icteric stages*. The onset may be acute or insidious, with fever, malaise, anorexia, nausea, vomiting and liver tenderness. These usually subside with the onset of jaundice. The patient starts to feel better within the next week or so and the jaundice disappears within a month. Recovery is slow, over a period of 4 to 6 weeks.

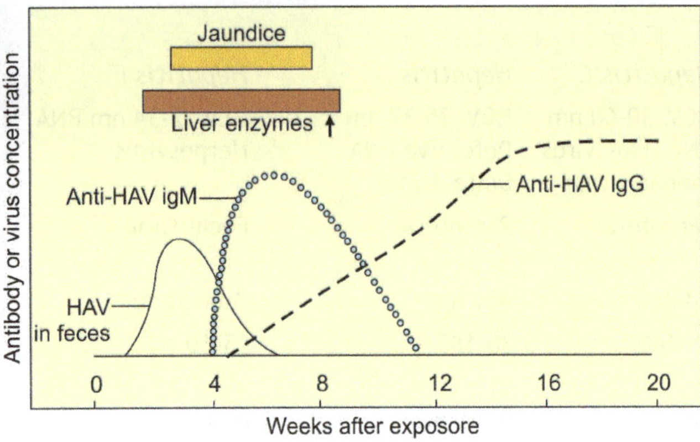

Fig. 67.2: Typical course of hepatitis type A

Hepatitis A is nearly always self-limiting, but relapses have been reported.The disease is milder in children, in whom many infections may be an icteric.

Complications such as fulminant hepatitis, fortunately rare, are seen mainly in older people. Unlike HBV, immune complex-related symptoms (e.g. arthritis, rash) rarely occur in people with HAV disease.

Laboratory Diagnosis

Etiological diagnosis of type A hepatitis may be made by demonstration of the virus or its antibody.

A. **Demonstration of the virus:** The virus can be visualized by immune electron microscopy (IEM) in fecal extracts during the late incubation period and the pre icteric phase, but seldom later.
B. **Demonstration of antibody:** The best way to demonstrate an acute HAV infection is by finding anti-HAV immunoglobulin M (IgM), as measured by an enzyme-linked immunosorbent assay (ELISA) or radioimmunoassay. Antibody IgM anti-HAV antibody appears during the late incubation period, reaches peak levels in 2 to 3 weeks and disappears after 3 to 4 months. IgG antibody appears at about the same time, peaks in 3 to 4 months and persists much longer, perhaps for life (Fig. 67.2). Demonstration of lgM antibody in serum indicates current or recent infection, while IgG antibody denotes recent or remote infection and immunity.
C. **Virus isolation:** The virus is grown in human simian cell cultures. It is notroutinely performed because efficient tissue culture systems for growing the virus are not available.

Prophylaxis

A. General Measures

The spread of HAV is reduced by interrupting the fecal-oral spread of the virus. This is accomplished by avoiding potentially contaminated water or food, especially uncooked shellfish. Chlorine treatment of drinking water is generally sufficient to kill the virus.

B. Immunization

There is only one serotype of HAV and HAV infects only humans, factors that help ensure the success of an immunization program. Natural infection with HAV, clinical or subclinical, leads to lifelong immunity. There is no cross-immunity between HAV and any of the other hepatitis viruses:

1. **Passive protection:** Specific passive prophylaxis by pooled normal human immunoglobulin (16% solution in a dose of 0.2-0.12 ml/kg body weight) intramuscular (IM), before exposure or in early incubation period, can prevent or attenuate clinical illness, while not necessarily preventing infection and virus excretion.
2. **Hepatitis A vaccine:**
   i. *Formalin inactivated, alum conjugated vaccine*—A safe and effective formalin inactivated, alum conjugaged vaccine containing HAV grown in human diploid cell culture is available for use in children and adults at high risk for infection, especially travelers to endemic regions. A full course consists of two intramuscular injections of the vaccine. Protection begins 4 weeks after injection and lasts for 10 to 20 years.
   ii. *Live HAV vaccine*—A live HAV vaccine has been developed in China.

Treatment

Treatment is symptomatic. No specific antiviral drug is available.

HEPATITIS B VIRUS (HBV)—SERUM HEPATITIS

Type B hepatitis is the most widespread and the most important type of viral hepatitis. Hepatits B virus (HBV) infects the liver and, to a lesser extent, the; kidneys and pancreas of only humans and chimpanzees. HBV establishes chronic infections, especially in those infected as infants. It is a major factor in the eventual development of liver disease and hepatocellular carcinoma in those individuals. As there is an effective vaccine against HBV, hepatocellular carcinoma becomes the only human cancer which is vaccine preventable.

Classification

HBV is assigned to a separate family *Hepadnaviridae* (Hepatitis DNA viruses) which consists of two genera:

i. **Orthohepadnavirus:** Containing HBV as well as the woodchuck and ground squirrel hepatitis viruses. HBV is *Hepadnavirus* type 1.
ii. **Avihepadnavirus:** Containing the Pekin duck and grey heron hepatitis viruses.

Structure

HBV is a 42 nm DNA virus with an outer envelope and an inner nucleocapsid core, 27 nm in diameter, enclosing the viral genome and a DNA polymerase (Fig. 67.3).
1. Which is circular double stranded DNA.

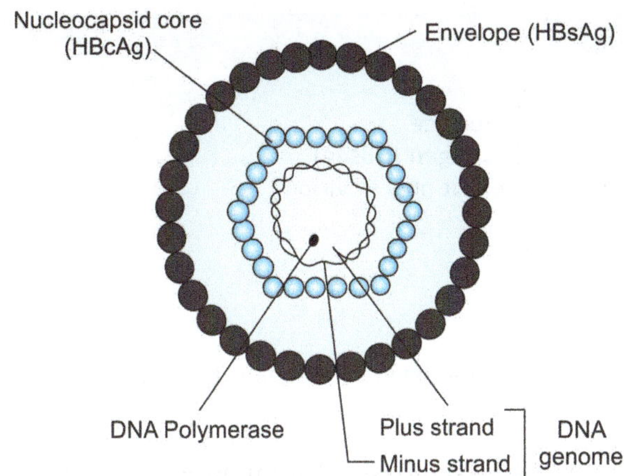

Fig. 67.3: Hepatitis B virus structure

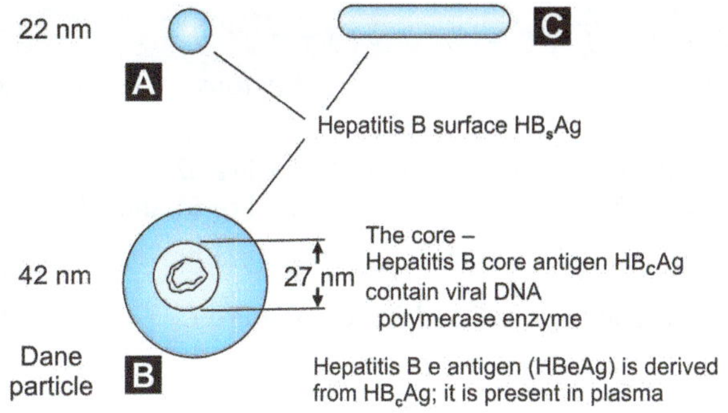

Fig. 67.4A to C: Different types of particles of HBV: A. Spherical 22 nm particle. B. Double shelled 42 nm particle (Dane particle). C. Tubular 22 nm particle

Australia Antigen

In 1965, Blumberg, studying human serum lipoprotein allotypes, observed in the serum of an Australian aborigine, a new antigen which gave a clearly defined line of precipitation with sera from two hemophiliacs who had received multiple blood transfusions. This was named the Australia antigen. By 1968 the 'Australia antigen' was found to be associated with serum hepatitis. It was subsequently shown to be the surface component of HBV. Therefore, the name Australia antigen was changed to hepatitis B surface antigen (HBsAg).

Types of Particles

Under the electron microscope, sera from type B hepatitis patients show three types of particles (Fig. 67.4A to C).

i. *Spherical particle:* The predominant form is a small, spherical particle of (22 nm diameter).
ii. *Tubular particle:* The second type of particle is filamentous or tubular (22 nm diameter) of varying length. Both types of particle consist solely of surplus virion envelope. The particles carry the hepatitis B surface antigen (HBsAg).
iii. *Dane particle:* The third type of particle, is a double walled spherical structure, (42 nm in diameter). This particle is the complete hepatitis B virus. It was first described by Dane in 1970 and so is known as the Dane particle.The outer surface, or envelope, contains HBsAg and surrounds a 27 nm inner nucleocapsid core that contains HBcAg.

Antigenic Structure

1. **Hepatitis B surface antigen (HBsAg):** The envelope proteins expressed on the surface of the virion and the surplus 22 nm diameter spherical and filamentous particles constitute the hepatitis B surface antigen (HBsAg). HBsAg consists of two major polypeptides, one of which is glycosylated.
2. **Hepatitis B core antigen (HBcAg):** The antigen expressed on the core is called the hepatitis B core antigen (HBcAg).
3. **Hepatitis B e antigen (HBeAg):** A third antigen called the hepatitis B e antigen (HBeAg) is a soluble nonparticulate nucleocapsid protein. The HBeAg and HBcAg proteins share most of their protein sequence.
4. **Viral genes and antigens**

Genome

The nucleocapsid encloses the viral genome consisting of two linear strands of DNA held in a circular configuration. One of the strands (the plus strand) is incomplete, so that the DNA appears partially double stranded and partially single stranded. Associated with the plus strand is a viral DNA polymerase, which has both DNA-dependent DNA polymerase and RNA-dependent reverse transcriptase functions. Although a DNA virus, it encodes a reverse transcriptase and replicates through an RNA intermediate. This polymerase can repair the gap in the plus strand and render the genome fully double stranded (Fig. 67.3).

The genome has a compact structure with four overlapping genes. These include structural proteins of the virion surface and core, a small transcriptional transactivator (X), and a large polymerase (P) protein that includes DNA polymerase, reverse transcriptase, and RNase H activities (Table 67.2, Fig. 67.5).

HBV Subtypes

The particles containing HBsAg are antigenically complex. It contains two different antigenic components—the common group reactive antigen a, and two pairs of type specific antigens d-y and w-r, only one member of each pair being present at a time. HBsAg can thus be divided into four major antigenic subtypes: adw, adr, ayw and ayr. The subtypes do not seem to be important in immunity because of the dominant antigen is shared by all. The subtypes breed true, and the index case and contacts in an outbreaks have the same subtype. The finding of identical subtypes would, of course, not confirm the possibility, but differing subtypes would rule it out.

Table 67.2: Genes coding for antigens of HBV

Gene	Regions	Antigen
S (Having three regions S, Pre-S1 and Pre-S2)	S S + Pre-S2 S + Pre-S1 and S2	Major protein (S) } Surface Middle protein (M) } antigen (HBsAg) Large protein (L)—Present only in virion
C (Having two regions C and Pre-C)	C C + Pre-C	Core antigen (HBcAg) HBeAg
P		DNA polymerase
X		HBx Ag (Nonparticulate antigen which leads to enhanced replication of HBV)

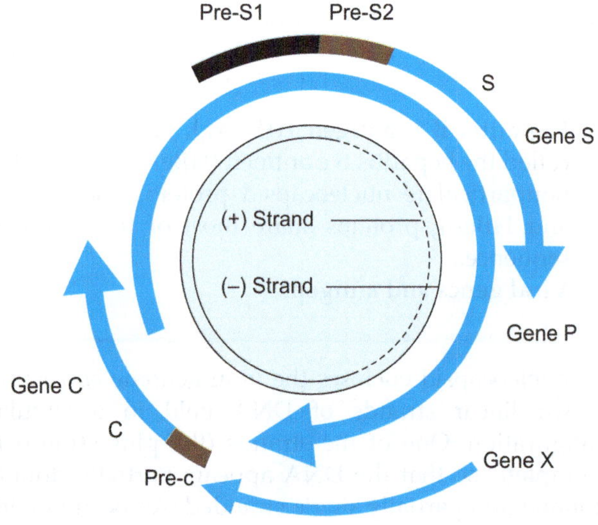

Fig. 67.5: HBV genes and gene products

They show a distinct geographical distribution. Subtype adw is common in Europe, Australia and the America; adr is prevalent in South and East India and the Far East ayw is common from West Asia through the Middle East, to Western and Northern India; ayr is very rare (Table 67.3). A number of other surface antigenic reactivities (a, x, f, t, j, n, g) have been reported, but not adequately studied.

Replication

HBV replicates within hepatocytes. Replication of viral nucleic acid starts within the hepatocyte nucleus where viral DNA can be free—, extrachromosomal, or integrated at various sites within the host chromosomes. However, integration is not essential for viral replication. Replication resembles that seen in retroviruses, in that DNA is synthesized from an RNA template by reverse transcription. To replicate hepadnavirus DNA, a full-length RNA copy is enclosed in core protein in the hepatocyte nucleus. This is copied to DNA by the polymerase, the RNA is destroyed and the DNA copied to form double-stranded DNA as the virion matures.

HBV DNA and protein have also been identified in extrahepatic sites such as bone marrow, spleen, lymph nodes and circulating lymphocytes, but apparently no damage is produced in these locations. The significance of this extrahepatic presence is not understood.

Table 67.3: Antigenic types of HBsAg

Antigenic types	Distribution
adw	Worldwide
adr	Asia
ayw	India, Africa, Russia
ayr	India, Africa, Russia

Cultivation

HBV does not grow in any conventional culture system. However, limited production of the virus and its proteins can be obtained from several cell lines transfected with HBV DNA. HBV proteins have been cloned in bacteria and yeast. The chimpanzee is susceptible to experimental infection and can be used as a laboratory model.

Stability

HBV is a relatively heat stable virus. It remains viable at room temperature for long periods. Heating to 60°C for 10 hours inactivates virus by a factor of 100-1000-fold. It is susceptible to chemical agents. Exposure to hypochlorite (10,000 ppm available chlorine) or 2 percent glutaraldehyde for 10 min will inactivate virus 100000-fold , though HBsAg may not be destroyed by such treatment. The stability of HBsAg does not always coincide with that of the infectious agent. HBsAg is not destroyed by ultraviolet irradiation of plasma or other blood products, and viral infectivity may also resist such treatment.

Clinical Syndromes

Acute Infection

The clinical presentation of HBV in children is less severe than that in adults, and infection may even be asymptomatic. Clinically apparent illness occurs in as many as 25 percent of those infected with HBV.

1. *Preicteric phase:* HBV infection is characterized by a long incubation period (about 1-6 months) and an insidious onset. Symptoms during the prodromal period may include fever, malaise, and anorexia, followed by nausea, vomiting, abdominal discomfort, and chills.

2. *Icteric phase:* The classic icteric symptoms of liver damage (e.g. jaundice, dark urine, pale stools) follow soon thereafter. Recovery is indicated by a decline in the fever and renewed appetite. Fulminant hepatitis occurs in approximately 1 percent of icteric patients and may be fatal.

   HBV infection can promote hypersensitivity reactions that are due to immune complexes of HBsAg and antibody. These may produce rash, polyarthritis, fever, acute necrotizing vasculitis, and glomerulonephritis.
3. *Convalecent phse:* About 90 to 95 percent of adults with acute hepatitis B infection recover within 1 to 2 months of onset. Mortality is about 0.5 to 2 percent, but may be more in post-transfusion cases. About 1 percent of patients, particularly those having simultaneous delta virus infection develop fatal fulminant hepatitis.

Chronic Infection

A proportion of cases (1-10 percent) remain chronically infected. They may be asymptomatic carriers or may progress to recurrent or chronic liver disease or cirrhosis. A few of them may develop hepatocellular carcinoma after many decades (Fig. 67.6).

Pathogenesis

The pathogenesis of hepatitis appears to be immunemediated. HBV replicates in the hepatocytes, reflected in the detection of viral DNA and HBcAg in the nucleus and HBsAg in the cytoplasm and at the hepatocyte membrane. During the incubation period, high levels of virus are present before the host immune response develops and controls the virus. During replication HBcAg and HBeAg are also present at the cytoplasmic membrane. These antigens induce both B and T cell responses. Damage to the hepatocyte can result from antibody-dependent, NK and cytotoxic T cell action.

Hepatocytes carry viral antigens and are subject to antibody-dependent NK cell and cytotoxic T cell attack. In the absence of adequate immune response, HBV infection may not cause hepatitis, but may lead to carrier state. Therefore infants and immunodeficient persons are more likely to become asymptomatic carriers following infection.

Epidemiology

HBV is worldwide in distribution. There is no seasonal distribution. The infection is usually sporadic, though occasional outbreaks have occurred in hospitals, orphanages and institutions for the mentally handicapped. Natural infection occurs only in humans. There is no animal reservoir. The virus is maintained in the large pool of carriers whose blood contains circulating virus for long periods, in some even lifelong.

The prevalence of HBV infection varies widely in different parts of the world. The high prevalence areas (10-20 percent) are East and South-East Asia, the Pacific Islands, and tropical Africa. The CIS (ex-USSR), the Indian subcontinent, parts of Africa, eastern and southeastern Europe and parts of Latin America are areas of medium prevalence (2-20 percent). The prevalence is low < 1 percent) in the rest of Europe, Australia and New Zealand, Canada, and the USA.

India falls in the intermediate group, with higher carrier rates in the southern part of the country and lower rates in the northern part.

The rich and the poor countries also differ in the age and modes of infection. In the rich countries, infection occurs mostly in adolescents and young adults through contaminated syringes and needles "typically among drug addicts, and through sex, particularly by homosexual intercourse". In the poor countries, infection occurs usually at younger ages, either perinatally from mother to baby, or horizontally among children. Perinatal and horizontal infection in infants and neonates generally leads to asymptomatic infection, with circulating HBeAg and HBV DNA, without any rise in transaminase levels. This is due to their inability to mount an immune response against the virus. Such cases become chronic

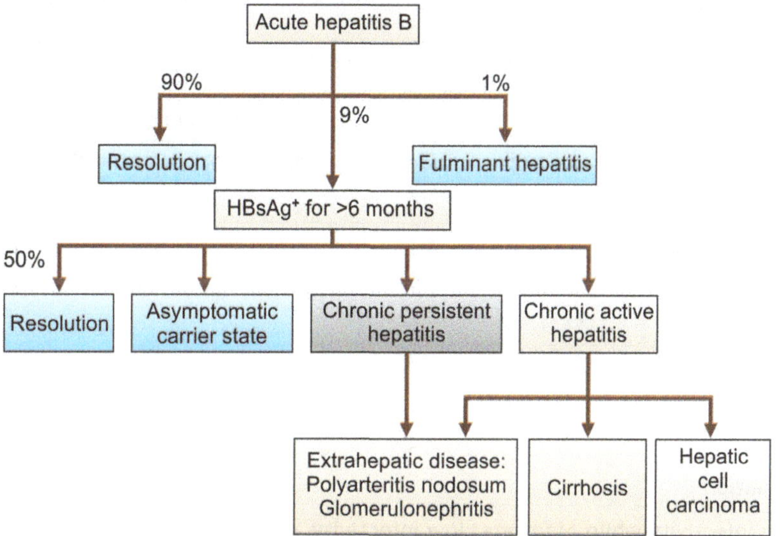

Fig.67. 6: Clinical outcomes of acute hepatitis B infection

carriers, with an enhanced risk of developing hepatocellular carcinoma in later life. It is estimated that there are 350 million HBV carriers; of these, 75 percent were infected at birth. The global death rate from hepatocellular carcinoma is estimated at 250 000 per annum.

Mode of Transmission

HBV is a blood-borne virus and there are three important modes of transmission:

1. Parenteral transmission
2. Perinatal transmission
3. Sexual transmission

1. Parenteral Transmission

HBV is transmitted only in blood and other body fluids, including cervical secretions, semen, and breast milk. Many other therapeutic, diagnostic, prophylactic and even nonmedical procedures are now the main modes of infection.

HBV is very highly infectious far more that HIV. Because the titers of virus are so high in body fluids (10^6-10^8 per ml), invisibly small quantities—0. 00001 ml or even lesscan transmit the infection. It is, therefore, easy to understand that minor abrasions or cuts can serve as portals of entry. These include shared syringes, needles and other sharp items or endoscopes, personal articles such as razors, nail clippers or combs, and practices such as acupuncture, tattooing, ritual circumcision, ear or nose piercing, and field camps for surgery or diseas'e detection by blood testing where separate sterile articles may not be available. Professionals using sharp articles like barbers, dentists and doctors may unwittingly transmit the virus if great care is not taken.

2. Perinatal Transmission

Vertical transmission from mother to child is one of the most important routes. HBV can be transmitted to babies through contact with the mother's blood at birth and in mother's milk. Babies born to chronic HBV-positive mothers are at highest risk for infection.

3. Sexual Transmission

Since HBV is present in semen and vaginal secretions, therefore, it can be transmitted by sexual contact. The risk of transmission by heterosexual and homosexual contact increases with the number of partners and the duration of such relationships. HBV infection has occurred after artificial insemination.

Hepatitis B Carriers

Carriers are of two types:

1. **Super carriers**: They have HBeAg, high titers of HBsAg and DNA polymerase in their blood. HBV may also be demonstrable in their blood. Very minute amount of serum or blood from such carriers can transmit the infection. About a quarter of the carriers in India are. HBeAg-positive.
2. **Simple carriers:** These are more common types of carriers who have low titer of HBsAg in blood, with negative HBeAg, HBV and DNA polymerase. They transmit the infection only when large volumes of blood are transferred as in blood transfusion. Many super carriers in time become simple carriers.

HBV Markers

The main antigens HBsAg, HBcAg, and HBeAg each induce corresponding antibodies. With the exception of HBcAg, all these antigens and antibodies, together with the viral DNA polymerase, can be detected in the blood at various times after infection and are referred to as 'markers', because their presence or absence in an individual patient marks the course of the disease and also gives a good idea of the degree of infectivity for others (Table 67.4). HBcAg is readily detectable only in the hepatocyte nuclei.

Laboratory Diagnosis

Specific Diagnosis

Specific diagnosis of hepatitis B rests on serological demonstration of the viral markers and can be carried out by detection of HBsAg, anti-HBs, HBeAg, anti-

Table 67.4: Interpretation of serological markers in HBV infection

Clinical condition	*Serological tests*					
	HBsAg	*HBeAg*	*Anti-HBS*	*Anti-HBe*	*Anti-HBc*	
					IgM	*IgG*
Late incubation period or early hepatitis	+	+	-	-	-	-
Acute hepatitis	+	+	-	-	+	-
Late/chronic HBV infection	+	±*	-	-	-	+
Simple carrier	+	-	-	-	-	+
Super carrier	+	+	-	-	-	+
Past infection	-	-	+	+	_	+
Immunity following vaccination	-	-	+	-	_	_

*When +, it indicates high infectivity while - indicates low infectivity.

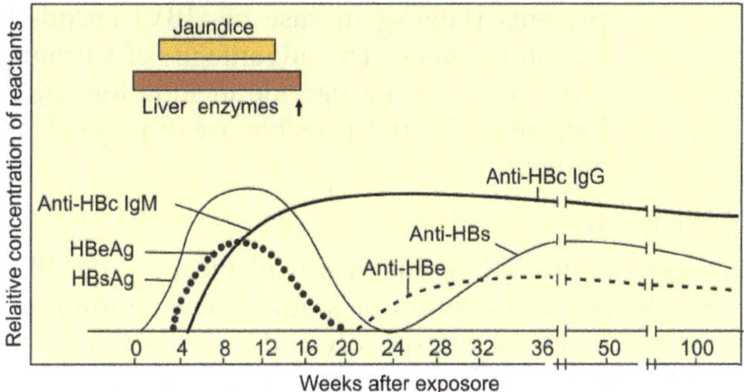

Fig. 67.7: Hepatitis antigens, antibodies and DNA in a patient recovering from acute HBV infection

HBe, IgM anti-HBc, IgG anti-HBc and HBV DNA in the serum. The sequence of appearance of viral markers in the blood is shown in Fig. 67.7. These can be detected by sensitive and specific tests like ELISA and RIA.

1. Detection of Viral Markers

i. *HBsAg:* HBsAg is the first marker to appear in blood after infection, being detectable even before elevation of transaminases and onset of clinical illness. HBsAg is usually detectable 2-6 weeks in advance of clinical and biochemical evidence of hepatitis and persists throughout the clinical course of the disease. In the typical case, it disappears within about 2 months of the start of clinical disease, but may sometimes last for 6 months and even beyond, but typically disappears by the sixth month after exposure.

ii. *HBcAg:* High levels of IgM-specific anti-HBc are frequently detected at the onset of clinical illness. Because this antibody is directed against the 27 nm internal core component of HBV, its appearance in the serum is indicative of viral replication. HBcAg is not demonstrable in circulation because it is enclosed within the HBsAg coat, but its antibody, anti-HBc appears in serum a week or two after the appearance of HBsAg. It is therefore the earliest antibody marker to be seen in blood, long before anti-HBe or anti-HBs. As anti-HBc remains lifelong, it serves as a useful indicator of prior infection with HBV; even after all the other viral markers become undetectable. Initially, antiHBc is predominantly IgM, but after about 6 months, it is mainly IgG. Selective tests for IgM or IgG anti-HBc therefore enable distinction between recent or remote infection respectively.

iii. *HBeAg:* HBeAg provides information about relative infectivity. Its presence denotes high infectivity and its absence, along with the presence of anti-HBe, indicates low infectivity. As it is invariably present during acute hepatitis, its testing is indicated only in chronic infection and carriers.

HBeAg appears in blood concurrently with HBsAg, or soon afterwards. Circulating HBeAg is an indicator of active intrahepatic viral replication, and the presence in blood of DNA polymerase, HBV DNA and virions, reflecting high infectivity. Before HBsAg disappears, HBeAg is replaced by anti-HBe, signaling the start of resolution of the disease. The disappearance of HBeAg coincides with the fall of transaminase levels in blood. Anti-HBe levels often are no longer detectable after 6 months.

The most useful detection methods are ELISA for HBV antigens and antibodies and PCR for viral DNA.

2. Viral DNA Polymerase

DNA polymerase activity, HBV DNA, and HBeAg, which are representative of the viremic stage of hepatitis B, occur early in the incubation period, concurrently or shortly after the first appearance of HBsAg.

3. Polymerase Chain Reaction (PCR)

Molecular methods such as DNA:DNA hybridization and PCR, at present used for HBV DNA testing are highly sensitive and quantitative. HBV DNA level in serum reflects the degree of viral replication in the liver and so helps to assess the progress of patients with chronic hepatitis under antiviral chemotherapy.

4. Biochemical Tests

In acute viral hepatitis caused by various hepatitis viruses, levels of serum transaminases (aminotransferases) are increased 5- to 100-fold. Both alanine and aminotransferase and aspartate aminotransferase, rise together late in the incubation period. Peak level is obtained about the time jaundice appears and reverts to normal in next 2 months. Serum bilirubin levels may rise up to 25-fold.

Prophylaxis

Measures for the control of HBV infection are the same as those for HIV infection.

A. **General prophylaxis:** General prophylaxis consists in avoiding risky practices like promiscuous sex, injectable drug abuse and direct or indirect contact with blood, semen or other body fluids of patients and carriers. Healthcare staff must take the obvious personal precautions, such as keeping cuts and abrasions covered and wearing gloves when injecting or operating upon actual and potential high-risk patients.

B. **Immunization:** Both passive and active methods of immunization are available.

   1. *Passive immunization:* Hyperimmune hepatitis B immune globulin (HBIG) prepared from human volunteers with high titer anti-HBs, administered IM in a dose of 300-500 IU soon after exposure to infection constitutes passive immunization. It may not prevent infection, but protects against illness and the carrier state.

HBIG must be given as soon as possible after an accident and preferably within 48 hours. A second dose is given 4 weeks later to those who do not respond to current vaccines. If the victim has not been vaccinated, HBIG should be used and a course of active immunization started, injecting the two materials into different body sites.

2. *Active immunization:* Active immunization is more effective.
   i. *Plasma-derived hepatitis B vaccine:* A vaccine for hepatitis B has been available since 1982. The initial vaccine was prepared by purifying HBsAg associated with the 22 nm particles from healthy HBsAg-positive carriers and treating the particles with virus-inactivating agents (formalin, urea, heat). This was immunogenic, but became unacceptable because its source was human plasma, limited in availability and not totally free from possible risk of unknown pathogens.
   ii. *Recombinant yeast hepatitis B vaccine:* The currently preferred vaccine is genetically engineered by cloning the S gene of HBV in baker's yeast. It consists of nonglycosylated HBsAg particles alone. This vaccine is safe, antigenic, free from side effects and as immunogenic as plasma-derived vaccine. It is given with alum adjuvant, IM into the deltoid or, in infants into the anterolateral aspect of the thigh. Three doses given at 0, 1 and 6 months constitute the full course. Seroconversion occurs in about 90 percent of the vaccinees.

      A special vaccine containing all antigenic components of HBsAg (Pre-S1, Pre-S2 and S) has been developed, which gives greater seroconversion. Clinical protection is believed to last much longer. Booster doses are needed only for those at high risk.
   iii. *Recombinant chinese hamster ovary (CHO) cell hepatitis vaccine* Expression system of CHO cells has been successfully used and the product is commercially available. This is the first vaccine using mammalian cell expression system.
   iv. *Synthetic peptide vaccines:* As the name indicates, these are chemically synthesized polypeptide vaccines. These are safe and cheap. These are still under experimental stage.
   v. *Hybrid virus vaccine:* Potential live vaccines using recombinant vaccinia virus have been prepared for hepatitis B, influenza, rabies, Epstein-Barr and human immunodeficiency viruses. Recombinant vaccines can be generated by incorporating foreign genes (HBsAg sequences in case of HBV) into vaccinia virus DNA. Recombinant vaccinia virus expresses proteins (HBsAg in case of HBV) encoded by foreign gene. The advantages of vaccinia virus recombinant vaccine include low cost, long shelf-life and possible use of polyvalent antigens.

Treatment

No specific antiviral treatment is available for acute HBV infection. Hepatitis B immune globulin may be administered within a week of exposure and to newborn infants of HBsAg-positive mothers. Interferon alpha, alone or in combination, with other antiviral agents such as lamivudine and famcyclovir, has been beneficial in some cases of chronic hepatitis. There is no effective treatment for the carrier state, though spontaneous resolution takes place in some of them.

HEPATITIS C VIRUS (HCV)

HCV resembles flaviviruses in structure and organization, and has been classified as a new genus *Hepacivirus* in the family Flaviviridae. HCV is a 50-60 nm virus with a linear single stranded RNA genome, enclosed within a core and surrounded by an envelope, carrying glycoprotein spikes.

The virus shows considerable genetic and antigenic diversity. Various viruses can be differentiated by RNA sequence analysis into at least six major genotypes (clades) and more than 70 subtypes. The genome of HCV encodes 10 proteins, including 2 glycoproteins (E1, E2). Some genotypes are seen worldwide, while others are localized. Because of this diversity there is little heterologous or even homologous postinfection immunity in hepatitis C.

The virus has not been grown in culture, but has been cloned in Escherichia coli.

Mode of infection: Infection is mainly by blood transfusion and other modes of contact with infected blood or blood products. Injectable drug abusers, transplant recipients and immunocompromised persons are at high risk. Sexual transmission is probably less important. The virus can be transmitted from mother to infant, though not as frequently as for HBV. In some countries, HCV infection has been associated with folk medicine practices.

Infections by HCV are extensive throughout the world. HCV infection is seen only in humans. The groups at risk are broadly similar to those listed for hepatitis B but their relative proportions are different.

Clinical Features

The incubation period is long, 15-160 days, with a mean of 50 days. HCV causes three types of disease:

1. **Acute hepatitis** with resolution of the infection and recovery in 15 percent of cases;
2. **Chronic persistent infection** with possible progression to disease much later in life for 70 percent.

3. **Severe rapid progression to cirrhosis** in 15 percent of patients. Many patients (20-50%) develop cirrhosis and are at high risk for hepatocellular carcinoma (5-25%) decades later. HCV promotes the development of hepatocellular carcinoma after 30 years in up to 5 percent of chronically infected patients.

Laboratory Diagnosis

A. **Antibody detection:** The diagnosis and detection of HCV infection are based on ELISA recognition of antibody. The antigens used are various structural and nonstructural proteins cloned in *E. coli.* Antibodies are directed against core, envelope, and NS3 and NS4 proteins and tend to be relatively low in titer. There have been three successive generations of such antigens, introduced to improve sensitivity and specificity of serological diagnosis. Even the third generation ELISA currently in use, employing NS-5 region protein and synthetic peptides becomes positive only months after the infection and shows nonspecific reactions. Confirmation by immunoblot assay is therefore recommended. In HCV infection, antibodies appear irregularly and late, limiting their diagnostic utility.

B. **HCV RNA identification:** Identification of HCV RNA in blood provides more sensitive and specific results within a few days of exposure to HCV. Reverse transcriptase—polymerase chain reaction, branched-chain DNA, and other genetic techniques can detect HCV RNA in seronegative people and have become key tools in the diagnosis of HCV infection.

Prophylaxis

Only general prophylaxis, such as screening of blood and blood products prior to transfusion, is possible. No specific active or passive immunizing agent is available.

Treatment

Recombinant interferon-alpha, alone or with ribavarin is the only known treatment for HCV.

HEPATITIS D VIRUS (HDV)

This curious little agent was first detected in people undergoing exacerbations of chronic HBV infections. In 1977, Rizzetto and colleagues in Italy identified a new viral antigen in the liver cell nuclei of patients infected with hepatitis B virus. This has been shown to be due to the hepatotropic virus Delta or Hepatitis D virus (HDV) is unique in that it uses HBV and target cell proteins to replicate and produce its one protein. Delta is a defective RNA virus dependent on the helper function of HBV for its replication and expression. It acquires an HBsAg coat for transmission. Therefore, it has no independent existence and can survive and replicate only as long as HBV infection persists in the host. It is often associated with the most severe forms of hepatitis in HBsAg-positive patients. It is a viral parasite, proving that "even fleas have fleas."

Morphology

HDV is enclosed within the hepatitis B surface antigen, HBsAg, and has no recognizable morphology of its own.

The HDV is a defective satellite virus requiring HBV as helper virus. HDV is a spherical, 36 nm particle with an outer coat composed of the hepatitis B surface antigen surrounding the circular single stranded RNA genome. The HDV RNA genome is very small (approximately 1700 nucleotides), and unlike other viruses, the singlestranded RNA is circular. Delta agent is thus an incomplete virus, reminiscent of the Dependoviruses (Fig. 67.8). It has been proposed to be classified in a new genus Deltavirus, because of its special features.

Pathogenesis

Its mode of transmission is the same as for HBV. Similar to HBV, the delta agent is spread in blood, semen, and vaginal secretions. However, it can replicate and cause disease only in people with active HBV infections.

Types of infection: Two types of infection are recognized: Coinfection and superinfection.

1. **Coinfection:** In coinfection, delta and HBV are transmitted together at the same time. Coinfection clinically presents as acute hepatitis B, ranging from mild to fulminant disease.
2. **Superinfection:** In superinfection, delta infection occurs in a person already harbouring HBV. More rapid, severe progression occurs in HBV carriers superinfected with HDV than in people co-infected with HBV and the delta agent.

No association has been noted between HDV and hepatocellular carcinoma. In simultaneous acute HBV and HDV infections, IgM anti-HBc will be detectable,

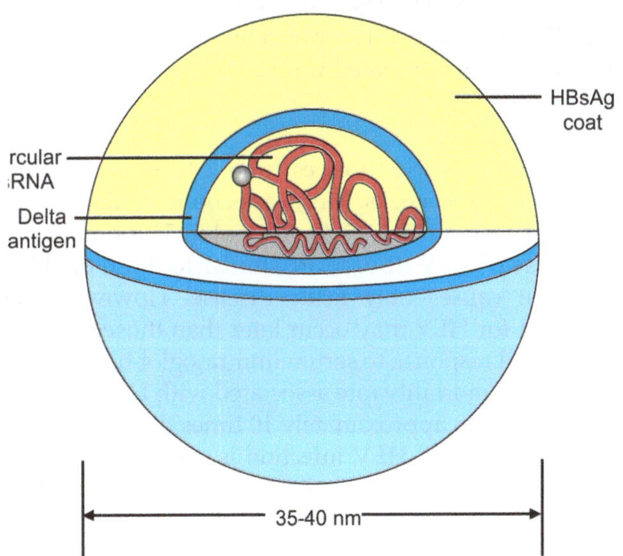

Fig. 67.8: The delta hepatitis virion

while in acute HDV infection superimposed on chronic HBV infections, anti-HBc will be of IgG class.

Laboratory Diagnosis

The only way to determine the presence of the agent is by detecting the delta antigen or antibodies. ELISA and radioimmunoassay procedures are available for doing this. Anti-delta antibodies appear in serum and can be identified by ELISA. The IgM antibody appears 2-3 weeks after infection and is soon replaced by the IgG antibody in acute delta infection. However, in chronic infection, the IgM antibody persists for years.

For the rapid identification of delta particles in circulation, RNA sequences have been cloned and DNA probes have been developed. The woodchuck has been found to be a suitable experimental model for the study of HDV infection.

Treatment

There is no known specific treatment for HDV hepatitis.

Prophylaxis

Because the delta agent depends on HBV for replication and is spread by the same routes, prevention of infection with HBV prevents HDV infection. Immunization with HBV vaccine protects against subsequent deltavirus infection.

HEPATITIS E VIRUSV (HEV) (ENTERICALLY TRANSMITTED NANB OR EPIDEMIC NANB HEPATITIS)

Hepatitis E Virus (HEV) has been provisionally classified in the genus Hepativirus under the family Caliciviridae. HEV is a spherical nonenveloped virus, 32-34 nm in diameter, with a single stranded RNA genome. The surface of the virion shows indentation and spikes. Comparison of virus strains from different areas indicates that only one serotype of the virus exists.

HEV (E-NANBH) (The *E* stands for "Enteric" or "epidemic") is predominantly spread by the feco-oral route, especially in contaminated water.

Clinical Features

The incubation period ranges from 2 to 9 weeks with an average of six weeks. Most cases occur in the young to middle aged adults (15-40 years old). The symptoms and course of HEV disease are similar to those of HAV disease; it causes only acute disease. However, the symptoms for HEV may occur later than those of HAV disease, and response to serum immunoglobulin G may be poor. The mortality rate associated with HEV disease is 1 to 2 percent, approximately 10 times that associated with HAV disease. HEV infection is especially serious in pregnant women (mortality rate of approximately 20%). HEV infection during pregnancy may cause a high rate of abortion and intrauterine death and increased perinatal mortality in babies born to women with fulminant hepatitis. Secondary attack rate among household contacts is very low in type E hepatitis, 2-3 percent as against 10-20 percent in HA V infection.

Epidemiology

HEV is predominantly spread by the fecal-oral route, especially in contaminated water. Hepatitis E has been shown to occur in epidemics, endemics and sporadic forms almost exclusively in the less developed parts of the world. A substantial proportion of cases of acute viral hepatitis occurring in young to middle-aged adults in Asia and the Indian subcontinent appear to be caused by HEV.

The largest such epidemic occurred in Delhi during the winter of 1955-56, following a breakdown in the water supply and sewage systems of Delhi caused by floods, affecting over 30,000 persons within six weeks, which affected pregnant women particularly severely. HAV was blamed at first, but 20 years later, the cause was eventually identified as a new agent, HEV. A similar epidemic of hepatitis E occurred between December 1975 and January 1976 in Ahmedabad city, India, again due to contaminated water supplies. In India, HEV is responsible for the majority of epidemic and sporadic hepatitis in adults.

Laboratory Diagnosis

Several antibody assays are being developed, mostly based on ELISA methods.

1. **Exclusion of hepatitis A and hepatitis B**—Exclusion of hepatitis A by IgM serology and hepatitis B by absence of HBsAg and IgM anti-HBc.
2. **Immunoelectron microscopy**—Immunoelectron microscopic examination of patient feces for aggregated calicivirus-like particles using monoclonal antibodies.
3. **ELISA tests**—for IgM and IgG anti-HEV.
4. **Western blot assay**—A Western blot assay for IgM and/or IgG antiHEV.
5. **Polymerase chain reaction (PCR)**—Polymerase chain reaction (PCR) assay for the detection of HEV RNA (as cDNA) in patient feces or in acute-phase sera.

Prophylaxis

General Measures

These depend on the maintenance of a clean water supply, and generally resemble those used to control HAV.

Immunization

Vaccines based on recombinant antigens are under development, and show some promise.

HEPATITIS G VIRUS

So far, the alphabetic designations of the various hepatitis viruses has been reasonably simple. Two flavivirus-like isolates were obtained in 1995 from Tamarin monkeys inoculated with blood from a young surgeon (GB) with acute hepatitis. It was termed GB, the patient's initials. A similar virus was isolated from another human specimen the same year. These isolates were called **GB viruses A, B and C** respectively.

In 1996, an isolate closely resembling GBV-C was obtained from a patient with **chronic hepatitis**. This has been called **hepatitis G virus (HGV)**.

Hepatitis G virus resembles HCV in many ways. HGV is a flavivirus, is transmitted in blood, and has a predilection for chronic hepatitis disease. It has not been grown, but its RNA genome has been cloned.

HGV RNA has been found in patients with acute, chronic and fulminant hepatitis, hemophiliacs, patients with multiple transfusions and hemodialysis, intravenous drug addicts and blood donors. HGV appears to be a blood-borne virus resembling HCV. Its role in hepatitis is yet to be clarified.

The virus is present worldwide. Majority of the individuals with HGV infection have no detectable evidence of liver disease. There have been, however, cases of **acute, fulminant and chronic hepatitis** where HGV is presently the only explanation for their liver disease. There is no evidence of a causal relationship between HGV infection and hepato-cellular carcinoma. HGV infection results frequently in **chronic viremia**. It often subsides after several years and anti-HG env antibody develops.

Laboratory Diagnosis

1. HGV is identified by detection of the genome by reverse transcriptase polymerase chain reaction (RT-PCR) or other RNA detection methods.
2. Recently, an immunoassay has been developed to detect anti-HG env. Serum HGV RNA indicates viremia, whereas anti-HG env is associated with recovery.

KNOW MORE

INDICATIONS FOR VACCINATION

Vaccination is recommended for infants, children, and especially people in high-risk groups. Vaccination is useful even after exposure for new-borns of HBsAg-positive mothers and people accidentally exposed either percutaneously or permucosally to blood or secretions from an HBsAg-positive person. Immunization of mothers should decrease the incidence of transmission to babies and older children, thus also reducing the number of chronic HBV carriers. Prevention of chronic HBV will reduce the incidence of PHC.

KEY POINTS

- At least six viruses, A through G (A, B, C, D, E and G) are hepatitis viruses.
- All the hepatitis viruses are RNA viruses, except the hepatitis B virus (HBV), which is a DNA virus belonging to the family Hepadnaviridae.
- Mode of transmission is *enteric* in hepatitis A and hepatitis E viruses while it is *parenteral* and *sexual* in both hepatitis B and hepatitis C viruses.
- Hepatitis A virus (HAV) can be demonstrated in the stool by immunoelectron microscopy (IEM). ELISA is the method of choice for detection of IgM and IgG antibodies in the serum.
- **Prevention of HAV infection depends on**: (a) vaccines containing formalin-inactivated HAV. (b) Prophylaxis with hepatitis A immunoglobulin to contacts within 2 weeks of exposure.
- Hepatitis B virus (HBV) is a double-walled spherical structure and measures 42 nm in diameter *(Dane particle)*. The outer surface or envelope of virus contains hepatitis B surface antigen (HBsAg). It encloses an inner icosahedral 27 nm nucleocapsid (core), which contains hepatitis B core antigen (HBcAg). Inside the core is the genome, a circular double stranded DNA and a DNA polymerase.
- The virion envelope consists of: 1. Hepatitis B surface antigen (HBsAg); 2. Hepatitis B core antigen (HBcAg); 3. Hepatitis Be antigen (HBeAg). Hepatitis B surface antigen (HBsAg) is also named as Australia antigen.
- HBV is a blood borne virus and there are three important modes of transmission: 1. Parenteral transmission; 2. Perinatal transmission; 3. Sexual transmission.
- **Laboratory diagnosis**: Specific diagnosis of hepatitis B rests on serological demonsration of the viral markers and can be carried out by detection of HBsAg, anti-HBs, HBeAg, anti-HBe, IgM anti-HBc, IgG anti-HBc and HBV DNA in the serum. The sequence of appearance of viral markers in the blood is important. These can be detected by sensitive and specific tests like ELISA and RIA. HBV DNA is also an indicator of viral replication and infectivity. Molecular methods such as DNA:DNA hybridization and PCR are used for HBV DNA testing are highly sensitive and quantitative.

Prophylaxis

Measures for the control of HBV infection are—General prophylaxis and immunization.General prophylaxis consists in avoiding risky practices like promiscuous sex, injectable drug abuse and direct or indirect contact with blood, semen or other body fluids of patients and carriers.

1. Passive immunization: Hyperimmune hepatitis B immune globulin (HBIG) administered IM in a dose of 300-500 IU soon after exposure to infection.
2. Active immunization: Active immunization is more effective such as plasma-derived hepatitis B vaccine, recombinant yeast hepatitis B vaccine

(Three doses given at 0, 1 and 6 months constitute the full course), recombinant chinese hamster ovary (CHO) cell hepatitis vaccine, synthetic peptide vaccines and Hybrid virus vaccine.

Hepatitis C Virus

- HCV can cause acute HCV infection, chronic HCV infection, and cirrhosis and other complications induced by hepatitis.
- Blood or blood products and also organs of infected patients are the major sources of infection.
- For diagnosis, antibody to HCV antigen can be detected by ELISA. The HCV RNA can be amplified by RT-PCR.

Hepatitis D Virus

- The hepatitis D virus (HDV) is an unusual, single-stranded, circular RNA virus and is unique in being an incomplete virus, that requires hepadnavirus helper functions for propagation in hepatocytes.
- Transmission of HDV occurs parenterally.

Hepatitis E Virus (HEV)

- Hepatitis E virus is the primary cause of enterically transmitted non-A, non-B hepatitis.
- It usually causes an acute, self-limiting disease similar to HAV.
- Specific diagnostic tests for infection due to HEV include PCR to detect HEV RNA, and ELISA, which detects both IgG and IgM anti-HEV antibodies.
- General measures for preventing HEV infection are by improved standards of sanitation and chlorinat. A vaccine may soon be available.
- Hepatitis G virus (HGV) is a flavivirus, is transmitted in blood, and has a predilection for chronic hepatitis disease.

IMPORTANT QUESTIONS

1. Name the hepatitis viruses. Describe the morphology and antigenic structure of hepatitis B virus.
2. Classify hepatitis viruses. Discuss the laboratory diagnosis of infections caused by hepatitis B virus.
3. Draw a neat labelled diagram of hepatitis B virus.
4. Write short notes on:
   Hepatitis A virus (HAV).
   Hepatitis B virus or Dane's particle.
   Hepatitis B surface antigen (HBsAg) or Australia antigen.
   Hepatitis B virus markers.
   Hepatitis C virus or (HCV).
   Hepatitis D virus or Delta agent.
   Non-A, Non-B hepatits.
   Hepatitis E virus.
   Hepatitis G virus.
   Prophylaxis of hepatitis B or hepatitis B vaccine.

FURTHER READING

Bradley OW, Krawczynski K, Kane MA. Hepatitis E. In: Belshe RB, editor. Textbook of human virology, ed 2, St Louis 1991, Mosby.

British Medical Bulletin 1990;Hepatitis 2:46.

Hadler SC, Fields HA. Hepatitis delta virus. In Belshe RB, editor. Textbook of human virology, ed 2, St Louis, 1991, Mosby.

Hagedorn CH, Rice CM. The hepatitis C viruses. Curr Top Microbiol Immuno l 2000;242:1-380.

Hepatitis NIAIO fact sheet. Available at http://www.niaid.nih.gov/publications/hepatitis.htm.

Hollinger FB, Emerson SU. Hepatitis A virus. In: Fields Virology. 4th ed. Knipe OM et al (editors). Lippincott Williams and Wilkins, 2001.

Hollinger FB and TJ Liary 2002. Hepatitis B virus. In. Field's Virology, Vol 2. 4th ed. Philadelphia: Williams and WIlkins.

Reyes GR, Baroudy BM. Molecular biology of NANBH agents: hepatitis C and hepatitis E viruses. Adv Virus Res 1991; 40:57-102.

Maillard ME and JR Gollan. Suppressing Hepatitis B. New Eng J Med 2003;348:848.

CHAPTER 68

Retroviruses-Human Immunodeficiency Virus (HIV)

Learning Objectives

After reading and studying this chapter, you should be able to:

- Classify retoviruses.
- Describe morphology of human immunodeficiency virus.
- Describe the antigenic structure of human immunodeficiency virus (HIV) and draw labelled diagram of HIV-1.
- Describe modes of transmission of human immunodeficiency virus (HIV).
- Describe types of exposure and their relative risks.
- Discuss opportunistic infections associated with human immunodeficiency virus (HIV) infection.
- Describe the laboratory diagnosis of human immunodeficiency virus (HIV) infection.
- Discuss the laboratory tests for detection of specific antibodies in human immunodeficiency virus (HIV) infection.
- Describe strategies of human immunodeficiency virus (HIV) testing.
- Describe the following: Antiviral therapy for HIV; postexposure prophylaxis.

RETROVIRUSES

These are RNA viruses that belong to family *Retroviridae* (Re=Reverse, tr=transcriptase). Members of this family possess an RNA genome and the characteristic biochemical feature is the presence of RNA dependent DNA polymerase (reverse transcriptase) within the virus

Classification

The family *Retroviridae* is classified into three sub-families (Table 68.1):

1. Oncovirinae

Oncovirinae comprising all oncogenic RNA viruses (formerly called oncornavirus).

2. Lentivirinae

Lentivirinae including the viruses causing 'slow infections' (lentus = slow) in animals as well as the human and related animal immunodeficiency viru (Table 68.2).

3. Spumavirinae

Spumavirinae containing the nononcogenic 'foamy viruses' (spuma = foam) causing asymptomatic infection in several animal species, and presenting as contaminants of primary cell cultures in which they induce foamy degeneration.

HUMAN IMMUNODEFICIENCY VIRUS (HIV)

Human immunodeficiency virus (HIV) types, derived from primate lentiviruses, are the etiologic agents of *acquired immunodeficiency syndrome* (AIDS). The illness was first described in 1981, and HIV-1 was isolated by the end of 1983. Since then, AIDS has become a worldwide epidemic, expanding in scope and magnitude as HIV infections have affected different populations and geographic regions.

The first indication of this new syndrome came in the summer of 1981, with reports from New York and Los Angeles (USA), of a sudden unexplained outbreak of two very rare diseases—Kaposi's sarcoma and *Pneumocystis carinii* pneumonia (PCP) in young homosexuals or addicted to injected narcotics. They appeared to have lost their immune competence. This condition was given the name **acquired immunodeficiency syndrome (AIDS).**

In **1983,** Luc Montagnier and colleagues from the Pasteur Institute, Paris isolated a retrovirus from a West African patient with persistent generalized lymphadenopathy, which is a manifestation of AIDS, and called it **lymphadenopathy associated virus (LAV).** In **1984,** Robert Gallo and colleagues from the National Institutes of Health, USA, reported isolation of a retrovirus from AIDS patients and called it human T cell lymphotropic virus-III (HTLV-III). Retroviruses HLTV-I and II had already been described earlier in association with human T cell leukemia (Table 68.1). Other similar isolates were reported from AIDS cases under different names. To resolve this nomenclatural confusion, the International Committee on Virus Nomenclature in

Table 68.1: Human retroviruses

Subfamily	*Genus*	*Virus*	*Disease*
1. Oncovirinae	*Retrovirus*	HTLV-1	Adult T cell leukemia/lymphoma
		HTLV-2	Prevalent in intravenous drug users, not associated with disease
2. Lentivirinae	*Lentivirus*	HIV-1 HIV-2	AIDS AIDS
3. Spumavirinae	*Spumavirus*	Human foamy virus	Nil

Table 68.2: Lentiviruses

A. Causing slow virus diseases in animals
   1. Visna/Maedi in sheep
   2. Caprine arthritis/equine infectious anemia.

B. Causing immunodeficiency
   1. In primates
      i. Human immunodeficiency viruses (HIV) types 1,2
      ii. Simian immunodeficiency viruses (SIV) causing Simian AIDS (SAIDS):
         a. Isolated from sooty mangabeys (SIV-SM) and from rhesus macaque (SIV-MAC) closely related to HIV type 2
         b. Isolated from chimpanzee—closely related to HIV type 1.
   2. In nonprimates
      i. Feline T lymphotropic virus (FTLV) causing Feline AIDS (FAIDS)

1986 decided on the generic name—**human immunodeficiency virus (HIV)** for these viruses. HIV occurs in two types: HIV-1 and HIV-2. Luc Montagnier was awarded Nobel prize in 2009 for the discovery of HIV.

Human Immunodeficiency Virus-2 (HIV-2)

A fourth human retrovirus, HIV-2, was isolated from mildly immunosuppressed patients in West Africa and appears to be less pathogenic. Fewer people succumb to HIV-2 than to HIV-1 and prior infection with HIV-2 may even help to prevent infection with HIV-1.

Structure

Enveloped Virus: HIV is a spherical **enveloped virus,** about 90-120 nm in size (Fig. 68.1).

Nucleocapsid: The **nucleocapsid** has an outer icosahedral shell and an inner coneshaped core, enclosing the ribonucleoproteins.

Genome: There are two identical copies of the positive sense, single-stranded RNA genome in the capsid (retroviruses are diploid). Also found within the capsid are the enzymes reverse transcriptase, (which is a characteristic feature of retroviruses). When the virus infects a cell, the viral RNA is transcribed by the enzyme, first

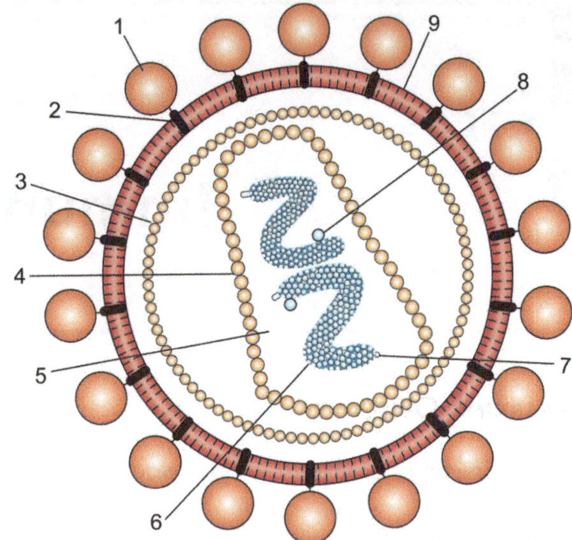

Fig. 68.1: Structure of HIV (diagrammatic representation) 1. Envelope glycoprotein spike (gp120), 2. Transmembrane pedicle glycoprotein (gp41), 3. Outer icosahedral shell of nucleocapsid (p18), 4. Cone shaped core of nucleocapsid (p24), 5. Inner core, 6. Viral proteins associated with RNA, 7. Viral RNA, 8. Reverse transcriptase, 9. Envelope lipid bilayer

into single stranded DNA and then to double stranded DNA (provirus) which is integrated into the host cell chromosome. The provirus can remain latent for long periods, though it influences host cell functions. At times, in response to viral promoters, the provirus initiates viral replication by directing synthesis of viral RNA and other components

Lipoprotein Envelope: During viral replication, when the naked virus buds out through the host cell surface membrane, it acquires a **lipoprotein envelope,** which consists of **lipid** derived from the host cell membrane and **glycoproteins** which are virus coded. The major virus coded envelope proteins are the projecting knob-like spikes on the surface and the anchoring transmembrane pedicles. The env polypeptide is composed of two subunits, the outer glycoprotein knob (gp120) and a transmembrane portion (gp41) which joins the knob to the virus lipid envelope. The receptor binding site for CD4 is present on gp120. Transmembrane pedicles cause cell fusion.

Viral Genes and Antigens

The genome organization is similar for all retroviruses in that their genomes contain in the same order the genes *gag, pol* and *env,* which code for the three groups of structural proteins (Figs 68.2 and 68.3). The genome of HIV contains the three structural genes *(gag, pol* and *env)* characteristic of all retroviruses, as well as other nonstructural and regulatory genes specific for the virus (Fig. 68.3). The products of these genes, both structural and- nonstructural, act as antigens (Table 68.2). Sera of infected persons contain antibodies to them. Detection of these antigens and antibodies is useful in the diagnosis and prognosis of HIV infections.

A. Genes Coding for Structural Proteins (Figs 68.2 and 68.3; Tables 68.3)

1. The Gag (Group-specific Antigen) Gene

The *gag* (group-specific antigen) gene determines the core and shell of the virus. It is expressed as a precursor protein, p55. This precursor protein is cleaved into three proteins, p15, p18 and p24, which make up the viral core and shell. The four proteins coded for by the *gag* gene of HIV are all found in the virion, including p24. The major core antigen is p24 which can be detected in serum during the early stages of HIV infection before antibodies appear. Late in the course of infection, the decline of free anti-p24 antibody and reappearance of p24 antigen in circulation point to exacerbation of the illness.

2. The Env

The *env* determines the synthesis of envelope glycoprotein gp160, which is cleaved into the two envelope components—gp120 which forms the surface spikes and gp41, the transmembrane anchoring protein. The spike glycoprotein gp120 is the major envelope antigen, and antibodies to gp120 are present in circulation till the terminal stage of the infection. The designation "gp41" refers to the structure of the molecule, which is a glycoprotein with an originally determined molecular weight of 41,000. "p24" refers to a protein (nonglycosylated) with a molecular weight of 24,000.

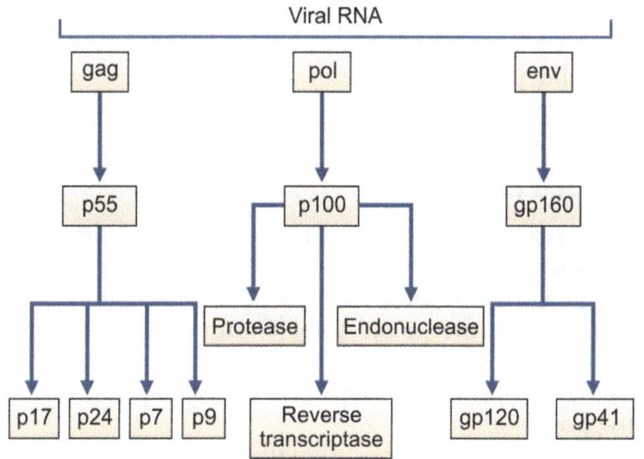

Fig. 68.2: The genomic organization of HIV structural genes and their protein products

Table 68.3: Major antigens of HIV

A. Envelope antigens:
   1. Spike antigen–gp120
      (Principal envelope antigen)
   2. Transmembrane pedicle protein–gp41
B. Shell antigen
   1. Nucleocapsid protein–p18
C. Core antigens
   1. Principal core antigen–p24
   2. Other core antigens–pl5, p55
D. Polymerase antigens–p31, p51, p66

3. The Pol Gene

The *pol* gene codes for the protease, endonuclease, integrase and reverse transcriptase. It is expressed as a precursor protein, which is cleaved into proteins p31, p51 and p66.

B. Nonstructural and Regulatory Genes

There are at least six regulatory genes in HIV and at least two in HTLV-1 (Fig. 68.3).

1. *tat* (trans activating gene) enhancing the expression of all viral genes. The *nef* gene is necessary for simian immunodeficiency virus (SIV) to be pathogenic in monkeys.
2. *nef* (negative factor gene) down regulating viral replication.
3. *rev* (regulator of virus gene) enhancing expression of structural proteins.
4. *vif* (viral infectivity factor gene) influencing infectivity of viral particles.
5. *vpu* (only in HIV-1) and *vpx* (only in *HIV-2*) enhancing maturation and release of progeny virus from cells. (Detection of the type-specific sequences *vpu* and *vpx* is useful in distinguishing between infection by HIV-1 and 2).
6. *vpr* stimulating promoter region of the virus.
7. LTR (long terminal repeat) sequences, one at either end, containing sequences giving promoter, enhancer and integration signals.

Classification

Lentiviruses have been isolated from many species (Table 68.2), including at least 26 different African nonhuman primate species. There are two distinct types of human AIDS viruses: HIV-1 and HIV-2. The two types are distinguished on the basis of genome organization and phylogenetic (evolutionary) relationships with other primate lentiviruses.

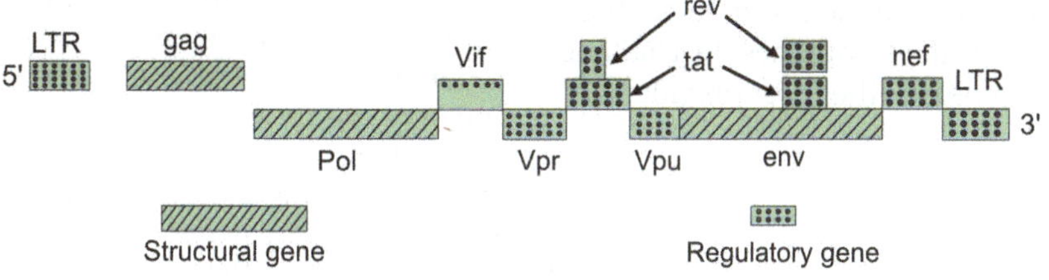

Fig. 68.3: HIV genome—diagrammatic representation

HIV-1 Groups

Based on *env* gene sequences, HIV-1 comprises three distinct virus groups—**M (for 'major'), O (for 'outlier')** and **N (for new).**

M Group

The predominant M group, which cause the large majority of HIV-1 infections worldwide, contains nine subtypes or "clades" (A-K, omitting E and I).

O Group

A few HIV-1 strains isolated from West Africa (Cameroon, Gabon) do not fall within the Group M and have been designated Group O.

N Group

Some recent isolates of HIV-1 from Cameroon, distinict from M and O groups have been called Group N. Recombinant forms of virus are also found in circulation in humans in different geographic regions.

Human Immunodeficiency Virus-2 (HIV-2)

HIV strains, first isolated from West Africa in 1986, which react with HIV type 1 antiserum very weakly or not at all have been termed HIV type 2. The envelope antigens of the two types are different, though their core polypeptides show some cross reactivity. HIV-1 and the chimpanzee virus carry a vpu gene, whereas HIV-2 and most SIVs have a *vpx gene*. The sequences of the *gag* and *pol* genes are highly conserved. It is much less virulent than HIV-1. It is largely confined to West Africa, though isolations have been reported from some other areas, including western and southern India.

HIV-2 Subtypes

Similarly, six subtypes of HIV-2 (A-F) have been identified. Within each subtype there is extensive variability. The genetic clades do not seem to correspond to neutralization serotype groups, and there is currently no evidence that subtypes differ in biology or pathogenesis. HIV-2 has only 40 percent genetic identity with HIV-I. It is more closely related to simian immunodeficiency virus than to HIV-1.

HIV-1 subtypes show a geographical distribution, though this is often blurred by viral trafficking. All known HIV virus groups and subtypes are present in Cameroon, West Africa, which may perhaps be the site of origin of the virus. Subtype A is the most prevalent, being found worldwide, while B is the most common in the Americas and Europe. The common subtypes in Africa are A, C and D, while in Asia the common subtypes are E, C and B. Subtype E is the commonest in Thailand. In India and China, subtype C is the most prevalent.

Resistance

1. Temperature

HIV is thermolabile, being inactivated in 10 minutes at 60°C and in seconds at 100°C. Over 60°C virus is inactivated 100-fold each hour. At room temperature (20-25°C), in dried blood it may survive for upto seven days. HIV is readily inactivated in liquids or 10 percent serum by heating at 56°C for 10 minutes, but dried proteinaceous material affords marked protection.

2. Lyophilization

It withstands lyophilisation. The virus in lyophilized blood products can be inactivated by heating at 68°C for 72 hours and in liquid plasma at 60°C for 10 hours.

3. Disinfectants

HIV is completely inactivated ($\geq 10^5$ units of infectivity) by treatment for 10 minutes at room temperature with any of the following: 10 percent household bleach, 50 percent ethanol, 35 percent isopropanol, 1 percent Nonidet p40, 0.5 percent Lysol, 0.5 percent paraformaldehyde, or 0.3 percent hydrogen peroxide. The virus is also inactivated by extremes of pH (pH 1.0, pH 13.0). However, when HIV is present in clotted or unclotted blood in a needle or syringe, exposure to undiluted bleach for at least 30 seconds is necessary for inactivation. Bleaching powder or household bleach are effective for surface decontamination. For treatment of contaminated medical instruments, a 2 percent solution of glutaraldehyde is useful.

Routes of Transmission

Virus is present in the blood, semen, and cervical and vaginal secretions, and these sources are important in transmission. HIV is spread only by three modes:

1. Sexual contact with infected persons (heterosexual or homosexual);
2. By blood and blood products;
3. From infected mother to babies (intrapartum, perinatal, postnatal).

The modes of transmission of HIV and their relative risks are shown in Tables 68.4 and 68.5.

1. Sexual Intercourse

HIV is primarily a sexually transmitted infection. Heterosexual transfer of virus is the route by which the great majority of infections are spread, accounting for 90 percent of the global total. Both sexes are affected equally. Transmission in the developing countries is almost always heterosexual and can take place in both directions.

The presence of other sexually transmitted diseases such as syphilis, gonorrhea, or herpes simplex type 2 increases the risk of sexual HIV transmission as much as a hundred-fold. Sex workers are at high risk due to their large number of partners; they are often an important reservoir. Transmission may be more likely from male to female. Asymptomatic virus positive individuals can transmit the virus.

Most early studies established that unprotected anal intercourse was a particular risk, especially to the passive, receptive partner (Table 68.6). The risk increases in proportion to the number of sexual encounters with different partners.

Table 68.4: Transmission of HIV Infection	
Routes	*Specific Transmission*
Known Routes of Transmission	
1. Inoculation in blood	Transfusion of blood and blood products Needle sharing among intravenous drug abusers Needlestick, open wound, and mucous membrane exposure in health care workers Tattoo needles
2. Sexual transmission	Anal and vaginal intercourse
3. Mother to baby	Intrauterine transmission Peripartum transmission Breast milk

Table 68.5: Efficiency of different routes of transmission of HIV	
Route	*Efficiency*
Blood transfusion	> 90%
Perinatal	13–40%
Sexual intercourse	
• Anal intercourse	1% per episode
• Vaginal intercourse	0.1% per episode
Intravenous drug use	0.5–1%

2. Blood and Blood Products

Transfusion of infectious blood or blood products is an effective route for viral transmission.

Contaminated Needles

This is particularly relevant in drug addicts who share syringes and needles. Drug and sexual routes merge when misusers support their habit by prostitution.

The use of unsterile syringes and needles by qualified and unqualified health workers makes iatrogenic infection likely. Injection users of illicit drugs are commonly infected through the use of contaminated needles.

Occupational exposure of health care workers to infected patients has resulted in transmission in a relatively small number of cases. The danger of needle-stick injury is present in medical and paramedical personnel, though the chances of infection are much less than with hepatitis B virus.

Contamination of eyes and **mucous membranes** is another possible route, but this is seldom confirmed. **Tattoo needles** and **contaminated inks** are other potential means by which HIV can be transmitted.

3. Mother to Child Transmission

Transmission of infection from mother to baby can take place before, during or after birth. Mother-to-infant transmission rates vary from 13 to 40 percent in untreated women (Table 68.5). Infants can become infected *in utero*, during the birth process, or, more commonly, through breast feeding. Transmission during breast feeding usually occurs early (by 6 months).

Pathogenesis

Infection is transmitted when the virus enters the blood or tissues of a person and comes into contact with a suitable host cell, principally the CD4 lymphocyte. The major determinant in the pathogenesis and disease caused by HIV is the virus tropism for CD4-expressing T cells and cells of the macrophage lineage (e.g., monocytes, macrophages, alveolar macrophages of the lung, dendritic cells of the skin, and microglial cells of the brain).

Specific binding of the virus to CD4 receptor is by the envelope glycoprotein gp120. However, for infection to take place, cell fusion is essential. This is brought about by the transmembrane gp41. The virus reaches the lymph node within 2 days of infection, and there the CD4 T cells are infected. Macrophages are persistently infected with HIV and are probably the major reservoirs and means of distribution of HIV.

HIV genome is uncoated and internalized into the cell after fusion of the virus with the host cell membrane. Viral reverse transcriptase mediates transcription of its RNA into double stranded DNA, which is integrated into the genome of the infected cell through the action of the viral enzyme integrase, causing a **latent infection.** HIV causes lytic and latent infection of CD4 T cells and persistent infection of cells of the monocyte macrophage family and disrupts neurons. The outcomes of these actions are **immunodeficiency** and **acquired immunodeficiency syndrome (AIDS) dementia.** From time to time, lytic infection is initiated, releasing progeny virions which infect other cells. The long and variable incubation period of HIV infection is because of the latency. HIV can be isolated from the blood, lymphocytes, cell-free plasma, semen, cervical secretions, saliva, tears, urine and breast milk in an infected individual.

The primary pathogenic mechanism in HIV infection is the damage caused to the CD4+ T lymphocyte. The T4 cells decrease in numbers and the T4:T8 (helper: suppressor) cell ratio is reversed. An important feature in HIV infection is the polyclonal activation of B lymphocytes leading to hypergammaglobulinemia. The hypergammaglobulinemia is more a hindrance than a help because it is composed mainly of useless immunoglobulin to irrelevant antigens and also autoantibodies. This may also be responsible for allergic reactions due to immune complexes (type 3 hypersensitivity). The loss of the CD4 T cells responsible for producing DTH allows the outgrowth of many of the opportunistic intracellular infections characteristic of AIDS (e.g., fungi and intracellular bacteria).

In peripheral blood, lymphoid tissue and other tissues such as brain where HIV replication occurs, HIV targets CD4 positive (CD4+) cells and cells of the monocyte -macrophage lineage. The latter may act as an important reservoir of virus. Macrophages are also important in carrying the virus into the central nervous system across the blood-brain barrier. The principal immuno-

Table 68.6: Immunological abnormalities in HIV infection
I. Features that characterize AIDS 1. Lymphopenia. 2. Selective T cell deficiency—Reduction in number of T4 (CD4) cells, Inversion of T4:T8 ratio. 3. Decreased delayed hypersensitivity on skin testing. 4. Hypergammaglobulinemia—predominantly IgG and IgA; and IgM also in children. 5. Polyclonal activation of B cells and increased spontaneous secretion of Ig. II. Other consistently observed features: 1. Decreased *in vitro* lymphocyte proliferative response to mitogens and antigens. 2. Decreased cytotoxic responses by T cells and NK cells. 3. Decreased antibody response to new antigens. 4. Altered monocyte/macrophage function. 5. Elevated levels of immune complexes in serum.

logical abnormalities seen in HIV infection are listed in Table 68.6.

Clinical manifestations in HIV infections are due not primarily to viral cytopathology but are: secondary to the failure of immune responses. This renders the patient susceptible to opportunistic infections and malignancies. An exception to this may be the dementia and other degenerative neurological lesions seen in AIDS. These may be due to the direct effect of HIV on the central nervous system.

Clinical Features of HIV Infection

The Centers for Disease Control (USA) have classified the clinical course of HIV infection under various groups (Table 68.7). The natural evolution of HIV infection can be considered in the following stages (Fig. 68.4):

Group I—Acute HIV Infection

The **acute seroconversion illness** resembles glandular fever, with adenopathy and flu-like symptoms.Within 3-6 weeks of infection with HIV, about 50 percent of persons experience low grade fever, malaise, headache, lymphadenopathy, sometimes with rash and arthropathy resembling glandular fever. Even fewer have the rare encephalitic presentation. Spontaneous resolution occurs within weeks. During this period there is a very high level of virus replication occurring in CD4+ cells.

Tests for HIV antibodies are usually negative at the onset of the illness but become positive during its course. Hence this syndrome has been called 'seroconversion illness', though in many of those infected, seroconversion occurs without any apparent illness. HIV antigenemia (p24 antigen) can be demonstrated at the beginning of this phase.

Group II—Asymptomatic or Latent Infection

A clinically asymptomatic or "latent" period follows the acute infection. During this time, there is a high level of ongoing viral replication. They show positive HIV antibody tests during this phase and are infectious. The infection progresses in course of time through various stages, CD4 lymphocytopenia, minor opportunistic infections, persistent generalized lymphadenopathy, AIDS-related complex (ARC), ultimately terminating in full blown AIDS, with its characteristic infections and malignancies.

The time from infection to death may be as long as 10 years and is inevitable in 70 percent of infected persons. The remainder may live as long as 17 years and form the 'long-term survivors' or 'non-progressors' group. The mechanisms for such prolonged survival are not clear, though many viral and host determinants may be responsible.

Group III—Persistent Generalized Lymphadenopathy (PGL)

Persistent generalized lymphadenopathy (PGL) *is* present in 25-30 percent of patients who are otherwise asymptomatic (Fig. 68.4). This has been defined as the presence of enlarged lymph nodes, at least 1 cm in diameter, in two or more noncontiguous extrainguinal sites, that persist for at least three months, in the absence of any

Table 68.7: Summary of classification system for HIV infection (Centers for Disease Control, USA)

Group I		Acute HIV syndrome
Group II		Asymptomatic infection
Group III		Persistent generalized lymphadenopathy
Group IV		Other diseases:
	Subgroup A	Constitutional disease—ARC
	Subgroup B	Neurologic diseases
	Subgroup C	Secondary infectious diseases
	Categary C1	Specified infectious diseases listed in the CDC surveillance definition for AIDS, such as *P. carinii* pneumonia, cryptosporidiosis, toxoplasmosis, generalized strongyloidiasis, cryptococcosis CMV or herpes diseases.
	Category C2	Other specified secondary diseases, such as oral hairy leukoplakia, salmonella bacteremia, nocardiosis, tuberculosis, thrush
	Subgroup D	Secondary cancers, such as Kaposi's sarcoma, lymphomas
	Subgroup E	Other conditions.

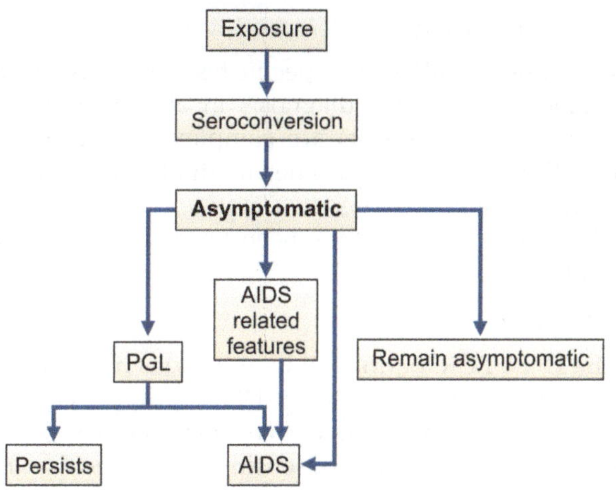

Fig. 68.4: Stages of infection with HIV

current illness or medication that may cause lymphadenopathy. The rate of progression of patients with PGL to AIDS is no greater than in those without adenopathy. This by itself is benign but the cases may progress to ARC or AIDS.

Group IV—AIDS Related Complex (ARC)

This group includes patients with considerable immunodeficiency, suffering from various constitutional symptoms or minor opportunistic infections. The typical constitutional symptoms are fatigue, unexplained fever, persistent diarrhea and marked weight loss of more than 10 percent of body weight. The common opportunistic infections are oral candidiasis, herpes zoster, hairy cell leukoplakia, salmonellosis or tuberculosis. Generalized lymphadenopathy and splenomegaly are usually present. ARC patients are usually severely ill and many of them progress to AIDS in a few months. With no treatment, the interval between primary infection with HIV and the first appearance of clinical disease is usually long in adults, averaging about 8-10 years. Death occurs about 2 years later.

The acquired immune deficiency syndrome (AIDS) presents in many ways, all due to the underlying severe loss of the ability to respond to infectious agents and to control tumors. The features classified as group IV include what was known as the AIDS-related complex or ARC.

AIDS

This is the end-stage disease representing the irreversible breakdown of immune defence mechanisms, leaving the patient prey to progressive opportunistic infections and malignancies. AIDS may be manifested in several different ways, including lymphadenopathy and fever, opportunistic infections, malignancies, and AIDS-related dementia (Table 68.8).

Pediatric AIDS

About a third to half the number of babies born to infected mothers are infected with HIV. In most cases, infection is transmitted to the baby in the perinatal period when the child's immune system is immature. It is probable that transmission can also take place during delivery or from breast milk. Pediatric AIDS—acquired from infected mothers—usually presents with clinical symptoms by 2 years of age. Death follows in another 2 years. Children may also acquire the infection from blood transfusion or blood products.

Laboratory Diagnosis

Laboratory procedures for the diagnosis of HIV infection include specific tests for HIV and tests for immunodeficiency as well. Evidence of infection by HIV can be detected in three ways:

A. Specific tests for HIV infection
B. Non-specific tests
C. Tests for opportunistic infections and tumor.

A. Specific Tests for HIV Infection

1. Antigen Detection

The virus antigens may be detectable in blood after about two weeks following a single massive infection, as by blood transfusion. The major core antigen p24 is the earliest virus marker to appear in blood and is the one tested for IgM antibodies appear in about 4-6 weeks, to be followed by IgG antibodies (Fig. 68.5).

Table 68.8: Opportunistic infections and malignancies commonly associated with HIV infection
I. BACTERIAL
1. *M. avium* complex
2. *Mycobacterium tuberculosis*—disseminated or extrapulmonary
3. *Salmonella*—recurrent septicemia
II. VIRAL
1. Cytomegalovirus
2. Herpes simplex virus
3. Varicella-zoster virus
4. Epstein-Barr virus
5. Human herpesvirus 6
6. Human herpesvirus 8
III. FUNGAL
1. Candidiasis
2. Cryptococcosis
3. Aspergillosis
4. *Pneumocystis carinii* pneumonia
5. Histoplasmosis
6. Coccidioidomycosis
IV. PARASITIC
1. Toxoplasomosis
2. Cryptosporidiosis
3. Isosporiasis
4. Microsporidiosis
5. Generalized strongyloidiasis
V. MALIGNANCIES
1. Kaposi's sarcoma
2. B cell lymphoma or non-Hodgkin's lymphoma
VI. SLIM DISEASE

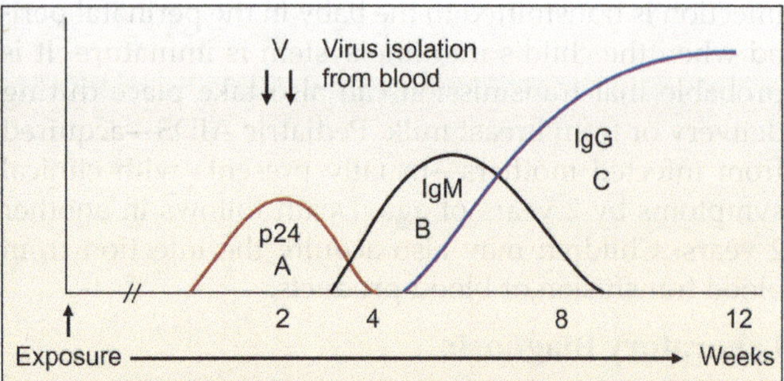

Fig. 68.5: Sequence of appearance of p24 antigen and antibodies after a massive HIV infection (as by blood transfusion) ↑ exposure: A = p24 antigen; B = IgM antibody; C = IgG antibody; 2, 4, 8, 12 = week after exposure.
↓ v ↓ = virus readily isoloated from blood.

The appearance of p24 antigenemia and viremia, followed by IgM antibody response coincides with the acute or seroconversion illness. Afterwards, free p24 antigen disappears from circulation and remains absent during the long asymptomatic phase, to reappear only when severe clinical disease sets in. The HIV window period is the time after infection has occurred but before evidence of HIV infection is detectable. The HIV diagnostic window represents a vulnerable period particularly from a blood safety standpoint, and several major efforts have been launched to minimize its duration. The p24 antigen capture assay (ELISA) which uses anti-p24 antibody as the solid phase can be used for this. The test is most useful in persons recently exposed to risk of infection, in whom antibody test is negative. The antigen often becomes undetectable after antibodies develop (because the p24 protein is complexed with p24 antibodies) but may reappear late in the course of infection, indicating a poor prognosis. HIV p24 antigen or HIV RNA is often detectable prior to antibody responses, and use of sensitive HIV antigen and nucleic acid testing has reduced the window period to the current minimum.

2. Virus Isolation

Once infected with HIV, a person remains infected for life. The virus is present in circulation and body fluids, within lymphocytes or cell-free. It can be isolated from CD4 lymphocytes of peripheral blood, bone marrow and serum. The technique of isolation is by co-cultivation of the patient's lymphocytes with uninfected lymphocytes in the presence of interleukin-2. Virus presence is detected by assays for reverse transcriptase and p24 antigen in the culture fluids.

Virus titers parallel p24 titers, being high soon after infection, low and antibody bound during the asymptomatic period, and again high towards the end.

Virus isolation is to be attempted only in laboratories with adequate containment facilities because of the risk involved.

With the advent of PCR, there are now few, if any, diagnostic uses of virus isolation.

3. Detection of Viral Nucleic Acid

As the most sensitive and specific test, PCR has become the gold standard for diagnosis in all stages of HIV infection. The PCR tests are complex and costly and are indicated only when other methods cannot give a definitive result.

Two forms of PCR have been used, DNA PCR and RNA PCR.

i. DNA PCR

In the DNA PCR, peripheral lymphocytes from the subject are lysed and the proviral DNA is amplified using primer pairs from relatively constant regions of HIV genome. The test is highly sensitive and specific when done with proper controls.

ii. RNA PCR

A related test, HIV RNA PCR can be used for diagnosis as well as for monitoring the level of viremia. Levels of RNA can be assayed as copy numbers and indicate the extent of virus replication in the patient. Measurement of plasma virus load is now essential for monitoring disease progression and response to antiviral therapy.

4. Antibody Detection

Demonstration of antibodies is the simplest and most widely employed technique for the diagnosis of HIV infection. The mean time to seroconversion after HIV infection is 3-4 weeks. Most individuals will have detectable antibodies within 6-12 weeks after infection, whereas virtually all will be positive within 6 months. Following sexual exposure to HIV, antibodies may take 2 months to appear, if infection has taken place. Therefore antibody testing will have to be done after 2-6 months to ascertain whether infection has occurred or not, after a single sexual exposure.

Serological tests for anti-HIV antibodies are of two type—screening and confirmatory tests (Tables 68.9, 68.10).

a. Screening (E/R/S) tests (Table 68.10)

i. Enzyme-linked Immunosorbent Assays (ELISA) Tests

HIV-I ELISA tests were the earliest approved serologic tests for HIV infection and remain the most sensitive approved commercial assays for infection. Direct solid phase antiglobulin ELISA is the method most commonly used. The antigen is obtained from HIV grown in continuous T lymphocyte cell line or by recombinant techniques and should represent all groups and subtypes of HIV-1 and HIV-2. The antigen is coated on microtiter wells or other suitable solid surface. The test serum is added, and if the antibody is present, it binds to the antigen. After washing away the unbound serum, antihuman immunoglobulin linked to a suitable enzyme is added, followed by a color-forming substrate. If the test serum contains anti-HIV antibody, a photometrically

Table 68.9: Laboratory tests for detection of specific antibodies in HIV infection
a. Screening (E/R/S) tests i. ELISA ii. Rapid tests – Dot blot assays – Particle agglutination (gelatin, RBC, latex, microbeads) – HIV spot and comb tests – Fluorometric microparticle technologies. iii. Simple tests. These are also based on ELISA principle but take 1-2 hours b. Supplemental tests i. Western blot assay ii. Immunofluorescence test.

detectable colour is formed which can be read by special ELISA readers. Conversion of substrate to product is quantitated by spectrophotometry.

Modifications of ELISA in which the antibody in test serum either competes with enzyme conjugated anti-HIV antibody, or is captured by antihuman immunoglobulin onto solid phase are more specific. Third generation assays employ a sandwich technique using enzyme-coupled HIV antigens and take advantage of the bi- or multi-valent nature of antibodies to improve specificity.

ELISA specific for IgM antibody is also available. Immunometric assays are highly sensitive and specific.

ii. Rapid Tests

Simple, rapid tests for detecting HIV antibodies are available for use in laboratories ill-equipped to perform EIA tests. These tests take less than 30 minutes and do not require expensive equipment. A number of 'rapid tests' have been introduced for this purpose such as: dot blot assays, particle agglutination (gelatin, RBC, latex, microbeads), HIV spot and comb tests, fluorometric microparticle technologies.

Tests using finger-prick blood, saliva and urine have also been developed.

iii. Simple Tests

These tests are not as fast as rapid tests. They take 1-2 hours and also do not require expensive equipment. These tests are also based on ELISA principle

b. Supplemental Tests

i. Western Blot Test

When EIA-based antibody tests are used for screening populations with a low prevalence of HIV infections (e.g., blood donors), a positive test in a serum sample must be confirmed by a repeat test. If the repeat EIA test is reactive, a confirmation test is performed to rule out false-positive EIA results. The most widely used confirmation assay is the Western blot technique, in which antibodies to HIV proteins of specific molecular weights can be detected.

In this test, HIV proteins, separated according to their electrophoretic mobility (and molecular weight) by polyacrylamide gel electrophoresis, are blotted onto strips of nitrocellulose paper. They retain their relative positions achieved on separation. The antigen impregnated nitrocellulose is then cut into strips, each strip having the full complement of vital proteins which were separated in the gel. Each strip is then incubated with a dilution of patient serum. Antibodies which attach to the separated viral antigens on the strip are detected by anti-human immunoglobulin antibody to which enzyme has been attached. The binding of this tracer antibody to the human immunoglobulin is detected by the addition of the enzyme conjugate followed by application of a substrate. The substrate changes color in the presence of enzyme and permanently stains the strip. The location or position on the strip at which a patient's antibodies attach to viral antigens indicates whether antibody is specific for viral antigens or directed against non-viral material from the cells in which the virus was grown.

Interpretation of Westron Blot (WB) Results

WB results are scored as **negative, positive,** or **indeterminate** (Fig. 68.6).

Negative Result

The WHO has suggested that a weakly reactive p17 band may be considered negative. Antibodies to viral

Table 68.10: Evolution of serological markers during HIV infection

	Markers			
State of infection	*p24 Ag (free)*	*Anti-HIV IgG*	*Anti-HIV IgM*	*Western blot pattern*
Early infection	–	–	–	–
Acute (seroconversion) illness	+ → –	– → +	+	Partial: p24 andl or gp120
Carrier asymptomatic	–	+	–	Full pattern
PGL	+	+	–	Loss of p24/p55
AIDS	+			Absence of p24: loss of other reactivities

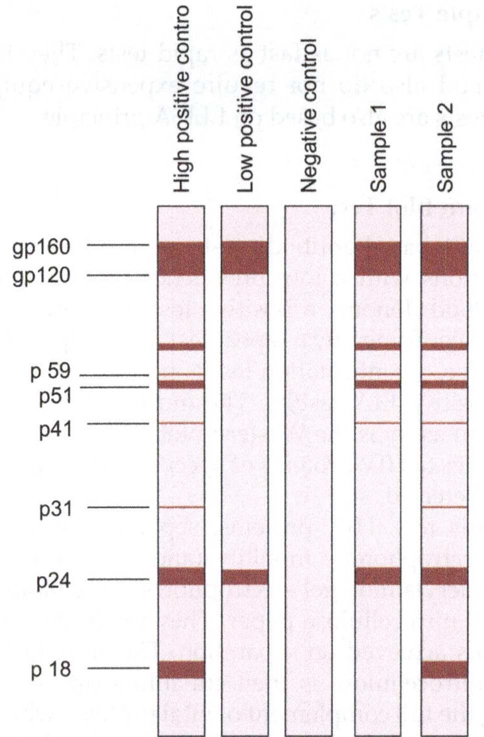

Fig. 68.6: Diagrammatic representation of Western blot test for HIV

core protein p24 or envelope glycoproteins gp41, gp120, or gp160 are most commonly detected.

Positive Result

In a positive serum, bands will be seen with multiple proteins, typically with p24 (*gag* gene, core protein), p31 (*pol* gene, reverse transcriptase) and gp41, gp120 or gp160 (*env* gene, surface antigens). A positive reaction with proteins representing the three *genes gag, pol, env* is conclusive evidence of HIV infection. The test may be considered positive even if it shows bands against at least two of the following gene products: p24, gp41, gp120/160. However, interpretation becomes difficult when bands that appear do not satisfy these criteria. This may happen in early infection but may also be nonspecific.

Indeterminate Result

Indeterminate results are not uncommon. Indeterminate results may arise from either insensitive detection of true reactivity (window period) or false reactivity with principally single-band reactivity. In such cases, the Western blot may be repeated, later. If no definitive result can be given even then, it may be necessary to have p24 assay done.

ii. Immunofluorescence Test

In this test, HIV infected cells are acetone fixed on to glass slides and then reacted with test serum followed by fluorescein conjugated anti-human gammaglobulin. A positive reaction appears as apple-green fluorescence of cell membrane under fluorescence microscope.

Many workers have shown that saliva is an acceptable and often favorable alternative to serum for HIV antibody testing. Blood of HIV-infected individuals is a hazardous substance that occasionally leads to HIV infection among health care workers

Strategies of HIV Testing

As the Western blot technique is costly, the practice now is to perform either two different types of ELISA or an ELISA with any of the rapid tests. A serum positive in both tests is considered positive. When in doubt, retesting after 1 or 2 months may be useful.

There are three strategies of HIV testing:

Strategy I

The serum is tested with one E/R/S test and if reactive, sample is considered positive and if non-reactive it is considered negative. This strategy is used for **transfusion safety.** For this purpose, a highly sensitive and very reliable test kit must be used.

Strategy II

The serum reactive with one E/R/S test is retested with a second E/R/S test with higher specificity, based on a different antigen preparation and/or different test principle. If found reactive on second E/R/S test also, it is reported as positive, otherwise as negative. This strategy is used for **HIV surveillance.** The majority of individuals seroconvert within 2 months after viral exposure. HIV infection for longer than 6 months without a detectable antibody response is very uncommon.

Strategy III

The serum reactive with two E/R/S tests is retested with a third E/R/S test. The third test should again be based on different antigen preparation or test principle. A serum testing reactive with all three E/R/S tests is reported positive. A serum sample non-reactive in third E/R/S is considered equivocal/borderline. Such individuals should be retested after three weeks. If this sample also provides an equivocal result, the person is considered to be HIV antibody negative. For asymptomatic HIV infection, it is necessary to confirm the diagnosis with three tests. Symptomatic infections with opportunistic infections, however, may be subjected to two tests.

The first test selected for any of the strategies should be of highest sensitivity and second and third E/R/S test selected should be of highest specificity.

Non-specific Tests

Immunological Tests

The following parameters help to establish the immunodeficiency in HIV infection:

a. Total leukocyte and lymphocyte count to demonstrate leukopenia and a lymphocyte count usually below 2000/mm^3.
b. T cell subset assays. Absolute CD4+ T cell count will be usually less than 200/mm^3. T4:T8 cell ratio is reversed.

c. Platelet count will show thrombocytopenia.
d. Raised IgG and IgA levels.
e. Diminished CMI as indicated by skin tests.
f. Lymph node biopsy showing profound abnormalities.

Tests for Opportunistic Infections and Tumors

Apart from diagnosing HIV infection, the laboratory would be called upon to identify the opportunistic infections that are a feature of AIDS. Routine microbiological methods would suffice for this. However, serological diagnosis of infections may not always be reliable in AIDS as antibody formation may be affected by the immune deficiency.

Applications of Serological Tests

Serological tests for HIV infection are employed in the following situations:

1. Screening
2. Seroepidemiology
3. Diagnosis
4. Prognosis.

1. Screening

Screening is defined as the systematic application of HIV testing, whether voluntary or mandatory, to entire populations or selected target groups. Screening of entire populations is neither feasible nor practicable. However, screening of a target population is valuable. As iatrogenic transfer of HIV is an important mode of spread of the infection to unsuspecting recipients, it should be mandatory that all donors of blood, blood products, semen, cells, tissues and organs be screened. As antibody tests are negative during the early stage of HIV infection when the individual is infectious, screening may not detect all dangerous donors but can still eliminate a large majority of them. Screening for p24 antigen can detect those in the window period also.

A person found positive for HIV antigen or antibody should never donate blood or other biological materials. As the infection can be transmitted from mother to baby before, during, or after birth, antenatal screening is useful. Some countries have laws requiring screening of incoming foreigners.

2. Seroepidemiology

Antibody surveys have been most useful in identifying the geographical extent of HIV infection and in other epidemiological studies such as spread of the infection from identified sources.

3. Diagnosis

Serology is almost always positive in persons with clinical features of AIDS. It may, however, be negative in acute illness and sometimes in the very late cases where the immune system is nonreactive. Routine serology may also be negative when the infection is with a different AIDS virus. For example, HIV-2 infections are likely to be missed if antibody testing is done with the HIV-1 antigen alone.

Antibody testing may also help to check whether infection has taken place following an exposure, such as sexual contact, blood transfusion or needlestick injury. Serology after two months and, if negative, after six months would be sufficient.

4. Prognosis

In a person infected with HIV, loss of detectable anti-p24 antibody indicates clinical deterioration. This is also associated with HIV antigenemia and increased virus titer in circulation.

Laboratory Monitoring of HIV Infection

Some laboratory tests are important in monitoring the course of HIV infection.

1. CD4+ T Cell Count

The most important of these is CD4+ T cell count which reflects the current immunological competence of the patient. HIV positive persons should have frequent CD+ T cell counts. When the count falls below 500 per mm^3, it is an indication of disease progression and the need for antiretroviral therapy. Counts below 200 denote risk of serious infections.

2. Direct Measurement of HIV RNA

Direct measurement of HIV RNA becomes necessary, particularly in the course of treatment. This is done usually by two methods, the reverse transcriptase PCR (RT-PCR) assay and the branched DNA (bDNA) assay.

3. Beta-2-Microglobulin and Neopterin

Beta-2-microglobulin and Neopterin are two substances which have a predictive value on the progression of HIV disease. They can be measured in serum or urine. Their concentrations are low in asymptomatic infection and rise with advancing disease.

Epidemiology

Routes of Transmission

HIV is spread only by three modes--sexual contact with infected persons (heterosexual or homosexual); by blood and blood products; and from infected mother to babies (intrapartum, perinatal, postnatal). HIV is not, however, transmitted by casual contact, touching, hugging, kissing, coughing, sneezing, insect bites, water, food, utensils, toilets, swimming pools, or public baths. The modes of transmission of HIV and their relative risks are shown in Table 62.7.

Geographic Distribution

HIV-1 infections are spreading worldwide, with the largest number of AIDS cases in sub-Saharan Africa but with a growing number of cases in Asia, the United States, and the rest of the world. HIV-2 is more prevalent in Africa (especially West Africa) than in the United States. Heterosexual transmission is the major means of spread of HIV-1 and HIV-2 in Africa, and both men and women are equally affected by these viruses. HIV-2 produces a disease similar to but less severe than AIDS.

AIDS in the developing countries differs from the disease in the Western countries clinically too. In Africa, the major manifestation is pronounced wasting so that it has been called the **'slim disease'**. The high prevalence of tuberculosis and parasitic infections complicate the clinical picture.

HIV Infection in India

HIV infection was detected rather late in India, the first cases having been found in female sex workers in Madras (Chennai) in 1986 and the first AIDS patient the same year from Bombay (Mumbai). Since then in every high risk group, the rate of infection has been mounting. HIV infection has spread throughout the country, though information is available from some parts only. By the end of 2003, India is believed to have about 5 million HIV-infected people, the second largest such population after South Africa.

Control

1. Sexual Transmission

The best method of checking sexual transmission of infection is health education regarding the danger of promiscuity and other high risk activities. The use of condoms and vaginal antiseptics could have an impact, but they need to be available and acceptable to the local population. The use of condoms offers considerable, though not complete, protection.

2. Exposure to Blood

Drug injectors can avoid risk by not injecting, or can reduce risk by using only clean equipment. Screening of all blood donors should eliminate almost all possibility of transmission. Factor VIII and other blood products are heat-treated, if possible, to inactivate HIV. All organ donors must be screened.

Occupational risk in the health care setting can be controlled by the implementation of safe working practices to prevent accidental injury and contamination with blood and body fluids.

3. Mother to Child Transmission

This can be reduced by identifying infected mothers and giving specific therapy in the later stages of pregnancy and to the baby after birth. All women who have been potentially exposed should seek HIV antibody testing before becoming pregnant and, if the test is positive, should consider avoiding pregnancy. HIV infected mothers should avoid breast feeding to reduce transmission of the virus to their children if safe alternative feeding options are available.

Prophylaxis

The prevention of AIDS rests at present on general measures such as health education, identification of sources and elimination of high risk activities. No specific vaccine is available.

Vaccine Development

No vaccine against HIV is available despite several trials. The high mutability, diverse antigenic types and subtypes, long latency and persistence in infected cells as provirus pose serious problems in the development of vaccines.

Strategies for Vaccine Production

Several possible strategies have been explored for vaccine production. These include immunization with:

a. Modified whole virus;
b. Subunits, based on envelope glycoproteins expressed in animal cells, bacteria, viruses-or as synthetic epitopes on adjuvant carriers; and
c. Target cell protection by anti-CD4 antibody or genetically engineered CD4. A number of candidate vaccines are being tested in clinical trials in humans.

Treatment

The anti-HIV drugs approved can be classified as nucleoside analogue reverse transcriptase inhibitors, non-nucleoside reverse transcriptase inhibitors, or protease inhibitors (Table 68.11). Other anti-HIV drugs being developed include different nucleotide analogues and other inhibitors of reverse transcriptase, receptor antagonists (CD4 and gp120 analogues), inhibitors of tat function (Ro247429), glycoprotein glycosylation inhibitors, interferon and interferon inducers, and antisense DNA to essential genome sequences.

In the current guidelines, AZT is recommended for the treatment of asymptomatic or mildly symptomatic people with CD4 counts of less than 500/μL and for the treatment of infected pregnant women to reduce the likelihood of transmission of the virus to the fetus. Unfortunately, the high mutation rate of HIV promotes the development of resistance to these drugs.

A cocktail of several antiviral drugs (e.g., AZT, 3TC, protease inhibitor) termed **highly active antiretroviral treatment (HAART),** each with different mechanisms of action, has less potential to breed resistance and has become a recommended therapy. Multidrug therapy can reduce blood levels of virus to nearly zero and reduce morbidity and mortality in many patients with advanced AIDS. Although HAART is a difficult drug regimen, many patients return to nearly normal on this therapy.

Apart from specific antiretroviral therapy, other measures in the treatment of AIDS include: (i) treatment and prophylaxis of opportunistic infections and tumors (ii) general management and (iii) immuno-restorative measures.

Postexposure Prophylaxis (PEP)

If an accidental exposure occurs, any wound should be washed with soap and water, or mucous membranes flushed with water. The accident must be reported so that, if necessary, prophylaxis can be started as soon as

Table 68.11: Potential autiviral therapies for HIV infection
Nucleoside analogue reverse transcriptase inhibitors
Azidothymidine (AZT) (Zidovudine)
Dideoxycytidine (ddC) (Zalciabine)
Dideoxyinosine (ddl)(Didanosine)
d4T (Stavudine)
3TC (Lamivudine)
ABC (Abacavir)
Non-nucleoside Reverse Transcriprase Inhibitors
Nevirapine (Viranune)
Delavirdine (Rescriptor)
Efavirenz (Sustiva)
Protease Inhibitors
Saquinavir (Invirase/Fortovase)
Ritonavir (Norvir)
Indinavir (Crixivan)
Nelfinavir (Viracept)
Amprenavit (Agenerase)
Highly Active Antiretroviral Therapy (HAART) (Combination)
Indinavir/AZT/3TC
Ritonavir/AZT/3TC
Nelfinavir/AZT/3TC
Nevirapine/AZT/ddl
Nevirapine/indinavir/3TC

possible. Knowledge of the status of the source patient is essential.

Exposure to blood, body fluid, other potentially infected material or an instrument contaminated with one of these materials may lead to risk of acquiring HIV infection. The risk of infection varies with the type of exposure and other factors. Most exposures do not result in infection. Health workers are normally at very low risk of acquiring infection during management of infected patients. Following exposure, postexposure prophylaxis (PEP) may be required depending upon the category of exposure and HIV status of exposure source (Table 68.12).

Basic PEP regimen consists of two drug combination while expanded PEP regimen is a combination of three drugs. In basic two drug regimen, Zidovudine 300 mg BD and Lamivudine 150 mg BD are used. A protease inhibitor is added to this combination of drugs in expanded three drug PEP regimen. Among protease inhibitors, lopinavir 400 mg BD or 800 mg OD or ritonavir 100 mg BD or 200 mg OD are preferred as third drug. These drugs must be started within the first 72 hours and ideally within 2 hours to be effective. The PEP should be continued for a period of four weeks. Both risk of infection and possible side-effects of antiretroviral drugs should be carefully considered when deciding to start PEP. Besides PEP, injured site on the wound should be thoroughly washed with soap and water. Antiseptics may also be used.

Therapy should be continued for 4 weeks and the victim followed with testing for virus for the next 6 months. Exposed persons should have post PEP HIV testing, at three months and at 6 months. If the test at six months is negative, no further testing is required.

KNOW MORE

Origin of AIDS

HIV in humans originated from cross-species infections by simian viruses in rural Africa, probably due to direct human contact with infected primate blood.

Acquired Immune Deficiency Syndrome (AIDS)

AIDS is one of the most devastating epidemics ever recorded. AIDS is only the last stage in the wide spectrum of clinical features in HIV infection. HIV disease progresses from an asymptomatic infection to profound immunosuppression, referred to as full-blown AIDS.

KEY POINTS

- Human immunodeficiency virus (HIV) causes AIDS, belongs to retroviruses.
- HIV is a spherical enveloped virus measuring up to 120 nm in diameter and consists of two identical copies of single-stranded positive sense RNA genome.
  The three important enzymes contained in the virion are reverse transcriptase, protease and integrase. In association with viral RNA is the reverse transcriptase enzyme.

Table 68.12: PEP regimen according to exposure and status of source

Category of exposure	*HIV positive and asymptomatic*	*HIV positive and clinically symptomatic*	*HIV status not known*
i. Mild exposure (Mucous membrane/non-intact skin with small volumes).	Consider two drug PEP regimen	Start two drug PEP regimen	Usually no PEP or consider two drug PEP regimen
ii. Moderate exposure (Mucous membrane/non-intact skin with large volume or percutaneous superficial exposure with solid needle)	Start two drug PEP regimen	Start three drug PEP regimen	Usually no PEP or consider two drug PEP regimen
iii. Severe exposure (Percutaneous with large volume)	Start two drug PEP regimen	Start three drug PEP regimen	Usually no PEP or consider two drug PEP regimen

- Important structural components of the virus include the surface antigen gp120, the transmembrane antigen gp41, the matrix protein p17 and the capsid antigen p25.
- The genome of HIV contains the three structural genes (*gag, pol* and *env*), as well as other nonstructural and regulatory genes specific for the virus (*tat, rev, nef, vif, vpr,* and *vpu*). The products of these genes, both structural and nonstructural, act as antigens
- Sera of infected persons contain antibodies to them. Detection of these antigens and antibodies is useful in the diagnosis and prognosis of HIV infections.
- HIV shows two distinct antigenic types—HIV-1 and HIV-2.
- There are three modes of transmission of HIV infection: *Sexual contact, parenteral* and *perinatal.*
- HIV virus shows trophism for CD4-expressing T-cells and macrophages.
- HIV infects principally the *CD4 lymphocytes.* The infection causes damage to T helper (T4) lymphocytes. *T4 cells* are *depleted* in numbers and the *T4:TB* (helper: suppressor) *ratio* is *reversed.*
- HIV causes acute infection, AIDS related complex (ARC), and AIDS. AIDS is the end-stage disease of HIV infection associated with opportunistic infections, malignancies, and neurologic diseases.
- When CD4+ cells fall below 200 per mm^3, the titer of virus increases markedly and there is irreversible breakdown of immune defencse mechanisms. Most of the patients with HIV disease die of infections other than HIV. e.g. opportunistic infections and malignancies. AIDS is the end stage of HIV infection.
- Laboratory diagnosis of HIV infection includes specific tests for HIV as well as tests for immunodeficiency. Specific tests include *antigen (P24) detection, virus isolation,* detection of *viral nucleic acid* and *antibody detection.*
- The p24 antigen is the earliest virus marker to appear in the blood. Viral isolation, detection of viral nucleic acid by *polymerase chain reaction* (PCR) and p24 antigen detection are useful for diagnosis *in window period.*
- HIV can be cultured by co-cultivation of lymphocytes with potentially infected and uninfected mononuclear cells.
- The detection of specific antibodies to HIV in the serum is the most commonly used method of serodiagnosis of patients with HIV and AIDS.
- The diagnosis of HIV infection is made by detecting serum antibodies to viral proteins, both core *(P24)* or envelope (*gp120, gp41*). There are two types of serological tests for anti-HIV antibodies: screening tests and supplementary tests.
- Screening tests—ELISA, rapid test, and simple test (E/R/S) are usually highly sensitive tests.
- Supplementary tests (Western blot and indirect immunofluorescence assay) are used as confirmatory tests for detection of HIV antibodies.
- The molecular methods include reverse transcriptase polymerase chain reaction (RT-PCR), nucleic acid based amplification (NASBA), and transcription mediated amplification (TMA) and branched chain DNA (BDNA).
- There are three strategies (strategy I to III) for HIV testing in India.
- Serological tests for HIV infection are employed for *screening, seroepidemiology, diagnosis* and *prognosis.*
- A safe and effective vaccine is yet to be available against HIV.
- Antiretroviral treatment (ART) is the mainstay in HIV treatment.
- Post exposure prophylaxis (PEP) may be required when there is exposure to blood, body fluid, other potentially infected material or an instrument contaminated with HIV.

IMPORTANT QUESTIONS

1. Describe the structure and laboratory diagnosis of human immunodeficiency virus.
2. Describe the antigenic structure of human immunodeficiency virus. (HIV) and draw labelled diagram of HIV-1 .
3. Discuss the modes of transmission and pathogenesis of human immunodeficiency virus (HIV).
4. Draw a diagram of human immunodeficiency virus (HIV) and label its parts.
5. Write short notes on:
   Antigens of human immunodeficiency virus (HIV).
   Opportunistic infections associated with human immunodeficiency virus (HIV) infection.
   Strategies of human immunodeficiency virus (HIV) testing.
   Control of HIV.
   Antiviral therapy for HIV.
   Postexposure prophylaxis.

FURTHER READING

Anderson RM. Understanding the AIDS pandemic. Sci Am 1992;266:58-66.

Centers for Disease Control and Prevention. Revised classification system for HIV infection. Morbidity Mortality Weekly Report 1993;41(RR-17).

Dewhurst S, da Cruz RL W, Whetter L. 2000. Bioscience. Pathogenesis and treatment of HIV-1 infection: recent developments (Y2K update). Available at http://bioscience.org/2000/v5/d/dewhurstlfulltext.htm

Fauci AS. The human immunodeficiency virus: infectivity and mechanisms of pathogenesis, Science 1988;239:617-622.

Levy IA. Pathogenesis of human immunodeficiency virus infection. Microbiol Rev 1993;57:183-289.

CHAPTER 69

Slow Virus and Prion Diseases

Learning Objectives

After reading and studying this chapter, you should be able to:
- Describe slow virus and prion diseases.
- Describe the following: Creutzfeldt-Jakob Disease (CJD); Kuru; Bovine spongiform encephalopathy (BSE) or mad cow disease; subacute sclerosing panencephalitis (SSPE); progressive multifocal leukoencephalopathy (PML).

INTRODUCTION

The term 'slow virus disease' is applied to a group of infections in animals and human beings, characterized by a very long incubation period and a slow but relentless course, terminating fatally. This usually ends months later in disability or death.

The concept of 'slow infection' was originally proposed by Sigurdsson (1954), a veterinary pathologist for slowly progressing infections of sheep, such as scrapie, visna and maedi. The recognition in recent years, that some chronic degenerative neurological diseases of human beings may have a similar pathogenesis, has led to considerable interest in this concept.

CHARACTERISTICS OF SLOW VIRUSES

1. *Incubation periods*-Long incubation period ranging from months to years,
2. *Course of illness*-Lasting for months or years, with remissions and exacerbations,
3. *Predilection for involvement of the central nervous system,*
4. *Absence of immune response* or an immune response that does not arrest the disease, but may actually contribute to pathogenesis,
5. *A genetic predisposition*
6. *Invariable fatal termination.*

CLASSIFICATION (TABLE 69.1)

Slow virus diseases may be classified into **three groups:** Group A, Group B, Group C.

Group A

Group A consisting of slowly progressive infections of sheep, caused by serologically related nononcogenic retroviruses called **lentiviruses** (from Latin. *lentus,* meaning slow). Examples: **Visna; Maedi (progressive pneumonia**).

Group B

Group B comprising **prion** diseases of the central nervous system (CNS) collectively known as the **subacute spongiform viral encephalopathies.** The unconventional slow viruses cause slow neurodegenerative diseases.

Animal Diseases

i. Scrapie,
ii. Bovine spongiform encephalopathy (BSE) (mad cow disease),
iii. Chronic wasting disease (in mule, deer and elk),
iv. Transmissible mink encephalopathy.

Human Diseases

i. Kuru
ii. Creutzfeldt-Jakob disease (CJD)
iii. Gerstmann-Striiussler-Scheinker (GSS) disease
iv. Fatal familial insomnia (FFI).

Group C

Some chronic degenerative diseases of the central nervous system in humans are caused by "slow" or chronic, persistent infections by **classic viruses.** Group C consists of two unrelated CNS diseases of human beings:

i. Sub-acute sclerosing panencephalitis
ii. Progressive multifocal leukoencephalopathy.

Group A

Visna

Visna, a demyelinating disease of sheep was originaly recognized in Iceland in 1935 where it was eradicated in 1951 by slaughter of all affected animals. The

Table 69.1: Slow virus and prion diseases				
Disease	*Agent*	*Hosts*	*Incubation period*	*Nature of disease*
Group A				
Visna	Retrovirus	Sheep	Months to years	Central nervous system demyelination
Group B				
a. Animal diseases				
i. Scrapie	Prion	Sheep, goat, mice	Months to years	Spongiform encephalopathy
ii. Bovine spongiform encephalopathy (BSE)	Prion	Cattle	Months to years	Spongiform encephalopathy
iii. Chronic wasting disease	Prion	Mule, deer, elk	Months to years	Spongiform encephalopathy
iv. Transmissible mink encephalopathy	Prion	Mink. Other animals	Months	Spongiform encephalopathy
b. Human diseases				
i. Kuru	Prion	Humans, chimpanzees, monkeys	Months to years	Spongiform encephalopathy
ii. Creutzfeldt-Jakob disease (CJD),	Prion	Humans, chimpanzees, monkeys	Months to years	Spongiform encephalopathy
iii. Gerstmann-Straussler-Scheinker (GSS) disease	Prion			
iv. Fatal familial insomnia (FFI)	Prion			
Group C				
i. Subacute sclerosing panencephalitis	Measles virus variant	Humans	2 to 20 years	Chronic sclerosing panencephalitis
ii. Progressive multifocal leukoencephalopathy	Polyomavirus JCV	Humans	Years	Central nervous system demyelination

disease has an incubation period of about two years. It has an insidious onset with pareses, progressing to total paralysis and death. Disease progression can be either rapid (weeks) or slow (years).

The virus can be grown in sheep choroid plexus tissue cultures, from the CSF, blood and saliva of affected animals. Infected animals develop antibodies to the virus. High levels of neutralizing antibody can be detected in circulation, but they do not protect the host. Instead, the CNS lesions may represent an immune disease, due to an antigen-antibody reaction on the surface of infected glial cells.

Maedi (Progressive Pneumonia)

Maedi (progressive pneumonia) is a slowly progressive fatal hemorrhagic pneumonia of sheep, with an incubation period of 2 to 3 years.

Visna and maedi (progressive pneumonia) viruses are closely related and are classified as **retroviruses (genus *Lentivirus*).** Human immunodeficiency virus, the causative agent of AIDS, also belongs to this group of lentiviruses. AIDS shows many features of a slow virus disease.

Group B (Prion Diseases)

Transmissible Spongiform Encephalopathies (Prion Diseases)

Degenerative central nervous system diseases have similar pathologic features. These diseases are described as transmissible spongiform encephalopathies. The causative agents are not conventional viruses; infectiviry is associated with proteinaceous material devoid of detectable amounts of nucleic acid. The term "prion" is used to designate this novel class of agents.

a. Prion Diseases of Animals

i. Scrapie

Scrapie is the prototype prion disease and has been known for some 200 years as a clinical entity in Europe. Scrapie shows marked differences in susceptibility of different breeds of animal. **The incubation period** is about two years. The affected animals are irritable and develop intense pruitus, scraping themselves against trees and rocks, hence the name scrapie. Emaciation and paralysis set in, leading to death.

The disease can be transmitted to sheep, mice and many other experimental animals by injection of suspensions of brain and spinal cord from affected animals. No immune response has been demonstrated in natural or experimental scrapie. Immunosuppression and interferon do not effect the course of the disease. The causative agent has been maintained in brain tissue explant cultures through several serial passages.

ii. Transmissible Mink Encephalopathy

Transmissible mink encephalopathy is a scrapie-like disease of farm- raised mink. It is believed to have spread to mink by feeding them on scrapie infected sheep meat.

iii. Bovine Spongiform Encephalopathy (BSE) or "Mad Cow Disease"

A disease similar to scrapie, designated **bovine spongiform encephalopathy (BSE**), or "**mad cow disease**," emerged in cattle in Great Britain in 1986. This outbreak was traced to the use of cattle feed that contained contaminated bone meal from scrapie-infected sheep and BSE-infected cattle carcasses. In 1988, the use of such cattle feed was prohibited.

The epidemic of "mad cow disease" peaked in Great Britain in 1993. It is estimated that over 1 million cattle were infected. BSE has also been found in other European Countries. In 1996, a new variant form of Creutzfeldt-Jakob disease (CJD) was recognized in the United Kingdom that occurred in younger people and had distinctive pathologic characteristics similar to those of BSE.

It is now accepted that the **new variant forms of CJD and BSE** are caused by a common agent, indicating that the BSE agent had infected humans. Through 2002, of 129 people who had been diagnosed with new variant CJD in England, 121 had died.

iv. Chronic Wasting Disease

A scrapie-like disease, designated chronic wasting disease, is found in mule deer and elk in the United States. There is no evidence that it has been transmitted to humans.

b. Human Prion Diseases

i. Kuru

Kuru was a fatal neurologic disease with cerebellar signs exclusive to the Fore tribes in Papua New Guinea. Kuru (means "**shivering" or "trembling")** was first described by Gajdusek and Zigas in 1957 as a mysterious disease seen only in the Fore tribe inhabiting the eastern highlands of New Guinea. The disease had an **incubation period** is 5 to 10 years and led to progressive cerebellar ataxia and tremors, ending fatally in 3 to 6 months. A total of 3700 cases occurred in a population of 35,000.

Transmission from human to human was associated with **ritual cannibalism** involving the consumption of the body of a dead member of the family after ritual nonsterilizing cooking. When Gajdusek began his study, he noted that **women and children in particular were the most susceptible to the disease**, and he deduced that the reasons were that the women and children prepared the food and they were given the less desirable viscera and brains to eat. Their risk for infection was higher because they handled the contaminated tissue, making it possible for the agent to be introduced through the conjunctiva or cuts in the skin, and they ingested the neural tissue, which contains the highest concentrations of the kuru agent. No cases have been seen in those born since 1957, when cannibalism ceased.

Carlton Gajdusek was awarded the Nobel prize for Medicine in 1976 for his important contributions on Kuru.

ii. Creutzfeldt-Jakob Disease (CJD)

Creutzfeldt-Jakob disease is a rare chronic encephalopathy of humans with associated dementia. CJD in humans develops gradually, with progressive dementia, ataxia, and myoclonus, and leads to death in 5 to 12 months. The disease was known as both sporadic and inherited forms.

The natural mode of transmission of Creutzfeldt-Jakob disease (GJD) is unknown, and it does not appear to be acquired from sheep. Iatrogenic CJD has been transmitted accidentally by contaminated growth hormone preparations from human cadaver pituitary glands, by corneal transplant, by contaminated surgical instruments, and by cadaveric human dura mater grafts used for surgical repair of head injury. There is no suggestion of CJD transmission by blood or blood products.

A protein very similar to scrapie PrP^{sc} is present in brain tissue infected with classic CJD. It has been speculated that the agent of CJD was derived originally from scrapie-infected sheep and transmitted to humans by ingestion of poorly cooked sheep brains.

An epidemic of mad cow disease in the United Kingdom and the unusual incidence of CJD in younger people (younger than 45 years) prompted concern that contaminated beef was the source of this new variant of CJD. The association between the bovine and human diseases has not been proved, however.

iii. Gerstmann-Straussler-Scheinker (GSS) Disease

It is rare neurologic diseases of humans due to prion-type agent.

iv. Fatal Familial Insomnia (FFI)

It is also a rare neurologic diseases of humans are due to prion-type agent. In fatal familial insomnia there is loss of ability to sleep and death within 1 to 2 years.

v. Alzheimer's Disease

There are some neuropathologic similarities between CJD and Alzheimer's disease, including the appearance of amyloid plaques. However, the disease has not been transmitted experimentally to primates or rodents, and the amyloid material in the brains of Alzheimer's patients does not contain Prpsc protein.

Group C

i. Subacute Sclerosing Panencephalitis (SSPE)

Subacute sclerosing panencephalitis (SSPE) is an extremely serious, very late neurologic sequela of measles. This disease occurs when **a defective measles virus** persists in the brain and acts as a slow virus. It has been reported that SSPE may also develop as a very rare late complication of **live measles virus vaccination**. A similar picture has also been described as a rare complication of **rubella infection.**

SSPE is most prevalent in children who were initially infected when younger than 2 years. The disease begins insidiously 5 to 15 years after a case of measles. It is characterized by progressive mental deterioration, involuntary movements, muscular rigidity, and coma. Death occurs 1 to 3 years after onset of symptoms.

Brain cells from patients show serological and electron microscopic evidence of measles virus infection. The virus cannot be isolated in routine cultures and the initial evidence linking SSPE with measles virus was based on the isolation of virus from infected brain tissue by co-cultivation of infected cells with cells permissive for measles virus growth. Unusually high levels of measles antibodies are found in the blood and cerebrospinal fluid of patients with SSPE and defective measles virus in brain cells. Antibody is regularly found in CSF and is pathognomonic. CMI to measles virus is absent in SSPE.

ii. Progressive Multifocal Leukoencephalopathy (PML)

Progressive multifocal leukoencephalopathy (PML) is a rare, degenerative CNS infection that usually occurs in persons with severe immunodeficiency disorders. The disease is characterized by memory loss, difficulty in speaking, and a loss of coordination. Death occurs in 3 to 4 months. Demyelination in the central nervous system of patients with progressive multifocal leukoencephalopathy results from oligodendrocyte infection.

Causative Agent

The **causative agent** is a **human polyomavirus (JC virus),** a member of the family polyomaviridae. PML has been seen in several patients with AIDS-associated neurologic complications.

Laboratory Diagnosis

The diagnosis must be made on clinical grounds, with confirmation by the characteristic histologic changes in brain tissue-described in the section on pathogenesis. There are no methods for directly detecting virus in tissue through the use of electron microscopy, antigen detection, or nucleic acid probes. Also, no serologic tests can detect viral antibody. Demonstration of a proteinase K: resistant form of PrP in a Western blot using antibody to PrP can confirm a case of CJD.

☞ KEY POINTS

- The term 'slow virus disease' is applied to a group of infections in animals and human beings, characterized by a very long incubation period and a slow but relentless course, terminating fatally.
- Slow virus diseases are classified into:
  - Group A; slowly progressive infections of sheep due to serologically related lentiviruses (nononcogenic retroviruses) such as visna and maedi.
  - Group B; prion diseases of the CNS (subacute spongiform viral encephalopathies). **Human diseases** include. Kuru, Creutzfeldt-Jakob disease (CJD), Gerstmann-Striiussler-Scheinker (GSS) disease and fatal familial insomnia (FFI) Scrapie, bovine spongiform encephalopathy (BSE) (mad cow disease), chronic wasting disease (in mule, deer, and elk), and transmissible mink encephalopathy occur in animals.
  - Group C diseases include subacute sclerosing panencephalitis and progressive multifocal leucoencephalopathy.
  - SSPE is a very rare delayed sequel to infection with measles virus, many years after the initial infection. *Papova virus* is responsible for PML.

KNOW MORE

Kuru

Before Gajdusek intervened, it was the custom of these people to eat the bodies of their deceased kinsmen. A suggested origin of kuru was from the cannibalistic consumption of a missionary who was dying from CJD.

IMPORTANT QUESTIONS

1. Write briefly on slow virus diseases.
2. Write short notes on:
   i. Creutzfeldt-Jakob Disease
   ii. Kuru
   iii. Bovine spongiform encephalopathy (BSE) or mad cow disease
   iv. Subacute sclerosing panencephalitis (SSPE)
   v. Progressive multi focal leukoencephalopathy (PML)

FURTHER READING

Prusiner SB: Molecular biology and genetics of neurodegenerative'diseases caused by prions, Adv Viros Res 1992;41:241-80.

Prusiner SB. Curr Top Microbial Immuno 1996;I207:1-162.

Prusiner SB. Prions of humans and animals. In Topleyand Wilson's Microbiology and Microbial Infections, 9th edn. London: Arnold 1988;805.

CHAPTER 70

Miscellaneous Viruses

Learning Objectives

After reading and studying this chapter, you should be able to:

- Describe Rubella virus.
- Describe the following: Lassa fever; severe acute respiratory syndrome (SARS).
- Describe structure of rotavirus.
- Discuss laboratory diagnosis of rotavirus infections
- List the viruses causing diarrhea.
- Discuss prophylaxis of hepatitis B infections of hepatitis B vaccine.
- Describe the following: Hepatitis C virus or Type C hepatitis; Hepatitis D virus or Delta agent; Hepatitis E virus; Hepatitis G virus.

RUBIVIRUS

Rubella virus has been classified in the family Togaviridae as the only member of the genus *Rubivirus*. It resembles togaviruses structurally and in many other features.

RUBELLA (GERMAN MEASLES)

Rubella or German measles is a mild exanthematous fever characterized by transient macular rash and lymphadenopathy.

Morphology: Rubella virus is a pleomorphic, roughly spherical particle, 50-70 nm in diameter, with single stranded RNA genome and surrounded by an envelope carrying hemagglutinin peplomers. The single-stranded RNA is infective, and replication occurs in the cytoplasm of infected cells.

Properties: It agglutinates goose, pigeon, day-old chick and human erythrocytes at 4°C, a characteristic which is utilized in the hemagglutination inhibition test for specific antibodies.

Resistance: The virus is inactivated by ether, chloroform, formaldehyde, beta propiolactone and desoxycholate. It is destroyed by heating at 56°C, but survives for several years at–60°C.

Clinical features: Humans are the only host for rubella. Infection is transmitted by the respiratory route. Incubation period is 2-3 weeks. The characteristic clinical features are:

1. *Rash:* A generalized rash develops first on the face and then spreads to the trunk and limbs, sparing the palms and soles. The rash is generally discrete and ordinarily disappears by the third day.
2. *Lymphadenopathy:* There is nontender enlargement of posterior cervical glands. The disease occurs principally in children but may affect all ages.
3. *Common complications*—are arthralgia and arthritis, commoner in women and with increasing age.
4. *Rare complications*—include thrombocytopenic purpura and encephalitis.

Congenital Rubella

Based on the results of *in vitro* studies, rubella infection is presumed to cause chromosomal breakages and inhibition of mitoses in infected embryonic cells. Fetal damage caused by maternal rubella is related to the stage of pregnancy.

Classical congenital rubella syndrome: The virus may spread to the fetus through the bloodstream, causing death due to infection in early pregnancy, congenital malformations in infection during the first trimester and more subtle damage in later infections. The commonest malformations caused by rubella are *cardiac defects, cataract* and *deafness*, which constitute the *classical congenital rubella syndrome.*

Epidemiology

Rubella has a worldwide distribution. Outbreaks usually occur in spring and early summer. Humans are the only host for rubella Infection is transmitted by the respiratory route. Serological surveys in different countries have shown that 80-90 percent are immune by the age of 15 years. About 10-20 percent of mothers are nonimmune and therefore vulnerable Currently, only 2-3 percent of young adult females are susceptible.

Laboratory Diagnosis

A. **Virus Isolation:** The virus may be isolated from blood during the early stage or more successfully from throat swabs in rabbit kidney or vero cells. Various tissue culture cell lines of monkey (BSC-1, Vero) or rabbit (RK-13, SIRC) origin, as well as primary African green monkey kidney cultures, may be used. Rubella produces a rather inconspicuous cytopathic effect in most of the cell lines. Therefore, rubella virus, in cell culture, is detected by interference with the CPE of a challenge virus (coxsackievirus A19) and by immunofluorescence or immunoperoxidase staining for detection of antigen. The virus grows better if cultures are incubated at a lower temperature, 33-35°C. Rubella virus isolation is not commonly employed for diagnosis because of the difficulties and delay involved.

B. **Serology:** Serological diagnosis is the method in routine use. ELISA for IgM and IgG antibodies gives valuable information. IgM antibody alone, without IgG means current acute infection. In case of rubella IgG antibody, four-fold or more rise in titer in paired sera has a diagnostic value. IgG antibody alone, without IgM means past infection or vaccination and denotes immunity, as there is only one serotype of rubella virus.

Prophylaxis

Passive prophylaxis: There is little evidence that administering normal human immunoglobulin after contact reduces the risk of maternal rubella and fetal infection, although it may attenuate the illness.

Active Prophylaxis

Rubella vaccines

Several **live attenuated vaccines** have been developed by serial passage of the virus in tissue culture. The vaccine in use now is the **RA 27/3 strain** grown in human diploid cell culture and administered by subcutaneous injection in a single dose of 0.5 ml. The vaccine is available as a single antigen or combined with measles and mumps components as MMR vaccine. Vaccine-induced immunity persists in most vaccinees for at least 14 to 16 years and probably is lifelong.

The vaccine virus is apparently not teratogenic. Inadvertent administration of the vaccine to a pregnant women may not therefore lead to congenital defects in the baby.

VIRAL HEMORRHAGIC FEVERS

Hemorrhagic manifestations are sometimes seen in many viral fevers such as exanthematous fevers—smallpox, chickenpox and measles; mosquito borne diseases—yellow fever, dengue and chikungunya; tick-borne fevers—Kyasanur Forest disease, Omsk hemorrhagic fever and the Crimean Congo hemorrhagic fever. However, the term **hemorrhagic viral fevers** is not applied to them, but only to a group of diseases, apparently zoonotic in nature, with typical hemorrhage features caused by viruses belonging to two families: **Arenavirus** and **Filovirus**.

ARENAVIRUSES

Arenaviruses are typified by pleomorphic particles that contain a segmented RNA genome; are surrounded by an envelope with large, club-shaped peplomers; and measure 50-300 nm in diameter (mean, 110-130 nm). Electron microscopy of thin sections shows characteristic electron-dense granules resembling grains of sand within virus particles. Hence the name *arena* (L), meaning **sand**. These particles are cellular ribosomes picked up by the virus presumably during maturation by budding from host cells.

Arenaviruses establish chronic infections in rodents. Each virus is generally associated with a single rodent species. Humans are infected when they come in contact with rodent excreta. Some viruses cause severe hemorrhagic fever.

Arenaviruses Causing Human Diseases

At least seven arenaviruses cause human disease—lymphocytic choriomeningitis (LCM), Lassa, Junin, Machupo, Guanarito, Sabia, Whitewater Arroyo.

1. **Lymphocytic choriomeningitis virus (LCM):** LCM virus has a worldwide distribution. It is endemic in mice and other animals (dogs, monkeys, guinea pigs, hamsters) and is occasionally transmitted to humans. It may chronically infect mouse or hamster colonies.

   Human infections: Most human infections are acquired by contact with laboratory mice or hamsters, presumably via mouse droppings.

   Lymphocytic choriomeningitis in humans is an acute disease manifested by **aseptic meningitis** or a **mild systemic influenza-like illness**. Rarely there is a severe encephalomyelitis or a fatal systemic disease. Many infections are subclinical. LCM has been reported to account for 5-10 percent of sporadic viral meningitis in human beings. The incubation period is usually 1-2 weeks.

2. **Lassa Fever:** The first recognized cases of Lassa fever occurred in 1969 among Americans stationed in the Nigerian village of Lassa.

   Natural reservoir is the multimammate rat, *Mastomys natalensis*. Rodent excreta probably act as the source of infection. The incubation period is 3-16 days. The virus is present in the throat, urine and blood of patients. Nosocomial infection has occurred frequently.

The antiviral drug ribavirin is the drug of choice for Lassa fever and is most effective if given early in the disease process. A potential vaccine is under study.

3. **South American hemorrhagic fevers:** The South American arenaviruses are all considered to be members of the Tacaribe group of arenaviruses. Rodents act as reservoirs and transmission is believed to occur through rodent excreta. Serious human pathogens are the closely related Junin, Machupo, Guanarito, and Sabia viruses.
   a. **Junin hemorrhagic fever (Argentine hemorrhagic fever)**
   b. **Machupo hemorrhagic fever (Bolivian hemorrhagic fever)**
   c. **Guanarito virus (the agent of Venezuelan hemorrhagic fever)**
   d. Both Guanarito virus and Sabia virus induce a clinical disease resembling that of Argentine hemorrhagic fever and probably have similar mortality rates.

FILOVIRUSES

They have been classified as Filoviridae. These are long threadlike viruses, hence the name (*filum* means thread). They range in size from 80 to 800-1000 nm. *Marburg* and *Ebolaviruses* causing hemorrhagic fever belong to the genus *Filovirus*.

Morphology: Marburg and Ebola viruses are enveloped negative sense single stranded RNA viruses with long tubular or filamentous forms (Fig. 70.1).

Marburg Virus

Marburg disease is a hemorrhagic fever that occurred simultaneously in laboratory workers in Marburg, Frankfurt (Germany) and Belgrade (Yugoslavia) in 1967. In each case, the infection arose from tissues of African green monkeys to which the laboratory workers had been exposed, imported to laboratories from Uganda via London.

Ebola Virus

In 1976, two severe epidemics of hemorrhagic fever occurred in Sudan and Zaire with high fatality. The causative virus was morphologically identical to the Marburg virus but antigenically distinct. The virus responsible was named Ebola virus after the Ebola river, beside which the first cases were noticed. In 1979, Ebola re-emerged in Sudan, with serial person-to-person spread. Another epidemic occurred in Kikwit, Zaire, in 1995.

Three distinct strains of Ebola virus have been recognized:

1. Zaire strain (EBOZ)
2. Sudan strain (EBO-S)
3. Reston strain (EBO-R)

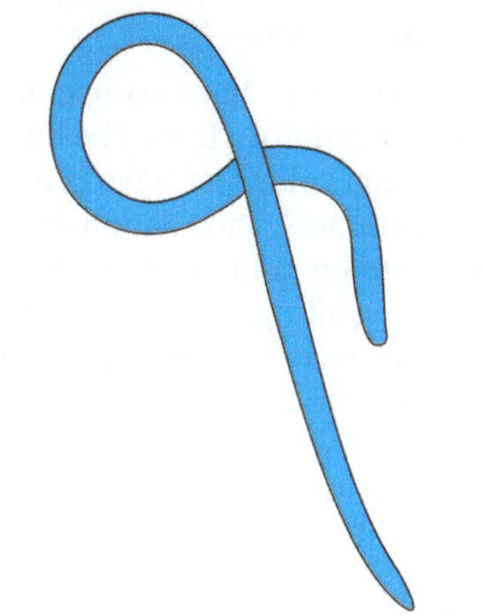

Fig. 70.1: Ebola virus (long thin filamentous form)

Pathogenesis

It is probable that Marburg and Ebola viruses have a reservoir host, perhaps a rodent or a bat, and become transmitted to humans only accidentally. Transmission among humans generally requires direct contact with blood or body fluids, although droplet infections can occur.

Laboratory Diagnosis

1. **Electron Microscopy:** In the blood and in the cytoplasm of the affected cells filamentous viruses can be seen by electron microscopy.
2. **Isolation of Virus:** Virus can be cultured in Vero cells from the blood during the febrile phase.Virus isolate is identified by electron microscopy and direct immunofluorescence. Virus culture must only be attempted in laboratories with required biosafety level.
3. **Serology:** Indirect immunofluorescence can be used for detection of antibodies in the serum.

CORONAVIRUSES

A group of spherical or pleomorphic enveloped RNA viruses, carrying petal or club shaped peplomers on their surface has been classified as coronaviruses (*Corona*, meaning **crown**) resembling the solar corona (Fig. 70.2). It has been classified in the family Coronaviridae with two genera: *Torovirus* and *Coronavirus*.

Torovirus: The toroviruses are widespread in ungulates and appear to be associated with diarrheas.

Coronavirus: There seem to be two serogroups of human coronaviruses. Coronaviruses of domestic animals and rodents are included in these two groups. There is a third distinct antigenic group which contains the **avian infectious bronchitis virus of chickens**.

The novel coronavirus recovered in 2003 from patients with severe acute respiratory syndrome (SARS) appears to represent a new (fourth) group of viruses.

Human Coronavuiruses

Coronavirus infections in humans usually remain limited to the upper respiratory tract. The human coronaviruses cause common colds and have been implicated in gastroenteritis in infants. Inoculation in human volunteers induces common cold after an incubation period of 2-5 days. The resulting immunity is poor and reinfections can occur even with the same serotype. They appear to be the second most common cause of common cold, particularly in winter, next only to rhinoviruses. Coronaviruses are suspected of causing some gastroenteritis in humans, but the agents have not been isolated.

Laboratory Diagnosis

i. Virus isolation—Nasopharyngeal washings are used for the isolation of virus. Human coronaviruses can be cultured in human fetal tracheal organ culture. Some strains may grow on monolayers of diploid human embryonic fibroblasts, with minimal cytopathic effects.
ii. Demonstration of antigen—ELISA test is used for detection of coronavirus antigens in respiratory secretions.
iii. Antibody detection—Antibodies in serum may be detected by ELISA.

Animal coronaviruses: Coronaviruses are established agents of diarrhea in calves, piglets and dogs. Coronaviruses have been observed in human feces also, some time in large number, but their significance is not known.

Severe Acute Respiratory Syndrome (SARS)

In November 2002, Guangdong province in South China experienced an outbreak of an unusual respiratory infection, with many deaths. The world outside knew about it only in February 2003, when a physician from Guangdong visited Hong Kong, fell ill and died there, after infecting twelve persons who had stayed in the same hotel. They, in turn, went to their separate countries to fall ill and initiate outbreaks there.

In February in Hanoi, Vietnam, a private hospital sought the help of a WHO local office about an unusual case of pneumonia. Dr Carlo Urbani, the nearest WHO infectious disease specialist volunteered. As a former President of Medesins sans Frontieres, he had ample experience in tackling epidemics. Sensing the danger, he immediately arranged for isolation and quarantine, but by then outbreaks had begun in many countries: Canada, USA, Ireland and some European countries, Hong Kong, China, Taiwan and most countries in South East Asia. The new disease was named **severe acute respiratory syndrome (SARS)**. It had affected over 30 countries, with many thousand cases and over 800 deaths by July, when the pandemic was controlled. India escaped the SARS epidemic; though a few suspect cases were detected and quarantined. Dr Carlo Urbani identified this epidemic, initiated early steps for its control and died of it within a month.

A coronavirus was found in the respiratory secretions of patients, identified by electron microscopy, and confirmed by growth in Vero cell culture, animal inoculation, cloning, sequencing and histology. Molecular and serological tests for rapid diagnosis were developed. This appears to be a new virus distinct from other coronaviruses, which had been classified into three types: **Mammalian viruses in types 1 and 2** and **avian viruses in type 3**. The new SARS virus becomes **coronavirus type 4**. It may be a recombinant of some animal and human virus.

Causative agent: It has been suggested that the causative agent isolated be designated '**Urbani SARS associated coronavirus**' in memory of Dr Carlo Urbani who identified the new epidemic, initiated early steps for its control and died of it within a month.

Mode of infection: SARS spreads by inhalation of the virus present in droplets or aerosols of respiratory secretions of patients. Fecal aerosols also may be infectious.

Clinical features: The incubation period is under 10 days. The disease starts as fever with cough or other respiratory symptoms. The outbreak of SARS was characterized by serious respiratory illness, including pneumonia and progressive respiratory failure. Diarrhea is sometimes seen. The chest radiograph shows pneumonic changes. Death is due to respiratory failure.

Laboratory Diagnosis

Specimen: Nasopharyngeal swab or aspirate, throat swab or stool specimens may be collected for laboratory diagnosis. Serum is used for antibody detection.

1. **Polymerase Chain Reaction (PCR):** Reverse transcription PCR (RT-PCR) has been used for early diagnosis.
2. **Virus culture:** Virus in clinical specimens can be cultured on Vero cell lines.
3. **Serology:** Demonstration of rise in titer of antibodies by ELISA or indirect immunofluorescent test in paired serum samples are useful later. However, for early diagnosis, PCR is preferred.

Treatment and Prophylaxis

No specific therapy or prophylaxis has been identified. The virus is highly mutable and so vaccine prophylaxis may not be easy. Control has been achieved by strict isolation and quarantine. However, ribavirin and steroids have been shown to be useful in treatment of critical patients.

REOVIRIDAE

The *Reoviridae* constitute a diverse family of viruses that infect humans, many mammals, other vertebrates (including birds), plants and insects. The family Reoviridae derives its name from the prototype virus which was known as the respiratory enteric orphan virus, because it could be isolated frequently from the respiratory and enteric tracts, and it was considered 'orphan' because it was not associated with any disease.

Classification

The family Reoviridae is divided into nine genera. Four of the genera are able to infect humans and animals: Orthoreovirus, Coltivirus, Orbivirus and Rotavirus. Four other genera infect only plants and insects, and one infects fish.

Morphology: All reoviruses have a double-shelled capsid, no envelope and measure 75-80 nm in diameter. The genome consists of double stranded RNA in 10-12 pieces, a feature unique among animal viruses. They are nonenveloped and resistant to lipid solvents.

Rotaviruses

Morphology: Morphologically, rotaviruses are polyhedrons of 75 nm diameter displaying characteristic sharp-edged double shelled capsids, which in electron micrographs look like spokes grouped around the hub of a wheel. The name is derived from *rota*, in Latin, meaning **wheel**. Both 'complete' and 'incomplete' particles are seen. The **complete** or '**double shelled**' virus measures about 70 nm in diameter and has a smooth surface. The **incomplete** or '**single shelled**" virus is smaller, about 60 nm, with a rough surface and is rota virus that has lost the outer shell. 'Empty' particles without the RNA core are also seen. The genome of rotaviruses is located inside the inner core and consists of 11 segments of double-stranded RNA. All segments, except one, code for only **one virus-specific protein (VP)**.

Classification

Rotaviruses have been classified into at least seven antigenic groups (A to G) based on antigenic epitopes on the internal structural protein VP6. These can be detected by immunofluorescence, ELISA, and immune electron microscopy (IEM). Multiple serotypes have been identified among human and animal rotaviruses.

Group A strains: Cause the majority of human infections and have been classified into subgroups (I and II) by ELISA, *CFT* or immune adherence agglutination, and into many serotypes (1, 2, 3 etc) by neutralization tests.

Group B strain: The Chinese adult diarrhoea rotavirus (ADRV) is of group B.

Group C rotaviruses: Cause occasional outbreaks in humans.

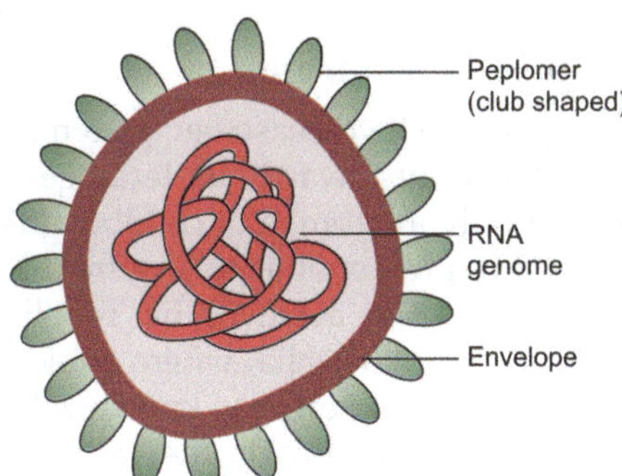

Fig. 70.2: Coronavirus

Animal Susceptibility

Rotaviruses have a wide host range and are a class of viruses causing diarrhea in the young of many animals and some birds.

The human rotavirus is related to the viruses of *epidemic diarrhea of infant mice (EDIM), Nebraska calf diarrhea* and the simian virus SAI1. Cross-species infections can occur in experimental inoculations. Human rotavirus infection has been transferred to piglets, calves and monkeys. It is not known whether human infection can be caused by animal rotaviruses.

Swine rotavirus infects both newborn and weanling piglets.

Propagation in Cell Culture

Rotaviruses are fastidious agents to culture. Most group A human rotaviruses can be cultivated if pretreated with the proteolytic enzyme trypsin and if low levels of trypsin are included in the tissue culture medium. As calf and simian viruses grow readily in cell cultures, they have been used as antigens for serological studies.

Epidemiology

Rotaviruses are the commonest cause of diarrhea in infants and children the world over and typically account for about half the cases of children hospitalized for diarrhea. Rotavirus infections usually predominate during the winter season. In tropical areas, infections occur evenly throughout the year.

Rotavirus diarrhea is usually seen in children below the age of five years but symptomatic infections are most common in children between ages 6 months and 2 years. By the age of five years, most children have had clinical or subclinical infection, so rotavirus diarrhea is very uncommon in older children and adults. Transmission appears to be by the fecal-oral route. Nosocomial infections are frequent.

Clinical Features

Infection is by the fecal-oral route. The incubation period is 2-3 days. Vomiting and diarrhea occur with little or no fever. Stools are usually greenish yellow or pale, with no blood or mucus. The disease is self-limited and recovery occurs within 5-10 days. Mortality is low. Group B rotaviruses have been implicated in large outbreaks of severe gastroenteritis in adults in China.

Laboratory Diagnosis

Laboratory diagnosis rests on demonstration of virus in stool collected early in the illness and on a rise in antibody titer.

A. **Demonstration of virus:** At the peak of the disease, as many as 10^{11} virus particles per milliliter of feces are present. Therefore, the diagnosis is not difficult. Virus in stool is demonstrated by immune electron microscopy (IEM), latex agglutination tests, or ELISA.

B. **Genotyping:** Genotyping of rotavirus nucleic acid from stool specimens by the polymerase chain reaction is the most sensitive detection method.
C. **Virus isolation:** Rotaviruses can be propagated in secondary or continuous cultures of monkey kidney cells. Cell culture, however, is not used for routine diagnosis
D. **Serologic tests:** Can be used to detect after antibody titer rise, particularly ELISA.

Treatment

Management consists of replacement of fluids and restoration of electrolyte balance either intravenously or orally, as feasible.

Control

In view of the fecal-oral route of transmission, wastewater treatment and sanitation are significant control measures.

Rotavirus Vaccine

An oral live attenuated rhesus-based rotavirus vaccine was licensed in the United States in 1998 for vaccination of infants. It was withdrawn a year later because of reports of intussusception (bowel blockages) as an uncommon but serious side effect associated with the vaccine.

A number of different candidate vaccines of live attenuated rotavirus of bovine or human origin, including bovine rotavirus monoreassortants carrying human VP7 genes of different serotypes, are currently being evaluated. Baculovirus-expressed virus-like particles, DNA-based vaccines and micro-encapsidated viral proteins or cDNAs are also being explored. It is hoped that (an) efficient rotavirus vaccine(s) will become available in the not too distant future.

Other Viruses Causing Diarrhea

In addition to rotaviruses and noncultivable adenoviruses, members of the family Caliciviridae are important agents of viral gastroenteritis in humans. The most significant member is Norwalk virus.

A. Norwalk Virus

Norwalk virus has been included in the family Caliciviridae. Caliciviruses are similar to picornaviruses but are slightly larger (27-40 nm). They exhibit a distinctive morphology in the electron microscope. The name calicivirus is derived from the presence of 32 cup shaped depressions on the virus surface (from *calyx*, meaning **cup**).

Historically, the Norwalk viruses were referred to as "small round structured viruses" based on their detection by electron microscopy.

Norwalk virus is the most important cause of epidemic viral gastroenteritis in adults. A 27 nm virus was shown to be responsible for an epidemic of gastroenteritis affecting school children and teachers in Norwalk, Ohio, in 1972. The virus induced diarrhea in human volunteers. Serological surveys have shown that infection with Norwalk virus is widespread in many countries. The viruses are most often associated with epidemic outbreaks of waterborne, food-borne, and shellfish-associated gastroenteritis.

Laboratory Diagnosis

1. **Electron microscopy:** The virus can be demonstrated in feces by electron microscopy.
2. **Antigen detection:** Radioimmunoassay (RIA) and ELISA can detect the virus and viral antigen.
3. **Serology:** Radioimmunoassay (RIA) and ELISA can detect antibody to Norwalk agent.

B. Adenoviruses

Several outbreaks of diarrhea in children have been associated with the presence of large numbers of adenoviruses in feces. Serotypes 40 and 41 are most commonly associated with acute diarrhea in infants.

The clinical syndrome is similar to that caused by rota-viruses, except that the infants tend to be older. Adenovirus gastroenteritis is sometimes complicated by intussusception.

C. Astrovirus

These star shaped (Greek, *astron* = **a star**, 28 nm isometric particles which have been associated with some epidemics of diarrhea in children. They are recognized as pathogens for infants and children, elderly institutionalized patients, and immunocompromised persons. Similar viruses have also been identified in lamb and calf diarrhea.

Laboratory Diagnosis

Demonstration of virus: The virus in the faeces can be demonstrated by immunoelectron microscopy and ELISA tests.

D. Coronavirus

These are well established causes of acute diarrhea in calves, piglets and dogs. They have been observed in human feces also but their relation to diarrhea is uncertain.

KNOW MORE

Expanded Rubella Syndrome

Active prophylaxis: The disease being so mild, prophylaxis is directed only towards its teratogenic hazard and so relevant only in women of childbearing age. An obvious method of protection is to acquire the infection before puberty. This was achieved by 'rubella parties', formerly practised in Australia, where adolescent girls voluntarily exposed themselves to known rubella cases.

Rubella was first described in the 18th century and was considered a mild illness with only occasional complications. However, in 1941, an astute Australian ophthalmologist, Sir Norman McAlister Gregg observed a sudden increase in congenital cataract in infants and related to it maternal rubella. Observations from different countries soon confirmed that maternal rubella induces congenital malformations of different kinds, the commonest being the triad of cataract, deafness and cardiac defects. The consequences of rubella *in utero* are referred to as the congenital rubella syndrome. Further progress had to wait till rubella virus was isolated in tissue culture in 1962.

☞ KEY POINTS

- Rubella virus has been classified in the family Togaviridae as the only member of the genus Rubivirus.
- Rubella or German measles is a mild exanthematous fever characterized by transient macular rash and lymphadenopathy. Infection is acquired by inhalation.
- Prophylaxis is relevant only in women of childbearing age and is best carried out by immunization with a live attenuated vaccine. (RA 27/3 strain grown in human diploid cell culture) is administered by subcutaneous injection to children 15 months of age as such or in combination (mumps-measles-rubella vaccine).
- Hemorrhagic viral fevers term is applied only to a group of diseases, apparently zoonotic in nature, with typical hemorrhage features caused by viruses belonging to two families:
- **Arenavirus:** At least seven arenaviruses cause human disease—lymphocytic choriomeningitis (LCM), Lassa, Junin, Machupo, Guanarito, Sabia, Whitewater Arroyo, and these are zoonotic diseases transmitted primarily from rodents to humans.
- **Filoviruses:** Marburg and Ebolaviruses causing hemorrhagic fever belong to the genus Filovirus.
- **Coronaviruses:** There are two genera in the Coronaviridae family: Torovirus and Coronavirus.
- **Torovirus:** It appears to be associated with diarrheas.
- **Coronavirus:** There seem to be two serogroups of human coronaviruses—of domestic animals and rodents. There is a third distinct antigenic group which contains the **avian infectious bronchitis virus of chickens**. The novel coronavirus recovered in 2003 from patients with **severe acute respiratory syndrome (SARS)** appears to represent a new (fourth) group of viruses.
- **Severe acute respiratory syndrome (SARS)** refers to a severe atypical pneumonia that assumed pandemic proportions in 2003. A novel coronavirus (the SARS-associated coronavirus [SARS-CoV]) was identified as the cause of the syndrome.
- **Reoviridae:** There are four genera in this family: Reovirus, Coltivirus, Orbivirus and Rotavirus.
- Rotaviruses are the commonest cause of diarrhea in infants and children. The human rotavirus is related to the viruses of **epidemic diarrhea of infant mice (EDIM)**, **Nebraska calf diarrhea** and the **simian virus SAI1**. Swine rotavirus infects both newborn and weanling piglets.
- Laboratory diagnosis is by demonstrating of virus by IEM, latex agglutination tests, or ELISA on clarified fecal samples. Polymerase chain reaction is the most sensitive detection method. Development of a rotavirus vaccine has been difficult.

Other Viruses Causing Diarrhea

- Viruses that are important causes of diarrhea include rotaviruses, the Norwalk virus, certain adenoviruses and coronaviruses, and astroviruses:

Severe Acute Respiratory Syndrome (SARS)

A novel coronavirus was identified as the cause of a worldwide outbreak of a severe acute respiratory syndrome (SARS) in 2003. In contrast, the outbreak of SARS was characterized by serious respiratory illness, including pneumonia and progressive respiratory failure. In all likelihood, the SARS virus originated in a nonhuman host and acquired the ability to infect humans. In rural regions of southern China, where the outbreak began, people, pigs, and domestic fowl live close together and there is widespread use of wild species for food and traditional medicine-conditions that promote the emergence of new viral strains.

IMPORTANT QUESTIONS

1. Write short notes on:
   Rubella or German measles
   Viral hemorrhagic fevers
   Lymphocytic choriomeningitis virus (LCM)
   Ebola virus
   Coronaviruses
   Severe acute respiratory syndrome (SARS)
   Viruses causing diarrhea
   Rotaviruses.

FURTHER READING

Burke B and Desselberger U. Rotavirus pathogenesis. Lancet 1996;218,299.

Holmes KV. SARS. New EngJ Med 2003;348:1948.

Kapikian AZ. Viral infections of the gastrointestinal tract New York: Marcel-Dekker 1994.

Lee HW and Van der Groen G. Hemorrhagic fever with renal syndrome. Progr Med Virol 1989;36:62.

Munro ND. et al. Temporal relations between maternal rubella and congenital defects. Lancet 1987;2:201.

CHAPTER 71

Oncogenic Viruses

Learning Objectives

After reading and studying this chapter, you should be able to:

- Describe oncogenic viruses.
- Classify oncogenic viruses.
- Describe oncogenes and mechanism of viral oncogenesis.
- List of viruses associated with human cancers.

INTRODUCTION

The association of viruses with malignancy dates from the observation by Ellerman and Bang (1908) that the mode of transmission in leukemia in fowls resembled that of an infectious disease. Peyton Rous (1911) showed that a fowl sarcoma could be transmitted to normal chickens by injecting them with cell-free filtrates of the tumor, a discovery for which he was awarded the Nobel prize belatedly in 1966. Shope isolated the rabbit fibroma virus in 1932 and the papilloma virus in 1933. Bittner (1936) proposed that breast cancer in mice could be caused by a virus transmitted from mother to offspring through breast milk. Considerable interest was aroused by the discovery by Stewart and Eddy (1957) of the polyoma virus which could produce a wide variety of neoplasms when injected into newborn rodents. Injection of certain types of human adenovirus into newborn hamsters was shown by Trentin (1962) to cause sarcomas.

It is now acknowledged that at least 15 percent of all human tumors worldwide have a viral cause. These include hepatocellular carcinoma caused by Hepatitis B or C viruses, uterine cervical cancer by certain types of papilloma viruses, anaplastic nasopharyngeal carcinomas by EB virus and adult cutaneous T cell lymphoma/ leukemia by HTLV-1.

ONCOGENIC VIRUSES

Oncogenic viruses are viruses that produce tumors in their natural hosts or in experimental animals, or induce malignant transformation of cells on culture. Transformation represents the various changes that accompany the conversion of a normal cell into the malignant cell.

PROPERTIES OF CELLS TRANSFORMED BY VIRUSES

Transformation from normal to malignant cell is a multistep process and may be partial or complete. Table 71.1 showes properties of cells transformed by viruses.

TYPES OF TUMOR VIRUSES

Both DNA and RNA viruses are oncogenic.

DNA Tumor Viruses

All known tumor viruses either have a DNA genome or generate a DNA provirus after infection of cells. DNA

Table 71.1: Properties of cells transformed by viruses

1. Altered cell morphology: Fibroblasts become shorter, parallel orientation is lost, chromosomal aberrations appear.
2. Altered cell metabolism: Increased growth rate, increased production of organic acids and acid mucopolysaccharides.
3. Altered growth characteristics: Transformed cells are altered in shape and lose the property of 'contact inhibition' so that, instead of growing as monolayer, they grow piled up, one over another, forming 'microtumors', capacity to divide indefinitely in serial culture, capacity to grow in suspension or in semisolid agar.
4. Antigenic alterations: Appearance of new virus specified antigens (T antigen-TSTA). Loss of surface antigens, cells become agglutinable by lectins.
5. Tumorogenicity: Capacity to induce tumors in susceptible animals.

tumor viruses are classified among the papilloma-, polyoma-, adeno-, herpes-, hepadna-, and pox-virus groups.

RNA Tumor Viruses

Most RNA tumor viruses belong to the retrovirus family. Retroviruses carry an RNA-directed polymerase (reverse transcriptase) that constructs a DNA copy of the RNA genome of the virus. The DNA copy (provirus) becomes integrated into the DNA of the infected host cell, and it is from this integrated DNA copy that all proteins of the virus are translated. Retroviruses are responsible for naturally occurring leukemia and sarcoma in several species of animals.

ONCOGENIC VIRUSES

A. Oncogenic DNA Viruses

DNA tumor viruses encode viral oncoproteins that are important for viral replication but also affect cellular growth control pathways.

I. Papovaviruses (See chapter 60)

1. **Papilloma viruses:** Papilloma viruses cause benign tumors in their natural hosts but some of them (e.g. condyloma acuminatum in humans, rabbit papilloma) may turn malignant. The association between human papilloma virus (HPV) infection and cancer of cervix uteri, particularly HPV types 16 and 18 has been established. In recent years, papillomaviruses have superseded herpes simplex as the prime suspect for causing cervical cancer.
2. **Polyoma virus:** The polyoma virus causes natural latent infection in laboratory and domestic mice. However, it induces a wide variety of histologically diverse tumors, when injected into infant mice or other rodents.
3. **Simian virus 40 (SV 40):** (See chapter 60).
4. **BK and JC virus:** (See chapter 60)

II. Poxvirus

Three members of the poxvirus group induce benign tumors. **Yaba virus** produces benign tumors (histiocytomas) in its natural host, monkeys. **Shope fibroma virus** produces fibromas in some rabbits and is able to alter cells in culture. **Molluscum contagiosum virus** produces small benign growths in human.

III. Adenovirus

No association of adenoviruses with human neoplasms has been found. Though some types (12, 19, 21) of human adenovirus may produce sarcomas in newborn rodents after experimental inoculation, they do not appear to have any association with human cancer.

IV. Herpesvirus

1. **Marek's disease:** Marek's disease is a highly contagious lymphoproliferative disease of chickens. Marek's disease can be induced in young chicken by the injection of the virus.

Table 71.2: List of oncogenic viruses

A. DNA viruses
- I. Papovavirus
  1. Papillomaviruses of human beings, rabbits and other animals
  2. Polyomavirus
  3. Simian virus 40
  4. BK and JC viruses
- II. Poxvirus
  1. Yaba virus
  2. Shope fibroma
  3. Molluscum contagiosum
- III. Adenovirus
  Many human and nonhuman types
- IV. Herpes virus
  1. Marek's disease virus
  2. Lucke's frog tumor virus
  3. Herpes virus pan, papio, ateles and saimiri
  4. Epstein-Barr virus
  5. Herpes simplex virus types 1 and 2
  6. Cytomegalovirus
- V. Hepatitis B virus

B. RNA viruses
- a. Retroviruses
  1. Avian leukosis viruses
  2. Murine leukosis viruses
  3. Murine mammary tumor virus
  4. Leukosis-sarcoma viruses of various animals
  5. Human T cell leukemia viruses
- b. Hepatitis C virus

Marek's disease can be prevented by vaccination with an attenuated strain of Marek's disease virus. This is the first instance of a malignant disease being controlled by a viral vaccine.

Other examples of herpesvirus-induced tumors in animals include lymphomas of certain types of monkeys and adenocarcinomas of frogs.

2. **Lucke's tumor of frogs:** A herpesvirus is considered to be the etiological agent of a renal adenocarcinoma in frogs.
3. **Herpesvirus saimiri:** It causes fatal lymphoma or reticulum cell sarcoma when injected into owl monkeys or rabbits.
4. **Epstein Barr virus:** Epstein-Barr (EB) herpesvirus causes acute infectious mononucleosis when it infects B lymphocytes of susceptible humans. EB virus is etiologically linked to Burkitt's lymphoma; to nasopharyngeal carcinoma (NPC); to post-transplant lymphomas; and to Hodgkin's disease. These tumors usually contain EB viral DNA (both integrated and episomal forms) and viral antigens. Malaria may be a cofactor of African Burkitt's lymphoma.
5. **Herpes simplex and cancer cervix:** An association has been proposed between herpes simplex type 2 infection and cancer of the uterine cervix, though not proved. It has also been suggested that herpes

simplex type I infection may be associated with cancer of the lip.

Kaposi's sarcoma-associated herpesvirus, also known as human herpesvirus 8 (KSHV/HHV8). It is suspected of being the cause of Kaposi's sarcoma, primary effusion lymphoma, and a particular lymphoproliferative disorder.

6. **Cytomegalovirus:** Cytomegalovirus infection has been associated with carcinoma of the prostate and Kaposi's sarcoma.

Some herpesviruses are associated with tumors in lower animals.

V. Hepatitis B virus

Hepatocellular carcinoma (HCC) following chronic infection with hepatitis B virus is one of the most prevalent forms of cancer. HBV has been claimed to be directly or indirectly involved in the etiology of hepatocellular carcinoma. It is also the only malignant disease of humans preventable by immunization against the causal agent. Unfor-tunately, however, the pathogenesis of this major cause of cancer is still poorly understood.

Chronic infection with hepatitis C virus is also considered to be a causative factor in hepatocellular carcinoma. Most probably, hepatitis C virus acts indirectly in the development of hepatocellular carcinoma.

B. Oncogenic RNA Viruses

Retroviruses

Retroviruses contain an RNA genome and an RNA directed DNA polymerase (reverse transcriptase). RNA tumor viruses in this family mainly cause tumors of the reticuloendothelial and hematopoietic systems (leukemias, lymphomas) or of connective tissue (sarcomas).

Oncogenic Retroviruses

Retroviruses are widely distributed, being found in nearly all vertebrates, including animals, birds and reptiles. Based on the host range and types of disease caused, oncogenic retroviruses can be considered under the following groups:

1. **Avian Leukosis Complex:** A group of antigenically related viruses which induce avian leukosis (lymphomatosis, myeloblastosis and erythroblastosis viruses) or sarcoma in fowls (Rous sarcoma virus, RSV).
2. **Murine Leukosis Viruses:** This group consists of several strains of murine leukemia and sarcoma viruses.
3. **Mammary tumor virus of mice:** This virus occurs in certain strains of mice having a high natural incidence of breast cancer. It is used to be known as the 'milk factor' or 'Bittner virus'. It multiplies in the mammary gland and is transmitted from mother to offspring through breast milk. Mammary cancer occurs only in susceptible strains of mice, after a latent period of 6-12 months.
4. **Leukosis-sarcoma viruses of other animals:** A number of viruses have been isolated from leukosis and sarcomas in various species of animals—cat, hamster, rat, guinea pig and monkey.
5. **Human Retroviruses:** Human T cell leukemia (lymphotropic) viruses (HTLV): Retroviruses named human T cell leukemia viruses were isolated in 1980 from cell cultures from adult patients with cutaneous T cell lymphoma (mycosis fungoides) and leukemia (Sezary syndrome) in the USA. HTLV-1 has been established as the causative agent of adult T cell leukemia-lymphomas (ATL) as well as a nervous system degenerative disorder called **tropical spastic paraparesis**. A related human virus, HTLV-2, has been isolated but has not been conclusively associated with a specific disease.

VIRUSES ASSOCIATED WITH HUMAN CANCER

Viruses associated with cancers in human beings are shown in Table 71.3.

Table 71.3: Viruses implicated in human cancers

Virus Family	*Virus*	*Tumour*
DNA viruses		
Papovaviridae	Human papilloma virus (HPV)	Cervical, vulvar, penile cancers Squamous cell carcinoma
Herpesviridae	Epstein-Barr virus (EBV)	Nasopharyngeal carcinoma African Burkitt's lymphoma B-cell lymphoma
	HSV type 2	Cervical carcinoma ?
	Human herpesvirus 8	Kaposi's sarcoma
Hepadnaviridae	Hepatitis B virus	Hepatocellular carcinoma
RNA viruses		
Flaviviridae	Hepatitis C virus	Primary liver cancer
Retroviridae	Human T-cell lymphotropic virus (HTLV-1)	Adult T cell leukemia/lymphoma

ONCOGENES

"Oncogene" is the general term given to genes that cause cancer. Viral oncogenes (V-onc), commonly known as 'cancer genes' are genes which encode proteins triggering transformation of normal cells into cancer cells. Normal versions of these transforming genes are present in normal cells and have been designated **proto-oncogenes** or **c-onc** (Table 71.4) . They are not of viral origin. These **c-onc** genes can also be activated by mutagenic stimuli to induce malignancy. Oncogenes are not essential for the replication of the virus and mutants lacking them occur, which replicate normally without being oncogenic.

Cellular oncogenes contain introns characteristic of eukaryotic genes, whereas viral oncogenes do not. Apparently viral oncogenes originated at some distant past from proto-oncogenes by recombination between retroviral and cellular genes. Transduction of the cellular genes was probably an accident, as the presence of the cellular sequences is of no benefit to the viruses.

Proto-oncogenes are widespread in vertebrates and metazoa—from human beings to fruitflies. They are well conserved in their genomes, suggesting that they serve some essential functions in normal cells. They have been found to code for proteins involved in regulating cell growth and differentiation. Incorrect expression of any component might interrupt that regulation, resulting in uncontrolled growth of cells (cancer). The presumed functions of many oncogenes have been identified. For example, the oncogene *src* is related to tyrosine-specific protein kinases, growth factors (*sis* is similar to human platelet-derived growth factor, a potent mitogen for cells of connective tissue origin), GTP-binding proteins (*Ha-ras*), and nuclear transcription factors (*myc, jun*), all concerned with regulation of normal cell growth (Table 71.4).

Transfection

Transfection is a useful method for the study of oncogenes. Certain mouse fibroblast cell lines, such as NIH 3T3, can take up foreign DNA, incorporate them into their genome and express them. This method of gene transfer is called **transfection**.

ANTI-ONCOGENES

A class of genes has been identified in normal human cells, whose function appears to be inhibition of malignant transformation. They have been termed **antioncogenes, tumor suppressor** or **growth suppressor genes**. The inactivation or functional loss of both alleles of such a gene is required for tumor formation—in contrast to the activation that occurs with cellular oncogenes.

Examples

i. **Retinoblastoma *(Rb)* gene:** The prototype of this inhibitory class of genes is the retinoblastoma *(Rb)* gene. The loss of *Rb* gene function is causally related to the development of retinoblastoma—a rare ocular tumor of children and other human tumors.
ii. p53 **gene:** The *p53* gene is mutated in over half of all human cancers.Specific chromosomal deletions, recognized in association with certain types of human cancers may reflect the loss of tumor suppressor genes.

MECHANISMS OF VIRAL ONCOGENESIS

The exact mechanisms of viral oncogenesis are not well understood. The molecular mechanisms responsible for activating a benign proto-oncogene and converting it into a cancer gene vary—but all involve genetic damage. Malignancy is a stable' heritable change and, as such, should be the result of a modification of the host cell genome.

DNA Genome

The viral DNA (or a portion of it) is integrated with the host cell genome in the case of oncogenic DNA viruses. The viral DNA being incomplete or 'defective', no infectious virus is produced. However, under its influence, the host cell undergoes malignant transformation. A virus transformed cancer cell is in many ways analogous to a bacterium lysogenized by a defective phage. The cell is not destroyed and no virus is produced in both cases. Acquisition of new characteristics by the transformed cell resembles lysogenic conversion in bacteria.

Retroviruses

In general, retroviruses induce tumors by two mechanisms—either by introducing into the cellular genome a new transforming gene (oncogene) or by inducing or altering the expression of a pre-existing cellular gene. Several molecular mechanisms have been suggested for the conversion of benign proto-oncogenes to cancer genes.

KNOW MORE

RNA tumor viruses are of two general types with respect to tumor induction. The highly oncogenic (direct-transforming) viruses carry an oncogene of cellular origin. The weakly oncogenic (slowly transforming) viruses do not contain an oncogene and induce leukemias after long incubation periods by indirect mechanisms. The two known cancer-causing retroviruses in humans act indirectly. Hepatitis C virus, a flavivirus, does not generate a provirus and appears to induce cancer indirectly.

☞ KEY POINTS

- Oncogenic viruses are viruses that produce tumors in their natural hosts or in experimental animals, or induce malignant transformation of cells on culture.

Table 71.4: Some oncogenes				
Viral oncogene	*Origin*	*Natural tumor*	*Human gene*	*Chromosomal location in human beings*
V-src	chicken	Sarcoma	C-src	20
V-sis	monkey	Sarcoma	C-sis	22
V-ras	rat	Sarcoma	C-ras	11
V-myc	chicken	Leukemia	C-myc	8
V-fes	cat	Sarcoma	C-fes	15
V-mos	mouse	Sarcoma	C-mos	8

V- Viral; C- Cellular

src = sarcoma of chicken, sis = simian sarcoma; ras = rat sarcoma; myc = myelomatosis of chicken; fes = feline sarcoma; mos = mouse sarcoma

- Transformation represents the various changes that accompany the conversion of a normal cell into the malignant cell.
- All known tumor viruses either have a DNA genome or generate a DNA provirus after infection of cells. In case of oncogenic DNA viruses, the viral DNA (or portion of it) is integrated with the host cell genome as prophage. As the integrated viral DNA is incomplete or 'defective', no infectious virus is produced. However, under its influence, the host cell undergoes neoplastic transformation.
- Most RNA tumor viruses belong to the retrovirus family. Retroviruses carry an RNA-directed polymerase (reverse transcriptase) that constructs a DNA copy of the RNA genome of the virus. The DNA copy (provirus) becomes integrated into the DNA of the infected host cell, and it is from this integrated DNA copy that all proteins of the virus are translated. Retroviruses are responsible for naturally occurring leukemia and sarcoma in several species of animals.
- Viral oncogenes (*V-onc*), commonly known as *'cancer genes'* are genes which encode proteins triggering transformation of normal cells into cancer cells Viral oncogenes (*v-on*) possess almost identical cellular counterparts (*c-on*) from which they were derived in the distant past and are not of viral origin.These c-on genes can also be activated by mutagenic stimuli to induce malignancy.
- The main oncogenic DNA viruses are papillomaviruses, Epstein-Barr virus, and hepatitis B. Most of the oncogenic RNA viruses are retroviruses; long-term infection with hepatitis C, a flavivirus, may result in liver cancer, but the mechanism is probably indirect.

IMPORTANT QUESTIONS

1. Define and classify oncogenic viruses.
2. Enumerate RNA and DNA oncogenic viruses. Describe mechanism of viral carcinogenesis.
3. Write short notes on:
   Viral oncogenes.
   Viruses associated with human cancer.

FURTHER READING

Burel S. Viral carcinogenesis: Revelation of molecular mechanisms and etiology of human disease. Carcinogenesis 2000; 21:405.

Friend SH et al. Oncogenes and tumor suppressor genes. New Engl J Med 1989;318,618.

Khalili K, Raab-Traub N (editors). Cancer viruses. Oncogene Rev 2003;22 (No. 2). [Entire issue].

SECTION FIVE

Medical Mycology

CHAPTER 72

General Properties, Classification and Laboratory Diagnosis of Fungi

LEARNING OBJECTIVES

After reading and studying this chapter, you should be able to:
- Differentiate between fungi and bacteria.
- Classify fungi.
- Describe laboratory diagnosis of fungal infections.
- Discuss diseases caused by fungi.

INTRODUCTION

Mycology is the study of fungi.

Beneficial Effects of Fungi

1. They reside in nature and are essential in breaking down and recycling organic matter.
2. Some fungi greatly enhance our quality of life by contributing to the production of food and spirits.
3. Other fungi have served medicine by providing useful bioactive secondary metabolites such as antibiotics (e.g., penicillin) and immunosuppressive drugs (e.g., cyclosporine).
4. Fungi have been exploited by geneticists and molecular biologists as model systems for the invest gation of a variety of eukaryotic processes.

Harmful Effects of Fungi

In addition to their disease-producing potential in humans, fungi are directly or indirectly harmful in many other ways. Fortunately, only a few hundred species of fungi have been implicated in human disease, and 90% of human infections by fungi can be attributed to a few dozen species. Fungus infections have assumed greater importance with the control of most bacterial infections in the developed countries. For instance, it has been stated that in the USA, fungus infections cause as many fatalities today as whooping cough, diphtheria, scarlet fever, typhoid, dysentery and malaria put together.

DIFFERENCES OF FUNGI FROM BACTERIA

All fungi are eukaryotic protista that differ from bacteria and other prokaryotes in many ways:

1. They possess rigid cell walls containing chitin, mannan and other polysaccharides.
2. The cytoplasmic membrane contains sterols.
3. Cytoplasmic contents include mitochondria and endoplasmic reticulum.
4. They possess true nuclei with nuclear membrane and paired chromosomes.
5. They may be unicellular or multicellular.
6. They divide asexually, sexually or by both processes.
7. Most fungi are obligate or facultative aerobes.
8. They are chemotrophic, secreting enzymes that degrade a wide variety of organic substrates into soluble nutrients which are then passively absorbed or taken into the cell by active transport.
9. The cells show various degrees of specialization.

GENERAL PROPERTIES OF FUNGI

Fungi grow in two basic forms, as yeasts and molds.

Yeast

The simplest type of fungus is the unicellular budding yeast.

Hypha

Elongation of the cell produces a tubular, thread like structure called **hypha**.

Hyphae may be **septate** or **nonseptate**. Some hyphae are divided into cells by cross-walls or septa, typically forming at regular intervals during hyphal growth. The septa, when present, have holes through which free flow of cytoplasmic material can take place. One class of medically important molds, the zygomycetes, produces hyphae that are rarely septated.

Mycelium

A tangled mass of hyphae constitutes the mycelium. Fungi which form mycelia are called molds or filamentous fungi (Fig. 72.1). Under standardized growth conditions in the laboratory, molds produce colonies with characteristic features such as rates of growth, texture, and pigmentation.

In a growing colony of filamentous fungus, the mycelium can be divided into the **vegetative mycelium** and the **aerial mycelium**.

Vegetative Mycelium

Hyphae that penetrate the supporting medium and absorb nutrients are the vegetative or substrate hyphae.

Aerial Hyphae

In contrast, **aerial hyphae** project above the surface of the mycelium and usually bear the reproductive structures of the mold.

CLASSIFICATION OF FUNGI

Fungi are placed in the phylum Thallophyta. It is divisible into two groups, algae and fungi.

Algae

Algae produce their own food by means of chlorophyll possessed by them.

Fungi

Fungi do not possess chlorophyll and are, therefore, saprophytes or parasites.

A. Morphological classification.
B. Systematic classification.

A. Morphological Classification

On the basis of morphology, there are four groups of fungi:

1. **Yeasts:** Yeasts are round, oval or elongated unicellular fungi. Most reproduce by an asexual process called budding in which the cell develops a protuberance, which enlarges and eventually separates from the parent cell. (Figs 72.1A and B). Some reproduce by fission. On culture, they form smooth, creamy colonies.

   Examples: *Saccharomyces cerevisiae*—nonpathogenic yeast and *Cryptococcus neoformans*—pathogenic yeast.

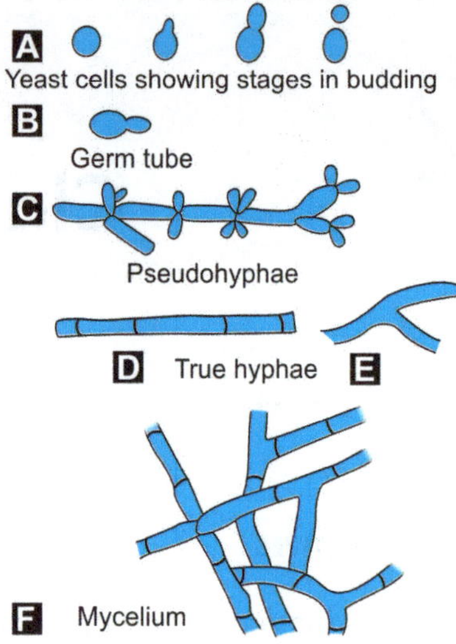

Fig. 72.1: Basic fungal morphology

2. **Yeast-like fungi:** Yeast-like fungi grow partly as yeast and partly as elongated cells resembling hyphae. The latter form a pseudomycelium.

   Example: *Candida albicans* is a pathogenic yeast-like fungus.

3. **Molds or filamentous fungi:** Molds or filamentous fungi form true mycelia and reproduce by the formation of different types of spores.

   Examples of mold: *Dermatophytes, Aspegillus, Penicillium, Mucor* and *Rhizopus*.

4. **Dimorphic fungi:** Many fungi pathogenic to man have a yeast form in the host tissue and in vitro at 37°C on enriched media and hyphal (mycelial) form in vitro at 25°C.

   Examples: *Histoplasma capsulatum, Sporothrix schenckii, Blastomyces dermatitidis, Coccidioides immitis, Paracoccidioides brasiliensis*, and *Penicillium marneffei*.

B. Systematic Classification

On the basis of formation of sexual spores, fungi have been divided into 4 classes (See Flow chart 72.1).

Flow chart 72.1

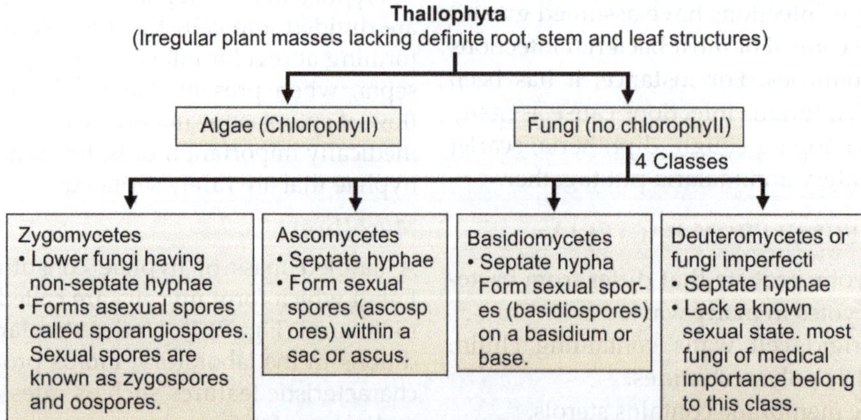

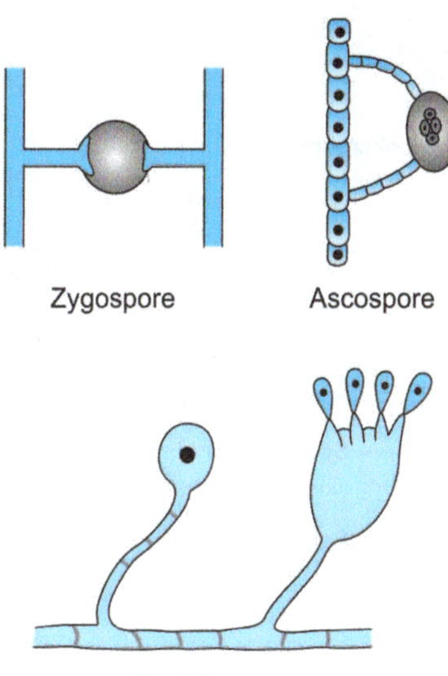

Fig. 72.2: Sexual spores

1. **Zygomycetes:** Zygomycetes are lower fungi which have nonseptate hyphae and form asexual spores called sporangiospores contained within swollen oospores.
   Examples: *Rhizopus, Absidia, Mucor, Pilobolus.*
   Other three classes Ascomyc`etes, Basidiomycetes and Deuteromycetes or Fungi Imperfecti possess septate hyphae.
2. **Ascomycetes:** The Ascomycetes form sexual spores (ascospores) within a sac or ascus. Ascomycetes include both yeasts and filamentous fungi.
3. **Basidiomycetes:** The Basidiomycetes form sexual spores (basidiospores) on a basidium or base.
   Examples: Mushrooms, *Filobasidiella neoformans* (anamorph), *Cryptococcus neoformans*.
4. **Deuteromycetes (Fungi Imperfecti):** Fungi that lack a known sexual state are placed in the class Deuteromycetes (Fungi Imperfecti). The members of this class may have lost the ability to reproduce sexually or their sexual states may have not been discovered. Most fungi of medical importance belong to this class.
   Examples: *Coccidioides immitis, Paracoccidioides brasiliensis, Candida albicans.*

REPRODUCTION AND SPORULATION

Types of Fungal Spores

Fungal spores are of two types: **sexual and asexual spores.**

1. **Sexual spores:** Sexual spore is formed by fusion of cells and meiosis as in all forms of higher life. Sexual spores are of four types—oospore, ascospore, zygospore and basidiospore (Fig. 72.2).
2. **Asexual spores:** These spores are produced by mitosis. These may be vegetative spores or aerial spores.

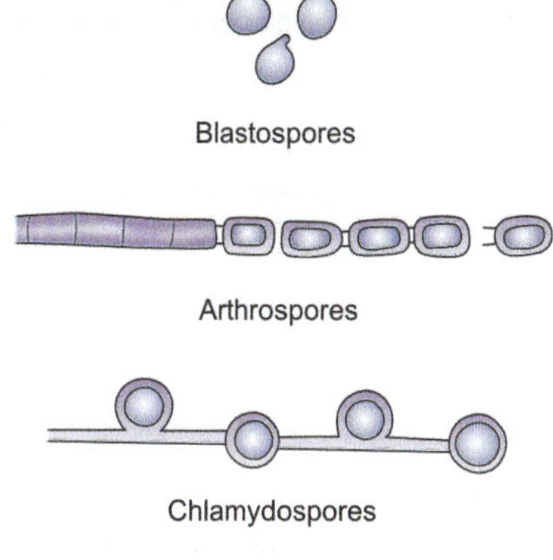

Fig. 72.3: Vegetative spores

a. Vegetative spores (Fig. 72.3).
   i. Blastospores: These are formed by budding from parent cell, as in yeasts.
   ii. Arthrospores: These are formed by the production of cross-septa into hyphae resulting in rectangular thick-walled spores.
   iii. Chlamydospores: These are thick-walled resting spores developed by rounding up and thickening of hyphal segments.
b. Aerial spores (Fig. 72.4)
   i. Conidiospores: Spores borne externally on sides or tips of hyphae are called conidiospores or simply conidia.
   ii. Microconidia: When conidia are small and single, these are called microconidia.
   iii. Macroconidia: These are large and septate conidia and are often multicellular.
   iv. Sporangiospores: These are spores formed within the sporangium. They develop on the ends of hyphae called sporangiophores.
   Examples: Mucor and **Rhizopus**.

LABORATORY DIAGNOSIS

A. Collection and Processing of Specimens

The sampling procedures vary according to the area and type of tissue involved. Specimens for the diagnosis of mycoses include: skin scrapings, oral scrapings, vaginal scrapings, corneal scrapings, hair and nails, blood, cerebrospinal fluid, urine, sputum, bronchial lavage specimens, pus and tissues. Causative agents of mycoses can be identified by following methods:

B. Direct Microscopy

1. **Potassium hydroxide (KOH) preparation:** A portion of the treated specimen should be examined microscopically to determine whether hyphal elements are present. Most specimens can be examined satisfactorily in wet mounts after partial

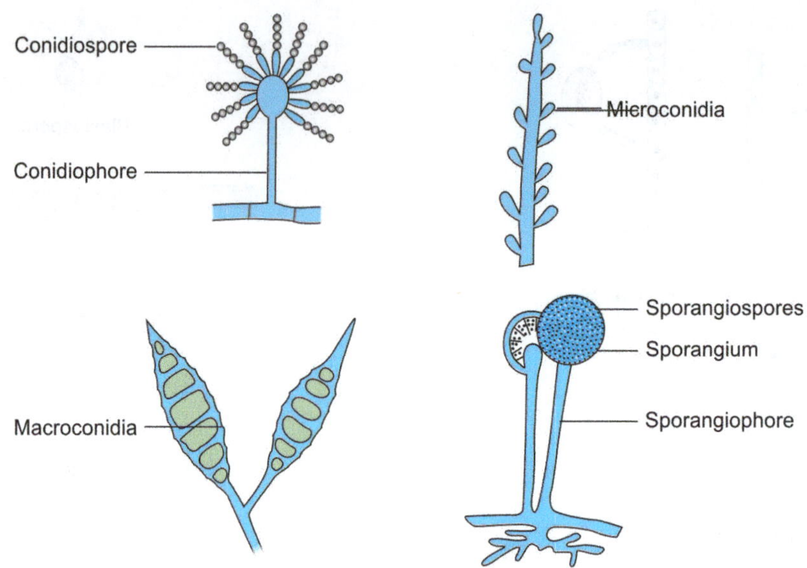

Fig. 72.4: Aerial spores

digestion of the tissue with 10-20% potassium hydroxide. The alkali digests cells and other tissue materials, enabling the fungus elements to be seen clearly.

Newer preparations for the KOH test may provide easier and more reliable results. These preparations incorporate dimethyl sulfoxide (DMSO) and a stain into the KOH solution. The DMSO facilitates more rapid breakdown of cellular debris without requiring heat while the stain is taken up by fungal elements, making them readily visible upon microscopic examination of the slide preparation.

2. **Potassium hydroxide (KOH) with Calcofluor white:** Addition of Calcofluor white and subsequent examination by fluorescence microscopy enhances the detection of most fungi since the fluorescent hydroxide-calcofluor binds to the fungal cell walls.
3. **Gram staining:** Gram films may also be used for the diagnosis of yeast infections of mucous membranes.
4. **India ink preparations:** India ink preparations may be used for detecting encapsulated yeast *Cryptococcus neoformans* in cerebrospinal fluid (CSF).
5. **Histology:** Common tissue stains used for detection of fungal elements are the periodic acid-Schiff (PAS), Grocott-Gomori methenamine-silver (GMS), hematoxylin and eosin (H and E), Giemsa, and the Fontana-Masson stains, are based on the presence of chitin and polysaccharides in their cell wall. The Giemsa stain is used primarily to detect *Histoplasma capsulatum* in blood or bone marrow. PAS attaches to polysaccharides in the fungal wall and stains fungi pink. The Fontana-Masson method stains melanin in the cell wall and identifies the presence of dematiaceous fungi.

C. Culture

1. **Culture media:** The commonest culture media used in mycology, Sabouraud's dextrose agar (SDA, pH 5.4), SDA with antibiotics, potato dextrose or the slightly modified potato flakes agar (PFA), and brain heart infusion (BHI) agar with blood and antibiotics. The antimicrobials usually included in SDA with antibiotics are chloramphenicol and gentamicin to inhibit bacterial growth and cycloheximide (actidione) to inhibit saprophytic fungi. The pH of Emmons' modification of SDA is close to neutral and is more efficient medium for primary isolation than the original formulation.
2. **Incubation:** Cultures are routinely incubated in parallel at room temperature 25°C (room temperature for weeks) and at 37°C for days. Many fungi develop relatively slowly and cultures should be retained for at least 2-3 weeks (in some cases up to 6 weeks) before being discarded. Yeasts usually grow within 1-5 days.

D. Identification of Fungi

Once an organism has grown, it is examined for characteristic gross and microscopic structures, so that identification can be made.

1. **Gross or macroscopic examination of cultures:** Growth characteristics useful for identification are the rapidity of growth, color and morphology of the colony on the obverse and pigmentation on the reverse. Molds are identified by their macroscopic and microscopic morphology. Yeasts are identified by sugar fermentation and their ability to assimilate carbon and nitrogen sources. Commercial kits are

available for the identification of medically important yeasts.

2. **Microscopic examination:** Microscopic characteristics that should be observed are the following:
   i. Septate versus sparsely septate hyphae
   ii. The types, size, shape, and arrangement of conidia.
   iii. Hyaline or dematiaceous hyphae.
   iv. Fruiting structures.

i. **Tease Mount**: For microscopic examination, slide mounts should be made in lactophenol cotton blue (LCB) from fungal colony (in teased mounts or slide cultures) to study the morphology of hyphae, spores and other structures. Teased mounts are prepared in lactophenol cotton blue (LCB) which contains lactic acid, phenol and cotton blue.

   Microscopic characteristics that should be observed are the following:

   a. **Septate versus sparsely septate hyphae:** Hyphal diameter, presence or absence of septa and of special structures are of diagnostic importance. Special hyphal structures frequently found are spring-like helical coils (spiral hyphae), localized swellings formed by tightly twisted hyphae resembling tennis racquets (racquet hyphae) and numerous short branches appearing at the ends of hyphae (favic chandelier) (Fig. 72.5).

   b. **Spores or Conidia:** The morphology of asexual spores or conidia is of diagnostic importance. They may be small, single celled 'microconidia' or large, single or multicelled 'macroconidia'. The type of spore formation is distinctive for different fungi. Blastospores are formed by budding, as in yeasts. Arthrospores are formed along the mycelium by segmentation and condensation of hyphae. Chlamydospores are thick walled resting spores formed by rounding up and thickening of hyphal segments (Fig. 72.3).

ii. **Cellophane tape preparation:** Cellophane tape preparations involve gently touching a piece of clear tape, sticky side down, to the surface of the colony and then removing it. The tape is placed onto a drop of LPCB on a slide and examined. An advantage of this procedure is that the conidial arrangement is retained. Tape preps should be read within 30 minutes and then discarded.

iii. **Slide culture:** Slide culture provides a useful technique for demonstrating the natural morphology of fungal structures and for encouraging conidiation in some poorly fruiting fungi. Another advantage to this type of slide is that it can be preserved indefinitely.

3. Miscellaneous Tests for the Identification of Molds:

i. Hair perforation test: It is done to differentiate *T. rubrum* from *T. mentagrophytes*.

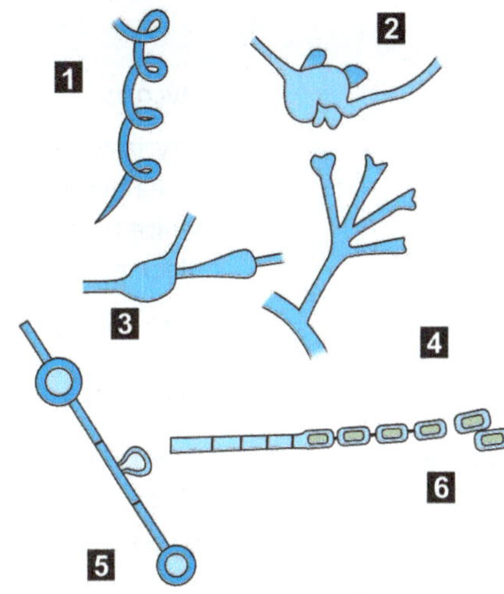

Fig. 72.5: Mycelial forms and asexual spores of fungi. 1. Spiral hypha, 2. Nodular organ, 3. Racquet mycelium, 4. Favi chandelier, 5. Chlamydo-spores along hypha, 6. Arthrospores

ii. Urease test: This test is done to differentiate *T. mentagrophytes* from *T. rubrum*.

iii. Thiamine requirement.

iv. Trichophyton agars.

v. Growth on rice grains.

4. Miscellaneous Tests for the Identification of Yeasts: Germ tube production, carbohydrate assimilation, chromogenic *substrates*, corneal agar, potassium nitrate assimilation, temperature studies, and urease.

Germ Tube Production: The germ tube test is probably the most important and easiest test to perform for identification of yeasts. Both *Candida albicans* and *Candida dubliniensis* are identified by germ-tube production.

E. Serologic Tests

The most common tests for:

a. **Fungal antibodies**:
   1. Immunodiffusion;
   2. Countercurrent immuno-electrophoresis (CIE);
   3. Whole cell agglutination;
   4. Complement fixation;
   5. Enzyme-linked immunosorbent assay (ELISA).

b. **Antigen detection:**
   1. Latex particle agglutination;
   2. ELISA.

F. Polymerase Chain Reaction (PCR)

Detection of fungal DNA in clinical material, principally blood, serum, broncho-alveolar lavage fluid and sputum, is increasingly used for diagnosis.

Mycoses (Fungus Infections)

Infection caused by fungus is known as mycosis (Plural mycoses).

Table 72.1: The major mycoses and causative fungi

Type of mycosis	*Mycosis*	*Causative Fungal Agents*
A. Superficial	Pityriasis versicolor	*Malassezia* species
	Tinea nigra	*Hortaea werneckii*
	White piedra	*Trichosporon* species
	Black piedra	*Piedraia hortae*
B. Cutaneous	Dermatophytosis	*Microsporum* species, *Trichophyton* species, and *Epidermophyton floccosum*
	Candidiasis of skin, mucosa, or nails	*Candida albicans* and other candida species
C. Subcutaneous	Sporotrichosis	*Sporothrix schencki*
	Chromoblastomycosis	*Phialophora verrucosa*, *Fonsecaea pedrosoi*, others
	Mycetoma	*Pseudallescheria boydii*, *Madurella mycetomatis*, others
	Phaeohyphomycosis	*Exophiala*, bipolaris, *exserohilum*, and others
D. Systemic (prlmary, endemic)	Coccidioidomycosis	*Coccidioides immitis*, *C. posadasii*
	Histoplasmosis	*Histoplasma capsulatum*
	Blastomycosis	*Blastomyces dermatitidis*
	Paracoccidioidomycosis	*Paracoccidioides brasiliensis*
E. Opportunistic	Systemic candidiasis	*Candida albicans* and other candida species
	Cryptococcosis	Cryptococcus neoformans
	Aspergillosis	*Aspergillus fumigatus* and other aspergillus species
	Mucormycosis (zygomycosis)	Species of *Rhizopus*, *Absidia*, *Mucor*, and other zygomycetes
	Penicilliosis	*Penicillium marneffei*

CLASSIFICATION OF MYCOSES (TABLE 72.1)

Classification of fungal disease according to primary sites of infections:

A. Superficial mycoses
B. Cutaneous mycoses
C. Subcutaneous mycoses
D. Systemic mycoses
E. Opportunistic mycoses.

A. Superficial Mycoses

These infections are limited to the outermost layers of the skin and hair. These include:

a. Infection of skin caused by *Malassezia furfur* (pityriasis versicolor) and *Exophiala werneckii* (tinea nigra), and
b. Infection of hair caused by *Piedraia hortae* (black piedra) and *Trichosporon beigelii* (white piedra).

B. Cutaneous Mycoses

Infections that extend deeper into the epidermis as well as invasive hair and nail diseases.

Examples:

a. Infection of skin, hair and nail caused by dermat phytes.
b. Infection of skin, nail and mucous membrane caused by *C. albicans* and other candida specicies.

C. Subcutaneous Mycoses

These infections involve the dermis, subcutaneous tissues, muscle, and fascia.

Examples: The principal subcutaneous mycoses are mycetoma, chromomycosis, sporotrichosis and rhinosporidiosis.

D. Systemic Mycoses

Infections that originate primarily in the lung but that may spread to many organ systems.

Examples: Systemic mycoses include blastomycosis, histoplasmosis, coccidioidomycosis and paracoccidioidomycosis.

E. Opportunistic Mycoses

Opportunistic infection occurrs in patients with debilitating diseases such as cancer or diabetes, or in whom the physiological state has been upset by immunosuppressive drugs, steroids, X-rays or broad spectrum antibiotics.

Opportunistic infections are caused mainly by fungi that are normally avirulent.

Examples: These include aspergillosis, penicilliosis, zygomycosis or mucormycosis, candidiasis and cryptococcosis.

F. Miscellaneous Mycoses

These include penicilliosis, otomycosis and keratomycosis.

KEY POINTS

- Mycology is the study of fungi.
- The cell wall of fungi possesses two characteristic cell structures: **chitin** and **ergosterol**.
- Fungi grow in two basic forms, as **yeasts** and **molds**. Elongation of the cell produces a tubular, thread like structure called **hypha** (septate or nonseptate). A tangled mass of hyphae constitutes the **mycelium**. Fungi which form mycelia are called molds or filamentous fungi. In a growing colony of filamentous fungus, the mycelium can be divided into the *vegetative mycelium* and the *aerial mycelium*.
- The fungi are classified in the phylum Thallophyta. The phylum consists of four classes of fungi—Zygomycetes, Ascomycetes, Basidiomycetes, and Deuteromycetes or Fungi Imperfecti.
- The fungi can also be classified as yeast, yeast-like fungi, molds and dimorphic fungi depending on their morphology.
- Laboratory diagnosis of fungal infections depends on: Direct microscopy, culture, serological tests, nonculture methods, and molecular methods.
- Mycoses (fungus infections): Infection caused by fungus is known as mycosis (Pl. mycoses).

IMPORTANT QUESTIONS

1. Describe the laboratory diagnosis of fungal diseases.
2. Write short notes on:
   Classification of fungi

FURTHER READING

Ajello L, Hay RJ, eds. Topley and Wilson's Microbiology and Microbial injections, 9th ed, vol 4, Arnold, London, Sydney, Auckland.

Casadevall A, Perfect JR: Cohen J, Powderly WG (editors), Infectious Diseases, 2nd ed. Volume 2, Chapter 237-241. Mosby, 2004.

Fothergill AW. Identification of dematiaceous fungi and their role in human disease. Clin Infect Dis 1996;22:S179.

CHAPTER

73 Superficial, Cutaneous and Subcutaneous Mycoses

Learning Objectives

After reading and studying this chapter, you should be able to:

- List superficial, cutaneous and subcutaneous mycoses.
- Describe causative agents of ectothrix and endothrix.
- Describe the following: Mycetoma; chromoblastomycosis; sporotrichosis; rhinosporidiosis.

A. SUPERFICIAL MYCOSES

These infections are limited to the outermost layers of the skin and hair. These include:

a. **Infection of skin**-Caused by *Malassezia furfur* (pityriasis versicolor) and *Exophiala werne ckii* (tinea nigra), and
b. **Infection of hair**-Caused by *Piedraia hortae* (black piedra) and *Trichosporon beigelii* (white piedra).

a. Infection of Skin

*Pityriasis Versicolor (**Tinea versicolor**)*

Pityriasis versicolor (Tinea versicolor) is a chronic, usually asymptomatic, involvement of the stratum corneum, characterized by discrete or confluent macular areas of discolozration or depigmentation of the skin. The areas involved are mainly the chest, abdomen, upper limbs and back. Disease is probably related to host or environmental factors. The disease is worldwide in distribution but is particularly prevalent in the tropics. It occurs mainly in young adults.

Causative Agent: Causative agent is a lipophilic, yeast-like fungus *Pityrosporum orbiculare (Malassezia furfur).* Malassezia species are lipophilic yeasts, and most required lipid in the medium for growth.The organism is found in areas of the body that are rich in sebaceous glands and it is a part of the normal flora of the skin.

Laboratory Diagnosis

1. Direct Microscopy

The diagnosis is confirmed by direct microscopic examination of scrapings of infected skin, treated with 10 to 20 percent KOH or stained with calcofluor white. Demonstration of clusters of the characteristic round yeast cells (5 to 8 µm in diameter) with short, stout hyphae, which may be curved and occasionally branched, is diagnostic (Fig. 73.1).

2. Culture

The fungus can be grown on Sabouraud's agar, covered with a layer of olive oil. Cultures are not routinely performed to confirm the diagnosis, because M. *furfur* requires a special medium containing fatty acids.Creamy colony develop within 5 to 7 days at 37°C. Lactophenol cotton blue mount of the colonies show budding yeast cells along with a number of bottle shaped cells. Hyphae are occasionally seen in culture.

Tinea Nigra

Tinea nigra (or tinea nigra palmaris) is a localized infection of the stratum corneum, particularly of the palms, producing black or brownish macular lesions. This condition is more prevalent in warm coastal regions and among young women. It is caused by the dematiaceous fungus ***Hortaea (Exophiala) werneckii****.*

Laboratory Diagnosis

1. Direct Microscopy

Microscopic examination of skin scrapings from the periphery of the lesion will reveal brownish, branched, septate hyphae and budding cells (Fig. 73.2).

2. Culture

On Sabouraud agar the fungus develops as grey, yeast-like colonies, which gradually become more mycelial and darker colored with age.

b. Infection of Hair

Piedra

Piedra is a fungal infection of the hair, characterized by the appearance of firm, irregular nodules along the hair

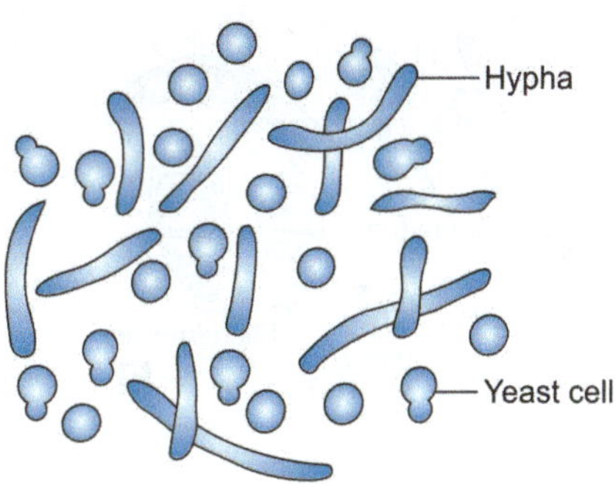

Fig. 73.1: Morphology of Pityriasis versicolor

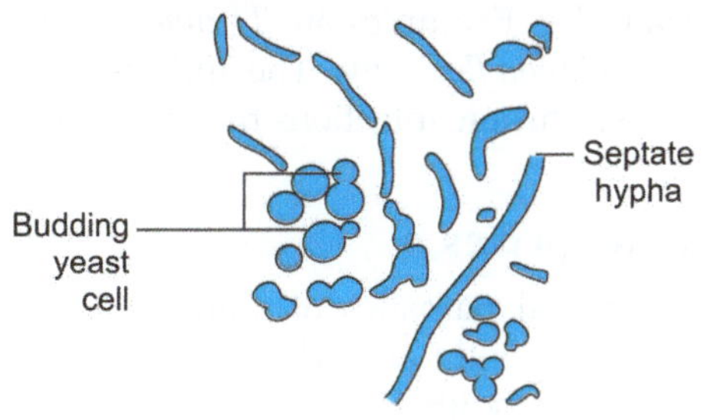

Fig. 73.2: *Hortaea wernickii*

shaft. The nodules are composed of fungus elements cemented together on the hair. Two varieties of piedra are recognized-black piedra caused by *Piedraia hortae* and white piedra caused by *Trichosporon beigelii.*

Black Piedra

This condition caused by *Piedraia hortae,* an organism that exists in the perfect (teleomorphic) state when it colonizes the hair shaft. It is characterized by the presence of black, hard nodules up to 1 mm in diameter, mainly on the hairs of the scalp. It occurs in humid, tropical climates.

Laboratory Diagnosis

1. Direct Microscopy

Crushing the nodules reveals the sexual reproductive phase, club-shaped asci each with eight ascospores. The differential diagnosis includes the nits of pediculosis and abnormal hair growth.

2. Culture

Culture is not necessary: Shaving to remove infected hairs is a satisfactory treatment.

White Piedra

White piedra, an infection of the hair, is caused by the yeastlike organism *Trichosporon beigelii*. Axillary, pubic, beard and scalp hair may be infected. This disease results in soft, white, greyish or light-brown nodules on the hair shafts, mainly in the axillae. The hair often breaks at the point of infection, leaving hairs with a clubbed or swollen end. Shaving of the affected area is usually sufficient to affect a cure.

B. CUTANEOUS MYCOSES

Infections that extend deeper into the epidermis as well as invasive hair and nail diseases.

Examples:

a. Infection of skin, hair and nail caused by dermatophytes.
b. Infection of skin, nail and mucous membrane caused by C. *albicans* and other candida species.

a. Dermatophytes

The dermatophytes are a group of closely related filamentous fungi that infect only superficial keratinised tissues-the skin, hair and nails. The term dermatomycosis, sometimes used as a synonym, would also include skin lesions produced by other fungi such as *Candida albicans* and also the cutaneous manifestations of systemic mycoses.

Dermatophytoses are among the most prevalent infections in the world. Although they can be persistent as a troublesome, they are not debilitating or life-threatening-yet millions of dollars are expended annually in their treatment. Being superficial, dermatophyte (ringworm), infections have been recognized since antiquity.

Classification

Dermatophytes have been classified into three genera–

1. *Trichophyton*: Trichophyton species infect hair, skin, or nails.
2. *Microsporum*: Microsporum species infect only hair and skin.
3. *Epidermophyton*: Epidermophyton attacks the skin and nails but not the hair.

About 40 species of dermatophytes are known to cause infection in humans and animals. Dermatophytes are probably restricted to the nonviable skin because most are unable to grow at 37°C or in the presence of serum.

Classification Depeding on Habitat

Dermatophytes are classified as anthropophilic, geophilic, zoophilic, or depending on whether their usual habitat is soil, animals, or humans.

1. Anthropophilic Species

Human beings are the main or only hosts for anthropophilic dermatophytes. Anthropophilic species may be transmitted by direct contact or through fomites, such as contaminated towels, clothing, shared shower stalls and similar examples. Examples are *T. rubrum, M. audouinii* and *Epidermophyton floccosum.* The anthropophilic group tends to cause chronic infections that may be difficult to cure.

2. Zoophilic Species

These are natural parasites of animals. Examples are *T. verrucosum* in cattle and *M. canis* in dogs and cats. Human infections with zoophilic dermatophytes cause severe inflammation but are more readily curable.

3. Geophilic Species

They occur naturally in soil, are relatively less pathogenic for human beings. Examples are *M. gypseum* and *T. ajelloi.*

The anthropophilic group tends to cause chronic infections that may be difficult to cure. The zoophilic and geophilic dermatophytes tend to cause inflammatory lesions that respond well to therapy and may occasionally heal spontaneously.

Identification

In skin, they are diagnosed by the presence of hyaline, septate, branching hyphae or chains of arthroconidia. In cultures on Sabouraud's agar, they form characteristic colonies consisting of septate hyphae and two types of asexual spores, microconidia and macroconidia. Differentiation into the three genera is based mainly on the nature of macroconidia (Fig 73.3). In culture, the many species are closely related and often difficult to identify. For some species of dermatophytes, a sexual reproductive state has been discovered and all dermatophytes with a sexual form produce ascospores and belong to the teleomorphic genus arrhroderma.

Morphology and Identification

Dermatophytes are identified by their colonial appearance and microscopic morphology after growth for 2 weeks at 25°C on Sabouraud's dextrose agar.

Trichophyton

Colonies may be powdery, velvety or waxy, with pigmentation characteristic of different species (Table 73.2). Microconidia are abundant and are arranged in clusters along the hyphae or borne on conidiophores. Macroconidia are relatively scanty. They are generally elongated, with blunt ends (Fig, 73.3, 73.4). Macroconidia have distinctive shapes in different species and are of importance in species identification. Some species possess special hyphal characters, such as spiral hyphae, racquet mycelium and favic chandeliers (Fig 73.4).

Important Species

T. rubrum, T. mentagrophyte, T tonsurans, T. schoenleinii, T. violaceum (Table 73.1).

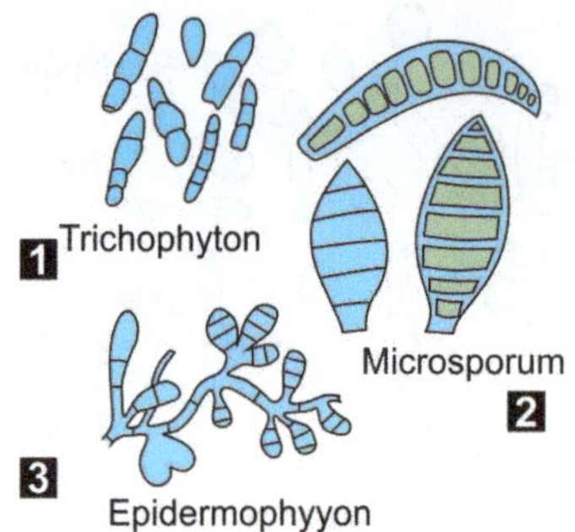

Fig. 73.3: Characterstic macroconidia of dermatophytes. (1) Cylindrical in Trichopyton; (2) Fusiform in Microsporum and; (3) Club shaped in Epidermophyton

T. rubrum

The typical colony of *T. rubrum* has a white, cottony surface and a deep red, nondiffusible pigment when viewed from the reverse side of the colony.The microconidia are small and piriform (pear-shaped) and club shaped thin walled macroconidia (scanty). *T. rubrum* is the most common species infecting human beings. It often causes chronic, treatment resistant lesions.

T. mentagrophyte

T. mentagrophyte has grape like clusters of microconidia (Fig. 73.4) and Cigar shaped macroconidia. It has no red pigment and urease positive. Hair perforation test is positive.

T. tonsurans

T. tonsurans produces a flat, powdery to velvety colony on the obverse surface that becomes reddish-brown on reverse. The microconidia are mostly elongate (Fig. 73.4).

Microsporum

Colonies are cotton-like, velvety or powdery, with white to brown pigmentation. Microconidia are relatively scanty and are not distinctive. Macroconidia are the predominant spore form. They are large, multicellular, spindle shaped structures, borne singly on the ends of hyphae. Both types of conidia are borne singly in these genera.Microsporum species infect only hair and skin but usually not the nail.

Important species are *M audouinii, M canis, M gypseum.*

Examples of pathogenic Microsporum species are: *M. audouinii, M. gypseum, M. canis, M. nanum, M. ferrugineum, M. equinum, M. fulvum, M. persic lor, M. gallinae, M. vanbreuseghemii, M. racemosum, M. praecox* etc.

M. audouinii

M audouinii rarely forms conidia in the culture but many thick walled chlamydospores (chlamydoconidia) may

Table 73.1: Salient characteristics of common dermatophytes		
Species	*Colony character*	*Morphology*
T. rubrum	Cottony which later becomes velvety with a blood-red pigment on the reverse side.	Few, long, narrow, pencil-shaped macroconidia.
T. mentagrophytes	Flat with a cream to buff or tan powdery surface. The reverse side of colonies on SDA may be brown, yellow or dark-red. However, dark-red colour is not produced on potato dextrose agar with 1 percent glucose.	Numerous microconidia, Thin-walled, club-shaped, spindle-shaped or long pencil-shaped macroconidia.
T. tonsurans	White, tan, yellow or reddish brown with central furrows.	Abundant clavate to tear-shaped microconidia and few irregular thick-walled macroconidia.
T schoenleinii	White to tan, glabrous, waxy, heaped and folded.	Microconidia very rare, macroconidia absent. Hyphae tend to become knobby and clubbed at ends (favic chandeliers).
T. vioaceum	Heaped and verrucose, glabrous with a deep violet color.	Irregularly branched hyphae. Microconidia usually not found. Macroconidia resembling those of *T. rubrum* may be formed.
M. audouinii	Slow growing, velvety, brownish.	Terminal chlamydospores, conidia rarely produced. Some isolates show large, irregular, spindle-shaped, thick-walled with smooth or echinulate surface macroconidia. Some Isolates produce pectinate hyphae.
M. canis	White to bright yellowish colony with bright yellow to orange-brown on reverse.	Abundant, thick-walled, spindle-shaped macroconidia and a few clavate, sessile microconidia.
M. gypseum	Powdery, buff colored.	Abundant, spindle-shaped macroconidia with thick and finely rough wall. Microconidia are clavate, sessue and rare.
E. floccosum	Powdery, greenish-yellow, radially folded. Reverse of the colony is colorless to yellowish-brown.	Macroconidia are numerous, club-shaped and are arranged in groups of two or three. Microconidia absent.

be present. On rice grain medium *M. audouinii* does not grow, while other *Microsporum* species grow rapidly. It is anthropophilic.

M. canis

M. canis forms numerous thick-walled, 8 to 15 celled macroconidia with curved or hooked spiny tips. A lemon-yellow or yellow-orange pigment develops on the reverse of the colony. It is zoophilic.

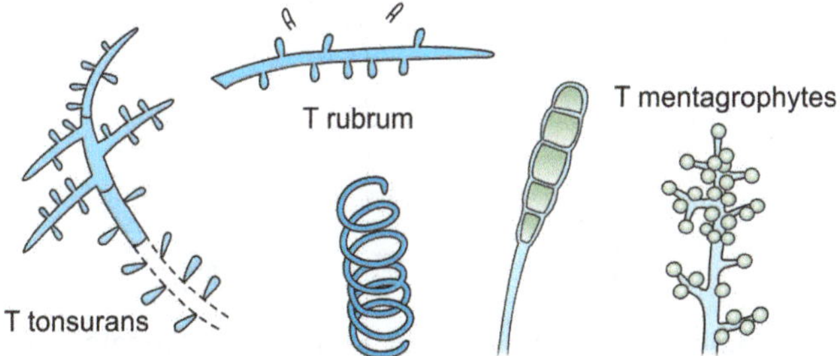

Fig. 73.4: Trichophyton species. Macroconidium, spiral hypha and typical microconidia

Table 73.2: Clinical types of dermatophytoses and their common causative agents	
Disease	*Common causative agents*
Tinea capitis	*Microsporum any species, Trichophyton most species*
Favus	*T. schoenleinii, T. violaceum, M. gypseum*
Tinea barbae	*T. rubnun, T.mentagrophytes, T. verrucosum*
Tinea imbricata	*T. concentricum*
Tinea corporis	*T. rubrum and any other dermatophyte*
T.cruris	*E.floccosum, T. rubrum*
T. pedis	*T. rubrum, E. Boccosum*
Ectothrix hair infection	*Microsporum species, T. rubrum, T.menrngrophytes*
Endothrix hair infection	*T. schoenieinii, T. tonsurans, T. violaceum*

M. gypseum

M gypseum colonies grow rapidly producing a flat, spreading, powdery surface that is cinnamonbuff to brown which consists of abundant macroconidia. They are boat-shaped, rough walled with 4 to 6 septa. It is geophilic.

Epidermophyton

Colonies are powdery and greenish-brown (khaki-colored). Microconidia are not produced. Macroconidia are abundant and multicellular (one to nine celled) smooth, thin walled, pear or c1ubshaped arranged in clusters, e.g. *E. floccosum, E. stockdaleae.*

Pathogenicity

Dermatophytes grow only on the keratinized layers of the skin and its appendages and do not ordinarily penetrate the living tissues. Lesions vary considerably according to the site of the infection and the species of fungus involved.

The mechanisms of pathogenesis in dermatophytosis are not clear. Sometimes, there is only dry scaling or hyperkeratosis, but more commonly there is irritation, erythema, edema and some vesiculation.

Dermatophytids (or 'id' reaction-Hypersensitivity to fungus antigens may play a role in pathogenesis and is probably responsible for the sterile vesicular lesions sometimes seen in sites distant from the ringworm. These lesions are called **dermatophytids** (or **'id'** reaction). In patients with foot ringworm this takes the form of a vesicular eczema of the hands, whereas patients with scalp ringworm (especially kerion) develop a follicular rash, usually on the trunk or limbs. These secondary lesions do not contain viable fungus and they disappear spontaneously when the infection subsides. Hypersensitivity can be demonstrated by skin testing with the fungus antigen, trichophytin.

Types of Hair Infection

In lesions, dermatophtes appear as hyphae and arthrospores. In scalp infection, the fungus invades the hair shaft and then the hyphae break up into chains of arthroconidia.Three types of hair infection can be seen in 10 percent KOH wet mounts:

1. **Ectothrix-**In this, the hyphae are sparsely distributed within hair shaft with a sheath of arthrospores on the surface of hair shaft. It is caused by *M audouinii, M. canis* and *T. mentagrophytes*-(Fig. 73.5A). In enctothrix infection the hair breaks off at, or just below, the mouth of the follicle to give what is described as *black dot* ringworm.
2. **Endothrix-**In this, the arthrospore formation occurs entirely within the hair completely filling the hair shaft. This is caused by *T. tonsurans* and *T. violaceum* (Fig. 73.5B). In ectothrix infection, the hair usually breaks off 2 to 3 mm above the mouth of the follicle, leaving short stumps of hair.
3. **Favus-**In this, there is sparse hyphal growth and formation of air spaces within the hair shaft. This is caused by *T. schoenleinii* (Fig. 73.5C). The hair remains intact but intense fungal growth within and around the hair follicle produces a waxy, honeycomb-like crust on the scalp.

Clinical Findings

Dermatophyte infections were mistakenly termed **ringworm or tinea** because of the raised circular lesions. *Tinea* comes from Latin and means "worm" or "moth. Table 73.3 lists the clinical types of dermatophytoses and their common causative agents.

The clinical forms are based on the site of involvement.

1. **Tinea corporis (Tinea glabrosa)-** Tinea corporis is ringworm of the smooth or nonhairy skin of the body. A special type is **Tinea imbricata** which is found in the tropics and is characterized by extensive concentric rings of papulosquamous scaly patches.
2. **Tinea cruris-**When the infection occurs in the groin and the perineum, it is called **tinea cruris** or **jock itch.**
3. **Tinea manus-**Tinea manus refers to ringworm of the hands or fingers. Dry scaly lesions may involve one or both hands, single fingers or two or more fingers.
4. **Tinea barbae or barber's itch-**It is involvement of the bearded areas of the face and neck.
5. **Tinea pedis or athletes' foot-**
   It is ringworm of the foot.

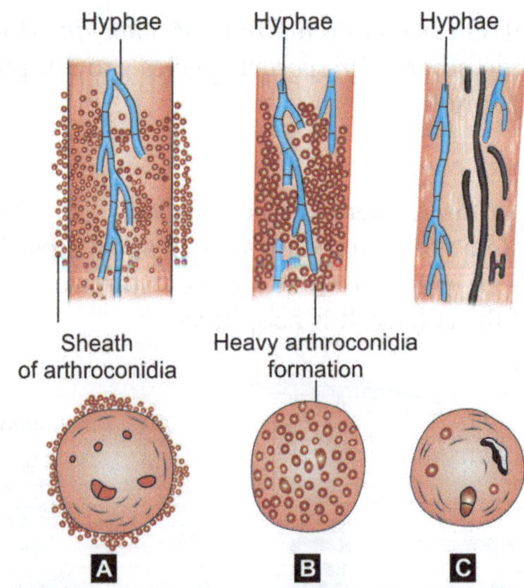

Fig. 73.5: Various forms of hair invasion by dermatophytes as seen in longitudinal and transverse sections of hair shaft; (A) ectothrix with hyphae sparsely distributed within hair shaft and a sheath of arthroconidia on the outside; (B) endothrix with heavy arthroconidia formation completely filling hair shaft; (C) favus showing sparse hyphal growth and formation of air spaces

6. **Tinea unguium (onychomycosis)**-Nail infection may follow prolonged tinea pedis.
7. **Tinea capitis**-Tinea capitis is dermatophytosis or ringworm of the scalp and hair.
8. **Favus-Favus,** is a chronic type of ringworm in which dense crusts (scutula) develop in the hair follicles, which lead to alopecia and scarring.
9. **Kerion**-Scalp infection sometimes produces severe boggy lesions with marked inflammatory reaction called **kerion**.

Laboratory Diagnosis

A. Specimens

Specimens consist of scrapings from borh rhe skin and rhe nails plus hairs plucked from involved areas. Microsporum-infecred hairs fluoresce under Wood's lighr in a darkened room.

B. Microscopic Examination

The routine method of diagnosis is by the examination of KOH mounts. Specimens are placed on a slide in a drop of 10 to 20 percent potassium hydroxide, with or without calcofluor white, which is a nonspecilic fungal cell wall stain viewed with a fluorescent microscope. A coverslip is added, and the specimen is examined immediately and again afrer 20 minutes. In skin or nails, regardless of the infecting species, branched septate hyphae or chains of arthroconidia (arthrospores) are seen (Fig. 73.6).

Selection of infected hair for examination is facilitated by these exposure to UV light (Wood's lamp). Infected hair will be fluorescent. Two types of hair infection may be distinguished in wet mounts, **'ectothrix'** in which arthrospores are seen as a sheath surrounding the hair and **'endothrix'** in which the spores are inside the hair shaft (Fig 73.5). Demonstration of the fungus in nails may be difficult and may be possible only after clearing with KOH for a day or two.

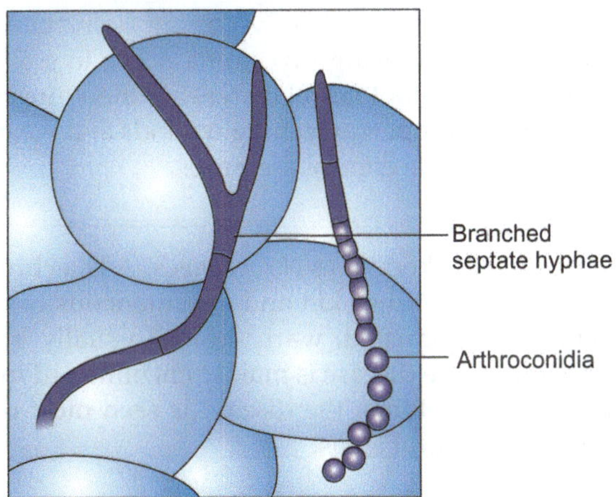

Fig. 73.6: Dermatophyte in potassium hydroxide mount of skin or nail scraping with branched septate hyphae and arthroconidia

C. Culture

Species identification is possible only by culture examination. Specimens are inoculated onto inhibitory mold agar or Sabouraud's agar slants conraining cycloheximide and chloramphenicol to suppress mold and bacterial growth, incubated for 1 to 3 weeks at room temperature, and further examined in slide cultures if necessary. Species are identified on the basis of colonial morphology (growth rate, surface texture and any pigmentation), microscopic morphology (macroconidia, microconidia), and, in some cases, nutritional requirements.

Epidemiology

Dermatophytosis occurs throughout the world but certain types of disease and some species of fungi show geographically restricted distribution. Reports estimate that they are the third most common skin disorder in children younger than 12 years and the second most common in older populations. The occurrence of these diseases varies with age, gender, ethnic group, and cultural and social habits of the population. Wearing shoes provides warmth and moisture, a setting for infections of the feet. *Tinea pedis,* so common in the temperate climates where all wear shoes is rare in the tropics where most walk barefoot. In developing countries, particularly in warm climates, scalp, body and groin infections predominate, with *T. rubrum* and *T. violaceum* among the most common causes. In India, tinea capitis occurs more often in the native children than in Europeans, whereas the Europeans have a higher incidence of tinea pedis. Many factors, such as age, hormones and intercurrent diseases, affect the susceptibility to dermatophytosis.

Treatment and Prevention

Topical therapy is satisfactory for most skin infections, but oral antifungals are required to treat infections of the nail and scalp, and severe or extensive skin infections.Topical agents include azole compounds, terbinafine, amorol fine and ciclopirox olamine. Oral griseofulvin is useful for scalp, skin and fingernail infections, but gives poor results in toenail infections, even after 18 months' therapy. Terbinafine and itraconazole have largely replaced griseofulvin for the treatment of nail infections.

Relatively little has been done to control the spread of ringworm. The prophylactic use of antifungal footpowder after bathing helps to reduce the spread of infection among swimmers. Foot-bath containing antiseptic solutions, which are commonplace in swimming pools, are of no value.

C. SUBCUTANEOUS MYCOSES

These infections involve the dermis, subcutaneous tissues, muscle, and fascia. They result from the traumatic inoculation of saprophytic fungi from soil or decaying vegetation into the subcutaneous tissue.

Examples

The principal subcutaneous mycoses are **mycetoma, chromomycosis, sporotrichosis and rhinosporidiosis.**

1. Mycetoma

Mycetoma is a chronic, granulomatous infection of the skin, subcutaneous tissues, fascia and bone, which most often affects the foot or the hand. The disease was originally reported by Gill (1842) from Madurai, south India, and Carter (1860) established its fungal etiology. It is therefore commonly known as Maduramycosis or Madura foot. However, this condition had been referred to in the Atharva Veda as *Padavalmika* (foot anthill).

Mycetoma occurs worldwide but more often among impoverished people who do not wear shoes. The disease is most prevalent in tropical and subtropical regions of Africa, Asia and Central America. Its incidence varies markedly from one place to another; for instance, in India, it is quite common in Tamil Nadu but rare in Kerala.

Etiology

It can be divided into three types, eumycetomas, actinomycetomas and botryomycosis. It may be caused by one of a number of different actinomycetes *(actinomycetoma)* or moulds *(eumycetoma)*. A similar condition called 'botryomycosis' is caused by *Staphylococcus aureus* and some other bacteria. Etiological diagnosis, therefore, is of importance in treatment. Important causative agents of these are given in Table 73.4. Eumycetoma is more prevalent in North India whereas actinomycetoma is more common in South India.

Pathogenesis

Triad of Symptoms

Infection follows traumatic inoculation of the organism into the subcutaneous tissue from soil or vegetable sources, usually on thorns or splinters and results in tumefactions, deformities and draining sinuses discharging fungal colonies called grains or granules (triad of symptoms). Subcutaneous tissues of the feet, lower extremities, hands, and exposed areas are most often involved. This process may spread to contiguous muscle and bone. Untreated lesions persist for years and extend deeper and peripherally causing deformation and loss of function.

Grains

Within host tissues, the organisms develop to form compacted colonies (grains) 0.5 to 2 mm in diameter, the color of which depends on the organism responsible; for example, *Madurella* grains are black and *Actinomadura pelletieri* grains are red. These 'granules' or 'grains' are microcolonies of the etiological agents and their demonstration is of diagnostic value. Observation of the color of the grains and the size and septations of the hyphae allow differentiation between actinomycetoma and eumycetoma. The color and consistency of the grains vary with the different agents causing the disease (Table 73.4). In actinomycotic mycetoma, the grains will be composed of very thin (less than 1 *μm* in diameter) filaments, while in mycotic lesions, they will be broader and often show septae and chlamydospores.

Laboratory Diagnosis

A. Direct Examination

The presence of grains in pus collected from draining sinuses or in biopsy material is diagnostic. The grains are visible to the naked eye and their color may help to identify the causal agent. Grains should be crushed in KOH (potassium hydroxide) and examined microscopically to differentiate between actinomycetoma and eumycetoma. Material from actinomycetoma grains may be gram-stained to demonstrate the gram-positive filaments.

B. Culture

Samples should also be cultured, at both 25 to 30°C and 37°C, on brain-heart infusion agar or blood agar for actinomycetes and on Sabouraud agar (without cycloheximide) for fungi. The fungi that cause eumycetoma are all septate moulds that appear in culture within 1 to 4 weeks.

Serological Tests

Serological precipitin tests are of little value for diagnosis and are not in routine use.

Treatment

The prognosis varies according to the causal agent, so it is important that the identity is established.The management of eumycetoma is difficult involving surgical debridement or excision and chemotherapy. Actinomycetoma responds well to rifampicin in combination with sulonamides or co-trimoxazole, but an average of 9 months therapy is required.

Epidemiology and Control of Mycetoma

The organisms producing mycetoma occur in soil and on vegetation. Barefoot farm laborers are therefore commonly exposed. Properly cleaning wounds and wearing shoes are reasonable control measures.

2. Chromoblastomycosis

This disease, also known as *chromomycosis,* is a chronic, localized disease of the skin and subcutaneous tissues, characterized by crusted, warty lesions usually involving the limbs. The disease is mainly encountered in the tropics. Like mycetoma, the disease is seen most often among males in rural areas.

Etiological Agents

The etiological agents are soil inhabiting fungi of the family Dematiaceae.

All are dematiaceous fungi, having melaninized cell walls. These include:

Table 73.4: Important causative agents and colour of the grains of various types of mycetoma	
Causative agent	*Color of the grains*
A. Eumycetoma	
• *Madurella mycetomatis*	Black
• *M. grisea*	Black
• *Exophiala jeanselmei*	Black
Curvularia geniculata	Black
Pseudallescheria boydii	White-yellow
• *Acremonium kifiense*	White-yellow
• *Aspergillus nidulans*	White
• *Fusarium species*	White
B. Actinomycetoma	
• *Actinomadura madurae*	White-yellow
• *A. pellitieri*	Red Yellow
• *Nocardia asteroids*	White-yellow
• *N. brasiliensis*	White
• *Nocardiopsis dassonvillei*	Cream
• *Streptomyces somaliensis*	Yellow
C. Botryomycosis	
• *Staphylococcus species*	White
• *Streptococcus species*	White
• *Escherichia coli*	White
• *Proteus species*	White
• *Pseudomonas aeruginosa.*	White
Actinobacillus lignieresi	Yellow

Phialophora verrucosa, Fonsecaea pedrosoi, Rhinocladiella aquaspersa, Fonsecaea compacta and *Cladophialophora carrionii.*

They enter the skin by traumatic implantation. The lesion develops slowly around the site of implantation. The infection is chronic and characterized by the slow development of progressive granulomatous lesions that in time induce hyperplasia of the epidermal tissue.

Laboratory Diagnosis

A. Microscopic Examination

The dark-colored fungal elements are relatively easy to see on microscopical examination of skin scrapings, crusts and pus.

The agents of chromoblastomycosis are identified by their modes of conidiation. Histologically, the lesions show the presence of the fungus as round or irregular, dark brown, yeast like bodies with septae, called *sclerotic cells* (Fig 73.7). Diagnosis can be established by demonstration of these sclerotic bodies in KOH mounts or in sections and by culture on Sabouraud's agar.

B. Culture

Culture on Sabouraud agar at 25 to 30°C yields slow-growing, greenish grey to black, compact, folded colonies. Cultures should be incubated for 4 to 6 weeks.

C. Serological Tests

Serological tests are not used routinely.

Phaeohyphomycosis

Phaeohyphomycosis is a term applied to infections characterized by the presence of darkly pigmented septate hyphae in tissue. Some of the more common causes of subcutaneous phaeohyphomycosis are *Exophiala jeanselmei., Phialophora richardsiae, Bipolaris spicifera,* and *Wangiella dermatitidis.* These species and others (e.g. *Exserohilum rostratum,* alternaria species, and curvularia species) may be implicated also in systemic phaeohyphomycosis

The sites of lesions may be cutaneous, subcutaneous, deeper tissues, or organs like the brain or lung. The tissue reactions and morphology of the fungus in lesions differ from those seen in chromoblastomycosis. Sclerotic cells or granules are not found.

The fungi appear in lesions as distorted hyphal strands.

Clinical Types of Phaeohyphomycosis

Phaeohyphomycosis is generally seen in debilitated or immunodeficient hosts. Some of the clinical types are:

a. **Brain abscess** caused by *Cladosporium bantianum.*
b. **Subcutaneous or intramuscular lesions with abscesses or cysts containing masses of brown hyphae** (formerly known as *phaeosporotrichose)* caused by *Phiaophora jeanselmei, P.spinifera, P. dermatitidis* or *P. richardsiae.*

3. Sporotrichosis

Sporotrichosis is a chronic, pyogenic granulomatous infection of the skin and subcutaneous tissues which may remain localized or show lymphatic spread. The disease is worldwide and occurs mainly in Central and South America, parts of the USA and Africa and Australia. It is rare in Europe.

Causative Agent

It is caused by *Sporothrix schenckii,* a saprophyte in nature.

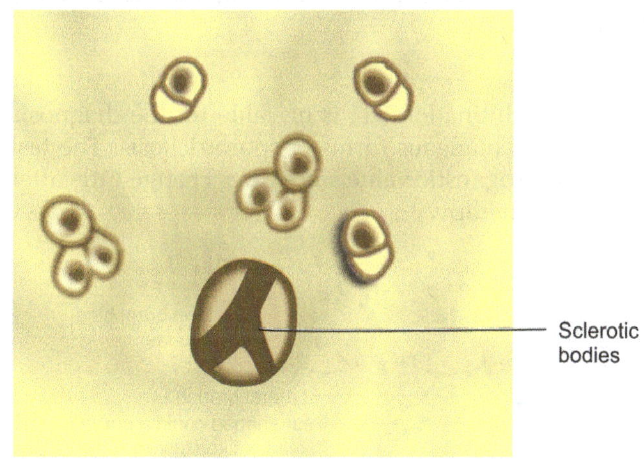

Fig. 73.7: Chromblastomycosis: KOH mount of lesion large septate 'scierotic bodies'

Morphology

S. schenckii is a dimorphic fungus. In nature and in culture at 25 to 30°C, it develops as a mould with very thin (1-2 μm) septate hyphae; spore-bearing hyphae carry clusters of oval spores. The yeast phase is formed in tissue and in culture at 37°C, and is composed of spherical or cigar-shaped cells (1-3 × 3-10 μm).

In infected tissues, the fungus is seen as cigar shaped yeast cells, without mycelia. Sometimes **'asteroid bodies'** are seen in the lesion, composed of a central fungus cell with eosinophilic material radiating from it.

Pathogenesis

The fungus is a saprophyte found widely on plants, thorns and timber. Infection is acquired through thorn pricks or other minor injuries. Rare instances of transmission from patients and infected horses and rats have been recorded.

Sporotrichosis most frequently presents as a nodular, ulcerating disease of the skin and subcutaneous tissues, with spread along local lymphatic channels but seldom extends beyond the regional lymph nodes. Most cases occur in the upper limb.

Laboratory Diagnosis

Diagnosis is made by culture as frequently the fungus may not be demonstrable in pus or tissues.

A. Micrscopic Examination

Direct microscopy is of little.

B. Culture

SDA or blood agar are the media used *S. schenckii* is a dimorphic fungus occurring in the yeast phase in tissues and in cultures at 37°C, and in the mycelial phase in nature and in cultures at room temperature. The septate hyphae are very thin (l-2 μm diameter) and carry flower-like clusters of small conidia borne on delicate sterigmata (Fig. 73.8).

Conidia are also produced along the sides of the hyphae.

C. Serology

A latex agglutination test is of value for the diagnosis of the extracutaneous forms of sporotrichosis. The test has poor prognostic value since titres change little after successful therapy.

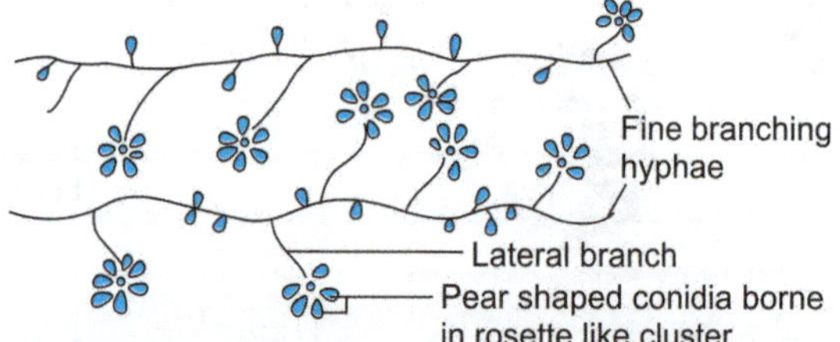

Fig. 73.8: *Sporothrix (Sporotrichum) schenkii*: culture mount showing fine branching hyphae and pear shaped conidle borne in rosette clusters at tips of lateral branches and singly along sides of hyphae

D. Skin Test

A skin test with sporotrichin antigen is 'positive in almost all patients with cutaneous sporotrichosis.

E. Animal Inoculation

Rats are highly susceptible and can be infected by intraperitoneal or intratesticular inoculation.

4. Rhinosporidiosis

Rhinosporidiosis is a chronic granulomatous disease characterized by the development of large polyps or wart-like lesions in the nose, conjunctiva and occasionally in ears, larynx, bronchus, penile urethra, vagina, rectum and skin. Though the disease was first identified in Argentina, the large majority of cases come from India and Sri Lanka. More than 90 percent of cases have been reported from India, Sri Lanka and South America. In India, sporadic cases occur all over but endemic foci exist in parts of Orissa, Andhra Pradesh, Kerala, Chennai and Raipur (Madhya Pradesh). Very rarely, there is hematogenous spread with metastatic lesions in the lungs, brain and bones.

Etiology

The causative fungus *Rhinosporidium seeberi.*

Mode of Infection

The mode of infection is not known though infection is believed to originate from stagnant water or aquatic life. It is believed that fish may be the natural hosts of *R. seeberi.*

Laboratory Diagnosis

It has not been cultured and animal inoculation is also not successful. Successful cultivation of the organism in epithelial cell culture has been reported.

Demonsration of Sporangia

Diagnosis depends on the demonstion of sporangia. Direct examination of the surface of the polypoid growth and histologic examination are the only ways to make

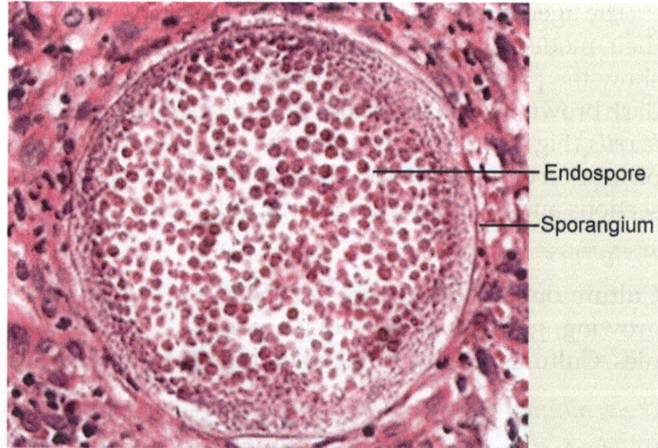

Fig. 73.9: Rhinosporidiosis: Sporangium with numerous endospores

a diagnosis. *R. seeberi* can be identified in hematoxylin and eosin stained sections, but sometimes one may need special stains. Histologically, the lesion is composed of large numbers of fungal spherules embedded in a stroma of connective tissue and capillaries. The spherules are 10 to 200 m in diameter and contain thousands of endospores (6-7 µm in diameter) (Fig 73.12). These spores develop into new sporangia when released

5. Subcutaneous Phycomycosis

In this condition, a painless subcutaneous nodule develops which enlarges to involve a whole limb or large areas of the body. It was originally reported from Indonesia and subsequently identified in many Asian and African countries.

Causative agent-is ***Basiodiobolus haptosporus***, a saprophytic phycomycete found in decaying vegetation and in the intestines of many reptiles and amphibians. It has been suggested that the infection may be acquired by insect bites.

KNOW MORE

Subcutaneous Mycoses

The fungi that cause subcutaneous mycoses normally reside in soil or on vegetation. They enter the skin or subcutaneous tissue by traumatic inoculation with contaminated material. In general, the lesions become granulomatous and expand slowly from the area of implantation. Extension via the lymphatics draining the lesion is slow except in sporotrichosis. These mycoses are usually confined to the subcutaneous tissues but in rare cases they become systemic and produce life-threatening disease.

☞ KEY POINTS

SUPERFICIAL MYCOSES

- Infection of skin is caused by *Malassezia furfur*. (pityraiasis versicolor).
- *Exophiala werneckii* (tinea nigra).
- *Piedroiohortae*, (black piedra).
- *Trichosporon beigelii* and *yeast-like organism* (white piedra).

CUTANEOUS MYCOSES

- The dermatophytes infect only superficial keratinized structure such as skin, hair, and nail but not deeper tissues.
- Dermatophytes belong to three genera: *Trichophyton, Microsporum,* and *Epidermophyton.*

SUBCUTANEUS MYCOSIS

- **Mycetoma** is a chronic, granulomatous infection of the skin, subcutaneous tissues, fascia and bone, which most often affects the foot or the hand. The disease was originally reported from Madurai, south India, and is commonly known as Maduramycosis or Madura foot.
- It can be divided into three types-eumycetomas, actinomycetomas and botryomycosis.
- **Chromoblastomycosis** is a chronic, localized disease of the skin and subcutaneous tissues, characterized by crusted, warty lesions usually involving the limbs. It is caused by *Phialophora verrucosa, Fonsecaea pedrosoi, Rhinocladiella aquaspersa, Fonsecaea compacta* and *Cladophialophora carrioni,* collectively called dematiaceous fungi.
- **Phaeohyphomycosis** is a heterogeneous group of cutaneous diseases caused by various dematiaceous fungi.
- **Sporotrichosis** is a chronic, pyogenic granulomatous infection of the skin and subcutaneous tissues which may remain localized or show lymphatic spread. It is caused by *Sporothrix schenckii,* a saprophyte in nature.
- **Rhinosporidiosis-**is a chronic granulomatous disease characterized by the development of large polyps or wart-like lesions in the nose, conjunctiva and occasionally in ears, larynx, bronchus, penile urethra, vagina, rectum and skin. The causative fungus *Rhinosporidium seeberi.*
- **Subcutaneous phycomycosis-**In this condition, a painless subcutaneous nodule develops which enlarges to involve a whole limb or large areas of the body. The causative agent is ***Basiodiobolus haptosporus***, a saprophytic phycomycete

IMPORTANT QUESTIONS

1. Discuss the morphology, cultural characters and pathogenicity of dermatophytes.
2. Describe the etiology, pathogenesis and laboratory diagnosis of mycetoma.
3. Discuss different kinds of subcutaneous mycoses.
4. Write short riotes on:
   Superficial mycoses
   Dermatophytes
   Mycetoma or Madura foot or Maduramycosis
   Chromoblastomycosis or verrucous dermatitis
   Sporotrichosis
   Rhinosporidiosis.

FURTHER READING

Fothergill AW. Identification of dematiaceous fungi and their role in human disease. Clin Infect Dis 1996;22:S179.

Kauffman CA, Hajeh RA, Chapman SW. Practice guidelines for the management of patients with sporotrichosis. Clin Infect Dis 2000;30:684.

McGinnis MR, Fader RC. Myceroma. A contemporary concept. Infect Dis Clin North Am 1988;2:939.

Revankar SG, et al. Disseminated phaeohyphomycosis: Review of an emerging mycosis. Clin Infect Dis 2002;34:467.

Weitzman I, Summerbell RC. The dermatophytes. Clin Microbiol Rev 1995;8:240.

CHAPTER 74

Systemic Mycoses

LEARNING OBJECTIVES

After reading and studying this chapter, you should be able to:

- List of fungi causing systemic infections.
- Descrbe histoplasmosis.

INTRODUCTION

Systemic mycoses-infections that originate primarily in the lung but that may spread to many organ systems. Systemic mycoses are caused by inhalation of airborne spores produced by the fungi which are present as saprophytes in soil and on plant material. From the lungs the fungus may disseminate to central nervous system (CNS), bone and other internal organs. Systemic mycoses include blastomycosis, histoplasmosis, coccidioidomycosis and paracoccidioidomycosis.

1. BLASTOMYCOSIS

Blastomycosis is a chronic infection of the lungs which may spread to other tissues, particularly skin, bone and genitourinary tract.

It is caused by *Blastomyces dermatitidis,* a dimorphic fungus. The disease has been called **North American blastomycosis** because it is endemic and in most cases occur in the United States and Canada. Despite this high prevalence in North America, blastomycosis has been documented in Africa, South America, and Asia. It is endemic for humans and dogs in the eastern United States. Khan et al (1982), from Delhi, isolated *B. dermatitidis* from a bat. Subsequently, Randhawa et at (1983), isolated the fungus from the bronchial aspirates of a patient. Jambhekhar et al (1988), reported a case of disseminated blastomycosis from Madhya Pradesh.

Pathogenesis

Soil is considered to be the source of infection, which is acquired by inhalation. Human infection is initiated in the lungs. Primary infection of the lung may resemble tuberculosis or histoplasmosis. Mild and self-limited cases have been documented, but their frequency is unknown.

1. **Asymptomatic:** The most common clinical presentation is a pulmonary infiltrate in association with a variety of symptoms indistinguishable from other acute lower respiratory infections.
2. **Chronic pneumonia.**
3. **Disseminated diseases:** The fungus may spread from the lungs through the bloodstream and form multiple abscesses in various parts of the body. Case fatality is high in the generalized disease.
4. **Cutaneous blastomycosis:** When dissemination occurs, skin lesions on exposed surfaces are most common. The cutaneous disease is usually on the skin of the face or other exposed parts of the body. The initial lesion is ative lesions. Lesions of bone, the genitalia (prostate, epididymis, and testis), and central nervous system also occur.

Morphology

Blastomyces dermatitidis is a dimorphic fungus. In tissue and in cultures at 37°C, the fungus appears as budding yeast cells, which are large (7-20 µm) and spherical, with thick, double contored walls. Each cell carries only a single broadbased bud (Fig. 74.1). The bud and the parent yeast are attached with a broad base, and the bud often enlarges to the same size as the parent yeast before they become detached. At room temperature, the culture is filamentous with septate hyphae and many round or oval conidia, and in older cultures, chlamydospores also.

Laboratory Diagnosis

Diagnosis can be made by direct microscopy, culture of sputum, pus and scrapings from skin lesions and serology.

A. Specimens

Specimens consist of sputum, pus, exudates, urine, and biopsies from lesions.

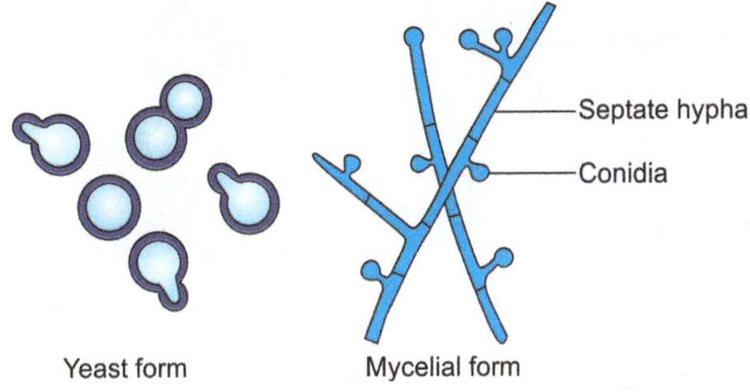

Fig. 74.1: *Blastomyces dermatitidis:* yeast and mycelial forms

B. Microscopic Examination

Potassium hydroxide (10%) mount of specimens may show characteristic thick-walled yeast cells with a single broadbased bud. These may also be apparent in histologic sections.

C. Culture

Colonies usually develop on Sabouraud's dextrose agar or enriched blood agar at 30°C within 2 weeks. The identification is confirmed by conversion to the yeast form after cultivation on a rich medium at 37°C, or by the exoantigen test or by a specific DNA probe.

D. Serology

Antibodies can be measured by the complement fixation (CF), immunodiffision (ID) tests and enzyme immunoassay (EIA). Overall, serologic tests are not as useful for the diagnosis of blastomycosis.

Treatment

Severe cases of blastomycosis are treated with amphotericin B. Uncomplicated pulmonary disease may respond to fluconazole.

2. PARACOCCIDIOIDOMYCOSIS

This is a chronic granulomatous disease of the skin, mucosa, lymph nodes and internal organs. It is confined to endemic regions of Central and South America. As the disease is confined to South America, it is called '**South American blastomycosis**'.

Causative Fungus

It is caused by *Paracoccidioides brasiliensis*, a dimorphic fungus.

Pathogenesis

It is characterized by primary pulmonary infection that spreads, by hematogenous route to mucosa of the nose, mouth and the gastrointestinal tract, skin, lymphatic system, and the internal organs producing chronic granulomatous reaction.

P. brasiliensis is inhaled, and initial lesions occur in the lung. After a period of dormancy that may last for decades, the pulmonary granulomas may become active, leading to chronic, progressive pulmonary disease or dissemination. In the usual case of chronic paracoccidioidomycosis, the yeasts spread from the lung to other organs, particularly the skin and mucocutaneous tissue, lymph nodes, spleen, liver, adrenals, and other sites. Many patients present with painful sores involving the oral mucosa.

Laboratory Diagnosis

1. Specimens

Sputum or pus, crusts and biopsies from granulomatous lesions

2. Direct Microscopy

Microscopy of sputum or pus, crusts and biopsies from granulomatous lesions usually reveals numerous yeast cells (10-60 μm) with multiple buds, which is diagnostic. (Figs 74.2A and B). Multiple buds attached to the mother cell (resembling 'pilot wheel') by very narrow necks are the hallmark of this fungus. Tissue sections should be stained with H & E, PAS and GMS.

3. Culture

P. brasiliensis grows in the mycelial phase in culture at 25-30°C, and in the yeast phase in tissue or at 37°C. Blood agar (without cycloheximide) and incubation at 37°C is recommended for isolation of the yeast phase (tissue form). Mycelial (mold) phase of the fungus develops on SDA incubated at 25-30°C. Identification depends on conversion from the mycelial to the yeast phase.

3. COCCIDIOIDOMYCOSIS

This is primarily an infection of the lungs caused by *Coccidioides immitis,* a dimorphic fungus found in the soil of semi-arid areas, mainly in the south-west USA and northern Mexico.

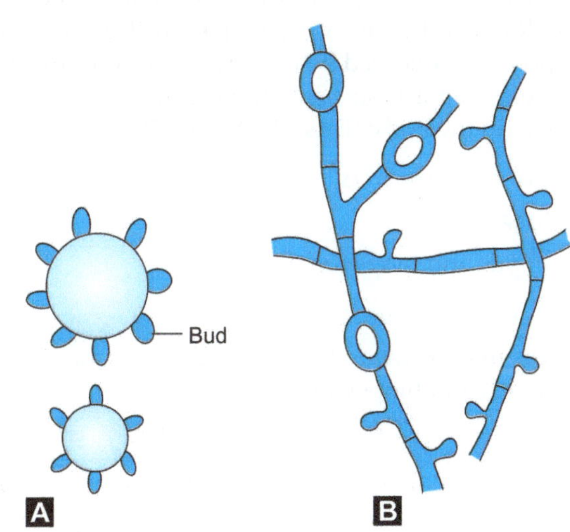

Figs 74.2A and B: *Paracoccidioides brasiliensis* (A) Yeast phase; and (B) Mycelial phase

The infection may be inapparent, benign, severe or even fatal. Agricultural workers with a higher exposure risk and dark-skinned people are especially prone to the disease. Recovery usually confers lifetime immunity.

Clinical Features

i. Asymptomatic

Infection is acquired by inhalation of dust containing arthrospores of the fungus. *C. immitis* usually causes an asymptomatic or self-limiting pulmonary illness, but a progressive and sometimes fatal secondary disease occasionally develops.

ii. Primary Pulmonary Disease

The 40 percent of individuals develop a self-limited influenza-like illness with fever, malaise, cough, arthralgia, and headache. This condition is called **Valley fever, San Joaquin Valley fever, or desert rheumatism.**

iii. Disseminated Disease

Less than one percent of infected persons develop chronic progressive disseminated disease (coccidioidal granuloma) which is highly fatal. It resembles clinically and histologically disseminated tuberculosis.

Morphology

The fungus is dimorphic, occurring in the tissue as a yeast and in culture as the mycelial form. The mycelial phase consists of hyphae which fragment into arthrospores which are highly infectious. In culture and in soil, *C. immitis* grows as a mold, producing large numbers of barrel-shaped arthrospores (4 × 6 µm diameter) which are highly infectious. They characteristically alternate with smaller intervening empty cells (Figs 74.3 A and B). *C. immitis* is probably the most virulent of all human mycotic agents. The inhalation of only a few arthroconidia produces primary coccidioidomycosis.

The yeast form is a spherule (15-75 µm diameter) with a thick, doubly refractile wall and filled with endospores. Endospores are released by rupture of the spherule wall and develop to form new spherules in adjacent tissue or elsewhere in the body (Figs 74.3A and B).

Laboratory Diagnosis

A. Specimens

Sputum, pus and biopsy material.

B. Microscopic Examination

Diagnosis may be made by microscopic examination of sputum, pus and biopsy material.

C. Culture

Specimen is inoculated on SDA medium in the test tube and incubated at 25-30°C. The arthroconidia are highly infectious and are a serious danger to laboratory staff. Consequently, Petri dishes should *never* be used for isolation of the organism and all procedures should be carried out in Category 3 containment facilities.

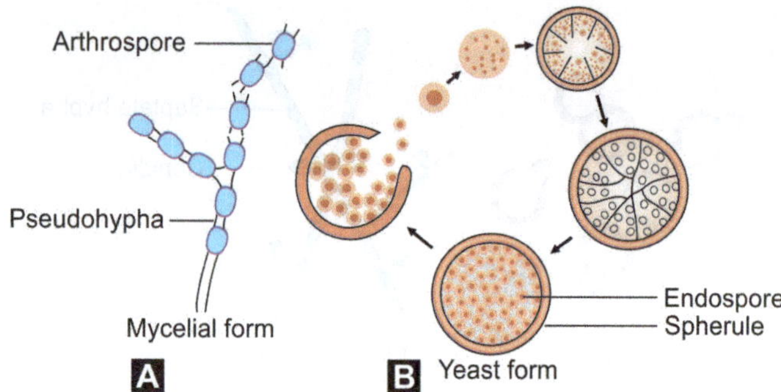

Figs 74.3A and B: *Coccidioides immitis:* (A) Arthrospores formation; (B) Spherule formation with endospores

D. Serological Tests

Serological tests such as precipitin test, latex agglutination test and complement fixation test play an important part in diagnosis.

E. Skin Test

Skin tests with coccidioidin, a culture filtrate antigen from the mycelial phase of *C. immitis,* or spherulin, an extract of the spherules are useful. The test becomes positive (5 mm diameter of induration at 48 hours) between 3-21 days of symptoms. About 90 percent of inhabitants in endemic areas exhibit positive skin test.

4. HISTOPLASMOSIS

Histoplasmosis is an intracellular infection of the reticuloendothelial system caused by the dimorphic fungus *Histoplasma capsulatum.* The disease was originally described by Darling (1905) who believed the causative agent to be a protozoon related to *Leishmania donovani*. Histoplasmosis is also known as **Darling's disease, reticuloendothelial cytomycosis, cave disease,** and **spelunker's disease.**

Morphology

H. capsulatum is dimorphic and grows in soil and in culture at 25-30°C as a mold and as an intracellular yeast in animal tissues. The yeast phase cells (2-3 × 3-4 µm) can also be produced *in vitro* by culture at 37°C on blood agar or other enriched media containing cysteine. In culture the mould colonies are fluffy, white or buff-brown. The mycelium is septate and two types of unicellular asexual spores are usually produced:

1. **Macroconidia:** Large, round, tuberculate macroconidia (8-14 µm in diameter) are most prominent and are diagnostic.
2. **Microconidia:** But smaller broadly elliptical, smooth-walled microconidia (2-4 µm in diameter) are also present in primary isolates.

In the laboratory, with appropriate mating strains, a sexual cycle can be demonstrated, yielding *Ajellomyces capsulatus*, a teleomorph that produces ascospores.

Epidemiology

The disease has a worldwide distribution but is most common in the USA. Twenty, five authentic cases of histoplasmosis have been reported from India. The etiologic agent of histoplasmosis, *H. capsulatum* var. *capsulatum,* grows in soil with a high nitrogen content, especially in areas contaminated with the excreta of bats and birds (starlings and chickens, in particular). Birds are not infected, whereas natural infection does occur in bats. The apparent immunity of birds to systemic histoplasmosis may be dependent directly upon their body temperature which is higher than the temperature at which the fungus can grow.

Cases of histoplasmosis have also been reported in Europe and Asia. A variant form of histoplasmosis occurs in Africa. The etiologic agent of this disease has been designated *H. capsulatum* var. *duboisii.*

Pathogenesis

Infection is acquired by inhalation. Most infections are asymptomatic and are detected only when individuals develop a positive skin test reaction. Some infected persons develop pulmonary disease which resembles tuberculosis.

1. Pulmonary Infection

A chronic form of histoplasmosis occurs mainly in adults. Large cavities develop directly from primary lesions in the lung or by reactivation of old lesions. The clinical picture closely resembles tuberculosis. In some cases, the infection may disseminate to give acute generalized disease.

2. Disseminated Histoplasmosis

Disseminated infection occurs most often in old age and infancy, or in individuals with impaired immune responses. The reticuloendothelial system is involved with resultant lymphadenopathy, hepatosplenomegaly, fever, anemia and a high rate of fatality.

3. Skin and Mucosa

Granulomatous and ulcerative lesions may develop on the skin and mucosa.

Laboratory Diagnosis

A. Specimens

Blood films, bone marrow slides, and biopsy specimens may be examined microscopically. Specimens for culture include sputum, urine, scrapings from superficial lesions, bone marrow aspirates, and buffy coat blood cells. In disseminated histoplasmosis, bone marrow cultures are often positive.

B. Microscopic Examination

Microscopy of smears of sputum or pus should be stained by the Wright or Giemsa procedure. Liver or lung biopsies stained with PAS or methenamine-silver may provide a rapid diagnosis of disseminated histoplasmosis in some patients. *H. capsulatum* is seen as small, oval yeast cells (2-4 *μm* in diameter), typically packed within the cytoplasm of macrophages or monocytes (Fig. 74.4).

C. Culture

Specimens are cultured in rich media, such as glucose cysteine blood agar at 37°C and on Sabouraud's agar or inhibitory mold agar at 25-30°C. On SDA, it forms white to tan fluffy colony with septate branching hyphae with two types of unicellular, asexual spores: (1) Large round, tuberculate macroconidia (8-14 μm) are most prominent and are diagnostic. (2) Small spores or microconidia are sessile or stalked, smooth-walled, round to pyriform, 2-4 μm in diameter (Fig. 74.5). On blood agar or other enriched media containing cysteine, small, round or oval budding yeast cells are produced.

D. Serological Tests

Serological tests like latex agglutination, precipitation and complement fixation become positive two weeks after infection. Serological tests are useful, but cross-reactions can occur, mainly with *C. immitis.*

Antibody tests fail to detect antibodies in up to 50 percent of immunosuppressed individuals. Tests for antigen detection by radio-immunoassay or ELISA are useful, but are not widely available.

E. Histoplasmin Skin Test

Delayed hypersensitivity to the fungus can be demonstrated by histoplasmin skin test. The test is similar to tuberculin test but antigen used is histoplasmin. Histoplasmin is a culture filtrate antigen of mycelial phase of *H. capsulatum.*

The histoplasmin skin test becomes positive soon after infection and remains positive for years. A positive 'histoplasmin skin test' indicates past or present infection, but does not differentiate active and past infections.

Treatment

Intravenous amphotericin B for up to 3 months is the treatment of choice for most forms of disseminated histoplasmosis; this is followed by oral itraconazole for 6-24 months in immunocompromised patients. Ketoconazole and itraconazole give good results in less ill cases.

African Histoplasmosis

This disease is caused by *H. capsulatum* var. *duboisii. H. capsulatum* var'. *duboisii* is morphologically identical to *H. capsulaturn* in its mycelial phase but differs in the

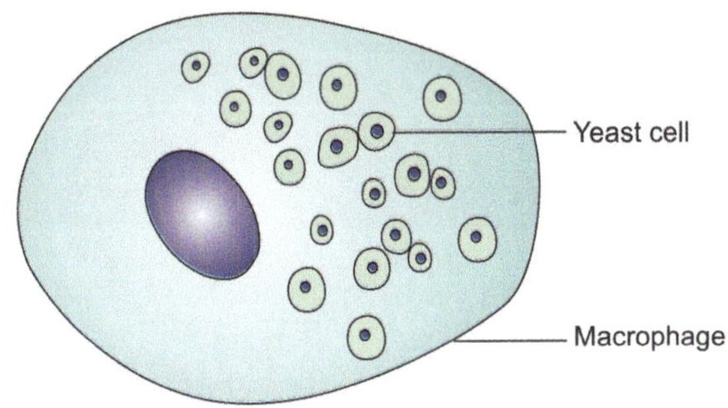

Fig. 74.4: *H. capsulatum*: Yeast cells in macrophage

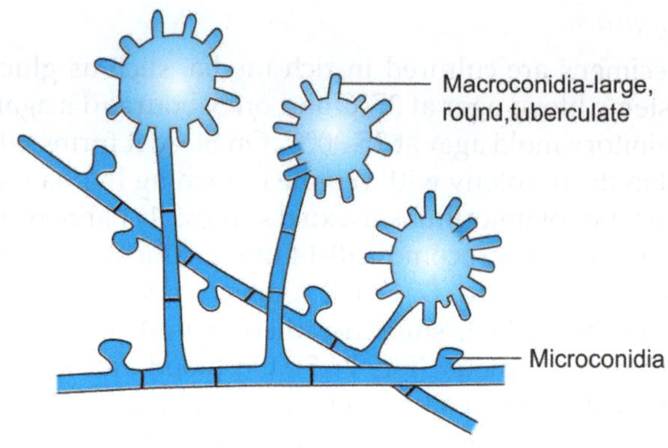

Fig. 74.5: *H. capsulatum*: Mycelial form

yeast phase both *in vivo* and *in vitro* and *H. duboisii* is morphologically identical to *H. capsulatum* in its mycelial phase but the yeast phase has larger cells (12-15 µm diameter).

African histoplasmosis is mainly restricted to the continent of Africa. It involves mainly the skin, subcutaneous tissues and bones. The lungs are not commonly affected and disseminated disease is infrequent.

KNOW MORE

Histoplasma capsulatum

Histoplasma capsulatum is a dimorphic soil saprophyte that causes histoplasmosis, the most prevalent pulmonary mycotic infection in humans and animals. In nature, *H. capsulatum* grows as a mold in association with soil and avian habitats, being enriched by alkaline nitrogenous substrates in guano. *H. capsulatum* received its name from the appearance of the yeast cells in histopathologic sections; however, it is neither a protozoan nor does it have a capsule.

KEY POINTS

- **Blastomycosis:** *Blastomycosis* is a chronic infection of the lungs which may spread to other tissues caused by *Blastomyces dermatitidis*, a dimorphic fungus. It is also known as *North American blastomycosis*. The fungus appears as budding yeast cells, large and spherical, with thick, double contored walls carrying only a single broadbased bud.
- **Coccidioidomycosis:** is primarily an infection of the lungs caused by *Coccidioides immitis*, a dimorphic fungus. Infection is acquired by inhalation of dust containing arthrospores of the fungus. In culture and in soil, *C. immitis* grows as a mold, producing large numbers of barrel-shaped arthrospores which are highly infectious. The yeast form is a spherule with a thick, doubly refractile wall and filled with endospores.
- **Histoplasmosis:** Histoplasmosis is primarily a disease of reticuloendothelial system caused by an intracellular fungus *Histoplasma capsulatum*, a dimorphic fungus.

  The fungal colony is characterized by thin, branching, septate hyphae that produce **tuberculate macroconidia** and **microconidia**.

  The macroconidia are thick-walled spherical spores and have finger-like projections. These are diagnostic form of the fungus.
- *H. capsulatum* causes acute pulmonary histoplasmosis, chronic pulmonary histoplasmosis, and progressive disseminated histoplasmosis. The infection is endemic in some parts of the world.

IMPORTANT QUESTIONS

Write short notes on:

Blastomycosis
Coccidioidomycosis (or) *Coccidioides immitis*
Histoplasmosis (or) Darling's disease
Paracoccidioidomycosis (or) Paracoccidioides brasiliensis

FURTHER READING

Franco M et al (editors). *Paracoccidioidomycosis*. CRC Press, 1994.

Kirkland TN, Fierer J. Coccidioidomycosis: A reemerging infectious disease. Emerg Infect Dis 1996;2:192.

Wheat LJ. Endemic mycoses in AIDS: A clinical review. Clin Microbiol Rev 1995;8:146.

Wheat LJ. Histoplasmosis: Recognition and treatment. Clin Infect Dis 1994;19:S19.

CHAPTER 75

Opportunistic Mycoses

Learning Objectives

After reading and studying this chapter, you should be able to:

- Describe diseases caused by *Candida albicans*.
- Describe the following: Thrush or oral thrush; Germ tube test or Reynlds-Braude phenomenon.
- Discuss laboratory diagnosis of candidiasis.
- List opportunistic fungi.
- Discuss cryptococcosis.
- Discuss laboratory diagnosis of *Cryptococcus neoformans*.
- Describe species of Asperillus.
- Describe the following: Aspergillosis; mucormycosis; pneumocystosis; otomycosis; mycotic keratitis.

OPPORTUNISTIC FUNGI

Patients with compromised host defenses are susceptible to ubiquitous fungi, are referred to as opportunistic fungi. Healthy people, if exposed to ubiquitous fungi, are usually resistant.

Causative Fungal Agents

A. Yeast like fungi (*Candida* spp., *Torulopsis, Cryptococcus*),
B. Filamentous fungi (*Aspergillus, Mucor, Absidia, Rhizopus, Cephalosporium, Fusarium, Penicillium, Geotrichum, Scoulariopsis*)
C. Others (*Pneumocystis carinii*).

A. YEAST LIKE FUNGI

Candidiasis

Candidosis (candidiasis, moniliasis) is an infection of the skin, mucosa, and rarely of the internal organs, caused by a yeast-like fungus *Candida albicans*, and occasionally by other Candida species. Several species of the yeast genus candida are capable of causing candidiasis. *Candidosis* is an opportunistic endogenous infection, the commonest predisposing factor being diabetes. They are members of the normal flora of the skin, mucous membranes, and gastrointestinal tract. Candida species colonize the mucosal surfaces of all humans during or soon after birth, and the risk of endogenous infection is ever-present.

Morphology

In culture or tissue, candida species grow as oval, budding yeast cells (3-6 µm in size). They also form pseudohyphae when the buds continue to grow but fail to detach, producing chains of elongated cells that are pinched or constricted at the septations between cells (Figs 75.1 and 75.2). *C. albicans* is dimorphic. In addition to yeasts and pseudohyphae, it can also produce true hyphae.

Species of Candida

Important species of *Candida* found in man are: (i) *C. albicans*; (ii) *C. stellatoidea*; (iii) *C. tropicalis*; (iv) *C. krusei*; (v) *C. guilliermondii*; (vi) *C. parapsilosis*; (vii) *C. glabrata*, (viii) *C. viswanathii*

Pathogenesis

A. Superficial (cutaneous or mucosal) candidiasis.
B. Systemic candidiasis.

A. Superficial (cutaneous or mucosal) Candidiasis

It is established by an increase in the local census of candida and damage to the skin or epithelium that permits local invasion by the yeasts and pseudohyphae. The risk factors associated with superficial candidiasis include AIDS, pregnancy, diabetes, young or old age, birth control pills, and trauma (burns, maceration of the skin). Lesions caused by candida are as follows:

a. Mucocutaneous Lesions

1. **Oral thrush:** Oral thrush can occur on the tongue, lips, gums, or palate. It is found commonly in bottle fed infants and the aged and debilitated. Creamy white patches appear on the tongue or buccal mucosa, that leave a red oozing surface on removal.
2. **Vulvovaginitis:** Yeast invasion of the vaginal mucosa leads to vulvovaginitis, characterized by

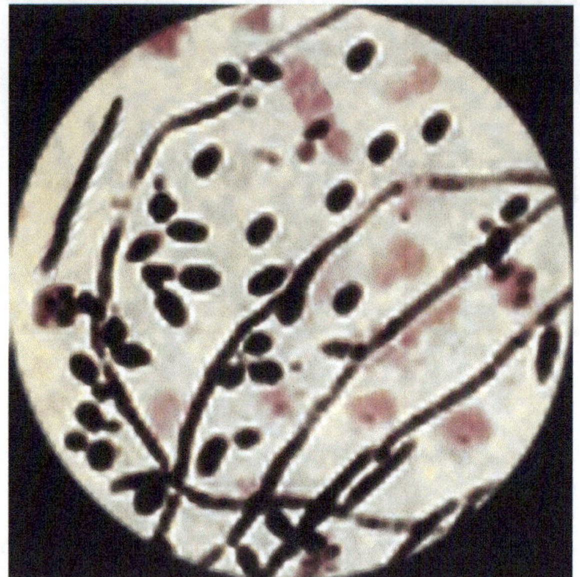

Fig. 75.1: Sputum specimen illustrating budding yeast cells and pseudohyphae of *Candida* species

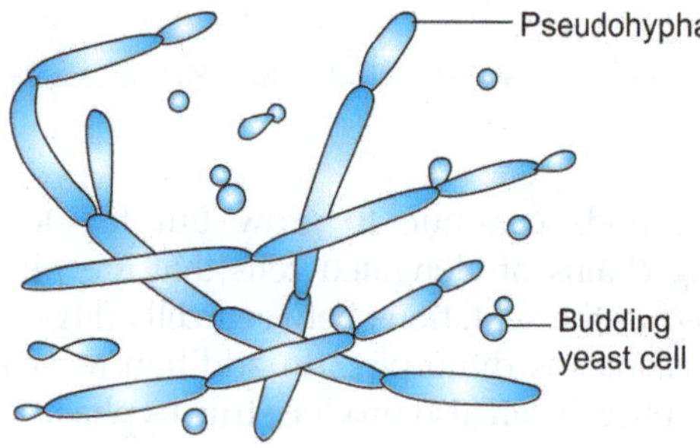

Fig. 75.2: *Candida albicans:* yeast form and pseudohyphae

irritation, pruritus, and vaginal discharge. This condition is often preceded by factors such as diabetes, pregnancy, or antibacterial drugs that alter the microbial flora, local acidity, or secretions.

3. Balanitis.
4. Conjunctivitis.
5. Keratitis.

b. Skin and Nail Infections

Cutaneous candidosis may be intertriginous or paronychial.

Intertriginous Infection

Intertriginous infection occurs in moist, warm parts of the body such as the axillae, groin, and intergluteal or inframammary folds. The sites affected are those where the skin is macerated by perspiration. It is most common in obese and diabetic individuals. The infected areas become red and moist and may develop vesicles.

Interdigital Involvement

Interdigital involvement between the fingers follows repeated prolonged immersion in water. It is most common in homemakers, bartenders, cooks, and vegetable and fish handlers.

Onychomycosis

Candidal invasion of the nails and around the nail plate causes onychomycosis, a painful, erythematous swelling of the nail fold resembling a pyogenic paronychia, which may eventually destroy the nail. Paronychia and onychia are seen in occupations that lead to frequent immersion of the hands in water.

Napkin Dermatitis

In infants it may lead to napkin dermatitis.

B. Systemic Candidiasis

It occurs when candida enters the bloodstream and the phagocytic host defenses are inadequate to contain the growth and dissemination of the yeasts. Systemic can-didiasis is most often associated with chronic adminis- tration of corticosteroids or other immunosuppressive agents; with hematologic diseases such as leukemia, lymphoma, and aplastic anemia; or with chronic granu- lomatous disease.

1. **Intestinal candidosis:** It is a frequent sequel to oral antibiotic therapy and may present as diarrhea not responding to treatment.
2. **Bronchopulmonary candidosis:** It is seen as a rare complication of pre-existing pulmonary or systemic disease.
3. Septicemia
4. Endocarditis
5. Meningitis
6. Kidney infections
7. **Urinary tract infections:** These are often associated with Foley catheters, diabetes, pregnancy, and antibacte-rial antibiotics.

C. Chronic Mucocutaneous Candidiasis

Most forms of this disease have onset in early childhood, are associated with cellular immunodeficiencies and endocrinopathies, and result in chronic superficial disfiguring infections of any or all areas of skin or mucosa.

Laboratory Tests

Diagnosis can be established by microscopy and culture.

A. **Specimens:** Specimens include swabs and scrapings from superficial lesions, blood, spinal fluid, tissue biopsies, urine, exudates, and material from removed intravenous catheters.

B. **Direct microscopy:** Tissue biopsies, centrifuged spinal fluid, and other specimens may be examined in Gram stained smears for pseudohyphae and budding cells. Wet films or gram-stained smears from lesions or exudates show budding gram-positive cells. As Candida can be seen on normal skin or mucosa as well, only its abundant presence is of significance. Demonstration of mycelial forms indicates colonisa tion and tissue invasion and is, therefore, of greater significance. Skin or nail scrapings are first placed in a drop of 10 percent potassium hydroxide (KOH) and calcofluor white.

C. **Culture:** Cultures are obtained on Sabouraud's dextrose agar (SDA) and on ordinary bacteriological culture media, e.g. blood agar at room temperature or at 37°C. Colonies are creamy white, smooth and with a yeast odour. Gram stained smear from colonies shows gram-positive budding yeast cells (Fig. 75.3).

D. **Identification:** The following tests are done to differentiate *C. albicans* from other species. *C. albicans* is identified by the production of germ tubes or chlamydospores. Other candida isolates are speciated with a battery of biochemical reactions.

a. *Germ tube test: C. albicans* has ability to form germ tubes within two hours when incubated in human serum at 37°C (Reynolds-Braude phenomenon) (Fig. 75.4).

b. Chlamydospores: Chlamydospores develop in a nutritionally deficient medium such as cornmeal agar at 20°C. They can be seen at the end of pseudohyphae (Fig. 75.5).

c. Carbohydrate fermentation and carbohydrate assimilation tests

These are used in identification of *C. albicans* and other species of candida. Species of candida can be identified depending on fermentation and assimilation of various carbohydrates. Sugar fermentation and assimilation tests can be used ro confirm the identification and speciate the more common candida isolates, such as *C tropiealis, C parapsilosis, C guillier mondii, C kefyr, C krusei, and C lusitaniae. C glabrata* is unique among these pathogens because it produces only yeast cells and no pseudohyphal forms.

E. Serology: Agglutinins appear in the sera of patients but as they are frequent in normal persons also, they are not helpful in diagnosis. The detection of circulating cell wall mannan, using a latex agglutination test or an enzyme immunoassay, is much more specific, but the test lacks sensitivity.

F. Skin Test: Delayed hypersensitivity to *Candida* is so universal that skin testing with *Candida* extracts is used as an indicator of the functional integrity of cell mediated immunity.

Treatment

Management of candidosis is mainly by removing the predisposing causes. All *Candida* strains are sensitive to nystatin but as it is poorly absorbed from the gut, it is not useful in systemic diseases. Amphotericin B, 5-fluorocytosine and clotrimazole may be used for disseminated candidosis.

Cryptococcosis

Cryptococcosis (torulosis, European blastomycosis, Busse-Buschke disease) is subacute or chronic infection caused by the capsulate yeast *Cryptococcus neoformans*. It is most frequently recognized as a disease of the central nervous system (CNS), although the primary site of infection is the lungs. The disease occurs sporadically throughout the world but it is now seen most often in patients with AIDS.

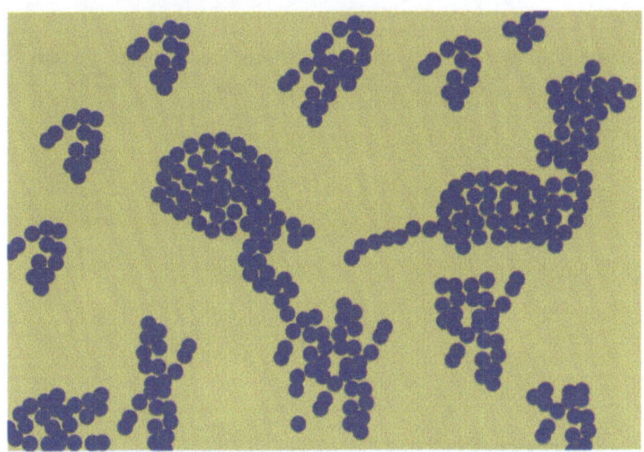

Fig. 75.3: Gram stained smear of *Candida albiocans*

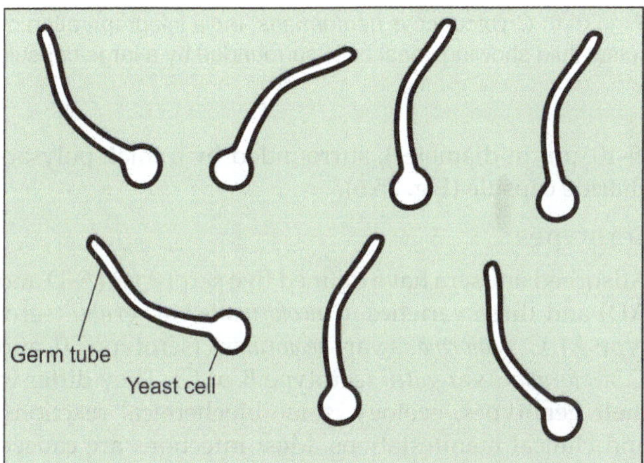

Fig. 75.4: Candida albicans showing Germ tubes

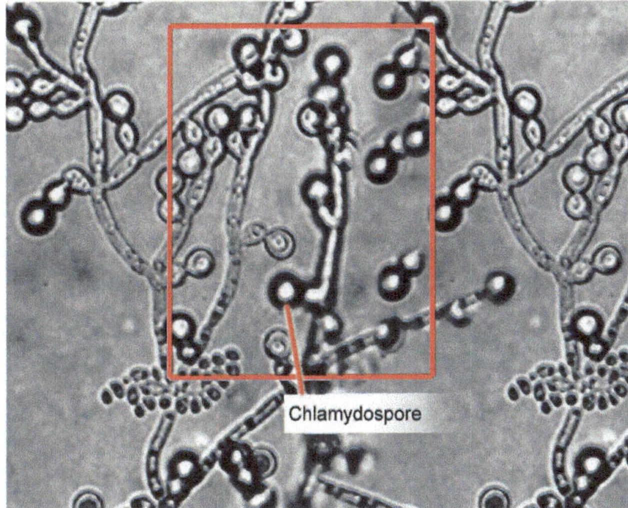

Fig. 75.5: Formation of Chlamydospores by Candida albicans when cultured on cornmeal agar at 25°C

Morphology

In culture, *C. neoformans* produces a whitish mucoid colony in 2-3 days. Microscopically, in culture or clinical material, *C. neoformans* is a spherical budding yeast

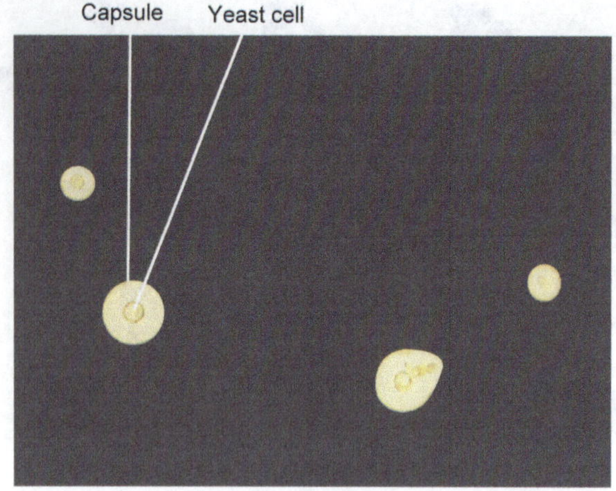

Fig. 75.6: *Cryptococcus neoformans:* India ink preparation of spinal fluid showing yeast cells surrounded by a large capsule

(5-10 µm in diameter), surrounded by a thick polysaccharide capsule (Fig. 75.6).

Serotypes

Adsorbed antisera have defined five serotypes (A-D and AD) and three varieties. *C neoformans* var. *grubii* (serotype A), *C. neoformans* var. *neoformans* (serotype D), and *C. neoformans* var *gattii* (serotype B or C). They differ in their genotypes, ecology, some biochemical reactions, and clinical manifestations. Most infections are caused by *C. neoformans* var. *neoformans*, which is commonly found in the excreta of wild and domesticated birds throughout the world. The birds themselves do not appear to become infected, probably because of their high body temperature. *C. neoformans* var. *gattii* is associated with the flowers of *Eucalyptus camaldulensis* (red river gum tree) and infections coincide with the distribution of the tree.

Sexual Reproduction

Sexual reproduction can be demon strated in the laboratory, and successful mating results in the production of mycelia and basidiospores. Two perfect stages of the fungus have been discovered. They belong to the class Basidiomycetes and have been termed *Filoba sidiella neoformans* var. *neoformans* (serotypes A and D) and *Filobasidiella neoformans* var. *bacillispora* (seroctypes B and C).

Epidemiology

Cryptococcosis is worldwide in distribution. As it was originally reported from Europe, it used to be known as 'European blastomycosis'. Several cases of cryptococcosis have been identified in India, this being the only deep mycosis common in this country. *C. neoformans* can produce disease in animals, particularly mastitis in cattle. The role of animal cryptococ cosis in causing human disease is not known.

Bird droppings (particularly pigeon droppings) enrich for the growth of *C. neoformans* and serve as a reservoir of infection. The organism grows luxuriantly in pigeon excreta, but the birds are not infected. In addition to patients with AIDS or hematologic malignancies, patients being maintained on corticosteroids are highly susceptible to cryptococcosis.

Pathogenesis

Infection is usually acquired by inhalation but may sometimes be through skin or mucosa. The primary pulmonary infection may be asymptomatic or may mimic an influenza-like respiratory infection, often resolving spontaneously. Pulmonary cryptococcosisis may lead to a mild pneumonitis. In patients who are compromised, the yeasts may multiply and disseminate to other parts of the body but preferentially to the central nervous system, causing cryptococcal meningoencephalitis. Other common sites of dissemination include the skin, eye, and prostate gland.

Cryptococcal Meningitis

Cryptococcal meningitis is the most serious type of infection and can resemble tuberculous or other chronic types of meningitis. Its onset is insidious and the course slow and progressive. It is often seen in AIDS. The meningeal form of cryptococcosis can occur in apparently healthy individuals, but occurs most frequently in patients with abnormalities of T lymphocyte function, including those with Hodgkin's disease, sarcoidosis, collagen disease and neoplasms.

Other Sites of Dissemination

Although predominantly a disease of the central nervous system (CNS), lesions of the skin, mucosa, viscera and bones may also occur. In its disseminated form, the disease may resemble tuberculosis. Visceral forms simulate tuberculosis and cancer clinically. Bones and joints may be involved. Cutaneous cryptococcosis varies from small ulcers to large granulomas. Rarely, lesions of skin and bones may occur without any evidence of infection elsewhere.

Laboratory Diagnosis

A. **Specimens:** Specimens include spinal fluid, tissue, exudates, sputum, blood, and urine. Diagnosis is established by demonstration of capsule, budding yeast cells in the lesions and by culture.
B. **Microscopic Examination—India ink or nigrosine preparation:** Specimens are examined in wet mounts, both directly and after mixing with India ink, which delineates the capsule. In unstained, wet preparations of CSF mixed with a drop of India ink or nigrosine, the capsule can be seen as a clear halo around the yeast cells (Fig. 75.6). The yeast cells of *C. neoformans* are round, 5-10 µm in diameter, and are surrounded by a mucopolysaccharide capsule.
Tissue sections: For examination of tissue sections it is best to use a specific fungal stain such as PAS. Alciari blue and mucicarmine stain the capsular

material, enabling the organisms to be differentiated from *H. capsulatum* and *B. dermatitidis*.

C. **Culture:** On Sabouraud agar (without cycloheximide) cultured at 25-30°C and 37°C, colonies normally appear within 2-3 days. Media with cycloheximide inhibit *C. neoformans* and should be avoided.In culture, *C. neoformans* appears as smooth, mucoid, cream coloured colonies. Cultures can be identified by growth at 37°C and detection of urease. Alternatively, on an appropriate. diphenolic sub strate, the phenol oxidase (or laccase) of *C. neoformans* produces melanin in the cell walls and colonies develop a brown pigment.

   Niger seed (bird seed) agar is a differential medium for presumptive identification of *C. neoformans*. It produces brown colonies on this medium within one week when incubated at 30°C. *C. neoformans* produces phenoloxidase, which oxidizes the caffeic acid in the niger seed into melanin.

D. **Serological tests:** Cryptococcal capsular polysaccharide antigen can be detected in CSF and blood by latex agglutination and ELISA test. A whole-cell agglutination test for serum antibody is positive in less than 50 percent of proven cases of cyptococcal meningitis.

E. **Animal inoculation test:** Intracerebral or intraperitoneal inoculation into mice leads to a fatal infection in case of *C. neoformans*. Capsulated budding yeast cells can be demonstrated in the brain of the infected mice.

Differentiation of pathogenic (*C. neoformans*) from other non-pathogenic cryptococci:

Pathogenic *C. neoformans* can be differentiated from nonpathogenic species by its ability to:

1. Grow at 37°C;
2. Hydrolyze urea;
3. Produce phenol oxidase—produce black colonies on niger seed agar, bird seed agar and caffeic acid agar, and;
4. Produce disease in mice on intracerebral and intraperitoneal inoculation (animal inoculation test positive). Capsulated budding yeast cells can be demonstrated in the brain of infected mice.

Treatment

Combination therapy of amphotericin B and flucyto sine has been considered the standard treatment for cryptococcal meningitis, though the benefit from adding flucytosine remains controversial. Fluconazole offers excellent penetration of the central nervous system.

B. FILAMENTAUS FUNGI

Aspergillosis

Aspergillosis is a spectrum of diseases that may be caused by a number of *Aspergillus* species. Aspergillus species are ubiquitous saprophytes in nature and aspergillosis occurs worldwide. There are more than 100 species of *Aspergillus* but only a few have been implicated in human disease. The most important are: *A. fumigatus, A. niger, A. flavus, A. terreusand A. nidulans*.

Pathogenesis

This mold produces abundant small conidia that are easily aerosolized. Following inhalation of these conidia, atopic individuals often develop severe allergic reactions to the conidial antigens. In immunocompromised patients—especially those with leukemia, bone marrow transplant patients, and individuals taking corticosteroids—the conidia may germinate to produce hyphae that invade the lungs and other tissues.

A. Localized Infections

Localized, noninvasive infections (colonization) by *Aspergillus* species may involve the nasal sinuses, the ear canal, the cornea, or the nails.

Examples:

Sinusitis—*A. flavus* and *A. fumigatus*.

Mycotic keratits—*A. flavus* and *A. fumigatus*.

Otomycosis—Species of *Aspergillus* particularly *A niger*.

B. Systemic Aspergillosis

1. Pulmonary Aspergillosis

a. **Allergic asthma:** In some atopic individuals, development of IgE antibodies to the surface antigens of Aspergillus conidia elicits an immediate asthmatic reaction upon subsequent exposure. About 10-20 percent of asthmatics react to *A. fumigatus*.

b. **Bronchopulmonary aspergillosis:** In others, the conidia germinate and hyphae colonize the bronchial tree without invading the lung parenchyma. This phenomenon is characteristic of allergic bronchopulmonary aspergillosis, which is clinically defined as asthma, recurrent chest infiltrates, eosinophilia, and both type I (immediate) and type III (Arthus) skin test hypersensitivity to Aspergillus antigen. They have difficulty breathing and may develop permanent lung scarring.

   Normal hosts exposed to massive doses of conidia can develop extrinsic allergic alveolitis which follows particularly heavy and repeated exposure to large numbers of spores. A well known example of this form of the disease is Maltsters lung, which occurs in workers who handle barley on which *A. clavatus* has sporulated during the malting process.

c. **Colonising aspergillosis (Aspergilloma):** Colonizing aspergillosis usually develops in preexisting pulmonary cavities, such as in tuberculosis or cystic disease. It is also referred to as fungus ball. The fungus grows into large 'balls' (*Aspergilloma*). Cases of aspergilloma rarely become invasive. Surgical removal becomes necessary as the disease commonly causes massive hemoptysis.

2. Invasive Aspergillosis

In invasive aspergillosis, the fungus actively invades the lung tissue. This form occurs in severely immunocompromised individuals who have a serious underlying illness. Neutropenia is the most common predisposing factor and *A. fumigatus* is the species most frequently involved.

The lung is the sole site of infection in 70 percent of patients. Disseminated aspergillosis involving the brain, kidney and other organs is a fatal complication sometimes seen in debilitated patients on prolonged treatment with antibiotics, steroids and cytotoxic drugs.

3. Endocarditis

Aspergillus species may rarely cause endocarditis in immunosuppressed patients and those who have undergone open heart surgery.

4. Paranasal Granuloma

A. flavus and *A. fumigatus* may colonize and invade the paranasal sinuses and the infection may spread through the bone to the orbit of the eye and brain. This condition is seen most often in warm dry climates and is common in parts of the Sudan and Northern India.

Laboratory Diagnosis

A. **Specimens:** Sputum, other respiratory specimens, or lung biopsy tissue provide good specimens. Blood samples are rarely positive.

B. **Microscopic Examination:** On direct examination of sputum with KOH or calcofluor white or in histologic sections, the fungus appears as nonpigmented septate mycelium, 3-5 µm in diameter, with characteristic dichotomous branching and an irregular outline. Rarely the characteristic sporing heads of Aspergillus species are present.

   In tissue sections, *Aspergillus* species are best seen after staining with PAS or methenamine-silver.

C. **Culture:** *Aspergillus* species grow readily on Sabouraud agar without cycloheximide at 25-37°C. Colonies appear after 1-2 days. Species are identified according to the morphology of their conidial structures. Asexual conidia are arranged in chains, carried on elongated cells called 'sterigmata', borne on the expanded ends (vesicles) of conidiophores (Fig. 75.7).

   As aspergilli are such common contaminants, their demonstration in exudates and isolation in cultures have to be interpreted with care.

D. **Skin tests:** Skin tests with *A. fumigatus* antigen are useful for the diagnosis of allergic aspergillosis.

E. **Serological tests:** Immunodiffusion, counterimmunoelectrophoresis (CIE) and ELISA are widely used for the detection of antibodies in the diagnosis of all forms of aspergillosis, particularly aspergilloma and allergic bronchopulmonary aspergillosis.

   For diagnosis of invasive aspergillosis, antigen detection has also been used successfully by techniques such as ELISA and latex agglutination.

F. **Polymerase chain reaction(PCR):** This is now increasingly used for diagnosis of invasive aspergillosis but its precise value is still being assessed.

Mucormycoses (Zygomycosis, Systemic Phycomycosis)

Mucormycosis is an opportunistic mycosis caused by a number of molds classified in the order Mucorales of the class Zygomycetes. The leading pathogens among this group of fungi are species of the genera *Rhizopus, Rhizomucor, Absidia, Cunninghamella,* and *Mucor*. These fungi are ubiq uitous thermotolerant saprophytes.

The conditions that place patients at risk include acidosis—especially that associated with diabetes mellitus—leukemias, lymphoma, corticosteroid treatment, severe burns, immunodeficiencies, and other debilitating diseases as well as dialysis with the iron chelator deferoxamine.

Clinical Manifestations

There are a number of different clinical varieties of mucormycosis:

1. Rhinocerebral Mucormycosis

The major clinical form is rhinocerebral mucormycosis, a rapidly fulminating infection which is almost invariably associated with acute diabetes mellitus, or with debilitating diseases such as leukemia or lymphoma. It results from germination of the sporangiospores in the nasal passages and invasion of the hyphae into the blood vessels, causing thrombosis, infarction, and necrosis. The disease can progress rapidly with invasion of the sinuses, eyes, cranial bones, and brain. Blood vessels and nerves are damaged, and patients develop edema of the involved facial area, a bloody nasal exudate, and orbital cellulitis.

The condition is rapidly fatal if untreated and, although the prognosis has improved over recent years, most diagnoses are still made at necropsy.

2. Thoracic Mucormycosis

Thoracic mucormycosis follows inhalation of the sporangiospores with invasion of the lung parenchyma and vasculature. In both locations, ischemic necrosis causes massive tissue destruction. Less frequently, this process has been associated with contaminated wound dressings and other situations.

3. Other Site of Invasion

Primary cutaneous infections have also been reported, but these are extremely rare. Subcutaneous forms of zygomycosis are less serious.

Laboratory Diagnosis

1. Specimens

Recognition of the fungus in tissue by microscopy is considerably more reliable than culture, but material

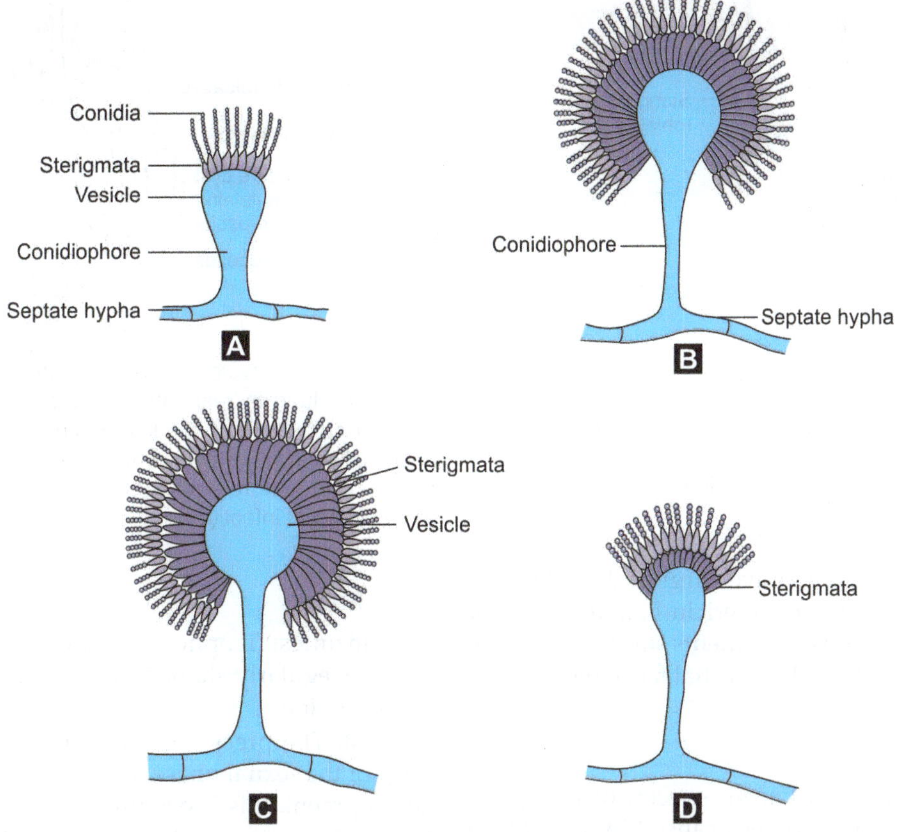

Figs 75.7A to D: *Aspergillus* spp. (A) *A. fumigatus*; (B) *A. flavus*; (C) *A. niger* and (D) *A. terreus*

such as nasal discharge or sputum seldom contains much fungal material and examination of a biopsy is usually necessary for a firm diagnosis.

2. Microscopy

Direct examination of curetted or biopsy material in potassium hydroxide (KOH) may reveal the characteristic broad, aseptate, branched mycelium and sometimes distorted hyphae. However, they are seen much more clearly when stained with methenamine-silver. The hyphae of these fungi do not stain with PAS. Biopsy is normally the best method of establishing the diagnosis and should be performed early in the course of the infection.

3. Culture

The fungi are readily isolated on Sabouraud agar without cycloheximide at 37°C, producing abundant cottony colonies. Isolation is of little diagnostic significance in the absence of strong supporting clinical evidence of infection.

4. Identification

Identification is based on the sporangial structures (Figs 75.8A to C).

i. **Mucor:** Shows nonseptate mycelium without rhizoids (root like structures). Sporangiophores, which may be branched, terminate in large globose sporangia containing numerous spores.

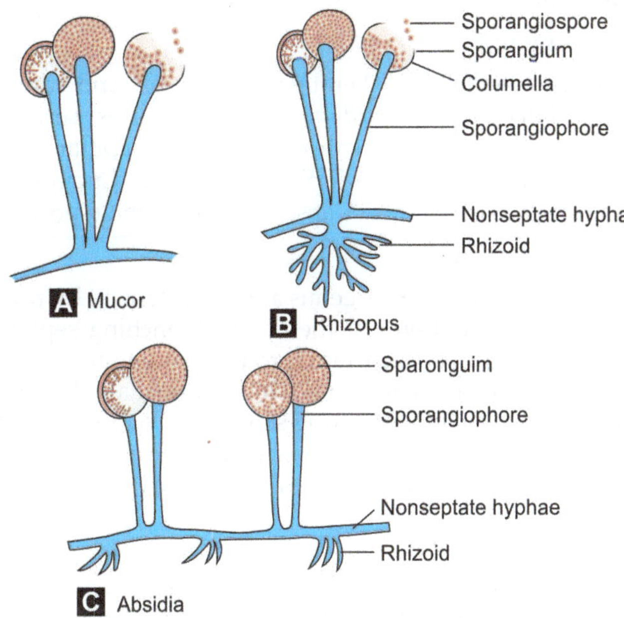

Figs 75.8A to C: Zygomycetes: (A) *Mucor* (B) *Rhizopus* (C) *Absidia*

ii. **Rhizopus:** shows nonseptate mycelium with rhizoids. Unbranched sporangiophores arise in groups directly above the rhizoids.

iii. **Absidia:** has also rhizoids but sporangiophores arise from the aerial mycelium inbetween the rhizoids .

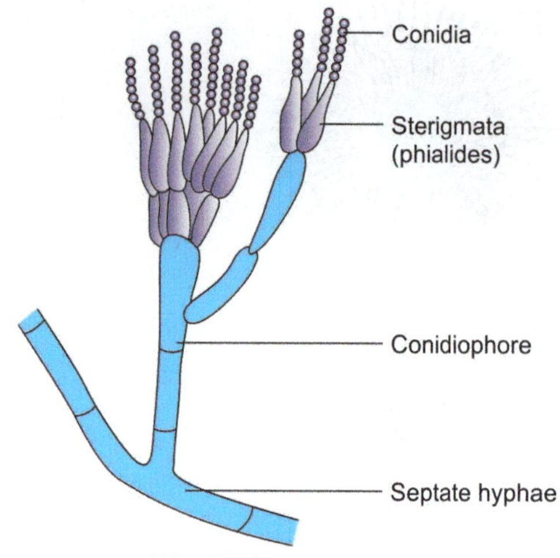

Fig. 75.9: Penicillium

Treatment

Treatment consists of aggressive surgical debridement, rapid administration of amphotericin B, and control of the underlying disease. Many patients survive, but there may be residual effects such as partial facial paralysis or loss of an eye.

Penicilliosis

There are more than 150 known species of the genus *Penicillium*. Except for infections caused by *Penicillium marnefeii*, the role other species of *Penicillium* have in infections of the clinical entity penicilliosis is difficult to confirm.

Pathogenesis

It causes penicillosis, keratitis, otomycosis and rarely deep infections. *Penicillium marneffei* causes serious disseminated disease with characteristic papular skin lesions in AIDS patients in South-East Asia. Cutaneous lesions and subcutaneous abscesses have been reported.

Identification

Fungi belonging to this genus are characterized by producing conidiophores at the tips of branching septate hyphae, which in turn may produce secondary structures termed metulae, from which flask-shaped structures called phialides bearing smooth or rough- shaped conidia are produced in chains, giving the en tire structure a brushlike or broom like appearance (Fig. 75.9).

C. OTHER FUNGAL AGENTS

Pneumocystis Jiroveci

Pneumocystis can inhabit the lungs of many mammals. Until recently, *P. carinii* was thought to be a protozoan, but nucleic acid sequencing showed conclusively that the organism was a fungus with a close relationship to Ascomycetes. It is now apparent from nucleic acid studies that *Pneumocystis* is not a single species. *P. carinii* is the species most commonly found in rats, and *P. jiroveci* is the species most often recovered from humans. Future studies will likely lead to the naming of many other species.

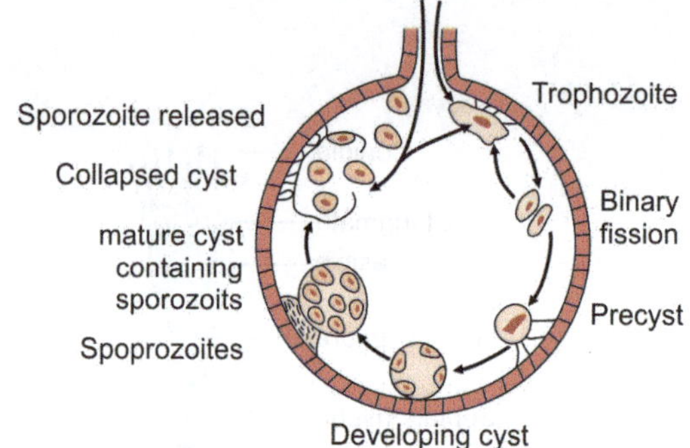

Fig. 75.10: Life cycle of *Pneumocystis jiroveci*. The parasite enters the lung in respiratory droplets and gets attached to alveolar epithelium. It divides by binary fission. Some form a thick-walled cyst withing which sporozoites develop. When mature cyst ruptures, sporozoites are released which initiate fresh cycles of infection

Morphology

Trophozoites: Trophozoites are thin walled (1-5 µm) and is irregularly shaped. It is thought to reproduce by binary fission.

Pre-cyst: The precyst is recognized as an intermediate stage of the sexual phase of reproduction leading to cyst development. It is 5 to 8 µm.

Cysts: Cysts are thick walled and spherical (about 8 µm), contain up to eight intracystic bodies. The cyst collapses, releasing trophozoites which initiate another cycle of multiplication, either in the same host, or in another if they have been spread by coughing. The collapsed cysts can be seen as irregular crescentic bodies (Fig. 75.10).

Cysts can be stained with Gomori methenamine silver (GMS) stain, toluidine blue, Giemsa, and calcofluor white. With GMS stain, the organisms appear deep blue-black.

Pathogenesis

P. carinii is normally a commensal in the lung, spread by respiratory droplets. Upto ten percent of healthy persons have been reported to carry the organisms in the lungs. *Pneumocystis* infection is acquired early in life. In immunocompetent individuals, infection is asymptomatic. However, in inimunocompromised patients, serious life-threatening pneumonia can develop.

Until the AIDS epidemic, human disease was confined to interstitial plasma cell pneumonia in malnourished or premature infants and immunosuppressed patients. Since the early 1980s, it has remained one of the primary opportunistic infections found in patients with AIDS. Prior to the introduction of effective chemoprophylactic regimens, it was a major cause of death among AIDS patients. A high incidence of disease also results from the use of immunosuppressive drugs in patients with malignancies or organ transplants. Underlying defects in cellular immunity apparently make patients susceptible to clinical infection with the organism.

Clinical Manifestations

Patients infected with *pneumocystis* may have fever, nonproductive cough, difficulty in breathing, and a low-grade fever. *P. carinii* pneumonia has been reported to be the most common life-threatening opportunistic infection in AIDS patients.

The multiplication of the parasite in the lungs induces a hyaline or foamy alveolar exudate containing numerous lymphocytes, macrophages and plasma cells, but no polymorphs. In stained sections, the exudate filling the alveoli shows a characteristic honeycomb pattern. Chest radiographs may be normal or show a diffuse interstitial infiltrate. The immune response to the organism after it attaches to and destroys alveolar cells is partly responsible for this radiographic pattern. When the infiltrate is examined, it is found to contain cells from the alveoli and plasma cells.

Laboratory Diagnosis

1. **Specimens:** To establish the diagnosis of *P. carinii* pneumonia, specimens of bronchoalveolar lavage, lung biopsy, or induced sputum are stained and examined for the presence of cysts or trophozoites.
2. **Staining:** Appropriate stains include Giemsa, toluidine blue, methenamine silver, and calcofluor white. A specific monoclonal antibody is available for direct fluorescent examination of specimens. Cyst wall stains black with methenamine silver staining. With the Giemsa stain, the organism appears round, and the cyst wall is barely visible. Intracystic bodies are seen around the interior of the organism. Fluorescent monoclonal antibody staining shows 'honeycomb' appearance of the cyst. *P. carinii* cannot be cultured.
3. **Serology:** While not clinically useful, serology has been used to establish the prevalence of infection. Serological tests can be used for diagnosis in suspected cases. Complement fixation titers of 1:4 or more is indicative of active disease. Latex agglutination test is also used.
4. **Polymerase chain reaction (PCR):** PCR for amplification of *P. carinii* DNA is a rapid method for detection of early infection.

Treatment

Acute cases of pneumocystis pneumonia are treated with trimethoprim-sulfamethoxazole (TMP-SMZ) or pentamidine isethionate. Prophylaxis can be achieved with daily TMP-SMZ or aerosolized pentamidine.

OTHER OPPORTUNISTIC FUNGI

Almost any fungus may invade a severely immunocompromised host and infections with many common fungi, including *Fusarium* species, *Trichosporon beigelii* and *Pseudallescheria boydii*, have been reported. Diagnosis is made by culture of the causative organism from clinical specimens and serological tests play little part. Tissue sections are often not very helpful since the causal fungi either have no special features to enable identification, or they resemble other fungal pathogens.

Infections are usually treated speculatively, and sometimes successfully, with amphotericin B.

OTOMYCOSIS

Otomycosis is a fungal infection of the external ear. About 10-20 percent of chronic ear infections are due to fungi. It is a very common disease and is usually caused by species of *A. niger, A. fumigatus, Penicillium, Candida albicans, C. tropicalis* and *C. krusei.*

The symptoms are itching, pain and deafness. Secondary bacterial infection, commonly due to *Pseu domonas* and *Proteus*, causes suppuration.

Diagnosis can be made by demonstration of the fungi in scrapings and by culture.

MYCOTIC KERATITIS

Keratomycosis or mycotic or fungal keratitis is an invasive fungal infection of the cornea secondary to injury, bacterial infection and treatment with antibacterial agents and steroids. They occur most often in hot climates and are caused by common saprophytic molds.

Causes

It is most frequently caused by *A. fumigatus, A. flavus, A. glaucus* and *A. niger*. In addition, species of *Fusarium, Curvularia, Candida, Acremonium, Paecilomyces, Penicillium, Alternaria, Fonsecea, Pseudallescheria, Drechslera and Aureobasidium* may also cause keratomycosis.

Pathogenesis

It usually follows trauma. Fungal spores colonize the injured tissue and initiate an inflammatory reaction leading to hypopyon, ulcer and endophthalmitis. Increased incidence of keratomycosis is due to widespread use of corticosteroids.

Laboratory Diagnosis

Diagnosis can be made by microscopic examination and culture of scrapings taken from the base or edge of the ulcer. Local application of amphotericin B, Nystatin and Pimaricin (Natamycin) may be useful.

KNOW MORE

- As members of the normal microbial flora, Candida and related yeasts are endogenous opportunists. Other opportunistic mycoses are caused by exogenous fungi that are globally present in soil, water, and air.
- Aspergillosis and mucormycosis are important opportunistic systemic mycoses

☞ KEY POINTS

- Patients with compromised host defenses are susceptible to ubiquitous fungi , are referred to as opportunistic fungi. *Candida albicans, Aspergillus fumigatus, Aspergillus niger, Penicillium* sp., *Rhizopus* and *Mucor* are some examples of opportunistic fungi.

Candidiasis

- Candidosis (candidiasis, moniliasis) is an infection of the skin, mucosa, and rarely of the internal organs, caused by a yeast-like fungus *Candida albicans*, and occasionally by other *Candida* species. It causes (a) Mucocutaneous lesions (Oral thrush); (b) Vulvovaginitis, conjunctivitis keratitis; Skin and nail infections—Intertriginous infection, interdigital involvement, onychomycosis, napkin dermatitis; Systemic candidiasis (intestinal candidosis, bronchopulmonary candidosis) septicemia, endocarditis, meningitis, kidney infections and urinary tract infections.
- Gram-stained smear of the exudates or tissue shows gram-positive, oval, budding yeast and pseudohyphae. Germ tube is a rapid method for identification of *C. albicans*.

Cryptococcosis

- Cryptococcosis is subacute or chronic infection caused by the capsulate yeast *Cryptococcus neoformans*.
- *C. neoformans* causes: Pulmonary cryptococcosis in immunocompromised hosts central nervous system (CNS) cryptococcosis; disseminated non-pulmonary non-CNS cryptococcosis.
- *Laboratory diagnosis*: In unstained, wet preparations of CSF mixed with a drop of India ink or nigrosine, the capsule can be seen as a clear halo around the yeast cells.
- A specific fungal stain such as PAS, Alciari blue and mucicarmine, stain the capsular material of *C. neoformans* in tissue specimens.

  The culture of centrifuged CSF specimens confirms diagnosis of the condition.
- Cryptococcal capsular polysaccharide antigen can be detected in CSF and blood by latex agglutination and ELISA test.

Aspergillosis

- *Aspergillus* is a filamentous fungus producing characteristic conidiophores and a typical colonial appearance. The species most frequently involved in human infections are *A. fumigatus*, *A. flavus* and *A. niger*.
- In immunocompetent hosts, *Aspergillus* species may primarily affect the lungs causing four main syndromes including allergic bronchopulmonary aspergillosis, chronic necrotizing aspergillus pneumonia, aspergilloma, and invasive aspergillosis.
- In immunocompromised host, *Aspergillus* cause a disseminated disease causing endophthalmitis, endocarditis, and abscesses in the viscera such as liver, spleen, kidney, soft tissues, and bone.

Zygomycosis

- Zygomycosis also known as mucormycosis or phycomycosis is an infection caused by saprophytic molds of the class Zygomycetes (mainly Mucor, Rhizopus and Absidia).
- Zygomycetes can cause rhinocerebral zygomycosis, pulmonary zygomycosis and gastrointestinal zygomycosis.

Pneumexystis Jiroveci

- Pneumexystis jiroveci is the causative agent of Pneumocystis carinii pneumonia (PCP). Transmission of infection occurs by inhalation.
- PCP is the most common opportunistic infection in HIV-patients. It also causes PCP in other patients with primary immune deficiencies.
- *Otomycosis* is a fungal infection of the external ear and is usually caused by species of *A. niger, A. fumigatus, Penicillium, Candida albicans, C. tropicalis* and *C. krusei*.
- *Keratomycosis or mycotic or fungal keratitis* is an invasive fungal infection of the cornea, secondary to injury, bacterial infection and treatment with antibacterial agents and steroids. It is most frequently caused by *A. fumigatus, A. flavus, A. glaucus* and *A. niger*.

IMPORTANT QUESTIONS

1. Describe the morphology, pathogenicity and laboratory diagnosis of *Candida albicans*.
2. Describe the morphology, cultural characters and laboratory diagnosis of *Cryptococcus neoformans*.
3. Write short notes on:
   Candidiasis, Candidiosis or Moniliasis
   Cryptococcosis
   Opportunistic systemic mycoses
   Aspergillosis
   Opportunistic fungi
   Mucormycosis or Zygomycosis
   Pneumocystis jiroveci.
   Keratomycosis.

FURTHER READING

Calderone RA (editors): Candida and Candidiasis. ASM Press, 2002.

Casadevall A, Perfect JR. Cohen J, Powderly WG (editors): Infectious Diseases, 2nd ed. Volume 2, Chapter 237-241. Mosby, 2004.

Cryptococcus neoformans. ASM Press, 1998. Denning DW: Invasive aspergillosis. Clin Infect Dis 1998;26:781.

Kumar S, Sirohiwal D, et al. Identification and strain differentiation of *Candida albicans* in women with recurrent vulvovaginal candidiasis. Asian J Obs Gynaec Pract 1997;1:56-58.

Lee FY, Mossad SB, Adal KA. Pulmonary mucormycosis: The last 30 years. Arch Intern Med 1999;159:6.

Ribes JA, Vanover-Sams CL, Baker DJ. Zygomycetes in human disease. Clin Microbiol Rev 2000;13:236.

Su TH, Martin WJ II. Pathogenesis and host response in *Pneumocystis carinii* pneumonia. Annu Rev Med 1994;45:261.

CHAPTER 76

Mycotoxicosis

Learning Objectives

After reading and studying this chapter, you should be able to:

- Describe mycotic poisoning.
- Describe the following: mycotoxicosis; aflatoxin; morphology and culture characteristics of *Staphylococcus aureus*.

Many fungi form poisonous substances. Mycotic poisoning is of two types:

MYCETISM

A fungus which is eaten for itself causes toxic effects. Mycetism may cause gastrointestinal disease, dermatitis or death. Mycetism has been known from ancient times, several varieties of poisonous mushrooms having been identified as inedible.

MYCOTOXICOSIS

The diseases that result from ingestion of food or feed that contains mycotoxins are known as **mycotoxicoses**.

In mycotoxicosis, fungal toxins contaminate some article of food. Many fungi produce poisonous substances called mycotoxins that can cause acute or, chronic intoxication and damage. The mycotoxins are secondary metabolites and their effects are not dependent on fungal infection or viability. The mycotoxicoses are most often the result of the accidental or recreational ingestion of fungi that produce these compounds.

A variety of mycotoxins are produced by mushrooms (e.g., *Amanita* species), and their ingestion results in a dose-related disease called **mycetismus**. Cooking has little effect on the potency of these toxins, which may cause severe or fatal damage to the liver and kidney.

Aflatoxin

Some fungi produce mutagenic and carcinogenic compounds that can be extremely toxic for experimental animals. One of the most potent is **aflatoxin**, which is elaborated by *Aspergillus flavus* and related molds and is a frequent contaminant of peanuts, corn, grains, and other foods.

Effect of Aflatoxin

It is highly toxic to animals and birds, and probably to human beings as well. It can cause hepatomas in ducklings and rats, and its possible carcinogenic effect in human beings has caused great concern. There have been several reports of aflatoxicosis from India, involving human beings and animals.

Ergot Alkaloids

Ergotoxicosis (ergotism) is due to the toxic alkaloids produced by the fungus *Claviceps purpurea,* while growing on the fruiting heads of rye. During the Middle Ages, epidemics known as St. Anthony's fire were associated with the consumption of bread and other bakery products made with contaminated rye and other grains. Ergot alkaloids have been used as oxytocic agents, promoting labor during childbirth by increasing the force and frequency of uterine contractions.

Other Mycotoxicoses

Several other mycotoxicoses that have caused human illness have also been described. Classic examples of human diseases caused by *Fusarium* mycotoxins include alimentary toxic aleukia, Urov or Kashin-Beck disease' and Akakabibye (scabby grain intoxication). Two of these are yellow rice toxicosis in Japan and alimentary toxic aleukia in the former Soviet Union.

There are also other fungi responsible for mycotoxicosis (Table 76.1).

PSYCHOTROPIC AGENTS

Toxic metabolites produced by fungi have been used by primitive tribes for religious, magical, and social purposes. The hallucinogenic agents (d-lysergic acid, psilocybin) produced by the *Psilocybe* species and other fungi have attracted much attention in recent years. In the 20th century, problems involving toxins of fungi were seen with the recreational use of psychotropic agents such as psilocybin and psilocin, as well as the semisynthetic derivative lysergic acid diethylamide (LSD).

Table 76.1: List of some mycotoxins and mycotoxin-producing fungi

Mycotoxin	*Mycotoxin-producing fungi*
Aflatoxin	*A. flavus, A. parsiticus, etc*
Ascladiol	*A. clavatus*
Butenolide	*F. tricinctum, F. nivale, F. equiseti*
Ergot alkaloid	*Claviceps* sp.
Fumigatin	*A. fumigatus*
Chlorine-containing peptide	*P. islandicum*
Muscarine, etc.	*Amanita muscaria, etc*
Ochratoxin A	*A. ochraceus*
Patulin	*P. urticae, A. clavatus, P. claviforme, P. expansum, A. giganteus, etc.*
Penicillic acid	*P. puberculum, P. cyclopium, P. thomii, etc*
Phalloidine	*Amanita phalloides*
Psilocybine	*Psilocybe* sp.
Psoralens	*Sclerotinia sclerotiorum*
Rubratoxin B	*P. rubrum, P. purpurogenum*
Scirpenols (nivalenol, fusarenon)	*F. nivale, F. tricinctum*

KEY POINTS

- **Mycotoxicoses:** Diseases caused by toxic metabolic products released by fungi.
- Mycotic poisoning is of two types:
  Mycetism: A fungus which is eaten for itself causes toxic effects.
  Mycotoxicosis: The diseases that result from ingestion of food or feed that contains mycotoxins are known as mycotoxicoses.
- **Aflatoxin:** It is elaborated by *Aspergillus flavus* and related molds. It is highly toxic to animals and birds, and probably to human beings as well.
- Ergot alkaloids: Ergotoxicosis (ergotism) is due to the toxic alkaloids produced by the fungus *Claviceps purpurea,* while growing on the fruiting heads of rye.
- **Psychotropic agents:** Toxic metabolites produced by fungi such as psilocybin and psilocin, as well as the semisynthetic derivative lysergic acid diethylamide (LSD).

IMPORTANT QUESTIONS

Write short notes on:

a. Mycotoxicosis.
b. Aflatoxin.

FURTHER READING

Casadevall A, Perfect JR, Cohen J, Powderly WG (editors): Infectious Diseases, 2nd ed. Volume 2, Chapter 237-241. Mosby 2004.

Kwon-Chung KJ, Bennet JE. Medical Mycology, Lea & Febiger 1992.

SECTION SIX

Miscellaneous

CHAPTER

77 Normal Microbial Flora of the Human Body

Learning Objectives

After reading and studying this chapter, you should be able to:

- Describe role of normal flora in human body.
- List organisms present in normal flora of upper respiratory tract and gastrointestinal tract.

INTRODUCTION

The term **"normal microbial flora"** denotes the population of microorganisms that inhabit the skin and mucous membranes of healthy normal persons. The normal microbial flora are more or less constant for each species. The skin and mucous membranes always harbor a variety of microorganisms that can be arranged into two groups.

Resident Flora

The **resident flora** consists of relatively fixed types of microorganisms regularly found in a given area at a given age. If disturbed, it promptly reestablishes itself. A knowledge of the normal flora of the body is essential to an understanding of the interaction of human beings and their pathogen laden environment.

Transient Flora

The transient flora consists of nonpathogenic or potentially pathogenic microorganisms that inhabit the skin or mucous membranes for hours, days, or weeks. It is derived from the environment, does not produce disease, and does not establish itself permanently on the surface. Members of the transient flora are generally of little significance so long as the normal resident flora remains intact. However, if the resident flora is disturbed, transient microorganisms may colonize, proliferate, and produce disease.

ROLE OF NORMAL MICROBIAL FLORA

The microorganisms that are constantly present on body surfaces are **commensals**. Their flourishing in a given area depends upon physiologic factors of temperature, moisture, and the presence of certain nutrients and inhibitory substances. Their presence is not essential to life, because "germ-free" animals can be reared in the complete absence of a normal microbial flora.

Beneficial Functions of Normal Flora

Yet the resident flora of certain areas plays a definite role in maintaining health and normal function:

1. **Prevent colonization by pathogens:** The sheer number of harmless bacteria in the lower bowel and the mouth make it unlikely that in a healthy person, an invading pathogen could compete for nutrients and receptor sites. On mucous membranes and skin, the resident flora may **prevent colonization by pathogens** and possible disease **through "bacterial interference".**
2. **Antimicrobial substances production:** Some bacteria of the bowel produce **antimicrobial substances** to which the producers themselves are immune. The antibiotic substances produced by some, for example, colicins, have a harmful effect on pathogens.
3. **Stimulus for the development of the immune system:** Bacterial colonization of the newborn infant acts as a **powerful stimulus for the development of the immune system;** raise the overall immune status of the host against pathogens having related or shared antigens.
4. The microflora of the intestinal tract synthesize vitamin K and several B vitamins which supply on occasion the body's needs, and also aid in the digestion and absorption of nutrients.
5. Normal flora may liberate endotoxins, which may activate alternate complement pathway and help the defense mechanisms of the body.

Harmful Effects of Normal Flora

Members of the normal flora may themselves produce disease under certain circumstances. Clinical problems caused by the normal flora arise under the following conditions:

1. **When the organisms are displaced from their normal site in the body to an abnormal site, e.g.** the introduction of the normal skin bacterium, *Staphylococcus epidermidis,* into the bloodstream where it can colonize catheters and heart valves, resulting in bacterial endocarditis.
2. **When potential pathogens gain a competitive advantage due to diminished populations of harmless competitors, e.g.** when normal bowel flora is depleted by antibiotic therapy leading to overgrowth by the resistant *Clostridium difficile,* which can cause a **severe colitis.**
3. **When some harmless, commonly ingested food substances are converted into carcinogenic derivatives by bacteria in the colon.** A well-known example is the conversion by bacterial sulfatases of the sweetener cyclamate into the bladder carcinogen cyclohexamine.
4. **Immunocompromised individuals:** Normal flora can overgrow and become pathogenic when individuals are immunocompromised.
5. **Drug resistance:** Penicillinase producing microorganisms can aggravate infection by conferring drug resistance against antibiotics, for example, increase in carriage of antibiotic resistant staphylococci.
6. **Confusion in diagnosis:** They cause confusion in diagnosis due to their ubiquitous presence in the body, and their morphological similarity with some pathogens.

NORMAL MICROBIAL FLORA OF THE HUMAN BODY

Normal Flora of the Skin

The predominant resident microorganisms of the skin are aerobic and anaerobic diphtheroid bacilli (e.g. *Corynebacterium, Propionibacterium*); nonhemolytic aerobic and anaerobic staphylococci *(Staphylococcus epidermidis,* occasionally *S. aureus,* and *Peptostreptococcus* species); gram-positive, aerobic, spore-forming bacilli that are ubiquitous in air, water, and soil; alphahemolytic streptococci (viridans streptococci) and enterococci (*Enterococcus* species); and gram-negative coliform bacilli and *Acinetobacter*. Fungi and yeasts are often present in skin folds; acid-fast, nonpathogenic mycobacteria occur in areas rich in sebaceous secretions (genitalia, external ear).

Although some organisms occur only superficially on the skin surface, much of the bacterial flora is located in the openings of the hair follicles. Hair frequently harbors *Staph. aureus* and forms a reservoir for cross-infection. Penicillin resistant staphylococci are seen in individuals working in hospitals.

Among the factors that may be important in eliminating nonresident microorganisms from the skin are the low pH, the fatty acids in sebaceous secretions, and the presence of lysozyme. Neither profuse sweating, nor washing and bathing can eliminate or significantly modify the normal resident flora.

Normal Flora of the Conjunctiva

The predominant organisms of the conjunctiva are diphtheroids *(Corynebacterium xerosis), S. epidermidis,* and nonhemolytic streptococci. Neisseriae and gram-negative bacilli resembling haemophili (*Moraxella* species) are also frequently present. The conjunctival flora is normally held in check by the flow of tears, which contain antibacterial lysozyme.

Normal Flora of the Nose, Nasopharynx and Accessory Sinuses

The floor of the nose harbors corynebacteria, staphylococci and streptococci. *Haemophilus* species and *Moraxella lacunata* may also be seen.

The nasopharynx of the infant is sterile at birth but, within 2-3 days after birth, acquires the common commensal flora and the pathogenic flora carried by the mother and the attendants. The nasopharynx can be considered the natural habitat of the common pathogenic bacteria which cause infections of the nose, throat, bronchi and lungs. Certain gram-negative organisms from the intestinal tract such as *Pseudomonas aeruginosa, E. coli,* paracolons and *Proteus* are also occasionally found in normal persons. After penicillin therapy, they may be the predominant flora.

Normal Flora of the Mouth

The mouth contains a plethora of organisms—pigmented and nonpigmented micrococci, some of which aerobic, gram-positive aerobic spore bearing bacilli, coliforms, *Proteus* and lactobacilli. The mucous membranes of the mouth and pharynx are often sterile at birth but may be contaminated by passage through the birth canal. Within 4-12 hours after birth, viridans streptococci become established as the most prominent members of the resident flora and remain so for life. They probably originate in the respiratory tracts of the mother and attendants. Early in life, aerobic and anaerobic staphylococci, gram-negative diplococci (neissertiae, *Moraxella catarrhalis),* diphtheroids, and occasional lactobacilli are added. When teeth begin to erupt, the anaerobic spirochetes, prevotella species (especially *P. melaninogenica), Fusobacterium* species, *Rothia* species, and *Capnocytophaga* species establish themselves, along with some anaerobic vibrios and lactobacilli. *Actinomyces* species are normally present in tonsillar tissue and on the gingivae in adults and various protozoa may also be present. Yeasts (*Candida* species) occur in the mouth.

Normal Flora of the Upper Respiratory Tract

Within 12 hours after birth, alpha-hemolytic streptococci are found in the upper respiratory tract and become the dominant organisms of the oropharynx and remain so for life. Small bronchi and alveoli are normally sterile.

In the pharynx and trachea, flora similar to that of the mouth establish themselves. The predominant organisms in the upper respiratory tract, particularly the pharynx, are nonhemolytic-and-alphahemolytic, streptococci and neisseriae. Staphylococci, diphtheroids, haemophili, pneumococci, mycoplasmas, and prevotellae are also encountered.

Normal Flora of the Gastrointestinal Tract

At Birth

At birth the intestine is sterile, but organisms are soon introduced with food. In 80-90 percent newborn infants, the meconium is sterile but in 10-20 percent, a few organisms, probably acquired during labor, may be present. In all cases, within 4-24 hours of birth, an intestinal flora is established partly from below and partly by invasion from above.

Breastfed Children

In breastfed children, the intestine contains lactobacilli (*L. bifidus* constituting 99 percent of total organisms in the feces), enterococci, colon bacilli and staphylococci. In artificially fed (bottle fed) children, intestine contains *L. acidophilus* and colon bacilli and in part enterococci, gram-positive aerobic and anaerobic bacilli. With the change of food to the adult pattern, the flora change. Diet has a marked influence on the relative composition of the intestinal and fecal flora.

Normal Adult

In the normal adult, the microorganisms on the surface of the esophageal wall are those swallowed with saliva and food. Because of the low pH of the stomach, it is virtually sterile except soon after eating. The stomach's acidity keeps the number of microorganisms at a minimum (10^3-10^5/g of contents) unless obstruction at the pylorus favors the proliferation of gram-positive cocci and bacilli. The normal acid pH of the stomach markedly protects against infection with some enteric pathogens, e.g. cholera.

It has been established, however, that *Helicobacter pylori*, a curved, gram-negative bacterium, can colonize the mucus-secreting epithelial cells of the human stomach and cause peptic ulcers or chronic gastritis.

As the pH of intestinal contents becomes alkaline, the resident flora gradually increases. The number of bacteria increases progressively beyond the duodenum to the colon, being comparatively low in the small intestine. In the adult duodenum, there are 10^3-10^6 bacteria per gram of contents; in the jejunum and ileum, 10^5-10^8 bacteria per gram; and in the cecum and transverse colon, 10^8-10^{10} bacteria per gram. In the upper intestine, lactobacilli and enterococci predominate, but in the lower ileum and cecum, the flora is fecal. In the sigmoid colon and rectum, there are about 10^{11} bacteria per gram of contents, constituting 10-30 percent of the fecal mass. Anaerobes outnumber facultative organisms by 1000-fold. In diarrhea, the bacterial content may diminish greatly, whereas in intestinal stasis the count rises.

In the normal adult colon, 96-99 percent of the resident bacterial flora consists of anaerobes: *Bacteroides* species, especially *B. fragilis; Fusobacterium* species; anaerobic lactobacilli, e.g. bifidobacteria; clostridia (*C. perfringens*, 10^3-10^5/g); and anaerobic gram-positive cocci (*Peptostreptococcus* species). Only 1-4 percent are facultative aerobes (gram-negative coliform bacteria, enterococci, and small numbers of *Proteus, Pseudomonas*, lactobacilli, *Candida*, and other organisms). More than 100 distinct types of organisms occur regularly in normal fecal flora. Minor trauma (e.g. sigmoidoscopy, barium enema) may induce transient bacteremia in about 10 percent of procedures.

Normal Flora of the Genitourinary Tract

The urinary tract, except for the external urethra of the male, is normally sterile, although a few nonpathogenic cocci can be present in the female urethra.

Mycobacterium smegmatis, a harmless commensal, is found in the smegma of the genitalia of both men and women. This may, by its presence in the voided specimens of urine, cause confusion. This organism is morphologically similar to the tubercle bacillus, but can be differentiated easily on the basis of its rapid growth on laboratory media. From apparently normal men, aerobic and anaerobic bacteria can be cultured from a high proportion, including lactobacilli, *Gard. vaginalis*, alpha-hemolytic streptococci and *Bacteroides species. Chlam. trachomatis* and *Ureaplasma urealyticum* may also be present. The female urethra is either sterile or contains a few gram-positive cocci.

At Birth

At birth the vagina is sterile. In the first 24 hours, it is invaded by micrococci, enterococci and diphtheroids. In 2-3 days, the maternal estrin induces glycogen deposition in the vaginal epithelium. This facilitates the growth of a lactobacillus (Doderlien's bacillus) which produces acid from glycogen, and the flora for a few weeks is similar to that of the adult. After the passively transferred estrin has been eliminated in the urine, the glycogen disappears, along with Doderlien's bacillus and the pH of the vagina becomes alkaline. This brings about a change in the flora to micrococci, alpha- and non-hemolytic streptococci, coliforms and diphtheroids.

At Puberty

At puberty, the glycogen reappears and the pH changes to acid due to the metabolic activity of Doderlien's bacilli, *E. coli* and yeasts. This appears to be an important mechanism in preventing the establishment of other, possibly harmful microorganisms in the vagina. If lactobacilli are suppressed by the administration of antimicrobial drugs, yeasts or various bacteria increase in numbers and cause irritation and inflammation. During pregnancy there is an increase in *Staphylococcus epidermidis*, Doderlien's bacilli and yeasts. Occasionally other members of the intestinal flora may be present. After menopause, lactobacilli again diminish in number and a mixed flora returns and the flora resembles that found before puberty.

KNOW MORE

Bacteria in the Blood and Tissues

As a rule, in the absence of disease, microorganisms are not found in blood or healthy tissue. Bacteria can get into the blood through cuts or abrasions, dental manipulations, or even food in the intestine (situations sometimes referred to as transient bacteremias). The commensals from the normal flora of the mouth, nasopharynx and intestinal tract may get into the blood and tissues. They are usually quickly eliminated by the normal defense mechanisms of the body. Occasional isolation of diphtheroids or nonhemolytic streptococci from normal and abnormal lymph nodes may be those which escaped elimination. They have little significance unless the organisms of doubtful pathogenicity are isolated more than once in serial blood cultures. There is always a risk, however, that a blood donor is unknowingly incubating some pathogen that could cause serious disease in a recipient. This is especially true of the virus that causes AIDS, but also applies to several other organisms.

☞ KEY POINTS

- The term 'normal microbial flora' denotes the population of microorganisms that inhabit the skin and mucous membrane of normal healthy individuals.
- The **resident flora** consists of relatively fixed types of microorganisms regularly found in a given area at a given age and cannot be removed permanently.
- **The transient flora** inhabit the skin or mucous membranes and does not produce disease, and does not establish itself permanently on the surface. The normal flora is composed principally of bacteria
- A resident, actively proliferating, microbial flora occurs in the skin, nose and oropharynx, the mouth and large intestine, the anterior parts of the urethra and the vagina. Other parts of the body, such as the rest of the respiratory and digestive tracts, the urinary bladder and the uterus, may contain small numbers of transient microorganisms.

IMPORTANT QUESTIONS

1. What is normal microbial flora of the human body? Write briefly on beneficial and harmful effects of normal flora.
2. Write short notes on:
   Normal bacterial flora.
   Difference between resident and transient flora.
   Normal flora of the skin.
   Normal flora of the intestine.
   Normal flora in the mouth and upper respiratory tract.
   Normal flora of the genitourinary tract.

FURTHER READING

Hentges DJ. Human Intestinal Microflora in Health and Disease, New York: Academic Press 1983.

Murray PR. 1998. Human Microbiota. In: Topley & Wilson's Microbiology and Microbial Infections, 9th edn, vol. 2: London.

Noble WC. Microbiology of Human Skin, London: Lloyd-Luke 1981.

CHAPTER

78

Infective Syndrome*

Learning Objectives

After reading and studying this chapter, you should be able to:

- Define bacteremia, septicemia, pyemia and endotoxemia.
- List causative organisms of septicemia, infective endocarditis and subacute endocarditis.
- Discuss laboratory diagnosis of subacute endocarditis.
- List causative organisms of meningitis.
- Discuss the laboratory diagnosis of acute pyogenic meningitis.
- Describe characteristics features of cerebrospinal fluid (CSF) in different forms of meningitis.
- Describe the following: Aseptic meningitis; tuberculous meningitis.
- List causative organisms of urinary tract infection.
- Discuss the laboratory diagnosis of urinary tract infection.
- List causative organisms of sore throat.
- Discuss the laboratory diagnosis of sore throat.
- List microbial pathogens that cause pneumonia.
- Define diarrhea and dysentery.
- List causative organisms of diarrhea and dysentery.
- Discuss laboratory diagnosis of diarrhea.
- Discuss laboratory diagnosis of dysentery.
- List causative organisms of food poisoning.
- Discuss the laboratory diagnosis of food poisoning.
- List microbial pathogens that cause pneumonia.
- List causative organisms of (i) Sexually transmitted diseases (STDs); (ii) Painless genital ulcer; painful genital ulcer; urethral discharge.
- Discuss the laboratory diagnosis of gonorrhea.
- Describe the following; Nongonococcal urethritis (NGU).
- Discuss the laboratory diagnosis of syphilis.
- List causative organisms of wound infection.
- Discuss the laboratory diagnosis of wound infection.
- List causative organisms of pyrexia of unknown origin (PUO).
- Discuss the laboratory diagnosis of pyrexia of unknown origin (PUO).

1. BACTEREMIA AND SEPTICEMIA

1. **Bacteremia:** Bacteremia may be defined as presence of bacteria in blood without any multiplication.
2. **Septicemia:** Septicemia is a condition in which bacteria circulate and actively multiply within the blood stream. Microorganisms causing septicemia are given in Table 78.1
3. **Pyemia:** Pyemia is essentially septicemia with metastatic infection.
4. **Endotoxemia:** Endotoxemia is a condition in which bacterial endotoxin circulates in the blood.

Bacteria gain entry into blood stream:

i. From breakages of blood vessels adjacent to skin or mucosal surfaces, or
ii. By phagocytic cells carrying organisms into capillaries or the lymphatic system.

Infective Endocarditis (Table 78.2)

Infective endocarditis denotes a condition of proliferation of microorganisms on the endothelium of the heart. Vegetation in the heart is the prototype lesion in endocarditis. Normal healthy endocardium of an immunocompetent host is generally not at risk when challenged transiently by a small number of organisms in the circulation. However, the heavy challenges that accompany intravenous drug use or the use of intravascular catheters in debilitated host or damage to the natural heart valve by rheumatic fever or atheroma and cardiac surgery, in particular, prosthetic valve replacement may lead to infections of the endocardium. Damaged endocardium provides a site for the aggregation of platelets

*This chapter was contributed by Dr. Savita Kumari, Professor, Department of Internal Medicine, Post Graduate Institute of Medical Education and Research (PGIMER), Chandigarh - 160 012.

Table 78.1: Causative organisms of septicemia

A. Gram negative bacilli (60-70% cases)
Salmonella typhi
S. paratyphi A
S. paratyphi B
S. paratyphi C
Brucella sp.
Haemophilus influenzae
Escherichia coli
Klebsiella pneumoniae
Proteus sp.
Enterobacter sp.
Bacteroides sp.
Pseudomonas sp.
B. Gram-positive bacilli
Listeria monocytogenes
C. Gram-negative cocci
Neisseria meningitidis
D. Gram-positive cocci (20-40% cases)
Staphylococcus aureus
Staph. epidermidis
Streptococcus pyogenes
Str. pneumoniae

Table 78.2: Causative agents of infective endocarditis

A. Bacteria
- *Viridans group of streptococci*
- *Streptococcus sanguis*
- *S. mutans*
- *S. intermedius*
- *S. mitis*
- *Groups F and G streptococci*
- *Staphylococcus epidermidis*
- *Enterococcus faecalis*
- *S. aureus*
- *Diphtheroids*
- *Haemophilus* sp.
- *Coliforms*
- *Pseudomonas*
- *Coxiella burnetii*
- *Chlamydia psittaci*

B. Fungi
- *Candida* sp.
- *Aspergillus* sp.

and fibrin is deposited to build up a sterile vegetation. This is then liable to be colonized by blood-borne organisms. It may occur as an acute rapidly progressive disease or in a subacute form. The earlier division of illness into acute and subacute forms is no longer considered useful.

Acute Endocarditis

Acute bacterial endocarditis starts abruptly with fever and usually an infection such as pneumonia is present somewhere else in the body or there is evidence of injected-drug abuse. Virulent species such as *Staphylococcus aureus* and *Streptococcus pneumoniae* are usually the cause and they infect both normal and abnormal heart valves. They can often produce a rapidly progressive disease, often with valve destruction and formation of abscesses in the heart muscle, leading to heart failure.

Subacute Endocarditis

By contrast, subacute bacterial endocarditis is usually caused by organisms with little virulence. These organisms of relatively low virulence cause infection on damaged or defective valve cusps and large firm vegetations comprising of dense fibrin, platelet aggregates with bacterial colonies are formed. It runs a chronic course. It is more common and comprises of almost 70 percent cases of bacterial endocarditis. Causative agents of subacute bacterial endocarditis are shown in Table 78.3.

Pathogenesis

Endocarditis is an endogenous infection acquired when organisms entering the bloodstream establish themselves on the heart valves. Therefore, any bacteremia can potentially result in endocarditis. Although patients at risk are those who have pre-existing cardiac disease, but about one-third of cases occur in previously normal hearts. About 70 percent patients have one or other predisposing cardiac lesion and other abnormalities that predispose to endocarditis. These include:

1. Rheumatic valvular disease
2. Congenital valve deformities
3. Cardiac valve prosthesis
4. Degenerative cardiac disease
5. Drug abuse.

Laboratory Diagnosis

Diagnosis of infective endocarditis depends on isolation of the causative agent from blood. Blood culture is the most important test for diagnosing infective endocarditis.

1. Specimen

The blood culture is the single most important laboratory test. Three to six samples of blood, 10 ml each should be collected from antecubital vein under all aseptic conditions using sterile disposable syringe over 24 hours. Samples should be collected before antimicrobial therapy is administered. Ideally three separate samples of blood should be collected within a 24-hour period and before antimicrobial therapy is administered. Each sample should be inoculated into 50-100 ml of glucose broth.

Large amount of blood is required because the number of organisms in the blood may be very few. In addition, blood provides nutrition for the growth of organisms. Repeated blood cultures are made because the bacteremia is intermittent.

2. Culture

Cultures are incubated at 37°C for at least 3 weeks. Subcultures are made on solid media such as blood agar and MacConkey agar after 24 hours, 48 hours and once a week thereafter. These solid media are incubated at 37°C for 24 hours.

Table 78.3: Causative agents of subacute endocarditis

A. Bacteria
Viridans group of streptococci—responsible for 60-80% of cases
Streptococcus sanguis
Str. mutans
Str. mitis
Enterococcus faecalis
Staph. epidermidis
Coxiella burnetii
Chlamydia psittaci
B. Fungi
Candida albicans
Aspergillus sp.

Coxiella burnetii and *Chlamydia* spp. cannot grow on cell free media. For the diagnosis of these agents, refer to corresponding chapters.

3. Identification

The isolated organism is identified by colony morphology, Gram staining, biochemical reactions and serological tests. Refer to the corresponding chapters for details of these agents.

4. Antibiotic Sensitivity Tests

Minimum inhibitory concentration (MIC) and minimum bactericidal concentration (MBC) of the antimicrobial agent for the isolated organism must be determined because the routine disk diffusion method is not adequate for guiding treatment of infective endocarditis. The measurement of MIC and MBC helps to determine the adequate dose of the antibiotic to be used for ensuring the serum levels that can penetrate the valves and kill the organisms.

5. Culture Negative Endocarditis

Blood cultures are persistently negative in about 10 to 20 percent cases. It may be due to following reasons:

i. *Recent antibiotic therapy:* Repeated blood cultures done after the stoppage of antibiotics may give positive results in such cases.
ii. *Inadequate number of specimens:* Repeated blood cultures are necessary as bacteremia is intermittent.
iii. *Infection with Coxiella burnetii or Chlamydia* sp. Blood cultures may be negative because *Coxiella burnetii* and *Chlamydia* sp. cannot grow on cell free media.

6. Other Tests for Diagnosis

i. Erythcyte sedimentation rate (ESR): It is elevated.
ii. Normocytic normochromic anemia.
iii. Total leukocyte count: Leukocytosis is common.
iv. C-reactive protein(CRP): More reliable than ESR in assessing progress.
v. Proteinuria
vi. Microscopic hematuria—usually present.
vii. Echocardiography: Key investigation for detecting and following the progress of vegetations, for assessing valve damage and for detecting abscess formation.

Treatment

The antibiotic treatment regimen for infective endocarditis depends upon the infecting organism.

For penicillin-susceptible **streptococci**, high dose penicillin is the treatment of choice. Patients with a good history of penicillin allergy can be treated with clindamycin or with a macrolide, although the latter may be less bactericidal. However, MIC (minimum inhibitory concentration) and MBC (minimum bactericidal concentration) tests should be performed to detect organisms that are less susceptible or tolerant to penicillin. These organisms and enterococci, which are always more resistant to penicillin, should be treated with a combination of penicillin (or ampicillin) and an aminoglycoside. Combinations such as these act synergistically against streptococci and enterococci.

Staphylococcal endocarditis, particularly in prosthetic valve endocarditis when the organisms may be hospital acquired and consequently often resistant to many antibiotics, often presents a more difficult therapeutic challenge. A β-lactamase stable penicillin such as cloxacillin is often suitable and may be given in combination with an aminoglycoside, rifampicin or fusidic acid. Vancomycin or teicoplanin should be used tor penicillin-allergic patients and for treating methicillin-resistant staphylococci.

Prevention

No scientifically proven preventive methods are available in people with known or suspected defects in their heart valves:

i. Importance of maintaining good dental hygiene because causative organisms of infective endocarditis are part of normal flora of mouth.
ii. Prophylactic antibiotic: The accepted practice is to give prophylactic antibiotic before dental extraction and other surgical procedures.

KNOW MORE

- The distinction between bacteremia and septicemia is essentially clinical but there is a quantitative implication. Thus, septicemia is thought of as a life-threatening emergency that must be dealt with urgently.

KEY POINTS

- Bacteremia may be defined as presence of bacteria in blood without any multiplication.
- Septicemia is a condition in which bacteria circulate and actively multiply within the blood stream.

- Pyemia is essentially septicemia with metastatic infection.
- Endotoxemia is a condition in which bacterial endotoxin circulates in the blood.
- Infective endocarditis denotes a condition of proliferation of microorganisms on the endothelium of the heart. Vegetation in the heart is the prototype lesion in endocarditis.
- Acute endocarditis: Virulent species such as *Staphylococcus aureus* and *Streptococcus pneumoniae* are usually the cause, and they infect both normal and abnormal heart valves.
- Subacute endocarditis is usually caused by organisms with little virulence. Viridans group of streptococci are responsible for 60 to 80 percent of cases.
- Diagnosis of infective endocarditis depends on isolation of the causative agent from blood. Blood culture is the most important test for diagnosing infective endocarditis. Three to six samples of blood, 10 ml each should be collected over 24 hours.

IMPORTANT QUESTIONS

1. Define bacteremia, septicemia, pyemia and endotoxemia. Name various organisms causing septicemia. How will you diagnose it in the laboratory?
2. Enumerate causative agents of infective endocarditis. How will you proceed to diagnose it in the laboratory?
3. Write short note on subacute endocarditis.

FURTHER READING

Deal WB. Fever of unknown origin: An analysis of 34 patients. PostgradMed J 1971;50:182.

Jacoby GA, Swartz MN. Fever of unexplained origin. New Engl J Med 1973;289:1407.

Larson EB, Featherstone HJ, Petersdorf RG. Fever of undetermined origin: Diagnosis and follow up of 105 cases. 1970-1980. Medicine (Baltimore) 1982;61:269.

Petersdorf RG, Beeson PB. Fever of unexplained origin. Report on 100 cases. Medicine (Baltimore) 1961;40:1.

Reese RE, Douglas RG (Eds). A Practical Approach to Infectious Diseases. Boston Toronto: Little, Brown & Co, 1986.

2. MENINGITIS

Meningitis is an inflammation of the membranes surrounding the brain and spinal cord. Most cases of meningitis fall into one of two categories: purulent meningitis and aseptic meningitis. The causative agents of these types are given in Table 78.4.

A. Purulent Meningitis (Acute Pyogenic Meningitis)

In purulent meningitis, the CSF is typically turbid due to the presence of large numbers of leukocytes, e.g. from 100 to several thousands/mm^3, most of which are polymorphs. The majority of cases are caused by one or other of three bacteria: meningococcus, pneumococcus and *Haemophilus influenzae*, which generally pass to the meninges from the respiratory tract via the bloodstream. In neonates and infants, coliform bacilli, group B streptococci and, less commonly, pseudomonads, salmonellae and *Listeria monocytogenes* may be the cause. Infections acquired through a carelessly performed lumbar puncture, an accidental wound or an infected neurosurgical wound may be due to pyogenic staphylococci or streptococci, coliform bacilli, anaerobic cocci or bacteroides. In patients with CSF-venous shunts, infection may be caused by *Staphylococcus epidermidis* (Table 78.4).

Laboratory Diagnosis

A. **Collection of Specimens**

   The principal specimen to be examined is of CSF collected by lumbar puncture under strict aseptic conditions to prevent the introduction of infection. Only 3-5 ml of fluid should be collected in a fresh sterile screw-capped containers. The specimen must be dispatched to the laboratory as quickly as possible, for delay may result in the death of delicate pathogens, such as meningococci, and the disintegration of leukocytes. It should not be kept in a refrigerator, which tends to kill *H. influenzae*. If delay for a few hours is unavoidable, the specimen is best kept in an incubator at 37°C.

B. **Laboratory Examination of CSF for Cells and Microorganisms**

1. Naked Eye Examination

Naked eye examination is done for the presence of turbidity and any sign of contamination with blood from the puncture wound.

Normal CSF is clear and colorless like water. A yellow color (xanthochromic) may result from a previous cerebral hemorrhage.

The specimen should then be examined by cell count, Gram film, culture and, if facilities are available, for its glucose and protein contents and the presence of hemophilus, meningococcal or pneumococcal antigens.

2. Cell Count

The leukocytes in the CSF are counted by microscopical observation of well-mixed, uncentrifuged fluid in a slide counting chamber. The relative numbers of polymorphs and lymphocytes should be noted, and the number of erythrocytes in specimens contaminated with blood (Table 78.5).

A wet film of centrifuged CSF deposit mixed with India ink will, when examined under oil-immersion, demonstrate the characteristic capsulate yeast cells of

Table 78.4: Causative agents of purulent and aseptic meningitis

Purulent meningitis	*Aseptic meningitis*
• *Neisseria meningitidis*	**A. Viruses**
• *Streptococcus pneumoniae*	• Enteroviruses (echoviruses, polioviruses, coxsackieviruses)
• *Haemophilus influenzae*	• Mumps
• *Escherichia coli*	• Herpes simplex
• *Group B streptococci*	• Varicella-zoster
• *Pseudomonas*	• Measles
• *Salmonellae*	• Adenoviruses
• *Staphylococcus aureus*	• Arboviruses
• *S. epidermidis*	**B. Spiral Bacteria**
• *Listeria monocytogenes*	• Syphilis (*Treponema pallidum*)
• *Klebsiella*	*Leptospira interrogans serovars. icterohaemorrhagiae* and *canicola*
• Anaerobic cocci	**C. Other bacteria**
• Bacteroides	Tuberculosis *(Mycobacterium tuberculosis)*
In neonates and infants	Partially treated with antibiotics
Esch. coli	Brain abscess
Group B streptococci	**D. Fungi**
Pseudomonads	*Cryptococcus neoformans*
salmonellae	*Candida albicans*
Staph. aureus	**E. Protozoa**
H. influenzae	• *Acanthamoeba*
Listeria monocytogenes	• *Naegleria*
Streptococcus pneumoniae	• *Toxoplasma gondii*
Klebsiella sp.	**F. Noninfective**
	Lymphoma
	Leukemias
	Metastatic and primary neoplasms
	Collagen-vascular disease

Cryptococcus neoformans and a wet film examined on a warm stage will show the slowly motile trophozoites of *Acanthamoeba (Hartmannella)* or *Naegleria*.

Dilution

CSF that is clear or only slightly turbid should be examined undiluted but when the specimen is highly turbid and its cell count very high, it may be necessary to dilute it 1 in 10 or 1 in 100 before examination. When separate counts are to be made of the leukocytes and erythrocytes, 0.85 percent NaCl solution should be used as diluent. If, however, the presence of large numbers of erythrocytes makes the recognition and counting of the leucocytes difficult, the dilution should be done with a counting fluid which lyses the erythrocytes and stains the nuclei of the leucocytes. A suitable fluid contains acetic acid and crystal violet.

Differential Leukocyte Count

If there is any difficulty in differentiating polymorphs and lymphocytes in the counting chamber, to assess the relative numbers of the two types of leukocytes.

3. Gram Film of CSF

After taking some CSF for the cell count, the remainder should be centrifuged to deposit any cells and bacteria and a film of the deposit should be stained by Gram's method. The finding of bacterial forms resembling meningococci, pneumococci, haemophili, coliform bacilli, streptococci or listeriae should at once be reported to the physician, for different antibiotics are preferred for treatment of the different infections:

The supernatant from the centrifuged CSF should be tested for its content of glucose and protein.

4. Culture

i. CSF Culture

Immediately after centrifugation of the CSF and the removal of some of the deposit for the Gram film, the remainder of the deposit should be seeded heavily on to culture media, blood agar and chocolate agar for incubation in humid air plus 5 to 10 percent, CO_2 and a tube of cooked-meat broth. A further blood agar plate should be seeded for incubation for 2 to 5 days in an anaerobic atmosphere with 5 to 10 percent CO_2.

When organisms are sufficiently numerous to be seen in the film, another blood agar plate should be seeded confluently and antibiotic disks applied, including disks with benzylpenicillin and chloramphenicol, so that sensitivity results may be obtained with minimal delay.

The cultures should be inspected after overnight incubation. The isolated organisms are identified by colony morphology, Gram staining from colonies, biochemical reactions and/or serological tests. Tests with appropriate antibiotics should be done on the isolate.

If no growth is apparent after overnight incubation, the plates should at once be reincubated for another day and then again inspected for growth.

If the plate cultures remain free from growth, and turbidity develops in the cooked-meat broth, the broth should be filmed and subcultured on to blood agar and heated-blood agar plates, incubated aerobically and anaerobically.

ii. Blood Culture

Blood culture is particularly useful in meningitis due to *N. meningitidis, H. influenzae* and *Str. pneumoniae*. About 50 percent of these cases have positive blood culture.

5. Biochemical Tests

The supernatant from the centrifuged CSF should be tested for its content of glucose and protein (Table 78.5).

6. Antigen Detection

The supernatant part of CSF contains antigen, which may be demonstrated by the performance of a latex agglutination test or coagglutination or counter-immunoelectrophoresis (CIEP) test. These tests are used for rapid diagnosis of meningitis and are particularly useful in

Table 78.5: Typical CSF findings in different types of meningitis					
		CSF in			
Characteristic	Noraml CSF	Acute pyogenic meningitis	Tuberculous meningitis	Viral meningitis
I. Pressure	Normal	Highly increased	Moderately increased	Slightly increased
II. Direct examination				
A. Cell count				
1. Total cell (per cu.mm	1-3	1,000-20,000	50-500	10-500
2. Predominant	Lymphocytes	Neutrophils (90-95%)	Lymphocytes (90%)	Lymphocytes
B. Biochemical analysis				
1. Total proteins (mg%)	30-45	100-600 (Highly increased)	80-120 (moderately increased)	60-80 (Slightly increased)
2. Sugars (mg%)	40-80	Diminished or absent (10-20)	Diminished (30-50)	Normal
III. Bacteriological examination				
A. Microsocopy				
1. Gram staining	Nil	Gram-negative cocci, Gram-positive cocci, Gram-negative bacilli or gram-positive bacilli may be found depending upon the causative agent responsible.	–	–
2. Ziehl-Neelsen staining	Nil	Nil	Acid-fast bacilli (AFB) may be found	–
B. Culture	Nil	According to the causative agent, specific organism may grow on appropriate media.	*M. tuberculosis* may grow on LJ media,	Viruses may be grown on cell cultures

partially treated patients in whom smear and culture may be negative. These are available for *N. meningitidis, Str. pneumoniae, H. influenzae* type b and Group B *Streptococcus.*

7. Agglutination

The isolated organisms may be grouped by agglutination with appropriate antisera.

8. Demonstration of Bacterial Endotoxin

This is especially useful when a patient has been partially treated and culture shows no growth. Bacterial endotoxin in blood can be detected by the limulus lysate test. The principle of the test is that extract prepared from the amoebocytes (blood cells) of the horse shoe crab *(Limulus polyphemus)* is coagulated when mixed with blood containing endotoxin. It is extremely sensitive test to detect bacterial endotoxin.

For detail identification of different organisms, refer to the respective chapters.

B. Aseptic Meningitis

In aseptic meningitis, the CSF is clear or only slightly turbid and contains only moderate numbers of leucocytes, e.g. 10-500/mm^3, most of which are lymphocytes, except in the earliest stage. The great majority of cases are due to viruses (viral meningitis), particularly enteroviruses of the echo, coxsackie and polio groups. Mumps virus is a moderately common cause and a few cases are due to herpes simplex, varicella-zoster, measles and adenoviruses. Arboviruses cause cases in countries where these viruses are common.

A few cases with CSF findings resembling those of viral meningitis are caused by leptospires (serovars canicola and icterohaemorrhagiae), fungi *(Cryptococcus neoformans* or *Candida albicans*) and amoebae *(Naegleria* or *Hartmannella)* and an underlying viral encephalitis may give a moderate lymphocytic exudate in the CSF.

It should also be noted that when antibiotic therapy is started at an early stage in a bacterial meningitis, the CSF findings may be like those of aseptic meningitis (Table 78.4, 78.5).

Laboratory Diagnosis

CSF is used for: Cell count and biochemical tests, microscopy, culture and other tests according to suspected causative agents (viruses, fungi or protozoa).

Tuberculous Meningitis

The CSF findings often resemble those of aseptic meningitis, but the cell count is usually slightly higher, e.g. 100 to 500 leukocytes/mm^3, mostly lymphocytes, and a veil clot (fibrin web) often develops when the CSF is allowed to stand undisturbed (Table 78.5).

Laboratory Diagnosis

1. **Specimen:** CSF is collected by lumbar puncture in a sterile container under aseptic conditions. When CSF is allowed to stand, a fibrin web (cobweb) often develops. Cell count and biochemical analysis can be done as described earlier.
2. **Microscopy:** The centrifuged deposit of the CSF should be examined in an auramine or Ziehl-Neelsen stained film for acid-fast bacilli.
3. **Culture:** Centrifuged deposit of CSF is inoculated on Lowenstein-Jensen (LJ) media and incubated at 37°C for 6-8 weeks. Identification of *M. tuberculosis* depends on colony morphology, ZN staining from colonies and biochemical reactions.

If facilities are available, some of the CSF should be inoculated into a guinea pig. As in purulent meningitis, the glucose content of the CSF is reduced and the protein content increased.

KEY POINTS

- Meningitis is an inflammation of the membranes surrounding the brain and spinal cord.
- Most cases of meningitis fall into one of two categories: purulent meningitis and aseptic meningitis.
- Acute pyogenic or purulent meningitis is characterized by the presence of acute inflammatory exudates with a large number of polymorphonuclear cells in the CSF. This condition is caused mostly by bacteria. The majority of cases are caused by one or other of three bacteria: meningococcus, pneumococcus and *Haemophilus influenzae*, which generally pass to the meninges from the respiratory tract via the bloodstream.
- Aseptic meningitis is characterized by the presence of increased lymphocytes and other mononuclear cells in the CSF, but absence of either bacteria or fungi by culture. The great majority of cases are due to viruses (viral meningitis), a few cases are caused by leptospires, fungi and amoebae.
- **Tuberculous meningitis:** The CSF findings often resemble those of aseptic meningitis, but the cell count is usually slightly higher, mostly lymphocytes, and a veil clot (fibrin web) often develops when the CSF is allowed to stand undisturbed.
- **Laboratory examination:** CSF is the specimen of choice. **Direct microscopy by Gram staining, antigen detection** and **culture** are key methods in laboratory diagnosis of acute pyogenic meningitis.
- When a tuberculous infection is suspected, the CSF should be examined in an auramine or Ziehl-Neelsen stained film for acid-fast bacilli and cultured Lowenstein-Jensen medium.

IMPORTANT QUESTIONS

1. Name various organisms causing meningitis. Discuss the laboratory diagnosis of acute pyogenic meningitis.
2. Write short notes on:
   Aseptic meningitis
   Tuberculous meningitis

FURTHER READING

Schneider Schaulies S, Dunster LM, ter Meulen V. Infections of central nervous system. In: Mahy BWJ, Collier L Arnold (Eds). Topley & Wilsons Microbiology and Microbial Infections, 9th edn. London, Sydney, Auckland 1998;835-72.

3. URINARY TRACT INFECTIONS

Urinary tract infection (UTI) may be defined as the presence of bacteria undergoing multiplication in urine within the urinary drainage system. Urinary tract infection is the second most common infection after respiratory tract for bacterial infection. Microorganisms causing UTI are shown in Table 78.6.

Types of UTI

Acute infections of UTI can be subdivided into two anatomic categories:

1. Lower UTI
   i. Urethritis
   ii. Cystitis
   iii. Prostatitis

   Lower UTI is due to ascending infection.
2. Upper UTI
   i. Acute pyelitis—infection of pelvis of kidney.
   ii. Acute pyelonephritis—infection of parenchyma of kidney.

Pyelonephritis is probably due to hematogenous infection.

Predisposing Factors

1. **Gender and sexual activity:** The females are more frequently affected by UTI because of its proximity to the anus, its short length (about 4 cm), and its termination beneath the labia.

   Sexual intercourse causes the introduction of bacteria into the bladder and is temporally associated with the onset of cystitis; it thus appears to be important in the pathogenesis of UTIs in younger women.

Table 78.6: Causes of urinary tract infection

A. **Gram-negative bacilli**
Gram-negative bacilli are by far the most common infecting agents.
1. *Escherichia coli* causes approximately 80% of acute infections in patients without catheters
2. *Proteus mirabilis*
3. *Klebsiella*
4. *Enterobacter* (occasionally)
5. *Serratia*
6. *Pseudomonas aeruginosa*

B. **Gram-positive cocci—lesser role in UTIs**
1. *Staphylococcus saprophyticus (10-15%)*
2. *S. epidermidis (1-5%)*
3. *S. aureus (1-5%)*
4. *Enterococcus* sp. *(15%)*

C. **Other organisms which may occasionally cause UTI**
1. *Mycobacterium tuberculosis*
2. *Enterobacter*
3. *Citrobacter*
4. *Salmonellae*
5. *Streptococcus pyogenes*
6. *S. agalactiae*
7. *Gardnerella vaginalis*

D. **Fungus**
Candida albicans may cause UTI in diabetics and immunocompromised patients.

2. **Pregnancy:** Pregnancy UTIs are detected in 2 to 8 percent of pregnant women.
3. **Obstruction:** Any impediment to the free flow of urine-tumor, stricture, stone, or prostatic hypertrophy—results in hydronephrosis and a greatly increased frequency of UTI.
4. **Neurogenic bladder dysfunction:** Interference with the nerve supply to the bladder, as in spinal cord injury, tabes dorsalis, multiple sclerosis, diabetes, and other diseases, may be associated with UTI.
5. **Vesicoureteral reflux:** Defined as reflux of urine from the bladder cavity up into the ureters and sometimes into the renal pelvis, occurs during voiding or with elevation of pressure in the bladder.
6. **Bacterial virulence factors:** Bacterial virulence factors markedly influence the likelihood that a given strain, once introduced into the bladder, will cause UTI.
7. **Genetic factors:** The number and type of receptors on uroepithelial cells to which bacteria may attach are at least in part genetically determined.

Clinical Features

1. **Asymptomatic bacteriuria:** Symptomless urinary tract infection is not uncommon and can be detected only by urine culture.
2. **Symptomatic:** The common symptoms of urinary tract infection are urgency and frequency of micturition, with associated discomfort or pain. The commonest condition is cystitis, due to infection of the bladder with a uropathogenic bacterium.

Laboratory Diagnosis

1. Specimen Collection

i. **Midstream specimen of urine (MSU):** Specimens of urine are generally collected in plastic universal containers, but midstream specimens from females are more conveniently collected in a widemouthed container such as a 12 oz (350 ml) glass jar or a sterile waxed cardboard container.
In male: From male patients, a midstream specimen of urine (MSU, the middle of the urine flow) is collected. Before collecting a sample, retract the prepuce, clean it with sterile normal saline and collect midstream specimen.
In female: In case of female, anogenital toilet is more important. Remove underclothing completely, sit comfortably on the seat and swing one leg to the side as far off as possible. Wash perineum and periurethral area with soap and water. Then clean with nonirritant antiseptic such as chlorhexidine. Separate apart labia with fingers of one hand and collect midstream urine.

ii. **Catheter specimen of urine (CSU):** Catheter specimen of urine (CSU) was commonly collected in the past from females, but catheterization for this purpose is no longer considered justifiable because it carries a 2 to 6 percent risk of introducing and initiating infection. A CSU is nowadays taken only if there are special indications for its requirement or in the course of a cystoscopic investigation.

iii. **Suprapubic stab:** Urine may be aspirated from the bladder into a syringe with a needle introduced aseptically through the skin and abdominal wall just above the pubis (suprapubic stab) from children and young infants. This procedure may also be used in adult women when uncontaminated specimen cannot be obtained by other methods.

iv. **Noninvasive method:** A noninvasive method of stimulating urine flow in a baby is by tapping just above the pubis with two fingers at 1 hour after a feed: one tap/second is given for 1 minute, an interval of 1 min is allowed, then tapping is resumed in this cycle.

v. **Early morning urine (EMU):** If tuberculosis of the urinary tract is suspected, the first urine passed in the day (early morning urine; EMU) is the most suitable specimen.

vi. **Initial flow of urine:** In the investigation of urethritis and prostatitis, the initial flow of urine, rather than a midstream specimen, should be examined.

2. Transport of Specimen

Once collected, a specimen of urine must be transported to the laboratory without delay, for urine is an excellent culture medium and contaminating bacteria can readily multiply to reach apparently significant numbers. If a delay of more than 1 to 2 hours is unavoidable, the multiplication of bacteria in the urine should be prevented by storage in a refrigerator at 4°C, or by transport in some form of refrigerated container, or by collection and transport in a container with boric acid at a final, bacteriostatic concentration of 1.8 percent.

3. Microscopy

Part of the specimen is used for bacteriological culture and the rest is examined immediately under the microscope.

The deposit of the centrifuged urine can be examined under microscope to find out the presence of pus cells, red blood cells and bacteria in it. Presence of more than 3 pus cells per high power field is suggestive of infection.

Some polymorphs are usually present in the urine of healthy, uninfected persons. The normal excretion of leucocytes in the urine varies from very few to up to about 10^6/24 hour and at times up to a few thousand may be present per ml of uncentrifuged urine. Generally, it is accepted that the leucocytes should be found in numbers at least as great as 10^4/ml before the presence of pyuria is accepted.

In the past, the microscopical examination was commonly done on a wet film or Gram-stained film of deposit centrifuged from the urine. Nowadays centrifugation is not recommended. Examination of uncentrifuged urine is more reliable.

Wet film examination: A leukocyte count sufficiently accurate for general purposes may be obtained from examination of a wet film of uncentrifuged urine. Under these conditions the finding of 1 leukocyte per 7 high power fields corresponds with 10^4 leukocytes per ml and the larger numbers than this indicates significant pyuria.

4. Culture

A. **Quantitative:** It is too laborious for routine use and not used routinely. The methods used are:
   i. Pour plate method
   ii. Surface viable count.
B. **Semiquantitative methods:** These are quicker methods.

i. Standard Loop Method

Measured quantity of urine (0.004 ml or 1/250 ml) with the help of standardized loop (internal diameter 3.26 mm) of nichrome or platinum wire of SWG 28 is inoculated on blood agar and MacConkey media and incubated overnight at 37°C for 24 hours. The number of colonies is counted and multiplied by 250 to get the bacterial count per ml. On the basis of this result, it can be reported whether the patient has significant bacteriuria or not. The identification of the isolate is carried out by cultural characteristics and biochemical reactions. Antibiotic sensitivity is then determined and treatment given accordingly.

Interpretation of Results

Kass (1957) gave a criterion of active bacterial infection of urinary tract as follows:

a. Count more than 10^5 bacteria of single species per ml: significant bacteriuria which indicates active UTI (Significant bacteria).
b. Count between 10^4 to 10^5 bacteria/ml is of doubtful significance and specimen should be repeated for culture.
c. Less than 10^4 organisms/ml and usually less than 10^3/ml accounts contamination.

ii. Filter paper Method

This method of semiquantitative culture is rapid and very economical in the use of culture medium, but growths are often confluent and, if mixed, require to be plated out to obtain pure subcultures for identifying and sensitivity tests.

iii. Dip-slide Method

Commercially prepared dip-slides or spoons can be bought. The dip-slide is small plastic tray carrying a layer of an appropriate agar culture medium. Opposite sides of the tray may carry different media, e.g. CLED agar medium on one side and MacConkey, brain heart infusion or *Pseudomonas* selective agar on the other. A midstream specimen of urine is collected in a clean container. The cap of the dip-slide container is unscrewed and held while the dip-slide is withdrawn from the container and briefly immersed in the urine. It is then incubated at 37°C overnight and examined for a growth of colonies. This method is relatively expensive.

5. Screening Techniques

Because urinary tract infection is such a common problem and bacteriological facilities are not always available, several screening techniques have been introduced for the presumptive diagnosis of significant bacteriuria. These include the following:

i. **Griess nitrite test:** It is based on nitrate-reducing enzymes produced by bacteria present in urine. The presence of nitrite, detectable by a simple test, indicates the presence of nitrate reducing bacteria in urine. Normal urine does not contain nitrite.
ii. **Catalase test:** The presence of catalase as evidenced by frothing on addition of hydrogen peroxide indicates bacteriuria, though a positive result is obtained also in hematuria.
iii. **Triphenyltetrazolium chloride (TIC) test:** It is based on the production of a pink-red precipitate in the reagent caused by the respiratory activity of growing bacteria.
iv. **Glucose test paper:** It is based on the utilization of the minute amounts of glucose present in normal urine, by bacteria causing the infection. Hence, it indicates bacteriuria.

v. **Polymorphonuclear neutrophils (PMNs):** PMNs are counted in uncentrifuged urine specimen with help of hemacytometer. 8 PMN/mm^3 is indicative of infection.
vi. **Leukocyte esterase:** Presence of this enzyme in urine is indication of bacteriuria.
vii. **Gram staining:** Presence of at least one bacteria per oil immersion field (examining 20 fields) correlates with significant bacteriuria (>10^5 bacteria/ml).
viii. **Dip slide culture methods:** Agar coated slides are immersed in urine or even exposed to the stream of urine during voiding, incubated and the growth estimated by colony counting or by color change of indicators.

None of the screening methods is as sensitive or reliable as a culture.

6. Identification and Sensitivity Tests

If similar colonies are found in numbers suggesting significant bacteriuria, a separate colony or a portion of apparently pure growth should be subcultured for identification and testing of its sensitivity to antibiotics.

7. Differentiation of Upper UTI and Lower UTI

The antibody coated bacteria test has been employed for the localization of the site of urinary infection. This is based on the assumption that bacteria coated with specific antibodies are present in the urine only when the kidneys are infected (upper UTI) and not when the infection is confined to the bladder (lower UTI). Antibody coated bacteria are detected by immunofluorescence using fluorescent tagged antihuman globulin or by staphylococcal coagglutination.

Tuberculosis of Kidney and Urinary Tract

Tuberculosis of kidney is a blood-borne infection. The patient presents with frequency and painless hematuria and routine urine culture does not show any pathogen. Tuberculosis must be considered in cases where pyuria is present without bacteriuria.

Laboratory Diagnosis

1. Specimen

Midstream urine specimen is not useful because excretion of *M. tuberculosis* from kidney is intermittent, hence, Early morning urine specimens should be collected in sterile container on three consecutive days.

2. Direct Ziehl-Neelsen Staining

Smear made from centrifuged deposit of urine is stained with Ziehl-Neelsen staining and may show acid-fast bacilli. Saprophytic mycobacteria (e.g. *M. smegmatis*) may be present in normal urine and may be confused with tubercle bacilli which may be excluded by using acid-alcohol as decolorizing agent in staining procedure. *M. tuberculosis* is acid-alcohol-fast while *M. smegmatis* is only acid-fast.

3. Culture

Culture is performed on Lowenstein-Jensen medium and incubated for 6-8 weeks. Growth is identified by Ziehl-Neelsen staining and biochemical tests.

KNOW MORE

Patients with signs or symptoms of urinary tract infection sometimes produce samples of urine that show pus cells but do not yield a significant growth of bacteria on routine culture. The explanation may be:

i. That the patient has been taking antibiotics prescribed on a previous occasion.
ii. Infection with an organism that does not grow on the media normally used for routine investigations, genitourinary tuberculosis or gonococcal infection, and infection with nutritionally exacting or anaerobic bacteria.

Nonbacterial urethritis or cystitis, or the urethral syndrome, the cause of which may be urethral or bladder infection with a chlamydia, ureaplasma, trichomonas or virus, which often remains unrecognized.

☞ KEY POINTS

- Urinary tract infection (UTI) may be defined as the presence of bacteria undergoing multiplication in urine within the urinary drainage system.
- Lower UTI include: urethritis, cystitis and prostatitis and is due to ascending infection.
- Upper UTI includes acute pyelitis and acute pyelonephritis.
- Laboratory diagnosis: For specimen collection, various methods used are: (1) Midstream Specimen of Urine (MSU); (2) Catheter specimen of urine; (3) Sup-rapubic stab; (4) Noninvasive method; (5)Early morning urine (EMU); (6) Initial flow of urine.
- Part of the specimen is used for bacteriological culture and the rest is examined immediately under the microscope.
- Culture: Standard loop technique is used for culture.
- Standard loop method: An inoculating loop of standard dimensions is used to take up a small, approximately fixed and known volume of mixed uncentrifuged urine and spread it over a plate of agar culture medium.
- Count more than 10^5 bacteria of single species per ml is called significant bacteriuria. It indicates active UTI.

IMPORTANT QUESTIONS

Name various organisms causing urinary tract infection. Discuss the laboratory diagnosis of this condition.

FURTHER READING

Clarridge JP, Pezzlo M, Vosti KL, Cumitech A. In: Wessfield AL, (Ed). Laboratory diagnosis of urinary tract infections, Washington DC American Society for Microbiology, 1987.

Pezzlo M. Urine and culture procedure. In: Isenberg HD, et al. (Eds) Clinical microbiology procedures handbook, Washington, DC American Society for Microbiology, 1992.

Murray PR, et al (Eds). Manual of clinical microbiology, 8th edn, Washington, DC ASM Press, 2003.

Winn W. Diagnosis of urinary tract infection: a modern procrustean bed. Am J Clin Pathol 1993;99:117.

4. SORE THROAT AND PNEUMONIA

Sore throat is essentially an acute tonsillitis or pharyngitis. It is characterized by redness and edema of mucosa, exudation of tonsils, pseudomembrane formation, edema of uvula, gray coating of tongue and enlargement of cervical lymph nodes. Causative agents of sore throat are given in Table 78.7.

Pseudomembrane Formation

Corynebacterium diphtheriae, *Candida albicans*, β-hemolytic group A Streptococcus, *Treponema vincentii* and *Leptotrichia buccalis* may lead to pseudomembrane formation.

Laboratory Diagnosis

The signs and symptoms of sore throat (pharyngitis) caused by streptococci and viruses are similar. These include pain on swallowing, congested tonsils and pharynx; enlarged lymph nodes and pyrexia. Viral pharyngitis or other causes of pharyngitis/tonsillitis must be differentiated from that caused by *Streptococcus pyogenes* since it is treatable with penicillin whereas viral infections are not. If untreated, streptococcal tonsillitis may give rise to complications such as peritonsillar abscess, sinusitis or immune complex diseases (rheumatic fever, glomerulonephritis).

A. Specimen

For bacteriological sampling, a plain, albumen-coated or charcoal-coated cotton-wool swab should be used to collect as much exudate as possible from the tonsils, posterior pharyngeal wall and any other area that is inflamed or bears exudate. Two throat swabs are collected. If it cannot be delivered to the laboratory within about 1 hour, it should be placed in a refrigerator at 4°C until delivery or, preferably, it should be submitted in a tube of transport medium for bacteriological specimens.

B. Direct Microscopy

From one swab, make two smears for Gram staining and Albert staining.

1. **Gram Staining:** It is not helpful unless Vincent's organisms or *Candida albicans* are suspected. Vincent's infection shows gram-negative spirochetes *(Borrelia vincentii)* and gram-negative fusiform bacilli *(Fusobacterium* sp.*)*. When *Candida albicans* is suspected, it appears as gram-positive oval budding yeast cells.
2. **Albert Staining:** It shows green colored, V and L shaped (Chinese letter pattern) bacilli with bluish-black metachromatic granules in infection due to *C. diphtheriae*.

C. Culture

Culture media are selected according to the organism suspected to be the causative agent of sore throat. Following media may be used for culture:

Blood agar: All the organisms will grow on this medium:

Crystal violet blood agar: It is selective for *Str. pyogenes* especially when incubated anaerobically.

Loeffler's serum slope: For isolation of *C. diphtheriae*, grow very rapidly (in 6-8 hours).

Potassium tellurite blood agar: Selective media for isolation of *C. diphtheriae*.

Sabouraud's dextrose agar (SDA): When suspecting *Candida albicans*, SDA should be included.

These culture media should be incubated at 37°C for overnight. In case of potassium tellurite blood agar, it should be incubated for 48 hours. A subculture should be made from Loeffler's serum slope onto potassium tellurite blood agar after 6-8 hours which is then incubated at 37°C for 48 hours.

D. Identification

1. Str. pyogenes

i. Morphology: Colonies are small (pin-point), circular, semitransparent, low convex discs having β-hemolysis.
ii. Gram staining: Small gram-positive spherical cocci occurring in chains is characteristic of *Str. pyogenes*
iii. Other tests for confirmation:
   - For β-hemolytic streptococci:
   - Bacitracin sensitivity test—for *Str. pyogenes*
   - Lancefield grouping—for all β-hemolytic streptococci.

2. C. diphtheriae

i. Morphology: On potassium tellurite blood agar, gray or black colored round colonies are seen.
ii. Gram staining: *C. diphtheriae* is seen as gram-positive bacilli.
iii. Albert staining: *C. diphtheriae* shows green coloured V or L shaped bacilli with bluish-black metachromatic granules.
iv. Other tests for confirmation:
   a. Biochemical tests.
   b. Toxigenicity tests: Elek's gel precipitation test; animal inoculation test.

Table 78.7: Causative agents of sore throat

A. Bacteria
- Streptococcus β-hemolytic group A and occasionally groups C and G
- *Corynebacterium diphtheriae*
- *Haemophilus influenzae*
- *Bordetella pertussis*
- *Neisseria gonorrhoeae*
- *Treponema vincentii*
- *Leptotrichia buccalis*

B. Fungi
- *Candida albicans*

C. Viruses
- Epstein-Barr virus
- Adenoviruses
- Coxsackievirus A

3. Candida albicans

i. Morphology: White or cream colored colonies may be seen.
ii. Gram staining: Candida albicans reveals gram positive budding yeast cells.
iii. Other tests for confirmation
   a. Germ tube test
   b. Carbohydrate fermentation and assimilation tests.

4. For other Causative Agents

Special culture media and different biochemical reactions or serological tests may be required as described in respective chapters. For details of tests mentioned above, refer to the corresponding chapters.

E. Antibiotic Sensitivity

All β-hemolytic group A streptococci are sensitive to penicillin G, and most are sensitive to erythromycin. *C. diphtheriae* is sensitive to penicillin.

PNEUMONIA

Pneumonia may be defined as inflammation and consolidation of the lung substance. Bacterial causes for pneumonia are listed in Table 78.8.

Classification

A. Community-acquired pneumonia.
B. Hospital-acquired pneumonia.
C. Pneumonia in immunocompromised patients.

A. Community-acquired Pneumonia

Community-acquired pneumonia has been thought to present as either of two syndrome—Typical or atypical.

Typical Pneumonia Syndrome

Typical pneumonia syndrome is usually caused by the most bacterial pathogen in community-acquired pneumonia, *Streptococcus pneumoniae,* but can also be due to other bacterial pathogens.

Atypical Pneumonia Syndrome

Historically, the term 'atypical pneumonia' was used for a primary pneumonia which was not demonstrably due to an accepted pathogen such as the pneumococcus. It is now recognized that a range of agents cause the syndrome and that these differ in various age and occupational groups. Atypical pneumonia is classically produced by *Mycoplasma pneumoniae* but can also be caused by *Legionella pneumophila, Chlamydia pneumoniae,* oral anaerobe, *P. carinii* and less frequently encountered pathogens. There is patchy consolidation of lungs.

1. **Lobar pneumonia:** It is an acute inflammation characterized by homogenous consolidation of one or more lobes. It is caused by *Streptococcus pneumoniae.*
2. **Bronchopneumonia:** It is almost always a secondary infection and generally follows viral infections of the respiratory tract. It is an acute inflammation of bronchi and the consolidation is scattered. It is caused by *Streptococcus pneumoniae Haemophilus influenzae* (rarely *Kleb. pneumoniae, Staph. aureus*).

Respiratory syncytial virus, *Mycoplasma pneumoniae, Chlamydia pneumoniae* and *Bordetella pertussis* cause bronchitis and bronchiolitis.

B. Hospital-acquired Pneumonia

Hospital-acquired or nosocomial pneumonia refers to a new episode of pneumonia occurring at least 2 days after admission to hospital.

Table 78.8: Microbial pathogens that cause pneumonia
A. Community-acquired pneumonia
Common organisms
Streptococcus pneumoniae
Chlamydophila pneumoniae
Mycoplasma pneumoniae
Legionella pneumophila
Uncommon organisms
Haemophilus influenzae
Staphylococcus aureus
Chlamydia psittaci
Coxiella burnetii
Kleb. Pneumoniae
Actinomyces israelii
Primary viral pneumonia
Influenza, parainfluenza virus and measles
Respiratory syncytial virus in infancy
Varicella (chickenpox)
B. Hospital-acquired pneumonia
Gram-negative bacilli (60%)
Escherichia, Pseudomonas, Klebsiella species.
Gram-positive organisms (16%)
Staph. aureus,
Anaerobes
Legionella sp.
C. Pneumonia in immunocompromised patients
Pnemocystis carinii
Gram-negative bacteria: Ps. aeruginosa
Mycobacteruim tuberculosis
Streptococcus pneumoniae
Haemophilus influenzae
Candida albicans
Aspergillus fumigatus
Viruses: Cytomegalovirus

C. Pneumonia in Immunocompromised Patients

The common causative agents include *Pneumocystis carinii, Staph. aureus, Ps. aeruginosa,* viral infections (CMV and herpes) and *M. tuberculosis.*

Laboratory Diagnosis

A. **Specimens:** Sputum, blood, pleural fluid, blood for serological tests.

B. **Direct microscopy**
   1. **Gram staining:** A clue to the probable pathogen is given by adequate number of pus cells alongwith presence of predominant organisms.
   2. **Ziehl-Neelsen staining:** Presence of acid-fast bacilli (AFB) gives a presumptive diagnosis of tuberculosis.
   3. **Giemsa staining:** For detection of cysts and trophozoites of Pneumocystis carini, Giemsa stain of sputum is useful.

C. **Culture:** Specimens such as sputum, blood and pleural fluid can be cultured on blood agar and chocolate agar. For culture, purulent portion of sputum is best. Sputum may be homogenized if the sputum is too viscid.
   1. **Blood agar:** It is incubated aerobically at 37°C under 5-10 percent CO_2.
   2. **Chocolate agar:** Incubation is also done at 37°C with 5-10 percent CO_2.
   3. **Lowenstein-Jensen (LJ) medium:** Three specimens of sputum are collected on three successive days. Culture is done on LJ media which is then incubated at 37°C for 6-8 weeks.
   4. Selective media are required for culture of *L. pneumophila.*
   5. Isolation of chlamydia can be done on cell-lines.

D. **Detection of bacterial antigens:** Direct immunofluorescence test for detection of *L. pneumophila* antigens, direct immunofluorescence examination of sputum is done.

E. **Serology:** Demonstration of antibody in patient's serum is done for diagnosis of some causative organisms. Most often these are diagnosed by high titers in a single sample, but it is better to demonstrate a four fold rise in titer of antibody.
   1. *Mycoplasma pneumoniae*
      - Complement fixation test (CFT)
      - Cold agglutinin test.
   2. *Chlamydia pneumoniae*
      Microimmunofluorescence with TWAR antigen
      - Complement fixation test (CFT)
      - Immunofluorescent antibody test.
   3. *Coxiella burnetii*
      - Complement fixation test (CFT).

KNOW MORE

Throat swabs for virological examination are collected in the same way but the heads are broken off into small vials of virus transport medium with penicillin and gentamicin for virus examination or with ampicillin only for *Mycoplasma pnemoniae.*

KEY POINTS

- Sore throat is essentially an acute tonsillitis or pharyngitis. It is characterized by redness and edema of mucosa, exudation of tonsils, pseudomembrane formation, edema of uvula, gray coating of tongue and enlargement of cervical lymph nodes.
- Laboratory diagnosis of sore throat caused by bacteria depends on direct microscopy and culture. For bacteriological sampling, two throat swabs are collected.
- Pneumonia may be defined as inflammation and consolidation of the lung substance which may be classified as community-acquired pneumonia, hospital-acquired pneumonia and pneumonia in immunocompromised patients. Common organisms causing community-acquired pneumonia are *Streptococcus pneumoniae, Chlamydia pneumoniae, Mycoplasma pneumoniae* and *Legionella pneumophila.*
- Laboratory diagnosis of pneumonia depends on direct microscopy and culture.

IMPORTANT QUESTIONS

1. Name various organisms causing sore throat. How will you diagnose it in the laboratory?
2. Name various bacterial causes of pneumonia. Discuss the laboratory diagnosis of pneumococcal pneumonia.

FURTHER READING

Bisno AL. Acute pharyngitis. N Engl J Med 2001;344:205.

Bisno AL, et al. Practice guidelines for the diagnosis and management of group A streptococcal pharyngitis. Clin Infect Dis 2002;35:113.

Finegold SM. Aspiration pneumonia. Rev Infect Dis 1991;13:(Suppl 9):S737.

Gwaity JM. Acute community-acquired sinusitis. Clin Infect-Dis 1996;23:1209.

Mcintosh K. Community-acquired pneumonia in children. N Engl J Med 2002;346:429.

5. DIARRHEA AND DYSENTERY

A. Diarrhea

Diarrhea is defined as the passage of loose, liquid or watery stools which are usually passed more than three times a day. Patient has bowel movements in excess of normal. Infective diarrhea may be caused by viruses, bacteria, protozoa and fungi (Table 78.9).

B. Gastroenteritis

Gastroenteritis (*gastro-*,"stomach," *entero-*, "intestine", *-itis*, "inflammation") may be defined as inflammation of the mucous membranae of stomach and intestine resulting in frequent loose motions with or without mucous and with or without blood, pain in abdomen and with or without fever. It is often used as synonym for acute diarrhea, especially when associated with vomiting.

C. Dysentery

Dysentery is a disease marked by frequent watery stools, often with blood and mucus, and characterized clinically by cramping abdominal pain, tenesmus, fever, and dehydration.

For all practical purposes, the terms diarrhea, gastroenteritis and dysentery are collectively included as *diarrheal diseases*.

D. Traveller's Diarrhea

Persons from developed countries visiting endemic areas may acquire various exotic intestinal pathogens and cause diarrheal illness soon after the traveller has returned by air—a condition known as "Traveller's Diarrhea". Important pathogens include Enteotoxigenic *Esch. coli* (ETEC) and pa rasites such as *Giardia lamblia* and *Entamoeba histolytica*.

DIARRHEA

A. Types of Bacterial Diarrhea

Etiology of diarrheal diseases is shown in table Table 78.10. Bacterial diarrhoea may be divided into two groups, those caused by invasive bacterial pathogens and those caused by noninvasive pathogens (Table 78.10).

Laboratory Diagnosis

1. Collection of Specimens

In most cases the stool is sent for bacterial culture. Because there are many other potential pathogens, the laboratory must be informed which tests to perform.

2. Direct Microscopy

Microscopic examination of the stool may reveal white blood cells if the patient has an inflammatory diarrhea. The bacterial pathogen may be visible on direct microscopic examination of the stool. In general, a wet film of a concentrate of the feces should be examined for protozoa, protozoal cysts and helminthes ova, and a stained film for the oocysts of cryptosporidium.

3. Culture

Selective and differential culture media are commonly used to attempt to identify bacterial pathogens in stool. The differential aspect of the media often allows differentiation of bacterial species based on colony morphology; the differences in colony appearance are usually due to different biochemical characteristics of the organisms.

a. Vibrios

Culture: Selective media such as TCBS or bile salt agar are used. Culture plates are incubated at 37°C for 24-48 hours. *Vibrio parahaemolyticus* is a halophilic vibrio—in media containing sodium chloride.

Identification: Identification of isolates is done by colony morphology, biochemical reactions and slide agglutination test.

b. Esch. coli

(i) ETEC; (ii) EPEC; (iii) EIEC; (iv) EAEC

i. **Culture:** Cuture is done on blood agar and MacConkey's agar. These media are incubated at 37°C for 24 hours.

Table. 78.9: Causative agents of infective diarrhea

A. Bacteria	**C. Protozoa**
• *Vibrio cholerae*	• *Entamoeba histolytica*
• *V. parahaemolyticus*	• *Giardia lamblia*
• *Escherichia coli* (ETEC, EPEC)	• *Cryptosporidium parvum*
• *Salmonella Enteritidis*	• *Isospora belli*
• S. Typhimurium	**D. Cestodes**
• Other Salmonella sp.	• *Hymenolepis nana*
• *Campylobacter* sp.	**E. Nematodes**
• *Yersinia enterocolitica*	• *Trichuris trichiura*
• *Shigella* sp.	• *Strongyloides stercoralis*
• *Clostridium perfringens*	• *Ascaris lumbricoides*
• *C. difficile*	• *Hookworms*
• *Staphylococcus aureus*	**F. Trematodes**
• *Bacillus cereus*	*Schistosoma mansoni*
• *Aeromonas hydrophilia*	
• *Plesiomonas shigelloides*	
B. Viruses	
• Rotavirus	
• Astrovirus	
• Calicivirus	
• Norwalk virus	
• Adenovirus	

Table 78.10: Types of bacterial diarrhea

i. Invasive bacterial pathogens
Salmonella species
Shigella species
Enteroinvasive *Esch. coli* (EIEC)
Enterohemorrhagic *Esch. coli* (EHEC)
Vibrio parahaemolyticus
Campylobacter jejuni
Yersinia enterocolitica

ii. Noninvasive bacterial pathogens
These organisms cause gastroenteritis or food poisoning by production of toxin.
Enterotoxigenic *Esch. coli* (ETEC)
Enteropathogenic *Esch. coli* (EPEC)
Vibrio cholerae
Shigella dysentriae type 1
Staphylococcus aureus
Bacillus cereus
Clostridium perfringens
Clostridium difficile

ii. **Identification:** Identification of isolates is done by colony morphology, biochemical reactions, slide agglutination with antisera. For the identification of EIEC strains Sereny test is used. Another method for identification of these strains is invasion of cultured HeLa cells. Production of verocytotoxin (VT) is confirmed by testing the strains on Vero cells, in which they cause cytopathic effects.

c. Salmonellae

S. typhimurium, S. enteritidis, S. dublin, S. choleraesuis, S. heidelberg and *S. thompson.*

i. **Culture:** Enrichment medium and the selective media such as MacConkey's agar, deoxycholate agar (DCA) and XLD agar. Wilson and Blair's medium may also be used.

Selective media are incubated at 37°C for 24 hours. Subculture is made onto selective medium after 6 hours incubation of enrichment medium. On MacConkey's agar and DCA pale nonlactose fermenting colonies develop while black shiny colonies are seen on Wilson and Blair's medium. On XLD medium, colonies are red in color.

ii. **Identification:** Colony morphology and biochemical reactions. The species is identified by agglutination test with polyvalent antisera using that of H and O.

d. Shigellae

All four serogroups of shigellae (*Shigella dysenteriae, S. flexneri, S. boydii* and *S. sonnei*) cause bacillary dysentery.

i. **Culture:** Culture is done on MacConkey's agar and DCA medium. *S. sonnei* is a late lactose

ii. **Identification:** It depends on colony morphology, biochemical reactions and agglutination test by group specific polyvalent antisera.

e. Campylobacter jejuni

i. **Culture Feces is on** inoculated on a selective medium containing vancomycin and polymyxin and is incubated at 43°C under microaerophilic conditions. *C. jejuni* and *E. coli* selectively grow against other fecal bacteria.

ii. **Identification:** Identification is by a characteristic morphology and gram-negative curved rods exhibiting darting motility, It is oxidase positive.

Yersinia species grow well at cooler temperatures of 25°C. This may be used in the microbiology laboratory by plating and incubating specimens at this temperature to enhance recovery of this organism. The use of certain selective media, such as cefsulodin-irgasan-novobiocin (CIN), can allow ready isolation of these species. Recovery of the organism can be increased by placing fecal samples in isotonic saline and keeping them at 40ºC before inoculation onto the selective medium.

f. Yersinia enterocolitica

i. **Specimens:** Feces, blood

ii. **Culture:** It can be isolated from feces or blood cultures. It is nonlactose fermenter (NLF).

g. Clostridium perfringens

i. **Culture:** Feces and food specimens are inoculated on blood agar for culture and incubated anaerobically at 37°C for 24 hours.

ii. **Identification is done by** colonies are either beta-hemolytic or nonhemolytic. These are gram-positive bacillus with subterminal spore. Further identified by Nagler reaction and serotyped by agglutination test.

h. Clostridium difficile

i. **Culture:** Feces is cultured on selective media with subsequent toxigenicity test.

ii. **Demonstration of toxin:** Toxin-characteristic effects on HEp-2 and human diploid cell cultures, or by ELISA. Toxin is specifically neutralized by antiserum

i. Staphylococcus aureus

i. **Culture:** Vomit, feces, suspected food are used for culture on selective medium (mannitol salt agar) or on ordinary media.

ii. **Identification:** The isolates are by colony morphology, Gram staining, catalase test and coagulase test. Phage typing may be done for epidemiological purposes.

iii. **Demonstration of enterotoxin:** Reverse passive latex agglutination assay (RPLA) or ELISA.

j. Bacillus cereus

i. **Culture:** Specimens such as vomit, feces and suspected food are used for culture on ordinary media (nutrient agar or blood agar) and incubated at 37°C for 24 hours. A special mannitol-egg yolk-phenol red-polymyxin agar (MYPA) medium may also be used.

ii. **Identification:** Colonies have curled hair appearance. Gram staining shows gram-positive spore bearing bacilli.

k. Other Bacteria

Aeromonas hydrophila and Plesiomonas shigelloides have been reported to cause diarrheal diseases.

B. Viruses

i. **Specimen:** Feces
ii. **Electron microscopy:** Electron microscopy of feces may demonstrate virus particles as in rotavirus and Adenovirus.
iii. **Fluorescent antibody test and ELISA:** These tests can detect viral antigens in feces.

C. Protozoa

i. Microscopy

- Saline preparation and iodine mount: In saline preparation motility of trophozoites can be observed while cysts take up iodine and appear distinct in iodine mount. Cysts and motile trophozoites of *E. histolytica* can be observed in feces of amoebic dysentery. *Giardia lamblia* cysts in formed stools, or trophozoites in fresh diarrheal stools can be seen in giardiasis. Trophozoites of *Balantidium coli* are found in liquid stool of this parasitic infection. Rarely cysts may be seen in formed stools.
- Acid-fast staining: Feces smear shows acid-fast oocyst of *Cryptosporidium parvum*.

ii. Serological Tests

Indirect hemagglutination assay (IHA) and ELISA are used to detect antibody titer in sera of patients with amoebiasis.

D. Fungus

There have been reports of diarrhea associated with *Candida albicans*.

DYSENTERY

Dysentery is a disease marked by frequent watery stools, often with blood and mucus, and characterized clinically by cramping abdominal pain, tenesmus (painful straining when passing the stools), fever, and dehydration. Dysentery results from 'enteroinvasive' microorganisms (Table 78.11). that penetrate through the mucosa and cause inflammation of the intestinal wall. Microorganisms causing dysentery are listed as shown in Table 78.11. For labratory diagnosis see above under "Diarehea".

Table.78.11: Microorganisms causing dysentery

A. Bacteria

1. *Shigella* sp, *Shigella dysenteriae, S. flexneri, S. boydii, S. sonnei*
2. *Escherichia coli: Enteroinvasive Esch. coli (EIEC), Enteropathogenic Esch. coli (EPEC), Enterohemorrhagic Esch. coli (EHEC)*

B. Protozoa

1. *Entamoeba histolytica*
2. *Balantidium coli*

KEY POINTS

- Diarrhea is defined as the passage of loose, liquid or watery stools. These liquid stools are usually passed more than three times a day.
- Gastroenteritis may be defined as inflammation of the mucous membrane of stomach and intestine resulting in frequent loose motions with or without mucous and with or without blood, pain abdomen and with or without fever. It is often used as synonym for acute diarrhea, especially when associated with vomiting.
- Bacterial diarrhea may be divided into two groups, those caused by invasive bacterial pathogens and those caused by noninvasive pathogens. *Vibrio cholerae, Esch. coli, Salmonellae* are some important bacterial causes of diarrheal diseases. Rotavirus is the most important viral etiology of diarrhea.
- Dysentery is a disease marked by frequent watery stools, often with blood and mucus, and characterized clinically by cramping abdominal pain, tenesmus, fever, and dehydration.
- Dysentery results from 'enteroinvasive' microorganisms. *Shigella* sp. bacillary cause dysentery and *Entamoeba histolytica* cause amoebic dysentery.
- Laboratory diagnosis of diarrhea, and dysentery depends on isolation of organism from the relevant specimen.

IMPORTANT QUESTIONS

1. Enumerate the different causes of diarrhea. How will you diagnose a case of diarrhea in the laboratory.
2. Enumerate the etiological agents of dysentery. Discuss in detail the laboratory diagnosis of dysentery.
3. Write short notes on:
   a. Traveller's diarrhea
   b. Viral diarrhea

FURTHER READING

Gilligan PH. *Escherichia coli.* EAEC, EHEC, EIEC, ETEC. Clin Lab Med 1999;19(3):505.

Thielman NM, Guerrant RL. Clinical practice. Acute infectious diarrhea. N Engl J Med 2004;350:38.

6. FOOD POISONING

The term bacterial food poisoning is restricted to acute gastroenteritis due to the presence of bacteria, usually in large numbers, or their products in food.

It is of three types (Table 78.12).

A. **Infective type:** In this type, multiplication of bacteria occurs *in vivo* when infective doses of microorganisms are ingested with food. Incubation period is generally 8 to 24 hours. The typical example of this type of food poisoning is by *Salmonellae*.

B. **Toxic type:** In this type, the disease follows ingestion of food with preformed toxin. Incubation period is short (2 to 6 hours). Example is staphylococcal food poisoning.

C. **Infective-toxic type:** In this type, bacteria release the toxin in the bowel. The incubation period is 6 to 12 hours. The typical example is *C. perfringens* food poisoning.

For the laboratory diagnosis, refer to the corresponding chapters. It has also been described earlier under "Laboratory Diagnosis of Diarrhea".

Table 78.12: Causative agents of food poisoning

1. **Infective type**
   - *Salmonella typhimurium*
   - *S. Enteritidis*
   - *S. Heidelberg*
   - *S. Indiana*
   - *S. Newport*
   - *S. Dublin*
   - *Vibrio parahaemolyticus*
   - *Campylobacter jejuni*
2. **Toxic type**
   - *Staphylococcus aureus.* *Bacillus cereus*
   - *Clostridm botulinum*
3. **Infective–toxic type**
   Clostridium perfringens

KEY POINTS

- The term bacterial food poisoning is restricted to acute gastroenteritis due to the presence of bacteria, usually in large numbers, or their products in food. It is of three types: (A) Infective type, (B) Toxic type, and infective-toxic type.
- Laboratory diagnosis of food poisoning depends on isolation of organism from the relevant specimen.

IMPORTANT QUESTIONS

1. Name various organisms causing food poisoning. Describe briefly the laboratory diagnosis of this condition.
2. Write short note on:
   Food-borne botulism

FURTHER READING

Butzler JP. Campylobacter, from obscurity to celebrity, Clin Microbiol Infect 2004;10:868.

Centers for Disease Control and Prevention. Multistate outbreak of Salmonella serotype typhimurium infections associated with drinking unpasteurized milk-Illinois, Indiana, Ohio, and Tennessee, 2002-2003, MMWR 26:613, 2003.

7. SEXUALLY TRANSMITTED DISEASES (STDs)

The sexually transmitted diseases (STDs) are a group of communicable diseases which are transmitted predominantly or entirely by sexual contact. The causative organisms include a wide range of bacterial, viral protozoal and fungal agents STD may present as genital ulcers, genital discharge without any genital lesion or only as systemic manifestations (Table 78.13).

Table 78.14. shows classification of sexually transmitted disease agents.

Etiology

The causative agents of STDs are summarized in Table 78.13. STDs can be differentiated into two types: (a) those causing ulcerative lesions and (b) those causing nonulcerative lesions.

Laboratory Diagnosis

Laboratory diagnosis of these diseases have already been described in corresponding chapters. However, salient features for laboratory diagnosis of important STDs are mentioned here.

A. Syphilis

Syphilis is caused by *Treponema pallidum*.

1. Specimens

(i) Fluid from chancre; (ii) Scrapings from ulcerated secondary lesions; (iii) Blood for serology.

2. Microscopy

Dark ground microscopy or phase contrast microscopy is used for demonstration of *T. pallidum* in exudate. The direct fluorescent antibody test for *T. pallidum* (DF ATP) is a better and safer method for microscopic diagnosis.

3. Serological tests

i. Nonspecific tests
   a. VDRL test
   b. Rapid plasma reagin (RPR) test

Table 78.13: Organisms causing sexually transmitted diseases

STDs	*Organisms*
A. Painless genital ulcers	
• Syphilis	*Treponema pallidum*
• Lymphogranuloma venereum (LGV)	*Chlamydia trachomatis*
• Donovanosis	*Calymmatobacterium granulomatis*
B. Painful genital ulcers	
• Chancroid	*Haemophilus ducreyi*
• Herpes genitalis	Herpes simplex viruses (HSV) type 2 and 1
C. Urethral discharge	
• Gonorrhea	*Neisseria gonorrhoeae*
• Nongonococcal urethritis (NGU)	*Chlamydia trachomatis* (types D–K) *Ureaplasma urealyticum* *Mycoplasma genitalium* *M. hominis*
D. Vaginal discharge	
• Gonorrhea	*N. gonorrhoeae*
• NGU	*C. trachomatis* *M. hominis*
• Trichomoniasis	*Trichomonas vaginalis*
• Vaginitis	*Gardnerella vaginalis* Mobiluncus sp.
• Vulva-vaginal candidiasis	*Candida albicans*
E. Genital warts	Human papilloma viruses
F. No genital lesions but only systemic manifestations	*HIV-1* and *HIV-2* Hepatitis B virus (HBV) Hepatitis C virus (HCV)
G. Miscellaneous	Group B streptococci Molluscum contagiosum virus Cytomegalovirus *Phthirus pubis* *Sarcoptes scabei* *Shigella* sp. *Campylobacter* sp. *Giardia lamblia* *Entamoeba histolytica*

ii. Specific tests
   a. TPHA (*Treponema pallidum* hemagglutination assay).
   b. FTA-ABS (Fluorescent treponemal antibody absorption test).
   c. TPI (*Treponema pallidum* immobilization test) VDRL and TPHA are two most commonly used tests.

B. Lymphogranuloma Venereum (LGV)

It is caused by *C. trachomatis* serotypes L1, L2 and L3.

1. **Direct microscopy:** Smears of material aspirated from the bubos may show the elementary bodies (Miyagawa's granulocorpuscles). The sensitivity of microscopy is very low.
2. **Isolation:** Isolation of the chlamydia by intracerebral inoculation into mice and into yolk sac of eggs has been replaced by cell cultures.
3. **Serological tests:** LGV patients develop high titers of circulating antibodies, with titers of 1:64 or more in CF test and 1:512 or more in micro-IF. Serological diagnosis is therefore feasible.
4. **Frei Test:** It is a skin test using LGV antigen and shows delayed type of hypersensitivity.

Table 78.14: Classification of sexually transmitted disease agents
A. Bacterial agents
• *Neisseria gonorrhoeae*
• *Chlamydia trachomatis*
• *Treponema pallidum*
• *Haemophilus ducreyi*
• *Mycoplasma hominis*
• *Ureaplasma urealyticum*
• *Calymmatobacterium granulomatis*
• *Shigella* spp.
• *Campylobacter sp.*
• *Group B streptococci*
• Bacterial vaginosis-associated organisms
B. Viral agents
Human (alpha) herpesvirus 1 or 2 (herpes simplex viruses (HSV). Human (beta) herpesvirus 5 (formerly cytomegalovirus):
• Hepatitis B virus (HBV)
• Human papilloma viruses
• Molluscum contagiosum virus
• Human immunodeficiency virus (HIV)
C. Protozoal agents
• *Entamoeba histolytica*
• *Giardia Lamblia*
• *Trichomonas vaginalis*
D. Fungal agents
• *Candida albicans*
E. Ectoparasite
• *Phthirus pubis*
• *Sarcoptes scabiei*

C. Donovanosis

Donovanosis is caused by *Calymmatobacterium granulomatis*.

1. **Specimen:** Tissue smear from the ulcer
2. **Staining:** Diagnosis can be made by demonstration of Donovan bodies in Wright-Giemsa stained impression smears from the lesions. They appear as rounded coccobacilli, 1-2 µm, within cystic spaces in large mononuclear cells. They show bipolar condensation of chromatin, giving a closed safety pin appearance in stained smears. Capsules are usually seen as dense acidophilic areas around the bacilli.

D. Chancroid

Chancroid or soft chancre is caused by *H. ducreyi*.

1. **Specimen:** Exudate
2. **Gram Staining:** Gram-negative coccobacilli are seen in Gram staining.
3. **Culture:** Exudate is cultured onto chocolate agar enriched with isovitalex and fetal calf serum, and containing vancomycin as selective agent. Culture plates are incubated at 35°C under 10 percent CO_2 and in high humidity in 2 to 8 days.
4. **Identification:** For identification of the organism colony morphology and Gram staining are useful.
   i. *Colony morphology*: *H. ducreyi* forms small, gray, translucent colonies.
   ii. *Gram staining:* It shows gram-negative coccobacilli.

E. *Herpes genitalis*

Herpes simplex virus (HSV), types 1 and 2 is the etiological agent but type 2 strains are more commonly associated.

1. **Specimens**
   i. Scrapings from base of the lesions
   ii. Blood for serology
2. **Microscopy:** Intranuclear type A inclusion bodies may be seen in Giemsa stained smears. The virus particle may also be demonstrated under the electron microscope.
3. **Virus isolation:** Diagnosis is confirmed by tissue culture in human diploid fibroblast cells. Typical cytopathic changes may appear as early as in 24 to 48 hours.
4. **Serology:** ELISA, neutralisation or complement fixation tests are used for antibody detection and is useful in diagnosis of primary infection.

F. Gonorrhea

It is caused by *Neisseria gonorrhoeae*.

1. **Specimens:** Specimens used are (i) Urethral discharge; (ii) An endocervical swab; (iii) In chronic cases: Massage of the prostate per rectum may express some exudate for examination when prostatitis is suspected and there is no spontaneous discharge from the urethra; (iv) Rectal swab.
2. **Direct Gram staining:** Gram stain of the penile exudate shows characteristic and diagnostic gram-negative intracellular diplococci (GNID) in granulocytes. The Gram stain is not sensitive or specific enough to use as detection of gonorrhea in women due to the presence of mixed normal flora.
3. **Culture:** Specimen transport is extremely important because *N. gonorrhoeae* dies rapidly in transport media, such as modified Stuart's medium. Commonly used selective media are modified Thayer-Martin and Martin-Lewis plates. Inoculated plates should immediately be placed in an incubator at 37°C for 24-48 hours in the presence of CO_2. Cultures with no visible growth must be held for 72 hours before discarding.
4. **Identification:** Identification of organism is based on colony morphology, Gram staining from colonies and biochemical reactions.
   i. **Colony morphology:** Small, gray, translucent, raised colonies.
   ii. **Gram staining:** Gram-negative diplococci
   iii. **Biochemical reactions:** They are oxidase positive and produce acid from glucose but not lactose, maltose or sucrose.
   iv. Direct detection methods, such as fluorescent antibody and agglutination tests, are also available.

v. Nucleic acid probes for the direct detection of *N. gonorrhoeae* from clinical specimens.

G. Nongonococcal Genital Infection

Symptoms of discharge and dysuria clinically indistinguishable from gonorrhoea caused by organisms other than *N. gonorrhoeae* is called nongonococcal urethritis (NGU). Causative agents are shown in Table 78.13. A significant proportion of nongonococcal genital infection in women is generally due to chlamydia trachomatis.

Laboratory Diagnosis

1. **Specimens**
   (i) Swabs from exudate of urethra; (ii) Cervical discharge.
2. **Direct examination**
   i. Giemsa stain: It shows intracytoplasmic inclusion bodies suggestive of *C. trachomatis*.
   ii. *Antigen detection:* For detection of elementary bodies of *C. trachomatis*, smears are made from exudate and examined by immunofluorescence test with a monoclonal antibody or by ELISA.
3. Culture: The exudate is inoculated on McCoy or HeLa cell cultures treated with cycloheximide. Intracytoplasmic glycogen-rich inclusions are detected by Giemsa stain or by immunofluorescence. These are suggestive of *C. trachomatis*.
4. Serology
   i. Complement fixation test (CFT)
   ii. Microimmunofluorescence or ELISA is useful for detection of serovar-specific antibody.

H. Trichomoniasis

It is caused by *Trichomonas vaginalis*.

1. **Specimen:** Swab of vaginal discharge is examined freshly. Specimen should be collected in stuart's transport medium if delay in transport is inevitable.
2. **Direct Microscopy:** Direct wet film shows motile trichomonads. Direct microscopy is at least 80 percent as positive as culture.
3. **Culture:** Fineberg's medium is used for culture of specimen and it is incubated for 5 days and examined for motile protozoa.

I. Bacterial Vaginosis-associated Organisms

The diagnosis of bacterial vaginosis does not depend on the isolation of a particular microorganism, e.g. Gardnerella vaginalis or *Mycoplasma hominis*, but on the replacement of the predominantly lactobacillary flora with a mixture of aerobes; and anaerobic organisms, a shift of pH to neutral or alkaline, and the presence of 'clue' cells. These changes are most easily assessed from the Gram stained film.

For diagnosis of Shigella sp, Campylobacter sp, Group B streptococci refer to corresponding chapters.

J. Vulvovaginal Candidiasis

It is caused by various species of *Candida* but *C. albicans* accounts for 80 percent of cases.

1. **Specimen:** Swab from vaginal secretions
2. **Direct microscopy**
   i. KOH mount: It shows yeast cells.
   ii. Gram staining: Gram staining shows characteristic gram-positive budding yeast cells and pseudohyphae.
3. **Culture:** Sabouraud's dextrose agar (SDA) is inoculated with the specimen and incubated at 37°C for 48 hours.
4. **Identification**
   i. **Colony morphology:** Colonies are creamy white and smooth.
   ii. **Gram staining:** Gram stained smear shows budding gram-positive yeast cells.
   iii. **Germ tube formation:** *C. albicans* forms germ tube within two hours when incubated in human serum at 37°C.
   iv. **Chlamydospores formation:** On cornmeal agar *C. albicans* forms chlamydospores.

K. Genital Warts

Genital warts, also known as condyloma acuminata are common in sexually active adults. These are usually due to human papillomavirus (HPV) types 6 and 11.

For detection of inclusion bodies of HPV, cytological or histological examination of cells in urine is used.

For diagnosis of *Shigella* sp, *Campylobacter* spp, Group B streptococci refer to corresponding chapters.ww

KNOW MORE

Bacterial Vaginosis-associated Organisms

The diagnosis of bacterial vaginosis does not depend on the isolation of a particular microorganism, e.g. *Gardnerella vaginalis* or *Mycoplasma hominis*, but on the replacement of the predominantly lactobacillary flora with a mixture of aerobes; and anaerobic organisms, a shift of pH to neutral or alkaline, and the presence of 'clue' cells. These changes are most easily assessed from the Gram stained film.

KEY POINTS

- The sexually transmitted diseases (STDs) are a group of communicable diseases which are transmitted predominantly or entirely by sexual contact. The causative organisms include a wide range of bacterial, viral protozoal and fungal agents.
- STDs can be differentiated into two types: (a) those

causing ulcerative lesions and (b) those causing non-ulcerative lesions.

- STDs with ulcerative genital ulcers are caused by *Treponema pallidum, Haemophilus ducreyi, Calymmatobacterium granulomatis, C. trachomtis* (LGV, or lymphogranuloma venereum strain), and Herpes simplex virus (HSV).
- STDs with nonulcerative lesions—Men: These include urethritis caused by *N. gonorrhoeae*; Nongonococcal urethritis by *C. trachomatis, Mycoplasma genitalium, U. urealyticum, Trichomonas vaginalis,* and HSV; Prostatitis/epididymitis by *C. trachomatis, N. gonorrhoeae,* HSV, *C. trachomatis,* and *T. pallidum.*
  Women: These include cystitis/urethritis caused by *C. trachomatis, N. gonorrhoeae,* HSV; Vulvitis by *Candida albicans,* HSV; Vaginitis by *C. albicans, T. vaginalis;* Cervicitis by *C. trachomatis, N. gonorrhoeae, M. genitalium;* and bacterial vaginosis by *Gardnerella vaginalis,* Mobiluncus sp, and *Mycoplasma hominis.*
- For laboratory diagnosis and treatment according to the suspected organism responsible for the manifestations of that particular STD.

IMPORTANT QUESTIONS

1. Name various organisms causing sexually transmitted diseases. Discuss the laboratory diagnosis of syphilis.
2. Write short notes on:
   a. Laboratory diagnosis of gonorrhoea.
   b. Nongonococcal urethritis (NGU).
   c. Chancroid.
   d. Lymphogranulum venereum (LGV).
   e. Donovanosis.
   f. Vulvovaginal candidiasis.

FURTHER READING

Morse SA, Morse SA. New tests for bacterial sexually transmitted diseases. Curr Opin Infect Dis 2001;14:51.

Schneede P, et al. Sexually transmitted diseases (STDs)—a synoptic overview for urologists Eur Uro 2003;144:7.

Sobel JD, Sobel JD. Bacterial vaginosis. Annu Rev Med 2000; 51:356.

Wasserheit IN, Wasserheit IN. HIV infection and other STDs: so close and yet so far. Sex Transm Dis 1999;26:550.

8. WOUND INFECTION

Wound infection occurs when one or more microorganisms evade the clearing effect of the host's defences, replicate in large numbers and attack and harm the host's tissues. Wound infections may be endogenous or exogenous. Infection of a wound is difficult to define and no clear rules can be given to distinguish it from contamination and colonization. It may be caused by a variety of aerobic and anaerobic species of bacteria (Table 78.15).

Table 78.15: Microorganisms causing wound infection

A. Aerobes

Staphylococcus aureus

β-hemolytic streptococci

- *Enterococcus* sp.
- *Streptococcus pneumoniae*
- *Escherichia coli*
- Other coliform bacilli
- *Proteus* sp.
- *Pseudomonas aeruginosa*
- *Mycobacterium marinum*
- *Nocardia* sp.

B. Anaerobes

- Peptostreptococci
- *Bacteroides* sp.
- *Clostridium perfringens* and other clostridia.

Actinomyces israelii

Laboratory Diagnosis

1. Specimens

Pus or *exudate* is often submitted on a swab for laboratory investigation. Whenever possible, pus or exudate should be submitted in a small screw-capped bottle, a firmly stoppered tube or syringe, or a sealed capillary tube. If possible, send two swabs taken from the depths of the wound or lesion, so that one can be used for the preparation of a smear for microscopy and the other for the seeding of cultures.

2. Naked-eye Examination

i. The pus of a staphylococcal lesion is typically creamy and thick in consistency, with pus cells evident on microscopy.
ii. The pus of a *Streptococcus pyogenes* infection is generally straw-colored and watery, with lysis of pus cells seen on microscopy.
iii. The pus of proteus infection has a fishy smell
iv. The pus of pseudomonas infection a sweet, musty odor and often a blue pigmentation.
v. Pus containing anaerobic organisms often has an offensive putrid smell, and that of actinomycosis may contain small microcolonies that appear as 'sulfur granules'. In some fungal infections such as mycetoma, black or brown granules may be present. The pus of an amoebic abscess is said to resemble anchovy sauce.

3. Microscopy

Smear stained by Gram's Method:

Gram-positive cocci: In typical clusters may suggest a staphylococcal infection,

In chains streptococcal infection:

Gram-positive diplococci may be given by either pneumococci or enterococci.

Gram-variable filaments of actinomyces may appear like chains of cocci and their fragments as diphtheroid bacilli.

Examination of a wet film may reveal the presence of fungi or motile bacteria.

Darkground microscopy of a wet film is useful in the diagnosis of primary syphilis.

A smear stained by the auramine or Ziehl-Neelsen method—for tubercle bacillus, another mycobacterium or a nocardia.

4. Culture

The specimen should be inoculated on to two plates of blood agar, the one for incubation at 37°C aerobically, preferably in air plus 5 to 10 percent CO_2, the other for incubation anaerobically in nitrogen hydrogen plus 5-10 percent CO_2, It should also be plated for aerobic incubation on MacConkey agar or CLED agar and be inoculated into a tube of cooked-meat broth for the enrichment of exacting aerobes and anaerobes. The culture plates are examined after overnight incubation at 37°C for 18 to 24 hours. If there is no growth, the aerobic and anaerobic blood agars should be reincubated for another 24 hours. If there is still no growth, the plates may be discarded unless there is an indication for longer incubation.

If tuberculous or fungal infection is suspected, the specimen should be cultured by the appropriate methods on the appropriate special media.

5. Identification of Isolates

After the bacteria cultured have been obtained in pure subcultures, any further necessary tests for their identification should be done, e.g. the coagulase test on staphylococci, Lancefield's grouping of β-hemolytic streptococci, and biochemical tests on coliform bacilli and anaerobes. For detailed laboratory diagnosis, refer to the corresponding chapters.

6. Antibiotic Sensitivity

At the same time the pure cultures should be tested for sensitivity to an extended range of antibiotics useful in therapy.

SKIN AND SOFT-TISSUE INFECTIONS

Various skin and soft-tissue infections include impetigo, folliculitis, furuncles, carbuncles, cellulitis, erysipelas, paronychia, erysipeloid and erythrasma. Impetigo is a common pyoderma. Initially, lesions of impetigo begin as small vesicles that pustulate and rupture, creating a thick, yellow, encrusted appearance. The lesions are superficial and painless but pruritic and easily spread by scratching. Folliculitis is inflammation and infection of hair follicles. Lesions of folliculitis may develop into 'a deeper inflammatory' nodule called a furuncle. A carbuncle is an abscess that extends even more deeply into the subcutaneous fat and may have multiple draining sites. Cellulitis is a diffuse inflammation and infection of the superficial skin layers. In contrast, erysipelas is a deeper form of cellulitis that involves not only the superficial epidermis but also the underlying dermis and lymphatic channels. Paronychia is an infection of the cuticle surrounding the nail bed. Erysipeloid is a superficial soft-tissue infection. Erythrasma is a chronic, pruritic, reddish-brown, macular infection found most commonly in men and obese patients diabetes mellitus. The causative agents are given in Table 78.16.

Table 78.16: Causative agents of skin and soft-tissue infections

Infection	*Causative agents*
Impetigo	*Streptococcus pyogenes*, *Staphylococcus aureus*, occasionally group B streptococci
Folliculitis	*S. aureus*, gram-negative bacilli, *Candida*
Furuncles	*S. aureus*
Carbuncles	*S. aureus*
Callulitis	*S. pyogenes*, *S. aureus*, *Haemophilus influenzae* in children
Erysipelas	*S. pyogenes*, occasionally other β-hemolytic streptococci or *S. aureus*
Paronychia	*S. aureus*, gram-negative bacilli, *Candida*
Erysipeloild	*Erysipelothrix rhusiopathiae*
Erythrasma	*Corynebacterium minutissimum*

KEY POINTS

- Wound infections may be endogenous or exogenous. It may be caused by a variety of aerobic and anaerobic species of bacteria.
- Laboratory diagnosis: Pus or exudate is often submitted on a swab for laboratory investigation. The basic procedures usually include a naked-eye examination of the specimen, microscopical examination of a Gram film, and culture on aerobic and anaerobic blood agar plates, on MacConkey agar and in cooked-meat broth. Gas chromatography may be performed directly on liquid specimens to indicate the presence of anaerobes.

IMPORTANT QUESTIONS

1. Name various microorganisms causing wound infection. Write briefly on laboratory diagnosis of wound infection.

FURTHER READING

Collee JC, et al. Mackie & Mc Cartney Practical Medical Microbilology,14th edn. London: Churchill Livingstone 1996; 53-94.

9. PYREXIA OF UNKNOWN ORIGIN (PUO)

Pyrexia of unknown origin (PUO) may be defined (a) any febrile illness (body temperature greater than 38°C) on several occasions, (b) duration of fever of more than 3 weeks, and (c) failure to reach a diagnosis despite 1 week of inpatient investigation. It is also known as 'fever of unknown origin' (FUO).

Causes of PUO

Infection is the most common cause of PUO. However, there are important noninfectious causes of fever. The causes of PUO are given in Table 78.17.

Laboratory Diagnosis of PUO

Tests should first be done for the more likely infections and then, if these are negative, tests for the less likely should be done.

A. Bacterial Infections

1. Specimens

Blood: For blood culture, peripheral blood smear, haematology, serology and other tests
Urine: Midstream urine specimen for UTI
Sputum: In cases of lung infections
Pus: In localized abscesses

2. Collection

These specimens must be collected in sterile containers under aseptic conditions. Blood is collected in blood culture bottles for culture and in a sterile vial for serology. Blood culture should be collected before antibiotics are given. Midstream urine specimen of urine (MSU) should be collected in a sterile universal container.

3. Culture

i. **Blood culture:** For blood culture, 5 ml of blood is collected in each bottle of 50 ml glucose broth and 50 ml taurocholate broth. These broths are incubated at 37°C for 24 hours and then subcultures are made on blood agar (from glucose broth) and MacConkey agar (from taurocholate broth). Blood agar and MacConkey agar plates are incubated at 37°C for 24 hours.
ii. **Urine culture:** A calibrated volume of midstream urine specimen is inoculated on blood agar and

Table 78.17: Causes of PUO

A. **Infective causes**
  a. *Bacterial*
    - Urinary tract infections
    - Lung, subdiaphragmatic, appendix and other deep abscesses
    - Septicemia associated with cryptic abscesses, pneumonia, pyelonephritis, biliary tract infection, infective endocarditis and immunodeficiencies;

  Enteric fever
    - Tuberculosis
    - Brucellosis
    - Syphilis
    - Relapsing fever
    - Rheumatic fever
    - Leptospirosis without jaundice or meningitis
    - Typhus fever
    - Nonmeningitic meningococcal infection
    - Q fever
  b. Parasitic
    - Malaria
    - Hepatic amoebiasis
    - leishmaniasis
    - Trypanosomiasis
    - Toxoplasmosis
    - Filariasis
  c. *Viral*
    - EBV infection
    - CMV infection
    - HIV infection
    - Rubella and other infectious fevers without typical rash

B. **Noninfective causes**
  a. *Neoplasms*
    - Hodgkin's lymphoma
    - Non-Hodgkin's lymphoma
    - Leukemia
    - Hypernephroma
    - Hepatoma
    - Disseminated malignancy
  b. *Connective tissue disorders*
    - Systemic lupus erythematosus (SLE)
    - Polyarteritis nodosa
    - Temporal arteritis
  c. *Granulomatous diseases*
    - Sarcoidosis
    - Crohn's disease
    - Granulomatous hepatitis
  d. *Drug Reactions*
    - Drug induced fevers

Very careful history taking is an essential preliminary to diagnosis. Enquiry must be made about foreign travel, occupation, contact with cases of infectious diseases and contact with animals. The physician must inform the microbiologist of any such relevant facts.

MacConkey agar. These media are incubated at 37°C for 24 hours. Culture should be performed on Lowenstein-Jensen (LJ) medium in case of renal tuberculosis.

iii. **Sputum culture:** Specimen is inoculated on blood agar and MacConkey agar plates and incubated at 37°C for 24 hours. Specimen should be cultured on Lowenstein-Jensen (LJ) medium in case of tuberculosis and incubated at 37°C for 6 weeks.

iv. **Pus culture:** Pus is inoculated in glucose broth, blood agar and MacConkey agar. These media are incubated at 37°C for 24 hours. Pus should be cultured on LJ media for *M. tuberculosis*. Culture of pus should be performed under anaerobic conditions when suspecting anaerobic organisms.

4. Identification

Organisms are identified on the basis of colony morphology, Gram staining, biochemical reactions and agglutination, etc. Ziehl-Neelsen (ZN) staining is performed to detect acid-fast bacilli (AFB) for *M. tuberculosis*. This is further confirmed by culture and biochemical reactions. For details of individual organism, refer to corresponding chapters.

5. Serology

Paired sera should be collected for serological tests for antibody responses to a range of possible pathogens, e.g. cytomegalovirus, hepatitis B virus, influenza virus, infectious mononucleosis virus, chlamydia, coxiella, rickettsia, mycoplasma, salmonella, brucella, legionella, leptospira, borrelia, treponema, toxoplasma, aspergillus and other fungi, and entamoeba. HIV infection should also be considered. The first specimen should be taken as early in the illness as possible and the second 2-4 weeks later. The antistreptolysin-O (ASO) test should be done for cryptic *Streptococcus pyogenes* infection.

B. Parasitic Infections

Stained peripheral blood films smears (thin and thick) will help in diagnosis of malaria, leishmaniasis, trypanosomiasis and filariasis. Wet blood film may show microfilaria in cases of filariasis. Serology is useful in amoebiasis.

C. Viral Infections

Viral infections may be detected either by tissue culture or by serology. Peripheral blood smear may be helpful in infectious mononucleosis. Paul-Bunnell test is useful in infectious mononucleosis.

D. Fungal Infections

Specimens may be cultured on Sabouraud's dextrose agar or Brain-heart infusion agar.

E. Other Tests for Diagnosis

1. *Skin tests*
   - **Mantoux test:** A tuberculin test and a chest X-ray should be done to detect tuberculosis.
   - Skin tests for histoplasmosis, coccidioidomycosis, sarcoidosis
2. Hematological investigations should be done to detect leukocytosis, suggestive of a cryptic abscess; eosinophilia, suggestive of helminthiasis; and atypical lymphocytes, suggestive of infectious mononucleosis.
3. *Immunologic tests*
   - LE cell phenomenon and antinuclear antibody test in SLE
4. **Biopsy:** Biopsy of liver and bone marrow should always be considered in the investigation of classical cases of PUO, but other tissues such as skin, lymph nodes and kidney may also be sampled.

☞ KEY POINTS

- Pyrexia of unknown origin (PUO) may be defined (a) any febrile illness (body temperature greater than 38°C) on several occasions, (b) duration of fever of more than 3 weeks, and (c) failure to reach a diagnosis despite 1 week of inpatient investigation. It is also known as 'fever of unknown origin' (FUO).
- The causes of PUO include infections bacterial, parasitic and viral) neoplasms, connective tissue disorders, granulomatous diseases and drug reactions.
- Tests should first be done for the more likely infections and then, if these are negative, tests for the less likely should be done.

IMPORTANT QUESTIONS

1. Define and enumerate the causes of pyrexia of unknown origin (PUO). Discuss the laboratory diagnosis of PUO.

FURTHER READING

Collee JC, et al. Mackie & Mc Cartney Practical Medical Microbilology. 14th edn. London: Churchill Livingstone 1996; 53-94.

CHAPTER 79

Hospital-Aquired Infection

Learning Objectives

After reading and studying this chapter, you should be able to:

- Define hospital-associated infection.
- List of routes of transmission of hospital-associated infections.
- Describe common hospital-associated infections and causative organisms responsible for these conditions.
- Describe diagnosis and control of hospital-associated infections.

INTRODUCTION

A *nosocomial infection is any infection acquired while in hospital.* The terms *hospital infection, hospital-acquired infection* or *nosocomial infection* (from *nosocomeion*, meaning hospital) are applied to infections developing in hospitalized patients, not present or in incubation at the time of their admission. Such infections may become evident during their stay in hospital or, sometimes, only after their discharge. An infection that is incubating in a patient when he or she is admitted into hospital is not a hospital infection. However, community-acquired infections brought into hospital by the patient may subsequently become hospital infections for other patients and hospital staff.

Incidence of Hospital Infection

The incidence of hospital infection has been reported to be 2-12 percent in the advanced countries. It is much higher in the crowded hospitals in the developing countries. On average, around 10 percent of all hospital patients will develop an infection as a result of their stay in hospital. Urinary, respiratory and wound infections are the most common.

SOURCES OF INFECTIONS

Hospital infection may be exogenous or endogenous in origin.

1. Exogenous

Exogenous source may be another person in the hospital *(crossinfection)* or a contaminated item of equipment or building service *(environmental infection).*

i. **Contact with other patients and staff:** Patients and hospital personnel suffering from infection, or asymptomatic carriers are the most important sources.

ii. **Environmental sources:** These include inanimate objects, air, water and food in the hospital.

   a. **Inanimate objects:** Equipments, materials, such as sanitary installation (bed pans, urinals), lights, table, blankets, medical equipment (endoscopes, catheters', needles, spatula and other instruments), floors, food. and water (contaminated by kitchen or other hospital staff or visitor).

2. Endogenous

A high proportion of clinically apparent hospital infections are endogenous *(self-infection),* the infecting organism being derived from the patient's own skin, gastrointestinal or upper respiratory flora.

FACTORS INFLUENCING HOSPITAL-ASSOCIATED INFECTIONS

A large number of factors contribute towards the development of hospital associate infection. These include the following:

1. **Age-**Natural resistance to infection is lower in infants and the elderly, who often constitute the majority of hospital patients.
2. **Susceptibility to infection:** Pre-existing disease, such as diabetes, or other conditions for which the patient was admitted to hospital, and the medical or surgical treatment, including immunosuppressive drugs, radiotherapy or splenectomy, may also reduce the patient's natural resistance to disease. Moreover, the natural defence mechanisms of the body surfaces may be bypassed either by injury or by procedures such as surgery, insertion of an indwelling catheter, tracheostomy or ventilatory support.
3. **Hospital environment:** The hospital environment is heavily laden with a wide variety of pathogens.

Patients shed them from their bodies; hospital personnel spread them through their hands and clothes. Bedding, linen and utensils act as fomites. Equipment may be contaminated. Pathogens are present in the hospital dust and air, and sometimes even in antiseptic lotions and ointments. Contamination of hospital food or water may cause outbreaks of infections.

4. **Diagnostic or therapeutic procedures:** During diagnostic or therapeutic procedures such as insertion of urethral or intravenous catheters, the slightest lapse in asepsis may lead to infection.
5. **Drug-resistance:** The hospital microbial flora is usually multidrug resistant due to injudicious use of antibiotics, thus limiting the choice of therapy.
6. **Transfusion:** Blood, blood products and "intravenous fluids used for transfusion, if not properly screened, can transmit many infections.
7. **Advances in medical progress:** Advances in treatment of cancer, organ transplantation, implanted prostheses and other sophisticated medical technologies enhance the risk of infection to patients.

MICROORGANISMS CAUSING HOSPITAL INFECTION

Almost any microbe can cause a hospital-acquired infection, but those that are able to survive in the hospital environment for long periods and develop resistance to antibiotics and disinfectants are particularly important in this respect. Though protozoal infections are rare. The pattern of hospital infection has changed over the years, reflecting advances in medicine and the development of antimicrobial agents. In the pre-antibiotic era the majority of infections were caused by gram-positive organisms, particularly *Streptococcus pyogenes* and *Staphylococcus aureus. Strep. pyogenes* was, perhaps, the most important cause of hospital infection formerly but is hardly ever encountered now as it is highly susceptible to antibiotics.

1. ***Staph. aureus: Staph. aureus*** strains, resistant to multiple antibiotics and belonging to phage type 80/81, spread globally in the 1950s and 1960s, colonising hospitals and causing nosocomial infection with such frequency that they came to be called 'hospital staphylococci'. Subsequently, epidemic or pandemic strains characterized by resistance to methicillin (MRSA) have been found in many hospitals worldwide, presenting a daunting challenge. The original phage types have since been replaced by others belonging to group III but staphylococci continue to be very common agents in hospital infection. *Staph. epidemlidis* and Group D streptococci also are sometimes responsible for hospital infections.
2. **Pseudomonas species:** *P. aeruginosa* and other *Pseudomonas* species have always been important causes of hospital infection because of their intrinsic resistance to most antibiotics and ability to survive and even multiply at low temperatures and in disinfectant solutions. They may also carry drug resistant plasmids. They are a common cause of hospital-acquired pneumonia; and infections of the urinary tract and burn wounds.
3. **Tetanus spores: Tetanus spores** can survive in dust and may sometimes contaminate items used in hospitals. Hospital tetanus is usually due to faulty sterilization techniques or other lapses in asepsis.
4. **Viral infections:** Viral infections probably account for more hospital-acquired infections than previously realized.
   i. **HIV and hepatitis B and C viruses:** are transmitted by contaminated blood or blood products. Screening of blood donors has reduced the risk to a large extent. However, HIV escapes detection during the window period. Screening for HCV antibody as practized now is not very satisfactory. In view of the inherent risk of transmission of known and unknown pathogens, it is safer to limit blood transfusion to the absolute minimum and employ autologous transfusion instead, wherever possible. The use of shared syringes and needles also carries the risk of transmission of these viruses.
   ii. **Viral diarrhea and chickenpox** are other viral infections that spread in hospitals.
   iii. **Cytomegalovirus, herpesvirus, influenza, enteroviruses and arenaviruses** may also cause hospital infection.
5. **Fungus and parasites:** The range of hospital pathogens also includes yeasts *(Candida albicans)*, moulds, (Aspergillus, Mucor) and protozoa *(Entamoeba histoLytica, Plasmodia, Pneumocystis carinii, ToxopLasma gondii)*.

ROUTES OF TRANSMISSION

1. **Contact:** The most common routes of transmission for hospital infection are:
   Direct contact: spread from person to person *(staphylococcal* and *streptococcal sepsis)*.
   Indirect contact: spread via contaminated hands or equipment (enterobacterial diarrhea, *Pseudomonas aeruginosa* sepsis).
2. **Airborne spread**
   Infections may be spread:
   i. **Droplets:** Droplets of respiratory transmitted by inhalation
   ii. **Dust:** Dust from bedding, floors; exudate dispersed from a wound during dressing and from the skin by natural shedding of skin scales, spread to the susceptible site, e.g., *Ps. aeruginosa, Staph. aureus.*
   iii. **Aerosols:** Aerosols produced by nebulizers, humidifiers and air conditioning apparatus

transmit certain pathogens like *Legionella* to the respiratory tract. Occurrence of *legionellae* in hospital water supply has led to outbreaks of infection mainly with *Legionella pneumophila.*

3. **Oral route-**Hospital food contains gram-negative bacilli which are most often antibiotic resistant *(P. aeruginosa, E. coli, Klebsiella* spp. and others), which may colonize the gut and later cause infection in susceptible patients.
4. **Parenteral route (inoculation):** With the use of single-use disposable needles, syringes and other devices and satisfactory procedure for the sterilization of surgical instruments, transmission of infection by parenteral route has been infrequent. However, certain infections may be transmitted by blood transfusion or tissue donation, contaminated blood-products (factor VIII), contaminated infusion fluids and from accidental injury with contaminated sharp instruments (HIV, hepatitis B and C). In areas of high prevalence of malaria, syphilis and these viruses stringent precautions should be taken to minimize transmission between patients by strict use of single-use items and from health care workers to patients and vice versa by minimizing needles tick episodes.
5. **Self-infection and cross-infection:** *Self-infection* may occur due to transfer into the wound of staphylococci (or occasionally streptococci) carried in the patient's nose and distributed over the skin, or of coliform bacilli and anaerobes released from the bowel during surgery, Alternatively, *cross-infection* may result from staphylococci or coliform bacilli derived from other patients or healthy staff' carriers.

COMMON HOSPITAL-ACQUIRED INFECTION

1. Urinary Tract Infection

Most hospital-acquired infections of the urinary tract are associated with urethral catheterization. Urinary tract infection is caused by *Esch. coli, Klebsiella, Proteus, Serratia, Pseudomonas, Providencia,* coagulase negative staphylococci, *enterococci* and *Candida albicans.*

2. Respiratory Infections

Aspiration in unconscious patients and pulmonary ventilation or instrumentation may lead to nosocomial pneumonia, particularly in those with pre-existing cardiopulmonary disease. The major pathogens include *Staph. aureus, Klebsiella spp., Enterobacter, Serratia, Proteus, Esch. coli, Pseudomonas aeruginosa, Acinetobacter, Legionella pneumophila* and respiratory viruses. Respiratory tract infection accounted for approximately 25 percent of hospital infection and was as common as that of urinary tract.

3. Wound and Skin Sepsis

The incidence of postoperative infection is higher in elderly patients. Most wound infections manifest within a week of surgery. *Staph. aureus* is the predominant pathogen, followed by *Pseudomonas aeruginosa* and then *Esch. coli, Proteus, enterococci* and coagulase negative staphylococci.

Nonsurgical sites of wound infections include infection 'cut-downs', umbilical stumps, ulcers and burns. *Ps. aeritginosa* is the most important cause of infection in burns.

4. Gastrointestinal Infections

Food poisoning and neonatal septicaemia in hospital have been reported. These infections are mainly associated with *salmonella* and *shigella* sonnei.

5. Burns

Staph aureus, Pseudomonas aeruginosa, Acinetobacter and *strept pyogenes* are responsible for hospital acquired infections in cases of burns.

6. Bacteremia and Septicemia

These may be consequences of infections at any site but are commonly caused by infected intravenous cannulae. gram-negative bacilli are the common pathogens. 'Cut-downs' are safer on the arms than on legs. Intravenous rehydration in diarrhea should be restricted to emergencies and should be replaced by oral fluids as early as possible. Infection can be prevented by proper skin toilet before 'cut-down' and the use of stainless steel needles instead of plastic cannulae.

Staph. epidermidis bacteremia is seen commonly in patients with artificial heart valves. Bacteremia in those with valvular defects may lead to endocarditis.

DIAGNOSIS AND CONTROL OF HOSPITAL INFECTION

The most important steps in preventing nosocomial infections are to first recognize their occurrence and then establish policies to prevent their development. Hospital infection may occur sporadically or as outbreaks.

Etiological diagnosis is by the routine bacteriological methods of smear, culture, identification and sensitivity testing. When an outbreak occurs, the source should be identified and eliminated. This requires the sampling of possible sources of infection such as hospital personnel, inanimate objects, water, air or food. Typing of isolate - phage, bacteriocin, antibiogram or biotyping—from cases and sites may indicate a causal connection. Obvious examples of sources of hospital outbreaks are nasal carriage of staphylococci by surgeons or pseudomonas growing in hand lotions. Carriers should be suitably treated.

The provision of sterile instruments, dressings and fluids is of fundamental importance in hospital practice. The cause of infection may be a defective autoclave or improper techniques such as boiling infusion sets in ward sterilizers. A careful analysis of the pattern of infection may often reveal the source but sometimes it eneludes the most diligent search.

Hospital infection	Causative organisms
1. Urinary tract infection	*Esch. coli, Klebsiella, Proteus, Serratia, Pseudomonas, Providencia,* coagulase negative staphylococci, enterococci and *Candida albicans.*
2. Respiratory infection	*Staph. aureus, Klebsiella, Enterobacter, Serratia, Proteus, Esch. coli, Pseudomonas aeruginosa, Acinetobacter, Legionella pneumophila* and respiratory viruses
3. Wound and skin sepsis	*Staph. aureus, Pseudomonas aeruginosa, Esch. coli, Proteus,* enterococci and coagulase negative staphylococci
4. Gastrointestinal infection	*Salmonella, Shigella sonnei* and viruses
5. Bums	*Staph. aureus, Pseudomonas aeruginosa, Acinetobacter and Str. pyogenes*

Table 79.1: Commonly occurring microorganisms in hospital infection

INFECTION CONTROL POLICY

The establishment of an effective infection control organization is the responsibility of good management of any hospital. There will normally be two parts:

Infection Control Committee (ICC)

Every hospital should have an infection control committee (ICC) with responsibilities that include the production and implementation of a disinfection policy. The committee will consist of a medical microbiologist who will usually serve as chairman, a physician, a surgeon, nurse teachers, nurse representatives for surgery, obstetrics, gynecology and medicine, and sterile service manager. Where hospitals obtain disinfectants through their purchasing department then the purchasing officer should be invited to attend relevant meetings. The ICC should meet regularly to formulate and update policy for the whole hospital matters having implications for infection control and to manage outbreaks of no nosocomial infection.

Roles of the Infection Control Committee

i. The surveillance of hospital infection.
ii. The establishment and monitoring of policies and procedures designed to prevent infection (e.g. catheter care policy, antibiotic policy, disinfectant policy).
iii. The investigation of outbreaks.

Infection Control Team

An infection control team of workers, headed by the **infection control doctor** (usually the microbilogists), to take up day-to-day responsibility for this policy. The functions of this team include **surveillance and control of infection and monitoring of hygiene practices,** advising the infection control committee on matters of policy relating to the prevention of infection and the education of all staff in the microbiologically safe performance of procedures. The *infection control nurse* is a key member of this team. Close working links between the microbiology laboratory, infection control nurse and the different clinical specialties and support services (including sterile services, laundry, pharmacy and engineering) are important to establish and maintain the infection control policy, and to ensure that it is rationally based and that the recommended procedures are practicable.

PREVENTION

The hospital-acquired infections can be prevented by following means:

1. **Sterilization:** The provision of sterile instruments, dressings, surgical gloves, facemasks, theater clothing and fluids.
2. **Cleaning and disinfection:** The general hospital environment can be kept in good order by attention to basic cleaning, waste disposal and laundry.
3. **Skin disinfection and antiseptics:** Thorough hand washing after any procedure—involving nursing care or close contact with the patient is essential.
   Procedures for preoperative disinfection of the patient's skin and for surgical scrubs are mandatory within the operating theater.
4. Rational antibiotic prophylaxis.
5. **Protective clothing.**
6. **Isolation:** To prevent the spread of specific infections to other patients *(source isolation)* and to protect susceptible or immunocompromised patients *(protective isolation).*
7. **Hospital building and design:** The routine maintenance of the hospital building to ensure that surfaces wherever possible are smooth, impervious and easy to clean.
8. **Equipment:** Any object or item of equipment for clinical use should be assessed to determine the appropriate method, frequency and site of decontamination.
9. **Personnel:** An occupational health service in hospitals should screen staff before employment and offer appropriate immunization. Hepatitis B vaccine should be given to all health care workers.
10. **Monitoring:** Monitoring of the physical performance of air-conditioning plants and machinery used for disinfection and sterilization is essential. In the event of an outbreak of hospital infection, more specific monitoring targeted at the known or likely causative microorganism should be considered.
11. **Surveillance and the role of the laboratory:** The detection and characterization of hospital infection incidents or outbreaks rely on laboratory data.

EFFICACY OF INFECTION CONTROL

These include sterilization, handwashing, closed drainage systems for urinary catheters, intravenous catheter care, perioperative antibiotic prophylaxis for contaminated wounds and techniques for the care of equipment used in respiratory therapy. Isolation techniques are assumed to be reasonable as suggested by experience or inference. Measures which are now considered to be ineffective include the chemical disinfection of floors, walls, sinks and routine environmental monitoring.

Effective surveillance and action by the infection control team have been shown to reduce infection rates. One important role of the team is to monitor compliance with practices known to be effective and to eliminate the many rituals or less effective practices which may even increase the incidence or cost of cross-infection. Innovations in infection control will need to be evaluated for efficacy and cost-effectiveness as further advances occur in medical care and limited health care resources are spread across hospital and community needs.

Unfortunately, in many hospitals, infection control is attempted by resorting to more and more of antibiotics. This is not only futile but may even be positively harmful by encouraging selective colonization by multiresistant pathogens. In the final analysis, prevention of hospital infection rests upon a proper understanding of aseptic practices and meticulous attention to hygienic principles. Sir William Osler's aphorism that 'soap, water and commonsense are the best disinfectants' applies even today in the context of hospital infection.

With this understanding it is possible that hospital infection can be controlled and largely prevented. The dictum of Florence Nightingale, made over a century ago, that 'the very first requirement in a hospital is that it should do the sick no harm', remains the goal.

KNOW MORE

Bacteremia and septicemia: The infecting organism may arise from a focus of infection in another system (secondary bacteremia) or no source may be identified clinically or after additional investigations (primary bactermia). The longer the cannulae are kept *in situ,* the greater the risk of infection. 'Cut-downs' on the leg veins in infants or children with diarrhea generally get left in place for long periods, the site being bathed in diarrheal stools. Phlebitis sets in with consequent bacteremia. Many a child admitted with diarrhea thus dies of septicemia.

KEY POINTS

- Hospital-acquired infections or nosocomial infections are infections occurring in hospitalized patients who were neither infected nor were in inclubation at the time of their admission.
- The sources of hospital-acquired infection may be exogenous or endogenous.
- Almost any microbe can cause a hospital-acquired infection, but those that are able to survive in the hospital environment for long periods and develop resistance to antibiotics and disinfectants are particularly important in this respect.
- Nosocomial infections are transmitted by air, direct contact, oral route, and parenteral route. UTIs, nosocomial pneumonia, and surgical wound infections are the common examples of hospital infections.
- Diagnosis of hospital-acquired infections is made by routine bacteriological methods.
- The source of infection may be traced by performing phage typing, bacteriocin typing, biotyping, or molecular typing.
- Active hospital surveillance is the key for successful hospital infection control.
- Every hospital must have an effective hospital-acquired infection control committee (ICC) which should be with responsibitis for the control of hospital-acquired infection (HAI) and monitoring of hygienic practices in the hospital.

IMPORTANT QUESTIONS

1. Define hospital-acquired infection. Enumerate organisms causing it. What are the factors which influence development of this infection?
2. Write short notes on:
   a. Routes of transmission of hospital-associated infections
   b. Diagnosis and control of hospital-associated infections
   c. Prevention of hospital-associated infections
   d. Disinfection policy
   d. Infection control committee

FURTHER READING

Ayliffe GAJ, Fraise A, Mitchell K, Geddes AM, Control of Hospitol Infection, 4th edn. Arnold, london 2000.

Bennett JV, Brachman PS. Hospital Infections 4th edn. Boston: Little Brown 1996.

Wenzel RP. Prevention and Control of Nosocomial Infection, 3rd edn. lippincott, Williams and Wilkins, Baltimore 1997.

CHAPTER 80

Laboratory Control of Antimicrobial Therapy

Learning Objectives

After reading and studying this chapter, you should be able to:

- List different methods of antibiotic sensitivity testing.
- Describe disk diffusion methods.
- Explain Stokes disk diffusion method and its reporting.
- Differentiate between Stokes disk diffusion and modified Stokes disk diffusion method.
- Explain the meaning of sensitive, intermediate and resistant applied to antimicrobial susceptibility test results.
- Describe the following: Epsilometer or E-Test; Minimum inhibitory concentration of antimicrobial agents; Minimum bactericidal concentration of antimicrobial agents.

INTRODUCTION

Clinical microbiologists have a major role to play in advising on the prescription of antibiotics for either treatment or prophylaxis of infection and in monitoring antibiotic use and clinical efficacy. Apart from rare exceptions like *Strep. pyogenes*, pathogenic bacteria exhibit very great strain variations in susceptibility to antibiotics and chemotherapeutic agents. This is particularly marked in the case of Staph. aureus and gram-negative bacilli. Therefore, it is essential to determine the susceptibility of isolates of pathogenic bacteria to antibiotics that are likely to be used in treatment. As strains of most pathogenic organisms differ from one another within their species in their antibiotic sensitivities, sensitivity tests are required as a routine.

ANTIBIOTIC SENSITIVITY TESTS

Antibiotic sensitivity tests are of two types:

A. Diffusion methods
   1. Kirby-Bauer disk diffusion method
   2. Stokes disk diffusion method

B. Dilution methods
   1. Broth dilution method
   2. Agar dilution method

Diffusion Methods

Here the drug is allowed to diffuse through a solid medium so that a gradient is established, the concentration being highest near the site of application of the drug and decreasing with distance (Fig. 80.1). The test bacterium is seeded on the medium and its sensitivity to the drug determined from the inhibition of its growth. Several methods have been used for the application of the drug. It may be added to ditches or holes cut in the medium or to hollow cylinders (Heady cups) placed on it. The method most commonly employed is to use filter paper disks, impregnated with antibiotics.

Disk Diffusion Methods

These methods are suitable for organisms that grow rapidly overnight at 35°C. These methods have, however, been found to be unsuitable for slow growing microbes.The various tech niques are technically simple, cheap and reliable.

Medium

The medium must support good overnight growth of test and control organisms; slow growth can result in the inhibition zones being abnormally large. Mueller-Hinton broth and agar may be used for testing aerobic

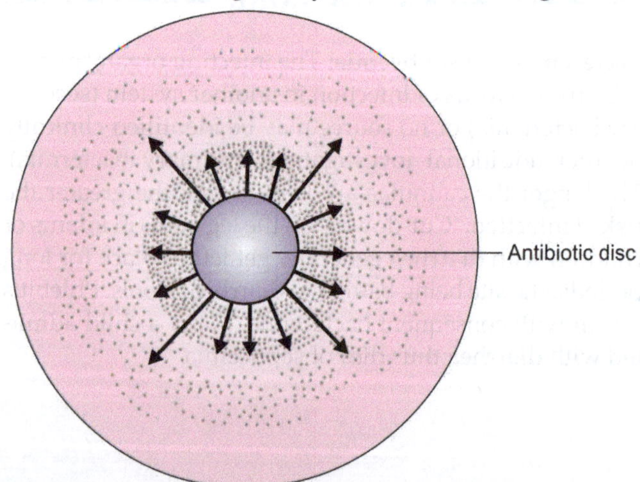

Fig. 80.1: The principle of antibiotic diffusion in agar. The concentration of antibiotic decreases as the distance from the disk increases

and facultative anaerobic isolates. The medium is and poured to a depth of 4 mm (25 ml medium) in flat-bottomed 9 cm Petridishes on a level surface. When set, the plates may be stored for up to a week at 4°C and their surfaces should be dried with their lids ajar before use. The pH of the medium must be close to 7.3. A more acid reaction decreases the activity of aminoglycoside and macrolide antibiotics. Incubation of the test plates in a CO_2-enriched atmosphere, or fermentation of sugars in the medium, may lower the pH sufficiently to produce this effect. A more alkaline pH favors the action of tetracyclines, novobiocin and fusidic acid, but interferes seriously with the activity of nitrofurantoin.

These media have minimal inhibitory effect on sulphonamides and trimethoprim. Large quantities of thymidine are present in some lots of media. Some organisms can use thymidine to bypass the mechanism of action of trimethoprim and grow, even though they are innately sensi tive to the antibiotic.

The addition of 5 percent lysed horse blood is needed to support the growth of fastidious species such as *Haemophilus influenzae*. Lysed horse blood should also be added for tests with sulphonamides and trimethoprim; its content of thymidine phosphorylase is needed to neutralize the inhibitory effect of thymidine in the medium on the action of these drugs. Low levels of free Ca^{2+} and Mg_2^+ ions in the medium increase the action of aminoglycoside antibiotics against *Pseudomonas aeruginosa* because when these organisms are grown in media deficient in these cations, cell wall permeability to *aminoglycoside* antibiotics is increased, while high levels decrease it.

The addition of 5 percent NaCI to the medium is needed in one of the methods for detecting resistance to methicillin in strains of staphylococci . The combined effects of medium chosen, added NaCI and incubation temperature all affect expression of resistance to methicillin. The pH of the medium must be close to 7.3.

Inoculum

Isolation of organisms in pure culture is of utmost importance. Prepare the inoculum from material picked up with a loop from five to ten colonies of the species to be tested. Inoculate them in a suitable broth medium. Incubate at 35-37°C for 4-6 hours when the growth is considered to be in logarithmic phase. The density of the organisms is adjusted to approximately 10^8 colony forming units (cfu)/ml by comparing its turbidity with that of 0.5 McFarland opacity standard.

If time does not permit incubation for 4-6 hours, young colonies may be removed from the surface of an agar plate that has been incubated overnight and diluted to proper density. Inoculum of standard strain should also be prepared as in case of the test strain.

Control Strains

Control strains for Kirby-Bauer and Stokes disk diffusion methods are given in Table 80 .1. For determination of minimum inhibitory concentration of antimicrobial agents against various bacteria, the same control strains given in Table 80.1 are used. However, in case of *S. aureus*, control strain *S. aureus* ATCC 29213 is used in place of ATCC 25923.

Antibiotic Disks

Commercially prepared disks 6 mm in diameter should be used. Manufacturers produce disks with accurate antibiotic content. If disks are prepared locally in the laboratory, then pure antimicrobial agents obtained from the manufacturers and not the ones for clinical use should be used. Proper diluents should be used. Distilled water serves to dissolve most antibiotic powders, but *chloramphenicol*, rifampicin and erythromycin must first be dis solved in a small amount of ethanol, nitrofurantoin and sulfonamides in small volume of NaOH solution, trimethoprim in weak acid (acetic or lac tic), and amoxicillin and ceftazidime in a small vol ume of saturated $NaHCO_3$ (Table 80.2).

Disks and disk dispensers should be stored in sealed containers with a desiccant, bulk stock being kept at -20°C if possible, otherwise at less than 8°C. Working stock, also kept in sealed containers with desiccant, should be stored at less than 8°C. Before they are opened for use, the containers should be allowed to warm up

Table 80.1: Control strains for Kirby-Bauer and stokes disc diffusion methods

Test bacteria	*Control strain*	
	Kirby-Bauer	*Stokes*
Coliform organisms	*E. coli* ATCC 25922	*C. coli* NCTC 10418
Pseudomonas	*P. aeruginosa* ATCC 27853	*P. aeruginosa* NCTC 10662
Haemophilus spp.	*H. influenzae* ATCC 49247	*H. influenzae* NCTC 11931
Gonococci	*N. gonorrhoeae* ATCC 49226	*N. gonorrhoeae* (sensitive strain)
Enterococci	*E. faecalis* ATCC 29212	*E. faecalis* NCTC 12697
Other organisms that can grow aerobically	*S. aureus* ATCC 25923	*S.aureus* NCTC 6571

Table 80.2: Diluents used for various antibiotics

Antibiotic	*Diluent*
Chloramphenicol	Ethanol
Rifampicin	Ethanol
Erythromycin	Ethanol
Nitrofurantoin	NaOH solution
Sulfonamides	NaOH solution
Trimethoprim	Acetic acid
Amoxycillin	NaHCOs (Saturated)
Ceftazidime	NaHCOs (Saturated)

slowly at room temperature to minimize condensation of moisture, which may lead to hydrolysis of the antibiotic. Therefore, they should be taken out from refrigerator 1-2 hours before applying on the culture medium. Disk contents of antimicrobial agents are given in Table 80.3.

Drugs to be tested against each species of bacteria should be grouped in sets of six or seven, the maximum number that can be accommodated on a single 100 mm diameter plate by Stokes and Kirby Bauer methods respectively. First line tests include those antibiotics that are locally available and commonly prescribed and reserve for second line tests those antibiotics for which prescription is restricted to special circumstances.

1. Kirby-Bauer Disk Diffusion Method

Procedure

A. **Preparation of inoculum (Growth Method):** Dip a sterile nontoxic cotton swab into the inoculum suspension and rotate the swab several times with firm pressure on he inside wall of the tube to remove excess fluid. Inoculate the dried surface of a Mueller-Hinton agar plate that has been 'brought to room temperature by streaking the swab three times over the entire agar surface. Replace the lid of the dish. Allow at least 3 to 5 minutes but no longer than 15 minutes for the surface of the agar to dry before adding the antibiotic disks.

B. **Testing of antibiotics:** The appropriate antimicrobial-impregnated disks are placed on the surface of the agar, using either sterile forceps or multidisk dispenser. Disks must be evenly distributed on the agar so that they are no closer than 24 mm from center to center. On a plate of 100 mm diameter, seven disks may be applied, one in the center and six in the periphery (Fig. 80.2). The plates are then incubated at 35°C for 16-18 hours (24 hours when testing staphylococci against methicillin or oxacillin or enterococci against vancomycin).

C. **Interpretation:** Examine the plates after overnight incubation. With the use of sliding calipers, a ruler, or a template, the zones of complete growth inhibition around each of the disks are carefully measured to within the nearest millimeter. The diameter of the disk is included in this measurement.

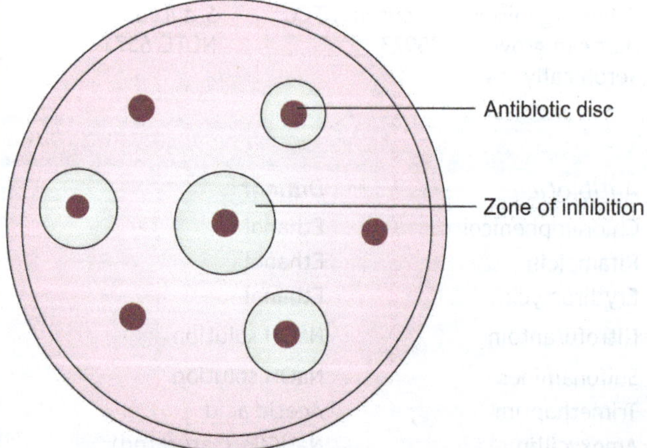

Fig. 80.2: Kirby-Bauer disk diffusion method

Table 80.3: Disk contents of various antimicrobial agents for the comparative methods

Antimicrobial agent (and test option)	*Disk content*
Benzylpenicillin	
staphylococci	2 IU (1.2 µg)
penumococci and meningococci	0.25 IU (0.15 µg)
Ampicillin	
Enterobacteriaceae and enterococci	10 µg
Haemophilus and Moraxella spp	2 µg
Amoxycillin/clavulanate	
Enterobacteriaceae	20 µg/10 µg
Haemophilus spp, Moraxella spp and staphylococci	2 µg/1 µg
Piperacillin	30 µg
Mezlocillin	30 µg
Azlocillin	30 µg
Cephalothin	30 µg
Cephalexin	30 µg
Cephadroxil	30 µg
Cephradine	30 µg
Cefuroxime	30 µg
Ceftazidime	10 µg
Cefotaxime	10 µg
Cefsulodin	30 µg
Methicillin	5 µg
Carbenicillin	100 µg
Ticarcillin	75 µg
Imipenem	10 µg
Gentamicin	10 µg
Amikacin	30 µg
Tobramycin	10 µg
Neomycin	30 µg
Netilmicin	10 µg
Erythromycin	5 µg
Clindamycin	2 µg
Tetracycline	10 µg
Fusidic acid	10 µg
Chloramphenicol	
Enterobacteriaceae	30 µg
haemophili, pneumococci and meningococci	10 µg
Colistin	10 µg
Nalidixic acid	30 µg
Nitrofurantoin	50 µg
Sulfafurazole	
Enterobacteriaceae and enterococci	100 µg
meningococci	25 µg
Trimethoprim	2.5 µg
Trimethoprim/sulphamethoxazole	1.2 µg/23.8 µg
Spectinomycin	100 µg
Vancomycin	30 µg
Rifampicin	5 µg
Ciprofloxacin	1 µg
Mupirocin	5 µg

Section 6 ◆ Miscellaneous

Table 80.4: Interpretation chart used in kirby-sauer disk diffusion method

*Antibiotic**	*Diameter of zone inhibiticn (in mm)*		
	Resistant	*Intermediate sensitive*	*Sensitive*
Benzylpenicillin	≤ 28	—	≥ 29
Ampicillin			
Enterobacteriaceae	≤ 13	14-16	≥ 17
Methicillin	≤ 9	10-13	≥ 14
Carbenicillin			
Esch. coli and *Proteus sp.*	≤ 17	18-20	≥ 23
Ps. aerugirwsa	≤ 13	14-17	≥ 18
Gentamicin	≤ 12	—	≥ 13
Amikacin	≤ 14	15-16	≥ 17
Erythromycin	≤ 13	14-17	≥ 18
Tetracycline	≤ 14	15-18	≥ 19
Chloramphenicol	≤ 12	13-17	≥ 18
Nalidixic acid	≤ 13	14-18	≥ 19
Trimethoprim	≤ 10	11-15	≥ 16
Ciprofloxacin	≤ 15	16-20	≥ 21

* Only limited antibiotics have been shown in the table.

The interpretation of zone size into susceptible (infection treatable with nonnal dosage), moderately susceptible (infection that may respond to therapy with higher dosage) or resistant (not treatable with this agent) is based on the inter pretation chart (Table 80.4). Reference strains of *S. aureus*, *E. coli*, *P. aerugiriosa*, etc. should be tested each time a new batch of disks or agar is used.

Kirby Bauer test results are interpreted using a table that relates zone diameter to the degree of microbial resistance.

2. Stokes Disk Diffusion Method

For Stokes disk diffusion method, the plate is divided into three parts. The control inoculum should be spread in two bands on either side of the plate, leaving a central band uninoculated. The test organism is inoculated on central one third and control on upper and lower thirds of the plate. However, in modified Stokes disk diffusion method, the test organism is inoculated in the upper and lower thirds and control on central one third. An uninoculated gap 2-3 mm wide should separate the test and control areas on which antibiotic disks are applied (Fig. 80.3). A maximum of six antibiotic disks can be accommodated on a single 100 mm diameter plate. Plates should be incubated in air at 35-37°C overnight (ideally for 16-18 h). Tests should not be read earlier.

Reading and Reporting Results

Diskard any plates on which growth is not semi confluent and repeat the tests. Measure the inhibition zones of the control strain, i.e. the distance in millimeters from the edge of the disk to the zone edge if that is obvious; if it is not, measure to the point of 80 percent inhibition of growth. The measurement should be made with callipers, a millimetre rule or a ruled template, and on the agar surface, not through the glass or plastic bottom or lid of the dish.

Each zone size is interpreted as follows:

Categories of sensitivity: Three categories of sensitivity can be recognized:

1. Sensitive: The zone size of the test strain is larger than, equal to or not more than 3 mm smaller than that of the control strain.
2. Intermediate: The zone size of the test strain is at least 2 mm, but also 3 mm smaller than that of the control strain.
3. Resistant: The zone size of the test strain is smaller than 2 mm.

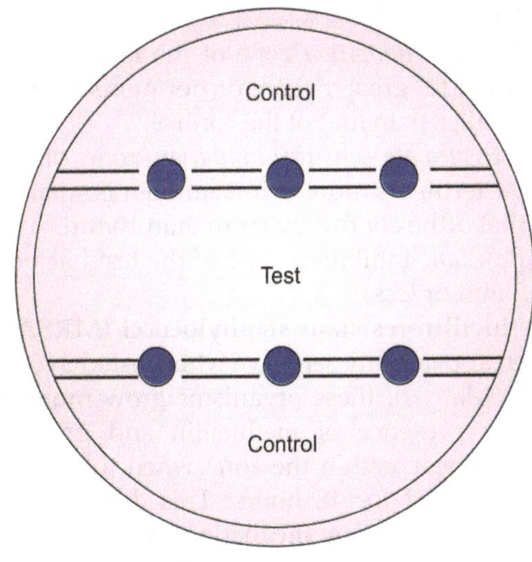

Fig. 80.3: Stokes disk diffusion method

Exceptions

1. **Penicillinase:** Penicillinase producing strain of *Staphylo coccus*-fails to form enough of the enzyme to neutralize penicillin close to the disk, it will show an inhibition zone but, it will be smaller and colonies at the edge are large and well developed and there is no gradual fading of growth towards the disk. These penicillinase-producing staphylococci show heaped up zone edges in tests of penicillins; accordingly they should be reported as resistant to penicillin irrespective of zone size.
2. **Motile organisms:** Motile organisms such as *Proteus mirabilis* and *P. vulgaris* may swarm when growing on agar surface resulting in a thin veil that may penetrate into the zones of inhibition around antibiotic disks. The zones of swarming should be ignored and the outer margin, which is usually clearly outlined, should be measured.
3. **Polymyxins:** Polymyxins diffuse poorly in agar so that zones are small. In this case, by Stokes method, report as:
   Sensitive: Zone radius equal to, wider than, or not more than 3 mm smaller than the control.
   Resistant: Zone radius more than 3 mm smaller than the control.
4. **Zones around ciprofloxacin:** Zones around ciprofloxacin discs are large with some control strains. These are interptreted as follows:
   (a) When sensitive control used is *Staph. aureus* or *Ps. aeruginosa*.
   Sensitive-Inhibition zone of the test bacterium is equal to, greater than, or not more than 7 mm smaller than that of control.
   Intermediate sensitive-Inhibition zone of the test bacterium is more than 2 mm but is smaller than that of the control by more than 7mm.
   Resistant-Inhibition zone of the test bacterium is 2 mm or less.
   (b) When sensitive control used is *Esch. coli* or *H. Influenzae*
   Sensitive-Inhibition zone of the test bacterium is equal to, greater than, or not more than 10 mm smaller than that of the control.
   Intermediate sensitive-Inhibition zone of the test bacterium is more than 2 mm but is smaller than that ofthe control by more than 10mm.
   Resistant-Inhibition zone of the test bacterium is 2 mm or less.
5. **Methicillin-resistant staphylococci (MRSA):** Will often appear fully sensitive when tested in ordinary way. Many of these organisms grow more slowly in the presence of methicillin and growth will only appear within the zone when the incubation is continued for 48 hours. This difficulty can be overcome either by incubating the culture at 30°C or test may be reliable when incubated at 37°C if 5 percent NaCI has been added to the medium.

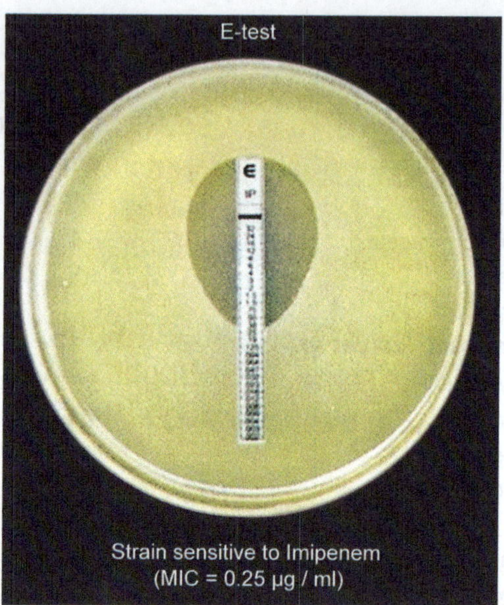

Fig. 80.4: The E-test (*Courtesy: Dr Krishna Prakash, Director Professor, Department of Microbiology, Maulana Azad Medical College, New Delhi -110002*)

6. **Trimethoprim and sulphamethoxazole disks:** For sensitivity tests trimethoprim and sulphamethoxazole disks containing both drugs are widely used. Such disks may be misleading., it is impossible to know whether the organism is sensitive to both or only to one of them when both drugs are present. To overcome this problem, each drug should, therefore, be tested separately.

3. Primary Sensitivity Tests

The disk diffusion methods, as described above, are done after the pathogenic bacteria are isolated from the clinical specimens. In these tests the specimen serves as the inoculum. A portion of it is spread uniformly over part or whole of the primary culture plate and antibiotic disks are applied.

In practice, it appears that primary sensitivity tests are of most value for specimens of urine, of some value for swabs or pus from patients attending accident and emergency departments, but of little value for specimens from patients already receiving antibiotics or for specimens from sites likely to be heavily contaminated, e.g. bedsores, varicose ulcers, vaginal swabs and wounds infected with mixed intestinal bacteria.

Epsilometer or E-Test

The E-test, a modification of the disk diffusion test, uti lizes a strip impregnated with a gradient of concentrations of an antimicrobial drug. Multiple strips, each containing a different drug, are placed on the surface of an agar medium that has been uniformly inoculated with the test organism so that they extend out radially from the center (Fig. 80.4). Each strip contains a gradient of an antibiotic and is labeled with a scale of minimal inhibitory concentration values. The lowest concen tration in the strip lies at the center of the plate. After 24 to 48 hours of incubation, an elliptical zone of inhibition appears. The MIC is determined by reading a number of the numerical scale printed on the strip at the point where the bacterial growth intersects it.

Minimum Inhibitory and Bactericidal Concentrations

The minimum inhibitory concentration (MIC) is the least amount of antimicrobial that will inhibit visible growth of an organism after overnight incubation. The minimum bactericidal concentration (MBC) is the amount of agent that will prevent growth after sub culture of the organism to antibiotic-free medium.

Principal uses of MIC

1. In the determination of antibiotic sensitivities of organisms from patients with serious infections, e.g. infective endocarditis.
2. In measuring the antimicrobial sensitivities of slow-growing organisms, e.g. *Mycobacterium tubercu-losis)*.
3. Reference laboratories determine MICs as reference points in the evaluation and comparison of new and existing antimicrobial agents.

Determination of MBC is not routinely performed in clinical laboratories, the exception being in the management of endocarditis when the dose and combination of antibiotics may be adjusted according to MBC.

Dilution Methods

Here, serial dilutions of the drug are prepared and inoculated with the test bacterium. This is too laborious for routine use. Dilution tests may be done by the tube dilution or agar dilution methods.

1. Broth Dilution Method

Serial dilutions of the drug in Mueller-Hinton broth are taken in tubes and a standardized suspension of the test bacterium inoculated. The inoculum is prepared as in case of disk diffusion methods by comparing with 0.5 McFarland opacity standard. An organism of known sensitivity should also be titrated. Incubate at 35-37°C for 16-18 hours and read the results. Incubate at 30°C for determination of MIC of methicillin (Fig. 80.5).

MIC is the lowest concentration of antimicrobial agent at which there is no visible growth. For deter mination of MBC, subculture from each tube showing no growth over a quarter of a nutrient medium free from antimicrobial agent. Incubate and examine them for growth. The tube containing lowest concentration of the antimicrobial agent that fails to yield growth, on subculture, is the MBC of the antimicrobial agent for the test strain. MIC inhibits the bacterial growth while MBC kills the bacterium.

Determining the MIC and MBC using these conventional methods gives precise information regarding an organism's susceptibility. The techniques, however, are labor-intensive and consequently expensive. In addition, individual sets of tubes must be inoculated to determine susceptibility to each different antimicrobial tested.

2. Agar Dilution Method

Here, serial dilutions of the drug are prepared in agar (Mueller-Hinton agar) and poured into plates. The 'agar dilution' method is more convenient when several strains are to be tested at the same time. The advantage is that many strains can be inoculated on each plate containing an antibiotic dilution.

Automated versions of sensitivity tests are available and are in use in large laboratories.

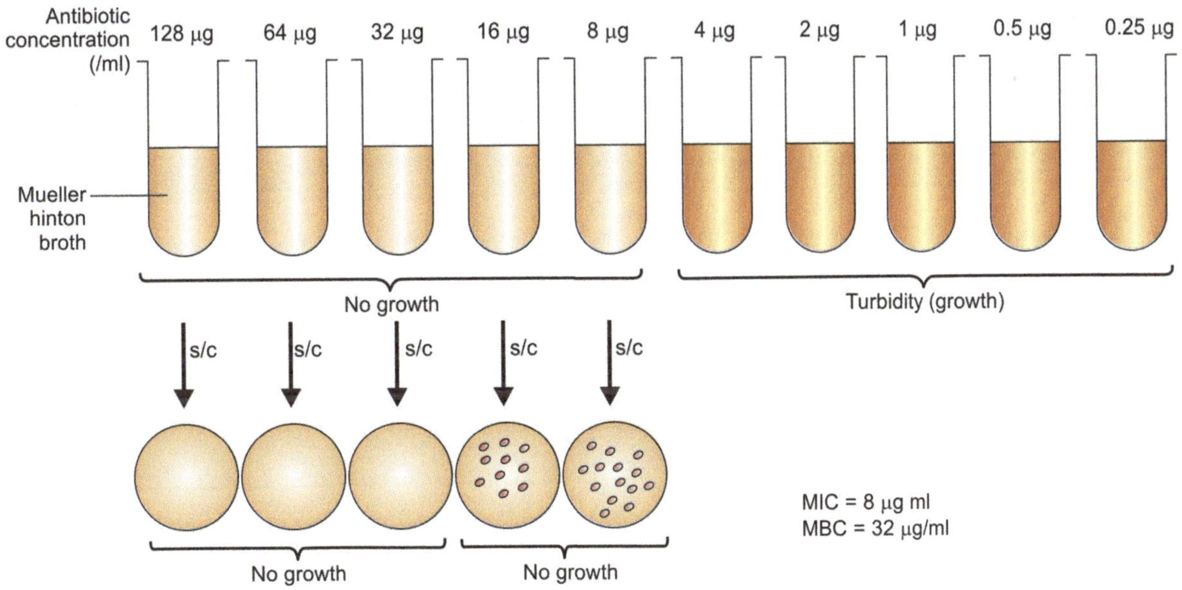

Fig. 80.5: Broth dilution methods showing MIC and MBC

ANTIBIOTIC ASSAYS IN BODY FLUIDS

These are required to verify whether adequate drug concentrations are achieved in blood and other body fluids, and to guard against excessive blood levels of potentially toxic drugs. Likewise, new drugs must be tested to determine achievable levels in the blood, urine, or other body fluids.

The assays are generally done by making serial dilutions of the specimen and inoculating standard suspensions of bacteria of known MIC. Assays by the agar diffusion method can also be done. A technique called the diffusion assay is used to measure the concentration of an antimicrobial in a fluid specimen. The test relies on the same principie as the Kirby Bauer test, except in this case it is the concentration of drug, not the sensitivity of organism, being assayed. This depends on the direct relationship between antibiotic concentration and the diameter of the zone of inhibition with a standard sensitive strain of bacterium.

KNOW MORE

Kirby: Bauer disk diffusion method.

The values in table were derived by finding the MIC values and zone diameters for many different microbial strains. A plot of MIC (on a logarithmic scale) versus zone inhibition diameter (arithmetic scale) is prepared for each antibiotic. These plots are then used to find the zone diameters corresponding to the drug concentrations actually reached in the body.

KEY POINTS

- Antibiatic susceptibility tests are of two types: diffusion tests and dilution tests.
- Diffusion tests consisists of Kirby-Bauer and Stokes' disk method. Stokes' disk method incorporates built-in controls against many variables and therefore provides dependable results.
- Epsilometer or E-test is a modification of the disk diffusion test. It uses an strip imptregnated with with a gradient of concentration antimicrobial drug.
- **Dilution tests:** There are two types of dilution tests: Broth dilution method and Agar dilution method.

IMPORTANT QUESTIONS

1. Name different methods of antibiotic sensitivity testing. Diskuss in detail Kirby-Bauer disk diffusion method for antimicrobial sensitivity testing.
2. Write short notes on:
   a. Kirby-Bauer disk diffusion method
   b. Stokes disk diffusion method
3. Write short notes on: Minimum inhibitory concentration of antimicrobial agents.
   Minimum bactericidal concentration of antimicrobial agents.

FURTHER READING

Amyes SGB, Miles RS. Laboratory control of anti microbial therapy. In: Mackie and McCartney

Practical Medical Microbiology, 14th ed. Collee. JG, Fraser AG, Marmion BP, Simmons A (eds), Churchill Livingstone, New York, Edinburg, Lon don, Madrid, Melbourne, San Francisco, Tokyo 1996:151-78.

Koletar SL. Concepts in Antimicrobial Therapy. In: Diagnostic Microbiology, 2nd ed. Mahon Cr, Manuselis G (eds) WB Saunders Company Phila delphia, London, Toronto, Montreal, Sydney, Tokyo 2000:52-104.

Winn WC Jr, , Allen SD,Janda WM, Koneman EW, Procop G W, Schrecken berger PC, Woods GL. Color Atlas and Textbook of Diagnostic Microbiology, 6h ed, Lippincott Philadelphia New York 2006:945-1021.

CHAPTER 81

Antimicrobial Chemotherapy

LEARNING OBJECTIVES

After reading and studying this chapter, you should be able to:

- Describe mechanism of action of antibacterial drugs.
- Describe mechanism of drug resistance.
- List chephalopsporins of first, second, third and fourth generation.

DISCOVERY OF ANTIMICROBIAL DRUGS

The modern era of chemotherapy began with the work of the German physician Paul Ehrlich (1854–1915). In 1932, the German chemist Gerhard Domagk, using the same dogged persistence demonstrated by Ehrlich, discovered that a red dye called prontosil was dramatically effective in treating streptococcal infections in animals. Prontosil molecule, producing a smaller molecule called sulfanilamide. This breakdown product acted against the infecting streptococci. Thus, the discovery of sulfanilamide, the first sulfa drug, was based on luck as well as scientific effort.

Antimicrobial agent: Antimicrobial agent is a chemical substance inhibiting the growth or causing the death of a microorganism. Many chemicals have this property if sufficiently high concentrations are used. An ideal antimicrobial agent exhibits selective toxicity. This term implies that a drug is harmful to a parasite without being harmful to the host.

ANTIBIOTIC

Antibiotic as originally defined was a chemical substance produced by various species of microorganisms that was capable of inhibiting the growth or causing death of other microorganisms in low concentration. However, with the advent of synthetic methods, this definition has been modified.

CHEMOTHERAPEUTIC AGENTS

Chemotherapeutic agents are the chemical substances used to kill or inhibit the growth of microorganisms already established in the tissues of the body. Synthetic compounds such as sulfonamides, quinolones, nitrofurans and imidazoles should strictly be referred to as chemotherapeutic agents. However, since some antibiotics can be manufactured synthetically while others are the products of chemical manipulation of naturally occurring compounds (semisynthetic antibiotics) the distinction is now ill defined. Nowadays the term antibiotic is used loosely to describe agents (mainly, but not exclusively, antibacterial agents) used to treat systemic infection.

Antimicrobial agent (AMA): It would be more meaningful to use the term antimicrobial agent (AMA) to designate synthetic as well as naturally obtained drugs that attenuate microorganisms.

ANTIBACTERIAL AGENTS

The principal types of antibacterial agents are listed in Table 81.1. These have been grouped according to their site of action.

MECHANISMS OF ACTION OF ANTIBACTERIAL DRUGS

Several microbial processes, including the synthesis of bacterial cell walls, proteins, and nucleic acids, metabolic pathways, and the integrity of the cytoplasmic membrane, are the targets of most antimicrobial drugs (Fig. 81.1). Mechanisms of action' of antibacterial agents can be placed under the headings:

1. Inhibition of bacterial cell wall synthesis;
2. Inhibition of bacterial cytoplasmic membrane function.
3. Inhibition of bacterial nucleic acid synthesis;
4. Inhibition of bacterial protein synthesis.

1. Inhibiton of Bacterial Cell Wall Synthesis

Bacterial cell walls are unique in that they contain peptidoglycan. Inhibitors of bacterial cell wall synthesis act on the formation of the peptidoglycan layer (Fig. 81.2). Bacteria that lack peptidoglycan, such as mycoplasmas, are resistant to these agents.

Table 81.1: Mechanisms of antibacterial drug action	
1. Inhibitors of bacterial cell wall synthesis	
Penicillins	Cephalosporins
Vancomycin	Bacitracin
Cycloserine	Fosfomycin
2. Inhibitors of bacterial cytoplasmic membrane function	
Polymyxins	Gramicidin
Tyrocidine	
3. Inhibition of bacterial nucleic acid synthesis	
Quinolones	Rifamycins
Nitroimidazoles	Nitrofurans
Novobiocin	
4. Inhibition of bacterial protein synthesis	
Aminoglycosides	Chloramphenicol
Tetracyclines	Macrolides
Lincosamides	Fusidic acid
Streptogramins	Mupirocins
5. Metabolic antagonism	
Trimethoprim	
Sulfonamides	Dapsone
Isoniazid	

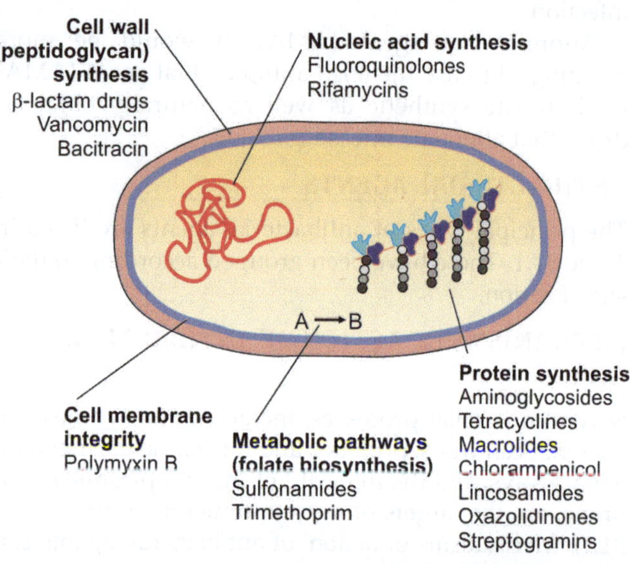

Fig. 81.1: Targets of antibacterial drugs

The antibiotics which inhibit cell wall synthesis are β-lactam antibiotics (penicillins and cephalosporins), glycopeptides, bacitracin, cycloserine, fosfomycin and isoniazid.

A. β-Lactam Agents

This group includes penicillins, cephalosporins and other compounds that feature a β-lactam ring in their structure (Fig. 81.2). All these compounds bind to proteins situated at the cell wall-cell membrane interface.

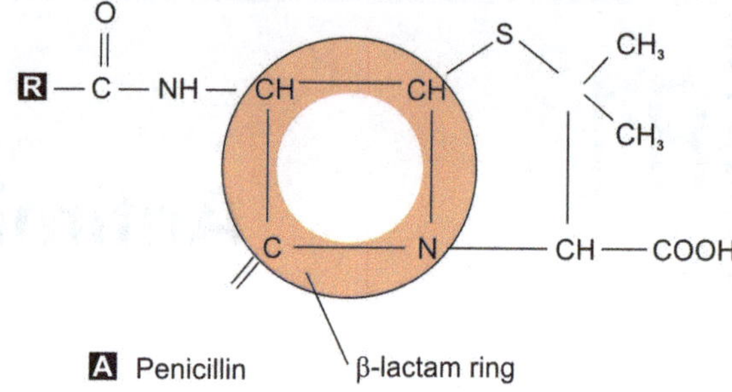

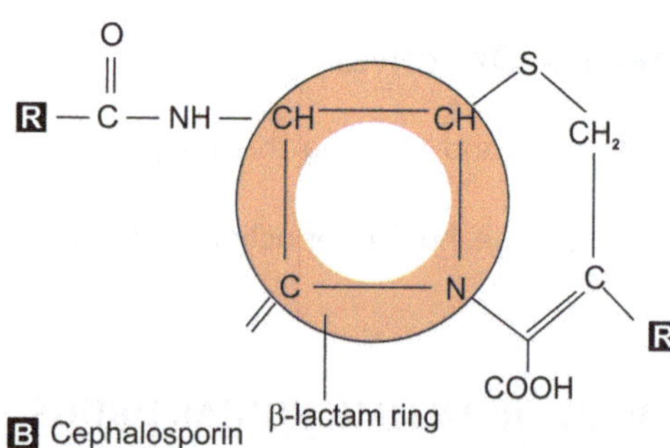

Fig. 81.2: The β-lactam ring of penicillins and cephalosporins the core chemical structure of (A) a penicillin (B) a cephalosporin. The β-lactam rings are marked by an orange circle. The R groups vary among different penicillins and cephalosporins

These penicillin-binding proteins (PBPs) are involved in cell wall construction, including the cross-linking of the peptidoglycan strands that gives the wall its strength. Opening of the β-lactam ring by hydrolytic enzymes, collectively called β-lactamases, abolishes antibacterial activity. Many such enzymes are found in bacteria.

Penicillins

Each member of the family of penicillins shares a common basic structure. Penicillins are a group of antimicrobial substances, all of which possess a common chemical nucleus (6-aminopenicillanic acid) which contains a β-lactam ring essential to their biologic activity. A side chain is attached to the β-lactam ring which determines many of the antibacterial and pharmacological characteristics of a particular type of penicillin. Only the side chain has been modified in the laboratory to create penicillin derivatives, each with unique characteristics.

All β-lactam antibiotics inhibit formation of bacterial cell wall. They particularly block the final transpeptidation reaction in the synthesis of cell wall peptidoglycan and also activate autolytic enzymes in the cell wall. These reactions lead to cell death. All β-lactam antibiotics must cross the bacterial cell wall, resist degradation by β-lactamases, and bind to penicillin-binding proteins situated at the cell wall-cell membrane interface.

Currently the family of penicillins can be loosely grouped into several categories, each of which consists of several different drugs:

Penicillins can be divided into several groups:

1. Those with highest activity against gram-positive organisms but susceptible to hydrolysis by β-lactamases:
   - Penicillin-G
   - Penicillin- V
2. Those with high activity against both gram-positive and gram-negative organisms but destroyed by β-lactamases:
   - Ampicillin
   - Amoxicillin
   - Carbenicillin
   - Ticarcillin
   - Piperacillin
3. Those stable in gastric acid and suitable for oral administration:
   - Penicillin- V
   - Cloxacillin
   - Ampicillin
4. β-lactamase resistant penicillins:
   - Methicillin
   - Nafcillin
   - Oxacillin
   - Cloxacillin
   - Dicloxacillin
   - Flucloxacillin
5. Penicillins active against pseudomonas:
   - Ampilcillin
   - Carbenicillin
   - Ticarcillin
   - Azlocillin
   - Mezlocillin
   - Piperacillin

Cephalosporins

Cephalosporins are a family of antibiotics originally isolated in 1948 from the fungus *Cephalosporium*, and their β-lactam structure is very similar to that of the penicillins (Fig. 81.2). In place of 6-aminopenicillanic acid, they have a nucleus of 7-aminocephalosporanic acid. As might be expected from their structural similarities, cephalosporins resemble penicillins in inhibiting the transpeptidation reaction during peptidoglycan synthesis. They are broad-spectrum drugs frequently given to patients with penicillin allergies.

Like the penicillins, the cephalosporins have been chemically modified to produce a family of various related antibiotics. They are grouped as the first-, second-, third-, and fourth generation cephalosporins (Table 81.2).

First-generation cephalosporins are more effective against Gram-positive than Gram-negative pathogens. Second-generation drugs act against many gram-negative as well as gram-positive pathogens. Third-generation drugs are particularly effective against gram-negative pathogens, and often also reach the central nervous system.

Other β-Lactam Antiblotics

Various agents with diverse properties share the structural feature of a β-lactam ring with penicillins and cephalosporins (Fig. 81.2). Two other groups of β-lactam drugs, carbapenems and monobactams, are very resistant to β-lactamases.

Carbapenems

The carbapenems are effective against a wide range of Gram-negative and Gram-positive bacteria. Two types are available, imipenem and meropenem.

i. Imipenem: Imipenem is inactivated by a kidney enzyme—dehydropeptidase in the human kidney, and is co-administered with a dehydropeptidase inhibitor, cilastatin.
ii. Monobactam: The only monobactam used therapeutically, aztreonam, is primarily effective against members of the family *Enterobacteriaceae*, which are Gram-negative rods.

Oxacephems (e.g. latamoxef): are broad-spectrum β-lactamase-stable compounds.

The clavam, clavulanic acid, exhibits poor antibacterial activity, but has proved useful as a β-lactamase inhibitor when used in combination with β-lactamase-susceptible compounds (e.g. co-amoxiclav, the combination of amoxycillin and clavulanic acid).

The sulfhones, sulbactam and tazobactam, also act as β-lactamase inhibitors and are marketed combined with ampicillin (or cefoperazone) and piperacillin, respectively.

Glycopeptides

Two glycopeptides, vancomycin and teicoplanin, are in clinical use. Their chief importance resides in their action against gram-positive cocci with multiple resistance to other drugs. They are mainly used in serious infections with staphylococci and enterococci that are resistant to other drugs.

Table 81.2: Important cephalosporins and their antibacterial spectrum

Class	*Compounds*	*Antibacterial spectrum*
First generation	Cephalothin	*S. aureus, streptococci* (other than enterococci), *E. coli, Klebsiella,*
	Cephalexin	*H. influenzae* and *P.mirabilis*
Second generation	Cefamandole Cefotaxime	First generation spectrum expanded to indole positive *Proteus, enterobacter, citrobacter, serratia* and many gram-negative anaerobes.
Third generation	Cefotaxime	Spectrum of second generation expanded to give high activity against
	Cefoperazone	*H.influenzae*, gonococci including b-lactamase producing strains and activity against *P.aeruginosa* and many gram-negative anaerobes.
Fourth generation	Cefepime Cefpirome	Spectrum similar to that of third generation compounds, but highly resistant to b-lactamases, hence active aainst many bacteria resistant to the earlier brugs. *P.aeruginosa* is also inhibited by cefepime.

Other Inhibitors of Bacterial Cell Wall Synthesis

These include bacitracin, cycloserine, fosfomycin and isoniazid.

Bacitracin

Bacitracin is active against Gram-positive bacteria, but is too toxic for systemic use. It is found in many topical preparations, and is also used in the laboratory in the presumptive identification of hemolytic streptococci of Lancefield group A.

Cycloserine

Cycloserine is used only as a second-line agent in infections with multiresistant strains of *Mycobacterium tuberculosis*.

Isoniazid

Isoniazid and some other compounds are used in tuberculosis

Fosfomycin

Fosfomycin exhibits a fairly broad-spectrum, notably against Gram-negative bacilli. It is mainly used for the treatment of urinary tract infection.

2. Inhibition of Bacterial Cytoplasmic Membrane Function

Only polymyxins have been regularly used systemically among membrane active agents used in human medicine. Polymyxins are a family of antibiotics produced by species of *Bacillus*. Two members of the family are in therapeutic use: polymyxin B and colistin (polymyxin E).

Polymyxins combine with the cytoplasmic membrane and alters their permeability, leading to leakage of cellular contents and eventual death of the cells.

Polymyxin B and colistin (polymyxin E) exhibit potent antipseudomonal activity, but toxicity has limited their usefulness, except in topical preparations and bowel decontamination regimens. They have serious nephrotoxicity and neurotoxicity which limits their use in clinical practice and are usually employed only as lifesaving measure.

3. Inhibitors of Nucleic Acid Synthesis

Quinolones

The quinolones are synthetic drugs that contain the 4.-quinolone ring. Quinolones act on the α subunit of DNA gyrase, an enzyme that engineers the breaking and rejoining of super coiled DNA. The first quinolone, nalidixic acid was synthesized in 1962. Their properties allow them to be roughly categorized into three groups. Nalidixic acid and its early congeners are narrow-spectrum agents active only against Gram negative bacteria. Their use is virtually restricted to urinary tract infection, although they have also been used in enteric infections and, in the case of acrosoxacin, in gonorrhea.

Newer quinolones like ciprofloxacin, norfloxacin, ofloxacin, pefloxacin and lomefloxacin are broad-spectrum quinolones. These have been successfully used in a wide variety of infections, but resistance is becoming more prevalent.

Rifamycins

Rifamycins inhibit bacterial growth by binding strongly to the DNA-dependent RNA polymerase of bacteria thus, inhibiting transcription of RNA from DNA. This group of antibiotics is characterized by excellent activity against mycobacteria, although other bacteria are also susceptible. Staphylococci in particular are exquisitely sensitive. Rifampicin and rifabutin are most widely used.

Rifampicin

The best known member of the group, is used in tuberculosis and leprosy. Rifampicin resistance results from a change in RNA polymerase due to a chromosomal mutation that occurs with high frequency.

Rifabutin

Rifabutin (ansamycin) is used in infections caused by atypical mycobacteria of the avium-intracellulare group.

Nitroimidazoles

Azole derivatives have wide-ranging antimicrobial activity against fungi, protozoa, helminths, as well as bacteria. Those that exhibit antibacterial activity are 5-nitroimidazoles. The antibacterial effect of 5-nitroimidazole is dependent on reduction of nitro group under anaerobic conditions, produced intracellularly in anaerobic organisms, to a short-lived intermediate product which kills the cell probably by inducing break in DNA strands. Because of the requirement for low Eh values, 5-nitroimidazoles are active only against anaerobic (and certain' microaerophilic) bacteria and anaerobic protozoa. The representative of the group most commonly used clinically is metronidazole, but similar derivatives include tinidazole, ornidazole and nimorazole. They are primarily antiprotozoal agents, but they exhibit potent activity against anaerobic bacteria.

Metronidazole is a useful alternative to vancomycin in the treatment of *Clostridium dificile* associated colitis. However, the susceptibility of certain microaerophilic organisms remains unexplained.

Nitrofurans

Various nitrofuran derivatives are in use around the world as antibacterial agents. These include nitrofurantoin and furazolidone.

Nitrofurantoin

Nitrofurantoin is the most familiar nitrofuran derivative, an agent used exclusively in urinary tract infection. It is bactericidal to most urinary pathogens at concentrations achievable in urine.

Furazolidone

Furazolidone is used in enteric infections. The mode of action of nitrofurans has not been elucidated, but it is probable that a reduced metabolite acts on DNA in a manner analogs to that of the nitroimidazoles.

Novobiocin

Novobiocin acts on the β-subunit of bacterial DNA gyrase. It is quite active against staphylococci and streptococci, but is no longer favored because of problems of resistance and toxicity. *Staphylococcus saprophyticus* is novobiocin resistant.

4. Inhibition of Bacterial Protein Synthesis

Several types of antibacterial drugs inhibit prokaryotic protein synthesis (Fig. 81.1). Bacteria have 70S ribosomes, whereas mammalian cells have 80S ribosomes. While all cells synthesize proteins, the structure of the prokaryotic 70S ribosome, which is composed of a 30S and a 50S subunit, is different enough from the eukaryotic 80S ribosome to make it a suitable target for selective toxicity. The mitochondria of eukaryotic cells also have 70S ribosomes, however, which may partially account for the toxicity of some of these drugs. They interfere with different stages of the process of protein synthesis.

The major classes of antibiotics that inhibit protein synthesis are the aminoglycosides, the tetracyclines, and the macrolides. Others include the lincosamides and chloramphenicol. Of these, only the aminoglycosides are bactericidal; the others are all bacteriostatic. Two classes of drugs that have recently been approved for use are the oxazolidinones and the streptogramins. A synergistic combination of two streptogramins is bacteriocidal against some organisms.

Aminoglycosides

The aminoglycosides irreversibly bind to the 30S ribosomal subunit, causing it to distort and malfunction. This blocks the initiation of translation and causes misreading of mRNA by ribosomes that have already passed the initiation step. Examples of aminoglycosides include streptomycin, kanamycin, neomycin, gentamicin, tobramycin, amikacin, etc.

Use

They are bactericidal compounds, and some, notably gentamicin and tobramycin, exhibit good activity against *Pseudomonas aeruginosa*. Such compounds have been widely used, often in combination with β-lactam antibiotics, with which they interact synergically, in the 'blind' treatment of sepsis in immunocompromised patients. Unfortunately, these all can cause severe side effects. The group also has in common a tendency to damage the eighth cranial nerve (ototoxicity) and the kidney (nephrotoxicity). Consequently, they are generally used only when other alternatives are not available. Amikacin is resistant to most of the common enzymes. Another aminoglycoside, neomycin, is too toxic for systemic use. However, it is a common ingredient in nonprescription topical ointments.

Chloramphenicol

Chloramphenicol binds to the 50S ribosomal subunit. It is mainly bacteriostatic. This compound, and the related thiamphenicol, also possess a very broad antibacterial spectrum.

Use

Use of chloramphenicol has been limited to typhoid fever, meningitis and a few other clinical indications because of the occurrence of a rare but fatal side-effect, aplastic anemia. Microorganisms resistant to chloramphenicol produce the enzyme chloramphenicol acetyltransferase, which destroys the drug activity.

Tetracyclines

The tetracyclines reversibly bind to the 30S ribosomal subunit, blocking the attachment of tRNA to the ribosome and preventing the continuation of protein synthesis. These drugs are actively transported into prokaryotic but not animal cells, which effectively concentrates them inside bacteria. This, in part, accounts for their selective toxicity. The various members of the tetracycline group are closely related and differ more in their pharmacological behavior than in antibacterial activity. Doxycycline and minocycline are in most common use.

Tetracyclines are broad-spectrum agents with important activity against chlamydiae, rickettsiae, mycoplasmas and, surprisingly, malaria parasites, as well as most conventional Gram-positive and Gram-negative bacteria. Tetracyclines can cause discoloration in teeth when used by young children.

Macrolides

The macrolides reversibly bind to the 50S ribosomal subunit and prevent the continuation of protein synthesis. Macrolides include erythromycin, azithromycin, clarithromycin, dirithromycin and spiramycin. Both clarithromycin and azithromycin have a longer half-life than erythromycin, so that they can be taken less frequently.

Macrolides as a group are effective against a variety of bacteria, including many Gram-positive organisms as well as the most common causes of atypical pneumonia "walking pneumonia". They often serve as the drug of choice for patients who are allergic to penicillin.

Lincosamides

Lincosamides bind to the 50S ribosomal subunit and resemble macrolides in binding site, antibacterial activity, and mode of action. The original lincosamide antibiotic, lincomycin, has been superseded by the 7-deoxy-7-chloro derivatives, clindamycin, which is better absorbed after oral administration and is more active against the organisms within its spectrum. These include staphylococci, streptococci and most anaerobic bacteria, against which clindamycin exhibits outstanding activity. However, it may lead to severe diarrhea which sometimes progresses to a life-threatening pseudomembranous colitis.

Fusidic Acid

The structure of fusidic acid is related to that of steroids, but the antibiotic is devoid of steroid-like activity. It blocks factor G, which is involved in peptide elongation. It has excellent activity against staphylococci, good activity against *Corynebacterium diphtheriae* and modest activity against streptococci, Gram-negative anaerobes, *Nocardia asteroides*, and *Mycobacterium tuberculosis*. It penetrates well into bone and has been widely used (generally in combination with a β-lactam antibiotic to prevent the selection of resistant variants) in the treatment of staphylococcal osteomyelitis.

Streptogramins

This is the collective name for a family of antibiotics that occur naturally as two synergic components, a peptolide and a depsipeptide. Derivatives suitable for parenteral administration, quinupristin and dalfopristin, have been developed as a combination product. The combination is effective against a variety of Gram-positive bacteria, including some of those that are resistant to β-lactam drugs and vancomycin.

Mupirocin

This is an antibiotic, produced by *Pseudomonas fluorescens*. It blocks incorporation of isoleucine into proteins. Its useful activity is restricted to staphylococci and streptococci. It is used only in topical preparations since it is inactivated when given systemically.

5. Metabolic Antagonism

Sulfonamides and Diaminopyrimidines

These agents affect DNA synthesis because of their role in folic acid metabolism. Relatively few antibacterial medications interfere with metabolic pathways. Among the most useful are the folate inhibitors-sulfonamides and trimethoprim. p-aminobenzoic acid (PABA) is an essential metabolite for many microorganisms. They convert it into dihydrofolic acid and then to tetrahydrofolic acid. This is an essential factor in the synthesis of certain amino acids, purines and pyrimidines needed for synthesis of proteins and nucleic acids, and thus for cell growth. Animal cells lack the enzymes in the folic acid synthesis portion of the pathway, which is why folic acid is a dietary requirement.

Mechanism of Action of Sulfonamides

Sulfonamides are analogs of para-aminobenzoic acid (PABA). Because of their similar structures (Fig. 81.3), there occurs competition between the sulfonamides and the PABA for the active site on the surface of the enzyme initiating the conversion of PABA to dihydrofolic acid. When sulfonamides enter into reaction in place of PABA, nonfunctional analgs of folic acid are formed, preventing further growth of bacterial cell. Human tissue cells and sulphonamide resistant bacteria also require folic acid for synthesis of nucleic acids but are capable of taking

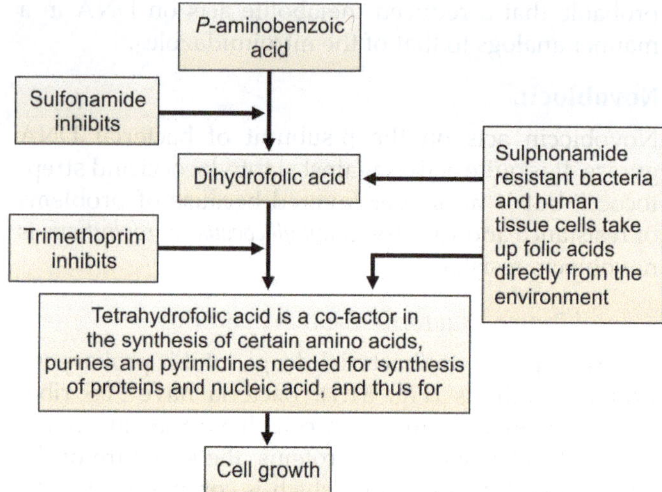

Fig. 81.3: Mechanism of action of sulfonamides

up preformed folic acid from the environment and their growth is independent of the conversion of PABA to folic acid. Other analogues of PABA are diami-nodiphenylsulfone (dapsone) and p-aminosalicylic acid (PAS) which are' active against lepra and tubercle bacilli respectively.

Diaminopyrimidines, which include the broad-spectrum antibacterial agent trimethoprim and the antimalarial compounds pyrimethamine and cycloguanil (the metabolic product of proguanil), prevent the reduction of dihydrofolate to tetrahydrofolate. Sulfonamides and diaminopyrimidines thus, act at sequential stages of the same metabolic pathway and interact synergically.

Sulfonamidesarebroad-spectrumantibacterialagents, predominantly bacteriostatic, but resistance is common and the group also suffers from problems of toxicity. They are now little used alone, but the combination of sulfamethoxazole with trimethoprim (co-trimoxazole) is still widely used, notably in the prophylaxis and treatment of *Pneumocystis carinii* pneumonia. They have some activity against protozoa, including *Plasmodium spp.* and *Toxoplasma gondii*. Sulfadoxine or sulfadiazine combined with pyrimethamine are used in malaria and toxoplasmosis respectively.

ANTIBIOTIC RESISTANCE

The spread of drug-resistant pathogens is one of the most serious threats to the successful treatment of microbial disease. It is now recognized, however that drug resistance limits the usefulness of all known antimicrobials. Understanding the mechanisms and the spread of antimicrobial resistance is an important step in curtailing the problem.

A. Mechanisms of Drug Resistance

Bacteria can resist the effects of antimicrobials through a variety of mechanisms. In some cases this resistance is innate, but in many others it is acquired. The most common mechanisms of acquired antimicrobial resistance are as follows:

1. Drug-Inactivating Enzymes

Some organisms produce enzymes that chemically modify a specific drug in such a way as to render it ineffective.

i. **Penicillinase:** The best-known example, is the hydrolysis of the β-lactam ring of many penicillins by the enzyme penicillinase.
ii. **Chloramphenicol acetyl transferase:** The enzyme chloramphenicol acetyl transferase chemically alters the antibiotic chloramphenicol.

2. Alteration in the Target Molecule

Because each chemotherapeutic agent acts on a specific target, resistance arises when the target enzyme or organelle is modified, so that it is no longer susceptible to the drug.

Examples

i. Alterations in the penicillin-binding proteins prevent β-lactam drugs from binding to them.
ii. Similarly, a change in the ribosomal RNA, the target for the macrolides, prevents those drugs from interfering with ribosome function.

3. Decreased Uptake of the Drug

The porin proteins in the outer membrane of Gram-negative bacteria selectively permit small hydrophobic molecules to enter a cell. Alterations in these proteins can, therefore, alter permeability and prevent certain drugs from entering the cell. By excluding entry of a drug, an organism avoids its effects.

Many gram-negative bacteria are unaffected by penicillin G because it cannot penetrate the envelope's outer membrane. Changes in penicillin binding proteins also render a cell resistant. A decrease in permeability can lead to sulfonamide resistance. Mycobacteria resist many drugs because of the high content of mycolic acids in a complex lipid layer outside their peptidoglycan. This layer is impermeable to most drugs.

4. Increased Elimination of the Drug

The systems that bacteria use to transport detrimental compounds out of a cell are called efflux pumps. This resistance 'strategy is to pump the drug out of the cell after it has entered.

B. Genetic Basis of Antibiotic Resistance

Antimicrobial resistance can be due to either spontaneous mutation, which alters existing genes, or acquisition of new genes.

1. Spontaneous Mutation

As cells replicate, spontaneous mutations occur at a relatively low rate. Even at a low rate, however, such mutations can ultimately have a profound effect on the resistance of a bacterial population to an antimicrobial drug. Usually such mutations result in a change in the drug receptor.

Drugs such as streptomycin used in tuberculosis treatment to which single point mutations can confer resistance are sometimes used in combination with another drug to prevent survival of resistant mutants. If any organism spontaneously develops resistance to one drug, the other drug will still kill it. This is the rational behind multiple drug therapy used in tuberculosis.

2. Gene Transfer

i. Conjugation

The most common mechanism of transfer of resistance is through the conjugative transfer of R plasmids (resistance plasmids).

R plasmids frequently carry several different resistance genes, each one mediating resistance to a specific antimicrobial drug. Thus, when an organism acquires an R plasmid, it acquires resistance to several different medications simultaneously.

ii. Transduction

Acquisition of resistance by transduction is common in staphylococci. The penicillin plasmids (carrying gene for β lactamase production) enclosed in a bacteriophage is transferred from a penicillin-resistant staphylococcus to a susceptible staphylococcus.

iii. Transformation

Resistance transfer can be demponstated experimentally but its significance is not known.

iv. Transposons

Many of the resistance genes on R plasmids are carried on transposons that can move from a plasmid to the chromosome, from one plasmid to another, or from the chromosome to a plasmid. Thus, if one organism has two different plasmids, an antibiotic-resistance gene can move from one to the other.

KNOW MORE

ACQUISITION OF RESISTANCE

Acquisition of resistance through spontaneous mutation is called *vertical evolution,* because it affects only the progeny of the altered cell. In contrast, acquisition of resistance through gene transfer is called **horizontal evolution**; even entirely unrelated organisms can gain new traits this way.

KEY POINTS

- Antimicrobial agent is a chemical substance inhibiting the growth or causing the death of a microorganism.
- Antibiotic as originally defined was a chemical substance produced by various species of microorganisms that was capable of inhibiting the growth or

causing death of other microorganisms in low concentration.

- Chemotherapeutic agents are the chemical substances used to kill or inhibit the growth of microorganisms already established in the tissues of the body.
- Mechanisms of action' of antibacterial agents can be: (1) Inhibition of bacterial cell wall synthesis; (2) Inhibition of bacterial cytoplasmic membrane function; (3) Inhibition of bacterial nucleic acid synthesis; (4) Inhibition of bacterial protein synthesis and (5) Metabolic antagonism.
- Mechanisms of drug resistance are (1) Drug-inactivating enzymes; (2) Alteration in the target molecule; (3) Decreased uptake of the drug; (4) Increased elimination of the drug.
- Antimicrobial resistance can be due to either spontaneous mutation, or acquisition of new genes.
- The most common mechanism of transfer of resistance is through the conjugative transfer of R plasmids (resistance plasmids).

IMPORTANT QUESTIONS

1. Define the terms antimicrobial agent, chemotherapeutic agent and antibiotic. Name various mechanisms of action of antibiotics giving examples.
2. Describe the structure and functions of bacterial cell wall. Name various antibiotics which affect cell wall synthesis.
3. Write short notes on:
   a. Antibiotics inhibiting protein synthesis.
   b. Antibiotics inhibiting bacterial cytoplasmic membrane function.
   c. Antibiotics inhibiting bacterial nucleic acid synthesis.
   d. Antibiotics that act as metabolic antagonists.
4. Discuss genetic basis of drug resistance in bacteria.

FURTHER READING

Finch RG, Greenwood D, Norrby SR, Whitley RJ. Antibiotic and chemotherapy: anti-infective agents and their use in therapy, 8th (Edn) Churchill Livingstone, Edinburgh 2002.

Greenwood D (ed.) Antimicrobial chemotherapy, 4th edn. Oxford University Press, Oxford 2000.

Kucers A, Crowe SM, Grayson Ml, Hoy JF. The use of antibiotics: a clinical review of antibacterial, antifungal and antiviral drugs, 4th (edn). Heinemann, london 1997.

Russell AD, Chopra I. Understanding Antibacterial Action and Resistance, 2nd edn. Ellis Horwood, Chichester 1996.

CHAPTER 82

Immunoprophylaxis

Learning Objectives

After reading and studying this chapter, you should be able to:

- List of immunizing agents.
- Describe the following: live attenuated vaccines; killed vaccines; toxoids.
- Explain immunization schedule.
- Discuss national immunization schedule.
- Describe passive immunization.

INTRODUCTION

An important contribution of microbiology to medicine has been immunization, which is one of the most effective methods of controlling infectious diseases. By systematic active immunisation, many developed countries have virtually eliminated 'vaccine preventable diseases' (VPD) such as diphtheria, pertussis, tetanus, measles, mumps, rubella and poliomyelitis. Major headway has been made against influenza, hepatitis B, pneumococci and *Haemophilus influenzae* type b infections, at least in some parts of the world. Of the major scourges of mankind, malaria, leprosy, helminthic diseases and human immunodeficiency virus infections have remained without useful vaccines.

Immunizing Agents

The immunizing agents may be classified as:

A. Vaccines
B. Immunoglobulins

A. VACCINES

A vaccine [Latin *vacca,* cow] is a preparation from an infectious agent that is administered to humans and other animals to induce protective immunity against a given disease. It stimulates the production of protective antibody and other immune mechanisms. Vaccines may be prepared from live modified organisms, inactivated or killed organisms, extracted cellular fractions, toxoids or combination of these. More recent preparations are sub-unit vaccines and recombinant vaccines.

Types of Vaccines

1. Live Vaccines

Live vaccines (e.g., BCG, measles, oral polio) are prepared from live (generally attenuated) organisms. These organisms have been passed repeatedly in the laboratory in tissue culture or chick embryos and have lost their capacity to induce full blown disease but retain their immunogenicity. In general, live vaccines are more potent immunizing agents than killed vaccines.

Live vaccines should not be administered to persons with immune deficiency diseases or to persons whose immune response may be suppressed because of leukemia, lymphoma or malignancy or because of therapy with corticosteroids, alkylating agents, antimetabolic agents or radiation. Pregnancy is another contraindication unless the risk of infection exceeds the risk of harm to the fetus of some live vaccines.

When two live vaccines are required they should be given either simultaneously at different sites or with an interval of at least 3 weeks. In the case of live vaccines, immunization is generally achieved with a single dose. The exception is polio vaccine which needs three or more doses to be given at spaced intervals to produce effective immunity. Live vaccines usually produce a durable immunity, but not always as long as that of the natural infection.

2. Killed (Inactivated) Vaccines

Organisms killed by heat or chemicals, when infected into the body stimulate active immunity. They are usually safe but generally less efficacious than live vaccines.

Killed vaccines usually require a primary series of 2 or 3 doses of vaccine to produce an adequate antibody response, and in most cases "booster" injections are required. The duration of immunity following the use of inactivated vaccines varies from months to many years. Inactivated polio vaccine has been quite an effective vaccine. Killed vaccines are usually administered by subcutaneous or intramuscular route.

3. Toxoids

Certain organisms produce exotoxins, e.g., diphtheria and tetanus bacilli. The toxins produced by these organisms are detoxicated and used in the preparation of vaccines. The antibodies produced neutralize the toxic moiety produced during infection, rather than act upon the organisms. In general, toxoid preparations are highly efficacious and safe immunizing agents.

4. Cellular Fractions

Vaccines, in certain instances, are prepared from extracted cellular fractions, e.g., meningococcal vaccine from the polysaccharide antigen of the cell wall, the pneumococcal vaccine from the polysaccharide contained in the capsule of the organism and hepatitis B polypeptide vaccines. Although the duration of experience with these vaccines is limited, their efficacy and safety appear to be high.

5. Mixed or Combined Vaccine

If more than one kind of immunizing agent is included in the vaccine, it is called a mixed or combined vaccine. The following are some of the well-known combinations:

DPT (Diphtheria-pertussis-tetanus)
DT (Diphtheria-tetanus)
DP (Diphtheria-pertussis)
DPT and typhoid vaccine
MMR (Measles, mumps and rubella)
DPTP (DPT plus inactivated polio)

6. Recombinant-Vector Vaccines

It is now possible to isolate genes that encode major antigens from a pathogen and insert them into nonvirulent viruses or bacteria. Recently-several micro-organisms have been used in the production of these recombinant-vector vaccines.

Examples: Adenovirus, vaccinia virus, canarypox virus, attenuated poliovirus, and attenuated strains of *Salmonella* and *Mycobacterium.*

7. DNA Vaccines

A DNA vaccine elicits protective immunity against a microbial pathogen by activating both branches of the immune system: humoral and cellular. Long-lasting memory are cells also are generated.

Examples: At present, there are human trials under way with several different DNA vaccines against malaria, AIDS, influenza, hepatitis B, and herpesvirus. Vaccines against a number of cancers (lymphomas, prostate, colon) are also being tested.

IMMUNIZATION

Immunization is of three types: active immunization, passive immunization and combined passive and active immunization.

A. Active Immunization

Active immunization, is one of the most powerful and cost-effective weapons of modern medicine. It is the protection of susceptible humans from communicable diseases by the administration of vaccines (vaccination).

Immunoprophylaxis may be in the form of:

1. *Routine immunization:* which forms part of basic health care. There are some infectious diseases whose control is solely based on active immunization, e.g., polio, tetanus, diphtheria and measles. Vaccination against these diseases is given as a routine during infancy and early childhood, with periodic boosters to maintain adequate levels of immunity.
2. *Immunization of individuals or selected groups:* There are immunizations against certain diseases which are offered to high-risk groups or restricted to definite geographic areas where the disease is endemic or a public health problem (e.g., yellow fever). Diseases for which improved or less costly vaccines are needed include tuberculosis, pertussis, meningococcal meningitis, hepatitis B, rabies, Japanese encephalitis, etc.

Immunization Schedules

National Immunization Schedule

The National Immunization Schedule is given in Table 82.1. The first visit may be made when the infant is 6 weeks old; the second and third visits, at intervals of 1 to 2 months. Oral polio vaccine may be given concurrently with DPT. BCG can be given with any of the three doses but the site for the injection should be different. The schedule also covers immunization of women during pregnancy against tetanus.

Expanded Programme on Immunization (EPI)

In May 1974, the WHO officially launched a global immunization programme, known as Expanded Programme on Immunization (EPI) to protect all children of the world against six vaccine-preventable diseases, namely - diphtheria, whooping cough, tetanus, polio, tuberculosis and measles by the year 2000. EPI was launched in India in January 1978. This is given in Table 82.2. The Programme is now called Universal Child Immunization, 1990-that's the name given to a declaration sponsored by UNICEF as part of the United Nations' 40th anniversary in October 1985. It is aimed at adding impetus to the global programme of EPI.

Table 82.1: National immunization schedule		
a. For infants		
At birth (for institutional deliveries	-	BCG and OPV-O dose
At 6 weeks	-	BCG (if not given at birth)
	-	DPT-1 and OPV-1
At 10 weeks	-	DPT-2 and OPV-2
At 14 weeks	-	DPT-3 and OPV-3
At 9 months	-	Measles
b. At 16–24 months	-	DPT and OPV
c. At 5–6 years	-	DT-the second dose of DT should be given at an interval of one month if there is no clear history or documented evidence of previous immunization with DPT
d. At 10 and 16 years	-	Tetanus Toxoid - The second dose of TT vaccine should be given at an interval of one month if there is no clear history or documented evidence of provious immunization with DPT, DT or TT vaccines
e. For pregnant women		
Early in pregnancy	-	TT-1 or Booster
One month after TT-1–TT-2		

Note: i. Intreval between 2 dose should not be less than one month.
II. Minor cough, colds and mild fever are not a contraindication to vaccination.
iii. In some states, Hepatitis B vaccine is given as routine immunization.

Table 82.2: WHO EPI immunization schedule (when early protection is a must)

Age	*Vaccine*
Birth	BCG, oral polio
6 weeks	DPT, oral polio
10 weeks	DPT, oral polio
14 weeks	DPT, oral polio
9 months	Measles

The Indian version, the Universal Immunization Programme, was launched on November 19, 1985 and was dedicated to the memory of Smt Indira Gandhi. The National Health Policy aimed at achieving universal immunization coverage of the eligible population by 1990.

B. PASSIVE IMMUNIZATION

Passive immunization is used, when it is considered necessary to protect a patient at short notice and for a limited period. Antitoxic, antibacterial or antiviral antibodies in human (homologous) or animal (heterologous) serum are injected to give temporary protection. Human (homologous) sera are much less likely to give rise to the adverse reactions occasionally associated with the animal (hetrologous) sera. Homologous sera confer protection for 3 to 6 months, whereas protection afforded by a heterologous serum is likely to last for only a few weeks.

Preparations for passive immunization: Three types of preparations are available for passive immunization:

a. Normal human immunoglobulin,
b. Specific (hyperimmune) human immunoglobulin
c. Antisera or antitoxins.

a. Normal Human Immunoglobulin

Normal human Ig is used to prevent measles in highly susceptible individuals and to provide temporary protection (upto 12 weeks) against hepatitis A infection for travellers to endemic areas and to control institutional and household outbreaks of hepatitis A infection.

Live vaccines should not normally be given for 12 weeks after an injection of normal human Ig, and if a live vaccine has already been given. NHIg injection should be deferred for 2 weeks.

b. Specific Human Immunoglobulin

These preparations are made from the plasma of patients who have recently recovered from an infection or are obtained from individuals who have been immunized against a specific infection. Therefore they have a high antibody content against an individual infection and provide immediate protection e.g., specific human Igs are used for chickenpox prophylaxis of highly susceptible individuals and for passive immunization against tetanus (human tetanus immunoglobulin, HTIG), hepatitis B (HBIG), rabies (HRIG), varicella-zoster (ZIG) and vaccinia (AVIG).

c. Antisera

The term *antiserum* is applied to materials prepared in animals. Originally, passive immunization was achieved by the administration of antisera or antitoxins prepared from nonhuman sources such as horses. Since human immunoglobulin preparations exist only for a small number of diseases, antitoxins prepared from non-human sources (against tetanus, diphtheria, botulism, gas gangrene and snake bite) are still the mainstay of passive immunization.

Administration of antisera may occasionally give rise to serum sickness and anaphylactic shock due to abnormal sensitivity of the recipient. The current trend is in favor of using immunoglobulins wherever possible.

C. Combined Passive and Active Immunization

In some diseases (e.g., tetanus, diphtheria, rabies) passive immunization is often undertaken in conjunction with inactivated vaccine products, to provide both immediate (but temporary) passive immunity and slowly developing active immunity. If the injections are given at separate sites, the immune response to the active agent, mayor may not be impaired by immunoglobulin.

INDIVIDUAL IMMUNIZATION

Vaccines offered under national programmes are limited by economic considerations and so some important vaccines may be omitted because they are costly. These may be supplemented by individual initiative, whenever possible.

Hepatitis B vaccine: Many developing countries, including India, have high endemicity for this virus. Inclusion of the hepatitis B vaccine in routine childhood immunization will therefore be beneficial. The recent reduction in cost of the vaccine as a result of indigenous manufacture, has made mass vaccination more feasible.

Three doses of killed vaccine are given at 0, 1 and 6 months intramuscularly into the deltoid or, in infants into the anterolateral aspect of thigh. Gluteal injection is not recommended because it may result in poor response.

MMR vaccine: A dose of MMR vaccine may be beneficial at 16 to 24 months or later, not only to reinforce immunity against measles but also to protect against mumps and rubella. It is live attenuated vaccine. It is contraindicated in pregnancy.

Typhoid vaccine: The original typhoid vaccine is not widely used because of its uncertain benefit and frequent adverse reactions. Two recent typhoid vaccines, the live oral Gal-E mutant vaccine and the injectable purified Vi polysaccharide vaccine may be acceptable because they offer prolonged protection and are free from reactions. They are recommended for immunization of those five-years-old or above and so may be employed at school entry.

KNOW MORE

- The modern era of vaccines and vaccination began in 1798 with Edward Jenner's use of cowpox as a vaccine against smallpox and in 1881 with Louis Pasteur's anthrax vaccine.
- The term "polyvalent" is applied to vaccines (e.g., polio, influenza vaccines) which are prepared from two or more strains of the same species. The term "auto" or "autogenous" vaccine is applied when the organism in the vaccine is obtained from the same patient.
- In India, the Expanded Programme on Immunization (EPI) and the Universal Immunization Programme (UIP) have been able to afford protection for much of the target population against VPDs and have led to a significant decline in the recorded incidence of VPDs, as well as of infant and child mortality.

KEY POINTS

- Immunoprophylaxis is the prevention of disease by the production of active or passive immunity.
- The immunizing agents are vaccines and immunoglobulins.
- Active immunization can be achieved by natural infection with a micro-organism administration of a vaccine.
- Vaccines may be live attenuated, killed, or in the form of toxoids.
- Live attenuated vaccines-BCG, smallpox vaccine, oral polio vaccine (OPV), mumps, measles, and rubella (MMR) vaccine, and yellow fever vaccine are some of the examples of live vaccines.
- Killed inactivated vaccines: Typhoid, cholera, pertussis, pneumococcal, rabies, hepatitis B, and influenza vaccines are the examples of killed inactivated vaccines.
- **Toxoids:** Tetanus toxoid and diphtheria toxoids are two most widely used toxoids for immunization.
- Subunit vaccines: Hepatitis B subunit vaccine.
- Passive immunization is carried out by administration of human and animal sera.
- Combined active and passive immunization is often carried out to confer slowly developing immunity and immediate passive immunity, respectively, against certain diseases such as diphtheria, tetanus, and rabies.

IMPORTANT QUESTIONS

Write short notes on:

Live attenuated vaccines

Killed vaccines

Toxoids

National immunization schedule

WHO EPI immunization schedule

FURTHER READING

Park K. Immunization schedules. In: Park's Textbook of Preventive and Social Medicine, 17th edn. Banarsidas Bhanot Publishers 1167, Prem Nagar, Jabalpur 2002;99.

CHAPTER

83 Bacteriology of Water, Milk and Air

Learning Objectives

After reading and studying this chapter, you should be able to:

- Describe bacteriological examination of water.
- Discuss bacteriological examination of milk.
- Describe the following: Water-borne pathogens; presumptive coliform count; Eijkman test.
- Describe settle plate method and slit samplere method for bacteriology of air.

BACTERIOLOGY OF WATER

Introduction

Safe and wholesome water is defined as water that is free from pathogenic agents, free from harmful chemical substances, pleasant to the taste and usable for domestic purposes. Many major human diseases, for example, typhoid fever, cholera and other diarrheal diseases, poliomyelitis and viral hepatitis A and B are waterborne. Both chemical and bacteriological examination of water supply from the source to the consumer should be regularily and systematically done, though in several cases chemists and bacteriologists may disagree. The aim of microbiological examination of water supplies is to detect whether pollution of the water by pathogenic organisms has occurred or not. It is impracticable to attempt directly to detect the presence of all the different kinds of water-borne pathogens, any of which may be present only intermittently. Instead, reliance is placed on testing the supply for microorganisms which indicate that fecal pollution has taken place.

BACTERIAL FLORA IN WATER

Bacterial flora can be divided into three groups:

1. **Natural water bacteria:** This group includes those organisms that are commonly found in water free from gross pollution.
2. **Soil bacteria:** These organisms are frequently washed into water during heavy rains and are not normal inhabitants of water.
3. **Sewage bacteria:** In this group many of the bacteria are normal inhabitants of the intestine of man and animals. Others live mainly on decomposing organic matter of either animal or vegetable origin. Bacterial flora in water is given in Table 83.1.

FACTORS DETERMINING THE NUMBER OF BACTERIA IN WATER

1. **Surface or deep water:** Surface water is more likely to be contaminated
2. **Salinity:** In saline water number of bacteria is less as compared to fresh water. However, halophilic bacteria can survive in saline water.
3. **Mineral springs:** These are usually pure and most of the organisms found in water derived from them come from imperfectly sterilized bottles.
4. **Nutrition:** When organic matter is plentiful, organisms are abundant, when it is scarce they are few, and tend to die out.
5. **Temperature:** When nutrition is available, rise in temperature leads to multiplication, otherwise, the number decreases. Low temperature favors survival of bacteria.
6. **Light:** Day light with the wavelength of 300400 nm is bactericidal. However, this effect is reduced by opacity and movement in water.
7. **Acidity:** Acidity of water has a bactericidal action, thus purifying water.
8. **Dissolved oxygen:** It is essential for survival of aerobes.
9. **Protozoal content:** Certain flagellates exterminate bacteria in water and bring down their number.
10. **Rain:** Early rain washes large number of bacteria from the soil which may contaminate water sources. Subsequent rains dilute the bacterial population.
11. **Storage:** Storage of water decreases bacterial count due to sedimentation and devitalization.

Table 83.1: Bacterial flora in water	
Natural water bacteria	*Micrococcus, Pseudomonas, Serratia, Flavobacterium, Chromobacterium, Acinetobacter* and *Alcaligenes*
Soil bacteria	*Bacillus subtilis, B. megaterium, B. mycoides, Enterobacter aerogenes* and *E. cloacae*
Sewage bacteria	
• Intestinal bacteria	*Escherichia coli, Enterococcus faecalis, Clostridium perfringens, Salmonella* Typhi and *Vibrio cholerae*
• Sewage bacteria proper	*Proteus vulgaris, Clostridium sporogenes, Zoogloea ramigera, Sphaerotilus natans, Haliscomenobacter hydrossis, Nostocoida limicola, Microthrix parvicella, Hexiaecter, Microscilla* and *Nocardia*

WATER-BORNE PATHOGENS

Water-borne Diseases include

1. Those caused by the presence of an infective agent
2. Those due to the presence of an aquatic host (Table 83.2).

INDICATOR ORGANISMS

Microorganisms for use as indicators of fecal pollution should satisfy several criteria. They should be present in feces in greater numbers than any pathogen yet be unable to proliferate in water to any extent. Moreover, they should be more resistant than pathogens to the stresses of the aquatic environment and disinfection processes. Usually a number of indicator organisms are sought. Such as coliforms and *E. coli* other bacteria also sometimes used as indicators of fecal pollution such as *streptococcus fecalis* and clostridium perfringers.

COLLECTION OF WATER SAMPLES

For collection, use heat-sterilized bottles containing a sufficient volume of sodium thiosulphate to neutralize the bactericidal effect of any chlorine or chloramine in the water which may lower bacterial counts by continued activity. Each bottle of 100 ml capacity should contain 0.1 ml of a fresh 1.8 percent (w/v) aqueous solution of sodium thiosulphate.

1. **Sampling from a tap or pump outlet:** When collecting the sample from taps, exercise extreme care to avoid contaminating it with bacteria from the environment. Allow water to run to waste for 2-3 min before running it into the bottle.
2. **Sampling from reservoir (streams, rivers, lakes and tanks):** When sampling from streams or lakes, open the bottle at a depth of about 30 cm with its mouth facing the current and ensure that water entering the bottle has not been in contact with the hand. Sample wells with weighted bottles. Collect at least 100 ml in each bottle.
3. **Sampling from a dug well:** A stone of suitable size is tied with the bottle. Then a clean cord of suitable length is tied with the bottle and lowered into the well. Immerse the bottle completely in the water. When the bottle is filled, pull it out, stopper it and wrap it in a kraft paper. The bottle should not touch the sides of the well any time.

Table 83.2: Biological (Water-borne Diseases)
1. Those caused by the presence of an infective agent: a. Viral: Viral hepatitis A, hepatitis E, Poliomyelitis, rotavirus diarrhea in infants b. Bacterial: Typhoid and paratyphoid fever, bacillary dysentery, *Esch. coli* diarrhea, cholera c. Protozoal: Amoebiasis, giardiasis d. Helminthic: Roundworm, threadworm, hydatid disease. e. Leptospiral: Weil's disease 2. Those due to the presence of an aquatic host a. Snail: Schistosomiasis b. Cyclops: Guineaworm, Fish tape worm

Transport

Stopper the bottle, label it with full details of the source of the water and time and date of collection, and deliver it to the laboratory as quickly as possible, at least within 6 hours, keeping it in a cool container and protected from light.

BACTERIOLOGICAL EXAMINATION OF WATER

The following tests are generally done for routine bacteriological analysis of water:

A. **Plate count**

B. **Counting of indicator organisms**

   a. Multiple tube test
      1. Total coliform count or Presumptive coliform count
      2. Eijkman test: Fecal coliform and confirmed *Escherichia coli* count.
      3. Count of fecal streptococci
      4. Count of Clostridium perfringens

   b. Membrane filtration tests.

A. Plate Count

The plate count expresses the number of all colony forming bacteria in 1 ml water. It is of limited value by itself, but as a supplementary test it provides information about the amount and type of organic matter in the water which may be useful in indicating the efficiency of the processes used for water treatment or the suitability of the water for large-scale production of food and drink.

B. Counting of Indicator Organisms

As the number of indicator bacteria in the water may be small, large volumes of the water have to be cultured. Two methods are available for this purpose, the multiple tube method and the membrane filtration method.

a. Multiple Tube Test

'Measured volumes of water and dilutions of water are added to a series of tubes or bottles containing a liquid indicator growth medium. The media receiving one or more of the indicator bacteria show growth and a characteristic color change which are absent in those receiving an inoculum of water without indicator bacteria. From the number and distribution of positive and negative reactions, the **most probable number (MPN)** of indicator organisms in the sample may be estimated by reference to statistical tables.

Advantages

The multiple tube method has the advantages that it can show gas formation by the bacteria and is suitable for the examination of turbid waters containing small numbers of the indicator bacteria, e.g. waters containing numerous saprophytic bacteria that might suppress growth of the coliforms.

1. Presumptive Coliform Count (Total Coliform Count)

The test is called presumptive because the reaction observed may occasionally be due to the presence of some other organisms and the presumption that the reaction is due to coliform organisms has to be confirmed.

Indicator Medium

The indicator medium used most has been MacConkey broth containing bromocresol purple to indicate by its color change to yellow the formation of acid from the lactose in the broth. An inverted Durham tube is placed in each bottle or tube of the medium. Bacteria capable of growth and the production of acid and gas in MacConkey broth are assumed to be coliform bacilli, i.e. 'presumptive coliforms'.

The following range is put up:

One 50 ml quantity of water added to 50 ml double strength medium.

Five 10 ml quantities each to 10 ml double strength medium.

Five 1 ml quantities each to 5 ml single strength medium.

Five 0.1 ml quantities each to 5 ml single strength medium.

MacConkey's fluid medium (modified) is used. The range of quantities depends on the likely strength of contamination. For highly contaminated waters, smaller volumes are tested. Incubate the seeded media aerobically at 37°C. After 24 hours and 48 hours of incubation, inspect the media and note the number of cultures of each volume of water that show the production of acid (color change) and gas. These acid- and gas-producing cultures are considered 'presumptive positive' growths of coliform bacilli, e.g. *escherichia, klebsiella* or *citrobacter*. Cultures not showing production of both acid and gas at 48 hours are considered negative.

The probable number of coliforms per 100 ml are read off from the probability tables of McCrady. This is known as the 'presumptive coliform count' or the most probable number of coliforms (MPN).

2. Eijkman Test : Fecal Coliform and Confirmed Escherichia Coli Count

Some spore-bearing bacteria give false-positive reactions in the presumptive coliform test. It is necessary, therefore, to confirm the presence of true ('fecal') coliform bacilli.

The Eijkman test is usually employed to find out whether the coliform bacilli detected in the presumptive test are *E. coli*. After the usual presumptive test, subcultures are made from all the tubes/ bottles showing acid and gas to fresh tubes of single strength MacConkey's medium already warmed to 44°C. Incubation at 44°C should be carried out in thermostatically controlled water baths that do not deviate more than 0.5°C from 44°C. Those showing gas in Durham's tubes contain *E coli*. Further confirmation of the presence of *E. coli* can be obtained by testing for indole production and citrate utilization.

From the combination of positive and negative results for gas and indole production at 44°C, read off the most probable number of *E. coli* per 100 ml of water. This latter value is known as the confirmed *E. coli* count.

3. Count of Fecal Streptococci

If there is difficulty in interpreting the results of the presumptive coliform and confirmed *E. coli* tests, as when presumptive coliforms are present but *E. coli* is absent, a demonstration of the presence of fecal streptococci will confirm the fecal origin of the coliform bacilli. Subcultures are made from all the positive bottles in the presumptive coliform test into tubes containing 5 ml of glucose azide broth. The presence of *Enterococcus. fecalis* is indicated by the production of acid in the medium within 18 hours at 45°C. The positive tubes should be plated onto MacConkey's agar for confirmation. Millipore membrane technique can also be adopted for this purpose.

4. Count of Clostridium Perfringens

This is tested by incubating varying quantities of the water in litmus milk medium (anaerobically) at 37°C for five days and looking for stormy fermentation.

b. Membrane Filtration Tests

In this method, a measured volume of the water sample is filtered through a membrane with a pore size small enough to retain the indicator bacteria to be counted on its surface. The membrane is then placed and incubated on a selective indicator medium at the appropriate temperature, so that the indicator bacteria grow into colonies on its upper surface. These colonies, which are recognized by their color, morphology and ability to grow on the selective medium, are counted and the bacteriological content of water calculated.

Tests for Pathogenic Bacteria

Specific pathogens such as typhoid bacilli or cholera vibrios may have to be looked for in water by employing enrichment and selective media under special circumstances. This used to be done by adding the water samples to tenfold concentrated liquid media, incubating and subculturing onto appropriate solid media.

Isolation of S. Typhi: For isolation of S. Typhi, equal volume of water is added to double strength selenite broth followed by incubation and subculture on Wilson and Blair's medium.

Isolation of V. cholerae: For isolation of V. cholerae, alkaline peptone water (10^{x}) is mixed with nine times its volume of water, incubated and subcultured on bile salt agar. A simpler and more sensitive method is to filter the water sample through membrane filters and incubate the filters on appropriate solid media.

Viruses in Water

It is recommended that, to be acceptable, drinking-water should be free from any viruses infections for man. Methods are available for the isolation of enteroviruses and other cytopathogenic viruses from water but they do not form part of routine testing. As a general rule, it is assumed that the viruses in water are destroyed by chlorination, when the concentration of free residual chlorine is at least 0.5 mg per liter, for a minimum contact period of 30 minutes at pH below 8 and a turbidity of 1 nephalometric turbidity unit or less

Protozoa in Water

Species of protozoa known to have been transmitted by the ingestion of contaminated drinking-water include *Entamoeba histolytica, Giardia spp.* and *rarely,* Balantidium coli. However, there is no good indicator for protozoal contamination of water. Coliform counts are not reliable as indicators of protozoal contamination of chlorinated water as they are more resistant to chlorine than are coliforms.

BACTERIOLOGY OF MILK

Human infections may be caused by the ingestion of animal milk which contains microorganisms derived either from the animal, e.g. by contamination with its feces, or from the environment or from milk handlers such as dairy workers.

Table 83.3: Milk borne diseases

1. Infections of animals that can be transmitted to man:
   Primary importance:
   Tuberculosis
   Brucellosis
   Streptococcal infections
   Staphylococcal enterotoxin poisoning
   Salmonellosis
   Q fever
   Lesser importance:
   Cowpox
   Foot and mouth disease
   Anthrax
   Leptospirosis
   Tick-borne encephalitis-transmitted through goat milk
2. Infections primary to man that can be transmitted through milk:
   Typhoid and paratyphoid fevers
   Shigellosis
   Cholera
   Enteropathogenic *Escherichi coli* (EEC)
   Nondiarrheal diseases
   a. Streptococcal infections
   b. Staphylococcal food poisoning
   c. Diphtheria
   d. Tuberculosis
   e. Enteroviruses
   f. Viral hepatitis

Milk Borne Diseases

Milk is a good medium for bacteria and a good vehicle for many pathogens. Milk bone diseases are as follows (Table 83.3).

1. **Infections of animals that can be transmitted to human beings (Table 83.4):** Occasionally, milk may be contaminated with *Streptobacillus moniliformis* from the nasal secretion of rats and with *Campylobacter jejuni* from animal feces. *Yersinia enterocolitica* is not uncommon in milk and may give rise to gastroenteritis if present in large numbers.

   The organisms that cause all the diseases mentioned above are destroyed by adequate pasteurization.
2. **Infections primary to man that can be transmitted through milk:**
   a. **Enteric infections:** These are caused by consumption of milk which has been mixed with water contaminated by human excreta. A less common source are the human carriers of enteric infections employed in the dairies. The diseases caused are typhoid and paratyphoid fevers, shigellosis, cholera (rarely) and diarrhea due to *E. coli.*

b. **Streptococcal infections:** Cows may have udder or teat infections and the organisms get into the milk. Milk handlers may be carriers and may contaminate the milk.

c. **Staphylococcal food poisoning:** Milk from cows suffering from staphylococcal mastitis is contaminated with the organism. If the milk is consumed after being allowed to remain at temperatures favorable for its multiplication the enterotoxin is produced, which causes food poisoning. Many such outbreaks have been reported.

d. **Diphtheria:** Milk contaminated either from a human carrier or more usually through diphtheritic lesions on the teats, when consumed unpasteurised, causes disease.

e. **Tuberculosis:** Milk contaminated by excretions from persons suffering from tuberculosis, when consumed, leads to the disease.

Bacteriological Examination of Milk

1. **Viable count:** This is estimated by doing plate counts with serial dilutions of the milk sample. Raw milk always contains bacteria, varying in number from about 500 to several million per ml.
2. **Test for coliform bacilli:** This is tested by inoculating varying dilutions of milk into 3 tubes of MacConkey's fluid medium with Durham tube and noting the production of acid and gas after incubation at 37°C for 48 hours. Contamination with coliforms comes mainly from dust, dirty utensils and dairy workers. All coliforms are killed by adequate pasteurization and their presence in pasteurized milk indicates faults in pasteurizer or postpasteurization contamination.
3. **Methylene blue reduction test:** It depends on the reduction of methylene blue by bacteria in milk when incubated at 37°C in complete darkness. The rate of reduction is related to the degree of bacterial contamination. Raw milk is considered satisfactory if it fails to reduce the dye in 30 minutes under standard conditions.

   Procedure: The test is performed by adding I ml of standard methylene blue solution to 10 ml of milk in a test tube. The tube is incubated in the dark at 37°C. The milk is considered satisfactory, if it fails to reduce the dye in 30 minutes.

   Resazurin test: The Resazurin test is similar but the dye resazurin, on reduction, passes through a series of color changes-from blue to pink to colourless the shade of color after incubation with milk for a particular period of time, depending on the degree of contamination. Generally the 10-minute resazurin test is done, in which the shade of color is noted after incubation with the milk for ten minutes.
4. **Phosphatase test:** Disodium p-nitrophenyl phosphate $\xrightarrow{\text{PHOSPHTASE}}$ p-nitrophenol-Yellow color: The enzyme phosphatase normally present in milk is inactivated if pasteurization has been carried out properly. This is a check on the pasteurizsation of milk. The test depends upon the ahility of the enzyme to liberate p-nitrophenol from disodium p-nitrophenyl phosphate and thereby produce a yellow color that can be quantitated by a colorimeter. Residual phosphatase activity indicates that pasteurization has not been adequate.
5. **Turbidity test:** This is a check on the 'sterilization' of milk. If milk has been boiled or heated to the temperature prescribed at least 100°C for 5 minutes for 'sterilization', all heat coagulable proteins are precipitated. If ammonium sulfate is then added to the milk, filtered and boiled for five minutes, no turbidity results. Absence of turbidity indicates that the milk has been heated to at least 100°C for at least 5 min.
6. **Examination for specific pathogens**

   a. **Tubercle bacillus:** Centrifuge 100 ml of milk at 3,000 rpm for 30 minutes and inoculate two guinea pigs. Keep the animals under observation for signs of tuberculous lesions. Kill one at 4 weeks, perform a necropsy and examine suspect lesions for tuber bacilli. If there are no tuberculous lesions, the other animal is kept for a further 4 weeks, then it is killed and examined. Tubercle bacilli may also be isolated in culture. Microscopic examination for tubercle bacilli is unsatisfactory.

   b. **Brucella:** *Brucella* may be isolated by inoculating cream from the milk sample on serum dextrose agar. or by injecting centrifuged deposit of the milk sample intramuscularly into guineapigs. The animals are sacrificed after 6 weeks and the serum tested for agglutinins and the spleen inoculated in culture media for brucellae. Brucellosis in animals can be detected also by demonstrating the antibodies in milk, by the milk-ring or the whey agglutination tests.

BACTERIOLOGY OF AIR

A person inhales over 15 cubic meters of air in the course of a day. Hence the bacterial content of the air one breathes is important, particularly when it contains pathogens. The bacterial content of air depends on the location, i.e. whether it is outdoor air or indoor air.

Outdoor Air

The bacterial content of outdoor air depends on many factors such as the density of human and animal populations, the nature of the soil, the amount of vegetation, the atmospheric conditions such as humidity, temperature and wind conditions, rainfall and sunlight.

Indoor Air

On the other hand, in the case of indoor air, the bacteria may be distributed through gross droplets and droplet nuclei from nose and mouth and through dust particles.

Observations of the number of bacteria-carrying particles in air may be required in premises where safe working depends on the air's content of bacteria being kept at a very low level.

Essential conditions for bacteriological examination of air

1. Surgical operation theater.
2. In hospital wards in which there is an outbreak of crossinfection
3. Premises where food articles are prepared and packed.
4. Premises where pharmaceutical materials are prepared.

Measurement of Air Contamination

The methods for bacteriological examination of air are of two types:

1. Settle plate method
2. Slit sampler method

1. *Settle plate method:* Petri dishes containing an agar medium of known surface area are left open for a measured period of time. Large bacteria-carrying dust particles settle on to the medium. The plates are incubated at 37°C for 24 hours and a count of the colonies formed shows the number of settled panicles. Blood agar medium may be used for the count of the pathogenic, commensal and saprophytic bacteria in the air. The method has the advantage of simplicity and is specially used for testing the air in surgical theaters and hospital wards.

   The optimal duration of exposure is that which will give a significant and readily countable number of well separated colonies, e.g. 30-100, and will depend on the dustiness of the air. In occupied rooms and hospital wards it is generally between 10 and 60 min.
2. **Slit sampler method:** The most efficient and convenient of the devices for counting the bacteria-carrying particles suspended in a unit volume of air is the slit sampler introduced by Bourdillon, et.al. (1941). By this method, the number of bacteria in a measured volume of air is determined. The efficiency of collection, even for the smallest bacterial particles, is very high.

Procedure

In this, a known volume of air is directed onto a plate through a slit 0.33 mm wide and 27.5 mm long with vertical parallel sides about 3 mm deep. At the correct negative pressure, air will enter through a slit of these dimensions at the rate of 1 ft^3/min. The culture medium is incubated and colonies counted which gives the number of bacteria present in the air.

Air Contamination Standards

Bacterial count should not exceed;

i. 50/ft^3 in factories, offices, homes and such places
ii. 10/ft^3 (353/m^3) in surgical theaters providing for most forms of surgery
iii. 1/ft^3 in theaters for operations on the central nervous system or dressing of burns.

BACTERIOLOGICAL EXAMINATION OF ENVIRONMENTAL DUST

A. Sweep Plate

Personal clothing, bed clothes and domestic furnishing material such as curtains may contain bacteria laden dust. When a Petri dish containing a suitable culture medium is removed from its lid and rubbed over the surface of the fabric, with the medium facing the fabric, dust settles on the medium. Colonies can be identified and counted after incubation.

B. Dust Sampling

Moistured cottonwool swabs may be used for collecting dust from the floor, wall, furniture and other surfaces. The swabs are placed in broth and anaerobic media such as Robertson's cooked meat and incubated. After subculturing on plates, the isolates can be identified. This is routinely employed for assessing the level of asepsis in surgical theatres, particularly for the detection of spores of tetanus bacilli and other clostridia in theater dust.

KNOW MORE

'Coliforms' (presumptive coliforms): The primary test employed as an indicator of fecal pollution of water is the presence of coliform bacteria because they are invariably present in the feces of human beings and other warm blooded animals in large numbers and can be easily detected in water, even in high dilutions. Though coliform bacteria are not exclusively of fecal origin, they serve as presumptive evidence, to be confirmed by the detection of thermotolerant *E. coli,* which provides definite proof of fecal pollution.

KEY POINTS

- Indicator organisms are coliform (e.g. *Escherichia coli*), fecal streptococci (e.g. *Enterococcus* species), and sulfite-reducing clostridia (e.g. *Clostridium perfringens*).
- Methods of water analysis include presumptive coliform count, differential coliform count, membrane filtration method, and detection offecal streptococci and C. *perfringens*.

The Eijkmen test is done to confirm that the caliform bacilli detected in the presumptive test are *E. Coli.*

- Pathogenic microorganisms in milk can transmit milk borne diseases such as tuberculosis, and typhoid fever.
- Bacteriological examination of milk can be carried out by colony counts, coliform counts; chemical tests such as methylene blue reduction test, phosphatase test, and turbidity test; and detection of specific pathogens.
- Settle plate method and slit sampler methods are used for bacteriological examination of air.

IMPORTANT QUESTIONS

1. Write short notes on:
   a. Bacterial flora in water
   b. Water-borne pathogens
   c. Bacteriological examination of water
   d. Bacteriology of milk
   e. Presumptive coliform count
   f. Eijkman test
   g. Bacteriology of air

FURTHER READING

Senior BW. Examination of water, milk, food and air. Mackie & McCartney Practical Medical Microbiology. Collee JG, Fraser AG, Marmion BP, Simmons A (Eds) 14th ed, 1996;883-926.

World Health Organization, Geneva: Guidelines for Drinking Water Quality. Vol. I. Recommendations (1983); Vol II. Health Criteria (1984);Vol. III. Drinking water quality control in small community supplies 1985.

CHAPTER 84

Hospital Waste Management

Learning Objectives

After reading and studying this chapter, you should be able to:

- Describe universal precautions.
- Describe the following: Categories of biomedical waste; waste segregation; waste treatment and disposal.
- Describe treatment and technologies for health care waste.

"Let the wastes of "the sick" not contaminate the lives of the healthy"

INTRODUCTION

Hospitals regularly generate waste which may be a potential health hazard to health care workers, the general public and the environment. Therefore, adequate management and disposal of waste is essential. Inadequate and inappropriate handling of health care waste may have serious public health consequences and a significant impact on the environment. This becomes all the more important in a situation peculiar to a developing country like India where poverty and ignorance induce many persons to sift and sort through dumped waste to make a living out of recyclables thus being exposed to danger of contracting diseases from hazardous components of waste.

UNIVERSAL PRECAUTIONS

Concerns about transmission of the hepatitis B virus (HBV) and human immunodeficiency virus (HIV) led to the introduction of 'universal precautions', to minimize the infections in medical laboratory workers and health care personnel. These universal precautions include:

1. Assume that all specimens/patients are potentially infectious for HIV and other blood-borne pathogens.
2. All blood specimens or body fluids should be placed in a leak-proof impervious bags for transportation to the laboratory.
3. Use gloves while handling blood and body fluid specimens and other objects exposed to them. If there is a likelihood of spattering, use face masks with glasses or goggles.
4. Wear laboratory coats or gowns while working in the laboratory. Wrap-around gowns should be preferred. These should not be taken outside.
5. Never pipette by mouth. Mechanical pipetting devices should be used.
6. Decontaminate the laboratory work surfaces with an appropriate disinfectant after the spillage of blood or other body fluids when the procedures are completed.
7. Limit use of needles and syringes to situations for which there are no other alternatives.
8. Biological safety hoods should be used for laboratory work.
9. All the potentially contaminated materials of the laboratory should be decontaminated before disposal or reprocessing.
10. Always wash hands after completing laboratory work and remove all protective clothings before leaving the laboratory.

Agents which are associated with laboratory acquired infections: Most common agents which are associated with laboratory acquired infections include hepatitis B virus, *Coccidioides immitis, Bacillus anthracis, Brucella* species, *Mycobacterium tuberculosis, Francisella tularensis* and shigella species.

DEFINITION OF BIOMEDICAL WASTE (BMW)

According to Biomedical waste (Management and Handling) Rules, 1998 of India, "Biomedical waste" means any waste, which is generated during the diagnosis, treatment or immunization of human beings or animals or in research activities pertaining thereto or in the production or testing of biologicals.

Between 75 to 90 percent of the waste produced by the health care providers is **non risk or "general"** health care waste, comparable to domestic waste. The remaining 10 to 25 percent health care waste is regarded as **hazardous** and may create a variety of health risk. However, if the infectious component gets mixed with the general noninfectious waste, the entire bulk of waste becomes potentially infectious.

Infectious wastes include all those medical wastes, which have the potential to transmit viral, bacterial or parasitic diseases. It includes both human and animal infectious waste and waste generated in laboratories, and veterinary practice. Infectious waste is hazardous in nature. Any waste with a potential to pose a threat to human health and life is called **hazardous waste**. The persons most at risk are the staff of hospitals particularly nurses and waste handlers. In countries such as India, scavengers and ragpickers are at serious risk.

Noninfectious hazardous waste may be chemical (toxic, corrosive, inflammable, reactive and otherwise injurious), radioactive, and pharmacolocal (surplus or time expired drugs).

CATEGORIES OF BIOMEDICAL WASTE

The detail categories of bio-medical waste as given in schedule 1 of BMW' 98 annexure A (Table 84.1).

WASTE SEGREGATION

Waste should be segregated at source, since 80 percent of the waste is non-hazardous and can be disposed off easily into the municipal bin. It is important that hazardous waste component is separated from nonhazardous waste and to collect these in appropriate receptacles. Mixing of waste will render the entire waste potentially hazardous. The wastes are segregated preferably at the point of generation. This is the most important step to safeguard the occupational health of health care personnel. Waste should be segregated in bags of different colors to facilitate appropriate treatment and disposal (Table 84.2).

Yellow Bags

Infectious nonsharp waste should be put in yellow bags. This includes: soiled bandages, dressings, soiled cotton and other soiled waste. Human and animal tissues and body parts should be placed in **double bags of yellow color.**

Red Bags

Red bags may be used for nonsharp waste (except anatomical tissues) if microwaving/autoclaving/chemical treatment followed by landfill is the disposal option. However, red bags used for infectious nonsharp waste must not be incinerated as red color contains cadmium, which causes toxic emissions. If colored bags are not available then a clearly visible label of appropriate color is acceptable.

Blue/White Translucent Bags

Plastic disposable items such as used gloves, catheters and IV sets should be put into **blue/white translucent bags** for shredding/maceration and disinfection before disposal (recycling or landfill).

Sharps (syringes and their attached needles, scalpel blades, any cutting or piercing articles, glass vials, used slides and small pieces of broken glass) should be discarded in **blue/white translucent puncture proof container.** Needles should not be recapped, bent or mutilated by hand. Needle should be destroyed in a needle destroying machine and tip of the needle should be broken. Sharps are then subjected to autoclaving/microwaving/chemical treatment/shredding. Sharps are the most hazardous component of hospital waste, because they have the ability to puncture and injure the skin thus causing direct access of pathogenic organisms into the tissues.

Black Bags

Incineration ash and solid chemical waste such as discarded medicines should be collected in **black bags** for disposal in secured landfill.

Proper segregation will minimize the waste stream needing special treatment, i.e. infectious waste. This practice also helps in safe handling and transporation of waste.

TREATMENT AND DISPOSAL TECHNOLOGIES FOR HEALTH CARE WASTE (TABLE 84.2)

Waste Treatment

Following techniques are in use for treatment of infected material.

1. Incineration
2. Autoclaving
3. Chemical disinfection
4. Wet and dry thermal treatment
5. Microwave irradiation
6. Inertization

1. Incineration

Incineration is a high temperature dry oxidation process, that reduces organic and combustible waste to inorganic incombustible matter and results in a very significant reduction of waste—volume and weight. This is a safe method of treating large solid infectious waste, particularly anatomy waste and amputated limbs, animal carcasses and the like. The incinerator subjects them to very high heat, converting them to ash, which would be only about a tenth of original volume. However, it is expensive and is generally used only by very large establishments.

Types of Incinerators: Three basic kinds of incineration technology are of interest for treating health care waste:

a. Double-chamber pyrolytic incinerators which may be especially designed to burn infectious health care waste.

Table 84.1: Schedule 1: Categories of biomedical waste in India

Option	*Waste category*	*Treatment and disposal*
Category No. 1	Human anatomical waste (human tissues, organs, body parts)	Incineration[2]/deep burial
Category No. 2	Animal waste (animal tissues, organs, body parts carcasses, bleeding parts, fluids, blood and experimental animals used in research, waste generated by veterinary hospitals colleges, discharge from hospital, animal house)	Incineration[2]/deep burial
Category No. 3	Microbiology and biotechnology waste (waste from laboratory cultures, stocks or specimens of micro-organisms, live or attenuated vaccines, human and animal cell culture used in research and infectious agents from research and industrial laboratories, waste from production biologicals, toxins, dishes and devices and for transfer of cultures)	Local autoclaving/microwaving/incineration[2]
Category No. 4	Waste sharps (needles, syringes, scalpels, blades, glass, etc. that may cause puncture and cuts. This includes both used and unused sharps)	Disinfection (chemical treatment@/autoclaving/microwaving and mutilation/shredding)
Category No. 5	Discarded medicines and cytotoxic drugs (wastes comprising of outdated, contaminated and discarded medicines)	Incineration®destruction and drugs disposal in secured landfills
Category No. 6	Solid waste (Items contaminated with blood, and fluids including cotton, dressing, soiled plaster casts, lines, beddings, other material contaminated with blood)	Incineration®autoclaving/microwaving
Category No. 7	Solid waste (wastes generated from disposable items other than the waste sharps such as tubings, catheters, intravenous sets, etc.	Disinfection by chemical treatment@@ autoclaving/microwaving and mutilation/shredding##
Categroy No. 8	Liquid waste (waste generated from laboratory and washing, cleaning, housekeeping and disinefecting activities)	Disinfection by chemical treatment@@ and discharge into drains
Categroy No. 9	Incineration ash (ash from incineration of any biomedical waste)	Disposal in municipal landfill
Categroy No. 10	Chemicals used in production of biologicals, chemicals used in disinfection, as insecticitdes, etc.	Chemical treatment@@ and discharge into drains for liquids and secured landfill for solids

@@ Chemical treatment using at least 1 percent hypochlorite solution or any other equipment chemical reagent. It must be ensured that chemical treatment ensures disinfection.

\## Multilation/shredding must be such so as to prevent unauthorized reuse.

® There will be no chemical pretreatment before incineration. Chlorinated plastics shall not be incinerated.

2 Deep burial shall be an option available only in towns with population less than five lakhs and in rural areas.

b. Single-chamber furnaces with static grate, which should be used only if pyrolytic incinerators are not affordable.
c. Rotary kilns operating at high temperatures, capable of causing decomposition of genotoxic substances and heat-resistant chemicals.

Double-Chambered incineration: An incinerator should consist of 2 chambers, primary and secondary. The temperature of primary chamber should be 750 to 850°C while the temperature in the secondary chamber should be 1000 to 1100°C. Waste is burnt in one chamber (primary chamber) at 800°C. Combustion of gases emitted from the first chamber, occurs in the second or secondary chamber which has a high temperature of 1000°C. The negative pressure is maintained inside the incinerator by the system, thereby forcing the end-gases out of the chimney.

The chimneys of incinerators should be 30 meters high and combustion efficiency (CE) of the incinerator should be at least 99 percent. It is computed as follows:

Table 84.2: Color coding and type of container for disposal of biomedical wastes			
Color coding	*Type of container*	*Waste category*	*Treatment option as per schedule 1*
Yellow	Plastic bag	Cat. 1, Cat. 2, and Cat. 3, Cat. 6	Incineration/deep burial
Red	Disinfected container/plastic bag	Cat. 3, Cat. 6, Cat. 7	Autoclaving/microwaving/ chemical treatement
Blue/white translucent	Plastic bag/puncture proof container	Cat. 4, Cat. 7	Autoclaving/microwaving/ chemical Treatment and destruction/shredding
Black	Plastic bag	Cat. 5 and Cat. 9 and Cat. 10 (solid)	Disposal in secured landfill

Notes:
1. Color coding of waste categories with multiple treatment options as defined in Schedule 1, shall be selected depending on the treatment option chosen, which shall be as specified in Schedule 1.
2. Waste collection bags for waste types needing incieration shall not be made of chlorinated plastics.
3. Categories 8 and 10 (liquid) do not require containers/bags.
4. Category 3 if disinfected locally need not be in containers/bags.

Advantage of incinerator: The incinerator has an advantage of dealing with all pathological and cytotoxic wastes. Body parts, animal waste, microbiological waste and soiled dressings can be treated with this technique.

Disadvantage of incinerator:
1. It generates highly toxic gases (e.g. dioxins and furans, if PVC plastics are present).
2. It adversely affects the health of the community.
3. Recycling and reprocessing of materials cannot be done.
4. Burning of plastic waste or sharps is also not recommended.

2. Autoclaving

Autoclaving, at 121°C for 60 minutes, is an effective method for treating infectious waste before disposal. A separate autoclave dedicated for waste treatment should be used. Waste arising from microbiology and biotechnology laboratories including cultures and stocks must be autoclaved before disposal by incineration. Plastic disposables including blood bags, urine bags, etc. should be autoclaved followed by shredding. It may then be considered for recycling. It is not recommended for pathological waste. Autoclaved material is typically landfilled, therefore, it has a large strain on land fill capacity.

Types of autoclaves: There are two kinds of autoclaves: (i) the prevacuum type and (ii) the gravity autoclave.

i. **Prevacuum type:** In the Prevacuum type, steam is created outside the chamber loaded with waste. Air in the chamber is then gradually removed and steam is injected in. This type of autoclave eliminate 'cold spots' and 'air pockets' (where the steam is unable to penetrate) by creating this vacuum. This ensures quicker heating. A temperature of 121°C and pressure of 15 pounds per square inch is used.
ii. **Gravity autoclave:** For gravity autoclave, the waste material should be subjected to autoclave residence time of not less than 60 minutes, while in a vacuum autoclave time period should not be less than 45 minutes. Biological *(Bacillus stearothermophilus spores)* or chemical indicators (strips/tapes) should be used for validation test of autoclave.

3. Chemical Disinfection

Chemicals are added to waste to kill or inactivate the pathogens it contains. Chemical disinfection is most suitable for treating liquid waste such as blood, urine, stools or hospital sewage. However, solid wastes including microbiological cultures, sharps, etc. may also be disinfected chemically with certain limitations.

Hypochlorite: The most economical and effective disinfectant is hypochlorite. For clean conditions available chlorine required is 0.1 percent and for dirty conditions (where there is presence of organic matter such as blood, etc.) available chlorine should be 1.0 percent. Various hypochlorites available are:

Sharp decontaminating unit (SDU) for syringes and needles: It consists of blue/white translucent outer plastic puncture proof and inner perforated container with handles. It is filled one-third with hypochlorite (1 percent available chlorine). Immediately after use the syringe with needle is filled with disinfectant. It is then dropped into the SDU, so that it is completely immersed in disinfectant. After 30 minutes, the inner perforated container is lifted and contents drained and put into puncture proof container for transferring to shredding machine.

4. *Wet and Dry Thermal Treatment*

Wet thermal treatment: Wet thermal treatment or steam disinfection is based on exposure of shredded infectious waste to high temperature, high pressure steam, and is similar to the autoclave sterilization process. The process is inappropriate for the treatment of anatomical waste and animal carcasses, and will not efficiently treat chemical and pharmaceutical waste.

Screw-feed technology: Screw-feed technology is the basis of a non-burn, dry thermal disinfection process in which waste is shredded and heated in a rotating auger. The waste is reduced by 80 percent in volume and by 20-35 percent in weight. This process is suitable for treating infectious waste and sharps, but it should not be used to process pathological, cytotoxic or radioactive waste.

5. *Microwave Irradiation*

This is another useful method of sterilization of small volume waste at the point of generation. Most microorganisms are destroyed by the action of microwave of a frequency of about 2450 MHz and a wave-length of 12.24 cm. The water contained within the waste is rapidly heated by the microwaves and the infectious components are destroyed by heat conduction. This cannot be used for animal or human body parts, metal items or toxic or radioactive material.

6. *Inertization*

The process of "inertization" involves mixing waste with cement and other substances before disposal, in order to minimize the risk of toxic substances contained in the wastes migrating into the surface water or ground water.

DISPOSAL

Landfilling, deep burial and sewage are used for disposal. Infectious waste after treatment can be disposed of by landfilling or deep burial. Liquid waste can be disposed in sewage drains. Besides treatment, incineration is also a method of disposal.

Treatment methods used for different types of infectious wastes are shown in Table 81.1.

BIOMEDICAL WASTE MANAGEMENT IN INDIA

National legislation is the basis for improving health care waste disposal practices in any country. It establishes legal control, and permits the national agency responsible for the disposal of health care waste, usually the Ministry of Health, to apply pressure for their implementation. The Ministry of Environment may also be involved. There should be a clear designation of responsibilities before the law is enacted.

The United Nations Conference on the Environment and Development (UNCED) in 1992 recommended the following measures:

a. Prevent and minimize waste production.
b. Reuse or recycle the waste to the extent possible.
c. Treat waste by safe and environmentally sound methods.
d. Dispose off the final residue by landfill in confined and carefully designed sites.

Biomedical Waste (Management and Handling) Rule 1998, prescribed by the Ministry of Environment and Forests, Government of India, came into force on 28th July 1998. This rule applies to those who generate, collect, receive, store, dispose, treat or handle biomedical waste in any manner. Table 84.1 shows the categories of biomedical waste, types of waste and treatment and disposal options under Rule 1998.

The biomedical waste should be segregated into containers bags at the point of generation of the waste. The color coding and the type of containers used for disposal of waste are as shown in Table 84.2.

WASTE MANAGEMENT PROGRAM

All laboratories should develop waste management program according to the specific needs of the individual laboratory. The policies and procedures should be incorporated in the laboratory's operating manuals. Emphasis should be on waste minimization (by reducing waste, reuse and recycling), proper segregation, and health and safety of the workers. All personnel generating, collecting, transporting and storing infectious waste must be trained under the program.

KNOW MORE

- It is estimated that the quantity of waste generated from hospitals in our country ranges from 1 to 2 Kg/bed/day.

Land Disposal

If a municipality or medical authority genuinely lacks the means to treat waste before disposal, the use of a landfill has to be regarded as an acceptable disposal route. There are two types of disposal land-**open dumps** and **sanitary landfills**. Health care waste should not be deposited on or around open dumps. The risk of either people or animals coming into contact with infectious pathogens is obvious.

KEY POINTS

- **Biomedical waste (BMW)** means any waste, which is generated during the diagnosis, treatment or immunization of human beings or animals or in research activities pertaining thereto or in the production or testing of biologicals.
- Handling of biomedical wastes is increasingly associated with the risk of acquiring infections caused by hepatitis B virus, *Bacillus anthracis, Mycobacterium tuberculosis, Francisella tularensis, Shigella* species, and *Brucella* species.

- Categories of biomedical wastes are: human anatomical waste; animal waste; microbiology and biotechnology waste; waste sharps; cytotoxic waste; solid waste; liquid waste, incineration ash human; chemicals used in production of bilogicals.
- Waste should be segregated in bags of different colors to facilitate appropriate treatment and disposal.
- Black plastic bags are used for incineration and solid chemical noninfectious wastes. Yellow plastic bags are used for infectious nonsharp substances. Blue or white bags are used for disposal of plastics and sharp instruments. Red bags are used for infectious nonsharp wastes.
- There are various techniques are of biomedical wastes treatment. These include incineration, autoclaving, chemical treatment wet and dry thermal treatment, microwave irradiation and inertization.

IMPORTANT QUESTIONS

1. Describe various techniques used for the treatment and disposal of hospital waste.
2. Write short notes on:
   a. Universal precautions
   b. Segregation of waste
   c. Hospital waste management of biomedical waste.

FURTHER READING

Acharya DB, Singh M, The Book of Hospital Waste Management; 2000.

Pruss, A, Ciroiut, E, Rushbrook P. Safe management of wastes from health care activities, WHO 1999.

Sharma AK, Biomedical Waste (Management and Handling) Rules, 1998, Suvidha Law House, Bhopal.

CHAPTER 85

Vehicles and Vectors

Learning Objectives

After reading and studying this chapter, you should be able to:

- List of diseases transmitted by (i) water and food; (ii) blood.

INTRODUCTION

To maintain an active infectious disease in a human population, the pathogen must be transmitted from one host or source to another. Transmission is the third link in the infectious disease cycle and occurs by four main routes: airborne, contact, vehicle, and vector-borne.

VEHICLES AND VECTORS

The agents of transmission that bring the microorganism from the reservoir to the host may be a living entity, in which case they are called *vectors*, or they may be a nonliving entity referred to as a *vehicle or fomite*.

Modes of transmission: The human host may acquire microbial agents by various means referred to as the modes of transmission. The mode of transmission is

A. Direct
B. Indirect.

A. **Direct:** Transmitted by direct contact between reservoir and host.
B. **Indirect:** Transmitted to host (human host) via **intervening agent(s)** such as vectors animals, insects, other humans and vehicles-water, food, air, medical devices, various other inanimate objects).

Vehicle-borne Transmission

Vehicle-borne transmission implies transmission of the infectious agent through the agency of water, food ice, blood, serum plasma or other biological products such as tissues and organs. Of these water and food are the most frequent vehicles of transmission, because they are used by everyone.

A. Diseases Transmitted by Water and Food

- Acute diarrheas
- Typhoid fever
- Cholera
- Poliomyelitis
- Hepatitis A virus infection
- Food poisoning
- Intestinal parasitic infestation.

B. Diseases Transmitted by Blood

Those transmitted by blood include:

1. Viruses

- Hepatitis B
- Human immunodeficiency viruses (HIV).
- Human T cell lymphotropic viruses.
- Infectious mononucleosis.
- Cytomegalovirus.

2. Bacteria

- Syphilis.
- Brucellosis.

3. Parasites

- Malaria
- Trypansomiasis (Chaga's disease)
- Trypanosoma cruzi.

Vector-borne Transmission

In infectious disease epidemiology, *vector* is defined as an arthropod or any living carrier (e.g. snail) that transports an infectious agent to a susceptible individual. Transmission by a vector may be mechanical or biological. In the latter case, the disease agent passes through a developmental cycle multiplication in the vector.

Mechanical Transmission

A crawling or flying arthropod through soiling of its feet or proboscis mechanically transports the infectious agent; or by passage organisms through its gastrointestinal tract and pass excreted. There is no development or multiplication of the infectious agent on or within the vector.

Biological Transmission

The infectious agent undergoing replication or development or both in vector and require incubation period before vector can transmit, biological transmission is of three types:

a. **Propagative:** When the disease agent undergoes no cyclical change, but multiplies in the body of the vector, transmission is said to be propagative, e.g. plague bacilli in rat fleas
b. **Cyclo-propagative:** The disease agent undergoes cyclical change, and multiplies in the body of the arthropod, e.g. malaria parasite in anopheline mosquito.
c. **Cyclo-developmental:** When the disease agent undergoes cyclical change but does not multiply in the body of the arthropod, e.g. filarial parasite in culex mosquito and guineaworm embryo in Cyclops.

Medical Entomology

A study of the arthropods of medical importance is known as medical entomology which is an important branch of preventive medicine. Arthropods act as vectors or carriers of diseases (See Chaper 11, Table 11.1).

KNOW MORE

Infestation

By infestation is meant the lodgement, development and reproduction of arthropods on the surface of the body or in the clothing e.g. louse infestation.

☞ KEY POINT

- Transmission of the infectious disease occurs by four main routes: airborne, contact, vehicle, and vector-borne.
- The agents of transmission that bring the microorganism from the reservoir to the host may be a living entity, in which case they are called vectors, or they may be a nonliving entity referred to as a vehicle or fomite.
- Modes of transmission -The mode of transmission is:
  A. Direct; B. Indirect.
- **Vehicle-borne:** Vehicle-borne transmission implies transmission of the infectious agent through the agency of water, food (including raw vegetables, fruits, milk and milk products), ice, blood, serum plasma or other biological products such as tissues and organs.
- **Vector-borne:** Vector is defined as an arthropod or any living carrier that transports an infectious agent to a susceptible individual. Transmission by a vector may be mechanical or biological.

IMPORTANT QUESTIONS

1. Discuss the role of vehicles and vectors in transmission of infectious agents.
2. Write short notes on:
   a. Water-borne diseases.
   b. Diseases transmitted by blood and blood products.
   c. Diseases transmitted arthropods as vectors.

FURTHER READING

Park K. In: Park's Textbook of Preventive and Social Medicine, 17th edn. Banarsidas Bhanot Publishers 1167, Prem Nagar Jabalpur 2002;99.

CHAPTER 86 Emerging and Re-emerging Infectious Diseases

Learning Objectives

After reading and studying this chapter, you should be able to:

- Describe the following: emerging infectious diseases; re-emerging infectious diseases.

INTRODUCTION

Sometimes infectious agents that have not been previously recognized appear. **Emerging infectious diseases** are those whose incidence in humans has increased during the last two decades or which threaten to increase in the near future. The term also refers to newly-appearing infectious diseases, or diseases that are spreading to new geographical areas-such as cholera in South America and yellow fever in Kenya. During the past 20 years, at least 30 new diseases have emerged to threaten the health of hundreds of millions of people. For many of these diseases there is no treatment, cure or vaccine and the possibility of preventing or controlling them is limited.

Examples of Emerging Pathogens

Table 86.1 summarises the etiological agents and infectious diseases in humans and/or animals recognized since 1973. The year may differ from first appearance and first identification of cases.

Approximately 75 percent of emerging pathogens are *zoonotic*, that is, communicated by animals to humans. When humans encroach upon a rainforest, they become exposed to viruses and other microbes that they otherwise would not have encountered. HIV/AIDS, avian influenza, monkeypox, Nipah, SARS, and Ebola are all the result, to a greater or lesser extent, of interactions with animals that led to the emergence and re-emergence of deadly diseases.

The diseases in question involve all the major modes of transmission—they are spread either from person to person, by insects or animals, or through contaminated water or food.

1. **Human immunodeficiency virus (HIV):** The most dramatic example of a new disease is AIDS, caused by the human immunodeficiency virus (HIV).
2. A new breed of deadly hemorrhagic fevers, of which Ebola is the most notorious, has struck in Africa, Asia, the United States and Latin America.
3. **Hantavirus pulmonary syndrome:** The United States has seen the emergence of **hantavirus pulmonary syndrome.** Other hantaviruses have been recognized for many years in Asia, where they cause hemorrhagic fever with renal involvement in humans.
4. **Foodborne and waterborne diseases:** Epidemics of foodborne and waterborne diseases due to new organisms such as **cryptosporidium** or new strains of bacteria such as *Escherichia coli* (0157:H7 strain of *E.coli)* have hit industrialized and developing countries alike. A completely new strain of cholera, 0139, appeared in south-eastern India in 1992 and has since spread north and west to other areas of India, into western China, Thailand and other parts of South-East Asia.
5. **Influenza pandemic:** The threat of a new **global influenza pandemic** is increasing. Epidemic strains of influenza viruses originate from China. The influenza virus is carried by ducks, chickens and pigs raised in close proximity to one another on farms. The exchange of genetic material between these viruses produces new strains, leading to epidemics of human influenza, each epidemic being due to a different strain.

RE-EMERGING, OR RESURGING DISEASES

Re-emerging, or *resurging*, diseases *are those that* have been around for decades or centuries, but have come back in a different form or a different location. Some infectious diseases once thought to be all but conquered have returned with a vengeance. Others have developed stubborn resistance to antibiotic drugs. Re-emerging pathogens are anthrax *(Bacillus anthracis)*, Botulism *(Clostridium botulinum*, Plague *(Yersinia pestis)*, Smallpox virus, Tularemia *(Francisella tularensis).*

Year	Agent	Type	Disease/Comments
1973	Rotavirus	Virus	Major cause of infantile diarrhea worldwide
1975	Parvovirus B19	Virus	Aplastic crisis in chronic hemolytic anemia
1976	*Cryptosporidium parvum*	Parasite	Acute and chronic diarrhea
1977	*Ebola virus*	Virus	Ebola hemorrhagic fever
1977	*Legionella pneumophila*	Bacterium	Legionnaires' disease
1977	Hantaan virus	Virus	Haemorrhagic fever with renal syndrome (HRFS)
1977	*Campylobacter jejuni*	Bacterium	Enteric pathogen distributed globally
1980	Human T-lymphotropic virus 1 (HTLV-1)	Virus	T-cell lymphoma-leukemia
1981	Toxin-producing strains of *Staphylococcus aureus*	Bacterium	Toxic shock syndrome
1982	*Escherichia coli 0157:H7*	Bacterium	Hemorrhagic colitis; hemolytic uremic syndrome
1982	*Borrelia burgdoferi*	Bacterium	Lyme disease
1982	HTLV-2	Virus	Hairy cell leukemia
1983	Human immunodeficiency virus (HIV)	Virus	Acquired immunodeficiency syndrome (AIDS)
1983	*Helicobacter pylori*	Bacterium	Peptic ulcer disease
1985	*Enterocytozoon bieneusi*	Parasite	Persistent diarrhea
1986	*Cyclospara cayetanensis*	Parasite	Persistent diarrhea
1986	BSE agent?	Non-conventional agent	Bovine spongiform encephalopathy in cattle (Mad cow disease)
1988	Human herpesvirus 6 (HHV-6)	Virus	Exanthem subitum
1988	*Hepatitis E virus*	Virus	Enterically transmitted non-A, non-B hepatitis
1989	Ehrlichia chaffeensis	Bacterium	Human ehrlichiosis
1989	*Hepatitis C virus*	Virus	Parenterally transmitted non-A, non-B liver hepatitis
1991	Guanarito virus	Virus	Venezuelan hemorrhagic fever
1991	*Encephalitozoon hellem*	Parasite	Conjunctivitis, disseminated disease
1991	*New species of Babesia*	Parasite	Atypical babesiosis
1992	*Vibrio cholerae 0139*	Bacterium	New strain associated with epidemic cholera
1992	*Bartonella henselae*	Bacterium	Cat-scratch disease; bacillary angiomatosis
1993	Sin Nambre virus	Virus	Hantavirus pulmonary syndrome
1993	*Encephalitozoon cuniculi*	Parasite	Disseminated disease
1994	Sabia virus	Virus	Brazilian hemorrhagic fever
1995	Human herpesvirus 8	Virus	Associated with Kaposi's sarcoma in AIDS patients
1996	nvCJD Australian bat lyssavirus	Virus	–
1997	H5N1	Virus	Avian flu (Bird flu)
1999	Nipah virus	Virus	–
2003	Corona virus	Virus	SARS

Table 86.1: New infectious diseases recognized since 1973

Important Examples of Re-emerging Infections in India

Appearance of plague in an explosive formin 1994 after a period-of quiescence of almost 27 years, cholera in 1995 and dengue hemorrhagic fever in 1996. New and previously unknown diseases continue to emerge (Table 86.2).

The emergence of drug-resistant strains of micro-organisms or parasites is promoted by treatments that do not result in cure. The reasons for outbreaks of new diseases, or sharp increases in those once believed to be under control, are complex and still not fully understood. The fact is however that national health has become an international challenge. An outbreak anywhere must now be seen as a threat to virtually all countries, especially those that serve as m.

Antimicrobial Resistance

Resistance by disease—causing organisms to antimicrobial drugs and other agents is a major public health problem worldwide. It is making a growing number of infections virtually untreatable, both in hospitals and in the general community. It is having a deadly impact on the control of diseases such as *Mycobacterium tuberculosis*, Malaria, Enterococci, Staphylococci, Streptococci, Pneumococci and *Haemophilus influenzae*, *Neisseria gonorrahoeae*, *Shigella dysenteriae*, *Salmonella Typhi*.

FACTORS RESPONSIBLE FOR EMERGENCE AND RE-EMERGENCE OF INFECTIOUS DISEASES

1. Economic development and land use, unplanned and under-planned urbanization;
2. Over-crowding and rapid population growth;
3. Poor sanitation;
4. Inadequate public health infrastructure;
5. Resistance to antibiotics;
6. Increased exposure of humans to disease vectors and reservoirs of infection in nature; and
7. Rapid and intense international travel.

Responding to Epidemics

The strategy for controlling re-emerging diseases is through available cost-effective interventions such as early diagnosis and prompt treatment, vector control measures and the prevention of epidemics, for malaria; and DOTS-directly observed treatment, short-course—for tuberculosis; by launching research initiatives for treatment regimens and improved diagnostics, drugs and vaccines; and above all by strengthening epidemiological surveillance and drug-resistance surveillance mechanisms and procedures with appropriate laboratory support for early detection, confirmation and communication. The laboratory technologist must relearn information once thought to be out of date. Media selection, identification techniques, and safety precautions must all be re-examined and implemented. Interaction with the infection control program strengthens the establishment of prevention and control strategies.

The category of diseases-"new diseases-new problems"-such as Ebola and other viral hemorrhagic fevers, is probably the most frightening. Much of this already applies to HIV/AIDS, one of the most serious diseases to emerge in recent decades.

Table 86.2: Re-emerging infectious diseases

Disease	*Causative agent*
A. Bacterial	
1. Tuberculosis	Multidrug resistant *Mycobacterium tuberculosis*
2. Typhoid fever	Multidrug resistant *Salmonella* Typhi
3. Leptospirosis	*Leptospira interrogans*
4. Melioidosis	*Burkholderia (Pseudomonas) pseudomallei*
5. Anthrax	*Bacillus anthracis*
6. Plague	*Yersinia pestis*
B. Parasitic	
1. Malaria	Drug resistant *Plasmodium falciparum*
2. Leishmaniasis	*Leishmania donovani*
3. Lymphatic filariasis	*Wuchereria bancrofti*, *Brugia malayi* and *B timori*

KNOW MORE

Examples of Bacterial Resistance

1. *Mycobacterium tuberculosis;* 2. *Malaria;* 3. *Enterococci;* 4. *Staphylococci;* 5. *Streptococci;* 6. *Pneumococci and Haemophilus influenzae;* 7. *Neisseria gonorrhoeae;* 8. *Shigella dysenteriae* 9. *Salmonella Typhi.*

☞ KEY POINTS

- Emerging infectious diseases are those whose incidence in humans has increased during the last two decades or which threaten to increase in the near future e.g. Coronavirus (SARS),Viral hemorrhagic fever viruses,West Nile virus
- Approximately 75 percent of emerging pathogens are zoonotic that is, communicated by animals to humans. HIV/AIDS, avian influenza, monkeypox, Nipah, SARS and Ebola are all the result, to a greater or lesser extent, of interactions with animals that led to the emergence and re-emergence of deadly diseases.
- Re-emerging, or resurging diseases are those that have been around for decades or centuries, but have come back in a different form or a different location. Such as anthrax (Bacillus anthracis), Botulism (Clostridium botulinum, Plague (Yersinia pestis), Smallpox virus, Tularemia (Francisella tularensis)
- Resistance by disease-causing organisms to antimicrobial drugs and other agents is a major public health problem worldwide.

IMPORTANT QUESTIONS

1. Write an essay on emerging and re-emerging infections.
2. Write short notes on
   i. Emerging infectious diseases
   ii. Re-emerging infectious diseases

FURTHER READING

Banerjee K. Emerging viral infections with reference to India; Indian J Med Res 1996;103:177-200.

Bloom BR, Murray CJL. Tuberculosis: commentary on a re-emergent killer. Science 1992;257:1055-64.

John TJ. Emerging and re-emerging bacterial pathogen in India. Indian J Med Res 1996;103:4-18.

World Health Organization. Emerging infectious diseases. Wkly Epidemiol Rec 1994;69:234-6.

SECTION SEVEN

Diagnostic Medical Microbiology

CHAPTER

87 Staining Methods

Learning Objectives

After reading and studying this chapter, you should be able to:

- Describe common staining techniques.
- Describe the following: Simple stains; differential stains; Gram stain; Acid fast stain (Ziehl-Neelsen staining of acid fast bacilli); Albert's stain.

INTRODUCTION

Because most microorganisms appear almost colorless when viewed through a standard light microscope, we often must prepare them for observation. Live bacteria do not show much structural detail under the light microscope due to lack of contrast. Hence it is customary to use staining techniques to produce color contrast.

PREPARING FILM OR SMEAR FOR STAINING

Slides

Film preparations are made either on cover-slips or on 3 × 1 in glass slides, usually the latter. It is essential that the cover-slips or slides be perfectly clean and free from grease, otherwise films will be uneven.

Smear Preparation

A thin film of material containing the microorganisms is spread over the surface of the slide. This film, called a **smear**, is allowed to air dry.

Fixation

Before the microorganisms can be stained, however, they must be fixed (attached) to the microscope slide. In most staining procedures the slide is fixed by passing it through the flame of a Bunsen burner several times, smear side up, or by covering the slide with methyl alcohol for 1 minute. Air drying and flaming fix the microorganisms to the slide. Fixing simultaneously kills the microorganisms and attaches them to the slide. It also preserves various parts of microbes in their natural state with only minimal distortion.

Staining

Stain is applied and then washed off with water; then the slide is blotted with absorbent paper. Without fixing, the stain might wash the microbes off the slide. The stained microorganisms are now ready for microscopic examination.

TYPES OF STAIN

Basic dyes: Stains are salts composed of a positive and a negative ion, one of which is colored and is known as the **chromophore.** The color of so-called **basic dyes** is in the positive ion; in **acidic dyes,** it is in the negative ion. Bacteria are slightly negatively charged at pH 7. Thus, the colored positive ion in a basic dye is attracted to the negatively charged bacterial cell.

Basic dyes, which include crystal violet, methylene blue, malachite green, and safranin, are more commonly used than acidic dyes.

Acidic Dyes: Acidic dyes are not attracted to most types of bacteria because the dye's negative ions are repelled by the negatively charged bacterial surface, so the stain colors the background instead. Examples of acidic dyes are **eosin, acid fuchsin, and nigrosin.**

Negative Staining

Preparing colorless bacteria against a colored background is called **negative staining**. It is valuable in the observation of overall cell shapes, sizes, and capsules because the cells are made highly visible against a contrasting dark background. Distortions of cell size and shape are minimized because heat fixing is not necessary and the cells do not pkk up the stain.

To apply acidic or basic dyes, microbiologists use three kinds of staining techniques: simple, differential, and special.

STAINED PREPARATIONS

Staining simply means coloring the microorganisms with a dye that emphasizes certain structures. Bacteria may be stained in the living state, but this type of staining is employed only for special purposes. Routine methods for staining of bacteria involve drying and fIxing smears, procedures that kill them. Fixing simultaneously kills the microorganisms and attaches them to the slide. It also preserves various parts of microbes in their natural state with only minimal distortion.

Common Staining Techniques

The following are staining techniques commonly used in bacteriology.

A. Simple stains
B. Differential stains
   a. Gram stain
   b. Acid fast stain (Ziehl-Neelsen staining of acid fast bacilli)
C. Special stains
   a. Negative staining
   b. Impregnation methods:

A. SIMPLE STAINS

A single stain is used in simple staining. A simple stain is an aqueous or alcohol solution of a single basic dye.

Procedure

The bacterial suspension is smeared on the surface of the slide, fixed by gentle heating and flooded with a dye solution (*e.g.* methylene blue) for about one minute. The dye is washed off with water and the slide blotted dry and examined. They provide color contrast, but impart the same color to all bacteria. Occasionally, a chemical is added to the solution to intensify the stain; such an additive is called a **mordant.** One function of a mordant is to increase the affinity of a stain for a biological specimen; another is to coat a structure (such as a flagellum) to make it thicker and easier to see after it is stained with a dye.

Some of the simple stains commonly used in the laboratory are **methylene blue, carbol fuchsin, crystal violet, and safranin.**

1. **Loeffler's methylene blue:** The films are stained for 3 min. then are washed with water. This preparation does not readily over-stain. Sections are stained for 5 min or longer.

   Loeftler's methylene blue is generally the most useful of the many preparations of this dye. It shows the characteristic morphology of polymorphs, lymphocytes and other cells more clearly than do stronger stains such as the Gram stain or ZN staining (dilute carbol fuchsin.)
2. **Polychrome methylene blue:** This is made by allowing Loeftler's methylene blue to 'ripen' slowly for 12 months or more in half filled bottles and shaken at intervals to aerate thoroughly the contents. The slow oxidation of the methylene blue forms a violet compound that gives the stain its polychrome properties. It may be ripened quickly by the addition of 1 percent potassium carbonate (K2C03). The preparation is used in a manner similar to Loeftler's methylene blue. It is also employed in McFadyean's reaction for the identification of anthrax bacilli in blood films.
3. **Dilute carbol fuchsin:** Dilute carbol fuchsin is made by diluting Ziehl-Neelsen's stain with 10-20 times its volume of water. Stain for 10-25 seconds and wash well with water.

B. DIFFERENTIAL STAINS

These stains impart different colors to different bacteria or bacterial structures. **Gram stain** and the **acid fast stain** are two most widely used differential stains.

Gram Stain (Figs 87.1A and B)

The Gram stain is the principle stain used for microscopic examination of bacteria. It was first devised by the histologist Hans Christian Gram (1884) as a method of staining bacteria in tissues. It is one of the most useful staining procedures because it classifies bacteria into two large groups: gram-positive and gram-negative.

REAGENTS

Violet dye

Crystal violet or methyl violet is used at concentrations of 0.5-2 percent. Solution is facilitated if the dye is first dissolved in alcohol and then added to the water.

1.	Crystal violet or methyl violet 6B	10 gm
2.	Absolute alcohol (100% ethanol)	100 ml
3.	Distilled water	1 Liter

Gram's iodine

1.	Iodine	10 gm
2.	Potassium iodide	20 gm
3.	Distilled water	1 Liter

Decolorizer

i. Acetone.
ii. Absolute alcohol (100% ethanol).
iii. Acetone-alcohol: This is a mixture of 1 volume of acetone with 1 volume of 95 percent ethanol. It requires application for about 10 seconds.

Safranine Counterstain

Safranine 0.5 percent in distilled water.

Procedure

1. A heat-fixed smear is covered with a basic purple dye, usually crystal violet (primary stain) for **one minute.**
2. After a short time, the purple dye is washed off, and the smear is covered with iodine, **a mordant** for **one minute.** (When the iodine is washed off, both gram-positive and gram-negative bacteria appear dark violet or purple).
3. Next, the slide is washed with alcohol or an alcohol acetone solution. This solution is a decolorizing

agent, which removes the purple from the cells of some species but not from others.

4. The alcohol is rinsed off, and the slide is then stained with safranin, a basic red dye for **one minute.**
5. The smear is washed again, blotted dry, and examined microscopically.

   Interpretation of gram-staining see chapter 90.

Acid Fast Stain (Ziehl-Neelsen Staining of Acid Fast Bacilli)

Acid fast stain was discovered by Ehrlich (1882), who found that after staining with aniline dyes, tubercle bacilli resist decolorisation with acids. The method, as modified by Ziehl and Neelsen, is in common use now.

Principle

Some bacteria, such as mycobacteria are resistant to aniline dyes and do not readily penetrate the substance of the tubercle bacillus and are therefore unsuitable for staining it. The dye can be made to penetrate the bacillus by the use of a powerful staining solution that contains **phenol,** and the **application of heat.** Once stained the tubercle bacillus cannot be decolorised even with powerful decolorizing agents for a considerable time and thus still retains the stain when everything else in the microscopic preparation has been decolorized. Hence, they are called **Acid-fast Bacilli (AFB).**

The stain used consists of **basic fuchsin**, with **phenol** (acts as a-mordant) added. The dye is basic and its combination with a mineral acid used as decolorizer produces a compound that is **yellowish brown in color** which is readily dissolved out of all structures except acid-fast bacteria. Any strong acid can be used as a decolorizing agent, but 20 percent sulphuric acid (by volume) is usually employed. In order to show structures and cells, including non-acid-fast bacteria, that have been decolorized, and to form a contrast with the red-stained bacilli, the preparation is counterstained with **methylene blue or malachite green.**

Acid fastness has been ascribed to the high content and variety of lipids, fatty acids and higher alcohols found in tubercle bacilli. A lipid peculiar to acid fast bacilli, a high molecular weight hydroxy acid wax containing carboxyl groups **(mycolic acid)** is acid fast in the free state. Acid-fastness depends also on the **integrity of the cell wall** besides lipid contents.

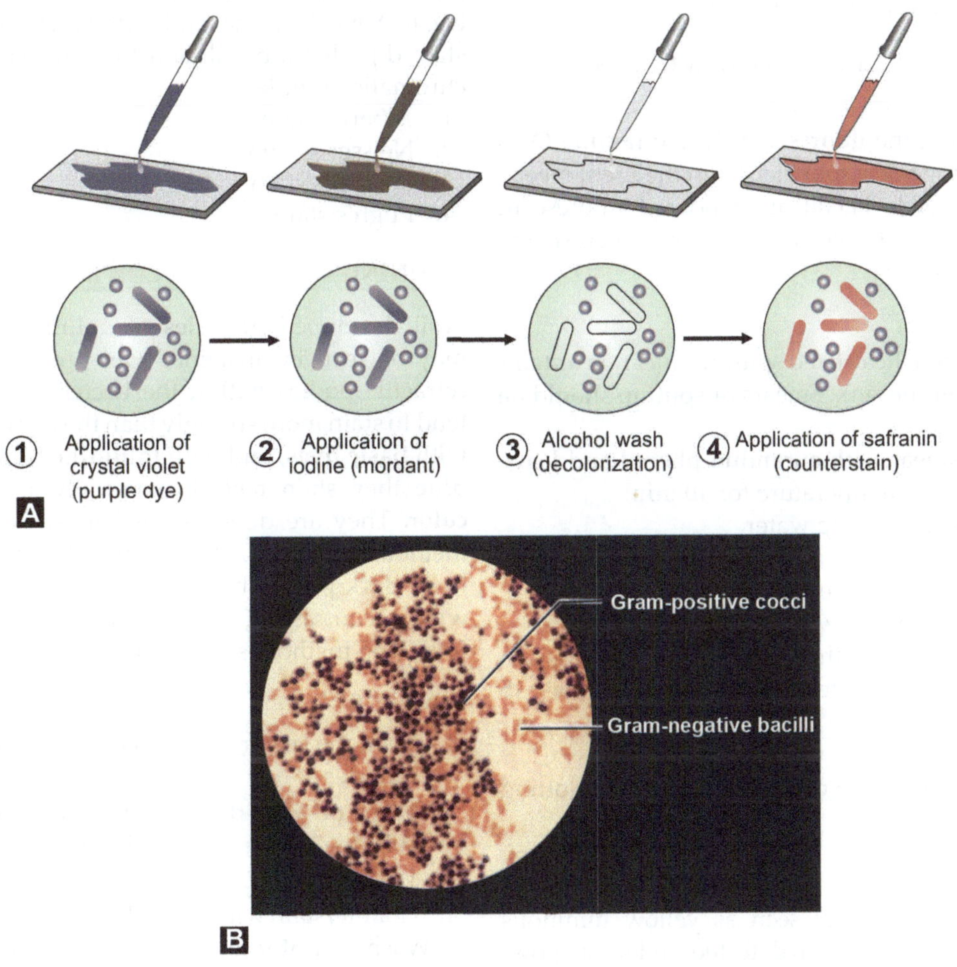

Figs 87.1A and B: Gram stain; (A) Steps in the Gram stain procedure; (B) Results of a Gram stain. The Gram-positive ccocci (purple) the Gram-negative bacilli (reddish-pink)

Procedure

1. Make a smear on a numbered slide, dry and fix by flaming.
   See chapter 90 for procedure.

ZN reagents

1. ZN carbol fuchsin

Basic fuchsin (powder)	5 g
Phenol (crystalline)	25 g
Alcohol (95% or 100% ethanol)	50 ml
Distilled water	500 ml

2. **Sulphuric acid (20%) decolorizer**

Concentrated sulphuric acid (98%, 1.835 g/ml)	250 ml
Distilled water	1 litre

3. **Alcohol 95 percent:** Ethanol 95 ml plus water to 100 ml, *or* Industrial methylated spirit.
4. Acid-alcohol decolorizer

Concentrated hydrochloric acid	75 ml
Industrial methylated spirit	2425 ml

5. Methylene blue counterstain
   Loeffler's methylene blue (see above).
6. Saturated solution of methylene blue in alcohol 300 ml
   KOH, 0.01 percent in water 1 liter

FLUOROCHROME STAINING FOR ACID FAST BACTERIA

A stain containing the **fluorescent dye, auramine O,** is substituted for the hot carbol fuchsin in the Ziehl-Neelsen method. Tubercle bacilli are rendered fluorescent and becomes easy to detect by fluorescence microscopy. Heating is unnecessary.

Procedure

1. Place the dried, heat-fixed sputum smear on a staining rack over the sink. Smears of sputum should be thin.
2. Cover the smear with **auramine phenol** and leave to stain at room temperature for **10 min.**
3. Wash off stain with tap water.
4. Cover the slide with an excess of 1 percent acid alcohol and leave to decolorize for 5 min.
5. Wash off decolorizer with tap water.
6. Cover the smear with the 0.1 percent potassium permanganate counterstain and leave for 15 seconds.
7. Wash off counterstain with tap water.
8. Dry on heated drier or dry in air. Do not use blotting paper.
9. Examine the film dry by fluorescence microscopy with a 4 mm objective.

Tubercle bacilli are seen as yellow luminous rods in a dark field. When detected under low power, the morphology of the bacilli is confirmed with an oil-immersion objective.

C. Special Stains

Special stains are used to stain specific structures inside or outside of a cell color and isolate specific parts of microorganisms, such as capsule stain, endospore stain and flagella stain.

i. **Negative staining:** Here, bacteria are mixed with dyes such as **Indian ink or nigrosin** that provide a uniformly colored background against which the unstained bacteria stand out in contrast. This is particularly useful in the demonstration of **bacterial capsules** which do not take simple stains.
ii. **Impregnation methods:** Cells and structures too thin to be seen under the ordinary microscope may be rendered visible if they are thickened by impregnation of silver on the surface, so. that they become visible under ordinary microscope. For example for the demonstration of spirochetes and bacterial flagella.

SPECIAL STAINS FOR CORYNEBACTERIUM DIPHTHERIAE, (STAINS TO DEMONSTRATE METACHROMATIC GRANULES)

Certain bacilli possess volutin granules in the protoplasm which aid in the identification of diphtheria bacilli. Several special and differential stains are used to stain diphtheria bacilli and to demonstrate their metachromatic granules.

1. Albert's stain
2. Neisser's stain
3. Ponder's stain
4. Pugh's stain.

STAINING OF VOLUTIN-CONTAINING ORGANISMS

Well developed granules of volutin (polyphosphate) may be seen in unstained wet preparations as round refractile bodies within the bacterial cytoplasm. They tend to stain more strongly than the rest of the bacterium with **basic dyes,** and with **toluidine blue or methylene blue** they stain metachromatically, a **reddish-purple color.** They are demonstrated most clearly by special methods, such as Albert's and Neisser's, which stain them dark purple but the remainder of the bacterium with a contrasting counterstain. For routine use the following method is recommended:

Neisser's Stain

1. Stain the smear with Neisser's methylene blue for 3 minutes.
2. Wash off with dilute iodine and leave some of this solution on the slide for 1 minute.
3. Wash in water.
4. Counterstain with neutral red solution for 3 minutes.

Wash in water and dry. Observe under oil immersion lens. By this method the bacilli exhibit deep blue granules and the bacterial protoplasm takes pink color.

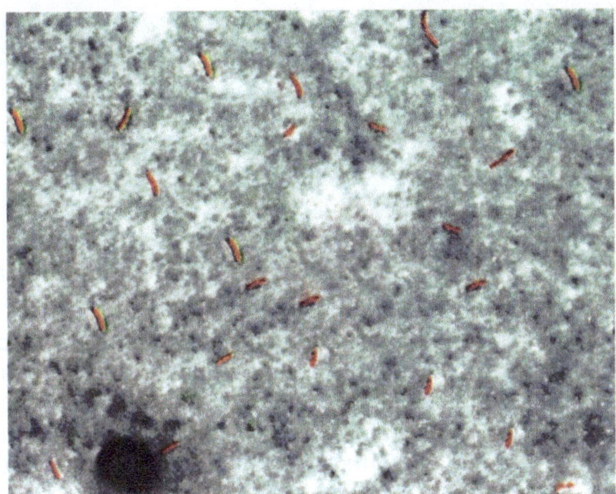

Fig. 87.2: Acid-fast bacteria. *Mycobacterium tuberculosis* in Ziehl-Neelsen stained smear

Ponder's Stain

1. Stain the smear with Ponder's stain for 1-2 minutes.
2. Wash with water and blot dry. Examine under oil immersion objective. The volutin granules stain dark blue, whereas the bacillus pale blue.

Pugh's Stain

1. Stain the smear with Pugh's stain (toluidine blue) for 2 to 3 minutes.
2. Wash with water and blot dry. Examine under oil immersion objective. The volutin granules stain red dish purple and the remainder of the organism light blue.

VITAL STAINING

Microorganisms may be stained in the living state. Many dyes are toxic and kill the cells on staining. Non-toxic staining, during which the cells retain their viability, is known as vital or intravital staining, e.g. methylene blue dye test for *Toxoplasma*.

SUPERVITAL STAINING

The staining of living cells after their removal from the host has been referred to as supervital or supravital staining, where the cells remain viable during the staining process.

Excess stain, if any can be removed by blotting paper.

For permanent preparations, the edges of coverslips may be sealed with nail polish. Lactic acid is a preservative, phenol is a disinfectant and cotton blue provides blue color to fungal filaments.

A. Special Stains to Observe Cell Structures

Special stains are used to color and isolate specific parts of microorganisms, such as endospores and flagella, and to reveal the presence of capsules.

CAPSULE STAIN

Capsules stain poorly, a characteristic exploited with a capsule stain, an example of a negative stain. It colors the background, allowing the capsule to stand out as a halo around an organism.

To observe capsules, a liquid specimen is placed on a slide next to a drop of India ink. A thin glass coverslip is then placed over the two drops, causing them to flow together. Unlike the stains discussed previously, the capsule stain is done as a wet mount-a drop of liquid on which a coverslip has been placed-rather than as a smear. At the optimum concentration of India ink, the fine dark particles of the stain color the background enough to allow the capsule to be visible.

2. Endospore (Spore) Staining

Endospores cannot be stained by ordinary methods, such as simple staining and Gram staining, because the dyes do not penetrate the wall of the endospore. A spore stain is used to make endospores more readily noticeable. This stain, like the classic acid-fast staining procedure, uses heat to facilitate staining. Generally, malachite green is used as a primary stain. Its uptake by the endospore is facilitated by gentle heat. When water is then used to rinse the smear, only endospores retain the malachite green. The smear is then counterstained, most often with the red dye safranin. The spores appear green amid a background of pink cells.

3. Flagella Staining

Bacterial flagella are too small to be seen with a light microscope without staining. The flagella stain overcomes this limitation by using a mordant that allows the staining agent to adhere to and coat these thin structures, effectively increasing their diameter. A tedious and delicate staining procedure uses a mordant and the stain carbolfuchsin to build up the diameters of the flagella until they become visible under the light microscope. Unfortunately, this staining procedure is difficult and requires patience and expertise.

Table 87.1 shows a summery of stain and their characteristics.

☞ KEY POINTS

- Staining means coloring a microorganism with a dye to make some structures more visible.
- Fixing uses heat or alcohol to kill and attach microorganisms to a slide.
- A smear is a thin film of material used for microscopic examination.
- Bacteria are negatively charged, and the colored positive ion of a basic dye will stain bacterial cells.
- The colored negative ion of an acidic dye will stain the background of a bacterial smear; a negative stain is produced.

Table 87.1: A summary of stains and their characteristics

Stain	Characteristics
Simple stains	Employ a basic dye to impart a color to a cell. Used to highlight microorganisms to determine cellular shapes and arrangements.
Differential	
Gram Stain	Classifies bacteria into two large groups: gram-positive and gram-negative. Gram-positive bacteria retain the crystal violet stain and appear purple. Gram-negative bacteria do not retain the crystal violet stain and remain colorless until counterstained with safranin and then appear pink.
Acid-fast stain	Used to distinguish *Mycobacterium* species and some species of *Nocardia*. Due to the lipid composition of their cell walls, these organisms do not readily take up stains. Acid.fast bacteria remain red, once stained with carbolfuchsin and treated with acid-alcohol, because they retain the carbolfuchsin stain. Non-acid.fast bacteria, appear. blue, when stained and treated the same way and then stained with methylene blue because they lose the carbolfuchsin stain and are then able to accept the methylene blue stain.
Special Stain	Used to color and isolate various structures, such as capsules, endospores, and flagella.
Capsule stain	Used to demonstrate the presence of capsules. The capsules appear as unstained halos around bacterial cells and stand out against a dark background because capsules do not accept most stains. This is an example of a negative stain.
Endospore stain	Used to detect the presence of endospores in bacteria. When malachite green is applied to a heat-fixed smear of bacterial cells, the stain penetrates the endospores and stains them green. When safranin (red) is then applied, it stains the remainder of the cells red or pink.
Flagella stain	The staining agent adheres to and coats the otherwise thin flagella, enabling them to be seen with the light microscope.

- A simple stain is an aqueous or alcohol solution of a single basic dye.
- It is used to make cellular shapes and arrangements visible.
- A mordant may be used to improve bonding between the stain and the specimen.
- Differential stains, such as the Gram stain and acid-fast stain, differentiate bacteria according to their reactions to the stains.
- The Gram stain procedure uses a purple stain (crystal violet), iodine as a mordant, an alcohol decolorizer, and a red counterstain. Gram-positive bacteria stain purple and Gram-negative bacteria stain pink.
- The acid-fast stain is used to stain organisms such as *Mycobacteria*, which do not take up stains readily; acid-fast organisms stain pink and all other organisms stain blue.
- **Special Stains**
  1. The endospore stain and flagella stain are special stains that color only certain parts of bacteria.
  2. Negative staining is used to make microbial capsules visible.

KNOW MORE

- Bacteria may be stained in the living state, but this type of staining is employed only for special purposes. Routine methods for staining of bacteria involve **drying** and **fixing smears**, procedures that kill them. Without fixing, the stain might wash the microbes off the slide.
- **Gram stain**

  Method: The staining technique consists of four steps:
  1. **Primary staining** with a basic pararosaniline (triphenyl methane) violet dye, namely crystal violet, methyl violet or gentian violet (a mixture of the two preceding dyes);
  2. Application of a dilute solution of iodine;
  3. **Decolorisation with** an organic solvent such as acetone, alcohol or aniline;
  4. **Counterstaining** with a dye of contrasting color, such as carbol fuchsin, safranine or neutral red.

IMPORTANT QUESTIONS

1. Describe in detail the Gram's staining. Descibe Gram staining mechanism
2. Give an account of differential stains.
3. Write Short Notes
   i. Simple staining.
   ii. Acid-fast stain or Ziehl-Neelsen's stain.
   iii. Negative staining.
   iv. Impregnation methods.
   v. Albert's stain.

FURTHER READING

Collee JC, et al. Mackie and Mc Cartney Practical *Medical Microbilology*. 14th edn. London: Churchill Livingstone 1996; 14th (Edn):113-129.

CHAPTER

88 Molecular Detection of Microorganisms

Learning Objectives

After reading and studying this chapter, you should be able to:

- Describe molecular methods for microbial identification.
- Describe staphylococcal diseases.
- Describe the following: Nucleic acid probes; polymerase chain reaction (PCR); PCR in diagnosis of infections agents.

INTRODUCTION

The DNA, RNA, or proteins of an infectious agent in a clinical sample can be used to help identify the agent like the evidence left at the scene of a crime. In many cases, the agent can be detected and identified in this way even if it cannot be isolated or detected by immunologic means. New techniques and applications of the techniques are being developed for the analysis of infectious agents. It also helps in disease prognosis and monitoring the response to treatment. However, in most cases the new molecular methods supplement rather than replace the conventional laboratory tests for diagnosis. The advantages of molecular techniques are their sensitivity, specificity, and safety. Molecular methods have been found to be advantageous in situations in which conventional methods are slow, insensitive, expensive or not available. Additionally, non-nucleic acid-based analytic methods that detect phenotypic traits not detectable by conventional strategies (e.g., cell wall components) have been developed to enhance bacterial detection, identification, and characterization.

MOLECULAR METHODS

Molecular methods are classified into three categories :

A. Hybridization
B. Amplication
C. Sequencing and enzymatic digestion of nucleic acids.

A. Hybridization (See Chapter 10 for detail)

DNA probes can be used like antibodies as sensitive and specific tools to detect, locate, and quantitate specific nucleic acid sequences in clinical specimens. Nucleic acid probes are segments of DNA or RNA labelled with radioisotopes or enzymes that can hybridize to complementary nucleic acid with high degree of specificity.

Principle of nucleic acid probes is described in Chapter 10 under heading 'DNA probes'. A number of DNA probes have been developed for direct detection of micro-organisms in clinical specimens and for identification of organisms after isolation of culture. Applications of DNA probe technology in microbiology are:

i. Nucleic acid probes for direct detection of groupA streptococci, *Chlamydia trachomatis* and *Neisseria gonorrhoeae* are available.
ii. Probes for identification of group A streptococci, group B streptococci, enterococci, *Haemophilus influenzae*, mycobacteria, *N. gonorrhoeae*, *Staphylococcus aureus*, *Streptococcus pneumoniae*, *Campylobacter sp.*, *Histoplasma capsulatum*, *Blastomyces dermatidis* and *Coccidioides immitis* isolated in culture are also available.
iii. DNA probes for detection of LT and ST toxins of *Esch. coli* are available.

B. Amplified Methods

1. Polymerase chain reaction (PCR)
2. Transcription mediated amplification (TMA)
3. Nucleic acid sequence amplification (NASBA)
4. Ligase chain reaction (LCR)

1. Polymerase Chain Reaction (PCR)

It is the target amplification system. The polymerase chain reaction (PCR) can detect single copies of viral DNA by amplifying the DNA many million-fold and is one of the newest techniques of genetic analysis. The basic principle of Polymerase chain reaction (PCR) has been described in chapter 10.

In this technique, a sample is incubated with two short DNA oligomers, termed primers, that are complementary to the ends of a known genetic sequence of the viral DNA, a heatstable DNA polymerase (Taq or other polymerases obtained trom thermophilic bacteria), nucleotides, and buffers. The oligomers hybridize to the appropriate sequence of DNA and act as primers for the polymerase, which copies that segment of the DNA. The sample is then heated to denature the DNA (separating the strands of the double helix) and cooled to allow hybridization of the primers to the new DNA. Each copy of DNA becomes a new template. The process is repeated many (30–50) times to amplify the original DNA sequence in an exponential manner. A target sequence can be amplified a million-fold in a few hours using this method. PCR has been applied in clinical laboratory for diagnosis of various infectious agents **(see chapter 10, Table 10.5).**

Besides originally described PCR, other types of PCR include reverse-transcriptase PCR (RT-PCR), nested PCR and multiplex PCR.

i. **RT-PCR:** Reverse transcription PCR (RT-PCR) amplifies an RNA target. In this technique, target is RNA instead of DNA. The unique step to this procedure is the use of the enzyme reverse transcriptase that directs synthesis of DNA from the viral RNA template. Once the DNA has been produced, relatively routine PCR technology is applied to obtain amplification.

ii. **Nested PCR:** Nested PCR involves the sequential use of two primer sets. The first set is used to amplify a target sequence. The amplicon obtained is then used as the target sequence for a second amplification using primers internal to those of the first amplicon. Essentially, this is an amplification of a sequence internal to an amplicon.

   The advantage of this approach is extreme sensitivity and confirmed specificity without the need for using probes.

iii. **Multiplex PCR:** Multiplex PCR is a method by which more than one primer pair is included in the PCR mixture. This will help in amplification of more than one target sequence in a clinical specimen. The control amplicon should always be detectable after PCR. Mutliplex PCRs are usually less sensitive than PCRs with single set of primers.

iv. **Arbitrary primed PCR:** Arbitrary primed PCR uses short primers that are not specifically complementary to a particular sequence of a target DNA. By comparing fragment migration patterns following agarose gel electrophoresis, strains or isolates can be judged to be the same, similar, or unrelated.

v. **Quantitative PCR:** Quantitative PCR is an approach that combines the power of PCR for the detection and identification of infectious agents with the ability to quantitate the actual number of targets originally in the clinical specimen. It can be used for studying and understanding the disease state (e.g., acquired immunodeficiency syndrome [AIDS]), the prognosis of certain infections, and the effectiveness of antimicrobial therapy.

vi. **Real time PCR:** Real time PCR combines rapid thermo cycling with the ability to detect target by fluorescently labeled probes as the hybrids are formed, i.e. in real time. This technology allows for high throughput of samples, multiplexing reactions, quantitation of target, and on-line monitoring.

2. Transcription Mediated Amplification (TMA)

Transcription mediated amplification (TMA) is an isothermal RNA amplification method and use three enzymes; reverse transcriptase (RT), RNAase H, and T7 DNA dependent RNA polymerase. RNA target is reverse transcribed into cDNA and then RNA copies are synthesized with the help of RNA polymerase. A 10^9 fold amplification of the target RNA can be achieved in about 2 hours.

Advantages of TMA include no requirement for a thermal cycler, and contamination risk is minimized. TMA based assays are available for detection of *M. tuberculosis, C trachomatis, N gonorrhoeae, Hepatitis Virus* (HCV) and human immunodeficiency virus 1 (HIV1).

3. Nucleic Acid Sequence-Based Amplification (NASBA)

Both TMA and NASBA are examples of transcription-mediated amplication. These isothermal assays use three enzymes: transcriptase (RT), RNAase H, and T7 DNA dependent RNA polymerase. Like TMA, it is also an isothermal RNA amplification method. The method is similar to TMA. RNA target is reverse transcribed into cDNA and then RNA copies are synthesized with the help of RNA polymerase. It also does not require thermal cycler. NASBA based kits for detection and quantitation of HIV-1 RNA and CMV RNA are available.

4. Ligase Chain Reaction (LCR)

Ligase chain reaction (LCR) is an amplification of probe nucleic acid rather than target nucleic acid. By this approach, an amplified probe is the final reaction product to be detected, while the target sequence is neither amplified nor incorporated into this product.

LCR uses two pairs of probes that span the target sequence of interest. Once annealed to the target sequence, a space remains between the probes that is enzymatically closed using a ligase (i.e., a ligation reaction). On heating, the joined probes are released as a single strand that is complementary to the target nucleic acid. These newly synthesized strands are then used as the template for subsequent cycles of probe annealing and ligations. Through the process, probe DNA is amplified to a level readily detectable using assays similar to those described for the biotin-avidin system. Like PCR, LCR also requires thermal cycler.

LCR based amplification has been used to detect *Chlamydia trachomatis* and *Neisseria gonorrhoeae*.

Applications of Molecular Methods in Clinical Laboratory

Molecular methods have a significant role in the following situations in clinical microbiology laboratory.

1. Detection of uncultivable and slow growing microorganisms.
2. Role in clinical virology.
3. Disease prognosis.
4. Response to treatment.

1. Detection of Uncultivable and Slow Growing Micro-organisms

The greatest advantage of molecular methods has been in the discovery of previously unrecognised or uncultivable organisms. Molecular methods have been used to detect previously unknown agents directly in clinical specimens by using broad-range primers for a number of micro-organisms. HCV, Sin nobre virus and Human herpes virus 8 (HHV-8), *Bartonella henselae* are some examples of human pathogens first identified from clinical specimens using molecular methods.

Molecular methods have the ability to detect nonviable organisms that are not retrievable by cultivation-based methods. These methods are also useful for fastidious micro-organisms which may die in transit or may be overgrown by contaminants when cultured. *N. gonorrhoeae* is one such example whose nucleic acid can be detected under circumstances in which it cannot be cultured. The use of improper collection, inappropriate transport conditions or delay in transport can reduce the viability of the organism but do not affect the nucleic acid detection.

These can detect and identify organisms that cannot be grown in culture or are extremely difficult to grow (e.g., hepatitis B virus and the agent of Whipple's disease) and also more rapid detection and identification of organisms that grow slowly (e.g., mycobacteria, certain fungi)

2. Role in Clinical Virology

Molecular methods to replace culture for detection of bacteria in routine practice are limited because of need to isolate the organisms for antibiotic sensitivity testing. These methods can actually replace the culture only in those micro-organisms which have predictable antibiotic susceptibility, and consequently, routine susceptibility testing is not performed.

Culture-based methods in virology are costly and antiviral susceptibility testing is not routinely done in clinical virology. Molecular approaches are often faster, more sensitive, and more cost-effective than the conventional approaches. Enteroviral meningitis, HSV encephalitis and CMV infections in immunocompromised patients are examples for which nucleic acid based tests are relevant and cost-effective for diagnosis.

3. Disease Prognosis

Molecular methods are able to quantitate infectious agent burden directly in patient specimens, an application that has particular importance for managing Human immunodeficiency virus (HIV) infections. Thus, it provides important information which may predict disease progression.

Molecular methods can be used for subtyping of certain viruses which may provide information about the severity of infection. HPV causes dysplasia, neoplasia and carcinoma of cervix in women. HPV types 16 and 18 are associated with a high risk of progression to neoplasia, whereas HPV types 6 and 11 have a low risk.

4. Response to Treatment and Drug Resistance

Molecular methods have been developed to detect the genes responsible for drug resistance that may not always readily be detected by phenotypic methods. Examples include detection of the *van* genes, which mediate vancomycin resistance among enterococci, and the *mec* gene, which encodes resistance among staphylococci and rifampicin resistance in *Mycobacterium tuberculosis*.

Molecular techniques have a significant role in predicting and monitoring patient responses to antiviral therapy. HIV-1 viral load assays have been developed to monitor the response of antiretroviral therapy. Viral load assays have also been used in monitoring the response to therapy in patients who are chronically infected with HBV and HCV.

Molecular methods can be used to detect drug resistance mutations in RT and protease genes of HIV-1. These mutations lead to lower levels of sensitivity to antiretroviral drugs and are important causes of treatment failure. This helps to determine an appropriate treatment in patients who do not respond to therapy.

KNOW MORE

NON-NUCLEIC ACID-BASED ANALYTIC METHODS

1. *Chromatography:* It may be:

1. Gas-Liquid Chromatography; 2. High-Performance Liquid Chromatography (HPLC).

2. *Electrophoretic Protein Analysis:* Can be used to generate microbial protein profiles.

KEY POINTS

- Molecular methods are classified into three categories: A. Hybridization; B. Amplication; C. Sequencing and enzymatic digestion of nucleic acids
- **Amplified methods**
  1. Polymerase chain reaction (PCR); 2. Transcription mediated amplification (TMA); 3. Nucleic acid sequence amplification (NASBA); 4. Ligase chain reaction (LCR).
- Molecular methods have a significant role in:
  1. Detection of uncultivable growing microorganisms; 2. Role in clinical virology; 3. Disease prognosis; 4. Response to treatment

IMPORTANT QUESTIONS

1. Write short notes on:
   Detection of micro-organisms by molecular methods
   Nucleic acid probes
   Polymerase chain reaction (PCR) in diagnosis of infectious agents.
   RT-PCR
   Applicaions of molecular methods in clinical microbiology

FURTHER READING

Forbes BA, Weissfeld AS, Sahm DF. Baily and Scott's diagnostic microbiology, ed 10, St. Louis 1998 Mosby.

Fredericks DN, Relman DA. Application of polymerase chain reaction to the diagnosis of infectious diseases, Clin Infect Dis 1999;29:475-86.

1. *Detection of Unculturable or Slow-growing Micro-organisms*

The greatest advantage of molecular methods has been in the discovery of previously unrecognised or unculturable organisms. Molecular methods have been used to detect previously unknown agents directly in clinical specimens by using broad-range primers for a number of microorganisms. Hepatitis C virus, Sin Nombre hantavirus, human herpes virus 8 (HHV-8), *Bartonella henselae* are some examples of human pathogens first identified from clinical specimens using molecular methods.

Molecular methods have the ability to detect nonviable organisms that are not retrievable by culture-based methods. These methods are also useful for fastidious microorganisms which may die in transport or may be overgrown by contaminants when cultured. *Neisseria gonorrhoeae* is one such example whose nucleic acid can be detected under circumstances in which it cannot be cultured. The use of improper collection, inappropriate transport conditions or delay in transport can reduce the viability of the organism but do not affect the nucleic acid detection.

These can detect and identify organisms that cannot be grown in culture or are extremely difficult to grow (e.g., hepatitis B virus and the agent of Whipple's disease and also more rapid detection and identification of organisms that grow slowly (e.g., mycobacteria, certain fungi).

Molecular methods to replace culture for detection of bacteria in routine practice are limited because of need to isolate the organisms for antibiotic sensitivity testing. These methods can actually replace the culture only in those micro-organisms which have predictable antibiotic susceptibility, and consequently, routine susceptibility testing is not performed.

Culture-based methods in virology are costly and antiviral susceptibility testing is not routinely done in clinical virology. Molecular approaches are often faster, more sensitive, and more cost-effective than the conventional diagnostic [illegible] methods. HSV encephalitis and CMV infections in immunocompromised patients are examples for which nucleic acid based tests are relevant and cost-effective for diagnosis.

Molecular methods are able to quantitate infectious agent burden directly in patient specimens, an application that has particular importance for managing Human immunodeficiency virus (HIV) infections. Thus, it provides important information which may predict disease progression.

Molecular methods can be used for subtyping of certain viruses which may provide information about the severity of infection. HPV causes dysplasia, neoplasia and carcinoma of cervix in women. HPV types 16 and 18 are associated with a high risk of progression to neoplasia whereas HPV types 6 and 11 have a low risk.

Response to Treatment and Drug Resistance

Molecular methods have been developed to detect the genes responsible for drug resistance that may not always readily be detected by phenotypic methods. Examples include detection of the *van* genes, which mediate vancomycin resistance among enterococci, and the *mec* gene, which encodes resistance among staphylococci and *rifampicin* resistance in *Mycobacterium tuberculosis*.

Molecular techniques have a significant role in predicting and monitoring patient responses to antiviral therapy. HIV-1 RNA load assays have been developed to monitor the response to antiretroviral therapy. Viral load measurements [illegible] determine the response to therapy [illegible] HIV infected individuals.

Molecular methods [illegible] resistance [illegible] mutations in RT and protease genes of HIV-1. These mutations lead to low levels of sensitivity to antiretroviral drugs and are important causes of treatment failure. This helps to determine an appropriate treatment for patients who do not respond to therapy.

KNOW MORE

NON-NUCLEIC ACID BASED ANALYTIC METHODS

1. *Chromatography:* It may be:
   1. Gas-Liquid Chromatography 2. High Performance Liquid Chromatography (HPLC).
2. *SDS-PAGE Protein Analysis:* Can be used to generate microbial protein profiles.

KEY POINTS

- Molecular methods are classified into three categories: A. Hybridization; B. Amplification; C. Sequencing and enzymatic digestion of nucleic acids.
- Amplified methods
  1. Polymerase chain reaction (PCR); 2. Transcription mediated amplification (TMA); 3. Nucleic acid sequence amplification (NASBA); 4. Ligase chain reaction (LCR).
- Molecular methods have a significant role in:
  1. Detection of unculturable, fastidious micro-organisms; 2. Quantitation of viral load; 3. Disease prognosis; 4. Response to treatment.

IMPORTANT QUESTIONS

1. Write short notes on:
   - Detection of micro-organisms by molecular methods
   - Nucleic acid probes
   - Polymerase chain reaction (PCR) in diagnosis of infectious agents.
   - RT-PCR
   - Applications of molecular methods in clinical microbiology.

FURTHER READING

Forbes BA, Weissfeld AS, Sahm DF. Baily and Scott's diagnostic microbiology, ed 10, St. Louis 1998 Mosby.

Fredericks DN, Relman DA. Application of polymerase chain reaction to the diagnosis of infectious diseases. Clin Infect Dis 1999;29:475-86.

Index

Page numbers followed by *f* refer to figure and *t* refer to table

C

D

F

I

J

K

L

M

P

Q

T

U

V

W

X

Y

Z